D0321788

NUCLEIC ACIDS

STRUCTURES, PROPERTIES, AND FUNCTIONS

NUCLEIC ACIDS

STRUCTURES, PROPERTIES, AND FUNCTIONS

Victor A. Bloomfield
UNIVERSITY OF MINNESOTA

Donald M. Crothers
YALE UNIVERSITY

Ignacio Tinoco, Jr.

WITH CONTRIBUTIONS FROM

John E. Hearst
UNIVERSITY OF CALIFORNIA, BERKELEY

David E. Wemmer
UNIVERSITY OF CALIFORNIA, BERKELEY

Peter A. Kollman
UNIVERSITY OF CALIFORNIA, SAN FRANCISCO

Douglas H. Turner
UNIVERSITY OF ROCHESTER

University Science Books
Sausalito, California

University Science Books
55D Gate Five Road
Sausalito, CA 94965

Fax: (415) 332-5393
www.uscibooks.com

Production manager: *Susanna Tadlock*
Manuscript editor: *Jeannette Stiefel*
Designer: *Robert Ishi*
Compositor: *Eigentype*
Printer & Binder: *Maple-Vail Book Manufacturing Group*

This book is printed on acid-free paper.

Library of Congress Cataloging-in-Publication Data

Bloomfield, Victor A.
Nucleic acids : structures, properties, and functions / by Victor A. Bloomfield, Donald M. Crothers, Ignacio Tinoco, Jr. ; with contributions from John E. Hearst ... [et al.].
p. cm.
Includes bibliographical references and index.
ISBN 0-935702-49-0 (hardcover)
1. Nucleic Acids. I. Crothers, Donald M. II. Tinoco, Ignacio. III. Title.
QP620.B64 1999
572.8—dc21 98-45268
CIP

Printed in the United States of America
10 9 8 7 6 5 4 3 2 1

Contents

Preface

In this book we present a comprehensive account of the structures and physical chemical properties of the nucleic acids, with emphasis on implications for biological function. The level of presentation assumes that the reader has knowledge of physical chemistry and molecular biology that would be obtained from introductory courses in these subjects.

We have three intended audiences: molecular biologists, physical biochemists, and physical chemists. Molecular biologists and nucleic acid biochemists have made remarkable strides in defining the major classes of nucleic acids, determining their sequences, and proposing connections between sequence and biological function. Such connections pertain not just to the direct coding of amino acid sequences in proteins, but also to regulation of transcription, translation, replication, and molecular evolution. The next step, one that requires physical chemical approaches, is to understand the mechanism by which a sequence of nucleotides exerts its coding or regulatory function. For this, one must know how the structure and dynamics of that sequence differ from those of its neighbors, and how those properties are affected by interaction with other molecules. One of our goals is to present to the molecular biologist the results that have been obtained on these issues, the experimental basis of these results, and some idea of what may be expected from even closer collaboration between biologists and physical chemists in the future.

For those physical biochemists already working on nucleic acids, or beginning a research career in this area, our goal is to provide a comprehensive treatment of the major experimental and theoretical approaches to the structures and physical properties of the nucleic acids.

Physical chemists have developed a powerful set of experimental and theoretical techniques for determining the structural and dynamical properties of molecules. New methods in spectroscopy and diffraction, and new theoretical approaches to complex molecular systems are producing ever more sensitive and detailed insights into nucleic acid behavior. We hope in this book to give physical chemists an overview of the biological context in which their contributions are being used, and some perspectives on the challenges that remain.

To reach three such different audiences with the same book requires careful organization. We have adopted a uniform plan for each chapter, leading off with a statement of the biological significance of each topic, and following with a clear presentation of the basic physical ideas and major results. The quantitative details, which are important to the physical chemists and physical biochemists, are developed in special sections or appendices.

The main logic of organization is through a systematic consideration of techniques used in the study of nucleic acid structure and properties. This should maximize the book's utility as a textbook in physical biochemistry, and as a reference for techniques that may be encountered in the literature or in one's own work. Examples of the uses of the various techniques go from the simpler techniques to the more complex. Readers interested in applications to particular molecules will find cross-references in the index.

We hope that the book will serve a variety of functions: as a textbook in physical biochemistry and biophysical chemistry classes; as an aid to molecular biologists or biochemists trying to gain a better understanding of physical techniques they use or encounter in the literature; and as a reference work for physical chemists and physical biochemists.

We are grateful to our colleagues who have sent us reprints and original figures for reproduction, and who have commented on the various versions of this book. We give special thanks to David Draper, whose detailed comments notably improved our efforts.

NUCLEIC ACIDS

STRUCTURES, PROPERTIES, AND FUNCTIONS

CHAPTER 1

Introduction

1. BIOLOGICAL ROLES OF THE NUCLEIC ACIDS

Nucleic acids are central molecules in the transmission, expression, and conservation of genetic information. The role of DNA as the carrier of genetic information has been amply demonstrated beginning with the classic experiments of Avery et al. (1944) and Hershey and Chase (1952). The classic example of how biological function follows from biomolecular structure comes from the elucidation by Watson and Crick (1953) of the structure of DNA as a double helix, using the X-ray fiber diffraction patterns generated by Franklin, Wilkins, and their associates (Franklin and Gosling, 1953; Wilkins et al., 1953) and the chemical evidence on base complementarity of Chargaff (1950). It was immediately obvious how information could be passed from one generation to the next by synthesizing a complementary strand for each of the parent strands and pairing with the parental complement. This mode of semiconservative replication was verified by Meselson and Stahl (1957) using the newly developed technique of density gradient ultracentrifugation.

Chemical analysis showed that there was a great deal of RNA in cells; most of the various types of RNA and their biological functions were understood in the 1950s and 1960s. Messenger RNA (mRNA) is the product of transcription of DNA into the carrier of the genetic code. It, in turn, is translated into proteins. Ribosomal RNA (rRNA) is complexed with ribosomal proteins to form ribosomes, the organelles on which protein synthesis occurs. Transfer RNA (tRNA) serves as the adaptor molecule between the amino acid and the genetic code triplet on the mRNA. Recently, we have learned that RNA plays even wider and unexpected roles. Genetic information can be carried in RNA viruses and retroviruses, and copied into DNA by reverse transcriptase. Complexes with proteins called small nuclear ribonucleoproteins (snRNPs) have been discovered that splice out intron sequences from transcripts of genomic DNA (Sharp, 1987). Ribonucleic acid itself can have enzymatic activity (such molecules are termed ribozymes) (Cech, 1987) and may have been the primordial enzyme, a key idea in considerations of prebiotic evolution (Gesteland and Atkins, 1993). A laboratory demonstration of RNA-catalyzed RNA polymerization has been achieved by Doudna and Szostak (1989).

The monomers of nucleic acids, the nucleotides and nucleosides, serve a diversity of roles. Some of these have been known over much of the span of modern biochemistry, such as the role of ATP as the energy currency in cells, and of the role of adenosine-containing cofactors nicotinamide adenine dinucleotide (NAD) and NADH as coenzymes in oxidoreduction reactions. More recently, cyclic adenosine - monophosphate (AMP) and cyclic guanosine monophosphate (GMP) have been shown to be crucial second messengers in controlling a wide variety of cellular processes.

The reader is referred to some of the basic texts and historical monographs listed at the end of this chapter for more information on these topics.

2. PHYSICAL CHEMISTRY AND NUCLEIC ACID FUNCTION

When the precursor to this book (Bloomfield et al., 1974) was written, most of these biological functions were known (but not all—splicing and RNA enzymatic activity were yet to come). What was not nearly so well known then as now was the detailed structural basis for these biological activities. A few of the crucial developments in the last 15 years are listed below (references will be given in later chapters).

- Discovery of Z-DNA, and determination of the X-ray structure of a B-DNA dodecamer, both of which demonstrated that DNA structure was highly variable at the atomic level, providing a concrete foundation for thinking about the structural basis of recognition and regulation.
- Discovery of naturally bent DNAs, which also demonstrated that particular sequences had defined and unique molecular structures.
- Recognition of the ubiquity and importance of supercoiling as a way of exerting long-range control of DNA.
- Recognition that ions and water can greatly influence the structure and interactions of DNA.
- Determination of the many non-Watson–Crick base pairings, and specific ribose-base–phosphate interactions present in folded RNA molecules.

Much of this progress has come from the development of new methods and new approaches to data analysis.

- Development of gel electrophoresis as a way to analyze the size and structure of nucleic acids and their complexes with proteins, permitting rapid and inexpensive analysis of complex molecules, and likely leading to sequencing of the human genome within a few years after this book is published.

- Ability to prepare synthetic nucleic acids, or purify natural ones, in adequate quantity for physical studies.

- Multidimensional NMR as a way to determine molecular structure in solution.

- Accumulation of data on a wide variety of nucleic acids, permitting recognition of regularities and suggesting structural generalizations.

- Random synthesis of nucleic acid sequences and selection for binding ability or catalytic activity, allowing identification of novel properties of nucleic acids, and revealing the range of sequences that can perform the same function.

- Computer analysis methods that have pointed to similarities between functionally and evolutionarily related molecules, suggesting common structures in RNAs that may regulate protein synthesis. (This topic, because of its mathematical and computer basis, may have considerable attraction for quantitatively oriented scientists. We do not discuss it in this book, but interested readers are directed to the references on computer analysis listed at the end of this chapter).

- Thermodynamic data bases for predicting secondary structure, loops, bulges, and so forth.

As an example of the important biological questions that physical chemistry may help us to answer, consider protein binding to a regulatory sequence of DNA. What structural features cause this specific binding site to be recognized? How can the protein search so much intracellular DNA to find that particular sequence? How much energy is required to deform the DNA and protein for optimal binding? How can the effects of sequence mutations on binding strength be understood? How does change in ionic strength affect binding? How does protein binding affect the structure and activity of distant DNA sequences, including the binding of other proteins? In this book, we have tried to show how answers to such questions may be obtained.

3. OUTLINE OF THIS BOOK

The first part, Chapters 2–8, treats the properties of the nucleic acids mainly at the level determined by atomic and molecular structure.

Chapter 2 (Bases, Nucleosides, and Nucleotides) discusses the physical and chemical properties of the monomeric building blocks of nucleic acid structure: the bases, nucleosides and nucleotides. Seemingly recondite monomer structural features such as sugar pucker and glycosidic bond conformation are shown later to be key determinants of helix geometry. Electron distributions in the bases influence their hydrogen-bonding and base stacking capabilities, as well as their spectroscopic behavior.

Chapter 3 (Chemical and Enzymatic Methods) discusses chemical and photochemical reactivity, important for structural determinations and mutagenesis. Much of the recent rapid progress in our understanding of nucleic acid sequence and structure, and of the binding of ligands, comes from increasingly sensitive and specific reactions. It also notes some of the naturally occurring base modifications that influence specificity and recognition.

Chapter 4 (Nucleic Acid Structures from Diffraction Methods) presents the basic ideas and results of diffraction techniques applied to nucleic acids. It begins with fiber diffraction, which gave the earliest insights into double helix structure and whose results are still important in delineating helical parameters for long sequences that cannot be crystallized. It continues with high-resolution X-ray crystallography of DNA oligonucleotides and tRNA, whose structures can be determined at atomic resolution. Results on A, B, and Z forms of DNA have had enormous impact on our thinking about the variability and flexibility of nucleic acid conformation. This chapter concludes with results of crystallographic studies on complexes with proteins, metal ions, water, and drugs—the sorts of complexes through which DNA exerts its biological functions.

Chapters 5 (Structure and Dynamics by NMR) and 6 (Electronic and Vibrational Spectroscopy) approach structure and dynamics through the NMR and optical spectroscopic techniques that have yielded so much of our current information about nucleic acids in solution. Nuclear magnetic resonance (NMR) has been the tool that provided the basic understanding of the structure of mononucleotides in solution. Now it is being used to determine the three-dimensional solution structures of tRNAs, rRNAs, and ribozymes. Atomic resolution studies of oligonucleotides are being used to probe the structure and dynamics of loops and bulges caused by base mismatches, and to determine the rates at which bases fluctuate between paired and unpaired states. Ultraviolet (UV) spectroscopy has for decades been one of the standard techniques for determining nucleic acid concentration and helix–coil transition behavior. In more sophisticated forms, it can elucidate the orientation of the bases along the helix, and give information on DNA conformation in packaged intracellular forms such as chromatin. Circular dichroism (CD) is the prime method for distinguishing the main helical forms of double-stranded nucleic acids, and was the first technique to show the existence of left-handed Z-DNA. Infrared (IR) and Raman spectroscopy are sensitive measures of backbone and base geometry; they can thus provide critical information on the conformation of DNA inside viruses.

Chapter 7 (Theoretical Methods) provides a theoretical reprise of many of the preceding topics. Developments in computer hardware and software, and in understanding of intermolecular forces, are making possible highly detailed modeling of the structural, energetic, and reactive properties of the nucleic acids and their complexes. Reliable predictions of the structural and dynamic effects of base substitutions, of conformational changes induced by ligand binding, and of the relative affinities of intercalating drugs are clearly on the horizon.

Chapter 8 (Conformational Changes) deals with conformational transitions, helix–coil, and helix–helix, which are crucial to understanding the forces stabilizing helices, to calculating the likelihood of finding biologically significant alternative structures, and to assessing the complexity of genomic sequences. It has strong emphasis on base stacking and double helix formation in single-stranded polynucleotides, matters that are crucial to understanding the way in which folding of RNA influences protein binding, and transcription and translation.

There follow two chapters on the size and shape of nucleic acids considered as macromolecules.

Chapter 9 (Size and Shape of Nucleic Acids in Solution) surveys the major experimental methods for characterizing the molecular weight, size, and shape of nucleic acids chains. These include the gel electrophoretic and density gradient centrifugation techniques that are so widely used in molecular biology laboratories, as well as other hydrodynamic, scattering, and microscopy techniques. It also considers size and shape from the theoretical point of view, including such matters as conformations of single-stranded molecules, the rigidity of double-stranded ones, and the occurrence of naturally bent DNA.

Chapter 10 (Supercoiled DNA) deals with the key topic of closed circular, supercoiled DNA. It discusses how the various facets of supercoiling—linking, twisting, and writhing—can be described and interrelated. It then shows how the considerable energy stored in supercoiled DNA can be utilized to drive reactions and conformational transitions.

The next four chapters (Chapters 11–14) consider the noncovalent interactions of nucleic acids with the many types of molecules that affect stability or regulate function.

Chapter 11 (Interaction of Nucleic Acids with Water and Ions) considers the most basic interactions with water and ions. Water is a fundamental determinant of nucleic acid structure, as recognized since the 1950s when the A, B, and C forms of DNA were found to be interconvertible as a function of relative humidity. The basis for water stabilization of nucleic acid structure is now being revealed by X-ray and computer modeling studies. Since DNA and RNA are highly charged, their properties are strongly influenced by ions in solution. Ions affect the relative stability of double and single helices, the bending rigidity of DNA, and the condensation of DNA into compact particles. Perhaps the most dramatic effect of ions is on the binding of proteins. A two-fold increase in salt can produce a 100-fold or more decrease in binding constant.

Chapter 12 (Interactions and Reactions with Drugs) surveys binding of small ligands such as drugs. Drug binding is not a passive event; it requires changes in nucleic acid structure, and can thus be used as a probe of structural changes. The structural, thermodynamic, and kinetic principles governing binding affinity are discussed here.

Chapter 13 (Protein–Nucleic Acid Interactions) deals with the important interactions between nucleic acids and proteins, both regulatory and structural. These complex systems stretch the limits of biophysical approaches. But a variety of experimental methods, coupled with computer analysis of sequences and recognition of motifs characterizing binding domains (such as zinc fingers and leucine zippers) have given remarkable insight.

Chapter 14 (Higher Order Structure) concludes the book with a discussion of the higher order structures and mechanisms in the packaging of DNA in viruses and chromatin. These, it has become increasingly apparent, must be understood if we are to comprehend much of the biological functioning of the nucleic acids.

4. OBTAINING NUCLEIC ACIDS FOR PHYSICAL CHEMICAL STUDY

There are reciprocal influences between physical biochemistry and molecular biology. Not only does molecular biology provide the problems for the physical chemistry of nucleic acids, it also may help to provide the nucleic acids themselves. The need in physical studies is for macromolecules to be as pure in size and composition, and in as large quantities, as possible. In addition, there should be a range of sizes, compositions, and sequences.

Traditionally, DNA has been isolated from readily available sources such as calf thymus, or from bacterial viruses. Once the importance of avoiding shear breakage was recognized, and delicate handling techniques were devised, bacteriophages gave a good source of monodisperse DNA. However, viral DNA is sufficiently large (2–110 million molecular weight) that it gives information mainly on average and long-range conformational properties.

For many types of studies, it is desirable to have short segments of DNA. Sequence and compositional variations can be more extreme and better controlled than in longer molecules. Local stiffness and bending can be analyzed more directly, since the DNA conformation approximates a rigid rod rather than a random coil. And small nucleic acids, with only a few dozen base pairs (bp), are most suitable for high-resolution NMR and X-ray structural studies. Large DNA molecules can be broken into shorter ones in a variety of ways. Unless extreme care is taken, high molecular weight DNA will be broken by the shear generated by standard laboratory manipulations, such as pouring and pipeting, into fragments of about 5 million molecular weight. Further breakage, down to about 1 million (1500 bp) or somewhat below, can be achieved by more vigorous mechanical shearing, for example, by repeated passage through a hypodermic syringe, or processing in a homogenizer or blender. Even smaller fragments, down to 50–100 bp, can be obtained by sonication for several hours. All of these procedures, however, produce a broad range of molecular sizes, and separation to obtain narrow length fractions is difficult.

To obtain more monodisperse preparations, many workers have turned to nuclease cleavage of chromatin. As described in Chapter 14, the central structural element of chromatin is the nucleosome, consisting of about 150 bp of DNA wrapped around a histone octamer. An additional several dozen bp of DNA, complexed with histone H1, constitute a spacer region between nucleosomes. By a carefully regulated sequence of enzymatic digestions and deproteinizations (Wang et al., 1990), it is possible to obtain hundreds of milligrams, or even grams, of 150 bp mononucleosomal DNA from conveniently available tissues rich in chromatin, such as calf thymus or chicken erythrocytes. The range of lengths in these preparations is quite narrow, about 150 ± 10 bp. This length is useful for many physical studies, since it corresponds to about one persistence length of DNA, and therefore exhibits measurable but not overwhelming flexibility (Chapter 9). Unfortunately, multiples of this fundamental length are much harder to purify in useful quantities, and the mononucleosomal DNA is quite random in base composition and sequence.

If molecules with different though defined lengths or with known sequences are desired, workers generally resort either to large-scale production of plasmids for DNAs in the range around 1 kbp, or to chemical synthesis for the range 10–50 bp. Plasmids may be engineered by recombinant DNA technology to contain desired sequences, and then overexpressed in *Escherichia coli* to yield a few milligrams of pure DNA (Lewis et al., 1985). Density gradient ultracentrifugation, which is discussed in Chapter 9, is an important technique for plasmid purification. The plasmids may either be used intact, or the sequences of interest may be excised by digestion with a restriction enzyme. In the latter case, though, it is not always straightforward to separate the desired sequence from the other fragments.

To obtain relatively short sequences, such as those used in X-ray crystallographic or NMR studies, it is generally most direct to use phosphoramidite solid-phase oligonucleotide synthesis (Itakura et al., 1984). A reactive phosphoramidite group on the 3′ position of deoxyribose is coupled to the 5′-OH of the next monomer unit. The amino groups of the bases are protected. At the end of the synthesis the protected bases must be deblocked and the products purified. Using commercial synthesizers, it is easy to produce milligram quantities of oligonucleotides one hundred bp long.

The polymerase chain reaction (PCR) is beginning to be used to produce physical quantities of DNA. This technique was originally developed to amplify genomic sequences for detailed analysis (Mullis and Faloona, 1987). Specific synthesis of DNA *in vitro,* via a polymerase catalyzed chain reaction, can amplify a particular sequence up to 1 million-fold. It employs two synthetic oligonucleotides as primers for enzyme-catalyzed DNA synthesis, using as templates the complementary strands in the region between the priming sites. Exponential increase of the sequence of interest is achieved by repeated cycles of thermal denaturation of the duplex, annealing of the primers, and polymerase-catalyzed synthesis. The key to practical success of the method is a highly processive and thermostable DNA polymerase from *Thermus aquaticus,* known as the *Taq* polymerase. It can produce microgram amounts from a single copy. To scale up to milligrams requires substantial development, including use of many reaction vessels in parallel; but the scale-up is technically feasible.

To obtain physical quantities of RNA is either easier or harder than for DNA, depending on the material desired. Certain naturally occurring RNAs, such as rRNAs and the various tRNAs, can be readily isolated in large quantities and are commercially available. To produce oligonucleotides, the same phosphoramidite solid-state synthetic chemistry is employed as for DNA, though the protection is more difficult because of the 2′-OH. Synthetic homopolymers with a random length distribution may be synthesized using polynucleotide phosphorylase. Large quantities of monodisperse RNAs with a desired sequence in the size range from a dozen to a few hundred bases can be obtained by RNA polymerase transcription from a synthetic DNA template (Milligan et al., 1987).

5. THE PERIODICAL LITERATURE OF NUCLEIC ACIDS

In such an active area, keeping up with the literature is difficult but necessary. The major journals that carry work related to physical chemistry of nucleic acids are *Biochemistry, Biophysical Journal, Biopolymers, Journal of Biological Chemistry, Journal of*

Biomolecular Structure and Dynamics, Journal of Chemical Physics, Journal of Molecular Biology, Journal of the American Chemical Society, Nature, Nature Structural Biology, Nucleic Acids Research, Proceedings of the National Academy of Sciences USA, RNA, Science, and *Structure.*

Also useful are the review series *Annual Review of Biophysics and Biophysical Chemistry* (now changed to *Annual Review of Biophysics and Biomolecular Structure*), *Annual Review of Biochemistry, Nucleic Acids and Molecular Biology, Current Opinion in Structural Biology, Progress in Nucleic Acid Research and Molecular Biology,* and *Quarterly Reviews of Biophysics.*

6. BOOKS AND MONOGRAPHS

Although research on nucleic acids advances at a very rapid rate, modern textbooks and monographs still provide solid background information. Some of the most useful are listed in Section 6.1.

6.1 Basic Texts—Molecular Biology and Biochemistry

Adams, R. L. P., Knowler, J. T., and Leader, D. P. (1992) *The Biochemistry of the Nucleic Acids*, 11th ed. Chapman and Hall, London.

Albert, B., Bray, D., Lewis, J., Raff, M., Roberts, K., and Watson, J.D. (1994). *Molecular Biology of the Cell*, Garland, New York.

Kornberg, A. and Baker, T. (1991). *DNA Replication*, 2nd ed. Freeman, New York.

Lewin, B. (1994). *Genes VI*, Oxford University Press, Oxford, UK.

Lodish, H., Baltimore, D., Berk, A., Zipursky, S.L., Matsudaira, P. and Darnell, J. (1995). *Molecular Cell Biology*, Scientific American Books, New York.

Sambrook, J., Fritsch, E. F., and Maniatis, T. (1989). *Molecular Cloning: A Laboratory Manual*, Cold Spring Harbor Press, Cold Spring Harbor, NY.

Singer, M. and Berg, P. (1991). *Genes and Genomes*, University Science Books, Sausalito, CA.

Singer, M. and Berg, P. (1997). *Exploring Genetic Mechanisms*, University Science Books, Sausalito, CA.

Stryer, L. (1994). *Biochemistry*, 4th ed., Freeman, New York.

Watson, J. D., Hopkins, N. H., Roberts, J. W., Steitz, J. A. and Weiner, A. M. (1987). Molecular Biology of the Gene, 4th Edition. Benjamin/Cummings Publishing Company, Menlo Park, CA.

6.2 Monographs on Physical Chemistry of Nucleic Acids

Blackburn, G. M. and Gait, M. J., Eds. (1990). *Nucleic Acids in Chemistry and Biology*. IRL Press, Oxford, UK.

Bloomfield, V. A., Crothers, D. M., and Tinoco, I., Jr. (1974). *Physical Chemistry of Nucleic Acids*. Harper & Row, New York.

Neidle, S. (Ed.) (1999). *Oxford Handbook of Nucleic Acid Structure*. Oxford University Press, UK.

Saenger, W. (1984). *Principles of Nucleic Acid Structure*. Springer-Verlag, New York.

Sinden, R. R. (1994). *DNA Structure and Function*. Academic, San Diego.

6.3 Physical Biochemistry, and Physical Chemistry with Biochemical Emphasis

Cantor, C. R. and Schimmel, P. R. (1980). *Biophysical Chemistry*. Freeman, San Francisco, CA.

Eisenberg, D. and Crothers, D. (1979). *Physical Chemistry with Applications to the Life Sciences*. Benjamin/Cummings Publishing Company, Menlo Park, CA.

Freifelder, D. (1982). *Physical Biochemistry: Applications to Biochemistry and Molecular Biology*, 2nd ed. Freeman, San Francisco, CA.

Tanford, C. (1961). *Physical Chemistry of Macromolecules*. Wiley, New York.

Tinoco, I., Jr., Sauer, K., and Wang, J. C. (1995). *Physical Chemistry: Principles and Applications in Biological Sciences*. 3rd ed., Prentice-Hall, Englewood Cliffs, NJ.

Van Holde, K. E., Johnson, W. C., and Ho, P. S. (1998). *Principles of Physical Biochemistry*. Prentice-Hall, Englewood Cliffs, NJ.

6.4 Historical Monographs

Chambers, D. A., Ed. (1995). *DNA: The Double Helix*, New York Academy of Sciences, New York.

Freifelder, D. (1978). *The DNA Molecule: Structure and Properties. Original Papers, Analyses, and Problems*. Freeman, San Francisco, CA.

Judson, H. F. (1979). *The Eighth Day of Creation: Makers of the Revolution in Biology*. Simon and Schuster, New York.

Olby, R. (1974). *The Path to the Double Helix*. University of Washington Press, Seattle, WA.

Watson, J. D. (1968). *The Double Helix*. Atheneum, Boston, MA.

7. COMPUTER ANALYSIS OF SEQUENCES AND STRUCTURES

Large data bases of sequences and structures, and powerful computer programs for analyzing them, have had major impact on research on the structure and function of nucleic acids. Some useful books and review articles follow:

Bishop, M. J. and Rawlings, C. J., Eds. (1987). *Nucleic acid and protein sequence determination: a practical approach.* IRL Press, Oxford, UK.

Karlin, S., Bucher, P., Brendel, V., and Altschul, S. F. (1991). Statistical methods and insights for protein and DNA sequences, *Annu. Rev. Biophys. Biophys. Chem.* **20**, 175–203.

Pearson, W. R. and Lipman, D. J. (1988). Improved Tools for Biological Sequence Comparison. *Proc. Natl. Acad. Sci. USA.* **85**: 2444–2448.

Sankoff D. and Kruskal J. B., eds. (1983). *Time Warps, String Edits and Macromolecules: The Theory and Practice of Sequence Comparison.* Addison-Wesley, Reading, MA, 382 pp.

Stormo, G. D. (1988). Computer methods for analyzing sequence recognition of nucleic acids. *Annu. Rev. Biophys. Biophys. Chem.* **17**, 241–263.

Also noteworthy is the *Nucleic Acids Database* (Berman et al., 1992), which assembles and makes available structural information on nucleic acids through an interactive on-line data base. Structures are selected by making choices based on a large variety of structural and experimental characteristics. The query is processed, and a results list is presented. From this list, the user can obtain the structural coordinates in several formats, view the structure itself online or remotely, and obtain other information. The World Wide Web address of the Nucleic Acids Database is http://ndbserver.rutgers.edu:80/.

References

Avery, O., MacLeod, C., and McCarty, M. (1944). Studies on the Chemical Nature of the Substance Inducing Transformation of Pneumococcal Types. *J. Expt. Med.* **79**, 137–157.

Berman, H. M., Olson, W. K., Beveridge, E. L., Westbrook, J., Gelbin, A., Demeny, T., Hsieh, S.-H., Srinivasan, A. R., and Schneider, B. (1992). The nucleic acid database. A comprehensive relational database of three-dimensional structures of nucleic acids, *Biophys. J.* **63**, 751–759.

Bloomfield, V. A., Crothers, D. M., and Tinoco, I., Jr. (1974). *Physical Chemistry of Nucleic Acids*. Harper & Row, New York.

Cech, T. R. (1987). The chemistry of self-splicing RNA and RNA enzymes. *Science* **236**, 1532–1539.

Chargaff, E. (1950). Chemical Specificity of Nucleic Acids and Mechanisms of their Enzymatic Degradation. *Experientia* **6**, 201–209.

Doudna, J. A. and Szostak, J. W. (1989). RNA-catalyzed synthesis of complementary-strand RNA. *Nature* **339**, 519–522.

Franklin, R. E. andGosling, R. (1953). Molecular Configuration in Sodium Thymonucleate. *Nature* **171**, 740–741.

Gesteland, R. F. and Atkins, J. F. (1993). *The RNA World*, Cold Spring Harbor Press, Cold Spring Harbor, NY.

Hershey, A. D. and Chase, M. (1952). Independent Functions of Viral Protein and Nucleic Acid in Growth of Bacteria. *J. Gen. Physiol.* **36**, 39–56.

Itakura, K., Rossi, J. J., and Wallace, R. B. (1984). Synthesis and Use of Synthetic Oligonucleotides, *Annu. Rev. Biochem.* **53**, 323–356.

Lewis, R. J., Huang, J. H., and Pecora, R. (1985). Three Plasmids Constructed for the Production of Monodisperse Semistiff DNA Samples. *Macromolecules.* **18**, 1530–1534.

Meselson, M. and Stahl, F. W. (1957). The Replication of DNA in *Escherichia Coli. Proc. Natl. Acad. Sci. USA* **44**, 671–682.

Milligan, J. F., Groebe, D. R., Witherell, G. W., and Uhlenbeck, O. C. (1987). Oligoribonucleotide synthesis using T7 RNA polymerase and synthetic DNA templates. *Nucleic Acids Res.* **15**, 8783–8798.

Mullis, K. B. and Faloona, F. (1987). Specific synthesis of DNA in vitro via a polymerase-catalyzed chain reaction. *Methods Enzymol.* **155**, 335–350.

Sharp, P. A. (1987). Splicing of messenger RNA precursors. *Science* **235**, 766–771.

Wang, L., Ferrari, M., and Bloomfield, V. A. (1990). Large-Scale Preparation of Mononucleosomal DNA from Calf Thymus for Biophysical Studies. *BioTechniques* **9**, 24–27.

Watson, J. D. and Crick, F. H. C. (1953). Molecular Structure of Nucleic Acids. A Structure for Deoxyribose Nucleic Acid. *Nature (London)* 171, 737–738.

Wilkins, M. H. F., Stocker, A. R., and Wilson, H. R. (1953). Molecular Structure of Desoxypentose Nucleic Acids. *Nature (London)* **171**, 738–740.

CHAPTER 2

Bases, Nucleosides, and Nucleotides

1. STRUCTURES

1.1 Configuration and Conformation

Configuration refers to the covalent bonding in a molecule; conformation refers to the three-dimensional (3D) structure of a molecule. The configuration is constant for a molecule; the conformation depends on the environment (temperature, pH, salt concentration, interactions with other molecules, etc.). The conformation is dynamic; there is always a range of different structures that the molecule samples at equilibrium. The primary structure is the configuration; the secondary structure and tertiary structure are the conformation.

The bases in DNA and RNA that are responsible for coding the genetic information are adenine, cytosine, guanine, and thymine for DNA and adenine, cytosine, guanine, and uracil for RNA. The configurations for the bases are given in Figure 2-1. Each base is essentially planar and its conformations are limited. Of course, all of the bonds undergo bending and stretching vibrations, which leads to in-plane and out-of-plane ring breathing modes and amino group umbrella motions. There is rotation about the bond joining each amino group to a base. In aqueous solution nuclear magnetic resonance (NMR) studies show that the cytosine amino group rotates at a rate of about 35 revolutions per second (rps) ($35\ s^{-1}$) at 0° C; adenine and guanine amino groups rotate much faster (McConnell and Seawell, 1973). Hydrogen bonding of an amino hydrogen slows rotation of the amino group. The methyl group of thymine rotates at the much faster rate of about $10^{10}\ s^{-1}$ (Jardetsky and Roberts, 1981).

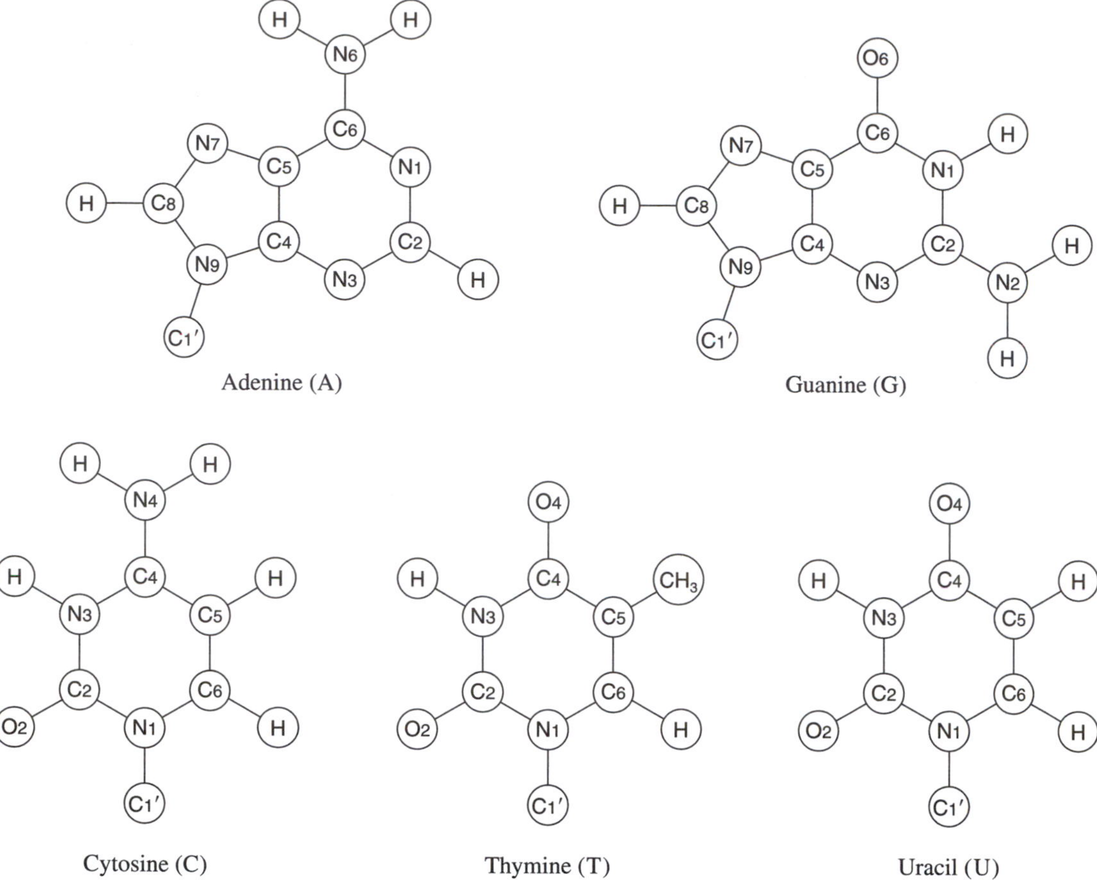

Figure 2-1
The coding bases in DNA and RNA. Adenine (A) is 6-amino purine. Guanine (G) is 2-amino, 6-keto purine; it has an imino group at N1. Purines are connected to DNA or RNA sugars at N9 in the imidazole rings. Cytosine (C) is 2-keto, 4-amino pyrimidine. Uracil (U) (found in RNA) is 2,4-diketo pyrimidine; it contains an imino group at N3. Thymine (T) (found in DNA) is 5-methyl uracil. The pyrimidines are connected to DNA or RNA sugars at N1.

A nucleoside is formed by attaching the base to a sugar. The configurations of the ribonucleosides found in RNA and of the deoxyribonucleosides found in DNA are shown in Figure 2-2. The pyrimidines are linked at N1 to the sugar; the purines are linked at N9. The four asymmetric carbon atoms (C1′, C2′, C3′, C4′) in a ribonucleoside can produce 16 stereoisomers. The only one found in RNA is β-D-ribosyl nucleoside. The configurations at C2′ and C3′ relative to C4′ shown in Figure 2-2 specify the sugar as ribose. (The other possibilities are arabinose, xylose, and lyxose.) The base is attached in the β configuration at C1′, which means it is on the same side of the ribose ring as the 5′ carbon. Each sugar has two mirror images; only the D form occurs naturally in RNA. The four β-D-ribosyl nucleosides (ribonucleosides) of the coding bases are adenosine, cytidine, guanosine and uridine. When thymidine occurs as a ribonucleoside it is called ribosylthymine or 5-methyluridine.

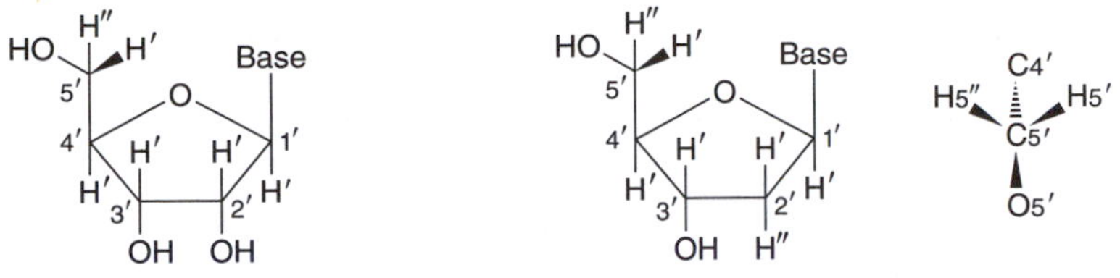

Ribonucleoside
(β-D-ribosyl nucleoside)

Deoxyribonucleoside
(β-D-2′-deoxyribosyl nucleoside)

Figure 2-2
The ribonucleosides found in RNA and deoxyribonucleosides found in DNA. The numbers for the sugar atoms are primed to distinguish them from base atoms. When there are two protons on one carbon, they are identified by primes and double primes. H2″ on deoxyribose replaces the 2′–OH group of ribose. H5″ and H5′ are designated as shown in the figure. The H5″, H5′ and O5′ are in clockwise order when one looks from C5′ toward C4′.

Deoxyribonucleic acid contains 2′ deoxyribonucleosides with the same configuration as in RNA; they are β-D-2′-deoxyribosyl nucleosides. The coding bases form deoxyadenosine, deoxycytidine, deoxyguanosine, and thymidine. Replacement of the 2′–OH of RNA by a hydrogen in DNA has major ramifications. For example, DNA is much less susceptible to hydrolysis than RNA and cannot have branched polynucleotides involving the 2′-, 3′- and 5′ -OH groups. Consequently, DNA should have more conformational flexibility than RNA because it lacks the steric hindrance of the hydroxyl in RNA, but it also lacks the extra hydrogen-bonding site, which gives RNA greater possibilities for specific interactions.

Nucleotides are formed by adding phosphate groups to nucleosides. In ribonucleosides phosphate esters can be made at the 2′-, 3′- and 5′ -OH groups; in deoxyribonucleosides, only 3′- and 5′-phosphate esters can be made. The common ribonucleotides are adenylic acid, cytidylic acid, guanylic acid, and uridylic acid; the common deoxyribonucleotides are deoxyadenylic acid, deoxycytidylic acid, deoxyguanylic acid, and thymidylic acid. The position of the phosphate group is specified by a 2′, 3′, or 5′ before the name of the compound. Alternatively, the molecules can be named as derivatives of nucleosides, such as adenosine-3′-phosphate. The abbreviation for 5′-nucleotides is pN; for 3′ nucleotides it is Np. Nucleosides with more than one phosphate are also found. Examples are 5′-diphosphates (ppN), 5′-triphosphates (pppN), 2′, 5′-diphosphates, and so on. Ribonucleosides can form 2′, 3′-diphosphates and 2′, 3′-cyclic monophosphates; both ribo- and deoxyribo-nucleosides can form 3′, 5′-cyclic monophosphates.

Exonucleases are phosphodiesterases that hydrolyze nucleic acids starting at the 5′- or the 3′-end of the chain to produce nucleotides. Some exonucleases are specific for DNA, some for RNA, and some (e.g., snake venom phosphodiesterase) attack both DNA and RNA. Most exonucleases produce 5′-nucleotides, but some yield 3′-nucleotides. Complete hydrolysis of RNA by OH^- ion leaves both 2′- and 3′-nucleotides. The reaction proceeds by first forming the 2′, 3′-cyclic phosphate, then cleaving either bond to produce a mixture of the two nucleotides.

1.2 Naturally Occurring Modified Bases in Transfer RNA

Modified bases occur in many nucleic acids, but by far the largest number of different modified bases occur in transfer RNAs (tRNAs). Some are slight modifications of the normal bases; others are hypermodified. All appear to be modified by specific enzymes after the tRNA is synthesized (Kline and Söll, 1982). The simple derivatives found in tRNA include

1. Adenine derivatives: 1-methyladenosine, N^6-methyladenosine, inosine (deaminated adenosine), 1-methylinosine.
2. Cytosine derivatives: 3-methylcytidine, 5-methylcytidine, 2-thiocytidine, N^4-acetylcytidine.
3. Guanine derivatives: 1-methylguanosine, N^2-methylguanosine, N^2, N^2-dimethylguanosine, 7-methylguanosine.
4. Uracil derivatives: ribosylthymine (5-methyluridine), 5-methoxyuridine, 5,6-dihydrouridine, 4-thiouridine, 5-methyl-2-thiouridine, pseudouridine (uracil is attached to ribose at the C5).

Much more exotic modified bases also occur (Kline and Söll, 1982). The adenosine derivatives include N^6 - (Δ^2-isopentenyl) adenosine (i^6A) and N^6 - threoninocarbonyl adenosine (t^6A). Guanosine derivatives are fluorescent and occur at the 3′-end of the anticodon in tRNA. The roles of the modified bases in tRNA function are generally not known; those that are known are reviewed by Agris (1995). Phenylalanine tRNA transcribed from DNA *in vitro* and containing no modified bases is recognized by tRNA synthetase and charged with phenylalanine as well as the natural tRNA is (Sampson and Uhlenbeck, 1988).

1.3 DNA Methylation

Natural DNA contains 5-methylcytosine, N^6-methyladenine, N^4-methylcytosine and 5-hydroxymethyluracil as minor components; the first two compounds are the most common. The methylation is done sequence-specifically after the DNA is synthesized. The amount of methylation varies among organisms; some have no methylation, but some plants have 30% of cytosines methylated. Some bacteriophages have 100% of their bases replaced by modified bases. For example, the T-even phages have cytosine completely replaced by 5-hydroxymethylcytosine. Here the modified nucleotide is synthesized first and then incorporated into the DNA by DNA polymerase. The methyl groups attached to the DNA bases come from *S*-adenosylmethionine; this compound has the S of methionine replacing the 5′-OH of adenosine. The enzymes that catalyze the reactions are called DNA methyl transferases or DNA methylases. Books by Adams and Burdon (1985) and Razin et al. (1984) describe the biochemistry and molecular biology of methylation in great detail.

The structural effects of methylation have not been studied extensively. 5-Methylcytosine favors formation of a left-handed DNA double helix (Z-DNA) in alternating CG sequences. N^6-methyladenine can exist in two isomers, only one of which can base pair. However, NMR studies show that rotation of the methylamino group is rapid and that the DNA double helix is only slightly destabilized (Engel and von Hippel, 1978).

The biological functions of methylation are many and are not well understood. In bacteria, methylation is used to distinguish self-DNA from foreign-DNA. Immediately after DNA replication, the new strand is methylated; any duplex DNA that is not methylated at the appropriate sequences is cut by restriction enzymes. Bacterial methylases are part of a restriction-modification system that takes the place of the immune system in higher organisms. Methylation is also part of base–base mismatch repair. When DNA polymerase makes a mistake in replicating DNA, DNA repair enzymes recognize the nonmethylated strand as the new strand that must be changed to produce the proper sequence. 5-Methylcytosine is the only modified base in the DNA of higher eukaryotes; in animals, it is predominantly in the sequence CpG. The role of methylation in the control of gene expression is actively being studied. There is evidence that undermethylation increases transcription of the DNA, but the mechanism is not known and many genes do not have this type of control.

1.4 Torsion Angles

The conformations of nucleosides and nucleotides depend on the torsion angles for rotation around each bond; their definitions, following IUPAC recommendations (IUPAC-IUB, 1983), are shown in Figure 2-3. Bond stretching and bond-angle bending occur and change with conformation, but the main differences among conformations is provided by differences in the torsion angles. There are seven torsion angles per nucleotide that must be specified to characterize the conformation (secondary structure) of a nucleic acid (see Fig. 2-3). The convention for defining a torsion angle is given in Figure 2-4. The adjacent backbone atoms A–B–C–D are in 5′ to 3′ order (Fig. 2-3); for example, angle β refers to chain atoms P5′–O5′–C5′–C4′. A positive angle is obtained for a clockwise rotation of D when one looks from B toward C. The zero angle is taken

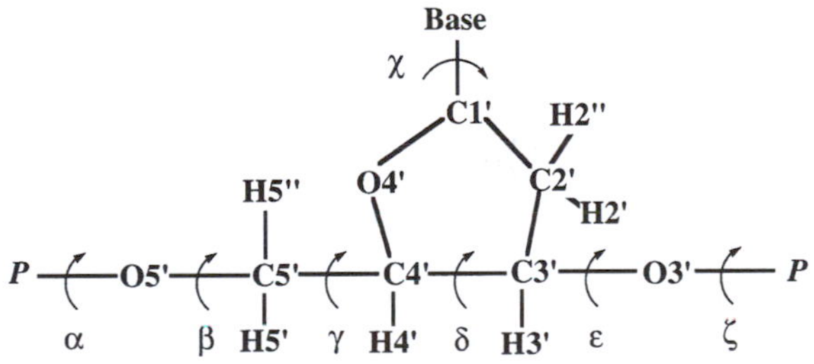

Figure 2-3
The seven torsion angles that specify the conformation of each nucleotide in a polynucleotide chain. The sugar–phosphate backbone is characterized by six torsion angles (α, β, γ, δ, ε, ζ). The orientation of the base relative to the sugar is specified by the glycosidic torsion angle, χ. The conformation of the sugar ring must also be described (see Fig. 2-5).

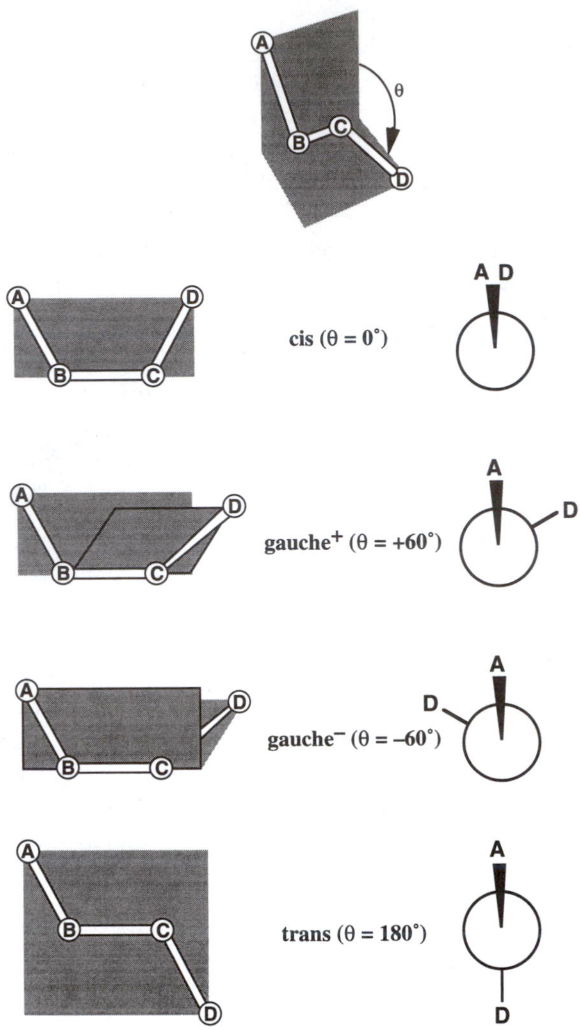

Figure 2-4
Each torsion angle is defined by the angle between the two planes specified by atoms A–B–C and B–C–D. An angle of 0° specifies all atoms in the same plane with A and D eclipsed (cis conformation). If one rotates D clockwise while looking from B to C, a positive torsion angle results; the trans planar conformation has torsion angles of $\pm 180°$. Gauche conformations have torsion angles of $+60°$ or $-60°$. The Newman projections for tetrahedrally bonded atoms in these conformations are shown on the right side of the figure.

as the cis conformation; trans (t) is $\pm 180°$. Gauche forms are labeled gauche$^+$ (g$^+$, $\theta = +60°$) and gauche$^-$ (g$^-$, $\theta = -60°$). The sugar–phosphate backbone conformation is quite variable, but conformations with the bases stacked tend to have $\alpha(g^-)$, β(t), γ(g$^+$), ε(t), ζ(g$^-$). The torsion angle δ is directly related to the sugar conformation and will be discussed separately.

Ribose and deoxyribose are five-membered rings, therefore five torsion angles can be specified. The convention is that torsion angle C4′–O4′–C1′–C2′ is ν_0, and that ν_1, ν_2,

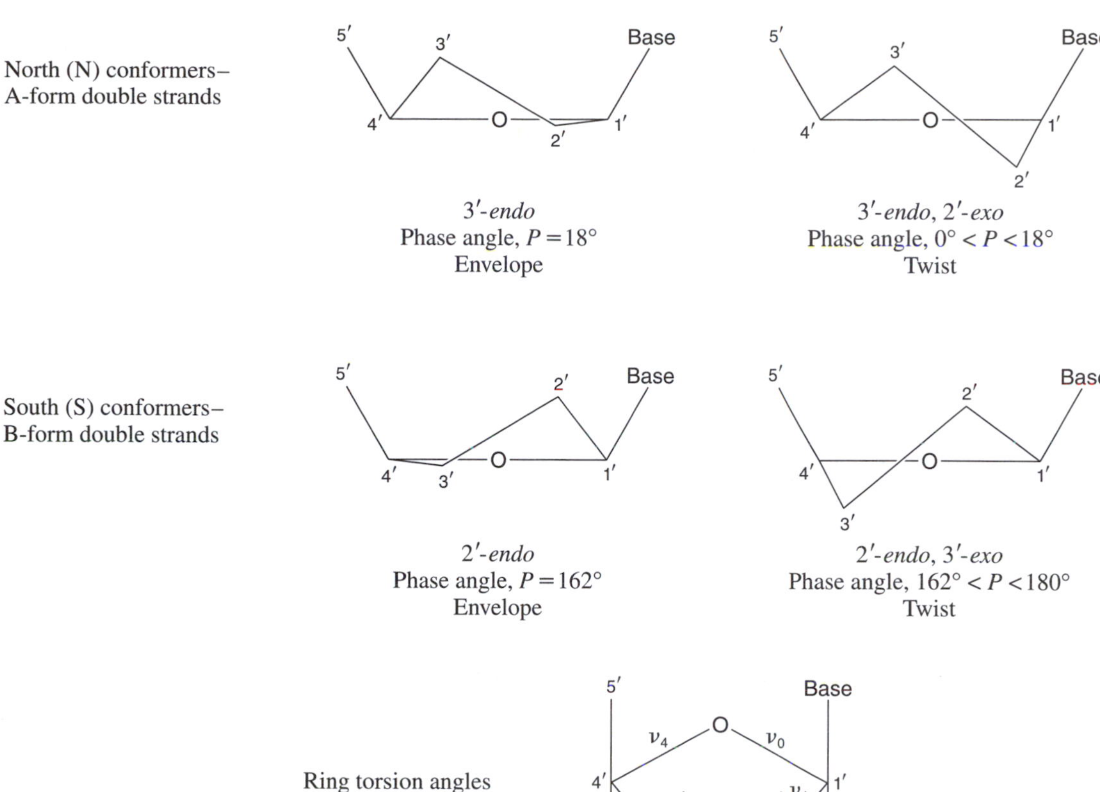

Figure 2-5
The conformations of ribose and deoxyribose found in nucleotides. The conformation can be specified by the atoms located out of the plane defined by C4′–O4′–C1′ (e.g., 3′-*endo*-2′-*exo* or 2′-*endo*-3′-*exo*), by the pseudorotation phase angle, P (N conformers have P near 0°, S conformers have P near 180°). Five torsion angles (ν_j) specify the conformation of a five-membered ring, but for equal bond angles and bond lengths two parameters are independent: pseudorotation phase angle, P, and maximum torsion angle, Φ_m. Note that ν_3 specifies the same torsion angle as δ of Figure 2-3, but that it has a different magnitude because δ depends on the chain atoms C5′–C4′–C3′–03′, whereas ν_3 depends on C2′–C3′–C4′–04′; $\delta = \nu_3 + 125°$.

ν_3, ν_4 continue clockwise around the ring (see Fig. 2-5). A planar five-membered sugar ring (all torsion angles equal to zero) is sterically and energetically very unfavorable; it is not found. By pulling one atom out of the plane, we release the strain, lower the energy, and produce a stable conformation. The conformation with the 3′ carbon out of the plane of C4′–O4′–C1′–C2′ ($\nu_0 = 0$) and on the same side of the plane as the base is called C3′-*endo* (a C3′-*exo* conformation would have the 3′ carbon on the opposite side of the plane). The conformation with the 2′ carbon out of the plane, and on the same side as the base, is C2′-*endo*. These two conformations approximate those found in DNA and RNA. In double-stranded B-form DNA, the deoxyribose sugars are approximately C2′-*endo*; in double-stranded A-form RNA, or A-form DNA, the ribose sugars are approximately C3′-*endo*. The different forms of double-stranded nucleic acids (A, B, . . . Z) are described in Chapter 3. In general, both the 2′ and 3′ carbons will not be in the plane formed by C4′–O4′–C1′; furthermore, other sugar pucker conformations (such as O4′-*endo* and C1′-*exo*) are also found in B-DNA crystal structures (Fratini et al., 1982).

Figure 2-5 illustrates the types of conformations found in polynucleotides; either the 3′ carbon (C3′-*endo* conformers) or the 2′ carbon (C2′-*endo* conformers) is on the same side of the C4′–O4′–C1′ plane as the base and the 5′ carbon. The displacement from the plane is 0.5 Å or less. The other carbon is either in the plane (a pure endo conformation), or on the other side of the plane (a mixed *endo*, *exo* conformation). Other conformations are possible, such as pure C2′ or C3′-*exo*, or C4′-, or O4′-, or C1′ -*endo* or -*exo*.

The conformation of a five-membered ring can be described (Altona, 1982, 1986) by only two parameters because of geometrical constraints in a five-membered ring of identical bond angles and bond lengths. All five torsion angles can be related to these two parameters. Spectroscopists studying cyclopentane realized that different conformations were equivalent to rotation about an axis perpendicular to the pentagon. For example, if one carbon was out of plane, rotation by $360°/5 = 72°$ was the same as moving that carbon back into the plane, and moving the next carbon out of plane. Thus a pseudorotation phase angle (P) and a maximum torsion angle (Φ_m, amplitude of pucker) were chosen to specify five torsion angles in five-membered rings. The actual torsion angles, ν_j, are given by

$$\nu_j = \Phi_m \cos[P + 144°(j - 2)], \qquad (j = 0, 1, 2, 3, 4) \tag{2-1}$$

Trigonometric identities show that the sum of the five torsion angles is equal to zero, as required for a ring. The pseudorotation phase angle, P, is a positive angle between 0° and 360°; it specifies which torsion angle is the maximum and which two atoms are out of the plane of the other three atoms. Figure 2-6 shows the conformations that occur as a function of the pseudorotation angle P. Phase angle $P = 0°$ (called north, N) produces the C3′-*endo*, C2′-*exo* conformation with the two atoms equidistant from the plane. The ring is twisted symmetrically. As P increases, the 2′-*exo* atom approaches the plane. For P between 0° and 18°, the conformations are twisted and favor C3′-*endo*. For $P = 18°$ the conformation is pure C3′-*endo*; the ribose looks like a letter envelope with the flap up. For P near 180° (south, S), the conformations are 2′-*endo* type; $P = 162°$ is pure C2′-*endo*. Crystal structures mainly, but not exclusively, reveal two ranges for the pseudorotation angle (deLeeuw et al., 1980): N-type, 3′-*endo* type ($P \cong 0° - 18°$), and S-type, 2′-*endo* type ($P \cong 162° - 180°$). The maximum torsion angle, Φ_m, (pucker amplitude) varies between 34° and 42°. In a crystalline B-DNA dodecanucleotide duplex with complementary sequence d(CGCGAATTCGCG)$_2$, there are 24 independent nucleotides. Eleven were found to be C2′-*endo*, six were O4′-*endo*,

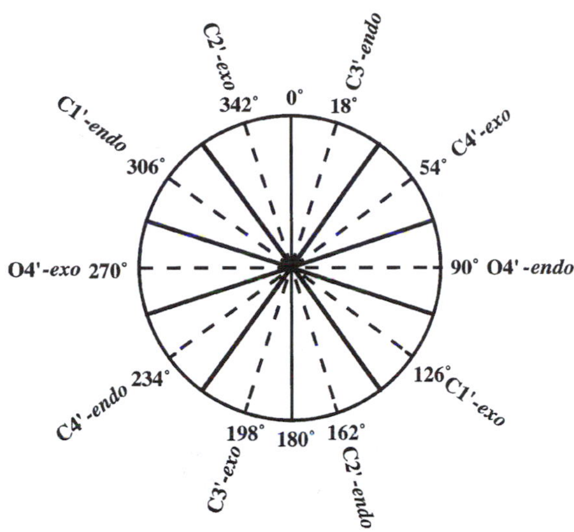

Figure 2-6
The pseudorotation phase angle, P, characterizes the pucker in a five-membered ring. Nearly the entire range of P values have been found in nucleic acids, but ribose and deoxyribose nucleotides are mainly found with a range near $P = 18°$ (C3′-*endo*) and $P = 162°$ (C2′-*endo*). As the pseudorotation angle increases from 0° to 360°, the sugar conformation alternates between symmetrical twist forms, with two adjacent atoms displaced equally above and below the plane of the other three atoms, and envelope forms with only one atom out of the plane of the other four. $P = 0°$ is the symmetrical twist with C3′ on the same side of the plane as the base and C5′, and with C2′ on the opposite side (3′-*endo*-2′-*exo*); $P = 18°$ is C3′-*endo*; $P = 36°$ is C3′-*endo*-C4′-*exo*; $P = 54°$ is C4′-*exo*, etc.

four were C1′-*exo*, and three were C4′-*exo* (Fratini et al., 1982). This corresponds to a range of P from 162° (C2′-*endo*) to 54° (C4′-*exo*). Nuclear magnetic resonance data on double-stranded RNA indicate nearly exclusively C3′-*endo* conformations. Both double-stranded Z-form DNA and RNA have alternating C2′-*endo* and C3′-*endo* sugar conformations.

Sugar conformations in nucleic acids are not static. A plot of energy versus pseudorotation phase angle has broad minima near C2′-*endo* and C3′-*endo* (Olson, 1982). Which means that a range of structures can exist near these minima. Furthermore, NMR measurements in solution show that there is a rapid (ns) equilibrium between sugar conformers in single-stranded oligonucleotides and at the ends of duplex oligonucleotides. Ribonucleotides in aqueous solution are 55–85% 2′-*endo* (depending on the base and whether the phosphate is 5′ or 3′); deoxyribonucleotides are 65–85% C2′-*endo* (Altona, 1986). Stacking of the nucleotides in single strands changes the ribonucleotides to mostly C3′-*endo* conformations and leaves the deoxyribonucleotides C2′-*endo*. For the usual values of the backbone torsion angles, the distance between adjacent phosphates in a polynucleotide is about 1 Å larger with C2′-*endo* conformations (Saenger, 1984, p. 237)

In a nucleoside, each base is attached to the 1′ carbon of the sugar by a glycosidic bond (to N1 in pyrimidines and to N9 in purines). The torsion angle about this bond is specified by the angle χ (Fig. 2-3). The two ranges found for this angle are designated syn and anti (Fig. 2-7). The anti conformation is usually more stable; it is found in mononucleotides and in right-handed double-stranded polynucleotides. The syn conformation requires some external stabilizing force; it is found for guanosine and deoxyguanosine in left-handed double-stranded polynucleotides [Z-DNA (Rich et al., 1984), Z-RNA (Hall et al., 1984)]. In the syn conformation the six-membered ring of the purines or the carbonyl at C2 of the pyrimidines is over the sugar; this causes steric interference. The syn conformation can be favored by attaching a bulky group at the 8 position of purines or the 6 position of pyrimidines. This substitution produces even more steric hindrance in the anti conformation and favors the syn conformation. Figure 2-7 illustrates this effect.

The definition of the glycosidic torsion angle χ is illustrated in Figure 2-8. Rotation about the glycosidic bond is hindered, with purines less hindered than pyrimidines. Purine nucleosides thus have a much wider range of χ values than pyrimidine nucleosides. For purines, $\chi = 0°$ corresponds to O4′–C1′ eclipsed by N9–C4; the position of C4 determines χ. For pyrimidines, $\chi = 0°$ corresponds to O4′–C1′ eclipsed by N1–C2; the position of C2 determines χ. The anti range corresponds to χ near $-135°$. Which places C4 and the six-membered ring of the purine away from the sugar; in pyrimidines, the C2 carbonyl is away from the sugar. The syn range centers around $+45°$; this places C4 of purines or C2 of pyrimidines above the sugar. The syn conformations have higher energies than the anti conformations in pyrimidine nucleosides. The sugar conformation also has a large effect on glycosidic torsion angles. The C2′-*endo* conformation provides less steric hindrance than C3′-*endo*; this means a wider range of allowed χ values and less energy cost for a syn conformation. All 5′-nucleotides are more hindered and have a smaller range of χ angles. In B-DNA, the bases are all anti with glycosidic torsion angles of about $-160°$; in A-DNA and A-RNA the bases are anti with glycosidic angles near $-100°$. In Z-DNA and Z-RNA, the bases alternate from syn ($\chi = +65°$) to anti ($\chi = -155°$).

Detailed discussion of the correlations found among all the torsion angles in a nucleotide from X-ray data and from theoretical calculations is given in Saenger (1984, Chapter 4). As might be expected, certain torsion angles favor or exclude others. For example, there is a strong correlation between pseudorotation phase angle P and glycosidic angle χ. Bond angles and bond lengths for the bases and derivatives are also given, with extensive references to the literature where coordinates can be found.

2. ELECTRON DISTRIBUTIONS

2.1 Charge Distributions

The electronic charge distribution for each base in its ground and excited states determines its properties. In principle, we can calculate the charge distribution for each state from the Schrödinger equation, but this is not very practical. Furthermore, we are interested in properties in aqueous solutions or other solvents, which greatly increases the difficulty of accurate calculations. However, useful approximate calculations of

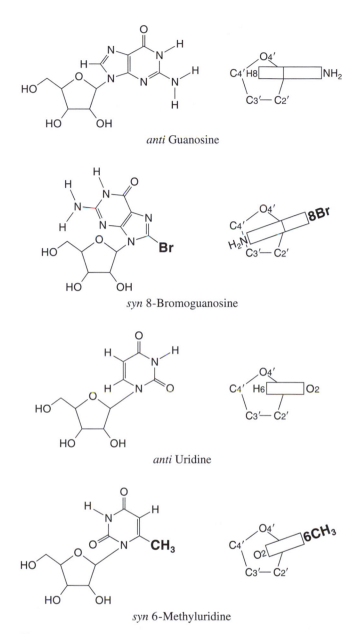

anti Guanosine

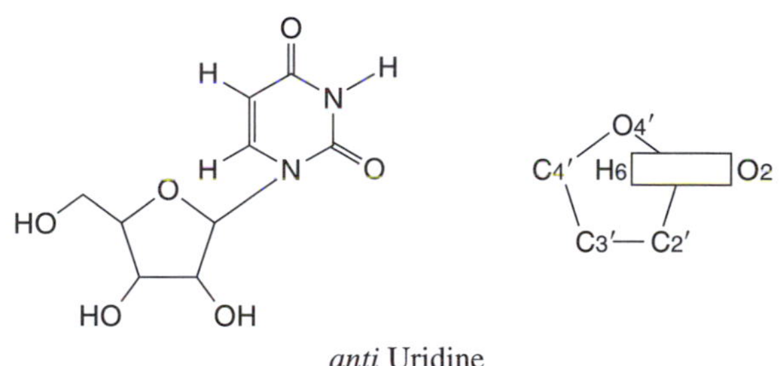

syn 8-Bromoguanosine

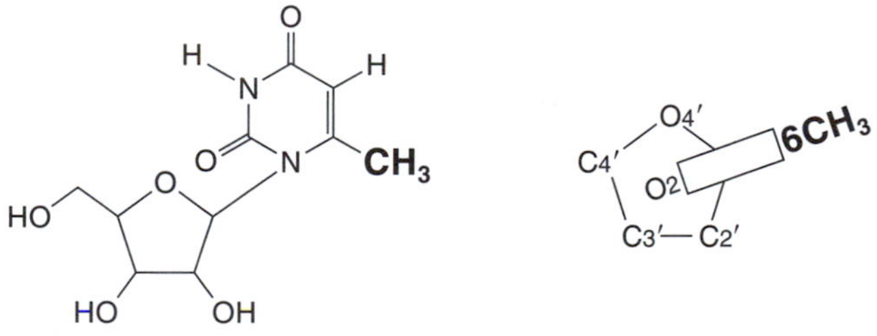

anti Uridine

syn 6-Methyluridine

Figure 2-7
Nucleosides and nucleotides are usually found in the anti conformation because of steric hindrance between the sugar and either N3 in purines or the carbonyl oxygen (O2) in pyrimidines. However, a bulky substituent at the 8 position of purines or the 6 position of pyrimidines will favor the syn conformation. A syn conformation is found for guanosine in left-handed Z-RNA and for deoxyguanosine in left-handed Z-DNA.

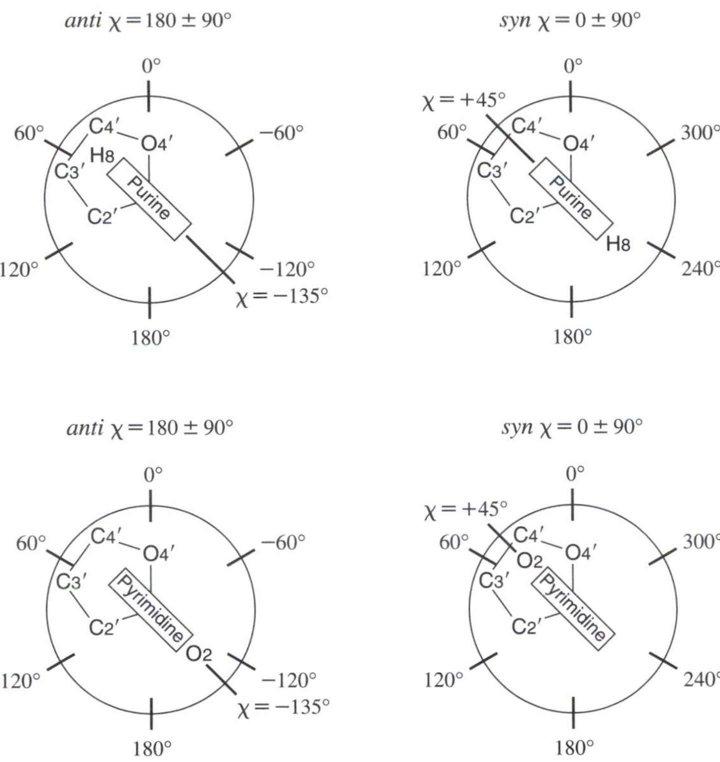

Figure 2-8
The glycosidic torsion angle χ is defined by O4′– C1′–N9–C4 for purines and O4′–C1′–N1–C2 for pyrimidines. When $\chi = 0^\circ$ the O4′–C1′ is eclipsed by the N9–C4 bond in purines and by the N1–C2 bond in pyrimidines. The syn conformations correspond to $0^\circ \pm 90^\circ$; anti conformations correspond to $180^\circ \pm 90^\circ$. In nucleotides steric hindrance limits the conformations actually found to a much narrower range of angles that depend on sugar pucker and base. The syn conformations are usually found with $\chi = 45^\circ \pm 45^\circ$; anti conformations are usually found with $\chi = -135^\circ \pm 45^\circ$.

charge distributions and resulting properties make it possible to understand experimental results, to suggest other experiments, and to use extrapolation to make quantitative predictions (see Chapter 7 and Singh and Kollman, 1984; Weiner et al., 1984; Bash et al., 1987).

The ground state charge distribution for a molecule is specified by the positions of the nuclei and by the electron density in space around the nuclei. For some purposes, the net charge of each atom in the molecule is sufficient, which can be obtained experimentally from high resolution (better than 1 Å) X-ray diffraction data on crystals of the molecules (Pearlman and Kim, 1990). However, calculations are most often used to obtain net charges (Singh and Kollman, 1984).

The attraction or repulsion for an external charge (such as a hydrogen ion) can be calculated from the net charges on the atoms of a molecule. One simply uses Coulomb's law. In this approximation, the energy of a proton (a unit charge) interacting with a molecule is

$$\text{Energy (kJ mol}^{-1}) = \frac{1389}{\varepsilon} \sum_i \frac{q_i}{R_i} \tag{2-2}$$

or

$$\text{Energy (kcal mol}^{-1}) = \frac{331.8}{\varepsilon} \sum_i \frac{q_i}{R_i} \tag{2-3}$$

where q_i is the net charge at each atom (in units of the elementary charge, such as -0.4 on a nitrogen or $+0.1$ on a carbon) and R_i is the distance in angstroms from each charge to the proton. The static dielectric constant of the medium surrounding the charges is ε; it characterizes how the medium shields the interaction of the charges. Its value is 1 for a vacuum, about 80 for liquid water at room temperature, and it is equal to the square of the refractive index ($\varepsilon \approx 2$–4) for nonpolar materials. In aqueous solution, ε will vary from about 1, if there is no solvent between the charges, to a value approaching 80 at large distances. A common procedure is to use a variable ε set equal to the magnitude of the distance between charges in angstroms.

The equation ignores the polarization of the electron distribution caused by the attractive force of the proton; this approximation worsens as the proton gets close to the molecule, but it can provide a useful picture of where protons, or other cations will tend to bond to a molecule. Figure 2-9 shows the potential energy of a unit charge in the presence of each of the bases.

Coulomb's law can also be used to calculate the electrostatic contribution to the interaction energies of two molecules (Weiner et al., 1984). One sums over the products of all net atomic charges on different molecules; each product is divided by the distance between the interacting atoms. A positive result represents a net repulsion, whereas a negative result is an attraction. In addition to electrostatic forces (charge–charge) one must add London–van der Waals forces; these are charge–polarizability and polarizability–polarizability attractions, and short-range electron overlap repulsions. Interaction energies have also been calculated using ground-state electric dipole moments to represent the charge distribution of a molecule. For two stacked bases, or coplanar hydrogen-bonded bases, the bases are too close together for the electric dipole approximation to be useful, whereas the net atomic charge (monopole) approximation is informative. Realistic energy calculations also require the incorporation of solvent effects (Bash et al., 1987). Theoretical calculations of interaction energies are discussed in Chapter 7.

2.2 Hydrogen-Bond Donors and Acceptors

The oxygen and nitrogen atoms of each base act as hydrogen-bond donors or acceptors; the sites are illustrated in Figure 2-10. Each base has donor and acceptor sites, so if there are no constraints on the geometry of the interaction each base can pair with itself or any other base through hydrogen bonding. The bases also hydrogen bond to polar amino acids in nucleic acid–protein interactions (see Chapter 12). Nucleosides

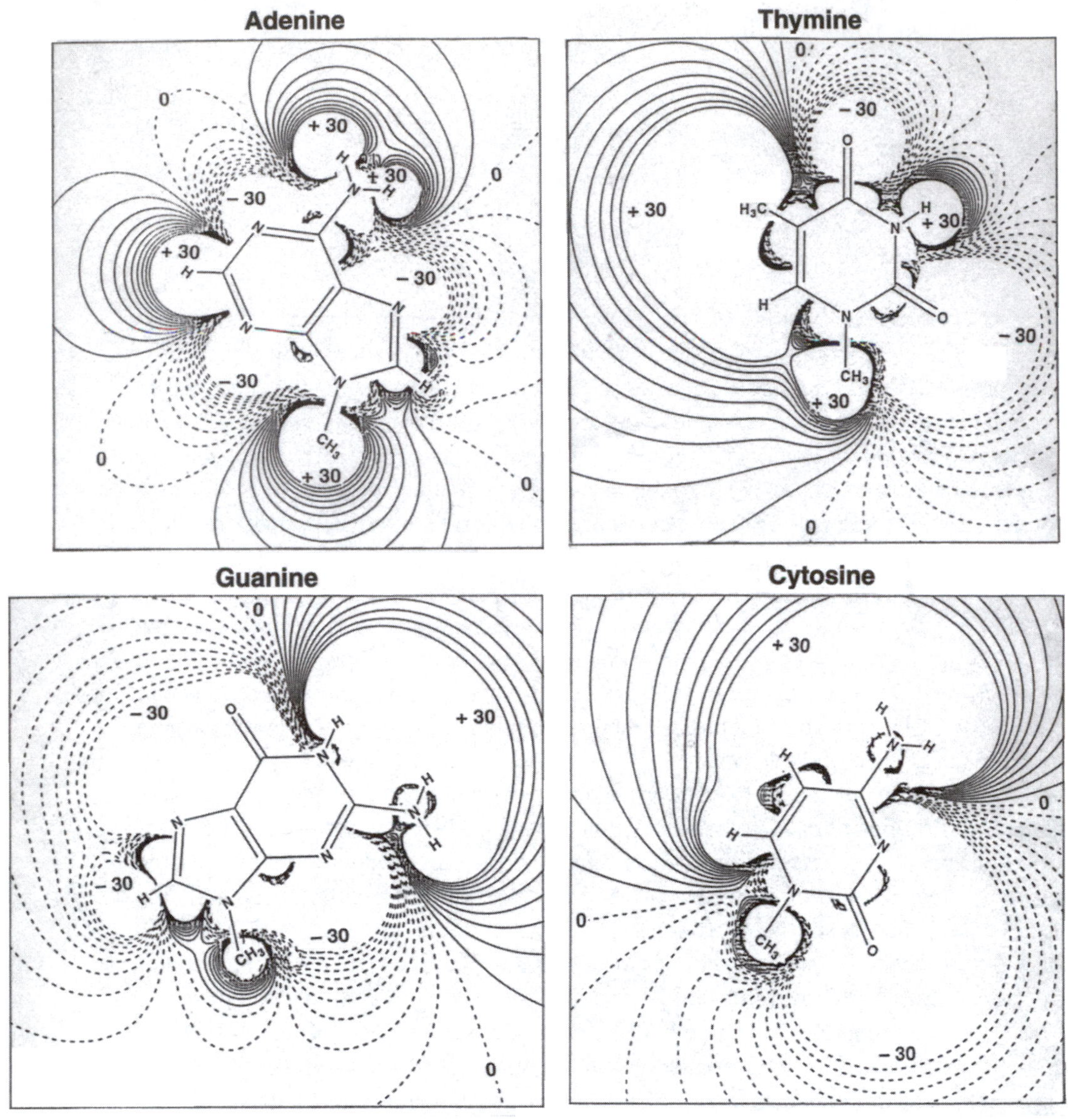

Figure 2-9
The potential energy of a unit positive charge in the electrostatic field in the plane of each base. A dielectric constant of 1 was used; the contours are drawn between +30 and −30 kcal mol^{-1} in increments of 3 kcal mol^{-1}. Solid contours are positive (repulsive to a positive charge); dotted contours are negative (attractive to a positive charge). A methyl group replaces the C1′ of the sugar group. The figures were kindly provided by Dr. David Pearlman, University of California, San Francisco, based on atomic charges determined by X-ray diffraction (see Pearlman and Kim, 1990).

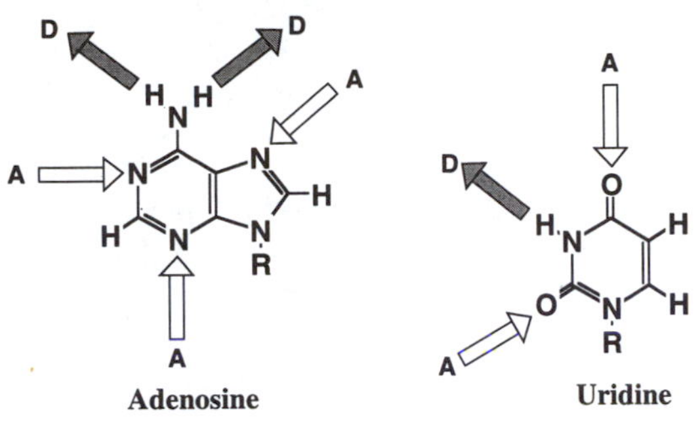

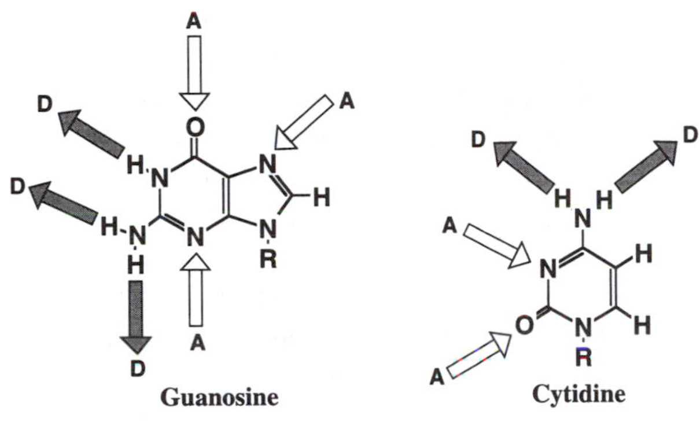

Figure 2-10
Potential hydrogen bonding sites in the bases of the nucleosides. Hydrogen-bonding donor sites (amino or imino protons) are labeled with dark arrows; hydrogen-bonding acceptor sites (carbonyl oxygens or aromatic nitrogens) are labeled with light arrows.

and nucleotides provide further hydrogen-bonding opportunities through the 2′-OH group and the phosphate group. In dilute aqueous solution, all the donor and acceptor sites of the nucleosides are hydrogen bonded to water, so the actual reaction that occurs when two bases pair, or a base hydrogen bonds to a polar amino acid, is the exchange of base–water hydrogen bonds for base–base or base–amino acid hydrogen bonds. Hydrogen bonding of bases in solution can be inferred from NMR chemical shifts and exchange lifetimes of amino or imino protons, or by infrared (IR) or Raman studies of carbonyl frequencies. In X-ray crystallography, hydrogen bonding is deduced from the distances between donor and acceptor atoms (N–N, N–O, O–O). The hydrogen-bonded atoms are within 2.8–3.0 Å; non-hydrogen-bonded atoms remain at least 3.4 Å apart. Hydrogen atoms can be located directly in high-resolution NMR structures, in very high resolution X-ray studies, or in neutron diffraction experiments.

2.3 Ionization

The sites of ionization and their pK_a values for ribonucleosides and thymidine are given in Figure 2-11; deoxyribonucleosides have pK_a values that are 0.1–0.3 pH units larger than ribonucleosides. The pK_a for dissociation of an acid (HA) is defined as

$$\mathrm{p}K_a = -\log K_a = -\log \frac{[a_H+]\,[a_A-]}{[a_{HA}]} \tag{2-4}$$

where $[a_H+]$, $[a_A-]$ and $[a_{HA}]$ are the equilibrium activities. Usually, the activity of the hydrogen ions is measured with a pH meter, and the ratio of salt and acid activities is replaced by the ratio of molarities. This replacement gives an apparent pK_a that depends on ionic strength; all pK_a values depend on temperature. An extensive collection of pK_a values is given in the Handbook of Biochemistry and Molecular Biology (Fasman, 1975). We quote values at 25°C and 0.1 ionic strength.

The addition of a phosphate to form nucleotides raises the nucleoside pK_a values by 0.2–0.6 pH units. The negatively charged phosphate group attracts the ionizable proton and thus raises the pK_a. In a polynucleotide the electrostatic field of all the neighboring phosphates can further increase the base pK_a values. An ionizable proton that is hydrogen bonded is stabilized by the interaction and its pK_a is thus raised. Tertiary conformations that concentrate many phosphates around a base can also produce large increases in pK_a. A shift in pK_a of one unit requires a free energy at 25°C of $\Delta G^\circ = RT \ln 10 = 1363 \text{ cal mol}^{-1} = 5706 \text{ J mol}^{-1}$. This value means that a favorable interaction of a base with another group that requires the presence (or absence) of an ionizable proton can also shift the pK_a. For example, the reaction of adenine or cytosine with a proton will not occur significantly at pH 6 (pK_a of adenosine is 3.5; pK_a of cytidine is 4.2). However, the reaction of adenine plus cytosine plus a proton can occur at pH 6, if the protonated base is stabilized by an $\mathrm{AH^+ \cdot C}$ base pair. The free energy decrease from binding $\mathrm{AH^+}$ to the cytosine must be greater than the free energy increase required to add a proton to adenine at the unfavorable pH.

The nucleotide phosphate group has two more ionizable protons whose pK_a values depend only slightly on the nature of the base and on whether the phosphate is on the 2′, 3′, or 5′ position. The equilibria are

$$\mathrm{R{-}O{-}P({=}O)(OH){-}OH} \xrightarrow{\mathrm{p}K_a \approx 1} \mathrm{R{-}O{-}P({=}O)(OH){-}O^-} \xrightarrow{\mathrm{p}K_a \approx 6\text{–}7} \mathrm{R{-}O{-}P({=}O)(O^-){-}O^-} \tag{2-5}$$

Thus, above pH 7 a nucleotide phosphate primarily carries a double negative charge; below pH 6 it has a −1 charge. The phosphodiester in an oligonucleotide or polynucleotide has only one ionizable proton; its pK_a is about 1. Thus, above pH 1 each phosphate group in a polynucleotide (except for terminal phosphates) is singly charged.

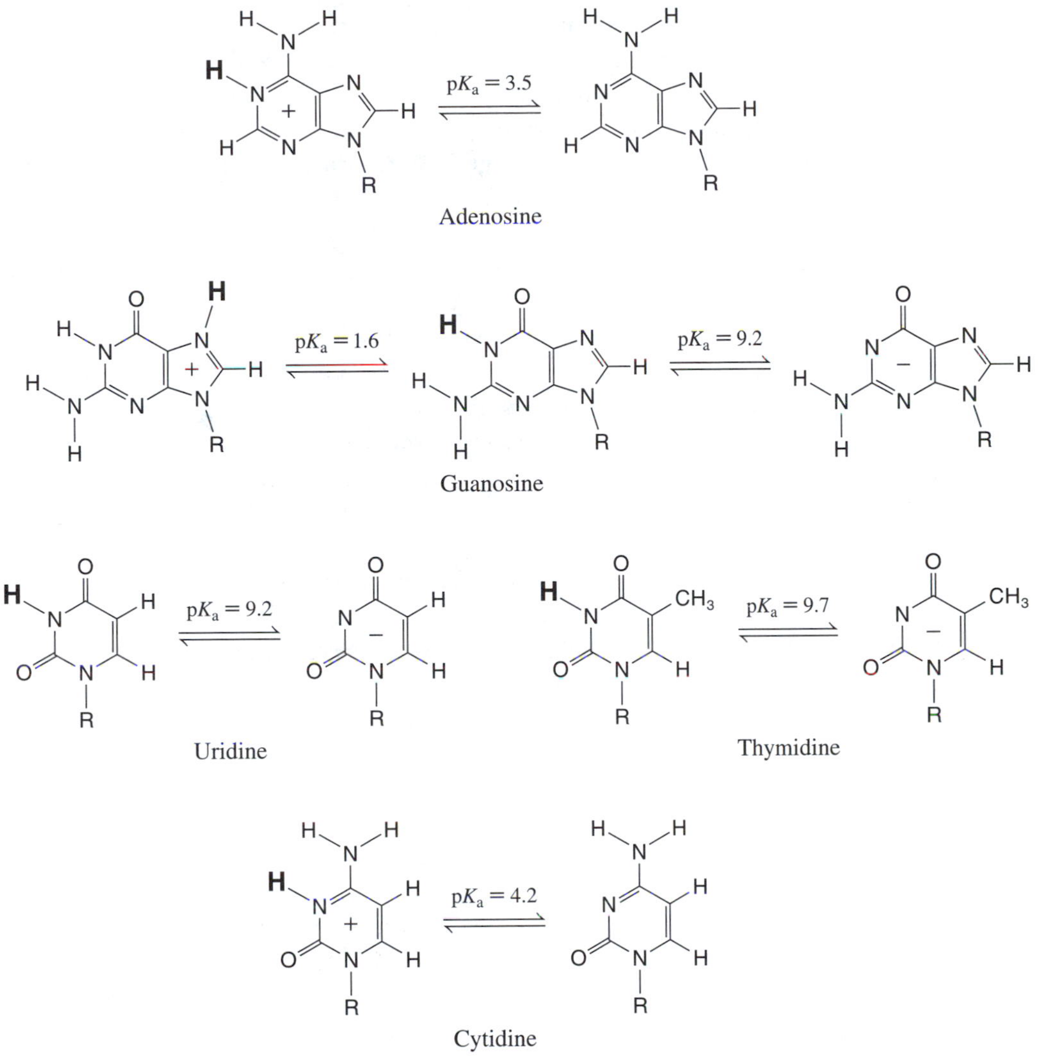

$pK_a = 3.5$

Adenosine

$pK_a = 1.6$ $pK_a = 9.2$

Guanosine

$pK_a = 9.2$

Uridine

$pK_a = 9.7$

Thymidine

$pK_a = 4.2$

Cytidine

Figure 2-11
Sites and pK_a values for protonation and ionization of five nucleosides. Note that at pH 7 all bases are neutral.

2.4 Rates of Proton Exchange

The rate of exchange with solvent of protons in nucleic acids (Englander and Kallenbach, 1984) and double-stranded oligonucleotides (Early et al., 1981) has been used to determine the stability and mobility (duplex breathing modes) of helices. Hydrogen bonding, shifts of pK_a values due to folding of a polynucleotide chain, or shielding of ionizable groups by binding to a ligand (such as a protein or drug), for example, will influence the rate of exchange. The rate of exchange of imino protons in guanine and thymine or uracil have been used to measure the kinetics of individual base pair opening in double strands, and the effect of bound drugs on these kinetics (Pardi et al., 1983). Thus proton exchange rates can probe structure and dynamics in nucleic acids (Gueron et al., 1987). Knowledge of exchange mechanisms for mononucleotides is important for understanding the data on polynucleotides.

The nucleic acid bases have exchangeable protons on the amino groups (N6 of adenine, N4 of cytosine, N2 of guanine) and the imino groups (N1 of guanine, N3 of uracil or thymine). In addition, adenine and guanine have slow but measurable exchange of the carbon-linked proton at the 8 position. The exchange rates have been measured by NMR (McConnell and Seawell, 1972; McConnell, 1978; McConnell et al., 1983; Fritzche et al., 1981; Raszka, 1974) and stopped-flow optical spectroscopy (Cross, 1975) for the N-linked protons and by tritium incorporation for the C-linked protons (Gamble et al., 1976). The general mechanism for proton exchange involves donation of a proton to a base (H_2O, OH^-, buffer), or acceptance of a proton from an acid (H_2O, H^+, buffer). Thus the rates will depend markedly on pH, concentration, and the type of buffer and solute concentration. Here we will discuss only the qualitative rates and mechanisms.

The lifetimes of amino or imino protons have been measured mainly by NMR. Each proton peak in an NMR spectrum has a line width that depends on its magnetic environment. There is an additional line broadening due to exchange processes such as proton exchange with solvent, or conformational changes, or chemical reactions. This excess line width, $\Delta\nu_e$, in reciprocal seconds s^{-1}, is directly proportional to the proton lifetime, τ, in a specific chemical state:

$$\Delta\nu_e = (\pi\tau)^{-1} \tag{2-6}$$

Excess line widths of a few hertz (Hz) to a few hundred hertz can be measured, therefore proton lifetimes ranging from about 1–100 ms can be measured directly. All imino and amino proton lifetimes for nucleotides occur in this range; they differ in their pH dependence and in the specific effects of each solvent component, such as phosphate, which is a powerful catalyst for exchange.

Cyclic 2′, 3′ mononucleotides, with singly charged phosphodiesters as found in polynucleotides, have been used to study exchange rates. Proton exchange rates for the amino groups of adenine and cytosine have a broad pH minimum from pH 6–7; guanine has a sharper minimum near pH 5.7 (McConnell and Seawell, 1972; McConnell, 1978; McConnell et al., 1983). The minimum proton exchange rate for the imino proton of cyclic 2′, 3′ guanylic acid is near pH 5 (McConnell and Seawell, 1972; McConnell, 1978; McConnell et al., 1983); for uridine it is pH 3.4 to 4.2 (Fritzche et al., 1981). Conditions can be chosen to minimize the rate of exchange of a particular proton, which is important, because the exchange rate with solvent can determine whether a proton is observable in an NMR spectrum (rapid exchange broadens the peak to invisibility). Each cytosine amino proton provides a separate NMR peak, and the rate of exchange is faster for the proton near N3 than that for the proton near C5 (Raszka, 1974). In adenine and guanine, the amino groups rotate too fast to give two separate NMR peaks, but base pairing, or other hydrogen bonding, can slow the rotation enough to provide distinct peaks.

Proton exchange at the 8 position of adenine or guanine takes place over a period ranging from hours to days; at room temperature and below, a proton at C8 will not exchange in D_2O for weeks. However, at higher temperatures exchange is faster and the rate is dependent on conformation. The C8 protons on different bases in tRNA exchange with tritiated H_2O at different rates (Gamble et al., 1976), and the C8 protons on syn guanines in Z-DNA, or Z-RNA, exchange at different rates than the C8 protons on anti guanines in A-RNA, or B-DNA. Thus the individual exchange rates for each G or A nucleotide can provide information about accessibility to solvent of the 8 position; this method is much less disruptive than chemical reactions as probes. For the 5′ mononucleotides, the H8 of guanylic acid exchanges about twice as fast as the H8 of adenylic acid at all temperatures. The lifetime for the exchange of tritium into the H8 of guanine in pG is about 1000 h at 40°C; the activation energy for both adenylic and guanylic nucleotides is about 22 kcal mol^{-1} (Gamble et al., 1976).

3. BASE PAIRING AND BASE STACKING

3.1 Base Pairing

The Watson–Crick base pairs that are found in the usual double-stranded DNA and RNA are shown in Figure 2-12. Their geometry is such that any sequence of base pairs can fit into the nucleic acid helix without distortion; the distances between C1′ atoms of sugars on opposite strands is essentially the same for A•T and G•C base pairs, (Donohue and Trueblood, 1960). The definition of major and minor grooves is shown in Figure 2.12. The minor groove is partly shielded by the sugar–phosphate backbones, therefore many of the interactions with the nucleic acids are on the major groove side. The characteristic groups in the two grooves are

Major Groove		Minor Groove	
Adenine	N6 amino, N7	Adenine	N3
Cytosine	N4 amino	Cytosine	C2 carbonyl
Guanine	C6 carbonyl, N7	Guanine	N2 amino, N3
Thymine	C4 carbonyl, C5 methyl	Thymine	C2 carbonyl
Uracil	C4 carbonyl	Uracil	C2 carbonyl

Many other hydrogen-bonded base pairs are possible if a continuous helix of arbitrary sequence is not required. Base pairs with two hydrogen bonds are shown in Figure 2-13; a good reference for structures of base pairs as determined by X-ray crystallography is given by Kennard and Hunter (1989). References to the occurrence of non-Watson–Crick base pairs studied by NMR and X-ray diffraction are listed in Table 2.1. One should always remember that bases can pair in a multitude of ways in addition to the Watson–Crick constraints. Ionization of a base can provide further opportunities for base pairing as shown in Figure 2-14.

Three-base coplanar combinations of A•(A•U) and G•(G•C) occur in $tRNA^{Phe}$ (Saenger, 1984, Chapter 15); the third base is in the major groove of the RNA duplex.

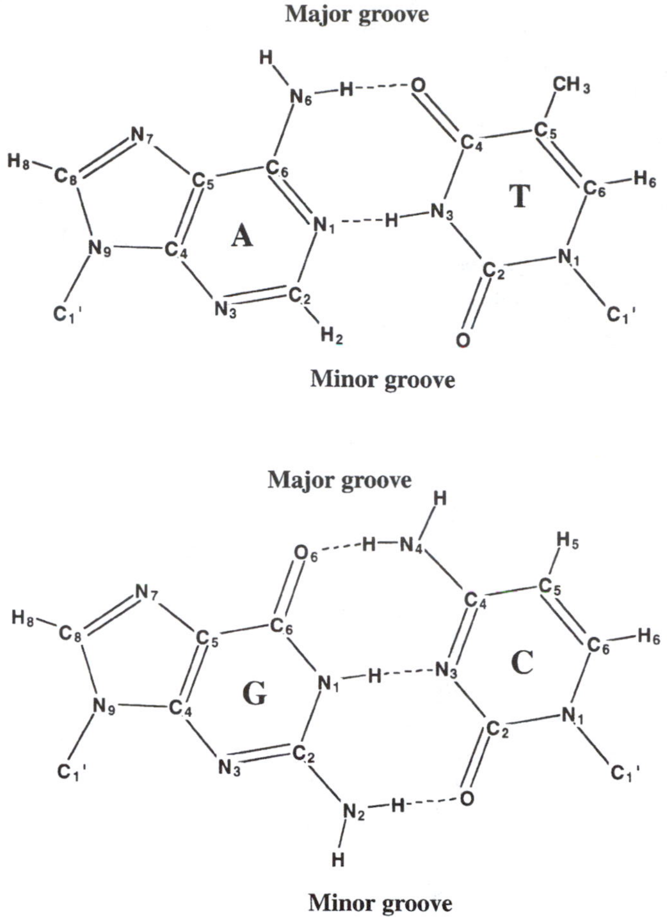

Major groove

Minor groove

Major groove

Minor groove

Figure 2-12
Watson–Crick base pairs found in the usual double-stranded DNA and RNA. The A • T and G • C base pairs have the same distance between the C1′ atoms of their sugars and can form a regular helix of any sequence. Each nucleic acid double helix has a major groove and a minor groove; the minor groove is on the side of the base pair where the sugars are attached.

Phylogenetic data indicates an A • (G • C) triplet in group I catalytic introns (Michel and Westhof, 1990); here the third base is in the minor groove. Three-base coplanar interactions occur in triple-stranded DNA polynucleotides with the third base in the major groove (Moser and Dervan, 1987; Htun and Dahlberg, 1989; Macaya et al., 1991; Roberts and Crothers, 1992; Van Meervelt et al., 1995). Hydrogen bonding for the base triples is shown in Figure 2-15; we see that protonated bases can be involved. The pK_a values of the bases are raised significantly by the attractive force of the polynucleotide phosphates for the protons, and by the stabilization caused by hydrogen bonding. Thus, protonated base pairs can occur near pH 7.

A minor groove A • (G • C) triple in an oligonucleotide model of part of the group I intron forms a nucleoside triple with hydrogen bonding from the adenine to the 2′-OH

Table 2.1
Non-Watson–Crick Base Pairs

Base Pair	Occurrence	Reference
A•G	tRNA, rRNA, ribozymes, oligos	Saenger (1984 p. 336) Noller (1984); Brown et al. (1986), Privé et al. (1987), Heus and Pardi (1991), Li et al. (1991), Cheng et. al. (1992), Wimberly et al. (1993), SantaLucia and Turner (1993), Pley et al. (1994), Greene et al. (1994), Katahira et al. (1994)
A•U Hoogsteen	rRNA	Wimberly et al. (1993)
A•T Hoogsteen	tRNA, oligos, crystal	Saenger (1984 p. 336), Quigley et al. (1986), Hoogsteen (1963)
G•C reverse Watson–Crick	tRNA	Saenger (1984 p. 336)
G•U, G•T wobble	tRNA, oligos	Crick (1966), Kalnik et al. (1988b), Hunter et al. (1986b), Holbrook et al. (1991), Hare et al. (1988)
C•U	rRNA, oligos	Patel et al. (1984), Wu and Marshall (1990), Holbrook et al. (1991)
G•G	Telomeres, mononucleotides, oligos	Sasisekharan et al. (1975), Borden et al. (1992), Kang et al. (1992), Smith et al. (1992)
U•U	Oligos	SantaLucia et al. (1991), Baeyens et al. (1995)
A^+•A^+	Polynucleotides	Rich et al. (1961)
A^+•C	Oligonucleotides	Hunter et al. (1986a, 1987), Kalnik et al. (1988a), Puglisi et al. (1990)
C^+•C	Polynucleotides, oligos	Borah et al. (1976), Pilch and Shafer, (1993)
G^+•C	Oligo–Antibiotic complex	Quigley et al. (1986)

of a ribose in the G•C pair (Chastain and Tinoco, 1992). Hydrogen bonds from bases to phosphates and to ribose groups may often contribute to nucleic acid structures other than double helices of Watson–Crick base pairs.

3.2 Base–Amino Acid Hydrogen Bonding

Nucleic acid–protein interactions include covalent bonds (such as the addition of the sulfhydryl of cysteine across the 5–6 double bond of pyrimidines), hydrogen bonds, ionic interactions, and London–van der Waals interactions. Hydrogen bonding can involve the side chains of polar amino acids and all the amide linkages. The amino

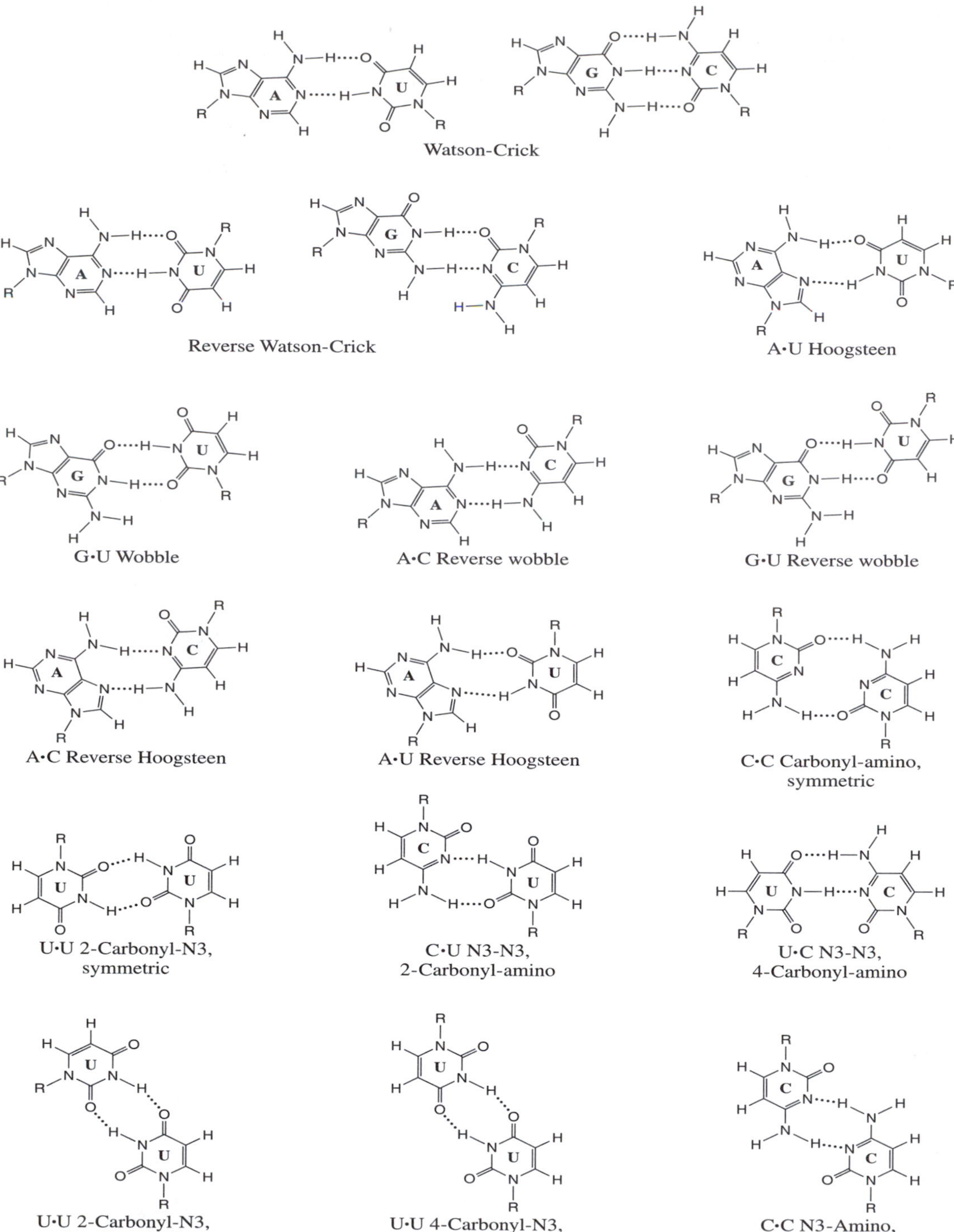

Figure 2-13
Base pairs that can be formed containing at least two hydrogen bonds (Saenger, 1984 p. 336). [This figure is used with permission from I. Tinoco. Jr. (1993). In *The RNA World*, Gesteland, R. and Atkins, J. Eds, Cold Spring Harbor Press, Cold Springs Harbor, NY, pp. 604–607].

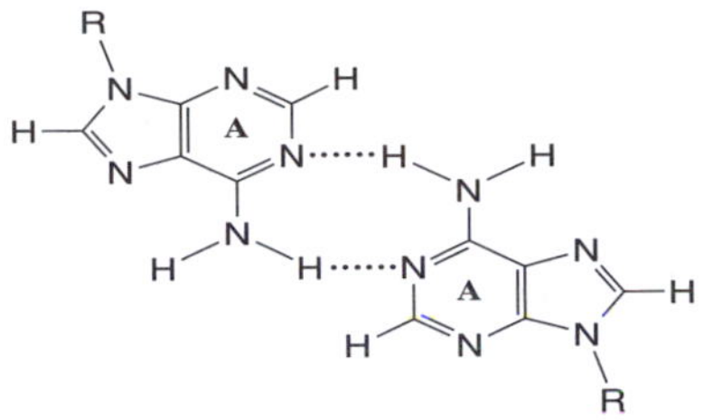

A•A N1-Amino,
symmetric

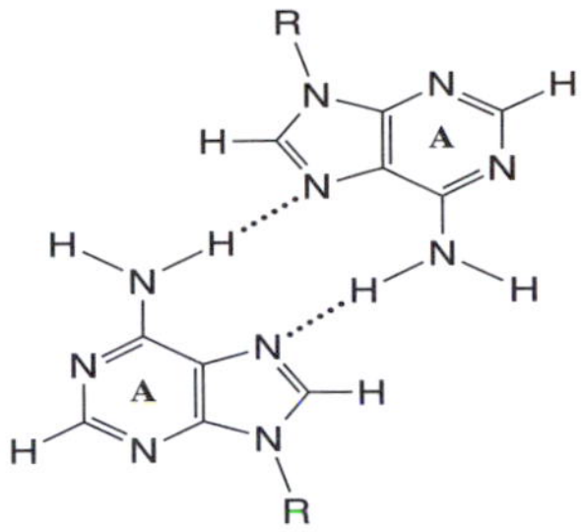

A•A N7-Amino,
symmetric

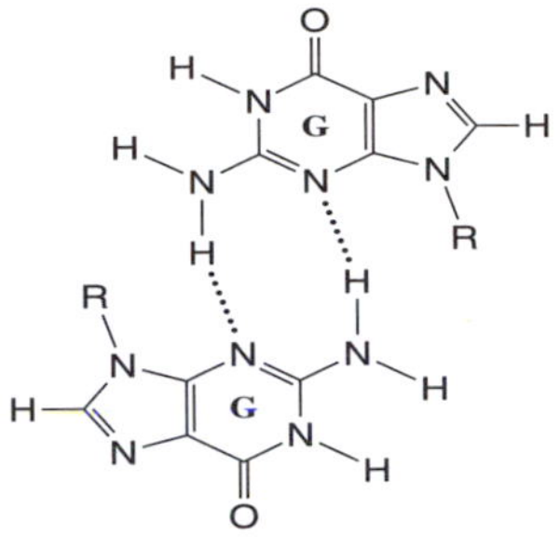

G•G N3-Amino,
symmetric

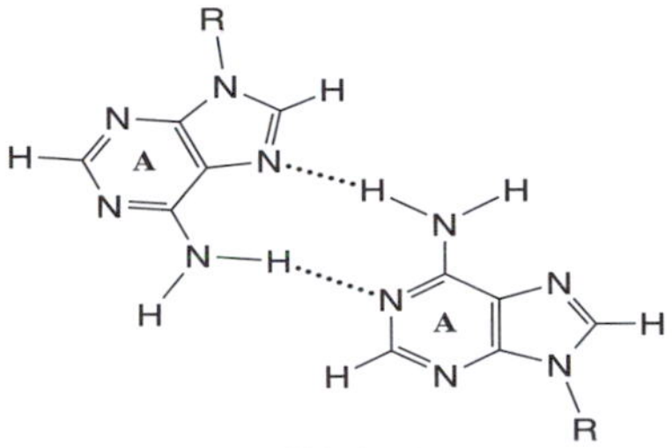

A•A N1-Amino,
N7-amino

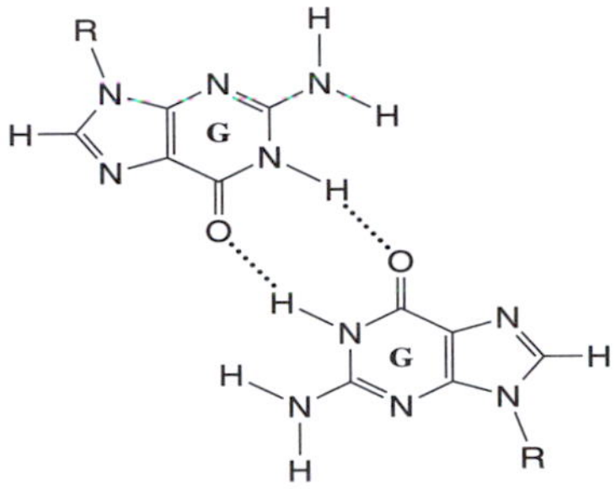

G•G N1-Carbonyl,
symmetric

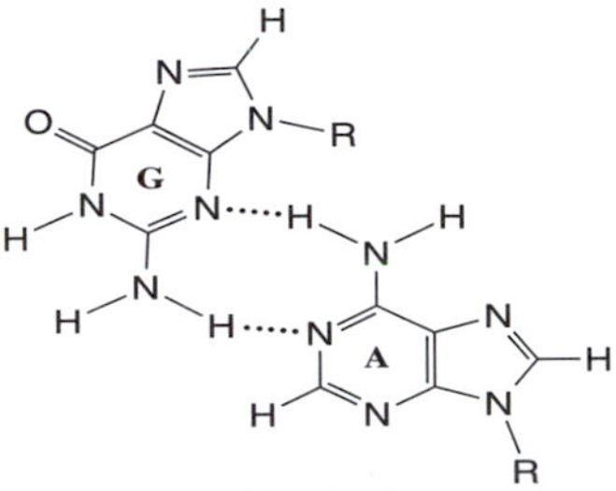

G•A N3-Amino,
amino-N1

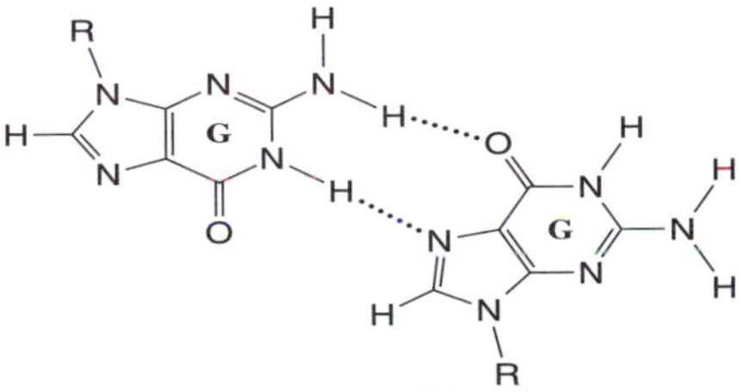

G•G N7-N1,
Carbonyl-amino

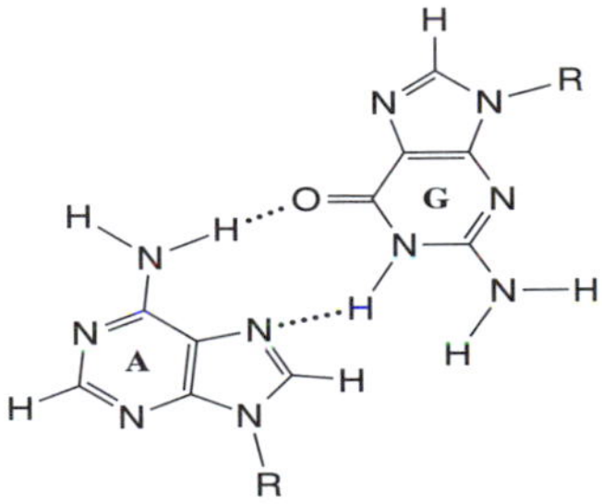

A•G N7-N1,
Amino-carbonyl

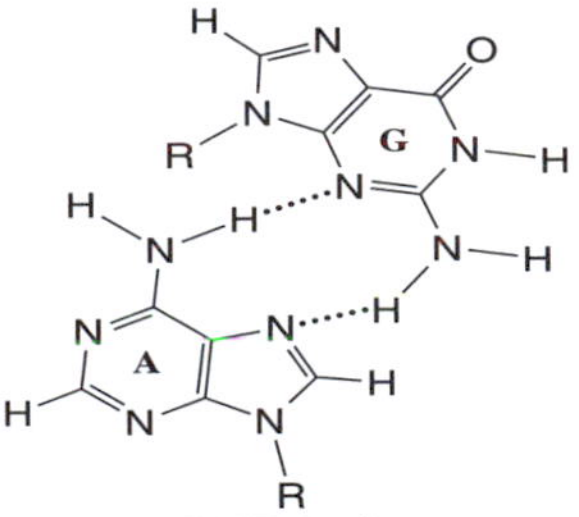

A•G N7-Amino,
amino-N3
(sheared)

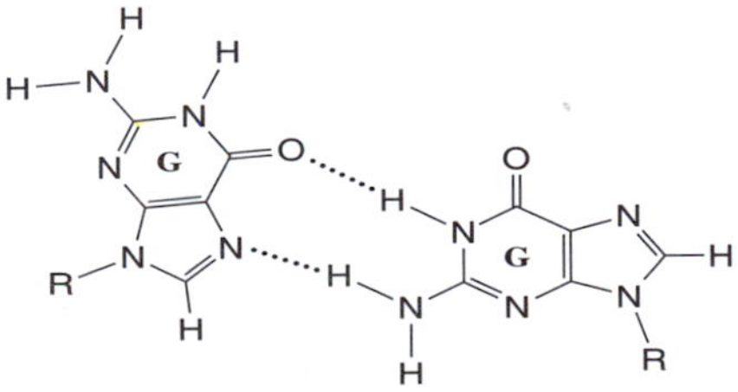

G•G N1-Carbonyl,
N7-amino

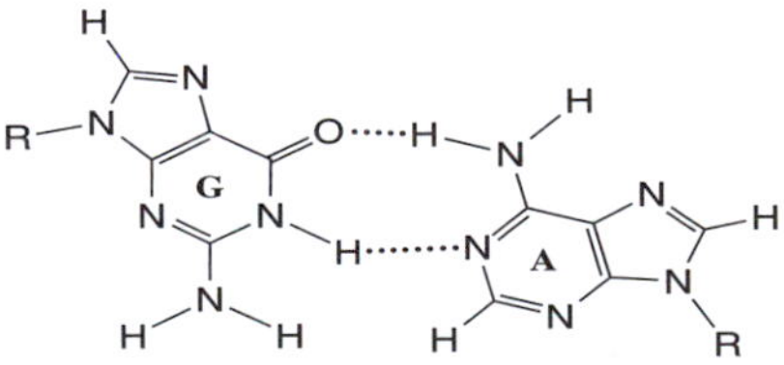

G•A N1-N1,
Carbonyl-amino
(imino)

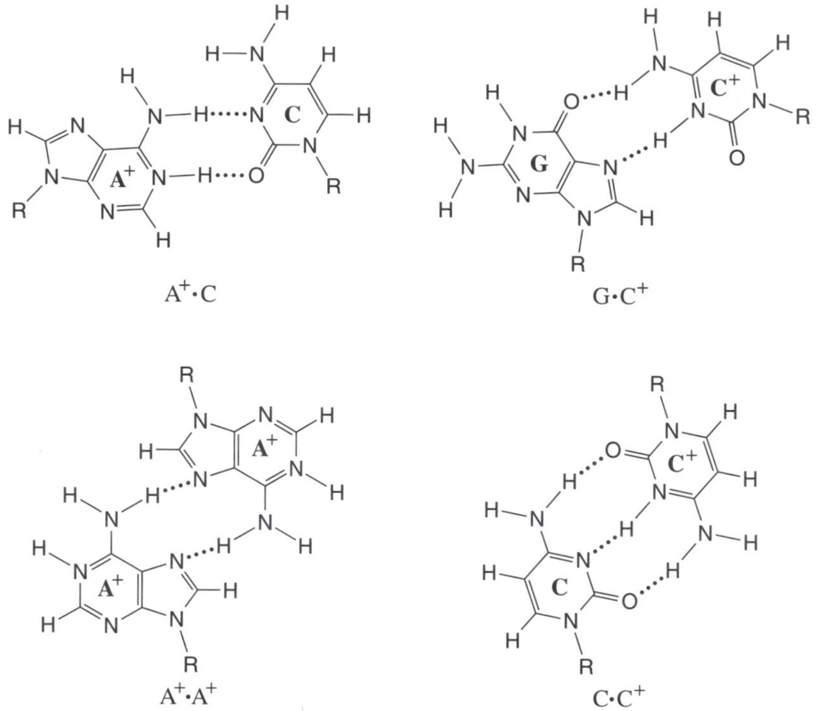

A$^+$·C

G·C$^+$

A$^+$·A$^+$

C·C$^+$

Figure 2-14
Some non-Watson–Crick base pairs containing at least two hydrogen bonds that can be formed from ionized bases. Note that the added protons do not take part in the hydrogen bonds in the A$^+$ · A$^+$ pair.

acid–base hydrogen bonding is characterized by whether it disrupts Watson–Crick base pairing or not. Figure 2-16 shows amino acid–base interactions with two hydrogen bonds that do not disrupt base pairing; all occur through the major groove of the double helix. The amino acids shown are glutamine (Gln, Q), asparagine (Asn, N), threonine (Thr, T), serine (Ser, S) and arginine (Arg, R). The NH groups in arginine can bond in a number of different ways with the N7 and carbonyl of guanine; only two are shown in the figure. The hydrogen bonding shown for glutamine or asparagine can also occur with any peptide bond of the protein backbone. Figure 2-17 shows amino acid–base hydrogen bonding that substitutes for Watson–Crick base pairing. Interactions with single-stranded DNA or RNA can take place at any of the base donor and acceptor sites.

3.3 Base Stacking

Soon after the Watson–Crick DNA structure was proposed, physical chemists looked for specific A · T and G · C base pairing in aqueous solution. They put the bases or nucleosides in water and tried to detect hydrogen bonding or specific 1:1 pairs. Instead, they found that all bases or nucleosides prefer to stack in aqueous solution; the flat

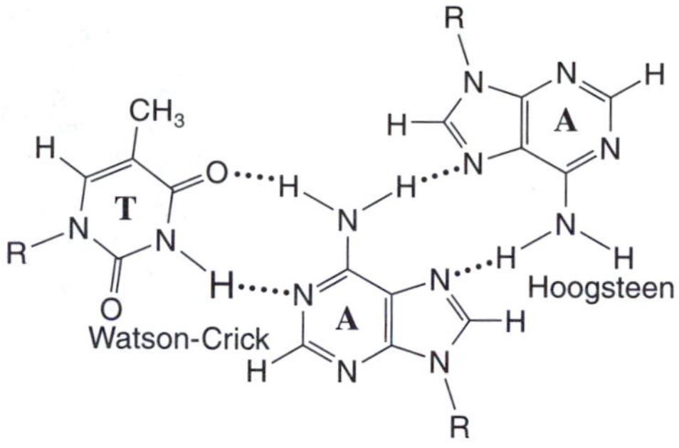

(T·A)A

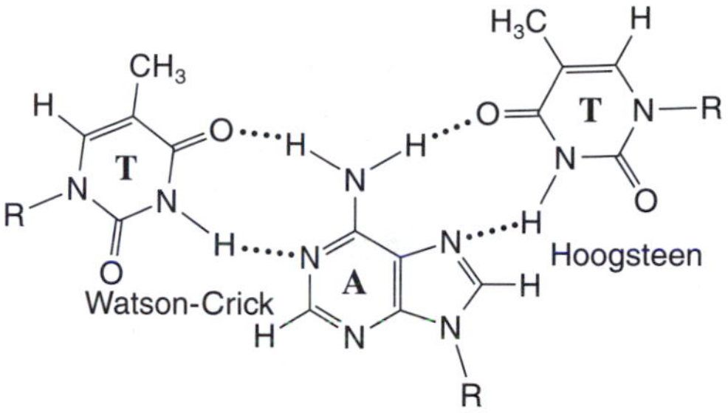

(T·A)T

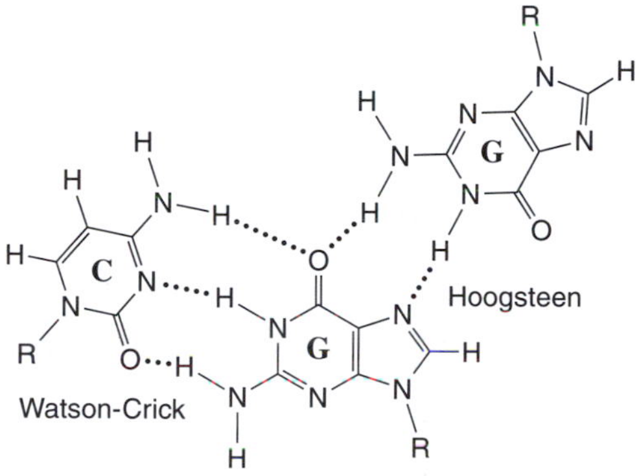

(C·G)G

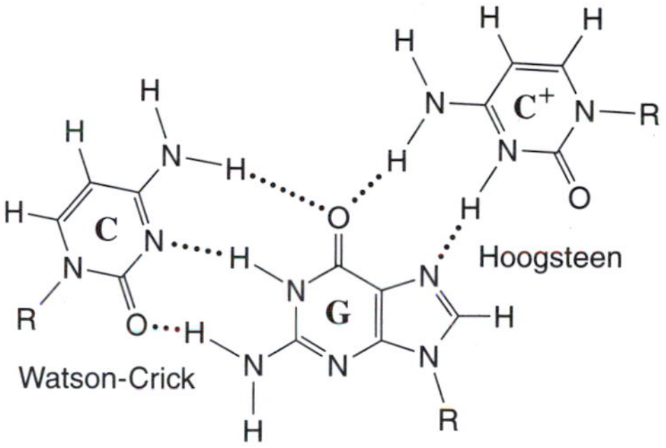

(C·G)C^+

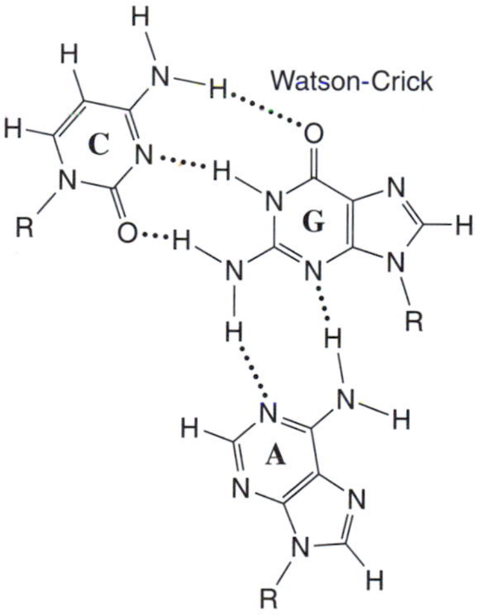

A(C·G)

Figure 2-15
Base triplets that occur in DNA, RNA and DNA–RNA hybrids. The top four triples involve formation of Hoogsteen base pairs in the major groove of the double helix; the bottom triple has the third base in the minor groove.

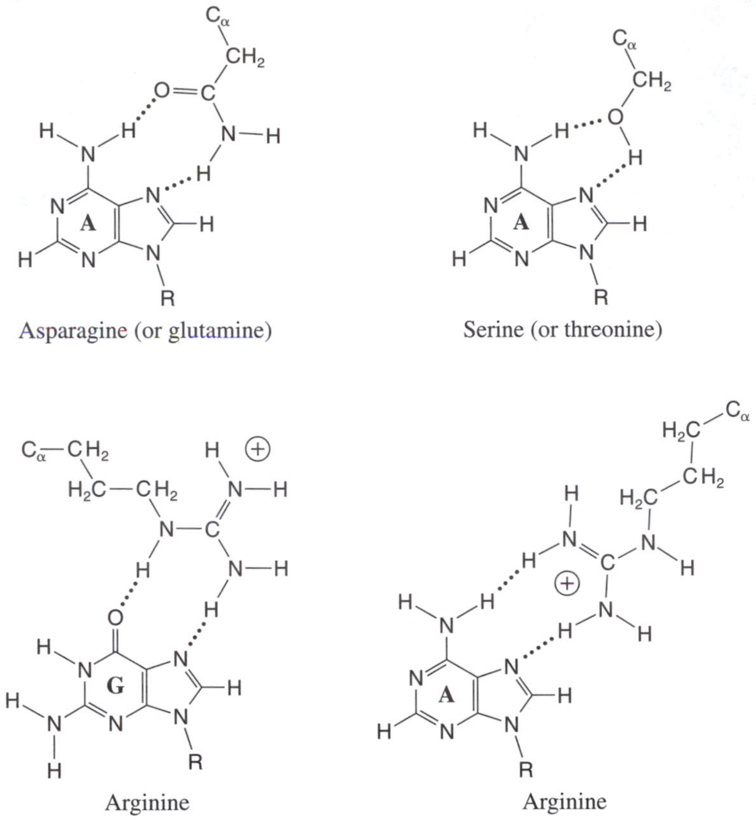

Figure 2-16
Interactions involving two hydrogen bonds between amino acids and bases that can occur through the major groove of a double helix.

bases aggregate to form columns containing 2, 3, or more bases. The distribution of bases and the average number of bases in a stack depend on the concentration of bases and the equilibrium constants for forming stacks, which illustrates that in aqueous solution interaction of the flat sides of the bases (without base–base hydrogen bonding) is favored over coplanar interaction of the edges (with base–base hydrogen bonding). In nonaqueous and weak hydrogen-bonding solvents such as chloroform or dimethyl sulfoxide (DMSO), and in the gas phase, coplanar interaction with hydrogen bonding is favored. The relative free energies of the paired and stacked species in solution will depend on all of the intermolecular interactions between the bases and the solvent, the base–base interactions in the presence of solvent, and the solvent–solvent interactions in the presence of bases. Thus, realistic theoretical calculations of free energies of pairing and stacking are very difficult. However, experimental measurements of thermodynamic values can give clues about the types of interactions responsible for stacking.

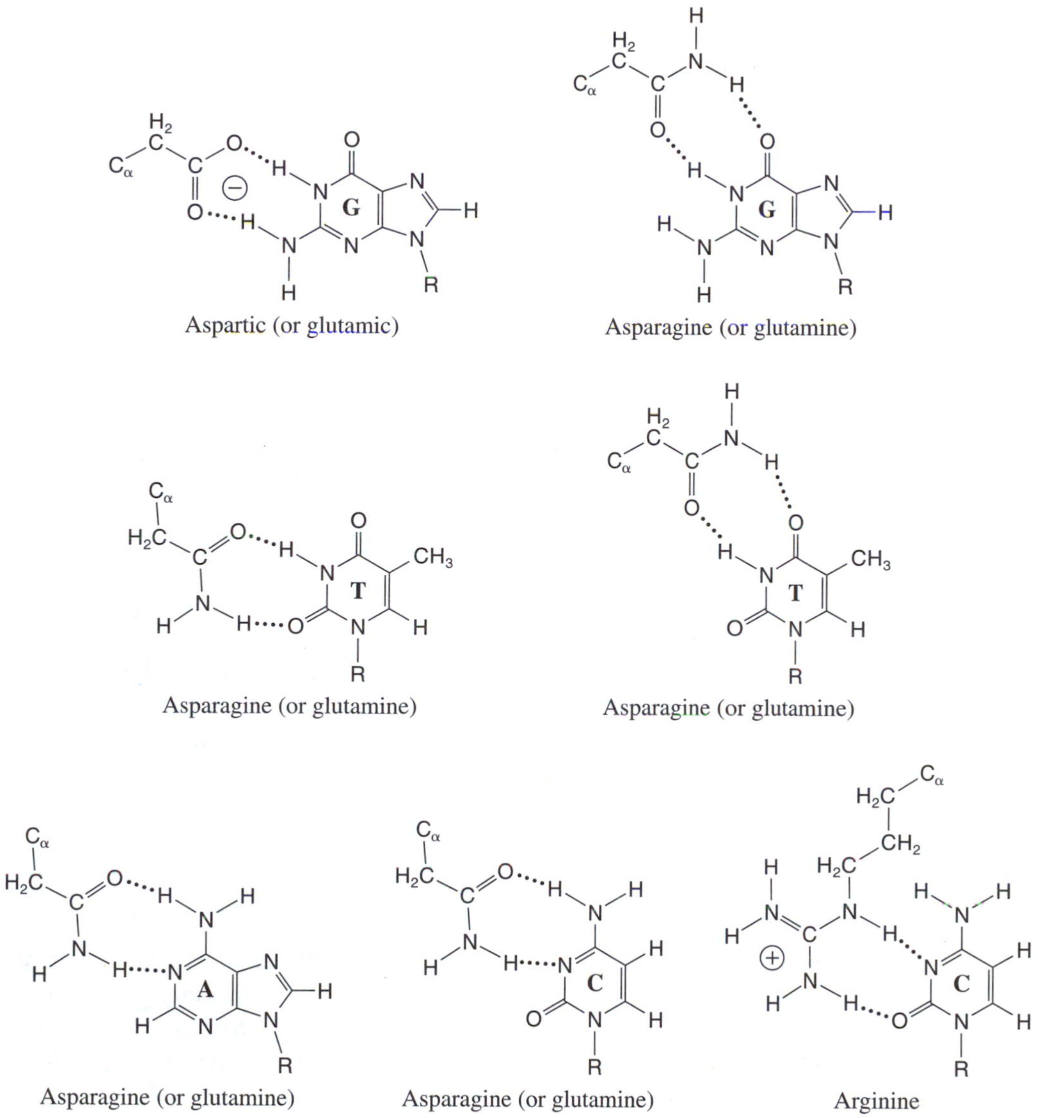

Aspartic (or glutamic)

Asparagine (or glutamine)

Asparagine (or glutamine)

Asparagine (or glutamine)

Asparagine (or glutamine)

Asparagine (or glutamine)

Arginine

Figure 2-17
Interactions involving two hydrogen bonds between amino acids and bases that take the place of Watson–Crick base pairing.

Experimental thermodynamic data are available for base pairing in nonaqueous solvents (Kyoguku et al., 1967) and in the gas phase (Sukhodub, 1987). In aqueous solution the thermodynamics of stacking of nucleosides has been measured by vapor phase osmometry (Ts'o, 1974), ultracentrifugation (Solie and Schellman, 1968), calorimetry (Gill et al., 1967), and NMR (Tribolet and Sigel, 1987). The pyrimidine nucleosides stack less than purine nucleosides, and the equilibrium constants K for stacking ($A + A = A_2$) range from 1 to 10 M^{-1}. Methylation of the bases increases stacking (Ts'o, 1974; Solie and Schellman, 1968; Gill et al., 1967). Table 2-2 gives

Table 2.2
Equilibrium Constants and Standard Thermodynamic Values for the Association of Nucleosides in Aqueous Solution[a]

Nucleoside	$K(25^\circ)$ (M^{-1})	$\Delta G^\circ(25^\circ)$ (kcal mol^{-1})	ΔH° (kcal mol^{-1})	ΔS° (eu)
Uridine	0.6	+0.3	−2.7	−10
Thymidine	0.9	+0.1	−2.4	−9
Cytidine	0.9	+0.1	−2.8	−10
Deoxycytidine	0.9	+0.1		
Adenosine	5	−1.0		
Deoxyadenosine	5–8	−1 to −1.2	−6.5	−18

Data are from Ts'o, (1974). Guanosine could not be studied because of its low solubility.

values for the measured equilibrium constants and the standard thermodynamic values. The standard free energy, enthalpy and entropy are obtained from the equilibrium constant and its temperature dependence using the van't Hoff equation.

$$\Delta G^\circ = -RT \ln K \tag{2-8a}$$

$$\Delta H^\circ = -R\partial(\ln K)/\partial(1/T) \tag{2-8b}$$

$$\Delta S^\circ = (\Delta H^\circ - \Delta G^\circ)/T \tag{2-8c}$$

The negative values of ΔS° immediately rule out hydrophobic interactions as dominant; hydrophobic interactions require release of water and an increase of entropy. The net base–base attractive forces provide a negative ΔH°, which essentially balances the unfavorable entropy loss in stacking the pyrimidine nucleosides and provides a small favorable free energy for stacking adenine nucleosides.

Intercalation of a base between two adjacent bases in a polynucleotide involves base stacking; this intercalation occurs in a tertiary interaction in tRNA^{Phe} (Saenger, 1984, p. 336). Base stacking is involved when nucleosides such as guanosine line up on a polynucleotide template of polycytidylic acid (Orgel, 1988). This type of interaction may have been important in the first prebiotic synthesis of complementary strands. It is clear that both base stacking and base pairing interactions are important for the formation of double strands. The stacking of bases with the side chains of the aromatic amino acids phenylalanine and tyrosine contribute to protein–nucleic acid interaction.

References

Adams, R. L. P. and Burdon, R. H. (1985). *Molecular Biology of DNA Methylation*, Springer-Verlag, New York.

Altona, C. (1982). Conformational Analysis of Nucleic Acids. Determination of Backbone geometry of Single-Helical RNA and DNA in Aqueous Solution, *Recl. Trav. Chim. Pays-Bas* **101**, 413–433.

Altona, C. (1986). DNA, the Versatile Vector of Life: Two-dimensional NMR Studies, *J. Mol. Struct.* **141**, 109–125.

Baeyens, K. J., De Bondt, H. L., and Holbrook, S. R. (1995). Structure of an RNA Double Helix Including Uracil–Uracil Base Pairs in an Internal Loop, *Struct. Biol.* **2**, 1–7.

Bash, P. A., Singh, U. C., Langridge, R., and Kollman, P. A. (1987). Free Energy Calculations by Computer Simulation, *Science* **236**, 564–568.

Borah, B., and Wood, J. L. (1976). The Cytidinium-cytidine Complex: Infrared and Raman Spectroscopic Studies, *J. Mol. Struct.* **30**, 13–30.

Borden, K. L., Jenkins, T. C., Skelly, J. V., Brown, T., and Lane, A. N. (1992). Conformational Properties of the G.G Mismatch in $d(CGCGAATTGGCG)_2$ Determined by NMR, *Biochemistry* **31**, 5411–5422.

Brown, T., Hunter, W. N., Kneale, G., and Kennard, O. (1986). Molecular Structure of the G • A Base Pair in DNA and its Implications for the Mechanism of Transversion Mutations, *Proc. Natl. Acad. Sci. USA* **83**, 2402–2406.

Chastain, M., and Tinoco, I. Jr. (1992). A Base-Triple Structural Domain in RNA, *Biochemistry*, **31**, 12733–12741.

Cheng, J.-W., Chou, S.-H., and Reid, B. R. (1992). Base Pairing Geometry in GA Mismatches Depends Entirely on the Neighboring Sequence, *J. Mol. Biol.* **228**, 1037–1041.

Crick, F. H. C. (1966). Codon–anticodon Pairing: The Wobble Hypothesis, *J. Mol. Biol.* **19**, 548–555.

Cross, D. G. (1975). Hydrogen Exchange in Nucleosides and Nucleotides. Measurement of Hydrogen Exchange by Stopped-Flow and Ultraviolet Difference Spectroscopy, *Biochemistry* **14**, 357–362.

deLeeuw, H. P. M., Haasnoot, C. A. G., and Altona, C. (1980). Empirical Correlations Between Conformational Parameters in β -D- Furanoside Fragments Derived from a Statistical Survey of Crystal Structures of Nucleic Acid Constituents. Full Description of Nucleoside Molecular Parameters, *Isr. J. Chem.* **20**, 108–126.

Donohue, J. and K. N. Trueblood (1960). Base Pairing in DNA, *J. Mol. Biol.* **2**, 363–371.

Early, T. A., Kearns, D. R., Hillen, W., and Wells, R. D. (1981). A 300- and 600- MHz proton NMR Study of a 12 Base Pair Restriction Fragment: Relaxation Behavior of the Low-field Resonances in Water, *Biochemistry* **20**, 3756–3764; A 300-MHz proton Nuclear Magnetic Resonance Investigation of Deoxyribonucleic Acid Restriction Fragments: Dynamic Properties, *Biochemistry* **20**, 3764–3769.

Engel, J. D. and P. H. von Hippel (1978). Effects of Methylation on the Stability of Nucleic Acid Conformations, *J. Biol. Chem.* **253**, 927–934.

Englander, S. W., and N. R. Kallenbach (1984). Hydrogen Exchange and Structural Dynamics of Proteins and Nucleic Acids, *Q. Rev. Biophys.* **16**, 521–655.

Fasman, G. D., Ed. (1975). *Handbook of Biochemistry and Molecular Biology, 3rd ed., Nucleic Acids,* Vol. 1, CRC Press, Cleveland, OH. (1975).

Fratini, A. V., Kopka, M. L., Drew, H. R., and Dickerson, R. E. (1982). Reversible Bending and Helix Geometry in a B-DNA Dodecamer: $CGCGAATT^{Br}CGCG$, *J. Biol. Chem.* **257**, 14686–14707.

Fritzche, H., Kan, L.-S., and Ts'o, P. O. P. (1981). Proton Nuclear Magnetic Resonance Study on Uridine Imido Proton Exchange, *Biochemistry* **20**, 6118–6122.

Gamble, R. C., Schoemaker, H. J. P., Jekowsky, E., and Schimmel, P. R. (1976). Rate of Tritium Labeling of Specific Purines in Relation to Nucleic Acid and Particularly Transfer RNA Conformation, *Biochemistry* **15**, 2791–1799.

Gill, S. J., Downing, M., and Sheats, G. F. (1967). The Enthalpy of Self-Association of Purine Derivatives in Water, *Biochemistry* **6**, 272–276.

Greene, K. L., Jones, R. L., Li, Y., Robinson, H., Wang, A., Zon, G., and Wilson, W. D. (1994). Solution Structure of a GA Mismatch DNA Sequence, $d(CCATGAATGG)_2$, Determined by 2D NMR and Structural Refinement Methods, *Biochemistry* **33**, 1053–1062.

Gueron, M., Kochoyan, M., and Leroy, J. L. (1987). A single mode of DNA base-pair opening drives imino proton exchange, *Nature (London)* **328**, 579–581.

Hall, K., Cruz, P., Tinoco, I. Jr., Jovin, T. M., and van de Sande, J. H. (1984). Z-RNA—A Left-Handed RNA Double Helix, *Nature (London)* **311**, 584–586.

Hare, D., Shapiro, L., and Patel, D. J. (1988). Wobble dG • dT Pairing in Right-Handed DNA: Solution Conformation of the d(CGTGAATTCGCG) Duplex Deduced from Distance Geometry Analysis of Nuclear Overhauser Effect Spectra, *Biochemistry* **25**, 7445–7456.

Heus, H. A. and A. Pardi (1991). Structural Features that Give Rise to the Unusual Stability of RNA Hairpins Containing GNRA Loops, *Science* **253**, 191–194.

Holbrook, S., Cheong, C., Tinoco, I. Jr., and Kim, S.-H. (1991) Crystal Structure of an RNA Double Helix Incorporating a Track of Non-Watson–Crick Base Pairs, *Nature (London)* **353**, 579–581.

Hoogsteen, K. (1963). The Crystal and Molecular Structure of a Hydrogen-bonded Complex Between 1-Methylthymine and 9-Methyladenine, *Acta Crystallogr.* **16**, 907–916.

Htun, H., and J. E. Dahlberg (1989). Topology and Formation of Triple-Stranded H-DNA, *Science* **243**, 1571–1576.

Hunter, W. N., Brown, T., Anand, N. K., and Kennard, O. (1986a). Structure of an Adenine.Cytosine Base Pair in DNA and Its Implications for Mismatch Repair, *Nature (London)* **320**, 552–555.

Hunter, W. N., Brown, T., and Kennard, O. (1987). Structural Features and Hydration of a Dodecamer Duplex Containing Two C•A Mispairs, *Nucleic Acids Res.* **15**, 6589–6606.

Hunter, W. N., Kneale, G., Brown, T., Rabinovich, D., and Kennard, O. (1986b). Refined Crystal Structure of an Octanucleotide Duplex with G•T Mismatched Base-pairs, *J. Mol. Biol.* **190**, 605–818.

IUPAC–IUB (1983). Recommended nomenclature and definitions are given in Abbreviations and Symbols for the description of Conformation of Polynucleotide Chains, *Eur. J. Biochem.* **131**, 9–15.

Jardetsky, O. and G. C. K. Roberts (1981). *NMR in Molecular Biology*, Academic, New York, p. 534.

Kalnik, M. W., Kouchakdjian, M., Li, B. F. L., Swann, P. F., and Patel, D. J.(1988a). Base Pair Mismatches and Carcinogen-Modified Bases in DNA: An NMR Study of A•C and A•O^4meT Pairing in Dodecanucleotide Duplexes, *Biochemistry* **27**, 100–108.

Kalnik, M. W., Kouchakdjian, M., Li, B. F. L., Swann, P. F., and Patel, D. J. (1988b). Base Pair Mismatches and Carcinogen-Modified Bases in DNA: An NMR Study of G•T and G•O^4meT Pairing in Dodecanucleotide Duplexes, *Biochemistry* **27**, 108–115.

Kang, C. H., Zhang, X., Ratliff, R., Moyzis, R., Rich, A. (1992). Crystal Structure of Four-Stranded *Oxytricha* Telomeric DNA, *Nature (London)* **356**, 126–131.

Katahira, M., Kanagawa, M., Sato, H., Uesugi, S., Fujii, S., Kohno, T., and Maeda, T. (1994). Formation of Sheared G:A Base Pairs in an RNA Duplex Modelled after Ribozymes, as Revealed by NMR, *Nucleic Acids Res* **22**, 2752–2759.

Kennard, O. and Hunter, W. N. (1989). Oligonucleotide Structure—A Decade Of Results From Single Crystal X-Ray Diffraction Studies, Q. Rev. Biophys. **22**, 327–379.

Kline, L. K. and Söll D. (1982). Nucleotide Modification in RNA, in *The Enzymes,* Vol. XV, 3rd Ed. Boyer, P. D., Ed., Academic, New York, pp. 567–582.

Kyoguku, Y., Lord, R. C., and Rich, A. (1967). The Effect of Substituents on the Hydrogen Bonding of Adenine and Uracil Derivatives, *Proc. Natl. Acad. Sci. USA* **57**, 250–257.

Li, Y., Zon, G., and Wilson, W. D. (1991). NMR and Molecular Modeling Evidence for a G•A Mismatch Base Pair in a Purine-rich DNA Duplex, *Proc. Natl. Acad. Sci. USA* **88**, 26–30.

Macaya, R. F., Gilbert, E. E., Malek, S., Sinsheimer, J. S., and Feigon J. (1991). Structure and Stability of X•G•C Mismatches in the Third Strand of Intermolecular Triplexes. *Science* **254**, 270–274.

McConnell, B. and Seawell, P. C. (1972). Proton Exchange of Nucleic Acids. Amino Protons of Mononucleotides, *Biochemistry* **11**, 4382–4392.

McConnell, B. (1978). Exchange Mechanisms for Hydrogen Bonding Protons of Cytidylic and Guanylic Acids, *Biochemistry* **17**, 3168–3176.

McConnell, B., Rice, D. J., and Uchima, F.-D. A. (1983). Exceptional Characteristics of Amino Proton Exchange in Guanosine Compounds, *Biochemistry* **22**, 3033–3037.

McConnell, B. and Seawell, P. C. (1973). Amino Protons of Cytosine. Chemical Exchange, Rotational Exchange, and Salt-Induced Proton Magnetic Resonance Chemical Shifts of Aqueous 2′, 3′ Cyclic Cytidine Monophosphate, *Biochemistry* **12**, 4426–4434.

Michel. F. and Westhof, E. (1990). Modelling of the Three-dimensional Architecture of Group I Catalytic Introns Based on Comparative Sequence Analysis, *J. Mol. Biol.* **216**, 585–610.

Moser, H. and Dervan, P. B. (1987). Sequence-Specific Cleavage of Double Helical DNA by Triple Helix Formation, *Science* **238**, 645–650.

Noller, H. F. (1984). Structure of Ribosomal RNA, *Annu. Rev. Biochem.* **53**, 119–162.

Olson, W. K. (1982). How Flexible is the Furanose Ring? An Updated Potential Energy Estimate, *J. Am. Chem. Soc.* **104**, 278–286.

Orgel, L. (1988). Evolution of the Genetic Apparatus: A Review. *Symp. Quant. Biol.* **52**, 9–16.

Pardi, A., Morden, K. M., Patel, D. J., and Tinoco, I. Jr. (1983). Kinetics for Exchange of the Imino Protons of the d(C-G-C-G-A-A-T-T-C-G-C-G) Double Helix in Complexes with the Antibiotics Netropsin and/or Actinomycin, *Biochemistry* **22**, 1107–1113.

Patel, D. J., Kozlowski, S. A., Ikuta, S., and Itakura, K. (1984). Dynamics of DNA Duplexes Containing Internal G•T, G•A, A•C, and T•C Pairs: Hydrogen Exchange at and Adjacent to Mismatch Sites, *FASEB J.* **48**, 2663–2670.

Pearlman, D. A., and Kim, S.-H. (1990). Atomic charges for DNA constituents derived from single-crystal X-ray diffraction data, *J. Mol. Biol.* **211**, 171–87.

Pilch, D.S. and Shafer, R.H. (1993). Structural and thermodynamic studies of d($C_3T_4C_3$) in acid solution—evidence for formation of the hemiprotonated CH^+ • C base pair, *J. Am. Chem. Soc.* **115**, 2565–2571.

Pley, H. W., Flaherty, K. M., and McKay, D. B. (1994) Three-dimensional Structure of a Hammerhead Ribozyme. *Nature (London)* **372**, 68–74

Prive', G. G., Heinemann, U., Chandrasekaran, S., Kan, L.-S., Kopka, M. L., and Dickerson, R. E. (1987). Helix geometry, Hydration, and G•A Mismatch in a B-DNA Decamer, *Science* **238**, 498–504.

Puglisi, J. D., Wyatt, J. R., and Tinoco, I., Jr. (1990). Solution Conformation of an RNA Hairpin Loop, *Biochemistry* **29**, 4215–4226.

Quigley, G. J., Ughetto, G., van der Marel, G. A., van Boom, J. H., Wang, A. H.-J. and Rich, A. (1986). Non-Watson–Crick G•C and A•T Base Pairs in a DNA-Antibiotic Complex, *Science* **232**, 1255–1258.

Raszka, M. (1974). Mononucleotides in Aqueous Solution: Proton Magnetic Resonance Studies of Amino Groups, *Biochemistry* **13**, 4616–4622.

Razin, A., Cedar, H., and Riggs, A. D. Eds. (1984). *DNA Methylation. Biochemistry and Biological Significance,* Springer-Verlag, New York.

Rich, A., Davies, D. R., Crick, F. H., and Watson, J. D. (1961). The Molecular Structure of Polyadenylic Acid, *J. Mol. Biol.* **3**, 71–86.

Rich, A., Nordheim, A., and Wang, A. H.-J. (1984). The Chemistry and Biology of Left-Handed Z-DNA, *Annu. Rev. Biochem.* **53**, 791–84.

Roberts, R. W., and Crothers, D. M. (1992). Stability and Properties of Double and Triple Helices: Dramatic Effects of RNA or DNA Backbone Composition, *Science* **258**, 1463–1466.

Saenger, W. (1984). *Principles of Nucleic Acid Structure*, Springer-Verlag, New York.

Sampson, J. R. and Uhlenbeck, O. C. (1988). Biochemical and Physical Characterization of an Unmodified Yeast Phenylalanine Transfer RNA Transcribed *in vitro, Proc. Natl Acad. Sci. USA* **85**, 1033–1037.

SantaLucia, J., Jr., Kierzek, R., and Turner, D. H. (1991). Stabilities of Consecutive A•C, C•C, G•G, U•C, and U•U Mismatches in RNA Internal Loops: Evidence for Stable Hydrogen-bonded U•U and C•C^+ Pairs, Biochemistry **30**, 8242–51.

SantaLucia, J., Jr., and Turner, D. H. (1993). Structure of $r(GGCGAGCC)_2$ in solution from NMR and restrained molecular dynamics, *Biochemistry* **32**, 12612–23.

Sasisekharan, V., Zimmerman, S., and Davis, D. R. (1975). The Structure of Helical 5′-guanosine Monophosphate, *J. Mol. Biol.* **92**, 171–179.

Singh, U. C. and P. A. Kollman (1984). An Approach to Computing Electrostatic Charges for Molecules, *J. Comp. Chem.* **5**, 129–145.

Smith, F. W. and J. Feigon (1992). Quadruplex Structure of Oxytricha Telomeric DNA, *Nature (London)* **356**, 164–168.

Solie, T. N. and Schellman, J. A. (1968). The Interaction of Nucleosides in Aqueous Solution, *J. Mol. Biol.* **33**, 61–77.

Sukhodub, L. F. (1987). Interaction and Hydration of Nucleic Acid Bases in a Vacuum. Experimental Study, *Chem. Rev.* **87**, 589–606.

Tinoco, I., Jr. (1993) in *The RNA World*, Gesteland R. and Atkins, J. Eds. Cold Spring Harbor Press, Cold Springs Harbor, NY, pp. 604–607.

Tribolet, R. and Sigel, H. (1987). Self-association of Adenosine 5′-monophosphate (5′ AMP) as a Function of pH and in Comparison with Adenosine, 2′-AMP and 3′-AMP, *Biophysical Chemistry* **27**, 119–130.

Ts'o, P. O. P. (1974). Bases, Nucleosides and Nucleotides, in *Basic Principles in Nucleic Acid Chemistry*, Ts'o, P. O. P., Ed., Vol. I, Academic, New York, pp. 453–584.

Van Meervelt, L., Vlieghe, D., Dautant, A., Gallois, B., Précigoux, G., and Kennard, O. (1995). High-resolution Structure of a DNA Helix Forming (C•G)*G Base Triplets, *Nature (London)* **374**, 742–744.

Weiner, S. J., Kollman, P. A., Case, D. A., Singh, U. C., Ghio, C., Alagona, G., Profeta, S., and Weiner, P. (1984). A New Force Field for Molecular Mechanical Simulations of Nucleic Acids and Proteins, *J. Am. Chem. Soc.* **106**, 765–784.

Wimberly, B., Varani, G., Tinoco, I., Jr. (1993). The Conformation of Loop E of Eukaryotic 5S Ribosomal RNA, *Biochemistry* **32**, 1078–1087

Wu, J. and Marshall, A. G. (1990). 500-MHz Proton NMR Evidence for Two Solution Strucutures of the Common Arm Base Paired Segment of Wheat Germ 5S Ribosomal RNA, *Biochemistry* **29**, 1772–1730.

CHAPTER 3

Chemical and Enzymatic Methods

with contributions by John Hearst

1. INTRODUCTION

The ability to determine and construct DNA and RNA sequences, and ultimately to relate sequence to function, depends on the chemical and enzymatic methods that are the subject of this chapter. The first general method for determining DNA sequence—developed by Maxam and Gilbert—relies on selective chemical cleavage of the DNA chain with reagents whose efficiency depends on the nature of the base at a particular position. For example, a reaction that selectively cleaves ^{32}P-labeled chains at guanosine residues will display an electrophoresis pattern with bands of increasing size (decreasing mobility) corresponding to increasing distance of each guanosine from the labeled end of the chain. An important feature of the cleavage reactions for sequencing, in addition to their nucleotide specificity, is that the chains are cut only once or less. This cutting produces a population of cleaved chains in which the reacted nucleotide can be hundreds of residues from the end carrying the label. As we will see, it is also important that the reactions produce uniform chemistry at the site of cleavage, because heterogeneous fragments of the destroyed nucleotide would yield multiple bands on the gel for each site of reaction.

The Sanger method for DNA sequence determination relies on the related principle of nucleotide-selective blockage of enzymatic synthesis of DNA. For example, if a small amount of dideoxyguanosine triphosphate is included in the mixture of dNTPs used to copy a DNA template, the growing chains eventually will be blocked at G residues (C residues in the template) because the lack of a 3′-OH will prevent the next base in the sequence to be added. Electrophoresis of an end-labeled mixture of the partially blocked synthesis reaction reveals a series of bands of increasing length terminated at each G in the sequence.

Once DNA sequences became known, methods for facile variation of sequence took on added importance. Chemical synthesis of DNA, supplemented by cloning methods to allow tests of function both *in vitro* and *in vivo,* has greatly accelerated the pace of progress in molecular biology. Years of effort by chemists have led to automated procedures for synthesis of DNA and RNA molecules of defined sequence. Chemical and enzymatic methods have not been limited to determination of sequence: the primary structure. Reactivity of DNA and RNA molecules to chemical and enzymatic probes depends on whether or not a nucleotide is in a double helix (the secondary structure) and sometimes also depends on features of the higher order folding of the molecule (the tertiary structure). In this chapter, we examine the chemical basis for these powerful methods.

2. SYNTHESIS AND HYDROLYSIS

2.1 Synthesis of DNA and RNA

Chemical synthesis of DNA and RNA (Gait, 1984; Sinha et al., 1984; Itakura et al., 1984) or RNA (Kierzek et al., 1986; Ogilvie et al., 1988; Wang et al., 1990) is usually done on a commercial apparatus using phosphoramidite chemistry on a solid support. The 3′ position of the sugar has a reactive phosphoramidite group—a trivalent phosphorus with two esters and an amide linkage

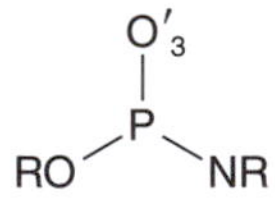

$$\mathrm{RO}-\overset{\overset{\displaystyle O'_3}{|}}{\mathrm{P}}-\mathrm{NR}$$

The trivalent phosphorus atom is subsequently oxidized to the pentavalent state found in natural nucleic acids.

Synthesis starts (see Fig. 3-1.) with the 3′-terminal monomer attached to a silica support with its 5′-OH available for reaction, but with the amino group on its base protected and, for ribose, its 2′-OH protected. The next monomer (with base, 5′-OH and if necessary, 2′-OH protected) is added so that coupling of its 3′-phosphoramidite to the 5′-OH occurs to produce a phosphite diester. The phosphorus is then oxidized to its pentavalent state to form the normal phosphate diester. After the last monomer (the 5′ terminus) is added, the molecule is removed from the solid support and is deprotected and purified. DNA molecules of approximately 150 nucleotides can be synthesized

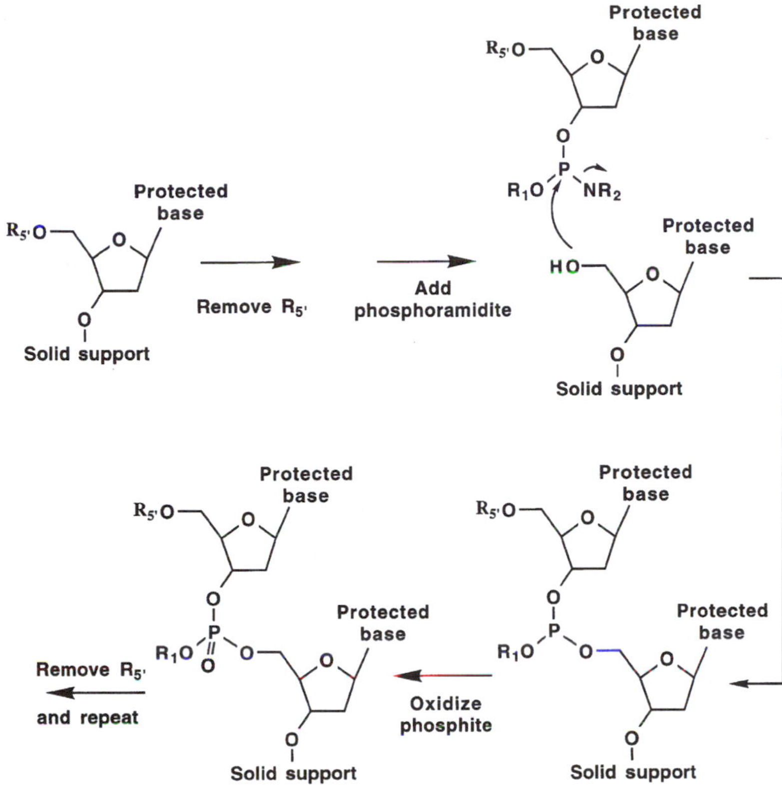

Figure 3-1
The chemical synthesis of DNA and RNA by the solid phase phosphoramidite method. The crux of the problem is to protect all the reactive groups (the amino groups of the bases, the phosphates, the 5′-OH groups and for RNA the 2′-OH groups) except the one 5′-OH group that the activated 3′-phosphoramidite attacks.

in microgram (μg) quantities. It is easier to synthesize DNA than RNA because the 2′-OH group must be specifically protected. Reaction at the 2′-OH can produce 2′–5′-phosphate diester linkages, or even branched chains during RNA synthesis. Chemical synthesis of DNA or RNA of course allows incorporation of unnatural and isotopically labeled nucleotides. Chemical reactions involving nucleic acids are described in *The Organic Chemistry of Nucleic Acids* (1986) by Mizuno.

Enzymatic synthesis of RNA oligonucleotides is easily done using T7 RNA polymerase and a chemically synthesized DNA template (Milligan et al., 1987; Wyatt et al., 1991). The DNA templates for longer RNAs are usually obtained by cloning the template and a promoter region; the DNA is transcribed by the corresponding T7, T3, or SP6 bacteriophage RNA polymerase (Sambrook et al., 1989, p. 5.58). To incorporate labeled or modified nucleotides into a long RNA strand, a combined chemical and enzymatic method can be used. A short RNA oligonucleotide is chemically synthesized containing the modified nucleotide. The chemically synthesized fragment is then linked to the long RNA using T4 DNA ligase (which requires a double-stranded substrate) and a DNA template that is complementary to the ends of the RNA molecules to be ligated (Moore and Sharp, 1992).

To amplify a natural DNA for sequencing or other study, the DNA can be cloned and replicated *in vivo.* A very useful collection of protocols for sequencing, mapping and characterizing natural nucleic acids is given in *Molecular Cloning, A Laboratory Manual* by Sambrook et al. (1989). The DNA can also be amplified in vitro using a polymerase chain reaction (PCR), which uses a heat-stable DNA polymerase to make many multiple copies of the DNA (Saiki et al., 1988). Two oligonucleotide primers (one for each strand) are used to determine which region of the duplex is amplified. After each round of synthesis the solution is heated to dissociate the DNA polymerase from the new duplex and to allow a new cycle of synthesis. As each cycle doubles the number of DNA templates, rapid multiplication results.

Synthesis of random sequences of DNA or RNA may seem like a useless endeavor, but if the random synthesis is followed by a highly specific procedure that selects molecules with a chosen property, very useful information can be obtained. The idea is simple. All possible sequences of a chosen length are synthesized by using a mixture of all four phosphoramidites in each synthesis step. Then a selection procedure is used to separate a small number of sequences with the desired property from all others. For example, Tuerk and Gold (1990) synthesized all $65{,}536(= 4^8)$ RNA hairpins with loops of eight nucleotides and a constant stem. Selection was done by binding to bacteriophage T4 DNA polymerase. The tightest binding fraction was amplified by PCR after the RNA was reverse transcribed to DNA. These sequences were selected again for the tightest binders to T4 DNA polymerase. After several cycles of selection and amplification only a few different loop sequences remain. Hairpin loops of eight nucleotides were used because translation of T4 DNA polymerase is controlled by binding to an eight-nucleotide hairpin in its messenger RNA. The selection experiment found the naturally occurring hairpin, as expected, but also discovered a quadruple mutant that bound equally tightly.

Randomizing a short oligonucleotide can produce all possible sequences, but for longer molecules the total amount of material synthesized limits the number of possible sequences. A random 50-mer can have 4^{50} different sequences; this is about 10^{30} different molecules. However, this number of molecules is 2×10^6 mol and has a mass of about 3×10^7 kg. Micromole amounts of oligonucleotides is more practical, which corresponds to 6×10^{17} molecules. Thus randomizing sequences longer than about 30 nucleotides ($4^{30} = 10^{18}$) means that only a fraction of the possible sequence space is investigated. However, the method of random synthesis and selection—also called SELEX, in vitro evolution, and directed molecular evolution—is discovering unexpected new binding and catalytic activities of RNA and DNA sequences (Dai et al., 1995; Gold et al., 1995; Huizenga and Szostak, 1995).

It is important, both in natural biosynthesis and after chemical synthesis, to be able to interconvert nucleosides and nucleotides. Nucleosides (either ribo- or deoxyribo-) can be phosphorylated to 5′-phosphates, 5′-diphosphates, and 5′-triphosphates using kinases; these enzymes catalyze the transfer of the terminal phosphate of adenosyl triphosphate (ATP) to a nucleoside, a phosphate, or a diphosphate (Anderson, 1973;

Richardson, 1981). Phosphatases, also called phosphomonoesterases, catalyze the hydrolysis of 2′, 3′, and 5′ nucleotides to nucleosides (Schmidt, 1961; Stadtman, 1961). Phosphodiesterases are nucleases; they catalyze the hydrolysis of polynucleotides to nucleotides. The phosphatases do not attack 2′, 3′-cyclic phosphates; nucleases are needed to hydrolyze these phosphodiesters. A good review of the mechanisms of kinases, phosphatases, and nucleases is given by Knowles (1980). A kinase can be used to radioactively label the 5′-ends of polynucleotides with ^{32}P; if necessary, a phosphatase is first used to remove the nonradioactive phosphate.

2.2 Hydrolysis of DNA and RNA

Making or breaking phosphodiester bonds is obviously central to the synthesis and sequencing of nucleic acids. It is also crucial in understanding RNA processing, catalytic RNA, and the action of RNAses and DNAses.

The mechanism of hydrolysis of a 3′, 5′-phosphate diester is shown very generally in Figure 3-2. The phosphorus in a phosphate is at the center of a tetrahedron of four oxygen atoms, but in the transition state to hydrolysis it becomes the center of a trigonal bipyramid with the attacking group at the fifth site. The leaving group is on a line through the phosphorus and the attacking group; there is an in-line displacement. In RNA, the attacking group is often the 2′-OH of ribose, leading to a 2′, 3′-cyclic phosphate that is opened subsequently by water to produce a mixture of 2′- and 3′- phosphates. This attack by the 2′-OH of ribose is involved in the mechanism of hydroxide-ion catalyzed RNA hydrolysis and most ribonuclease catalyzed reactions. Self-cleaving hammerhead ribozymes leave 2′, 3′-cyclic phosphates (Forster and Symons, 1987), and thus utilize the same mechanism. Figure 3-2(*a*) shows a base, B, removing the hydrogen from the 2′-OH group; this base can be OH^- or a side chain group on an enzyme. The reaction is also favored by acid catalysis, HA, at the leaving oxygen and by a metal ion, or other positive charges, neutralizing the negative phosphate charge in the transition state. This mechanism predicts that there is an inversion of configuration at the phosphorus on hydrolysis. Inversion has been found by using a chiral phosphorus in the reaction; either oxygen isotopes (Knowles, 1980) or sulfur replacing an oxygen (Eckstein, 1985) produced the chiral group.

The same type of mechanism, but with attack by an exogenous hydroxyl group [Fig. 3-2(*b*)], can produce a 5′-phosphate and a free 3′-hydroxyl after hydrolysis. This reaction occurs in DNA, but it also can occur in RNA. The *Tetrahymena* ribozyme, which catalytically cleaves itself and other RNAs, uses an exogenous guanosine for the attacking hydroxyl and leaves a 5′-phosphate (with inversion of configuration) and a free 3′-hydroxyl group (McSwiggen and Cech, 1989).

2.2.1 Nucleases

DNAses can leave either 5′- or 3′-phosphates after cleavage; most, but not all, RNAses leave cyclic 2′, 3′ phosphates that slowly open to give mixed 2′ and 3′ phosphates. Table 3-1 lists properties of some of the more useful nucleases; further discussion of their use will be given in the following sections.

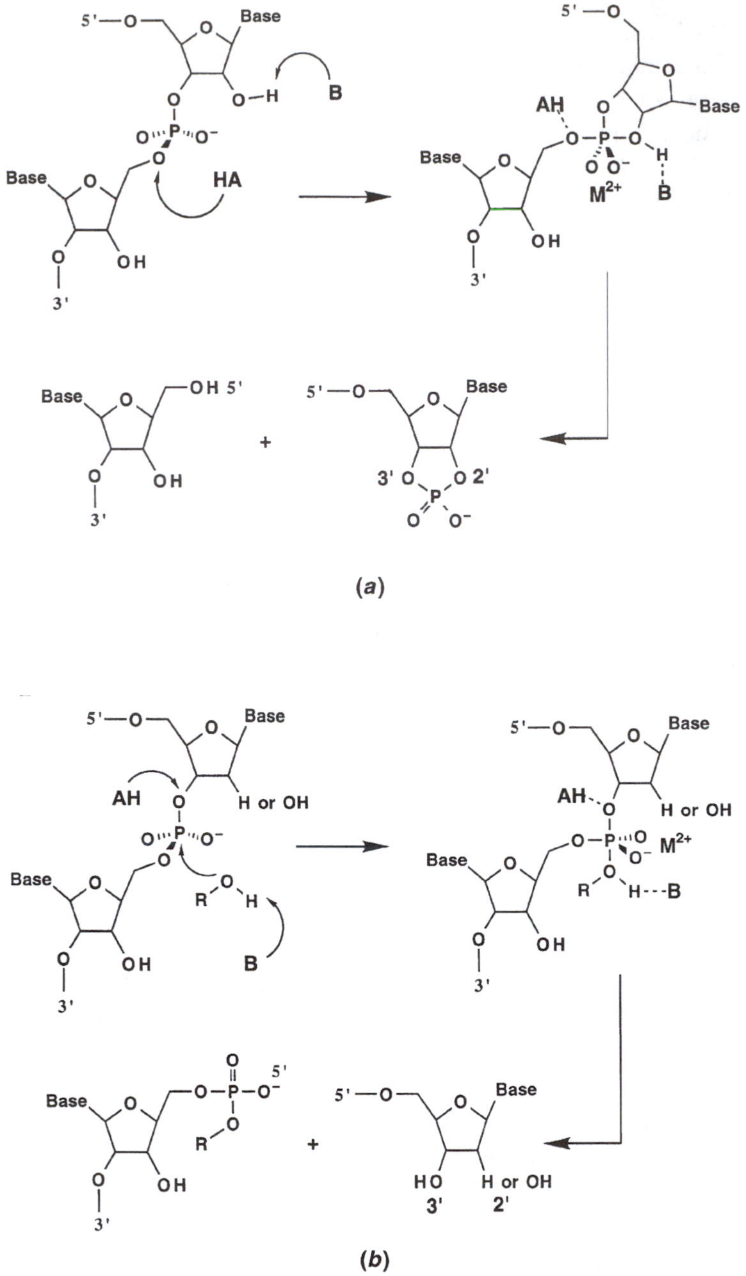

(*a*)

(*b*)

Figure 3-2
A general mechanism for the hydrolysis of phosphodiesters catalyzed by acids (AH), bases (B) and metal ions (M^{2+}). An in-line attack of an oxygen on phosphorus produces a trigonal bipyramide transition state. (*a*) In RNA, the attacking group can be the 2′-OH of the adjoing ribose thus leaving a cyclic 2′,3′-phosphate and a 5′-hydroxyl. (*b*) An external hydroxyl can be the attacking group in DNA or RNA; this leaves a 5′-phosphate and a 3′-hydroxyl.

Table 3.1
Properties of Some Useful Nucleases

Enzyme	Specificity	Phosphates	Reference
RNase T1	G-specific RNA endonuclease	2′, 3′	Ehresmann et al. (1987)
RNase CL 3	C-specific RNA endonuclease	2′, 3′	Ehresmann et al. (1987)
RNase U2	A>G-specific RNA endonuclease	2′, 3′	Ehresmann et al. (1987)
RNase A	C, U-specific RNA endonuclease	2′, 3′	Silberklang et al. (1977)
RNase Phy M	A, U-specific RNA endonuclease	2′, 3′	Donis-Keller et al. (1980)
RNase T2	Single-strand-specific RNA endonuclease	2′, 3′	Ehresmann et al. (1987)
RNAse V1	Double-strand-specific RNA endonuclease	5′	Auron et al. (1982)
RNAse H	RNA endonuclease of RNA–DNA hybrid duplexes	5′	Berkower et al. (1973)
Nuclease S1	Single-strand-specific DNA or RNA endonuclease	5′	Sambrook et al. (1989) p. 5.78
Mung bean nuclease	Single-strand-specific DNA or RNA endonuclease	5′	Laskowski (1980)
Micrococcal nuclease	Single-strand-specific DNA or RNA endonuclease	3′	Sulkowski et al. (1969)
Venom exonuclease (phosphodiesterase I)	3′ DNA or RNA exonuclease	5′	Laskowski (1971)
Spleen exonuclease (phosphodiesterase II)	5′ DNA or RNA exonuclease	3′	Bernardi and Bernardi (1971)
DNAse I	Nonspecific DNA endonuclease	5′	Sambrook et al. (1989) p. 5.83
DNA Exonuclease III	3′ DNA exonuclease	5′	Sambrook et al. (1989) p. 5.85
Nuclease Bal 31	3′ DNA exonuclease, single-strand-selective endonuclease	5′	Sambrook et al. (1989) p. 5.73

2.2.2 *Restriction Enzymes*

Restriction enzymes are endonucleases that cleave DNA double strands at specific sites; their use has led to many of the recent advances in molecular biology. There are several hundred restriction enzymes known; commercial catalogs and updated computer data bases are the best ways to keep up with the increasing variety available. Sambrook et al. (1989) has a useful list (p. 5.3) and a discussion of their use. The recognition sequences of restriction enzymes are usually palindromes; this means the sequence is self complementary. The lengths of the recognition sequence vary from four to eight nucleotides. The longer the recognition sequence the fewer the cuts made in a DNA. An enzyme specific for a sequence of 8 nucleotides would cut about once in 4^8 or 65,536 base pairs. For ease in mapping genomic DNA, enzymes that cut a chromosome at only a few sites would be ideal. There have been attempts to increase the length of the recognition sequence by mutations or by chemical modifications of the enzymes. An alternative method uses a specific binding protein (such as the lac repressor) that can recognize 10–20 nucleotides. The repressor protects nucleotides at the repressor binding site from methylation by a methyl transferase. As restriction enzymes do not cut at methylated sequences all restriction sites not protected by repressor are blocked. When the repressor is removed only this site is available for cleavage; thus a specific cut in 4^{10}–4^{20} or 10^6–10^{12} nucleotides is made (Koob and Szybalski, 1988). Dervan (1992) describes methods based on triple-strand formation to specifically cleave DNA into megabase-sized pieces.

Restriction enzymes can produce blunt ends in the DNA by cleavage at the twofold axis of symmetry of the palindrome, or they can produce cohesive ends by cleaving at sites at either side of the twofold palindrome symmetry axis. The cohesive ends can have either 5′ or 3′ single-strand overhangs. Some restriction enzymes leave terminal 5′-phosphates; others leave terminal 3′-phosphates.

3. DETERMINATION OF SEQUENCE

The first step in characterizing a nucleic acid is to determine the sequence of its nucleotides. A naturally occurring double-stranded DNA is usually too long to sequence completely, so the positions of fragments within the large DNA are determined by mapping. Our emphasis will be on sequencing of pieces of DNA or RNA by enzymatic and chemical methods. The oligonucleotides can be naturally occurring pieces, or be synthesized chemically or enzymatically.

3.1 DNA Sequencing

3.1.1 *Maxam–Gilbert Chemical Sequencing*

The Maxam–Gilbert DNA sequencing method depends on base-specific chemical reactions to break polynucleotide chains specifically at A, G, C, or T (Maxam and Gilbert, 1977, 1980). The logic of sequence determination by chemical reactions is simple and elegant. A polynucleotide is labeled at either its 5′- or 3′-end with ^{32}P then treated with a base-specific reagent (e.g., specific for G), so that less than 1% of the

G's react. Further treatment of the chemically modified polynucleotide causes strand breakage at each modified G. This breakage produces a set of end-labeled fragments whose lengths reveal the positions of G's within the sequence; the lengths can be determined to a resolution of one nucleotide by gel electrophoresis. Of course, many other fragments result, but only the end-labeled ones are detected. Use of four base-specific reactions in separate experiments provides the position of each of the bases relative to the end label; this is the sequence.

The reactions devised by Maxam and Gilbert share important common features: (1) A base-specific reaction either cleaves the aromatic ring of the base or breaks the glycosidic linkage that holds it to the deoxyribose sugar. (2) The first reaction step destabilizes the deoxyribose ring to attack by agents such as piperidine, allowing cleavage of the ring between O and C1′. (3) Conversion of the resulting aldehyde (C=O) function at C1′ to C=N destabilizes the deoxyribose ring fragment by a process called β elimination, which cleaves the C5′–O and C3′–O bonds that hold the backbone of the DNA chain together. (4) The result is conversion of phosphodiesters to phosphomonoesters: the 5′ fragment of the chain terminates in a 3′-phosphate, and the 3′ fragment of the chain begins with a 5′-terminal phosphate.

The Maxam–Gilbert reactions for specific bases are (see Fig. 3-3):

G + A	Piperidine in acid
G	Dimethyl sulfate + piperidine in base
C + T	Hydrazine + piperidine in base
C	Hydrazine, 1.5 *M* NaCl + piperidine in base

The *G + A specific reaction* [illustrated in Fig. 3-3(*a*)] begins with acid-catalyzed depurination (pH 2), a reaction which pyrimidines do not undergo appreciably. Protonation at N7 yields a resonance form with positive charge at N9. The electron-withdrawing character of N9 then destabilizes the glycosidic bond to attack by water (Step 1) at C1′. Once the base is gone, the ribose ring can open to expose the aldehyde function at C1′ to reaction with piperidine (Step 2). The resulting Schiff base compound makes the 2′ proton acidic enough to be removed by piperidine acting as a base. A double bond forms between C2′ and C3′, and the 3′-$ROPO_3^{2-}$ group is released as an anion by a process called β elimination (Step 3). Next, the 4′ proton is abstracted and the β elimination process is repeated, with loss of the 5′ $ROPO_3^{2-}$ group (Step 4). The result is release of the two strands with 3′ and 5′-phosphate groups attached to their ends at the site of cleavage, plus the deoxyribose fragment from the depurinated G or A nucleotide. Note that in this mechanism of DNA cleavage there is no breakage of carbon-carbon bonds in the backbone.

Production of terminal phosphates is an important feature of the Maxam–Gilbert method. Failure of one of the β elimination reactions, for example, would yield a phosphodiester at the end of one of the chains, whose electrophoretic mobility would be substantially smaller than that of a corresponding chain terminated by a phosphomonoester. The reduced mobility results because the phosphomonoester has two negative charges whereas the phosphodiester only has one, and because of the added frictional drag of the ribose ring fragment that forms the diester.

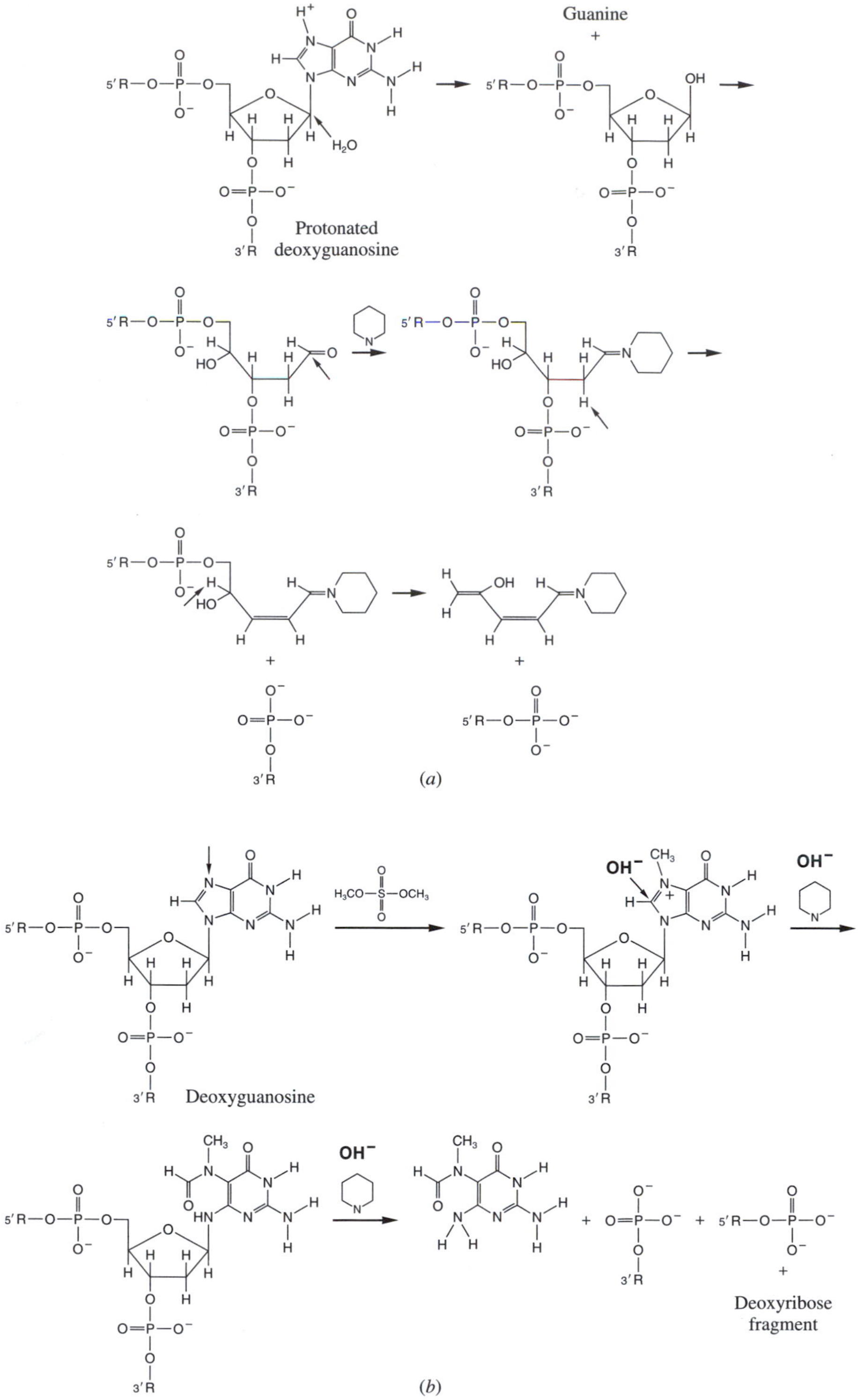

Figure 3-3
The chemistry involved in the sequencing of DNA by the Maxam–Gilbert method. (*a*) The G + A specific reaction; (*b*) G specific reaction; (*c*) C + T specific reaction.

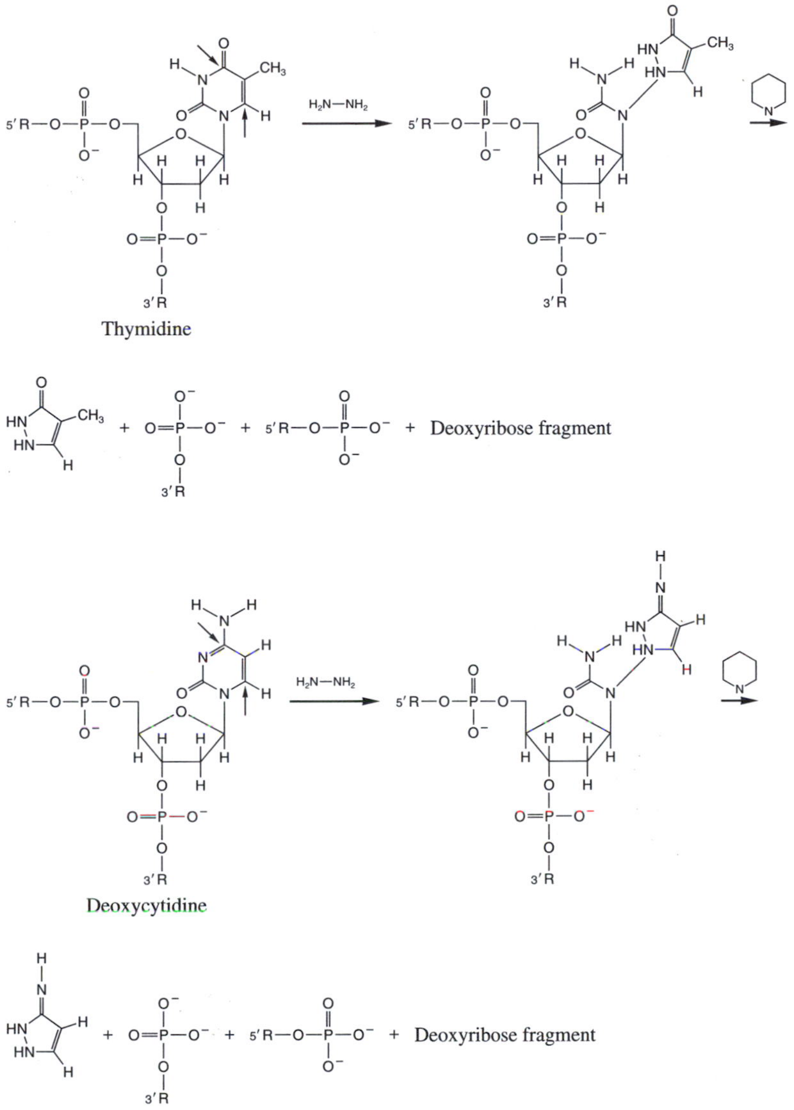

Thymidine

$H_2N{-}NH_2$

+ Deoxyribose fragment

Deoxycytidine

$H_2N{-}NH_2$

+ Deoxyribose fragment

(*c*)

The *G specific reaction* involves methylation of guanine at N7 with dimethyl sulfate [Step 1 in Figure 3-3(*b*)]. Other bases are also methylated; but the next step, addition of piperidine at pH 8, cleaves the chain only at methylated G residues. Methylated G is subject to ring opening by attack of OH^- at C8 (Step 2), which is followed by piperidine displacement of the glycosidic linkage, breakage of the C1′–O bond and β elimination to produce 5′- and 3′- phosphate chains.

The *C + T specific reaction* is illustrated in Figure 3-3(*c*). In the composite Step 1, hydrazine adds across the 5,6 double bond, and then attacks the 4-carbonyl function to yield a new five-membered ring. This structure degrades spontaneously, opening the original pyrimidine ring. In step 2, elimination of the 5-membered ring is accompanied by breakage of the ribose ring at the C1′–O bond. As before β elimination produces 5′- and 3′- phosphate chains.

The *C specific reaction* differs in that NaCl is added to 1.5 *M*, conditions that strongly favor the reaction of C over T.

The aliquots from the four reactions are placed in adjacent wells in a polyacrylamide gel. After electrophoresis of the oligonucletide chains and detection of the radioactive bands with photographic film or luminescent plates, the sequence is read from the sequencing ladders.

3.1.2 Sanger Dideoxy Sequencing

The logic of sequencing in the Sanger method is the same as that of Maxam and Gilbert—the length of an end-labeled oligonucleotide that terminates at one type of base reveals the position of that base. Instead of using chemical reagents to cause specific strand scission, a complementary DNA strand is synthesized. A small fraction of a single type of dideoxynucleoside triphosphate [e.g., dideoxy guanosine 5′-triphosphate (ddGTP)] is used to cause chain termination during synthesis (Sanger et al., 1977). Gel electrophoresis is used to determine the chain lengths of the nested set of end-labeled oligonucleotides.

Synthesis of DNA is done by DNA polymerase that requires a primer (a complementary oligonucleotide with a free 3′-end) to start synthesis. Thus, either single- or double-stranded DNA can be used as a template; the primer determines which strand and what region of the strand will be sequenced. The four triphosphates (ATP, CTP, GTP, TTP) are added plus a few percent of one dideoxy NTP. The dideoxy nucleoside lacks the 3′-OH necessary to continue the chain, so when it is incorporated by DNA polymerase the growing chain stops.

The newly synthesized chains can be radioactively labeled so that gel electrophoresis of four separate reactions produces sequencing ladders. Alternatively, a different fluorescent group can be covalently attached to each different dideoxynucleotide. Now the four reactions are all done in one solution and only one electrophoresis ladder is made. Each rung of the ladder is identified by the characteristic fluorescence maximum of the dideoxynucleotide (Smith et al., 1986). The ladder is scanned with a laser and multiple detectors; software provides a printed sequence complete with question marks at uncertain positions.

Dideoxy sequencing is the most commonly used method, however, regions of high GC content cause problems. Base pairing and other self-structure of the nucleotide chains alter their electrophoretic mobility and cause compression and overlap of the sequencing ladder. Two cures that have been used are replacement of inosine (ITP) for guanosine (GTP) in the polymerization reaction to reduce base pairing, and using higher temperaures for the electrophoresis (Sambrook et al., 1989).

3.2 RNA Sequencing

3.2.1 Reverse Transcriptase Sequencing

One way to determine the sequence of an RNA strand is to first convert it to a DNA strand. Reverse transcriptase is a DNA polymerase that uses either single- stranded

DNA or RNA as a template (Verma, 1981; Hahn et al., 1989). It can thus be used to sequence RNA (or DNA) by the Sanger dideoxy method. Reverse transcriptases are also very useful in determining secondary structure; this is discussed in Section 4.2.

3.2.2 Peattie Chemical Sequencing

A method for sequencing RNA with chemical reagents adapted from the Maxam–Gilbert DNA method was devised by Peattie (1979). Aniline at pH 4.6 (instead of piperidine used for DNA) cleaves the RNA chain specifically at the site of each chemically modified base. (RNA is too sensitive to base-catalyzed hydrolysis to allow use of piperidine.) The reaction mechanism is analogous to the piperidine cleavage illustrated in Figure 3-3. However, the second β elimination step does not occur. Hence aniline cleavage leaves a 5′-phosphate and a heterogenous 3′-end (Kochetkov and Budovskii, 1972). Thus 3′-labeled RNA strands are used for the sequence analysis. The base-specific chemical reactions are similar to those for DNA: G $-$ dimethyl sulfate $+$ aniline, U $-$ hydrazine $+$ aniline, C $>$ U $-$ hydrazine $+ 3\,M$ NaCl $+$ aniline, A $>$ G $-$ diethylpyrocarbonate $+$ aniline. The diethylpyrocarbonate reaction (shown in Fig. 3-4) replaces the acid-catalyzed depurination step used in DNA sequencing. Depurination and depyrimidination occur 100–1000 times slower in RNA than in DNA (Kotchekov and Budowskii, 1972).

3.2.3 Ribonuclease Sequencing

Base-specific or base-selective ribonucleases are very useful in the sequence characterization of RNA. For short oligonucleotides, partial digestion with nucleases may suffice to give the entire sequence. Base-specific ribonucleases are listed in Table 3-1. The ribonucleases cut on the 3′-side of the specific nucleotide to leave an oligonucleotide, or nucleotide, containing the specific base at the 3′-terminus. There is a mixture of 2′, 3′ cyclic phosphates, 2′-phosphates and 3′-phosphates on the 3′-end. To determine the sequence, RNA molecules labeled at the 3′- or 5′- end are partially hydrolyzed with the base-specific nucleases in Table 3-1 and the fragments separated by gel electrophoresis. Comparison with a hydrolysis ladder produced by hydroxide ion, which is not base specific but also leaves 3′-phosphates, provides the sequence. As some modified nucleotides are not recognized by ribonuclease, the OH^- hydrolysis insures cleavage after every nucleotide.

Complete digestion of RNAs with base-specific nucleases can also be used to characterize their sequences (Kuchino and Nishimura, 1989). In fingerprinting, two or more nucleases are used to hydrolyze an RNA completely. The fragments from each reaction are separated in two dimensions using paper electrophoresis in the first dimension and thin-layer chromatography (TLC) in the second (Branch et al., 1989a). This fingerprinting method can locate modified nucleotides including branched 2′- and 3′-linked ribose groups; it is especially helpful in characterizing small amounts of natural RNAs.

4. DETERMINATION OF SECONDARY STRUCTURE

Reaction with chemical reagents is an excellent method to probe the accessibility of each reactive site on every nucleotide. The reactivity of each base is unique, but the specific environment of each nucleotide also modulates the reactivity of each atom. For double-stranded DNA or RNA, the conformation (A or B or Z) affects the reactivity. The junctions between different conformations may be most reactive. Small differences in structure, such as widening or narrowing of major or minor gooves caused by specific sequences, will also change the reaction rates. The precise locations of binding of proteins or drugs can be determined by the differences in reactivities of the bound and naked nucleic acid. For RNA, covalent reactions can reveal which bases are single-stranded (unbonded and usually most reactive), which are in base-paired secondary structures, and which are involved in tertiary structures. Single-stranded regions can also occur in double-stranded DNA when the sequence is a palindrome—an inverted repeat, which can lead to hairpin structures on opposite strands called a cruciform; the hairpin loops can be detected by single-strand-specific reagents.

4.1 End-Labeling Nucleic Acids

It is usually necessary to end-label an RNA or DNA before studying its structure. Labeling at the 5′-end is done with [$\gamma-{}^{32}$P]ATP and T4 polynucleotide kinase (Richardson, 1981). A convenient 3′- labeling method uses T4 RNA ligase to add radioactively labeled pNp to the 3′-OH end of RNA or DNA (Uhlenbeck and Gumport, 1982).

A classical method of labeling RNA at the 3′-end involves oxidation of the vicinal 2′, 3′-OH groups with periodate (IO_4^-) to 2′, 3′-dialdehydes (Winter and Brownlee, 1978). The dialdehyde is not stable, but will react further with amines, for example; this reaction can be used to attach the 3′-end of an RNA to a column. The dialdehyde can be reduced to the stable dialcohol with borohydride (BH_2^-); this method can be used for 3′-end labeling RNA with tritium. If the dialdehyde is treated with base, β elimination occurs to remove the terminal ribose and to release a 5′-phosphate. Periodate treatment of the 3′-end of an RNA followed by β elimination can thus be used to remove the 3′-terminal nucleoside; this reaction has been used to remove the reactive terminal guanosine in the catalytic intron of *Tetrahymena* ribosomal RNA (Tanner and Cech, 1987).

4.2 Chemical Reactivity of Nucleic Acids

The two methods commonly used to determine relative reactivity of nucleotides in a nucleic acid are chemical cleavage of an end-labeled chain at the site of the modified base and chain termination by reverse transcriptase at this site. The chemical cleavage reactions using aniline for RNA and piperidine for DNA were described in Sections 3.3.1 and 3.3.2 on sequencing. Reverse transcriptase can make complementary copies of RNA or DNA single strands, but it stops or pauses at modified bases, or at the end of the strand where cleavage has occurred (Ehresmann et al., 1987; Stern et al., 1988a). A primer is added, which is complementary to the region to be analyzed; the primer is the start of the copy made by reverse transcriptase. The end of the copy occurs at the reacted base, therefore the length of the copy identifies the location of this base. A standard sequencing reaction is run in parallel to provide the exact position. The primer extension method can be used for large molecules, but is less practical for oligonucleotides of less than 20 nucleotides.

Table 3-2 lists reagents commonly used for probing secondary and tertiary structure; a useful review of some the chemical reactions is given by Sigman and Chen (1990). Double-stranded regions react more slowly with these reagents than do single-stranded regions. Reactions are shown in Figures 3–4 through 3–6.

Dimethyl sulfate DMS [$CH_3O(SO_2)OCH_3$] is an example of alkylating agents that have been used on nucleic acids; others include methyl methane sulfonate [$CH_3(SO_2)OCH_3$] and *N*-ethyl *N*-nitrosourea, ENU [$CH_3CH_2N(NO)CONH_2$]. Essentially all the nitrogen and oxygen atoms in the bases, sugars and phosphates can be alkylated in aqueous solution (Singer and Kusmierek, 1982; Singer and Grunberger, 1983). However, the main sites of alkylation on the bases are illustrated for DMS in Figure 3-4. Reaction occurs at N1 of adenine, N3 of cytosine, and N7 of guanine. The reactions with adenine and cytosine are greatly reduced in double-stranded regions; the reaction with guanine is affected by stacking and by tertiary interactions. Alkylation of guanines makes the nucleotides susceptible to depurination; further treatment can cause strand scission. Cleavage at the 3-methylcytosine is done after mild hydrazinolysis. The N1 methyladenine is detected by its stopping of reverse transcriptase.

N-Ethyl *N*-nitrosourea preferentially alkylates phosphates in nucleic acids to form triesters that cause cleavage of the chain in mild alkali (Vlassov et al., 1981). Clearly, there is no specificity to the reactivity of each phosphate except for its conformation and its interactions. Thus, ENU provides a useful complement to the set of reagents that react with the bases.

Diethylpyrocarbonate reacts with N7 of adenine (Fig. 3-4) and less strongly with N7 of guanine and with the amino group of cytosine (Ehrenberg et al., 1976; Peattie and Gilbert, 1980). The modification is detected by strand scission at the reacted base. In duplexes, the N7 of adenine is in the major groove, which slows the reaction with DEP. Stacking is also important in the reaction.

The reagent CMCT (Moazed et al., 1986) is a carbodiimide that reacts at guanine N1 and uracil or thymine N3; kethoxal (Moazed et al., 1986) reacts as other glyoxals do at guanine N1–N2; and bromacetaldehyde (Lilley, 1983) adds across N1–N6 of adenine (see Fig. 3-5). Thus these reagents monitor Watson–Crick hydrogen-bonding sites and are good probes of duplex formation. The reagents do not lead to cleavage of the chain so detection is done by using reverse transcriptase.

Reagents that react specifically with pyrimidines (see Fig. 3-6), and react faster in single-stranded regions include hydroxylamine (H_2NOH), bisulfite ion (HSO_3^-), osmium tetroxide (OsO_4), permanganate ion (MnO_4^-) and hydrazine (H_2NNH_2). Hydrazine is mainly used in sequencing and was described earlier (Fig. 3-3). The hydroxylamine reaction with uridine is similar to that of hydrazine; with cytosine hydroxylamine forms hydroxyaminocytosine (Phillips, 1967). Bisulfite adds across the 5–6 double bond of pyrimidines (Hayatsu, 1976); the modified bases can be detected by treatment with hydrazine, then cleavage with aniline. Bisulfite catalyzes the exchange of deuterium or tritium for hydrogen in the 5 position, thus it is useful for deuteriating U, T, or C in oligonucleotides to simplify NMR spectra (Brush et al., 1988).

Table 3.2
Chemical Probes Used for Nucleic Acid Structure Determination[a]

Probe	Specifity	Cleavage	Reference
Dimethyl sulfate (DMS)	Adenine N1	No	Singer and Grunberger (1983)
Dimethyl sulfate (DMS)	Cytosine N3	Yes	Singer and Grunberger (1983)
Dimethyl sulfate (DMS)	Guanine N7	Yes	Singer and Grunberger (1983)
Ethylnitrosourea (ENU)	Phosphates	Yes	Vlassov et al. (1981)
Diethylpyrocarbonate (DEP)	Adenine N7	Yes	Ehrenberger et al. (1976)
Diethylpyrocarbonate (DEP)	Guanine N7	Yes	Ehrenberger et al. (1976)
CMCT[b]	Guanine N1	No	Moazad et al. (1986)
CMCT[b]	Uracil or thymine N3	No	Moazad et al. (1986)
Kethoxal[c]	Guanine N1–N2	No	Moazad et al. (1986)
Bromoacetaldehyde	Adenine N1–N6	No	Lilley (1983)
Hydroxylamine	Uracil or thymine C4–C6	Yes	Phillips (1967)
Hydroxylamine	Cytosine C5–C6	No	Phillips (1967)
Bisulfite	Pyrimidine C5–C6	No	Hayatsu (1976)
Osmium tetroxide	Pyrimidine C5–C6	Yes	Lilley and Palacek (1984)
Permanganate	Pyrimidine C5–C6	Yes	Hayatsu and Ukita (1967)
Hydrazine	Pyrimidine C4–C6	Yes	Maxam and Gilbert (1980)
Ni^{2+}-Complex[d]	Guanine N7	Yes	Chen et al. (1993)
2:1 1,10-Phenanthroline: Cu^{2+}	Single strand	Yes	Sigman (1986)
Ethylenedianinetetraacetic acid EDTA–Fe^{2+}	Solvent accessible regions	Yes	Dervan (1986)
Methidiumpropyl–EDTA–Fe^{2+}	Double strand	Yes	Hertzberg and Dervan (1984)
Uranyl acetate	Nonselective (light activated)	Yes	Gaynor et al. (1989)
$Rh(phen)_2phe^{3+}$[e]	Tertiary interaction sites (light activated)	Yes	Barton (1986)

[a]Jaeger et al. (1993).

[b]1-Cyclohexyl-3-(2-morpholinoethyl) carbodiimide methane-p-toluene sulfonate.

[c]β -Ethoxy - α - ketobutyraldehyde.

[d][2,12-Dimethyl-2,7,11,17-tetraazabicyclo[11.3.1]heptadeca-1(17),2,11,13,15-pentaene]nickel(II).

[e]Bis(phenathroline)(phenanthrenequinonediimine)-rhodium(III).

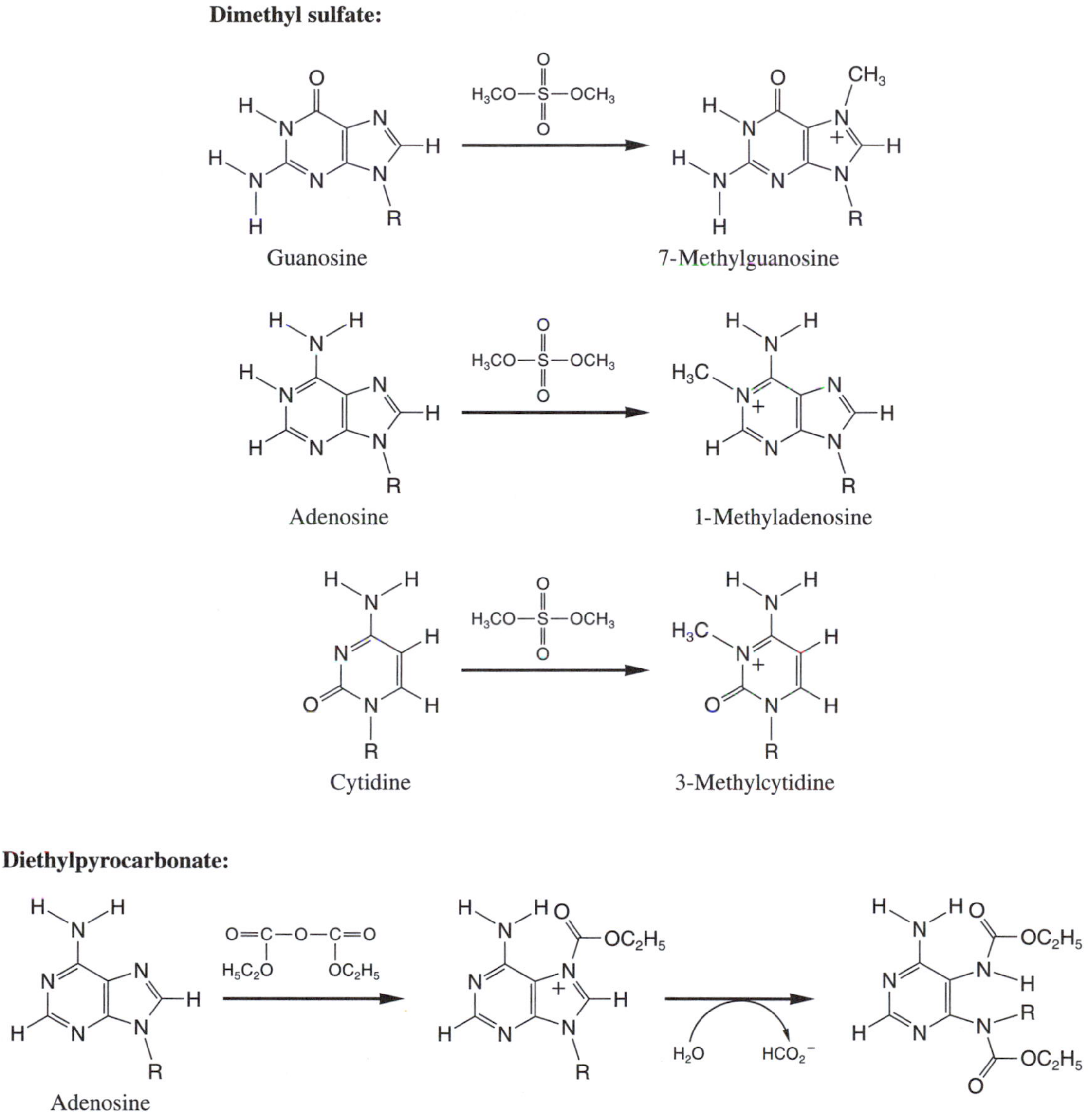

Dimethyl sulfate:

Guanosine → 7-Methylguanosine

Adenosine → 1-Methyladenosine

Cytidine → 3-Methylcytidine

Diethylpyrocarbonate:

Adenosine

Figure 3-4
Dimethyl sulfate and other alkylating reagents react with adenine, guanine, and cytosine to give *N*-methyl derivatives. Diethylpyrocarbonate reacts most rapidly with N7 of adenine as illustrated. It also reacts slowly with N7 of guanine and the amino group of cytosine.

Osmium tetroxide in the presence of pyrimidine adds across the 5–6 double bond of single-stranded pyrimidines (Lilley and Palacek, 1984). After further treatment, both osmium tetroxide and hydroxylamine can lead to strand cleavage at the reactive site. Permanganate ion oxidizes the 5–6 double bond of pyrimidines (with preference for thymine and uracil over cytosine) to the 5,6-dihydroxy compound (Hayatsu and Ukita,

Kethoxal (β-ethoxy-α-ketobutyraldehyde):

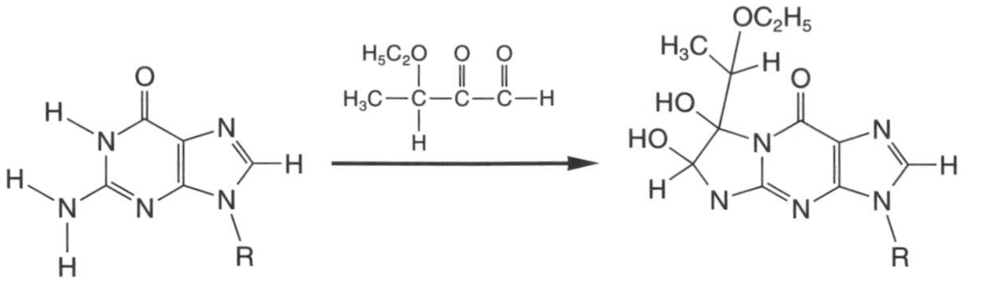

Guanosine

Carbodiimide (CMCT):

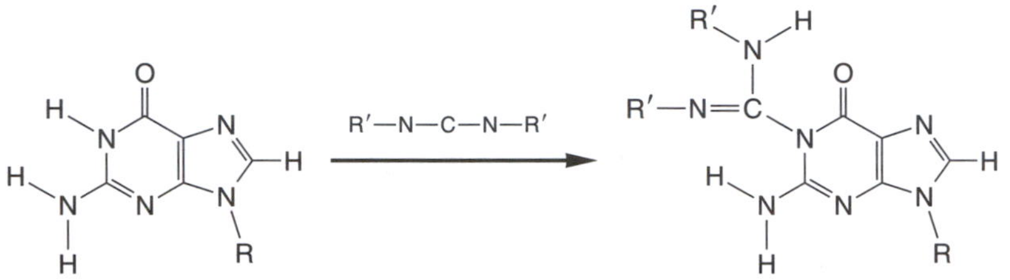

R′—N—C—N—R′

Guanosine

R′—N—C—N—R′

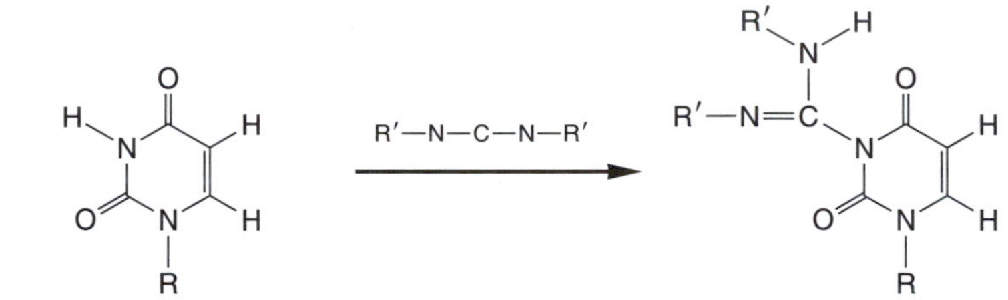

Uridine

Bromoacetaldehyde:

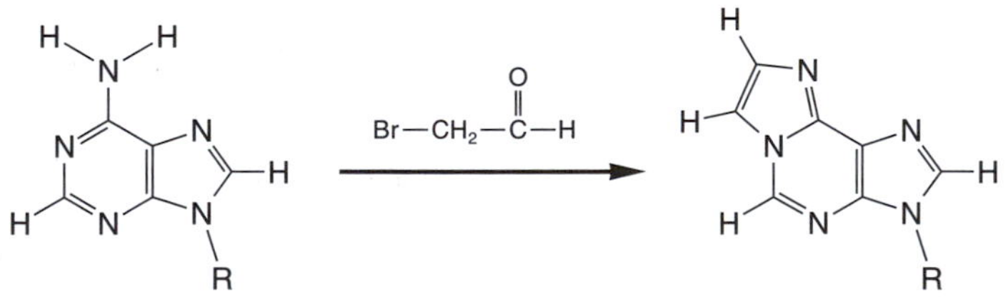

Adenosine

1-N6-Ethenoadenosine

Figure 3-5
Glyoxals (kethoxal), carbodiimides (CMCT) and bromoacetaldehyde all react at the Watson–Crick hydrogen-bonding sites of bases. They directly monitor base pairing.

Bisulfite:

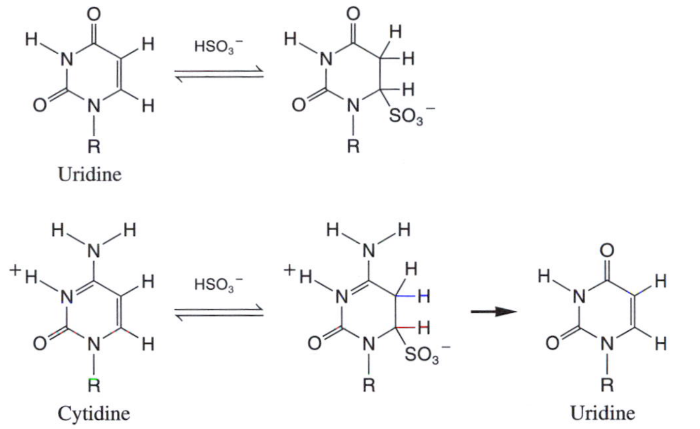

HSO_3^-

Uridine

HSO_3^-

Cytidine

Uridine

Osmium tetroxide:

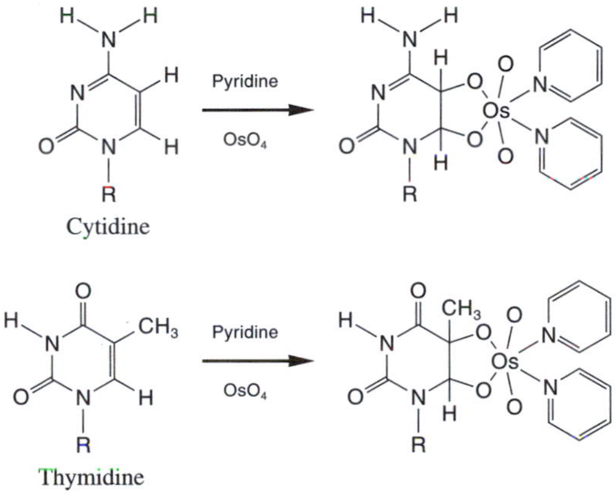

Pyridine

OsO_4

Cytidine

Pyridine

OsO_4

Thymidine

Hydroxylamine:

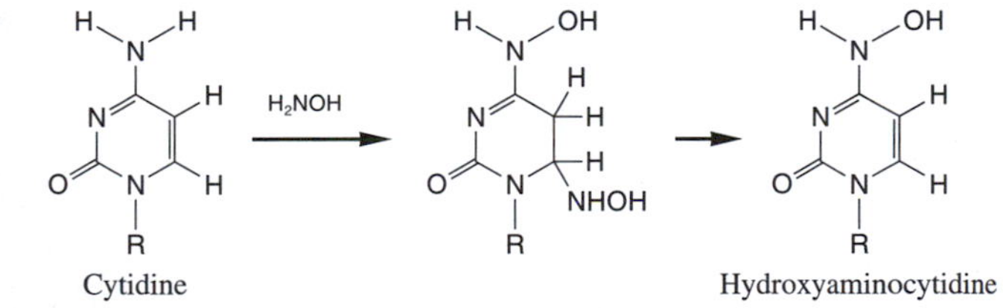

H_2NOH

Cytidine

Hydroxyaminocytidine

Permanganate (MnO_4^-):

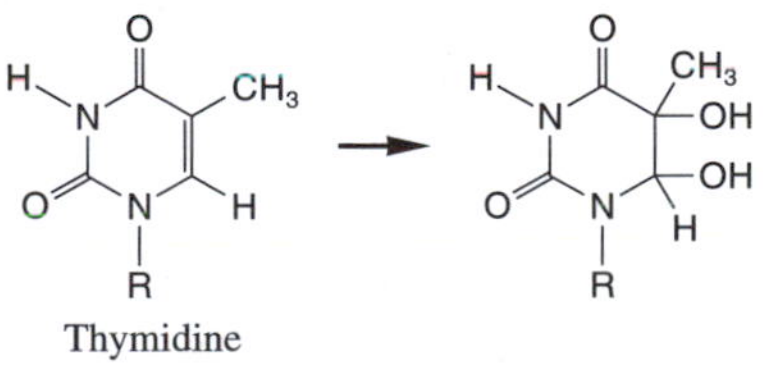

Thymidine

Figure 3-6
Bisulfite ion, osmium tetroxide and hydroxylamine all add across the 5–6 double bond of pyrimidines. Permanganate ion oxidizes this bond to form the dihydroxy derivative.

1967; Rubin and Schmid, 1980). The reaction can be monitored by isolation of the modified nucleotide or by chemical cleavage at the site.

A Ni^{II} complex that reacts at N7 of guanine and leads to chain scission after oxidation with persulfate (Chen et al., 1992; Chen at al., 1993) has the specificity of T1 ribonuclease, but reacts with both DNA and RNA. It preferentially reacts with single-stranded guanines, but the small size of the reagent apparently makes it more reactive than T1 at partially solvent-accessible regions, such as ends of double helices and small loops.

4.3 RNA Secondary Structure

4.3.1 Chemical Probes

Nearly all the reagents in Table 3-2 have been used to study secondary structure in RNA. Ribosomal RNAs (Noller, 1984; Moazed et al., 1986; Romby et al.,1987) and tRNAs (Peattie and Gilbert, 1980; Dock-Bregeon et al., 1989) have been particularly popular, but all types of RNAs have been probed. Applications to RNA are reviewed in Ehresmann et al. (1987), and useful articles are found in Dahlberg and Abelson (1989).

4.3.2 Nucleases

The base-specific nucleases listed in Table 3.1 are all more active on single- than double-stranded regions; they can thus be used to learn about conformation. In fact, extra stable secondary structures can interfere with their use in sequence determination. Nucleases often used to probe RNA secondary structure are nuclease S_1, which hydrolyzes single-stranded RNA or DNA, and RNAse V_1, which is specific for double-stranded sequences of RNA. A review of the applications of nucleases in RNA structure determination is given by Rajbhandary et al. (1982), Ehresmann et al. (1987), and Knapp (1989).

Nuclease S_1 from *Aspergillus oryzae* (Vogt, 1973) hydrolyzes single strands independent of sequence to leave 5′-phosphates. Nuclease S_1 digestion can be done at 85°C to produce a relatively uniform hydrolysis ladder to compare with other enzymes that leave 5′-phosphates, such as RNase V_1. Nuclease digestion is better than a OH^- hydrolysis ladder that leaves oligonucleotides with 3′-phosphates which have different mobilities in electrophoresis. Nuclease S_1 hydrolysis is inhibited by double-strand formation and by tertiary folding of RNA; in $tRNA^{Phe}$ only the anticodon loop is cut (Wrede et al., 1979). In general, S_1 will cut hairpin loops, internal loops, and bulges (of more than one nucleotide) and is very useful in assessing RNA folding (Ehresmann et al., 1987).

Ribonuclease V_1 is an endonuclease from cobra venom (Lockard and Kumar, 1981) that hydrolyses RNA phosphodiesters in double strands and in stacked single strands. It leaves 5′-phosphates but unlike S_1 it does not cut DNA efficiently. Cleavage is not sequence-specific but it does vary within a duplex region. Each strand of the duplex is cut independently; in Escherichia coli $tRNA^{Phe}$ the 5′ side of the acceptor stem is cut strongly but the 3′ side is not (Auron et al., 1982). RNAse V_1 cuts in stacked regions of internal loops or hairpin loops (Romby et al., 1988; Wyatt et al., 1990). It seems to require a sequence of four to six duplex or stacked nucleotides to provide a site for hydrolysis (Lowman and Draper, 1986).

A combination of enzymes and chemical reagents makes a powerful tool for determining RNA secondary and tertiary structure. Only radioactive amounts of material are needed for analysis, and the differences in sizes of the different probes will monitor a wide range of steric barriors. The enzymes work over a surprising range of temperatures and solvent conditions; high concentrations of enzymes produce useful results even in environments where they have only marginal activity. The chemical reagents can of course cover an even wider range of conditions.

In interpreting the data, it is important to remember the limitations of the methods. The size of an enzyme may limit its access to a susceptible site. The specificities of the enzymes and the reactivities of the probes are not always well understood. Because a kinetic effect is being measured, not an equilibrium one, a conformation present in small concentration but with a very high reactivity will be greatly over represented in the concentrations of products formed. Furthermore, binding of an enzyme can perturb equilibria.

4.4 DNA Secondary Structure

Many of the reagents and enzymes discussed above are also useful to probe DNA secondary structure. Here the main questions are how does the sequence affect the double helix conformation and where do the rare single-stranded regions occur. The formation of unusual structures, such as cruciforms, triple strands, junctions, and so on in DNA (sometimes induced by negative supercoiling) are one important cause of unpaired bases (see a review by Tullius, 1991). The use of chemical reagents to determine where proteins bind—footprinting—is described in Chapter 12.

4.4.1 Chemical Probes

Negative supercoiled DNA favors structures with decreased winding in the Watson–Crick duplex and thus induces cruciform formation. The strands of the DNA unwind and form hairpins on opposite strands at sequences that are palindromes: inverted repeats. The bases in the loops of the hairpins are hyperreactive to single-strand-selective reagents, such as bromoacetaldehyde, glyoxal, osmium tetroxide, and bisulfite (Lilley, 1983; Furlong et al., 1989). Osmium tetroxide OsO_4 has been very useful in detecting crucifrom formation *in vitro* and *in vivo* (McClellan et al., 1990). The junctions between the B- and Z-forms of DNA act like loop regions and are especially reactive with single-strand-selective reagents (Johnston and Rich, 1985).

Triple-strand formation is induced in polypurine-rich regions of DNA by negative supercoiling and by low pH. The formation of the triplex releases a single strand that is hyperreactive to single-strand-selective chemical reagents (Wells et al, 1988). The guanines in the triplex are protected from alkylation at N7 by DMS (Voloshin et al., 1988).

Both chemical probes and some naturally occurring drugs can cleave DNA by an oxidative mechanism that begins by abstraction of a hydrogen atom from the deoxyribose ring. The resulting radical reacts further with oxidants such as O_2 or H_2O_2, leading to chain cleavage. The products of the reaction depend on which hydrogen is removed. The products of $1'$ and $4'$ hydrogen abstraction have been characterized by

Sigman and co-workers (Pope et al., 1982; Spassky and Sigman, 1985; Kuwabara et al., 1986; Thederahn et al., 1989). Abstraction of a 4′ hydrogen leads to oxidative cleavage of the sugar ring, yielding a 5′ chain fragment that has a phosphoglycolate residue on its 3′-end. Because of the extra charge from the carboxylic acid, this product migrates slightly faster on gel electrophoresis than the corresponding 3′ phosphate marker from the Maxam–Gilbert cleavage reaction. The 3′-chain fragment undergoes a β-elimination reaction to yield a 5′-terminal phosphate and a base propenal. Abstraction of hydrogen from C1′ produces a 3′ chain fragment that has a 5′-terminal phosphate as a result of a β-elimination reaction. A transient product can be detected in which the residue of the sugar ring remains attached to the 5′-chain fragment, but this reacts further in a β-elimination reaction to yield a 3′-terminal phosphate and 5-methylene-2-furanone as the remnant of the deoxyribose ring.

Cleavage that begins by abstraction of the 5′ hydrogen in the presence of O_2 releases the 5′-terminal chain fragment with a terminal 3′-phosphate attached. The 3′-chain fragment retains the oxidized deoxyribose ring, with a mixture of products corresponding to aldehyde and carboxylic acid functionalities at the 5′ carbon (Kappen and Goldberg, 1983; Kappen et al., 1987). The chain terminated by the carboxylic acid residue migrates approximately one nucleotide slower than the marker resulting from Maxam–Gilbert cleavage of the same nucleotide; whereas the aldehyde-terminated chain, being uncharged, migrates approximately two nucleotides slower. If the oxidative cleavage is carried out in strong base (OH^-), H4′ in the aldehyde (but not the carboxylic acid) is sufficiently acidic to enable β elimination to take place. Therefore the aldehyde product is replaced in this case by a 3′-terminal chain fragment carrying a 5′-phosphate.

Hydroxyl radicals are oxidative agents that can be formed by ferrous ions in the presence of oxygen and a reducing agent; they cleave the sugar-phosphate backbone of polynucleotides. An iron(II) complex of ethylenediammine tetraacetate, $Fe^{II}\cdot$(EDTA), is an effective cleaving reagent. Sequence-specific cleavage of duplex DNA is attained by connecting Fe(EDTA) to an oligonucleotide that forms a triple helix at the desired cutting site (Moser and Dervan, 1987). An iron ethidium-propyl-ethylenediaminetetra-acetic acid (methidiumpropyl EDTA or MPE) complex is specific for double-stranded DNA because it intercalates into the duplex (Hertzberg and Dervan, 1982, 1984). The (MPE) Fe^{II} complex plus O_2 is proposed to produce hydroxyl radicals (HO$\cdot$) near the sugar residues in the minor groove of the DNA. The first step of the chain cleavage reactions is thought to be hydrogen abstraction by the HO$\cdot$ at the C1′ or C4′ positions of the deoxyribose ring.

A novel application of hydroxyl radical cleavage is in determining the number of base pairs per turn of duplex DNA in solution. The DNA is adsorbed on calcium phosphate and Fe^{II}(EDTA) is added. In solution, the DNA is cleaved uniformly at every nucleotide, but cleavage of the adsorbed DNA is decreased at nucleotides in contact with the solid salt surface. A modulation of the cutting pattern is obtained (see Fig 3.7) with a periodicity of 10.5 bp turn (Tullius and Dombroski, 1985). This study also provides evidence for sequence variability in DNA since 5′-pyrimidine-purine-3′ sequences were observed to cleave more slowly than other dinucleotide sequences.

Iron(II) complexes have been mainly applied to structural studies on DNA (Dervan, 1986; Jezewska et al., 1989), but they have also been used to probe solvent accessibility of RNA (Latham and Cech, 1989; Wang and Cech, 1992).

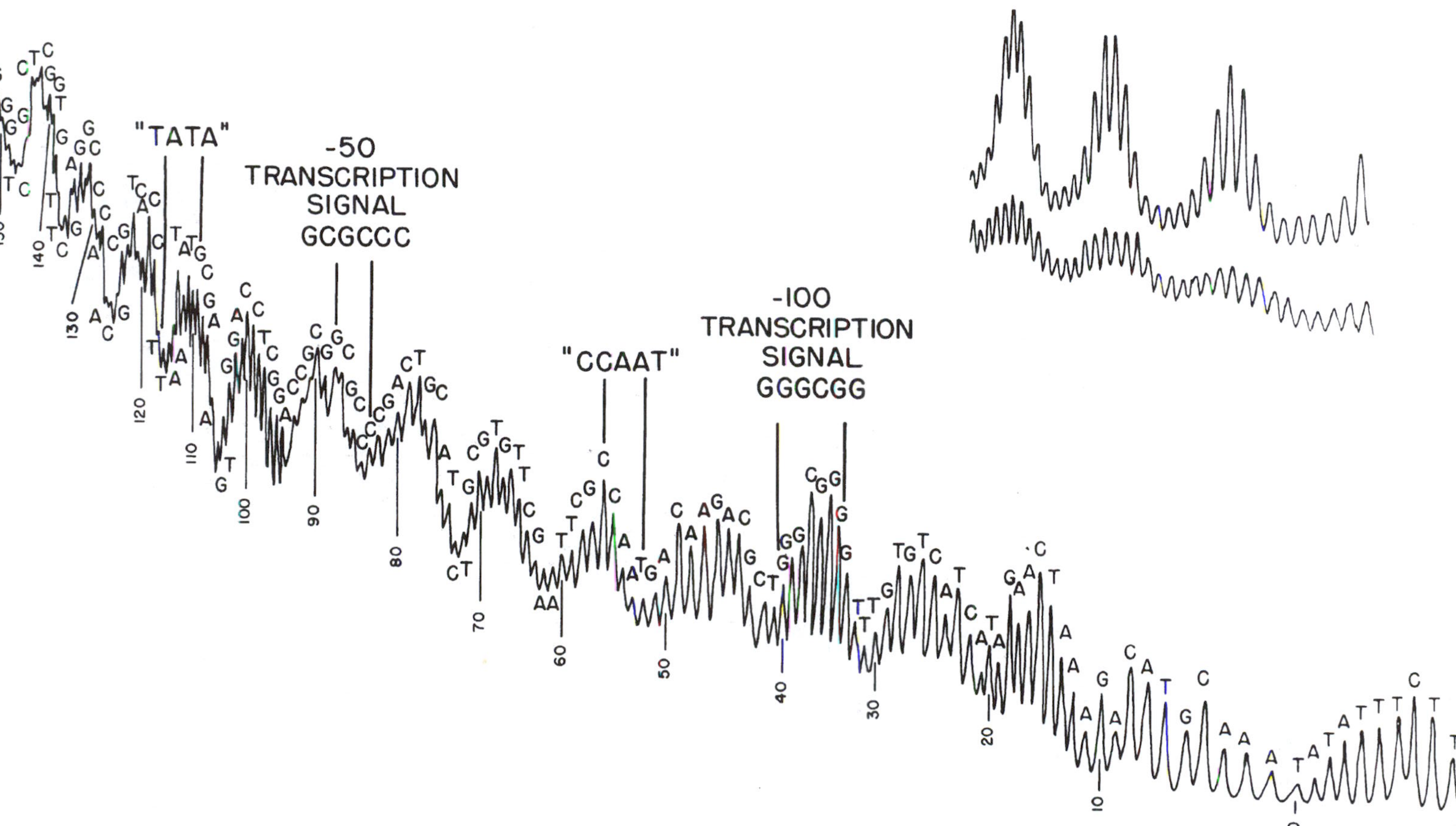

Figure 3-7
The Fe • EDTA generated hydroxyl radicals cleave a DNA adsorbed to a crystal suface. The modulation of the reactivity shows that this duplex has an average of 10.5 bp/turn, but that there is also sequence-specific variability. [Reprinted with permission from Tullius, T. D., and Dombroski, B. A. (1985) Iron (II) EDTA used to measure the helical twist along any DNA molecule, *Science* **230**, 679–681. Copyright©1985, American Association for the Advancement of Science].

Light-activated uranyl acetate has similar properties to Fe^{II} (EDTA) (Gaynor et al., 1989). It cleaves both DNA and RNA nonsequence specifically at solvent accessible nucleotides.

An oxygen-dependent cleavage of DNA by a 1,10-phenanthroline cuprous complex has useful structural specificity. The $[Cu^{II}(phen)_2]^{+2}$ cleavage has mechanistic similarities to the hydroxyl radical cleavage (Sigman et al., 1979; Sigman, 1986). The mechanism involves attack at the C1′ and C4′ sugar ring positions, but the metal–ion complexed active species $[CuO]^{+1}$ and H_2O_2 as cooxidant rather than freely diffusing HO• are likely to be the oxidizing species (Goyne and Sigman, 1987; Thederahn et al., 1989). A-form poly(rA)•poly(dT) helix is cleaved 10–20% as rapidly as the B-form poly(dA)•poly(dT) (Pope and Sigman, 1984). However, copper-phenanthroline cleavage is also useful for studying RNA structure (Wang et al., 1991). Under similar conditions in 3*M* NaCl, poly(dG-dC) is not cleaved, presumably because this helix is in Z form. Preferred cleavage has also been demonstrated in the Pribnow boxes of the Ps and L8-UV-5 lac promoters (Spassky and Sigman, 1985) suggesting an unusual helical structure of the Pribnow box. Sequence-specific cleavage of single-stranded DNA has been accomplished by linking 1,10-phenanthroline to the 5′-end of an oligonucleotide. After hybridization, the cleavage is activated by the addition of H_2O_2 (Chen and Sigman, 1986).

Light induces DNA strand scission *in vitro* by tris(1,10-phenanthroline)cobalt(III), $[Co^{III}(phen)3]^{3+}$, and tris(4,7-diphenylphenanthroline)cobalt(II), $[Co^{II}(dip)3]^{2+}$ complexes (Barton and Raphael, 1984). A number of low-spin d^6 transition metal complexes including $[Co(NH_3)_6]^{3+}$, $[Rh(phen)_3]^{3+}$, and $[Ru(phen)_3]^{2+}$ also cleave DNA (Fleisher et al., 1986). The mechanisms for these reactions typically do not require O_2. These observations have led to a set of relatively specific probes for DNA structure that bring about photoinduced cleavage *in vitro* and *in vivo.* While the complete specificities of these agents are not yet understood, the tris phenanthrolines are chiral, and the resolved compounds are capable in some cases of differentiating right- and left-handed helices as well as other more subtle structural modifications in DNA (Barton, 1986). Figure 3-8 shows the structure of L-tris(4,7-diphenylphenanthroline) cobalt(III), $[L\text{-}Co(dip)3]^{3+}$, a complex incapable of binding to B-form DNA but that

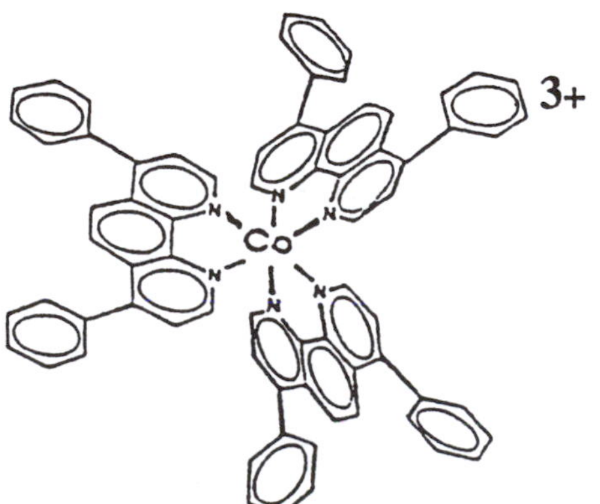

Figure 3-8
A chiral metal complex that specifically binds to nucleic acids and leads to light-induced cleavage (Barton, 1986). The molecule shown is tris(4,7-diphenylphenanthroline)cobalt(III), $[Co^{(III)})(dip)3]^{3+}$.

does bind to Z-DNA. Photocleavage induced by this compound has been demonstrated in the promoter and enhancer regions as well as downstream of the 3′-termini of genes in SV40. In one case, there was correspondence between the site of cleavage and the site of binding of anti-Z-DNA antibodies (Muller et al., 1987). Analogous ruthenium complexes with phenanthroline derivatives can distinguish between A- and B-forms of DNA; they show selective binding to different regions of pBR322 (Mei and Barton, 1988). These observations suggests that control DNA sequences, coding genes, and other regions have different conformations.

4.4.2 Nucleases

The enzyme S_1 nuclease hydrolyzes both single-stranded DNA and RNA as described earlier (Vogt, 1973; Ehresmann et al., 1987). It cleaves the single-stranded regions in loops of cruciforms, in the strand released after triple-strand formation (Johnston, 1988; Htun and Dahlberg, 1989) and in internal loops. It does not cleave single base–base mismatches (Dodgson and Wells, 1977).

The location of introns and exons in a genomic DNA can be mapped by hybridizing its mRNA with the DNA. The introns are not complementary to the RNA, therefore the DNA will form single-stranded loops that are cut by S_1 nuclease (Sambrook et al., 1989, p. 7.58).

The numbers of base pairs per turn of DNAs adsorbed on flat surfaces have been determined by cleavage with nucleases. The nucleic acids were added to calcium phosphate crystals or to powdered mica and treated with DNase I or micrococcal nuclease. Random sequence DNA and alternating poly(dA-dT) showed 10.5–10.6 bp turn, whereas the homopolynucleotide poly(dA)•poly(dT) showed 10.0 bp turn (Rhodes and Klug, 1981).

5. DETERMINATION OF TERTIARY STRUCTURE

All the chemical and enzymatic methods described above can obviously be used for probing tertiary as well as secondary structure. Transfer RNA with its well-characterized three-dimensional (3D) structure has been an excellent model for testing the effect of tertiary interactions on chemical reactivity (Holbrook and Kim, 1983). In addition to the reagents discussed previously, measurements of the rates of tritium exchange with the H8 of purines have been used as nonperturbing probes of RNA tertiary structure (Gamble and Schoemaker, 1976; Schimmel and Redfield, 1980). Tertiary interactions in a DNA can protect DNA from nuclease, or chemically-induced, cleavage. Selective protection of four-way junctions in DNA has been determined (Churchill et al., 1988; Murchie et al., 1990)

The rates of reactions with the chemical reagents, with enzymes, and with the tritium-labeled water depend on the accessibility of the site. Kinetics controls the reactivity. A method that depends on equilibrium—thermodynamics—is the binding of oligonucleotides to nucleic acids. The extent of (reversible) binding depends on the sequence and on the competition between intermolecular binding to the added oligonucleotide and intramolecular binding with other sequences of the target (Uhlenbeck, 1972; Freier and Tinoco, 1975). Equilibrium dialysis of a radioactively labeled oligonucleotide is most often used to measure the binding, but fluorescence measurements can also be used. Binding equilibrium constants for a set of oligonucleotides are determined and their relative values are interpreted in terms of binding to single-stranded regions or to structured regions.

5.1 Cross-Linking

5.1.1 Chemical Cross-Linking

Reagents with two reactive groups can be used to covalently link nucleotides that are far apart in the sequence. The interaction that brings the two nucleotides close enough to cross-link may be part of the secondary structure or the tertiary structure. Whatever the cause, a cross-link provides an upper bound to the distance between the nucleotides. The localization of the cross-link can be done approximately by electron microscopy, or precisely by partial hydrolysis of the nucleic acid and identification of the reacted nucleotides.

Chemical cross-linking reagents that have been used to determine the 3D structure of 16S rRNA include *N*-acetyl-*N*′-(*p*-glyoxylbenzoyl)cystamine (Wollenzien et al., 1985), bis(2-chloroethyl)-methylamine or "nitrogen mustard", *p*-azidophenyl acetimidate, and iminothiolane (Brimacombe et al., 1988). When the chemical cross-linking results are combined with psoralen photocross-links, ultraviolet (UV) cross-links, chemical probing, and phylogenetic comparisons (Gutell et al., 1985) a highly constrained 3D structure for the RNA is obtained (Stern et al., 1988b; Brimacombe et al., 1990; Hubbard and Hearst, 1991).

5.1.2 Psoralen Photocross-Linking

Psoralen is a three ring heterocyclic compound with the structure shown below:

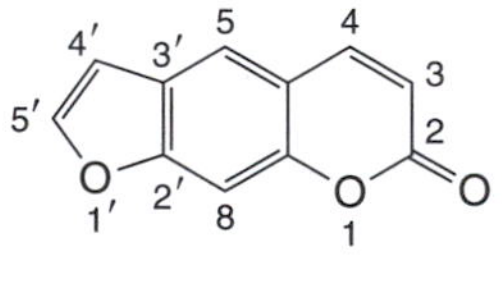

It is one of a class of compounds that photoreacts specifically with nucleic acids; proteins do not react and no nucleic acid–protein cross-links occur. In the absence of UV irradiation, no reaction occurs. The photocross-linking can be done *in vivo* or *in vitro* to probe conformations in a wide range of environments. The most useful compounds are water soluble derivatives such as 4′-hydroxymethyl-4, 5′, 8-trimethylpsoralen (HMT) and 4′-aminomethyl-4, 5′, 8-trimethylpsoralen (AMT) (Isaacs et al., 1977). Both DNA and RNA photoreact, but DNA is much more reactive. For a review of applications of psoralen to nucleic acids see Shi et al. (1990).

The mechanism of the photoaddition of a psoralen to a nucleic acid helix involves several steps (see Fig. 3-9). The initial step is the intercalation of the psoralen

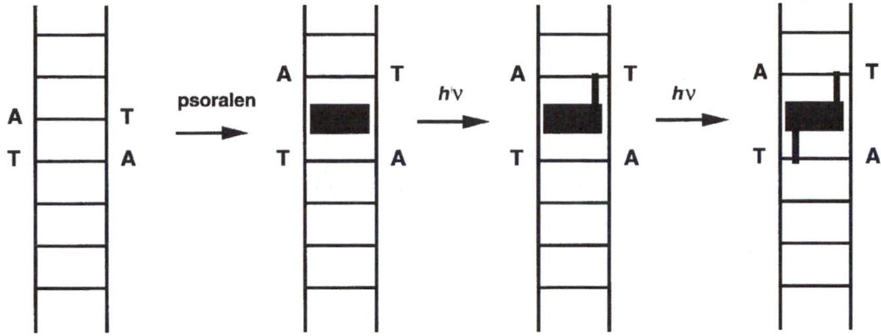

Figure 3-9
Mechanism of the cross-linking photoreactions of psoralens with the 5–6 double bonds of neighboring pyrimidines in a DNA helix. The first step is a thermal reaction involving the intercalation of the psoralen in the helix. The second step is cyclobutane formation at the 4′–5′ bond of the furan ring after the absorption of a photon. The final step is the formation of the second cyclobutane ring at the 3–4 bond of the pyrone ring upon absorption of a second photon, thus creating a covalent cross-link between the two strands of the DNA helix.

into the nucleic acid helix in a dark reaction. When an intercalated psoralen absorbs a photon of wavelength between 300 and 400 nm, either the 3–4 double bond of the pyrone ring or the 4′–5′ double bond of the furan ring is sensitized to react by cycloaddition to the 5–6 double bond of an adjacent pyrimidine. The second cycloaddition, to the opposite nucleic acid strand, can only form if two conditions are met. First, the monoadduct formed must be a furan-side monoadduct (the first cycloaddition must have occurred at the 4′–5′ double bond of the psoralen). The adduct is then a coumarin derivative that can still absorb a photon of wavelength between 300 and 380 nm, making the cross-link possible. Second, a pyrimidine has to be adjacent to the psoralen monoadduct on the opposite strand. Thus, for cross-link to form, the original intercalation had to occur in either a 5′-purine–pyrimidine-3′ site or in a 5′-pyrimidine–purine-3′ site in the helix. The rates of reactivity are base and sequence dependent; thymines and uracils are more reactive than cytosines. Furthermore, 5′TpA3′ sites are far more photoreactive than the 5′ApT3′ sites (Hearst et al., 1984; Cimino et al., 1985). X-ray crystallographic (Peckler et al., 1982) studies established the configuration and stereochemistry of the psoralen–pyrimidine product. A detailed structure of the DNA–psoralen cross-link has been obtained by NMR studies (Spielmann et al., 1995).

The photocross-linking reaction has been applied to the investigation of DNA structure in virus particles. The best example relates to the fd bacteriophage that contains a circular single-stranded DNA genome of 6408 bases. The origin of DNA replication is known from sequencing studies to contain a complex of four hairpins that when cross-linked is readily visualized by electron microscopy. The phage particle is a long cylindrical filament with the DNA at its center. Cross-linkage established that the origin complex is located at the end of the filamentous phage (Huang and Hearst, 1981).

Extensive studies were carried out on 16S rRNA by psoralen photocross-linking (Thompson and Hearst, 1983; Wollenzein et al., 1985). A cross-linkage map was first generated using electron microscopy; later studies identified which uracil nucleotides were linked. Maps were generated both for the isolated 16S rRNA *in vitro* and for the same rRNA in the 30S subunit of the ribosome. It was concluded that the RNA has equivalent regions of secondary structure in both environments.

Photocross-linkable oligonucleotides provide a unique tool for the study of the kinetics and equilibria of hybridization between a small probe and a large nucleic acid. Oligonucleotides prepared with a photoreactive psoralen covalently attached (Gamper et al., 1984; Shi and Hearst, 1987) may be used as hybridization probes that can be irreversibly bonded to their target sequence by near UV light. The hybridization of an oligonucleotide probe to a nucleic acid must often compete with self structure in the target, and optimum conditions for specificity of probe binding require rapid fixation. The ability to bind the probe irreversibly with a light pulse is thus very useful (Gamper et al., 1986, 1987).

5.1.3 Ultraviolet Photocross-Linking

Irradiation of nucleic acids with UV light in their absorption region of 200–300 nm causes cross-links to form. The chemical structures of the crosslinks are not generally known, but their formation can give very useful information about the 3D structure of a folded RNA or DNA (Prince et al., 1982; Branch et al., 1989b; Brimacombe et al., 1988, 1990). The cross-linking reactions include the formation of pyrimidine dimers linked by a cyclobutyl ring across their 5–6 double bonds (Fisher and Johns, 1976). Four isomers form; two are cis, with the bases facing each other; two are trans, with the bases on opposite sides of the cyclobutane ring. The N1–H bonds (the glycosidic bonds) can be parallel (syn) or antiparallel (anti). In duplex DNA only the cis–syn isomer forms, as expected from the orientation of adjacent bases in the polynucleotide. Photooxidation of purines in the presence of oxygen is another source of crosslinking in nucleic acids (Elad, 1976). Ultraviolet photocross-linking has the great advantage that no chemical reagent has to be added to the nucleic acid solution; thus equilibria among different secondary structures are not perturbed.

References

Anderson, E. P. (1973). Nucleoside and Nucleotide Kinases, in *The Enzymes,* Vol. IX, Part B, 3rd ed., Boyer, P. D., Ed., Academic Ps, New York pp. 49–96.

Auron, P. E., Weber, L. D., and Rich, A. (1982). Comparison of Transfer Ribonucleic Acid Structures Using Cobra Venom and S_1 Nucleases, *Biochemistry* **21**, 4700–4706.

Barton, J. K. and Raphael, A. L. (1984). Photoactivated Stereospecific Cleavage of Double-Helical DNA by Cobalt(III) Complexes, *J. Am. Chem. Soc.* **106**, 2466–1468.

Barton, J. K. (1986). Metals and DNA: Molecular Left-Handed Complements, *Science* **233**, 727–733.

Berkower, I., Leis, J., and Hurwitz, J. (1973). Isolation and Charaterization of an Endonuclease from *Escherichia coli* Specific for Ribonucleic Acid•Deoxyribonucleic Acid Hybrid Structures, *J. Biol. Chem.* **248**, 5914–5921.

Bernardi, A. and Bernardi, G. (1971) Spleen Acid Exonuclease, in *The Enzymes*, Vol. 4, 3rd ed. Boyer, P. D. Ed., Academic Ps, New York, pp.329–336.

Branch, A. D., Benenfeld, B. J., and Robertson, H. D. (1989a). RNA Fingerprinting, Methods Enzymol, **180**, 130–154.

Branch, A. D., Benenfeld, B. J., and Robertson, H. D. (1989). Ultraviolet-induced Cross-linking Reveals a Unique Region of Local Tertiary Structure in Potato Spindle Tuber Viroid and HeLa 5S RNA, *Proc. Natl. Acad. Sci. USA* **82**, 6590–6594.

Brimacombe, R., Atmadja, J., Stiege, W., and Schuler, D. (1988). A Detailed Model of the Three-Dimensional Structure of *Escherichia coli* 16S Ribosomal RNA *in situ* in the 30 S Subunit, *J. Mol. Biol.* **199**, 115–136.

Brimacombe, R., Gruer, B., Mitchell, P., Osswald, M., Rinke-Appel, J., Schüler, D., and Stade, K. (1990). Three-Dimensional Structure and Function of *Escherichia coli* 16S and 23S rRNA as Studied by Crosslinking Techniques, Hill, W., Dahlberg, A., Garrett, R. A., Moore, P. B., Schlessinger, D., and Warner, J. R., Eds., in *The Ribosome: Structure, Function, and Evolution* American Society of Microbiology, Washington, DC, pp. 93–106.

Brush, C. K., Stone, M. P., and Harris, T. M. (1988). Selective Reversible Deuteriation of Oligodeoxynucleotides: Simplification of Two-Dimensional Nuclear Overhauser Effect NMR Spectral Assignment of a Non-Self-Complementary Dodecamer Duplex, *Biochemistry* **27**, 115–122.

Chen, X., Burrows, C. J., and Rokita, S. E. (1992). Conformation-Specific Detection of Guanine in DNA: Ends, Mismatches, Bulges, and Loops, *J. Am. Chem. Soc.* **114**, 322–325.

Chen, C-H. B. and Sigman, D. S. (1986). Nuclease Activity of 1,10-Phenantroline-Copper: Sequence Specific Targeting, *Proc. Natl. Acad. Sci. USA* **83**, 7147–7151.

Chen, X., Woodson, S. A., Burrows, C. J., and Rokita, S. E. (1993). A Highly Sensitive Probe for Guanine N7 in Folded Structures of RNA: Application to tRNA and Terahymena Group I Intron, *Biochemistry* **32**, 7610–7619.

Churchill, M. E. A., Tullius, T. D., Kallenbach, N. R., and Seeman, N. C. (1988). A Holliday Recombination Intermediate is Twofold Symmetric, *Proc. Natl. Acad. Sci. USA* **85**, 4653–4656.

Cimino, G. D., Gamper, H. B., Isaacs, S. T., and Hearst, J. E. (1985). Psoralens as Photoactive Probes of Nucleic Acid Structure and Function: Organic Chemistry, Photochemistry, and Biochemistry, *Annu. Rev. Biochem.* **54**, 1151–1193.

Dahlberg, J. E. and Abelson, J. N., Eds. (1989). RNA Processing, *Methods Enzymol.,* **180**.

Dervan, P. (1986). Design of Sequence-Specific DNA-Binding Molecules, *Science* **232**, 464–471.

Dervan, P. (1992). Reagents for the Site-Specific Cleavage of Megabase DNA, *Nature (London)* **359**, 87–88.

Dock-Bregeon, A. C., Westhof, E., Giegé, R., and Moras, D. (1989). Solution Structure of a tRNA with a Large Variable Region: Yeast $tRNA^{Ser}$, *J. Mol. Biol.* **206**, 707–722.

Dodgson, J. B. and Wells, R. D. (1977). Action of Single-strand Specific Nucleases on Model DNA Heteroduplexes of Defined Size and Sequence, *Biochemistry* **16**, 2374–2379.

Donis-Keller, H. (1980). Phy M: an RNase Activity Specific for U and A Residues Useful in RNA Sequence Analysis, *Nucleic Acids Res.* **8**, 3133–3142.

Eckstein, F. (1985). Nucleoside Phosphorothioates, *Ann. Rev. Biochem.* **54**, 367–402.

Ehresmann, C., Baudin, F., Mougel, M., Romby, P., Ebel, J.-P. and Ehresmann, B. (1987). Probing the Structure of RNAs in Solution, *Nucleic Acids Res.* **15**, 9109–9128.

Ehrenberg, L., Fedorcsak, I., and Solymosy, F. (1976). Diethylpyrocarbonate in Nucleic Acid Research, *Prog. Nucleic Acid Res. Mol. Biol.* **16**, 189–262.

Elad, D. (1976). Photoproducts of Purines, Wang, S. Y., Ed., in *Photochemistry and Photobiology of Nucleic Acids*, Vol. I, Academic Ps, New York, pp. 357–380.

Fisher, G. J. and Johns, H. J. (1976). Pyrimidine Photohydrates, in *Photochemistry and Photobiology of Nucleic Acids*, Vol.I, Wang, S. Y., Ed., Academic Ps, New York. pp. 225–294, pp. 169–224.

Fleisher, M. B., Waterman, K. C., Turro, N. J., and Barton, J. K. (1986). Light Induced Cleavage of DNA by Metal Complexes, *Inorg. Chem.* **25**, 3549–3551.

Forster, A. C. and Symons, R. H. (1987). Self-Cleavage of Plus and Minus RNAs of a Virusoid and a Structural Model for the Active Site, *Cell* **49**, 211–220.

Freier, S. M. and Tinoco, I., Jr. (1975). The Binding of Complementary Oligoribonucleotides to Yeast Initiator Transfer RNA, *Biochemistry* **14**, 3310–3314.

Furlong, J. C., Sullivan, K. M., Murchie, A. I. H., Gough, G. W., and Lilley, D. M. J. (1989). Localized Chemical Hyperreactivity in Supercoiled DNA: Evidence for Base Unpairing in Sequences That Induce Low-Salt Cruciform Extrusion, *Biochemistry* **28**, 2009–2017.

Gait, M. J., Ed. (1984). *Oligonucleotide Synthesis: A Practical Approach*, IRL Press, Oxford, UK.

Gamble, R. C., and Schoemaker, J. P. (1976). Rate of tritium labeling of specific purines in relation to nucleic acid and particularly transfer RNA conformation. *Biochemistry* **15**, 2791–2799.

Gamper, H. B., Cimino, G. D., and Hearst, J. E. (1987). Solution Hybridization of Crosslinkable DNA Oligonucleotides to M13DNA: Effect of Secondary Structure on Hybridization Kinetics and Equilibria, *J. Mol. Biol.* **197**, 349–362.

Gamper, H. B., Cimino, G. D. Isaacs, S. T., Ferguson, M., and Hearst, J. E. (1986). Reverse Southern Hybridization, *Nucleic Acids Res.* **14**, 9943–9954.

Gamper, H.B., Piette, J., and Hearst, J. E., (1984). Efficient Formation of a Crosslinkable HMT Monoadduct at the KpnI Recognition Site, *Photochem. Photobiol.* **40**, 29–34.

Gaynor R., Soultanakis, E., Kuwabara, M., Garcia, J., and Sigman, D. (1989). Specific Binding of a Hela Cell Nuclear Protein to RNA Sequences in the Human Immunodeficiency Virus Transactivating Region, *Proc. Natl. Acad. Sciences USA* **86**, 4858–4862.

Goyne, T. E. and Sigman, D. S., (1987). Nuclease Activity of 1,10-Phenanthroline-Copper Ion. Chemistry of Deoxyribose Oxidation, *J. Am. Chem. Soc.* **109**, 2846–2848.

Gutell, R. R., Weiser, B., Woese, C. R., and Noller, H. F., (1985). Comparative Anatomy of 16-S-Like Ribosomal RNA, *Prog. Nucleic Acid Res. Mol. Biol.* **32**, 155–216.

Hahn, C. S., Strauss, E. G., and Strauss, J. H., (1989). Dideoxy Sequencing of RNA Using Reverse Transcriptase, *Methods Enzymol.* **180**, 121–130.

Hayatsu, H. (1976). Bisulfite Modification of Nucleic Acids and Their Constituents, *Prog. Nucleic Acid Res. Mol. Biol.* **16**, 75–124.

Hayatsu, H. and Ukita, T. (1967). The Selective Degradation of Pyrimidines in Nucleic Acids by Permanganate Oxidation, *Biochem. Biophys. Res. Commun.* **29**, 556–561.

Hearst, J. E., Isaacs, S. T., Kanne, D., Rapoport, H., and Straub, K. (1984). The Reaction of the Psoralens with Deoxyribonucleic Acid, *Q. Rev. Biophys.* **17**, 1–44.

Hertzberg, R. P. and Dervan, P. B., (1982). Cleavage of double-helical DNA by (Methidiumpropyl-EDTA)iron(II), *J. Am. Chem. Soc.* **104**, 313–315.

Hertzberg, R. P. and Dervan, P. B. (1984). Cleavage of DNA with Methidiumpropyl-EDTA-Iron(II): Reaction Conditions and Product Analyses, *Biochemistry* **23**, 3934–3945.

Holbrook, S. R. and Kim, S.-H. (1983). Correlation Between Chemical Modification and Surface Accessibility in Yeast Phenylalanine Transfer RNA, *Biopolymers* **22**, 1145–1166.

Htun, H. and Dahlberg, J. E. (1989). Topology and Formation of Triple-Stranded H-DNA, *Science* **243**, 1571–1576.

Huang, C.-C. and Hearst, J. E. (1981). Fine Mapping of Secondary Structures of fd Phage in the Region of the Replication Origin, *Nucleic Acids Res.* **9**, 5587–5599.

Hubbard, J. M and Hearst, J. E. (1991). Computer Modelling 16S Ribosomal RNA, *J. Mol. Biol.* **221**, 889–907.

Isaacs, S. T., Shen, C-K. J., Hearst, J. E., and Rapoport, H. (1977). Synthesis and Characterization of New Psoralen Derivatives with Superior Photoreactivity with DNA and RNA, *Biochemistry* **16**, 1058–1064.

Itakura, K., Rossi, J. J., and Wallace, R. B. (1984). Synthesis and Use of Synthetic Oligonucleotides, *Annu. Rev. Biochem.* **53**, 323–356.

Jaeger, J. A., SantaLucia, J., Jr., and Tinoco, I., Jr. (1993). Determination of RNA Structure and Thermodynamics, *Annu. Rev. Biochem.* **62**, 255–287.

Jezewska, M. J., Bujalowski, W., and Lowman, T. M. (1989). Iron(II)-ethylenediaminetatraacetic Acid Catalyzed Cleavage of DNA Is Highly Specific for Duplex DNA, *Biochemistry* **28**, 6161–6164.

Johnston, B. H. (1988). The S_1-Sensitive Form of $d(C\text{-}T)_n \cdot d(A\text{-}G)_n$: Chemical Evidence for a Three-Stranded Strucuture in Plasmids, *Science* **241**, 1800–1804.

Johnston, B. H., Kung, A. H., Moore, C. B., and Hearst, J. E. (1981). Kinetics of Formation of Deoxyribonucleic Acid Cross-Links by 4′-(Aminomethyl)-4,5′,8- trimethylpsoralen, *J. Am. Chem. Soc.* **20**, 735–738.

Johnston, B. H. and Rich, A. (1985). Chemical Probes of DNA Conformation: Detection of Z-DNA at Nucleotide Resolution, *Cell* **42**, 713–724.

Kappen, L. S., Ellenberger, T. E., and Goldberg, I. H. (1987). Mechanism and Base Specificity of DNA Breakage in Intact Cells by Neocarzinostatin, *Biochemistry* **26**, 384–390.

Kappen, L. S. and Goldberg, I. H. (1983). Deoxyribonucleic Acid Damage by Neocarzinostatin Chromophore: Strand Breaks Generated by Selective Oxidation of C-5′ of Deoxyribose, *Biochemistry* **22**, 4872–4878.

Kierzek, R., Caruthers, M. H., Longfellow, C. E., Swinton, D., Turner, D. H., Freier, S. M. (1986). Polymer-Supported RNA Synthesis and its Application to Test the Nearest-Neighbor Model for Duplex Stability, *Biochemistry* **25**, 7840–7846.

Knapp, G. (1989). Enzymatic Approaches to Probing of RNA Secondary and Tertiary Structure, *Methods Enzymol.* **180**, 192–212.

Knowles, J. R. (1980). Enzyme-Catalyzed Phosphoryl Transfer Reactions. *Annu. Rev. Biochem.* **49**, 877–919.

Kochetkov, N. K. and Budovskii, E. I., Eds. (1972). *Organic Chemistry of Nucleic Acids, Part B*, Plenum, New York.

Koob, M. and Szybalski, W. (1988). Conferring Operator Specificity on Restriction Endonucleases, *Science* **241**, 1084–1086.

Kuchino, Y. and Nishimura, S. (1989). Enzymatic RNA Sequencing, Processing, *Methods Enzymol.* **180**, 154–163.

Kuwabara, M., Yoon, C., Goyne, T., Thederahn, T., and Sigman, D. S. (1986). Nuclease Activity of 1,10-phenanthroline-copper Ion: Reaction with CGCGAATTCGCG and Its Complexes with Netropsin and EcoRI, *Biochemistry* **25**, 7401–7408.

Laskowski, M. (1971). Venom Exonuclease, Boyer, P. D. Ed., in *The Enzymes*, Vol. 4, 3rd ed. Academic, New York, 313–328.

Laskowski, M. (1980). Purification and Properties of the Mung Bean Nuclease, *Methods Enzymol.* **65**, 263–169.

Latham, J. A. and Cech, T. R. (1989). Defining the Inside and Outside of a Catalytic RNA Molecule, *Science* **245**, 276–245.

Lilley, D. M. J. (1983). Structural Perturbation in Supercoiled DNA: Hypersensitivity to Modification by a Single-strand-selective Chemical Reagent Conferred by Inverted Repeat Sequences, *Nucleic Acids Res.* **11**, 3097–3112

Lilley, D. M. J. and Palecek, E. (1984). The Supercoil-stabilized Cruciform of Col E1 is Hypersensitive to Osmium Tetroxide, *EMBO J.* **3**, 1187–1192.

Lockard, R. E. and Kumar, A. (1981). Mapping tRNA Structure in Solution Using Double-strand-Specific Ribonuclease V_1 from Cobra Venom. *Nucleic Acids Res.* **9**, 5125–5140.

Lowman, H. B. and Draper, D. E. (1986). On the Recognition of Helical RNA by Cobra Venom V_1 Nuclease *J. Biol. Chem.* **261**, 5396–5403.

Maxam, A. M. and Gilbert, W. (1977). A New Method for Sequencing DNA, *Proc. Natl. Acad. Sci. USA* **74**, 560–564.

Maxam, A. M. and Gilbert, W. (1980). Sequencing End-Labeled DNA with Base-Specific Chemical Cleavages, *Methods Enzymol.* **65**, 499–560.

Mc Clellan, J. A., Boublikova, P., Palecek, E., and Lilley, D. M. J. (1990). Superhelical Torsion in Cellular DNA Responds Directly to Environmental and Genetic Factors, *Proc. Natl. Acad. Sci. USA* **87**, 8373–8377.

McSwiggen, J. A. and Cech, T. R. (1989). Stereochemistry of RNA Cleavage by the *Tetrahymena* Ribozyme and Evidence That the Chemical Step Is Not Rate-Limiting, *Science* **244**, 679–694.

Mei, H.-Y. and Barton, J. K. (1988). Tris(tetramethylphenanthroline)ruthenium (II): A Chiral Probe That Cleaves A-DNA Conformations, *Proc. Natl. Acad. Sci. USA* **85**, 1339–1343.

Milligan, J. F., Groebe, D. R., Witherell, G. W., and Uhlenbeck, O. C. (1987). Oligoribonucleotide Synthesis using T7 RNA Polymerase and Synthetic DNA Templates. *Nucleic Acids Res.* **15**, 8783–8798.

Mizuno, Y. (1986). The Organic Chemistry of Nucleic Acids, Elsevier, The Netherlands.

Moazed, D., Stern, S., and Noller, H. F. (1986). Rapid Chemical Probing of Conformation in 16S Ribosomal RNA and 30 S Ribosomal Subunits Using Primer Extension, *J. Mol. Biol.* **187**, 399–416.

Moore, M. J. and Sharp, P. A. (1992). Site-specific Modification of Pre-messenger-RNA—The 2′-hydroxyl Groups at the Splice Sites, *Science* **256**, 992–997.

Moser, H. E. and Dervan, P. B. (1987). Sequence-Specific Cleavage of Double Helical DNA by Triple Helix Formation, *Science* **238**, 645–650.

Muller, B. C., Raphael, A. L., and Barton, J. K. (1987). Evidence for Altered DNA Conformations in the Simian Virus 40 Genome: Site-Specific DNA Cleavage by the Chiral Complex L-tris(4,7-diphenyl-1,10-phenanthroline)cobalt(III), *Proc. Natl. Acad. Sci. USA* **84**, 1764–1768.

Murchie, A. I. H., Carter, W. A., Portugal, J., and Lilley, D. M. J., (1990). The Tertiary Structure of the Four-Way DNA Junction Affords Protection Against DNase I Cleavage *Nucleic Acids Res.* **18**, 2599–2605.

Noller, H. F. (1984). Structure of Ribosomal RNA, *Annu. Rev. Biochem.* **53**, 119–162.

Ogilvie, K. K., Usman, N., Nicoghosian, K., and Cedergren, R. J. (1988). Total Chemical Synthesis of a 77-Nucleotide-Long RNA Sequence Having Methionine-Acceptance Activity, *Proc. Natl. Acad. Sci. USA* **85**, 5764–5768.

Peattie, D. A. (1979). Direct Chemical Method for Sequencing RNA, *Proc. Natl. Acad. Sci. USA* **76**, 1760–1764.

Peattie, D. A. and Gilbert, W. (1980). Chemical Probes for Higher-order Structure in RNA, *Proc. Natl. Acad. Sci. USA* **77**, 4679–4682.

Peckler, S., Graves, B., D., Kanne, Rapoport, H., Hearst, J. E., and Kim, S.-H. (1982). Structure of a Psoralen-Thymine Monoadduct Formed in Photoreaction with DNA, *J. Mol. Biol.* **162**, 157–172.

Phillips, J. H. (1967). The Reaction of Hydroxylamine with Pyrimidine Bases, *Methods Enzymol.,* **12**, 34–38.

Pope, L. M., Reich, K. A., Graham, D. R., and Sigman, D. S. (1982). Products of DNA Cleavage by the 1,10-phenanthroline-copper Complex. Inhibitors of *Escherichia coli* DNA Polymerase I, *J. Biol. Chem.* **257**, 12121–12128.

Pope, L. E. and Sigman, D. S. (1984). Secondary Structure Specificity of the Nuclease Activity of the 1,10-Phenanthroline–Copper Complex, *Proc. Natl. Acad. Sci. USA* **81**, 3–7.

Prince, J. B., Taylor, B. H., Thurlow, D. L. J., Ofengand and Zimmerman, R. A. (1982). Covalent Crosslinking of $tRNA^{val}$ to 16S RNA at the Ribosomal P Site: Identification of Crosslinked Residiues, *Proc. Natl. Acad. Sci. USA* **79**, 5450–5454.

Rajbhandary, U. L., Lockard, R. E., and Reilly, R. M. (1982). Use of Nucleases in RNA Sequence and Structural Analyses, Linn, S. M., and Roberts, R. J., Eds., *Nucleases*, Cold Spring Harbor Press, Cold Springs Harbor, New York, 275–289.

Rhodes, D. and Klug, A. (1981). Sequence-Dependent Helical Periodicity of DNA, *Nature London* **292**, 378–380.

Richardson, C. C. (1981). Bacteriophage T4 Polynucleotide Kinase, Boyer, P. D., Ed., in *The Enzymes*, 3rd Ed., Vol. 14, Academic, New York, 299–303.

Romby, P., Moras, D., Dumas P., Ebel, J. P., and Giegé, R. (1987). Comparison of the tertiary structure of yeast tRNA(Asp) and tRNA(Phe) in solution. chemical modification study of the bases. *J. Mol. Biol.* **195**, 193–204.

Romby, P., Westhof, E., Toukifimpa, R., Mache, R., Ebel, J. P., Ehresmann, C., and Ehresmann, B. (1988). Higher order structure of cholorplastic 5S ribosomal RNA from spinach. *Biochemistry* **27**, 4721–4730.

Rubin, C. M. and Schmid, C. (1980). Pyrimidine-Specific Chemical Reactions Useful for DNA Sequencing, *Nucleic Acids Res.* **8**, 4613–4619.

Saiki, R. K., Gelfand, D. H., Stoffel, S., Scharf, S. J., Higuchi, R., Horn, G. T., Mullis, K. B., and Erlich, H. A. (1988). Primer-directed Enzymatic Amplification of DNA with a Thermostable DNA Polymerase, *Science* **239**, 487–491.

Sambrook, J., Frisch, E. F., and Maniatis, T. (1989). *Molecular Cloming, A Laboratory Manual*, 2nd ed., Cold Spring Harbor Laboratory Press, Cold Springs Harbor, New York.

Sanger, F., Nicklen, S., and Coulson, A. R. (1977). DNA Sequencing with Chain-terminating Inhibitors, *Proc. Natl. Acad. Sci. USA* **74**, 5463–5467.

Schmidt, G. (1961). Nonspecific Acid Phosphomonoesterases, Boyer, P. D., Lardy, H., and Meyrback K., Eds., in *The Enzymes*, Vol. V, 2nd ed., Academic, New York, 37–48.

Shi, Y. and Hearst, J. E. (1987). Wavelength Dependence for the Photoreactions of DNA–Psoralen Monoadducts. 1. Photoreversal of Monoadducts. 2. Photo-Cross- Linking of Monoadducts, *Biochemistry* **26**, 3786–3798.

Shi, Y., Lipson, S. E., Chi, D. Y., Spielmann, H. P., Monforte, J. A., and Hearst, J. E. (1990). Applications of Psoralens as Probes of Nucleic Acid Structure and Function, in *Bioorganic Photochemistry: Photochemistry and the Nucleic Acids*, Morrison, H., Ed., Wiley, New York.

Sigman, D. S. (1986). Nuclease Activity of 1,10-Phenanthroline–Copper Ion, *Acc. Chem. Res.* **19**, 180–186.

Sigman, D. S., Graham, D. R., D'Aurora, V., and Stern, A. M. (1979). Oxygen- Dependent Cleavage of DNA by the 1,10-Phenanthroline • Cuprous Complex, *J. Biol. Chem.* **254**, 12269–12272.

Sigman, D. S. and Chen, C,-h. B. (1990). Chemical Nucleases–New Reagents In Molecular Biology, *Annu. Rev. Biochem.* **59**, 207–236.

Silberklang, M., Prochiantz, A., Haenni, A.-L., and Rajbhandary, U. L. (1977). Studies on the Sequence of the 3′-Terminal Region of Turnip-Yellow-Mosaic-Virus RNA, *Eur. J. Biochem.* **72**, 465–478.

Singer B. and Grunberger, D. (1983). *Molecular Biology of Mutagens and Carcinogens*, Plenum, New York, 360 pp.

Singer, B. and Kusmierek, J. T. (1982). Chemical Mutagenesis, *Annu. Rev. Biochem.* **51**, 655–693.

Sinha, N. D., Biernat, J., and Köster, H. (1984). Polymer Support Oligonucleotide Synthesis XVIII: Use of β-Cyanoethyl-N, N-Dialkylamino-N-Morpholino Phosphoramidite of Deoxynucleosides for the Synthesis of DNA Fragments Simplifying Deprotection and Isolation of the Final Product, *Nucleic Acids Res.* **12**, 4539–4557.

Smith, L. M., Sanders, J. Z., Kaiser, R. J., Hughes, P., Dodd, C., Connell, C. R., Heiner, C., Kent, S. B. H., and Hood, L. E. (1986). Fluorescence Detection in Automated DNA Sequence Analysis, *Nature (London)* **321**, 674–679.

Spassky, A. and Sigman, D. S. (1985). Nuclease Activity of 1,10-Phenanthroline-Copper Ion. Conformational Analysis and Footprinting of the lac Operon, *Biochemistry* **24**, 8050–8056.

Spielmann, H. P., Dwyer, T. J., Sastry, S. S., Hearst, J. E., and Wemmer, D. E. (1995). DNA Structural Reorganization Upon Conversion of a Psoralen Furan-side Monoadduct to an Interstrand Cross-link: Implications for DNA Repair, *Proc. Natl. Acad. Sci. USA* **92**, 2345–2349.

Stadtman, T. C. (1961). Alkaline Phosphatases, Boyer, P. D., Lardy, H. and Myrback, K., Eds., in *The Enzymes*, Vol. V, 2nd ed., Academic, New York 55–71.

Stern, S., Moazed, D., and Noller, H. F. (1988a). Structural analysis of RNA Using Chemical and Enzymatic Probing Monitored by Primer Extension, *Methods Enzymol.* **164**, 481–489.

Stern, S., Weiser, B., and Noller, H. F. (1988b). Model for the Three-dimensional Folding of 16S Ribosomal RNA, *J. Mol. Biol.* **204**, 447–481.

Sulkowski, E. and Laskowski, M. (1969). Action of Microccal Nuclease on Polymers of Deoxyadenylic and Deoxythymidylic Acids, *J. Biol. Chem.* **244**, 3818–3822

Tanner, K and Cech, T. R. (1987). Guanosine Binding Required for Cyclization of the Self-Splicing Intervening Sequence Ribonucleic Acid from *Tetrahymena Thermophyla*, *Biochemistry* **26**, 3330–3340.

Thederahn, T., Kuwabara, M., Larsen, T. A., and Sigman, D. S. (1989). Nuclease Activity of 1,10-phenanthroline-copper: Kinetic Mechanism, *J. Amer. Chem. Soc.* **111**, 4941–4946.

Thompson, J. F. and Hearst, J. E. (1983). Structure of *E.Coli* 16S RNA Elucidated by Psoralen Crosslinking, *Cell* **32**, 1355–1365.

Tullius, T. D. (1991). The Use of Chemical Probes to Analyze DNA and RNA Structures, *Curr. Opinion Struct. Biol.* **1**, 428–434.

Tullius, T. D. and Dombroski, B. A. (1985). Iron (II) EDTA Used to Measure the Helical Twist Along Any DNA Molecule, *Science* **230**, 679–681.

Uhlenbeck, O. C. (1972). Complementary Oligonucleotide Binding to Transfer RNA, *J. Mol. Biol.* **65**, 25–41.

Uhlenbeck, O. C. and Gumport, R. I. (1982). T4 RNA ligase, in *The Enzymes*, 3rd ed. Vol 15, Boyer, P. D., Ed., Academic, New York, pp. 31–58.

Verma, I. M. (1981). Reverse Transcriptase, in *The Enzymes*, 3rd ed., Vol. 14, Boyer, P. D., Ed., Academic, New York, pp. 87–96.

Vlassov, V. V., Giegé, R., and Ebel, J.-P. (1981). Tertiary Structure of tRNAs in Solution Monitored by Phosphodiester Modification with Ethylnitrosourea, *Eur. J. Biochem.* **119**, 51–59.

Vogt, V. M. (1973). Purfication and Further Properties of Single-Strand-Specific Nuclease from Aspergillus Oryzae. *Eur. J. Biochem.* **33**, 192–200.

Voloshin, O. N., Mirkin, S. M., Lyamichev, V. I., Belotserkovskii, B. P., and Frank-Kamenetskii, M. D. (1988). Chemical Probing of Homopurine-Homopyrimidine Mirror Repeats in Supercoiled DNA, *Nature (London)* **333**, 475–476.

Wang, J. F. and Cech, T. R. (1992). Tertiary Structure Around the Guanosine-Binding Site of the Tetrahymena Ribozyme, *Science* **256**, 526–529.

Wang, Y. H., Lin, P. N., Sczekan, S. R., McKenzie, R. A., and Theil, E. C. (1991). Ferritin mRNA Probed, Near the Iron Regulatory Region, with Protein and Chemical (1,10-phenanthroline-Cu) Nucleases. A Possible Role for Base-paired Flanking Regions, *Biol. Metals* **4**, 56–61.

Wang, Y.-Y., Lyttle, M., and Borer, P. N. (1990). Enzymayic and NMR Analysis of Oligoribonucleotides Synthesized with 2′-trialkylsilyl Protected Cyanoethylphosphoramidite Monomers, *Nucleic Acids Res.* **18**, 3347–3352.

Wells, R. D., Collier, D. A., Hanvey, J. C., Shimizu, M. and Wohlraub, F. (1988). The Chemistry and Biology of Unusual DNA Structures Adopted by Oligopurine • oligopyrimidine sequences, *FASEB J.* **2**, 2939–2949.

Winter,G. and Brownlee, G. G. (1978). 3′ End Labelling of RNA with ^{32}P Suitable for Rapid Gel Sequencing, *Nucleic Acids Res.* **5**, 3129–3139.

Wollenzien, P., Murphy, R. F., Cantor, C. R., Expert-Besançon, A., and Hayes, D. H. (1985). Structure of the *Escherichia coli* 16S Ribosomal RNA: Psoralen crosslinks and *N*-acetyl-N′-(p-glyoxylbenzoyl)-cystamine Crosslinks detected by Electron Microscopy, *J. Mol. Biol.* **184**, 67–80.

Wrede, P., Wurst, R., Vournakis, J., and Rich, A. (1979). Conformational changes of yeast tRNA^{Phe} and *E. coli* tRNA2^{Glu} as indicated by different nuclease digestion patterns. *J. Biol. Chem.* **254**, 9608–9616.

Wyatt, J. R., Puglisi, J. D., and Tinoco, I., Jr. (1990). RNA Pseudoknots: Stability and Loop Size Requirements, *J. Mol. Biol.* **213**, 455–470.

Wyatt, J. R., Chastain, M., and Puglisi, J. D. (1991). Synthesis and Purification of Large Amounts of RNA Oligonucleotides, *Biotechniques* **11**, 764–769.

CHAPTER 4

Nucleic Acid Structures from Diffraction Methods

1. INTRODUCTION

The biological functions of nucleic acids are ultimately coupled to their three-dimensional (3D) structures, either in the presence or absence of any interacting proteins. Diffraction methods are among the most powerful means of obtaining information about the 3D structures of nucleic acids and other biological molecules. Single-crystal (crystallographic) X-ray diffraction methods at high resolution can reveal the most complete 3D structure of nucleic acids, and most of the concrete and precise information about 3D structures of nucleic acids has been derived by this method. When a single crystal is not available, X-ray diffraction methods can still be applied to oriented fibers, although resolution is lower and numerous assumptions are required to derive a molecular structure from the fiber diffraction pattern. This approach was taken by Watson and Crick (1953) when they proposed the double-helical structure of DNA.

Since the double helix model was proposed nearly 40 years ago, its beauty and simplicity have dominated the thinking of biological scientists. However, it is now clear that DNA structures are not simple and in fact can assume many different forms and conformations. Although we know very little in atomic detail about the structures of single-stranded DNA, multistranded (more than double stranded) DNA, supercoiled DNA, hairpin structures, and others, we know in considerable detail the 3D structures of double helical DNA in three forms: B, A, and Z. For RNA structures, we know very little except A-form double helical RNA and tRNA. We expect much more complex structures for rRNA, small nuclear RNA, catalytic RNA, and others. This chapter concentrates on the B-, A-, and Z-double helices and tRNA.

The fundamental questions about DNA structure and its relation to function have been succinctly expressed by Dickerson and co-workers (Yanagi et al., 1991): "How does a particular base sequence in double helical B-DNA affect the local structure of that helix, and how does it affect the ability of the helix to be bent, twisted, or otherwise deformed by binding to another molecule such as a drug or protein? In brief, in what way does sequence influence the deformation and the deformability of the helix, and are either of these properties used in the recognition process?" If the DNA helix were completely rigid and uniform, the only information usable in recognition would be the pattern of hydrogen-bond donor and acceptor groups at the edges of the bases in the floors of the major and minor grooves of the helix (Seeman et al., 1976). While acidic and basic groups of proteins do make such contacts (see Chapter 13 on DNA–protein interactions) it seems clear that additional factors such as buried waters and localized bending are also crucial to binding specificity. It is important to know whether local distortions from average helical geometry preexist and help with the recognition process.

Indeed, it may be speculated (Yanagi et al., 1991) that consensus sequences are not truly consensus in their one-dimensional (1D) information content, since these sequences are often quite different, but rather in their 3D structure.

In the sections that follow, we begin with a general description of X-ray diffraction and its application to crystalline and fibrous specimens. We then present some general features of duplex DNA and RNA structure, many of which were apparent even from the earlier, lower resolution fiber diffraction studies. Next follows a discussion of the higher resolution structures of nucleic acids obtained from X-ray crystallography. These structures include regular double helices of DNA and RNA, and double helices with various imperfections. We describe how high-resolution crystallography enables determination of electronic structures of nucleic acid constituents, and how ions and water networks appear to influence relative stability of helices. The chapter ends with a description of the crystal structure of tRNA.

2. DIFFRACTION METHODS

Some characteristic X-ray diffraction patterns of nucleic acids are shown in Figures 4-1 through 4-3. These diffraction patterns are intended to show the nature of the data from which structures are deduced, and to display the dramatic differences in information

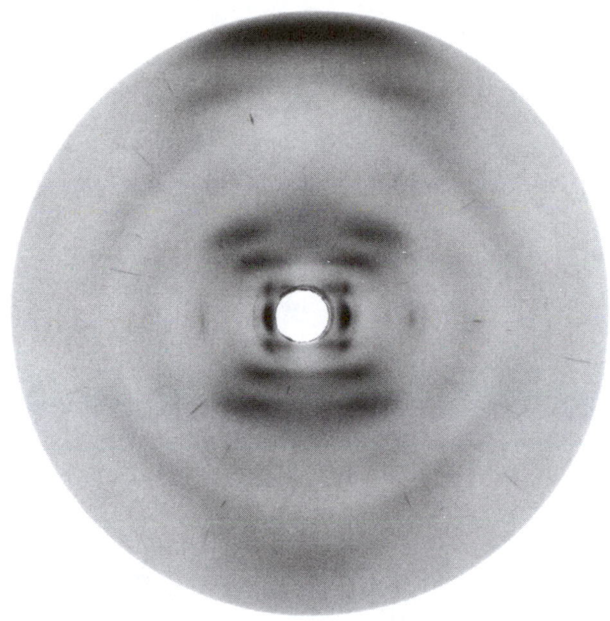

Figure 4-1
Low-resolution fiber diffraction pattern of DNA. [Courtesy of Dr. R. Chandrasekharan, Purdue University. Reprinted from Leslie, A. G. W., Arnott, R., Chandrasekharan, R., and Ratliff, R. L., Polymorphism of DNA double helices, *J. Mol. Biol.*, **143**, 49–72, copyright ©1980, by permission of the publisher Academic Press Limited London.]

content of different diffraction experiments. Figure 4-1 is similar to fiber diffraction patterns obtained by Franklin and Wilkins and co-workers in the early 1950s, from which Watson and Crick deduced the double helix structure of DNA. Figure 4-2 is a modern, state-of-the-art fiber diffraction photograph, showing many more spots and therefore allowing more detailed structural interpretation. Figure 4-3 is a very high-resolution pattern from a Z-DNA crystal. It contains thousands of measurable spots, and therefore allows direct determination of atomic-level detail.

2.1 Basic Principles of Diffraction from Crystals and Fibers

Some of the basic features of these fiber and crystal diffraction patterns can be understood from the familiar Bragg's law

$$n\lambda = 2\,d\,\sin\theta_b \tag{4-1}$$

where d is the distance between scattering planes (i.e., regularly repeating features in the structure), θ_b is one-half of the scattering angle, λ is the X-ray wavelength, and n is an integer. The $\sin\theta_b$ is proportional to d^{-1}, so that small distances in the structure correspond to large angles (i.e., large distances on the film recording the diffracted

Figure 4-2
High-resolution fiber diffraction pattern of DNA. [Courtesy of Dr. R. Chandrasekharan, Purdue University. From Arnott, S., Chandrasekharan, R., Puigjaner, I. H., Birdsall, D. L., and Ratliff, R. L. (1983). *Nucleic Acids Res.* **11**, 1457–1474, by permission of Oxford University Press.]

image) and vice versa. This reciprocal relation between "real space" and "diffraction space" is fundamental to diffraction experiments. Astbury's early studies of X-ray diffraction from DNA fibers showed a strong reflection at an angle corresponding to 3.4 Å, which he correctly interpreted as being the distance between the planar bases stacked one upon the other in the fiber. This 3.4 Å reflection is evident as the very intense horizontal streak at the top and bottom of the photograph in Figure 4.1.

More generally, the desired information about molecular structure is contained in the positions and intensities of the diffraction spots. The basic equations relating structure and diffraction are

$$F(hk\ell) = \sum f_j \exp[2\pi i(hx_j + ky_j + \ell z_j)] \tag{4-2}$$

$$I(hk\ell) = |F(hk\ell)|^2 \tag{4-3}$$

where $I(hk\ell)$ is the intensity of a diffracted X-ray beam coming out from the crystal at an angle described by three integers (Miller indices, or coordinates of the diffraction maxima in reciprocal space) h, k, ℓ; $F(hk\ell)$ is the structure factor; and f_j is the X-ray scattering factor of atom j whose Cartesian coordinates relative to the unit cell axes are (x_j, y_j, z_j). The summation runs over all atoms in the unit cell.

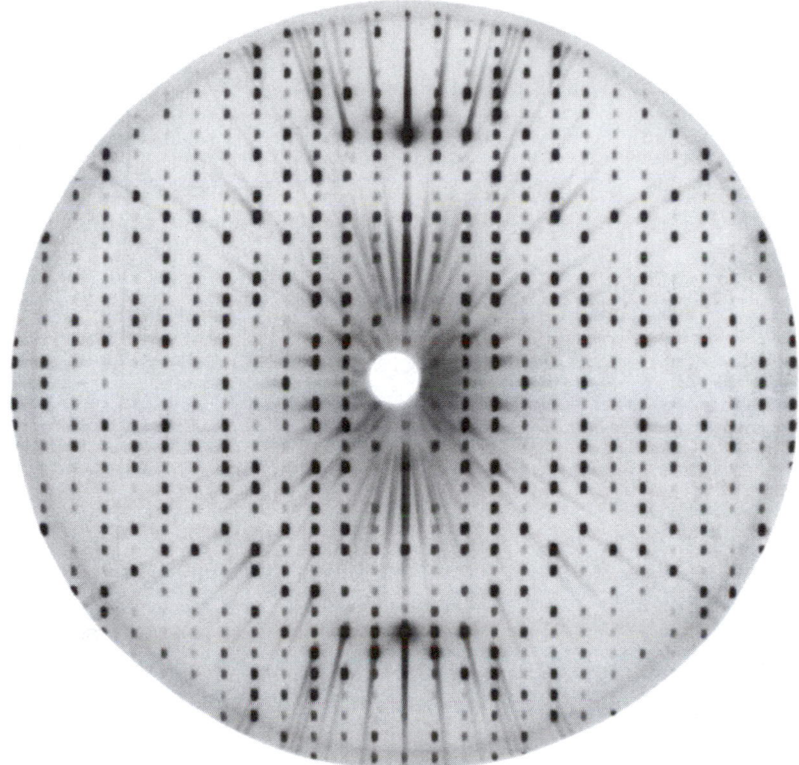

Figure 4-3
High-quality single-crystal diffraction pattern of Z-DNA. [Courtesy of Dr. A.H-J. Wang, University of Illinois.]

For fiber diffraction, the structure factor is adapted to cylindrical coordinates (r_j, f_j, z_j), yielding

$$G_{n,\ell}(R) = \sum f_j J_n(2\pi R r_j) \exp[i(2\pi z_j - n\phi_j)] \tag{4-4}$$

$$I(R, \ell) = \sum |G_{n,\ell}(R)|^2 \tag{4-5}$$

where $I(R, \ell)$ is the intensity of the diffraction spot located on the ℓ-th layer plane at radial distance R in reciprocal diffraction space; J_n is the n-th order Bessel function; and the sum in Eq. 4-5 runs over all n. Note that the fiber diffraction intensity does not depend on the angular positions of the scattering elements, because the fiber has rotational disorder about the molecular long axis.

2.2 Fiber Diffraction and Its Limitations

Most of the early studies of nucleic acid structures were done on fibers of nucleic acids. Any linear polymer can be pulled into a fiber forcing linear molecules to line up along the fiber axis. When such fibers are exposed to monochromatic X-ray radiation, characteristic diffraction patterns appear. The patterns can often be interpreted by inspection to obtain gross features of the average molecular structures in the fibers, such as the pitch, or distance along the axis per turn of the helix; the step height, or vertical distance between adjacent residues; the number of residues per turn; the pitch angle; the helix radius; and sometimes the number of strands per molecule.

The process of interpreting the pattern in more detail consists of proposing a model that is qualitatively consistent with the diffraction pattern, calculating the expected diffraction pattern by Eqs. 4-4 and 4-5, and adjusting the model to fit the expected diffraction pattern (locations and intensities of spots) to the observed pattern. The last two steps are repeated until the differences between the observed and calculated patterns are minimized. However the number of diffraction data are far fewer than necessary to uniquely define the 3D structure of the molecules in a fiber. Therefore, one has to make many simplifying assumptions to derive the structural models of nucleic acids, such as:

1. All the repeating units, whether mono-, di-, or multi-nucleotide, have the same conformation.
2. Connection between the repeating units is identical.
3. The backbone conformation is independent of the base sequence.
4. The structure has helical symmetry.

The most serious weakness in this procedure is that the correctness of the starting model is crucial. Even when the guessed model is wrong, it may still satisfy experimental data, which are very limited and few in number. This limitation is also true of other spectroscopic data that can suggest a model encumbered with many assumptions.

Although fiber diffraction cannot yield the level of detail and reliability obtained from single-crystal work, it is still capable of providing useful information, particularly on molecules with sequences that cannot be crystallized. As an example, Leslie et al. (1980) measured the diffraction patterns of DNA duplex fibers containing all four repeated dinucleotide sequences, 8 of the 12 possible repeated trinucleotide sequences, and 7 analogues in which G was replaced by hypoxanthine. They found evidence for at least six additional structural forms in addition to the canonical A, B, and C conformations. This finding was important early evidence for the extensive polymorphism of DNA, which has been examined in more detail by single-crystal work.

2.3 Single-Crystal Diffraction and Its Limitations

In contrast to fiber diffraction (cf. Figs. 4-1–4-3), single-crystal X-ray diffraction produces a large number of diffraction data, sometimes several thousand, thus its power. However, obtaining single crystals is very difficult and unpredictable, and has been the limiting step in X-ray crystallography of nucleic acids. Indeed, as Dickerson (1992) points out, the first single-crystal structure of a DNA molecule was solved in 1979 (Wang et al., 1979), 20 years after the first protein structure (myoglobin). The reason was that new chemical methods were needed to synthesize deoxyoligonucleotides of adequate length and sequence homogeneity. Automated synthesis techniques, using "gene machines," were not adequate, and tedious manual organic synthesis was required. More recent solid state methods based on the phosphoramidite technique can be used; purification of the product is the rate-limiting step (Kennard and Hunter, 1989).

The crystals usually contain cations and spermine. A dehydrating alcohol such as hexane-1,6-diol or 2-methyl-2,4-pentanediol is often used. The crystals usually contain about 50% nucleic acid and 50% solvent. This high solvent content often leads to disorder or low resolution (> 2 Å).

In contrast to protein crystallography, where the phase problem is generally solved by isomorphous replacement with heavy atoms (Blundell and Johnson, 1976), nucleic acid crystallography generally uses the molecular replacement method starting with a model based on idealized coordinates from fiber diffraction. This model is refined using various constraints on bond lengths and angles and energy minimization, until disagreement between calculated and observed diffraction pattern is minimized. For structures with base pair mismatches, the contributions from the atoms in the mismatch are initially omitted so as not to bias the refinement. The mismatch atoms are included when they begin to show up in the electron density difference map (Kennard and Hunter, 1989).

The goodness of fit of a proposed structure with the observed diffraction pattern is generally summarized in terms of the R factor

$$R = \left(\sum \left||F_o| - |F_c|\right|\right) / \sum |F_o| \tag{4-6}$$

where F_o and F_c are the observed and calculated structure factors, respectively. Generally, for a correct structure of a synthetic oligonucleotide solved to 2 Å resolution, R will be about 15%. Less well resolved, or incorrect, structures may have R about 25%. There is often confusion about resolution of an X-ray structure and the uncertainties in positions of the various atoms in the structure. In fact, because each atomic position is considerably overdetermined in a high-resolution structure, the position is much more accurately known than the resolution. For example, in a structure with diffraction spots extending to 2 Å, an R of 15% implies errors in atomic positions of about 0.1 Å.

Not all single-crystal structures are equally reliable. Pitfalls in various stages of large molecule crystallography can lead to badly mistaken structures (Brändén and Jones, 1990). Ultimately, the number of diffraction data per atom in the molecule is the parameter that determines the reliability of the structure determined. It is therefore essential to keep in mind that crystal structures determined from low-resolution data (smaller number of diffraction data) are less reliable than those from high resolution data (larger number of diffraction data). To determine the 3D structure of a molecule unambiguously, one needs to locate the position of each atom in the molecule, that is, x, y, and z coordinates for each atom. Therefore to determine the structure of a DNA duplex 12 bp long, one would require a minimum of about 1500 independent diffraction data points to identify approximately 500 non-hydrogen atoms, although constraints of standard bond lengths and angles reduce the number of independent coordinates to be determined. (Positions of hydrogen atoms cannot be determined by the method, but are inferred from positions of non-hydrogen atoms.) Fiber diffraction from DNA typically would give only a few tens of independent diffraction data. On the other hand, a single-crystal diffraction from a crystal of oligonucleotides may give 1000 to 10000 independent diffraction data points, depending on the resolution of the data.

It must also be realized that crystallography reveals the structure of only those atoms that are fixed at a given position in most of the molecules throughout the entire crystal. For example, if a part of the molecule is flexible or disordered, the method cannot provide an unambiguous structure for that portion; and the same is true for bound water molecules, metal ions, and other small molecules. If a water molecule or metal ion is bound to DNA at a specific site, it will be revealed by this method. However, if a water molecule is bound statistically in several places as a function of time or as a function of molecular population, it will not be revealed.

If more than one conformation is in equilibrium under a given condition, crystallography is not applicable. At best, it would give the structure of only one conformation.

In going from fiber to single-crystal diffraction, one observes local variations in structural parameters superimposed on the overall helix regularities. It is presumably these variations that lead to specificity of protein and drug binding and to other conformational subtleties that modulate nucleic acid function. However, it is important to realize that reported crystal structures have inevitable uncertainties, which must be critically taken into account. To quote Dickerson (1992) "Observable local features in an oligonucleotide crystal structure can have three origins: 1. Effects of base sequence, 2. Effects of intermolecular contacts within the crystal, and 3. Inadequacies in data and refinement, including a loss of detail because of low resolution of the analysis. The goal is to study No. 1, to monitor and control No. 2, and as much as possible to eliminate or at least minimize No. 3."

Dickerson (1992) has tabulated over 100 DNA crystal structures determined up to early 1991, grouping them according to A, B, Z or other forms, and within each class according to space group (i.e., mode of crystallization), giving unit cell dimensions and the number of unique base pairs in one asymmetric unit. For example, he lists 14 A-DNA structures that have crystallized in space group $P4_32_12_1$, and 10 B-DNA structures in space group $P2_12_12_1$. Within a given space group, all structures are isomorphous, with equivalent atoms in essentially the same places and with the same intermolecular crystal contacts. If structures with different sequences but in the same space group are compared, one obtains no information about the extent to which local features may be due to crystal packing perturbations. Comparison with identical sequences in other space groups is necessary.

In some space groups, the two halves of the duplex are symmetrically equivalent, so the asymmetric unit is only one-half of the molecule. In others, the two ends of the molecule are crystallographically distinct. In this latter case, if the molecule is chemically symmetrical about its midpoint, as is the case for most DNAs analyzed to date by X-ray methods, then structural differences between the ends can be attributed to crystal packing forces or to errors in the structural analysis.

One commonly listed indicator of the quality of a crystal structure is its resolution, that is, the distance in angstroms to which the diffraction pattern has been measured. Since crystals are three dimensional, a twofold increase in resolution, from 3.0 to 1.5 Å, implies eight times more data to be collected and analyzed, and eight times more information about the structure. In most cases, not all of the highest resolution diffraction spots are measurable, so a more reliable measure of structure quality may be the number of observed data per unique base pair (Dickerson, 1992). For the structures in the literature, this varies from under 50 to several thousand.

The R-factor can be improved (made smaller) by including water molecules in the structure. Such molecules are undoubtedly present, and make an important contribution to the structure, as discussed later. However, care must be taken that sufficient data support the assignment to immobilized solvent molecules of peaks in the electron density difference maps. A plausible measure of the weight of evidence in favor of including solvent molecules is the number of observed X-ray reflections per water molecule accepted (Dickerson, 1991). For some typical structures, this ratio ranges from 34 to 74.

Distinctions between influences of intermolecular crystal contacts, and errors in diffraction data and refinement, can be distinguished by "symmetry check plots" (Dickerson, 1992). In this procedure, oligomers with self-complementary sequences are examined in space groups in which the two halves of the molecule are not symmetrically equivalent. If helix parameters at the two ends of the molecule are different, this cannot be due to sequence effects. By examining different sequences in the same space groups, the maximum distortion due to crystal packing forces can be determined. The remainder must be due to errors in data and refinement. Systematic application of this approach gives an estimate of meaningful structural variation along the helix.

Comparison of the structures of crystals of the same sequence in different environments (space groups) can give information on the deformability in response to environment (Yanagi et al., 1991).

3. GENERAL FEATURES OF DUPLEX DNA AND RNA STRUCTURE

3.1 Watson–Crick Model of DNA

In 1953, Watson and Crick proposed a revolutionary, double helical model for DNA. The bases from the two strands were paired by complementary hydrogen bonding in the helix interior, the sugar–phosphate backbones extended along the outside, minimizing electrostatic repulsions; and the strands ran in antiparallel directions.

This model was derived from the interpretation of three types of experimental observations. (1) The diffraction pattern from highly hydrated DNA fibers obtained by Franklin and Wilkins was compatible with the helical model for DNA. (2) Biochemical studies of base composition ratio by Chargaff and co-workers (Zamenhof et al., 1952) suggested that the number of A bases was equal to T, and G bases to C. This constancy in ratio can be explained if A pairs with T and G with C in a double helix. (3) The crystal structure of a 5′ deoxycytidine (Furberg, 1951) suggested a favorable conformation for the sugar, for the connecting bond between the sugar and the base, and for the bond between the sugar and the phosphate; and only one of two tautomeric forms of the base was present at neutral pH. It was assumed that this conformation or some minor variant would hold for all the nucleotides in a DNA fiber.

This fundamental double helix model has since been abundantly confirmed and refined, using both improved fiber diffraction data (Arnott and Hukins, 1973), and single-crystal X-ray diffraction as discussed in detail below. One of the most important early findings is that fiber diffraction patterns of nucleic acids change depending on the environment—such as relative humidity, types of salt, ionic strength, and solvent—suggesting that nucleic acid structures change in response to their environment. These polymorphic forms probably reflect DNA polymorphism inside cells under various salt and ionic strength conditions, and in the presence of proteins that interact with DNA. The functional significance of DNA polymorphism is not yet understood, but the *in vitro* aspects of these changes are discussed in more detail in Chapter 11.

3.2 Models of DNA Forms A, B, and Z, and the A-RNA Form

Early X-ray diffraction studies of various forms of nucleic acid fibers resulted in several characteristic and distinct diffraction patterns. From these different patterns several classes of double helical models have been proposed. Among these, three have since been verified by single crystal diffraction studies of oligonucleotide duplexes. These are DNA forms A and B, and the A-RNA form. A fourth, DNA-Z form, was first discovered by single-crystal X-ray diffraction method and subsequently rediscovered from fiber diffraction studies. (Z-RNA has been demonstrated by NMR.)

To display and contrast the overall structural features of these forms—their relative size, length, depth and width of grooves—three different representations are given in Figure 4-4: (1) side-view, (2) end-view, and (3) cylindrical ribbon models. There are 20 bp in each model. For a more quantitative comparison, average structural features are listed in Table 4-1.

The minor groove is the side of the base pairs facing towards the sugar–phosphate backbone; the major groove is the side away. The width of either groove is calculated as the shortest distance between phosphates across the groove, minus 5.8 Å to account for the sum of the van der Waals radii of the two phosphates. The depth of the minor groove is the distance between P and N2 of G, N3 of A, or O2 of either C or T, minus the sum of the van der Waals radii (1.4 Å for O, 1.5 Å for N). The depth of the major groove is the distance between P and O6 of G, N6 of A, or O4 of T and N4 of C, again minus the sum of the van der Waals radii (Kennard and Hunter, 1989).

The major distinguishing features of the A, B, and Z forms are

- A and B are right-handed and can occur with any sequence, Z is left handed and occurs with alternating Pu-Py sequences, mainly GC.
- A is thick and compressed along the helix axis, Z is elongated and thin, B is intermediate.
- The displacement dx of the base pairs from the helix axis (see Figs. 4-5 and 4-7) is a characteristic feature. In B-DNA, the bases are on the helix axis, so dx is very small (average 0.8 Å). In the A-form, dx averages about −4 Å. In Z-DNA, the displacement is of opposite sense: $< dx > \approx$ 3–4 Å.

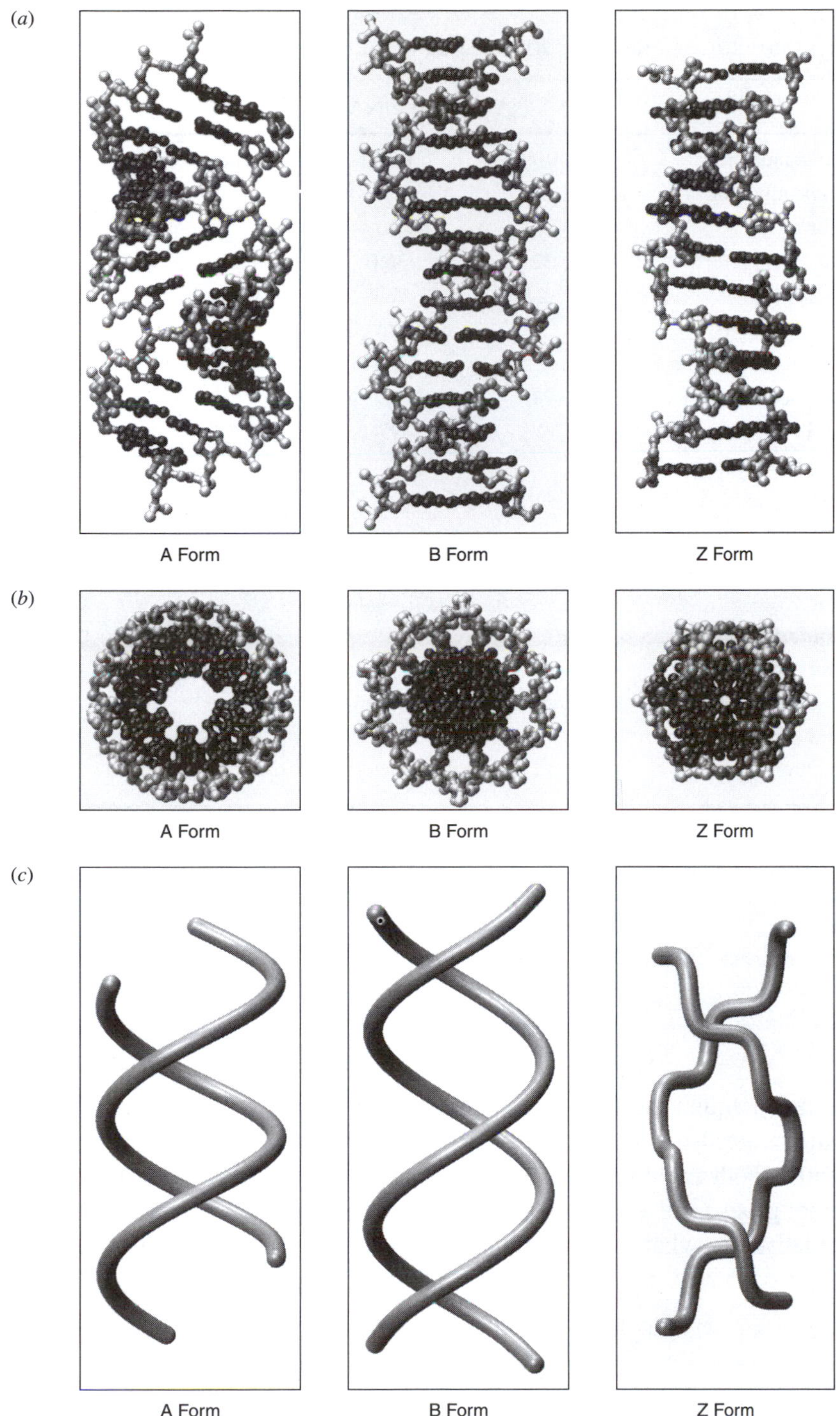

Figure 4-4
Three different representations of 3D structures of double helical DNA. From left to right, the A, B, and Z forms, each containing 20 bp. (*a*) Ball-and-stick side views. (*b*) Ball-and-stick end views. (*c*) Cylindrical ribbon representations. [Courtesy of Dr. Shri Jain and Dr. Helen Berman, Rutgers University.]

Table 4.1
Average Structural Parameters for Various Helical Forms

	A-DNA	B-DNA	Z-DNA	Z(WC)-DNA
Helix handedness	Right	Right	Left	Left
bp/repeating unit	1	1	2	2
bp/turn	11	10	12	12
Helix twist, (°)	32.7	36.0	−10[a], −50[b]	−68[a], +8[b]
Rise/bp, (Å)	2.9	3.4	−3.9[a], −3.5[b]	−3.9[a], −3.5[b]
Helix pitch, (Å)	32	34	45	45
Base pair inclination, (°)	12	2.4	−6.2	−5.8
P distance from helix axis, (Å)	9.5	9.4	6.2[a], 7.7[b]	5.6[a], 9.1[b]
X displacement from bp to helix axis, Å	−4.1	0.8	3.0	−1.6
Glycosidic bond orientation	anti	anti	anti[c], syn[d]	anti[c], syn[d]
Sugar conformation	C3′-*endo*	C2′-*endo*[e]	C2′-*endo*[c] C3′-*endo*[d]	C3′-*endo*[d] C2′-*endo*[c]
Major groove depth	13.5	8.5	Convex	Flat
width, (Å)	2.7	11.7		
Minor groove depth	2.8	7.5	9	Deep
width, (Å)	11.0	5.7	4	Narrow

[a]CpG step.
[b]GpC step.
[c]cytosine.
[d]guanine.
[e]There is a range of conformations.

- As a consequence of the *x*-displacement differences, the major and minor groove patterns are also different. The B-DNA has a wide major groove and a narrow minor groove; both are of similar depth. The A-form DNA and RNA have a narrow, deep major groove and a wide, shallow minor groove. In Z-DNA, the major "groove" is actually somewhat convex, and the minor groove is deep but narrow, and is lined by phosphates.
- The mean twist angle in B-DNA is 36° (though there is a wide range from 24° to 51°), giving 10 bp/turn. In A-DNA, the mean twist angle is 31°, yielding 11 bp/turn. In Z-DNA, there are 12 bp/turn, but the helix repeat is 2 bp (Pu-Pyr), so the mean twist angle is 60°, divided unequally between a G-C step of 50° and a C-G step of 10°.
- The pucker of the sugar ring tends to be C3′-*endo* in A, and C2′-*endo* in B though with broad distributions. In Z, the conformation is C2′-*endo* at C and C3′-*endo* at G.

- The glycosyl bond is anti in A and B, and alternates syn (at G) and anti (at C) in Z-DNA.
- The base pairs in A-DNA form are inclined at a large angle to the helix axis; those in B and Z forms are more nearly perpendicular.
- Survey of a large number of crystal structures shows that the Z-DNA is much more rigid than B or A form; and that B-DNA bends easily by collapsing the major groove.

For all these forms, the local mobility of the phosphate group is the highest, the sugars next, and the bases the least mobile (Holbrook and Kim, 1984).

3.3 Z-DNA and Its Puzzles

The first DNA sequences studied by crystallography were d(C-G)$_3$ (Wang et al., 1979) and d(C-G)$_2$ (Drew et al., 1980) because of the observation (Pohl and Jovin, 1972) that poly(dC-dG) could undergo a transition to a form with reversed circular dichroism (CD) that was provoked by high salt or alcohol. Surprisingly, these crystals had a left-handed twist, adding a third structural class, Z-DNA, to the right-handed A- and B-DNAs known from fiber diffraction. [For definition of slight differences between different crystal forms in the Z-DNA family, such as Z_I, Z_{II}, and Z', see Chapter 12 in Saenger (1984).]

The backbone chains in B- and Z-DNA run in opposite directions. In B and A, they run upward ($3' \rightarrow 5'$) at the left of the minor groove, and downward at the right. In Z, the overall sense is reversed. This reversed sense would require 180° reversal of glycosidic bonds, or all syn. Sterically, the six-membered C ring cannot accomodate this (though the five-membered ring in G connected to the sugar can), so conformation at C is anti but at G is syn, generating a local chain reversal that leads to the characteristic zigzag backbone path and two base repeating unit of Z-DNA.

How does DNA flip from B to Z, when the opposite direction of backbone chains implies that interconversion is not simply a matter of twisting the helix ends? There is no evidence for strand breakage and rejoining, nor for denaturation and renaturation. The original structure determination (Wang et al., 1979) proposed that the B-helix was first elongated, until the bases were able to rotate about their glycosidic bonds. Recompression to normal stacking distance then leads to Z-conformation. However, this process would be accompanied by a lot of steric hindrance, and should be slowed down by substitutions of increasing size on C5 of cytosine, which in fact accelerate the transition (Jovin et al., 1987).

Harvey (Harvey, 1983) proposed from molecular modeling a transition mechanism in which base pairs are flipped over one at a time, with maintenance of Watson–Crick pairing, in a cavity produced by longitudinal breathing. Since the cavity propagates down the helix after base pair flipping, the cooperativity of the transition is explained. Olson et al. (1982) showed that a postulated pathway that reversed the direction of the DNA backbone would drastically reduce the root-mean-square (rms) dimensions of the DNA chain, a process that seems unlikely.

The ready transformability from right- to left-handed DNA, with opposite sense of sugar–phosphate backbone, is therefore very difficult to understand if Z-DNA in solution is like Z-DNA in the crystal. Most solution measurements bear on only two characteristics: the left-handed helix, most commonly demonstrated by supercoiling experiments; and the dinucleotide repeat, which gives characteristic NMR spectra and chemical reactivity. It might be that other models could give equally good agreement with these experiments, but better account for the reversed chain direction and four other puzzling experimental results (Ansevin and Wang, 1990). (1) Some purine-pyrimidine repeats, such as $(dC\text{-}dG)_n$ or $(dA\text{-}dC)_n$ yield a left-handed helix, while others, such as $(dA\text{-}dT)_n$ or $(dC\text{-}dG)_2$-dT-dA-$(dC\text{-}dG)_2$, do not. (2) Amino proton exchange in G:C base pairs is an order of magnitude slower in left-handed than in right-handed helices. (3) The Z-DNA readily forms a precipitated Z* state. (4) The reversible B–Z transition is facilitated, rather than blocked, by bulky groups in the major groove of B-DNA.

These puzzling results have led to the proposal (Ansevin and Wang, 1990; Dickerson, 1992) of a new form of DNA, called Z(WC)-DNA, because it is like Z in being left handed with a zigzag backbone, but having Watson–Crick backbone directions. It is not proposed that the crystal structures of oligomeric Z-DNA are incorrect, but rather that an alternate form is energetically possible, has structural features compatible with observations (see Table 4-1), and rationalizes the puzzles noted in the previous paragraph. The structure of the Z(WC)-DNA helix was developed by energy minimization of a molecular model of an alternating G-C copolymer under the constraints that it be left handed and have 12 bp/turn, a 44-Å pitch, a dinucleotide repeat, and Watson–Crick chain directions. The resultant structure appears similar to Z-DNA in stereo drawings. Comparison of its structural features with those of the Z(II) crystal structure, B-DNA, and the fiber diffraction model of left handed DNA (Arnott et al., 1980) are shown in Table 4-1. The Z(WC)-DNA differs from Z(II) in having better defined hydrophobic patches and phosphate groups more closely spaced across the groove, features that may be important in protein–DNA recognition.

Since this model has Watson–Crick backbone directions, it immediately eliminates the major puzzle about how the B–Z transition is sterically possible. Indeed, it suggests that since a bulky group such as 2-acetylaminofluorene in the major groove induces zigzag distortions in a regular B-form helix, it should favor formation of left-handed DNA, thus dealing with puzzle (4) above. A notable feature of the proposed structure of Z(WC)-DNA is that a short, strong additional hydrogen bond is formed between the N2-amino group of guanine and a negative oxygen of the adjacent 5′-phosphate. This G-specific interaction explains both puzzle (1), why a sequence of two A-T base pairs is not stable, and puzzle (2), why the exchange of a G-amino proton is abnormally slow in left-handed DNA. Puzzle (3), the ready precipitation of left-handed DNA, is proposed to result from a four-stranded complex between two Z(WC)-DNA molecules with slightly modified helical parameters. Since this model has not been verified by crystallographic studies, its validity thus far rests on model-building and explanation of chemical results. However, it seems to merit consideration in explanation of a variety of *in vitro* and *in vivo* results.

4. DETAILED STRUCTURE AND SEQUENCE DEPENDENCE OF DOUBLE-STRANDED DNA

In an effort to find a level of description intermediate between average helix parameters from fiber diffraction or model building, and atomic coordinates from crystal structures, a set of helix parameters, shown in Figure 4-5, have been defined at an international workshop (Dickerson, 1989) to specify translational and rotational displacements within base pairs and from one base pair to the next.

While the mean values of local helix parameters as determined from crystal structures agree well with fiber diffraction results, the range of variation is unexpectedly large. This variation is shown in Figures 4-6 through 4-8. To give just two examples from 12 B-DNA structures (Dickerson, 1992; Yanagi et al., 1991): The mean and standard deviation of the helical twist angle Ω is $36.1° \pm 5.9°$, and the range is 24°–51°. The mean and standard deviation of the rise per base pair is 3.36 ± 0.46 Å, and the range is 2.5–4.4 Å. Such wide variations would have major impacts on specific recognition and binding by proteins and drugs, through their effects on the positioning of hydrogen bonding groups in the grooves of the helix.

A brief discussion of some of the most important of the helix structure parameters follows. The dependence of these parameters are ultimately to be understood from base stacking interactions and intrinsic local mobility of the bases within the constraints of the backbone conformation, which is dictated by the conformation of sugar moieties.

4.1 Base Stacking

The most prominent feature of all crystal structures of DNA is the extensive stacking of bases. It is fair to assume that base stacking plays one of the most important roles in stabilizing the helical structure of DNA. In Z-DNA form, because of the rigidity of the backbone, the base stacking pattern is well fixed. However, in right-handed B and A form of DNA, the way one base or base pair stacks on the neighbor bases or base pairs is substantially different and varies depending on the base sequence. In fiber diffraction studies, such differences were perforce assumed non-existent. The stacking mode can be specified by structural parameters (Fig. 4-5) that define the rotation of two bases in a pair, and of successive base pairs.

According to computer calculations (Haran et al., 1984), the relative displacement of the bases is largely governed by the energetics of base stacking. The stacking energy is sensitive to local changes in relative orientation of neighboring base pairs. Deviations of local crystal structure geometries of A- and B-form DNAs from idealized fiber diffraction coordinates are accountable by equipartition (i.e., smoothing out) of stacking energy along the double helix through adjustment of other local conformational variables.

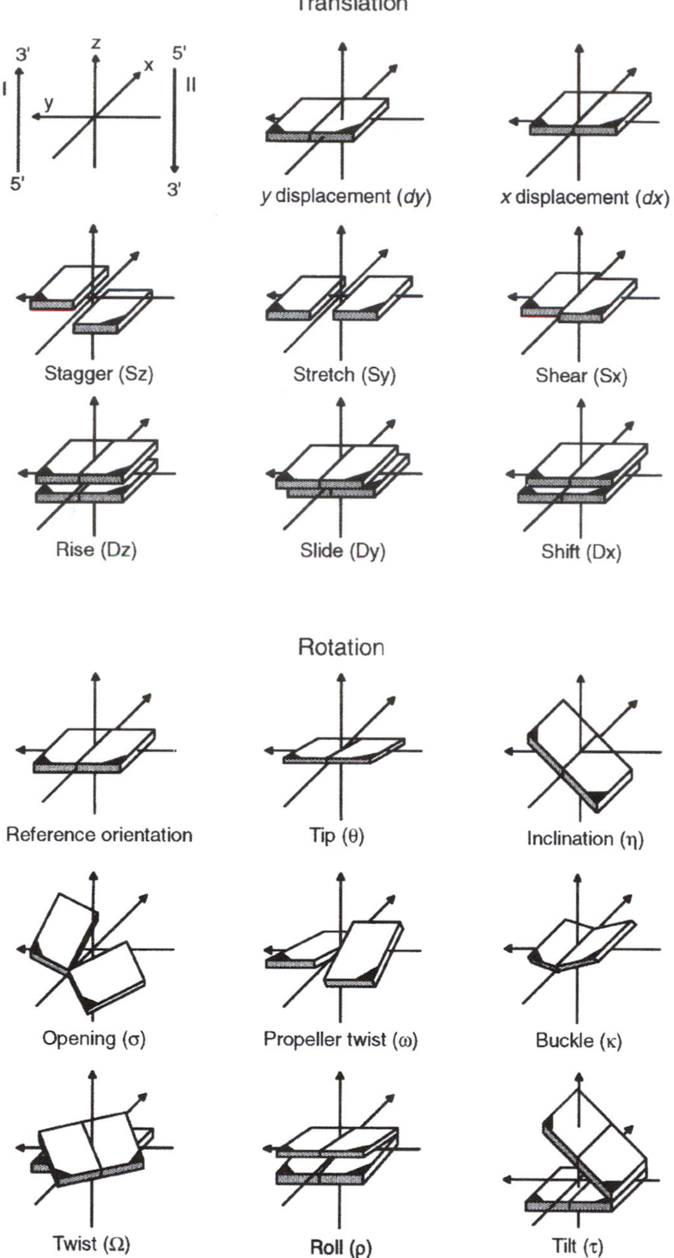

Figure 4-5
Definitions of translations and rotations involving bases and base pairs. The shaded edge represents the minor groove, and the black corners are the points of attachment of the glycosidic bonds to the C1′ sugar atoms. In each part of this figure, strand I of the double helix is at the left, running from 5′ to 3′ along the $+z$ axis, while strand II is at the right, running from 5′ to 3′ along the $-z$ axis. Positive roll widens the angle between base pairs on the side toward the minor groove. Positive tilt widens the angle between base pairs on the side toward strand I. Positive inclination rotates the base pair clockwise when viewed into the minor groove, along the $+x$ axis. Positive propeller twist rotates the base attached to strand II clockwise relative to the base attached to strand I, when viewed along the $-y$ axis.

4.2 Twist

Twist (Ω) is defined as the angle by which 1 bp has to rotate to match with the next base pair around the helical axis. It varies widely (Fig. 4-6) from 1 bp to the next, from 24° to 51° in B-DNA (see above) and from 23° to 44° in A-DNA (Shakked and Kennard, 1985; Heinemann et al., 1987). These variations are probably influenced, at

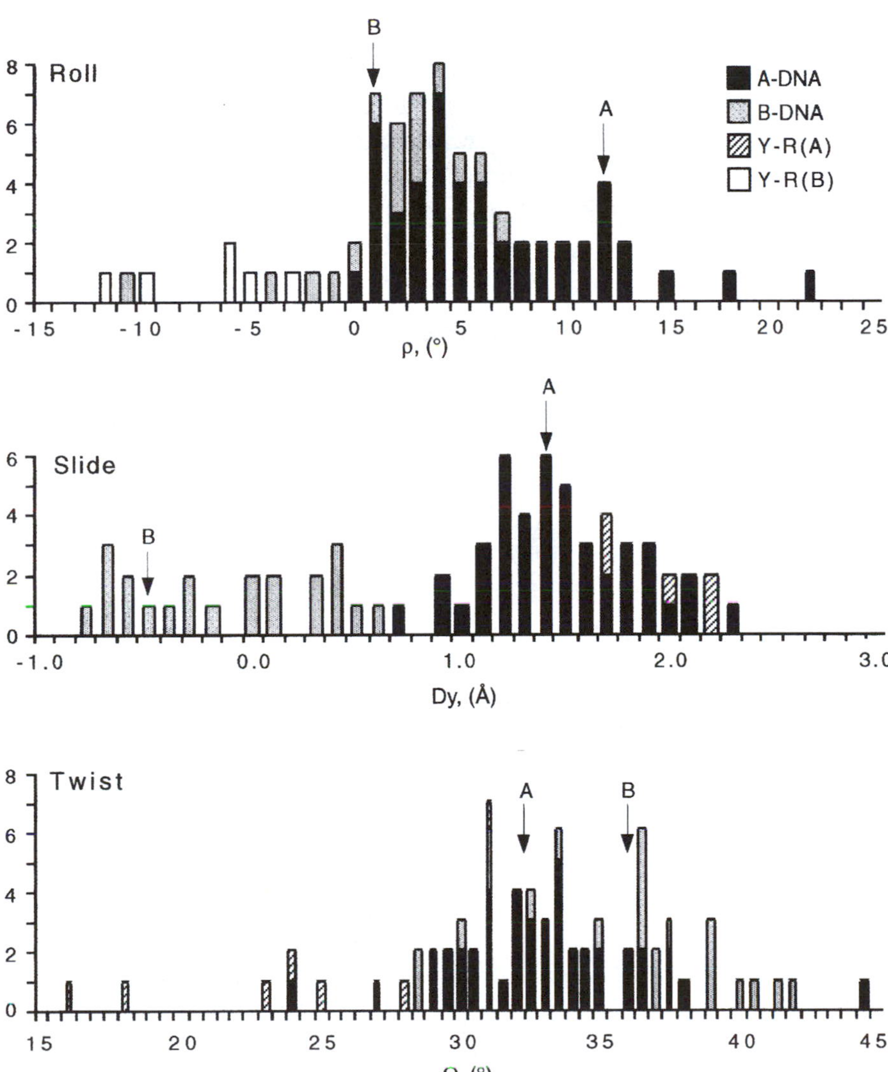

Figure 4-6
Distribution of roll, slide, and twist angles in A- and B-DNA helices. [Reprinted from *Prog. Biophys. Mol. Biol.* **47**, 159–195, Shakked, Z. and Rabinovich, D. The effect of the base sequence on the fine structure of the DNA double helix, Copyright ©(1986), with kind permission from Elsevier Science Ltd., The Boulevard, Langford Lane, Kedington 0X5 1GB, UK.]

least in part, by the attempt to maximize the base stacking overlap as 1 bp encounters the neighboring base pairs. Thus instead of having a constant value as assumed in deriving models from fiber diffraction data, the helical twist varies to compromise the best overlap or base stacking of individual base pairs within the constraints of backbone structure. When one follows the helical twist along the length of duplex, one sees an oscillating pattern around the average values, as if overtwisting in 1 bp is compensated by undertwisting in the next base pair, and so forth.

4.3 Inclination

Inclination (η) is the angle made by a base pair plane with a plane perpendicular to the helical axis. It varies from 9° to 22° in A-DNA form and is very small, about 2.4°, in B-DNA form (i.e., the bases are essentially perpendicular to the helix axis). This variation again presumably is a result of the tendency to maximize the base overlaps within the constraints of a given backbone structure. Inclination is usually anticorrelated with rise per base pair.

4.4 Propeller Twist

The propeller twist angle (ω) is formed between two base planes that form a base pair. In both A-DNA and B-DNA forms, substantial variations of propeller twist angles have been observed: between 6° and 16° in the A-DNA form, 13°–18° in the B-DNA form, but a little in Z-DNA form. In oligo(dA) • oligo(dT) tracts, propeller twist is so severe that each base forms bifurcated hydrogen bonds, one to its cognate base and the other to the base 5′ to that (Coll et al., 1987; Nelson et al., 1987).

4.5 Roll

Roll (ρ) is the angle between two successive base pair planes. This angle describes whether two adjacent base pairs are opened on the major groove or minor groove side. The roll angle in A-DNA form ranges from 6° to 9°, but in the B form it is almost zero on average, although there are variations (Fig. 4-6). Bending of DNA is the result of base pair rolls, as discussed in Chapter 9.

4.6 Buckle

The buckling angle (κ) is usually small, a few degrees, for all three forms of DNA and the A-RNA form.

4.7 X-Displacement from Helix Axis

The *x*-displacement (*dx*) is defined as the perpendicular distance from the long axis of the base pair (drawn from purine C8 to pyrimidine C6) to the helix axis. It is positive if the axis passes by the major groove side of the base pair, negative if it passes by the minor groove. Figure 4-7 shows that the displacements of the base pairs from the helix axis do not overlap among the A, B, and Z forms. This parameter is therefore a good definition of helix family (Dickerson, 1988).

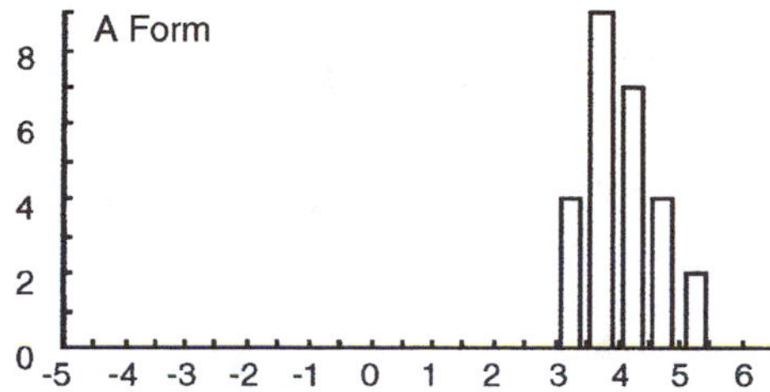

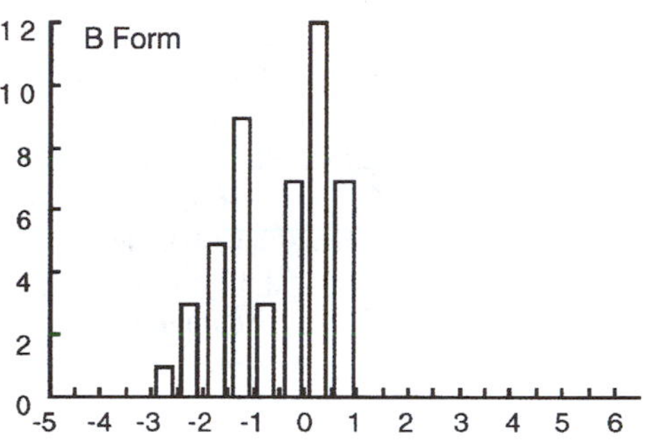

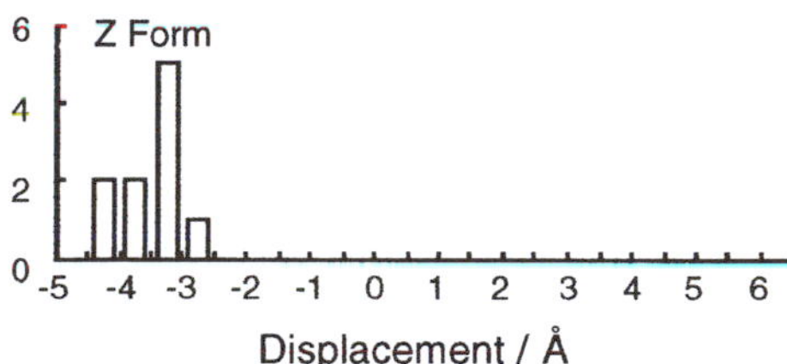

Figure 4-7
Frequency histograms of the *x* displacements of base pairs from the helix axis, as seen in published X-ray structures of the A, B, and Z types of helical oligomers. Displacement is positive if the axis drawn from purine C8 to pyrimidine C6 passes by the major groove side of the base pair, and negative if it passes by the minor groove side. [Adapted from Dickerson, 1988.]

4.8 Conformation of Sugar

Sugar conformations are discussed in detail in Chapter 2. The two most commonly found conformations are C3′-*endo*/C2′-*exo* ($P \approx 0°$) and C2′ *endo*/C3′-*exo* ($P \approx 180°$). The former is observed in RNA structures and A-DNA form, and the latter in B-DNA form. In Z-form DNA both conformations are found in alternating sequence; sugars in purine nucleotides are all C3′-*endo* and those in pyrimidine are C2′-*endo*. Here again wide range of variation around each conformation is observed. Figure 4-8 shows that there is substantial overlap in *P* among the three DNA helix forms, so that it is not as reliable a defining character between A- and B-DNA as is base pair *x* displacement from the helix axis.

4.9 Sequence Dependence of Structural Parameters

The question now arises, can patterns of dependence of these helical parameters on base sequence be discerned? If so, this implies that the base sequence is not just a linear code, but a structural code as well.

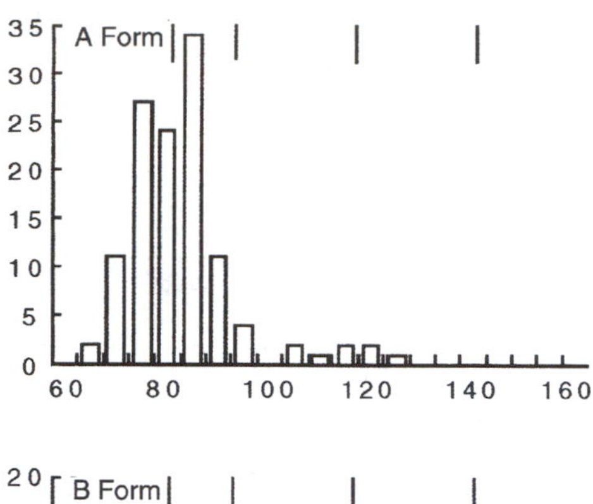

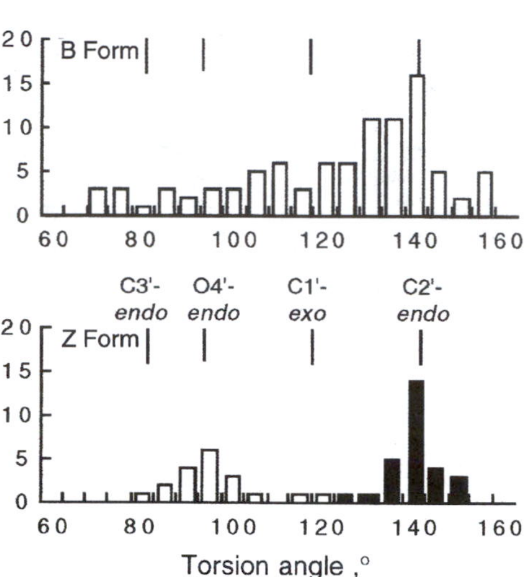

Figure 4-8
Frequency histograms of the occurrence of torsion angle δ (C5′-C4′-C3′-O3′) in published X-ray structures of A, B, and Z oligomers. For Z-DNA, the open bars are purines, and the black bars are pyrimidines. Atypical 3′-terminal purines are omitted. [Adapted from Dickerson, 1988.]

To look for such sequence-dependent effects, a fairly large data base of structures is needed. The first work on B-DNA was with variants of the dodecamer C-G-*x*-*x*-*x*-*x*-*x*-*x*-*x*-*x*-C-G (Drew et al., 1981). All of these crystallized in the same space group $P2_12_12_1$ and therefore shared the same microenvironment. They also only diffracted to relatively modest resolution, between 2.5 and 1.9 Å, which restricted the amount of information obtainable from sequence-structure comparisons, though useful correlations were made (Shakked and Rabinovich, 1986). More recently, B-DNA decamers have been crystallized in different space groups—monoclinic or orthorhombic. They diffract to higher resolution, 1.3–1.6 Å, and form essentially infinite helices in the crystal because of the 10 bp repeat.

Analysis of B-DNA structures discloses a number of important regularities (Kennard and Hunter, 1989; Dickerson, 1992; Grzeskowiak et al., 1991; Yanagi et al., 1991). The base step parameters twist, rise, and roll are highly correlated, which is not surprising given the finite extensibility of the backbone chain (high-twist requires low rise). Most of the observed base steps belong to one of two profiles. The high twist profile (HTP) also has low rise and negative roll. The low-twist profile (LTP) also has high rise and positive roll. An intermediate twist profile is seen in some cases, and a few others (C-G and C-A) are variable. There is a strong correlation between twist profile and base sequence:

Low twist profile	All R-R (and Y-Y) except G-A, which is HTP
Intermediate	All R-Y except G-C, which is HTP
Variable (VTP)	All Y-R except T-A, which is HTP

Clear examples of sequence-dependent twist profiles are seen in d(AT) tracts, where the helix twist angle is smaller at the A-T (R-Y) step than at the T-A (Y-R) step. The structural or energetic reasons for these regularities are not apparent.

The variable twist steps C-G and C-A are influenced by their flanking steps: Large twists on either side restrict a VTP step to a small twist, and small flanking twists allow a large central twist. This pattern suggests that the proper unit of analysis is three successive base steps, or four successive base pairs (Calladine, 1982; Dickerson, 1992). There are 136 such unique tetrads: 16 (4×4) self-complementary sequences (e.g. CCGG and TCGA) plus 120 ($[4^4 - 16]/2$) nonself-complementary sequences. Of these, only 35 were represented in data sets as of Dickerson's 1992 survey. Only when more sequences are crystallized and solved will it be possible to obtain a reliable correlation of structure with sequence.

From comparison of decamer and dodecamer structures, it appears that large base pair propeller requires two conditions (Yanagi et al., 1991). (1) The base pair must be A • T rather than G • C, and (2) the base pair must be involved in a three-center hydrogen bond with the following base pair in the major groove (e.g., (Nelson et al., 1987)). Such three-center hydrogen bonds occur frequently at C-C, C-A, A-A and A-C steps. Thus, large propeller is expected only at the leading base pair of A-A and A-C steps.

The width of the minor groove in the B-DNA decamers appears to correlate with base sequence. A wide minor groove (which can accomodate two ribbons of structured water down the groove) occurs when phosphates across the groove from each other are both in the BII conformation. This conformation occurs when the main chain torsion angles ε and ζ (see Fig. 2-3) are (gauche$^-$, trans) rather than the usual (trans, gauche$^-$). The phosphate conformation is influenced by base stacking and overlap, which as noted above have a dependence on sequence. B_{II} conformations occur only when the second base in a step is a purine: Y-R or R-R. The minor groove is especially narrow across the 2 bp of steps A-T, T-A, A-A, and G-A (Yanagi et al., 1991). The influence of groove width on hydration pattern is discussed in Chapter 11.

4.10 Environmental Dependence of Structural Parameters

While it is generally considered that right-handed duplex DNA is either in the A or B form, there is some evidence that intermediate conformations may sometimes be adopted depending on environment. For example, it is sometimes considered that the groove width is distinctive of DNA form. However, the minor groove width of A-DNA is observed to vary from 8.7 to 10.2 Å, and the major groove from 3.0 to 10.1 Å(Kennard and Hunter, 1989). Under solution conditions where the B form might be expected, certain sequences such as d(GGGGCCCC) (McCall et al., 1985) and d(GGATGGGAG) (McCall et al., 1986) have extensive stacking of purines that leads to a structure resembling the A form. The latter sequence is an essential part of the binding site of transcription factor IIIA, a protein that can bind both DNA and RNA. The DNA might therefore be expected to have a more RNA-like conformation. Circular dichroism studies of the complete 54 bp binding site as well as the nonamer sequence are consistent with an intermediate conformation (Fairall et al., 1989), although NMR shows that the nonamer is in the B form (Aboul-ela et al., 1988).

Another example is the octamer d(GGGCGCCCC) which, depending on the space group in which it crystallizes, can adopt one of three different A-form structures (Shakked et al., 1989). These differ in regard to tilting of the base pairs, rise per base pair, and width and depth of the major groove. These structural differences are attributable to hydration and molecular packing, which depend on the specific crystal form, and to temperature. The implications for DNA recognition by proteins are obvious.

Some sequences appear to be delicately balanced energetically between conformations, able to adopt one or the other depending on solution conditions or interactions with other molecules. A striking example of this is d(GG^{Br}UA^{Br}UAAC), which has a crystal structure containing both A and B forms (Doucet et al., 1989). The crystal lattice is formed of A-DNA molecules arranged with sixfold screw symmetry. These A-DNA molecules enclose a 26-Å diameter channel that contains a B-form molecule.

5. DOUBLE HELICAL RNA DUPLEX

Structure determination of a 14 bp RNA duplex (Dock-Bregeon et al., 1988) revealed that the duplex has two kinks dividing the molecule into three segments (Fig. 4-9). Each segment has a conformation very similar to the A-RNA form derived from fiber diffraction studies and to double helical portions of crystal structures of tRNA. As in RNA fiber (Arnott, 1970) and transfer RNA (Holbrook et al., 1978), the major groove is deep and narrow, while the minor groove is wide and shallow. The mean helical twist is 33.1°, mean rise per residue is 2.79 Å, and the base pairs are tilted 16.7° from the helical axis on the average. The structure shows a very extensive propeller twist (18.6°) similar to those found in tRNA and in an (AT) tract of DNA (Holbrook et al., 1978; Coll et al., 1987; Nelson et al., 1987). As expected, riboses have C3′-*endo* conformations in general. In about one-half of the cases, OH2′ appears to form a hydrogen-bond to O4′ of the following residue, similar to the finding in the tRNA structure (Holbrook et al., 1978).

The crystal structure of the RNA dodecamer r(GGAC**UUCG**GUCC)$_2$ containing the non-Watson–Crick base pairs G • U and U • C has been determined (Holbrook et al., 1991). These four noncomplementary nucleotides form a regular double helix, rather than the internal loop that might have been expected. The U • C pairs each form only one base–base hydrogen bond, but are stabilized by a water that bridges the ring nitrogens and four waters in the major groove which link bases and phosphates. The G • U wobble pair is also stabilized by a water molecule in the minor groove that bridges the unpaired guanine amino and the ribose OH of the uracil. This hydrogen-bonding to water shows that specific hydrogen bonds between groups other than just the bases are important in stabilizing double helical RNA.

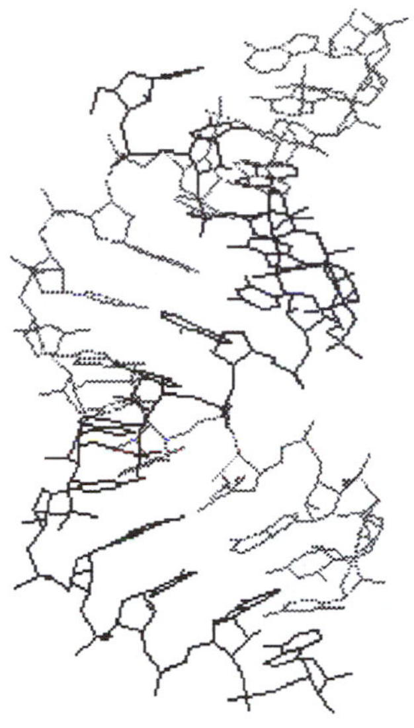

Figure 4-9
Wireframe view of an RNA duplex containing 14 bp of self-complementary sequence, $U(UA)_6A$. The two strands of the double helix are displayed in dark and light gray. The structure can be divided into three segments represented by noncolinear helical axes. [Structure determined by Dock-Bregeon et al., 1988.]

6. DNA HELICES WITH IMPERFECT BASE PAIRING

Several crystal structures with base pair mismatches have been determined. These include G•T in B-DNA form (Hunter et al., 1987), in A-DNA form (Kneale et al., 1985), and in Z-DNA form (Brown et al., 1986); G•A mismatch in B-DNA form with both bases in the anti-conformation (Privé et al., 1987) and with A in the syn conformation (Brown et al., 1986); and A•C mismatch in B-DNA form, where A is probably protonated at the N1 position (Hunter et al., 1986). In all cases, the perturbation of the duplex due to the mismatch is localized to the mismatch site and sometimes to the immediate neighbors, and the overall DNA conformation remains unchanged. The relative inextensibility of the sugar–phosphate backbone prevents G•A purine–purine anti–anti mismatches from causing bulges in the double helix. Instead, the mismatch is displaced toward the center of the cylindrical duplex (Yanagi et al., 1991).

Mistakes made by DNA polymerase during DNA replication or repair could lead to an unpaired extra base, which may loop out from or stack inside the DNA duplex. A tridecamer containing extra unpaired bases, $d(CGCAGAATTCGCG)_2$, shows interesting differences between crystal and solution. The 2.6-Å resolution crystal structure (Joshua-Tor and Sussman, 1991) shows that one of the extra A′s loops out of the duplex, while the other stacks into it. The looped-out A from one duplex intercalates into the stacked-in bulge site of a neighboring, symmetrically related duplex in the crystal. The two extra A’s form a reversed Hoogsteen base pair. In solution, this intermolecular interaction is not possible, and NMR shows that both A’s are stacked in the helix (Patel et al., 1982). The conformation of the looped-out bulge in the crystal structure is minimally disruptive of the helix, since the backbone makes a “loop-the-loop” and the extra A is flipped over with respect to the other bases in the strand.

A hairpin structure with a Z-DNA stem (Fig. 4-10) shows that as few as four nucleotides can form a loop (Chattopadhyaya et al., 1988). However, the loop conformation appears to be determined by the interaction with neighbor molecules. Thus, the loop conformation in solution is likely to be very different from that found in this crystal structure. Furthermore, the same sequence in a solution of moderate or low ionic strength was found to be in B-DNA form for the stem and had a different loop conformation according to two-dimensional(2D) NMR studies (Wolk et al., 1988).

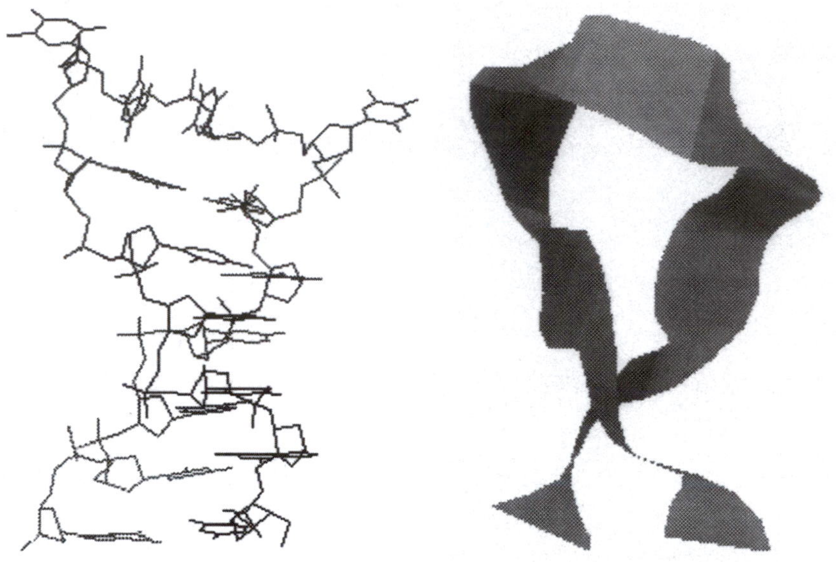

Figure 4-10
Wireframe and ribbon views of a DNA hairpin structure. The stem is in a Z-DNA conformation, and the loop is made of only four nucleotides. [Structure determined by Chattopadhyaya et al., 1988.]

7. LOCAL MOBILITY OF DOUBLE HELICAL NUCLEIC ACIDS

Nucleic acids do not have static structures; their dynamic mobility is essential for their function. One can divide nucleic acid motion into two classes: (1) large-scale motion, such as bending, supercoiling, or unwinding; and (2) local mobility, which is an intrinsic property of the constituents of nucleic acids, and in fact is a basis for understanding the larger scale motion as well as base stacking patterns described above. Single-crystal diffraction data can provide a quantity called the "thermal" parameter, which measures the sum of two types of motion. One arises from the motion or disorder of the entire molecule in the crystal, which is usually small and negligible for the purpose of discussion here. The other is local mobility which, in turn, is the sum of two kinds of motion: (1) the vibrational motion of molecules or the components of the molecule; and (2) conformational disorder. Although it is difficult to distinguish between these two classes of mobility, both reflect an ability on the part of the molecule to move and thus provide some measure of the local mobility of the nucleic acids in crystals. The types of local mobility observed from this line of analysis can be analyzed using helix structure variables like those diagrammed in Figures 4-6 and 4-7 (Holbrook and Kim, 1984). Various combinations of these local mobility elements may produce observed base stacking patterns and other features of structure–function relations in nucleic acids.

8. TRANSFER RNA

Transfer RNA is a class of small RNA molecules an average of 70–80 nucleotides long, that plays a key role in translating the genetic information coded in mRNA into proteins. During this process, a tRNA molecule interacts with numerous proteins, some specifically and others nonspecifically. Over 100 different cytoplasmic tRNAs have been sequenced and all can be arranged into a clover leaf secondary structure as originally proposed for yeast alanine tRNA (Holley et al., 1965). A generalized cloverleaf pattern is shown in Figure 4-11 together with the "L" pattern that is a rearrangement of the cloverleaf pattern to simulate the 3D structure of tRNA (Kim et al., 1974). Recently, it has been shown that tRNA from organelles of eukaryotic cells such as mitochondria and chloroplast often have substantially different secondary structures. Since the 3D structures of these unusual organelle tRNAs have not been determined yet, the following description of tRNA structure applies only to cytoplasmic tRNA.

The common features of the secondary structure of all the cytoplasmic tRNAs are as follows (Jack et al., 1976; Kim, 1979):

1. There are 7 bp in the amino acid (AA) stem, 3 or 4 bp in the dihydrouracil (D) stem, 5 bp in the anti-codon (AC) stem, and 5 bp in the TC (T) stem.
2. The base pairs in the stems are of the Watson–Crick type with an occasional G•U pair.
3. There are always two nucleotides between the AA and D stems, one between the D and AC stems and no nucleotide between the AA and T stems.
4. There are always 7 nucleotides in the AC and T loops, 7–10 nucleotides in the D loop, and 4–21 nucleotides in the variable arm (V arm).
5. Certain bases are conserved among all tRNAs except for the initiation tRNA: U between the AA and D stems; A at the beginning of the D loop; two G's at the D loop; U in the AC loop; a GC base pair in the T stem; a TUC sequence and A in the T loop; and CCA at the 3′-end.
6. Some positions are semiconserved. For example, the base after the sequence TUC in the T loop is always a purine, either G or A.

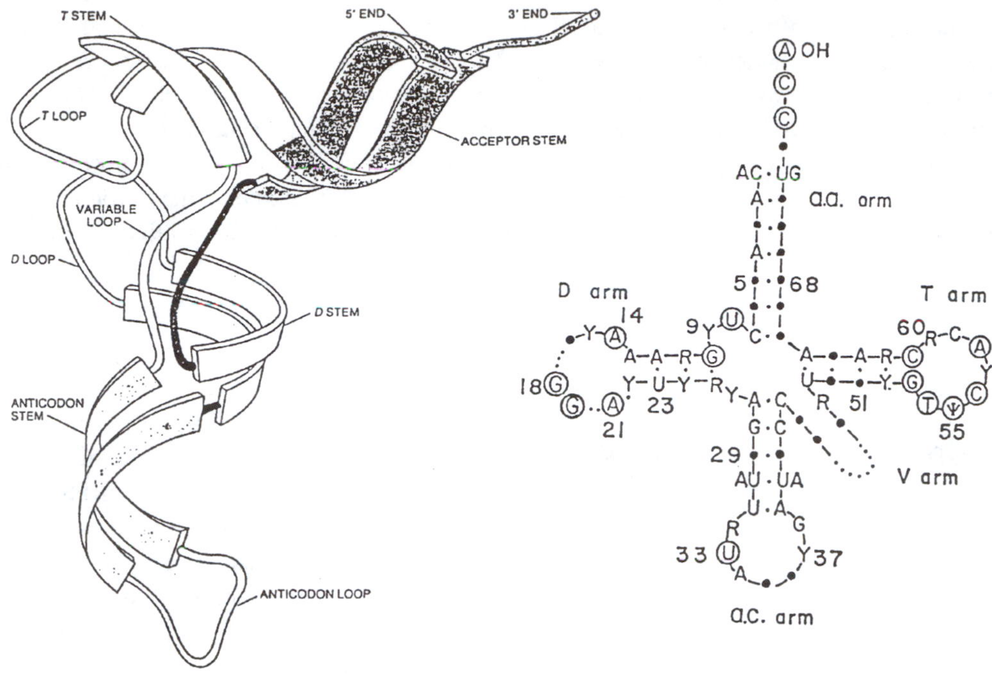

Figure 4-11
(*a*) Backbone structure of yeast phenylalanine tRNA. (*b*) Cloverleaf secondary structure of generalized cytoplasmic tRNA. Conserved and semiconserved bases are indicated by circled and uncircled letters, respectively. [Reprinted with permission from Kim, 1979.]

X-ray crystallographic studies (Kim et al., 1974) reveal that the majority of these conserved and semiconserved bases are involved in forming the tertiary structure of the tRNA; that is, they are utilized to form a common architectural frame for all tRNA structures. All the crystal structures so far determined are very similar to the yeast phenylalanine tRNA and therefore the following discussion will primarily focus on the yeast tRNA structures.

8.1 Overall Shape of "L"

As schematically shown in Figure 4-11, the overall shape of cytoplasmic tRNA is that of a letter "L" upside down. The amino acid stem and T stem form one continuous double helical arm, and the D stem and anti-codon stem form the other long double helical arm of the L. Each extended arm of the L is about 60 Å long and has a diameter of about 20 Å. Each stem is an antiparallel right-handed double helix similar to the A-RNA form. The 3′-end where peptide elongation occurs is at one extreme of the molecule, while the anti-codon triplet that recognizes the codon on the mRNA is at the other extreme. The T loop, which has a largely conserved base sequence in all cytoplasmic elongation tRNAs, is located at the corner of the L. Although the overall conformation of each double helix is that of A-RNA form, there are substantial variations from the canonical A-RNA conformation derived from fiber diffraction studies. Three views of the yeast phenylalanine tRNA structure, rotated about a vertical axis, are shown in Figure 4-12.

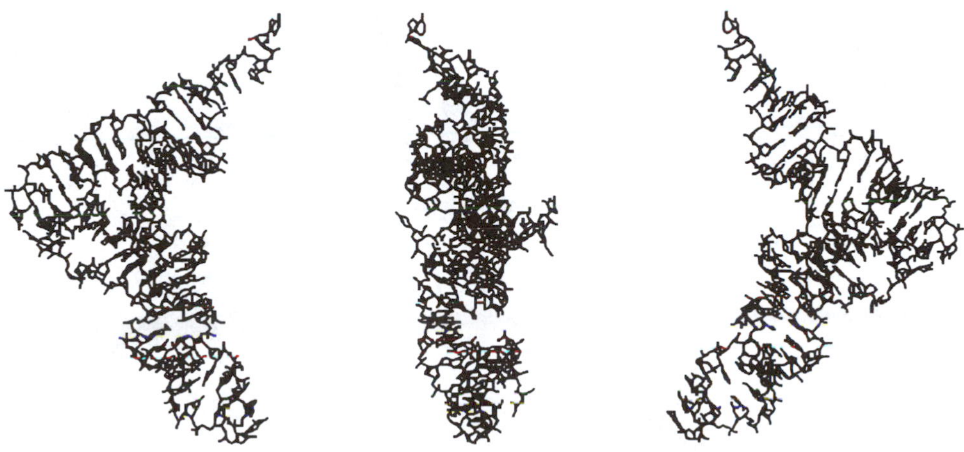

Figure 4-12
Three views of the crystal structure of yeast phenylalanine tRNA, rotated by 90° about the vertical axis. [Structure determined by Sussman et al., 1978. Reprinted with permission.]

8.2 Extensive Base Stacking

Although only about one-half of the bases are predicted to be in the double helical stem based on the "cloverleaf" model, the 3D structure reveals that more than 90% of the bases are stacked. Such extensive base stacking, including bases in the loop, must be one of the major stabilizing forces and is likely to be a universal feature of all free nucleic acids (see Fig. 4-13).

8.3 Tertiary Base Pairs Are Non-Watson–Crick Type

In yeast phenylalanine tRNA, there are nine tertiary base pairs (Fig. 4-13). Eight of these are non-Watson–Crick type and most of these involve bases that are either conserved or semiconserved (Rich and Kim, 1978) in all cytoplasmic elongation tRNAs. All of these tertiary base pairs are located at the intersection of the two arms of the L and are presumably essential for maintaining the L-shaped basic frame.

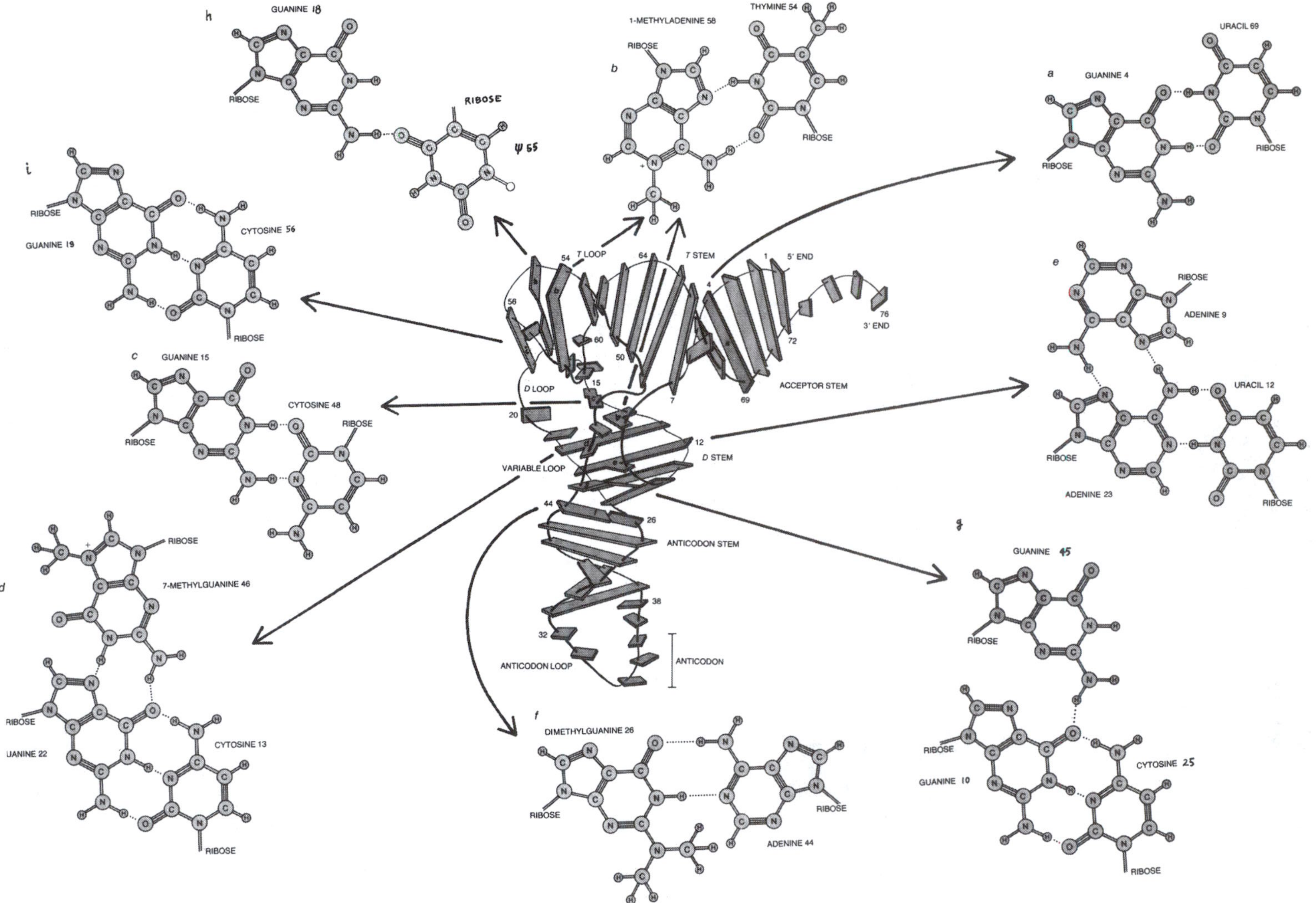

Figure 4-13
Tertiary base pairs and their locations observed in the crystal structure of yeast phenylalanine tRNA. [With permission from Rich and Kim, 1978.]

8.4 Flexibility of the Molecule

Refinement of the structure reveals that the two extremes of the molecules, that is, the 3′ end at one end of the L and the anti-codon loop at the other end, are more flexible than the center of the molecules by almost threefold (Sussman et al., 1978). This flexibility may be important for the biological function of tRNA molecules.

8.5 Tightly Bound Ions: Magnesium and Spermine

It has been known that divalent ions such as Mg^{2+} greatly stabilize tRNA molecules. In case of yeast phenylalanine tRNA, four tightly bound magnesium ions have been identified (Holbrook et al., 1977): three in the D loop, one in the sharp turn formed by the residues 8–12, and one in the anti-codon loop (see Fig. 4-14). These magnesium hydrates make many hydrogen bonds, mostly to the looped backbone, thus stabilizing the conformation of the loops to which they are bound. Polyamines such as spermine are known to stabilize the tRNA conformation as well. These polyamine molecules in general bind to the deep groove of double helical stems formed by the D stem and anti-codon stem and another in the deep groove of the double helix formed by the amino acid stem and the T stem (Holbrook et al., 1978; Quigley et al., 1978). The precise locations of their binding are not well established. Thus it appears that the spermine stabilizes the joint between two double helical stems to make a contiguous one, and the magnesium ions stabilize the conformation of tight turns or loops of the molecule.

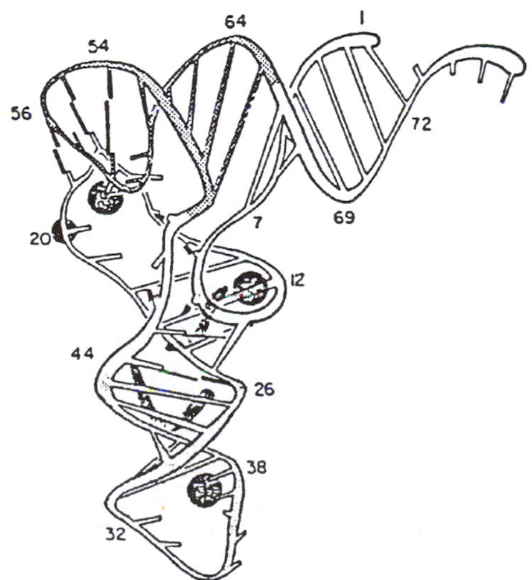

Figure 4-14
Positions of tightly bound magnesium ions in yeast phenylalanine tRNA. [From Holbrook, S. R., Sussman, J. L., Warrant, R. W., Church, G. M., and Kim, S. H. 1977. *Nucleic Acids Res.* **4**, 2811–2820 by permission of Oxford University Press.]

References

Aboul-ela, F., Varani, G., Walker, G. T., and Tinoco, I., Jr. (1988). The TFIIIA recognition fragment d(GGATGGGAG):d(CTCCCATCC) is B-form in solution *Nucleic Acids Res.* **16**, 3559–3572.

Ansevin, A. T. and Wang, A. H. (1990). Evidence for a new Z-type left-handed DNA helix: Properties of Z(WC)-DNA, *Nucleic Acids Res.* **18**, 6119–6126.

Arnott, S. (1970). The geometry of nucleic acids, *Prog. Biophys. Mol. Biol.* **21**, 267–319.

Arnott, S., Chandrasekaran, R., Birdsall, D. L., Leslie, A. G. W., and Ratliff, R. L. (1980). Left-Handed DNA Helices, *Nature (London)* **283**, 743–745.

Arnott, S., Chandrasekaran, R., Puigjaner, L. C., Walker, J. K., Hall, I. H., Birdsall, D. L., and Ratliff, R. L. (1983). Wrinkled DNA, *Nucleic Acids. Res.* **11**, 1457–1474.

Arnott, S. and Hukins, D. W. L. (1973). Refinement of the Structure of B-DNA and Implications for the Analysis of X-Ray Diffraction Data from Fibers of Biopolymers, *J. Mol. Biol.* **81**, 93–105.

Blundell, T. N. and Johnson, L. N. (1976). Protein Crystallography, Academic, New York.

Brändén, C.-I. and Jones, T. A. (1990). Between objectivity and subjectivity, *Nature (London)* **343**, 687–689.

Brown, T., Hunter, W. N., and Kennard, O. (1986). Structural characterization of the bromouracil-guanine base pair mismatch in a Z-DNA fragment, *Nucleic Acid Res.* **14**, 1801–1809.

Brown, T., Hunter, W. N., Kneale, G. and Kennard, O. (1986). Molecular structure of the G•A base pair in DNA and its implications for the mechanism of transversion mutations, *Proc. Natl. Acad. Sci. USA* **83**, 2402–2406.

Calladine, C. R. (1982). Mechanism of sequence-dependent stacking of bases in B-DNA, *J. Mol. Biol.* **161**, 343–352.

Chattopadhyaya, R., Ikuta, S., Crzeskowiak, K., and Dickerson, R. E. (1988). X-ray structure of a DNA hairpin molecule: d(C-G-C-G-G-G-T-T-T-C-G-C-G-C-G), *Nature (London)* **334**, 175–179.

Coll, M., Frederick, C. A., Wang, A. H. J., and Rich, A. (1987). A bifurcated hydrogen-bonded conformation in the d(A.T) base pairs of the DNA dodecamer d(CGCAAATTTGCG) and its complex with distamycin, *Proc. Natl. Acad. Sci. USA* **84**, 8385–8389.

Dickerson, R. E. (1988). Usual and Unusual DNA Structures: A Summing-Up. *Unusual DNA Structures*. Springer-Verlag, New York, 287–306.

Dickerson, R. E. (1989). Definitions and Nomenclature of Nucleic Acid Structure Parameters, *EMBO J.* **8**, 1–4.

Dickerson, R. E. (1992). DNA structure from A to Z, *Methods Enzymol.* **211**, 67–111.

Dock-Bregeon, A. C., Chevrier, B., Podjarny, A., Moras, D., de Bear, J. S., Gough, G. R., Gilham, P. T., and Johnson, J. E. (1988). High resolution structure of the RNA duplex $[U(U\text{-}A)_6A]_2$, *Nature (London)* **335**, 375–378.

Doucet, J., Benoit, J. P., Cruse, W. B., Prange, T., and Kennard, O. (1989). Coexistence of A- and B-form DNA in a single crystal lattice., *Nature (London)* **337**, 190–192.

Drew, H., Takano, T., Tanaka, S., Itakura, K. and Dickerson, R. E. (1980). High-Salt d(CpGpCpG), a Left-Handed Z' DNA Double Helix, *Nature (London)* **286**, 567–573.

Drew, H. R., Wing, R. M., Takano, T., Broka, C., Tanaka, S., Itakura, K. and Dickerson, R. (1981). Structure of a B-DNA dodecamer: configuration and dynamics, *Proc. Natl. Acad. Sci. USA* **78**, 2179–2183.

Fairall, L., Martin, S., and Rhodes, D. (1989). The DNA binding site of the Xenopus transcription factor IIIA has a non-B-form structure, *EMBO J.* **8**, 1809–1817.

Furberg, S. (1951). The crystal structure of cytidine, *Acta Crystallogs* **3**, 325–331.

Grzeskowiak, K., Yanagi, K., Privé, G. G., and Dickerson, R. E. (1991). The structure of B-helical C-G-A-T-C-G-A-T-C-G and comparison with C-C-A-A-C-G-T-T-G-G. The effect of base pair reversals, *J. Biol. Chem.* **266**, 8861–8883.

Haran, T. E., Berkovich-Yellin, Z., and Shakked, Z. (1984). Base-stacking interactions in double-helical DNA structures: experiment vs theory, *J. Biomol. Struct. Dyn.* **2**, 397–412.

Harvey, S. C. (1983). DNA Structural Dynamics: Longitudinal Breathing as a Possible Mechanism for the B-Z Transition, *Nucleic Acids Res.* **11**, 4867–4878.

Heinemann, U., Lauble, H., Franck, R., and Blöcker, H. (1987). Crystal structure analysis of an A-DNA fragment at 1.8~ resolution: d(GCCCGGGC), *Nucleic Acid Res.* **15**, 9531–9500.

Holbrook, S. R., Cheong, C., Tinoco, I., Jr., and Kim, S.-H. (1991). Crystal structure of an RNA double helix incorporating a track of non-Watson–Crick base-pairs. *Nature (London)* **353**, 579–581.

Holbrook, S. R. and Kim, S.-H. (1984). Local Mobility of Nucleic Acids as Determined from Crystallographic Data. I. RNA and B-Form DNA, *J. Mol. Biol.* **173**, 3651–3688.

Holbrook, S. R., Sussman, J. L., Warrant, R. W., Church, G. M., and Kim, S. H. (1977). RNA-ligand interactions. I: Magnesium binding sites in yeast phenylalanine tRNA, *Nucleic Acids Res.* **4**, 2811–2820.

Holbrook, S. R., Sussman, J. L., Warrant, R. W., and Kim, S. H. (1978). Crystal structure of yeast phenylalanine transfer RNA, II: Structural features and functional implications, *J. Mol. Biol.* **123**, 631–660.

Holley, R. W., Apgar, J., Everett, G. A., Madison, J. T., Marguisse, M., Merrill, S. H., Penwick, J. R., and Zamire, A. (1965). Structure of a ribonucleic acid, *Science* **147**, 1462–1465.

Hunter, W. N., Brown, T., Anand, N. N., and Kennard, O. (1986). Structure of an adenine-cytosine base pair in DNA and its implications for mismatch repair, *Nature (London)* **320**, 552–555.

Hunter, W. N., Brown, T., Kneale, G., Anand, N. N., Rabinovich, D., and Kennard, O. (1987). The structure of guanosine-thymidine mismatches in B-DNA at 2.5 Å resolution, *J. Biol. Chem.* **262**, 9962–9970.

Jack, A., Ladner, J. E., and Klug, A. (1976). Crystallographic refinement of yeast phenylalanine transfer RNA at 2.5 Å resolution, *J. Mol. Biol.* **108**, 619–649.

Joshua-Tor, L. and Sussman, J. L. (1991). The three-dimensional structure of a bulge-containing DNA fragment, *J. Biomol. Struct. Dynam.* **8**, a094.

Jovin, T. M., Soumpasis, D. M., and McIntosh, L. P. (1987). The transition between B-DNA and Z-DNA, *Annu. Rev. Phys. Chem.* **38**, 521–560.

Kennard, O. and Hunter, W. N. (1989). Oligonucleotide structure: a decade of results from single crystal X-ray diffraction studies, *Q. Revs. Biophys.* **22**, 327–379.

Kim, S. H. (1979). Crystal structure of yeast phenylalanine tRNA and general structural features of other tRNAs. in *Transfer RNA Structure, Properties and Recognition*. Schimmel, P. R., Söll, D., and Abelsonb, J. N., Eds. New York, Cold Spring Harbor, New York, pp. 83–100.

Kim, S. H., Suddath, F. L., Quigley, G. J., McPherson, A., Sussman, J. L., Wang, A. H., Seeman, N. C., and Rich, A. (1974). Three-dimensional tertiary structure of yeast phenylalanine transfer RNA, *Science* **185**, 435–440.

Kneale, G., Brown, T., Kennard, O., and Rabinovich, D. (1985). G•T base-pairs in a DNA helix: The crystal structure of d(G-G-G-G-T-C-C-C), *J. Mol. Biol.* **186**, 805–814.

Leslie, A. G. W., Arnott, S., Chandrasekharan, R., and Ratliff, R. L. (1980). Polymorphism of DNA double helices, *J. Mol. Biol.* **143**, 49–72.

McCall, M., Brown, T., Hunter, W. N., and Kennard, O. (1986). The crystal structure of d(GGATGGGAG): an essential part of the binding site for transcription factor IIIA., *Nature (London)* **322**, 661–664.

McCall, M., Brown, T., and Kennard, O. (1985). The crystal structure of d(GGGGCCCC)—a model for poly(dG).poly(dC), *J. Mol. Biol.* **183**, 385–396.

Nelson, H. C. M., Finch, J. T., Luisi, B. E., and Klug, A. (1987). The structure of an oligo(dA).oligo(dT) tract and its biological applications, *Nature (London)* **330**, 221–226.

Olson, W. K., Srinivasan, A. R., Marky, N. L., and Balaji, V. N. (1982). Theoretical probes of DNA conformation: Examining the B $\rightarrow$ Z conformational transition, *Cold Spring Harbor Symp. Quant. Biol.* **47**, 229-241.

Patel, D. J., Kozlowski, S. A., Marky, L. A., Rice, J. A., Broka, C., Itakura, K., and Breslauer, K. J. (1982). Extra adenosine stacks into the self-complementary d(CGCAGAATTCGCG) duplex in solution, *Biochemistry* **21**, 445–451.

Pohl, F. M. and Jovin, T. M. (1972). Salt-induced cooperative change of a synthetic DNA, equilibrium and kinetic studies with poly (dG-dC), *J. Mol. Biol.* **67**, 375–396.

Privé, G. G., Heinemann, U., Chandrasegaran, S., Kan, L.-S., Kopka, M. L., and Dickerson, R. E. (1987). Helix Geometry, Hydration, and G.A Mismatch in a B-DNA Decamer, *Science* **238**, 498–504.

Quigley, G. J., Teeter, M. M., and Rich, A. (1978). Structural analysis of spermine and magnesium ion binding to yeast phenylalanine transfer RNA, *Proc. Natl. Acad. Sci. USA* **75**, 64–68.

Rich, A. and Kim, S. H. (1978). The three-dimensional structure of transfer RNA, *Sci. Am.* **238**, 52–62.

Saenger, W. (1984). *Principles of Nucleic Acid Structure*, Springer-Verlag, New York.

Seeman, N. C., Rosenberg, J. M., and Rich, A. (1976). Sequence-specific recognition of double helical nucleic acids by proteins, *Proc. Natl. Acad. Sci. USA* **73**, 804–808.

Shakked, Z., Guerstein-Guzikevich, G., Eisenstein, M., Frolow, F., and Rabinovich, D. (1989). The conformation of the DNA double helix in the crystal is dependent on its environment., *Nature (London)* **342**, 456–460.

Shakked, Z. and Kennard, O. (1985). The A form of DNA. *Biological Molecules and Assemblies*. Wiley, New York, pp. 1–36.

Shakked, Z. and Rabinovich, D. (1986). The effect of the base sequence on the fine structure of the DNA double helix, *Prog. Biophys. Molec. Biol.* **47**, 159–195.

Sussman, J. L., Holbrook, S. R., Warrant, R. W., Church, G. M., and Kim, S.-H. (1978). Crystal structure of yeast phenylalanine transfer RNA. I: Crystallographic refinement, *J. Mol. Biol.* **123**, 607–630.

Wang, A. H.-J., Quigley, G. J., Kolpak, F. J., Crawford, J. L., van Boom, J. H., van der Marel, G., and Rich, A. (1979). Molecular Structure of a Left-Handed Double Helical DNA Fragment at Atomic Resolution, *Nature (London)* **282**, 680–686.

Watson, J. D. and Crick, F. H. C. (1953). A Structure for Deoxyribose Nucleic Acid, *Nature (London)* **171**, 737–788.

Wolk, S. K., Hardin, C. C., Germann, M. W., Van de Sande, J. H., and Tinoco, I., Jr. (1988). Comparison of the B- and Z-form hairpin loop structures formed by d$(CG)_5$ T_4 $(CG)_5$, *Biochemistry* **27**, 6960–6967.

Yanagi, K., Privé, G. G., and Dickerson, R. E. (1991). Analysis of local helix geometry in three B-DNA decamers and eight dodecamers, *J. Mol. Biol.* **217**, 201–214.

Zamenhof, S., Brawermann, G., and Chargaff, E. (1952). On the deoxypentose nucleic acids from several microorganisms, *Biochim. Biophys. Acta* **9**, 402–405.

CHAPTER 5

Structure and Dynamics by NMR

by David Wemmer

Many studies of nucleic acid function rely on having a knowledge of the structure of the molecule involved. In a majority of cases, these molecules carry out their function in solution, where in addition to the average structure, dynamic processes may be important. There are only two methods that are capable of determining structures to high resolution: X-ray diffraction and nuclear magnetic resonance (NMR) spectroscopy. Crystals can provide the most accurate and detailed structures, but it must be remembered that although the positions of atoms are being determined accurately, the molecule is ordered in a crystal lattice, hence the crystal structure may or may not fairly represent the solution structure. In addition, although it is possible to see some effects of dynamics in crystals, the molecule is basically trapped in a single conformation, while in solution it may interconvert among several possible structures. Over the past 15 years, the development of high-field NMR spectrometers and multidimensional methods have made it possible to extend detailed NMR analyses to biomolecules, including nucleic acids. With systematic application of these methods it is possible to derive good models for the solution structures of nucleic acids, and to determine features of conformational equilibria, when present. It is possible to adjust solution conditions to a wide range of ionic strengths, solvent compositions, and temperatures to investigate the effects of these parameters on the structures. It is also possible to examine interactions with specific ions, drugs, proteins, and other nucleic acids. The flexibility of this method makes it very attractive for examining solution structures of many small nucleic acids.

The basic principles of the NMR methods are presented in this chapter, along with a discussion of how the useful parameters for structure determination are derived, and of how NMR can be used to determine structures of nucleic acids in solution. Application of this approach is illustrated by a number of representative studies, though no attempt is made to be at all comprehensive.

1. INTRODUCTION TO NMR METHODS

1.1 Magnetically Active Nuclei

A number of different nuclei with spin $\frac{1}{2}$ can be detected sensitively by NMR, making ^{1}H, ^{31}P, ^{13}C, and ^{15}N the most important for the NMR of nucleic acids. Protons (^{1}H) are abundant (99.98% of natural hydrogen), occur at many positions in the molecule, and are the most sensitive of these nuclei to detect, hence they have been the most studied. The ^{31}P nucleus occurs only in the phosphate groups, but it is abundant (100%) and fairly sensitive. The ^{13}C and ^{15}N nuclei occur at important positions in nucleic acids, but their low natural abundance ($\sim$ 1% and 0.4% respectively) and low sensitivity has limited their utility. With new experiments, which are discussed below, and isotope labeling or enrichment, their use is steadily increasing.

1.2 Chemical Shifts

The NMR spectrum represents the frequencies at which transitions are induced between the two allowed states of the spin $+\frac{1}{2}$ and $-\frac{1}{2}$ (spin aligned with an external magnetic field = "up," or against it = "down" respectively). All spins of a given type have similar resonance frequencies, in a field of 11.5 T protons resonate at 500 MHz, ^{31}P at 202 MHz, ^{13}C at 125 MHz, and ^{15}N at 50 MHz. However, different nuclei of the same type, but in different chemical environments, have slightly different frequencies. These differences are referred to as chemical shifts, which arise from screening of the nuclear spin from the magnetic field by the electrons in the surrounding bonds (the rearrangement of the electrons in the external field induces a small field opposed to the applied field). Since the response is induced by the applied field, it scales linearly with it; doubling the applied field doubles the frequency shift. For this reason, it is conventional to give the size of the shift relative to a reference compound, such as the methyl groups in (trimethylsilyl)propionate (TSP) for ^{1}H and ^{13}C, H_3PO_4 for

^{31}P and HNO_3 for ^{15}N, as a fraction of the absolute resonance frequency. Thus a frequency difference of 750 Hz for a ^1H at 500 MHz corresponds to a chemical shift of $(750/500,000,000) = 1.50 \times 10^{-6}$, or 1.50 parts per million (ppm). These shifts are difficult to predict theoretically, but are extremely useful for identifying the chemical environments of the spins through empirical correlations. The dispersion in chemical shifts provided by variations in local environment is critical for assignment and interpretation of resonances. The ^{13}C chemical shifts are usually distinct for each chemically different type of carbon, and establishing connections to attached protons can be quite useful. The ability to interpret ^{13}C chemical shifts has been improving dramatically in the past few years (de Dios et al., 1993).

1.3 Spin–Spin Coupling

The spins are also sensitive to the presence of magnetic moments from other spins in their surroundings. This "coupling" between spins (termed spin–spin-, scalar-, or *J*-coupling) is transmitted through bonds between any pair of spins, but falls off rapidly with the number of bonds. The frequency shift induced by a spin j at a neighboring spin i depends on the spin state of j, up or down, and a constant representing how well the coupling is transmitted, J_{ij}. This effect is independent of the applied field strength, and so the coupling constants J_{ij} are given directly in frequency units, hertz (Hz). The coupling constants that are observable in nucleic acids range from about 140 Hz for direct (one bond separating spins) ^{13}C–^1H couplings, to about 1 Hz for the four bond coupling between methyl and aromatic protons, ^1H_3C–C_5–C_6–^1H, in thymine. Since the energy difference between the spin states is very small, essentially one-half of the spins at any particular position will be up, and one-half will be down, shifting neighboring spins in opposite directions. Thus each neighbor j will split the i resonance into a pair of lines, or a doublet, with peak separation J_{ij}. If several neighboring spins are present, each subsequently divides the resonance into further components. Thus neighboring spins can be counted by the number of splittings observed. The values of coupling constants can provide structural information as described in Section 5.3.2.

1.4 Relaxation Times

Finally, it is useful to introduce two time constants (called relaxation times), T_1 and T_2, to completely describe the behavior of spins. At equilibrium spins are aligned with the applied magnetic field, which defines the z axis of a coordinate system, with no component along the x or y axis. However, when a pulse of radio frequency (rf) radiation is applied to the spins, an $x - y$ component can be induced, and the z component can be altered from its equilibrium value. The spins return to their equilibrium condition by a first-order exponential decay. The "spin–lattice" relaxation time, T_1, is the time constant for the return of the z component of magnetization to its equilibrium value. The "spin–spin" relaxation time, T_2, is the time constant for the $x - y$ components to decay to their equilibrium value of zero. The relaxation time T_2 is directly related to the width of a resonance peak, the full width at half maximum (fwhm) height (in Hz) being given by $1/\pi T_2$.

1.5 Fourier Transform NMR Spectroscopy

Most NMR spectrometers now operate in the pulse-Fourier transform (FT) mode. With this method, the spins are excited by a short pulse of rf radiation near the spin's natural precession frequency. This pulse tips the magnetic moments of the spins from their equilibrium direction along the static magnetic field into the $x-y$ plane, where their frequencies are detected through the voltage they induce in a coil surrounding the sample. To analyze what frequencies and amplitudes are present, an FT of the voltage versus time signal is done. The result of this process is a spectrum (amplitude vs. frequency), that is exactly the same as would have been obtained in a conventional absorption experiment sweeping the frequency, Figure 5.1. The pulse-FT process is about 1000 times faster, making signal averaging practical to detect signals from dilute samples (Martin et al., 1980). In addition, the FT approach makes possible dynamic experiments in which the time dependence of magnetization can be followed.

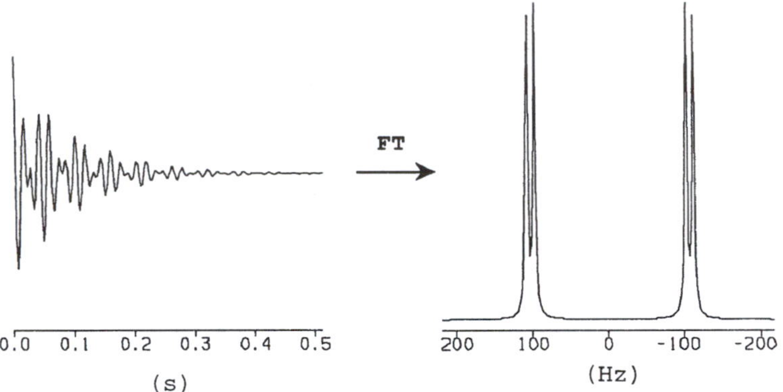

Figure 5-1
A schematic diagram of a time domain signal and its FT, the absorption spectrum, are shown. [*Concepts in Magnetic Resonance*, Wemmer, D. Copyright ©1989. Reprinted by permission of John Wiley & Sons, Inc.]

Experiments are often labeled as to whether they involve only one type of spin, homonuclear, (e.g., protons only); or more than one type, heteronuclear, (e.g., ^{1}H and ^{31}P or ^{13}C). In the heteronuclear case, the different types of spins have very different resonance frequencies, and hence pulses are applied separately to each of the spin species. One useful result of this is that one type of spin can be continuously irradiated during observation of the other, removing all splittings from the irradiated spins. This technique, termed heteronuclear decoupling, is particularly useful in heteronuclear correlation experiments (discussed below), in which the splittings just lead to lower sensitivity by splitting a resonance into many lines.

1.6 Two-Dimensional NMR

Two dimensional (2D) NMR experiments are also done with short pulses and FT, but rather than having one pulse and a detection time period, multiple pulses and delays are present. Any 2D experiment can be broken into preparation, evolution, mixing, and detection periods (Ernst et al., 1987), as shown schematically in Figure 5-2. A 2D experiment is carried out by systematically varying the evolution period, called t_1, in a stepwise fashion and detecting signals as a function of t_2 for each different t_1 value. These data constitute a matrix of signal amplitudes as a function of the two time variables. To obtain the spectrum (amplitude vs. two frequencies) two FTs must be done, one for rows and one for columns in the 2D matrix, as shown schematically in Figure 5-2. The interesting information comes when the spins exchange magnetization during the mixing period. In this case their evolution is at different frequencies during the t_1 and t_2 periods, giving rise to "cross" peaks in the 2D spectrum (peaks that are not on the diagonal). A number of different mechanisms exist for "correlating" spins (causing the magnetization transfer), and are discussed further below. It is important to know that all of the basic physics underlying these experiments is fully understood (Goldman, 1988). That is to say, given any set of chemical shifts and coupling constants for a groups of spins, the response to any pulse sequence of interest can be calculated. Descriptions of such calculations are available in both introductory (Martin et al., 1980) and advanced forms (Ernst et al., 1987).

1.6.1 Correlated Spectroscopy

One important step in analysing spectra is the identification of spins belonging to the same residue through their spin–spin couplings to one another, which can be done with a 2D experiment. Each nucleic acid base and each sugar forms a separate group of coupled protons, linked only by coupling through the ^{31}P in the phosphates. To correlate coupled spins of the same type (homonuclear, e.g., protons in nucleic acids), the 2D experiment used is COSY (for Correlated Spectroscopy), having the pulse sequence shown in Figure 5-3, experiment A. In this case just two pulses are required, with the evolution period t_1 between them, and the acquisition period t_2 immediately after the second pulse. The first pulse creates coherence (magnetization in the $x-y$ plane) on each spin, which then evolves at its natural frequency during t_1. If spins are coupled, then the second pulse (which is the entire mixing period) transfers some of the original coherence to each coupled spin. The cross-peaks occur at the intersections of two resonance frequencies only if the two spins are directly coupled. The transfer process creates lines that are intrinsically antiphase in character, that is one-half of the lines of each multiplet are up and the other one-half are down. No problems are caused when the couplings are large compared with the line width, but when they are smaller than the line width, the positive and negative components can cancel. When the couplings are quite small compared to the line width, the cancellation is severe and cross-peaks become weak, and hence difficult to detect. In this simplest COSY experiment, the peaks on the diagonal are out of phase with the cross-peaks (when cross-peaks are absorptive, diagonals are dispersive = derivative-like). Such line shapes are inconvenient for analysis of cross-peaks near the diagonal, and have led to a modification, called double quantum filtered (DQF) COSY, which gives all peaks the same phase. The DQF sequence has a third pulse inserted immediately after the second for which the direction in the $x-y$ plane is rotated between scans in a defined pattern. This process gives cross and diagonal peaks with the same phase, and also removes all singlets (spins not coupled to any others) from the diagonal. The filtering process

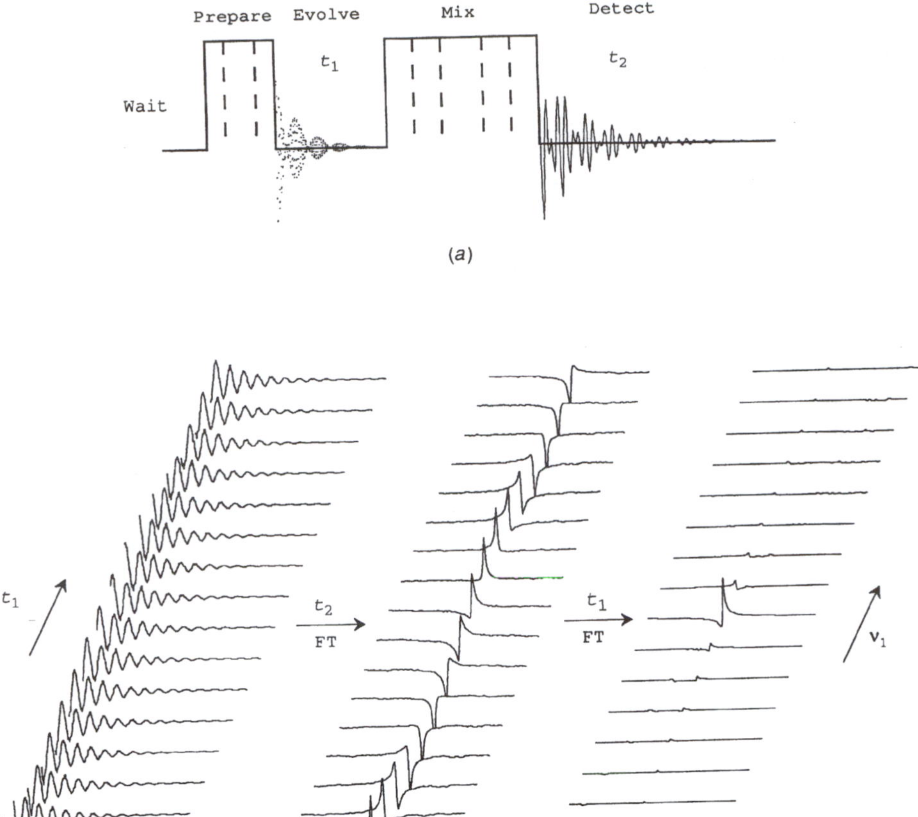

Figure 5-2
(*a*) A generalized diagram of a 2D pulse sequence is shown. The dashed lines in the Prepare and Mix periods indicate that these parts may actually be made up of several different pulses and/or delays. (*b*) Time domain signals from a 2D experiment are shown at the left, the data after transformation of t_2 into ν_2 are shown in the middle, and the fully transformed 2D spectrum is shown at the right. To make the pattern easy to see a single peak labeled by two frequencies ν_1 and ν_2 has been used for this example. [*Concepts in Magnetic Resonance* Wemmer, D., Copyright ©1989. Reprinted by permission of John Wiley & Sons, Inc.]

Figure 5-3 (*facing page*).
The pulse sequences for five commonly used multidimensional experiments are shown. Radio frequency pulses are indicated by vertical bars, with the angle of rotation of the spins indicated above it. The time period t_1 (and also t_2 in experiment E) is used parametrically in all experiments, t_2 (t_3 in experiment E) is the time during which actual data are collected, and the time period(s) τ and Δ are fixed during any given experiment. Experiments A, B, and C involve just protons, while D and E include ^{13}C, ^{15}N or $^{31}P = X$. Experiment E is a three dimensional (3D) experiment, in which it is easy to identify the nuclear Overhauser effect spectroscopy (NOESY) and heteronuclear multiple quantum coherence (HMQC) sections by comparison with C and D.

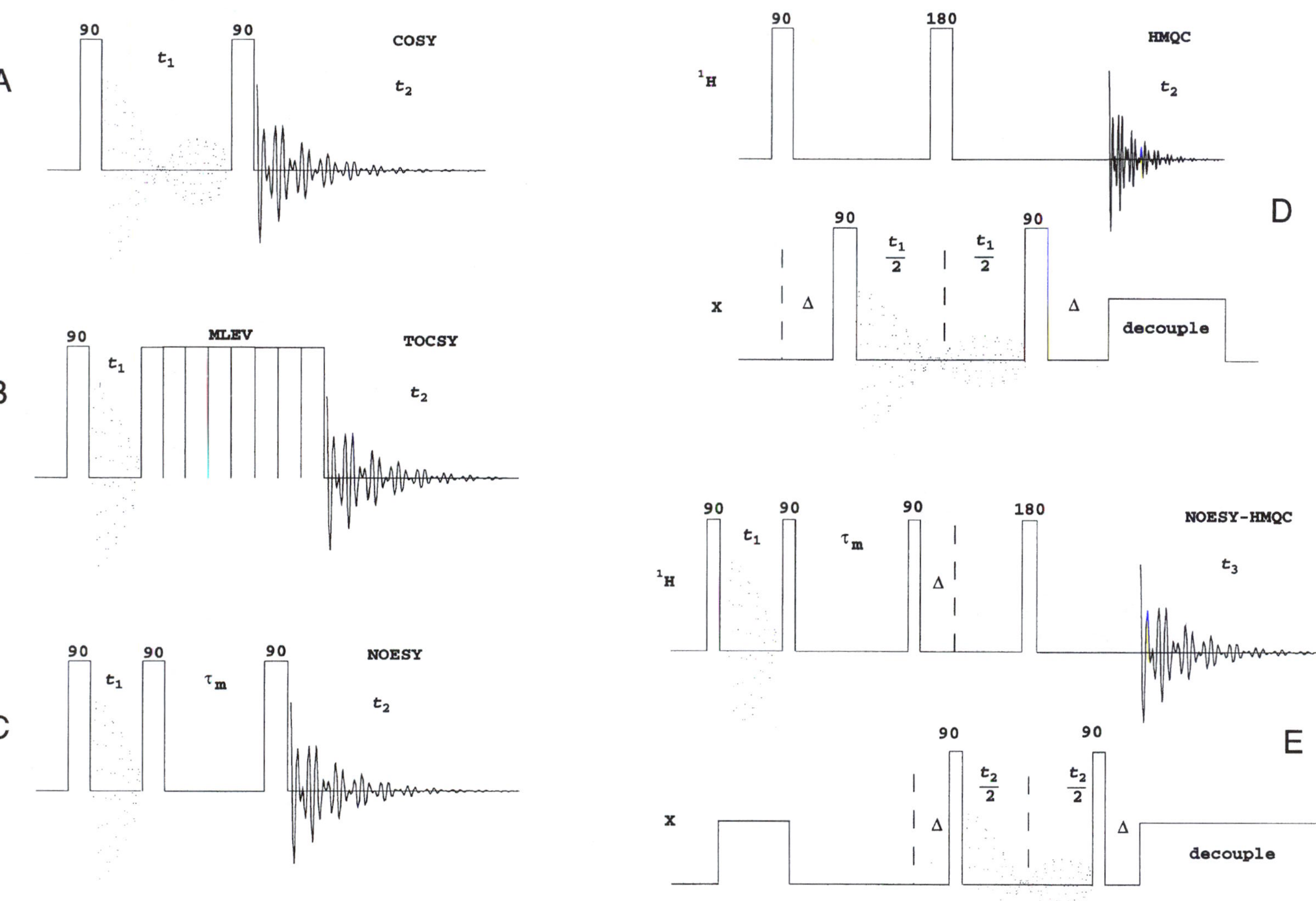
A
90
t_1
90
COSY
t_2
B
90
t_1
MLEV
TOCSY
t_2
C
90
t_1
90
τ_m
90
NOESY
t_2
D
^1H
90
180
HMQC
t_2
X
Δ
90
$\frac{t_1}{2}$
$\frac{t_1}{2}$
90
Δ
decouple
E
^1H
90
t_1
90
τ_m
90
Δ
180
NOESY-HMQC
t_3
X
Δ
90
$\frac{t_2}{2}$
$\frac{t_2}{2}$
90
Δ
decouple

leads to a small decrease in sensitivity, but the improved diagonal characteristics are generally worthwhile. Now this is the standard COSY experiment used.

1.6.2 Relayed Coherence Transfer Spectroscopy

In some cases, it is desirable to have extended coupling information, that is, to identify not only spins directly coupled to one another but also those more distant in the residue. Extended connectivities can be generated by repeating the COSY transfer step in the 2D experiment, either once or more, to make a RELAY or multiple RELAY coherence-transfer experiment. The time period required for each step except the first and last is fixed, so the experiment remains two dimensional. During a fixed delay τ, coherence transfer occurs in a way that depends on the coupling constants of the spins involved. Thus it is possible that intense RELAY peaks can be seen for one residue, but that none appear for another due to coupling constant differences. The dependence is predictable for specific coupling constants and spins systems (Bax and Drobny, 1985), making it possible to choose reasonable values before doing the experiment. Occasionally, more than one experiment is required to obtain complete information. Cross-peaks in the RELAY spectrum occur going from any given spin to its directly coupled neighbors and to the spins to which they are coupled, that is, from H1′ to H2′, H2″, and H3′ (though H1′ and H3′ are not directly coupled) in deoxyribose sugars.

1.6.3 Total Coherence Transfer Spectroscopy

A good alternative for the RELAY experiment has been developed, called TOtal Coherence transfer SpectroscopY (TOCSY). The preparation and evolution periods are the same as those in COSY or RELAY experiments, but the mixing period consists of a continuous sequence of pulses that is designed to remove all effects of chemical shifts, Figure 5-3 experiment B. The entire spin system then behaves as though it is strongly coupled, and evolves from the initial state, where magnetization is localized on each single spin, to one in which it is shared among all spins within the coupled group at long mixing times. Thus cross-peaks occur from each spin to all others in the coupled spin system, although cross-peaks tend to get weaker with more intervening spins. For small oligonucleotides, such as 12 bp DNAs, at least the H1′, H2′, H2″, H3′, and H4′ couplings can be identified (Flynn et al., 1988), in addition to cytosine H5–H6 and thymine CH_3–H6. The TOCSY experiment has the additional advantage that some of the magnetization transfer is "in phase" and does not suffer from the "antiphase" cancellation as in COSY.

During the evolution and mixing times of these experiments, spin–spin (T_2) relaxation dephases (destroys) x–y magnetization. This destruction competes with the transfer to other spins, and limits the number of transfer steps that can be carried out while still retaining sufficient signal to be detected at the end. The width of each part of a resonance increases essentially linearly with molecular weight, decreasing T_2 proportionately. Thus the problems of cancellation and inefficient transfer become increasingly severe with increasing molecular weight.

A transfer of coherence similar to that in the COSY experiment can also be carried out when the spins are of different types, for example, an ^{1}H and a ^{13}C, or an ^{1}H and a ^{31}P. In this case, pulses must be given to each of the spin species, such as in the sequence shown in Figure 5-3 experiment D. The mechanism for transfer can be basically the same as in COSY, or instead involve multiple quantum coherence, but there is no diagonal in the spectrum since coherence is prepared and detected on different types of spin. Peaks in the 2D spectrum occur at the intersection of the frequencies of the ^{1}H and the coupled spin. When single-bond couplings are used (^{1}H–^{13}C or ^{1}H–^{15}N) the coupling constants are quite large ($\geq$ 90 Hz), and transfer is very efficient. However, in some cases of interest (^{1}H–^{31}P) there are no directly attached protons, and longer range couplings must be used. Antiphase cancellation can be a problem, but many such experiments have been successfully carried out. The experiments can be done either with or without heteronuclear decoupling, depending on whether just chemical shift correlations or the coupling constants are of interest. Since the detection of spins with high-resonance frequencies is more sensitive than those with low frequencies, these correlation experiments are best done by transfer of magnetization starting on the X spin and going to the ^{1}H, termed inverse detection. For ^{13}C and ^{15}N at natural abundance ($\sim$ 1 and 0.3% of the nuclei, respectively), this requires eliminating signals from ^{1}H–^{12}C or ^{14}N pairs (with no couplings) which are greater than or equal to 100 times larger. Fortunately, a number of ways have been designed to do this, while maintaining the high sensitivity.

1.6.4 Nuclear Overhauser Effect Spectroscopy

A critical part of the analysis of biomolecular structures by NMR is determination of proximity of protons. Identifying proximal protons is done by detecting mutual spin-lattice relaxation, with a 2D experiment known as Nuclear Overhauser Effect SpectroscopY (NOESY). The pulse sequence is shown in Figure 5-3 experiment C. In this case exchange of information occurs via populations (z magnetization) rather than coherence (x–y magnetization). To achieve this, two pulses, together with the t_1 time period, are needed for preparation. The mixing period is simply a delay time, τ_m, during which relaxation occurs. Finally, the populations are probed by the final pulse and observation time t_2. A cross-peak is seen connecting the frequencies of two spins i and j if they are close enough ($\leq$ 5 Å) that mutual relaxation (cross-relaxation) occurs. The magnitude of the effect between the spins depends on the distance between the spins and on how fast the molecule, or part of the molecule containing the spins, is moving. The rate of motion of the vector connecting the two spins is characterized by a correlation time, τ_c. In NMR applications, the correlation time often can be taken as a single time constant related to the average rate for tumbling (rotating) of the molecule in solution (also described by a rotational diffusion coefficient; see Chapter 9). The correlation time will depend on the size and shape of the molecule, the viscosity of the solution, and the temperature. For molecules with molecular weights above about 2 kDa the amplitude of a cross-peak for an isolated pair of spins, at short mixing times, is directly proportional to the correlation time, τ_c, and inversely proportional to the sixth power of the distance between spins $(r_{ij})^6$. Thus one can obtain distances between protons by comparing two NOEs: one for a fixed interproton distance, such as H5–H6 in uracil or cytosine, and one for a distance that is dependent on conformation, such as H8 of a purine and H1′ of its sugar.

$$\frac{r_{ij}^6}{r_{ab}^6} = \frac{\text{NOE}_{ab}}{\text{NOE}_{ij}} \tag{5-1}$$

The distance dependence is sufficiently strong that direct cross-peaks are not seen when the protons are further than 5 Å apart. The range of the effect can be adjusted somewhat by altering the length of the mixing period. When one is interested in quantitative analysis of the cross-peak intensities, especially for moderately long mixing times, considerable care must be used. This topic will be addressed further in Section 3. For assignments, it is usually sufficient to know that NOESY cross-peaks can be expected for all spin pairs less than 4 Å apart.

The exchange of magnetization between two frequencies during the mixing period can also occur through chemical transfer of a spin (changing its frequency during the mixing time), for example, a deprotonation/protonation from solvent, or through alteration of a conformation. For such an exchange mechanism to give rise to a cross-peak, two conditions must be met. First, the lifetime in either state must be of the same order of magnitude as the mixing time, and must also be shorter than the time constant T_1 for the spins to return to their equilibrium populations. Second, both states must be significantly populated (i.e., both sets of peaks must be visible in the spectrum). Occasionally, cross-peaks from NOE and chemical exchange may be confused, since they both give rise to cross-peaks in a spectrum. However, chemical exchange increases with temperature (since it occurs as a thermally activated process), while NOEs do not. In addition, rotating frame NOE (ROESY—a full discussion is beyond the scope of this chapter) can be used to distinguish these.

1.7 Three-Dimensional and Four-Dimensional NMR

The principles described above for 2D experiments are easily extended to three (3D) or four (4D) dimensions. The basic idea is just to concatenate two experiments, end to end, giving two or three time periods that are incremented in a stepwise fashion. The mechanisms for information transfer in such experiments are exactly the same as in 2D. During data collection, each combination of time variables must be collected (i.e., each t_1, t_2 pair in a 3D experiment) such as seen in the HMQC–NOESY in Figure 5-3 experiment E. The reason for adding dimensions is to provide additional resolution (spreading a 2D spectrum along a third axis), which makes it easier to assign resonances and to quantify cross-peak intensities (since fewer overlap). Such experiments can be homonuclear, [e.g., the TOCSY–NOESY combination has been useful for DNA (Radhakrishnan et al., 1992)] or heteronuclear, [e.g., ^{13}C-correlated-NOESY, which has been applied to isotope labeled RNA (Nikonowicz and Pardi,

1993)]. Such experiments take a few days of spectrometer time (compared with ~ 1 day typical for a 2D experiment), and require a larger amount of disk space. However, the improved resolution frequently makes them worthwhile.

1.8 Solvent Suppression

For observation of most proton resonances in nucleic acids, samples can be dissolved in deuterated water D_2O (solvent), to avoid the large proton signal that occurs in normal water (H_2O). However, the more acidic protons in nucleic acids (those attached to nitrogen or oxygen) have a lifetime for exchange with solvent of a few seconds or less, the precise value depending on temperature, pH and buffer concentration (for more detail see Section 5.5). Thus, if a sample is dissolved in D_2O (solvent), such protons are replaced by deuterons before even a typical 1D NMR spectrum can be measured. Therefore, H_2O solvent must be used in order to characterize the resonances from these protons. Thus a dynamic range problem is created for the detector if the standard pulse sequence is used (1 m*M* solute signals would have to be detected in the presence of the 100 *M* solvent signal). To avoid this problem, specialized techniques are applied to suppress the solvent signal. Two common approaches are to saturate the solvent resonance using a weak rf pulse before beginning the normal pulse sequence (termed presaturation), or to simply design a sequence that excites resonances of interest, but not those of the solvent (selective excitation).

The presaturation method is simple and easy to use, and often works at low temperatures. However with increasing temperature the solvent exchange rates of the relevant protons increase to the point that chemical exchange transfers the saturated protons onto the nucleic acid rapidly, and the signals of interest are again lost. Presaturation has the advantage that once the solvent is saturated completely, nonselective pulses can be used for exciting the remainder of the signals, giving undistorted intensities throughout the spectrum. Presaturation is also very simple to add to any 2D experiment.

A number of pulse sequences can be used for selective excitation. In many of these sequences the resonances of interest are rotated downward toward the x–y plane by each of a number of pulses, due to the rotation about the z axis during the times between them. Signals at intermediate frequencies will only be tipped part way toward the x–y plane, and hence appear with reduced intensity, while the water ends up along the z axis. Consequently, its signal is greatly reduced. There are a number of variations of this approach, which may produce somewhat better suppression of the solvent signal (1:–3:3:–1, jump–return, or 1:1 echo) but they operate on the same basic principle (Hore, 1983). Any or all of the pulses in a 2D experiment can be replaced with these selective pulses, but it must be realized that the amplitudes in the resulting spectrum will be significantly distorted, giving reduced peak intensities near the solvent frequency. When many pulses are used, the distortions are multiplied.

The experiments described in this section are the basic tools for collecting information about nucleic acids, which can then be analyzed to determine assignments, structures, and information about the dynamics.

2. SYSTEMATIC ASSIGNMENT OF RESONANCES

Any thorough analysis of the NMR spectrum of a biopolymer requires assignment of the resonances; that is, association of each resonance with a specific spin in the molecule. The first systematic approach for assigning resonances was developed by Wüthrich (1986) for proteins, but was soon extended to nucleic acids as well. In both systems, a sequential approach is used, that is, one that takes advantage of the proximity of covalently bonded neighbors in the primary structure of the molecule. In general, this requires spin–spin coupling information (from COSY) to identify the groups of spins within each residue, and proximity information (from NOESY) to connect resonances from neighboring residues. In nucleic acids, exchangeable protons (those attached to nitrogen or oxygen) and nonexchangeable (those attached to carbon) can be seen to form somewhat separated groups, (Fig. 5-4). Since this is true, they are often assigned independently. For DNA and RNA, the assignment process for exchangeable resonances in simple duplexes is very similar, hence these will be discussed together. However, for the nonexchangeable protons there are a number of differences, and DNA will be discussed first, then RNA.

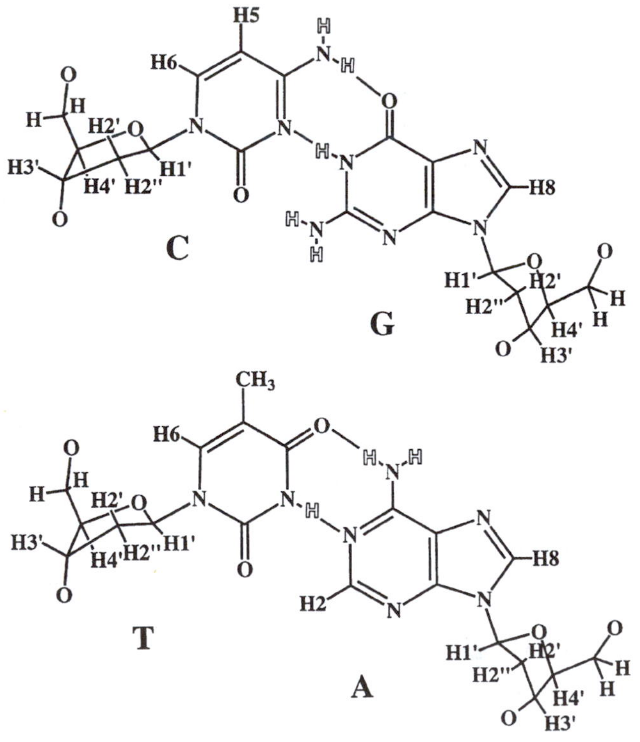

Figure 5-4
The structures of C • G and A • T base pairs are shown, with the nonexchangeable protons of the base and sugar labeled. The hydrogen bonded, exchangeable protons (imino = NH and amino = NH_2) are indicated by outlined letters.

2.1 Exchangeable Protons

The normal Watson–Crick base pairs are shown in Figure 5-4 with exchangeable protons indicated by outline letters. As can be seen, many of these are involved in the hydrogen bonding between base pairs, and in fact are good probes of the secondary and tertiary structure of the nucleic acid. The spatial separation of this group from most of the nonexchangeable protons (note the exception of the adenine H2) means that they can be assigned independently. The difference in chemical environment of the imino and amino protons leads to a large difference in chemical shift. The hydrogen-bonded iminos resonate in the 11–16 ppm region, while the hydrogen-bonded aminos are at 8–9 ppm, and the nonhydrogen bonded ones are in the 6.5–7.5 ppm range. These ranges can be seen in the spectrum of a DNA oligomer, (Fig. 5-5). In a NOESY spectrum taken in H_2O solution, it is possible to observe cross-peaks from each imino to the imino groups of the immediate neighbors, and also to the amino and H2 protons within the base pair. For DNA oligonucleotides, it has been found that the cytosine amino proton resonances are much sharper than those of the purines (usually so broad that they cannot be seen), due to slower rotation about the C–N bond. For each G•C base pair, a strong cross-peak is expected from the imino to one amino proton, which in turn has a strong cross-peak to the other amino of that cytosine. For an A•T base pair, there is expected a single strong cross-peak to the adenine H2 proton (even sharper than the cytosine aminos), which does not have further strong cross-peaks. Thus one can follow the chain of imino to imino NOEs along the sequence, and identify the type of base pair at each step through the NOEs to amino or H2 protons. In most cases, this information is sufficient to assign all of the imino protons. Loop regions are more difficult to deal with because nonsequential NOEs may be observed. The ability to assign resonances from nonbasepaired sequences (hairpin loops, internal loops, or bulge loops) depends on the particular sequence and structure involved, and must be considered for each individual case.

2.2 DNA

A qualitatively similar process can be used for assigning all of the nonexchangeable resonances of DNA (Scheek et al., 1983; Chazin et al., 1986). The distribution of resonances can be seen in Figure 5-5. Usually, assignments are done in D_2O solution to avoid problems with suppression of the water resonance. The aromatic protons of bases can be identified through a COSY experiment. The cytosine and uracil H5 and H6 are coupled (8 Hz) to one another, thymine shows a weak four bond coupling between the methyl and H6 (1 Hz), and the purine H8s are both singlets, making it impossible to distinguish A and G by couplings. The COSY can also be used to identify the groups of coupled protons within each sugar. In DNA, the duplex regions are normally in a B-type conformation, with the sugar near the C2′-*endo* pucker. The H1′ to H2′ and H2″ cross-peaks are normally quite well resolved, the H2′ to H3′ cross-peaks can usually be seen, as can some of the H3′ to H4′ cross-peaks. The H4′, H5′ and H5″ cross-peaks often have poor chemical shift dispersion, and only a few of the couplings can be followed in the COSY spectrum. The TOCSY spectra can provide remote connectivities, for

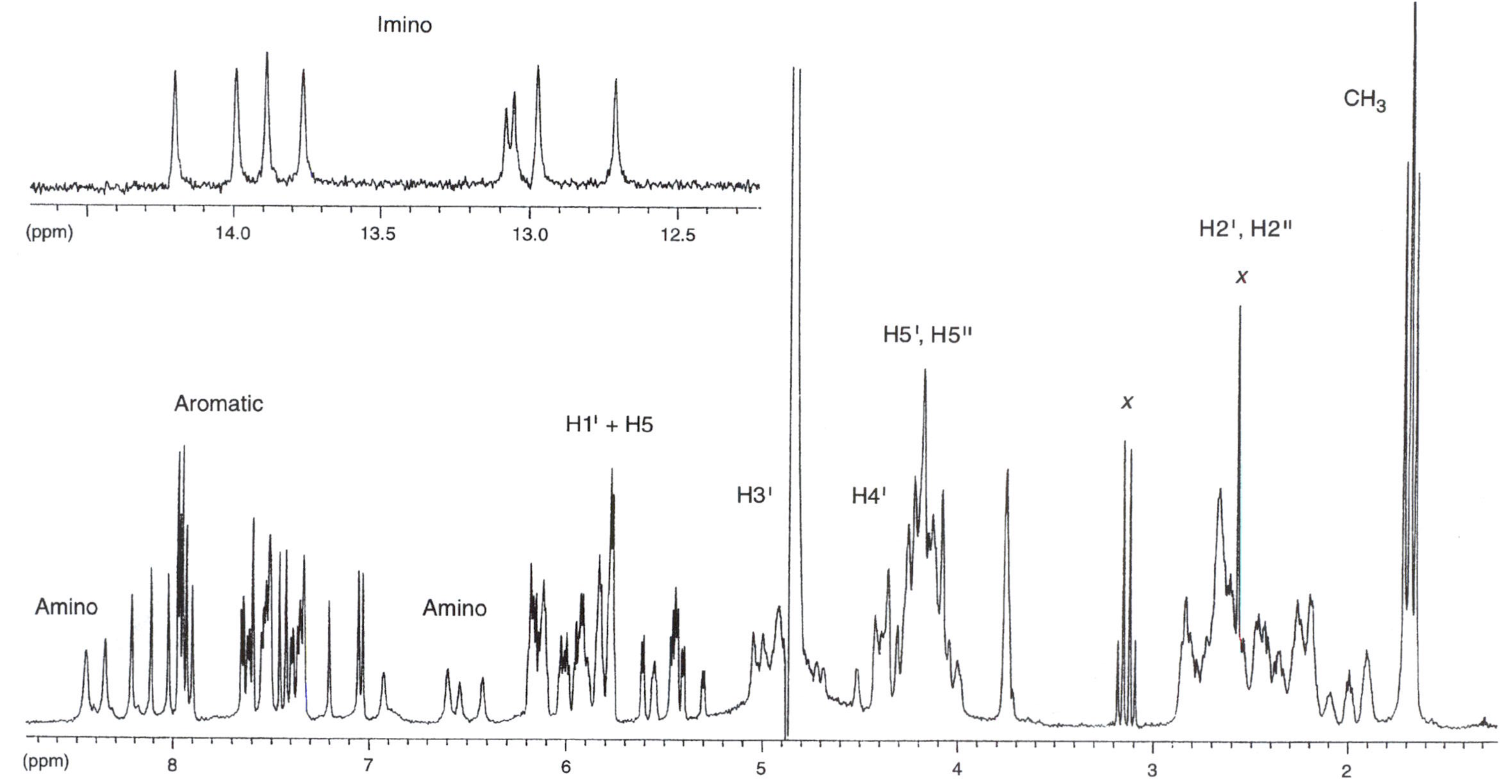

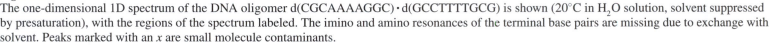

Figure 5-5
The one-dimensional 1D spectrum of the DNA oligomer d(CGCAAAAGGC)·d(GCCTTTTGCG) is shown (20°C in H_2O solution, solvent suppressed by presaturation), with the regions of the spectrum labeled. The imino and amino resonances of the terminal base pairs are missing due to exchange with solvent. Peaks marked with an *x* are small molecule contaminants.

example, from the H1′ directly to the H3′ and H4′, but couplings to the H5′ and H5″ are still weak.

In a right-handed DNA helix, near the B form in conformation, the base aromatic proton is stacked above the sugar of the preceding residue, (Fig. 5-6). In this geometry, there are *inter* residue NOEs visible between the aromatic proton (H6 or H8) and the sugar H1′ and H2″ of the residue to the 5′ side. The *intra* residue distances between the aromatic and the sugar H1′ and H2′ are similar to the interresidue ones, and hence NOEs between them will be visible also. Thus it is possible to step from base aromatic to the sugar H1′ of the same residue, to the neighboring aromatic for the whole length of each DNA strand, as shown in Figure 5-6. The only exceptions to this pattern occur at

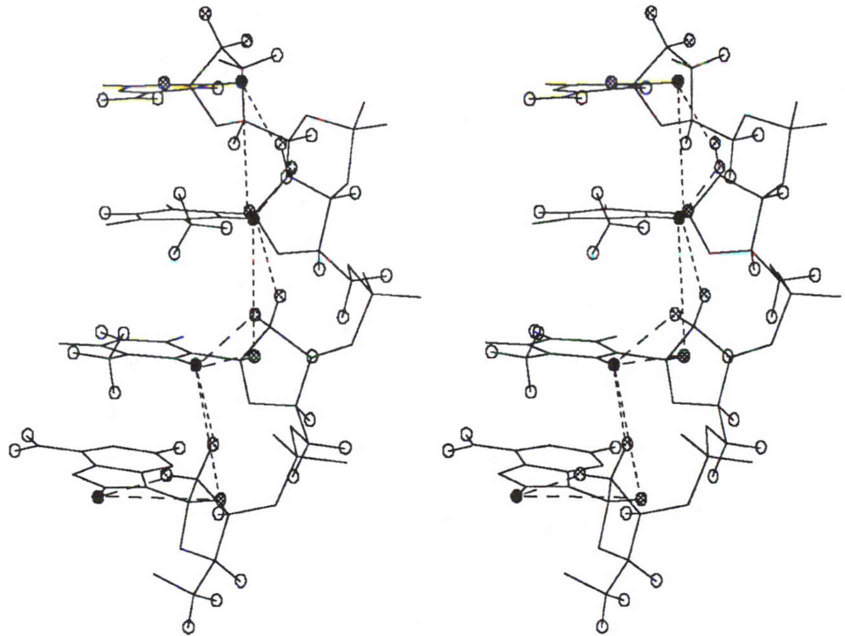

Figure 5-6
A stereopicture of a four base segment of DNA (sequence ATTC) is shown. The positions of protons are indicated by circles, the aromatic protons (H8 or H6) are solid circles, the sugar C1′ protons are shaded, and the sugar H2′ and H2 are cross-hatched. The short dashes indicate the sequential connectivites between each aromatic and the 5′ neighboring H1′ and H2, while the long dashes indicate the intraresidue aromatic to H1′ and H2′. Together these form the sequential steps used for assignment of right-handed helical DNA. Note that the methyl groups in each thymine or H5 of cytosine are immediately above the preceding base aromatic proton, providing an independent connectivity between bases. [Wemmer, D. (1992). NMR Studies of Nucleic Acids and Their Complexes, in *Biological Magnetic Resonance*, Vol. 10, Berliner, L. J. and Reuben, J., Eds., Plenum, New York, pp. 196–264. Reprinted with permission from Plenum Publishing Corporation.]

the ends of the DNA, for the 5′-end there is no preceding sugar, and hence the terminal base will have a cross-peak only to the H1′ of that residue, while for the 3′-end, the lack of a 3′ neighbor means that the H1′ proton will have a cross-peak only to its own base. If there are ambiguities due to chemical shift degeneracy of aromatic or H1′ protons, then an independent path involving sugar H2′, H2″ and aromatics can be followed instead, or these alternate connectivities can be used to confirm the assignments. Additional NOEs are present at each step involving a pyrimidine, from the protons at the 5 position of the base (either H5 or the CH_3, for C and T, respectively) to the 5′ neighbor's H1′. At each step, the COSY data can be used to identify the base type for the sequentially identified aromatic, checking to see that it corresponds with the primary sequence. The identification of sugar protons belonging to the same residue can also be compared in COSY and NOESY spectra.

For a self-complementary sequence, the two strands are completely equivalent. Hence, there will be just one set of assignments involving all resonances. However, for a duplex in which the two strands are different sequences, there will be two independent sequential walks, one for each of the strands. Such an example is shown in Figure 5-7 for an 11-mer DNA duplex. If the chemical shifts of all the protons are well resolved, the NOESY serves to sequence the oligonucleotide (although the sequence should be known beforehand). When degeneracies occur in complex spectra, then the sequence information can be used to resolve uncertainties. This method has proved straightforward and reliable for regular B-DNA duplexes up to about 40 nucleotides.

Note that the assumption of a right-handed helix (implicit in predicting which NOEs will be seen) is essentially self-confirming. If there are irregularities such as looped out bases, hairpin loops, or stretches of Z-DNA, then some connectivites described in this sequential assignment process will not be found. When such irregularities are present (which usually can be predicted from the sequence alone) there may be difficulty in assigning all of the bases and sugars with this approach. The COSY data can still be used to identify base types, and regardless of the conformation some of the intraresidue base-to-sugar NOEs must be visible. By eliminating assigned resonances from duplex regions, and taking the partial NOE information into consideration, it is sometimes possible to fully assign resonances in spite of structural irregularities. An example is shown in Figure 5-8 for a hairpin loop with sequence TAAT. Although several of the sequential NOEs are missing, those that are present are sufficient to clearly identify the resonances. For longer loop segments, or those with many repeats of the same base, it may not be possible to complete the assignments with the COSY–NOESY data alone. An alternative assignment approach, which may work in such cases, is described below.

2.3 RNA

In RNA, the assignment process is slightly altered by several factors. The bases C and U have the same coupling pattern, and hence cannot be distinguished from COSY data (although they are easily distinguished from purines). In addition, duplex RNA is typically in an A-type conformation, with a corresponding sugar pucker of C3′-*endo*. For this conformation, the sugar H1′ to H2′ coupling constant is small ($\geq$1 Hz), making the identification of the *intra*sugar H1′–H2′ couplings impossible. Also, the addition

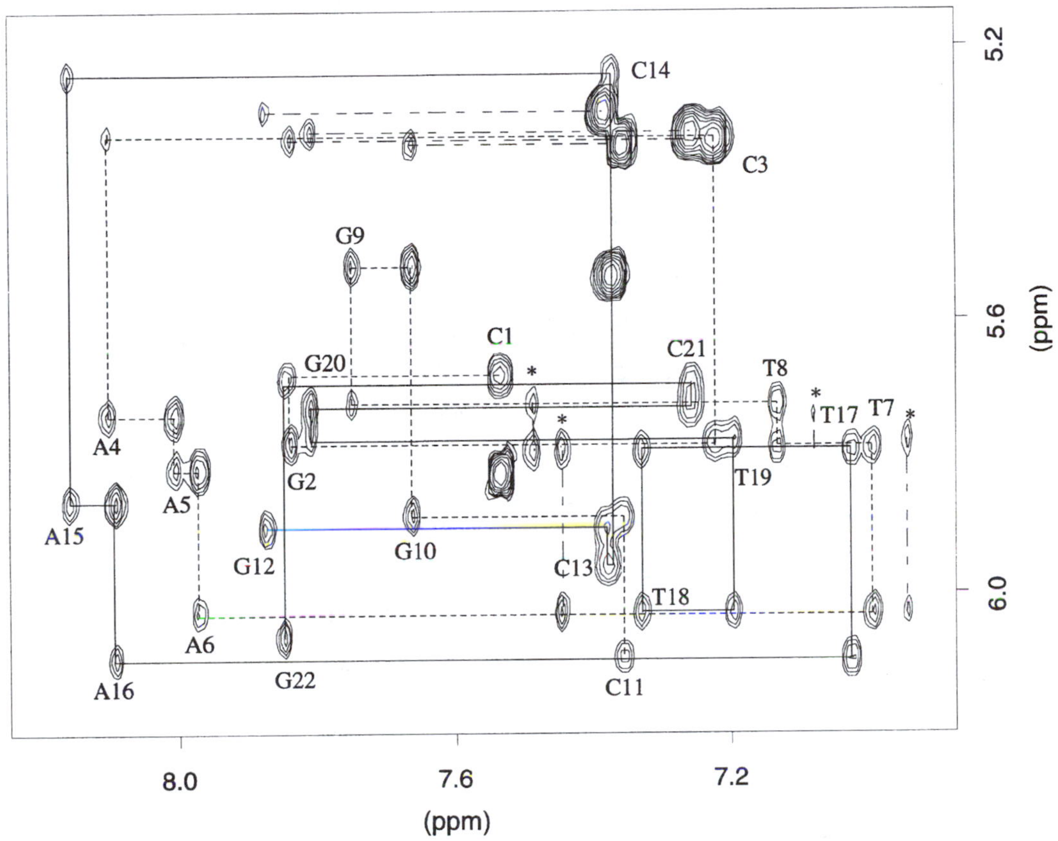

Figure 5-7
The aromatic proton (horizontal) to sugar H1′ and cytosine H5 (vertical) region of a NOESY spectrum of the DNA duplex: 5′-CGCAAATTGGC-3′ • 5′-GCCAATTTGCG-3′ is shown. The sequential connectivities traced with a solid line for one strand, and a dashed line for the other. The intraresidue connectivity (aromatic to H1′) of each base is labeled with the identity of that base. The cross-peaks with a vertical line and * connect adenosine H2 to sugar H1′, as mentioned in the text for A-rich sequences. The horizontal dashed lines connect the cytosine H5 to the sugar of the preceding residue.

of the 2′-OH group (relative to DNA) moves the H2′ resonance into the same chemical shift range as those for the H3′ and H4′ protons, making this region of the NOESY spectrum more crowded. Finally, the difference in the geometry of the base stacking in A versus B helices moves some of the protons apart, particularly the aromatic and neighboring H1′. The loss in cross-peak intensity is somewhat compensated by a shorter distance between the aromatic and the neighboring H2′. The same basic idea for sequential assignments applies as with DNA, but these differences make the process

(*a*)

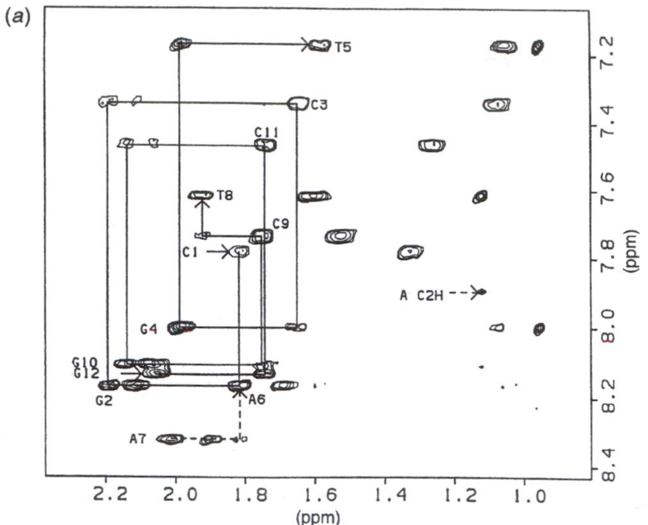

(*b*)

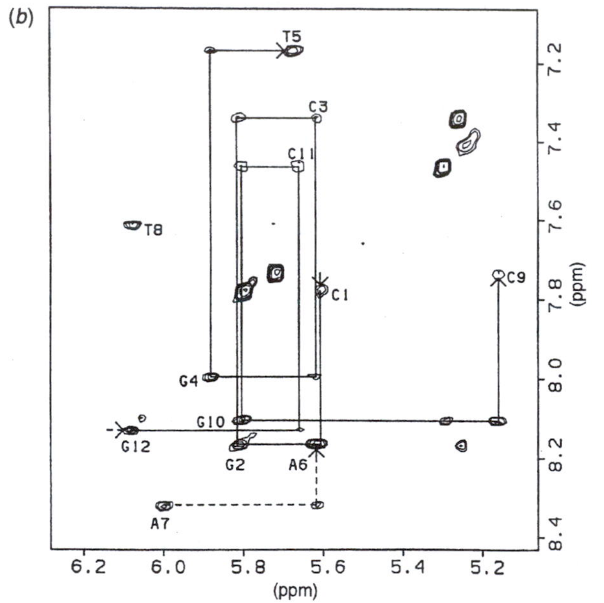

Figure 5-8
The aromatic proton to sugar H2′,H2″ (*a*) and aromatic proton to sugar H1′ regions of a NOESY spectrum of the DNA oligonucleotide d(CGCGTAATCGCG) are shown. The sequential connectivity path is traced in each region, and intraresidue cross-peaks are labeled with the residue. The center TAAT sequence forms a hairpin loop structure and some of the connectivites that would be expected in a duplex are missing, (e.g., T8 H6 to A7 H1′ and A6 H8 to T5 H1′). In spite of this, it is possible to unambiguously assign all resonances by using all of the information available. [Part *b* was reproduced from Wemmer, D. (1992) NMR Studies of Nucleic Acids and Their Complexes, in *Biological Magnetic Resonance*, Vol. 10, Berliner, L. J and Reuben, J., Eds., Plenum, New York. Reprinted with permission from Plenum Publishing Corporation.]

technically somewhat more difficult, but still quite feasible (Chou et al., 1989). An example of the aromatic to H1′ and other sugar cross-peak regions are shown for an RNA oligonucleotide in Figure 5-9.

2.4 ^{31}P Resonances

The phosphate serves to link the residues in all nucleic acids. The spin on each ^{31}P can serve as a source of spectral information about residues that are bridged by a particular phosphate, and about local conformation. For duplex DNAs, the ^{31}P resonances occur in a fairly narrow window, but for moderate size oligomers a single peak for each phosphate can often be seen. These can be assigned directly using a chemical approach, incorporating ^{17}O into a single phosphate group during synthesis. The presence of the rapidly relaxing spin $\frac{5}{2}$ ^{17}O causes a broadening of the ^{31}P due to unresolved coupling. When compared to the normal ^{31}P spectrum (with all ^{16}O, having no spin), the broadening is obvious, and leads to assignment of the single phosphate that was labeled (Schroeder et al., 1987). The labeling is time consuming and costly, but leads to direct assignment of the peak, and can be applied to fairly large molecules.

Each phosphorus is coupled to protons on the attached sugars, the H5′, H5″ and H4′ of the 5′ sugar, and the H3′ of the 3′ sugar (Marion and Lancelot, 1984). If the protons can be assigned using COSY–NOESY experiments, then the assignments can be transferred to the ^{31}P by application of a hetero–correlated spectrum, such as shown in Figure 5-10(*a*). On the other hand, the presence of these couplings can be used in a different way to actually establish the assignments without reference to NOESY data.

The heteronuclear coupling can provide a mechanism for generating internucleotide ^{1}H–^{1}H cross-peaks that do not occur in COSY spectra (Frey et al., 1985). This coupling is done by carrying out two steps of heteronuclear transfer, the first from the sugar H3′ to the 3′ ^{31}P, then from this same ^{31}P to the H4′ of the 5′ side residue. In the resulting hetero-RELAY-COSY experiment there will be *inter*nucleotide H3′ to H4′ cross-peaks. The RELAY spectrum is used in conjunction with the normal COSY in which *intra*residue peaks are identified. The advantage of either of these approaches is that they require no assumptions about the geometry of the helix (although it is possible that conformations could occur for which the necessary couplings for generating cross-peaks are nearly zero). The disadvantage is that with either method fairly crowded regions of the spectrum are used, and some of the couplings are moderately weak, limiting their applicability to systems of low molecular weight.

The most powerful experiments (helping with both ^{1}H and ^{31}P assignments) are the hetero-TOCSY based ones (Kellogg and Schweitzer, 1993). In this case, the hetero-TOCSY step generates correlations from the phosphates to many of the protons in each sugar, though there are differences in the patterns seen in DNA and RNA due to the different sugar conformations. Since transfer is through in-phase coherence, it minimizes cancellation effects at higher molecular weights. In addition, the hetero- and homonuclear transfers occur at the same time, increasing efficiency. The advantage of this relative to just heterocorrelation is easily seen in Figure 5-10(*b*). The hetero-TOCSY step can also be followed by a NOESY step, allowing transfer to protons near those initially undergoing coherence transfer. While this makes more cross-peaks in

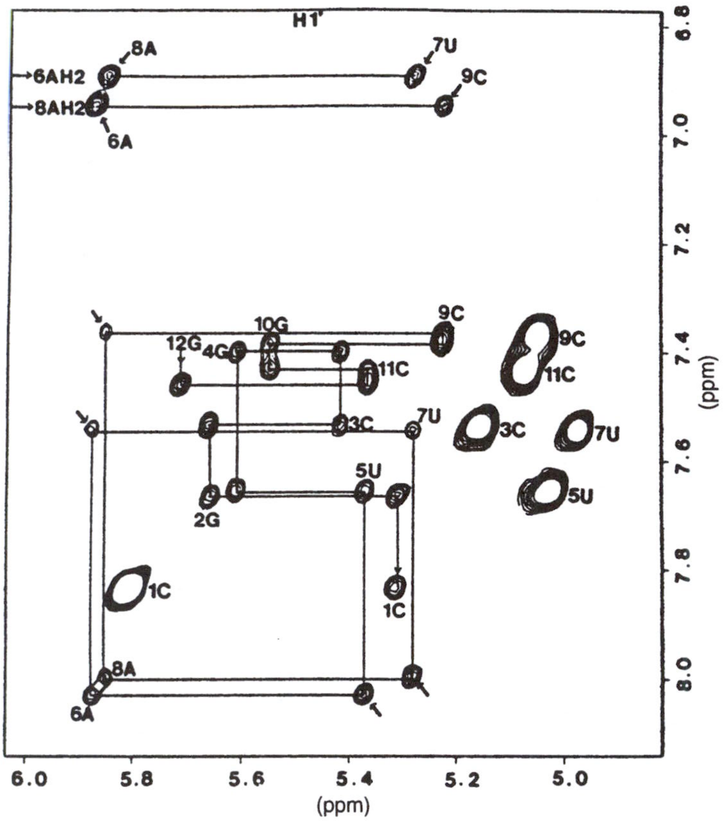

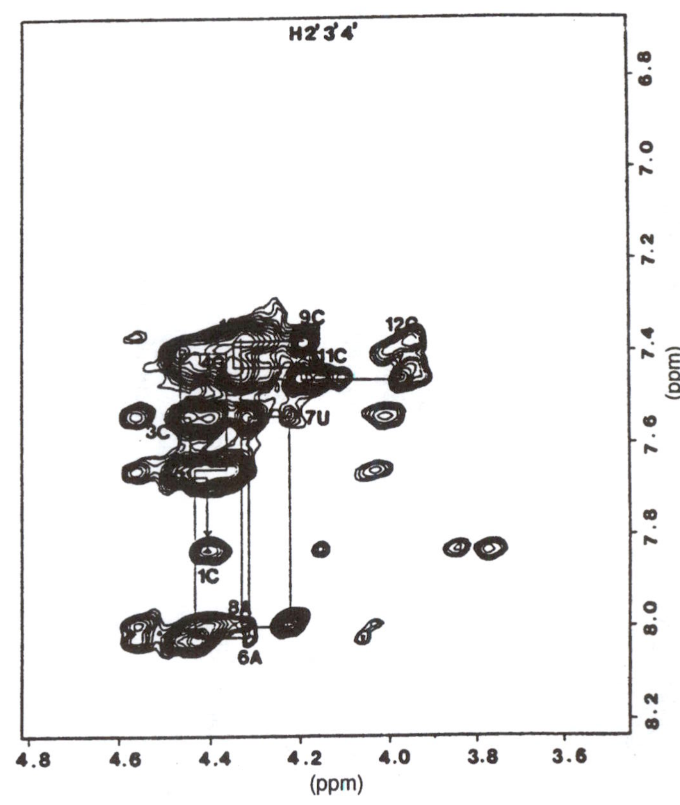

Figure 5-9
The aromatic to H1′ (lower left), aromatic to H2′, H3′ and H4′ (lower right) and H1′ to H2′, H3′ and H4′ regions (upper right) of a NOESY spectrum of the RNA oligomer r(CGCGUAUACGCG)$_2$ duplex are shown. The sequential assignment path is traced in the lower left segment, while intraresidue connectivities are indicated in the upper right. [Reprinted with permission from Chou, S.-H., Flynn, P., and Reid, B. R. (1989). *Biochemistry*, **28**,2422–2435.Copyright ©1989 American Chemical Society.]

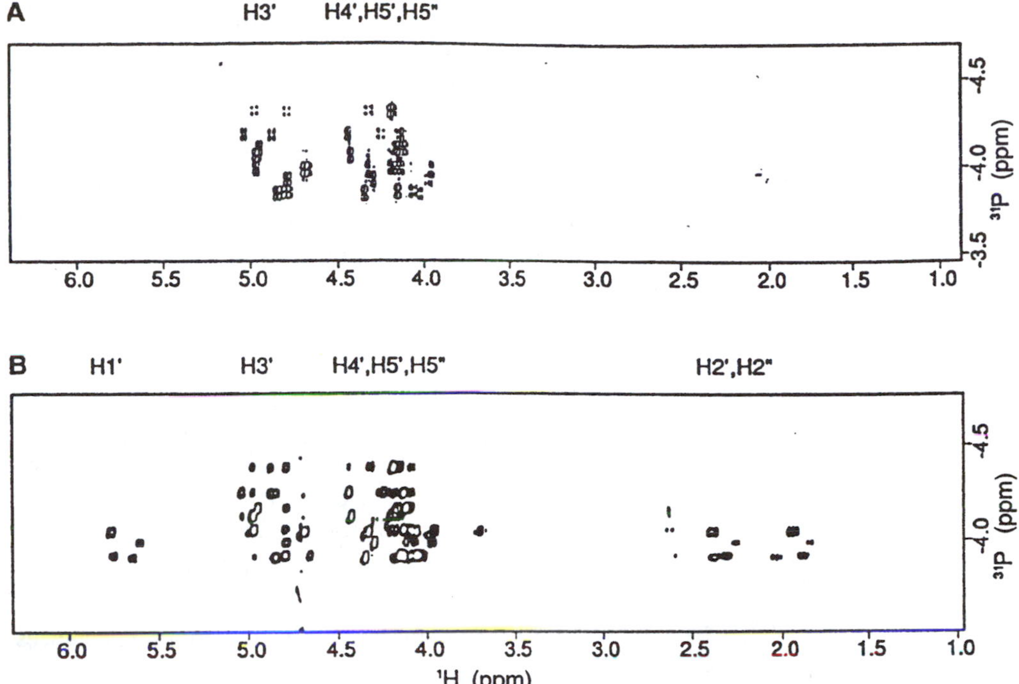

Figure 5-10
Examples of ^{1}H–^{31}P correlation spectra are shown. In (*A*), the normal heterocorrelation experiment was used, establishing connectivities between ^{31}P and directly coupled protons. In (*B*), a heteroTOCSY transfer was used, establishing correlations between the ^{31}P and many protons within each sugar, allowing use of less croweded regions for assignment. Note the antiphase character of the COSY-type transfer in (*A*), while the TOCSY transfer in (*B*) is in-phase. Kellogg, G. W. and Schweitzer, B. I., Fig. 2, *J. Biomol. NMR*, **3**, (1993), 577–593, Two- and three-dimensional ^{31}P-driven procedures for complete assignment of backbone resonances in oligodeoxyribonucleotides, Copyright ©, with kind permission from Kluwer Academic Publishers.]

the spectrum, it does provide new information. To recover resolution, the experiment can also be converted to a 3D format, allowing the protons to evolve to frequency label them, then doing the NOESY mixing, and finally detection. This extra transfer spreads the proton TOCSY–NOESY data out into separate ^{31}P planes, and is a powerful aid for assigning proton resonances as well as the ^{31}P.

2.5 Isotope Enrichment Based Methods

Spectral complexity is often a limiting factor in analysis of large nucleic acids. Hence, methods for simplifying / sorting resonances can be important. One approach has been to ^{2}H label DNA or RNA, which can be done through exchange reactions. Heating adenosine or guanosine in D_2O at alkaline pH exchanges the H8 position quite readily. With DNA, this can be done on an individual strand, eliminating resonances from both A and G from that strand. With RNA prepared by T7 polymerase transcription, the NMPs can be exchanged allowing elimination of A and/or G (depending on the mix used during RNA synthesis). Heating in D_2O in the presence of bisulfite will also exchange the H5 of C or U, through an addition–elimination reaction (Brush et al., 1988). By collecting NOESY spectra of various exchanged samples, prepared with or postlabeled with different deuterium patterns, it is possible to identify connectivities that might otherwise be too severely overlapped for clear interpretation. Some care must be used to match solvent conditions for the samples since resonances can shift with salt, pH, and temperature. It is also often important to compare different labeled samples.

Both ^{13}C and ^{15}N have been observed at natural abundance in small nucleic acids. With ^{1}H detected heterocorrelation experiments it has been straightforward to carry over proton assignments to the directly bonded heteronuclei (LaPlante et al., 1988). To assign nonprotonated atoms has been somewhat more difficult, but has been accomplished for DNA using long-range correlation experiments (Ashcroft et al., 1989). The natural abundance correlations are also useful for aiding with proton assignments. As noted in the discussion of chemical shifts, the range for each carbon type (in DNA, most carbon types in RNA) is distinct. Thus direct proton correlations to those carbons establish the proton type, which has been helpful for finding abnormally shifted proton resonances (e.g., an H1′ in among the H3′, H4′ cross-peaks of a small RNA), as can be seen in Figure 5-11 (Varani and Tinoco, 1991).

For both DNA and RNA, there have been a number of applications of ^{15}N and ^{13}C labeling (isotope enrichment) to enhance information for assignments. The first experiments used ^{15}N labeled bases and biosynthetic incorporation into tRNA to identify resonances. Identification was done by direct observation of the ^{15}N - ^{1}H splitting in the spectrum, and also by establishing ^{15}N - ^{1}H heteronuclear correlations (Griffey et al., 1983). By labeling a single type of base, such as uridine, it is possible to unambiguously identify U containing base pairs, including base triples for which the usual NOE method leaves some ambiguity. Qualitatively similar experiments using chemically synthesized bases labeled at a single site have also been done, but the extra effort required for full chemical synthesis of the labeled building blocks has limited applications. Obviously, this approach immediately provides residue-type assignments (based both on where the label is chemically inserted and on the chemical shift of the label). It has been particularly valuable in analysis of the nonprotonated nitrogens, which can hydrogen bond to ligands. The ^{15}N labeled sites provide a probe for detection of hydrogen bonds (Rhee et al., 1993).

A major advance in NMR studies of RNA has come through application of methods for uniform isotope enrichment. In this approach, cells are grown on labeled medium (^{15}N ammonium chloride and ^{13}C glucose, methanol, or acetate depending on the organism), and the total RNA is harvested. The RNA is enzymatically degraded to the mononucleotide monophosphates, which are purified and may be separated from one another if desired. Another set of enzymes is added to convert the monophosphates to triphosphates, which are then used in any desired combination for *in vitro* polymerase synthesis of RNA (Batey et al., 1992). With the isotopes at essentially 100% abundance, correlation experiments are very quick and easy. More importantly, two new classes of experiments come into play: heteronuclear coherence transfer and heteronuclear

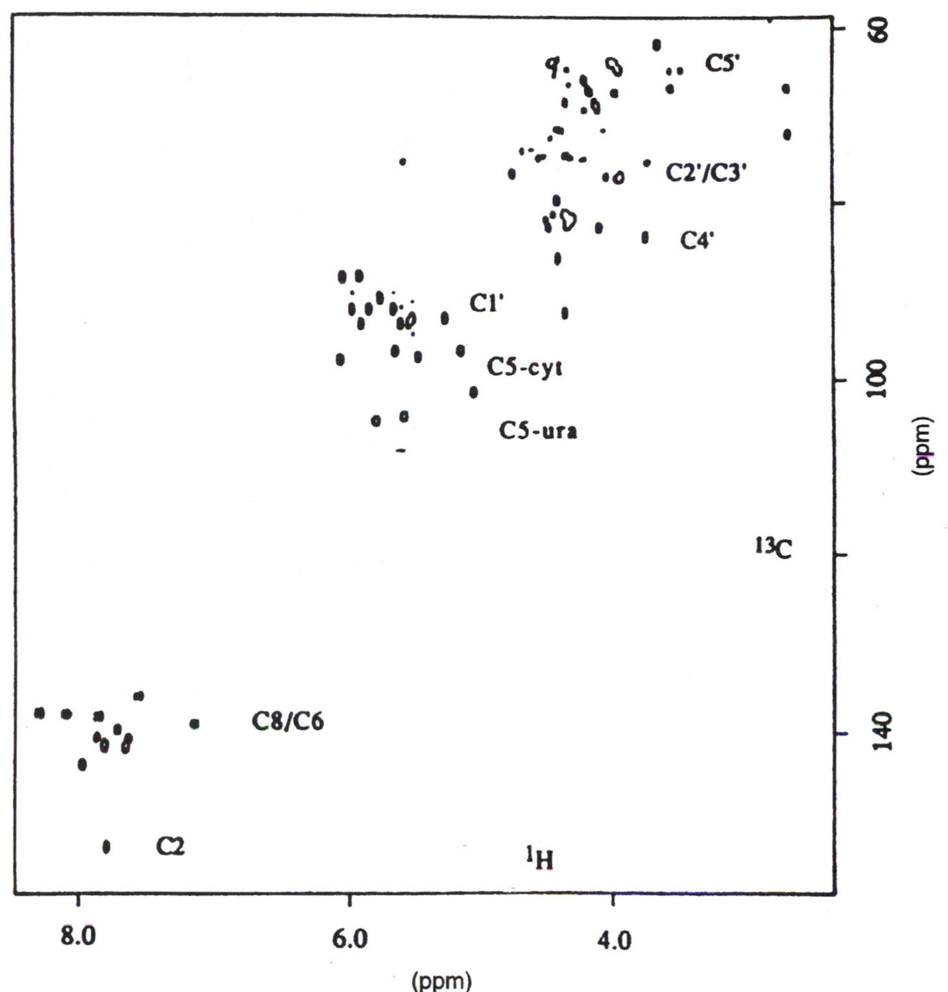

Figure 5-11
A natural abundance ^{13}C-^{1}H correlation spectrum is shown for an RNA tetraloop. Note the groupings in carbon chemical shift, which allows carbon-type identification even though some proton resonances are outside their normal chemical shift range. Reprinted with permission from G. Varani and I. Tinoco, Jr. (1991). *J. Am. Chem. Soc.*, **113**, 9349–9354. Copyright ©1991 American Chemical Society.]

filtering (Griffey and Redfield, 1987). In the coherence transfer experiments, a series of hetero-COSY-like steps is carried out, for example, from the H1′ to the C1′, from C1′ to the C6/C8 of the base, then to the H8/H6. Frequency labeling of any combination of these spins can be done, the important fact is that a correlation of the sugar and base protons can be done without the need for any NOE transfer step. Eliminating this step is important because the NOE transfers through space, and it is not possible to be sure whether a correlation is intranucleotide or internucleotide. Another application is to carry out ^{13}C-TOCSY transfers for correlating all of the resonances in sugar. The one bond ^{13}C–^{13}C couplings are fairly large ($\sim$ 40 Hz), giving fairly rapid coherence transfer, and are essentially independent of conformation. Thus the geometries that give small proton–proton couplings (H1′ to H2′ in C3′-*endo*) do not affect the carbon-transfer experiments. The second class of experiments uses the ^{13}C (or ^{15}N) as a spreading parameter, separating the proton resonances into separate planes in a 3D (or 4D) spectrum according to the chemical shift of the attached ^{13}C (or ^{15}N). There is some correlation between the ^{1}H and ^{13}C frequencies for each carbon type, so the resolution gain is not as large as if both shifts were randomly distributed over their range, but there is still an improvement in effective resolution (Nikonowicz and Pardi, 1993). Equivalent uniform labeling experiments have recently been done for DNA (Zimmer and Crothers, 1995), and while the needed nucleotides can be prepared, the enzymatic synthesis is more difficult (requiring an initial priming region). However, it is likely that clever new approaches can overcome this limitation, and futher use in DNA will soon follow.

3. PARAMETERS FOR STRUCTURE ANALYSIS

Usually, the central goal of an NMR study of an oligonucleotide is the characterization of its structure. There are two basic sources of structural information from NMR data: ^{1}H–^{1}H NOE intensities and spin–spin coupling constants. The NOEs can be used to obtain interproton distance estimates, while coupling constants can be used to determine dihedral angles—torsion angles for rotation around single bonds. Additional structural information may come from chemical shifts or exchange behavior, but these are always less directly interpretable in structural terms, and hence are subject to different possible interpretations.

3.1 Nuclear Overhauser Effect

The theory of the NOESY experiment has been treated in detail in a number of publications (Jeener et al., 1979; Macura and Ernst, 1980), here only the basic important features are summarized. As described in Section 1.6.4, NOESY cross-peak intensities (or more properly the volume integrals) initially grow as $\tau_m(\tau_c/r^6)$, where τ_m is the NOESY mixing time, τ_c is the tumbling correlation time for the pair of protons, and r is the distance between the protons involved. This formula assumes a single, isotropic correlation time, which is probably a good approximation for most small oligomers, but might break down for longer ones. For a long cylinder, in principle two correlation times are needed: one for rotation about the long axis and one for tumbling of that long axis. Such nonisotropic rotation can be taken into account during the analysis of structures (Wang et al., 1992a), but has not become routine. The overall rotational motions are best analyzed by other methods, such as fluorescence depolarization.

A more significant problem is that spins in nucleic acids occur in various groups, with few that can be considered isolated, as assumed implicitly in this equation. Proximity of several spins can be important because for large interproton distances transfer of magnetization through a third nearby spin (usually referred to as spin diffusion) may be faster than the direct transfer pathway (for an example of such effects see Olejniczak et al., 1986). At very short mixing times, the problem of spin diffusion is reduced, but the cross-peaks from proton pairs at long distances will be extremely weak. Rather than use the equation above, with its significant limitations, there are two better approaches that have been developed. The first takes advantage of the fact that the inverse problem, calculation of NOEs given all of the distances between protons, is fairly easy to solve (Borgias and James, 1988). A couple of groups developed approaches in which an initial guess is made at a possible geometry (e.g., a regular B-DNA), and then NOE peak intensities are calculated (Borgias and James, 1990). This set of intensities is merged with the real experimental intensities, and a matrix inversion is done to obtain new distance estimates. The process is repeated until all observed NOE intensities and the computed distances are compatible (*including* effects of spin diffusion). The process should converge to the same set of distances regardless of starting structure, hence the initial guess is only relevant to speed of convergence. The effects of local motions can also be included (Koning et al., 1991). The distances after convergence can then be used as restraints for structure calculations. This process can be carried out for a whole set of NOESY mixing times, from which it is possible (at least in principle) to determine both short (2 Å) and long (4+ Å) distances. The second approach is simply to use the NOE intensities during the structure calculation, varying coordinates until the predicted NOE intensities (again calculated taking spin diffusion into consideration) match the experimental ones (Nikonowicz et al., 1991, Robinson and Wang, 1992, Kim and Reid, 1992). A systematic comparision of these methods has not yet been done. Presently, both approaches seem to be satisfactory and either approach is much better than ignoring the problem of spin diffusion completely.

The equations normally used for the NOE calculations contain the assumption that all proton pairs tumble with the same correlation time, τ_c. There are some indications that there is at least some internal motion in duplex oligonucleotides, although experiments comparing the buildup rates of H2′–H2″ pairs (1.8 Å) and cytosine H5–H6 pairs (2.5 Å) suggest that such motion is limited (Reid et al., 1989). Since the nature of internal motion is not known, it is difficult to assess its effect on the NOE calculations. LeMaster et al., (1988) showed that for proteins fast (compared with τ_c), uncorrelated vibrational motions do not cause significant changes in the distance dependence of the NOE. The NOE predicted from the average position of the atoms is the same, to within a few percent, as that calculated more properly using a full analysis of the dynamics. However, slow motions, again relative to τ_c, could cause averaging (with a weighting of NOE buildup rates, going as $1/r^6$), which would affect distance determinations. These motions might be difficult to detect by NMR (see the section on dynamics below). In general, it is not possible to go from an averaged NMR parameter (such as NOE intensity or coupling constant) back to the combination of conformers and populations that gave rise to it. However, if several average parameters are measured, knowledge of the distribution of conformations can be obtained.

For NOESY data taken in H_2O solution, additional considerations may arise. In some cases, the intensity of peaks from different spectral regions are distorted by the excitation profile of selective pulses, and of course this must be considered in deriving distance estimates from cross-peak intensities. Although the pulse profile can, in principle, be calculated precisely, it is likely that some additional uncertainty is introduced in making the correction. A second factor is that the nitrogen-bound protons may exchange with solvent during the mixing time, leading again to reduced

NOE cross-peak intensities. Such behavior is commonly seen for imino resonances coming from base pairs near the ends of a helix. A second type of exchange is seen for amino protons, coming from rotation about the C–N bond, which is relatively fast for G and A aminos, usually leading to very broad lines, as pointed out above. Even for C amino groups, for which the lines are sharper, the rotation is probably sufficiently fast that exchange occurs during the NOESY mixing time, leading to a non-NOE transfer of magnetization. This transfer again complicates the quantitative analysis of cross-peaks involving these amino protons. To date, NOEs involving exchangeable protons have only been used as qualitative constraints.

It is also worth noting that care must be used in collecting and analyzing the NOE data. First, to obtain uniformly quantitative data the nonselective T_1 relaxation time should be measured, and the repetition rate should be slow enough to allow fairly complete relaxation of all spins. If different groups of spins have substantially different T_1 values (which often happens for the adenosine H2s), then their NOE intensities will be distorted relative to others. Similarly, it has been found that most protons in small RNA oligomers relax more slowly than equivalent protons in DNA (Wang et al., 1992b). Second, the cross-peaks vary significantly in line-shape, and integrations should be done to ensure that an appropriate region is included for each. Third, the digital resolution should be sufficiently high to allow accurate integration of the peaks. In general, the typically used values for recycle delay, integration regions, and digital resolution are sufficient for accurate analysis, but clearly the more quantitative the analysis, the more care is required in carrying through all phases of the measurement.

3.2 Spin–Spin Coupling

Spin–spin couplings provide structural information in a different way from the NOE intensity. The frequently seen three-bond vicinal coupling constants depend on the dihedral angle, as indicated in Figure 5-12, which is usually called a Karplus curve (J value vs. dihedral angle). These curves depend on both the types of atoms involved (^{1}H–^{1}H vs. ^{1}H–^{31}P e.g.) and the substituents neighboring the bond. The necessary parameters for describing the curves are obtained empirically by measuring couplings in a large number of compounds of known, fixed geometry (for oligonucleotides appropriate compounds include cyclic phosphodiesters, etc.). The required calibrations and analysis for nucleic acids have been done in great detail by Altona (1982). For the ^{1}H–^{1}H coupling constants, $^3J_{HH}$, the substituent effects can be included in one equation:

$$^3J_{HH} = 13.7\cos^2\phi - 0.73\cos\phi + \sum_i \Delta\chi_i[0.56 - 2.47\cos^2(z_i\phi + 16.9|\Delta\chi_i|)] \tag{5-2}$$

where ϕ is the dihedral angle, $\Delta\chi_i$ is the difference in electronegativity between a substituent and H (some $\Delta\chi$ values for O: 1.3; C: 0.4; N: 0.85; P: -0.05), and z_i is a relative orientation factor that is either $+1$ or -1 depending on the position of the substituent.

For the proton phosphorus couplings, a somewhat simpler relation can be used.

$$^3J(\mathbf{HCOP}) = 15.3cos^2\phi - 6.1\cos\phi + 1.6 \tag{5-3}$$

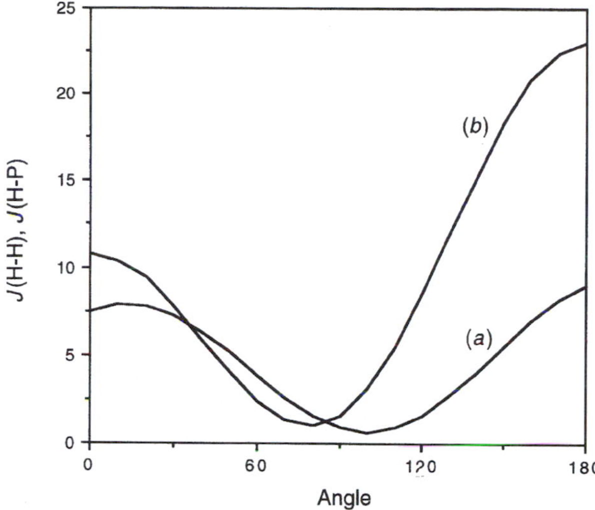

Figure 5-12
Two Karplus curves (coupling constant vs. dihedral angle between the spins involved) are shown. Curve (*a*) is for the **HCOP** coupling; curve (*b*) is for a **HCCH** coupling, such as that between the H1′ and H2′. Both curves were calculated from the equations given in the text, [Derived by Altona 1982.]

Regardless of the source of the calibration, if such a relation is known then determinations of spin–spin couplings can be converted to estimates of angles. Note that in many cases there is not 1:1 relationship between the value of J and the angle; that is, there are often two angles that are consistent with a given coupling value. When this occurs then other information (NOEs, steric hindrance, or couplings involving other spins) must be used to determine which is the correct value.

The spin–spin couplings can be determined in a number of different ways. First, in very small oligos the normal 1D spectrum may be reasonably well resolved and the splittings can be determined directly, although it is sometimes difficult to tell which lines are associated with which combination of coupling constants. It is usually beneficial to use a simulation program, fitting peaks to each component of the multiplet. When done carefully, even rather small couplings can be accurately determined. One-dimensional data can easily be recorded with high digital resolution, so that digital factors should not limit accuracy. Unfortunately, for most oligonucleotides of interest resonances overlap significantly (particularly those from the sugar), and it is not possible to identify components belonging to a specific multiplet. In such cases it is possible to use one of several available 2D experiments, including J-resolved, COSY, exclusive COSY (E.COSY), or heterocorrelated spectroscopy.

The multiplet fine structure of COSY cross-peaks can be used to determine couplings. As noted above, the cross-peaks are intrinsically antiphase in character with respect to the active coupling, but in phase with respect to passive couplings, shown schematically in Figure 5-13. The number of lines within a cross-peak can be quite large, and there is at least partial overlap of components, particularly at the moderate digital resolution used in typical experiments ($2k \times 2k$ data points at 2 Hz/pt). In spite of this, however, the values of many different coupling constants can be determined by simulation of the cross-peak fine structure (Widmer and Wüthrich, 1987). If particularly small couplings are sought, then it is important to work at higher digital resolution. In addition, for slowly tumbling molecules the central components involving couplings to geminal protons (e.g., H1′ to H2′,H2″) are affected by spin-flip processes distorting the couplings. Careful simulations can account for these affects, but ignoring them in the past may have contributed to misinterpreting conformational averaging in sugars (Harbison, 1993).

In a modification of the COSY experiment, called E.COSY, or a simpler version purged E. COSY (P.E.COSY) (Mueller, 1987), part of the multiplet is reduced in intensity in a way that removes some of the cancellation problem, making it possible to measure smaller couplings (Bax and Lerner, 1988). Such multiplets are shown in Figure 5-13. Any heteronuclear couplings to either of the spins involved in the multiplet will appear as passive couplings in the fine structure. These can be removed by insertion of a 180° pulse during the t_1 period, and by heteronuclear decoupling during t_2. Selective excitation during both parts of the COSY experiment can also lead to the E.COSY type line shape, and allows data collection at very high digital resolution (Emsley et al., 1993). Simulation of cross-peaks can be used to extract coupling constants, (Fig. 5-14).

Although the heteronuclear couplings can be determined in principle from the ^{1}H–^{1}H multiplet structure, it is often more desirable to measure them separately in heteronuclear experiments. Again there are several modifications that can be used, analogous to those discussed above. First, when resolution permits, heteronuclear *J*-resolved experiments are straightforward to carry out, and again have the advantages of high sensitivity and digital resolution. The heteronuclear correlation experiments also have cross-peak fine structure that can be analyzed. For ^{31}P the resonances usually cover a rather narrow band width, making it possible to collect high digital resolution data in a short time. For other heteroatoms, the chemical shift range is usually large, and the number of data points that can be acquired in a reasonable time limits the resolution. Folding permits high-resolution experiments to be carried out (Schmieder et al., 1992).

Conformational averaging affects coupling constants, in a fashion similar to that discussed above for the NOE. In the fast exchange limit, the measured value of couplings will be a simple population weighted average over all conformations sampled. In some cases, for example, in cyclic structures such as the ribose ring, couplings alone can be used to show that more than one conformation must contribute to a particular coupling value. Couplings alone suffice because specification of the sugar conformation only requires two parameters (see Chapter 2), the pseudorotation phase and amplitude, and there may be as many as five couplings that can be measured. Since the number of measured values exceeds the number of unknown parameters, it is possible to determine the weights of contributing conformers, assuming that a minimum number are present. In some cases if only a single averaged coupling constant is measured, then there may be an apparent contradiction with distances determined from NOEs (which are averaged in a different way). Such behavior sometimes occurs for terminal sugars in oligonucleotides.

(a)

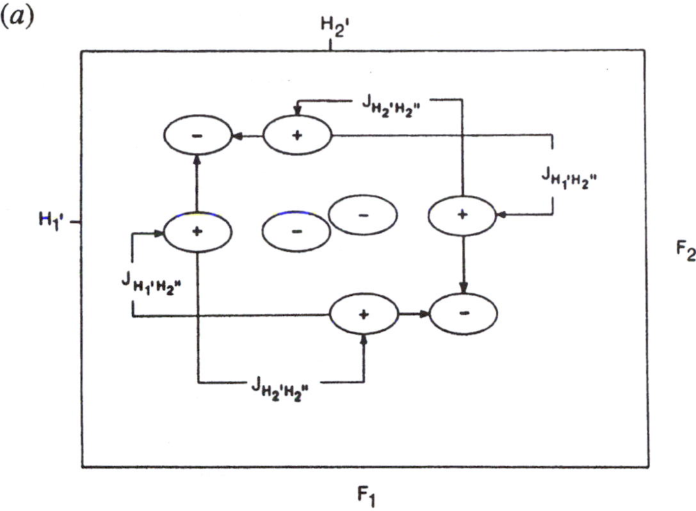

(b)

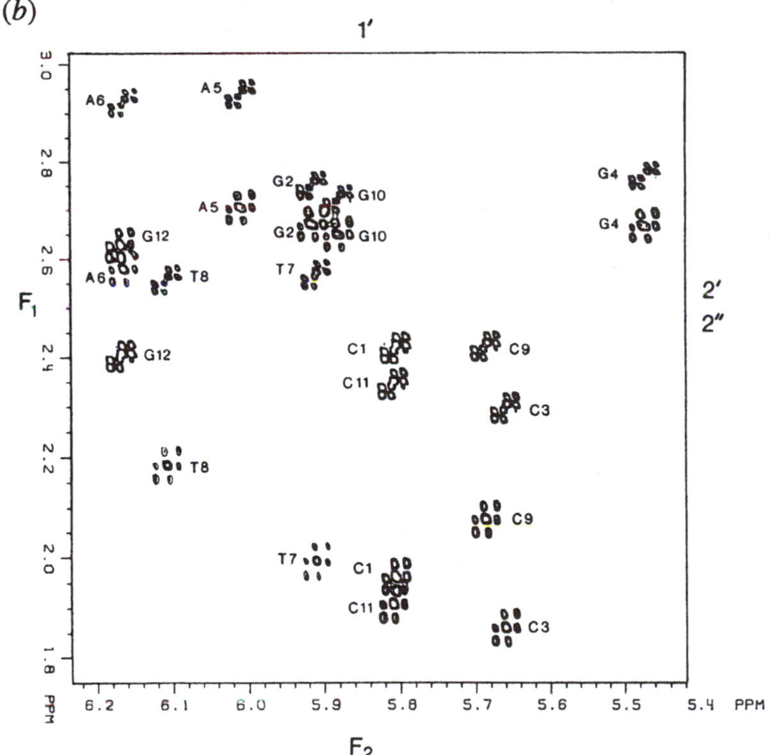

Figure 5-13
A schematic expansion of a multiplet from an exclusive E.COSY type spectrum is shown (bottom) together with a segment of an actual spectrum containing the cross-peaks between the H1′ and H2′, H2″ protons of the DNA dodecamer d(CGCGAATTCGCG)2. [Reprinted with permission from Bax and Lerner 1988.]

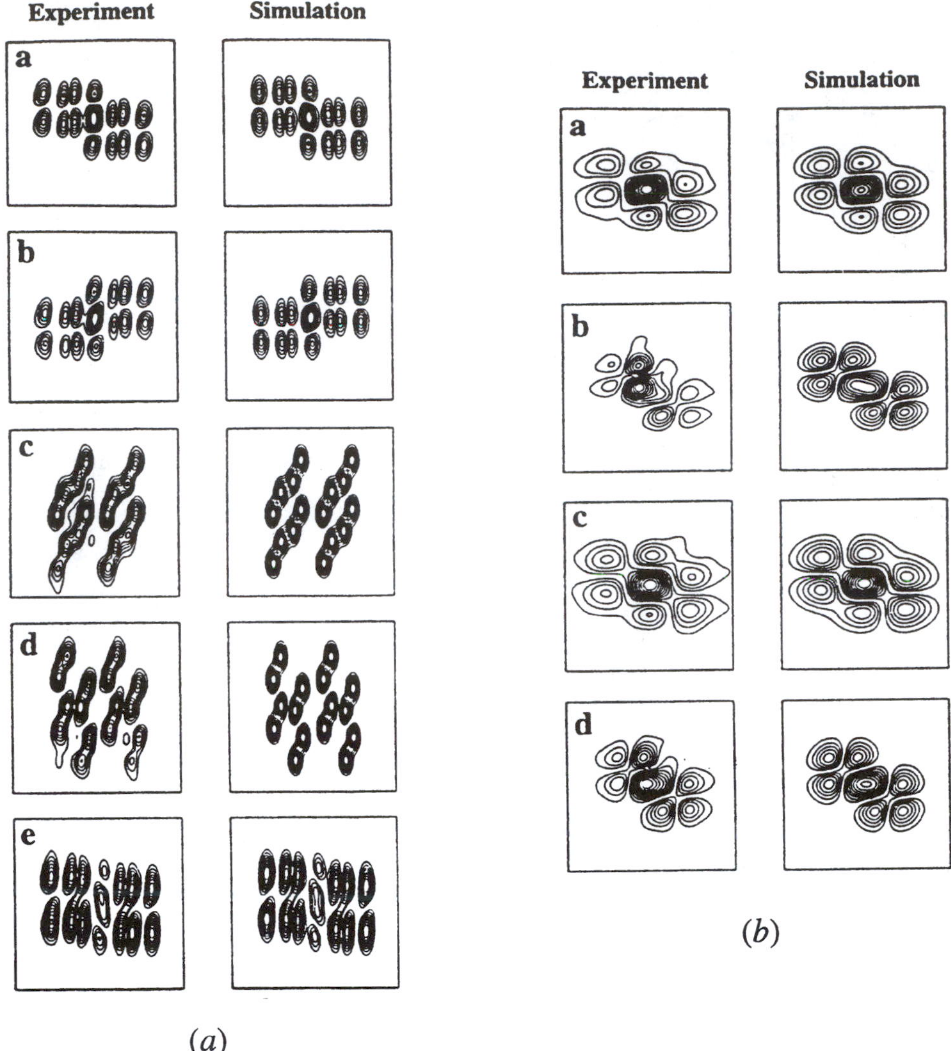

Figure 5-14
Small sections from selective COSY experiments, and the simulated spectra used to obtain coupling constants, on the DNA tetramer TATA (*a*), and the duplex oligomer d(GCGTACGC)$_2$ (right) are shown. In the left panel, a, c, and e are the H1′–H2′, H2′ - H2″, and H2′–H3′ cross-peaks respectively. Panels b and d are the same as a and c but with a selective 180° pulse to change the sign of a coupling. In the right panel, a and b are H1′ - H2′ and H1′ - H2″ cross-peaks for T4, and c and d are the same for A5 in the duplex oligomer. [Reprinted with permission from L. Emsley, T. J. Dwyer, H. P. Spielmann, and D. E. Wemmer (1993). *J. Am. chem. Soc.*, **115**, 7765–7771. Copyright ©1993 American Chemical Society.]

3.3 Chemical Shifts

The chemical shift contribution due to aromatic "ring currents" can be calculated reasonably accurately, and can be used in a limited sense for structural interpretation (Perkins, 1982). The most aromatic base, adenine, gives rise to the largest shifts on neighboring bases, followed by guanine, cytosine, and thymine (uracil) in decreasing order of effect. The major problem in interpreting chemical shifts is not in interpretation of the ring current shift itself, but rather in determination of other contributions to the chemical shift. In spite of this problem, qualitative patterns are often clear; for example, the resonance of a proton placed above or below an adenine or guanine plane will be shifted upfield. Thus, intercalation into a duplex will usually cause an upfield shift in the protons of the intercalator. It has been realized that ^{13}C chemical shifts are also sensitive to conformation, for example the ^{13}C shifts reflect sugar pucker (Santos et al., 1989). Although all of the factors affecting ^{13}C chemical shifts are not known, the empirical correlation observed thus far is good enough to make the carbon shifts valuable as conformational indicators.

4. CHARACTERIZATION OF STRUCTURES

Both the intensity of NOESY cross-peaks and the values of coupling constants are sensitive to the conformation of a nucleic acid. However, in spite of this it is difficult to determine the full structure of a molecule from these parameters. Some of the local structural features, such as the sugar conformation and the glycosidic angle, can be determined accurately with few assumptions. The global structure is more difficult to determine, and extensive NMR data are needed to accurately determine structural parameters, such as helix twist and rise, and groove width. The methods that have been applied to these problems will be discussed below.

4.1 Sugar Pucker

A qualitative analysis of the sugar pucker is usually possible just from the cross-peaks seen in a COSY spectrum. For C2′-*endo* and similar S(outh) conformers, the couplings between the H1′ and both the H2′ and H2″ are substantial, and give rise to cross-peaks. The H3′ to H2′ coupling is moderate in strength, sufficient to also give a cross-peak, but the H3′ to H2″ is near zero and cross-peaks between these protons are not seen. On the other hand for C3′-*endo* N(orth) conformers, the H1′ to H2′ coupling is near zero eliminating cross-peaks, while H1′ to H2″, and H3′ to H2′ and H2″ couplings are substantial and do give cross-peaks. Obviously, for RNA the H2″ is replaced by an OH, and hence all cross-peaks involving it are absent. For a more detailed analysis of the sugar conformation, Altona (1982) showed that the values of the coupling constants involving the sugar H1′, H2′, H2″, H3′, and H4′ are sufficient to determine both the pseudorotation angle of the ring and the amplitude. In some cases, it is not possible to measure all of the individual coupling constants, and an alternative approach using the widths of multiplets (sums of coupling constants) works well instead (Rinkel and Altona, 1987). Any particular conformation of the sugar will give rise to certain combinations of coupling constants, which may be used either directly in Karplus relations to determine angles, or through a graphical analysis to determine directly the pseudorotation angle.

If several of the couplings can be determined, then it is possible to determine whether the coupling values are consistent with a single C2′-*endo*, C3′-*endo*, or any other single conformation, or whether conformational averaging is occurring. If averaging is present, simple assumptions about the conformers being averaged can be made, and then relative populations may also be derived. In general, the same coupling values could arise through more complex averages of many conformations, but the simplest average may well be correct.

4.2 Glycosidic Angle

For determination of the glycosidic angle, χ, NOESY data must be used. The frequently seen conformers fall into two groups, anti and syn, with the aromatic proton far from the H1′ and close to it, respectively. The distances from the aromatic proton of the base (H6 or H8) to the sugar H1′ and H2′ are dependent both on the sugar pucker and the value of χ. If the sugar pucker can be determined from couplings, then the NOE intensities can be analyzed to give χ. The relations between the distances and the χ value are shown in Figure 5-15.

In a right-handed B-DNA helix there is a repeated pattern of NOEs, which was used for the sequential assignment of resonances. As noted above, the presence of such connectivities in fact verifies the right-handed nature of the helix, and when taken together with the determination of sugar pucker from coupling constants, can be used to identify regions of nucleic acids that are in B-type conformations. An analogous pattern, with different sugar pucker and different NOE intensities, is seen for the A form. For completely double-stranded nucleic acids, the overall conformation can be determined from circular dichroism (CD) studies as well (see Chapter 6). From NMR, it is possible to identify precisely where transitions between local conformations occur. For example, a number of different stem–loop DNA sequences have been studied. In all of these, the duplex stem region has been in the B form [excepting an alternating CG stem at very high salt, which goes to the Z conformation (Wolk et al., 1988)]. The B-sugar conformation and stacking are seen to continue through the first base of each loop. The second base in the loop is clearly not stacked normally, and the conformation seems to depend on the loop sequence (Blommers et al., 1989) (see Fig. 5-8). Similarly, when an extra base is present on one of the strands it has been possible to distinguish cases in which it stacks into the helix in a fairly normal B geometry from those in which it is bulged out (examples are discussed further below). Such qualitative analyses have been valuable in understanding other chemical and physical properties, other RNA hairpins, pseudoknots, and internal loops, discussed later.

Other structures may also have characteristic features that may be analyzed in a qualitative way. For example, in Z-DNA and Z-RNA, which occur most readily in alternating purine pyrimidine (usually GC) sequences, there is an alternation of successive base pairs between syn (purine) and anti (pyrimidine) conformations. Although the assignment of resonances in these forms cannot use the same rules as for the B or A form, the presence of the syn and anti forms is obvious from NOE intensities in the aromatic

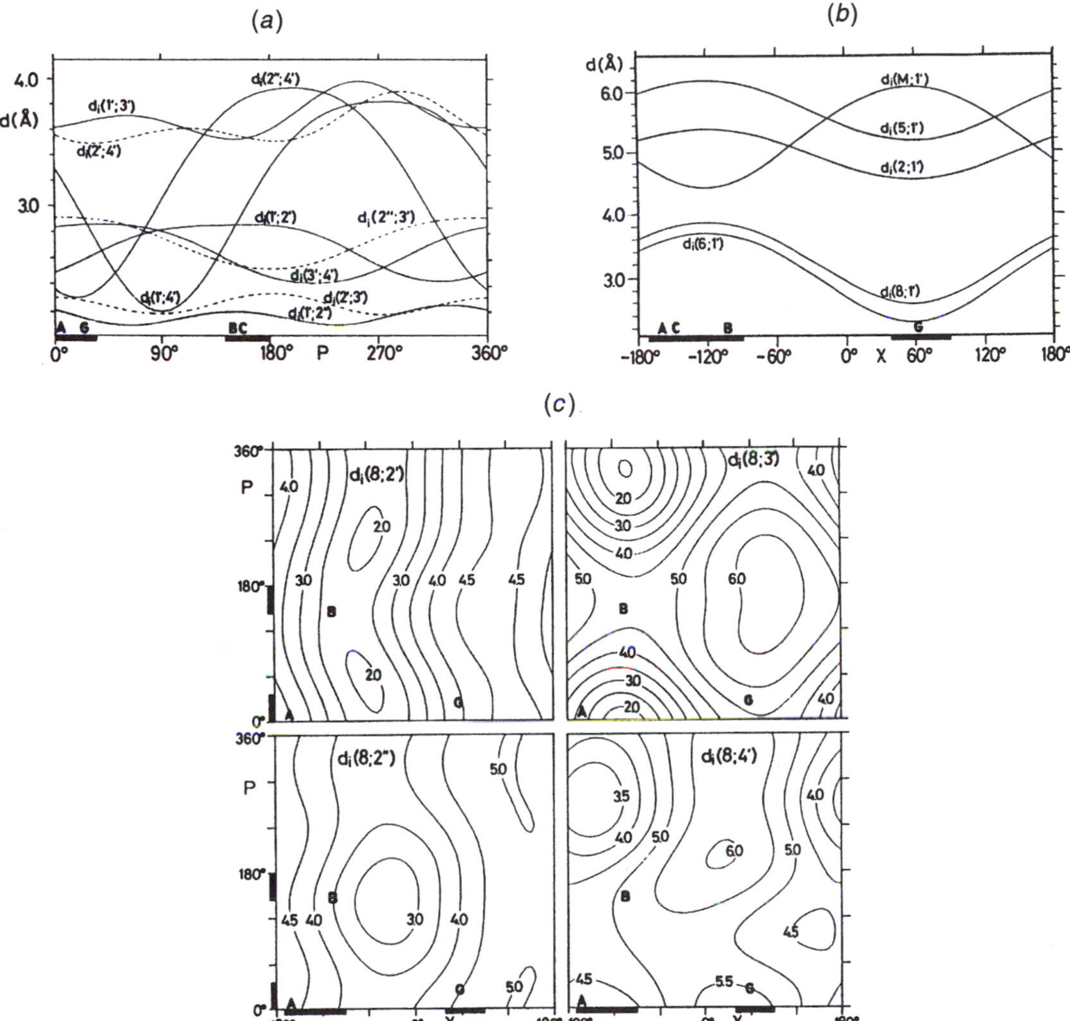

Figure 5-15
Plots are shown, which indicate distances between particular pairs of protons in DNA residues as a function of the conformation of that residue. In (*a*), the distances between indicated proton pairs is given as a function of the sugar pseudorotation angle *P*. The letters A and B indicate the *P* values of A- and B-form DNA, while G and C indicate the positions of G and C residues in Z-form DNA. In (*b*), the distances between indicated proton pairs are given as a function of the glycosidic angle χ annotated as in (*a*). In (*c*), several intraresidue distances between the aromatic proton and sugar protons are given as a function of both *P* and χ. These are for the H8 of a purine, but the values of an H6 for a pyrimidine are very similar. [NMR of Proteins and Nucleic Acids, Wüthrich, K., Copyright ©1986. Reprinted with permission of John Wiley & Sons, Inc.]

to H1′ region of the spectrum (Feigon et al., 1984). In telomer DNA sequences, there are G rich sequences, which also show alternating syn–anti patterns. Nuclear magnetic resonance has been used to verify the formation of G quartets, and the geometry of the loops in a number of telomeric sequences. It has also probed "intercalated cytosine" type structures involving $C \cdot C^+$ base pairs (Gehring et al., 1993), centromere sequence structures involving G-G stacks and G • A base pairs (Chou et al., 1994), and parallel stranded DNA structures (Zhou et al., 1993). For many of these, the 'low resolution' NMR structures gave the first clear structural picture, and gave insight into possible function of the sequences.

4.3 Detailed Analysis of Nucleic Acid Structures

4.3.1 Distance Constraints

Developing more detailed models—sets of coordinates that satisfy all of the NMR derived constraints as well as having appropriate covalent structures (bond lengths, angles, van der Waals contacts)—can be done in a number of different ways. In all of them, as many constraints on distances and angles as possible must be determined. The constraints usually take the form of distance estimates derived from the NOE, with some estimate of the uncertainty (to give both lower and upper bounds on the distance). In general, both the accuracy and precision of the constraints is important to the quality of the final structures, probably much more so than for the calculation of protein structures from NMR data, due to the lack of tertiary folding and a compact interior in nucleic acids. Different workers have used a variety of approaches for generating initial models to refine. For double-stranded helices, coordinates derived from fiber diffraction work (Nilges et al., 1987) representing regular A-form or B-form helices have been used. Since fiber diffraction gives an "average" structure, it is easy to create coordinates for specific sequences of DNA or RNA that have the correct average parameters. A wider range of starting structures (including duplexes and loop regions) can be generated through distance geometry "embedding" (Hare and Reid, 1986). This process takes all known distance information (covalent bond angles and bond distances, coupling constants, and NOEs), and carries through a set of coordinate transformations to obtain approximate structures for the molecule. Since the input information is incomplete, such structures are approximately correct globally, but have many problems with the local structure. There were concerns about sampling with this algorithm, but in the latest versions of the distance geometry programs these have been taken care of. The least defined starting structure that can be used is just a random set of atomic positions (Nilges et al., 1988) within a certain distance of the origin. Convergence takes longer with this approach, but a better sampling of structures consistent with the data may be obtained.

4.3.2 Energy Minimization

In any of these initial models, there are disagreements (from few to many depending on the type of starting model) between the initial model and the constraints, which must be removed through some optimization or minimization process. It is possible to carry out this optimization including just the NMR data, supplemented by constraints maintaining known covalent geometry, or with the addition of a full potential energy function including electrostatics, hydrogen bonds, dispersion forces, and so on (see Chapter 7.) Although minimization of an error function (which quantitatively assesses the violations of the input constraints) is sufficient in principle, in practice minimization algorithms suffer from getting trapped in local minima. Such trapping occurs equally whether or not a full potential energy function is used. The most satisfactory alternative to local minimization has been a molecular dynamics approach, generally called simulated annealing. In this method, random velocities are assigned to each atom, with the average velocity defining an effective temperature of the system. The atoms are then allowed to move following simple Newtonian mechanics, including "energy" terms that are quadratic in the distance violation (or error in NOE intensity). The error function is then treated just as any other potential energy term would be, with forces acting on atoms corresponding to derivatives of the function with respect to positions of the atoms. The system is allowed to evolve for some length of time, and then is "cooled" slowly by damping all velocities. The end of this process corresponds to a minimization of the error function. Unlike the simple minimization, however, the evolution at high temperature makes it much easier for the molecule to move between local minima, and eventually fall into the lowest minimum of the sampled region of conformational space. If the distance information being used is quite complete, this should also represent the global minimum of the error function. By using this method, it is fairly easy to remove essentially all disagreements between the set of coordinates and the input constraints, resulting in a final set of coordinates that agree (at least in a first-order sense) with all of the constraints, including the NMR data. Since this process uses random velocities in the initial stages, the calculation can be repeated, and will generally yield different coordinates at the end. The variation among the sets of coordinates gives information about the uncertainties in the determined structure. In a rough sense, the magnitude of the spread in structures is like the resolution in a crystal structure—when the spread is small (resolution is high) the coordinates of each atom are more precisely determined. However, unlike crystal structures there may be big differences in how well local and global parameters are determined.

When full energy terms are used during refinement, structures jointly satisfy the experimental NMR data and minimize potential energy (Gronenborn and Clore, 1989). It is often difficult to determine which features of a structure are really well determined from the NMR data, and which arise from the potential. With proteins, refinements are often done first with NMR data alone, then subsequently with potential terms as well. As methods for determination of nucleic acid structures continue to develop one can hope that standards for their description will also be established to make interpretation easier. Presently, it can only be said—read the paper describing a structure carefully to determine exactly how measurements and calculations were done. Simulations have shown that some helix parameters can be determined reasonably well, while others are underdetermined (Pardi et al., 1988; Metzler et al., 1990).

5. DYNAMIC ANALYSIS

Nuclear magnetic resonance is sensitive to dynamic processes over a wide range of frequencies (or equivalently lifetimes), however, there are also ranges in which both the NMR spectrum and relaxation parameters are remarkably insensitive. The effects of motion will be considered beginning from the lowest rates. When more than one conformation of a molecule is present in solution, separate resonances from each will be seen (with intensities reflecting populations) as long as the separation in chemical shifts (in frequency units) is large relative to the rate of interconversion. For example, a difference in chemical shifts of 1 ppm at 500 MHz corresponds to a frequency (and rate) of 500 s^{-1}. As the rate of exchange approaches the separation frequency the peaks become broadened, eventually collapsing into a single peak at the average (population weighted) chemical shift. In the slow exchange limit the line-width of each peak is related to the lifetime in that conformation. Once the peaks have coalesced, the line-width is related to both the rate of exchange, as well as the chemical shift separation. In many cases, it is possible to vary the rate of exchange by changing the sample temperature (rate $= A\exp[-E_a/RT]$). When line-shapes can be observed over a significant range of temperatures it is possible to determine the activation energy as well as the rate at each temperature. When the rate of exchange becomes very high, the average peak returns to a very narrow width, and there is no direct indication of the exchange process in the spectrum.

5.1 Slow Exchange

If the interconversion rate is very slow, saturation transfer can be used (with either the NOESY experiment, or a 1D pulse sequence) to detect exchange. In this case, the cross-peaks connect resonances from the spin in two different chemical environments. The intensity of the cross-peak is determined by the ratio of the exchange rate to the spin–lattice relaxation rate (the inverse of the spin–lattice relaxation time T_1). If both the cross-peak intensities and relaxation rates are measured, then the true T_1 value and the exchange rate can be determined. Since T_1 is usually on the order of 1 s for nucleic acid oligonucleotides, and intensity changes of less than 10% can be measured, it is possible to measure rates as low as 0.1 s^{-1}. With the 2D version of the experiment, it is possible to detect and measure the rates of several simultaneous exchange processes. An example of exchange of a drug between different binding sites on a DNA undecamer, determined by this methods, is discussed below, (Fig. 5-16).

5.2 Rapid Exchange

When exchange processes become very fast, they can be detected through their direct effect on the spin–lattice relaxation time, T_1. This time constant is for populations of spin states that have been perturbed, to return to equilibrium; it is determined by how often fluctuations in the magnetic environment occur with the correct frequency to stimulate transitions of the spins. The magnetic environment is determined by dipolar couplings to other spins [^1H–^1H, or ^1H–X (nonproton)], and through chemical shift anisotropy, which is usually important only for X $\neq$ ^1H nuclei (chemical shift anisotropy is the

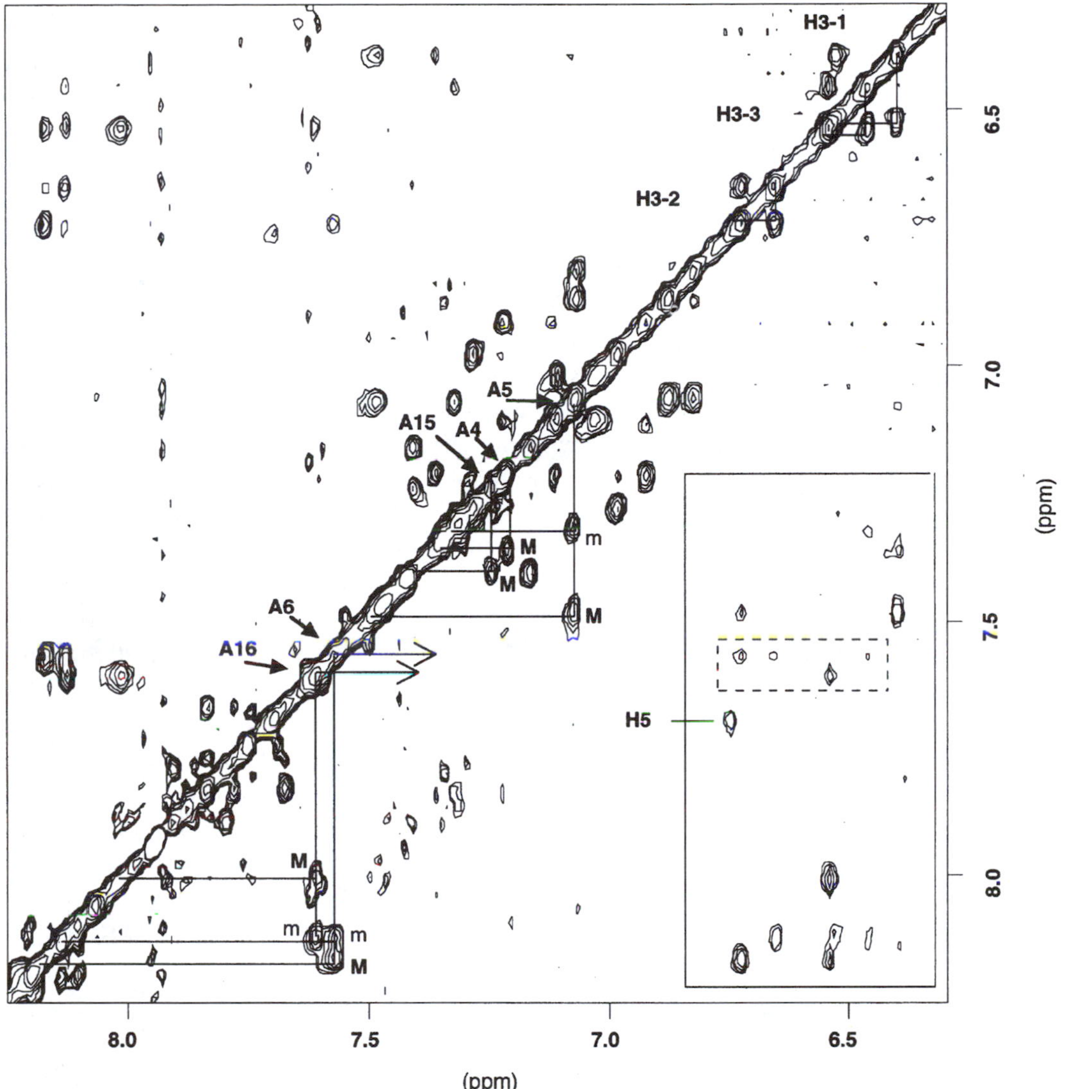

Figure 5-16
A section of a NOESY–exchange spectrum is shown, which contains the aromatic and sugar H1′ protons of a DNA oligonucleotide and the aromatic protons of the drug distamycin-A. Although the spectrum is complicated, and peaks are present both from chemical exchange of the drug between different binding sites and from NOEs between protons, it is possible to assign and interpret the spectrum. The parameters m and M label the minor and Major forms of complex with DNA (present in about a 3:2 ratio). Most of the peaks near the diagonal correspond to chemical exchange of protons between free DNA chemical shifts and the major and minor forms of complex, while those away from the diagonal are from NOEs. The identities of some of the adenosine H2 resonances are indicated. The box to the lower right contains NOEs used to assign the DNA and to determine intermolecular contacts. The exchange cross-peaks make it possible to determine rate constants for transfer of drug between each different kind of binding site.

difference in chemical shift along different directions in the molecule). Although only the average chemical shift is measured in solution, the variation in chemical shift with orientation as the molecule tumbles provides one of the mechanisms of spin–lattice relaxation (Levy, 1974). Any sort of motion of the spins, whether an internal conformational transition or the overall tumbling of the molecule in solution, can be described by a correlation function, $F(\tau)$. This correlation function can be thought of as the probability that a vector in some initial orientation is still in that orientation a time τ later. For most random processes, the correlation function is exponential, $F(\tau) = K\exp(-\tau/\tau_c)$, where K reflects the amplitude of the motion, and τ_c is the time constant reflecting the rate of the motion. The contribution to the relaxation is calculated through the FT of $F(\tau)$, which is known as the spectral density function $J(\omega)$. The relevant terms for dipolar spin lattice relaxation of protons are $J(0)$, $J(\omega_H)$ and $J(2\omega_H)$ where ω_H is the Larmor frequency of the spin. It is the difference between $J(0)$ and $J(2\omega_H)$ that gives rise to the NOE. For dipolar ^{1}H–X relaxation, the relevant spectral densities are $J(\omega_H + \omega_X)$ and $J(\omega_H - \omega_X)$, while for shift anisotropy relaxation it is just $J(\omega_X)$. Since the Larmor frequencies are high (the lowest we consider here being ^{15}N at 50 MHz when protons resonate at 500 MHz) relaxation is most sensitive to motions that are on the nanosecond time scale. In order to obtain the maximum information from relaxation, it is best to measure line widths (spin–spin or T_2 relaxation), spin–lattice relaxation (T_1) and NOEs, optimally at several field strengths to vary the Larmor frequency. It should be noted that because of the spin diffusion that occurs among protons, moving magnetization among spins, proton T_1 relaxation measurements can only provide an average behavior. For the rare X nuclei, spin diffusion does not occur, and localized differences in motion can be identified. Perhaps the most obvious case of internal motion is the rotation of the thymine methyl group about the C6–CH_3 bond, which normally has a correlation time much shorter than the overall rotation ($\sim \tau_c \approx 10^{-11}$ s). Further discussion of motions detected by NMR are given for specific systems below.

5.3 Intermediate Exchange Rates

From the previous discussion it should be clear that line-widths and relaxation times measured in solution are insensitive to motions of intermediate rate (in the range of 10^4–10^8 s^{-1}). Part of this gap can be filled through solid-state NMR experiments, particularly using samples enriched in deuterium, ^{2}H. This isotope of hydrogen has a spin of 1, and is affected by quadrupole couplings that do not occur for spin $\frac{1}{2}$ nuclei. In solids this coupling makes lines very broad (the order of 200,000 Hz!), but in turn this makes the line-shape sensitive to motions in the 10^4–10^6 s^{-1} range (Kintanar et al., 1989; Wang et al., 1994).

5.4 Imino Proton Exchange

A final method for detection of dynamic properties of nucleic acids is through the determination of imino proton-exchange rates. In all of the methods discussed above, the presence of a dynamic process is detected through a direct effect on some feature of a resonance. Exchange measurements serve as an indirect measure of infrequent fluctuations that normally would not contribute to the spectrum, or to relaxation parameters. It was noted previously that protons bonded to nitrogen are relatively acidic. Hence, these protons can exchange with the solvent. This process is acid and base catalyzed, and requires that the proton to be exchanged is accessible to both catalyst and solvent. When an imino proton is hydrogen bonded in the interior of a duplex region it is inaccessible, and hence cannot exchange. However, fluctuations in the structure occur to an "open" state exposing the imino proton, and allowing exchange to occur. This process is shown schematically in Figure 5-17. Thus the exchange rates reflect the probability that such a transition to the open-state occurs, even though the equilibrium constant for the open state is too small to have a measurable effect on the spectrum. Exchange rates are measured by determining saturation transfer rates to solvent, or through the effect of exchange on the line-width. It has been shown that in order to use the exchange rate to actually measure the opening rate, one must be certain that an exchange event occurs whenever a base pair opens. Experimentally, this is achieved by increasing the concentration of base catalyst in the solution, plotting the exchange rate versus base concentration, and extrapolating to infinite base (Leroy et al., 1988b). By comparing exchange rates for a number of different sequences, it has become clear that the fluctuations leading to the open state are quite localized (1 base pair can open without either neighbor opening). Unfortunately, the nature of the open state is not yet understood, although it is clear that it is different from the unstacked state, which leads to melting detected in UV absorbance studies (Benight et al., 1988). In principle, similar information might be obtained by observation of amino proton exchange, but spectral crowding and broad lines make this difficult.

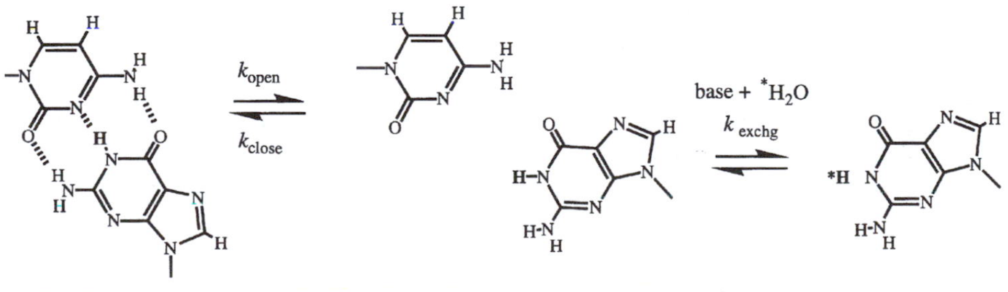

k_{open} k_{close} base + $^*H_2O$ k_{exchg}

Figure 5-17
Schematic drawing of the process of base pair opening and exchange. The proton that is derived from the solvent that is exchanged into the base pair is indicated with an asterisk. Although the equilibrium constant for the open state is quite small, the detection of the exchange events by NMR is straightforward.

6. EXAMPLES OF NMR STUDIES OF NUCLEIC ACIDS

6.1 Helix Conformational Analysis

A wide variety of different nucleic acids systems have now been examined by NMR; van de Ven and Hilbers (1988) reviewed all systems studied through mid-1988. Much of the initial NMR work was done on duplex oligonucleotides (all Watson–Crick base pairs) because their spectra are relatively simple. Although a seemingly simple application, NMR has been useful to examine differences in the behavior of particular sequences in solution and in crystals. Many short DNA oligonucleotides that have been crystallized have done so in an A conformation, for example, the TFIIIA recognition sequence d(GGATGGGAG)•d(CTCCCATCC) (McCall et al., 1986). In this case, nuclease mapping and protein binding characteristics, together with the crystallization in the A form, led to the suggestion that this DNA sequence is in the A form in solution, or at least has a higher propensity to convert to the A form than other sequences, which could be related to the binding activity (Rhodes and Klug, 1986). As discussed in previous sections there are a number of criteria by which the A and B forms can be distinguished, even without a detailed structural analysis. For this sequence in solution, both the NOE patterns and the coupling constants showed clearly that all residues are B-like in conformation (Aboul-ela et al., 1988). In addition the transition to an A conformation induced by addition of ethanol to the solvent, and monitored by CD, showed that the propensity to covert to A form was the same as other DNAs. All of the DNA oligonucleotides that crystallized in the A form and that have been studied by NMR have been clearly B form in normal water solutions. These observations indicate that it is crystal forces and/or solvent conditions for crystallization which lead to stabilization of the A form, rather than an intrinsic propensity inherent in these sequences. Subsequent work has shown that different parts of TFIIIA contribute to binding of DNA and RNA.

6.2 Refined DNA Structures

There are now many DNA sequences for which extensive NOE data have been collected. The refinement methodology, consideration of spin diffusion, and the number of restraints determined from experimental data have been increasing with time. It is likely that a number of the early structures determined (pre -1990) would change significantly if they were redone with current methods. In the more recent structures, there are local conformational variations within the B family of structure, which can be determined with reasonable precision (Cheng et al., 1992b; Weisz et al., 1994). As one might expect, it has been found that inclusion of ^{1}H–^{31}P coupling constants improves the precision in determining phosphate positions, which in turn improves the overall structure (Kim et al., 1992). A recent example of calculated structures for an alternating purine–pyrimidine sequence is shown in Figure 5-18.

Model calculations have been done using distances derived from a known set of coordinates (avoiding any experimental uncertainties in deriving distances from NOE data). The distances are then used to generate structures by the same algorithms that were used for real experimental data, and results were compared with the original

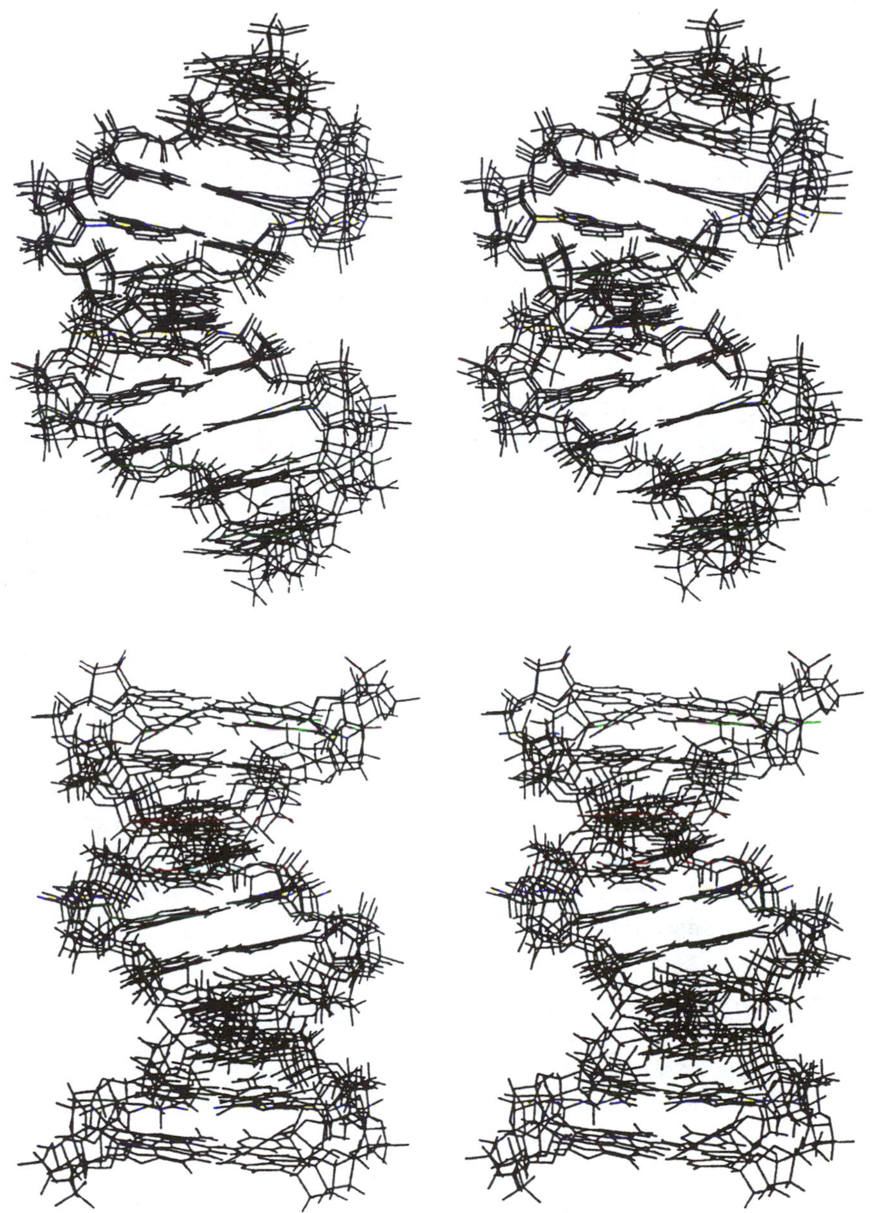

Figure 5-18
Structures of the DNA oligomer $(GCGTATACGC)_2$ are shown. These structures were calculated using RMD refinement with NOE intensities. [Reprinted from Cheng, J.-W., Chou, S.-H., Salazar, M. and Reid, B. R. (1992b). *J. Mol. Biol.* **228**, 118–137. Copyright ©1992, by permission of the publisher, Academic Press Limited London.]

structure. This approach has shown that there are some features of the local structure that can be determined well, but others are poorly defined when working with NOE data alone (Metzler et al., 1990). Extensive simulations including coupling constants in addition to NOEs have not yet been done. These would be helpful in again establishing the limits of this structure determination method.

6.2.1 Bent DNA

There has been considerable interest in "bent DNA", which results from the presence of several adenosine residues together, sometimes called a poly-A tract. For a number of different sequences containing such A tracts, interstrand NOEs from adenine H2 to sugar H1′ are seen, which are not seen in the absence of an A tract and are predicted not to occur in a normal B-form structure. The effect is particularly pronounced at the 3′-end of the string of A residues, correlating with the bending behavior (Kintanar et al., 1987). There are changes in chemical shifts seen as the length of the A tract increases (Katahira et al., 1988), and there is a correlation between NMR observable distances and groove width in these sequences (Chuprina et al., 1991). There is an abrupt change in the sugar pucker seen for the first bases beyond the end of the A tract as well. The pseudorotation phase angles are 100°–130°, versus 150°–180° in the A tract (Celda et al., 1989). When the imino proton exchange in these sequences is examined, there is a very good correlation between abnormally low opening rates for the A • T base pairs (Leroy et al., 1988a) (lifetimes increased by a factor of more than 10) and the presence of a bend-inducing sequence (determined by gel mobility). Although the structural basis for the slowing of opening is not clear, it appears that NMR can identify some features of the unusual structure that give rise to this behavior. Further quantitative NMR studies are likely to contribute to understanding of the local structure that determines DNA bending.

6.2.2 Nonstandard Pairing and Stacking, Hairpin Loops

Nuclear magnetic resonance has also been successful in characterization of DNAs containing non-Watson–Crick (or mismatched) base pairs and base insertions. Some mispaired bases can be accommodated in the double helical structure with little distortion, G • T wobble pairs, for example (Hare et al., 1986)) and others apparently just stabilized by stacking. In other mismatches, there are larger conformational changes. There have been a number of studies of G • A mispairs, and it has been found that the pairing depends on the flanking sequences (Cheng et al., 1992a, Chou et al., 1992). With a carcinogen-modified base, a syn conformation has been seen, which is necessary to fit the bulky modifying group into the duplex without disrupting neighboring base pairs (Norman et al., 1989). Similarly, it has been found that some single base insertions on one strand of a duplex are bulged, or looped out, allowing stacking of the base pairs on either side of the extra base (Morden et al., 1983). In other cases, the extra base is intercalated into the helical segment (Kalnik et al., 1989a), disrupting some of the stacking on the opposite strand, but maximizing the interaction of the extra base with the neighboring bases of the same strand. It appears that pyrimidine bases tend to be bulged out, while purines tend to be stacked in, although all combinations of extra bases and neighboring base contexts have not yet been examined. There is even evidence for conformational transitions between "in" and "out" in some cases (Kalnik et al., 1989b).

A variety of hairpin loops have also been examined, with a number of interesting findings. One of the first to be characterized structurally, a T_4 loop, showed that there might be pairing between the first and last T's of the sequence (Hare and Reid, 1986). More extensive studies have shown that there are a number of "loop sequences" which might have been thought to have four unpaired bases really have just two. The base pair flanking these loops sometimes takes on a reverse geometry to minimize the distance between phosphates connected by the loop (Blommers et al., 1991).

6.3 RNA Oligonucleotides

RNA oligonucleotides have provided us with a wider range of structures to study. Analysis of the secondary structures of hundreds of ribosomal RNAs identified two hairpin loop sequences that occurred much more frequently than would be predicted on a statistical basis (Tuerk et al., 1988). These loop sequences confer extra stability to hairpins, raising the melting temperature by as much as 20°C over the T_m values for other loop sequences with four bases. The sequence (UUCG) has been examined in detail by NMR (Varani et al., 1990; Allain andVarani, 1995), using both NOE data and coupling constants. A very compact loop structure was found containing interactions other than Watson–Crick base pairs. These include a reverse-wobble anti-U syn-G base pair between the first and fourth bases in the loop, C2′-*endo* sugars for the second and third residues, and a hydrogen bond between the cytosine amino and one of the loop phosphates, (Fig. 5-19). It has been noted previously that in order to span the ends of the duplex segment with a small loop the sugar residues of the central two loop residues take on the unusual (for RNA) C2′-*endo* conformation, which places the attached phosphates further apart than the usual C3′-*endo* conformer (Saenger, 1984). Another ultrastable loop family, with GNRA sequences (N = any nucleotide, R = purine) has also been characterized (Heus and Pardi, 1991). The novel and interesting structure called a pseudoknot has also been examined by NMR (Puglisi et al., 1990; van Belkum et al., 1989; Shen and Tinoco, 1995). A pseudoknot is formed when a single strand on either side of a hairpin folds back to form base pairs with the hairpin loop. A pseudoknot thus contains two stems and two loops (see Fig. 5-20). Detailed NMR studies were carried out to show that the two predicted, costacked sections of duplex are formed in solution, and that the connecting segments form appropriate connecting loops. The NMR studies have shown that in order to accommodate this structure there are sugars that are in the C2′-*endo* conformation both at the point where the two stems stack together, and in the loops. With some sequences the pseudoknot is in equilibrium with one of the constituent hairpins, in slow exchange on the NMR time scale leading to separate resonances for the two forms. This equilibrium is magnesium ion dependent, and can be followed in detail by NMR. It has been possible to see saturation transfer between alternate conformations, and from this to determine rate constants for the interconversion. The information about the pseudoknot itself was supplemented by studies of the constituent hairpin loops. It was shown that one of the hairpins was stabilized by two additional base pairs relative to the pseudoknot itself, a

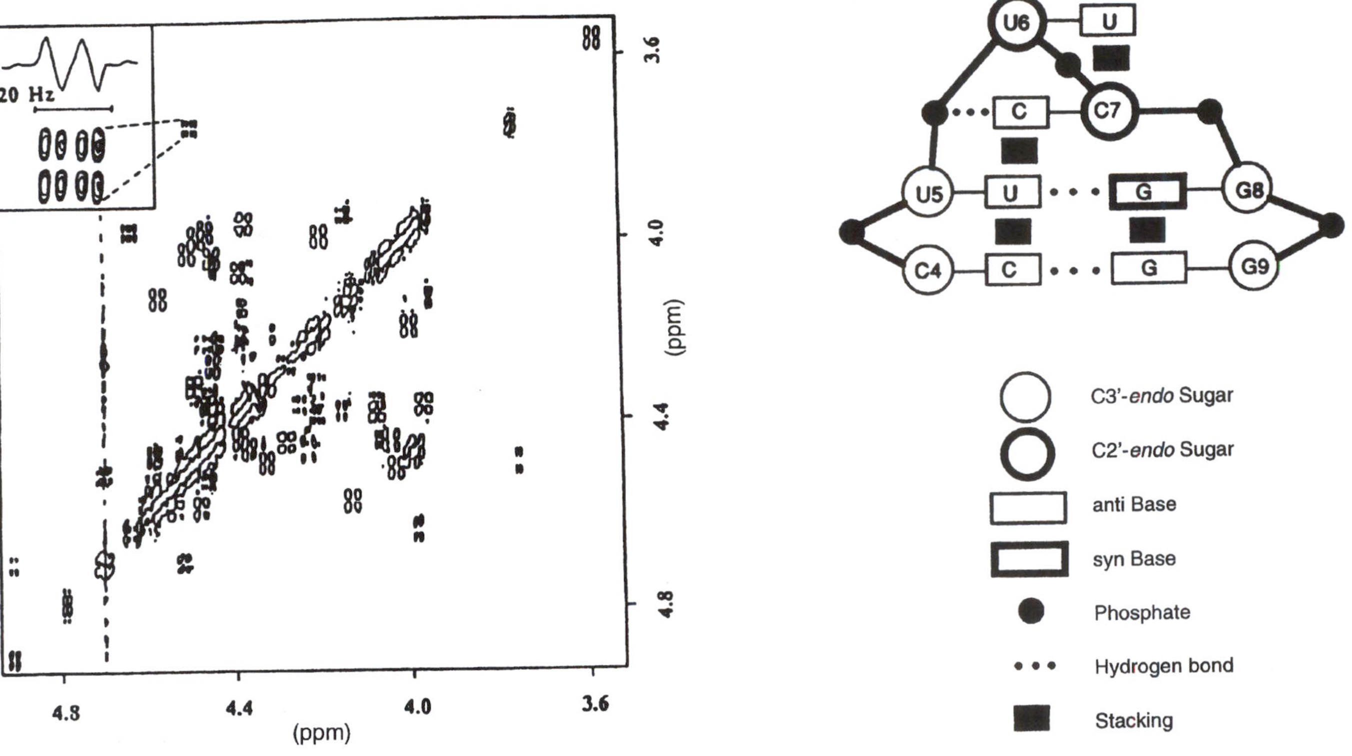

Figure 5-19
A part of a DQF COSY spectrum of an extra stable RNA hairpin loop is shown. The values of the coupling constants determined from this spectrum are used as constraints in analyzing the structure. A shematic drawing of the structural features of the loop are shown at the right, with conformations and interactions labeled. [Reprinted with permission from *Nature*, Cheong, C., Varani, G., and Tinoco, I., Jr. (1990). *Nature (London)* **346**, 680–682. Copyright ©1990 Macmillan Magazines Limited.]

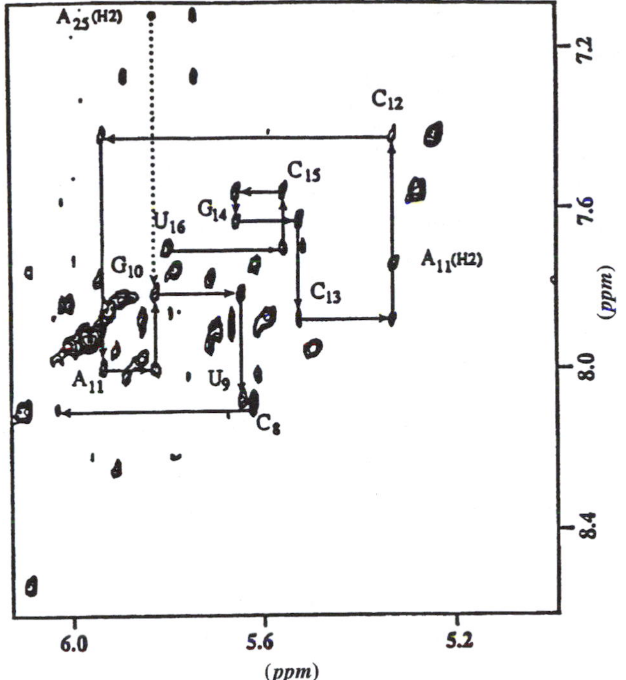

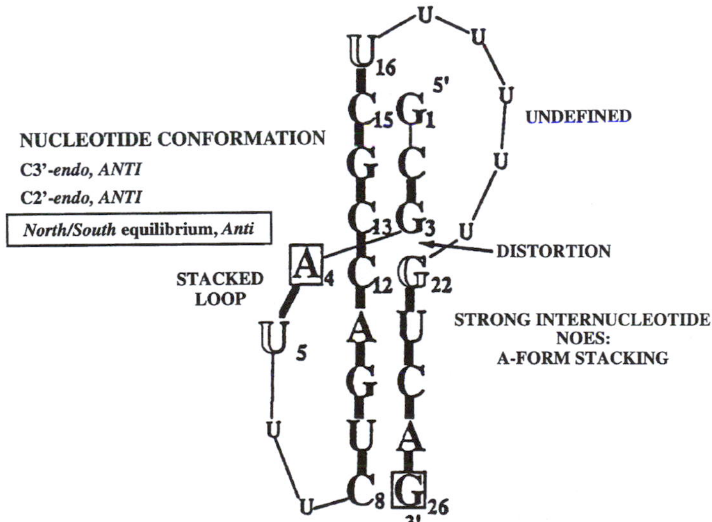

Figure 5-20
A part of a NOESY spectrum is shown from an RNA pseudoknot. The assignment pathway is indicated for one segment of the molecule by sequentially connecting the resonances from neighboring base pairs. With this procedure, it was possible to assign all of the duplex region resonances and some of those from the loops. A schematic drawing of the structure of the pseudoknot is shown with some of the conformational features labeled. [Reprinted from *J. Mol. Biol.*, Puglisi, J. D., Wyatt, J. R., and Tinoco, I., Jr. (1990), Conformation of an RNA Pseudoknot, 437–453, Copyright ©1990, by permission of the publisher Academic Press Limited London.]

U • A and G • U after a mismatched (A • C) (Saenger, 1984). This information, together with the structural, kinetic and thermodynamic data, gives a good picture of both the folded structure of the pseudoknot and its unfolding.

The NMR studies of fragments of RNA have also been carried out, structures of Loop E of eukaryotic 5S RNA (Wimberly et al., 1993) and the sarcin/ricin loop from 28S RNA (Szewczak et al., 1993) are good examples. Although these examples are not high resolution structure determinations, both provide new insight into the structures of complex natural RNAs. The 5S Loop E is an internal loop, flanked on both sides by duplex segments. Different base pairing schemes had been suggested based upon chemical and enzymatic modification studies. The NMR data showed that the structure was neither of those suggested, but rather that it contained G • A, A • A and U • U mismatches and a folded-back chain on one of the strands. The sarcin/ricin loop has 17 nucleotides that would be indicated as "unpaired" according to the sequence. However like the internal Loop E there is actually extensive pairing, again nonconventional A • A, U • A (reverse Hoogsteen), and A • G types. The backbone is also irregular, forming a looped G, which hydrogen bonds to the opposite backbone. These examples just begin to provide insight into how complicated an RNA structure can be in "single stranded" regions, the genetic data showing that the unusual structures formed are critical for proper activity of these sequences. With labeled RNAs now available, it can be anticipated that structures of larger, even more complicated RNAs may soon be solved.

6.4 Complexes with Other Molecules

6.4.1 Drugs

Complexes of nucleic acids with other molecules, ranging from water and simple ions through intermediate-size organic molecules (drugs) to proteins have been examined. A number of differences have been found in results obtained with NMR, in comparison to other methods. One example of this finding was that distamycin-A can bind to the sequence AAATTT in several different modes (Pelton and Wemmer, 1990), while crystals grown under the same conditions exhibited drug bound in only one site. A surprise from the NMR work was that one of the binding modes had two distamycins side by side within the minor groove of the DNA. Modeling work, and crystallography on DNA alone, have shown that the intrinsic groove width in A,T rich sequences is not large enough to accommodate two drugs, requiring that the binding of the second drug significantly expand the groove. This binding mode had not been seen previously with other techniques, and changes the consideration of forces that stabilize the binding of the drug. It has also lead to modified molecules that bind specifically to G, C rich sequences (Geierstanger et al., 1994). A dimer binding mode was also determined by NMR (Gao and Patel, 1989) for the drug chromomycin, which has been subsequently verified crystallographically. For these drugs, the symmetry in the complexes was evident from the small number of resonances observed. This finding, taken together with the stoichiometry (determined from titrations), force us to the conclusion that dimers of drug must form, even when it is not possible to assign all of the resonances. By carrying out detailed NOE studies of the complex in both of these cases, it has been possible to define the precise binding site of the drug, the interactions with the DNA oligomers, and changes in DNA structure. The most valuable information is generally provided by intermolecular NOEs. At intermediate drug/DNA ratios, it is also possible to study the exchange of the drug between binding sites, Figure 5-19, which can provide information about exchange mechanisms, and may also help in assignment of resonances. A great variety of complexes with other drugs have been similarly examined by NMR. It is often possible to address whether the drug interacts sequence specifically (causing shifts in some resonances but not others), what the stoichiometry of binding is, and qualitatively what the binding mode is (shifts caused by intercalation, NOEs to protons in the major or minor groove, etc.). Covalent adducts of a drug to a DNA oligomer have also been made, and the structure of the resulting modified molecule has been determined by NMR (Norman et al., 1990; Lin et al., 1991).

6.4.2 Proteins

Nuclear magnetic resonance studies have been undertaken for a number of different protein–DNA complexes, primarily of regulatory proteins. An interesting example is the binding of arginine to the TAR RNA sequence (Puglisi et al., 1992). It had been shown that one specific arginine in the Tat protein of human immunodeficiency virus (HIV), which binds to the TAR RNA, was responsible for much of the specificity. By studying the RNA in the presence and absence of arginine, it was shown that a substantial conformational rearrangement occurred in the RNA, making a new base triple, to form a binding pocket for the guanidinium end of the arginine. This structural interpretation also explained the data that were available regarding sequence conservation. The affinity of the arginine alone was low since there are many other basic residues in the real Tat sequence, but the specificity seems to be largely explained.

Since the spectra of even small proteins are quite complex, the presence of both protein and DNA in the complex leads to very complicated spectra. Early studies used just ^{1}H NMR, but in most of the recent cases the protein has been uniformly labeled with ^{13}C and ^{15}N, and spectral editing techniques can be applied to separate the spectra and to identify the intermolecular contacts (Otting and Wüthrich, 1990). For the *Escherechia coli lac* repressor DNA binding domain (the "headpiece"), NMR early data showed that the binding orientation on a half operator DNA is opposite to that which was predicted from models (Boelens et al., 1987). Subsequent studies with labeled protein have allowed identification of many contacts between the protein and DNA (Chuprina et al., 1993). It was found that there are relatively few direct hydrogen bonds from side chains to bases, instead there are extensive apolar contacts and many direct and water mediated hydrogen bonds to the phosphates. These observations fit well with genetic as well as biochemical data. The homeo domain *antennapedia* bound to its cognate DNA has also been extensively investigated (Qian et al., 1989, Billeter et al., 1993). Again full characterization relied on use of isotopes, in particular identification of many of the contacts of the protein to the DNA. This study also reinforces the importance of water in forming the interface between protein and DNA. Studies of GAL4 and GATA-1, both zinc-containing proteins with DNA, have also been undertaken (Omichinski et al., 1993; Baleja et al., 1994). To date, there have been fewer studies of protein–RNA complexes, though recent examples indicate that they will not be long in coming (Howe et al., 1994, Görlach et al., 1992). The further developments of labeling, filtering experiments and the use of magnetic field gradients will continue to lead to more detailed structures of new complexes.

7. CONCLUSIONS

In summary, it is clear that solution NMR provides a powerful tool for the analysis of the structures of nucleic acids. There are limitations both in the size of the molecules that can be studied, and the accuracy to which certain features of the structures can be determined. In spite of this, it seems clear that NMR is presently the most generally applicable method available for understanding the structures of small nucleic acids in solution, their structural interconversions, and their interactions with other small molecules. As more applications take advantage of the use of isotopes, the size and complexity range that can be studied will be significantly increased, paralleling past work in protein systems (Marion et al., 1989). The addition of further heteronuclear coupling constants should also improve structure determinations. Continued application of 3D and 4D experiments will expand the molecular weight range accessible to NMR analysis, making possible studies of many more nucleic acid systems of current biochemical interest.

References

Aboul-ela, F., Varani, G., Walker, G. T., and Tinoco, I., Jr. (1988). The TFIIIA recognition fragment d(GGATGGGAG):d(CTCCCATCC) is B-form in solution, *Nucleic Acids Res.* **16**, 3559–3572.

Altona, C. (1982). Conformation analysis of nucleic acids. Determination of backbone geometry of single-helical RNA and DNA in aqueous solution, *Rec. Trav. Chim. Pays-Bas* **101**, 413–433.

Ashcroft, J., LaPlante, S.R., Borer, P.N., and Cowburn, D. (1989). Sequence Specific ^{13}C NMR Assignment of Non-protonated Carbons in [d(TAGCGCTA]2 Using Proton Detection, *J. Am. Chem. Soc.* **111**, 363–365.

Baleja, J. D., Mau, T., and Wagner, G. (1994). Recognition of DNA by GAL4 in solution: Use of a monomeric protein–DNA complex for study by NMR, *Biochemistry* **33**, 3071–3078.

Batey, R. T., Inada, M., Kujawinski, E., Puglisi, J.D., and Williamson, J.R. (1992). Preparation of isotopically labeled ribonucleotides for multidimensional NMR spectroscopy of RNA, *Nucleic Acids Res.* **20**, 4515–4523.

Bax, A. and Drobny, G. (1985). Optimization of Two-Dimensional Homonuclear Relayed Coherence Transfer NMR Spectroscopy, *J. Magn.Res.* **61**, 306–320.

Bax, A. and Lerner, L. (1988). Measurement of ^{1}H–^{1}H Coupling Constants in DNA Fragments by 2D NMR, *J. Magn. Res.* **79**, 429–438.

Benight, A. S., Schurr, J. M., Flynn, P. F., Reid, B. R., and Wemmer, D. E. (1988). Melting of a Self-complementary DNA Minicircle. Comparison of Optical Melting Theory with Exchange Broadening of the NMR Spectrum, *J. Mol. Biol.* **200**, 377–399.

Billeter, M., Qian, Y-Q., Otting, G., Muller, M., Gehring, W., and Wüthrich, K. (1993). Determination of the NMR solution structure of an Antennapedia homeodomain–DNA complex, *J. Mol. Biol.* **234**, 1084–1093.

Blommers, M. J. J., Walters, J. A. L. I., Haasnoot, C. A. G., Aelen, J. M. A., van der Marel, G. A., van Boom, J. H., and Hilbers, C. W. (1989). Effects of Base Sequence on the Loop Folding in DNA Hairpins, *Biochemistry* **28**, 7491–7498.

Blommers, M. J. J., van de Ven, F. J. M., va der Marel, G. A., van Boom, J. H., and Hilbers, C. W. (1991). Three-dimenstional structure of a DNA hairpin in solution, *Eur. J. Biochem.* **201**, 33–51.

Boelens, R., Scheek, R. M., van Boom, J. H., and Kaptein, R. (1987). Complex of lac Repressor Headpiece with a 14 Base-pair *lac* Operator Fragment Studied by Two–dimensional NMR, *J. Mol. Biol.* **193**, 213–216.

Borgias, B. A. and James, T. L. (1988). COMATOSE, A Method for Constrained Refinement of Macromolecular Structure Based on Two-Dimensional Nuclear Overhauser Effect Spectra, *J. Magn. Reson.* **79**, 493–512.

Borgias, B. A. and James, T. L. (1990). MARDIGRAS–A Procedure for Matrix Analysis of Relaxation for Discerning Geometry of an Aqueous Structure, *J. Magn. Reson.* **87**, 475–87.

Brush, C. K., Stone, M. P., and Harris, T. M. (1988). Selective Reversible Deuteriation of Oligodeoxynucleotides: Simplification of Two-Dimensional Nuclear Overhauser Effect NMR Spectral Assignment of a Non-Self-Complementary Dodecamer Duplex, *Biochemistry* **27**, 115–122.

Celda, B., Widmer, H., Leupin, W., Chazin, W.J., Denny, W.A., and Wüthrich, K. (1989). Conformational Studies of d(AAAAATTTTT)$_2$ Using Constraints from Nuclear Overhauser Effects and from Quantitative Analysis of the Cross-Peak Fine Structure in Two-Dimensional ^{1}H NMR Spectra, *Biochemistry* **28**, 1462–1471.

Chazin, W. J., Wüthrich, K., Hyberts, S., Rance, M., Denny, W. A., and Leupin, W. (1986). ^{1}H NMR Assignments for d-(GCATTAATGC)$_2$ using Experimental Refinements of Established Procedures, *J. Mol. Biol.* **190**, 439–453.

Cheng, J-W., Chou, S-H., and Reid, B. R. (1992a). Base Pairing Geometry in GA Mismatches Depends Entirely on the Neighboring Sequence, *J. Mol. Biol.* **228**, 1037–1041.

Cheng, J-W., Chou, S-H., Salazar, M. and Reid, B.R. (1992b). Solution Structure of [d(GCGTATACGC)$_2$], *J. Mol. Biol.* **228**, 118–137.

Cheong, C., Varani, G., and Tinoco, I., Jr. (1990). Solution Structure of an Unusually Stable RNA Hairpin, 5′ GGAC(UUCG)GUCC, *Nature (London)* **346**, 680–682.

Chou, S-H., Cheng, J-W., and Reid, B. R. (1992). Solution Structure of [d(ATGAGCGAATA)]2, *J. Mol. Biol.* **228**, 138–155.

Chou, S-H., Flynn, P. and Reid, B.R. (1989). Solid-Phase Synthesis and High-Resolution NMR Studies of Two Synthetic Double-Helical RNA Dodecamers: r(CGCGAAUUCGCG) and r(CGCGUAUACGCG), *Biochemistry* **28**, 2422–2435.

Chou, S-H., Zhu, L. M., and Reid, B. R. (1994). The Unusual Structure of the Human Centromere (GGA)2 Motif—Unpaired Guanosine Residues Stacked Between Sheared G•A Pairs, *J. Mol. Biol.* **244**, 259–268.

Chuprina, V. P., Lipanov, A. A., Fedoroff, O. Y., Kim, S-G., Kintanar, A., and Reid, B. R. (1991). Sequence effects on local DNA topology, *Proc. Natl. Acad. Sci.* **USA 88**, 9087–9091.

Chuprina, V. P., Rullmann, J. A., Lamerichs, R. M., van Boom, J. H., Boelens, R., and Kaptein, R. (1993). Structure of the complex of lac repressor headpiece and and 11 base-pair half-operator determined by NMR spectroscopy and restrained molecular dynamics, *J. Mol. Biol.* **234**, 446–462.

de Dios, A. C., Pearson, J. G., and Oldfield, E. (1993). Secondary and tertiary structural effects on protein NMR chemical shifts: an *ab initio* approach, *Science* **260**, 1491–1496.

Emsley, L., Dwyer, T. J., Spielmann, H. P., and Wemmer, D. E. (1993). Determination of DNA Conformational Features from Selective Two-Dimensional NMR Experiments, *J. Am. Chem. Soc.* **115**, 7765–7771.

Ernst, R. R., Bodenhausen, G., and Wokaun, A. (1987). *Principles of Nuclear Magnetic Resonance in One and Two Dimensions*, Oxford University Press, Oxford, UK.

Feigon, J., Wang, A.H.-J., van der Marel, G. A., Van Boom, J. H., and Rich, A. (1984). A one- and two-dimensional NMR study of the B to Z transition of (m^{5}dC-dG)$_3$ in methanolic solution, *Nucleic Acids Res.* **12**, 1243–1263.

Flynn, P. F., Kintanar, A., Reid, B. R., and Drobny, G. (1988). Coherence Transfer in Deoxyribose Sugars Produces by Isotropic Mixing: an Improved Intraresidue Assignment Strategy for the Two-Dimensional NMR Spectra of DNA, *Biochemistry* **27**, 1191–1197.

Frey, M. H., Leupin, W., Sørensen, O. W., Denny, W. A., Ernst, R. R., and Wüthrich, K. (1985) Sequence-Specific Assignment of the Backbone ^{1}H and ^{31}P NMR Lines in a Short DNA Duplex with Homo- and Heteronuclear Correlated Spectroscopy, *Biopolymers* **24**, 2371–2380.

Gao, X. L. and Patel, D. J. (1989). Solution Structure of the Chromomycin–DNA Complex, *Biochemistry* **28**, 751–762.

Gehring, K., Leroy, J. L., and Gueron, M. (1993). A Tetrameric DNA Structure with Protonated Cytosine•Cytosine Base Pairs, *Nature (London)* **363**, 561–565.

Geierstanger, B. H., Mrksich, M., Dervan, P. B., and Wemmer, D. E. (1994). Design of a G•C Specific DNA Minor Groove-Binding Peptide, *Science* **266**, 646–650.

Goldman, M. (1988). *Quantum Description of High-Resolution NMR in Liquids*, Oxford University Press, Oxford, UK.

Görlach, M., Wittekind, M., Beckman, R. A., Mueller, L., and Dreyfuss, G. (1992). Interaction of the RNA-binding domain of the hnRNP C proteins with RNA, *EMBO J.* **11**, 3289–3295.

Griffey, R. H., Poulter, C. D., Bax, A., Hawkins, B. L., Yamaizumi, Z., and Nishimura, S. (1983). Multiple quantum two-dimensional ^{1}H–^{15}N nuclear magnetic resonance spectroscopy: Chemical shift correlation maps for exchangeable imino protons of *E.coli* tRNAfMet in water, *Proc. Natl. Acad. Sci. USA* **80**, 5895–5897.

Griffey, R. H. and Redfield, A. G. (1987). Proton-detected heteronuclear edited and correlated nuclear magnetic resonance and nuclear Overhauser effect in solution, *Q. Rev. Biophys.* **19**, 51–82.

Gronenborn, A. M. and Clore, G. M. (1989). Analysis of the Relative Contributions of the Nuclear Overhauser Interproton Distance Restraints and the Empirical Energy Function in the Calculation of Oligonucleotide Structures Using Restrained Molecular Dynamics, *Biochemistry* **28**, 5978–5984.

Harbison, G. S. (1993). Interference Between J-couplings and Cross-relaxation in Solution NMR Spectroscopy—Consequences for Macromolecular Structure Determination, *J. Am. Chem. Soc.* **115**, 3026–3027.

Hare, D. R. and Reid, B. R. (1986). Three-dimensional Structure of a DNA Hairpin in Solution: Two-dimensional NMR Studies and Distance Geometry Calculations on d(CGCGTTTTCGCG), *Biochemistry* **25**, 5341–5350.

Hare, D., Shapiro, L., and Patel, D. J. (1986). Wobble dG X dT Pairing in Right-Handed DNA: Solution Conformation of the d(CGTGAATTCGCG) Duplex Deduced from Distance Geometry Analysis of Nuclear Overhauser Effect Spectra, *Biochemistry* **25**, 7445–7456.

Hore, P. J. (1983). Solvent Suppression in Fourier Transform Nuclear Magentic Resonance, *J. Magn. Res.* **55**, 283.

Howe, P. W. A., Nagai, K., Neuhaus, D., and Varani, G. (1994). NMR studies of U1 snRNA recognition by the N-terminal RNP comain of the human U1A protein, *EMBO J.* **13**, 3873–3881.

Hues, H. A. and Pardi, A. (1991). Structural Features That Give Rise to the Unusual Stability of RNA Hairpins Containing GNRA Loops, *Science* **253**, 191–194.

Jeener, J., Meier, B. H., Bachmann, P., and Ernst, R. R. (1979). Investigation of Exchange Processes by Two-Dimensional NMR Spectroscopy, *J. Chem. Phys.* **71**, 4546–4553.

Kalnik, M. W., Norman, D. G., Swann, P. F., and Patel, D. J. (1989a). Conformation of Adenosine Bulge-containing deoxytridecanucleotide Duplexes in Solution. Extra Adenosine Stacks into Duplex Independent of Flanking Sequence and Temperature, *J. Biol. Chem.* **264**, 3702–3712.

Kalnik, M. W., Norman, D. G., Zagorski, M. G., Swann, P. F., and Patel, D. J. (1989b). Conformational Transitions in Cytidine Bulge-containing Deoxytridecanucleotide Duplexes: Extra Cytidine Equilibrates Between Looped Out (low temperature) and Stacked (elevated temperature) conformations in Solution, *Biochemistry* **28**, 294–303.

Katahira, M., Sugeta, H., Kyogoku, Y., Fujii, S., Fujisawa, R., and Tomita, K. (1988). One- and two-dimensional NMR studies on the confomration of DNA containing the oligo(dA)oligo(dT) tract, *Nuc. Acids Res.* **16**, 8619–8632.

Kellogg, G. W. and Schweitzer, B. I. (1993). Two- and Three-dimensional ^{31}P-driven RME procedures for complete assignment of backbone resonances in oligodeoxyribonucleotides, *J. Biomol. NMR* **3**, 577–95.

Kim, S-G., Lin, L-J., and Reid, B. R. (1992). Determination of Nucleic Acid Backbone Conformation by ^{1}H NMR, *Biochemistry*. **31**, 3564–3574.

Kim, S-G. and Reid, B. R. (1992). Automated NMR Structure Refinement via NOE Peak Volumes. Application to a Dodecamer DNA Duplex, *J. Magn. Res.* **100**, 382–390.

Kintanar, A., Huang, W.-C., Schindele, D. C., Wemmer, D. E., and Drobny, G. (1989). Dynamics of Bases in Hydrated $[d(CGCGAATTCGCG)]_2$ *Biochemistry* **28**, 282–293.

Kintanar, A., Klevit, R. E., and Reid, B. R. (1987). Two-dimensional NMR investigation of a bent DNA fragment: Assignment of the proton resonances and preliminary structre analysis, *Nucleic Acids Res.* **15**, 5845–5862.

Koning, T. M. G., Boelens, R., van der Marel, G. A., van Boom, J. H., and Kaptein, R. (1991). Structure Determination of a DNA Octamer in Solution by NMR Spectroscopy. Effect of Fast Local Motions, *Biochemistry* **30**, 3787-3797.

LaPlante, S. R., Ashcroft, J., Cowburn, D., Levy, G. C., and Borer, P. N. (1988). ^{13}C-NMR Assignments of the Protonated Carbons of $[d(TAGCGCTA)]_2$ by Two-dimensional Proton Detected Heteronuclear Correlation, *J. Biomol. Struct. Dyn.* **5**, 1089–1099.

LeMaster, D. M., Kay, L. E., Brünger, A. T., and Prestegard, J. H. (1988). Protein Dynamics and Distance Determination by NOE Measurements, *FEBS Lett.* **236**, 71–76.

Leroy, J. L., Charretier, E., Kochoyan, M., and Gueron, M. (1988a). Evidence from Base-pair Kinetics for Two Types of Adenine Tract Structures in Solution: The Relation to DNA Curvature, *Biochemistry* **27**, 8894–8898.

Leroy, J. L., Kochoyan, M., Huynh-Dinh, T., and Gueron, M. (1988b). Characterization of base-pair Opening in Deoxynucleotide Duplexes Using Catalyzed Exchange of the Imino Proton, *J. Mol. Biol.* **200**, 223–238.

Levy, G. C., Ed. (1974). *Topics in Carbon-13 NMR Spectrscopy* Vol. 1, Wiley, New York, pp. 79–149.

Lin, C. H., Beale, J. M., and Hurley, L. H. (1991). Structure of the (+)-CC-1065-DNA Adduct: Critical Role of Ordered Water Molecules and Implications for Involvement of Phosphate Catalysis in the Covalent Reaction, *Biochemistry* **30**, 3507–3602.

Macura, S. and Ernst, R. R. (1980). Elucidation of Cross Relaxation in Liquids by Two-Dimensional NMR Spectroscopy, *Mol. Physics* **41**, 95–117.

Marion, D. and Lancelot, G. (1984). Sequential Assignment of the ^{1}H and ^{31}P Resonances of the Double Stranded Deoxynucleotide d(ATGCAT)$_2$ by 2D-NMR Correlation Spectroscopy, *Biochem. Biophys. Res. Comm.* **124**, 774–783.

Marion, D., Driscoll, P. C., Kay, L. E., Wingfield, P. T., Bax, A., Gronenborn, A. M., and Clore, G. M. (1989). Overcoming the Overlap Problem in the Assignment of ^{1}H NMR Spectra of Larger Proteins by Use of Three-Dimensional Heteronuclear ^{1}H-^{15}N Hartmann–Hahn–Multiple Quantum Coherence and Nuclear Overhauser–Multiple Quantum Coherence Spectroscopy: Application to Interleukin 1β, *Biochemistry* **28**, 6150–6156.

Martin, M. L., Martin, G. J., and Delpuech, J.-J. (1980). *Practical NMR Spectroscopy*, Heyden and Son, London, UK.

McCall, M., Brown, T., Hunter, W. N., and Kennard, O. (1986). The crystal structure of d(GGATGGGAG) forms an essential part of the binding site for transcription factor IIIA, *Nature (London)* **322**, 661–664.

Metzler, W. J., Wang, C., Kitchen, D. B., Levy, R. M., and Pardi, A. (1990). Determining Local Conformational Variations in DNA, *J. Mol. Biol.* **214**, 711–736.

Morden, K. M., Chu, Martin and Tinoco, I., Jr. (1983). Unpaired Cytosine in the Deoxyoligonucleotide dCA$_3$CA$_3$ G•dCT$_6$G Is Outside the Helix, *Biochemistry* **22**, 5557–5563.

Mueller, L. b. (1987), P. E. COSY, a Simple Alternative to E.COSY, *J. Magn. Res.* **72**, 191–196.

Nikonowicz, E. P., Meadows, R. P., Fagan, P., and Gorenstein, D. G. (1991). NMR Structural Refinement of a Tandem G•A Mismatched Decamer d(CCAAGATTGG)2 via the Hybrid Matrix Procedure, *Biochemistry* **30**, 1323–1334.

Nikonowicz, E. P. and Pardi, A. (1993). An Efficient Procedure for Assignment of the Proton, Carbon and Nitrogen Resonances in ^{13}C/^{15}N Labeled Nucleic Acids, *J. Mol. Bio.* **232**, 1141–1156.

Nilges, M., Clore, G. M., and Gronenborn, A. M. (1988). Determination of Three-dimensional Structures of Proteins from Interproton Distance Data by Dynamical Simulated Annealing from a Random Array of Atoms. Circumventing Problems Associated With Folding, *FEBS. Lett.* **238**, 289–294.

Nilges, M., Clore, G. M., Gronenborn, A. M., Piel, N., and McLaughlin, L. W. (1987). Refinement of the Solution Structure of the DNA Decamer 5′d(CTGGATCCAG)$_2$: Combined Use of NMR and Restrained Molecular Dynamics, *Biochemistry* **26**, 3718–3733.

Norman, D., Abuaf, P., Hingerty, B. E., Live, D., Grunberger, D., Broyde, S., and Patel, D. J. (1989). NMR and Computational Characterization of the N-(deoxyguanosin-8-yl)aminofluorene adduct [(AF)G] Opposite Adenosine in DNA: (AF)G[syn].A[anti] Pair Formation and its pH Dependence, *Biochemistry* **28**, 7462–7476.

Norman, D., Live, D., Sastry, M., Lipman, R., Hingerty, B. E., Tomasz, M., Broyde, S., and Patel, D. J. (1990). NMR and Computational Characterization of Mitomycin Cross-Linked to Adjacent Deoxyguanosines in the Minor Groove of the d(TACGTA):d(TACGTA) Duplex, *Biochemistry* **29**, 2861–2875.

Olejniczak, E. T., Gampe, R. T., Jr. and Fesik, S. W. (1986). Accounting for Spin Diffusion in the Analysis of 2D NOE Data, *J. Magn. Res.* **67**, 28–41.

Omichinski, J. G., Clore, G. M., Schaad, O., Felsenfeld, G., Trainor, C., Appella, E., Stahl, S. J., and Gronenborn, A. M. (1993). NMR Structure of a specific DNA complex of Zn-containing DNA binding domain of GATA-1, *Science* **261**, 438–446.

Otting, G. and Wüthrich K. (1990). Heteronuclear Filters In 2-dimensional [H-1, H-1] NMR Spectroscopy—Combined Use With Isotope Labelling For Studies Of Macromolecular Conformation And Intermolecular Interactions, *Q. Rev. Biophys.* **23**, N1:39–96.

Pardi, A., Hare, D. R., and Wang, C. (1988). Determination of DNA structure by NMR and distance geometry techniques: A computer simulation, *Proc. Natl. Acad. Sci. USA* **85**, 8785–8789.

Pelton, J. G. and Wemmer, D. E. (1990). Binding Modes of Distamycin A with d(CGCAAATTTGCG)$_2$ Determined by Two-Dimensional NMR, *J. Am. Chem. Soc.* **112**, 1393–1399.

Perkins, S. J. (1982). Application of Ring Current Calcualtions to the Proton NMR of Proteins and Transfer RNA *Biological Magnetic Resonance*, Vol.4, Berliner, L. J. and Reuben, J., Eds., Plenum, New York, pp. 79–144.

Puglisi, J. D., Tan, R., Calnan, B. J., Frankel, A. D., and Williamson, J. R. (1992). Conformation of the TAR RNA–Arginine Complex by NMR Spectroscopy, *Science* **257**, 76–80.

Puglisi, J. D., Wyatt, J. R., and Tinoco, I., Jr. (1990). Conformation of an RNA Pseudoknot, *J. Mol. Biol.* **214**, 437–453.

Qian, Y. Q., Billeter, M., Otting, G., Müller, M., Gehring, W. J., and Wüthrich, K. (1989). The Structure of the Antennapedia Homeodomain Determined by NMR Spectroscopy in Solution: Comparison with Prokaryotic Repressors, *Cell* **59**, 573–580.

Radhakrishnan, I., Patel, D. J., and Gao, X. (1992). 3D Homonuclear NOESY–TOCSY of an Intramolecular Pyrimidine•Purine Pyriminde DNA Triplex Containing a Central G•TA Triple: Nonexchangeable Proton Assignments and Structural Implications, *Biochemistry* **31**, 2514–2523.

Reid, B. R., Banks, K., Flynn, P., and Nerdal, W. (1989). NMR Distance Measurements in DNA Duplexes: Sugars and Bases Have the Same Correlation Times, *Biochemistry* **28**, 10001–10007.

Rhee, Y. S., Wang, C., Gaffney, B. L., and Jones, R. A. (1993). ^{15}N-Labeled Oligo-deoxynucleotides. 6. Use of ^{15}N NMR to Probe Binding of Netropsin and Distamycin to (d[CGCGAATTCGCG]$_2$), *J. Am. Chem. Soc.* **115**, 8742–8746.

Rhodes, D. and Klug, A. (1986). An underlying repeat in some transcription control sequences corresponding to half a double helical trun of DNA, *Cell* **46**, 123–132.

Rinkel, L. J. and Altona, C. (1987). Conformational Analysis of the Deoxyribofuranose Ring in DNA by Means of Sums of Proton–proton Coupling Constants: A Graphical Method, *J. Biomol. Struct. Dyn.* **4**, 621–649.

Robinson, H. and Wang, A. H-J. (1992). A Simple Spectral-Driven Procedure for the Refinement of DNA Structures by NMR Spectroscopy, *Biochemistry* **31**, 3524–3533.

Saenger, W. (1984). *Principles of Nucleic Acid Structure*, Springer-Verlag, New York.

Santos, R. A., Tang, P., and Harbison, G. S. (1989). Determination of the DNA Sugar Pucker Using ^{13}C NMR Spectroscopy, *Biochemistry* **28**, 9372–9378.

Scheek, R. M., Russo, N., Boelens, R., and Kaptein, R. (1983). Sequential Resonance Assignments in DNA ^{1}H NMR Spectra by Two-Dimensional NOE Spectroscopy, *J. Am. Chem. Soc.* **105**, 2914–2916.

Schmieder, P., Ipple, J. H., van den Elst, H, van der Marel, G. A., van Boom, J. H., Altona, C., and Kessler, H. (1992). Heteronuclear NMR of DNA with the neteronucleus in natural abundance: facilitated assignment and extraction of coupling constants, *Nucleic Acids Res.* **20**, 4747–4751.

Schroeder, S. A., Fu, J. M., Jones, C. R., and Gorenstein, D. G. (1987). Assignment of Phosphorus-31 and Nonexchangeable Proton Resonances in a Symmetrical 14 Base Pair lac Pseudooperator DNA Fragment, *Biochemistry* **26**, 3812–3821.

Szewczak, A. A., Moore, P. B., Chan, Y-L., and Wool, I. G. (1993). The conformation of the sarcin/ricin loop from 28S ribosomal RNA, *Proc. Natl. Acad. Sci. USA* **90**, 9581–9585.

Tuerk, C., Gauss, P., Thermes, C., Groebe, D. R., Gayle, M., Guild, N., Stromo, G., d'Aubenton-Carafa, Y., Uhlenbeck, O. C., and Tinoco, I., Jr. (1988). CUUCGG Hairpins: Extraordinarily Stable RNA Secondary Structures Associated with Various Biochemical Processes, *Proc. Natl. Acad. Sci. USA* **85**, 1364–1368.

van Belkum, A., Wiersema, P. J., Joordens, J., Pleij, C., Hilbers, C. W., and Bosch, L. (1989). Biochemical and Biophysical Analysis of Pseudoknot-containing RNA Fragments. Melting Studies and NMR Spectroscopy, *Eur. J. Biochem.* **183**, 591–601.

van de Ven, F. J. M. and Hilbers, C. W. (1988). Nucleic acids and nuclear magnetic resonance, *Eur. J. Biochem.* **178**, 1–38.

Varani, G. and Tinoco, I. Jr. (1991). Carbon Assignments and Heteronuclear Coupling Constants for an RNA Oligonucleotide from Natural Abundance ^{13}C–^{1}H Correlated Experiments, *J. Am. Chem. Soc.* **113**, 9349–9354.

Wang, A. C., Kennedy, M. A., Reid, B. R., and Drobny, G. (1994). A Solid-state ^{2}H Investigation of Purine Motion in a 12 Base Bair RNA Duplex, *J. Magn. Res.* **105B**, 1–10.

Wang, A. C., Kim, S-G., Flynn, P. F., Chou, S-H., Orban, J., and Reid, B. R. (1992b). Errors in RNA NOESY Distance Measurements in Chimeric and Hybrid Duplexes: Differences in RNA and DNA Proton Relaxation, *Biochemistry* **31**, 3940–3946.

Wang, A. C., Kim, S-G., Flynn, P. F., Sletten, E., and Reid, B. R. (1992a). Considerations in the Application of Orientation-Dependent Analysis of NOE Intensities to DNA Oligonucleotides *J. Magn. Res.* **100**, 358–366.

Weisz, K., Shafer, R. H., Egan, W., and James, T. L. (1994). Solution Structure of the Octamer Motif in Immunoglobulin Genes via Restrained Molecular Dynamics Calculations, *Biochemistry* **33**, 354–366.

Wemmer, D., (1989). *Concepts in Magnetic Resonance* **1**, 59–72.

Wemmer, D. (1992). NMR Studies of Nucleic Acids and their Complexes, in *Biological Magnetic Resonance*, Vol. 10, Berliner, L. J. and Reuben, J., Plenum, New York.

Widmer, H. and Wüthrich, K. (1987). Simulated Two-Dimensional NMR Cross-Peak Fine Strcutures for ^{1}H Spin Systems in Polypeptides and Polydeoxynucleotides, *J. Magn. Res.* **74**, 316–336.

Wimberly, B., Varani, G., and Tinoco, I., Jr. (1993). The Conformation of Loop E of Eukaryotic 5S Ribosomal RNA, *Biochemistry* **32**, 1078–1087.

Wolk, S. K., Hardin, C. C., Germann, M. W., van de Sande, J. H., and Tinoco, I., Jr. (1988). Comparison of the B- and Z-Form Hairpin Loop Structures Formed by $d(CG)_5T_4(CG)_5$, *Biochemistry* **27**, 6960–6967.

Wüthrich, K. (1986). NMR of Proteins and Nucleic Acids, Wiley, New York.

Zhou, N., Germann, M.W ., van de Sande, J. H., Pattabiraman, N., and Vogel, H. J. (1993). Solution Structure of the parallel-stranded hairpin d(T8 link C4A8) as determined by Two-dimensional NMR, *Biochemistry* **32**, 646–656.

Zimmer, D. P. and Crothers, D. M. (1995). NMR of enzymatically synthesized uniformly ^{13}C ^{15}N-labeled DNA oligonucleotides. *Proc. Natl. Acad. Sci. USA* **92**, 3091–3095.

CHAPTER 6

Electronic and Vibrational Spectroscopy

Ultraviolet (UV) absorption and circular dichroism (CD) spectroscopies that probe electronic properties of the bases are very useful and general tools for characterizing nucleic acids, although they do not provide the atomic level detail of crystallography or NMR. Absorbance versus temperature curves (melting curves) can reveal GC content in double-stranded DNA, or provide an estimate of the amount of double-stranded regions in RNA. Circular dichroism gives even more information about nucleic acid helical structures. Optical anisotropy measurements on aligned samples give information on base alignment and backbone curvature. Vibrational spectroscopies—infrared (IR) and Raman—can give detailed information on backbone, sugar, and base orientations and interactions.

We measure the spectral properties of nucleic acids to learn about conformations, how they depend on their environment, and how fast they can change. We thus learn about structure, thermodynamics, and kinetics. The interpretation and application of optical and other spectroscopic techniques to biological macromolecules are discussed in great detail in the three volume text *Biophysical Chemistry* by Cantor and Schimmel

(1980). Chapters 7 and 8 of Part II, *Techniques for the Study of Biological Structure and Function*, are especially appropriate for this chapter. A useful review of a wide range of spectroscopic methods applicable to biological macromolecules is *Biochemical Spectroscopy, Methods Enzymol.*, **246** (1995) edited by K. Sauer.

1. ELECTRONIC ABSORPTION SPECTRA IN THE ULTRAVIOLET

1.1 Electronic Absorption Fundamentals

Three concepts are key to the use of absorption spectroscopy in nucleic acid studies: extinction coefficient, wavelength of maximum absorption, and hypochromicity. In addition, circular dichroism is closely related to electronic absorption and gives even more detail about nucleic acid conformations.

As light passes through an absorbing solution, its intensity decreases exponentially as described by the Beer–Lambert law,

$$I = I_0\, 10^{-\varepsilon c \ell} \tag{6-1}$$

where I_0 is the incident light intensity, I is the transmitted light intensity, ℓ is the path length (cm), c is the molar concentration, and ε is the molar extinction coefficient, $M^{-1}\text{cm}^{-1}$. The I/I_0 ratio is the transmittance of the solution and $A = \log(I_0/I)$ is the absorbance, sometimes called optical density (OD). Combining these definitions gives

$$A = \varepsilon \ell c \tag{6-2}$$

The extinction coefficient is a function of the wavelength of the exciting light. If the ground-state O has energy E_O and the excited state A has energy E_A, then according to Planck's law

$$E_\text{A} - E_\text{O} = h\nu_\text{OA} = hc/\lambda_\text{OA} \tag{6-3}$$

where h is Planck's constant (6.626×10^{-34} J s), c is the speed of light (2.997×10^{8} m s^{-1}), and ν_OA and λ_OA are the frequency and wavelength of the light that excite the transition. Because of vibrational substructure, optical absorption bands are broad; ν_OA and λ_OA generally refer to the frequency and wavelength of the absorption maximum. In electronic spectroscopy, it is more common to use wavelength than frequency to describe the transition.

A major application of absorbance is to determine concentrations of nucleic acids. The bases have absorption maxima near 260 nm, with extinction coefficients about 10,000 $M^{-1}\text{cm}^{-1}$, so concentrations of 10^{-4} molar nucleotides give an absorbance (A_{260}) of about 1 in a 1-cm pathlength cell. Use of concentrations determined by absorbance is so common that the amount of nucleic acid is often specified in OD units. One OD unit is the amount of nucleic acid in 1 mL of a solution that has an absorbance at 260 nm of 1.0 in a 1-cm cell. For a typical double-stranded nucleic acid with an extinction coefficient per nucleotide of 7000 $M^{-1}\text{cm}^{-1}$, 1 OD unit is approximately 0.15 μmol of nucleotides, which is approximately 50 μg of material.

The absorbance of a polynucleotide depends on the sum of the absorbances of the nucleotides plus the effect of the interactions among the nucleotides. The interactions cause a single strand to absorb less than the sum of its nucleotides, and a double strand to absorb less than its two component single strands. The effect is called hypochromism or hypochromicity; it depends on the stacking of the bases (Tinoco, 1960). The decrease in absorption of a double strand ranges up to 40% relative to the nucleotides and 25% relative to the single strands. The converse term, *hyperchromicity*, refers to the *increase* in absorption when a double-stranded nucleic acid is dissociated into single strands or nucleotides.

The percent hypochromicity, h_o, is defined as

$$\%h_o = \left[1 - \frac{\varepsilon(\text{polynucleotide})}{\varepsilon(\text{reference})}\right] \times 100 \tag{6-4}$$

The extinction coefficient of the polynucleotide at a chosen wavelength is divided by that of a reference compound such as the sum of the nucleotides, or the denatured single strands at a high temperature.

Circular dichroism is the difference in light absorption for incident left and right circularly polarized light. Only molecules that are not superimposable on their mirror images—chiral molecules—can have CD. Thus, the nucleic acid bases are not circularly dichroic; but nucleosides or nucleotides, with their chiral sugar groups, are. Molar CD is

$$\varepsilon_L - \varepsilon_R = (A_L - A_R)/\ell c \tag{6-5}$$

where A, ℓ, and c are defined as in absorption. The subscripts L and R refer to left and right circularly polarized light. We see that there can only be CD in a wavelength range where there is absorbance, and that CD can be positive or negative. Mirror-image molecules, such as D and L nucleotides, have the same CD spectrum, but with opposite signs. Circular dichroism can be measured with the same amount of material and in the same concentrations as absorbance, but it can provide more information than absorbance can about the conformations of nucleic acids, especially the stacking of the bases.

1.2 Electronic Transitions and Excited States

In order to make more than empirical correlations (e.g., higher absorbance and lower CD correlates with single strands; lower absorbance and higher CD correlates with double strands), we need to understand quantitatively how the optical properties relate to structure. Since electronic transitions take place between ground and excited states, we must understand the properties of both. The Schrödinger equation relates the nuclear positions and electronic distributions to the optical properties, but ab initio calculations of excited-state properties are even more impractical than for ground state properties. A practical approach is first to use the measured optical properties of the nucleotides to learn about their excited electronic states, then to use the interactions among the nucleotides to calculate the polynucleotide optical properties (Tinoco and Williams, 1984). Optical properties that depend on the excited electronic states include UV absorption spectra, CD and optical rotatory dispersion, and fluorescence and phosphorescence.

Separation of the absorption and CD spectra into bands allows assignment of the bands to electronic transitions. The shape of each electronic band depends on the vibrational substructure of each electronic level; two closely overlapping electronic bands may be difficult to separate experimentally. After a transition. the electronic state of the molecule has changed from the ground to an excited state. It remains in the excited state for a few picoseconds (ps) or nanoseconds (ns), then returns to the ground state. For a transition from ground-state O to excited-state A the excited state has a higher energy (calculated from Eq. 6-3) and a different electronic distribution. This different arrangement of electrons changes (for a short time) the net charge on each atom, and all the properties that depend on it—such as the pK_a value, the intermolecular interactions, and the chemical reactivity.

1.2.1 Electric Dipole Transition Moment

The important parameter for the optical properties is not the charge distribution in either the ground or excited state, but a combination of the two called the electric dipole transition moment, μ_{OA}. It represents the movement of charge density in going from the ground to the excited state. The larger this movement, the stronger the absorption (just as a radio signal is stronger, the larger the amplitude of electron motion in the antenna). The relation between μ_{OA} and absorption intensity can be obtained by integrating the absorption band corresponding to the electronic transition. For a transition from O to A, the dipole strength D_{OA}, which is equal to the square of the electric dipole transition moment, is

$$D_{OA} = \mu_{OA}^2 = \frac{6909hc}{8\pi^3 N_A} \int \frac{\varepsilon(\lambda) d\lambda}{\lambda} \tag{6-6}$$

The electric dipole transition moment has units of charge times distance; the centimeter-gram-second (cgs) unit for the charge is electrostatic units (esu) and for the distance it is centimeters. As the charge on an electron is 4.8030×10^{-10} esu and its displacement on excitation is on the order of 1 Å (10^{-8} cm), the transition moment magnitude is of the order 10^{-18} esu cm. One debye is defined as 10^{-18} esu cm, so that transition moment magnitudes are often given in units of debyes. For a molar extinction coefficient ε (units of M^{-1} cm^{-1}) and the dipole strength D_{OA} [in (esu cm)^2], the constant in front of the integral sign in Eq. 6-6 becomes 9.185×10^{-39}.

The electric dipole transition moment is a vector (it has both magnitude and direction). Its magnitude characterizes for each transition the total amount of light absorption in the band integrated; its direction characterizes the polarization of light that will produce maximum absorption. To measure the direction of an electric dipole transition moment, it is necessary to orient the molecules (e.g., in a crystal or in a stretched film) and to measure the absorption with linearly polarized light. The plane of polarization of the incident light is oriented along different directions of the molecule to obtain the three components of the electric dipole transition vector. The direction of polarization of fluorescence relative to the orientation of incident plane-polarized light provides information about the relative directions of absorption and fluorescence transition directions. The electric dipole transition vectors are crucial in calculating the optical properties of polynucleotides from nucleotides.

1.2.2 Transitions in the Nucleotide Bases

The absorption and CD spectra (Sprecher and Johnson, 1977) for the two purine deoxyribo-5′-nucleotides are given in Figure 6-1; the three pyrimidine deoxyribo-5′-

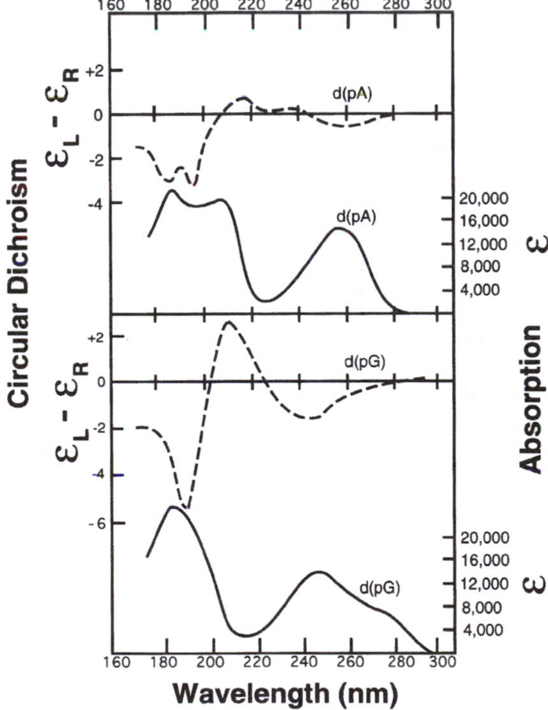

Figure 6-1
Circular dichroism (_ _ _ _) and absorption (—) spectra of purine deoxyribonucleotides; molar extinction coefficients in units of M^{-1} cm^{-1}, ε and $\varepsilon_L - \varepsilon_R$, are plotted versus wavelength. The upper curves are for 5′-deoxyriboadenylic acid; the lower curves are for 5′-deoxyguanylic acid. [Data are from Sprecher and Johnson, 1977.]

nucleotides are shown in Figure 6-2. Absorption and CD begin at about 300 nm in the ultraviolet and continue to lower wavelengths. The spectra shown stop at about 180 nm mainly because of increasing light absorption by water. The absorption and CD spectra for the ribonucleotides are very similar to those shown for the deoxyribonucleotides.

The transitions leading to the absorption and CD spectra shown in Figures 6-1 and 6-2 are transitions occurring in the bases of the nucleotides. The sugars and phosphates contribute to the spectra mainly below 190 nm. Electronic transitions in the bases are classified as being in the plane or perpendicular to the plane of the base; no transition is possible at any other angle in a planar molecule because of symmetry restrictions. The in-plane transitions are called π–π^* because these transitions involve movement of the π electrons in the plane of the base. (Of course, all the electrons are affected, but the π electrons are dominant.) The out-of-plane transitions are called n–π^* because nonbonding electrons on nitrogen or oxygen of a base are excited into the π electron system. The n–π^* transitions are usually much weaker in absorption than the π–π^* transitions and are easily hidden under the stronger transitions; however, sometimes they can be found in the CD spectra.

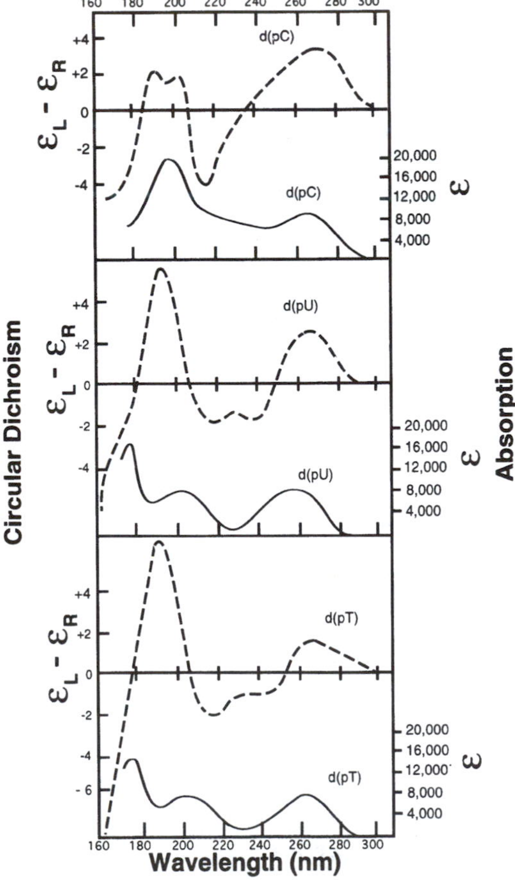

Figure 6-2
Circular dichroism and absorption spectra of pyrimidine deoxyribonucleotides plotted as in Figure 6-1. The curves are from top to bottom: 5′-deoxyribocytidylic acid, 5′-deoxyribouridylic acid, and thymidylic acid. [Data are from Sprecher and Johnson, 1977.]

The directions of the π–π^* transition dipoles obtained for the bases are given in Figure 6-3 (Chou and Johnson, 1993). The directions were obtained from crystal spectra, spectra of molecules oriented in stretched films, and polarization of fluorescence. Less than one-half of the bands have been assigned. Quantum mechanical calculations have been done to try to reproduce the measured values for the transition dipoles and to predict the others (Gueron et al., 1974), but much more definitive results are needed before these predicted results can be used with confidence to calculate spectra of nucleic acids as a function of conformation (Tinoco and Williams, 1984).

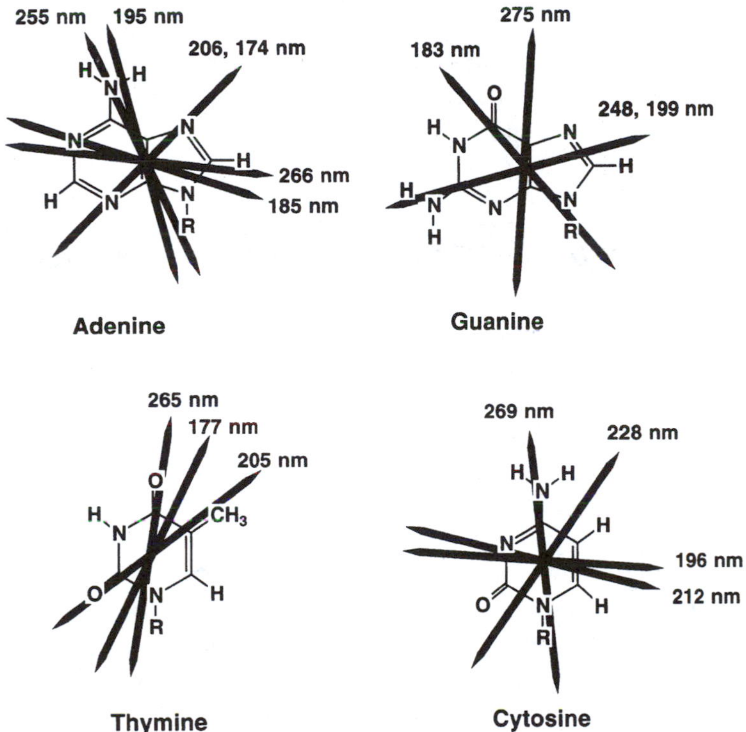

Figure 6-3
Directions of electric dipole transition moments for electronic transitions in nucleic acid bases. Directions for uracil are similar to those for thymine. The directions and magnitudes are used to interpret the optical properties of polynucleotides. [Data are from Chou and Johnson, 1993.]

1.2.3 Ionization Constants of Excited States

Essentially all properties of an excited state are different from those of the ground state, but one property is particularly easy to measure—the pK_a^* of the excited state. When a base ionizes, the absorption spectrum of the ionized form is different from the un-ionized form. Thus, the pH dependence of the absorption can provide the ratio of the two forms and the pK_a for the ground state. An identical experiment using the fluorescence (or phosphorescence) spectra gives the pK_a^* of the excited state. The fluorescence is emitted by the excited state and the ionized and un-ionized forms of the excited states will have different fluorescence spectra; therefore, the pH dependence of fluorescence gives pK_a^* (Weller, 1961). The pK_a shifts are large. For 5′-thymidylic acid pK_a^*- pK_a is +2 for the first excited singlet state (fluorescence) and +5.7 for the first excited triplet (phosphorescence). For 5′-adenylic acid pK_a^*- pK_a is +3.3 for the first excited singlet state and −5.3 for the first excited triplet state (Gueron et al., 1974). These large shifts in acidity of the ionizable groups on electronic excitation illustrate the large changes in chemical reactivity that can occur for the excited bases; obviously photochemical reactions can be very different from ground-state reactions.

1.3 Effects of Base–Base Interaction on Polynucleotide Extinction Coefficients

The absorption of a nucleic acid depends on the sum of the absorption of each nucleotide plus an effect due to their interactions. Thus the extinction coefficient at any wavelength of a nucleic acid depends on its sequence and conformation. The conformational dependence of the interactions means that absorption measurements can be used to learn whether a nucleic acid is single or double stranded, and to follow structural changes in each. The interactions among stacked bases cause a decrease in absorption—a hypochromicity. Therefore double strands absorb less than single strands, which absorb less than the component nucleotides.

The explanation for the change of absorption of interacting molecules is qualitatively simple. When a molecule absorbs light there is a transitory motion of electrons, which is characterized by an electric dipole transition moment. The magnitudes and directions of these transition dipoles have been determined for the nucleic acid bases as shown in Figure 6-3. For two interacting molecules, the transition dipole of each is influenced by the neighboring molecule; a classical polarizability mechanism is illustrated in Figure 6-4. A transition dipole induces a dipole in the neighboring molecule. The positive end of the dipole (the head) will induce a negative charge near it, and the negative end (the tail) will induce a positive charge. Thus in stacked bases the induced dipole is opposite to the transition dipole, no matter what the direction of the transition dipole is—as long as it is in the plane of the base. This means that the effective transition dipoles of all in-plane transitions (such as π–π^* transitions) in stacked bases are less than those of noninteracting bases. The sum of each transition dipole plus the dipoles induced in neighboring bases is less than the transition dipole of the isolated base. Since the extinction coefficient is proportional to the square of the magnitude of the transition dipole (Eq. 6-6), the absorbance is less. Thus, there is hypochromicity in stacked bases relative to the unstacked bases. For paired bases, the measured directions of transition dipoles also lead to hypochromicity as shown in the figure for the 280-nm transition of guanine in G•C. There are other contributions to the interaction between transitions in molecules, but the simple picture in Figure 6-4 provides a useful understanding.

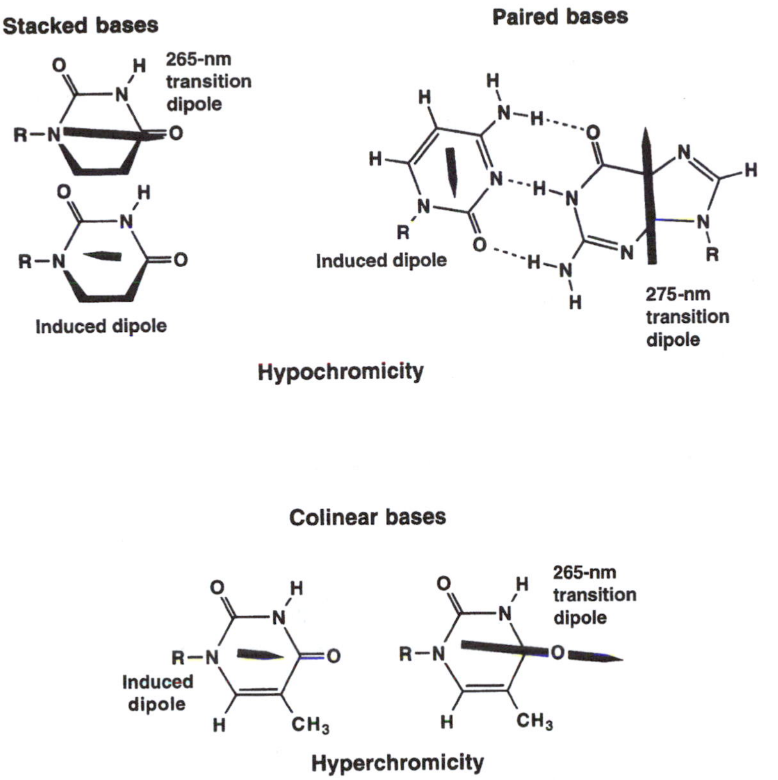

Figure 6-4
Orientations of bases that lead to hypochromicity (transition dipole induces an opposite sign dipole in neigboring base) and hyperchromicity (transition dipole induces same sign dipole in neigboring base). All transitions in the plane of the bases in stacked bases are hypochromic. In coplanar bases (either paired or colinear), hypochromicity or hyperchromicity depends on the direction of the electric dipole transition moment of the absorption band.

Other transition directions could lead to hyperchromicity—an increase in absorption. For example, a transition in guanine along the N–H bond of the imino group would induce the same sign dipole in the paired cytosine, and cause a hyperchromicity. This hyperchromicity does not occur for the known transition directions and for the geometry of Watson–Crick base pairs. However, other base–base orientations or other transition directions in modified bases could produce hyperchromicity. An example is illustrated in Figure 6-4 for two uracils oriented on a line. Hyperchromicity would also occur for stacked bases with transitions perpendicular to the plane of the bases (such as n–π^* transitions). Whether such transitions are significant in nucleic acid bases is not known. There is hyperchromicity at the extreme long wavelength end of the absorption bands of a nucleic acid, but the explanation for this is not clear. For nearly all the UV absorption spectra, nucleic acids are hypochromic relative to their nucleotides and to their less-ordered states. There are theoretical equations to calculate hypochromicity or hyperchromicity (Cantor and Schimmel, 1980; DeVoe, 1964); here we stress the more empirical and practical results.

The interactions between the transition dipoles are Coulombic. As there is no net change in charge for a transition, the interactions can be approximated by dipole–dipole interactions; they are of order r^{-3} with r the distance between bases. This r independence allows a nearest neighbor approximation to be useful, which means that we need to consider only interactions of a transition with the nucleotides on either side in a single strand, or the paired base and the neighboring base pairs in a double strand. The nearest neighbor approximation leads to a simple equation (Cantor and Tinoco, 1965) to estimate the extinction coefficient of a single strand of any sequence. The equation is accurate enough to use routinely:

$$\begin{aligned}\varepsilon(\text{ApCpGpUp}\cdots\text{ApG}) &= 2[\varepsilon(\text{ApC}) + \varepsilon(\text{CpG}) + \varepsilon(\text{GpU}) + \cdots \varepsilon(\text{ApG})] \\ &\quad -[\varepsilon(\text{Cp}) + \varepsilon(\text{Gp}) + \varepsilon(\text{Up}) \cdots + \varepsilon(\text{Ap})] \qquad (6\text{-}7)\end{aligned}$$

$\varepsilon(\text{ApCpGpUp}\cdots\text{ApG})$ is the extinction coefficient per mole of strand. The parameters $\varepsilon(\text{ApC})$, $\varepsilon(\text{CpG})$, and so on. are the extinction coefficients of the component dinucleoside phosphates per mole of nucleotide; that is why they are multiplied by 2. The parameters $\varepsilon(\text{pC})$, $\varepsilon(\text{pG})$, and so on. are the extinction coefficients of the nucleotides. The dinucleoside phosphates count each nucleotide twice, except for the two end ones, which is the reason the extinction coefficients of the nucleotides (except the two end ones) must be subtracted.

The equation assumes nearest neighbor only interactions, and requires that the conformations and optical properties of the dinucleoside phosphates represent the conformations and optical properties present in the single strand. An extinction coefficient for the single strand can be measured by hydrolysis of the single strand to nucleotides. From the measured absorbances before and after hydrolysis, and the known extinction coefficients of the nucleotides, the extinction coefficient of the single strand can be determined without assumptions. Alternatively, an absolute measure of concentration, such as a quantitative analysis of phosphorus, can be used. The calculated extinction coefficient usually agrees with the measured value to better than $\pm 10\%$. Table 6-1 gives values to use in the equation above.

An analogous equation would apply to double strands, but two base-pair fragments are not available for its implementation. Extinction coefficients for double-stranded polynucleotides can be calculated from their nearest neighbor frequencies and from extinction coefficients measured for synthetic polynucleotides of simple repeating sequences (Allen et al., 1984; Gray et al., 1978). However, an extinction coefficient for a naturally occurring double-stranded nucleic acid can be estimated from its A$\cdot$T base composition fraction, f^{AT}, alone. For a random sequence, the extinction coefficient depends on the square of the base composition (Cantor and Schimmel, 1980, p. 1154). It is well known that natural DNA sequences are not random, but except for highly repetitive sequences, this deviation from randomness is not important for the spectra.

Table 6.1
Extinction Coefficients per Nucleotide (M^{-1} cm^{-1} × 10^{-3}) at 260 nm, 25°C, 0.1 Ionic Strength, pH 7 for Nucleotides and Dinucleoside Phosphates to Calculate Extinction Coefficients of Single Strands[a]

	RNA	DNA		RNA	DNA
Ap	15.34	15.34	CpG	9.39	9.39
Cp	7.60	7.60	CpU (CpT)	8.37	7.66
Gp	12.16	12.16	GpA	12.92	12.92
Up (Tp)	10.21	8.70	GpC	9.19	9.19
ApA	13.65	13.65	GpG	11.43	11.43
ApC	10.67	10.67	GpU (GpT)	10.96	10.22
ApG	12.79	12.79	UpA (TpA)	12.52	11.78
ApU (ApT)	12.14	11.42	UpC (TpC)	8.90	8.15
CpA	10.67	10.67	UpG (TpG)	10.40	9.70
CpC	7.52	7.52	UpU (TpT)	10.11	8.61

[a]Table I from Gray et al., 1995. The extinction coefficients are estimated to be accurate to ±0.10 for the monomers and ±4% for the dimers.

The quadratic equation is

$$\varepsilon = \varepsilon_{AA} f_{AT}^2 + 2\varepsilon_{AG} f_{AT}(1 - f_{AT}) + \varepsilon_{GG}(1 - f_{AT})^2 \quad (6\text{-}8)$$

Here ε_{AA} is the contribution to the extinction coefficient of A•T base pairs interacting with other A•T base pairs; ε_{AG} is from the interaction of A•T base pairs with G•C base pairs; and ε_{GG} is the interaction of G•C base pairs with G•C base pairs. These parameters can be obtained from average extinction coefficients of synthetic polynucleotides, such as poly[d(A)•d(T)], poly[d(A-T)•d(A-T)], etc. The conformations of the polynucleotides are assumed to be similar to the conformations of those sequences in the nucleic acids (Allen et al., 1984; Gray et al., 1981). Parameters can also be determined from measured extinction coefficients for natural DNAs of known base composition (Felsenfeld and Hirschman, 1965); the results are similar. The quadratic equation can be rearranged so that only the last term is quadratic.

$$\varepsilon = \varepsilon_{GG} + (\varepsilon_{AA} - \varepsilon_{GG}) f_{AT} + (\varepsilon_{AA} + \varepsilon_{GG} - 2\varepsilon_{AG}) f_{AT}(f_{AT} - 1) \quad (6\text{-}9)$$

Using parameters from synthetic polynucleotides (Allen et al., 1984; Gray et al., 1981), one obtains for double-stranded DNA

$$\varepsilon(260) = 7585 - 1285 f_{AT} + 1685 f_{AT}(f_{AT} - 1) \quad (6\text{-}10)$$

and for double-stranded RNA

$$\varepsilon(260) = 7010 - 530 f_{AU} - 130 f_{AU}(f_{AU} - 1) \qquad (6\text{-}11)$$

Note that for RNA the equation is nearly linear in base composition, and that the extinction coefficients do not vary much with base composition. For 50% A•T, the calculated $\varepsilon(260)$ for DNA is 6521 M^{-1} cm^{-1}; for 50% A•U, the calculated $\varepsilon(260)$ for RNA is 6778 M^{-1} cm^{-1}. Base composition and concentration of a DNA can be obtained more easily from the hyperchromicity on melting; which is discussed in Section 6.1.4.2.

Successful use of these equations requires pure DNA or RNA free from protein. The ratio of absorbance at 260 nm to that at 280 nm is a measure of protein present in the sample, as most proteins contain aromatic amino acids that absorb strongly at 280 nm. For pure DNA, the A(260)/A(280) ratio is about 1.9; it is lower if protein contaminants are present.

1.4 Detection of Conformational Changes by UV Spectroscopy

Electronic absorption measurements have been crucial in studying conformational changes in nucleic acids. This topic is discussed fully in Chapter 8; here we present the spectroscopic fundamentals.

1.4.1 Helix–Helix Transitions

Absorption spectra have been used to detect changes in conformation of double-stranded DNA and RNA; the best example is the right-handed B to left-handed Z transition. When this occurs, the absorption increases above 270 nm and decreases below this wave length. The change in the ratio A(295)/A(260) has been used to follow the transition (Pohl and Jovin, 1972; Hall et al., 1984), although this ratio can also be increased by metal cation binding. A more sensitive criterion is the CD; which will be described in Section 3.2.

1.4.2 Melting Curves

When an oligonucleotide or a polynucleotide is heated in solution, the absorbance increases. (For all aqueous solutions, there is a slight decrease in absorption on heating due to dilution of the solution as the solvent expands. Between 4 and 100°C water expands about 4%; for quantitative work a correction is made for expansion at each temperature.) For single strands, stacking decreases with increasing temperature and there is a gentle rise in absorbance. For double-stranded oligonucleotides there is an S-shaped increase in absorbance as single strands are produced. This increase becomes very abrupt for double-stranded nucleic acids, because the double-strand–single strand equilibrium is essentially a phase change, like the melting of ice. All the absorbance versus temperature curves are called melting curves; the different types are illustrated in Figure 6-5. To analyze the melting curves, we need to know how the extinction coefficient of each species depends on temperature and conformation.

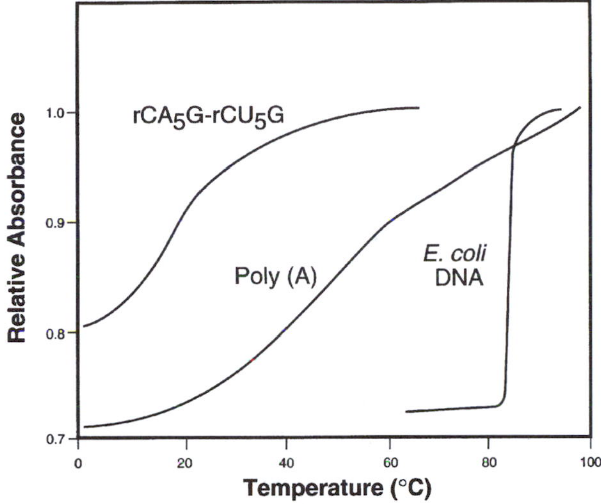

Figure 6-5
Absorbance melting curves for various polynucleotides and oligonucleotides. A double-stranded DNA dissociating to single strands produces a sharp cooperative transition. The unstacking of polyadenylic acid single strands yields a very broad transition. The double-strand to single-strand transition in a heptanucleotide can be fit by a single temperature-dependent equilibrium constant. The absorbance at 260 nm was measured for *Escherichia coli* DNA in 0.1 *M* KCl. [Data from Marmur and Doty, 1962]. The absorbance at 258 nm was measured for polyadenylic acid in 0.1 *M* LiCl. [Data from Leng and Felsenfeld, 1966.] The absorbance at 260 nm was measured for r CA_5G + r CU_5G in 1.0 *M* NaCl, pH 7, 10 m*M* phosphate, 0.1 m*M* ethylendiaminetetracetic acid (EDTA). [Data from the Aboul-ela, 1987.]

The abrupt melting of a nucleic acid is the easiest to analyze. Below the melting temperature, the absorbance of the double strand does not change much; at the transition a large increase in absorbance occurs over a 4°–8° temperature range, and above the melting temperature the absorbance increases slightly with temperature. Derivative melting curves reveal changes in absorbance over wider temperature ranges, as discussed in Section 1.4.3. The melting curve is characterized by the melting temperature t_m, the temperature range over which the transition occurs, and the increase of absorbance for the transition. The melting temperature is defined as the temperature at the midpoint of the absorbance transition. It depends on the base sequence of the DNA and on the solvent, but is independent of DNA concentration. The hyperchromicity, h_r, is defined in terms of absorbances, A, as

$$\%h_r = 100\frac{A(\text{melted species}) - A(\text{double strand})}{A(\text{double strand})} \tag{6-12}$$

For natural DNAs, the melting hyperchromicities at 260 nm are about 40% (Gray et al., 1978). The melting temperature and hyperchromicity for a DNA serve as criteria of purity and nativeness.

The entire absorption spectrum and its change with temperature is used to characterize the DNA. The hyperchromicity on melting DNA depends on wavelength and on base sequence (or base composition). For natural DNAs, equations can be obtained that are similar to those given earlier for the extinction coefficients, except that the parameters represent the change of extinction on melting. The wavelength-dependent parameters have been obtained by melting DNAs of known base composition (Felsenfeld and Hirschman, 1965) . These parameters plus measurements of the change of absorbance at 250, 260, 270, and 280 nm on melting a DNA sample provides the concentration of the DNA and its base composition to ± 0.03 in fraction of A • T (Hirschman and Felsenfeld, 1966). Only two wavelengths are needed to obtain two unknowns, but an over determination gives better precision.

Single strand melting curves have been used to learn about the thermodynamics and cooperativity of stacking in oligonucleotides and polynucleotides. Each base can be designated in a first approximation as being stacked or unstacked. At any temperature, the number of bases stacked will depend on the free energy of stacking. The distribution of stacked bases (in clusters or randomly distributed) depends on the cooperativity of stacking. As the temperature is raised, the bases unstack by an amount dependent on the enthalpy of stacking; the unstacking produces an increase in absorption. To interpret the melting curves, one must take into account the length dependence of the hypochromicity for stacked bases.

The percent hypochromicity of N stacked bases relative to the mononucleotides is:

$$\%h_0 = 100\frac{\varepsilon(\text{monomer}) - \varepsilon(N\text{-mer})}{\varepsilon(\text{monomer})} \tag{6-13}$$

The extinction coefficient of the N-mer is per nucleotide. We expect the hypochromicity to increase with the number of nucleotides in the stack, N, from the value for two stacked bases to the value for a polynucleotide. Applequist and Damle (1966) found a value of 25.3% hypochromicity at 257 nm for polyadenylic acid at 2°C. A dinucleoside phosphate, ApA, has a hypochromicity of only 13.3%; a 10-mer has a hypochromicity of 20%. As each N-mer has $N - 1$ nearest neighbor interactions the nearest neighbor only formula for hypochromicity of an N-mer, h_N, relative to the polymer, h_∞, is:

$$h_N = [(N-1)/N]h_\infty \tag{6-14}$$

Thus, for a hypochromicity model based on nearest-neighbor-only interactions, one would expect the dimer to have one-half the hypochromicity of the polymer (12.6%) and a 10-mer to have nine-tenths the hypochromicity of the polymer (22.8%). The data for polyadenylic acid oligonucleotides approach the polymer hypochromicity more slowly than in the equation above, but this simple equation is a good approximation for estimating hypochromicity from either dimer or polymer data.

Analysis (Applequist and Damle, 1966) of poly- and oligo-adenylic acid melting data in 0.15 M NaCl gives ΔH° of stacking of -9.4 kcal mol^{-1} and ΔS° of stacking of -29.3 cal deg^{-1} mol^{-1}. The cooperativity parameter, σ, which characterizes the probability of starting a new stack compared to continuing an old stack, is 0.6. Because a value of $\sigma = 1$ corresponds to no cooperativity, it is clear that stacking is not very cooperative. A highly cooperative reaction such as formation of an α-helix polypeptide has a σ of order 10^{-4}.

Melting of double-stranded polynucleotides to single strands involves loss of base pairs and a decrease in stacking. A low-melting temperature yields stacked single strands, whereas a high-melting temperature yields mainly unstacked single strands. The contributions of stacking and base pairing to the melting transition are complicated and it is difficult to obtain cooperativity parameters from the shape of the transition. The base composition heterogeneity is the main determinant of the shape of the transition. This shape is best analyzed from derivative curves discussed in Section 1.4.3.

The melting of double-stranded oligonucleotides can be conveniently analyzed using a two-state model. For a duplex in equilibrium with equimolar amounts of single strands, the fraction of strands in the duplex, f_D, is

$$f_D = \frac{A - A_S}{A_D - A_S} \tag{6-15}$$

Here A is the absorbance measured at any temperature, A_D is the absorbance of the duplex, and A_S is the absorbance of the single strands. Both A_S and A_D will depend on temperature, but the absorbance of the single strands will usually vary the most. The equilibrium constant for melting, K, can be calculated as a function of temperature from f_D and the total concentration of nonself-complementary oligonucleotides c_T according to the equation

$$K = \frac{[S]^2}{[D]} = \frac{(1 - f_D)^2}{2f_D} c_T \tag{6-16a}$$

For self-complementary oligonucleotides the equation is

$$K = \frac{[S]^2}{[D]} = \frac{2(1 - f_D)^2}{f_D} c_T \tag{6-16b}$$

From $K(T)$ and its temperature dependence, the thermodynamic parameters of the transition can be obtained by using the standard equations $\Delta G^\circ = -RT \ln K$, $\partial(\ln K)/\partial(1/T) = \Delta H^\circ/R$, and $\Delta G^\circ = \Delta H^\circ - T\Delta S^\circ$. The graph of $\ln K$ versus 1/T is a van't Hoff plot.

A two-state or, as it is also called, an all-or-none model may be reasonable for the melting of oligonucleotides of less than about 12–14 bp; for longer duplexes, intermediate species with melted ends become important (Werntges et al., 1986). If melting of a duplex involves equilibria with hairpins and single strands, then obviously a two-state model is not appropriate (Marky et al., 1983). Thermodynamic parameters can be obtained from the concentration dependence of melting temperatures, or from the temperature dependence of absorbance at constant oligonucleotide concentration. Alternatively, one can obtain the change in absorbance for a small change in temperature, $\Delta A/\Delta T$, from a temperature-jump experiment (Gralla and Crothers, 1973). The temperature-jump method has the advantage that the time dependence of ΔA can be seen. Thus changes in absorbance due to the double-strand to single-strand equilibrium can be distinguished from the much faster absorbance changes due to single-strand unstacking. A detailed description of oligonucleotide melting data and the concentration dependence of duplex and hairpin melting is given in Chapter 8.

The stoichiometry of complex formation between two different oligonucleotides, or polynucleotides, or indeed any two substances, can be determined from absorbance measurements. All that is required is that the complex have a different extinction coefficient from its components. Two solutions with equal concentrations in moles of nucleotides of the two components are prepared, then 10 or more mixtures are made containing a constant total concentration of nucleotides, but with different ratios of the components. The absorbance is measured after equilibrium is attained—this may take hours, particularly for guanine-containing species. The absorbance is plotted versus the mole fraction of one of the components. A minimum or maximum in the plot (called a Job plot) corresponds to the mole fraction of component in the stoichiometric complex. For stable complexes, where the reaction proceeds essentially to completion, the minimum or maximum is very sharp. For less stable complexes, the minimum or maximum is rounded, but still shows a distinct extremum. If more than one type of complex forms, an abrupt change of slope rather than an extremum can signal the formation of a distinct species. The stoichiometry of many polynucleotide systems has been determined by this method (Felsenfeld and Rich, 1957; Bloomfield et al., 1974).

1.4.3 Fine Structure Analysis

Differential melting (Wada et al., 1980) of natural DNA molecules can provide information about regions in the DNA that melt cooperatively over a narrow temperature range (a few tenths of a degree). The derivative of the absorbance melting curve is obtained either by measuring the difference in absorbance of two aliquots of the same solution whose changing temperatures differ by a constant amount (e.g., 0.25°C), or by numerically differentiating a melting curve. As an example, the melting of λDNA (Vizard and Ansevin, 1976) is shown in Figure 6-6. Note that λDNA, which is about 47 kbp in length, melts over a 15° range with at least 34 melting subcomponents.

The melting of a DNA revealed by a differential absorbance melting spectrum can be as narrow as 3°C for a small bacteriophage, or can be wider than 25°C for a eukaryotic DNA. The melting spectrum is made up of a sum of peaks, each peak representing the cooperative melting of a small region of the DNA over a range of about 0.5°C. The areas of the resolved peaks in Figure 6-6 have been estimated to represent the melting of regions containing from 300 to 4700 bp. The size and melting temperature of a cooperatively melting region depends on its base sequence, the base sequence of the surrounding DNA, and the length of the DNA (Wartell and Benight, 1985). Analysis of absorbance melting spectra can thus reveal specific characteristics, such as the presence of repetitive DNA sequences, regions of special stability or special sensitivity to specific ions, and the presence of mismatches. Differential melting spectra can be used as a rapid method to identify homology or differences (e.g., due to a deletion) among DNAs (Wada et al., 1980).

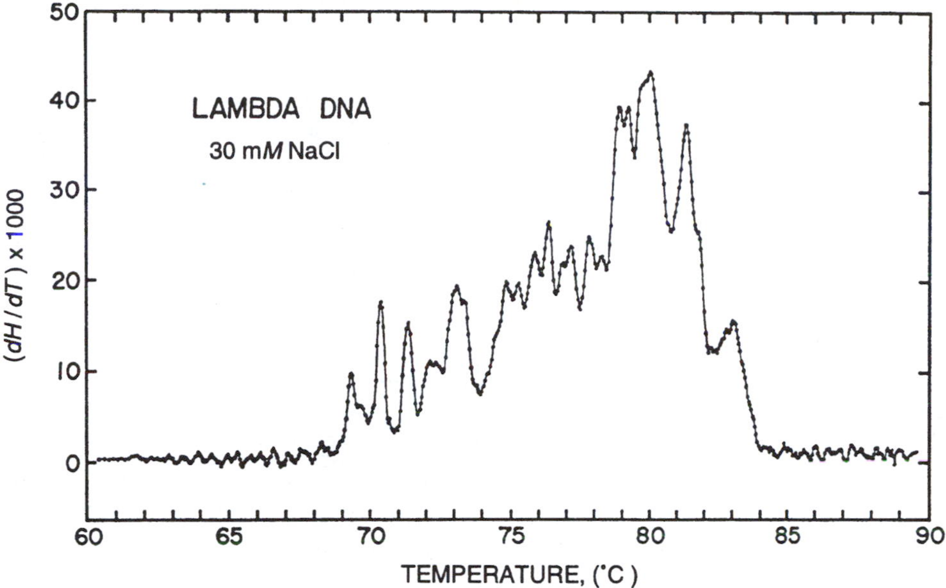

Figure 6-6
The thermal derivative of hyperchromicity at 270 nm versus temperature for lambda DNA in 30 m*M* NaCl. The wavelength of 270 nm was chosen because this is approximately the value that weights A • T and G • C melting equally. The heating rate was 0.1°C min^{-1} and absorbance data were obtained for thermal increments of 0.045°C. The hyperchromicity derivative was calculated numerically. [Reprinted with permission from D. L. Vizard and A. T. Ansevin, *Biochemistry*, **15**, 741 (1976). Copyright ©1976 American Chemical Society See Fig. 3.]

2. FLUORESCENCE AND PHOSPHORESCENCE

2.1 Fundamentals of Emission Spectroscopy

After light is absorbed by a molecule, the energy has various paths that it may take. Possible paths include loss as heat (which simply raises the temperature of the solution; this is the most common fate of absorbed photon energy), photochemical reactions (in which the light energy is partly converted to chemical energy), and emission of the light as fluorescence or phosphorescence. The fraction of the absorbed photons that are emitted as light is called the quantum yield. The intensity of emitted light decays exponentially after the exciting light is turned off; emission lifetimes (τ) can vary from picoseconds to seconds. The wavelength dependence of the emitted light (fluorescence or phosphorescence spectrum) is characteristic of the molecule, but may also depend on the wavelength λ of the exciting light. The excitation spectrum (the intensity of emitted light as a function of λ of the exciting light) is also characteristic. It is usually similar, but not identical, to the absorption spectrum. Finally, the emitted light may be polarized (although the exciting light is not). All these measurable properties can provide useful information about the configuration and conformation of nucleic acids. Furthermore, fluorescence energy-transfer experiments, in which one group is excited (the donor) while another group is monitored for emission (the acceptor) can reveal the distance between donor and acceptor (Cantor and Schimmel, 1980, pp. 433–465).

Fluorescence and phosphorescence are distinguishable both experimentally and theoretically. Nearly all molecules have their electrons paired in the ground state; they are in a singlet state (free radicals are exceptions). When light is absorbed, an excited singlet state (all electrons paired) is produced. Emission of light from an excited singlet state is called fluorescence. Transitions between singlet states can occur easily (the transitions are quantum mechanically allowed), so fluorescence lifetimes in solution are usually in the picosecond to microsecond range.

An excited electronic state can be produced, which has two unpaired electrons; this is a triplet state. Usually, an excited triplet state is attained via an excited singlet state. A singlet excited state is produced by the light, then a nonradiative transition to a lower energy triplet state occurs. Emission of light from an excited triplet state is called phosphorescence. Light-emitting transitions between triplets and singlets are not probable (they are quantum mechanically forbidden just as light absorbing transitions from singlet to triplet are forbidden), so phosphorescence lifetimes are longer than fluorescence lifetimes. In solution, they are usually in the millisecond to second range.

Both fluorescence and phosphorescence spectra are shifted to higher wavelength (lower energy, red-shifted) relative to the absorption spectra. Usually, the lowest wavelength region of the fluorescence overlaps the highest wavelength region of the absorption. The phosphorescence is red-shifted relative to the fluorescence because the excited triplet state is lower in energy than the excited singlet state. The energy levels for ground and excited states and the resulting fluorescence and phosphorescence spectra are illustrated in Figure 6-7. Note that the shapes of the absorption and the fluorescence curves depend on the vibrational levels of the electronic states.

2.2 Emission Properties of Nucleotides and Polynucleotides

The five common nucleotides all have a weak fluorescence at room temperature in aqueous solution. The quantum yields range from 0.3×10^{-4} to 1.2×10^{-4}, and the fluorescence lifetimes are in picoseconds (Callis, 1983). These very low quantum yields make measurement very difficult because small amounts of strongly fluorescing impurities can invalidate the results. The hypermodified Wye base, which occurs immediately adjacent to the 3′ side of the anticodon in $tRNA^{Phe}$ has a quantum yield of about 0.06 and a lifetime near 6 ns (Cantor and Schimmel, 1980, p. 443). Fluorescence has been used to study conformational changes in tRNA (Beardsley et al., 1970), kinetics of binding between codon and anticodon (Yoon et al., 1975), distances between fluorescent probes in ribosomes (Cantor, 1980), in ribozymes (Tuschl et al., 1994), and the ends of four-way junctions in DNA (Clegg et al., 1992, see below). A fluorescent base analog which has been widely used to substitute for adenylic acid for biochemical purposes, is 1, N^6-ethenoadenylic acid (Secrist et al., 1972); its quantum yield is about 0.4 and its lifetime is about 30 ns (Baker et al., 1978). Both naturally occurring and synthetic fluorescent molecules can be very useful probes of nucleic acid structures and interactions. Ethidium, an intercalating agent, is the most commonly used. It fluoresces very weakly when free in solution, but strongly when intercalated

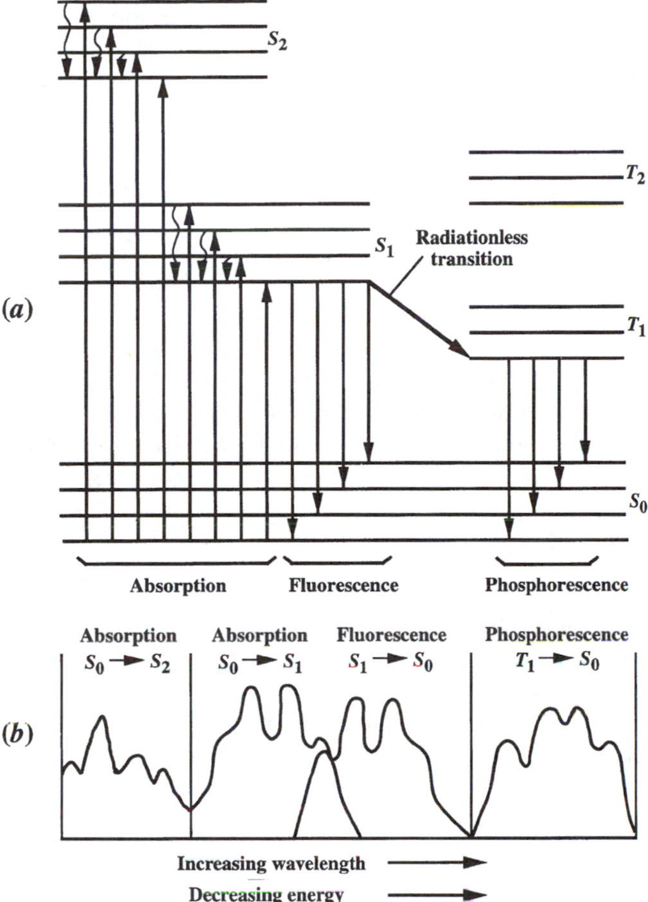

Figure 6-7
(*a*) Energy-level diagrams for absorption, fluorescence, and phosphorescence, and the corresponding spectra. Vibrational-electronic (vibronic) transitions from a ground-state singlet level, S_o, with all electrons paired, to an excited-state singlet level, S_1 or S_2, with all electrons paired, lead to absorption of light. Vibrationally excited states of the excited electronic state quickly decay to the ground vibrational level of the excited electronic state. Transitions from this level to the ground vibrational-electronic states lead to fluorescence. A radiationless transition can occur from the excited singlet, S_1, to an excited triplet state, T_1, with two unpaired electrons. Transitions from this state to the vibrational-electronic states of the ground-state singlet lead to phosphorescence. (*b*) The corresponding spectra show the absorption at the short wavelength end of the spectrum, the fluorescence in the middle overlapping the absorbance at one wavelength, and the phosphorescence at the long wavelength end. Radiation-linked singlet–triplet transitions are improbable, thus phosphorescence lifetimes are longer than fluorescence lifetimes.

between bases. Ethidium bromide is commonly used as a fluorescent stain for DNA in gel electrophoresis experiments.

Phosphorescence apparently does not occur for nucleotides or polynucleotides at room temperature; it is necessary to make measurements at 80 K or below. At these temperature the quantum yields for fluorescence increase to 0.01–0.1, and the quantum yields for phosphorescence are about 0.005 (Gueron et al., 1974).

2.3 Fluorescence Energy Transfer

Singlet excitation energy may not only be emitted as fluorescence or degraded non-radiatively by internal conversion. It may also be transferred from one chromophore to another. This transfer can occur by emission from one molecule and absorption by another, in which case the efficiency of energy transfer is usually negligible and varies with concentration in a predictable way. A more subtle and important mechanism of energy transfer, however, is by a resonance interaction proposed by Förster. Förster-transfer involves a dipole–dipole interaction, in which the oscillating transition moment in the *donor* molecule that has been excited induces an oscillating moment in the *acceptor.* Transfer of singlet excitation energy from donor to acceptor then occurs by direct resonance interaction, without emission of a photon. Since this is a dipole-induced dipole interaction, it has a characteristic R^{-6} dependence, where R is the distance between donor and acceptor. Fluorescence energy transfer can thus be used as a "spectroscopic ruler" to measure R (Stryer, 1978).

The efficiency of energy transfer, E, is defined as the probability that deexcitation of the donor occurs by resonance transfer to acceptor. The energy transfer efficiency is thus a number between 0 and 1. It can be shown that

$$E = \frac{1}{1 + (R/R_o)^6} \tag{6-17}$$

The parameter R_o is the distance at which transfer is half-efficient. In most cases, R_o is between 15 and 45 Å, leading to accessible R values between 10 and 80 Å. This range of distances is useful in probing nucleic acid structures and is hard to attain with other solution techniques. The parameter R_o increases with the spectral overlap integral between the fluorescence spectrum of the donor and the absorption spectrum of the acceptor (see Selvin and Hearst, 1994). It also depends on the angle between the transition dipole moments of donor and acceptor. Since this angle is usually unknown, a random distribution of orientations is generally assumed. However, this assumption can lead to inaccuracy if the donor and acceptor are both held in fixed conformations, as may happen if planar fluorophores stack on nucleic acid bases.

The efficiency of energy transfer, E, is measured either by the decrease in donor fluorescence in the presence of the acceptor, or the increase in acceptor fluorescence in the presence of the donor. If ϕ_D is the quantum yield for the donor fluorescence, and ϕ_{D+A} is the quantum yield for donor fluorescence in the presence of the acceptor, E is

$$E = 1 - \phi_D/\phi_{D+A} \tag{6-18a}$$

Since energy transfer to the acceptor decreases the singlet lifetime τ_{D+A} of the donor from its normal value τ_D in the absence of acceptor, fluorescence lifetime measurements can also be used to determine E according to the equation

$$E = 1 - \tau_{D+A}/\tau_D \qquad (6\text{-}18b)$$

An example of how resonance energy transfer can be used to analyze a complex nucleic acid structure is provided by the work of Clegg et al. (1992) on the four-way DNA junction that is postulated as the intermediate in homologous recombination. A model was proposed in which, if the Mg^{2+} or other cation concentration is high enough, the helical arms stack end-on-end in pairwise fashion to generate two quasicontinuous, coaxial helices, forming an X-shape. Two isomers are possible, depending on the stacking partners, which are illustrated in Figure 6-8. The junction was constructed from four 34-nucleotide strands (b, h, x, and r), so that each arm was 17 bp long. Fluorescein was used as the donor and rhodamine as the acceptor. The complete set of 12 doubly labeled derivatives was prepared, with donor conjugated to one of the 5′-termini and acceptor to another. If ends B and H are closest to each other in the junction, then the energy transfer effieicncy, E, should be greatest when these two ends are labeled with donor–acceptor pairs, and R—X labeling should be symmetrical with B—H. Figure 6-9 shows that these expectations are fulfilled in the presence of Mg^{2+}, but that all pairs of ends are approximately equidistant when Mg^{2+} is absent.

The dynamics of a DNA four-way junction has been measured by time-resolved fluorescence resonance energy transfer (Eis and Millar, 1993). In fluorescence energy transfer, each distinct conformational species has a unique distance between donor and acceptor, and thus has a unique efficiency of energy transfer. A flexible molecule produces a range of fluorescent lifetimes corresponding to the range of distances between donor and acceptor. A DNA four-way junction of 8 bp in each arm was found to have a Gaussian distribution of distances between the arms. A mean distance of 57.2 $\AA$ was measured with a full width at half-maximum for the Gaussian of 18.7 $\AA$. Clearly, the four-way junction is not rigid, but has a wide variation of angles between its arms.

2.4 Fluorescence Anisotropy Decay

If the exciting light is polarized, the fluorescence will also be polarized, or anisotropic. The rate of depolarization, or decay of anisotropy, of fluorescence emission gives information on the rotational, bending, and twisting motions of nucleic acids.

3. CIRCULAR DICHROISM

Polarized light can provide useful additional information about molecular structure and interactions. We have already seen in Figures 6-1 and 6-2 that the CD spectra of the mononucleotides are considerable more distinctive than their absorbance spectra. Note that the molar extinction coefficients of the nucleotides are at least 1000 times larger than the difference in extinction coefficients for left and right circularly polarized light.

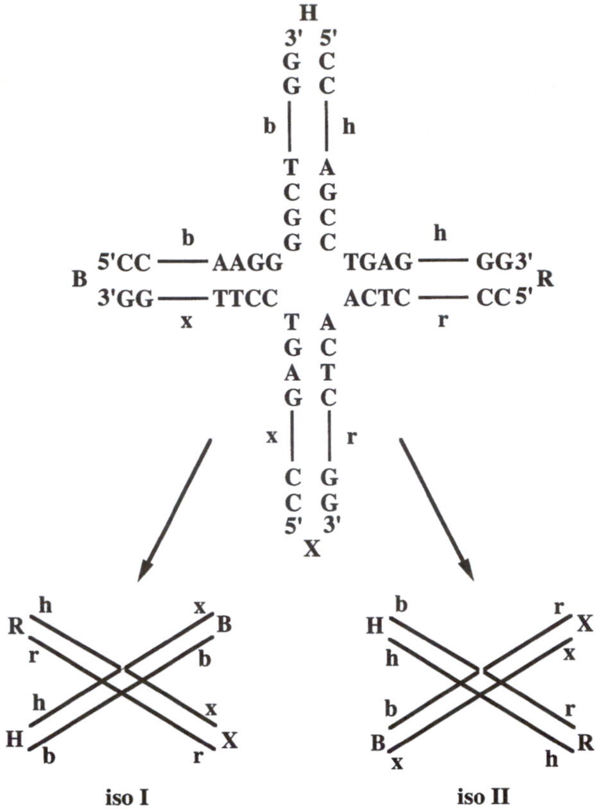

Figure 6-8
Possible folding isomers of a four-way DNA junction. The junction is composed of four 34-nucleotide oligonucleotides (b, h, r, and x), forming 17 bp arms. The fluorescent donor (fluorescein) and acceptor (rhodamine) are covalently attached in all pairwise combinations to the 5′-ends of the arms B, H, R, and X. Fluorescence resonance energy-transfer measurements show whether the pairs (B,X) and (R,H), or (B,H) and (R,X) are closer to each other, enabling discrimination between *isoI* and *isoII*. For the junction sequence shown, the data in Figure 6-9 indicate that isoII is predominant. [Reprinted with permission from R. Clegg, A. Murchie, A. Zechel, C. Carlberg, S. Diekmann and D. Lilley, *Biochemistry*, **31**, 4846 (1992). Copyright ©1992 American Chemical Society.]

However, the sensitivity of CD is as great as that of absorbance; the same concentrations are used to measure both.

The CD spectra of oligonucleotides and polynucleotides are even more characteristic of their structures. The molar CD of a polynucleotide is the sum of the molar CDs of its nucleotides, plus the effect of the interactions among the nucleotides. However, the effects of the interactions are much larger in CD spectroscopy than in absorption

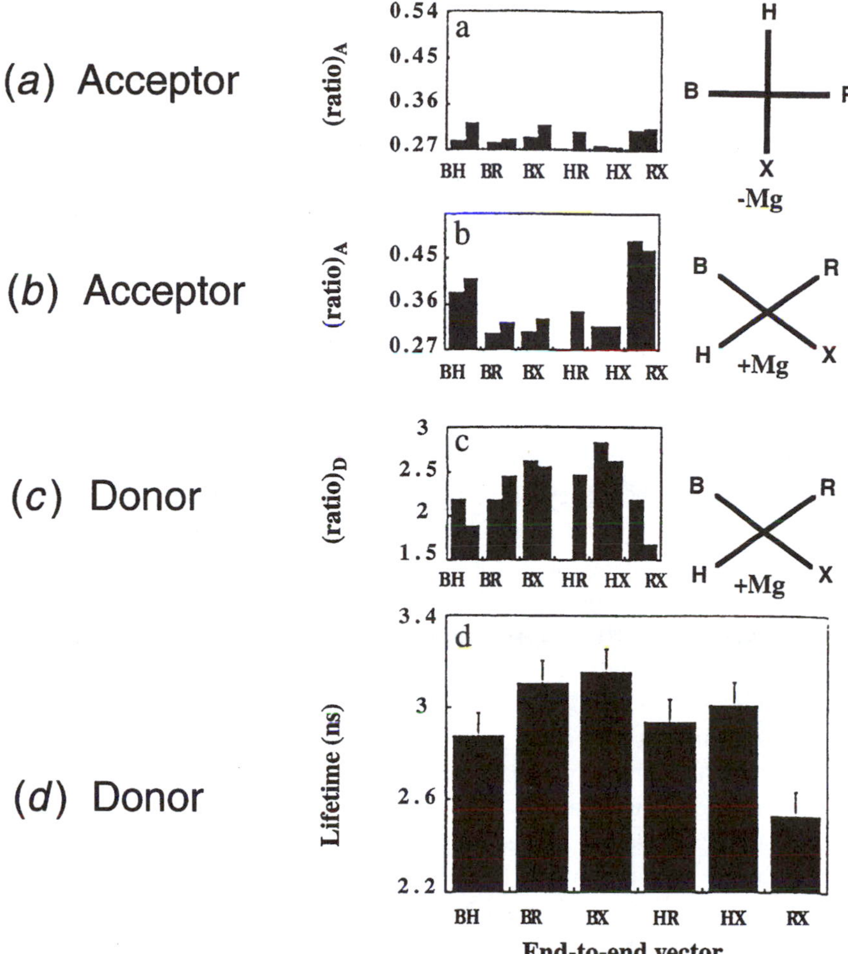

Figure 6-9
Fluorescence resonance energy-transfer experiments enable discrimination between the folding isomers of the four-way DNA junction shown in Figure 6-8. (*a*) In the absence of Mg^{2+}, enhancement of acceptor emission is similar for all pairs , indicating no preferential folding. In 5 m*M* Mg^{2+}, acceptor emission (*b*) is relatively enhanced, and donor emission (*c*) and lifetime (*d*) are relatively reduced for the pairs (BH) and (RX), indicating that these ends are closest together, as in *isoII*.

spectroscopy. The CD of double-stranded DNA or RNA can be a factor of 10 or more greater than the sum of the CDs of its nucleotides. Furthermore, because CD spectra have both sign and magnitude at each wavelength, they are more sensitive to conformation than absorption spectra are, e.g., double-stranded, right-handed DNA or RNA can easily be distinguished from left-handed DNA or RNA (Riazance et al., 1985; Williams et al., 1986). Circular dichroism spectra can also distinguish different forms of right-handed duplexes, such as the A and B form (Sprecher et al., 1979).

3.1 Fundamentals

We begin with a brief review of the nature of light and its polarization properties. Light is an oscillating electromagnetic field in which the electric and magnetic fields oscillate in a plane perpendicular to the direction of propagation of the light. For linearly polarized light (also called plane-polarized light), the electric field oscillates in only one direction—for example, along the y axis for light propagating along the z axis. Linearly polarized light would then be represented as a vector on the y axis whose length changes from $+1$ to 0 to -1 to 0 to $+1$ at the frequency of the light as shown in Figure 6-10. In general, linearly polarized light is represented as an oscillating vector oriented along any direction in the x–y plane (for light propagating along the z-axis). The length of the vector is proportional to the magnitude of the electric field of the light.

Circularly polarized light is represented by a vector of constant length that rotates in the x–y plane at the frequency of the light. The tip of the vector traces out a circle; it can move clockwise or counterclockwise to give right- or left-handed circularly polarized light. Because the light is propagating, the electric field (the tip of the vector) of circularly polarized light produces a right- or left-handed helix in space. In contrast, the electric field of linearly polarized light produces a sine wave in space. Vector addition of the electric fields of right and left circularly polarized beams propagating in the same direction with the same frequency and magnitude produces a linearly polarized beam. Circular dichroism, as defined in Eq. 6-5, is the differential absorption of the right and left circularly polarized beams.

Also, vector addition of the electric fields of x- and y-polarized light of the same frequency and magnitude but shifted in phase by $\pi/2$ rad (90°) produces circularly polarized light. Vector addition of fields of unequal magnitudes shows that elliptically polarized light will result. Thus, polarized light can be linearly polarized along any axis; it can be right or left circularly polarized, or it can be right or left elliptically polarized. The change in state of polarization of light after it is incident on a sample gives information about the structure and conformation of the molecules in the sample.

Another measure of optical activity is ellipticity. When linearly polarized light (which is equivalent to equal amplitudes of left and right circularly polarized light) is incident on an absorbing chiral substance, differential absorption of the circularly polarized light changes the linear polarization to elliptical polarization. The ellipse is characterized by an angle, θ, whose tangent is equal to the ratio of the minor to the major axis of the ellipse, and whose sign gives the handedness of the ellipse. The molar ellipticity $[\theta]$ in the standard units of deg M^{-1} cm^{-1} $\times$ 100 where the factor of 100 appears for historical reasons, is directly proportional to the CD by the relation

$$[\theta] = 3298(\varepsilon_L - \varepsilon_R) \tag{6-19}$$

Circular dichroism spectra are often given with units of molar ellipticity rather than molar extinction coefficient.

Optical rotatory dispersion is a third measurable property of a solution of optically active molecules. Rotation of the plane of linearly polarized light can occur at any wavelength in a chiral medium; it is not necessary for absorption to occur. The rotation

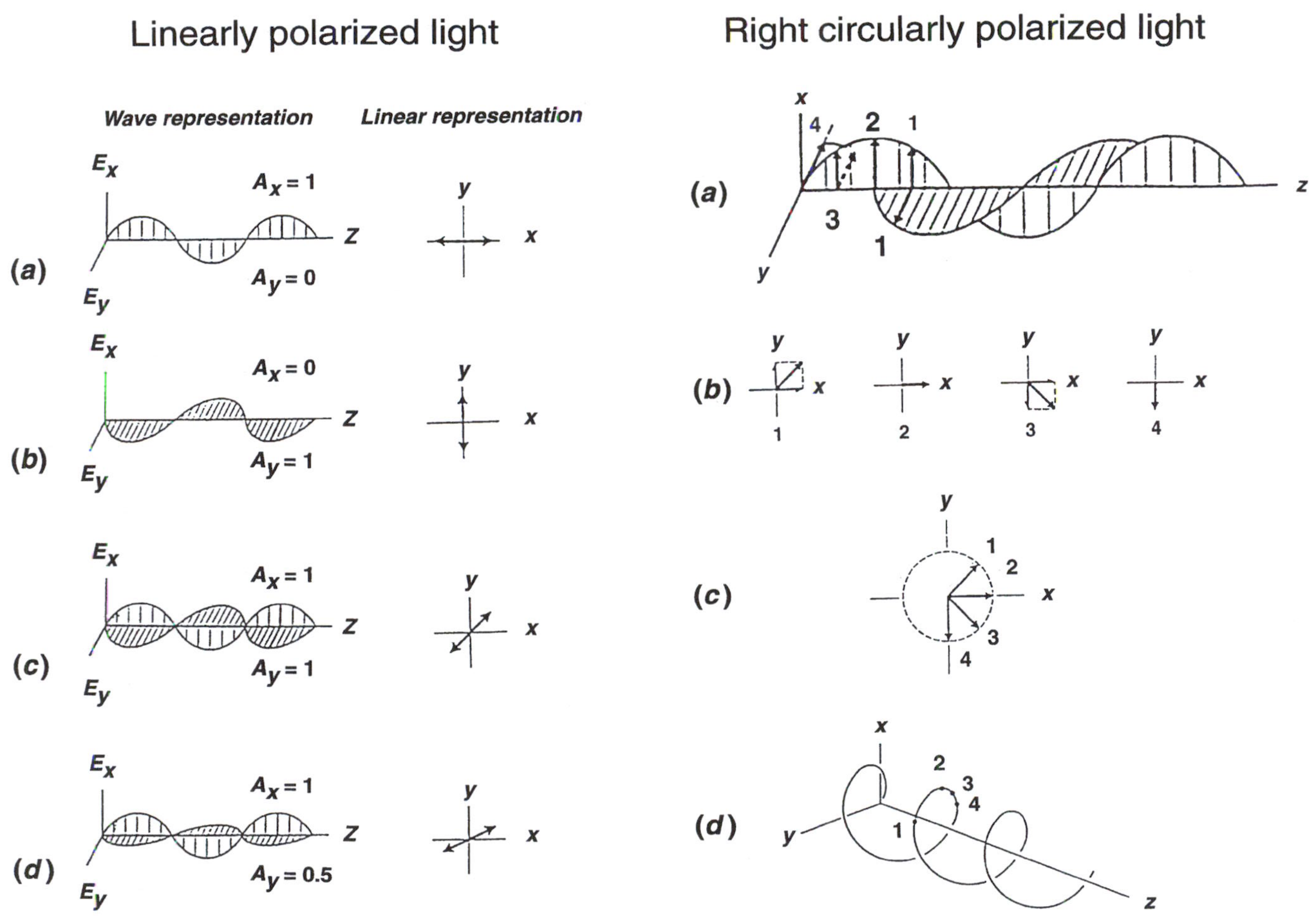

Figure 6-10
Representations of different types of polarized light by an oscillating vector. The tip of the vector indicates the magnitude and direction of the electric field of the light as a function of time and space. On the left is shown linearly polarized light propagating in the z direction. The polarization is along : the x axis (*a*), the y axis (*b*), 45° from the x axis (*x*), and 30° from the x axis (*d*). On the right is shown right circularly polarized light propagating in the z direction. Panels (*a*) and (*b*) illustrate that circularly polarized light can be considered as a sum of two perpendicular vectors oscillating 90° out of phase. The resulting tip of the electric vector of the light produces a circle as seen by an observer moving with the light (*c*); it produces a helix as seen by a stationary observer (*d*). [Reprinted with permission from Kliger et. al., 1990. This figure is from Figures 2-2 and 2-3.]

can be related to the different refractive indices of the medium for left and right circularly polarized light. Thus rotation can be measured outside absorption bands—such as in the visible region for polynucleotides. Clockwise rotation as viewed by an observer is called positive; anticlockwise is negative. The wavelength dependence of the rotation, the optical rotatory dispersion, gives much the same information as the CD and ellipticity when measured over the same wavelength range. If for instrumental reasons, or because of the nature of the sample, measurements cannot be made in the absorption region, then rotation measurements may be the only way to characterize the optical chirality. The molar rotation, $[\phi]$, and the molar ellipticity, $[\theta]$, have the same units. They are related to each other by an integral equation called the Kronig–Kramers transform. By integrating the ellipticity over one CD band, we obtain the contribution of that band to the rotation at any wavelength λ. Optical rotatory dispersion is more often seen in the older literature, but it has uses even today in special cases (such as when the absorption bands of the sample are outside the range of the instrumentation). A thorough discussion of polarized light and the basics of optical activity is given in Kliger et al. (1990).

For particles such as viruses, chromosomes, condensed DNA, membrane fragments, and other complexes that are large compared to the wavelength of the light, differential scattering of circularly polarized light can make an important contribution to the measured CD. The scattering can be recognized as an apparent CD outside the absorption wavelength region. Methods for correcting for these effects have been described by Dorman et al. (1973). For molecules small compared to the wavelength of light (with a dimension < 100 nm) the scattering effects are negligible; this includes nonaggregated nucleic acids and proteins. Scattering effects are considered in more detail in Section 3.4.

3.2 Relation of Circular Dichroism to Nucleic Acid Structure

The intensity of absorption depends on the magnitude of electron density displacement during the transition. Circular dichroism depends in addition on the shape of the path taken by the electron density during the transition. The light-induced electron motion must be along a segment of a helical path for CD to occur. By a helical path we mean one that involves circular motion plus linear motion perpendicular to the plane of the circle. Helices, in contrast to circles or straight lines, have a handedness; that is just what is required for chirality.

The helical electron motion may be due either to chiral molecular structure, or to perturbations induced by an asymmetric environment. Induced helices that are identical except for their handedness will have CD spectra opposite in sign but of equal magnitude. Obviously, mirror-image molecules (enantiomers) will have CD spectra that are identical, but of opposite sign. Non-chiral molecules (which are identical to their mirror images) and racemic mixtures (with equal amounts of mirror images) must have zero CD. In principle, the CD spectrum can be quantitatively related to the conformation, but as is often the case, in practice the calculations are not always definitive (Tinoco and Williams, 1984).

Circular dichroism of nucleic acids is mainly dependent on the stacking geometry of the bases. The difference in CD between a nucleoside, adenosine (A), and a dinucleoside phosphate, adenylyl-3′-5′-adenosine (ApA), is illustrated in Figure 6-11. Note that the CD of the dinucleoside phosphate per adenosine is about a factor of 10 larger than the CD of adenosine. In adenosine, the CD depends on the interaction of adenine with its ribose and phosphate groups; in the dinucleoside phosphate, the CD is mainly from the chiral adenine–adenine interaction. The combination of positive and negative extrema on either side of 260 nm is called an exciton band; there is another exciton band at 215 nm. The positive signs (long wavelength component positive, short wavelength component negative) of these two bands indicate that the two adenines are forming a right-handed stack in ApA (Bush and Tinoco, 1967).

Polynucleotides with more than one type of base and many overlapping bands cannot be analyzed so simply. However, for double-stranded nucleic acids, the CD below 220 nm is characteristic of the sense of the helix (Riazance et al., 1985; Gray et al., 1990); CD spectra for right-handed A and B forms, and left-handed Z form are shown in Figure 6-12. The right-handed nucleic acids (A-DNA, B-DNA, A-RNA) have an intense positive peak near 186 nm and negative CD below 180 nm; the left-handed molecules (Z-DNA, Z-RNA) have an intense negative peak at 190–195 nm, a crossover at 184 nm, and a positive peak below 180 nm. Calculations of CD spectra (Williams et al., 1986) for sequences other than those measured show that below 220-nm right-handed double helices have a positive CD couplet and left-handed duplexes have a negative couplet. So, a good method to establish the sense of a duplex helix is to measure the CD in the wavelength range from 170 to 220 nm.

At longer wavelengths—the 260-nm range—the spectra can be used to distinguish A type conformations from B type; although it is important to corroborate the CD

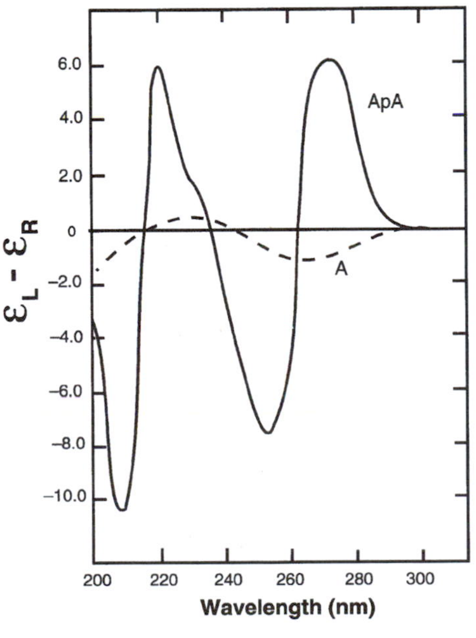

Figure 6-11
Circular dichroism spectra of adenylyl-3′-5′-adenosine (ApA) compared with adenosine (A); for both molecules, the spectra are given per mole of nucleoside. The spectra are for aqueous solutions at pH 7 and room temperature. [Data are from Warshaw and Cantor, 1970.]

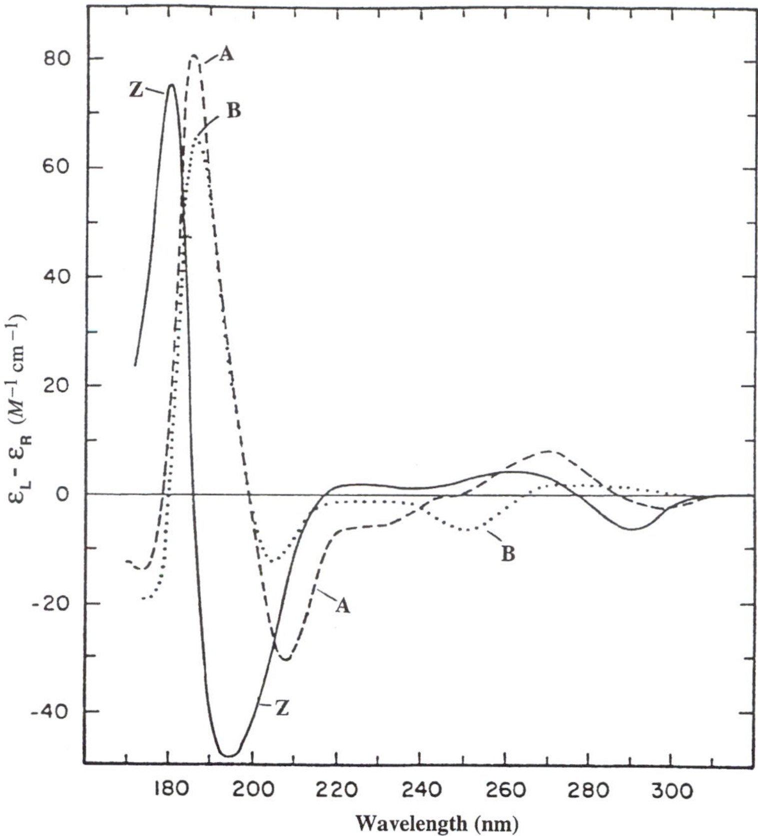

Figure 6-12
Circular dichroism spectra for poly[d(C-G)·d(C-G)] in right-handed A form, right-handed B form, and left-handed Z form (units are M^{-1} cm^{-1} mol^{-1} of nucleotide). Below 200 nm, the two right-handed forms are similar to each other and characteristically different from the left-handed form. The A-form spectrum was in 80% 1,1,1-trifluoroethanol and 0.67 m*M* phosphate; the B-form spectrum was in 10 m*M* phosphate; the Z-form spectrum was in 2 *M* $NaClO_4$ and 10 m*M* phosphate. All spectra were at 22°C. [Reprinted from Riazance, J. H., Baase, W. A., Johnson, W. C., Jr., Hall, K., Cruz, P., and Tinoco, I., Jr. (1985). *Nucleic Acids Res.* **13**, 4983–4989 by permission of Oxford University Press. See Fig. 1.]

results by other methods, such as Raman scattering or NMR. The CD signatures of A- and B- conformations were originally established (Tunis-Schneider and Maestre, 1970) by measuring the CD of films of DNA at relative humidities corresponding to conditions used to determine the X-ray structures of fibers. Since that time work has been done to correlate CD spectra and conformation; reviews appear periodically (Johnson, 1978; Tinoco et al., 1980; Tinoco and Williams, 1984; Gray et al., 1995). The CD spectra of A-RNA, A-DNA, B-DNA and Z-DNA in the 200–320-nm range are shown in Figure 6-13. Both A-RNA and A-DNA have spectra similar in shape. The A-RNA has a maximum near 260 nm, a minimum near 210 nm, and a small negative

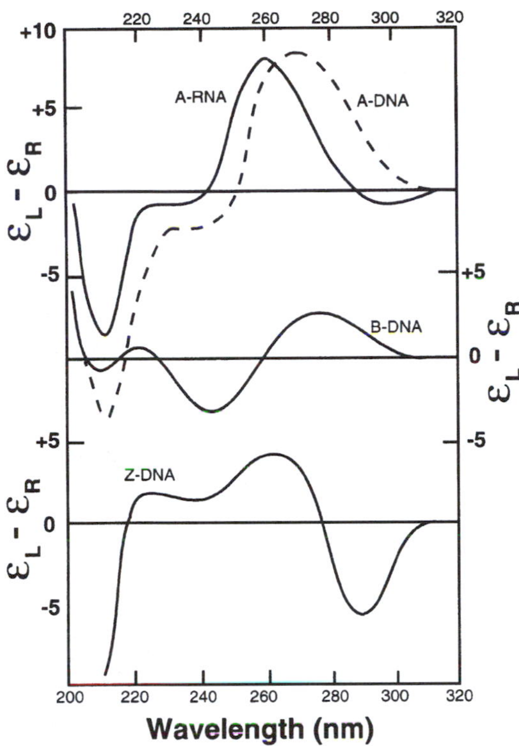

Figure 6-13
Circular dichroism spectra above 200 nm for right-handed A-RNA and A-DNA, right-handed B-DNA, and left-handed Z- DNA (units are M^{-1} cm^{-1} mol^{-1} of nucleotide). The A-RNA is *Penicillium chrysogenum* fungal virus double-stranded RNA with a G + C content of 54%; it is in 0.01 *M* Na^+, pH 7. [Data are from Gray et al., 1981.] The A-DNA is from *E. coli* with G + C content of 50%; it is in 80% trifluoroethanol, 0.667 *M* phosphate, pH 7. [Data are from Sprecher et al., 1979.] The B-DNA is from *E. coli* DNA in 0.02 *M* Na^+, pH 7. [Data are from Gray et al., 1978.] The Z-DNA is poly[d(CG) · d(CG)] in 2*M* $NaClO_4$, pH 7. [Data are from Riazance et al., 1985.]

CD between 290 and 300 nm. The A-DNA has a maximum at 270 nm, a minimum near 210 nm and zero CD at 300 nm and beyond. B-DNA has a conservative CD spectrum above 220 nm with approximately equal positive (275 nm) and negative (245 nm) components centered around 260 nm. The B-DNA maximum has less than one-half the magnitude of the A-DNA maximum. Of course, the exact shapes and magnitudes of the CD spectra will depend on the base sequences, but the overall patterns will remain constant.

The Z-DNA has a conservative spectrum above 240 nm with approximately equal negative (290 nm) and positive (260 nm) components centered around 280 nm. The approximate inversion of the CD spectrum of Z-DNA relative to B-DNA was the first evidence for left-handed DNA (Pohl and Jovin, 1972), but X-ray determination of a detailed structure was needed to establish the left-handed conformation (Wang et al., 1979a).

The characteristic differences in CD between A-DNA and B-DNA can be used to identify conformations in solution. The duplex d(GGATGGGAG) · d(CTCCCATCC) is a portion of the gene recognized by transcription factor IIIA (TFIIIA). In the crystal this oligonucleotide is A form (McCall et al., 1986); and as TFIIIA also binds the corresponding 5S RNA transcribed from the DNA, it was postulated that the DNA sequence was A form in solution (Rhodes and Klug, 1986). However, CD and NMR measurements show unequivocally that d(GGATGGGAG) · d(CTCCCATCC) is B form in solution (Aboul-ela et al., 1988). Furthermore, CD studies of a longer fragment of the same region show that the conformation does not change to A form when the TFIIIA protein binds to it (Gottesfeld et al., 1987). Thus qualitative CD studies can easily establish the type of conformation for a nucleic acid; more detailed structure determination requires other methods such as NMR or X-ray crystallography.

The CD spectrum in the 260-nm region of left-handed Z-RNA is very dependent on the solvent conditions. The original observation was that poly[r(C-G)] in 6*M* $NaClO_4$ underwent a transition with increasing temperature from a conformation with an A-RNA CD spectrum to one with a very different CD spectrum. The high-temperature spectrum was shown by NMR to correspond to a left-handed Z-type conformation (Hall et al., 1984); it is shown in Figure 6-14. Note that the CD spectrum is always positive above 220 nm and that it is very different from that of Z-DNA in this region, although below 220 nm the CD spectra of both Z-RNA and Z-DNA are very similar (Riazance et al., 1985). In general, high salt and high temperatures favor Z-RNA; and in 4 *M* $MgCl_2$, poly[r(C-G) • r(C-G)] has a CD spectrum in the 260-nm region (Figure 6-14) very similar to that of Z-DNA (Cruz et al., 1986). Below 200 nm, the high absorbance of the $MgCl_2$ precludes measurement of the CD. We do not know the reason for the differences in CD spectra for the two left-handed forms of poly[r(C-G) • r(C-G)]. Either there are differences in conformation such as the difference between the A and B form, or the high concentrations of Mg^{2+} or lower pH of the $MgCl_2$ cause changes in base spectra that change the CD.

Figures 6-12–6-14 show that CD spectra can be used to distinguish easily among the major conformations of nucleic acids. The correlations have been experimental. One measures the CD and then uses other methods to establish the conformation. When the solvent or temperature causes a change in CD, a change in conformation is assumed, but the nature of the change is not known. An attempt to interpret changes

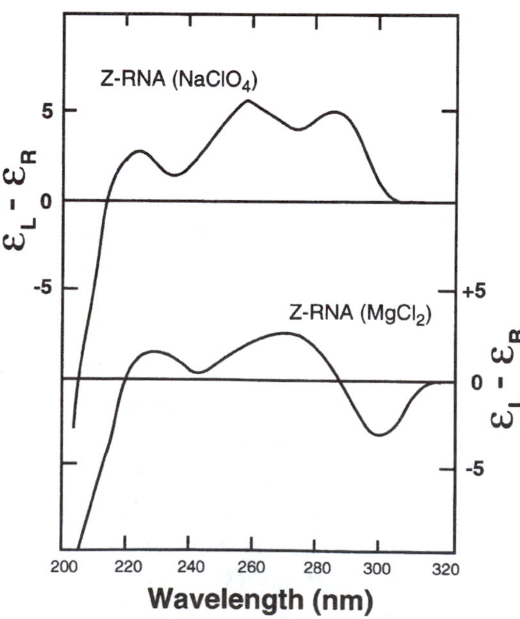

Figure 6-14
Circular dichroism spectra of two different forms of left-handed Z-RNA for poly[r(C-G) • r(C-G)] (units are M^{-1} cm^{-1} per mol of nucleotide). The top spectrum is in 6 *M* $NaClO_4$, 10 m*M* phosphate, pH 7, 45°C; the bottom spectrum is in 4 *M* $MgCl_2$, pH 4, 25°C. The spectrum in $NaClO_4$ is very similar to the Z-DNA spectrum below 200 nm, but is different above 240 nm. The spectrum in $MgCl_2$ is similar in shape to the Z-DNA spectrum above 210 nm; below this wavelength it cannot be measured because of the strong solvent absorption. [Data are from Cruz et al., 1986.]

in CD was made by Johnson et al. (1981). Calculations of the CD were made for a double-stranded deoxyoligonucleotide containing 11 bp with a sequence of all 10 nearest neighbors. The CD spectrum was calculated over the range from 200 to 350 nm for all reasonable right-handed conformations; both A-family ($3'$-*endo* sugar) and B-family ($2'$-*endo* sugar) structures were considered. The range of conformations with acceptable energies varied from 12 bp/turn for the A-family to 9 bp/turn for the B-family. The base pairs per turn are related to the winding angle by

$$\text{Winding angle} = 360^\circ/(\text{bp/turn}) \tag{6-20}$$

An approximate linear correlation was found between winding angle and propeller twist (the dihedral angle between bases in a base pair). The lowest winding angle of 30° (12 bp/turn) has a high propeller twist. As the winding angle increases the propeller twist decreases to zero, then increases in the opposite direction so that the highest winding angle of 45° (8 bp/turn) has the lowest (negative) value of propeller twist. A good linear correlation was found between the calculated CD at 275 nm and the winding angle, as well as the propeller twist. The calculation is approximate so the magnitudes cannot be trusted. However, the conclusion is probably valid that the CD can be used to determine changes in winding angle for right-handed duplexes. The data of Sprecher et al. (1979) can provide the needed information. The CD spectra of four different DNAs with $G + C$ content from 31 to 72% were measured with the DNAs in the A form (80% 1,1,1-trifluoroethanol), B form in moderate salt, and B form in 6 M NH_4F; all solvents contained a pH 7.0 phosphate buffer. At 270 nm, where both A and B-DNA have maxima, the values varied from a high of $\varepsilon_L - \varepsilon_R$ (270 nm) of $+9.44 M^{-1}\ \text{cm}^{-1}$ for A-form *microccus luteus* DNA to a low of $-0.55\ M^{-1}\ \text{cm}^{-1}$ for B-form calf thymus DNA in high salt. In order to correlate CD with winding angle, one must know the number of base pairs per turn in each conformation. If one assumes the following values: A (32.7° = 11 bp/turn), B in moderate salt (34.6° = 10.4 bp/turn), and B in high salt (35.3° = 10.2 bp/turn) from topological measurements of circular DNAs in solution (Wang, 1979b; Baase and Johnson, 1979), the average change of CD at 270 nm for the four DNAs was a decrease of $3.2 \pm 0.1\ M^{-1}\ \text{cm}^{-1}$ per 1° increase in winding angle. It is clear that the CD at 270 nm is very sensitive to winding angle.

Under normal solvent conditions RNA is expected to be A form and DNA is expected to be B form; the differences in CD among different RNAs or among different DNAs should then mirror differences in sequence. The CD differences will depend on the differences in interactions among the sequences and on the different local geometries the sequences may have. The first approximation is that the CD depends only on the nearest neighbor sequence; this has been found to be a useful approximation for interpreting optical properties. Gray and co-workers measured the CD spectra of synthetic double-stranded polynucleotides of repeating monomer, dimer, or trimer sequences for DNA (Allen et al., 1984; Gray et al., 1978) and for RNA (Gray et al., 1981). Although there are 10 Watson–Crick nearest neighbors, for a very long duplex (where end effects are negligible) there are only eight independent nearest neighbor frequencies. Gray et al. found that the CD of natural DNAs and the CD of synthetic RNAs with known nearest neighbor frequencies could be fit well using the CD of a small number of synthetic duplexes. However, in order to obtain the best fit for DNAs, the CD spectra of some duplexes were omitted. For example, poly[d(A)·d(T)] and poly[d(G)·d(C)] were replaced by poly[d(A-A-T)·d(A-T-T)] and poly[d(A-G-G)·d(C-C-T)]. The latter duplexes contain the (A-A)·(T-T) and (G-G)·(C-C) nearest neighbor sequences, but presumably have a local conformation more nearly the same as a natural DNA. From the synthetic DNA and RNA duplexes, sets of CD contributions of double-stranded Watson–Crick nearest neighbor sequences can be obtained for the purpose of nearest neighbor frequency analyses of polynucleotides. Except for the (A-A)·(T-T), (A-A)·(U-U), and (G-G)·(C-C) nearest neighbors, these CD contributions should not be confused with the actual contributions due to neighboring Watson–Crick pairs, since only eight independent contributions can be obtained from the polynucleotides. To obtain all 10 contributions needed to calculate the CD of oligonucleotide duplexes, measurements on additional oligonucleotides are needed.

Measured CD spectra of polynucleotides have been used to discover and characterize novel conformations other than Watson–Crick paired double helices (see Gray et al., 1995 for a review). The CD of a duplex is compared with the CD of known structures to determine the nearest neighbor sequences and conformations present. Double helices with bulged thymidines were proposed in simple repeating sequences of poly[d(G-G-A)·d(C-T)] (Evans and Morgan, 1982), and (C^+·C) pairs occur at low pH where the cytosines are half protonated (Gray et al., 1980, 1984; Edwards et al., 1990). The CD of double and triple helices involving dA, dT, rA, and rU have been measured to investigate the conformations of the individual strands in the helices (Steely et al., 1986). Simple repeating sequences occur often in the genomes of higher organisms; they may have unique structural roles.

To summarize, CD is most useful as a method to compare conformations and to detect changes when the solvent or temperature is changed. It is sensitive to interactions of neighboring bases and to any change in the spectrum of the bases, such as that caused by protonation.

3.3 Fluorescence-Detected Circular Dichroism

Fluorescence-detected CD (FDCD) has two main uses (Tinoco et al., 1987). One is the selective measurement of the CD of fluorescent chromophores to probe conformation near fluorophores in the sample. The other is the measurement of the differential absorption part of the CD, that is, without the effects of differential scattering. The measurement of FDCD uses alternately right and left circularly polarized light incident on the sample, but instead of detecting the transmitted light, the fluorescent light is measured (Lobenstein et al., 1985). The differential fluorescence intensity depends on the difference in absorption of right and left circularly polarized light by the fluorophore, or by any group that transfers energy to the fluorophore. The fluorescence quantum yields of the common nucleic acid bases are too small to be useful, so to probe conformation either naturally occurring fluorophores such as the Wye base in phenylalanine tRNA are used, or fluorescent dyes such as ethidium are added. Fluorescence-detected CD of the Wye base detected specific conformational changes in the anticodon loop of $tRNA^{Phe}$ with temperature and Mg^{2+} (Turner et al., 1975). These studies have the sensitivity of fluorescence measurements and the conformational information of CD; they monitor only the Wye base and any bases that transfer energy to the Wye base.

When ethidium binds to DNA, the chiral nature of the DNA induces a CD in the nonchiral ethidium. Fluorescence-detected CD separates the induced CD in the ethidium from the CD of all the base pairs far from the ethidium. Lamos et al. (1986) used FDCD to show that when ethidium binds to poly[d(G)-d(C)] in the Z form, the intercalated ethidium switches the Z-DNA at the binding site into the same conformation found for ethidium–B-DNA complexes. The ethidium binds cooperatively to the Z-DNA in clusters to give regions of DNA with high concentrations of bound ethidium and regions with no ethidium. In B-DNA, the ethidium binds randomly.

In an FDCD experiment the quantity measured is

$$\frac{F_L - F_R}{F_L + F_R} \tag{6-21}$$

where F_L and F_R are the fluorescence intensities when left or right circularly polarized light is incident. This result is analogous to the usual CD where $A_L - A_R$ is equal (for small values of $A_L - A_R$) to

$$\left(\frac{-2.303}{2}\right)\left[\frac{I_L - I_R}{I_L + I_R}\right] \tag{6-22}$$

and I_L, I_R are the transmitted intensities. The theory of FDCD (Tinoco and Turner, 1976) shows that if all the fluorescence comes from one chromophore, the FDCD measurement gives the Kuhn dissymmetry factor for the chromophore. This factor, $(\varepsilon_L - \varepsilon_R)/\varepsilon$, is the ratio of the CD to the extinction coefficient of the chromophore; it is a direct measure of the chirality of the chromophore. Because the FDCD measures the ratio of fluorescence intensities, it is independent of the quantum yield of fluoresence for a single fluorophore. As the quantum yield changes, the sensitivity and signal-to-noise (S/N) change, but the FDCD magnitude does not. For a mixture of fluorophores, or a fluorophore that exists in multiple conformations, the contribution of each species to the FDCD depends on its quantum yield (Tinoco and Turner, 1976). Measurement of fluorescence also introduces the problem of photoselection (Tinoco et al., 1977), where molecules that rotate slowly compared to the fluorescence lifetime (ns) act as if they were oriented. The measured values of $(\varepsilon_L - \varepsilon_R)$ and ε do not correspond to the randomly oriented molecule; they emphasize the CD and extinction along particular directions of the molecule. The directions depend on the directions of the absorption and fluorescence transition moments. For ethidium-DNA and tRNA, the molecular rotations are fast enough to avoid this problem. Special experimental arrangements can be used to obtain the desired average properties even when the molecules do not rotate fast enough (Tinoco et al., 1977).

Fluorescence-detected CD is very useful for obtaining the part of CD that comes solely from circular differential absorption. The difference in transmitted intensity when left and right circularly polarized light is incident on the sample has contributions from absorption and scattering (see Section 3.4). As only light absorbed by the fluorophore leads to fluorescence, FDCD is an excellent way to distinguish between differential absorption and differential scattering. One adds a non-chiral fluorophore to the solution of the (non-fluorescent) macromolecule to be studied. The fluorophore should not bind to the macromolecule; however, it completely surrounds the macromolecule in solution and detects the light not absorbed. Light is incident on the solution containing free non-chiral fluorophore and chiral non-fluorescent macromolecule. The non-chiral fluorophore absorbs all the light either transmitted or scattered by the macromolecule, thus its signal is a measure of the differential absorption of the light by the macromolecule. For samples with negligible scattering, the usual CD measurement will give the same result as the FDCD measured by adding a non-chiral fluorescent reporter molecule.

Maestre and Reich (1980) and Reich et al. (1980) used α-naphthylamine as a fluorescent reporter molecule to measure the FDCD of polylysine- and ethanol-condensed DNA. These DNA aggregates have very large values of CD and have large contributions from circular differential scattering as shown by large CD values outside the absorption bands of the DNA. This type of CD spectrum is called psi-type (for polymer and salt-induced type, see Chapter 14); it is very different from the CD of A-, B-, or Z-DNA (Lerman, 1973). The FDCD measurements showed that after removing the scattering contributions—removing the long-wavelength scattering tails—one still had a large and unusual CD. This result was interpreted in terms of a chiral side-by-side arrangement of the DNA molecules in the condensates, which led to cholesteric liquid crystal behavior. The CD spectra depend on chiral periodicities in the aggregate of the order of the wavelength of light. A general theory of these types of spectra has been developed (Keller and Bustamante, 1986).

3.4 Circular Differential Scattering

Circular intensity differential scattering (CIDS) of light is the preferential scattering of one sense of circularly polarized light relative to the other. In the simplest case, the state of polarization of the scattered light is not measured. The scattering occurs at all angles and the CIDS will be different for every angle. To obtain measurable values of circular differential scattering. the scattering particle must be at least one-tenth the wavelength of light. The largest effects occur for particle size similar to the wavelength of light. Large particle size and large scattering intensity are not sufficient to obtain circular differential scattering; it is necessary to have a chiral particle with a chiral size parameter similar to the wavelength. For example, for a helix the chiral parameters are radius and pitch—the pitch is the distance parallel to the helix axis between successive turns. Thus, the pitch/wavelength ratio is crucial, not the total helix length/wavelength ratio. When there is a match between a chiral size parameter and wavelength the CIDS becomes very large. For example, chiral liquid crystals can completely reflect one sense of circularly polarized light of the critical wavelength and not interact with the other sense of circularly polarized light.

The theory of the angular dependence of CIDS for randomly oriented macromolecules has been derived; it is reviewed by Tinoco et al. (1987). The derivation uses classical physics; it is analogous to the derivation of X-ray scattering except that the polarization of the incident light is taken into account. The angular dependence of the

CIDS is sensitive to the dimensions that govern the chirality of the particle, but is substantially less sensitive to the overall size of the particle.

The angular dependence of the CIDS has been measured for suspensions of octopus sperm cells (Maestre et al., 1982; Shapiro et al., 1994); the entire cell is a helix. The maximum CIDS signal occurs at $\pm 40°$ to the incident beam and has a value of $(I_L - I_R)/(I_L + I_R)$ equal to about 0.02; here I_L, I_R are scattered intensities. The scattering pattern is consistent with a left-handed helix of 650-nm pitch and 250-nm radius; the dimensions are in approximate agreement with electron microscopy images. The measurements are sensitive to scattering impurities and unless the sample has a strong signal, artifacts are difficult to avoid. The CIDS from uncondensed DNA has been too small to measure. Condensed DNA does provide measurable signals, but the CIDS is time dependent and the polydispersity of the samples makes the angular dependence difficult to interpret.

The total observed CD is a sum of absorbance and scattering contributions:

$$\varepsilon_L - \varepsilon_R = (a_L - a_R) + (s_L - s_R) \tag{6-23}$$

The contribution of CIDS to the CD can be calculated from integrating the polarization- and angular-dependent scattering cross-sections of the particle over all angles (Bustamante et al., 1983). This integration gives $(s_L - s_R)$, the scattering contribution to $(\varepsilon_L - \varepsilon_R)$. Values of $(s_L - s_R)$ as a function of wavelength can be obtained from the measured CD by subtracting the value of $(a_L - a_R)$ measured using FDCD. Values of CIDS integrated over a range of angles can be measured experimentally by using different size pinholes in front of the detector in a CD apparatus. The difference in measured CD using two different pinholes gives the CIDS integrated over the angle of acceptance of the smaller pinhole minus the CIDS integrated over the acceptance angle of the larger pinhole. The integrated CIDS measured by this indirect method agrees qualitatively with the direct measure of CIDS. The combination of $(s_L - s_R)$, which depends on the long-range organization (> 20 nm) of the particle, and $(a_L - a_R)$, which depends more on local, short-range interactions, makes a powerful combination of tools to understand conformation in complex systems.

4. LINEAR DICHROISM

4.1 Fundamentals

Additional information about the orientation of nucleic acid chromophores can be obtained from linear dichroism (LD), defined as the difference in absorption of linearly polarized light polarized parallel and perpendicular to some external axis of orientation:

$$\mathrm{LD} = A_\parallel - A_\perp \tag{6-24}$$

The average absorbance A, which would be measured in unoriented solutions, is

$$A = (A_\parallel + 2A_\perp)/3 \tag{6-25}$$

and the reduced LD is then defined as

$$\mathrm{LD}^r = \mathrm{LD}/A = (A_{\parallel} - A_{\perp})/A \tag{6-26}$$

The orientation is most commonly produced by application of a flow or electrical field, but other methods are also useful and are discussed below.

Linear dichroism gives information on the orientation of chromophores with respect to the orienting field. If the orientation of the molecule with respect to the field is known, then LD can be interpreted in terms of the orientation of the chromophores with respect to the molecular axes. The LD can then be used, for example, to determine the tilt of the bases with respect to the helix axis, the angle of bending of a DNA segment, the orientation of a bound ligand with respect to the bases, or the distortion of a polynucleotide produced by protein binding.

Comprehensive reviews of the theory and practice of LD, with discussions of many applications, have been written by Michl and Thulstrup (1986), Schellman and Jensen (1987), and Nordén et al. (1992).

The orientational dependence of absorption occurs because light is absorbed only when its electric vector has a component along the direction of a transition dipole moment in the molecule. The electric and magnetic fields of the light are perpendicular to its direction of propagation. Thus, a crystal with nucleic acid base planes all parallel to a face will not absorb light via out-of-plane transitions for light incident perpendicular to this face. The out-of-plane transitions are perpendicular to the electric field of the light and thus do not interact with the light. The in-plane π–π^* transitions will absorb light dependent on the direction of polarization of the light. The absorbance is a maximum when the direction of polarization is parallel to the direction of the transition dipole (Figure 6-3) of the chromophore. Thus, measurements of LD can be used to determine the orientation of molecules, or of chromophores within a molecule, when the transition dipole direction is known. If the molecular orientation is known, then transition dipole directions within a molecule can be determined, which is how spectra of crystals are used to determine directions of transition dipoles.

As noted above, the LD depends on both the orientation of the molecule in the field, and the orientation of the transition moment with respect to the molecular axes. The reduced LD can be written as the product of a molecular orientation factor S and an optical factor O:

$$\mathrm{LD}^r = S \times O \tag{6-27}$$

The most common case is that in which the molecule (such as a DNA double helix) has cylindrical average symmetry with respect to the orienting field. Then,

$$S = \frac{1}{2}(3 < \cos^2 \Theta > -1) \tag{6-28}$$

where Θ is the angle between the helix axis and the laboratory axis established by the orienting field. The angular brackets denote a thermal average over all values of Θ. If the molecule is perfectly aligned ($\Theta = 0$, $\cos \Theta = 1$), then $S = 1$. If there is no alignment at all, $S = 0$. If the nucleic acid has tertiary structure, then S is a product of Eq. 6-28 with similar terms corresponding to each level of folding with effective cylindrical symmetry. For example, if the molecule were a coiled supercoil, there would be a multiplicative term $\frac{1}{2}(3\cos^2\beta_1 - 1)$ where β_1 is the supercoil pitch angle, and another multiplicative term $\frac{1}{2}(3\cos^2\beta_2 - 1)$ where β_2 is the coil pitch angle (Nordén et al., 1992).

The optical factor O may be obtained by the following argument. Eliminating $A_\perp$ between Eqs. 6-24 and 6-25, we find

$$\text{LD} = \frac{3}{2}(A_\| - A) \tag{6-29}$$

For complete alignment of the helix axes, $A_\| = \varepsilon_0 \cos^2\phi$, where ϕ is the angle between the transition moment and the helix axis, and ε_0 is the absorbance when the transition moment is completely aligned with the electric vector of the light. The isotropic average absorbance $A = \varepsilon_0/3$, since $< \cos^2\phi >= \frac{1}{3}$ for random orientation. Substituting these results in Eq. 6-26 gives

$$\text{O} = \text{LD}/A(\text{perfect orientation}) = \frac{3}{2}(3\cos^2\phi - 1) \tag{6-30}$$

Trigonometric transformation to angles more descriptive of base geometry (Fig. 6-15) gives for each transition i (Nordén, 1978; Causley and Johnson, 1982; Chou and Johnson, 1993):

$$O_i = \frac{3}{2}[3\sin^2\alpha_j \sin^2(\chi_j - \delta_{ij}) - 1] \tag{6-31}$$

where α_j is the inclination angle between the perpendicular to the base j and the helix axis, χ_j is the angle between inclination axis of base j and the N3–C6 axis of purines (N1–C4 of pyrimidines), and δ_{ij} is the angle in the plane of the base j between the transition dipole i and the N3–C6 axis of purines (N1–C4 of pyrimidines). The observed optical factor is then an average over all transitions denoted by subscripts i:

$$O = \frac{\sum F_i \varepsilon_i(\lambda) O_i}{\sum F_i \varepsilon_i(\lambda)} \tag{6-32}$$

where the summation is taken over all bases and transitions, weighted by the fraction F_i contributed by each base to transition i. Equation 6-32 shows that if all of the transitions have the same O_i, the observed optical factor will be independent of wavelength. Conversely, if O varies with λ, the O_i values must be different, implying that the bases are not perpendicular to the helix axis. (Another possibility is that there are n–π^* transitions perpendicular to the plane of the bases, but this appears not to be the case.)

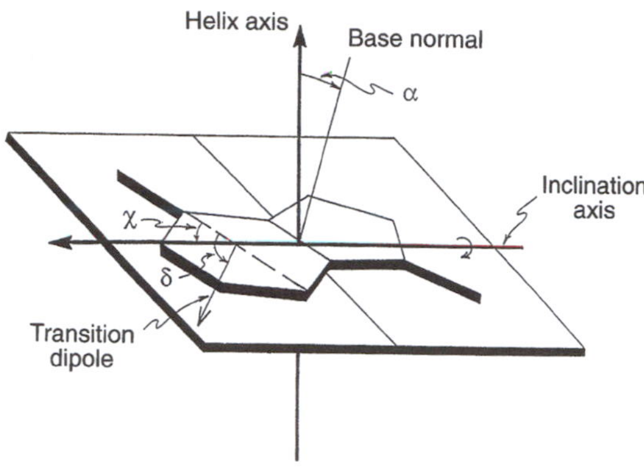

Figure 6-15
Diagram of angles used in the interpretation of LD meassurements on nucleic acids. Angles α_j and χ_j specify the orientation of each base. The inclination angle α_j is the angle between the plane of the base and the helix axis. The angle χ_j is the angle between the inclination axis and the N3–C6 axis of purines or the N1–C4 axis of pyrimidines. The angle δ_{ij} specifies the orientation of each transition dipole in the base plane; δ_{ij} is the angle between the transition dipole and the N3–C6 axis of purines or the N1–C4 axis of pyrimidines. [Reprinted with permission from R.-J. Chou and W. C. Johnson, Jr., *J. Am. Chem. Soc.*, **115**, 1205 (1993). Copyright ©1993 American Chemical Society. See Fig. 1.]

4.2 Methods of Aligning Samples

In general, the strategy of LD experiments is to define the orientation factor S, so that the optical factor O can be determined. Either very strong fields are applied, so that orientation can be extrapolated to perfect alignment ($S = 1$); or rather weak fields are used, for which an analytical theory of S is available.

Large nucleic acids, 1000 bp or longer, can be aligned in the velocity gradients produced by hydrodynamic flow to produce *flow dichroism.* The shear rate, G, is the velocity gradient dv/dx, where v is the local velocity of fluid flow and x is perpendicular to v. The parameter G has units of reciprocal seconds (s^{-1}). Velocity gradients can be produced by flowing the solution down a capillary tube or through a narrow slit, or by using a Couette apparatus in which the solution is held between two concentric cylinders, one of which is rotating. The Couette apparatus, with outer cylinder rotating, is the best of these alternatives because the velocity gradient is uniform between the cylinders, and hydrodynamic stability is greatest. For rigid particles, the orientation depends on the ratio $\beta = G/D_r$, where D_r is the rotational diffusion coefficient. For flexible polymers, the orientation parameter is $\beta = G\tau_1$, where τ_1 is the longest viscoelastic relaxation time (see Chapter 9 for detailed discussions of D_r and τ_1). In order to produce very highly oriented solutions, β must be on the order of 10^2, so that hydrodynamic forces greatly overbalance diffusion. At low shear rates, the linear dichroism generally depends on β^2, with proportionality constants depending on the optical arrangement of the experiment. Equations and graphs for various experimental configurations are given by Nordén et al. (1992).

If orientation is imposed by electric fields, S is given by Eq. 6-28 where Θ is the angle between the electric field and the molecular axis. The average is taken over a Boltzmann distribution, where the energy of the dipole μ in the field E is $U = -\mu E \cos\Theta$:

$$< \cos^2\Theta >= \frac{\int_0^\pi \cos^2\Theta e^{\mu \mathrm{E}\cos\Theta/kT}\sin\Theta d\Theta}{\int_0^\pi e^{\mu \mathrm{E}\cos\Theta/kT}\sin\Theta d\Theta}. \tag{6-33}$$

Unfortunately, the mechanism of interaction of nucleic acids with electric fields is complex, as elaborated in Chapter 11. The main interaction is a polarization of the ion atmosphere surrounding the polynucleotide, leading to an induced dipole moment proportional to the electric field and the polarizability α of the ion atmosphere: $\mu = \alpha E$. For medium-sized or large DNA molecules, α appears to be a function of E and of the duration of the orienting field. Thus, it is difficult to obtain an expression for $< \cos^2\Theta >$ as a function of E. Empirically, extrapolations of LD versus $1/E$ or $1/E^2$ to infinite field strength (apparent perfect orientation) are used. At low E, S is proportional to E^2 (the Kerr law).

Magnetic fields can also be used to orient DNA, since the motion of the π electrons in the plane of the bases produces a strong magnetic anisotropy. The strength of the magnetic dipole interaction is several orders of magnitude less than that of the electric dipole, so very high field superconducting magnets are required.

Very high degrees of orientation can be achieved in condensed phases. Large double-stranded polynucleotides can be very highly oriented in stretched or stroked films or fibers, or while undergoing gel electrophoresis. Monomers can be precisely oriented in crystals; polarized spectra of such crystals have been the main source of data on directions of transition moments, as in Figure 6-3.

4.3 Base Tilts in DNA

Flow and electric dichroism have produced some provocative results about the tilts of the bases in DNA in solution. Before considering these results, it is instructive to consider what is expected if the base pairs are perpendicular to the helix axis, and if the helix axis is perfectly aligned in the field ($S = 1$). If the helix axis is along the direction of polarization of the light (i.e., the orienting field and the polarization vector of the light are colinear), then the bases will be perpendicular to the field of the light and will not absorb: $A_{\parallel} = 0$. Then, Eqs. 6-26 and 6-29 yield $\mathrm{LD}^r = -\frac{3}{2}$. A value of the reduced LD less negative than this indicates either incomplete orientation or bases tilted from perpendicular to the helix axis. For example, Figure 6-16 shows reduced electric dichroism of a 250 bp fragment of calf thymus DNA at both low and high fields (Hogan et al., 1978). The inset shows an extrapolation to infinite field by using a plot of LD^r versus $1/E$. The intercept is -1.2, significantly less than the -1.5 expected.

The orientation of base pairs relative to the helix axis in synthetic polynucleotides and in double-stranded DNA in solution has been measured by flow dichroism and electric dichroism. Consider flow dichroism first. Johnson and coworkers (Causley and Johnson, 1982; Chou and Johnson, 1993; Kang and Johnson, 1993) oriented nucleic acids by pumping solutions between plates about 30 μm apart

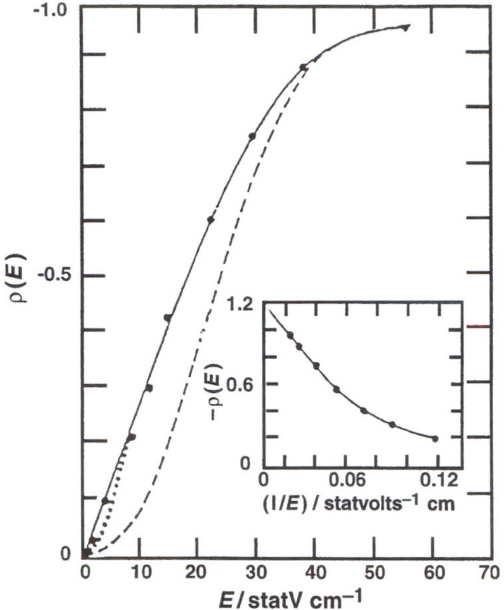

Figure 6-16
Reduced electric dichroism of a 250 bp fragment of calf thymus DNA at both low and high fields (Hogan et al., 1978). The inset shows an extrapolation to infinite field by using a plot of LD^r versus $1/E$.

to obtain shear rates in the range of 9000–62,000 s^{-1}. They made measurements on *E. coli* DNA in B and A form, and on synthetic polynucleotides: [d(A)·d(T)], [d(A-T)·d(A-T)], [d(C)·d(G)], [d(C-G)]·[d(C-G)], [d(A-C)]·[d(G-T)], [d(A-G)]·[d(C-T)], [r(A)·r(A^+)], and [r(C)·r(C^+)]. The wavelength range from 180 to 300 nm was studied. Measuring LD over this wide wavelength range, which encompasses three π–π^* absorption bands with known directions for the three transition dipoles, suffices to specify the plane of a base relative to the flow orientation axis. This method does not require knowledge about the extent of orientation in the flow; that is, the value of S need not be determined from a separate measurement.

For all wavelengths the LD was negative, thus all measurable transition dipoles make an angle greater than 55° to the helix axis ($3\cos^2\phi - 1$ is negative). This result is expected for π–π^* transitions in the plane of the bases; thus, the negative LD indicates that out-of-plane transitions are not significant. The reduced LD was wavelength-dependent; thus the base planes are not perpendicular to the orientation axis. If the base planes were perpendicular to the orientation axis, the reduced dichroism would remain constant independent of the directions of the transition dipoles in the base planes. Furthermore, the results were not shear dependent; this indicates a rigid molecule that is not distorted by the orientation. These conclusions are independent of quantitative results, but much more detailed conclusions can be reached from the measured values of reduced LD and knowledge of the directions of the transition dipoles. Table 6-2 gives measured inclination angles α (the angle between the normal to the plane of a base and the orientation axis of the DNA—presumably the helix axis) for natural and synthetic DNA molecules.

The most noteworthy result is the large inclination of the base planes for B-DNA in solution. The canonical B-form geometry from fiber diffraction data gives values

Table 6.2
The Orientation of the Base Planes Relative to the Flow Orientation Axis in DNA

DNA or RNA	Solvent	Base	α deg[a]
B-Form poly[d(A)•d(T)]	10 mM Na^+, pH 7	Adenine	23.2 ± 0.8
		Thymine	42.1 ± 2.5
B-Form poly[d(A-T)•d(A-T)]	10 mM Na^+, pH 7	Adenine	18.6 ± 0.6
		Thymine	34.8 ± 2.0
B-Form poly[d(C)•d(G)]	2 mM Na^+, pH 7	Guanine	20.1 ± 0.6
		Cytosine	33.8 ± 1.0
B-Form poly[d(C-G)]•[d(C-G)]	2 mM Na^+, pH 7	Guanine	21.4 ± 0.5
		Cytosine	34.0 ± 0.7
Z-Form poly[d(C-G)]•[d(C-G)]	70% TFE[b]	Guanine	27.1 ± 1.1
		Cytosine	32.1 ± 1.7
B-Form *E. coli* DNA	0.01–5.5 *M* salt	Adenine	16.1 ± 0.5
		Thymine	25.0 ± 0.9
		Guanine	18.0 ± 0.6
		Cytosine	25.1 ± 0.8
A-Form *E. coli* DNA	80% TFE[b]	Adenine	27.8 ± 1.0
		Thymine	34.7 ± 0.9
		Guanine	14.3 ± 1.0
		Cytosine	35.2 ± 0.5

[a]The inclination angle α is the angle between the normal to the base plane and the orientation axis of the DNA (see Fig. 6-15). Data are from Chou and Johnson (1993).

[b]TFE is trifluoroethanol.

near 7° for the base inclination (tilt) relative to the helix axis; also, data for Z-DNA crystals show a base tilt of about 7°. In contrast, the inclination angles in Table 6-2 are 16°–25° for B-DNA and 27° to 32° for Z-DNA Also, there is a large propeller twist in all the base pairs. Of course, the DNA conformation in solution need not be the same as in fibers or crystals, but such large discrepancies are striking.

Electric dichroism studies (Hogan et al., 1978; Wu et al., 1981; Charney and Yamaoka, 1982; Lee and Charney, 1982; Diekmann et al., 1982) of short fragments of B-form DNA extrapolated to infinite field gave similar results: They all yielded orientations of the 260-nm transition of 70°–75° with respect to the electrical orientation axis. This result corresponds to an inclination angle of the bases relative to the helix axis of at least 15°–20°. However, large DNA fragments such as linearized pBR322 with 4362 bp (Diekmann et al., 1982) and a 9200 bp calf thymus fragment (Lee and Charney, 1982) did give values of the minimum base inclination angle of about 6°

in agreement with canonical B form. Furthermore, short DNA fragments ($\sim$ 100 bp) have very sequence-dependent inclination angles, but they are always greater than the B form (Diekmann et al., 1982).

There are several possible interpretations of these large inclination angles. One is that, since DNA is not a rigid rod, all angles measured are averages. In LD the average over $\cos^2$ will increase the apparent inclination of the bases significantly for fluctuations of bases perpendicular to the rod axis. The fluctuations in a flexible rod that bends slightly will produce a nonzero apparent angle, even when the average angle is zero—the base planes are perpendicular to the helix axis.

Another possibility is that the double helix is not a rod, but rather is bent or has a superhelical shape (Lee and Charney, 1982). The local conformation is B form, but the bases may be tilted relative to the orientation axis because of the bending of the Watson–Crick helix. When molecules are oriented by a flow or electric field, the inclination angles are determined relative to the direction of orientation, which is not necessarily the local helix axis. Thus the measured angles depend on the global structure of the DNA (pitch of the superhelix or bend angles of the rod) as well as the local helix conformation. An elaboration of this idea is that the high electric field will tend to stretch the molecules and remove the bend or superhelix. This effect increases with increasing chain length, so long molecules will be straightened into linear rods, but shorter ones will not. As DNA bending occurs at runs of A•T base pairs and depends on the phasing of these sequences, the sequence-dependence of the electric dichroism for short fragments is not surprising.

It is interesting to note that when DNA forms side-by-side aggregates in solution that are similar to fibers used for X-ray diffraction, electric dichroism yields a base inclination of 7° in agreement with B form (Mandelkern et al., 1981).

In summary, LD measurements yield the angles between the transition moments and the axis of orientation of the molecules. These angles can be interpreted in terms of large base pair tilts and twists, and large propeller twists between bases in a base pair. Alternatively, there can be a sequence-dependent bending of the DNA helix or a super helical structure imposed on the synthetic polynucleotides or DNAs. It appears that the conformations of B- and Z-DNA duplexes are very sensitive to the environment and can be very different in solutions, in crystals, and in fibers. The DNA can differ in local secondary structure (characteristics of neighboring base pairs) and in global tertiary structure (bending or supercoiling).

4.4 Transient Electric Dichroism

When an electric field is applied to a solution of molecules, their orientation in the field is not instantaneous. Likewise, when the field is removed, the molecules disorient back to a random orientation in a finite time. The orientation and disorientation can be followed by measuring dichroism or birefringence. The disorientation is exponential in time and the relaxation time is equal to the rotational correlation time of the molecules, inversely proportional to the rotational diffusion coefficient. These hydrodynamic parameters have been used to determine the lengths of DNA molecules, and to determine whether they are bent. Details are given in Chapter 9.

5. INFRARED ABSORPTION AND RAMAN SCATTERING

5.1 Basic Ideas

Infrared absorption (IR) and Raman scattering both depend on the vibrational energy levels of the molecules. Infrared spectroscopy obeys the same principles as absorption spectroscopy at any other wavelength. Light with a frequency corresponding to the energy difference between two vibrational energy levels—usually of the ground electronic state—stimulates transitions between the two levels with a net absorption of the IR light. This is illustrated in Figure 6-17.

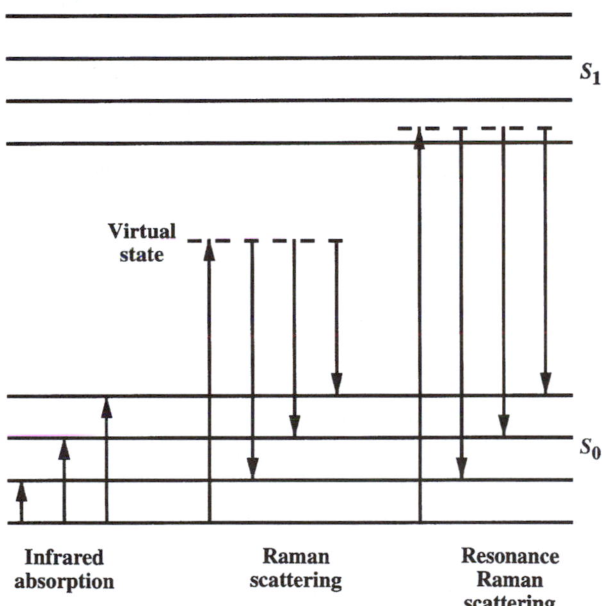

Figure 6-17
An energy diagram illustrating transitions in IR absorption, Raman scattering, and resonance Raman scattering. Infrared absorption occurs when there are transitions between vibrational energy levels; the molecule is usually in its ground electronic state (S_0). Scattering occurs whenever light is incident on a molecule. In Raman scattering, the scattered light has a different frequency from the incident light. The incident light excites the molecule from a particular vibrational state to a virtual state; the excited molecule returns to a different vibrational state. The difference in frequencies between incident and scattered light gives the vibrational energy. In resonance Raman, the virtual state is in or near an excited electronic state; the scattering is greatly enhanced.

The energy difference between vibrational levels is

$$\Delta E = h\nu_0 \tag{6-34}$$

where h is Planck's constant and ν_0 is the fundamental vibration frequency. For simple diatomic molecules consisting of atoms with masses m_1 and m_2,

$$\nu_0 = \frac{1}{2\pi}\sqrt{\frac{k}{\mu}} \tag{6-35}$$

where k is the Hooke's law force constant for stretching the bond and μ is the reduced mass $m_1 m_2/(m_1 + m_2)$. This equation shows that the vibrational frequency will be shifted to lower values if the bond is weakened and k is reduced (e.g., if some electron density is withdrawn from it by hydrogen bonding) or if the effective mass becomes larger (e.g., by isotopic substitution).

The motions probed by vibrational spectroscopy are the concerted vibrational motions (normal modes of vibration) of groups of atoms in the backbone and bases of DNA and RNA. These motions are influenced by the conformational state of the nucleic acid: whether it is double or single stranded, whether in A, B, or Z form, and so on. Vibrations of NH and C=O groups are especially sensitive to hydrogen bonding. The IR and Raman spectra of nucleic acids and their complexes have many absorption bands corresponding to these various motions, and are rich in structural information. The characteristic vibrational frequencies are still given by equations of the form of Eq. 6-35, but with k and μ now representing weighted averages of all the atoms participating in the group motion.

Although IR and Raman spectroscopy both probe vibrational motions, there are significant differences in experimental procedures and in the specific spectral bands that are observed. These differences are explained in detail below. Briefly, IR spectroscopy directly measures the absorption of IR radiation. It detects vibrations in which the net dipole moment of the molecule changes during the transition from ground to excited state. Since water also absorbs in the IR at wavelengths characteristic of C=O and N—H vibrations, it has been difficult to carry out IR spectroscopy on biopolymers in aqueous solutions. Raman spectroscopy, in contrast, is a form of scattering in which vibrational transitions are detected as frequency differences between incident and scattered light in the visible or UV region. It detects vibrations in which the net polarizability of the molecule changes during the transition. Raman spectroscopy does not suffer from interference by water absorption, and it can be applied to crystals and precipitates, but it has its own set of experimental difficulties.

There are several excellent surveys of experimental and theoretical aspects of Raman and IR spectroscopy with applications to nucleic acids and proteins (Hartman et al., 1973; Tsuboi and Takahashi, 1973; Tsuboi, 1974; Tu, 1982; Carey, 1982; Spiro, 1987; Peticolas et al., 1987; Tsuboi et al., 1987: Peticolas, 1995).

5.2 Infrared Absorption

The theoretical explanation for IR absorption is similar to that for electronic absorption. The intensity of absorption, when the frequency of incident light corresponds to an energy difference ν_0 in vibrational energy levels, depends on the square of a transition electric dipole moment. For a vibrational transition the transition dipole is directly proportional to the change in permanent dipole moment during the vibration. Infrared absorption thus requires the permanent dipole of the molecule to change as the nuclei vibrate.

The IR region of the spectrum corresponds to vibrational transitions; it is useful to assign the transitions to particular vibrations. A molecular vibration will involve motion of many nuclei, but some vibrations can be reasonably well localized. An illustration of the IR frequency range and the modes of vibrations identified is given in Figure 6-18 for an aromatic amino group such as in adenine, cytosine or guanine. The frequency range from 200 to 3400 cm^{-1} (wavelength range of 50–3 μm) covers the IR range used for nucleic acids.

Infrared absorption spectra for isolated nucleic acid bases and for bases in nucleosides and nucleotides have been measured in crystals, films, nonaqueous solvents, and D_2O solutions (Kyogoku et al., 1967; Tsuboi, 1974; Miles, 1975). The Fourier transform infrared (FTIR) method provides a great improvement over the conventional method in which a monochromator is used to obtain a spectrum of absorbance versus wavelength. In FTIR, a broad band of IR radiation is incident on the sample and the interference between different wavelengths transmitted by the sample is measured. Fourier transformation of the interferogram provides the conventional spectrum, but with greater sensitivity and S/N. The IR extinction coefficients are about a factor of 10 lower than the UV extinction coefficients. However, the large number of resolved IR bands with specific sensitivities to different conformations and interactions are very useful. Furthermore, the low extinction coefficients may be an advantage when studying equilibrium complexes that only form in high concentrations.

The region from 1450 to 1750 cm^{-1}, where the double bond (C=O, C=C, C=N) vibrations occur, is very sensitive to hydrogen bonding. There are significant frequency shifts of carbonyl groups and coupled ring vibrations when protons are added, are removed, or are hydrogen bonded. Infrared studies of the bases in $CHCl_3$, where the absence of water allows a significant amount of base–base hydrogen bonding, showed that Watson–Crick base pairs form more stable complexes than other base pairs (Kyogoku et al., 1967). In D_2O solution, tautomerism and ionization of the bases have been studied (Miles, 1961; Tsuboi, 1974;); the original determination that the bases were actually in the keto forms in solution, as postulated by Watson and Crick, was done by IR measurements. The shift in IR spectra with hydrogen bonding has been used to follow transitions from double strand to single strands in polynucleotides (Thomas, 1969). Figure 6-19 shows the IR spectrum of poly[r(A) • r(U)] compared with the averaged spectra of the two single strands, or the essentially identical spectrum of the sum of the nucleotides. Also shown is the spectrum of poly [r(G) • r(C)] compared with the sum of the nucleotide spectra. On melting the poly [r(A) • r(U)], the spectrum shifts to the sum of the single strands. As the IR shifts depend primarily on hydrogen bonding,

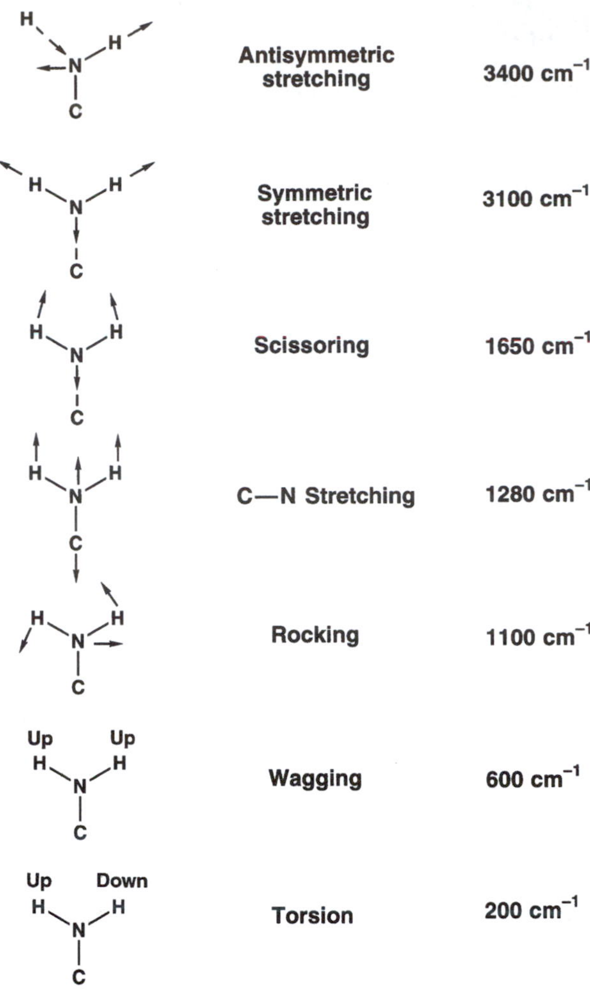

Figure 6-18
An illustration of the nomenclature and approximate frequencies for different types of vibrations in a planar C—NH_2 bond as found in adenine, cytosine, and guanine. Bond stretching frequencies are higher than frequencies for changing bond angles. The frequencies shown would all shift to lower values upon substitution of deuterium for hydrogen. [Reprinted with permission from Tsuboi, 1974.]

the absorbance should be directly proportional to the number of base pairs formed. In contrast, the UV absorbance depends on stacking and the interaction of more than one base. Infrared absorption in the 1500–1750-cm^{-1} frequency range can thus be used to count the number of A • U and G • C base pairs in RNAs (Thomas, 1969; Tsuboi, 1974).

Sugar–phosphate vibrations and some base vibrations are sensitive to conformation, which provides a useful measure of A to B to Z transitions (Taboury and Taillandier, 1985; Adam et al., 1986; Taillandier et al., 1987). The phosphate group shown

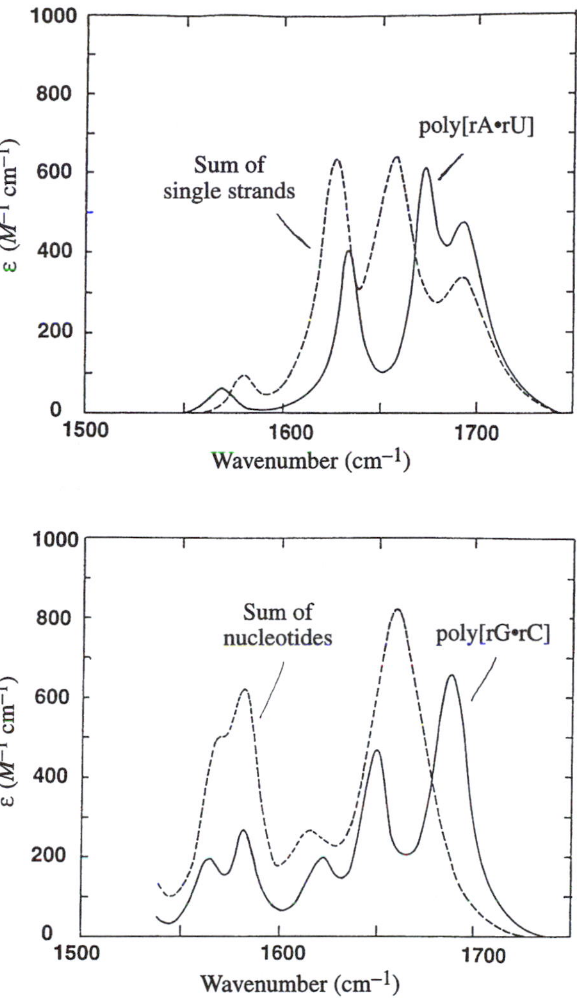

Figure 6-19
Comparison of the IR spectra in the double bond stretch region of base-paired poly [r(A) • r(U)] with its single strands (equivalent to its nucleotides) and of base-paired poly[r(G) • r(C)] with its nucleotides. The frequency shifts allow counting the number of A • U and G • C base pairs in an RNA. [Determination of the Base Pairing Content of Ribonucleic Acids by Infrared Spectroscopy, Thomas, G. J., Jr., *Biopolymers*, **7**, 325–334 (1969). Copyright ©1969. Reprinted by permission of John Wiley & Sons, Inc.]

below has four vibrations that depend on the conformation; they are the symmetric and antisymmetric stretches of the phosphodioxy group (OPO^-), and the symmetric and antisymmetric stretches of the phosphodiester group (ROPOR′).

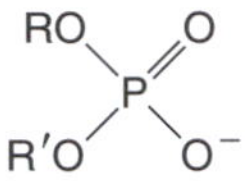

There are other marker bands that distinguish between C2′-*endo* sugars and C3′-*endo* sugars (Spiro, 1987). As the most extensive work on conformational transitions in nucleic acids has been done with Raman scattering, we will discuss the results in Section 6.5.4.

Infrared absorption spectra provide a wide range of bands to help characterize conformation, tautomerism, ionization, hydrogen–deuterium exchange, and hydrogen bonding in nucleic acids. Measurements can be made in D_2O solutions, films, or more complex environments such as viruses.

5.3 Infrared Linear Dichroism

Infrared LD measurements of oriented films of nucleic acids provides information about the orientation of particular bonds in the molecule. Interpretation of the data is done by the same methods used in UV linear dichroism; the inclination of a vibrational transition dipole is obtained relative to the orientation axis. To relate this measurement to molecular coordinates, one must orient the molecule completely in a known direction. Instead, the relative orientation of two bonds is usually determined from measured LD values of their two bands.

A study of the phosphate group orientation was done in oriented films of calf thymus and salmon sperm DNA; the C2—O2 bond in thymine was the reference bond (Pohle et al., 1984). The symmetric stretch of the phosphodioxy (1230 cm^{-1}) and its antisymmetric stretch (1090 cm^{-1}) suffice to fix the phosphate relative to the helix axis of a duplex. The measurements were done at high relative humidity (B form), at low relative humidity (A form) and in Li^+ salts (C form). The phosphate orientation is very different in these forms, particularly for A and B forms, as expected. The absolute orientation of the phosphate group agrees well in general with X-ray diffraction data on fibers and crystals. This finding lends credence to the IR results and establishes IR linear dichroism as a useful method to determine conformation in oriented systems.

5.4 Raman Scattering

In a scattering experiment, the incident light can have any frequency; there is no frequency matching requirement for light to be scattered. The intensity of the scattered light depends on the frequency of the light and on the properties of the molecule, but some light will be scattered at any frequency. A laser source is used to excite the sample, and the scattered light is collected and dispersed by a monochromator to obtain the frequency distribution of the scattered light. Most of the scattered light—the Rayleigh scattering—has nearly the same frequency as the incident light. But a very small fraction of the scattered light differs from the incident light by frequencies corresponding to energy levels in the molecule. The scattered light has gained energy from the molecule or lost energy to the molecule. The frequency shifted scattered light is the Raman scattering; the energy changes are shown in Figure 6-17 for a molecule that has gained vibrational energy from the incident light.

Raman spectra are presented in terms of frequency shifts (in cm^{-1}—called wavenumbers) relative to the incident light. The frequency (s^{-1}) is obtained from the frequency (cm^{-1}) by multiplication by the speed of light ($cm\ s^{-1}$). Strong absorption in the IR by water makes Raman scattering particularly useful for biological materials. Although much of the IR spectrum cannot be measured directly in aqueous solution, the Raman spectrum can be used to study the vibrational energy levels. In a Raman experiment, a wavelength can be chosen that is neither absorbed by the solvent nor by the sample. If a wavelength is chosen near the electronic absorption band of a chromophore, resonance Raman scattering occurs (see Fig. 6-17). Nonresonant Raman scattering reveals the vibrational levels of all the bonds in the sample. Resonant Raman preferentially excites bonds of the chromophore selected, which means that unique parts of a complex mixture can be studied.

Scattering of light is a two-photon interaction; a photon excites the molecule to a virtual state (a virtual state does not correspond to a single energy level of the molecule; it is not a stationary state) and a photon is emitted when the molecule is deexcited. Because the excited state is a virtual state, all the excited electronic states contribute to the probability of the transitions. The scattering of light depends on the fourth power of transition dipoles and on the fourth power of the frequency of the scattered light. The final result (Albrecht, 1961) is that the Raman scattering is proportional to the square of the transition polarizability tensor (the polarizability tensor characterizes the magnitude and direction of the electric dipole induced in a molecule by an electric field). Raman scattering requires that the polarizability tensor change as the nuclei vibrate. Both Raman scattering and IR absorption depend on vibrational transitions, but the intensities of the band for a particular transition will be different. The two spectra have different selection rules, and the intensities depend on different interactions.

Raman scattering can be measured for molecules in solution, in crystals, in fibers, or other complex mixtures. This wide range of environments is a great advantage over other optical methods. In solution, it requires a factor of 10 greater concentration than needed for UV absorbance or CD, but Raman scattering experiments can be done using only $5 \mu L$ of solution. Fluorescent impurities also present problems; they emit so much light that the Raman scattering signal may be lost.

Raman spectra can easily distinguish between the different conformations of DNA and RNA (Peticolas et al., 1987). Figure 6-20 compares A-type, low relative humidity fibers of DNA with B-type, high relative humidity fibers. The most striking difference is near 800 cm^{-1}, where the 807 cm^{-1} band, which has been assigned to the phosphodiester antisymmetric stretch in the A form, disappears in the B form (Peticolas et al., 1987).

Figure 6-21 shows the changes in Raman spectra when poly[d(G-C)] changes from B form to Z form. Also shown is the Raman spectrum of crystalline $d(C\text{-}G)_3$. The comparison of the Raman spectrum of the crystal, where X-ray diffraction provided

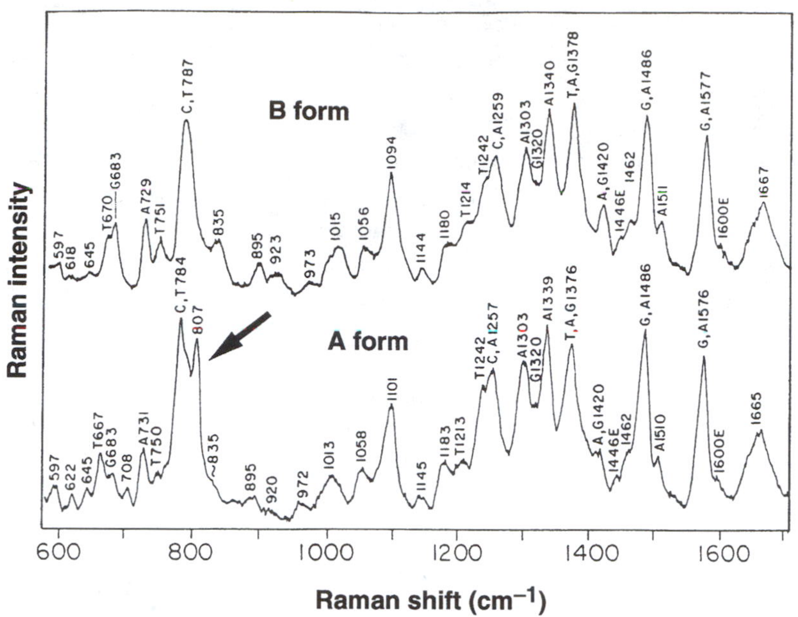

Figure 6-20
Comparison of the Raman spectra of A-type (92% relative humidity) and B-type (75% relative humidity) calf thymus DNA fibers. Note the disappearance of the phosphodiester antisymmetric stretch at 807 cm^{-1} when A-DNA converts to B-DNA. [Nucleic Acids, in *Biological Applications of Raman Spectroscopy*, Peticolas, W. L., Kubrasek, W. L., Thomas, G. A., and Tsuboi, M., Vol. 1, Spiro, T. G., Ed., Copyright ©1987. Reprinted by permission of John Wiley & Sons, Inc. See Fig. 3.]

a detailed structure, with the Raman spectrum of the solution proved that the form of poly[d(G-C)] in concentrated aqueous salt solution was indeed the left-handed Z form (Thamann et al., 1981). There are several significant differences seen in the spectra. The phosphodiester stretch near 800 cm^{-1} which distinguishes the A and B form, is also different in the Z form. A band at 675 cm^{-1} in B form shifts to 625 cm^{-1} in Z form; this band is assigned as a guanine breathing mode—in a ring breathing mode the ring expands and contracts. The frequency of this guanine breathing mode is different for *anti*-guanosine and for *syn*-guanosine. Similar changes occur in the Raman spectra of A-RNA on conversion to Z-RNA (Trulson et al., 1987). The Raman spectrum indicates that there are two left-handed forms of RNA; the one in 4 *M* $MgCl_2$ has a Raman and a CD spectrum similar to that of Z-DNA, the one in 6 *M* NaBr or $NaClO_4$ does not. Raman spectra of crystals and in solution have been used to deduce the propensity of different sequences to occur in different conformations (Wang et al., 1987). Details of a B-form crystal conformation have been compared with the same sequence in solution; the B form in the crystal showed a much greater heterogeneity than in solution (Benevides et al., 1988).

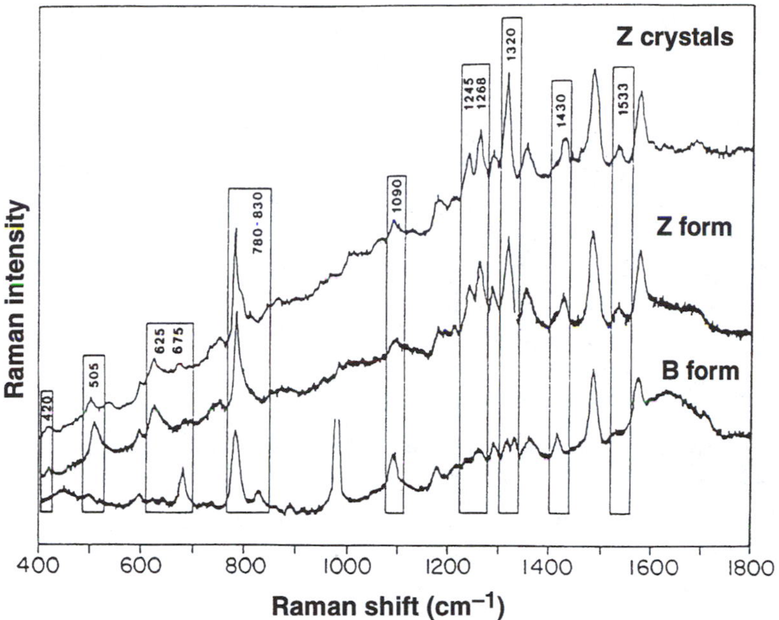

Figure 6-21
Comparison of the Raman spectra of poly[d(G-C)] in 4 *M* NaCl (Z form) and in 0.1 *M* NaCl (B form) with the spectrum of crystalline Z form d(CGCGCG). The identity of the crystal spectrum with the high salt spectrum defines the high salt form as left-handed Z. Nucleic Acids, in *Biological Applications of Raman Spectroscopy*, Petricolas, W. L., Kubasek, W. L., Thomas, G. A., and Tsuboi, M., Vol. 1, Spiro, T. G., Ed., Copyright ©1987. Reprinted by permission of John Wiley & Sons, Inc. See Fig. 18.

Raman scattering bands shift due to hydrogen bonding and conformational changes; there are also changes in intensity of Raman bands without significant shifts. Unstacking of bases leads to a Raman hyperchromism analogous to UV hyperchromism. As each type of base has different Raman bands, unstacking of each type of base in a nucleic acid can be monitored and compared with the loss of hydrogen bonding and the change of backbone conformation (Small and Peticolas, 1971). Transfer RNAs and 5S RNA have been analyzed this way (Chen et al., 1978).

RNA and DNA viruses (Fish et al., 1981; Thomas et al., 1983) and nucleosomes (Goodwin et al., 1979) have been studied by Raman scattering. Changes in the secondary structure of the RNA and DNA are clearly visible when the nucleic acid is encapsidated in a virus or wrapped around a histone. The ability to study nucleic acid bands and protein bands concurrently provides useful information about the protein–nucleic acid interaction (Thomas, 1987). Raman spectra have been obtained of single chromosomes in intact cells (Puppels et al., 1990).

Raman scattering and other optical spectroscopic methods respond to, and thus measure, all the species present in the sample. If each species has a distinct Raman band, the concentration of each species can be monitored, and the details of a transition can be followed. Nuclear magnetic resonance spectra depend on a much longer time scale than optical spectroscopy, where the kinetics of a transition becomes important. If equilibrium between two species is attained in milliseconds or faster, the NMR shows peaks that are the average for the two species. Only slow exchange between species gives individual NMR peaks for each species, whereas individual Raman peaks are seen for any rate of chemical exchange. Thus rapid exchange between syn and anti or 2′-*endo* and 3′-*endo*, which could not be distinguished from an intermediate conformation by NMR, might be resolved by Raman scattering. Samples with mixtures of conformations can thus be analyzed effectively (Taillandier et al., 1987).

5.5 Vibrational Circular Dichroism and Raman Optical Activity

The same principles that govern the CD of electronic transitions apply to vibrations as well. Vibrational (or IR) CD will be observed if the charge displacements associated with a vibrational transition have a helical character. In double-stranded DNA, the helical motion is thought to arise from dipolar coupling of transitions associated with the helically arranged bases. For example, in polynucleotides and oligonucleotides containing dC-dG, the vibrational CD in the region 1550–1750 cm^{-1} appears to be due to coupled stretching modes of the carbonyl residues (Gulotta et al., 1989; Zhong et al., 1990). Since the dipole change is nearly parallel to the C=O bond direction, vibrational CD can give information on the relative positions and orientations of the purine and pyrimidine bases.

An example is shown in Figure 6-22, for poly(dG-dC) • poly(dG-dC) in the B conformation (low salt) and the Z conformation (0.7 *M* $MgCl_2$). In the lower panel, the IR absorption spectra associated with the C=O stretch show some substructure due to exciton splitting of coupled vibrations. In the upper panel, the vibrational CD spectra reflect the change of handedness accompanying the B–Z transition. In the B

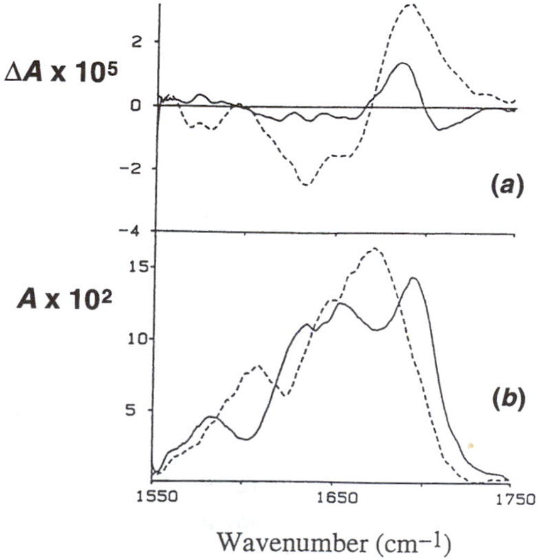

Figure 6-22
Infrared CD (*a*) and absorption (*b*) spectra of poly(dG-dC) • poly(dG-dC) in the low salt B-conformation (solid lines) and in 0.7 *M* $MgCl_2$ Z conformation (dashed lines). [IR Vibrational CD in Model Deoxyoligonucleotides: Observation of the B → Z Phase Transition and Extended Coupled Oscillator Intensity Calculations, M. Gulotta, D.J. Goss, M. Diem, *Biopolymers* **28**, 2047–2058 (1989). Copyright ©1989. Reprinted by permission of John Wiley & Sons, Inc.]

conformation, the positive lobe is at lower wavenumbers, while in the Z conformation, it is at higher wavenumbers. The vibrational CD spectrum is also more intense for Z-DNA, owing to changes in the distances and orientations of the C=O dipoles.

In analogy to the difference in scattered intensity for right and left circularly polarized light (CIDS), there exists a difference in Raman scattered intensity for chiral molecules in circularly polarized light . This difference is usually called Raman optical activity (ROA). The ROA characterizes chiral vibrations, just as vibrational CD does. Reviews of ROA are given by Hug (1994) and Nafie et al. (1995).

References

Aboul-ela, F., Ph. D. Thesis, Sequence-Dependent Structure and Thermodynamics of DNA Oligonucleotides and Polynucleotides: UV Melting and NMR Studies, University of California, Berkeley, CA, 1987.

Aboul-ela, F., Varani, G., Walker, G. T., and Tinoco, I., Jr. (1988). The TFIIIA Recognition Fragment d(GGATGGGAG): d(CTCCCATCC) is B-form in Solution, *Nucleic Acids Res.* **16**, 3559–3572.

Adam, S., Liquier, J., Taboury, J. A., and Taillandier, E. (1986). Right- and Left-Handed Helices of Poly[d(A-T)] • Poly[d(A-T)] Investigated by Infrared Spectroscopy, *Biochemistry* **25**, 3220–3225.

Albrecht, A. C. (1961). On the Theory of Raman Intensities, *J. Chem. Phys.* **34**, 1476–1484.

Allen, F. S., Gray, D. M., and Ratliff, R. L. (1984). On the First-Neighbor Analysis of Nucleic Acid CD Spectra: The Definitive T Matrix and Considerations of Various Methods, *Biopolymers* **23**, 2639–2659.

Applequist, J. and Damle, V. (1966). Thermodynamics of the One-Stranded Helix–Coil Equilibrium in Polyadenylic Acid, *J. Am. Chem. Soc.* **88**, 3895–4000.

Baase, W. A. and Johnson, W. C., Jr. (1979). Circular Dichroism and DNA Secondary Structure, *Nucleic Acids Res.* **6**, 797–814.

Baker, B. M., Vanderkoi, J., and Kallenbach, N. R. (1978). Base Stacking in a Fluorescent Dinucleoside Monophosphate: εApεA, *Biopolymers* **17**, 1362–1367.

Beardsley, K., Tao, T., and Cantor, C. R. (1970). Studies on the Conformation of the Anticodon Loop of Phenylalanine Transfer Ribonucleic Acid. Effect of Environment on the Fluorescence of the Y Base, *Biochemistry* **9**, 3524–3530.

Benevides, J. M., Wang, A. H.-J., van der Marel, G. A., van Boom, J. H., and Thomas, G. J., Jr. (1988). Crystal and Solution Structures of the B-DNA Dodecamer d(CGCAAATTTGCG) Probed by Raman Spectroscopy: Heterogeneity in the Crystal Structure Does Not Persist in the Solution Structure, *Biochemistry* **27**, 931–938.

Bloomfield, V. A., Crothers, D. M., and Tinoco, I., Jr. (1974). *Physical Chemistry of Nucleic Acids*, Harper and Row, New York, pp.322–328.

Bush, C. A. and Tinoco, I., Jr. (1967). Calculation of the Optical Rotatory Dispersion of Dinucleoside Phosphates. Appendix. Derivation of the ORD and CD Curves from Absorption and Rotational Strengths, *J. Mol. Biol.* **23**, 601–614.

Bustamante, C., Tinoco, I., Jr., and Maestre, M. F. (1983). Circular Differential Scattering Can Be an Important Part of the Circular Dichroism of Macromolecules. *Proc. Natl. Acad. Sci. USA* **80**, 3568–3572.

Callis, P. R. (1983). Electronic States and Luminescence of Nucleic Acid Systems, *Annu. Rev. Phys. Chem.* **34**, 329–357.

Cantor, C. R. and Tinoco, I., Jr. (1965). Absorption and Optical Rotatory Dispersion of Seven Trinucleoside Diphosphates, *J. Mol. Biol.* **13**, 65–77.

Cantor, C. R. and Schimmel, P. R. (1980). *Biophysical Chemistry. Part II. Techniques for the Study of Biological Structure and Function*, Freeman, San Francisco, CA.

Carey, P. R. (1982). *Biochemical Applications of Raman and Resonance Raman Spectroscopies*, Academic, New York, p.262.

Causley, G. C. and Johnson, W. C., Jr. (1982). Polynucleotide Conformation from Flow Dichroism Studies, *Biopolymers* **21**, 1763–1780.

Charney, E. and Yamaoka, K. (1982). Electric Dichroism of Deoxynucleic Acid in Aqueous Solutions: Electric Field Dependence, *Biochemistry* **21**, 834–842.

Chen, M. C., Giegé, R., Lord, R. C., and Rich, A. (1978). Raman Spectra of Ten Aqueous Transfer RNAs and 5S RNA. Conformational Comparison with Yeast Phenylalanine Transfer RNA, *Biochemistry* **17**, 3134–3138.

Chou, P.-J. and Johnson, W. C., Jr. (1993). Base Inclinations in Natural and Synthetic DNAs *J. Am Chem. Soc.* **115**, 1205–1214.

Clegg, R., Murchie, A., Zechel, A., Carlberg, C., Diekmann, S., and Lilley, D. (1992). Fluorescence Resonance Energy Transfer Analysis of the Structure of the Four-way DNA Junction, *Biochemistry* **31**, 4846–4856.

Cruz, P., Hall, K., Puglisi, J., Davis, P., Hardin, C. C., Trulson, M. O., Mathies, R. A., Tinoco, I., Jr., Johnson, W. C., Jr., and Neilson, T. (1986). The Left-Handed Z-form of Double-Stranded RNA, in *Biomolecular Stereodynamics* **IV**, Adenine Press, New York, pp. 179–200.

DeVoe, H. (1964). Optical Properties of Molecular Aggregates. I. Classical Model of Electronic Absorption and Refraction, *J. Chem. Phys.* **41**, 393–400.

Diekmann, S., Hillen, W., Jung, M., Wells, R. D., and Pörschke, D. (1982). Electric Properties and Structure of DNA Restriction Fragments from Measurements of the Electric Dichroism, *Biophys. Chem.* **15**, 157–167.

Dorman, B. P., Hearst, J. E., and Maestre, M. F. (1973). UV Absorption and Circular Dichroism Measurements on Light Scattering Biological Specimens; Fluorescent Cell and Related Large-angle Light Detection Techniques, *Methods Enzymol.* **27D**, 767–796.

Edwards, E. L., Patrick, M. H., Ratliff, R. L., and Gray, D. M. (1990). A•T and C^+•C Base Pairs Can Form Simultaneously in a Novel Multistranded DNA Complex, *Biochemistry* **29**, 828–836.

Eis, P. S. and Millar, D. P. (1993). Conformational Distribution of a Four-Way DNA Junction Revealed by Time-Resolved Fluorescence Resonance Energy Transfer, *Biochemistry* **32**, 13852–13860.

Evans, D. H. and Morgan, A. R. (1982). Extrahelical Bases in Duplex DNA, *J. Mol. Biol.* **160**, 117–122.

Felsenfeld, G. and Hirschman, S. Z. (1965). A Neighbor-Interaction Analysis of the Hypochromism and Spectra of DNA, *J. Mol. Biol.* **13**, 407–427.

Felsenfeld, G., and Rich, A. (1957). Studies on the Formation of Two- and Three-Stranded Polyribonucleotides, *Biochim. Biophys. Acta* **26**, 457–468.

Fish, S. R., Hartman, K. A., Stubbs, G. J., and Thomas, G. J., Jr. (1981). Structural Studies of Tobacco Mosaic Virus and its Components by Laser Raman Spectroscopy, *Biochemistry* **20**, 7449–7457.

Goodwin, D. C., Vergne, J., Brahms, J., Defer, N., and Kruh, J. (1979). Nucleosome Structure: Sites of Interaction of Proteins in the DNA Grooves as Determined by Raman Scattering, *Biochemistry* **18**, 2057–2064.

Gottesfeld, J. M., Blanco, J., and Tenant, L. L. (1987). The 5S Gene Internal Control Region is B-form Both Free in Solution and in a Complex with TFIIIA, *Nature (London)* **329**, 460–462.

Gralla, J. and Crothers, D. M. (1973). Free Energy of Imperfect Nucleic Acid Helices III. Small Internal Loops Resulting From Mismatches, *J. Mol. Biol.* **78**, 301–319.

Gray, D. M., Cui, T., and Ratliff, R. L. (1984). Circular Dichroism Measurements Show that C•C^+ Base Pairs Can Exist with A•T Base Pairs Between Antiparallel Strands of an Oligonucleotide Double-Helix, *Nucleic Acids Res.* **12**, 7565–7580.

Gray, D. M., Hamilton, F. D., and Vaughn, M. R. (1978). The Analysis of Circular Dichroism Spectra of Natural DNAs Using Spectral Components from Synthetic DNAs, *Biopolymers* **17**, 85–106.

Gray, D. M., Hung, S.-H., and Johnson, K. H. (1995) Absorption and Circular Dichroism Spectroscopy of Nucleic Acid Duplexes and Triplexes, *Methods Enzymol.* **246**, 19–34,

Gray, D. M., Liu, J.-J., Ratliff, R. L., and Allen, F. S. (1981). Sequence Dependence of the Circular Dichroism of Synthetic Double–Stranded RNAs, *Biopolymers* **20**, 1337–1382.

Gray, D. M., Johnson, K. H., Vaughan, M. E., Morris, P. A., Sutherland, J. C., and Ratliff, R. L. (1990). The Vacuum UV CD Bands of Repeating DNA Sequences Are Dependent on Sequence and Conformation, *Biopolymers* **29**, 317–323.

Gray, D. M., Vaughan, M., Ratliff, R. L., and Hayes, F. N. (1980). Circular Dichroism Spectra Show That Repeating Dinucleotide DNAs May Form Helices in Which Every Other Base is Looped Out, *Nucleic Acids Res.* **8**, 3695–3707.

Gueron, M., Eisinger, J., and Lamola, A. A. (1974). Excited States of Nucleic Acids, in *Basic Principles in Nucleic Acid Chemistry*, Vol. I, Ts'o, P. O. P., Ed., Academic, New York, pp. 312–398.

Gulotta, M., Goss, D. J., and Diem, M. (1989). IR Vibrational CD in Model Deoxyoligonucleotides: Observation of the B $\rightarrow$ Z Phase Transition and Extended Coupled Oscillator Intensity Calculations, *Biopolymers* **28**, 2047–2058.

Hall, K., Cruz, P., Tinoco, I., Jr., Jovin, T. M., and van de Sande, J. H. (1984). Z-RNA—A Left-handed RNA Double Helix, *Nature (London)* **311**, 584–586.

Hartman, K. A., Lord, R. C., and Thomas, G. J., Jr. (1973). Structural Studies of Nucleic Acids and Polynucleotides by Infrared and Raman Spectroscopy, in *Physico-Chemical Properties of Nucleic Acids*, Vol. 2, Duchesne, J., Ed. Academic, New York, pp. 1–89.

Hirschman, S. Z. and Felsenfeld, G. (1966). Determination of DNA Composition and Concentration by Spectral Analysis, *J. Mol. Biol.* **16**, 347–358.

Hogan, M., Dattagupta, N., and Crothers, D. M. (1978). Transient Electric Dichroism of Rod-Like Molecules, *Proc. Natl. Acad. Sci. USA* **75**, 195–199.

Hug, W. (1994). Vibrational Raman Optical Activity Comes of Age, *Chimia* **48**, 386–390.

Johnson, B. B., Dahl, K. S., Tinoco, I., Jr., Ivanov, V. I., Zhurkin, V. B. (1981). Correlations Between DNA Structural Parameters and Calculated Circular Dichroism, *Biochemistry* **20**, 73–78.

Johnson, W. C., Jr. (1978) Circular Dichroism Spectroscopy and the Vacuum Ultraviolet Region, *Annu. Rev. Phys. Chem.* **29**, 93–114.

Kang, H. and Johnson, W. C., Jr. (1993). Linear Dichroism Demonstrates that the Bases in Poly[d(AC)] • Poly[d(GT)] and Poly[d(AG)] • Poly[d(CT)] Are Inclined to the Helix Axis, *Biopolymers* **33**, 245–253.

Keller, D. and Bustamante, C. (1986). Theory of the Interaction of Light with Large Inhomogenous Molecular Aggregates. II. Psi-Type Circular Dichroism, *J. Chem. Phys.* **84**, 2972–2980.

Kliger, D. S., Lewis, J. W., and Randall, C. E. (1990). *Polarized Light in Optics and Spectroscopy*, Academic, Boston, p. 304.

Kyogoku, Y., Lord, R. C., and Rich, A. (1967). The Effect of Substituents on the Hydrogen Bonding of Adenine and Uracil Derivatives, *Proc. Natl. Acad. Sci. USA* **57**, 250–257.

Lamos, M. L., Walker, G. T., Krugh, T. R., and Turner, D. H. (1986). Fluorescence-Detected Circular Dichroism of Ethidium Bound to Poly(dG-dC) and Poly(dG-m^{5}dC) under B- and Z-Form Conditions, *Biochemistry* **25**, 687–691.

Lee, C.-H. and Charney, E. (1982). Solution Conformation of DNA, *J. Mol. Biol.* **161**, 289–303.

Leng, M. and Felsenfeld, G. (1966). A Study of Polyadenylic Acid at Neutral pH *J. Mol. Biol.* **15**, 455–466.

Lerman, L. S. (1973). The Polymer and Salt-Induced Condensation of DNA, in *Physico-Chemical Properties of Nucleic Acids*, Vol. 3, Duchesne, J., Ed. Academic, New York, pp. 59–76.

Lobenstein, E. W., Schaefer, W. C., and Turner, D. H. (1985). Fluorescent Detected Circular Dichroism of Proteins with Single Fluorescent Tryptophans, *J. Am. Chem. Soc.* **103**, 4936–4940.

Maestre, M. F. and Reich, C. (1980). Contribution of Light Scattering to the Circular Dichroism of Deoxyribonucleic Acid Films, Deoxyribonucleic Acid–Polylysine Complexes, and Deoxyribonucleic Acid Particles in Ethanolic Buffers, *Biochemistry* **19**, 5214–5223.

Maestre, M. F., Bustamante, C., Hayes, T. L., Subirana, J. A., and Tinoco, I., Jr. (1982). Differential Scattering of Circularly Polarized Light by the Helical Sperm Head from the Octopus Eledone cirrhosa. *Nature (London)* **298**, 773–774.

Mandelkern, M., Dattagupta, N., and Crothers, D. M. (1981). Conversion of B DNA Between Solution and Fiber Conformations, *Proc. Natl. Acad. Sci. USA* **78**, 4294–4298.

Marmur, J. and Doty, P. (1962). Determination of the Base Composition of Deoxyribonucleic Acid (DNA) *J. Mol. Biol.* **5**, 109–118.

Marky, L. A., Blumenfeld, K. S., Kozlowski, S., and Breslauer, K. J. (1983). Salt-Dependent Conformational Transitions in the Self-Complementary Deoxydodecanucleotide d(CGCGAATTCGCG): Evidence for Hairpin Formation, *Biopolymers* **22**, 1247–1257.

McCall, M., Brown, T., Hunter, W. N., and Kennard, O. (1986). The Crystal Structure of d(GGATGGGAG) Forms an Essential Part of the Binding Site for Transcriptional Factor IIIA, *Nature (London)* **322**, 661–664.

Michl, J. and Thulstrup, E. W. (1986). *Spectroscopy with Polarized Light*, VCH, New York, 573 pp.

Miles, H. T. (1961). Tautomeric Forms in a Polynucleotide Helix and Their Bearing on the Structure of DNA, *Proc. Natl. Acad. Sci. USA* **47**, 781–802.

Miles, H. T. (1975) Spectra and tables compiled in Handbook of Biochemistry and Molecular Biology, 3rd ed., *Nucleic Acids*, Vol. 1, Fasman, G. D., Ed., CRC Press, Cleveland, Oh, pp.604–623.

Nafie, L. A., Yu, G. S., and Freeman, T. B. (1995) Raman Optical Activity of Biological Molecules, *Vibrational Spectroscopy* **8**, 231–239.

Nordén, B. (1978). Applications of linear dichroism spectroscopy, *Appl. Spectrosc. Rev.* **14**, 157–248.

Nordén, B., Kubista, M., and Kurucsev, T. (1992). Linear dichroism spectroscopy of nucleic acids, *Q. Revs. Biophys.* **25**, 51–170.

Peticolas, W. L. (1995). Raman Spectroscopy of Proteins and Nucleic Acids, *Methods Enzymol.* **246**, 389–415.

Peticolas, W. L., Kubasek, W. L., Thomas, G. A., and Tsuboi, M. (1987). Nucleic Acids, in *Biological Applications of Raman Spectroscopy*, Vol. 1, Spiro, T. G., Ed., Wiley, New York, pp. 81–133.

Pohl, F. M. and Jovin, T. M., (1972). Salt-induced Co-operative Conformational Change of a Synthetic DNA: Equilibrium and Kinetic Studies with Poly(dG-dC), *J. Mol. Biol.* **67**, 375–396.

Pohle, W., Zhurkin, V. B., and Fritsche, H. (1984). The DNA Phosphate Orientation. Infrared Data and Energetically Favorable Structures, *Biopolymers* **23**, 2603–2622.

Puppels, G. J., de Mul, F. F. M., Otto, C., Greve, J., Robert-Nicoud, M., D. Ardt-Jovin, J., and Jovin, T. M. (1990). Studying Single Living Cells and Chromosomes by Confocal Raman Microspectroscopy, *Nature (London)* **347**, 301–303.

Reich, C., Maestre, M. F., Edmondson, S. and Gray, D. M. (1980). Circular Dichroism and Fluorescent-Detected Circular Dichroism of Deoxyribonucleic Acid and Poly[d(A-C) • d(G-T)] in Ethanolic Solutions: A New Method for Estimating Circular Intensity Differential Scattering. *Biochemistry* **19**, 5208–5213.

Rhodes, D. and Klug, A. (1986). An Underlying Repeat in Some Transcriptional Control Sequences Corresponding to Half a Double Helical Turn of DNA, *Cell* **46**, 123–132.

Riazance, J. H., Baase, W. A., Johnson, W. C., Jr., Hall, K., Cruz, P., and Tinoco, I., Jr. (1985). Evidence for Z-Form RNA by Vacuum UV Circular Dichroism, *Nucleic Acids Res.* **13**, 4983–4989.

Sauer, K., Ed. (1995). *Methods Enzymol.* **246**, 816.

Schellman, J. and Jensen, H. P. (1987). Optical Spectroscopy of Oriented Molecules, *Chem Rev.* **87**, 1359–1399.

Secrist, J. A., III, Barrio, J. R., Leonard, N. J., and Weber, G. (1972). Fluorescent Modification of Adenosine-Containing Coenzymes. Biological Activities and Spectrosopic Properties, *Biochemistry* **11**, 3499–3506.

Selvin, P. R. and Hearst, J. E. (1994). Luminescence Energy Transfer Using a Terbium Chelate—Improvements on Fluorescence Energy Transfer, *Proc. Natl. Acad. Sci. USA* **91**, 10024–10028.

Shapiro, D. B., Maestre, M. F., McClain, W. M., Hull, P. G., Shi, Y., Quinby-Hunt, M. S., Hearst, J. E., and Hunt, A. J. (1994). Determination of the Average Orientation of DNA in the Octopus Sperm *Eledone cirrhosa* Through Polarized Light Scattering, *Appl. Opt.* **33**, 5733–5744.

Small, E. W. and Peticolas, W. L. (1971). Conformational Dependence of the Raman Scattering Intensities from Polynucleotides. III. Order–Disorder Changes in Helical Structures, *Biopolymers* **10**, 1377–1416.

Spiro, T. G., Ed. (1987). In *Biological Applications of Raman Spectroscopy*, Vol. 1: Raman Spectra and the Conformations of Biological Macromolecules, Vol. 2: Resonance Raman Spectra of Polyenes and Aromatics, Wiley, New York.

Sprecher, C. A., Baase, W. A., and Johnson, W. C., Jr. (1979). Conformation and Circular Dichroism of DNA, *Biopolymers* **18**, 1009–1019.

Sprecher, C. A. and Johnson, W. C., Jr. (1977). Circular Dichroism of the Nucleic Acid Monomers, *Biopolymers* **16**, 2243–2264.

Steely, H. T., Jr., Gray, D. M., and Ratliff, R. L. (1986). CD of Homopolymer DNA.RNA Hybrid Duplexes and Triplexes Containing A • T or A • U Base Pairs, *Nucleic Acids Res.* **14**, 11071–10090.

Stryer, L. (1978) Fluorescence Energy Transfer as a Spectroscopic Ruler, *Annu. Rev. Biochem.* **47**, 819–846.

Taboury J. A. and Taillandier, E. (1985). Right-Handed and Left-Handed Helices of Poly(dA-dC) • (dG-dT), *Nucleic Acids Res.* **13**, 4469–4482.

Taillandier, E., Ridoux, J.-P., Leupin, W., Denny, W. A., Wang, Y., Thomas, G. A., and Peticolas, W. L. (1987). Infrared and Raman Studies Show that Poly(dA) • Poly(dT) and $d(AAAAATTTTT)_2$ Exhibit a Heteronomous Conformation in Films at 75% Relative Humidity and a B-Type Conformation at High Humidities and in Solution, *Biochemistry* **26**, 3361–3368.

Thamann, T. J., Lord, R. C., Wang, A. H.-J., and Rich, A. (1981). The High Salt Form of Poly (dG-dC)·Poly (dG-dC) is Left-Handed Z-DNA: Raman Spectra of Crystals and Solutions, *Nucleic Acids Res.* **9**, 5443–5457 .

Thomas, G. J., Jr. (1969). Determination of the Base Pairing Content of Ribonucleic Acids by Infrared Spectroscopy, *Biopolymers* **7**, 325–334.

Thomas, G. J., Jr. (1987). Viruses and Nucleoproteins, in *Biological Applications of Raman Spectroscopy*, Vol. 1, Spiro, T. G., Ed., Wiley, New York, pp. 81-133.

Thomas, G. J., Jr., Prescott, B., and Day, L. A. (1983). Structure Similarity, Difference and Variability in the Filamentous Viruses fd, If1, Pf1 and Xf. Investigations by Laser Raman Spectroscopy, *J. Mol. Biol.* **165**, 321–356.

Tinoco, I., Jr. (1960). Hypochromism in Polynucleotides, *J. Am. Chem. Soc.* **82**, 4785–4790.

Tinoco, I., Jr., Bustamante, C., and Maestre, M. F. (1980). The Optical Activity of Nucleic Acids and Their Aggregates, *Annu. Rev. Biophys. Bioeng.* **9**, 107–141.

Tinoco, I., Jr., Ehrenberg, B., and Steinberg, I. Z. (1977). Fluorescence Detected Circular Dichroism and Circular Polarization of Luminescence in Rigid Media: Direction Dependent Optical Acitivity Obtained by Photoselection, *J. Chem. Phys.* **66**, 916–920.

Tinoco, I., Jr., Mickols, W., Maestre, M. F., and Bustamante, C. (1987). Absorption, Scattering and Imaging of Biomolecular Structures with Polarized Light, *Annu. Rev. Biophysics Biophys. Chem.* **16**, 319–349.

Tinoco, I., Jr. and Williams, A. L.,, Jr. (1984). Differential Absorption and Differential Scattering of Circularly Polarized Light: Applications to Biological Macromolecules, *Annu. Rev. Phys. Chem.* **35**, 329–355.

Tinoco, I., Jr. and Turner, D. H. (1976). Fluorescence Detected Circular Dichroism. Theory, *J. Am. Chem. Soc.* **98**, 6453–6456.

Trulson, M. O., Cruz, P., Puglisi, J. D., Tinoco, I., Jr. and Mathies, R. A. (1987). Raman Spectroscopic Study of Left-Handed Z-RNA, *Biochemistry* **26**, 8624–8630.

Tsuboi, M. (1974). Infrared and Raman Spectroscopy, in *Basic Principles in Nucleic Acid Chemistry*, Vol. 1, Ts'o, P. O. P., Ed., Academic, New York, pp. 399–452.

Tsuboi, M. and Takahashi, S. (1973). Infrared and Raman Spectra of Nucleic Acids—Vibrations in the Base-residues, in *Physico-Chemical Properties of Nucleic Acids*, Vol. 2, Duchesne, J., Ed., Academic, New York, pp. 91–145.

Tsuboi, M., Nishimura, Y., Hirakawa, A. Y., and Peticolas, W. (1987). Resonance Raman Spectroscopy and Normal Modes of the Nucleic Acid Bases, in *Biological Applications of Raman Spectroscopy*, Vol. 2, Spiro, T. G., Ed., Wiley, New York, pp. 109–179.

Tu, A.T. (1982). *Raman Spectroscopy in Biology: Principles and Applications*, Wiley, New York.

Tunis-Schneider, M. J. B. and Maestre, M. F. (1970). Circular Dichroism Spectra of Oriented and Unoriented Deoxyribonucleic Acid Films, *J. Mol. Biol.* **52**, 521–541.

Turner, D. H., Tinoco, I., Jr. and Maestre, M. F. (1975). Fluorescence Detected Circular Dichroism Study of the Anticodon Loop of Yeast $tRNA^{Phe}$, *Biochemistry* **14**, 3794–3799.

Tuschl, T., Gohlke, C., Jovin, T. M.,, Westhof, E., Eckstein, F. (1994). Three-Dimensional Model for the Hammerhead Ribozyme Based on Fluorescence Measurements, *Science* **266**, 785–789.

Vizard, R. L. and Ansevin, A. T. (1976). High Resolution Thermal Denaturation of DNA: Thermalites of Bacteriophage DNA, *Biochemistry* **15**, 741–750.

Wada, A., Yabuki, S., and Husimi, Y. (1980). Fine Structure in the Thermal Denaturation of DNA: High Temperature-Resolution Spectrophotometric Studies, *CRC Crit. Rev. Biochem.* **9**, 87–144.

Wang, J. C. (1979b). Helical Repeat of DNA in Solution, *Proc. Natl. Acad. Sci. USA* **76**, 200–203.

Wang, A. H.-J., Quigley, G. J., Kolpak, F. J., Crawford, J. L., van Boom, J. H., van der Marel, G., and Rich, A. (1979a). Molecular Structure of a Left-handed Double Helical DNA Fragment at Atomic Resolution, *Nature (London)* **282**, 680–686.

Wang, Y., Thomas, G. A., and Peticolas, W. L. (1987). Sequence Dependence of the B to Z Transition in Crystals and Aqueous NaCl Solutions for Deoxyoligonucleotides Containing All Four Canonical DNA Bases, *Biochemistry* **26**, 5178–5186.

Warshaw, M. M. and Cantor, C. R. (1970) Oligonucleotide Interactions. IV. Conformational Differences Between Deoxy- and Ribodinucleoside Phosphates, *Biopolymers* **9**, 1079–1103.

Wartell, R. M. and Benight, A. S. (1985). Thermal Denaturation of DNA Molecules: A Comparison of Theory with Experiment, *Phys. Rep.* **126**, 67–107.

Weller, A. (1961). Fast Reactions of Excited Molecules. *Prog. Reac. Kinet.* **1**. 187–214.

Werntges, H., Steger, G., Riesner, D., and Fritz, H.-J. (1986). Mismatches in DNA Double Strands: Thermodynamic Parameters and Their Correlation to Repair Efficiencies, *Nucleic Acids Res.* **14**, 3773–3790.

Williams A. L., Jr., Cheong, C., Tinoco, I., Jr., and Clark, L. B., (1986). Vacuum Ultraviolet Circular Dichroism as an Indicator of Helical Handedness in Nucleic Acids, *Nucleic Acids Res.* **14**, 6649–6659.

Wu, H. M., N. Dattagupta and Crothers, D. M. (1981). Solution Structural Studies of the A and Z Forms of DNA, *Proc. Natl. Acad. Sci. USA* **78**, 6808–6811.

Yoon, K., Turner, D. H., and Tinoco, I., Jr. (1975). The Kinetics of Codon–Anticodon Interaction in Yeast Phenylalanine Transfer RNA, *J. Mol. Biol.* **99**, 507–518.

Zhong, W., Gulotta, M., Goss, D. J., and Diem, M. (1990). DNA Solution Conformation Via Infrared Circular Dichroism: Experimental and Theoretical Results for B-family Polymers, *Biochemistry* **29**, 7485–7491.

CHAPTER 7

Theoretical Methods

by Peter Kollman

1. THEORETICAL FOUNDATIONS

1.1 Overview

Theory can provide an understanding of the structures and interactions of nucleic acids. Theoretical studies include the application of quantum mechanics, molecular mechanics and dynamics, and statistical mechanics to rationalize, understand, and predict experimentally observable properties. If the theoretical foundations and parameters are sufficiently well established, theory can also be used to calculate aspects of molecular behavior that are unobservable because of experimental complexity, a use that is particularly important in nucleic acid and other biological molecules. An example, discussed later in Section 7.2, is the calculation of the distribution of torsional displacements, relative to perfect helical geometry, as a function of base sequence. Such information is difficult to extract unequivocably from NMR or X-ray diffraction experiments. If such information is not verifiable experimentally, it must be viewed with caution, but the calculations can provide considerable insight. The extensive use of molecular theory to understand biomolecular behavior has depended strongly on the development of supercomputers.

If our computational power were great enough, the laws of quantum mechanics could be solved directly to predict the behavior of any molecular system. These laws are expressed in the Schrödinger equation, proposed in 1926, which is assumed to contain the fundamental physics that enables all the properties of individual molecules to be derived. In principle, the electronic properties of a molecule can be calculated from the interactions among all the electrons and nuclei in a molecule. The fundamental conceptual basis for qualitative insights into chemical bonding and reactivity are contained in such a theory (Woodward and Hoffmann, 1970). However, it has only been since the 1960s with the advent of significant computing power, that this equation has been able to be used in an ab initio (first principles) rigorous approach to derive properties of molecules. Even so, the size of the molecules that ab initio approaches can treat accurately enough to compete with experiment is still small (Schaefer, 1986; Szabo and Ostlund, 1982; Parr and Yang, 1989). Semiempirical applications of the quantum mechanical equations have also been very useful (Dewar et al., 1985).

Unlike the direct quantum mechanical methods, molecular mechanics and molecular dynamics approaches (Burkert and Allinger, 1982; McCammon and Harvey, 1987; Goodfellow, 1990) assume an effective potential for atom–atom interactions, based either entirely on empirical data or on both empirical data and quantum mechanical calculations. This effective potential can be used in three different ways to obtain molecular conformations, or to calculate molecular interactions. Let us consider a single molecule, a molecule surrounded by solvent, or any collection of molecules; this is called a system. The arrangement of the atoms in the system is called a conformation (for a single molecule) or a configuration (for many molecules), although the terms are often used interchangeably.

In *energy minimization*, an initial conformation is varied so as to lower its energy; this is continued until an energy minimum is reached. This procedure finds the local energy minimum near the starting structure. It will provide the best structure not too different from the starting structure; for example, it can thus be used to improve a structure obtained from X-ray or NMR data, (Brünger, 1990).

In the *Monte Carlo* method, a statistical sample of the potential energy surface for many different configurations is obtained. Large changes in configuration are made, including those that raise the energy. Changes that lower the energy are favored, but maxima as well as minima are sampled. This protocol allows a global minimum to be found if the search of configuration space is wide enough, and it provides a free energy minimum, instead of an energy minimum. The free energy of a system depends on the number of different configurations that have nearly the same energy. Thus the free energy minimum depends not only on the depth of the energy minimum, but also on its width. The Monte Carlo method provides the shape of the energy surface and thus the free energy minimum.

Molecular dynamics uses the effective potential between atoms to calculate the forces between the atoms. Each atom is given an initial velocity and Newton's equations of motion are solved to calculate how the velocity (magnitude and direction) is affected by all the other atoms. Motion of all the molecules is thus simulated; the molecules can change conformations and can interact. The free energy of the system can be calculated from the molecular dynamics to obtain minimum free energies and free energies of activation for non-covalent processes.

In the remainder of this section, we describe the basic ideas behind the development of molecular mechanics and molecular dynamics approaches to simulating intramolecular and intermolecular interactions of molecules in general and nucleic acids in particular.

1.2 Intermolecular Interactions

Accurate ab initio calculations on small molecules and their intermolecular complexes can be done. We use the water molecule as an example, because of the importance of water as a solvent and because the structure and energy of the hydrogen bond between two water molecules has been accurately described by calculations on $(H_2O)_2$ and H_2O itself (Kollman, 1977a).

Water has a dipole moment of 1.85 D (Eisenberg and Kauzmann, 1969), which arises from the fact that the electrons are arranged so as to produce a net negative charge near the oxygen nucleus and a net positive charge near the hydrogen nuclei. (The dipole moment is the product of charge and separation distance. One negative electronic charge of 4.8×10^{-10} esu separated by 1 Å, or 10^{-8} cm, from an equal positive charge produces a dipole moment of 4.8×10^{-18} esu cm, or 4.8 D.) Water also has quadrupole and higher electric moments as well (Margenau and Kestner, 1970; Kollman, 1977b). For some analyses, it is useful to reduce the complete quantum mechanically calculated charge distribution to a series of classical mechanics point charges or multipoles, centered on the atoms or bonds of the molecule (Singh and Kollman, 1984). Figure 7-1 shows such a simple representation of the water molecule's charge distribution.

One can analyze the energy of interaction of two water molecules by quantum mechanical perturbation theory and divide the total energy of interaction between the two water molecules into the major components: (1) An electrostatic interaction energy between all the multipoles of each molecule, including a quantum mechanical "penetration term," which has no classical mechanical analogy; (2) Pauli exchange repulsion between the electrons of the two molecules, which also has no classical mechanical analogy and arises from the fact that the wave function for the system must be antisymmetric with respect to electron exchange; (3) a polarization of each of the charge distributions of the molecule by the presence of the other molecule, which can be described by a classical model with induceable dipoles on the atoms or bonds of the molecule; (4) charge transfer of electron density from one molecule to another,

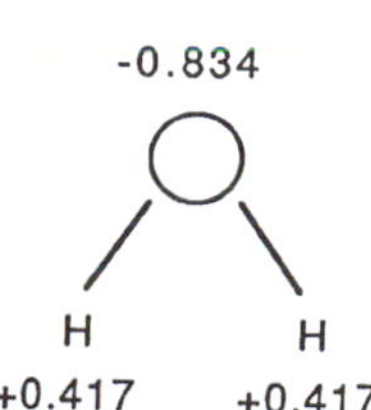

Figure 7-1
The TIP3P model of water with the atomic charges (q_i) on each atom. In simulations of many water molecules, the energy of the system is represented as $E = \sum q_i q_j / R_{ij} + A_{OO}/R_{OO}^{-12} - B_{OO}/R_{OO}^{-6}$ where the sum is over pairs of molecules. For each molecule pair, there are 9 (3 × 3) $q_i q_j / R_{ij}$ electrostatic terms, but a single van der Waals term (R^{-6} and R^{-12}) involving only the oxygen parameters and distances. (See nonbonded interaction in Eq. 7-1.)

and (5) induced dipole–induced dipole (dispersion) attraction between the molecules (Kollman, 1995). There are, in addition, other terms that tend to be smaller but not negligible.

For polar molecules, the most important attractive component is the electrostatic, whereas dispersion attraction is the most important attractive component for nonpolar molecules. Exchange repulsion is the component that is most critical in describing the intermolecular separation of molecules. These various simple components can be reasonably represented by analytical potential functions, which describe the energy of the system as a function of intermolecular geometry. (Weiner et al., 1984)

Most analytical potential functions include electrostatic, dispersion, and exchange repulsion effects, although there are a few that simulate polarization and charge transfer as well. Those that leave out these two and other higher order effects have included them implicitly if the potential function parameters are fit to experimental data. For example, water potentials, such as TIP3P (Jorgensen et al., 1983), that reproduce the enthalpy of the liquid have a dipole moment that is approximately 30% higher than the gas-phase value, thus presumably representing the polarization effect of surrounding water molecules on a molecule in the liquid.

In any case, one of the premises of molecular mechanics treatments of intermolecular interactions is that these interactions can be represented by analytical potential functions. The validity of this premise can be tested by comparing simulated results with those from experiments.

1.3 Intramolecular Interactions

The assumption usually made in analyzing the energy of molecules with conformational flexibility is that the relative energies of the different conformations can be represented by analytical potential functions that contain two features: (1) a Fourier series of energy terms representing the intrinsic energy of rotation around a given bond and (2) nonbonded (electrostatic, dispersion and exchange repulsion) interactions between the rest of the molecule whose relative orientations are affected by the bond rotation. The goal is to describe the energy of the system as a function of its various torsion angles (defined in Fig. 2-4). As an example of the results of such calculations in a small, model system, we consider ethane and its derivatives.

There is evidence from quantum mechanical calculations that the conformational energy of rotation around a given bond can be understood in terms of electron delocalization effects from bonding orbitals on one side of the bond to antibonding orbitals on the other side (Levine, 1991). One can represent the quantum mechanical wave function of a molecule as a combination of localized orbitals. These localized orbitals can be described as bonding (σ), antibonding (σ^*) and lone-pair (n) orbitals. A simple representation of the ground state of the molecule will have all electrons in purely bonding (σ) and lone-pair (n) orbitals.

However, a more complete representation will allow some mixture of antibonding orbitals into the wave function. Brunck and Weinhold (1979) note that this mixing of σ^* orbitals into the wave function is conformation dependent and is maximum for bonds attached to a central bond when these bonds are trans with respect to each other.

For example, ethane [Fig. 7-2(*a*)] prefers to be staggered instead of eclipsed because this conformation allows maximum trans σ–σ^* mixing. Nonbonded interactions between the hydrogens also favor the staggered over the eclipsed conformations, but the magnitude of this exchange repulsion–dispersion contribution is only 0.5 kcal mol^{-1} out of a total barrier height of 2.8 kcal mol^{-1}.

Such nonbonded (exchange repulsion, dispersion, and electrostatic) interactions modulate the intrinsic quantum mechanical torsion effects in an important way in such molecules as butane and 1,2-difluoroethane [Fig. 7-2(*b*)]. For example, in butane, the trans conformation is more stable than the gauche by approximately 0.7 kcal mol^{-1} because the exchange repulsion between the methyl group is minimized in that conformation. On the other hand, in 1,2-difluoroethane, the gauche conformation is more stable than the trans by 0.7 kcal mol^{-1} in the gas phase, and this conformation preference increases in solution. Why is this? It has been shown that this gauche preference comes about because a gauche conformation of the fluorines enables a trans orientation between fluorines and hydrogen. These trans F···H orientations maximize the C-H(σ)-C-F(σ^*) bond–antibond mixing and stabilize the gauche conformation by about 6 kcal mol^{-1} (Brunck and Weinhold, 1979). On the other hand, the nonbonded (mainly electrostatic) repulsions between fluorines is approximately 5 kcal mol^{-1} greater in the gauche conformation than the trans, leading to an approximately 1 kcal mol^{-1} net stabilization of the gauche conformation. This gauche preference increases in solution because the quantum mechanical electron delocalization effects are not very affected by solvent, but the solvation of the gauche conformer is more favorable than the trans conformer because the gauche conformer has a larger dipole moment.

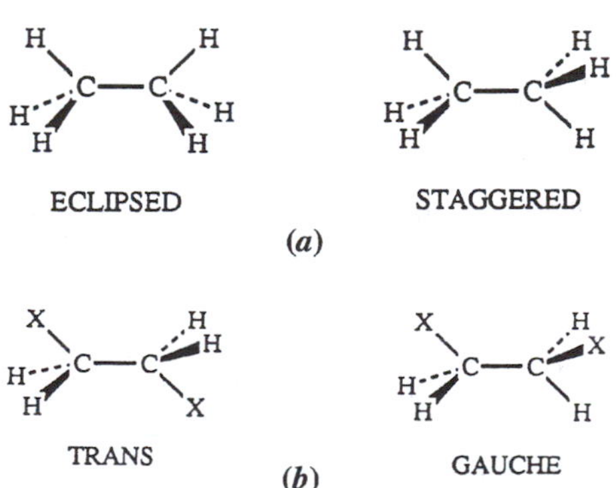

Figure 7-2
(*a*) Ethane in staggered and eclipsed conformations. In the staggered conformation H—C—C—H torsion angles are 60° (gauche$^+$), 180° (trans) or 300° (gauche$^-$); in the eclipsed conformation all angles are cis (0°, 120° or 240°). (*b*) The trans and gauche conformations of butane (X = CH_3) and 1,2-difluoroethane (X = F).

In summary, we can understand molecular conformational preferences in terms of electron delocalization, a quantum mechanical effect that can be empirically represented by a Fourier series, and nonbonded interactions that can be represented analogously to intermolecular nonbonded interactions.

1.4 An Analytical Model for Intermolecular and Intramolecular Effects: A Force Field

As described above, conformations and interactions can be most rigorously studied using quantum mechanical methods. Quantum mechanical methods solve for the electronic structure of molecules and thus derive the effective Born–Oppenheimer potential for nuclear motion from first principles (Levine, 1991). Ab initio methods solve for the energy and wave functions with the "correct" (typically nonrelativistic) Hamiltonian. Semiempirical quantum mechanical methods simplify this process by leaving out much of the most time-consuming part of the calculations, the evaluation of electron repulsion integrals, and making appropriate empirical adjustments to other terms in the Hamiltonian to compensate (Dewar et al., 1985).

However, none of the quantum mechanical methods are currently able to address many of the questions of interest to nucleic acid chemists because the calculations on systems of more than 10–20 non-hydrogen atoms are either prohibitively time consuming, or, if carried out at an approximate level of quantum mechanical theory, rather inaccurate. Instead, one can use the insights gained from the quantum mechanical calculations on small model systems and construct an analytical model that contains the most important physical effects expected from the model systems. The usefulness of such an approach derives from the accuracy of the Born–Oppenheimer approximation, in which one describes the motion of the nuclei of molecules on a potential surface caused by the electronic structure. The Born–Oppenheimer approximation works because electrons are so much lighter than nuclei that they respond rapidly to changes in nuclear positions (Levine, 1991).

A typical analytical model (Weiner et al., 1984) for the total potential energy of the system V_{total} is presented in Eq. 7-1; the equation and its associated parameters is called a force field.

$$V_{\text{total}} = \sum_{\text{bonds}} K_r(r - r_{\text{eq}})^2 + \sum_{\text{angles}} K_\theta(\theta - \theta_{\text{eq}})^2 + \sum_{\text{dihedrals}} \frac{V_n}{2}[1 + \cos(n\phi - \gamma)] + \sum_{i<j}\left(\frac{A_{ij}}{R_{ij}^{12}} - \frac{B_{ij}}{R_{ij}^{6}} + \frac{q_i q_j}{\varepsilon R_{ij}}\right) + \sum_{H\text{-bonds}}\sum_{i<j}\left(\frac{C_{ij}}{R_{ij}^{12}} - \frac{C_{ij}}{R_{ij}^{10}}\right) \quad (7\text{-}1)$$

There are analytical energy terms for bond stretching, bond bending, torsional rotation, and out-of-plane distortion (top line of Eq. 7-1). Nonbonded energies include exchange repulsion, dispersion, and electrostatic terms, and there is an additional hydrogen-bond term, which is not included in many analytical models in the literature, but which can be used to simulate "charge transfer" and other nonbonded interaction energies (bottom line of Eq. 7-1).

In the above model, the bond energy term represents the energy of distorting the bonds from their equilibrium values, r_{eq}. The stronger the bond, the larger the force constant K_r. A similar energy term represents bond angle distortions from their equilibrium values, θ_{eq} with force constants K_θ. A Fourier series represents the relative energy of the molecule as a function of torsion angles with periodicity n, barrier height V_n, and phase angle γ. For rotation around the bond linking two tetrahedral atoms

(sp^3 bonding), n is 3; for sp^2 bonding, n is 2. If $\gamma = 0°$ the staggered conformation is more stable (e.g., ethane); if $\gamma = 180°$, the eclipsed conformation is more stable. The nonbonded energy is represented by exchange repulsion A_{ij}/R_{ij}^{12}, dispersion attraction B_{ij}/R_{ij}^{6}, and electrostatic energy $q_i q_j/\varepsilon R_{ij}$, each being a sum over all atoms i and j separated by at least three bonds. There are parameters A_{ij} and B_{ij} for each pair of interacting atoms; q_i and q_j are net charges and ε is the dielectric constant. The final hydrogen-bonded term replaces the typical exchange repulsion–dispersion terms with C_{ij}/R_{ij}^{12} and D_{ij}/R_{ij}^{10} functions for hydrogen-bonded atoms.

One should appreciate that analytical models such as Eq. 7-1 have a large number of parameters that can come either from experiment or from quantum mechanical calculations on model systems. For example, Weiner et al. (1984) used quantum mechanical methods to derive the atomic charges for nucleic acid constituents and empirical methods to derive most of the rest of the potential parameters. Other potentials include those developed by Brooks et al. (1983) and van Gunsteren and Berendsen (1990). The validity of such models must be tested as widely as possible in comparison with experiment, in order to critically analyze the limitations and applicability of a function such as Eq. 7-1.

One can also go beyond the approximations inherent in Eq. 7-1 in a number of ways. For example, in both the force fields developed by Allinger et al. (1989) and by Maple et al. (1994) one includes a considerably more complex representation of intramolecular energies. For example, one can include higher order terms in the individual terms (e.g., quartic terms in bond stretching) or coupling between them (e.g., bond-torsional coupling). On the other hand, one can also use a more complex function for non-bonded interactions; for example, including induced dipoles in the model (Lybrand and Kollman, 1985; Caldwell et al. 1990; Dang et al. 1991). The importance of these two extensions in representing nucleic acid properties is yet to be established.

The force fields developed for nucleic acids fall into two general classes: one category allows only variation in dihedral angle space (with the possible exception of angle bending changes in the sugar ring) and thus does not consider the first two terms in Eq. 7-1. Such approaches have been developed by Sasisekharan (1973), Olson (1975), Broyde and Hingerty (1983), and Lavery and Hartmann (1994), among others. These approaches are inherently more efficient in terms of the number of independent conformational variables in the system than Cartesian coordinate approaches, but are more difficult to use in molecular dynamics. Cartesian coordinate approaches, in which all degrees of freedom are allowed to vary using Eq. 7-1, have been implemented for nucleic acids by Kollman et al. (1981), Levitt (1983), Weiner et al. (1984), and Nilsson and Karplus (1986), among others. This approach appears more general and flexible, although it is less efficient from the point of view of number of degrees of freedom.

The power of using a force field such as Eq. 7-1 lies in the fact that it is easy to calculate efficiently its analytical derivatives. The first derivative, or gradient, gives the (negative) force on each atom, which is used in molecular dynamics as discussed in Section 7.1.5.2. The analytical second derivatives at energy minima allow one to calculate normal mode frequencies and thermochemical properties (McCammon and Harvey, 1987).

2. SIMULATION METHODS

2.1 Computer Graphics and Distance Geometry

The advent of computer graphics hardware and software has allowed for real-time manipulation of molecules in stereo and color (Gund et al., 1987), and has included such features as molecular dot surfaces (Langridge et al., 1981). These hardware and software advances allow one to dock and to manipulate molecules and molecular complexes in order to understand intermolecular specificity. The role of electrostatic interactions, hydrophobic interactions, hydrogen bonding, exchange repulsion, and dispersion effects in DNA interactions with proteins and with drugs (Weiner et al., 1982) can be evaluated.

The application of qualitative computer graphics techniques requires the input of molecular structures, which for macromolecules typically come from X-ray crystallographic and fiber diffraction data, NMR spectroscopy, or from model building based on such data. Another model-building approach is that of distance geometry (Crippen, 1981), which uses stereochemical principles for describing bond distances and angles, then adds other constraints formulated as atom–atom distances. From upper and lower bound distance matrices one can construct a number of structures consistent with the atom–atom distances. The root-mean-square difference between atom positions in the various structures gives a quantitative measure of how well the input atom–atom distances have defined the three-dimensional (3D) structure. Distance geometry has been most powerful in creating stereochemically reasonable structures that cover as wide a range of structures as allowed by the data. This method has been used in combination with two-dimensional (2D) NMR methods to describe 3D structures of oligonucleotides in solution (see Chapter 5).

2.1.1 Molecular Mechanics Methods—Energy Minimization, Monte Carlo, Molecular Dynamics

The above methods do not use energies at all in describing molecules and depend only on the stereochemical principles and experimental data used as input. Energy-based, molecular mechanical methods use simple analytical expressions (such as Eq. 7-1) to describe the energy of the system. One can begin the system in a particular nuclear configuration and minimize the energy (energy minimization) or solve Newton's equation of motion on such a surface (molecular dynamics). Such methods can be applied to systems with thousands of atoms because the energy functions and their derivatives are so simple to evaluate.

The single most difficult problem in theoretical simulations of complex molecules is the local minimum problem, that is, the fact that it is very difficult and time consuming to evaluate all the low-energy structures for even dinucleoside phosphates. Thus, determining the global minimum as well as all the populated local minima of a complex molecule in solution is extremely difficult, despite the fact that, with molecular mechanics energy functions, the energy and its derivatives can be rapidly evaluated. Often, in simulations of complex molecules, solvent and counterions are not included explicitly because such inclusion increases the number of particles considerably. For example, one turn of a B-form double helix requires 20 Na^+ counterions and several thousand H_2O molecules for realistic modeling. However, the result of simulations on small molecules suggests that if one includes solvent and counterions, and uses equations such as 7-1, one can reproduce some experimental data well (Bash et al., 1987b).

In recent years, there have been some exciting developments in "implicit models" to represent the environment (solvent and counterions) around macromolecules. The solution of the electrostatic Poisson–Boltzmann equation (Sharp and Honig, 1990) has been used to describe counterion and electrostatic solvation effects. The GBSA model developed by Still et al. (1990) enables the representation of electrostatic and van der Waals solvent effects in an approximate way. The AMSOL models solves the semiempirical quantum mechanical electronic structure problem in the presence of continuous solvent, with the solvation terms analogous to those used by GBSA (Cramer and Truhlar, 1991).

As noted above, there are two molecular mechanics methods (McCammon and Harvey, 1987; Goodfellow, 1990), which are used to obtain average properties: *Monte Carlo* and *molecular dynamics*. In the Monte Carlo method as devised by Metropolis et al. (1953), one starts with a given configuration and changes it randomly. The energy of the initial configuration and the changed configuration is calculated. Then the Boltzmann factor P

$$P = e^{-(E_{\text{new}} - E_{\text{old}})/RT} \tag{7-2}$$

is computed to determine whether to accept the new configuration or stay at the old. Here E_{new} and E_{old} are the energies of the new and old configurations, R is the gas constant, and T is the absolute temperature. Note that if E_{new} is less than E_{old}, the Boltzmann factor is greater than 1. Changes in configuration that lower the energy of the system ($P > 1$) are always allowed. If E_{new} is greater than E_{old}, an increase in energy occurs, and P is less than 1. The larger the energy increase, the smaller is the value of the Boltzmann factor. Changes in configuration that raise the energy are accepted only if P is greater than a random number between 0 and 1. In this way, a sampling of the configurational space of the system is done. Averages of any desired property are calculated as a function of the number of steps; eventually equilibrium is reached so that the averages no longer change. The fact that steps that increase as well as decrease the energy are allowed, provides a path for the system to move out of a local minimum and attain a global minimum if enough steps are taken (assuming the process converges).

In molecular dynamics (Flores and Moss, 1990), one starts with a given configuration, velocities often chosen from a Maxwellian distribution and uses Newton's equations of motion to calculate how the positions and velocities of the atoms of the system changes with time. The starting configuration specifies a position $\boldsymbol{r}_i$ and a velocity $\boldsymbol{v}_i$ for each atom. The atoms move for 1 or 2 fs (10^{-15} s), then the force $\boldsymbol{F}_i$ on each atom i, which according to standard physics is the negative gradient of the potential energy, is calculated

$$\mathbf{F}_{\text{i}} = -\nabla_i V_{\text{tot}} \tag{7-3}$$

by taking the analytical derivative of the energy in Eq. 7-1. The gradient operator is

$$\nabla_{\mathrm{i}} = \frac{\partial}{\partial x}\mathbf{i} + \frac{\partial}{\partial y}\mathbf{j} + \frac{\partial}{\partial z}\mathbf{k} \tag{7-4}$$

where (**i**, **j**, **k**) are unit vectors along the (x, y, z) axes, and the derivatives are taken at the position of atom i. Newton's equations are then used to calculate the acceleration $d^2\mathbf{r}_{\mathrm{i}}/dt^2$ of each atom i from its mass m_i and the force acting on it:

$$\mathbf{F}_{\mathrm{i}} = m_{\mathrm{i}}\frac{d^2\mathbf{r}_{\mathrm{i}}}{dt^2} \tag{7-5}$$

The forces accelerate the atoms—change the velocities—and the system evolves for another one or two femtoseconds. The position and velocity of each atom can be calculated from knowledge of its previous position and velocity. Thus the system can change in time so as to sample a wide range of configurations. A movie can be made of the process, or snapshots can be taken, to see how the molecules move with time. Eventually, an equilibrium configuration should be reached so that only fluctuations around this configuration are seen. Average properties, including energy, can be calculated. The molecular dynamics can be simulated at different temperatures by adjusting the average velocities and thus the average kinetic energy of the atoms. The average kinetic energy of an atom is directly proportional to the absolute temperature T:

$$< \frac{1}{2}mv^2 >= \frac{3}{2}k_{\mathrm{B}}T \tag{7-6}$$

where k_{B} is the Boltzmann constant and m and v are the mass and velocity of an atom; the average $< >$ is over all atoms in the system.

Molecular dynamics can be used with artificial force terms to simulate many different situations. For example, potentials can be added to require close agreement with experimental values for distances or torsion angles determined by NMR, or with X-ray diffraction intensities (Brünger, 1990; Brünger and Karplus, 1991). A starting conformation with random torsion angles is used, then the conformation evolves with forces calculated from Eq. 7-1 plus experimental constraints. If similar equilibrium conformations are reached from many different starting conformations, one gains confidence that the conformation corresponds to a global minimum. Clearly, any constraints can be added (such as requiring certain hydrogen bonds to be present, or a specific cross-link to form) to simulate the configurations of a system with a particular characteristic.

2.1.2 Free Energy Perturbation Calculations

Although the internal energy of a system can be most directly calculated, the free energy is more valuable since it is more directly related to common experimental measurements. The free energy G of a system (such as a molecule in a solvent) can be calculated from the possible values of energies of the system and the probability that the system will have each possible energy (Mezei and Beveridge, 1986; Bash et al., 1987a; Beveridge and DiCapua, 1989). Only energy or free energy differences can be measured, so only changes in free energy are relevant. At constant temperature and pressure, a free energy change, ΔG, can be expressed in terms of changes in energy, E, pressure, P, volume, V, temperature, T, and entropy, S:

$$\Delta G = \Delta E + P\Delta V - T\Delta S \tag{7-7}$$

All the information needed to calculate the free energy is contained in the potential energy of interaction, the force field such as given in Eq. 7-1. The shape of the multidimensional potential energy surface as a function of movement of the molecules in the system characterizes changes in energy, volume, and entropy, and thus the change in free energy.

The free energy difference between two states A and B is

$$\Delta G_{AB} = -RT \ln < e^{-(E_B - E_A)/RT} >_A \tag{7-8}$$

where $E_B - E_A$ is the difference in the force field energy for the two states (calculated from Eq. 7-1), and the average $< >_A$ is taken over the configurations of the system in state A. The calculation is done by a series of small perturbations in the free energy, ΔG_i,

$$\Delta G_{AB} = \sum_i \Delta G_i \tag{7-9}$$

with each ΔG_i evaluated by using Eq. 7-8:

$$\Delta G_i = -RT \ln < e^{-[E(\lambda_i) - E(\lambda_{i-1})]/RT} > \lambda_{i-1} \tag{7-10}$$

The potential energy is written as a function of a variable λ which corresponds to some system parameter and varies from 0 to 1:

$$E(\lambda) = (1 - \lambda)E_A + \lambda E_B \tag{7-11}$$

Clearly when $\lambda = 0$, the potential energy corresponds to E_A, when $\lambda = 1$, the potential energy corresponds to E_B; the value of λ represents a normalized reaction coordinate.

Molecular dynamics or Monte Carlo calculations are done at each step to calculate the average over the configurations corresponding to energy $E(\lambda_{i-1})$. The key requirement is that each increment in energy, $E(\lambda_i) - E(\lambda_{i-1})$, be of the order of RT, so that the molecular dynamics at each step can sample the configurations corresponding to the energy differences. The number of λ_i values that need to be used and the size of each step depend on the reaction being simulated. It is not necessary to use equal size steps throughout the reaction; the size of the step should depend on how rapidly the energy changes as the reaction proceeds. If the increments in λ are small enough, the exponential in Eq. 7-10 can be expanded to give

$$\Delta G_i \approx < E(\lambda_i) - E(\lambda_{i-1}) > = < E(\lambda_i) > - < E(\lambda_{i-1}) > \qquad (7\text{-}12)$$

For example, consider the opening of a base pair at the end of a DNA helix of 6 bp surrounded by 2594 water molecules (Pohorille et al., 1990). The initial state is B-DNA form in water. The reaction coordinate was chosen as the distance between N1 of adenine and N3 of thymine of an end base pair. It was varied from 2.6 (the A•T hydrogen-bond distance is 2.8 Å) to 6.5 Å (which gives room for water molecules to bond to all donors and acceptors on the end bases). The free energy change relative to the B form was calculated after each small increment in distance; Figure 7-3 shows the results. A free energy difference of +1 kcal mol^{-1} was obtained in going from the A•T terminal base pair to an open form (water hydrogen-bonded bases). An activation free energy of +4 kcal mol^{-1} was also calculated, but this value is dependent on the reaction coordinate chosen. The difference in equilibrium free energy is independent of path, so it should not depend on the reaction coordinate. Despite their interest and plausibility, one must realize that the sampling problems are severe in such systems.

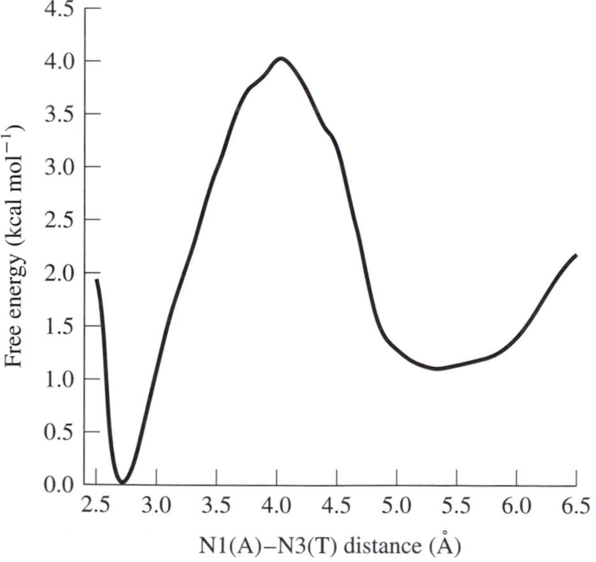

Figure 7-3
The calculated free energy change during the opening of a terminal A•T base pair in a self-complementary hexanucleotide, [d(TCGCGA)$_2$, surrounded by 2594 water molecules. The reaction coordinate was chosen as the distance between N1 of adenine and N3 of thymine. A total of 450 ps of molecular dynamics simulation was done. [Data from Pohorille et al., 1990.]

The simulation shown in Figure 7-3 was done for 450 ps; it should be considered illustrative only, until other reaction paths are tried and longer simulations are done.

An important application of free energy calculations is to compare the behavior of two similar molecules or molecular groupings, by simulating the conversion of one into the other (Bash et al., 1987b). The reaction coordinate for a reaction need not be an experimentally possible one; the free energy difference between two states is independent of path. For example to calculate the difference in solvation free energy for adenine and guanine, a molecular dynamics simulation is first done on adenine surrounded by water molecules. The molecular dynamics simulation is continued as adenine is slowly changed toward guanine; the parameters in Eq. 7-1 representing adenine are gradually changed to those for guanine. Thus the hydrogen atom on C2 of adenine begins to change to the amino group of guanine. The C—H bond distance is increased slowly as the charge and radius for hydrogen atom change from the values for hydrogen to the values for nitrogen. Two new hydrogen atoms are grown on the nitrogen atom to form the amino group. At C6 the adenine amino is changed to the guanine carbonyl oxygen. An imino proton is grown at the N1 of adenine. The energy change at each step should be small—of the order of RT. The free energy difference of converting adenine to guanine in aqueous solution is the sum of all the small free energy steps calculated from Eq. 7-10 or Eq. 7-12.

The conversion of one molecule into another by incrementally adding or removing groups seems bizarre, but it is often the easiest way to calculate a free energy difference. For example, the relative binding constants for two intercalating drugs can be calculated by simulating the conversion of one drug to the other both in the intercalated state, and as the free molecule in solution. Consider two drugs, A and B, which bind at the same site to macromolecule P with equilibrium binding constants K_A and K_B and standard free energies of binding ΔG°_A and ΔG°_B. The difference in standard free energies for binding A and B gives the relative binding constants

$$\Delta G^\circ_B - \Delta G^\circ_A = -RT \ln(K_B/K_A) \tag{7-13}$$

The diagram below shows that this difference is equal to the difference in standard free energies between that for converting bound A to bound B, ΔG°_{ABP}, and that for converting free A to free B, ΔG°_{AB}.

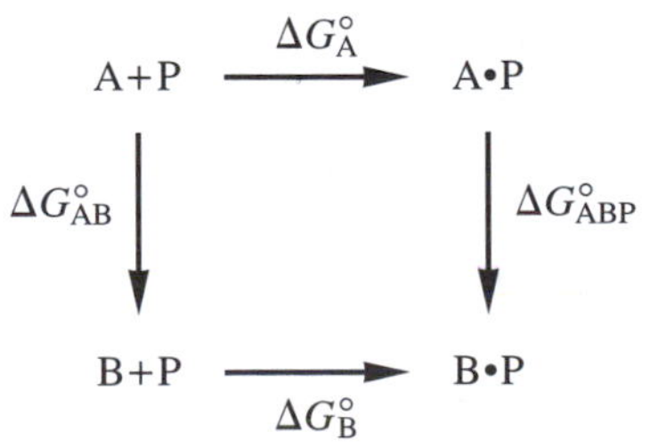

$$\begin{array}{ccc} A+P & \xrightarrow{\Delta G^\circ_A} & A\bullet P \\ \Delta G^\circ_{AB} \downarrow & & \downarrow \Delta G^\circ_{ABP} \\ B+P & \xrightarrow[\Delta G^\circ_B]{} & B\bullet P \end{array}$$

$$\Delta G^\circ_B - \Delta G^\circ_A = \Delta G^\circ_{ABP} - \Delta G^\circ_{AB} \tag{7-14}$$

In practice, it is often more accurate to calculate the binding free energy difference by simulating the conversions of A to B than to simulate the binding of each drug separately.

Of course, if one wishes to simulate covalent (bond-making or breaking) processes, or processes involving atoms other than H, C, N, O, P, etc., equations such as Eq. 7-1 as currently formulated will not work. Either explicit use of quantum mechanical methods or different analytical functions are necessary. One can handle some ions with Eq. 7-1 because the interactions with their neighbors are primarily electrostatic, but, for transition metal ions. where the electronic state is a sensitive function of ligand geometry, Eq. 7-1 cannot be used. Even ab initio approaches require considerable effort to get correct spin states in simple transition metal systems (Veillard, 1986), although density functional approaches appear particularly promising in this regard (Parr and Yang, 1989). Another possible solution to this problem, which might allow simulations of excited states and reactive processes of complex systems, is to develop combined quantum and molecular mechanical methodologies (Warshel and Levitt, 1976; Singh and Kollman, 1985; Bash et al., 1987b). These appear promising, but are so computer intensive that their potential has not begun to be tapped.

In summary, there is a considerable and powerful arsenal of computer-based methods that enable simulations of nucleic acids. Nonetheless, the ability of these methods to make meaningful contact with experiment, give mechanistic insight, and be predictive, is still limited by the fact that the systems have many degrees of freedom and the computer capabilities to scan these are still too limited.

2.1.3 Representation of Simulation Results

Computer simulations on nucleic acids face three major difficulties, compared to simulations of small molecules or even proteins. First, because of the charged phosphate groups, long-ranged Coulombic forces must be included. There is considerable uncertainty how to parameterize these forces in an equation such as Eq. 7-1. Second, the long-range (energy varies as r^{-1} rather than r^{-6} for dispersion forces) means that the model system must be large, with many molecules included, especially waters whose dipolar and hydrogen-bonding character is not easy to handle computationally. This fact makes computations very long and expensive, even on the most powerful supercomputers. Third, the large number of atoms in a DNA oligomer means that many structural variables must be monitored during the computation. In this sense as well, protein simulations are easier than those of nucleic acids, since it is often easier in a protein to identify an active site region or particular amino acid side-chain whose behavior should be monitored. Making sense of all the numbers in a simulation of a DNA oligomer is a daunting task. Very recently, exciting progress has been made in overcoming the first two of these difficulties, as described in Section 7.6, Concluding Perspective.

The last of these difficulties has been attacked by Beveridge, Lavery, and coworkers (Ravishankar et al., 1989; Beveridge et al., 1990) in an approach called "Dials and Windows". They have devised a graphical display that represents the values of the torsional angles on each strand (Fig. 7-4) as conformational wheels or dials and the range of linear and helicoidal parameters as windows. The values of each parameter for a given base or base pair is shown horizontally, while the sequence is represented vertically. The advantage of this representation in clearly displaying differences between structures is clear. The reader is referred to the original paper (Ravishankar et al., 1989) for a more extensive discussion of dials and windows figures not shown here.

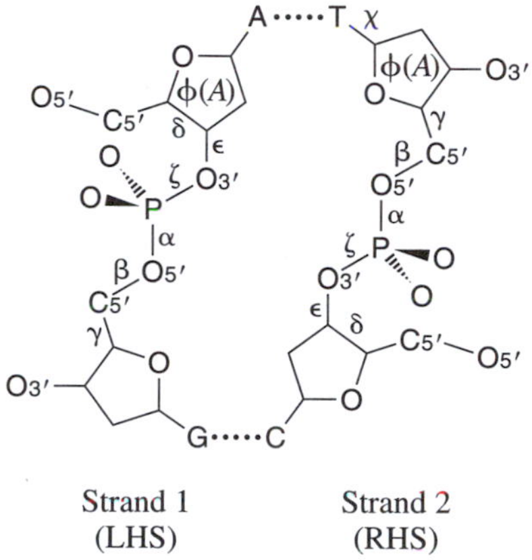

Strand 1 (LHS) Strand 2 (RHS)

Figure 7-4
IUPAC definition of backbone torsion angles in two chains of double helix, used in conformational dials analysis. α, β, γ, δ, ε, and ζ are the backbone dihedral angles, χ is the exocyclic sugar–base torsion angle, and $\phi(A)$ is the sugar–pucker angle, where A is the amplitude.

2.2 Thermodynamic Factors Determining the Free Energies of Molecular Association in Solution

Even if one had a function such as Eq. 7-1 that described the energic or enthalpic interaction between two solutes A and B accurately for gas-phase association, there are other critical factors to be considered before one can simulate the free energies for a solution-phase association. The first important factor is the translational, rotational, and vibrational entropy changes upon association (Page and Jencks, 1971). These entropy changes are invariably unfavorable for association; they are the reason that the equilibrium concentration of gas-phase H_2O dimers is very small compared to gas-phase H_2O monomers at room temperature (Kollman, 1995). To understand solution-phase association, let us consider the following thermodynamic cycle:

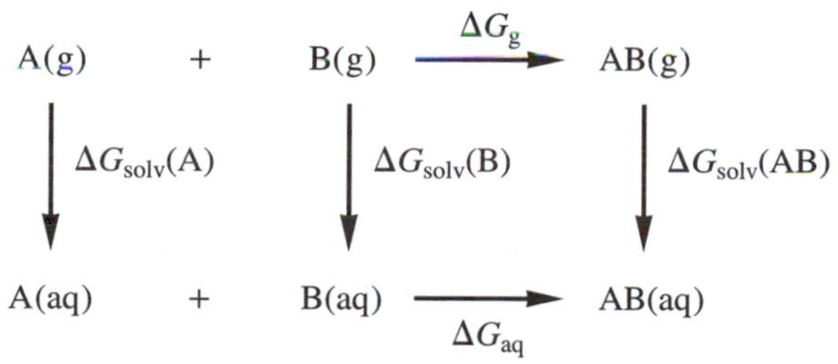

$$
\begin{array}{ccccc}
\mathrm{A(g)} & + & \mathrm{B(g)} & \xrightarrow{\Delta G_{\mathrm{g}}} & \mathrm{AB(g)} \\
\downarrow \Delta G_{\mathrm{solv}}(\mathrm{A}) & & \downarrow \Delta G_{\mathrm{solv}}(\mathrm{B}) & & \downarrow \Delta G_{\mathrm{solv}}(\mathrm{AB}) \\
\mathrm{A(aq)} & + & \mathrm{B(aq)} & \xrightarrow[\Delta G_{\mathrm{aq}}]{} & \mathrm{AB(aq)}
\end{array}
$$

The free energy of association of two molecules A and B in solution ΔG_{aq} differs from the corresponding free energy in the gas-phase ΔG_{g} by the various solvation free energies of A, B, and AB.

$$\Delta G_{aq} = \Delta G_{g} + \Delta G_{solv}(AB) - \Delta G_{solv}(A) - \Delta G_{solv}(B) \tag{7-15}$$

If A and B are more highly charged or have larger dipole moments than the complex AB and the solvent is water, then $\Delta G_{solv}(A) + \Delta G_{solv}$ (B) will likely be more negative than ΔG_{solv} (AB), and water will significantly retard the association of the molecule. On the other hand, if A and B are both nonpolar, then both ΔG_{solv}(A) + ΔG_{solv} (B) and ΔG_{solv} (AB) are unfavorable, because the water in the first coordination shell of non-polar molecules is entropically destablized. The fact that AB has a smaller surface area than the sum of the areas of A and B leads to the ΔG_{solv} (AB) being more favorable than ΔG_{solv}(A) + ΔG_{solv} (B) (Kollman, 1995). This is called hydrophobic association because the association is largely solvent driven and comes from water's stronger interactions with itself than with the solute.

When cationic drugs or proteins associate with anionic nucleic acids, there are strong solute–solute electrostatic interactions but also strong electrostatic interactions of both cations and anions with the solvent water. For example, both Na^+ and Cl^- have free energies of solvation of about 100 kcal mol^{-1} in water. Because of these strong interactions, there is little tendency for them to associate even though the ion–ion attraction is so strong. For example, calculations on the free energy of association of Na^+ and Cl^- in water suggest little if any tendency for ion pairing (Jorgensen et al., 1987), consistent with experiment. These contributions to the free energy of association may largely cancel, and the net favorable interactions may come from the entropy of counterion release from A and B when they associate (Anderson and Record, 1990). Thus, the Na^+ ions, which must surround DNA for electroneutrality, are released when a cationic drug interacts with DNA, which is favorable entropically. Record has made an analogy between the hydrophobic effect and counterion release effects (Anderson and Record, 1990). Thus, in summary, there are a number of subtle factors that make solution association much more complex and difficult to predict from first principle than gas-phase association, particularly for ionic solutes.

2.3 Formation of Double Helices

We attempt to make the ideas discussed above more concrete by describing the important factors that determine the double- to single-strand equilibrium of nucleic acids. The hydrogen-bond complementarity of adenine-thymine and guanine-cytosine is critical to the genetic code, but in solution, water also forms hydrogen bonds with the bases. Thus, it seems that hydrogen bonding between bases is not a strong driving force by itself for double-strand association in water (Cieplak and Kollman, 1988, Dang and Kollman, 1990). However, if there is a mismatch in hydrogen bonding, this will clearly further decrease the tendency for association. Base stacking, on the other hand, is likely to give more favorable dispersion attraction in the double-stranded form. Also, water is released when bases stack more effectively, leading to a hydrophobic contribution driving the single-strand $\rightleftharpoons$ double-strand equilibrium to the right (Cieplak and Kollman, 1988, Dang and Kollman, 1990).

Kollman and coworkers (Cieplak and Kollman, 1988, Dang and Kollman, 1990) have used computer simulation methods to show that the free energy for stacking between A and T or G and C is more favorable than Watson–Crick hydrogen bonding in water. In the gas phase, hydrogen bonding is more favorable. This result suggests that stacking, not hydrogen bonding per se, provides a driving force for double helix formation. Evidence that hydrogen bonds do contribute to double helix stability (Chapter 8) is deduced by comparing the thermodynamic stability of a complementary double helix to one with unpaired (dangling) bases at the end. For example, adding an A • U base pair decreases the free energy more than the sum of the effects of adding an A and adding a U separately. One interpretation of this is that the increased stability provided by the individual dangling bases is a measure of stacking, and the further increment in stability when a base pair is added is a measure of hydrogen bonding. However, one can speculate that when a dangling base stacks on the end of a double helix, some of its hydrogen-bonding sites are partially desolvated and occluded, which is an unfavorable interaction. When a second complementary base is added, these desolvation–occlusion effects are not a factor, because the two dangling bases can form hydrogen bonds to each other. The increased stability of the base pair over the two unpaired dangling bases is thus only indirectly caused by the base–base hydrogen bonding. This interpretation can be tested by the free energy perturbation methods described earlier.

In addition to stacking and hydrogen bonding, phosphate–phosphate repulsion and conformational entropy effects are important in double-strand formation. It is difficult for theoretical methods to simulate all of the above effects quantitatively. If one considers base pairing or stacking of isolated nucleic acid bases, recent free energy perturbation approaches are capable of semiquantitative simulation of these (Bash et al., 1987b, Cieplak and Kollman, 1988, Dang and Kollman, 1990). The polyanion effects could be effectively treated with nonlinear Poisson–Boltzmann methods (Sharp and Honig, 1990), but the conformational entropy effects that appear when one studies the equilibrium between nucleic acids, not just nucleic acid bases, make it currently impossible for theoretical simulation methods to calculate the thermodynamics of strand equilibrium from first principles.

3. THEORETICAL RATIONALE FOR CONFORMATIONAL PREFERENCES OF NUCLEIC ACID COMPONENTS

Above, we have described the physical effects critical in determining molecular conformation. Figure 7-5 illustrates the various bonds in a polynucleotide chain. Let us now review what theoretical calculations tell us about qualitatively and semiquantitatively conformational preferences for the various bonds involved in nucleic acids:

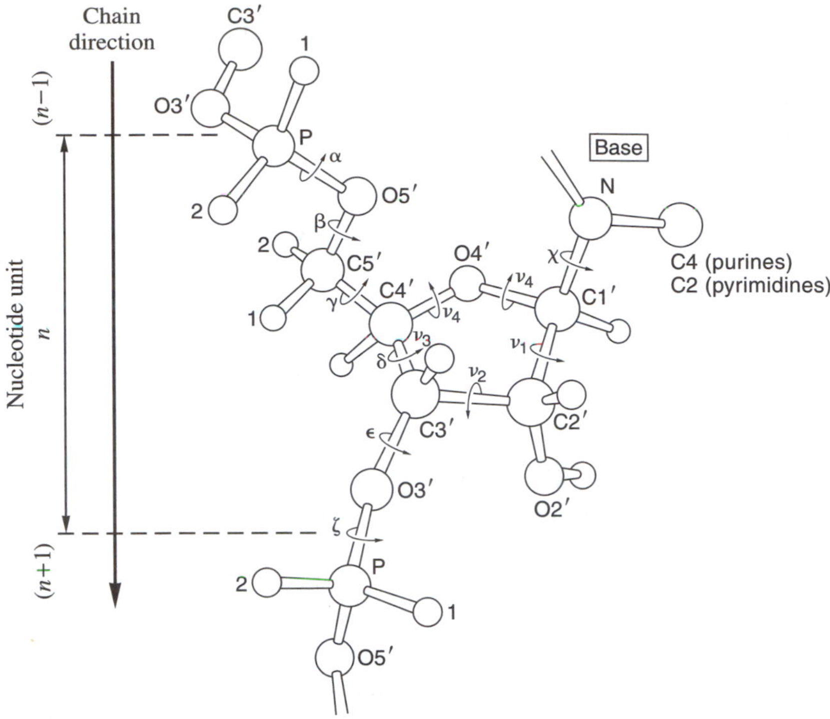

Figure 7-5
Atomic numbering scheme and definition of torsion angles for a polyribonucleotide chain. Counting of nucleotides is from top to bottom, in the direction of O5′ → O3′.

—C—C— Bonds: C5′—C4′ (γ); C4′—C3′ (ν3 or δ); C3′—C2′ (ν2); and C2′—C1′(ν1) are examples of such bonds in DNA and RNA. In molecules with only carbons and hydrogens attached to the center —C—C—bond, it is a reasonable theoretical model to use a three-fold torsion term to describe the barrier to rotation, modulated by nonbonded effects, as in *n*-butane. With one electronegative atom on an end of the C—C bond, it is not clear from model compounds whether an additional Fourier term should be added to the intrinsic torsional energy for this bond. However, for O—C—C—O units, it is clear (as in 1,2-difluoroethane above) that a twofold Fourier term is required to reproduce the gauche tendency of such bonds (Marsh et al., 1980; Olson, 1982a).

—C—O— Bonds: Examples of these bonds are C3′—O3′ (ε) and O5′—C5′ (β). The simplest model is to treat these like ethyl methyl ether, with a threefold torsion and appropriate nonbonded interactions between the atoms on each side of the bond.

—P—O— Bonds: Here, it is clear that both two- and threefold torsional terms are required to fit the known gauche tendency of C3′—O3′—P—O5′ (ζ) and

O3′—P—O5′—C5′ (α) units (Newton, 1973). As in the case of O—C—C—O units, the gauche tendency in these systems is clearly understandable in electronic structure terms, and is most simply described as a delocalization of lone-pair electrons on one ester oxygen into the antibonding orbital involving the phosphorus and the other ester oxygen. Such delocalization is maximized when the orbitals are trans (or cis), which happens when the O—C bond is gauche with respect to the P-O ester bond.

—C(sp3) - N(sp2)— Bond (χ): This bond is an example of a sixfold rotor. In CH_3NO_2, the barrier to rotation is nearly zero, so the simplest model is to make the Fourier terms zero for this bond, with nonbonded effects determining the conformational preferences.

—sp^2—sp^2— Bonds: These bonds, which occur inside the bases, are kept planar by twofold Fourier terms with phase $\gamma = 180°$, and the magnitude of the V_2 coefficient determined from the magnitude of this term in model compounds (Weiner et al., 1984).

The above choices for parameters lead to the following conformational preferences in nucleosides and nucleotides (gauche is g; trans is t): For rotations involving the phosphodiester linkage O3′—P—O5′ (ζ and α) the g^+, g^+ (or g^- ,g^-) conformation is calculated to be most favorable, followed by g, t and t, t. Quantum mechanical calculations (Newton, 1973) and simulations on dimethyl phosphate (Jayaram et al., 1988) also follow this tendency, as do X-ray data (Berman and Shieh, 1981). For the O5′—C5′ (β) bond, the preference is for t conformations, as observed by NMR (Patel et al., 1987) and crystallography (Saenger, 1984). Such a conformation minimizes the P $\cdots$ C4′ repulsions. The calculated preference for the rotamers around the C5′—C4′ (γ) bond in nucleosides and nucleotides is $g^+ > t > g^-$, which is consistent with NMR data (Patel et al., 1987) and comes partially from the fact that the g^+ and t conformations have the more favorable gauche O—C—C—O orientation. The C3′—O3′ (ε) bond is calculated to have only two low-energy conformations, with t more stable than g^-. The g^+ conformation is much higher in energy than the other two because it involves significant P $\cdots$ C2′, C4′ repulsions. X-ray and NMR studies on nucleotides suggest predominantly t conformations around the C3—O3′ bond (Patel et al., 1987, Saenger, 1984). These conformational preferences are preserved in both A- and B-form double helices: $\alpha = g^-, \beta = t, \gamma = g^+, \varepsilon = t; \zeta = g^-$.

Because of ring closure, only two independent dihedral angles are needed to describe the conformational preferences in the five-membered sugar ring: the pseudorotation phase (P) and the average out-of-plane distance of the atoms (Altona and Sundaralingam, 1972, Cremer and Pople, 1975). Sugar conformations and the pseudorotation phase angle are described in Chapter 2. The low-energy furanose sugar conformations predicted by the theory correspond to C2′-*endo* (P = 162°) and C3′-*endo* conformations (P = 18°), with a low barrier (1.5–2 kcal mol^{-1}) between them at an O1′-*endo* conformation and a higher (3-4 kcal mol^{-1}) barrier at O1′-*exo*. The average out-of-plane distance of the ring atoms is calculated to be approximately 0.35–0.40 Å. These results are quite consistent with available experiments (Patel et al., 1987;

Saenger, 1984). Olson (1982a) has suggested that the reason that theoretical calculations (and X-ray and NMR experiments) find C2′-*endo* to be more stable than C3′-*endo* for 2′ deoxyfuranoses is that this C2′-*endo* conformation allows a more gauche O3′—C3′—C4′—O1′ conformation than in the C3′ *endo* conformation. In ribofuranosides, with both O3′—C3′—C4′—O1′ and O2′—C2′—C1′—O1′ interactions, both C2′-*endo* and C3′-*endo* conformations have an equal number of gauche O—C—C—O interactions and X-ray and NMR data on ribofuranosides suggest that these conformations are closer in energy than for deoxyribonucleosides (Patel et al., 1987, Saenger, 1984). The sugar conformations are very different in A and B form. Both A-DNA and A-RNA have C3′- *endo* conformations, whereas B-DNA has C2′-*endo* conformations. In Z-DNA and Z-RNA, the sugar conformations alternate from C2′-*endo* for dC or rC to C3′-*endo* for dG or rG. The requirements of the helix geometries clearly are more significant than any differences in ribose and deoxyribose conformational preferences.

Theoretical calculations and experiments suggest that nucleosides and nucleotides have both anti and syn local minima in the glycosidic angle χ (see Fig. 2-7). For purines, the syn [χ (O1′—C1′—N—C8) $\approx 240°$] and anti ($\chi \approx 60°$) conformations are comparable in energy, which is consistent with experiment. For pyrimidines, the anti conformation [χ (O1′—C1′—N1—C6) $\approx 60°$] is calculated to be more stable than syn, also consistent with experiment. This difference between χ preferences in purines and pyrimidines comes from the C2=O2 group of the pyrimidines, but $\chi \approx 60°$ and 240° are local minima for both because such conformations minimize the exchange repulsion between atoms of the furanose ring and the C6(C2) and C8(C4) atoms of the base. In double helices, all bases are anti except for the guanines in Z-DNA and Z-RNA.

In summary, theory is in agreement with experimental data on conformational tendencies for the various torsion angles in nucleosides and nucleotides.

4. SEQUENCE-DEPENDENT EFFECTS WITHIN HELICES

Figure 4-4 shows the structures of A-, Z-, and B-DNA helices. Despite the fact that theoretical approaches are not able to calculate the free energy for the single-strand $\rightleftharpoons$ double-strand equilibrium in DNA for the reasons discussed above, calculations have been able to give interesting insight into the base sequence-dependent stability of the different double helices. For example, the homopolynucleotide poly dG • poly dC undergoes its double- to single-stranded transition 12° lower than poly d(G—C) • poly d(G—C), whereas poly dA • poly dT melts 6° higher than poly d (A-T) • poly d (A-T). Theoretical calculations can reproduce and rationalize the difference between the homo- versus heteropolynucleotide stabilities in GC versus AT DNA by making the assumption that all the stability differences come from the double-stranded form. The relative molecular mechanical energies of pairs of base pairs [d(G-G) • d(C-C) and the average of the energies of d(G-C) • d(G-C) and d(C-G) • d(C-G)] are then used to calculate the relative stability of the double-stranded form (Kollman et al., 1981). The assumptions are far from being accurate, but the fact that many sequence-dependent trends are reproduced in that manner gives one confidence that they are not grossly wrong. Given these assumptions, the theory can give a physical picture of why, in a

B-DNA-like model, the heteropolynucleotide is more stable in the GC series, whereas the homopolynucleotide is more stable in AT. It turns out that van der Waals interactions tend to favor the homopolynucleotide, because of more effective base stacking.

A key to understanding the electrostatic interactions lies in the dipole moments of the bases, which are about 6 D for G and C and about 3 D for A and T. Furthermore, the directions of the dipole moments in the bases are closer to parallel for A versus T, whereas they are pointed in almost antiparallel directions in G versus C. The poly d(G) • poly d(C) homopolynucleotide has large and unfavorable electrostatic interactions between intrastrand stacked bases, because the bases are aligned only 36° from a parallel direction (Fig. 7-6). In the poly d(G-C) • poly d(G-C) heteropolynucleotide, because the G and C dipoles are nearly antiparallel to each other, these intrastrand stacking interactions are large and favorable. Although other components contribute, it appears that the single most important energy term that leads to the GC heteropolynucleotide stability is the electrostatic term. In the AT polynucleotides, not only are these electrostatic interactions expected to be approximately four times smaller (roughly proportional to the product of the dipole moments), but the alignments of the AT dipole moments in the B-DNA helix are approximately parallel, leading to the relative helical stability inferred from van der Waals interactions alone. Although the physical picture is less clear, the simple molecular mechanics calculations can reproduce the relative stability of the homopolynucleotide versus heteropolynucleotide of alternating AT/GC, as well as having reasonable success with the relative stabilities of trinucleotide repeating polymers of two AT and one GC of different sequence order (Weiner et al., 1984).

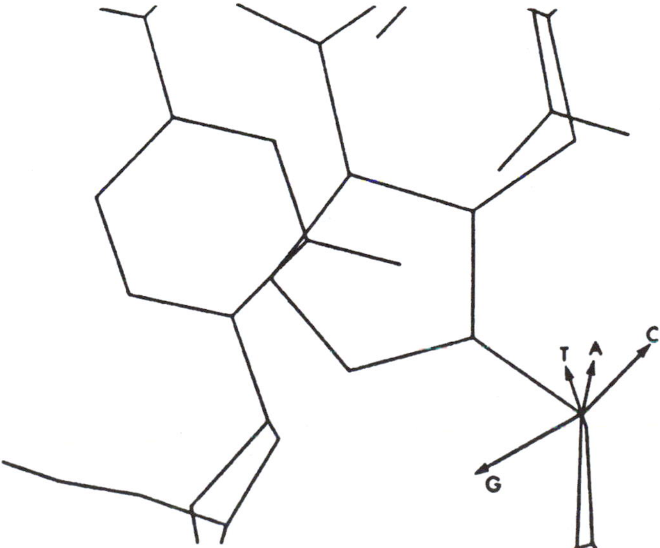

Figure 7-6
Projection of base dipoles on the glycosidic bond of B-DNA.

Calculations on A-DNA models suggest a reversal of stabilities, with the homopolynucleotide more stable in the GC series and the heteropolynucleotide more stable in the AT and AU series (Tilton et al., 1983). These are consistent with the observed reversal found in RNA polynucleotides, where polyr(A-U) • polyr(A-U) heteropolynucleotides are more stable than polyr(A) • polyr(U) homopolynucleotides, but polyr(G) • polyr(C) homopolynucleotides are more stable than polyr(G-C) • polyr(G-C) heteropolynucleotides. The different base orientations in the A-from helix clearly have a large effect on the van der Waals stacking and electrostatic energies in this helix.

Thus, there are interesting sequence-dependent stabilities in A- and B-DNA helices, which, somewhat surprisingly, can be qualitatively modeled with molecular mechanics minimization on fragments of double helices. These successes of the theory are encouraging and support the fact that van der Waals plus electrostatic nonbonded interactions between bases are the critical factors in determining the observed stabilities.

Analogous molecular mechanical calculations have been done on Z-DNA helices of various sequences (Kollman et al., 1982). By comparing the relative energies of polyd(G-C) and polyd(G-^{5Me}C) helices in B and Z conformations, one can rationalize why 5-methyl substitution selectively stabilizes Z-DNA. However, the calculations are less successful in predicting under what conditions polyd(G-C) will be in a B form and under what conditions it will be in the Z form. The reason is that such calculations include solvent only implicitly and none are very accurate. The various models to account for the solvent include those that employ a distance-dependent dielectric constant (Weiner et al., 1984), a dielectric constant of 4 (Olson, 1975), a Debye–Hückel electrostatic screening term (Broyde and Hingerty, 1983), or a reduction of phosphate partial charges (Nilsson and Karplus,1986). Thus, there is a difference in the ability of simple molecular mechanical theory to examine sequence-dependent stabilities within a given helical family and its ability to examine relative stabilities between different helical families. Furthermore, without including solvation effects explicitly, one cannot compare A • T versus G • C stabilities within a family.

One example in which the molecular mechanical theory was successful was the prediction that a nonalternating purine–pyrimidine sequence would be able to form a Z helix at only a modest energy cost. This prediction required a syn pyrimidine in the DNA helix, a conformation that had never been seen in small molecule models with a hydrogen substituent at the C6 position, but which the calculations showed would only cost about 1–2 kcal mol^{-1} of strain energy (Kollman et al., 1982). Thus, one can see that theoretical molecular mechanical calculations have often given interesting insight into the properties of A, B, and Z helices. Although the starting geometries for such helices come from experiment and model building, all degrees of freedom were allowed to change during the minimization. The typical calculations (Kollman et al., 1981) moved the atoms from their initial geometry by a root-mean-square average distance of only ≈ 0.5 Å. Additional consideration of calculations on energetics of the B–Z transition, including more detailed consideration of ion and solvent effects, is presented in Chapter 11.

It is clear from analyses of DNA X-ray structures (Chapter 4) that sequence-dependent effects may be much larger than suggested from simple minimization studies. Molecular dynamics simulations on different sequences of B-DNA helices have been done using various energy models. Particularly important energetically, but also particularly difficult to model, are electrostatic contributions. Some studies with explicit solvent have used no electrostatic charges at all, others have modified the phosphate charge, and others have put in explicit counterions and used a distance-dependent dielectric constant. Molecular dynamics studies with explicit solvent have been reported, using different energy models and periodic boundary conditions.

One such study (Rao et al., 1986) showed that, as suggested from molecular mechanics, sugar repuckering is the most flexible degree of freedom in the helices, in the sense that this variable undergoes the most transitions between the two local minima (C2′-*endo* and C3′-*endo*) in its potential surface (Fig. 7-7). In fact, many other types of transitions occur in these simulations: α (g → t), β (t → g^+), γ (g → t), ε (t → g^-), and ζ (g → t), with the helix remaining B-DNA-like throughout the simulation. Analyses of hydrogen bond distances, base pair tilts, base pair twists, and helix repeats are interesting in that most of the helix repeat angles for a given base pair step vary around the canonical value of about 36° (10 bp per helix repeat) with fluctuations up to 45°–50° and down to 20°–25°. Nonetheless, the average repeat stays near 10, close to the X-ray value of 10.0 and the solution value (depending on sequence) of 10.0–10.7.

An even more complete computational study of the "Dickerson dodecamer" d(CGCGAATTCGCG) was carried out by Swaminathan et al. (1991). It involved the dodecamer, 22 Na^+ counterions, and 1927 water molecules, enough to provide more than two complete hydration shells for the DNA. It was carried out for 140 ps of molecular dynamics simulation, which corresponds to 140,000 steps of 1 fs each. (100 ps took 40 h on one of the fastest supercomputers available in 1990). At each step, the interactions given by an equation similar to Eq. 7-1 had to be computed between all

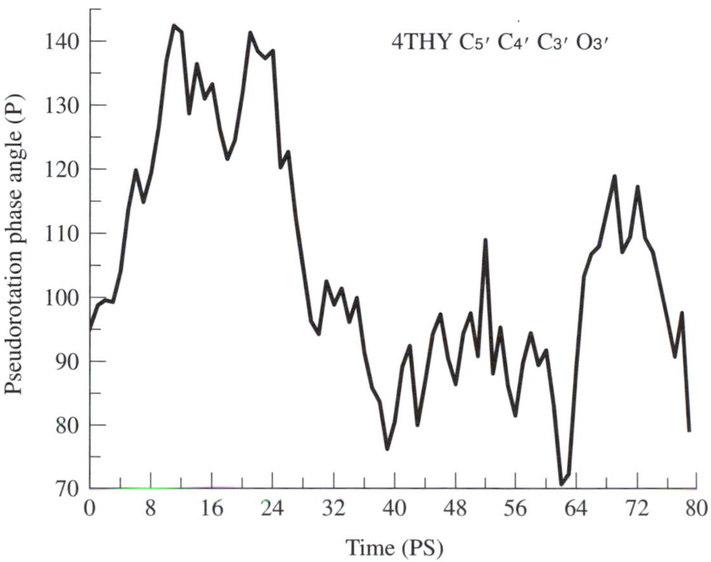

Figure 7-7
Sugar pucker phase angle (*P*) as a fuction of time during an 80-ps molecular dynamics simulation on a $d(ATGCAT)_2$ • actinomycin D complex. The phase angle for the second nucleotide (T_2) is depicted.

pairs of atoms. A detailed look at some of the calculational strategies and results may serve to give a sense of the nature of such calculations.

The force field used in the calculation was taken from GROMOS, but with several significant modifications. The first was addition of a Hooke's law constraining potential for hydrogen bonds involved in Watson–Crick base pairing, which served to maintain the structure in a basically B-form conformation. Without this constraint, rather large axis deformations and base pair openings are observed. A second modification was the use of switching functions, which make nonbonded interactions go smoothly to zero between 7.5 and 8.5 Å (otherwise the electrostatic interactions are too long-ranged for computational feasibility). A third stratagem was to group Na+ with phosphates, to assure electroneutrality in the list of nonbonded pair interactions. Extensive Monte Carlo equilibration of the solvent was then performed to get the structure into a state that yielded a stable dynamic trajectory.

Figure 7-8 shows three typical indexes of the progress of the calculation over its 140-ps duration: the total energy, which should decrease somewhat as the structure

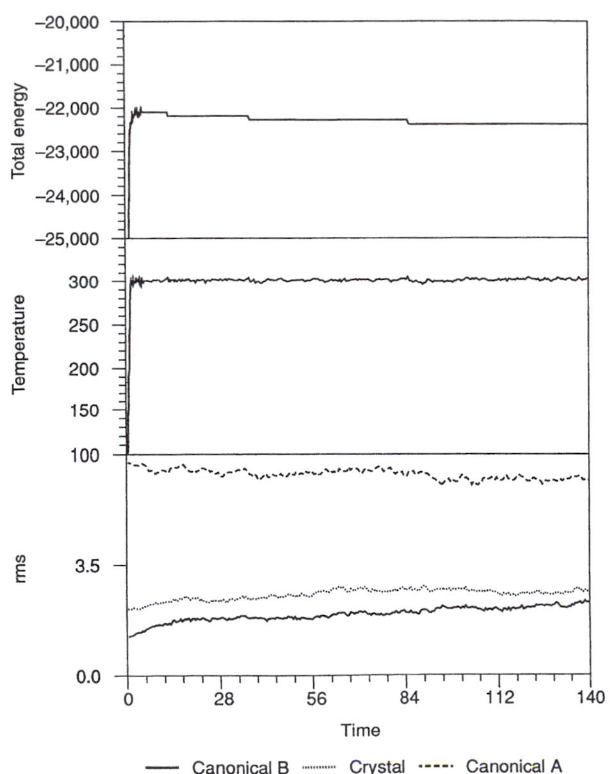

Figure 7-8
Time evolution of total energy, kinetic energy temperature, and root-mean-square (rms) deviation from standard structures for molecular dynamics simulation of dodecamer. [Reprinted with permission from Swaminathan, S., Ravishanker, G., and Beveridge, D., *J. Am. Chem. Soc.* **113**, 5027–5040 (1991). Copyright ©1991 American Chemical Society.]

"anneals" and then remain roughly constant; the temperature (proportional to the average kinetic energy of the atoms, Eq. 7-6), which should remain essentially constant; and the rms deviation of the atomic positions from three standard structures (canonical A, canonical B, and crystallographic B). As should be the case, the rms deviation is much greater with respect to the A structure than either of the B structures. Figure 7-9 shows how the radial coordinate of the conformational dials can be used to represent the time evolution of the simulated structure. The angular variable starts at the center of the circle at $t = 0$ ps, and reaches the circumference at $t = 140$ ps. Thus a sense of

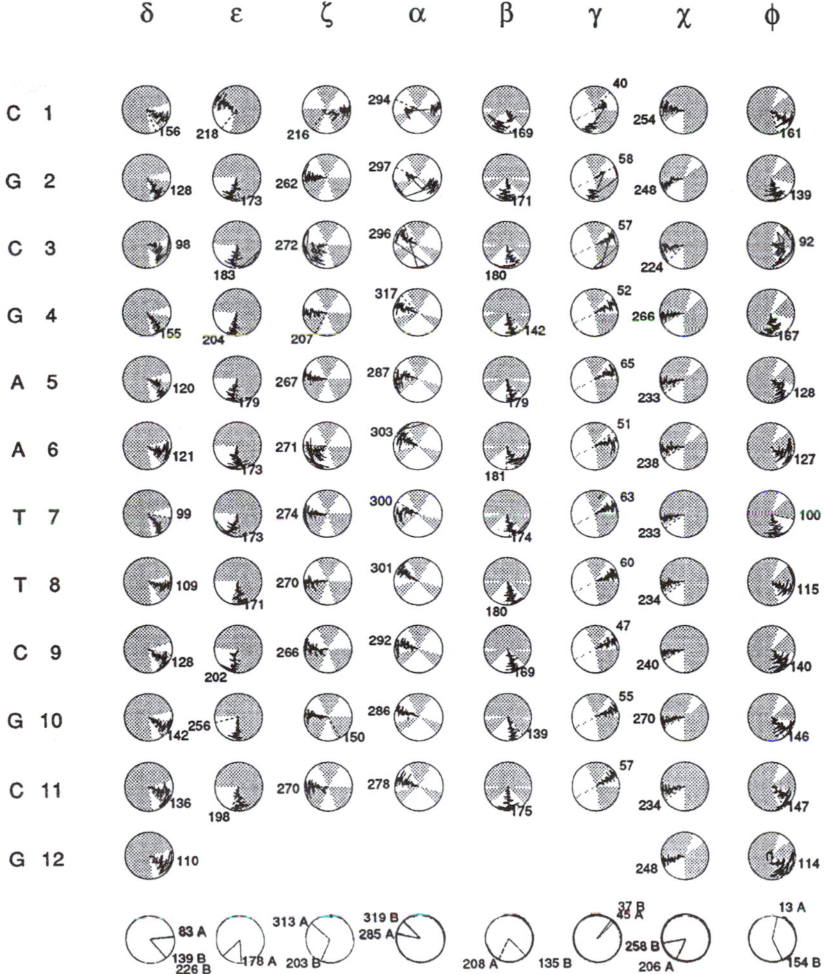

Figure 7-9
Time evolution of torsional variables for dodecamer in conformational dials representation. For reference, the crystallographic B structure (Drew et al., 1981) is indicated by the straight lines, and sterically forbidden, regions of conformational space (Olson, 1982b) are indicated by shaded areas. [Reprinted with permission from Swaminathan, S., Ravishanker, G., and Beveridge, D., *J. Am. Chem. Soc.* **113**, 5027–5040 1991. Copyright ©1991 American Chemical Society.]

structural fluctuation, and correlation between structural variables, can be conveyed as a function of time. For reference, the crystallographic B structure (Drew et al., 1981) is indicated by the straight lines, and sterically forbidden regions of conformational space (Olson, 1982b) are indicated by shaded areas. Similar time-evolution diagrams in the helicoidal windows representation are shown in Figure 7-10, where $t = 0$ is at the bottom of each window and $t = 140$ ps is at the top.

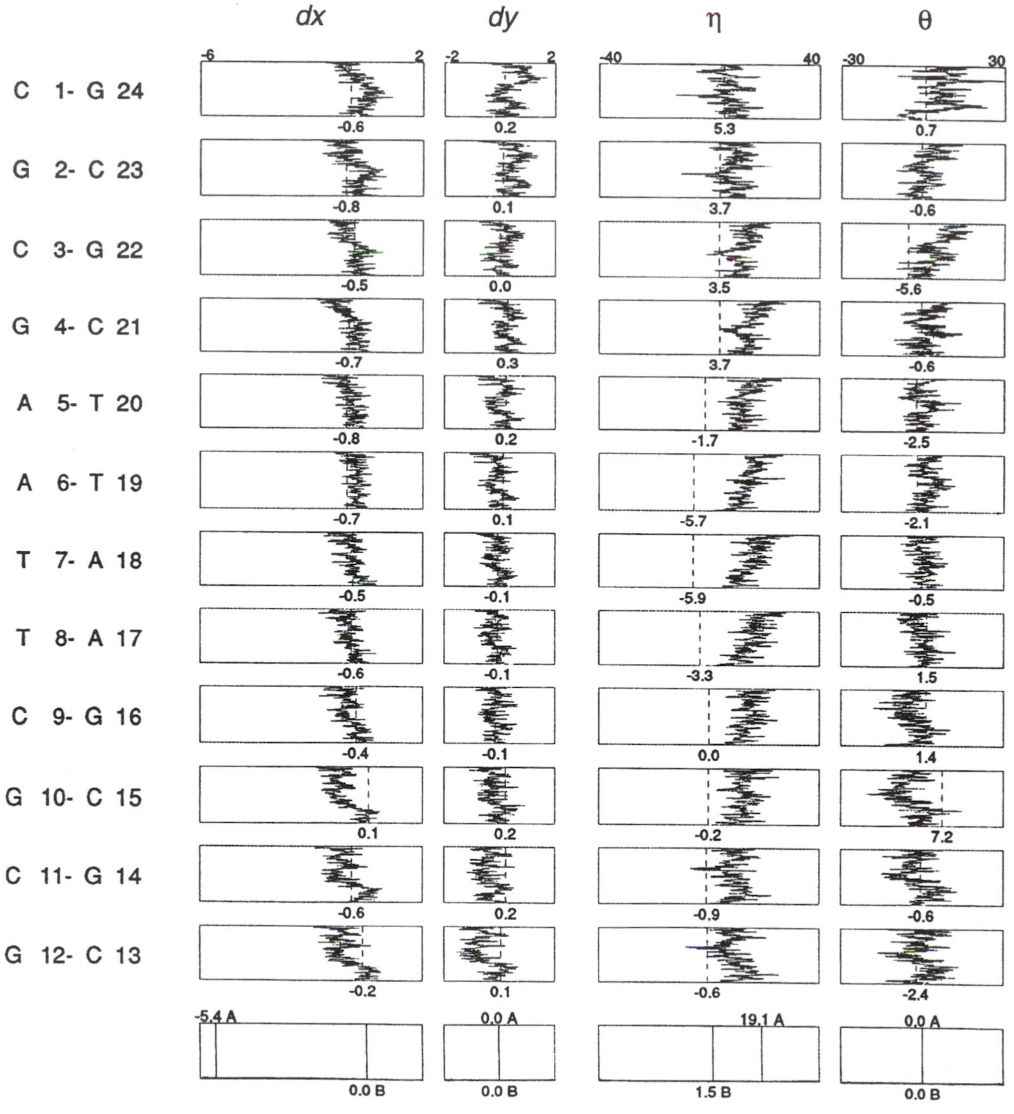

Figure 7-10
Time evolution of axis–base pair parameters for dodecamer in helicoidal windows representation. [Reprinted with permission from Swaminathan S., Ravishanker, G., and Beveridge, D., *J. Am. Chem. Soc.* **113**, 5027–5040 (1991). Copyright ©1991 American Chemical Society.]

The computed structure showed several important similarities to the crystallographic one, such as local axis deformation near the GC–AT interfaces in the sequence, and a large propeller twist in the base pairs (Swaminathan et al., 1991). However, one of the striking features of the X-ray structure, the narrowing of the minor groove in the AT central region, was not observed over the time course of the simulation and was conjectured to be a result of crystal packing forces. An important general result was the demonstration, by a parallel in vacuo simulation, that water molecules must be explicitly included to maintain the major and minor groove structure of the DNA. Further discussion of these calculations, with special emphasis on water and ionic interactions, is given in Chapter 11.

As supercomputers become faster, longer simulations become possible. A 1-ns molecular dynamics simulation of the dodecamer (McConnell et al, 1994) progressed through three substates, each with lifetimes of several hundred picoseconds. Even longer simulations may be necessary to reveal the full dynamic complexity of DNA.

In the simulation by McConnell et al., (1994) a different model was used to improve dynamic stability. No counterions were included and the charge on each phosphate was reduced to -0.24. This emphasizes the difficulty in dealing with large electrostatic effects with full solvent–counterion representations. A particularly promising solution to this problem has been presented by York et al. (1994), who have carried out simulations on DNA and protein crystals using the "particle mesh Ewald" method to describe the electrostatic interactions. This method appears to be superior in "maintaining" observed structures during molecular dynamics.

The role of sequence-dependent effects on DNA structure is extremely important in its ramifications for DNA bending and DNA–protein interactions. The root causes for it are likely to be contained in force field equations such as Eq. 7-1, but the sampling difficulties and the trajectory convergence and stability difficulties noted above make a definitive analysis and understanding difficult. Nonetheless, the studies by Beveridge and co-workers using molecular dynamics (Subramanian et al., 1990; McConnell et al., 1994) and Lavery and Hartmann (1994) (using molecular mechanical methods in helicoidal coordinates) have made significant inroads in coming to grips with sequence dependent structural effects.

5. MISMATCHES, DAMAGED DNA, AND DNA-DRUG INTERACTIONS

5.1 Mismatched Helices

One can use a combination of model building and molecular mechanics/dynamics to study nonself-complementary DNA helices. As an illustration of the fruitful interplay of experimental and theoretical approaches, we consider such studies on helices $d(CGCGAATTCGCG)_2$, $d(CGTGAATTCGCG)_2$ (G•T mismatch), $d(CGAGAATTCGCG)_2$ (G•A mismatch) $d(CGCGAATTCACG)_2$ (A•C mismatch), and $d(CGCGAATTCTCG)_2$ (C•T mismatch). The NMR studies (Patel et al., 1982; Patel et al., 1984) clearly showed that the G•T base pair formed a wobble structure, with T carbonyl—O2 • • • G imino-H1 and T imino—H3 • • • G carbonyl—O6 hydrogen bonds [see Fig. 7-11(*a*)]. Molecular mechanics energy minimization starting with the Watson–Crick base paired geometry found such a wobble structure with no barrier between Watson-Crick and wobble (Keepers et al., 1984). Subsequent X-ray studies have confirmed the G • T wobble base structure (Kennard, 1985). For the G • A mismatch, the NMR studies found a G *anti* • A *anti* structure [Fig. 7-11(*c*)] with a hint of less intense peaks for a G *anti* • A *syn* structure [Fig. 7-11(*b*)]. This finding was corroborated by theoretical studies, although the latter has to be corrected for exaggerated G 2NH2• • • phosphate hydrogen bonding in a G *syn* • A *anti* structure.

The conclusion from theoretical studies was a small energetic preference for G *anti* • A *anti*, with G *anti* • A *syn* close in energy (Keepers et al., 1984). These results have been corroborated by X-ray studies (Hunter et al., 1986a), with both types of structures being found in G • A mismatches in different sequences. The NMR and theoretical studies agreed on the most favorable A • C mismatch, but crystallographic studies (Hunter et al., 1986b) suggest a protonated A • CH^+ base pair not considered in either the NMR or theoretical studies. The data from the NMR studies were not clear enough to evaluate the C • T mismatch structure found in the theoretical calculations (Patel et al., 1982, 1984). NMR studies suggested that the sequence $d(CGCAGAATTCGCG)_2$ has

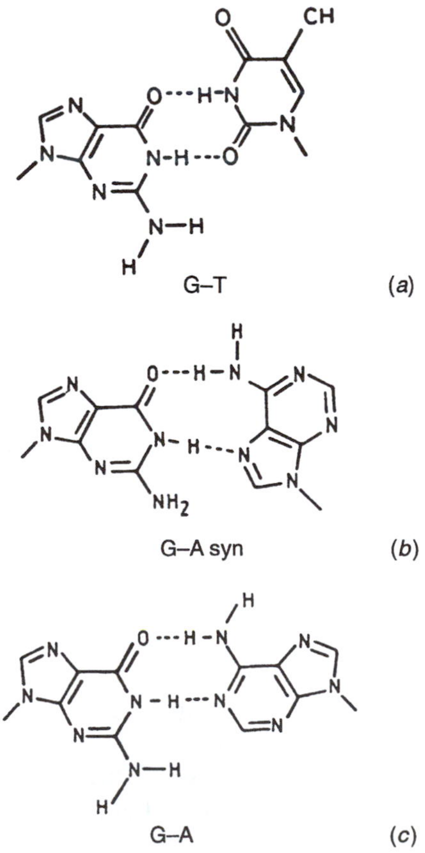

G–T (*a*)

G–A syn (*b*)

G–A (*c*)

Figure 7-11
Schematic diagram of G • T and G • A base pairs.

the extra adenine stacked into the DNA helix and model building–molecular mechanics suggested a detailed structure for this sequence (Patel et al., 1982, Patel et al., 1984). The above studies suggest a synergistic interaction between NMR studies and theoretical molecular mechanics, with each playing a role in the analysis of mismatched base pair DNA helices.

5.2 Radiation-Induced Dimers

A second interesting interplay between theory and experiment comes from studies on psoralen cross-linked and thymine dimer structures in DNA helices. Pearlman et al. (1985) built structural models of such covalently modified DNA and refined them with molecular mechanics. Independently, Rao et al. (1984) used the same molecular mechanics methods to study a thymine dimer structure. The structures resulting from these two studies were different from previous structures and different from each other. These two studies showed that hydrogen bonding was retained in cyclobutane thymine dimer structures, even though those structures had distorted A • T hydrogen bonds at the site of covalent modification. Subsequent NMR studies by Kemmik et al. (1987) on $d(CGAATTCG)_2$ supported the conclusion that the hydrogen bonds were weakened but not broken when the T_2 fragment was photochemically cross-linked. The structures presented by Pearlman et al. (1985) and Rao et al. (1984) (see color plate for Fig. 7-12) differed in the amount of kink induced in the DNA helix by the thymine dimer covalent modifications. Thus it was not a surprise that when Pearlman et al. (1985) built a large helix kink into their starting structure they found large kinks while Rao et al. (1984) started without one. and found small ones. Some experimental data on psoralen cross-links and thymine dimers favor small kinks (Sinden and Hagerman, 1984), others medium and some large (Hussain et al., 1988), so the experimental results are not clear.

The above case history shows that until more reliable force fields and more efficient search methods are developed, molecular mechanics cannot give definitive predictions on the optimum structure of large oligonucleotide fragments. Energy minimization is likely to remain close to the initial model structure and more extensive conformational searches are unlikely to have a small enough uncertainty in the energy to define one structure or set of structures as a global minimum.

5.3 DNA Interactions with Other Molecules

5.3.1 Non-Covalent Interactions

Double helical DNA can interact strongly and specifically with other molecules through its various functional groups. Typical examples of noncovalent interactions with DNA are those involving small molecules, such as netropsin, of the right shape and hydrogen-bonding (electrostatic) properties to interact with the DNA minor groove. Netropsin has strong electrostatic interactions with the minor groove of AT rich DNA (see color plate for Fig. 7-13). Using NMR NOE data, Caldwell and Kollman (1986) model-built and energy-minimized a netropsin–$d(CGCGAATTCGCG)_2$ complex. They subsequently found a correspondence of the netropsin atoms to within 0.6–1.0 Å of values determined from an X-ray structure (Kopka et al., 1985). This result illustrates again how a combination of model building–computer graphics based on NMR data, plus energy minimization to refine structures, can correspond reasonably well with experiment. As noted first by Seeman et al. (1976) and subsequently by Weiner et al. (1982) the hydrogen bonds in the major and minor groove of double helical DNA lead to characteristic patterns in electrostatic potential in the grooves. In the minor groove of A•T, only electron donors (hydrogen-bond acceptors N3 of adenine and O2 of thymine) are found. The resulting negative potential is complementary to the positive potential of the netropsin N—H group. The G•C polymers also have a guanine $2NH_2$ group that breaks the negative potential of the minor groove, and through its van der Waals properties, opens the minor groove and may interact repulsively with parts of netropsin. In the major groove, G•C pairs have a $- \; - \; +$ potential, due to G N7, G O6, and C $4NH_2$, whereas an A•T pair has $- \; + \; -$ from A N7, A $6NH_2$, C O4. Thus, a G•C base pair has a greater electrostatic potential gradient across the groove and $dG_n \cdot dC_n$ stretches of DNA have the largest possibilities for electrostatic complementarity. Nonetheless, the DNA–Eco R1 complex of McClarin et al. (1986) shows how major groove specificity can be achieved with various hydrogen-bonding schemes to specific A•T or G•C pairs. Caruthers (1980) has also shown how thymine methyl groups can play an important role in enhancing DNA-protein binding through dispersion attraction with protein side chains. Thus, it is clear that both small and large molecules can interact strongly and specifically with the hydrogen bonds (electrostatic potential) of the grooves of DNA. The minor groove has a narrow shape appropriate for binding a small molecule like netropsin with a snug fit between the drug and DNA. The major groove is larger and more open. Larger structures such as proteins can interact with it, but it is clear that a snug steric van der Waals fit to such a convex surface is more difficult than in the case of the concave minor groove.

Another motif for molecules to interact with DNA is through intercalation, where the DNA base pairs unwind and the drug slips between the bases. Ethidium, 9-aminoacridine and proflavin are examples of simple cationic intercalators that interact strongly with DNA. Other drugs such as actinomycin D, echinomycin/triostin, and daunomycin can be thought of as both intercalators and groove binders, because they have both intercalating groups and groups that interact with hydrogen bonding groups in the DNA minor groove. For example, the 5′-G-C-3′ sequence specificity of actinomycin comes from hydrogen-bonding interactions of the actinomycin peptide chromophore with the guanine N3 and $2NH_2$ groups in the minor groove of the DNA (see color plate for Fig. 7-14) (Lybrand et al., 1986). Triostin has a C-G preference as well, due to peptide hydrogen bonds with the guanine $2NH_2$ group (Singh et al., 1986). Triostin also has the ability to induce non-Watson–Crick (Hoogsteen) base pairing next to the site of intercalation. This finding has been interpreted as caused by effective van der Waals interactions between the sugar–phosphate backbone and the valyl side chain of the chromophore, an interaction that is enhanced by the narrow DNA helix at Hoogsteen base pairs. Figure 7-15 (see color plate) shows the Watson–Crick base paired geometry on the left and Hoogsteen on the right; note the "ideal" fit between triostin and nucleic acid in the latter case and the gap between groups in the former.

5.3.2 Covalent Interactions

DNA can form covalent interactions with various molecules. Such interactions with aromatic hydrocarbon diol epoxides (e.g., benzpyrenediol epoxide) have been implicated in the carcinogenic action of these molecules. However, anticancer drugs such as cis platinum, mitomycin and anthramycin may also exert their anticancer action through covalent adducts with DNA. Molecular modeling–molecular mechanics studies on such molecules have provided detailed models on the complexes. For example, both kinked and nonkinked cis platinum adducts have been constructed and shown to be possible (Kozelka et al., 1985, 1986), with the kinked adduct more consistent with NMR data. Rao et al. (1986a) correctly predicted stereochemistry and groove direction of anthramycin–DNA adducts prior to NMR analysis (see color plate for Fig. 7-16). Rao et al. (1986b) stated that their models could not determine whether G O6 or G $2NH_2$ adducts of mitomycin were more likely, based on molecular mechanics, at a time when O6 alkylation by this molecule was accepted as experimental fact. Subsequent experiments suggested instead that guanine 2NH2 was the site of both mono-adduct and covalent cross-link formation. Thus, the modeling has been useful in such covalent adduct DNA complexes in providing reasonable and detailed 3D models, but these methods cannot unequivocally establish the lowest energy structure in all but the simplest molecules.

6. CONCLUDING PERSPECTIVE

In the last few years, two exciting developments have made a tremendous impact on the ability to simulate nucleic acid systems realistically. First, the development of methods like particle mesh Ewald (PME), a computationally efficient way to include long range electrostatic effects) (Darden *et al*, 1993; Essmann *et al*, 1995) has allowed the simulation of stable nucleic acid trajectories. This was most dramatically illustrated by Cheatham *et al* (1995), where a DNA duplex and an RNA hairpin loop were stable and consistent with the experimentally derived structures for a nanosecond of molecular dynamics, whereas using non-bonded cutoffs, the solutes rapidly fell apart. Both sets of simulations used full representations of water and counterions. The second major development has been increased computer power, both in the form of faster, less expensive workstations and PC's and in the harnessing of powerful parallel computers. This has enabled simulations to be carried out into the multinanosecond range for nucleic acid systems, to further test simulation protocols and force fields. Recent reviews of the state of the art in such simulations of nucleic acids include those by Auffinger and Westhof (1998). Some of the highlights in the recent past have been: (1) The demonstration that simulations on DNA duplexes, whether starting in A or B forms, which differ by 6 Å root mean square deviation, converge to the same B family structure in approximately half a nanosecond of molecular dynamics (Cheatham and Kollman, 1996); (2) The ability of locally enhanced sampling to find the correct structure of an RNA hairpin loop when starting with the wrong structure (Simmerling *et al*, 1998); (3) The stabilization of DNA in the A form in 85% ethanol and in the presence of cobalt hexammine (Cheatham and Kollman, 1997a, 1997b); (4) Simulations of tRNA structural features (Auffinger and Westhof, 1997) and the hammerhead ribozyme (Auffinger and Westhof, 1997); (5) Simulations of sequence dependent DNA bending (Young and Beveridge, 1998, Pastor et al, 1997). It is clear that one is now in the position to simulate the structures and energies of nucleic acid systems in a much more realistic way than before, allowing more synergistic interactions with experimental studies.

In addition to the above simulations in which one is describing structure and structural transitions, recent studies which use the molecular dynamics trajectories in aqueous solvent, but calculate free energies using the solute molecular mechanical internal energies from those trajectories plus continuum solvatioin methods to estimate the solvent-solute and solvent-solvent free energies are showing promise. For example, Srinivasan et al. (1998), Jayaram et al. (1998) and Cheatham et al. (1998) show that one can effectively describe the preferences for A-DNA and B-DNA under various conditions.

References

Allinger, N. L., Yuk, Y. H., and Lii, J.-H. (1989). Molecular Mechanics. The MM3 Force Field for Hydrocarbons, *J. Amer. Chem. Soc.* **111**, 8551–8566.

Altona, C. and Sundaralingam, M. (1972). Conformational Analysis of the Sugar Ring in Nucleosides and Nucleotides. A New Description Using the Concept of Pseudorotation, *J. Amer. Chem. Soc.* **94**, 8205–8212.

Anderson, C. F., and Record, M. T., Jr. (1990). Ion Distributions around DNA and Other Cylindrical Polyions—Theoretical Descriptions and Physical Implications. *Annu. Rev. Biophys. Biophys. Chem.* **19**, 423–465.

Arnott, S., Campbell Smith, P., and Chandresekhar, R. (1976). *CRC Handbook of Biochemistry*, G. Fasman, Ed., CRC Press, Cleveland, OH, p. 11.

Arnott, S. and Hukins, D. W. L. (1972). Optimised Parameters for A-DNA and B-DNA. *Biochem. Biophys. Res. Commun.* **47**, 1504–1509.

Arnott, S. and Hukins, D. W. L. (1973). Hukins Refinement of the Structure of B-DNA and Implications for the Analysis of X-Ray Diffraction Data from Fibers of Biopolymers. *J. Mol. Biol.* **81**, 93–105.

Auffinger, P. and Westhof, E. (1997) RNA Hydration: Three Nanoseconds of Multiple Molecular Dynamics Simulations of the Solvated tRNA(Asp) Anticodon Hairpin. *J. Mol. Biol.*, **269**, 326–341.

Auffinger, P. and Westhof, E. (1998) Simulations of the Molecular Dynamics of Nucleic Acids. *Current Opinion In Structural Biology*, **8**, 227–236.

Bash, P., Field, M., and Karplus, M. (1987a) Free Energy Perturbation Method for Chemical Reactions in the Condensed Phase: A Dynamical Approach Based on a Combined Quantum and Molecular Mechanics Potential. *J. Amer. Chem. Soc.* **109**, 8092–8094.

Bash, P., Singh, U. C., Langridge, R., and Kollman, P.(1987b). Free Energy Calculations by Computer Simulation. *Science* **236**, 564–568.

Berendsen, H. J. C., Postma, J. P. M., van Gunsteren, W. F., DiNola, A., and Haak, J. R. (1984). *J. Chem. Phys.* **81**, 3684–3690.

Berman, H. M. and Shieh, H. S.(1981). *Topics in Nucleic Acid Structure*, Neidle, S., Ed., Macmillan, London, UK.

Beveridge, D. L. and DiCapua, F. M. (1989). Free Energy via Molecular Simulation: Applications to Chemical and Biomolecular Systems. *Annu. Rev. Biophys. Biophys. Chem.* **18**, 431–492.

Beveridge, D. L., Subramanian, P., Jayaram, B., Swaminathan, S., and Ravishankar, G. (1990). Molecular simulation studies on the d(CGCGAATTCGCG) duplex: Hydration, ion atmosphere, structure and dynamics. Sarma, R. H., and Sarma, M. H., Ed., in *DNA & RNA*. Adenine Press. Albany, NY, 79–112.

Brooks, B. R., and Karplus, M. (1983). CHARMM: A Program for Macromolecular Energy Minimization, and Dynamics Calculation. *J. Computational Chem.* **4**, 187–217.

Broyde, S. and B. Hingerty, (1983). Conformation of 2-Aminofluorene-Modified DNA. *Biopolymers* **22**, 2423–2441.

Brunck,T. K. and Weinhold, F.(1979). Quantum-Mechanical Studies on the Origin of Barriers to Internal Rotation about Single Bonds. *J. Amer. Chem. Soc.* **101**, 1700–1709.

Brünger, A. T. (1991). Refinement of Three-dimensional Structures of Proteins and Nucleic Acids, Chapter 5, in *Molecular Dynamics, Applications in Molecular Biology*, Goodfellow, J. M., Ed., CRC Press, Boca Raton, FL.

Brünger, A. T. and Karplus, M. (1991). Molecular Dynamics Simulations with Experimental Restraints, *Acc. Chem. Res.* **24**, 54–61.

Burkert, U. and Allinger, N. L., (1982). *Molecular Mechanics*, American Chemical Society, Washington, DC.

Caldwell, J. C., Dang, L. X., and Kollman, P. A. (1990). Implementation of Non-Additive Intermolecular Potentials by Use of Molecular Dynamics: Development of a Water–Water Potential and Water–Ion Cluster Interactions. *J. Am. Chem. Soc.* **112**, 9144–9148.

Caldwell, J. C. and Kollman, P. A. Molecular Mechanical Study of Nelsopsis-DNA Interactions, Biopolymers, **25**, 249–266.

Caruthers, M.(1980). Deciphering the Protein-DNA Recognition Code, *Acc. Chem. Res.* **13**, 155–160.

Cheatham, T. E., Crowley, M. F., Fox, T. and Kollman, P. A. (1997) A Molecular Level Picture of the Stabilizatiion of A-DNA in Mixed Ethanol-Water Solutions. *Proc. Natl. Acad. Sci. USA*, **94**, 9626–9630.

Cheatham, T. E. and Kollman, P. A. (1996). Observations of the A-DNA to B-DNA Transition During Unrestrained Molecular Dynamics in Aqueous Solution. *J. Mol. Biol.*, **259**, 434–444.

Cheatham, T. E. and Kollman, P. A. (1997). Insight into the Stabilization of A-DNA by Specific Ion Association: Spontaneous B-DNA to A-DNA Transition Observed in Molecular Dynamics Simulations of d[ACCCGCGGGT](2) in the Presence of Hexaamminecobalt(III). *Structure*, **5**, 1297–1311.

Cheatham, T. E., Miller, J. L., Fox, T., Darden, T. A. and P. A. Kollman. (1995) Molecular Dynamics Simulations on Solvated Biomolecular Systems—The Particle Mesh Ewald Method Leads to Stable Trajectories of DNAm RNA, and Proteins. *J. Am. Chem. Soc.*, **117**, 4193–4194.

Cheatham, T. E., J. Srinivasan, D. A. Case, and P. A. Kollman (1998). Molecular Dynamics and Continuum Solven Studies of the Stability of poly dG-poly dC and poly dA-poly dT Duplexes in Solution, *J. Biomol. Struct. Dyn.*, **16**, 2654–280.

Cieplak, P. and Kollman, P. (1988). Calculation of the Free Energy of Association of Nucleic Bases in Vacuo and Water Solution. *J. Am. Chem. Soc.* **110**, 3734–3739.

Cornell, W. D., Creplak, P., Bayly, C. I., Gould, I. R., Merz, K. M., Ferguson, D. M., Spellmeyer, D. C., Fox, T., Caldwell, J. W., and Killman, P. A. (1995). A Second Generation Force Field for the Simulation of Proteins, Nucleic Acids and Organic Molecules. *J. Am. Chem. Soc.* **117**, 5179–5197.

Cramer, C. J. and Truhlar, D. G. (1991). General Parameterized SCF Model for Free Energies of Solution in Aqueous Solution, *J. Am. Chem. Soc.* **113**, 8305–8311.

Cremer, D. and Pople, J. A. (1975). A General Definition of Ring Puckering Coordinates. *J. Am. Chem. Soc.* **97**, 1354–1358.

Crippen, G., (1981) Distance Geometry and Conformational Calculations, in Chemometrics Research Study Series, Vol. 1, Bawden, D., Ed., Research Studies Press, Wiley, New York.

Dang, L., Caldwell, J., Rice, J. E., and Kollman, P. (1991). Ion Solvation in Polarizible Water: Molecular Dynamics Simulations, *J. Am. Chem. Soc.* **113**, 2481–2486.

Dang, L. and Kollman, P. (1990). Molecular dynamics simulations study of the free energy of association of 9-methyladenine and 1-methylthymine in water. *J. Am. Chem. Soc.* **112**, 503–507.

Darden, T., York, D. and Pedersen, L. (1993) Particle Mesh Ewald - An N.Log (N) Method For Ewald Sums In Large Systems. *J. Chem. Phys.* **98**, 10089–10092.

Dewar, M., Zoebisch, F., Healy, E., and Stewart, J. (1985). AM1: A New General Purpose Quantum Mechanical Molecular Model. *J. Am. Chem. Soc.* **107**, 3902–3909.

Drew, H. R., Wing, R. M., Takano, T., Broka, C., Tanaka, S., Itakura, K., and Dickerson, R. (1981). Structure of a B-DNA dodecamer: configuration and dynamics. *Proc. Natl. Acad. Sci. USA* **78**, 2179–2183.

Essmann, U., Perera, L., Berkowitz, M. L., Darden, T., Lee, H. and Pedersen, L.G. (1995) A Smooth Particle Mesh Ewald Method. *J. Chem. Phys.*, **103**, 8577–8593.

Eisenberg, D. and Kauzmann, W. (1969). *The Structure and Properties of Liquid Water*. Oxford University Press, Oxford, UK.

Flatters, D., Young, M. L., Beveridge, D. L. and Lavery, R. (1997). Conformational Properties of the TATA-Box Binding Sequence of DNA. *J. Biomol. Struct. Dyn.*, **14**, 757–765.

Flores, T. P. and Moss, D. S. (1990). Simulating the Dynamics of Macromolecules, Chapter 1 in *Molecular Dynamics, Applications in Molecular Biology*, Goodfellow, J. M., Ed, CRC Press, Boca Raton, FL.

Gund, P., Halgren, T. A., and Smith, G. M. (1987). Molecular Modeling as an Aid to Drug Design and Discovery. *Annu. Rep. Med. Chem.* **22**, 269–279.

Goodfellow, J. M., Ed. (1990). *Molecular Dynamics, Applications in Molecular Biology*, CRC Press, Boca Raton, FL.

Hermann, T., Auffinger, P. and Westhof, E. (1998) Molecular Dynamics Investigations of Hammerhead Ribozyme RNA. *Euro. Biophys. J. with Biophys. Letts.*, **27**, 153–165.

Hunter, W., Brown, T., Anard, N., and Kennard, O. (1986b). Structure of an Adenine.Cytosine Base Pair in DNA and Its Implications for Mismatch Repair, *Nature (London)* **320**, 552–555 (1986b).

Hunter, W., Brown, T., and Kennard, O. (1986a). Structural Features and Hydration of d(C-G-C-G-A-A-T-T-A-G-C-G); a Double Helix Containg Two G.A Mispairs, *J. Biomol. Struct. Dyn.* **4**, 173–191.

Hussain, I., Griffith, J., and Sancar, A. (1988). Thymine Dimers Bend DNA. *Proc. Nat. Acad. Sci.* **85**, 2558–2562.

Jayaram, B., Mezei, M., and Beveridge, D. L. (1988). Conformational Stability of Dimethyl Phosphate Anion in Water: Liquid-State Free Energy Simulations. *J. Amer. Chem. Soc.* **110**, 1691–1694.

Jayaram, B., D. Sprous, M. A. Young and D. A. Beveridge (1998), Free Energy Analysis of the Conformational Preferences of A and B Forms of DNA in Solution, *J. Amer. Chem. Soc.*, 10629–10633.

Jorgensen, W. L., Buckner, J. K., Huston, S. E., and Rossky, P. J. (1987). Hydration and Energetics for $(CH_3)_3$ CCl Ion Pairs in Aqueous Solution, *J. Amer. Chem. Soc.* **109**, 1891–1899.

Jorgensen, W. L., Chandresekhar, J., Madura, J. D., Impey, R. W., and Klein, M. L. (1983). Comparison of Simple Potential Functions for Simulating Liquid Water. *J. Chem. Phys.* **79**, 926–935.

Keepers, J., Schmidt, P., James, T., and Kollman, P. (1984). Molecular-Mechanic Studies of the Mismatched Base Analogs of $d(CGCGAATTCGCG)_2$,$d(CGTGAATTCGCG)_2$, $d(CGAGAATTCGCG)_2$, $d(CGCGAATTCACG)_2$, $d(CGCGAATTCTCG)_2$, and d(CGCAGAATTCGCG) • d(CGCGAATTCGCG). *Biopolymers* **23**, 2901–2929.

Kemmik, J., Boelens, R., Koenig, T., Kaptein, R., van der Marel G., and van Boom, J. (1987). Conformational Changes in the Oligonucleotide Duplex d(GCGTTGCG) • d(CGCAACGC) Induced by Formation of a *cis–syn* Thymine Dimer, *Eur. J. Biochem.* **162**, 37–43.

Kennard, O. (1985). Structural Studies of DNA Fragments: The G • T Wobble Base Pair in A,B and Z DNA; The G • A Base Pair in B-DNA, *J. Biomol. Struct. Dyn.* **3**, 205–226.

Kollman, P. (1977a). *Modern Theoretical Chemistry Applications of Electronic Structure Theory*. Vol 4, Schaefer, H. F., Ed., Plenum, New York.

Kollman, P. (1977b). Noncovalent Interactions. *Acc. Chem. Res.* **10**, 365–371.

Kollman, P., Weiner, P., and Dearing, A. (1981) Studies of Nucleotide Conformations and Interactions. The Relative Stabilities of Double-Helical B-DNA Sequence Isomers. *Biopolymers* **20**, 2583–2621.

Kollman, P., (1995). Drug-Target Binding Forces in *Burger's Medicinal Chemistry and Drug Discovery*, Vol. 1, M. E. Wolff, Ed., Wiley-Interscience, New York, pp. 399–412.

Kollman, P., Weiner, P., Quigley, G., and Wang, A. (1982). Molecular-Mechanical Studies of Z-DNA: A Comparison of the Structural and Energeric Properties of Z- and B-DNA. *Biopolymers* **21**, 1945–1969.

Kopka, M., Yoon, C., Goodsell, D., Djusa, P., and Dickerson, R. (1985) The Molecular Origin of DNA-Drug Specificity in Netropsin and Distamycin, *Proc. Nat. Acad. Sci.* **82**, 1376–1380.

Kozelka, J., Petsko, G., Lippard, S., Quigley, G., and Lippard, S. (1985). Molecular Mechanics Calculations on *cis*-$[Pt(NH_3)_2d(GpG)]$ Adducts in Two Oligonucleotide Duplexes , *J. Am. Chem. Soc.* **107**, 4079–4081.

Kozelka, J., Petsko, G., Quigley, G., and Lippard, S. J. (1986). High-Salt and Low-Salt Models for Kinked Adducts of *cis*-Diamminedichloroplatinum (II) with Oligonucleotide Duplexes. *Inorg. Chem.* **25**, 1075–1077.

Langridge, R., Ferrin, T. E., Kuntz, I. D., and Connolly, M. L. (1981). Real-Time Color Graphics in Studies of Molecular Interactions, *Science* **211**, 661–666.

Lavery, R. and Hartmann, B. (1994) Modeling DNA Conformational Mechanics. *Biophys. Chem.* **50**, 33–45.

Levine, I. (1991) *Quantum Chemistry*. 4th ed., Prentice Hall, Englewood Cliffs, NJ.

Levitt, M. (1983). Computer Simulation of DNA Double-helix Dynamics, *Cold Spring Harbor Symp. Quant. Bio.* **47**, 251–262.

Lybrand, T., Brown, S., Shafer, R., and Kollman, P. (1986). Computer Modeling of Actinomycin D Interactions with Double-helical DNA. *J. Mol. Biol.* **191**, 495–507.

Lybrand, T. and Kollman, P. A. (1985). Water–Water and Water–Ion Potential Functions Including Terms for Many-Body Effects, *J. Chem. Phys.* **83**, 2923–2933.

Maple, J. R., Hwang, M. J., Stockfisch, T. P., Dinur, U., Waldman, M., Ewig, C. S., and Hagler, A. T. (1994). Derivation of Class II Force Fields. 1. Methodology and Quantum Force Field for the Alkyl Functional Group and Alkane Molecules, *J. Comp. Chem.* **15**, 162–182.

Mackerell, A. D., Wiorkiewicz-Kuczera, J., and Kaplus, M. (1995) An all-atom Empirical Energy Function for the Simulation of Nucleic Acids. *J. Am. Chem. Soc.* **117**, 11946–11975.

Margenau, H. and Kestner, N.(1970). *Theory of Intermolecular Forces*, Pergammon, Oxford, UK.

Marsh, F. J., Weiner, P., J. Douglas, E., P. Kollman, A., G. Kenyon, L., and Gerlt, J. A. (1980). Theoretical Calculations on the Geometric Destabilization of 3',5′- and 2′, 3′- Cyclic Nucleotides, *J. Amer. Chem. Soc.* **102**, 1660–1665.

McCammon, J. A. and Harvey, S. C. (1987). *Dynamics of Proteins and Nucleic Acids*, Cambridge University Press, New York.

McClarin, J., Fredrick, C., Wang, B. C., Greene, P., Boyer, H., Grable, J., and Rosenberg, J. (1986). Structure of the DNA-Eco RI Endonuclease Recognition Complex at 3 Å Resolution. *Science* **234**, 1526–1541.

McConnell, K., Nirmala, R., Young, M., Ravishankerand, G., and Beveridge, D. (1994). A nanosecond molecular dynamics trajectory for a B DNA double helix, Evidence for substates. *J. Am. Chem. Soc.* **116**, 4461–4462.

Metropolis, N., Rosenbluth, A. W., Rosenbluth, M. N., Teller, A. H., and Teller, E. J.,(1953). Equation of State Calculations by Fast Computing Machines. *J. Chem. Phys.* **21**, 1087–1092.

Mezei, M., and Beveridge, D. L. (1986). Free Energy Simulations. *Ann. New York Acad. Sci.* **482**, 1–23.

Newton, M. D. (1973). A Model Conformational Study of Nucleic Acid Phosphate Ester Bonds. The Torsional Potential of Dimethyl Phosphate Monoanion. *J. Amer. Chem. Soc.* **95**, 256–258.

Nilsson, L. and Karplus, M. (1986). Empirical Energy Functions for Energy Minimization and Dynamics of Nucleic Acids. *J. Comp. Chem.* 7, 591–616.

Olson, W., (1975). Configuration-Dependent Properties of Randomly Coiling Polynucleotide Chains. II. The Role of the Phosphodiester Linkage. *Biopolymers* **14**, 1797–1810.

Olson, W. (1992a). How Flexible Is the Furanose Ring? 2. An Updated Potential Energy Estimate, *J. Amer. Chem. Soc.* **104**, 278–286.

Olson, W. K. (1982b). Theoretical Studies of Nucleic Acid Conformation: Potential Energies, Chain Statistics, and Model Building, in *Topics in Nucleic Acid Structures*, Part 2, Neidle, S., Ed., Macmillan Press, London, 1–79.

Page, M. I. and Jencks, W. P. (1971). Entropic Contributions to Rate Accelerations in Enzymic and Intramolecular Reactions and the Chelate Effect. *Proc. Nat. Acad. Sci.* **68**, 1678–1683.

Parr, R. G. and Yang, W. (1989). *Density-Functional Theory of Atoms and Molecules*, Oxford University Press, New York,.

Pastor, N., Pardo, L. and Weinstein, H. Does TATA Matter? A Structural Exploration of the Selectivity Determinants in its Complexes with TATA Box-Binding Protein. (1997) *Biophys. J.*, **73**, 640–6652.

Patel, D., Kozlowski, S., Ikutura, S. (1984). Deoxyadenosine-Deoxycytidine Pairing in the d(C-G-C-G-A-A-T-T-C-A-C-G) Duplex: Conformation and Dynamics at and Adjacent to the dA • dC Mismatch Site. *Biochemistry* **23**, 3207–3217, 3218–3226.

Patel, D., Kozlowski, S., Marky, L., Rice, J., Broka, C., Dallas, J., Itakura, K., and Breslauer, K. (1982). Structure, Dynamics, and Energetics of Deoxyguanosine-Thymidine Wobble Base Pair Formation in the Self- Complementary d(CGTGAATTCGCG) Duplex in Solution. *Biochemistry* **21**, 437–444.

Patel, D. J., Shapiro, L., and Hare, D. (1987). DNA and RNA: NMR Studies of Conformations and Dynamics in Solution. *Q. Rev. Biophys.* **20**, 35–112.

Pearlman, D., Holbrook, S., Prokle, D., and Kim, S. H. (1985). Molecular Models for DNA Damaged by Photoreaction, *Science* **227**, 1304–1308.

Pohorille, A., Ross, W. S., and Tinoco, I., Jr. (1990). DNA Dynamics in Aqueous Solution: Opening the Double Helix. *Int. J. Supercomputer Appl.* **4**, 81–96.

Rao, S., Keepers, J., and Kollman, P. (1984). The Structure of d(CGCGAAT[]TCGCG)• d(CGCGAATTCGCG): The Incorporation of a Thymine Photodimer into a B-DNA Helix, *Nucleic Acids Res.* **12**, 4789–4807.

Rao, S. N., Singh, U. C., and Kollman, P. (1986a) Molecular Mechanics Simulations on Covalent Compexes between Anthramycin and B DNA. *J. Med. Chem.* **29**, 2484–2492.

Rao, S. N., U. Singh, C., and Kollman, P. (1986b). Conformations of the Noncovalent and Covalent Complexes between Mitomycins A and C and $d(GCGCGCGCGC)_2$, *J. Am. Chem. Soc.* **108**, 2058–2068.

Rao, S. N., Singh, U. C., and Kollman, P. A. (1986). Molecular Dynamics Simulations of DNA Double Helices: Studies of Sequence Dependence and the Role of Mismatch Pairs in the DNA Helix. *Israel J. Chem.* **27**, 189–197.

Ravishanker, G., Swaminathan, S., Beveridge, D. L., Lavery, R., and Sklenar, H. (1989). Conformational and Helicoidal Analysis of 30 PS of Molecular Dynamics on the d(CGCGAATTCGCG) Double Helix: Curves, Dials and Windows. *J. Biomol. Struct. Dyn.* **6**, 669–699.

Saenger, W., (1984). *Principles of Nucleic Acid Structure*, Springer-Verlag, New York.

Sasisekharan, V., (1973). *Conformation of Biological Molecules and Polymers*, Bergmann, E. and Pullman, B., Eds., Jerusalem, Israel.

Schaefer III, H. F. (1986). Methylene, A Paradigm for Computational Quantum Chemistry. *Science* **231**, 1100–1107.

Seeman, N., Rosenberg, J., and Rich, A. (1976). Sequence-Specific Recognition of Double Helical Nucleic Acids by Proteins. *Proc. Nat. Acad. Sci.* **73**, 804–808.

Sharp, K. A. and Honig, B. (1990). Electrostatic Interactions in Macromolecules: Theory and Experiment *Annu. Rev. Biophys. Biophys. Chem* **19**, 301–332 (1990).

Simmerling, C., Miller, J., and Kollman, P.A. Combined Locally Enhanced Sampling and Particle Mesh Ewald as a Strategy to Locate the Experimental Structure of a Non-Helical Nucleic Acid. *J. Am. Chem. Soc.* (in press).

Sinden, R. and Hagerman, P. (1984). Interstrand Psoralen Cross-Links Do Not Introduce Appreciable Bends in DNA, *Biochemistry* **23**, 6299–6303.

Singh, U. C. and Kollman, P. (1984). An Approach to Computing Electrostatic Charges for Molecules. *J. Comp. Chem.* **5**, 129–145.

Singh, U. C. and Kollman, P. (1985). A Combined A*b Initio* Quantum Mechanical and Molecular Mechanical Method for Carrying out Simulations on Complex Molecular Systems: Applications to the CH_3Cl^+Cl- Exchange Reaction and Gas Phase Protonation of Polyethers. *J. Comp. Chem.* **7**, 718–730.

Singh, U. C., Pattabiraman, N., Langridge, R., and Kollman, P. A. (1986). Molecular Mechanical Studies of $d(CGTACG)_2$: Complex of Triostin A with the Middle A·T Base Pairs In Either Hoogsteen or Watson–Crick Pairing. *Proc. Natl. Acad. Sci. USA* **83**, 6402–6406.

Singh, U. C., Weiner, P., Caldwell, J., and Kollman, P. (1986), *AMBER 3.0* University of California, San Francisco, CA.

Srinivasan, J., T. E. Cheatham, P. Cieplak, P. A. Kollman and D. A. Case (1998). Continuum Solvent Studies of the Stability of DNA, RNA and Phosphoramidate-DNA Helices, *J. Amer. Chem. Soc.*, **120**, 9401–9409

Still, W. C., Tempczyk, A., Hawley, R. C., and Henrickson, T. (1990). Semianalytical Treatment of Solvation for Molecular Mechanics and Dynamics. *J. Amer. Chem. Soc.* **112**, 6127–6120.

Subramanian, P. S., Swaminathan, S., and Beveridge, D. L. (1990). Theoretical account of the 'spine of hydration' in the minor groove of duplex d(CGCGAATTCGCG). *J. Biomol. Struct. Dyn.* **7**, 1161–1165.

Swaminathan, S., Ravishanker, G., and Beveridge, D. (1991). Molecular dynamics of B-DNA including water and counterions: A 140-ps trajectory for d(CGCGAATTCGCG) based on the GROMOS force field. *J. Am. Chem. Soc.* **113**, 5027–5040.

Szabo, A. and Ostlund, N. (1982). *Modern Quantum Chemistry*, Macmillan, New York, (1982).

Tilton R. F., Jr., Weiner, P. K., and Kollman, P. A. (1983). An Analysis of the Sequence Dependence of the Structure and Energy of A- and B-DNA Models Using Molecular Mechanics. *Biopolymers* **22**, 969–1002.

Van Gunsteren, W. F. and Berendsen, H. J. C. (1990). Molecular Dynamics Computer Simulation. Method, Application and Perspectives in Chemistry. *Angew. Chem* **102**, 1020–1055.

Veillard, A., Ed. *Quantum Chemistry: The Challenge of Transition Metals and Coordination Chemistry*, Vol. 176, NATO ASI Series C Dordecht, The Netherlands.

Warshel, A. and Levitt, M. (1976). Theoretical Studies of Enzymatic Reactions: Dielectric, Electrostatic and Steric Stabilization of the Carbonium Ion in the Reaction of Lysozyme. *J. Mol. Biol.* **103**, 227–249.

Weiner, S. J., Kollman, P. A., Case, D. A., Singh, U. C., Ghio, C., Alagona, G., Profeta, S., Jr., and Weiner, P. (1984). A New Force Field for Molecular Mechanical Simulation of Nucleic Acids and Proteins. *J. Am. Chem. Soc.* **106**, 765–784.

Weiner, P. K., Landridge, R., Blaney, J. M., Schaefer, R., and Kollman, P. A. (1982). Electrostatic Potential Molecular Surfaces. *Proc. Natl. Acad. Sci. USA* **79**, 3754–3758.

Woodward, R. B. and Hoffman, R. (1970). *The Conservation of Orbital Symmetry*, Verlag Chemie, Weinheim, Bergstrasse.

York, D. M., Wlodawer, A., Pedersen, L. G., and Darden, T. A. (1994). Atomic Level Accuracy in Simulations of Large Protein Crystals. *Proc. Natl. Acad. Sci.* **91**, 8715–8718.

CHAPTER 8

Conformational Changes

by Douglas H. Turner

To be functional, nucleic acids must adopt particular three dimensional (3D) conformations. For example, Figure 8-1(*a*) shows the sequence of a self-splicing intron from *Tetrahymena thermophila* (Cech and Bass, 1986). At high temperatures, this molecule has the conformation of a random coil, and is not active. At temperatures near 37°C, however, the interactions discussed in Chapter 7 force the molecule to fold on itself to give the base pairing, or secondary structure, shown in Figure 8-1(*b*) (Cech

(*a*) cucucuAAAUAGCAAUAUUUACCUUUGGAGGGAAAAGUUAUCA
GGCAUGCACCUGGUAGCUAGUCUUUAAACCAAUAGAUUGCA
UCGGUUUAAAAGGCAAGACCGUCAAAUUGCGGGAAAGGGGU
CAACAGCCGUUCAGUACCAAGUCUCAGGGGAAACUUUGAGA
UGGCCUUGCAAAGGGUAUGGUAAUAAGCUGACGGACAUGGU
CCUAACCACGCAGCCAAGUCCUAAGUCAACAGAUCUUCUGUU
GAUAUGGAUGCAGUUCACAGACUAAAUGUCGGUCGGGGAAG
AUGUAUUCUUCUCAUAAGAUAUAGUCGGACCUCUCCUUAAU
GGGAGCUAGCGGAUGAAGUGAUGCAACACUGGAGCCGCUGG
GAACUAAUUUGUAUGCGAAAGUAUAUUGAUUAGUUUUGGAG
UACUCGuaag

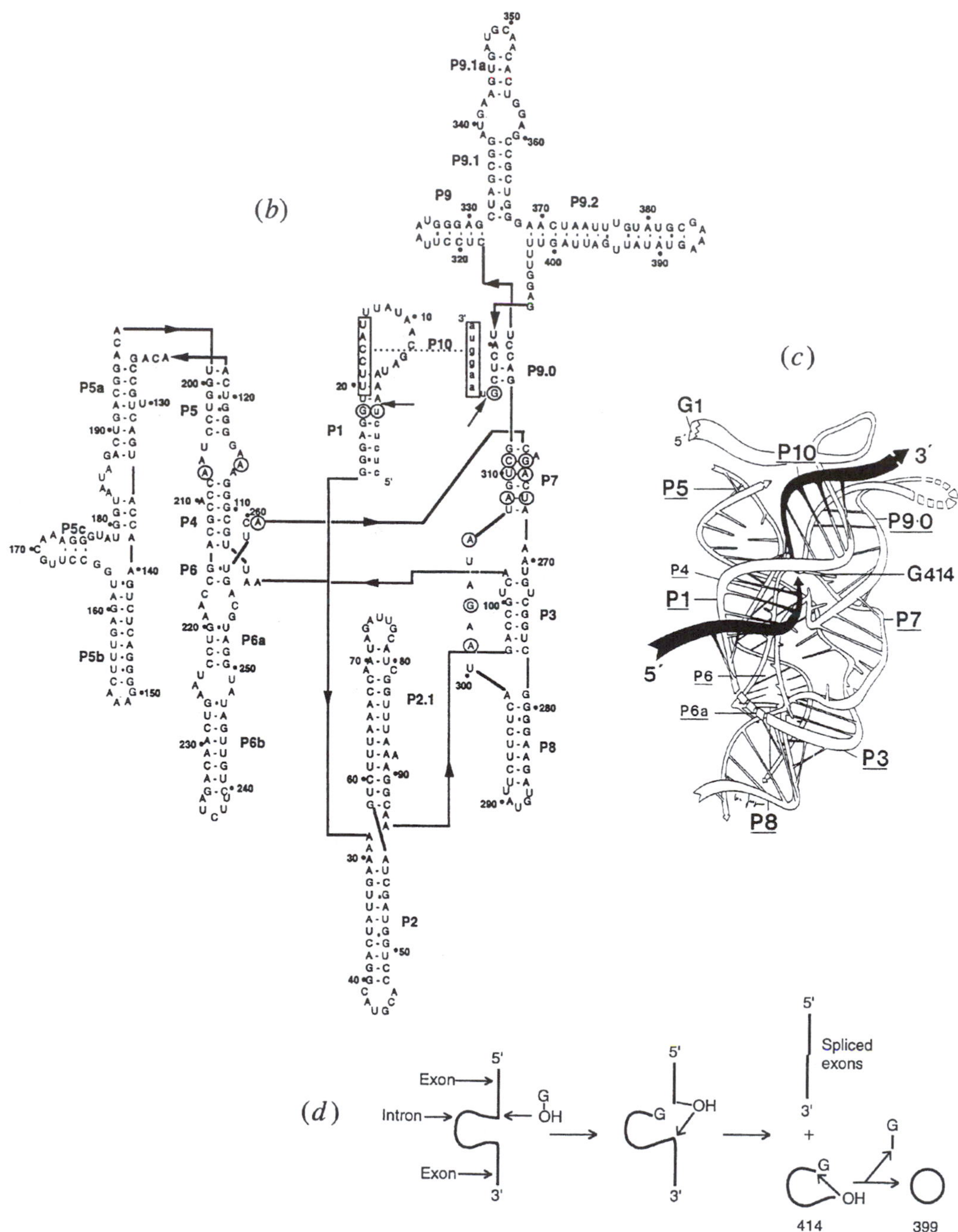

Figure 8-1 (*facing page*).
(*a*) The sequence of the self-splicing intron from the rRNA precursor of *T. thermophila* is shown in capital letters. Small letters give the adjacent sequences of the attached exons.
(*b*) Secondary structure formed by the sequence in (*a*). [Reprinted with permission from Cech, R. R., Damberger, S. H., and Gutell, R. R. (1994). *Nature Struct. Biol.* **1**, 273–280.] Nucleotides that are circled are conserved in other group I introns. (*c*) Proposed model for 3D structure of active site for self-splicing intron [Reprinted with permission from Michel and Westhof, 1990].
(*d*) Sequence of reactions mediated by structure in (*b*). [From Cech and Bass, 1986.]

et al., 1994). These same interactions also determine the 3D folding of the molecule, which has been suggested to look like Figure 8-1(*c*) at the active site (Michel and Westhof, 1990). When the molecule is in the necessary conformation, it can catalyze the reactions shown in Figure 8-1(*d*) (Cech and Bass, 1986; Cech, 1990). Since transcription of RNA occurs by stepwise addition of nucleotides, conformational changes are required for it to assume its active state.

Self-splicing introns are examples of nucleic acids with catalytic functions. The function of much nucleic acid, however, is to carry information, such as the sequence of a protein. In these cases, conformation is still important for interactions with other molecules that regulate expression and transmission of the information (Oxender et al., 1979; Shen et al., 1988; Tang and Draper, 1989).

As described in Chapter 3, it is now straightforward to determine the sequence of a nucleic acid. Sequences containing more than 1 billion nucleotides are known, and determination of the 3 billion nucleotides of the human genome is in sight. Determination of the secondary and 3D structures and of the dynamics of these structures is more difficult, however (see Chapters 4 and 5). This finding is particularly true for RNA. Thus, transfer RNA is the only complete natural RNA whose 3D structure is known. While DNA overwhelmingly occurs as a fully paired, right-handed double helix, X-ray and solution studies reveal considerable sequence-dependent differences in conformation (Saenger, 1984; Dickerson et al., 1982; Wang et al., 1982; Arnott et al., 1982; Calladine, 1982; Dickerson, 1983). Presumably, these differences are important for interactions of DNA with other molecules. Thus the sequence dependence of conformation and its functional effects are important, but relatively unexplored.

One of the goals of biophysical chemistry is to use the available information on the structure and dynamics of nucleic acids to deduce principles that allow accurate prediction of these properties from sequence alone. In this chapter, experimental results on conformational changes are described, and the disciplines of thermodynamics, statistical mechanics, and kinetics are used to derive general principles. The results can be used to test predictions made from our knowledge of intermolecular and intramolecular forces as discussed in Chapter 7. Knowledge of these forces will ultimately permit accurate predictions of properties from sequence.

1. SINGLE-STRAND STACKING

Much biological activity of nucleic acids depends on regions of single-stranded, unpaired nucleotides. This configuration maximizes the number of active groups available for recognition or reaction. For example, in tRNA (see Figs. 4-11–4-14), recognition of the codon depends on pairing with the single-stranded anticodon. Transfer RNA is charged with an amino acid at one of the hydroxyl groups of the 3′-terminal A in the single-stranded sequence NCCA. As shown in Figure 8-1, several single stranded nucleotides of the self-splicing intron are conserved. Presumably they are important for catalysis (Kim and Cech, 1987; Michel and Westhof, 1990).

One of the simplest conformational changes in nucleic acids involves the transition of an unpaired single strand from a random coil in which the bases are not stacked to an ordered, helical structure in which the bases are stacked. This transition is illustrated in Figure 8-2. The thermodynamics and dynamics of this transition have been studied for a number of cases, and provide insight into the interactions governing the properties of single strands. Most of the methods used in these studies are also applicable to other conformational changes.

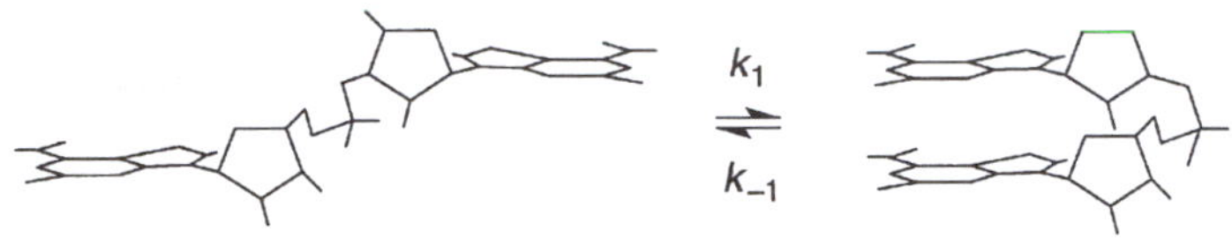

Figure 8-1
Transition from unstacked to stacked conformation in single-stranded dinucleoside monophosphate.

1.1 Thermodynamics of Single Strand Stacking

The main methods for determining thermodynamic parameters for nucleic acid conformational changes are calorimetry and the temperature dependence of spectroscopic properties. Both methods are easy to understand in principle. In practice, neither is easy to use with transitions for single-strand stacking because the transitions occur over a wide temperature range. For example, reported enthalpy changes, ΔH°, for single strand stacking in poly (A) range from -3 to -13 kcal mol^{-1}.

1.1.1 Calorimetry

Calorimetric methods measure the heat released or absorbed by a chemical reaction. The most popular calorimetric method applied to conformational changes of nucleic acids is differential scanning calorimetry (DSC) (Sturtevant, 1987; Breslauer et al., 1992). In this method, electrical energy is used to slowly increase the temperatures of two matched cells. One cell contains the sample of interest and the other contains a blank (e.g., buffer). If the temperature change induces a chemical reaction in the sample but not the blank, then the amount of electrical energy required to raise the temperature of the sample cell will be increased by the amount of heat absorbed by the reaction. Thus a differential scanning calorimeter measures the difference in heat required to raise the temperatures of the two cells. The data is reported as excess heat capacity, ϕC_p, versus temperature as shown in Figure 8-3.

One disadvantage of DSC is that any reaction occurring in the sample cell can affect the excess heat capacity. For example, if hydration of the polymer chain is temperature dependent, then ϕC_p will be affected over the temperature range where hydration is changing. This disadvantage is not serious for studies of conformational changes that occur over small temperature ranges. In these cases, baselines before and after the transition of interest can be extrapolated to subtract out from ϕC_p any effects from other reactions. This subtraction leaves the change in heat capacity, ΔC_p°, associated with the transition of interest. The area under the ΔC_p° curve for the temperature

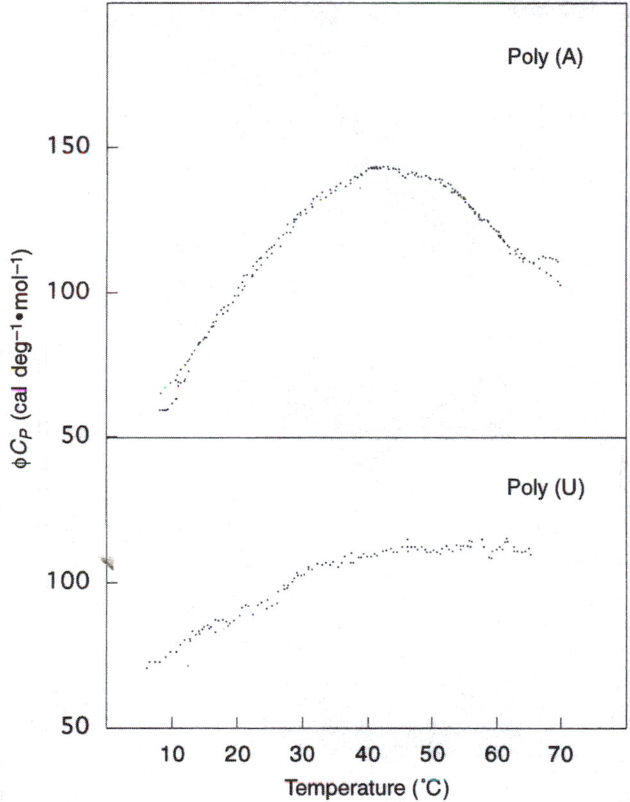

Figure 8-2
Excess heat capacity, ϕC_p, versus temperature for poly(A) and poly(U) as measured by differential scanning calorimetry. [Calorimetric Determination of the Heat capacity Changes Associated with the Conformational Transitions of Polyriboadenylic Acid and Polyribouridylic Acid, Suurkuusk, J., Alvarez, J., Freire, E., and Biltonan, R., *Biopolymers*, **16**, 2641–2652. Copyright ©1997. Reprinted by permission of John Wiley & Sons, Inc.]

interval, T1 to T2, in which the reaction is occurring is the ΔH° for the transition (see Figs. 8-3 and 8-4):

$$\Delta H^\circ = \int_{T_1}^{T_2} \Delta C_p^\circ dT \tag{8-1}$$

Unfortunately, the stacking reaction for a single stranded polynucleotide is incomplete in the normally accessible temperature range of 0–100°C. This limitation makes it difficult to determine the baseline relevant for measuring area under the excess heat capacity curve.

1.1.2 Temperature Dependence of Spectroscopic Properties

The ΔH° for a reaction is related to the temperature dependence of the equilibrium constant by the van't Hoff equation:

$$\frac{\partial \ln K}{\partial T} = \frac{\Delta H^\circ}{RT^2} \tag{8-2}$$

Thus any method that provides an equilibrium constant as a function of temperature can be used to determine ΔH°. Spectroscopic methods are most common. For example, the equilibrium constant for the reaction shown in Figure 8-2 is

$$U \rightleftharpoons S \qquad K = \frac{[S]}{[U]} = \frac{\alpha}{1-\alpha} \tag{8-3}$$

where [S] and [U] are concentrations of stacked and unstacked species, respectively, and $\alpha = [S]/([S] + [U])$, the fraction of strands in the stacked state. Therefore any property that allows determination of [S] and [U] can be used to determine K. For example, the extinction coefficients for stacked and unstacked conformations are usually different (see Chapter 6, Section 1.1). Thus the absorbance, A, of a sample containing dinucleoside monophosphate will be

$$A = \varepsilon_s[S]\ell + \varepsilon_u[U]\ell = C_T\ell[\alpha\varepsilon_s + (1-\alpha)\varepsilon_u] \tag{8-4}$$

Here ε_s and ε_u are extinction coefficients for the stacked and unstacked species, respectively, and ℓ is the pathlength of the cell. If ε_s and ε_u are known, then the relevant concentrations can be determined from the absorbance and knowledge of the total concentration, $C_T = [S] + [U]$. Unfortunately, it is often difficult to determine ε_s and ε_u, particularly when the transition occurs over a large temperature range so that the sample is never completely one species.

When ε_s and ε_u cannot be directly measured, plots of absorbance versus temperature must be fit to a model for the transition of interest. For the transition shown in Figure 8-2, the two-state model is the simplest and most common model. The dinucleoside monophosphate is assumed to be either stacked or unstacked, and the ΔH° and ΔS°

for the reaction are assumed to be independent of temperature. For this model, plots of absorbance versus temperature can be fit with four variables: ε_s, ε_u, ΔH°, and ΔS°, since

$$K = \exp(-\Delta H^\circ/RT + \Delta S^\circ/R) \qquad (8\text{-}5)$$

This treatment assumes the change in absorbance is entirely due to the stacking reaction and this assumption can be checked by measuring the time dependence of the absorbance change as described in Section 8.1.2.

When single-strand stacking occurs in an oligonucleotide or polynucleotide, there is an additional consideration. A new stack can either lengthen an existing region of stacked nucleotides, or it can occur in a region that was previously completely random coil. The equilibrium constants for these two processes can be different, indicating cooperativity for stacking. In this case, a somewhat more complex model, the one-dimensional (1D) Ising model, must be used for the analysis (Zimm and Bragg, 1959; Applequist, 1963). In this model, the equilibrium constants for propagating or initiating a stacked region are denoted s and βs, respectively. Thus β for a noncooperative transition is one. As a transition becomes more cooperative, β becomes smaller. While s has the temperature dependence given above for K, β is assumed to be temperature independent. For this model, α is given by (Applequist, 1963):

$$\alpha = 0.5 + 0.5(\mathrm{s} - 1)[(1 - \mathrm{s})^2 + 4\beta \mathrm{s}]^{-1/2} \qquad (8\text{-}6)$$

The addition of β means five parameters are required to fit data to the model. In practice, the transitions are too broad to reliably allow a five parameter fit. Simultaneous fitting of spectroscopic and calorimetric data can improve the situation, however (Freier et al., 1981).

Representative thermodynamic parameters for some dinucleoside monophosphates and polynucleotides are listed in Table 8-1. Included in the table are melting temperatures, t_m, the temperatures where half the bases are stacked ($K = 1$ and/or $\alpha = 0.5$). For any transition,

$$\Delta G^\circ = -RT \ln K = \Delta H^\circ - T\Delta S^\circ \qquad (8\text{-}7a)$$

Here T is the Kelvin temperature, $T = 273.15 + t$, where t is the temperature in degrees Celsius (°C). Thus, for a unimolecular transition, U $\rightleftharpoons$ S, at the T_m, $-RT_\mathrm{m} \ln(1) = 0 = \Delta H^\circ - T_\mathrm{m}\Delta S^\circ$, so

$$T_\mathrm{m} = \frac{\Delta H^\circ}{\Delta S^\circ} \quad \text{(unimolecular transition)} \qquad (8\text{-}7b)$$

All temperatures in the above equations are in K. When the units of ΔH° are kilocalories per mole, they must be multiplied by 1000 if ΔS° is in entropy units (eu, cal mol^{-1} deg^{-1}). Note that round off errors often affect T_m values by several degrees. While measured T_m's are reasonably reliable, caution is required when considering values of ΔH° and ΔS° for single-strand stacking. As noted above, all involve extrapolation

Table 8.1
Thermodynamic Parameters for Single-Strand Stacking

Molecule	Reference	Method[a]	[NaCl]	β	ΔH° (kcal mol^{-1})	ΔS° (eu)[c]	t_M °C
dApA	Olsthoorn et al. (1981).	CD			−7.3	−23	49
ApA	Olsthoorn et al. (1981).	CD			−7.2	−25	22
ApA	Powell et al. (1972).	A	0.01–1		−8.5	−28	26
ApU	Olsthoorn et al. (1981).	CD			−7.3	−25	22
CpC	Powell et al. (1972).	A	0.01–1		−8.5	−30	13
UpU	Simpkins and Richards (1967).	A,ORD	0.1		No stacking		< 0
Poly(A)	Filimonov and Privalov (1978).	C	0.01–0.1		−3.0	−10	40
Poly(A)	Suurkuusk et al. (1977).	C	0.02	0.6	−8.5	−27	46
Poly(A)	Freier et al. (1981).	A,C	0.05–0.1	0.5[b]	−6.8[b]	−22	38
Poly(C)	Freier et al. (1981).	A,C	0.2	1	−9.0	−28	53
Poly(U)	Richards et al. (1963); Inners and Felsenfeld (1970).	V,S	0.0002–2		No stacking		< 0

[a]Methods: A = absorbance; C = calorimetry; CD = circular dichroism; ORD = optical rotatory dispersion; S = sedimentation; V= viscosity.

[b]Average of two values.

[c]eu = cal k^{-1} mol^{-1}

of data outside experimentally accessible temperature limits. As discussed in Section 8.1.2, kinetic results indicate the two-state model also may not be adequate for certain conditions.

Despite the uncertainties, some features of single-strand stacking are clear. The cooperativity is small, with β typically between 0.5 and 1. The salt dependence is also negligible. The sequence dependence, however, is considerable. While poly(A) and poly(C) are largely stacked at 20°C, poly (U) is a random coil. Structural studies of oligomers by NMR and ORD have provided evidence that the negligible stacking of U is not restricted to UU sequences. For example, in the sequences AUG and UGUG, the purines stack together while the U bases remain unstacked (Lee and Tinoco, 1980; van der Hoogen et al., 1988b).

1.2 Kinetics of Single Strand Stacking

1.2.1 Temperature-Jump Relaxation Spectroscopy

The kinetics of single strand stacking are very fast. They can be measured, however, by temperature-jump relaxation spectroscopy (Eigen and De Maeyer, 1963; Bernasconi, 1976; Turner, 1986). In this method, the temperature of a solution is raised quickly, which changes the equilibrium constant for any reaction that has a nonzero ΔH° (see Eq. 8-2). If the temperature-jump occurs faster than the equilibrium can adjust, then the time dependence of the approach to the new equilibrium concentrations can be followed spectroscopically. The rate constants for the reaction can be derived from this time dependence. For example, for the two-state reaction shown in Figure 8-2, the time dependence of the change in concentration of unstacked species, $\Delta[\mathrm{U}]$, is a single exponential:

$$\Delta[\mathrm{U}] = \Delta[\mathrm{U}]_0 \exp(-t/\tau) \tag{8-8}$$

Here $\Delta[\mathrm{U}]_0$ is the displacement from the new equilibrium at the time of the temperature-jump, and t is time. For a unimolecular reaction, $\tau^{-1} = k_1 + k_{-1}$, the sum of the forward and reverse rate constants. The equilibrium constant is given by $K = k_1/k_{-1}$. Thus measurements of the relaxation time, τ, and K allow determination of k_1 and k_{-1}.

1.2.2 Results

Single relaxation times have been measured for CpC, CpA, ApC, poly(C), and poly(A) (Freier et al., 1981; Pörschke, 1978; Dewey and Turner, 1979). These results are listed in Table 8-2. At Na^+ concentrations above 0.05 M, the relaxation times for poly(A) and ApA depend on the monitoring wavelength (Pörschke, 1978), which means there is more than one conformational change in these cases. Studies by NMR indicate this results because two stacked conformations exist (Kondo and Danyluk, 1976). Thus the two-state model is not appropriate, and rate constants cannot be derived easily.

All the forward rate constants for single strand stacking are about 10^7 s^{-1}. While this value is large compared to many other reactions, it is slow for a simple stacking reaction. For example, the relaxation time for stacking of the two adenines joined by

Table 8.2
Kinetic Parameters for Single Strand Stacking

Molecule	Reference	$[Na^+]$ (M)	t °C	τ (ns)	$k_1 \times 10^6$ s^{-1}	$E_{a,1}$ (kcal mol^{-1})	$\Delta S_1^\ddagger$ (eu)	$k_{-1} \times 10^6$ s^{-1}	$E_{a,-1}$ (kcal mol^{-1})	$\Delta S_{-1}^\ddagger$ (eu)
ApA	Pörschke (1978).	1	4	50, >100	*a*			*a*		
ApC	Pörschke (1978).	1	4	42	13[b]			11[b]		
CpA	Pörschke (1978).	1	4	30	14[b]			19[b]		
CpC	Pörschke (1978).	1	4	30	19[b]	2.1	−19	14[b]	11	12
A^9—$(CH_2)_3$—A^9	Pörschke (1978).	1	4	<15						
Poly(A)	Dewey and Turner (1979).	0.05	10	174	6.2	6.2	−7.5	1.6	14	18
Poly(C)	Freier et al. (1981).	0.05	10	45	19	−0.8	−30	3.1	7.8	−2
Poly(C)	Freier et al. (1981)	1	15	67	13	1.9	−22	1.8	11	6
Poly(dA)	Dewey and Turner (1979).	0.05	10	48	22	3.3	−15	3.4	9.6	3

[a]Rate constants are not calculated because more than one relaxation time is observed.

[b]Equilibrium constants used to calculate k_1 and k_{-1} from τ were taken from Davis and Tinoco (1968) for ApC and CpA, and from Powell et al. (1972) for CpC.

a trimethylene bridge in 9,9′-trimethylenebisadenine is faster than 15 ns (Pörschke, 1978), indicating $k_1 > 7 \times 10^7 s^{-1}$. The forward rate constant for bimolecular stacking reactions of bases and of planar dye molecules is typically about $10^9 M^{-1} s^{-1}$ (Pörschke and Eggers, 1972; Dewey et al., 1979). Given the high local concentration of bases in a single-stranded polymer, the magnitude of the unimolecular rate constant would be at least as large if controlled by the same factors. These comparisons suggest the sugar–phosphate backbone somehow limits the stacking rate. Further insights into this are provided by the activation energies, E_a, and entropies, $\Delta S^\ddagger$, as derived from the temperature dependence of rate constants and the Eyring equation:

$$k = (eRT/hN)\exp(-E_a/RT + \Delta S^\ddagger/R) \tag{8-9}$$

Here e is the base for natural logarithms (2.72), h is Planck's constant, and N is Avogadro's number. For the stacking of poly(C) and poly(A), E_a is only a few kilocalories per mole (Freier et al., 1981; Dewey and Turner, 1979). Furthermore, stacking rate constants depend linearly on solvent viscosity. This finding is consistent with a diffusion controlled rate. The activation entropies, however, are about -10 to -30 eu. Thus the relatively slow rate constants are a consequence of a very ordered transition state, which could result from the necessity of constraining bonds in the backbone to single conformations before rotational diffusion to the stacked conformation. Another possible contributor to the large, unfavorable $\Delta S^\ddagger$ is specific solvation requirements for the transition state.

1.3 Interactions Determining Structure and Dynamics of Single-Strand Stacking

The results for the thermodynamics and kinetics of single-strand stacking provide insight into the interactions that determine these properties. In particular, conformational entropy favors the unstacked, random coil conformation. These unfavorable entropy effects must be overcome by bonding interactions, commonly called stacking interactions, in order to stabilize the ordered, stacked helix. Both effects are discussed below, along with some unexplained effects.

1.3.1 Conformational Entropy Effects

The covalent bonds in a dinucleoside monophosphate are shown in Figure 2-3. There are a limited number of conformations that can be adopted around each bond. For most bonds in polymers, the number of available conformations is typically three. X-ray and NMR data on nucleic acids suggest the appropriate number of available conformations for the bonds shown in Figure 2-3 are (Berman, 1981; Olson, 1982): $\alpha(1)$, $\beta(3)$, $\gamma(3)$, $\delta(2)$, $\varepsilon(3)$, $\zeta(2)$, $\chi(2)$. If the stacked state of a dinucleoside monophosphate is restricted to a single conformation, but the unstacked state can sample all available conformations for each bond, then initiation of stacking will be opposed by an unfavorable initiation conformational entropy term of

$\Delta S^\circ_{ic} = -R \ln(1 \times 3 \times 3 \times 2 \times 2 \times 3 \times 2 \times 2 \times 2) = -13$ eu. Notice that two glycosidic (χ) and two sugar (δ) bonds are included in this calculation, one for each nucleoside. If the stacked dimer in Figure 8-2 was part of a polynucleotide, then stacking an adjacent nucleotide on the pre-existing stack would require constraining rotation about one more glycosidic and one more sugar bond, rather than two of each. In this simple model, propagation of a helix is associated with a propagation conformational entropy change, ΔS°_{pc}, of -11 eu. The values measured for ΔS° and $\Delta S^\ddagger$ of stacking in dinucleoside monophosphates and in polymers range from about -10 to -30 eu (see Tables 8-1 and 8-2). Thus conformational effects can account for a large part of the measured equilibrium and activation entropy changes. They can also account for the small cooperativity observed. If the cooperativity parameter β arises only because two glycosidic and two sugar bonds are constrained for initiation and one each for propagation, then its value is predicted to be $\beta = \exp[(-R \ln 2 \times 2)/R] = 0.25$. Measured values range from about 0.5 to 1. Of course, this model may be oversimplified. For example, more than one configuration may contribute to the stacked state.

1.3.2 Stacking

The origin of stacking interactions has been the subject of much discussion. From the thermodynamic data, it is clear that stacking is driven by a favorable, modest ΔH°. This finding is consistent with noncovalent bonding due to dispersion forces as described in Section 7.1.

It has often been suggested that stacking is also driven by classical hydrophobic interactions. In this model, ordered water is released from around the bases upon stacking, and this provides a favorable entropy term. The magnitude expected for such an effect can be estimated from the dimerization of benzene in water. The ΔS° for benzene dimerization is a favorable 20 eu (Tucker et al., 1981). The ΔS° for single strand stacking, however, is unfavorable (see Table 8-1). Moreover, the magnitude is either that expected for conformational entropy effects, or is even more unfavorable. Thus the thermodynamic parameters provide no evidence that classical hydrophobic bonding is important for driving stacking.

1.3.3 Unexplained Effects

The results on single strand stacking also raise some interesting questions. Comparison of Tables 8-1 and 2-2 indicates the ΔH° values for stacking of ApA, poly(A), and deoxyadenosine monomers are similar. The ΔH° values for stacking of CpC and poly(C), however, are very different from the ΔH° for association of cytidines. Moreover, the order for strength of stacking in monomers, dimers, and polymers is different: U $\leq$ C $<$ A versus UpU $\ll$ CpC $<$ ApA versus poly(U) $\ll$ poly(A) $<$ poly(C). This result suggests some new interaction in poly(C). The activation parameters for stacking in poly(C) and poly(A) are also suggestive of a difference (Table 8.2). The $E_{a,1}$ and $\Delta S^\ddagger_1$ for poly(A) are reasonable for a diffusion controlled process. For poly(C), however, $E_{a,1}$ is lower and $\Delta S^\ddagger_1$ higher than expected for diffusion control. One possibility is a special, specific solvation of poly(C).

2. DOUBLE HELIX FORMATION BY OLIGONUCLEOTIDES WITHOUT LOOPS

The main structural motif for natural nucleic acids is the double helix with Watson–Crick base pairs. Cellular DNA is almost exclusively in this form. Known structures of RNA are more than 50% double helix. Thus an understanding of the principles governing double helix formation is essential for understanding and predicting the properties of nucleic acids. Oligonucleotides provide convenient model systems for discovering these principles.

2.1 Thermodynamics of Duplex Formation

2.1.1 Methods

The methods used for measuring the thermodynamics of double helix formation by oligonucleotides are similar to those used for single-strand stacking. Double helix formation, however, is a very cooperative process, so the transitions occur in smaller temperature intervals. For example, Figure 8-4 shows a DSC curve for duplex formation

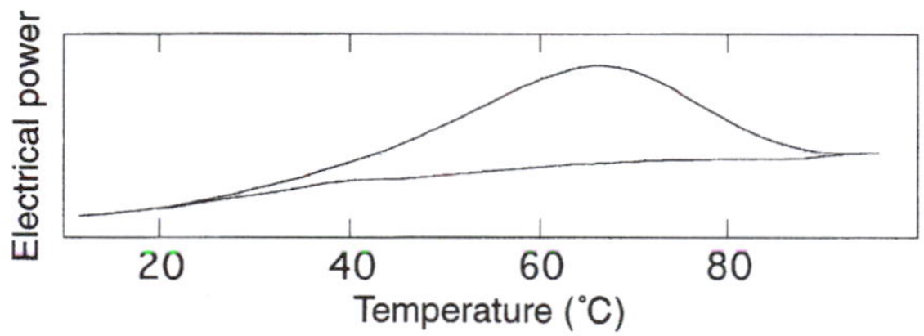

Figure 8-3
Raw data from a DSC experiment on dGCGCGC. Shown is electrical power (fed back to sample cell to maintain it at same temperature as reference cell) versus temperature. For the upper curve, the sample cell contained 1×10^{-3} *M* dGCGCGC in 1 *M* NaCl, 45 m*M* cacodylate, pH 7. For the lower curve, both sample and reference cells contained only buffer. [Reprinted with permission from Albergo, D. D., Marky, L. A., Breslauer, K. J., and Turner, D. H. (1981). *Biochemistry*, **20**, 1409–1413. Copyright ©American Chemical Society.]

by dGCGCGC (Albergo et al., 1981). The beginning and end of the transition and the baselines outside the transition region are apparent, which makes integration of the ΔC_p° curve to obtain ΔH° relatively reliable (see Eq. 8-1). To obtain ΔS°, the same data can be plotted as $\Delta C_p^\circ / T$ versus temperature and integrated:

$$\Delta S^\circ = \int_{T_1}^{T_2} \frac{\Delta C_p^\circ}{T}\, dT \tag{8-10}$$

An advantage of calorimetry is that the thermodynamic parameters obtained in this way do not depend on a theoretical model for the transition. As long as the baselines can be determined, the thermodynamic parameters can be obtained by taking the area under the transition curve.

The temperature dependence of spectroscopic properties is also used to obtain thermodynamic parameters for duplex formation. In this case, the data must be fit to a theoretical model to derive parameters. In practice, a simple two-state model is used most often (Martin et al., 1971; Borer et al., 1974; Turner et al., 1988; Petersheim and Turner, 1983). The simplifying assumption in the two-state case is that a given strand is either maximally base paired or completely not base paired (see Fig. 8-5). This assumption corresponds to a completely cooperative transition. More general statistical mechanical treatments are discussed in Appendix 8-11.

The equations used to fit spectroscopic data to the two-state model depend on whether the oligonucleotides are self- or nonself-complementary:

Self-complementary $$2A \rightleftharpoons A_2 \qquad K = \frac{[A_2]}{[A]^2} = \frac{\alpha}{2(1-\alpha)^2 C_T} \tag{8-11}$$

Nonself-complementary $[A]_0 = [B]_0 = C_T/2$ $$A + B \rightleftharpoons AB \qquad K = \frac{[AB]}{[A][B]} = \frac{2\alpha}{(1-\alpha)^2 C_T} \tag{8-12}$$

Here α is the fraction of total strand concentration, C_T, that is in duplex; and for the nonself-complementary case, it has been assumed that the total concentration of A and B strands is each $C_T/2$. If the spectroscopic property is absorbance, A, then at any temperature T

Self-complementary $$A = C_T \ell[\varepsilon_A(1-\alpha) + \varepsilon_{A_2}\alpha/2] \tag{8-13}$$

Nonself-complementary $$A = \frac{C_T \ell}{2}[(\varepsilon_A + \varepsilon_B)(1-\alpha) + \varepsilon_{AB}\alpha] \tag{8-14}$$

Here ε_A, ε_B, ε_{A_2}, and ε_{AB} are extinction coefficients for single-stranded A and B, and for duplexes A_2 and AB, respectively. If these extinction coefficients are known, then absorbance versus temperature data can be fit with Eqs. 8-11–8-14 and 8-7a to provide values of ΔH° and ΔS° for the transition. In practice, the extinction coefficients are usually temperature dependent. For example, the extinction for a single strand will vary

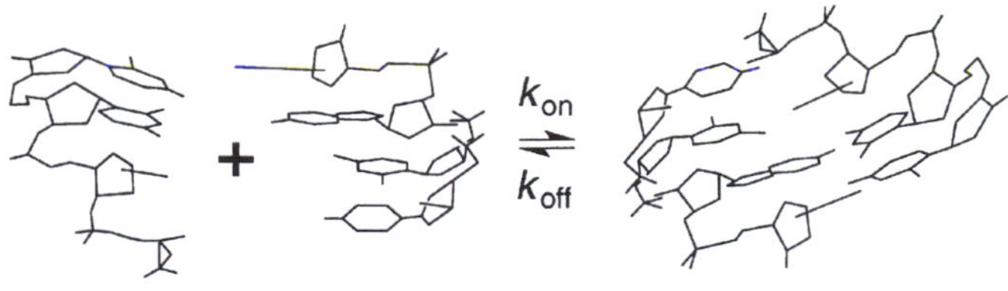

Figure 8-4
Transition from two single strands to a double helix.

with temperature because of the stacked-to-unstacked equilibrium discussed in Section 8.1. When nonself-complementary oligomers are studied, this temperature dependence can sometimes be measured independently with the individual single strands. In most cases, the temperature dependences of extinction coefficients are assumed to be linear, for example, $\varepsilon_A = m_A T + b_A$. The values for m_A and b_A are determined by fitting the absorbance versus temperature data in the regions where only single strands or only duplexes occur, or by including the linear dependence of extinction in fitting the shape of the entire duplex-to-single strand transition curve. Note that not every spectroscopic property is as simple as absorbance. For example, NMR chemical shifts cannot easily be used to derive thermodynamic properties (Pardi et al., 1981).

For a two-state transition, the concentration dependence of duplex formation provides another method for determining thermodynamic parameters. Defining the melting temperature, T_m, as the temperature at which $\alpha = 0.5$ and plugging into Eqs. 8-11 and 8-12 gives the results that at the T_m, K is $1/C_T$ for a self-complementary transition and $4/C_T$ for a nonself-complementary transition. Substituting these results into $\Delta G^\circ = -RT \ln K = \Delta H^\circ - T\Delta S^\circ$, and rearranging leads to

Bimolecular, self-complementary
$$\frac{1}{T_m} = \frac{R \ln C_T}{\Delta H^\circ} + \frac{\Delta S^\circ}{\Delta H^\circ} \quad (8\text{-}15a)$$

Bimolecular nonself-complementary
$[A]_0 = [B]_0 = C_T/2$
$$\frac{1}{T_m} = \frac{R \ln(C_T/4)}{\Delta H^\circ} + \frac{\Delta S^\circ}{\Delta H^\circ} \quad (8\text{-}15b)$$

Thus, a plot of $1/T_m$ versus $\ln(C_T)$ should be linear, and ΔH° and ΔS° can be determined from the slope and intercept. Equations 8-15a and 8-15b are special cases of the general equations for N strands associating to form an N-mer (Marky and Breslauer, 1987):

N-mer, Self-complementary
$$\frac{1}{T_m} = \frac{(N-1)R}{\Delta H^\circ} \ln C_T + \frac{[\Delta S^\circ - (N-1)R \ln 2 + R \ln N]}{\Delta H^\circ} \quad (8\text{-}16a)$$

N-mer, Nonself-complementary
$[A]_0 = [B]_0 = [C]_0 = \cdots$
$$\frac{1}{T_m} = \frac{(N-1)R}{\Delta H^\circ} \ln C_T + \frac{[\Delta S^\circ - (N-1)R \ln 2N]}{\Delta H^\circ} \quad (8\text{-}16b)$$

If a transition is two state, then all the methods described above should give the same thermodynamic parameters. If a transition is not two state, then the magnitude of ΔH° determined from calorimetry will be larger than that determined spectroscopically (Sturtevant, 1987). Moreover, the two spectroscopic methods may give different results, since non-two-state behavior typically affects the shape of a melting curve more than the T_m. These comparisons, therefore, provide tests for two-state behavior.

2.1.2 Results for Duplexes without Loops

Table 8-3 lists thermodynamic parameters determined for duplex formation by some oligonucleotides. The enthalpy and entropy changes are quite large relative to those for single-strand stacking. In contrast to single-strand stacking, these changes also increase substantially in magnitude as the sequence is made longer (e.g. GCCGGC vs. CCGG). These trends reflect the high cooperativity of the duplex transition.

Table 8-3 contains some results for sequences that have been studied in both deoxy and ribo forms, in parallel and antiparallel strands, and in 3′–5′and 2′–5′linked oligomers. The thermodynamic parameters are different in each case. For sequences that have been studied, duplexes with antiparallel strands are more stable than duplexes with parallel strands (Rippe and Jovin, 1992), and duplexes with 3′–5′phosphodiester bonds are more stable than duplexes with 2′–5′phosphodiester bonds (Kierzek et al., 1992; Jin et al., 1993). For 3′–5′linked duplexes, whether a deoxy or ribo duplex is more stable depends on the sequence. Different stabilities might be expected for sequences containing AU or AT pairs. The RNA and DNA sequences with only GC pairs also have different stabilities, however. Thus the 2′-OH affects duplex stability. This 2′-OH effect is even observed if a single 2′-OH is replaced by H in an oligonucleotide (Bevilacqua and Turner, 1991). This effect may be related to the different conformational preferences for RNA and DNA (see Chapter 4). For RNA–DNA hybrid duplexes, the available data suggest interesting sequence dependence (Martin and Tinoco, 1980; Sugimoto et al., 1995). For example, at total strand concentrations of $4 \times 10^{-4} M$, the melting temperatures of $rCA_5G \cdot dCT_5G$ and $dCA_5G \cdot rCU_5G$ are 19 and less than 0°C, respectively (Martin and Tinoco, 1980). It has been suggested that the relative stabilities of RNA • RNA, DNA • DNA, and RNA • DNA duplexes may determine the positions for factor-independent termination of transcription (Martin and Tinoco, 1980).

Table 8-3 also illustrates the large dependence of stability on base pair composition. At $10^{-4} M$ oligonucleotide strands, the duplex with 9 bp formed by CA_7G and CU_7G melts at 32°C, whereas the 4 bp duplex $(GGCC)_2$ melts at 34°C. Evidently, GC pairs are more stable than AU pairs. The sequence dependence of stability depends on more than composition, however. For example, another duplex with four GC pairs, $(CGCG)_2$, melts at about 19°C at $10^{-4} M$.

Nearest neighbor models (Borer et al., 1974; Turner et al., 1988; Goldstein and Benight, 1992; Gray 1997a,b; Allawi and SantaLucia, 1997; Xia et al., 1998) provide reasonable approximations of the sequence dependence of duplex stability. One such model, the Independent Nearest Neighbor-Hydrogen Bonding or INN-HB model, assumes that the stability of a given base pair depends on the identity of the adjacent base pair and that the stability of a helix depends on these nearest neighbor interactions and the base composition of the helix as reflected by the terminal base pairs (Xia et al., 1998). For example, the nearest neighbors 5′CG3′/3′GC5′, 5′GC3′/3′CG5′, and 5′GG3′/3′CC5′ are different. Thus $(GGCC)_2$ and $(CGCG)_2$ are composed of different nearest neighbors and can have different stabilities in the model. The duplexes $(GCCGGC)_2$ and $(GGCGCC)_2$, however, both have the same nearest neighbors and same base compositions, and therefore must have identical stabilities within the model. Inspection of the experimental results in Table 8-3 for GCCGGC ($t_m = 67.2$°C) and GGCGCC ($t_m = 65.2$°C) indicates the nearest neighbor model is a reasonable approximation for this case. A study of pairs of oligonucleotides with identical nearest neighbors and base compositions but different sequences indicates the thermodynamic parameters for such oligomers are generally within about 10% of each other (Kierzek et al., 1986; Xia et al., 1998). Thus the nearest neighbor model is better than a simple composition model, but is not perfect.

Table 8.3
Thermodynamic Parameters for Duplex Formation by Oligonucleotides

Sequence	Reference	Measured[a]				Predicted[b]			
		ΔH° (kcal mol^{-1})	ΔS° (eu)	ΔG°_{37} (kcal mol^{-1})	t_m (°C)	ΔH° (kcal mol^{-1})	ΔS° (eu)	ΔG°_{37} (kcal mol^{-1})	t_m (°C)
RNA, 3′-5′, Antiparallel Strands, 1 *M* NaCl									
CCGG	Petersheim and Turner (1983).	−34.2	−95.6	−4.6	27.1	−33.8	−95.0	−4.4	25.3
CGCG	Xia et al. (1998).	−33.3	−95.6	−3.7	19	−32.6	−93.2	−3.6	18.8
GCGC	Freier et al. (1985b).	−30.5	−83.4	−4.6	26.5	−36.8	−103.4	−4.7	29.1
GGCC	Freier et al. (1983).	−35.8	−98.1	−5.4	34.4	−38.1	−105.2	−5.4	34.9
CGCGCG	Freier et al. (1986b).	−54.5	−146.4	−9.1	57.9	−58.1	−156.8	−9.4	58.5
GCAUGC	Groebe, Cameron, Freier, Turner, and Uhlenbeck, unpublished results.	−62.3	−177.2	−7.4	45.7	−56.4	−157.2	−7.6	48.3
GCCGGC	Freier et al. (1985b).	−62.7	−166.0	−11.2	67.2	−63.6	−168.8	−11.2	66.6
GCGCGC	Freier et al. (1985b).	−66.0	−178.5	−10.6	62.1	−62.3	−167.0	−10.5	63.1
GGCGCC	Freier et al. (1986b).	−67.8	−182.0	−11.3	65.2	−63.6	−168.8	−11.2	66.6
$CA_7G + CU_7G$	Nelsen et al. (1981); Freier et al. (1986b).	−59.8	−175.1	−5.5	31.6	−58.2	−169.5	−5.7	32.4
RNA, 2′-5′, Antiparallel Strands, 1 *M* NaCl									
GCGCGC	Kierzek et al. (1992).	−22.2	−56.2	−4.8	24.4				
CGGCGCCG	Kierzek et al. (1992).	−44.3	−121.2	−6.8	45.3				
DNA, 3′-5′, Antiparallel Strands, 1 *M* NaCl									
CGCGCG	Senior et al., (1988b).	−46.4	−122.8	−8.3	55.7	−51.2	−137.4	−8.6	55.7
GCATGC	Williams et al. (1989.	−42.4	−118	−5.9	38.3	−43.6	−121.6	−5.9	38.5
GCGCGC	Albergo et al. (1981).	−59.6	−162.7	−9.1	56.1	−50.4	−134.6	−8.7	56.5
CA_7G+CT_7G	Aboulela et al. (1985)	−68.0	−196	−7.2	40.1	−63.5	−182.5	−6.8	43.1
DNA, 3′-5′, Antiparallel Strands, 0.1 M NaCl									
$5'A_{10}TA_2T_4A_3TAT_33'$ $3'T_{10}AT_2A_4T_3ATA_35'$	Rippe and Jovin (1992); Rentzeperis (1992).	−151	−457						
$5'A_5GA_3GTAGT_4A_2GTAT_33'$ $3'T_5CG_3CATCA_4T_2CATA_35'$	Rippe and Jovin (1992); Rentzeperis (1992).	−167	−500						
DNA, 3′-5′, Parallel Strands, 0.1 *M* NaCl									
$5'A_{10}TA_2T_4A_3TAT_33'$ $5'T_{10}AT_2A_4T_3ATA_33'$	Rippe and Jovin (1992); Rentzeperis (1992).	−116	−374			−116	−354	−6.2	
$5'A_5GA_3GTAGT_4A_2GTAT_33'$ $5'T_5CT_3CATCA_4T_2CATA_33'$	Rippe and Jovin (1992); Rentzeperis (1992).	−116	−381						

[a]Thermodynamic parameters determined from T_m^{-1} versus log C_T plots. The t_m is for $1 \times 10^{-4} M$ strand concentration.

[b]Thermodynamic parameters predicted from Table 8.4 for 1 *M* NaCl and from Rippe and Jovin (1992). Values for ΔG°_{37} calculated from the predicted ΔH° and ΔS° using $\Delta G^\circ_{37} = \Delta H^\circ - 310.15\Delta S^\circ$ may differ from the listed ΔG°_{37} due to round off errors. The t_m is for $1 \times 10^{-4} M$ strand concentration.

Table 8.4
Thermodynamic Parameters for Helix Initiation and Propagation in 1*M* NaCl

Propagation Sequence			RNA[a]			DNA[b]			RNA/DNA[c]		
			$\Delta H°$ (kcal mol^{-1})	$\Delta S°$ (eu)	$\Delta G°_{37}$ (kcal mol^{-1})	$\Delta H°$ (kcal mol^{-1})	$\Delta S°$ (eu)	$\Delta G°_{37}$ (kcal mol^{-1})	$\Delta H°$ (kcal mol^{-1})	$\Delta S°$ (eu)	$\Delta G°_{37}$ (kcal mol^{-1})
RNA		DNA									
	→ GC CG ←		−14.88	−36.9	−3.42	−9.8	−24.4	−2.24	−8.0	−17.1	−2.7
	→ GG CC ←		−13.39	−32.7	−3.26	−8.0	−19.9	−1.84	−12.8	−31.9	−2.9
									−9.3	−23.2	−2.1
	→ CG GC ←		−10.64	−26.7	−2.36	−10.6	−27.2	−2.17	−16.3	−47.1	−1.7
→ GA CU ←		→ GA CT ←	−12.44	−32.5	−2.35	−8.2	−22.2	−1.30	−5.5	−13.5	−1.3
									−8.6	−22.9	−1.5
→ GU CA ←		→ GT CA ←	−11.40	−29.5	−2.24	−8.4	−22.4	−1.44	−7.8	−21.6	−1.1
									−5.9	−12.3	−2.1
→ CA GU ←		→ CA GT ←	−10.44	−26.9	−2.11	−8.5	−22.7	−1.45	−9.0	−26.1	−0.9
									−10.4	−28.4	−1.6

Propagation Sequence (RNA)	Propagation Sequence (DNA)	RNA[a] $\Delta H°$ (kcal mol^{-1})	RNA[a] $\Delta S°$ (eu)	RNA[a] $\Delta G°_{37}$ (kcal mol^{-1})	DNA[b] $\Delta H°$ (kcal mol^{-1})	DNA[b] $\Delta S°$ (eu)	DNA[b] $\Delta G°_{37}$ (kcal mol^{-1})	RNA/DNA[c] $\Delta H°$ (kcal mol^{-1})	RNA/DNA[c] $\Delta S°$ (eu)	RNA/DNA[c] $\Delta G°_{37}$ (kcal mol^{-1})
→ CU / GA ←	→ CT / GA ←	−10.48	−27.1	−2.08	−7.8	−21.0	−1.28	−7.0	−19.7	−0.9
								−9.1	−23.5	−1.8
→ UA / AU ←	→ TA / AT ←	−7.69	−20.5	−1.33	−7.2	−21.3	−0.58	−7.8	−23.2	−0.6
→ AU / UA ←	→ AT / TA ←	−9.38	−26.7	−1.10	−7.2	−20.4	−0.88	−8.3	−23.9	−0.9
→ AA / UU ←	→ AA / TT ←	−6.82	−19.0	−0.93	−7.9	−22.2	−1.00	−7.8	−21.9	−1.0
								−11.5	−36.4	−0.2
For Bimolecular Associations										
Initiation		3.61	−1.5	4.09	0.2	−5.6	1.96	1.9	−3.9	3.1
Each terminal AU or AT		3.72	10.5	0.45	2.2	6.9	0.05			
Symmetry correction (self-complementary)		0	−1.4	0.43	0	−1.4	0.43			
Symmetry correction (nonself-complementary)		0	0	0	0	0	0			

[a]Xia et al., (1998).

[b]Allawi and SantaLucia (1997).

[c]Sugimoto et al., (1995). For RNA/DNA hybrids with two sets of parameters, the top set corresponds to the top strand as RNA and the bottom set corresponds to the bottom strand as RNA. For example, $\Delta H° \begin{matrix} 5'\text{rGG}3' \\ 3'\text{dCC}5' \end{matrix} = -12.8$ kcal mol^{-1} and $\Delta H° \begin{matrix} 5'\text{dGG}3' \\ 3'\text{rCC}5' \end{matrix} = -9.3$ kcal mol^{-1}. Alternative analyses of the RNA/DNA data base have also been presented (Gray, 1997b).

Application of the INN-HB nearest neighbor model to RNA and DNA duplexes containing only Watson–Crick pairs requires determining thermodynamic parameters for 10 nearest neighbors, helix initiation, and helix termination by AU or AT for each case. For RNA/DNA hybrids, more parameters are required (Sugimoto et al., 1995; Gray, 1997a,b). The parameters have been determined by fitting to optical melting data for oligonucleotides (Xia et al., 1998; Allawi and SantaLucia, 1997; Sugimoto et al., 1995; Gray, 1997b). Some results are listed in Table 8-4, and comparisons of predictions from the model with measurements are shown in Table 8-3. For DNA, calorimetric data for both oligomers and polymers have also been fit to a similar model (Breslauer et al., 1986). SantaLucia (1998) and Owczarzy et al. (1997) compare the thermodynamic parameters for DNA obtained by various methods. A nearest neighbor analysis has also been done for the available data on parallel stranded double helixes (Rippe and Jovin, 1992).

In deriving the parameters in Table 8-4, it was necessary to consider the fact that duplexes formed by self-complementary oligomers have a twofold rotational symmetry, whereas single-strands and nonself-complementary oligomers do not. To correct for this symmetry difference, ΔS° for self-complementary oligomers must be reduced by $R\ln 2 = 1.4$ eu. Including this correction and simply summing up appropriate parameters allows prediction of thermodynamic parameters for the melting of any RNA or DNA duplex containing only Watson–Crick base pairs. The melting temperature in degrees Celsius, t_m, for any concentration can then be calculated from a rearrangement of Eqs. 8-15a and 8-15b:

Bimolecular, self-complementary
$$t_m = \frac{\Delta H^\circ}{\Delta S^\circ + R\ln(C_T)} - 273.15 \tag{8-17}$$

Bimolecular nonself-complementary ($[A]_0 = [B]_0 = C_T/2$)
$$t_m = \frac{\Delta H^\circ}{\Delta S^\circ + R\ln(C_T/4)} - 273.15 \tag{8-18a}$$

Equation 8-18a holds if the concentrations of the two nonself-complementary strands are equal. In many applications, including probing with oligonucleotides for complementary sequences in polynucleotides, one strand is in large excess. In this case, when t_m is defined as the temperature where half of the less concentrated sequence is bound:

Bimolecular nonself-complementary $[B]_0 \gg [A]_0$
$$t_m = \frac{\Delta H^\circ}{\Delta S^\circ + R\ln(C_B - 0.5C_A)} - 273.15 \tag{8-18b}$$

Equations 8-17 and 8-18a can be combined into one by including all the changes between self- and nonself-complementary oligomers in a constant, A:

$$t_m = \frac{\Delta H^\circ}{A + \Delta S^\circ_{NN} + R\ln(C_T)} - 273.15 \tag{8-19}$$

Here, ΔS°_{NN} is the entropy change without any symmetry term, C_T is always the total strand concentration, and A is -1.4 and -2.8 eu for self- and nonself-complementary oligomers, respectively. Note that in Eqs. 8-17–8-19, if the units for ΔH° are in kilocalories per mole, they must be multiplied by 1000 if ΔS° is in entropy units. Sample calculations are shown in Figure 8-6.

The results in Table 8-4 allow prediction of the thermodynamic properties of fully paired duplexes. Many nucleic acid associations, however, include additional unpaired nucleotides on the ends of double helixes. Two examples are the associations of tRNA and mRNA and of hybrid probes with DNA or RNA. For RNA, these unpaired nucleotides or "dangling ends" add stability to the double helix in a manner that is very sequence dependent (Turner et al., 1988). For example, at 10^{-4} *M* strands, the melting temperatures of $(GGCC)_2$, $(AGGCC)_2$ and $(GGCCA)_2$ are 34.4, 38.1, and 57.9°C, respectively (Freier et al., 1983b, 1985a). The effects of at least the first unpaired nucleotide adjacent to a helix can be approximated by a nearest neighbor model. The available parameters are listed in Table 8-5 (Turner et al., 1988). One striking observation for RNA oligomers in 1 *M* NaCl is that all the free energy increments for terminal unpaired nucleotides on the 5′ side of helixes are very similar and small, averaging -0.2 kcal mol^{-1}. Similar free energy increments are measured for addition of a 5′-phosphate suggesting that 5′-dangling ends interact little with the adjacent helix (Freier et al., 1985). Stability increments for 3′ dangling ends are sequence dependent, however, ranging from -0.1 to -1.7 kcal mol^{-1}. Thus some 3′ dangling ends can stabilize a helix more than some base pairs (cf. Tables 8-5 and 8-4), indicating a strong interaction. This difference between 5′ and 3′ stacking can be rationalized from structural considerations. Figure 8-7 shows stereoviews of $(AGGCC)_2$ and $(GGCCA)_2$ in A- form geometry. Whereas the 3′-dangling A of $(GGCCA)_2$ stacks directly on the opposite strand G of the adjacent base pair, the 5′-dangling A of $(AGGCC)_2$ is not close to the opposite strand C of the adjacent base pair. Thus it is not surprising that interactions of the 3′A with the opposite strand help hold the duplex together, whereas interactions of the 5′A with the opposite strand are negligible. Fewer data are available for DNA, but it appears that 5′-dangling ends will add more stability than 3′-dangling ends (Senior et al., 1988b; Mellema et al., 1984).

A 5′ and 3′ dangling end opposite each other is called a terminal mismatch. Free energy increments associated with terminal mismatches are listed in Table 8-6 (Freier et al., 1986a; Hickey and Turner, 1985a; Sugimoto et al., 1987b; Serra et al., 1994). For pyrimidine–pyrimidine and CA mismatches, these stability increments are essentially the sum of the increments for the constituent dangling ends. For purine–purine mismatches, the increment may be less. Thus there is no thermodynamic evidence for interactions between the bases in most terminal mismatches. The exceptions are terminal GU mismatches. In certain cases, the pairing of a terminal G and U provides more stability than the sum of the equivalent dangling ends. This finding is consistent with hydrogen-bond formation within GU mismatches as suggested by Crick (1966) in the "wobble" hypothesis (see Chapter 2).

2.2 Interactions Determining Stability of the Double Helix

Conformational entropy and stacking interactions present in single-stranded nucleic acids are also present in duplexes. In addition, hydrogen bonds are formed between base pairs, and duplex formation leads to increased condensation of counterions around

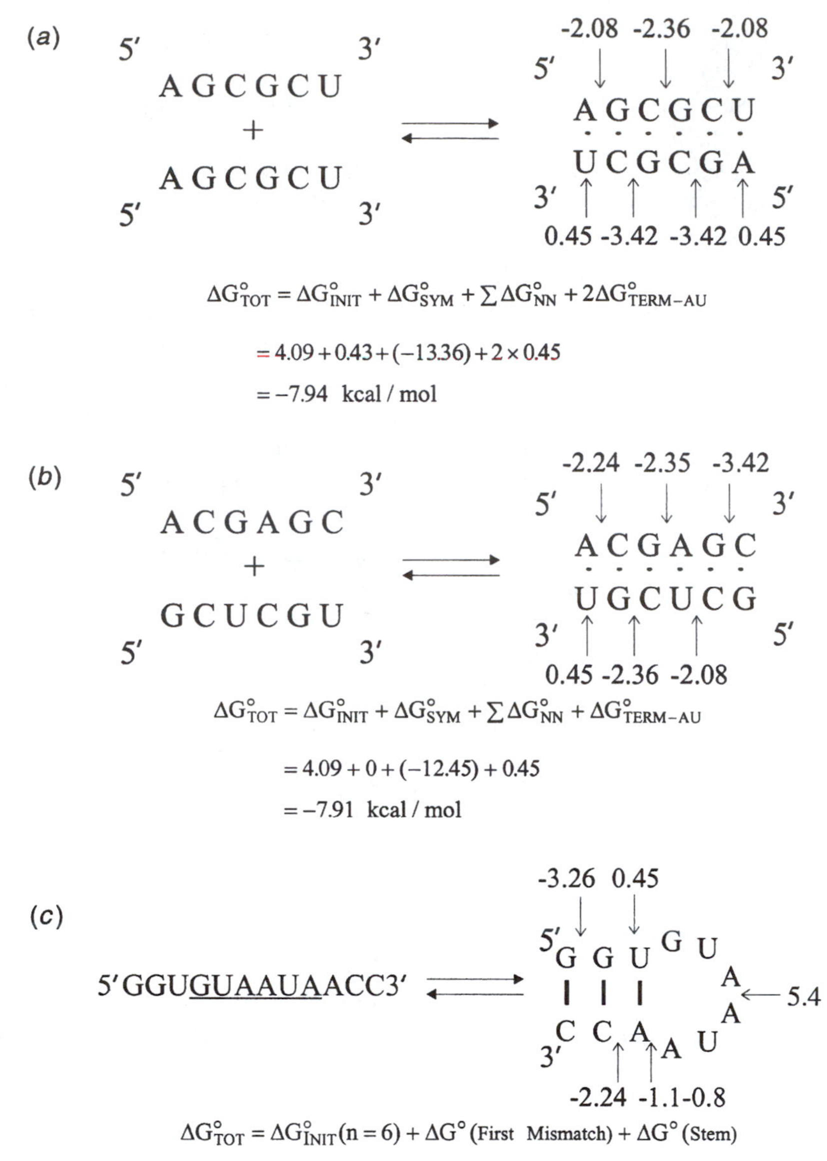

Figure 8-5

Predicting stability with the nearest neighbor model of Xia et al. (1998). (*a*) Calculation of free energy change for duplex formation by a self-complementary sequence. (*b*) Calculation for nonself-complementary sequence. Calculations of ΔH° and ΔS° are similar to those for ΔG°, except there is no symmetry term for ΔH°. Thus for the top sequence, $\Delta H^\circ = 3.61 + 2(-10.48) + 2(-14.88) + (-10.64) + 2(3.72) = -50.31$ kcal mol^{-1}, $\Delta S^\circ = -1.5 - 1.4 + 2(-27.1) + 2(-36.9) + (-26.7) + 2(10.5) = -136.6$ eu, and t_m at 10^{-4} *M* strands = 324.8 K = 51.6°C. For the sequence in (*b*), $\Delta H^\circ = 3.61 + (-11.40) + (-10.64) + (-12.44) + (-10.48) + (-14.88) + 3.72 = -52.51$ kcal mol^{-1}, $\Delta S^\circ = -1.5 + (-29.5) + (-26.7) + (-32.5) + (-27.1) + (-36.9) + 10.5 = -143.7$ eu, and t_m at 10^{-4} *M* strands = 318.7 K = 45.6°C. Calculations for intramolecular folding are illustrated in (*c*) and Figure 8-16. For intramolecular folding, ΔG°_{INIT} is replaced by ΔG°_{loops}, and there are no symmetry terms. Additional examples are given elsewhere (Xia et al., 1998).

Table 8.5
Thermodynamic Parameters for Unpaired Terminal Nucleotides in RNA at 1 *M* NaCl[a]

Propagation	X=A			X=C			X=G			X=U		
Sequence	ΔH°	ΔS°	ΔG°_{37}	ΔH°	ΔS°	ΔG°_{37}	ΔH°	ΔS°	ΔG°_{37}	ΔH°	ΔS°	ΔG°_{37}
3′ Unpaired Nucleotides												
→ AX / U ←	−4.9	−13.2	−0.8	−0.9	−1.2	−0.5	−5.5	−15.0	−0.8	−2.3	−5.4	−0.6
→ CX / G ←	−9.0	−23.4	−1.7	−4.1	−10.7	−0.8	−8.6	−22.2	−1.7	−7.5	−20.4	−1.2
→ GX / C ←	−7.4	−20.0	−1.1	−2.8	−7.9	−0.4	−6.4	−16.6	−1.3	−3.6	−9.7	−0.6
→ UX / A ←	−5.7	−16.4	−0.7	−0.7	−1.8	−0.1	−5.8	−16.4	−0.7	−2.2	−6.8	−0.1
5′ Unpaired Nucleotides												
→ XA / U ←	1.6	6.1	−0.3	2.2	7.9	−0.3	0.7	3.4	−0.4	3.1	10.6	−0.2
→ XC / G ←	−2.4	−6.0	−0.5	3.3	11.8	−0.3	0.8	3.4	−0.2	−1.4	−4.3	−0.1
→ XG / C ←	−1.6	−4.5	−0.2	0.7	3.1	−0.3	−4.6	−14.8	0.0	−0.4	−1.2	0.0
→ XU / A ←	−0.5	−0.7	−0.3	6.9	22.8	−0.1	(0.6)	(2.7)	(−0.2)	(0.6)	(2.7)	(−0.2)

[a]Turner et al., (1988). The parameters ΔH° and ΔG° (in kcal mol^{-1}); ΔS° (in eu). Values in parentheses are estimated.

Table 8.6
Thermodynamic Parameters for Terminal Mismatches in RNA[a]

ΔG°_{37} (kcal mol^{-1})

		$\overrightarrow{GX}$ / $\overleftarrow{CY}$					$\overrightarrow{CX}$ / $\overleftarrow{GY}$		
X/Y	A	C	G	U	X/Y	A	C	G	U
A	−1.1	(−1.5)	−1.3		A	−1.5	−1.5	−1.4	
C	−1.1	(−0.7)		(−0.5)	C	(−1.0)	(−1.1)		−0.8
G	−1.6		−1.4		G	−1.4		−1.6	
U		(−1.0)		(−0.7)	U		−1.4		−1.2

		$\overrightarrow{AX}$ / $\overleftarrow{UY}$					$\overrightarrow{UX}$ / $\overleftarrow{AY}$		
X/Y	A	C	G	U	X/Y	A	C	G	U
A	−0.8	(−1.0)	(−0.8)		A	−1.0	−0.8	−1.1	
C	−0.6	(−0.7)		(−0.7)	C	−0.7	−0.6		−0.5
G	−0.8		(−0.8)		G	−1.1		−1.2	
U		(−0.8)		(−0.8)	U		−0.6		−0.5

ΔH° (kcal mol^{-1})

		$\overrightarrow{GX}$ / $\overleftarrow{CY}$					$\overrightarrow{CX}$ / $\overleftarrow{GY}$		
X/Y	A	C	G	U	X/Y	A	C	G	U
A	−5.2	(−4.0)	−5.6		A	−9.1	−5.6	−5.6	
C	−7.2	(+0.5)		(−4.2)	C	(−5.7)	(−3.4)		−2.7
G	−7.1		−6.2		G	−8.2		−9.2	
U		(−0.3)		(−5.0)	U		−5.3		−8.6

		$\overrightarrow{AX}$ / $\overleftarrow{UY}$					$\overrightarrow{UX}$ / $\overleftarrow{AY}$		
X/Y	A	C	G	U	X/Y	A	C	G	U
A	−3.9	(+2.0)	(−3.5)		A	−4.0	−6.3	−8.9	
C	−2.3	(+6.0)		(−0.3)	C	−4.3	−5.1		−1.8
G	−3.1		(−3.5)		G	−3.8		−8.9	
U		(+4.6)		(−1.7)	U		−1.4		+1.4

ΔS° (eu)

		$\overrightarrow{GX}$ / $\overleftarrow{CY}$					$\overrightarrow{CX}$ / $\overleftarrow{GY}$		
X/Y	A	C	G	U	X/Y	A	C	G	U
A	−13.2	(−8.2)	−13.9		A	−24.5	−13.5	−13.4	
C	−19.6	(+3.9)		(−12.2)	C	(−15.2)	(−7.6)		−6.3
G	−17.8		−15.1		G	−21.8		−24.6	
U		(+2.1)		(−14.0)	U		−12.6		−23.9

		$\overrightarrow{AX}$ / $\overleftarrow{UY}$					$\overrightarrow{UX}$ / $\overleftarrow{AY}$		
X/Y	A	C	G	U	X/Y	A	C	G	U
A	−10.2	(+9.6)	(−8.7)		A	−9.7	−17.7	−25.2	
C	−5.3			(+1.5)	C	−11.6	−14.6		−4.2
G	−7.3		(−8.7)		G	−8.5		−25.0	
U		(+17.4)		(−2.7)	U		−2.5		+6.0

[a]Freier et al. (1986a); Hickey and Turner, (1985); Sugimoto et al. (1987b); Serra et al. (1994). Values in parentheses are estimated.

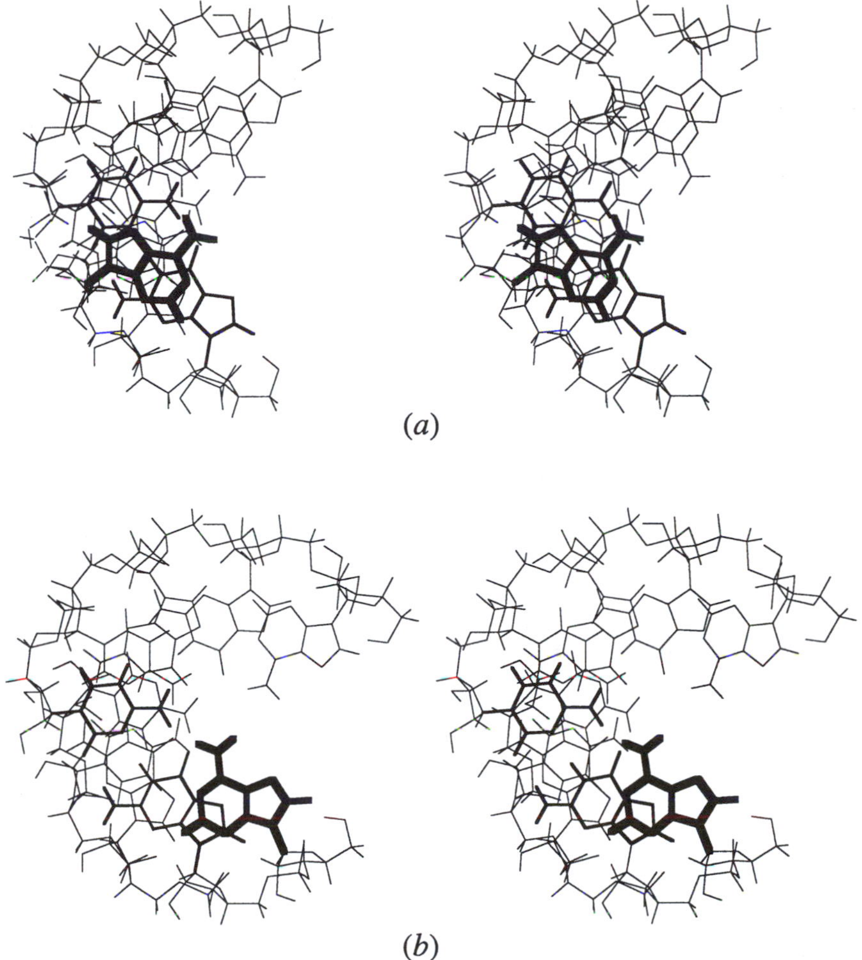

Figure 8-6
Stereoviews of $(GGCCA)_2$ (*a*) and $(AGGCC)_2$ (*b*) in A-form geometry. The terminal A closest to the reader is in boldface.

the backbone. All these factors affect the stability of the duplex. Solvent effects may also play a role. While exact partitioning of these effects on stability is not yet possible, a qualitative picture is emerging, and is discussed below.

2.2.1 Conformational Entropy

In Section 1.3.1, the conformational entropy associated with propagating a single-strand stacked helix by one additional nucleotide was estimated as -11 eu. Since propagation of a double helix by an additional base pair requires limiting the conformations accessible to two nucleotides, the conformational entropy associated with this process is estimated as $2 \times -11 = -22$ eu. Inspection of Table 8-4 indicates that 3/4 of the measured ΔS° values for duplex propagation are within 6 eu of this theoretical value. As with single strands, this suggests conformational entropy effects account for a large part of the observed ΔS°.

Conformational entropy effects may also account for the stabilization of RNA duplexes by 5′-dangling ends. As noted above, stabilization from a 5′-dangling end is about the same as from a 5′-phosphate. It has been suggested by Sundaralingam that addition of a 5′-phosphate restricts the conformations of the ribose group (Sundaralingam, 1969, 1973). This restriction could provide the small stabilization that is observed.

2.2.2 Stacking

An empirical indication of the contributions of stacking to duplex stability is provided by the stability increments for 3′-dangling ends in Table 8-5. Comparison of some selected increments with those for base pairs is provided in Figure 8-8. This comparison is deceptive when considering the favorable attractive forces of stacking interactions, however, because the favorable free energy increment for stacking of most 3′-dangling ends includes the unfavorable conformational entropy associated with that stacking. This unfavorable component of the stacking ΔG° has been empirically estimated at about 1.9 kcal mol^{-1} at 37°C (Freier et al., 1986c), somewhat less than the value of $-T\Delta S^\circ = -(-11 \times 310) = 3.4$ kcal mol^{-1} expected from the theoretical considerations given above. When the empirical estimate of conformational effects is factored out, the favorable stacking interactions are estimated to be as large as $-1.7 - 1.9 = -3.6$ kcal mol^{-1}. The magnitudes of these favorable interactions are very sequence dependent. In fact, the negligible ΔG° values measured for $\begin{matrix}\overrightarrow{\text{UU}}\\ \text{A}\end{matrix}$ and $\begin{matrix}\overrightarrow{\text{UC}}\\ \text{A}\end{matrix}$ may indicate no stacking at all for these sequences. This sequence dependence is consistent with the idea that electronic interactions between the bases are responsible for stacking.

2.2.3 Hydrogen Bonding

Inspection of Figure 8-8 indicates the stability increments for some base pairs are much larger than the increments for stacking of the constituent nucleotides. This suggests pairing between the bases contributes an additional interaction (Freier et al., 1986c; 1985a). Consistent with this are results with DNA oligomers showing that substitution of weakly hydrogen bonding diflourotoluene for T in an AT pair decreases duplex stability by 3.6 kcal mol^{-1} at 25°C (Moran et al., 1997). Presumably, the effect of hydrogen bonds is due to the difference between hydrogen bonding in a base pair and hydrogen bonding of the separated bases with water. Specific solvation effects could also be important. Estimates for the contributions of differential hydrogen bonding can be made by taking stability increments for base pairs, subtracting stability increments for stacking of the constituent nucleotides, and correcting for conformational entropy effects. For example (Fig. 8-8), in $(\text{GCCGGC})_2$, the ΔG°_{37} for adding each terminal GC pair is $[\Delta G^\circ_{37}(\text{GCCGGC}) - \Delta G^\circ_{37}(\text{CCGG})]/2 = -3.3$ kcal mol^{-1}. Stacking of each 3′-dangling end in $(\text{CCGGC})_2$ is associated with a ΔG°_{37} of

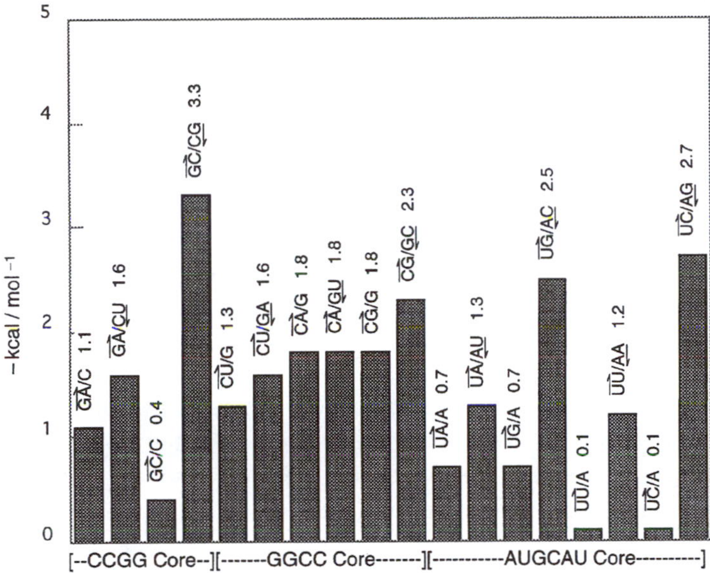

Figure 8-7
Free energy increments associated with duplex formation at 37°C from adding a terminal base pair or 3′ unpaired terminal nucleotide to CCGG, GGCC, or AUGCAU cores. For example, the first bar on the left is ΔG°_{37} for adding a 3′ unpaired A adjacent to a GC pair, ΔG°_{37} $\left(\begin{smallmatrix}5'\text{GA}3'\\ \text{C}\end{smallmatrix}\right) = 1/2\ [\Delta G^\circ_{37}(\text{CCGGA}) - \Delta G^\circ_{37}(\text{CCGG})] = -1.1$ kcal mol^{-1}. Note that the free energy increments for adding 5′ unpaired terminal nucleotides are not shown but average only -0.2 kcal mol^{-1}.

$[\Delta G^\circ_{37}(\text{CCGGC}) - \Delta G^\circ_{37}(\text{CCGG})]/2 = -0.4$ kcal mol^{-1}. The effect of the 5′-terminal G is $[\Delta G^\circ_{37}(\text{GCCGG}) - \Delta G^\circ_{37}(\text{CCGG})]/2 = -0.2$ kcal mol^{-1}. (This is no more than the contribution from adding a 5′ terminal phosphate suggesting an unpaired 5′-terminal G is not stacked.) Constraining the 5′-terminal G in a base pair, however, requires overcoming the estimated 1.9 kcal mol^{-1} of conformational free energy. Thus the free energy gained from pairing the terminal G and C is estimated as $\Delta G^\circ_{37,p} = -3.3 - (-0.4 - 0.2 + 1.9) = -4.6$ kcal mol^{-1}. Since this pairing involves three hydrogen bonds, the estimated free energy increment per hydrogen bond is -1.5 kcal mol^{-1} hydrogen bond. Similar calculations on other sequences give ΔG° values for a hydrogen bond that range from -0.5 to -1.5 kcal mol^{-1} hydrogen bond (Freier et al., 1986c; Turner et al., 1987). It has been suggested that this range is due to a sequence dependent competition between hydrogen bonding and stacking (Turner et al., 1987).

Another empirical estimate for the contributions of hydrogen bonds is provided by comparing duplex stabilities for sequences with different numbers of hydrogen bonds (Freier et al., 1986c; Turner et al., 1987). Such comparisons must also consider changes in stacking interactions that may result from changing the number of hydrogen bonding groups on the bases. Inspection of Table 8-5, however, indicates that at least for stacking of 3′-terminal unpaired bases, there is little effect of changing hydrogen-bonding groups. Both A and G have very similar thermodynamic parameters for 3′ stacking, and U and C are also similar. Thus any large changes in stacking energies between AU and GC pairs are probably a second-order effect resulting from redistribution of electrons on pairing. Even these effects can be minimized by comparing stabilities of base pairs with single changes in structure. For example, removing the 2-amino group from G in a GC pair leaves an IC pair with two hydrogen bonds. Table 8-7 contains several comparisons of oligomers with similar sequences, but different numbers of hydrogen bonds. The stability increments per hydrogen bond range from -0.5 to -2 kcal mol^{-1}.

2.2.4 Counterion Condensation

Repulsions of negatively charged backbone phosphates destabilize double helixes. Condensation of positively charged counterions around the helix favors duplex stability by neutralizing much of the negative charge on the phosphates. Experimental and theoretical aspects of this effect are discussed in Section 8.5, and in Chapter 11, respectively.

2.2.5 Solvent

By analogy to proteins, classical hydrophobic interactions are often invoked as a source of stability for double helixes. There is no evidence, however, to support the view that classical hydrophobic interactions stabilize duplexes or single-strand stacking. As discussed above, in most cases, the ΔS° of duplex formation is similar to or more unfavorable than expected from conformational terms. There is no indication of the large, positive ΔS° expected for classical hydrophobic interactions. Hydrophobic interactions are also associated with a large change in heat capacity, ΔC_p°. Values of ΔC_p° reported for single- to double-strand transitions of nucleic acids, however, are either modest or zero (Suurkuusk et al., 1977; Freier et al., 1983b, 1985a, 1986a,c; Kierzek et al., 1986; Breslauer et al., 1986). These generalizations even hold for duplexes containing thymine. Since thymine has a methyl group, some hydrophobic interaction might be expected.

One solvent effect that may be important for helix stability is specific solvation. This topic is discussed in detail in Chapters 4, 7, and 11. Such specific solvation could explain the excess unfavorable ΔS° observed for some helix transitions. The quantitative consequences of such solvation have not been determined, however.

2.3 Kinetics of Duplex Formation

2.3.1 Experimental Measurements

The kinetics of duplex formation by oligonucleotides have been measured primarily with the temperature-jump method. For the reaction shown in Figure 8-5, the time dependence of the change in concentration of single strands has the same form as Eq. 8-8. For duplex formation, the relaxation time, τ, is given by

$$\text{Self-complementary} \qquad \tau^{-1} = 4k_{on}[\mathrm{A}] + k_{off} \tag{8-20}$$

$$\text{Nonself-complementary} \qquad \tau^{-1} = k_{on}([\mathrm{A}] + [\mathrm{B}]) + k_{off} \tag{8-21}$$

Table 8.7
Comparison of Stabilities of Duplexes with Different Numbers of Hydrogen Bonds[a]

Reference Duplex	Duplex with Fewer H Bonds	Reference	Differences in No. of H Bonds	$\Delta\Delta t_m$ at $10^{-4}M$(°C)	$\Delta\Delta G^\circ_{37}$/H Bond (kcal mol^{-1} H Bond)
GCCGGC	ICCGGC	Turner et al. (1987)	2	15.9	−1.6
CGGCCG	CGGCCI	Turner et al. (1987)	2	8.5	−0.7
dCA_3GA_3G+dCT_3CT_3G	dCA_3IA_3G+dCT_3CT_3G	Martin et al. (1985)	1	2.8	−0.5
dCA_3CA_3G +dCT_3GT_3G	CA_3CA_3G +dCT_3IT_3G	Aboul−ela et al. (1985)	1	8.9	−1.8
$dG_3A_2GCT_2C_3$	$dG_3A_2ICT_2C_3$	Kawase et al. (1986)	2		−1.3
dCGTA′CG[d]	dCGTACG	Gaffney et al. (1984)	2		−0.5
CGGCCG	CAGCUG	Freier et al. (1986b)	2	20.1	−1.6
GCGCGC	GUGCAC	Freier et al. (1986b)	2	14.5	−1.5
GCGCGC	GCAUGC	Freier et al. (1986b)	2	16.4	−1.6
CGCGCG	UGCGCA	Freier et al. (1986b)	2	4.8	−0.5
GAUGCAUC	AAUGCAUU	Freier et al. (1986b)	2	12.2	−1.5
AUGCGCAU	AUACGUAU	Freier et al. (1986b)	2	18.4	−1.8
AUGCGCAU	AUGUACAU	Freier et al. (1986b)	2	18.7	−1.9

[a]I is inosine. A′ is 2-aminoadenine.

Here [A] and [B] are the equilibrium concentrations of the single strands at the higher temperature; k_{on} and k_{off} are the forward and reverse rate constants at the higher temperature as illustrated in Figure 8-5. Thus a plot of τ^{-1} versus single-strand concentration has an intercept of k_{off} and a slope of $4k_{on}$ or k_{on} depending on whether the sequence is self- or nonself-complementary. The activation energies for k_{on} and k_{off} can be determined from measurements as a function of temperature (see Eq. 8-9).

Typical results for the kinetics of duplex formation are compiled in Table 8-8. The forward rates, k_{on}, range from about 10^5–10^7 $M^{-1}s^{-1}$, and depend on both sequence and salt concentration. Higher salt concentrations increase the forward rate as expected for association of two like-charged species. The forward rate, however, is slower than the diffusion limit. Moreover, in several cases, the activation energy for k_{on} is negative or zero. That is, k_{on} can decrease with increasing temperature. Any elementary reaction step must have a positive activation energy. For example, diffusion controlled reactions in solution have activation energies of about 4 kcal mol^{-1}. Thus the mechanism of duplex formation must be more complicated than indicated in Figure 8-5.

Figure 8-9 illustrates a mechanism consistent with the kinetic experiments (Pörschke and Eigen, 1971; Craig et al., 1971; Williams et al., 1989; Wetmur and Davidson, 1968; Manning, 1976). In this mechanism, the rate-determining step is formation of a nucleus containing a small number of base pairs. This nucleus adds an additional base pair faster than it dissociates. Thus the double helix can "zip up" after formation of the nucleus. The stability of the intermediate preceding the nucleus, however, is temperature dependent. Like any double helix, its stability decreases as temperature increases. Thus the concentration of the intermediate preceding the nucleus decreases as temperature increases, and this can lead to a decrease in k_{on}. The measured activation energy for k_{on} depends on the $\Delta H°$ associated with formation of the intermediate preceding the nucleus and the activation energy for forming the last base pair in the nucleus (see Fig. 8-10): $E_{A,on} = \Delta H^\circ_{n-1} + E_{A;n-1\rightarrow n}$. Here n is the number of base pairs in the nucleus. The enthalpy change, ΔH°_{n-1} can be estimated from the values in Table 8-4. The activation energy for propagating the helix an additional base pair is expected to be similar to that for single strand stacking, roughly 5 kcal mol^{-1}. An estimate of the size of the nucleus can be made by adding appropriate parameters until the measured activation energy is approximated. For example, the measured activation energy for duplex formation by A_6U_6 is -3 kcal mol^{-1} (Craig et al., 1971). Formation

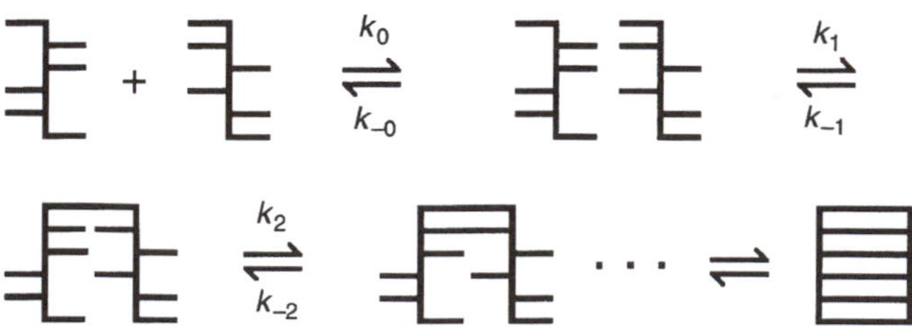

Figure 8-8
Mechanism for duplex formation. Step 0 is alignment of strands without inter-strand bonding. Subsequent steps involve formation of base pairs by hydrogen bonding.

Table 8.8
Kinetic Parameters for Duplex Formation by Oligonucleotides

Sequence	Reference	$[Na^+]$ (M)	t °C	k_{on} ($10^6 M^{-1}s^{-1}$)	k_{off} (s^{-1})	$E_{A,on}$ (kcal mol^{-1})	$E_{A,off}$ (kcal mol^{-1})
$A_9 + U_9$	Pörschke and Eigen (1971).	0.05	23.3	0.53	2300	−8	36
$A_{11} + U_{11}$	Pörschke and Eigen (1971).	0.05	23.3	0.50	210	−12	53
$A_{14} + U_{14}$	Pörschke and Eigen (1971).	0.05	23.5	0.61		−8	
A_4U_4	Craig et al. (1971).	0.25	21	1	3000	−6	37
A_5U_5	Craig et al. (1971).	0.25	21	2	2000	−4	50
A_6U_6	Craig et al. (1971).	0.25	21	1.5	15	−3	60
A_7U_7	Craig et al. (1971).	0.25	21	0.8	1	+5	65
	Breslauer and Bina–Stein (1977).	1	22.1	2.7	3.0	−6	45
A_2GCU_2	Pörschke et al. (1973).	0.05	23.3	1.6	450	3	33
	Pörschke et al. (1973).	1	23	10	200	8	33
A_3GCU_3	Pörschke et al. (1973).	0.05	23.3	0.75	3	7	50
A_4GCU_4	Pörschke et al. (1973).	0.05	23.3	0.13	1.5	8	26
	Pörschke et al. (1973).	1	23	0.9	1	9	26
$A_4G_2 + C_2U_4$	Pörschke et al. (1973).	0.05	16.8	11.4	320	13	34
$A_5G_2 + C_2U_5$	Pörschke et al. (1973).	0.05	23.3	4.4	340	7	43
$CA_5G + CU_5G$	Nelson and Tinoco (1982).	1	21.1	4.6	330	0	39
$dCA_5G + dCT_5G$	Nelson and Tinoco (1982).	1	20	9	70	−0.5	43
GGGC + GCCC	Podder (1971).	0.1	21.5	5.4	40	4.6	29
dGCGCGC	Freier et al. (1983a).	1	25	12.0	0.31	0.8	57
$A_9 + dT_9$	Hoggett and Maass (1971).	1	23	10	240	−2	43
dCGTGAATTCGCG	Chu and Tinoco (1983).	0.17	31.8	0.08	1.0	22	68
dCGCAGAATTCGCG	Chu and Tinoco (1983).	0.17	31.8	0.07	0.7	16	74
dGCATGC	Williams et al. (1989).	0.012	31.1	0.98	3.9	15	47
		0.042	31.1	1.6	2.1		
		1	31.1	9.9	2.2	−5	40
dGCATGC(0.01 M Mg^{2+})	Williams et al. (1989).	0.012	31.1	7.3	2.4	−3	43

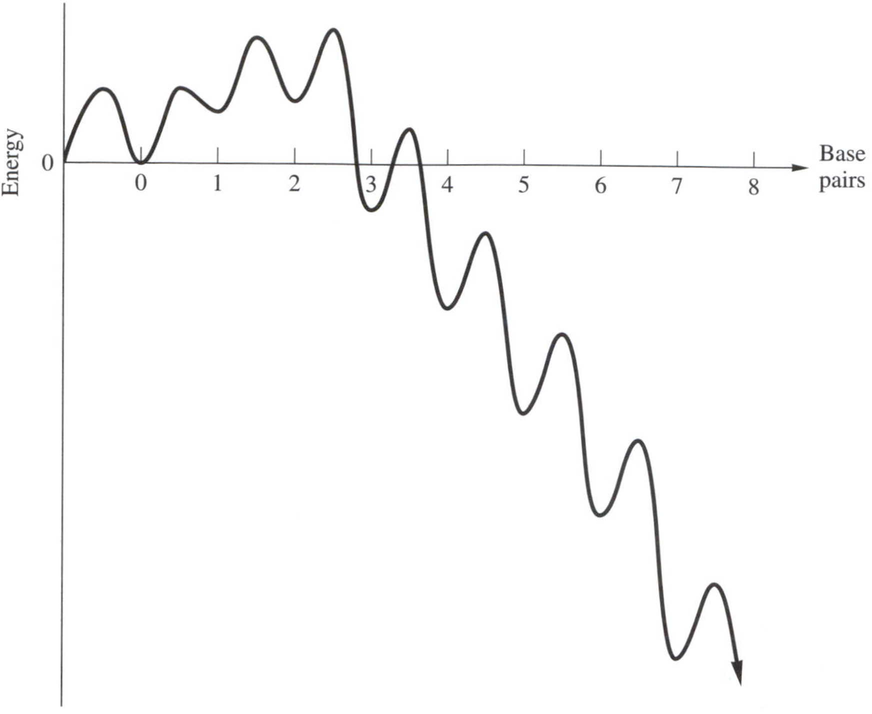

Figure 8-9
Energy profile corresponding to mechanism in Figure 8-9 for formation of duplex from separated single strands.

of four AU base pairs is associated with a $\Delta H^\circ \approx 3.6 + 2(3.7) + 3(-6.8) = -9.4$ kcal mol^{-1}. If addition of the next base pair requires an activation energy of 5 kcal mol^{-1}, then the overall activation energy would be $E_{\text{A,on}} = -9.4 + 5 = -4.4$ kcal mol^{-1}. Thus the results are consistent with a nucleus of 5 bp. Since several assumptions are required, the estimate for the size of the nucleus is rough. In general, the results in Table 8-8 are consistent with nuclei of 5 ± 1 bp for oligomers containing only AU pairs, and 2 ± 1 bp for oligomers containing at least two GC pairs.

The data in Table 8-8 can also be used to estimate a rate constant of about 10^7 s^{-1} for addition of the next base pair to the nucleus (Pörschke and Eigen, 1971; Craig et al., 1971). This rate constant is similar to that for single-strand stacking, which supports the choice of activation energy used above.

The activation energies reported for homo-duplex formation by dCGTGAATTCGCG and dCGCAGAATTCGCG are larger than expected from the above analysis (Chu and Tinoco, 1983). Both these oligomers are long, self-complementary, and contain non-Watson–Crick pairings. One possibility is that each oligomer has intramolecular base pairs that must be broken before duplexes can form. Another is that non-Watson–Crick pairings may introduce effects that are not easily predicted from available information. Further experiments are required to sort out these effects.

The reverse rates in Table 8-8 cover a wide range and have large activation energies. These results are consistent with the model described above. Dissociation of duplex requires breaking enough base pairs to return to the nucleus. The energy required for this will depend on sequence and length. If the size of the nucleus and the activation energy for losing a base pair from the nucleus are known, then the ΔH°values in Table 8-4 can be used to predict $E_{A,off}$. For example, for A_6U_6, 7 bp must be broken to return to the 5 bp nucleus. The activation energy for the reverse rate is predicted to be $E_{A,off} = 6 \times 6.8 + 9.4 + (5 + 6.8) = 62$ kcal mol^{-1}. The measured value is 60 kcal mol^{-1} (Craig et al., 1971).

2.3.2 Predicting Kinetics of Duplex Formation

The results described above suggest a way to roughly predict the kinetics of duplex formation between oligonucleotides that do not have intramolecular base pairing. At high Na^+ concentration or in the presence of Mg^{2+}, the forward rate is almost always $10^6 M^{-1}s^{-1}$ within an order of magnitude, and the temperature dependence is modest. The equilibrium constant for duplex formation can be predicted from the parameters in Table 8-4, since $K = \exp(-\Delta G^\circ/RT)$ and $\Delta G^\circ = \Delta H^\circ - T\Delta S^\circ$. The predicted values of K and k_{on} can then be used to predict k_{off} since $K = k_{on}/k_{off}$. For example, the predicted equilibrium constants at 37°C for formation of $(GGCC)_2$ and $(GGCCGGCC)_2$ are 6.6×10^3 and 3.1×10^{12}, respectively, giving predicted off rate constants of 1.5×10^2 and 3×10^{-7} s^{-1}, respectively. The half lives for dissociation of these helixes are predicted from $t_{1/2} = \ln(2)/k_{off} = 0.693/k_{off}$ to be about 5 ms and 27 days, respectively, at 37°C. Note that the off rate is very temperature dependent. For example, at 0°C, the half lives for $(GGCC)_2$ and $(GGCCGGCC)_2$ are predicted to be about 20 s and 30 million years. This type of rough approximation is often sufficient for designing experiments.

3. DOUBLE HELIX FORMATION BY OLIGONUCLEOTIDES WITH LOOPS

Loops are regions of non Watson-Crick paired nucleotides flanked by one or more double helixes. The known varieties of loops are illustrated in Figure 8-11. All occur in structures of RNA (e.g., see Fig. 8-1). While natural DNA is primarily double helix, loops can occur transiently or even be favored under certain conditions. Thus it is important to understand the properties of loops.

There are several experimental ways to determine the thermodynamic properties of loops. All involve measuring properties for structures containing the loop of interest and subtracting out other contributions. For example, ΔG° for the bulge loop in the duplex formed by GCG<u>A</u>GCG + CGCCGC is obtained by taking ΔG° for this bulge duplex and subtracting ΔG° for the fully paired duplex formed by GCGGCG + CGCCGC:

$$\Delta G^\circ_{bulge} = \Delta G^\circ(\text{GCG}\underline{\text{A}}\text{GCG} + \text{CGCCGC}) - \Delta G^\circ(\text{GCGGCG} + \text{CGCCGC})$$

The assumption is that the loop does not affect the stabilities of other regions so the free energies are additive. This assumption may not be very good (Longfellow et al., 1990). Often loops are studied in structures formed by single strands, for example the hairpin formed by $A_6C_6U_6$. If comparisons in such cases are made with helixes formed by bimolecular associations, it is necessary to correct for initiation of the bimolecular helix first. Unfortunately, fewer data are available for loops than for fully base paired helixes. Much of the available data is discussed below. Examples of calculations of thermodynamic properties for structures containing loops can be found in Xia et al. (1999).

3.1 Bulge Loops

Bulge loops are loops in which unpaired nucleotides occur on only one strand of a double helix (Fig. 8-11). In structures of natural RNA, the most common bulge contains one nucleotide, and bulges are known to be important for protein binding and tertiary folding (Peattie et al., 1981; Romaniuk et al., 1987; Flor et al., 1989; Cate et al., 1996, 1997). In DNA, bulges may be important in frameshift mutagenesis in sequences with repeating base pairs (Okada et al., 1972), and detection of bulges by gel electrophoresis can reveal mutations (Triggs-Raine and Gravel, 1990; Rommens et al., 1990). Chemical modification experiments on large RNA

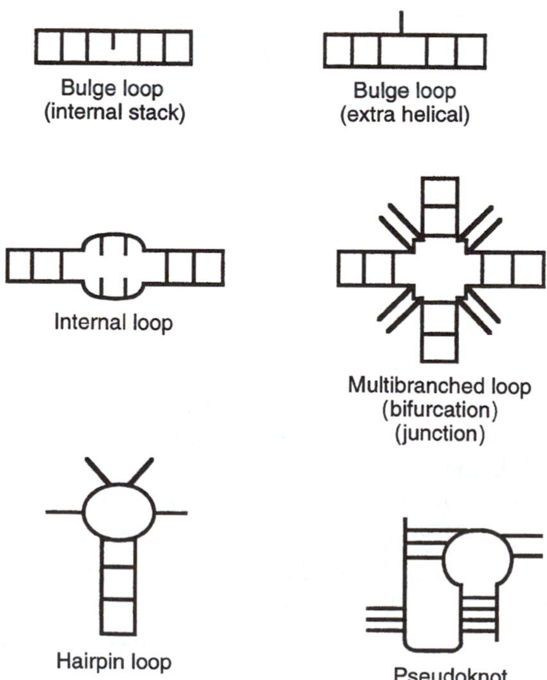

Figure 8-10
Schematic of various types of loops.

molecules (Moazed et al., 1986) suggest the structure of a bulge may be sequence dependent. Using R and Y to denote purines and pyrimidines, respectively, bulged nucleotides in N^RR sequences are usually moderately or strongly reactive. Bulged nucleotides in Y^RY sequences are usually protected from modification. NMR experiments indicate the bulged As in $(dCGC\underline{A}GAATTCGCG)_2$, $(dCGC\underline{A}GAGCTCGCG)_2$ and $(dCGCGAAATTT\underline{A}CGCG)_2$ are intercalated in the helix (Patel et al., 1982, 1986; Hare et al., 1986), while the bulged C in d(CAAA$\underline{\text{C}}$AAAG)·d(CTTTTTTG) and the bulged U in r(CUGG$\underline{\text{U}}$GCGG)·r(CCGCCCAG) are extrahelical (Morden et al., 1983; van den Hoogen et al., 1988) (see Chapter 4). Thermodynamic parameters for single nucleotide bulges indicate they destabilize the helix, but with little dependence on the identity of the bulged base (Longfellow et al., 1990; Groebe and Uhlenbeck, 1989). The destabilization, however, does depend on more than the adjacent base pairs (Longfellow et al., 1990). Thus the nearest neighbor model is oversimplified for bulges. A limited number of sequences have been studied, and representative data are listed in Table 8-9.

Some bulges occur in sequences that allow migration of the unpaired nucleotide. For example, any of the middle C nucleotides in the lower strand of $^{\text{dGATGGG-CAG}}_{\text{CTACCCCGTC}}$ could be bulged. The NMR experiments indicate such migration occurs at a rate of at least 100–1000 s^{-1} (Woodson and Crothers, 1987).

Theoretical considerations indicate bulge loops should be more destabilizing as the number of unpaired nucleotides increases. While only a few measurements have been made on bulges larger than 1, the results are mostly in agreement with this prediction (see Table 8-9) (Longfellow et al., 1990; Yuan et al., 1979; Weeks and Crothers, 1993).

3.2 Internal Loops

An internal loop is formed when a double helix is interrupted by non-Watson–Crick paired nucleotides on both strands (Fig. 8-11). The size of an internal loop is the total number of nucleotides in the loop. Internal loops can be symmetrical or asymmetrical with respect to the number of loop residues on each strand. Asymmetric internal loops are less stable than symmetric internal loops (Peritz et al., 1991). A particularly important subclass of internal loop, the mismatch, contains two nucleotides. Stabilities of mismatches are important for determining stabilities of duplexes formed between DNA hybridization probes and target sequences that are not completely complementary. Due to redundancy in the genetic code, this situation often occurs when the sequence of a probe is designed from the sequence of a gene's protein product. Modified bases, like inosine, that are less discriminating than A, C, G, and T, can be used in such cases to lessen the effect of the mismatch (Martin et al., 1985; Cheong et al., 1988; Nichols et al., 1994). Formation of large internal loops is the first step in melting of DNA (Blake and Decourt, 1987; Wada et al., 1980; Gotoh, 1983; Wartell and Benight, 1985). In RNA, internal loops of many sizes are known to occur (see Fig. 8-1). These loops allow structural bends (Murphy and Cech, 1993) and tertiary interactions (Costa and Michel, 1995), and can also form binding sites (Sassanfar and Szostak, 1993; Fan et al., 1996; Jiang et al., 1996; Dieckmann et al., 1996; Yang et al., 1996).

Considering the wide variety of possible internal loops, relatively few experimental data are available. Some measurements involve single mismatches in DNA

Table 8.9
Free Energy Increments at 37°C for Bulge Loops[a]

Bulge Sequence	Reference	$[Na^+]$ (M)	t_m(°C $10^{-4}M$)	Δt_m(°C $10^{-4}M$)	$\Delta\Delta G^\circ_{bulge,\,37}$ (kcal mol^{-1})
RNA					
poly(A,A°)·polyU	Fink and Crothers (1972)	0.2–0.5			2.9
$5'GCG^{A}GCG3'$ $3'CGC\text{-}CGC5'$	Longfellow et al. (1990)	1	39	20	3.5
$5'GCG^{U}GCG3'$ $3'CGC\text{-}CGC5'$	Longfellow et al. (1990)	1	38	21	3.7
$5'GCG\text{-}GCG3'$ $3'CGC_{A}CGC5'$	Longfellow et al. (1990)	1	39	20	3.5
$5'\,GCG^{A}GCGA3'$ $3'ACGC\text{-}CGC5'$	Longfellow et al. (1990)	1	48	22	5.1
$5'\,GCG^{A}GUCA3'$ $3'ACGC\text{-}CAG5'$	Longfellow et al. (1990)	1	49	16	2.0
$5'\,{}^{G}GGACUC^{A}CG^{AU}$ $3'_{UA}UCUGAG\text{-}GC_{A}{}^{U}$	Groebe and Uhlenbeck (1989)	1	79	16	3.4
$5'\,{}^{G}GGACUC^{G}CG^{AU}$ $3'_{UA}UCUGAG\text{-}GC_{A}{}^{U}$	Groebe and Uhlenbeck (1989)	1	79	16	3.5
$5'^{G}GGACUC^{U}CG^{AU}$ $'_{UA}UCUGAG\text{-}GC_{A}{}^{U}$	Groebe and Uhlenbeck (1989)	1	80	14	2.7
$5'GCG^{AA}GCG3'$ $3'CGC\text{—}CGC5'$	Longfellow et al. (1990)	1	24	35	5.2
$5'GCG^{AAA}GCG3'$ $3'CGC\text{—}CGC5'$	Longfellow et al. (1990)		14	45	5.7
$5'GCG\text{—}GCG3'$ $3'CGC_{AAA}CGC5'$	Longfellow et al. (1990)	1	38	21	3.7
DNA					
$5'dCA_3^{C}A_3G3'$ $3'\,GT_3\text{-}T_3C5'$	Morden et al. (1983)	1	18	14	2.6
$5'dCTG^{A}CCCATC3'$ $3'\,GAC\text{-}GGGTAG5'$	Woodson and Crothers (1987)	0.1	31	15	3.1
$5'dCTG^{C}CCCATC3'$ $3'\,GAC\ GGGTAG5'$	Woodson and Crothers (1987)	0.1	37	10	2.1

[a]The parameters Δt_m and $\Delta\Delta G^\circ_{bulge,37}$ are differences between the duplexes with and without the bulge.

oligonucleotides (Gaffney and Jones, 1989; Martin et al., 1985; Tibanyenda et al., 1984; Aboul-ela et al., 1985; Allawi and SantaLucia, 1997, 1998). Table 8-10 lists some of these results. Substituting a mismatch for a base pair always gives a less stable duplex. In general, the destabilization is associated with a less favorable ΔH° and more favorable ΔS° of duplex formation, which is consistent with the expectation of reduced bonding in a mismatch coupled with increased flexibility.

Data for RNA internal loops indicate that internal loops become more destabilizing as they increase in size (Weeks and Crothers, 1993; Peritz et al., 1991; Gralla and Crothers, 1973). For example, the melting temperatures at $10^{-4}\,M$ strand for

Table 8.10
Thermodynamic Parameters for Duplex Formation in 1 *M* NaCl by DNA Oligonucleotides Containing Mismatches[a]

Sequence	Mismatch	$-\Delta H^\circ$ (kcal mol^{-1})	$-\Delta S^\circ$ (eu)	$-\Delta G^\circ_{37}$ (kcal mol^{-1})	t_m(°C at $10^{-4}M$)
$CA_3CA_3G + CT_3GT_3G$		64.5	183	7.7	42.9
$CA_3GA_3G + CT_3CT_3G$		62.8	179	7.3	40.8
$CA_3AA_3G + CT_3TT_3G$		68.0	196	7.2	40.1
$CA_3TA_3G + CT_3AT_3G$		58.6	168	6.5	36.8
$A_3GA_3G + CT_3GT_3G$	GG	53.5	158	4.5	25.6
$CA_3TA_3G + CT_3GT_3G$	TG	55.6	165	4.4	25.7
$CA_3GA_3G + CT_3AT_3G$	GA	52.6	156	4.2	23.9
$CA_3GA_3G + CT_3TT_3G$	GT	46.7	137	4.2	22.3
$CA_3AA_3G + CT_3GT_3G$	AG	39.9	116	3.9	18.0
$CA_3AA_3G + CT_3AT_3G$	AA	36.9	107	3.7	15.0
$CA_3CA_3G + CT_3TT_3G$	CT	53.2	161	3.3	19.1
$CA_3TA_3G + CT_3CT_3G$	TC	50.0	151	3.2	17.5
$CA_3CA_3G + CT_3AT_3G$	CA	(40.3)	(120)	(3.1)	(13)
$CA_3TA_3G + CT_3TT_3G$	TT	(54.6)	(167)	(2.8)	(17)
$CA_3AA_3G + CT_3CT_3G$	AC	(35.8)	(106)	(2.9)	(9)
$CA_3CA_3G + CT_3CT_3G$	CC	(55.3)	(171)	(2.3)	(15)
5′CAACTTGATATTAATA 3′GTTGAACTATAATTAT		102.1	289	12.4	55.8
5′CAACTTGATATTAATA 3′GTTGAGCTATAATTAT	TG	92.6	266	10.1	49.4
5′CAACTTGATATTAATA 3′GTTGAACTATAGTTAT	TG	95.5	274	10.5	50.5
5′CAACTTGATATTAATA 3′GTTGAACTCTAATTAT	TC	98.4	286	9.7	47.3
5′CAACTTGATATTAATA 3′GTTGAATTATAATTAT	GT	91.3	264	9.4	47.1
5′CAACTTGATATTAATA 3′GTTGAACCATAATTAT	AC	90.9	265	8.7	44.6
5′CAACTTGATATTAATA 3′GTTGAACAATAATTAT	AA	92.0	267	9.26	46.2

[a]Data for $CA_3XA_3G + CT_3YT_3G$ sequences are from Aboul-ela et al. (1985). Data in parentheses are significantly less accurate. Other sequences are from Tibanyenda et al. (1984). Similar results for d(GGTTXTTGG) + d(CCAAYAACC) have been reported by Gaffney and Jones (1989).

$(CGCA_nGCG)_2$ with $n = 0, 1, 2$, and 3, are 58, 40, 36, and 32°C, respectively (Peritz et al., 1991). The stability of an internal loop is also dependent on the sequence in the loop (SantaLucia et al., 1991; Wu et al., 1995; Schroeder et al., 1996; Xia et al., 1997). For example, Table 8-11 lists ΔG° values for internal loops containing adjacent, identical mismatches. For this case, GU, GA, and UU double mismatches often stabilize a duplex, while other double mismatches destabilize a duplex. This sequence

Table 8.11
Free Energy Increments (ΔG°_{37} in kcal mol^{-1}) for Tandem Mismatches in RNA Oligonucleotides in 1 M NaCl[ab]

Mismatches / Closing bp	→ UG GU ←	→ GU UG ←	→ GA AG ←	→ AG GA ←	→ UU UU ←	→ GG GG ←	→ CA AC ←	→ CU UC ←	→ UC CU ←	→ CC CC ←	→ AC CA ←	→ AA AA ←
5′G 3′C	−4.9	−4.1	−2.6	−1.3	−0.5		1.0	1.1			0.9	1.5
5′C 3′G	−4.2	−1.1	−0.7	−0.7	−0.4	0.8	1.1	1.4	1.4	1.7	2.0	1.3
5′U 3′A	−2.6	−0.3	0.7		1.1		[1.9][c]	2.2	2.8			2.8
5′A 3′U	−1.9	0.2	0.3		0.6		2.3				2.5	2.8

[a]Free energy increments are calculated as in the following example:

$$\Delta G^\circ_{37,\mathrm{loop}}\begin{pmatrix}\mathrm{CAGG}\\ \bullet \quad \bullet\\ \mathrm{GGAC}\end{pmatrix} = \Delta G^\circ_{37}(\mathrm{CGC\underline{AG}GCG}) - \Delta G^\circ(\mathrm{CGCGCG}) + \Delta G^\circ_{37}\begin{pmatrix}\mathrm{CG}\\ \mathrm{GC}\end{pmatrix}$$

Most sequences had 3 bp on each side of the internal loop. Some values are averages from more than one sequence.

[b]He et al., (1991); SantaLucia et al., (1991); Wu et al., (1995); Mathews et al., (1999).

[c]Sequence with this tandem mismatch had either an unusual conformation or mixture of conformations.

dependence is thought to be due to hydrogen bonding in the stable loops (SantaLucia et al., 1991; Wu et al., 1995). The NMR studies of internal loops have revealed a variety of noncanonical interactions indicating that rules for thermodynamic stability may be complex (Wimberly et al., 1993; SantaLucia and Turner, 1993; Peterson et al., 1994; Battiste et al., 1994; Wu and Turner, 1996; Wu et al., 1997).

The most common mismatch in RNA is GU, and often GU mismatches are conserved (Gutell et al., 1994; Michel and Westhof, 1990). This conservation may imply importance in tertiary and functional interactions, and this has been shown for group I introns (Pyle et al., 1994; Knitt et al., 1994; Strobel and Cech, 1995). Table 8-12 lists thermodynamic parameters for GU mismatches (Mathews et al., 1999). Note that the nearest neighbor, 5′GU3′/5′GU3′, is destabilizing in most but not all contexts. This sequence dependence is a non-nearest neighbor effect. The destabilizing motif rarely occurs in secondary structures of natural RNA. Additional terms are applied for terminal GU pairs, and these terms are assumed to be the same as those listed for terminal AU pairs in Table 8.4.

Table 8.12
Thermodynamic Parameters for Helix Propagation by G:U Pairs in RNA Oligonucleotides in 1 *M* NaCl[a]

Propagation Sequence	ΔH° (kcal mol^{-1})	ΔS° (eu)	ΔG°_{37} (kcal mol^{-1})
$\overrightarrow{\text{GC}}$ / $\overleftarrow{\text{UG}}$	−12.59	−32.5	−2.51
$\overrightarrow{\text{GG}}$ / $\overleftarrow{\text{UC}}$	−12.11	−32.2	−2.11
$\overrightarrow{\text{GG}}$ / $\overleftarrow{\text{CU}}$	−8.33	−21.9	−1.53
$\overrightarrow{\text{GA}}$ / $\overleftarrow{\text{UU}}$	−12.83	−37.3	−1.27
$\overrightarrow{\text{GU}}$ / $\overleftarrow{\text{UA}}$	−8.81	−24.0	−1.36
$\overrightarrow{\text{CG}}$ / $\overleftarrow{\text{GU}}$	−5.61	−13.5	−1.41
$\overrightarrow{\text{UG}}$ / $\overleftarrow{\text{AU}}$	−6.99	−19.3	−1.00
$\overrightarrow{\text{AG}}$ / $\overleftarrow{\text{UU}}$	−3.21	−8.6	−0.55
$\overrightarrow{\text{UG}}$ / $\overleftarrow{\text{GU}}$	−9.26	−30.8	+0.30
$\overrightarrow{\text{GU}}$ / $\overleftarrow{\text{UG}}$	−14.59(−14.14)[b]	−51.2(−42.2)[b]	+1.29(−1.06)[b]
$\overrightarrow{\text{GG}}$ / $\overleftarrow{\text{UU}}$	−13.47	−44.9	+0.47

[a]He et al., (1991); Mathews et al. (1999). Additional terms are applied for terminal GU pairs, and these are assumed to be the same as those listed for terminal AU pairs in Table 8.4.

[b] $\overrightarrow{\text{GU}}$ / $\overleftarrow{\text{UG}}$ in the contexts $\overrightarrow{\text{AGUU}}$ / $\overleftarrow{\text{UUGA}}$, $\overrightarrow{\text{CGUG}}$ / $\overleftarrow{\text{GUGC}}$, and $\overrightarrow{\text{UGUA}}$ / $\overleftarrow{\text{AUGU}}$ has a ΔG°, ΔH°, and ΔS° of +1.29 kcal mol^{-1}, −14.59 kcal mol^{-1}, and −51.2 eu, respectively, but in the context $\overrightarrow{\text{GGUC}}$ / $\overleftarrow{\text{CUGG}}$ it has a ΔG°, ΔH°, and ΔS° of −1.06 kcal mol^{-1}, −14.14 kcal mol^{-1}, and −42.2 eu, respectively.

3.3 Hairpin Loops

Hairpin loops occur when nucleic acid strands fold back on themselves to make base pairs (Fig. 8-11). Hairpins are widespread in structures of natural RNA (see Fig. 8-1). In DNA, there are many instances of sequences capable of forming hairpins as alternates to the fully base paired double helix. There is evidence that such hairpins will form when the DNA is placed under superhelical tension (Gellert et al., 1979; Panayotatos and Wells, 1981; Sullivan and Lilley, 1986).

Conformational changes involving hairpins are intramolecular. Thus thermodynamic parameters can be determined from calorimetric and spectroscopic data in a manner similar to that described for single-strand stacking in Section 8.1.1. Fortunately, hairpin transitions are more cooperative than those of single strands, thus simplifying determination of baselines. For transitions considered two-state, ΔH° is simply derived from (Bloomfield et al., 1974):

$$\text{Unimolecular} \qquad \Delta H^\circ = 4RT_\mathrm{m}^2 \left(\frac{\partial \alpha}{\partial T}\right)_{T=T_\mathrm{m}} \tag{8-22a}$$

Here the partial derivative is the slope at the T_m of a plot of fraction of strands in single-strand state versus temperature. The value of ΔS° can then be determined from Eq. 8-7b. Equation 8-22a is a specific case of the general equation for an equilibrium with N strands (Marky and Breslauer, 1987):

$$N\text{-mer} \qquad \Delta H^\circ = (2 + 2N)RT_\mathrm{m}^2 \left(\frac{\partial \alpha}{\partial T}\right)_{T=T_\mathrm{m}} \tag{8-22b}$$

More data are available for hairpins than for other loops (Table 8-13; Hilbers et al., 1994). The free energy increment for loop formation is unfavorable, and the minimum loop size is 2 (Orbons et al., 1986, 1987; Blommers et al., 1989; Jucker and Pardi, 1995). The stability of hairpins is dependent on loop size, the sequence in the loop, and the base pair closing the loop (see Table 8-13). For example, the CG closed tetraloop formed by C<u>UUCG</u>G is unusually stable (Tuerk et al., 1988; Antao and Tinoco, 1992). The complete basis for the sequence dependence of hairpin stability has not yet been unraveled (Hilbers et al., 1994), although hydrogen bonding and stacking within the loop probably make contributions (Varani et al., 1991; SantaLucia et al., 1992; Serra et al., 1994). An equation that provides a reasonable prediction for the free energy increment in kilocalories per mole associated with RNA hairpin loops greater than or equal to 4 nucleotides is (Serra et al., 1994, 1997; Mathews et al., 1999):

$$\Delta G^\circ_{37,\mathrm{loop}} = \Delta G^\circ_{37,\mathrm{length}} + \Delta G^\circ_{37,\mathrm{mm}} - 0.8 \text{ if first mismatch is GA or UU} \tag{8-23}$$

Here $\Delta G^\circ_{37,\mathrm{length}}$ is the length dependence of hairpin ΔG° (see Table 8-14) and $\Delta G^\circ_{37,\mathrm{mm}}$ is the free energy increment for the first mismatch in the loop (Table 8-6). The sarcin/ricin loop from 28S ribosomal RNA has 17 nucleotides, many of which are involved in noncanonical interactions (Szewczak et al., 1993). This result suggests that the complete set of rules for hairpin stability may be complex.

Table 8.13
Thermodynamic Parameters for Hairpin Loops

Stem Sequence	Loop Sequence	Reference	Salt	ΔH° (kcal mol^{-1})	ΔS° eu	ΔG°_{37} (kcal mol^{-1})	t_m °C
RNA							
GGGAUAC •••••• ACCUAUG	A3	Groebe and Uhlenbeck (1988).	1*M* Na^+	−56.2	160.9	−6.3	76.4
	A4	Groebe and Uhlenbeck (1988).	1*M* Na^+	−63.8	−181.2	−7.6	78.9
	A5	Groebe and Uhlenbeck (1988).	1*M* Na^+	−64.6	−183.4	−7.7	79.1
	A7	Groebe and Uhlenbeck (1988).	1*M* Na^+	−61.5	−175.1	−7.2	78.1
	A9	Groebe and Uhlenbeck (1988).	1*M* Na^+	−52.6	−151.7	−5.6	73.3
	U3	Groebe and Uhlenbeck (1988).	1*M* Na^+	−60.1	−171.1	−7.0	78.1
	U4	Groebe and Uhlenbeck (1988).	1*M* Na^+	−67.4	−191.0	−8.2	79.7
	U5	Groebe and Uhlenbeck (1988).	1*M* Na^+	−67.4	−191.1	−8.1	79.5
	U7	Groebe and Uhlenbeck (1988).	1*M* Na^+	−66.0	−189.0	−7.4	76.1
	U9	Groebe and Uhlenbeck (1988).	1*M* Na^+	−73.8	−212.0	−8.0	74.3
GGU ••• CCA	AUAAUA	Serra et al. (1993).	1*M* Na^+	−25.9	−82.4	−0.3	40.7
GGA ••• CCU	AUAAUA	Serra et al. (1993).	1*M* Na^+	−29.2	−91.8	−0.7	44.2
GGG ••• CCC	AUAAUA	Serra et al. (1993).	1*M* Na^+	−33.5	−101.0	−2.2	58.7
GGC ••• CCG	AUAAUA	Serra et al. (1993).	1*M* Na^+	−34.3	−101.7	−2.7	64.0
GGC ••• CCG	GUAAUA	Serra et al. (1993).	1*M* Na^+	−36.9	−107.8	−3.4	68.5
GGAC •••• CCUG	UUCG	Antao and Tinoco (1992).	1*M* Na^+	−55.9	−159.9	−6.3	76.2
GGAC •••• CCUG	UUUG	Antao and Tinoco (1992).	1*M* Na^+	−44.0	−128.0	−4.3	70.3
GGAG •••• CCUC	UUCG	Antao and Tinoco (1992).	1*M* Na^+	−44.8	−131.4	−4.0	67.7
CA GA •• • $GUm^5\Psi$	$CmUG_m$ AYA^A	Clore et al. (1994).	0.5*M* K^+	−38	−118	−1.4	53
DNA							
GGAC •••• CCTG	TTCG	Antao and Tinoco (1992).	1*M* Na^+	−31.3	−93.9	−2.2	60.4
GGAC •••• CCTG	TTTG	Antao and Tinoco (1992).	1*M* Na^+	−31.0	−93.0	−2.2	59.8
GGAG •••• CCTC	TTCG	Antao and Tinoco (1992).	1*M* Na^+	−30.2	−92.5	−1.5	53.8
ATCCTA •••••• TAGGAT	T_4	Hilbers et al. (1985).	0.2*M* Na^+	−39	−119.2		54
	T_5	Hilbers et al. (1985).	0.2*M*Na^+	−43	−132.7		51
	T_6	Hilbers et al. (1985).	0.2*M* Na^+	−43	−133.9		48
	T_7	Hilbers et al. (1985).	0.2*M*Na^+	−43	136.0		44
CGAACG •••••• GCTTGC	A_4	Senior et al. (1988a).	0.1*M* Na^+	−33.7			63.4
	G_4	Senior et al. (1988a).	0.1*M* Na^+	−41.8			65.6
	C_4	Senior et al. (1988a).	0.1*M* Na^+	−46.3			65.8
	T_4	Senior et al. (1988a).	0.1*M* Na^+	−49.3			67.4
m^5C Gm^5C • • • Gm^5C G	GA	Orbons et al. (1987).		−28.9	87.8		56

Table 8.14
Free Energy Increments (kcal mol^{-1}) for RNA Loops at 37°C in 1 *M* NaCl[a]

Loop Size	Internal Loop	Bulge Loop	Hairpin Loop
1		+3.8	
2	+0.4[b]	+2.8	–
3	[c]	+3.2	+5.7
4	[c]	(+3.6)	+5.6
5	+1.8	(+4.0)	+5.6
6	+2.0	(+4.4)	+5.4
7			+5.9
8			+5.6
9			+6.4

[a]From Mathews et al. 1999. For larger loop sizes, n, use $\Delta G^\circ(n) = \Delta G^\circ(n_{max}) + 1.75RT \ln (n/n_{max})$, where n_{max} is 6, 6, and 9 for internal, bulge, and hairpin loops, respectively. Parameters not derived from experimental measurements are listed in parentheses. Parameters for hairpin loops are for closure by CG or GC, and for loops greater than or equal to 4 assume additional stability is conferred by terminal mismatches at helix ends (see Table 8-6). A reasonable approximation for hairpin loops greater than or equal to 4 is given by: $\Delta G^\circ_{37,loop} = \Delta G^\circ_{37,length} + \Delta G^\circ_{37,mm} - 0.8$ (if first mismatch is UU or GA (not AG)). Asymmetric internal loops with branches of N1 and N2 nucleotides must be penalized additionally by the minimum of 3.0 or 0.5 $|N1 - N2|$ kcal mol^{-1}. For internal loops larger than 4, each terminal GA and AG mismatch is given a favorable free energy increment of -1.1 kcal mol^{-1} and each terminal UU is given a bonus of -0.7 kcal mol^{-1}. For each AU or GU pair closing an internal loop, a penalty of 0.2 kcal mol^{-1} is applied in addition to the 0.45 kcal mol^{-1} penalty for terminating a helix with AU or GU. The parameter for bulge loops of one nucleotide is based on the assumption that additional stability is conferred by stacking of the adjacent base pairs as approximated by nearest neighbor parameters (see Table 8-4). It is assumed there is no stacking across bulges of two or more nucleotides.

[b]Rough average for single mismatches that are not GG and that have 2 adjacent GC pairs. The equivalent value for GG mismatches is -2 kcal mol^{-1}.

[c]Free energies of internal loops with 3 or 4 nucleotides are estimated with special rules (Xia et al., 1999; Mathews et al., 1999).

The kinetics of hairpin formation have been studied by temperature-jump methods (Pörschke, 1974; Riesner et al., 1973; Coutts, 1971; Coutts et al., 1974; Orbons et al., 1986, 1987). Forward rates are between 10^4 and 10^5 s^{-1} and change little with temperature. As with duplex formation, this result allows rough prediction of the rate of unfolding for a hairpin, if the equilibrium constant for hairpin formation is known or can be predicted. For example, at the melting temperature, $K = k_{for}/k_{rev} = 1$. Thus the rate of unfolding of a hairpin at the T_m is also roughly $10^4 - 10^5$ s^{-1}. In a temperature-jump experiment, the corresponding relaxation time is given by $\tau^{-1} = k_{for} + k_{rev}$, or about 5–50 μs at the T_m.

3.4 Multibranch Loops (Junctions)

Multibranch loops (junctions) occur when three or more helixes intersect (Fig. 8-11). The term loop is somewhat misleading since it is possible for these structures to exist with no unpaired nucleotides. Models for such a structure with four helixes have been studied with DNA oligonucleotides (Marky et al., 1987; Lu et al., 1992). The ΔH° for formation of one structure was simply the sum of the ΔH° values for formation of each separate helix in the structure. Thus the junction did not significantly perturb the bonding in attached helixes (Marky et al., 1987). The thermodynamic parameters for forming a four-arm junction structure from two paired duplexes are $\Delta G^\circ_{18} = 1.1$ kcal mol^{-1} and $\Delta H^\circ = 27.1$ kcal mol^{-1} (Lu et al., 1992). A DNA junction with three helixes is stabilized by adding unpaired nucleotides to the junction (Leontis et al., 1991). Melting of tRNA involves changes in the size of a multibranch loop (Riesner and Römer, 1973; Crothers and Cole, 1978; Crothers et al., 1974).

3.5 Knotted Loops

If two nucleotides i and j are base paired, a pseudo knot is formed when a nucleotide between i and j in the primary sequence is base paired with a nucleotide not between i and j (Fig. 8-11). These structures were first postulated to explain similar biochemical reactivities for tRNA and the 3′ ends of certain viral RNAs from plants (Pleij et al., 1985). They have subsequently been shown to occur in oligonucleotides of appropriate sequence (Wyatt et al., 1990). It was suggested that the minimum loop sizes for the 5′ and 3′ loops are two and three nucleotides, respectively, for A-form RNA, and perhaps even one or two for a distorted structure (Pleij et al., 1985). Studies of oligonucleotide models are consistent with the predicted minimum loop sizes for A form, and indicate pseudoknot stability is dependent on loop size and sequence (Wyatt et al., 1990). The presence of Mg^{2+} or high Na^+ concentrations also stabilizes pseudoknots (Wyatt et al., 1990).

3.6 Large Loops

Loops larger than 12 nucleotides are rarely seen in nucleic acid structures. Nevertheless, the properties of large loops are important for successful predictions of nucleic acid structure, and for predicting the melting behavior of nucleic acids. Unfortunately, it is difficult to experimentally determine the properties of large loops, especially given the number of possible permutations. Theoretical considerations, however, provide a reasonable approximation for the ΔG° for initiation of large loops (Jacobson and Stockmayer, 1950). Considerations of conformational entropy effects lead to the following equation for the ΔG° of initiation of a large loop with N nucleotides:

$$\Delta G^\circ_{\text{loop}}(N) = \Delta G^\circ_{\text{loop}}(n) + aRT \ \ln(N/n) \tag{8-24}$$

Here $\Delta G^\circ_{\text{loop}}(n)$ is the experimentally determined ΔG° for a loop of n nucleotides, and a is a constant. Theoretical considerations suggest the value of a is about 1.75 (Fisher, 1966). Experiments on bulge and internal loops are consistent with Eq. 8-24

(Weeks and Crothers, 1993). A set of loop parameters derived from experimental and theoretical results is given in Table 8-14.

Reactions involving covalently closed circular nucleic acids provide another situation where the free energies of large loops are probably important. For example, the melting temperature of a covalently closed circular DNA of 26 nucleotides that forms a dumbbell-shaped structure is more than 30°C higher than for the same sequence with a single break in the sugar–phosphate backbone (Erie et al., 1989). This difference corresponds to a more favorable ΔG°_{37} of 7 kcal mol^{-1} for folding into the dumbbell. A large fraction of this favorable free energy is expected from the conformational constraints placed on a melted loop when it is covalently closed (Jaeger et al., 1990). The same effect can also largely account for the enhanced binding observed between oligomers when one is a covalently closed circle (Prakash and Kool, 1992). While such a conformational effect might be expected to provide a more favorable entropy for folding, a more favorable enthalpy is actually observed (Erie et al., 1989; Prakash and Kool, 1992). This apparent anomaly is seen even in reactions of small molecules, however (Jencks, 1975).

4. DOUBLE HELIX FORMATION WITH POLYNUCLEOTIDES

An understanding of helix formation in oligonucleotides provides a basis for understanding the properties of nucleic acid polymers. It is likely that interactions important in oligomers will be major factors determining polynucleotide properties, but that new interactions will also be important. Tertiary interactions are expected to be weaker energetically than secondary structure interactions, and there is some experimental evidence to support this (Crothers et al., 1974; Banerjee et al., 1993; Jaeger et al., 1993; Laing and Draper, 1994). The most extensively studied cases of melting of nucleic acid polymers are tRNA and DNA. In both cases, melting typically starts with formation of large loops.

4.1 Transfer RNA

The experimentally determined pathway for melting of *Escherichia coli* formylmethionine tRNA in 0.17 M Na$^+$ is shown in Figure 8-12 (Crothers et al., 1974). First, tertiary interactions and the dihydrouridine stem melt, followed by the $T\Psi C$ stem, the anticodon stem, and finally the acceptor stem. The order of melting for other tRNAs may be different, indicating a sequence dependence (Riesner and Römer, 1973).

The kinetics of melting for tRNA are similar to expectations from oligomers (Riesner and Römer, 1973; Crothers et al., 1974). The $T\Psi C$ and anticodon hairpins typically melt with relaxation times of 10–100 μs. Melting of the acceptor stem is associated with a relaxation time of about 1 ms, presumably because it is the final stem to melt and therefore opens a large hairpin loop. Melting of the least stable helix, the dihydrouridine stem, is associated with relaxation times of about 1–10 ms. This time

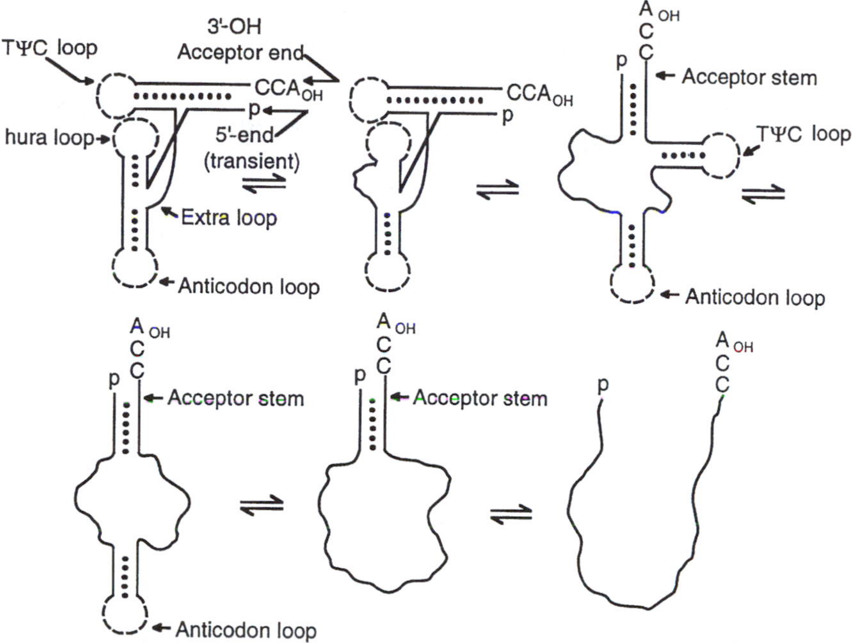

Figure 8-11
Pathway for melting of *E. coli* formylmethionine tRNA in 0.17 *M* Na^+ [Reprinted with permission from Crothers, D. M., Cole, P. E., Hilbers, C. W., and Shulman, R. G., The Molecular Mechanism of Thermal Unfolding of *Escherichia coli* Formylmethionine transfer RNA, *J. Mol. Biol.* **87**, 63–88, Copyright ©1974, by permission of the publisher Academic Press Limited London.] The dihydrouridine helix and tertiary interactions melt first, followed in succession by the $T\Psi C$, anticodon, and acceptor stem helixes.

is much longer than expected from model hairpins, and is associated with concomitant disruption of tertiary structure. This long relaxation time is one example of effects in polymers that are not expected from simple model systems.

4.2 DNA

A typical differentiated melting curve and pathway for melting of DNA are shown in Figure 8-13 (Blake and Decourt, 1987; Wada et al., 1980; Gotoh, 1983; Wartell and Benight, 1985; Lerman et al., 1984). Depending on sequence and length, DNA can begin melting from the ends or the middle. As temperature is raised, AT rich regions melt first, leaving internal loops in long DNA. Typically, loops with nucleotides from 100–350 bp are formed. A single large internal loop of *n* nucleotides is more favorable than two separate internal loops containing a total of *n* nucleotides (see Table 8-14). Thus loops coalesce. Occasionally, it is necessary to also consider intermediates with hairpin loops (Wartell and Benight, 1985). The result, as shown in Figure 8-13, is a series of transitions as the temperature is raised. The parameters in Table 8-4 coupled with parameters for internal and hairpin loops, should be able to predict the melting behavior for any DNA sequence. Alternative sets of parameters have also been suggested (Wada et al., 1980; Gotoh, 1983; Wartell and Benight, 1985; Breslauer et al., 1986; Lerman et al. 1984; Klump, 1990; Delcourt and Blake, 1991; Quartin and Wetmur, 1989; Doktycz et al., 1992). Parameters for loops have been determined by measuring the melting of DNAs that have AT rich regions inserted (Blake and Delcourt, 1987). For sequences longer than about 300 bp, separation of strands can be neglected in these predictions, but it must be included for shorter sequences (Wartell and Benight,

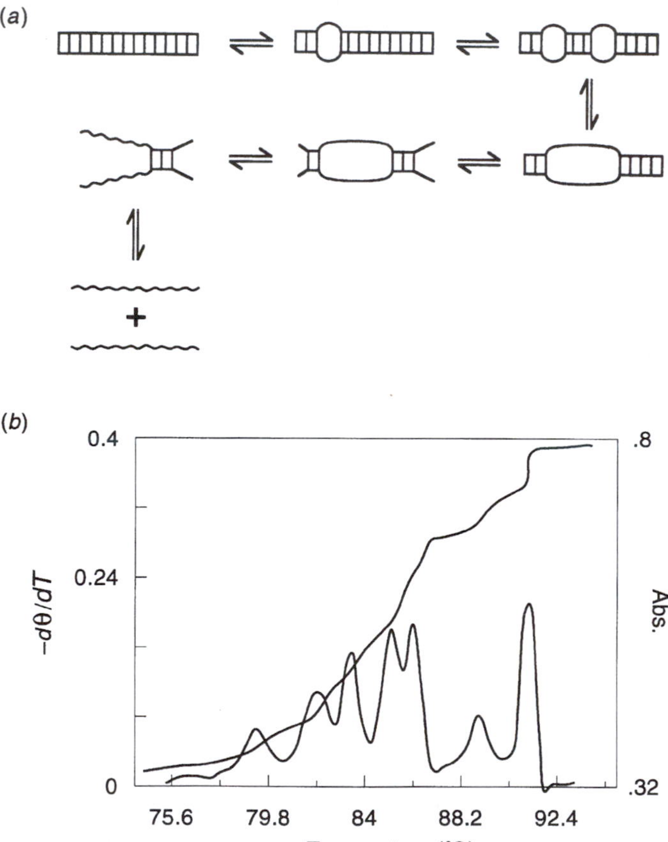

Figure 8-12
(*a*) Typical pathway for melting of DNA. (*b*) Absorbance versus temperature and differentiated melting curves for the 1630 bp Hinf I restriction endonuclease fragment of plasmid pBR322. [Reprinted from Wartell, R. M. and Benight, A. S., Thermal Denaturation of DNA Molecules: A Comparison of Theory with Experiment, *Phys. Rep.*, **126**, 67–107. Copyright ©1985 with permission of Elsevier Science-NL, Amsterdam, The Netherlands.]

1985). In several cases, predictions are reasonably successful (Wartell and Benight, 1985; Steger, 1994).

For long DNA sequences with relatively random distributions of nearest neighbors, simple equations have been developed empirically to approximately predict t_m from the fraction of GC content, F_{GC} (Marmur and Doty, 1962). One that is useful for long DNAs of quasirandom sequence with $0.3 < F_{GC} < 0.7$ and $0.02 \leq [Na^+] \leq 0.4M$ is (Blake, 1996)

$$t_m(°C) = 193.67 - (3.09 - F_{GC})(34.64 - 6.52 \log[Na^+]). \tag{8-25}$$

Another that includes the effect of duplex length, D, and the percentage of mismatches, P, but neglects the effect of GC content on the salt dependence is (Wetmur, 1991)

$$t_m(°C) = 81.5 + 41F_{GC} + 16.6 \log_{10}\left(\frac{[Na^+]}{1.0 + 0.7[Na^+]}\right) - \frac{500}{D} - P \tag{8-26}$$

This equation is valid up to 1 M Na^+. The analogous equation for RNA is

$$t_m(°C) = 78 + 70F_{GC} + 16.6 \log_{10}\left(\frac{[Na^+]}{1.0 + 0.7[Na^+]}\right) - \frac{500}{D} - P \tag{8-27}$$

An interesting application of DNA melting is the use of gels containing denaturant gradients to separate DNAs of similar length but different sequence, and to detect single base pair changes between DNAs (Lerman et al., 1984). These applications arise from a large decrease in gel mobility when denaturation induces an internal loop. Predictions for comparison with experiment are possible since denaturants such as urea and formamide mimic a temperature increase (Lerman et al., 1984; Klump and Burkart, 1977).

The kinetics of association and dissociation for large DNAs is quite different from that of oligomers (Wetmur and Davidson, 1968; Bloomfield et al., 1974; Cantor and Schimmel, 1980; Studier, 1969; Record and Zimm, 1972). This results from sequence complexity and large size. For example, association rates are affected by intramolecular helix formation and by incorrect intermolecular helix formation. Thus, when two separated strands of a large DNA are quickly cooled, they are kinetically trapped in nonnative structures and essentially never able to reform the perfectly matched helix. Under conditions where the perfectly matched helix can be formed, the rate constant k_2 for this association has been found experimentally to be given by (Wetmur and Davidson, 1968; Wetmur, 1991)

$$k_2 = \frac{k'_N \sqrt{L}}{N} \tag{8-28}$$

Here k'_N is the nucleation rate constant, L is the length of the shortest strand participating in duplex formation, and N is the complexity of the sequence. Complexity has been defined as "the total number of base pairs present in nonrepeating sequences" (Wetmur, 1991) or "the length of DNA needed to contain one copy of the entire sequence" (Cantor and Schimmel, 1980). The rate constant k'_N is a complicated function of temperature and other conditions, approaching zero at the T_m of a large DNA. It is maximal roughly 25°C below the T_m, where a typical value is $3.5 \times 10^5\ M^{-1}\ s^{-1}$ in 1 *M* NaCl (Wetmur, 1991). The kinetics of dissociation near the T_m for large DNAs occurs in the 100–1000 s time range and cannot be fit with a simple functional form. This observation is thought to be due to large frictional effects in the unwinding process and is avoided *in vivo* by the actions of nicking-closing enzymes (topoisomerases) and gyrases that break phosphodiester bonds to allow easier rotation.

4.3 Nucleic Acid Hybridization

Nucleic acid hybridization and methods that rely on it are the methods of choice for detecting specific sequences in complex mixtures. For example, the polymerase chain reaction (PCR) and various blotting techniques rely on specific binding of short or long nucleic acid primers or probes to target sequences (Wetmur, 1991). Oligonucleotide arrays on solid support can detect thousands of sequences simultaneously (Schera et al., 1995; Chee et al., 1996). The best conditions for specificity are dependent on many factors, including the T_m and kinetics for the binding and for the disruption of structure in individual sequences. The principles described above can often be used to predict optimum conditions for experiments requiring hybridization (Wetmur, 1991; Steger, 1994). The effects of solid supports, however, have not been studied in detail (Wetmur, 1991).

In applying the principles described above, it is important to recognize the parameter that is important for a given experiment. For example, PCR relies on dissociation of helixes, so the system should be brought to a temperature above T_m to insure complete dissociation. For the annealing step, specificity is important, so the temperature should be a little, typically 10°C, below T_m. For probing blots, the kinetics of dissociation is the important parameter, since unhybridized probe is removed by washing for a given time period. For this application, Wetmur (Wetmur, 1991) suggests defining a dissociation temperature T_d as the temperature at which one-half of the correctly matched hybrid is released in a time t_{wash}. The parameter T_d can be predicted from the expected activation energy for probe dissociation, $E_{A,off}$, the melting temperature T_m (in K) for the hybrid duplex at the probe concentration, and the half-life for probe dissociation at the T_m, $t_{1/2}$:

$$\ln(t_{wash}/t_{1/2}) = (E_{A,off}/R)(T_d^{-1} - T_m^{-1}) \qquad (8\text{-}29)$$

4.4 Tertiary Interactions

The term "tertiary interaction" is used in several different ways. Sometimes it refers to noncanonical interactions, for example a hydrogen bond to a phosphate group as sometimes seen in hairpin and other loops (Varani et al., 1991; Heus and Pardi, 1991; SantaLucia et al., 1992; Pley et al., 1994). Here we use a more restricted definition (Chastain and Tinoco, 1991). If two nucleotides i and j are paired, then a tertiary interaction is any interaction, direct or indirect, between a nucleotide k between i and j in the sequence, and another nucleotide ℓ which is not between i and j. Thus a pseudoknot (Fig. 8-11) is one form of tertiary interaction. Another is metal ion mediated interactions between nucleotides k and ℓ (Lu and Draper, 1994).

One approach to understanding tertiary interactions is to study binding of oligonucleotides to natural RNAs.The methods employed include kinetics (Sugimoto et al., 1988, 1989b; Herschlag and Cech, 1990; Bevilacqua et al., 1992), fluorescence (Bevilacqua et al., 1992; Sugimoto et al., 1989a), equilibrium dialysis (Bevilacqua and Turner, 1991), and gel retardation (Pyle et al., 1990). The last method is the fastest, but is not completely understood. When using gel retardation, it is important to allow sufficient incubation time before loading samples on the gel (Bevilacqua and Turner, 1991; Pyle et al., 1994). Another method for studying tertiary interactions is to monitor the effects of site directed mutation on RNA folding and function (Murphy and Cech, 1994; Costa and Michel, 1995).

Little is known about the thermodynamics of tertiary interactions. One interaction that has been identified involves hydrogen bonding to 2′-OH groups for recognition of an oligomer substrate by a group I ribozyme (Sugimoto et al., 1989b; Pyle and Cech, 1991; Bevilaqua and Turner, 1991; Strobel and Cech, 1993). Each such interaction can provide a free energy increment of about 1 kcal mol^{-1} (Bevilacqua and Turner, 1991). In one case, the interaction involves a hydrogen bond from the H of the 2′-OH to the N1 of an A (Pyle et al., 1992).

Unfavorable tertiary interactions are also possible. For example, UCdGU binds to a group I ribozyme at least 30-fold more weakly than UCdG (Moran et al., 1993). A model of the binding site (Michel and Westhof, 1990) suggests that it is designed to put strain on the phosphodiester bond between G and U. This bond is the site of cleavage in the second step of group I splicing, and such strain may provide a catalytic advantage (Moran et al., 1993). Such substrate destabilization has also been reported in a ribozyme model system for the first step of splicing (Bevilacqua et al., 1994; Narlikar et al., 1995).

Favorable and unfavorable tertiary interactions are presumably sensitive to the overall folding of an RNA. Thus they can give rise to cooperativity and anticooperativity in binding of substrates. Such effects have also been observed with a group I ribozyme (Bevilacqua et al., 1993; McConnell et al., 1993).

5. ENVIRONMENTAL EFFECTS ON HELIX STABILITY

5.1 Salt Concentration

A theoretical treatment of salt effects on helix stability for polyelectrolytes is given in Chapter 11. Here we focus on experimental results. In solutions containing only Na^+ or similar ions, increasing salt concentration up to about 1 M continuously increases helix stability. Up to about 0.2 M Na^+, the increase in T_m is linear with log $[Na^+]$ (see Eq. 11-20). The rate of increase depends on base composition. For example, from fitting data obtained on different, natural DNAs (Owen et al., 1969), Frank-Kamenetskii (1971) found $dT_m/d\log[Na^+] = 18.30 - 7.04F_{GC}$ (F_{GC} is the fractional GC content). From studies of subtransitions in the melting of lambda DNA, Blake and Haydock (1979) proposed a similar equation: $dT_m/d\log[Na^+] = 19.96 - 6.65F_{GC}$. Thus, the T_m of a deoxypolynucleotide with only AT base pairs increases almost 20°C for every factor of 10 increase in $[Na^+]$.

At salt concentrations above 1 *M*, addition of salt lowers the T_m of DNA. The lowering is relatively independent of cation, but is strongly dependent on anion with $CCl_3COO^- > SCN^- > ClO_4^- > CH_3COO^- > Br^-, Cl^-$ (Hamaguchi and Geiduschek, 1962). This order correlates with the effect of these ions on the solubility of the bases (Robinson and Grant, 1966). The better denaturants are most effective in enhancing base solubility.

There are few studies of the effect of Mg^{2+} concentration on polynucleotide melting (Riesner and Römer, 1973; Blagoi et al., 1978; Krakauer, 1974). As with Na^+, addition of Mg^{2+} initially increases T_m. The effect saturates at about $10^{-3} - 10^{-2}M$, however. Thereafter, addition of Mg^{2+} lowers T_m (Blagoi et al., 1978). Somewhat counterintuitively, addition of Na^+ in the presence of enough Mg^{2+} to neutralize the backbone charges results in a lowering of T_m. This lowering of T_m is expected from theoretical considerations (Record, 1975; Manning, 1978). In particular, at saturating concentrations of Mg^{2+}, the effective charge on single strands is larger than on double strands due to counterion condensation. Thus, single strands are stabilized more by increases in ionic strength because of Debye–Hückel screening effects.

Relatively little is known about salt effects on stabilities of oligonucleotide helixes. Qualitatively, the effects are similar to those observed for polynucleotides. For example, $dT_m/d\log[Na^+]$ for dGCATGC and dGGAATTCC are 11 and 12°C, respectively (Williams et al., 1989; Erie et al., 1987), only a little less than predicted for polymers with the same GC content. This result is somewhat surprising since theoretical considerations indicate less charge will be neutralized in oligonucleotides than in polynucleotides because of the reduced charge density at the ends (Record and Lohman, 1978; Olmsted et al., 1989). Relatively little experimental data is available, however. One interesting question is: At what length does an oligomer behave much like a polymer? This question has not been investigated in detail. The values of $dT_m/d\log[Na^+]$ for A_7U_7 and poly (A) poly (U) are similar, however: 17.4 and 19.6°C, respectively (Hickey and Turner, 1985b).

5.2 Solvent Effects

Addition of cosolvents to aqueous solutions of nucleic acids typically destabilizes the ordered form, and this is often used as a substitute for denaturation by temperature (Lerman et al., 1984). Typically, the T_m of a double helix will be a linear function of cosolvent concentration (Klump and Burkart, 1977; Hickey and Turner, 1985b). For example, the T_m values of calf thymus and salmon sperm DNA decrease 2.5°C M^{-1} urea (Klump and Burkart, 1977). The effects of cosolvents have been studied by measuring the concentration of cosolvent required to give 50% denaturation of T4 DNA at 73°C (Levine et al., 1963) and by measuring the effect of cosolvents on thermal denaturation curves of various oligonucleotides (Hickey and Turner, 1985b; Albergo and Turner, 1981). Some of these results are listed in Table 8-15. Somewhat surprisingly, there is no simple correlation between the concentration required for 50% denaturation at 73°C and the effect on T_m. This finding may reflect the very different conditions and assays

Table 8.15
Effects of Denaturants on Duplex Stability

	Molarity for 50% denaturation of T4 DNA at 73°C, 43 m*M* Ionic Strength		Δt_m(°C) at 10 mol % for 18.7μM Oligo in 1 *M* NaCl	
Cosolvent	Predicted[a]	Observed[b]	A_7U_7[c]	$(dGC)_3$[d]
Alcohols				
Methanol	3.9	3.5		6.6
Ethanol	1.3	1.2	8.3	11.1
1-Propanol	0.47	0.54	9.1	8.4
2-Propanol	0.64	0.90	8.4	
Ethyleneglycol	1.7	2.2	7.7	
Glycerol		1.8	7.9	
Cyclohexyl alcohol		0.22		
Phenol		0.08		
Other Compounds				
Pyridine		0.09		
1,4 Dioxane		0.64	18.7	
Formamide	1.5	1.9	14.8	12.0
N, N dimethylformamide	0.54	0.60	16.9	~22
Urea	1.1	1.0	17.8	13
Acetonitrile		1.2		
TritonX-100		> 10%		

[a]Herskovits and Harrington, (1972); Herskovits and Bowen, (1974).

[b]Levine et al., (1963).

[c]Hickey and Turner, (1985).

[d]Albergo and Turner, (1981).

used in the experiments. For example, the cosolvent required for 50% denaturation was measured for an ionic strength of 0.04 with an antibody assay, whereas the effect on T_m was measured at 1 *M* Na^+ with optical melting curves. The cosolvent concentrations required for 50% denaturation correlate well with enhancement of base solubility (Levine et al., 1963; Herskovits and Harrington, 1972; Herskovits and Bowen, 1974), and this has predictive value (see Table 8-15). This correlation seems reasonable since the bases are more exposed to solvent in the denatured form, and favorable interactions between bases and cosolvent favor denaturation.

5.3 pH

The stability of most double helixes is relatively insensitive to pH between 5 and 9. At lower and higher pH values, stability is decreased (Bloomfield et al., 1974; Blake, 1996). At low pH, bases in the single strand bind more protons than in the duplex, thus favoring single strand. At high pH, guanine, thymine, and uracil are deprotonated, thus precluding normal hydrogen bonding and increasing charge repulsion. At both extremes, the charge on the single strand is different from that around pH 7. This affects the salt dependence of the T_m (Manning, 1978; Record et al., 1976). In particular, $dT_m/d\log[Na^+]$ will be larger at low pH and smaller at high pH. For example, $dT_m/d\log[Na^+] \approx 0$ near pH 11 (Record, 1967).

Some sequences form double helixes only at low pH. For example, poly(A), poly(dA), poly(C), and poly(dC) form duplexes with transition midpoints between pH 4 and 8 depending on conditions of salt and temperature (Inman, 1964; Hartman and Rich, 1965; Holcomb and Timascheff, 1968; Adler et al., 1969; Guschlbauer, 1975). This results from hydrogen bonding schemes in which a proton is shared between two bases, for example $A \cdot A^+$. The protonated pairs do not have to be adjacent. Poly(dCT) forms a duplex with $C \cdot C^+$ pairs alternating with unpaired thymines (Gray et al., 1980). Duplex formation by $C \cdot C^+$ pairing has also been observed at the ends, but not the middle, of mixed deoxy sequences. For example, $d(C_4A_4T_4C_4)$ forms such duplexes (Gray et al., 1984), but $d(A_{10}C_4T_{10})$, $d(AACC)_5 + d(CCTT)_5$, and $d(A_6C_6A_6) + d(T_6C_6T_6)$ do not (Edwards et al., 1988). This context dependence indicates $C \cdot C^+$ pairs are hard to form between stretches of AT pairs. In RNA, however, $C \cdot C^+$ pairs have been observed in the middle of the duplex $(CGC\underline{CC}GCG)_2$ (SantaLucia et al., 1991). Triple-strand helixes involving $C \cdot G \cdot C^+$ pairing have also been observed to form upon protonation of C (Lee et al., 1979). Little is known about the pH dependence of structure for other sequences, but there are suggestions that interesting effects will be discovered (Topping et al., 1988; Kao and Crothers, 1980; Legault and Pardi, 1994). For example, it has been shown that 5S rRNA undergoes a tertiary structural switch between pH 7 and 8 (Kao and Crothers, 1980).

6. TRIPLE HELIXES

Hydrogen bonded base triples are found in tRNA (see Fig. 4.13). A triple-stranded structure consisting of $T \cdot AT$ and $C^+ \cdot GC$ base triples has been proposed for regions of DNA containing $(dT\text{-}dC)_n \cdot (dA\text{-}dG)_n$ repeats (Christophe et al., 1985; Mirkin et al., 1987; Voloshin et al., 1988; Htun and Dahlberg, 1988; Johnston, 1988; Wells et al., 1988; Mirkin and Frank-Kamenetskii, 1994). When this triple strand forms, the remaining $(dA\text{-}dG)_n$ strand is released from pairing as shown in Figure 8-14, and the overall structure is referred to as H-DNA. It has also been proposed that pairing to give triple strands could regulate transcription (Cooney et al., 1988; Miller and Sobell, 1966) and be used for site specific modification or cleavage of nucleic acid (Strobel et al., 1988; Praseuth et al., 1988). These ideas may provide the basis for therapeutics that bind as a third strand.

Only sequences that are largely polypurine and polypyrimidine have been observed to form triple strands (Lee et al., 1979; Wells et al., 1988; Felsenfeld and Miles, 1967; Michelson et al., 1967). Known triple-strand pairings include: $U \cdot AU$ (Felsenfeld et al., 1957; Stevens and Felsenfeld, 1964), $A \cdot UA$ (Broitman et al., 1987), $T \cdot AT$ (Riley et al., 1966), $C^+ \cdot GC$ (Lee et al., 1979; Lipsett, 1964), $T \cdot CG$ (Yoon et al., 1992), $U \cdot CG$ (Michel et al., 1990; Michel and Westhof, 1990), $G \cdot TA$ (Yoon et al., 1992; Griffin and Dervan, 1989), $G \cdot CG$ (Lipsett, 1964; Lipsett, 1963), and $A \cdot GC$ (Chastain and Tinoco, 1992). Triple-strand helixes have also been observed with 2′-5′ linked oligonucleotides (Jin et al., 1993). The most thoroughly studied triplex is poly(A)·2

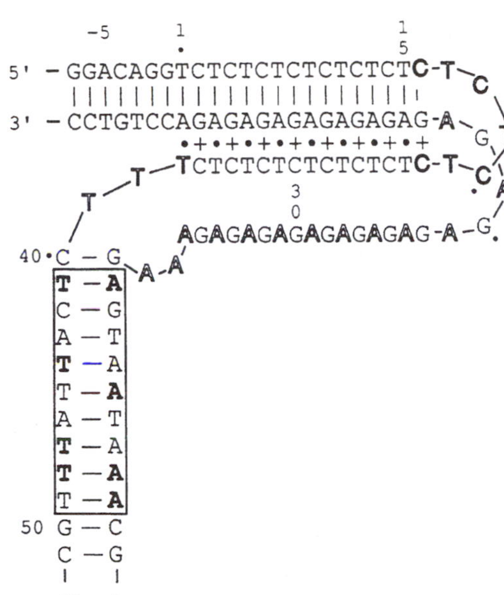

Figure 8-13
Schematic of structure for H-DNA [Reprinted with permission from Htun, H. and Dahlberg, J. E. (1988). *Science* **241**, 1791–1796. Copyright ©American Association for the Advancement of Science.] In this structure, Watson–Crick base pairs break to allow formation of a triple helix.

poly(U) (Felsenfeld et al., 1957; Stevens and Felsenfeld, 1964; Ross and Scruggs, 1965; Krakauer and Sturtevant, 1968). The stoichiometry of this complex was obtained from optical mixing curves (Felsenfeld et al., 1957; Stevens and Felsenfeld, 1964). The stability of the triplex relative to duplex and single strands depends on temperature and salt concentration as shown in the phase diagram in Figure 8-15 (Stevens and Felsenfeld, 1964; Krakauer and Sturtevant, 1968). As expected, high-salt favors triplex which is true for Mg^{2+} as well as for Na^+ (Felsenfeld et al., 1957). For example, in region 2 in Figure 8-15 the combination of poly(A)·2 poly(U) + poly(A) is more stable than 2 [poly(A)·poly(U)]. Thus any duplexes present will disproportionate to form triplex and single-strand poly(A):

$$2[\text{poly(A)poly(U)}] \rightleftharpoons [\text{poly(A)2 poly(U)}] + \text{poly(A)}$$

The ΔH° for this reaction is about 4 kcal mol^{-1} (Krakauer and Sturtevant, 1968). The ΔH° for formation of triplex from duplex and poly(U) is about -4 kcal mol^{-1} (Ross and Scruggs, 1965; Krakauer and Sturtevant, 1968). The time required for triple helix formation ranges from less than 1 to many minutes depending on length and conditions (Felsenfeld et al., 1957; Pörschke and Eigen, 1971).

7. G-QUARTETS

Hydrogen-bonded *G* quartets are known to occur in the telomeric regions at the ends of chromosomes (Williamson et al., 1989), and G tetraplexes can form from G monomers (Gellert et al., 1962; Pinnavaia et al., 1978), G rich DNA and RNA oligomers (Sen and Gilbert, 1988; Sundquist and Klug, 1989; Sen and Gilbert, 1990; Kim et al., 1991;

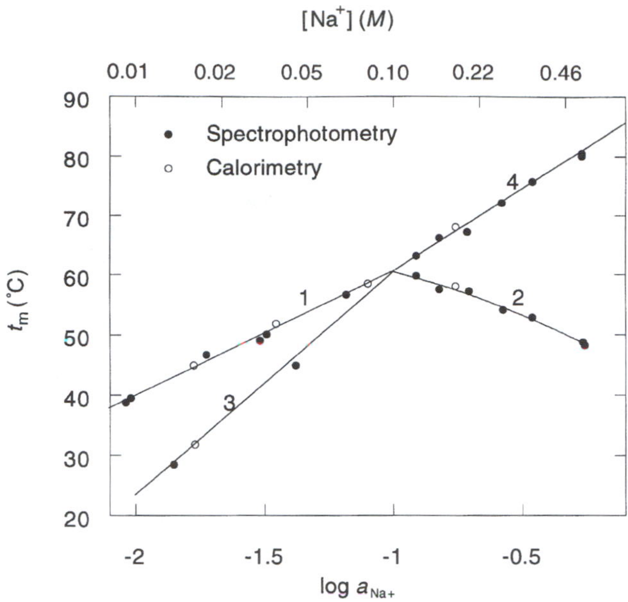

Figure 8-14
Dependence of t_m on the concentration of Na^+ at neutral pH in the absence of divalent ions for various reactions of polyA and polyU [Heats of Helix–Coil Transactions of PolyA- PolyU Complexes, Krakauer, H. and Sturtevant, J. M. *Biopolymers*, **6**, 491–512. Copyright ©1968. Reprinted by permission of John Wiley & Sons, Inc.] The lines correspond to the following reactions: (1) poly(A • U) → poly(A) + poly(U), (2) poly(A • U) → 1/2 poly(U • A • U) + 1/2 poly(A), (3) poly(U • A • U) → poly (A • U) + poly(U), (4) poly(U • A • U) → poly(A) + 2 poly(U). Note that the topmost region corresponds to single stranded poly(A) + poly(U).

Hardin et al., 1991; Jin et al., 1992; Lu et al., 1993), and polymers (Zimmerman et al., 1975). Formation of the G tetrad is favored by K^+ relative to Na^+ (Gellert et al., 1962; Pinnavaia et al., 1978; Sen and Gilbert, 1990; Hardin et al., 1991; Hud et al., 1996). It has been suggested that formation of the G tetrad may be important for several biological processes (Williamson et al., 1989; Sen and Gilbert, 1988; Sundquist and Klug, 1989; Sen and Gilbert, 1990; Williamson, 1994). The thermodynamics of tetraplex formation has been studied for several oligomer sequences (Jin et al., 1992; Lu et al., 1993). In the presence of K^+, ΔG°_{25} and ΔH° per tetrad appear to be -2 to -3 kcal mol^{-1} and -20 to -30 kcal mol^{-1}, respectively (Jin et al., 1992).

8. PREDICTING SECONDARY STRUCTURE

A major application of our knowledge of conformational changes is the prediction of nucleic acid structure from sequence. Most applications have involved RNA, but the methods are also applicable to DNA. One goal of these methods is to predict the base pairing, or secondary structure, for single-stranded chains. Experiments indicate that tertiary structure is less stable than most secondary structure (Crothers et al.,

1974; Banerjee et al., 1993; Jaeger et al., 1993; Laing and Draper, 1994). Thus as a first approximation, tertiary interactions can be neglected when predicting secondary structure. The conclusion that stabilities of small nucleic acid structures are largely determined by strong, local interactions such as stacking and hydrogen bonding suggests that summing the free energy changes for such interactions will provide an approximation for the stability of a given structure (Turner et al., 1988; Papanicolaou et al., 1984; Tinoco et al., 1971). The structure predicted to be most prevalent at equilibrium is then the one with the lowest free energy (see Eqs. 8-3 and 8-7a). It should be realized, however, that the relative concentrations predicted for various species is quite sensitive to the ΔG° values in the calculation. The predicted equilibrium constant between two species at 37°C changes by a factor of 10 for every 1.4 kcal mol^{-1} in ΔG°. Thus current predictions should be considered rough approximations.

As discussed in Section 2, stabilities of oligonucleotides without loops can be predicted well with a nearest neighbor model. Thus, this approximation provides a reasonable treatment for helical regions. Much less is known about the sequence dependence of stability for structures with loops, and current methods largely neglect this sequence dependence. This restriction can be eliminated when more experimental data become available. Even with these limitations, the loop parameters listed in Table 8-14 are useful for predicting RNA secondary structure.

If sequence dependent free energy parameters were available for every possible loop, it would still be difficult to completely predict the most stable conformation from sequence. The reason is that the time required to try every possibility is usually enormous. For example, for a sequence of N nucleotides with A, C, G, and U occurring randomly, the number of valid structures goes approximately as 1.8^N (Zuker and Sankoff, 1984). If ΔG° could be calculated for 1000 structures every second, it would take about 10^{10} years, roughly the age of the universe to try all valid possibilities for an 80 nucleotide sequence. To circumvent this problem, clever computer algorithms have been written that avoid trying every possibility (Papanicolaou et al., 1984; Zuker and Stiegler, 1981; Williams and Tinoco, 1986; Nussinov et al., 1982; Zuker, 1989; Mathews et al., 1999; Rivas and Eddy, 1999). These algorithms, however, require additional approximations. Two of the approaches are discussed below.

8.1 Combinatorial Algorithms

Combinatorial algorithms develop a list of all helixes that can be formed from a sequence (Papanicolaou et al., 1984; Gouy, 1986). These algorithms can include knotted structures. The algorithms then try combinations of these helixes in search of the lowest free energy. Various tricks are used to avoid computing combinations that are not likely to be low in free energy. Nevertheless, the time required is large because the number of helixes, L, and the number of combinations grow approximately as N^2 and 2^L, respectively.

8.2 Recursive Algorithms

Recursive (or dynamic) algorithms usually make the approximation that if two nucleotides pair, then any nucleotide between them in the primary sequence can only pair with other nucleotides between the originally paired nucleotides (Zuker and Sankoff, 1984; Williams and Tinoco, 1986; Nussinov et al., 1982). Thus knots are usually not allowed, but the other loops shown in Figure 8-11 are allowed. This approximation permits the lowest free energy structure to be found from consideration of the lowest free energy structure for each possible subfragment of the sequence. The computation for each new subfragment makes use of the computations for each smaller subfragment, which makes the algorithms recursive, and therefore fast. The time required typically grows as N^3 or N^4, depending on the generality of the free energy parameters. When knots are included in a recursive algorithm, the time grows as N^6 (Rivas and Eddy, 1999).

The most popular recursive algorithm can use more than thermodynamic parameters for deducing structure (Zuker and Stiegler, 1981; Zuker, 1989; Mathews et al., 1999). For example, if it is known that a given nucleotide is susceptible to nuclease cleavage, the predicted structure can be forced to include this. Thus experimental data can play a role in the process.

8.3 Suboptimal Structures

Except for short sequences, free energy minimization may never lead reliably to the exact secondary structure, because many approximations are mandated by experimental and computational considerations. Even if it were possible to predict the most stable structure, it is likely that other structures will be important for understanding function. For these reasons, both combinatorial and recursive algorithms have been designed to permit prediction of suboptimal structures (Steger et al., 1984; Williams and Tinoco, 1986; Zuker, 1989; Gouy, 1986). The output from these programs can then be tested against experimental data (Mathews et al., 1997) or used to design experiments aimed at testing various possibilities. If more than one sequence is known for molecules with similar functions, then comparison of suboptimal structures may help identify the true structure and features required for function (Lück et al., 1996; Mathews et al., 1997). Figure 8-16 provides an example of a sequence with two structures that are similar in free energy. Programs for folding RNA are now available on the WEB (e.g., http://rna.chem.rochester.edu, and http://www.ibc.wustl.edu/~zuker).

8.4 Evaluating Predictions

The performance of free energy minimization procedures can be tested by comparing predicted structures for various RNA sequences with those deduced from sequence comparisons. In tests of a recursive algorithm on various RNA sequences, roughly 70% of the known base pairs are present in the free energy minimized structures (Mathews et al., 1999). The best computer-generated suboptimal structure has roughly 85% of the known base pairs (Mathews et al., 1999). Considering the lack of experimental data for many free energy parameters, the required approximations, and the difficulty of the problem, the agreement between known and energy minimized structures is surprisingly good. This finding supports the assumption that secondary structure is largely determined by local interactions and is not very dependent on tertiary structure.

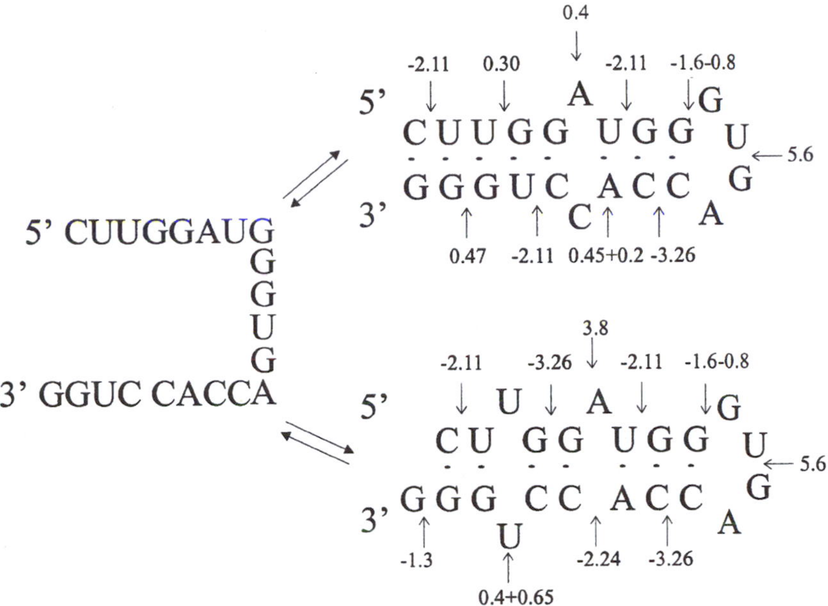

Figure 8-15
Example of an RNA sequence with two secondary structures of similar free energy. The sequence is a subfragment of 5S rRNA from *Philosamia cynthia ricini*. A program that gives suboptimal structures will predict both structures.

Presumably, performance will continue to improve as more parameters are measured experimentally.

It may also be possible to improve predictions of structure from sequence by adding considerations from the kinetics of folding. From the forward and reverse rate constants of hairpin formation, it is clear that a nucleic acid is similar to a computer algorithm in that it does not have time to try all possible pairings. Thus the pathway of folding must also be coded into the sequence. So far, only the folding of tRNA has been studied in detail (Riesner and Römer, 1973; Crothers and Cole, 1978; Crothers et al., 1974). Therefore general rules for folding are not clear yet. Preliminary attempts have been made, however, to introduce kinetics into algorithms for predicting secondary structure (Gultyaev et al., 1995; Schmitz and Steger, 1996).

9. PREDICTING THREE DIMENSIONAL STRUCTURE

Once the secondary structure of a nucleic acid is known, the next challenge is to predict the 3D structure. Good models are available for the 3D structures of double helixes (see Chapter 4), and determination of all the double helical regions greatly constrains possible 3D foldings. It is still necessary, however, to determine folding between helical regions. Presumably, this will be determined by the same factors that are important for secondary structure: conformational entropy, stacking, hydrogen bonding, counterion binding, and perhaps specific solvation. Contributions from all these factors can be seen in the crystal structures of tRNA and part of a group I intron (see Figs. 4.11–4.14 and color plate for 8-17) (Kim et al., 1974; Robertus et al., 1974; Cate et al., 1996). The limited data base available, however, makes it difficult to deduce general principles.

Since the factors governing 3D and secondary structure seem similar, it should be possible to use information from studies of conformational changes in oligonucleotides and polynucleotides to help predict 3D structure. The structure of yeast phenylalanine tRNA shown in Figure 8-17 illustrates three correlations of 3D structure with free energy changes measured for stacking in oligonucleotides. The first correlation involves single-strand stacking. Table 8-1 indicates that single-strand AA and CC sequences have considerable stacking. In the tRNA crystal structure, the sequences 35AA36 and 74CC75 are both stacked. Conversely, Table 8-1 indicates single-strand UU sequences do not stack. Dihydrouracil, D, is similar to uracil, except the pyrimidine ring is not planar. This lack of planarity should lead to even less tendency to stack. Inspection of Figure 8-17 shows that both dihydrouracils in the sequence 16DD17 are unstacked.

The second correlation involves coaxial stacking of adjacent helixes. In the tRNA crystal structure, the base pairs 7UA66 and 49CG65 are stacked on each other. This type of interaction is also associated with a large favorable free energy change in an oligonucleotide model system (Walter et al., 1994; Walter and Turner, 1994).

The third correlation is observed for stacking of unpaired nucleotides (dangling ends) adjacent to base pairs (Sugimoto et al., 1987a; Turner et al., 1988; Burkard et al., 1999). Figure 8-17 (*a*) illustrates dangling end sequences in yeast phenylalanine tRNA that have ΔG° values more favorable than -1 kcal mol in Table 8-5. These are strongly stacking sequences in oligonucleotides, and in the tRNA structure each such dangling end is stacked on its adjacent base pair. Figure 8-17 (*b*) illustrates dangling end sequences in tRNA that have ΔG° values less favorable than -0.4 kcal mol^{-1}. These weakly stacking sequences are not stacked on the adjacent base pairs in tRNA. In four out of five such cases, these weakly stacking sequences occur at places where there is a turn in the sugar-phosphate backbone. These are the only places in the structure where sharp turns occur at the ends of helical regions, which suggests weakly stacking sequences are favored at positions where the backbone turns. While limited, the results suggest free energies measured in oligonucleotides reflect the strengths of fundamental interactions that will help determine 3D as well as secondary structure.

Studies of oligonucleotides binding to natural RNAs are suggesting additional interactions that will be important for determining 3D structure. For example, interactions with 2′-OH groups, can be used for aligning helixes (Sugimoto et al., 1989; Pyle and Cech, 1991; Bevilacqua and Turner, 1991; Pyle et al., 1992). This type of helix alignment was suggested by a phylogenetic analysis of about 80 sequences (Michel and Westhof, 1990). This analysis also suggested several other tertiary interactions such as triple helix formation and interactions with tetraloops. Some have been confirmed by site directed mutagenesis (Michel et al., 1990; Costa and Michel, 1995), and by X-ray diffraction (Pley et al., 1994; Cate et al., 1996). Presumably, as more structures are determined, more patterns will be recognized. For example, pairing between complementary loops could also be important.

10. HELIX-HELIX TRANSITIONS

This chapter has focused on helix–coil transitions. Transitions are also known in which a helix changes conformation. Transitions between A, B, and Z forms have been studied. The Z form is most stable for alternating CG sequences, but variations are possible (Jovin and Soumpasis, 1987). The A and B forms can be stabilized for random sequence deoxy polymers, but the B form has not been observed for polyribonucleotides. Table 8-16 lists conditions under which various conformations are stable. Often unusual conformations are induced by conditions that affect the effective salt concentration and water activity. For example, A $\rightarrow$ Z, B $\rightarrow$ Z, and B $\rightarrow$ A transitions are all promoted by ethanol. High salt promotes A and B $\rightarrow$ Z transitions. Another similarity is that these transitions typically have small ΔH° values that are often positive (Klump and Jovin, 1987; Chaires and Sturtevant, 1986). Thus, for example, high temperature tends to promote the Z conformation. Despite extensive characterization of these transitions, there is no consensus on the quantitative contributions of various fundamental interactions. It is likely, however, that charge, solvation, and steric effects are important. There is also no clearly documented case in which these types of transitions serve a known physiological function. Discovery of functional significance is probably only a matter of time.

APPENDIX

A.1 Statistical thermodynamics of transitions

The two-state model for deriving thermodynamic parameters from spectroscopic data is not general. While short oligomers often exhibit two-state transitions, long oligomers

Table 8.16
Conditions Favoring Unusual Conformations

Transition	Reference	Cation or Salt	Solvent	T(°C)
A $\rightarrow$ Z	Cruz et al., (1986)	4.8–6 M $NaClO_4$	H_2O	35–45
[Poly(rCrG)]	Cruz et al., (1986)	4.8 M $NaClO_4$	20% EtOH	25
	Cruz et al., (1986)	5–7 M NaBr	H_2O	~20–65
	Trulson et al., (1987)	3.8 M $MgCl_2$	H_2O	~20
B $\rightarrow$ Z	Pohl and Jovin (1972)	4 M NaCl	H_2O	25
[Poly(dCdG)]	Pohl and Jovin (1972)	2.5 M $NaClO_4$	H_2O	25
	Pohl (1976); Hall and Maestre (1984)	$\sim 5 \times 10^{-4}$ M Na^+	10-50% EtOH	20-50
	Pohl and Jovin (1972)	1 M $MgCl_2$	H_2O	25
	Behe and Felsenfeld (1981)	2×10^{-5} M $Co(NH_3)_6^{3+}$	H_2O	20
B $\rightarrow$ Z	Behe and Felsenfeld (1981)	≥ 0.7 M NaCl	H_2O	~20
[Poly(dm5CdG)]	Behe and Felsenfeld (1981)	$\geq 6 \times 10^{-4} M$ $MgCl_2$	H_2O	~20
B $\rightarrow$ A, [DNA]	Ivanov et al. (1974)	$\sim 5 \times 10^{-4}$ M NaCl	80% EtOH	5–30

do not. Often terminal base pairs, base pairs adjacent to loops, or regions of weak base pairs melt first. Analysis of these non-two-state transitions requires a statistical thermodynamic model (Wada et al., 1980; Gotoh, 1983; Wartell and Benight, 1985; Bloomfield et al., 1974; Cantor and Schimmel, 1980; Poland, 1978).

In any statistical thermodynamic model, all the thermodynamic information is contained in the molecular partition function, q:

$$q = \sum_{j=0}^{n} \exp[-G_j/RT] \tag{8-A.1}$$

where G_j is the free energy of the jth configuration, and the summation is over all possible configurations. When only duplex formation is being considered, G_j for the completely single stranded state is set at 0, and the remaining G_j values are replaced by ΔG_i values, the difference in free energy between a given duplex configuration and the single-stranded state, to give

$$q = 1 + \sum_{i=1}^{n} \exp[-\Delta G_i/RT] \tag{8-A.2}$$

It is more convenient, however, to work with the conformational partition function, q_c:

$$q_c = q - 1 = \sum_{i=1}^{n} \exp[-\Delta G_i/RT] = \sum_{i=1}^{n} K_i \tag{8-A.3}$$

where K_i is the equilibrium constant for forming duplex configuration i from single strands. Any model can now be used to derive an expression for q_c. The trick is to use a model complicated enough to describe the experimental results, but simple enough to permit unambiguous extraction of the necessary parameters from data. This approach will be illustrated with the simplest model, the zipper model (Bloomfield et al., 1974; Cantor and Schimmel, 1980; Poland, 1978).

In the zipper model, only one helical region is allowed per duplex. To illustrate the process in its simplest form, we assume that the equilibrium constant for adding one base pair to an existing helix is always the same, s; that the equilibrium constant for initiating the helix is κ (often labeled σs in the literature); and that only perfectly aligned helixes are stable enough to contribute to q_c. These approximations are adequate for a sequence such as $G_3C + GC_3$ (see Figure 8-18). With these assumptions, the equilibrium constant for forming each duplex configuration with j base pairs from the single strands is κs^{j-1}. If there are $g_j(N)$ distinguishable duplexes with j base pairs that can be formed by a single strand with N bases, then Eq. 8-A.3 can be written

$$q_c = \kappa \sum_{j=1}^{N} g_j(N) s^{j-1} \tag{8-A.4}$$

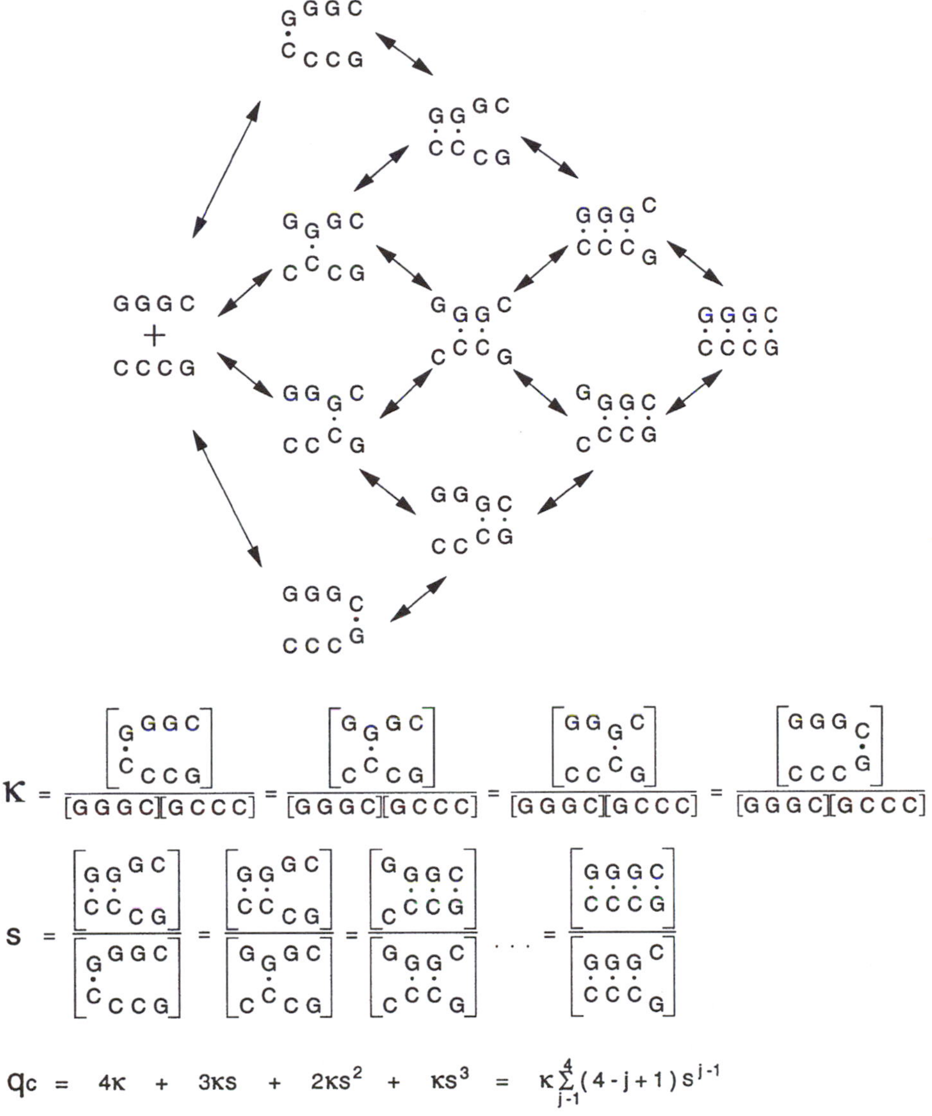

Figure 8-18
Application of the zipper model to calculate the conformational partition function, q_c, for duplex formation by G_3C+GC_3. The zipper model allows only one helical region per duplex.

When the two strands in the duplex have different sequence (i.e., are nonself-complementary), and only completely aligned duplexes are allowed (see Fig. 8-18), then $g_j(N) = N - j + 1$, giving

$$q_c = \kappa \sum_{j=1}^{N} (N-j+1)s^{j-1} = \kappa[(N+1)\sum_{j=1}^{N} s^{j-1} - \sum_{j=1}^{N} js^{j-1}] \qquad \text{(8-A.5)}$$

These are related to the finite geometric series, for which it can be shown that

$$\sum_{j=1}^{N} s^{j-1} = \frac{s^N - 1}{s - 1} \tag{8-A.6}$$

$$\sum_{j=1}^{N} js^{j-1} = \frac{Ns^{N+1} - (N+1)s^N + 1}{(s-1)^2} \tag{8-A.7}$$

so

$$q_c = \kappa \left(\frac{(N+1)(s^N - 1)}{s - 1} - \frac{Ns^{N+1} - (N+1)s^N + 1}{(s-1)^2} \right) = \kappa \frac{s^{N+1} - (N+1)s + N}{(s-1)^2} \tag{8-A.8}$$

Note that if $s > 1$ and N is large, $q_c \approx \kappa s^{N-1}$. Thus, in this limit q_c is simply the equilibrium constant for the two-state model.

To determine q_c it is necessary to measure κ and s. One way to do this is by analyzing optical melting curves. In the simplest case, the concentrations of the two complementary strands are equal, and the absorbance A of a sample depends only on the number of base pairs in the sample. For this case, the fraction X_b of bases paired is given by

$$X_b = \frac{A - A_s}{A_d - A_s} \tag{8-A.9}$$

Here A_s and A_d are absorbances when the sample is completely single strands or fully paired duplexes, respectively. (Note the similarity to Eq. 8-14, with α corresponding to X_b.) We need to express X_b in terms of q_c. For nonself-complementary strands, A and B, the equilibrium can be written

$$A + B \rightleftharpoons C_1 + C_2 + \cdots + C_n \tag{8-A.10}$$

where $C_1, C_2, \ldots, C_n$ are the possible configurations of duplex (see Fig. 8-18). Assuming the total concentrations of strands A and B are equal.,

$$q_c = \frac{[C_1] + [C_2] + \ldots + [C_n]}{[A][B]} = \frac{0.5XC_T}{[0.5(1-X)C_T]^2} = \frac{2X}{(1-X)^2 C_T} \tag{8-A.11}$$

where C_T is the total concentration of strands,

$$C_T = [A] + [B] + 2\sum_{i=1}^{n} [C_i] \tag{8-A.12}$$

and X is the fraction of strands in duplex,

$$X = 2([C_1] + [C_2] + \cdots + [C_n])/C_T \tag{8-A.13}$$

The average number of base pairs per duplex, $< n >$, can be calculated by realizing that the fraction of duplexes with j base pairs, f_j, is equal to $\kappa g_j(N)s^{j-1}$, normalized by the sum over all j, q_c. Thus

$$\begin{aligned} < n > &= \sum_{j=1}^{N} jf_j = \frac{1}{q_c}\sum_{j=1}^{N} j\kappa g_j(N)s^{j-1} \\ &= \frac{1}{q_c}\frac{d}{ds}\sum_{j=1}^{N}\kappa g_j(N)s^j = \frac{1}{q_c}\frac{d(sq_c)}{ds} \end{aligned} \tag{8-A.14}$$

This equation allows X_b to be expressed in terms of X and q_c:

$$X_b = \frac{X < n >}{N} = \frac{X}{Nq_c}\frac{d(sq_c)}{ds} \tag{8-A.15}$$

Solving Eq. 8-A.11 for X in terms of q_c leads to

$$X_b = \frac{1 + q_c C_T - \sqrt{1 + 2q_c C_T}}{Nq_c^2 C_T}\frac{d(sq_c)}{ds} \tag{8-A.16}$$

By using Eq. 8-A.9, X_b can be measured as a function of C_T and/or N, and the resulting simultaneous equations solved for κ and s.

While the above example illustrates the main features of the statistical approach, it is not general. For intramolecular helix formation with the same assumptions, the theory is similar (Bloomfield et al., 1974; Cantor and Schimmel, 1980). For most real cases, however, it is more complicated. Most sequences do not have the same s for each base pair (e.g., see Table 8-4). Many duplexes contain loops of various kinds so that more than one helical region is present. In this case, additional helix initiation parameters must be added that are different from κ. In general, the absorbance will not depend only on the number of base pairs (Poland, 1978). Thus X_b in Eq. 8-A.9 must be replaced by a more complicated function. Clearly, the number of parameters increases rapidly for real systems, which explains the relatively rare application of statistical models.

References

Aboul-ela, F., Koh, D., Tinoco, I., Jr., and Martin, F. H. (1985). Base–base Mismatches. Thermodynamics of Double Helix Formation for $dCA_3XA_3G + dCT_3YT_3G$ (X,Y = A,C,G,T), *Nucleic Acids Res.* **13**, 4811–4824.

Adler, A. J., Grossman, L., and Fasman, G. D. (1969). Polyriboadenylic and Polydeoxyriboadenylic Acids. Optical Rotatory Studies of pH-Dependent Conformations and Their Relative Stability, *Biochemistry* **8**, 3846–3859.

Albergo, D. D., Marky, L. A., Breslauer, K. J., and Turner, D. H. (1981). Thermodynamics of (dG-dC)$_3$ Double-Helix Formation in Water and Deuterium Oxide, *Biochemistry* **20**, 1409–1413.

Albergo, D. D. and Turner, D. H. (1981). Solvent Effects on Thermodynamics of Double-Helix Formation in (dG-dC)$_3$, *Biochemistry* **20**, 1413–1418.

Allawi, H. T. and SantaLucia, J., Jr. (1997). Thermodynamics and NMR of Internal G·T Mismatches in DNA, *Biochemistry* **36**, 10581–10594.

Allawi, H. T. and SantaLucia, J., Jr. (1998). Nearest Neighbor Thermodynamic Parameters for Internal GA Mismatches, *Biochemistry* **37**, 2170–2179.

Antao, V. P. and Tinoco, I., Jr. (1992). Thermodynamic parameters for loop formation in RNA and DNA hairpin tetraloops, *Nucleic Acids Res.* **20**, 819–824.

Applequist, J. (1963). On the Helix-Coil Equilibrium in Polypeptides, *J. Chem. Phys.* **38**, 934–941.

Arnott, S., Chandrasekaran, R., Hall, I. H., Puigjaner, L. C., Walker, J. K. and Wang, M. (1982). DNA Secondary Structures: Helices, Wrinkles, and Junctions, *Cold Spring Harbor Symp. Quant. Biol.* **47**, 53–65.

Banerjee, A. R., Jaeger, J. A., and Turner, D. H. (1993). Thermal unfolding of a group I ribozyme: The low-temperature transition is primarily disruption of tertiary structure, *Biochemistry* **32**, 153–163.

Battiste, J. L., Tan, R., Frankel, A. D., and Williamson, J. R. (1994). Binding of an HIV Rev peptide to Rev responsive element RNA induces formation of purine-purine base pairs, *Biochemistry* **33**, 2741–2747.

Behe, M. and Felsenfeld, G. (1981). Effects of methylation on a synthetic polynucleotide: The B–Z transition in poly(dG-m5dC):poly(dG-m5dC), *Proc. Natl. Acad. Sci. USA* **78**, 1619–1623.

Berman, H. M. (1981). Conformational Principles of Nucleic Acids, in *Topics in Nucleic Acid Structure*, Part 1, Chapter 1 Neidle, S., Ed., Macmillan, London, UK.

Bernasconi, C. F. (1976). *Relaxation Kinetics*, Academic Press, New York.

Bevilacqua, P. C., Johnson, K. A., and Turner, D. H. (1993). Cooperative and Anticooperative Binding to a Ribozyme, *Proc. Natl. Acad. Sci. USA* **90**, 8357–8361.

Bevilacqua, P. C., Kierzek, R., Johnson, K. A., and Turner, D. H. (1992). Dynamics of Ribozyme Binding of Substrate Revealed by Fluorescence-Detected Stopped-Flow Methods, *Science* **258**, 1355–1358.

Bevilacqua, P. C., Li, Y., and Turner, D. H. (1994). Fluorescence-Detected Stopped Flow with a Pyrene Labeled Substrate Reveals that Guanosine Facilitates Docking of the 5′ Cleavage Site into a High Free Energy Binding Mode in the *Tetrahymena* Ribozyme, *Biochemistry* **33**, 11340–11348.

Bevilacqua, P. C. and Turner, D. H. (1991). Comparison of Binding of Mixed Ribose-Deoxyribose Analogues of CUCU to a Ribozyme and to GGAGAA by Equilibrium Dialysis: Evidence for Ribozyme Specific Interactions with 2′ OH Groups, *Biochemistry* **30**, 10632–10640.

Blagoi, Y. P., Sorokin, V. A., Valeyev, V. A., Khomenko, S. A., and Gladchenko, G. O. (1978). Magnesium Ion Effect on the Helix–Coil Transition of DNA, *Biopolymers* **17**, 1103–1118.

Blake, R. D. (1996). Denaturation of DNA, in *Encyclopedia of Molecular Biology and Molecular Medicine*, Vol. 2, Meyers, R. A., Ed. VCH, New York, 1–19.

Blake, R. D. and Delcourt, S. G. (1987). Loop Energy in DNA, *Biopolymers* **26**, 2009– 2026.

Blake, R. D. and Haydock, P. V. (1979). Effect of Sodium Ion on the High-Resolution Melting of Lambda DNA, *Biopolymers* **18**, 3089–3109.

Bloomers, M. J. J., Walters, J. A. L. I., Haasnoot, C. A. G., Aelen, J. M. A., van der Marel, G. A., van Boom, J. J., and Hilbers, C. W. (1989). Effects of Base Sequence on the Loop Folding in DNA Hairpins, *Biochemistry* **28**, 7491–7498.

Bloomfield, V. A., Crothers, D. M. and Tinoco, I., Jr. (1974). *Physical Chemistry of Nucleic Acids*, Chapter 6, Harper and Row, New York.

Borer, P. N., Dengler, B., Tinoco, I., Jr., and Uhlenbeck, O. C. (1974). Stability of Ribonucleic Acid Double-Stranded Helices, *J. Mol. Biol.* **86**, 843–853.

Breslauer, K. J. and Bina-Stein, M. (1977). Relaxation Kinetics of the Helix–Coil Transition of a Self-Complementary Ribo-Oligonucleotide: A_7U_7, *Biophysical Chem.* **7**, 211–216.

Breslauer, K. J., Frank, R., Blöcker, H., and Marky, L. A. (1986). Predicting DNA Duplex Stability from the Base Sequence, *Proc. Natl. Acad. Sci. USA* **83**, 3746–3750.

Breslauer, K. J., Freire, E., and Straume, M. (1992). Calorimetry: A Tool for DNA and Ligand-DNA Studies, *Methods Enzymol.* **211**, 533–567.

Broitman, S. L., Im, D. D., and Fresco, J. R. (1987). Formation of the triple-stranded polynucleotide helix, poly (A:A:U), *Proc. Natl. Acad. Sci. USA* **84**, 5120–5124.

Burkard, M. E., Kierzek, R., and Turner, D. H. (1999). Thermodynamics of Unpaired Terminal Nucleotides on Short RNA Helixes Correlates with Stacking at Helix Termini in Larger RNAs, *J. Mol. Biol.* **290**, 967-982.

Calladine, C. R. (1982). Mechanics of Sequence-Dependent Stacking of Bases in B-DNA, *J. Mol. Biol.* **161**, 343–352.

Cantor, C. R. and Schimmel, P. R. (1980). *Biophysical Chemistry*, Part III, Chapter 23, Freeman, San Francisco, CA.

Cate, J. H., Gooding, A. R., Podell, E., Zhou, K., Golden, B. L., Kundrot, C. E., Cech, T. R., and Doudna, J. A. (1996). Crystal Structure of a Group I Ribozyme Domain: Principles of RNA Packing, *Science* **273**, 1678–1685.

Cate, J. H., Hanna, R. L., and Doudna, J. A. (1997). A Magnesium Ion Core at the Heart of a Ribozyme Domain, *Nature Struct. Biol.* **4**, 553–558.

Cech, T. R. (1990). Self-splicing of Group I Introns, *Annu. Rev. Biochem.* **59**, 543–568.

Cech, T. R. and Bass, B. L. (1986). Biological Catalysis By RNA, *Annu. Rev. Biochem.* **55**, 599–629.

Cech, T. R., Damberger, S. H., and Gutell, R. R. (1994). Representation of the Secondary and Tertiary Structure of Group I Introns, *Nature Struct. Biol.* **1**, 273–280.

Chaires, J. B. and Sturtevant, J. M. (1986). Thermodynamics of the B to Z transition in poly (m5dG-dC), *Proc. Natl. Acad. Sci. USA* **83**, 5479–5483.

Chastain, C. M. and Tinoco, I., Jr. (1991). Structural elements in RNA, *Prog. Nucleic Acid Res. Mol. Biol.* **41**, 131–177.

Chastain, C. M. and Tinoco, I., Jr. (1992). Poly(rA) Binds Poly(rG)·Poly(rC) to Form a Triple Helix, *Nucleic Acids Res.* **20**, 315–318.

Chee, M., Yang, R., Hubbell, E., Berno, A., Huang, X. C., Stern, D., Winkler, J., Lockhart, D. J., Morris, M. S., and Fodor, S. P. A. (1996). Accessing Genetic Information with High-Density DNA Arrays, *Science* **274**, 610–614.

Cheong, C., Tinoco, I., Jr., and Chollet, A. (1988). Thermodynamic Studies of Base Pairing Involving 2,6-Diaminopurine, *Nucleic Acids Res.* **16**, 5115–5122.

Christophe, D., Cabrer, B., Bacolla, A., Targovnik, H., Pohl, V., and Vassart, G. (1985). An unusually long poly(purine)-poly(pyrimidine) sequence is located upstream from the human thyroglobulin gene, *Nucleic Acids Res.* **13**, 5127–5144.

Chu, Y. G. and Tinoco, I., Jr. (1983). Temperature-Jump Kinetics of the dC-G-T-G-A-A-T-T-C-G-C-G Double Helix Containing a G:T Base Pair and the dC-G-C-A-G-A-A-T-T-C-G-C-G Double Helix Containing an Extra Adenine, *Biopolymers* **22**, 1235-1246.

Clore, G. M., Gronenborn, A. M., Piper, E. A., McLaughlin, L. W., Graeser, E., and Van Boom, J. H. (1984). The Solution Structure of a RNA Pentadecamer Comprising the Anticodon Loop and Stem of Yeast $tRNA^{Phe}$, *Biochem. J.* **221**, 737–751.

Cooney, M., Czernuszewicz, G., Postel, E. H., Flint, S. J., and Hogan, M. E. (1988). Site-Specific Oligonucleotide Binding Represses Transcription of the Human c-myc Gene in Vitro, *Science* **241**, 456–459.

Costa, M. and Michel, F. (1995). Frequent Use of the Same Tertiary Motif by Self-Folding RNAs, *EMBO J.* **14**, 1276–1285.

Coutts, S. M., Gangloff, J., Dirheimer, G. (1974). Conformational Transitions in $tRNA^{Asp}$ (Brewer's Yeast). Thermodynamic, Kinetic, and Enzymatic Measurements on Oligonucleotide Fragments and the Intact Molecule, *Biochemistry* **13**, 3938–3948.

Coutts, S. M., (1971). Thermodynamics and Kinetics of G:C Base Pairing in the Isolated Extra Arm of Serine-Specific Transfer RNA from Yeast, *Biochem. Biophys. Acta* **232**, 94–106.

Craig, M. E., Crothers, D. M., and Doty, P. (1971). Relaxation Kinetics of Dimer Formation by Self Complementary Oligonucleotides, *J. Mol. Biol.* **62**, 383–401.

Crick, F. H. C. (1966). Codon--Anticodon Pairing: The Wobble Hypothesis, *J. Mol. Biol.* **19**, 548–555.

Crothers, D. M. and Cole, P. E. (1978). In *Transfer RNA*, Altman, S., Ed., MIT Press, Cambridge, MA, pp. 196–247.

Crothers, D. M., Cole, P. E., Hilbers, C. W., and Shulman, R. G. (1974). The Molecular Mechanism of Thermal Unfolding of *Escherichia coli* Formylmethionine Transfer RNA, *J. Mol. Biol.* **87**, 63–88.

Cruz, P., Hall, K., Puglisi, J., Davis, P., Hardin, C. C., Trulson, M. O., Mathies, R. A., Tinoco, I., Jr., Johnson, W. C., Jr., and Neilson, T. (1986). The Left-Handed Z-form of Double Stranded RNA, in *Biomolecular Stereodynamics IV*, Sarma, R. H. and Sarma, M. H., Eds., Adenine Press, Albany, NY, pp. 179–200.

Davis, R. C. and Tinoco, I., Jr. (1968). Temperature-dependent Properties of Dinucleoside Phosphates, *Biopolymers* **6**, 223–242.

Delcourt, S. G. and Blake, R. D. (1991). Stacking Energies in DNA, *J. Biol. Chem.* **266**, 15160–15169.

Dewey, T. G., Raymond, D. A., and Turner, D. H. (1979). Laser Temperature Jump Study of Solvent Effects on Proflavin Stacking, *J. Am. Chem. Soc.* **101**, 5822– 5826.

Dewey, T. G. and Turner, D. H. (1979). Laser Temperature-Jump Study of Stacking in Adenylic Acid Polymers, *Biochemistry* **18**, 5757–5762.

Dickerson, R. E. (1983). Base Sequence and Helix Structure Variation in B and A DNA, *J. Mol. Biol.* **166**, 419–441.

Dickerson, R. E., Drew, H. R., Conner, B. N., Kopka, M. L., and Pjura, P. E. (1982). Helix Geometry and Hydration in A-DNA, B-DNA, and Z-DNA, *Cold Spring Harbor Symp. Quant. Biol.* **47**, 13–24.

Dieckmann, T., Suzuki, E., Nakamura, G. K., and Feigon, J. (1996). Solution Structure of an ATP-binding RNA Aptamer Reveals a Novel Fold, *RNA*, **2**, 628–640.

Doktycz, M. J., Goldstein, R. F., Paner, T. M., Gallo, F. J., and Benight, A. S. (1992). Studies of DNA Dumbbells. I. Melting Curves of 17 DNA Dumbbells with Different Duplex Stem Sequences Linked by T4 Endloops: Evaluation of the Nearest-Neighbor Stacking Interactions in DNA, *Biopolymers* **32**, 849–864.

Edwards, E. L., Ratliff, R. L., and Gray, D. M. (1988). Circular Dichroism Spectra of DNA Oligomers Show that Short Interior Stretches of $C:C^+$ Base Pairs Do Not Form in Duplexes with A:T Base Pairs, *Biochemistry* **27**, 5166–5174.

Eigen, M. and De Maeyer, L. (1963). in *Techniques of Organic Chemistry*, Vol. VIII, Part 2, Friess, S. L., Lewis, E. S. and Weissberger, A., Eds., Wiley-Interscience, New York, p. 895.

Erie, D. A., Jones, R. A., Olson, W. K., Sinha, N. K., and Breslauer, K. J. (1989). Melting Behavior of a Covalently Closed, Single-Stranded, Circular DNA, *Biochemistry* **28** , 268–273.

Erie, D., Sinha, N., Olson, W., Jones, R., and Breslauer, K. J. (1987). A Dumbbell-Shaped, Double-Hairpin Structure of DNA: A Thermodynamic Investigation, *Biochemistry* **26**, 7150–7159.

Fan, P., Suri, A. K., Fiala, R., Live, D., and Patel, D. J. (1996). Molecular Recognition in the FMN–RNA Aptamer Complex, *J. Mol. Biol.* **258**, 480–500.

Felsenfeld, G., Davies, D. R., and Rich, A. (1957). Formation of a Three-Stranded Polynucleotide Molecule, *J. Am. Chem. Soc.* **79**, 2023–2024.

Felsenfeld, G. and Todd Miles, H. (1967). The Physical and Chemical Properties of Nucleic Acids, *Annu. Rev. Biochem.* **36**, 407–448.

Filimonov, V. V. and Privalov, P. L. (1978). Thermodynamics of Base Interaction in $(A)_n$ and $(A:U)_n$, *J. Mol. Biol.* **122**, 465–470.

Fink, T.R. and Crothers, D. M. (1972). Free Energy of Imperfect Nucleic Acid Helices. I. The Bulge Defect, *J. Mol. Biol.* **66**, 1–12.

Fisher, M. E. (1966). Effect of Excluded Volume on Phase Transitions in Biopolymers, *J. Chem. Phys.* **45**, 1469–1473.

Flor, P., Flanegan, J. B., and Cech, T. R. (1989). A Conserved Base-pair within Helix P4 of the Tetrahymena Ribozyme Helps to Form the Tertiary Structure Required for Self-Splicing, *EMBO J.* **8**, 3391–3399.

Frank-Kamenetskii, M. D. (1971). Simplification of the Empirical Relationship between Melting Temperatures of DNA, Its GC Content and Concentration of Sodium Ions in Solution, *Biopolymers* **10**, 2623–2624.

Freier, S. M., Albergo, D. D., and Turner, D. H. (1983a). Solvent Effects on the Dynamics of $(dG\text{-}dC)_3$, *Biopolymers* **22**, 1107–1131.

Freier, S. M., Alkema, D., Sinclair, A., Neilson, T., and Turner, D. H. (1985). Contributions of Dangling End Stacking and Terminal Base-Pair Formation to the Stabilities of XGGCCp, XCCGGp, XGGCCYp, and XCCGGYp Helixes, *Biochemistry* **24**, 4533–4539.

Freier, S. M., Burger, B. J., Alkema, D., Neilson, T., and Turner, D. H. (1983). Effects of 3′ Dangling End Stacking on the Stability of GGCC and CCGG Double Helices, *Biochemistry* **22**, 6198–6206.

Freier, S. M., Hill, K. O., Dewey, T. G., Marky, L. A., Breslauer, K. J., and Turner, D. H. (1981). Solvent Effects on the Kinetics and Thermodynamics of Stacking in Polylcytidylic Acid, *Biochemistry* **20**, 1419–1426.

Freier, S. M., Kierzek, R., Caruthers, M. H., Neilson, T., and Turner, D. H. (1986). Free Energy Contributions of G:U and Other Terminal Mismatches to Helix Stability, *Biochemistry* **25**, 3209–3213.

Freier, S. M., Kierzek, R., Jaeger, J. A., Sugimoto, N., Caruthers, M. H., Neilson, T., and Turner, D. H. (1986b). Improved Free-Energy Parameters for Predictions of RNA Duplex Stability, *Proc. Natl. Acad. Sci. USA* **83**, 9373–9377.

Freier, S. M., Sinclair, A., Neilson, T., and Turner, D. H. (1985b). Improved Free Energies for G:C Base-Pairs, *J. Mol. Biol.* **185**, 645–647.

Freier, S. M., Sugimoto, N., Sinclair, A., Alkema, D., Neilson, T., Kierzek, R., Caruthers, M. H., and Turner, D. H. (1986c). Stability of XGCGCp, GCGCYp, and XGCGCYp Helixes: An Empirical Estimate of the Energetics of Hydrogen Bonds in Nucleic Acids, *Biochemistry* **25**, 3214–3219.

Gaffney, B. L. and Jones, R. A. (1989). Thermodynamic Comparison of the Base Pairs Formed by the Carcinogenic Lesion O6-Methylguanine with Reference both to Watson–Crick Pairs and to Mismatched Pairs, *Biochemistry* **28**, 5881–5889.

Gaffney, B. L., Marky, L. A., and Jones, R. A. (1984). The Influence of the Purine 2-Amino Group on DNA Conformation and Stability–II. Synthesis and Physical Characterization of d[CGT(2-NH_2)ACG], d[CGU(2-NH_2)ACG], and d[CGT(2-NH_2)AT(2-NH_2)ACG], *Tetrahedron* **40**, 3–13.

Gellert, M., Lipsett, M. N., and Davies, D. R. (1962). Helix Formation by Guanylic Acid, *Proc. Natl. Acad. Sci. USA* **48**, 2013–2018.

Gellert, M., Mizuuchi, K., O'Dea, M. H., Ohmori, H., and Tomizawa, J. (1979). DNA Gyrase and DNA Supercoiling, *Cold Spring Harbor Symp. Quant. Biol.* **43**, 35–40.

Goldstein, R. F. and Benight, A. S. (1992). How many Numbers are Required to Specify Sequence-Dependent Properties of Polynucleotides? *Biopolymers* **32**, 1679–1693.

Gotoh, O. (1983). Prediction of Melting Profiles and Local Helix Stability for Sequenced DNA, *Adv. Biophys.* **16**, 1–52.

Gouy, M. (1986). Secondary Stucture Prediction of RNA, in *Nucleic Acid and Protein Sequence Analysis: A Practical Approach*, Bishop, M. J. and Rawlings, C. J., Eds., IRL Press, Oxford, UK, pp. 259–284.

Gralla, J. and Crothers, D. M. (1973a). Free Energy of Imperfect Nucleic Acid Helices II. Small Hairpin Loops, *J. Mol. Biol.* **73**, 497–511.

Gralla, J. and Crothers, D. M. (1973b). Free Energy of Imperfect Nucleic Acid Helices, *J. Mol. Biol.* **78**, 301–319.

Gray, D. M. (1997a). Derivation of Nearest-Neighbor Properties from Data on Nucleic Acid Oligomers. I. Simple Sets of Independent Sequences and the Influence of Absent Nearest Neighbors, *Biopolymers* **42**, 783–793.

Gray, D. M. (1997b). Derivation of Nearest-Neighbor Properties from Data on Nucleic Acid Oligomers. II. Thermodynamic Parameters of DNA-RNA Hybrids and DNA Duplexes, *Biopolymers* **42**, 795–810.

Gray, D. M., Cui, T., and Ratliff, R. L. (1984). Circular dichroism measurements show that C:C^+ base pairs can coexist with A:T base pairs between antiparallel strands of an oligodeoxynucleotide double-helix, *Nucleic Acids Res.* **12**, 7565–7580.

Gray, D. M., Vaughn, M., Ratliff, R. L., and Hayes, F. N. (1980). Circular dichorism spectra show that repeating dinucleotide DNAs may form helices in which every other base is looped out, *Nucleic Acids Res.* **8**, 3695–3707.

Griffin, L. C. and Dervan, P. B. (1989). Recognition of Thymine•Adenine Base Pairs by Guanine in a Pyrimidine Triple Helix Motif, *Science* **245**, 967–971.

Groebe, D. R. and Uhlenbeck, O. C. (1988). Characterization of RNA Hairpin Loop Stability, *Nucleic Acids Res.* **16**, 11725–11735.

Groebe, D. R. and Uhlenbeck, O. C. (1989). Thermal Stability of RNA Hairpins Containing a Four-Membered Loop and a Bulge Nucleotide, *Biochemistry* **28**, 742–747.

Gultyaev, A. P., van Batenburg, F. H. D., and Pleij, C. W. A. (1995). The Computer Simulation of RNA Folding Pathways using a Genetic Algorithm *J. Mol. Biol.* **250**, 37–51.

Guschlbauer, W. (1975). Protonated polynucleotide structures. Thermodynamics of the melting of the acid form of polycytidylic acid, *Nucleic Acids Res.* **2**, 353–360.

Gutell, R. R., Larsen, N., and Woese, C. R. (1994). Lessons from an Evolving RNA: 16S and 23S rRNA Structures from a Comparative Perspective, *Microbiol. Rev*. **58**, 10–26.

Hall, K. B. and Maestre, M. F. (1984). Temperature-Dependent Reversible Transition of Poly(dCdG): Poly(dCdG) in Ethanolic and Methanolic Solutions, *Biopolymers* **23**, 2127–2139.

Hamaguchi, K. and Geiduschek, E. P. (1962). The Effect of Electrolytes on the Stability of the Deoxyribonucleate Helix, *J. Am. Chem. Soc.* **84**, 1329–1338.

Hardin, C. C., Henderson, E., Watson, T., and Prosser, J. R. (1991). Monovalent Cation Induced Structural Transitions in Telomeric DNAs: G-DNA Folding Intermediates, *Biochemistry* **30**, 4460–4472.

Hare, D., Shapiro, L. and Patel, D. J. (1986). Extrahelical Adenosine Stacks into Right-Handed DNA: Solution Conformation of the d(C-G-C-A-G-A-G-C-T-C-G-C-G) Duplex Deduced from Distance Geometry Analysis of Nuclear Overhauser Effect Spectra, *Biochemistry* **25**, 7456–7464.

Hartman, K. A., Jr. and Rich, A. (1965). The Tautomeric Form of Helical Polyribocytidylic Acid, *J. Am. Chem. Soc.* **87**, 2033–2039.

He, L., Kierzek, R., SantaLucia, J., Jr., Walter, A. E., and Turner, D. H. (1991). Nearest-Neighbor Parameters of G.U Mismatches: $\frac{5'GU3'}{3'UG5'}$ Is Destabilizing in the Contexts $\begin{matrix}\text{CGUG}\\ \bullet\ \ \bullet\\ \text{GUGC}\end{matrix}$, $\begin{matrix}\text{UGUA}\\ \bullet\ \ \bullet\\ \text{AUGU}\end{matrix}$, and $\begin{matrix}\text{AGUU}\\ \bullet\ \ \bullet\\ \text{UUGA}\end{matrix}$, but Stabilizing in $\begin{matrix}\text{GGUC}\\ \bullet\ \ \bullet\\ \text{CUGG}\end{matrix}$, *Biochemistry* **30**, 11124–11132.

Herschlag, D. and Cech, T. R. (1990). Catalysis of RNA Cleavage by the Tetrahymena thermophila Ribozyme. A. Kinetic Description of the Reaction of an RNA Substrate Complementary to the Active Site, *Biochemistry* **29**, 10159–10171.

Herskovits, T. T. and Bowen, J. J. (1974). Solution Studies of the Nucleic Acid Bases and Related Model Compounds. Solubility in Aqueous Urea and Amide Solutions, *Biochemistry* **13**, 5474–5483.

Herskovits, T. T. and Harrington, J. P. (1972). Solution Studies of the Nucleic Acid Bases and Related Model Compounds. Solubility in Aqueous Alcohol and Glycol Solutions, *Biochemistry* **11**, 4800–4811.

Heus, H. A. and Pardi, A. (1991). Structural Features that give Rise to the Unusual Stability of RNA Hairpins Containing GNRA Loops, *Science* **253**, 191–194.

Hickey, D. R. and Turner, D. H. (1985a). Effects of Terminal Mismatches on RNA Stability: Thermodynamics of Duplex Formation for ACCGGGp, ACCGGAp, and ACCGGCp, *Biochemistry* **24**, 3987–3991.

Hickey, D. R. and Turner, D. H. (1985b). Solvent Effects on the Stability of A_7U_7p, *Biochemistry* **24**, 2086–2094.

Hilbers, C. W., Haasnoot, C. A. G., de Bruin, S. H., Joordens, J. J. M., Van Der Marel, G. A., and Van Boom, H. H. (1985). Hairpin Formation in Synthetic Oligonucleotides, *Biochimie* **67**, 685–695.

Hilbers, C. W., Heus, H. A., van Dongen, M. J. P., and Wijmenga, S. S. (1994). The Hairpin Elements of Nucleic Acid Structure: DNA and RNA Folding, *Nucleic Acids Mol. Biol.* **8**, 56–104.

Hoggett, J. G. and Maass, G. (1971). Thermodynamics and Kinetics of the Formation of a Hybrid Double-Helix of Oligomers of Riboadenylic and Deoxyribothymidylic Acids of Definite Chain Lengths, *Ber. Bunsengers Phys. Chem.* **75**, 45–54.

Holcomb, D. N. and Timascheff, S. N. (1968). Temperature Dependence of the Hydrogen Ion Equilibria in Poly(riboadenylic Acid), *Biopolymers* **6**, 513–529.

Htun, H. and Dahlberg, J. E. (1988). Single Strands, Triple Strands, and Kinks in H-DNA, *Science* **241**, 1791–1796.

Hud, N. V., Smith, F. W., Aret, F. A. L., and Feigon, J. (1996). The Selectivity for K^+ versus Na^+ in DNA Quadruplexes is Dominated by Relative Free Energies of Hydration: A Thermodynamic Analysis by ^{1}H NMR, *Biochemistry* **35**, 15383–15390.

Inman, R. B. (1964). Transitions of DNA Homopolymers, *J. Mol. Biol.* **9**, 624–637.

Inners, L. D. and Felsenfeld, G. (1970). Conformation of Polyribouridylic Acid in Solution, *J. Mol. Biol.* **50**, 373–389.

Ivanov, V. I., Minchenkova, L. E., Minyat, E. E., Frank- Kamenetskii, M. D., and Schyolkina, A. K. (1974). The B to A Transition of DNA in Solution, *J. Mol. Biol.* **87**, 817–833.

Jacobson, H. and Stockmayer, W. H. (1950). Intramolecular Reaction in Polycondensations, I. The Theory of Linear Systems, *J. Chem. Phys.* **18**, 1600–1606.

Jaeger, J. A., Turner, D. H., and Zuker, M. (1989). Improved Predictions of Secondary Structures for RNA, *Proc. Natl. Acad. Sci. USA* **86**, 7706–7710.

Jaeger, J. A., Zuker, M., and Turner, D. H. (1990). Melting and Chemical Modification of a Cyclized Self-Splicing Group I Intron: Similarities of Structures in 1 *M* Na^+, in 10 m*M* Mg^{2+}, and in the Presence of Substrate, *Biochemistry*, **29**, 10148–10158.

Jaeger, J. A., Westhof, E. and Michel, F. (1993). Monitoring of the cooperative unfolding of the sunY group I intron of bacteriophage T4, *J. Mol. Biol.* **234**, 331–346.

Jencks, W. P. (1975). Binding Energy, Specificity, and Enzymic Catalysis: The Circe Effect, *Adv. Enzymol.* **43**, 219–410.

Jiang, F., Kumar, R. A., Jones, R. A., and Patel, D. J. (1996). Structural Basis of RNA Folding and Recognition in an AMP–RNA Aptamer Complex, *Nature (London)* **382**, 183–86.

Jin, R., Chapman, W. H., Jr., Srinivasan, A. R., Olson, W. K., Breslow, R., and Breslauer, K. J. (1993). Comparative Spectroscopic, Calorimetric, and Computational Studies of Nucleic Acid Complexes with 2′,5″ versus 3′,5″-Phosphodiester Linkages, *Proc. Natl. Acad. Sci. USA* **90**, 10568–10572.

Jin, R., Gaffney, B. L., Wang, C., Jones, R. A., and Breslauer, K. J. (1992). Thermodynamics and Structure of a DNA Tetraplex: A Spectroscopic and Calorimetric Study of the Tetramolecular Complexes of $d(TG_3T)$ and $d(TG_3T_2G_3T)$, *Proc. Natl. Acad. Sci. USA* **89**, 8832–8836.

Johnston, B. H. (1988). The S1-Sensitive Form of $d(C\text{-}T)_n{:}d(A\text{-}G)_n$: Chemical Evidence for a Three-Stranded Structure in Plasmids, *Science* **241**, 1800–1804.

Jovin, T. M. and Soumpasis, D. M. (1987). The Transition Between B-DNA and Z-DNA, *Annu. Rev. Phys. Chem.* **38**, 521–560.

Jucker, F. M. and Pardi, A. (1995). Solution Structure of the CUUG Hairpin Loop: A Novel RNA Tetraloop Motif, *Biochemistry* **34**, 14416–14427.

Kao, T. H. and Crothers, D. M. (1980). A proton-coupled conformational switch of Escherichia coli 5S ribosomal RNA, *Proc. Natl. Acad. Sci. USA* **77**, 3360–3364.

Kierzek, R., Caruthers, M. H., Longfellow, C. E., Swinton, D., Turner, D. H., and Freier, S. M. (1986). Polymer-Supported RNA Synthesis and Its Application to Test the Nearest-Neighbor Model for Duplex Stability, *Biochemistry* **25**, 7840–7846.

Kierzek, R., He, L., and Turner, D. H. (1992). Association of 2′-5′ Oligoribonucleotides, *Nucleic Acids Res.* **29**, 1685–1690.

Kim, S. H. and Cech, T. R. (1987). Three-dimensional Model of the Active Site of the Self-Splicing rRNA Precursor of Tetrahymena, *Proc. Natl. Acad. Sci. USA* **84**, 8788–8792.

Kim, J., Cheong, C., and Moore, P. B. (1991). Tetramerization of an RNA Oligonucleotide Containing a GGGG Sequence, *Nature (London)* **351**, 331-332.

Kim, S. H., Suddath, F. L., Quigley, G. J., McPherson, A., Sussman, J. L., Wang, A. H.-J., Seeman, N. C., and Rich, A. (1974). Three-Dimensional Tertiary Structure of Yeast Phenylalanine Transfer RNA, *Science* **185**, 435–440.

Klump, H. H. (1990). Saenger, W. Ed., Landolt-Börnstein, New Series, Nucleic Acids, Subvol. C, Spectroscopic and Kinetic Data, in VII Biophysics, Vol 1, Springer-Verlag, Berlin, Germany, pp. 241–256.

Klump, H. and Burkart, W. (1977). Calorimetric Measurements of the Transition Enthalpy of DNA in Aqueous Urea Solutions, *Biochim. Biophys. Acta* **475**, 601–604.

Klump, H. H. and Jovin, T. M. (1987). Formation of a Left-Handed RNA Double Helix: Energetics of the A-Z Transition of Poly[r(G-C)] in Concentrated $NaClO_4$ Solutions, *Biochemistry* **26**, 5186–5190.

Knitt, D. S., Narlikar, G. J., and Herschlag, D. (1994). Dissection of the Role of the Conserved G•U Pair in Group I RNA Self-Splicing, *Biochemistry* **33**, 13864–13879.

Kondo, N. S. and Danyluk, S. S. (1976). Conformational Properties of Adenylyl- 3′5′-adenosine in Aqueous Solution, *Biochemistry* **15**, 756–768.

Krakauer, H. (1974). A Thermodynamic Analysis of the Influence of Simple Mono- and Divalent Cations on the Conformational Transitions of Polynucleotide Complexes, *Biochemistry* **13**, 2579–2589.

Krakauer, H. and Sturtevant, J. M. (1968). Heats of the Helix–Coil Transitions of the Poly A-Poly U Complexes, *Biopolymers* **6**, 491–512.

Laing, L. G. and Draper, D. E. (1994). Thermodynamics of RNA Folding in a Conserved Ribosomal RNA Domain, *J. Mol. Biol.* **237**,560–576.

Lee, J. S., Johnson, D. A., and Morgan, A. R. (1979). Complexes formed by $(pyrimidine)_n$:$(purine)_n$ DNAs on lowering the pH are three-stranded, *Nucleic Acids Res.* **6**, 3073–3091.

Lee, C. H. and Tinoco, I., Jr. (1980). Conformation Studies of 13 Trinucleoside Diphosphates by 360 MHz PMR Spectroscopy. A Bulged Base Conformation. I. Base Protons and H1′ Protons, *Biophys. Chem.* **11**, 283–294.

Legault, P. and Pardi, A. (1994). In-Situ Probing of Adenine Protonation in RNA by C-13 NMR, *J. Am. Chem. Soc.* **116**, 8390–8391.

Leontis, N. B., Kwok, W. and Newman, J. S. (1991). Stability and Structure of Three-way DNA Junctions Containing Unpaired Nucleotides, *Nucleic Acids Res.* **19** , 759–766.

Lerman, L. S., Fischer, S. G., Hurley, I., Silverstein, K., and Lumelsky, N. (1984). Sequence-Determined DNA Separations, *Annu. Rev. Biophys. Bioeng.* **13**, 399–423.

Levine, L., Gordon, J. A., and Jencks, W. P. (1963). The Relationship of Structure to the Effectiveness of Denaturing Agents for Deoxyribonucleic Acid, *Biochemistry* **2**, 168–175.

Lipsett, M. N. (1963). The Interactions of Poly C and Guanine Trinucleotide, *Biochem. Biophys. Res. Commun.* **11**, 224–228.

Lipsett, M. N. (1964). Complex Formation between Polycytidylic Acid and Guanine Oligonucleotides, *J. Biol. Chem.* **239**, 1256–1260.

Longfellow, C. E., Kierzek, R. and Turner, D. H. (1990), Thermodynamic and spectroscopic study of bulge loops in oligoribonucleotides, *Biochemistry* **29**, 278–285.

Lu, M. and Draper, D. E. (1994). Bases defining an ammonium and magnesium ion dependent tertiary structure within the large subunit ribosomal RNA, *J. Mol. Biol.* **244**, 572–585.

Lu, M., Guo, Q. and Kallenbach, N. R. (1993). Thermodynamics of G-Tetraplex Formation by Telomeric DNAs, *Biochemistry* **32**, 598–601.

Lu, M., Guo, Q., Marky, C. A., Seeman, N. C., and Kallenbach, N. R. (1992). Thermodynamics of DNA Branching, *J. Mol. Biol.* **223** , 781–789.

Lück, R., Steger, G., and Riesner, D. (1996). Thermodynamic Prediction of Conserved Secondary Structure: Application to the RRE Element of HIV, the tRNA-like Element of CMV and the mRNA of Prion Protein, *J. Mol. Biol.* **258**, 813–826.

Manning, G. S. (1976). On the application of polyelectrolyte limiting laws to the helix–coil transition of DNA. V. Ionic effects on renaturation kinetics, *Biopolymers* **15**, 1333–1343.

Manning, G. S. (1978). The molecular theory of polyelectrolyte solutions with applications to the electrostatic properties of polynucleotides, *Q. Rev. Biophys.* **11**, 179–246.

Marky, L. A. and Breslauer, K. J. (1987). Calculating Thermodynamic Data for Transitions of any Molecularity from Equilibrium Melting Curves, *Biopolymers* **26**, 1601–1620.

Marky, L. A., Kallenbach, N. R., McDonough, K. A., Seeman, N. C., and Breslauer, K. J. (1987). The Melting Behavior of a DNA Junction Structure: A Calorimetric and Spectroscopic Study, *Biopolymers* **26**, 1621–1634.

Marmur, J. and Doty, P. (1962). Determination of the Base Composition of Deoxyribonucleic Acid from its Thermal Melting Temperature, *J. Mol. Biol.* **5**, 109–118.

Martin, F. H., Castro, M. M., Aboul-ela, F., and Tinoco, I., Jr. (1985). Base Pairing Involving Deoxyinosine: Implications for Probe Design, *Nucleic Acids Res.* **13**, 8927–8938.

Martin, F. H. and Tinoco, I., Jr. (1980). DNA–RNA Hybrid Duplexes Containing Oligo (dA:rU) Sequences Are Exceptionally Unstable and May Facilitate Termination of Transcription, *Nucleic Acids Res.* **8**, 2295–2299.

Martin, F. H., Uhlenbeck, O. C., and Doty, P. (1971). Self-Complementary Oligoribonucleotides: Adenylic Acid-Uridylic Acid Block Copolymers, *J. Mol. Biol.* **57**, 201–215.

Mathews, D. H., Banerjee, A. R., Luan, D. D., Eickbush, T. H., and Turner, D. H. (1997). Secondary Structure Model of the RNA Recognized by the Reverse Transcriptase from the R2 Retrotransposable Element, *RNA* **3**, 1–16.

Mathews, D. H., Sabina, J., Zuker, M., and Turner, D. H. (1999). Expanded Sequence Dependence of Thermodynamic Parameters Improves Predictions of RNA Secondary Structure, *J. Mol. Biol.* **288**, 911–940.

McConnell, T. S., Cech, T. R., and Herschlag, D. (1993). Guanosine Binding to the *Tetrahymena* Ribozyme: Thermodynamic Coupling with Oligonucleotide Binding, *Proc. Natl. Acad. Sci. USA* **90**, 8362–8366.

Mellema, R. J., van der Woerd, R., van der Marel, G. A., van Boom, J. H., and Altona, C. (1984). Proton NMR Study and Conformational Analysis of d(CGT), d(TCG), and d(CGTCG) in Aqueous Solution. The Effect of a Dangling Thymidine and of a Thymidine Mismatch on DNA Mini-duplexes, *Nucleic Acids Res.* **12**, 5061–5078.

Michel, F., Ellington, A. D., Couture, S., and Szostak, J. W. (1990). Phylogenetic and Genetic Evidence for Base-Triples in the Catalytic Domain of Group I Introns, *Nature (London)* **347**, 578–580.

Michel, F. and Westhof, E. (1990). Modelling of the Three-dimensional Architecture of Group I Catalytic Introns Based on Comparative Sequence Analysis, *J. Mol. Biol.* **216**, 585–610.

Michelson, A. M., Massoulie, J., and Guschlbauer, W. (1967). Synthetic Polynucleotides, *Prog. Nucleic Acids Res. Mol. Biol.* **6**, 83-141.

Miller, J. H. and Sobell, H. M. (1966). A Molecular Model for Gene Repression, *Proc. Natl. Acad. Sci. USA* **55**, 1201–1205.

Mirkin, S. M. and Frank-Kamenetskii, M. D. (1994). H-DNA and Related Structures, *Annu. Rev. Biophys. Biomolec. Struct.* **23**, 541–576.

Mirkin, S. M., Lyamichev, V. I., Drushlyak, K. N., Dobrynin, V. N., Filippov, S. A., and Frank-Kamenetskii, M. D. (1987). DNA H form requires a homopurine–homopyrimidine mirror repeat, *Nature (London)* **330**, 495–497.

Moazed, D., Stern, S., Noller, H. F. (1986). Rapid Chemical Probing of Conformation in 16S Ribosomal RNA and 30S Ribosomal Subunits Using Primer Extension, *J. Mol. Biol.* **187**, 399–416.

Moran, S., Ren, R. X.-F., and Kool, E. T. (1997). A Thymidine Triphosphate Shape Analog Lacking Watson–Crick Pairing Ability is Replicated with High Sequence Selectivity, *Proc. Natl. Acad. Sci. USA* **94**, 10506–10511.

Moran, S., Kierzek, R., and Turner, D. H. (1993). Binding of Guanosine and 3′ Splice Site Analogues to a Group I Ribozyme: Interactions with Functional Groups of Guanosine and with Additional Nucleotides, *Biochemistry* **32**, 5247–5256.

Morden, K. M., Chu, Y. G., Martin, F. H. and Tinoco, I., Jr. (1983). Unpaired Cytosine in the Deoxyoligonucleotide Duplex $dCA_3CA_3G{:}dCT_6G$ is Outside of the Helix, *Biochemistry* **22**, 5557–5563.

Murphy, F. L. and Cech, T. R. (1994). GAAA Tetraloop and Conserved Bulge Stabilize Tertiary Structure of a Group I Intron Domain, *J. Mol. Biol.* **236**, 49–63.

Murphy, F. L. and Cech, T. R. (1993). An Independently Folding Domain of RNA Tertiary Structure within the Tetrahymena Ribozyme, *Biochemistry* **32**, 5291–5300.

Narlikar, G. J., Gopalakrishnan, V., McConnell, T. S., Usman, N., and Herschlag, D. (1995). Use of binding energy by an RNA enzyme for catalysis by positioning and substrate destabilization, *Proc. Natl. Acad. Sci. USA* **92**, 3668–3672.

Nelson, J. W., Martin, F. H. and Tinoco, I., Jr. (1981). DNA and RNA Oligomer Thermodynamics: The Effect of Mismatched Bases on Double-Helix Stability, *Biopolymers* **20**, 2509–2531.

Nelson, J. W. and Tinoco, I., Jr. (1982). Comparison of the Kinetics of Ribooligonucleotide, Deoxyribooligonucleotide, and Hybrid Oligonucleotide Double-Strand Formation by Temperature-Jump Kinetics, *Biochemistry* **21**, 5289–5295.

Nichols, R., Andrews, P. C., Zhang, P., and Bergstrom, D. E. (1994). A Universal Nucleoside for Use at Ambiguous Sites in DNA Primers, *Nature (London)* **369**, 492–493.

Nussinov, R., Tinoco, I., Jr., and Jacobson, A. B. (1982). Secondary Structure Model for the Complete Simian Virus 40 Late Precursor mRNA, *Nucleic Acids Res.* **10**, 351–363.

Okada, Y., Streisinger, G., Owen, J., Newton, J., Tsugita, A., and Inouye, M. (1972). Molecular Basis of a Mutational Hot Spot in the Lysozyme Gene of Bacteriophage T4, *Nature (London)* **236**, 338–341.

Olmsted, M. C., Anderson, C. F., and Record, M. T., Jr. (1989). Monte Carlo description of oligoelectrolyte properties of DNA oligomers: Range of the end effect and the approach of molecular and thermodynamic properties to the polyelectrolyte limits, *Proc. Natl. Acad. Sci. USA* **86**, 7766–7770.

Olson, W. K. (1982). Theoretical Studies of Nucleic Acid Conformation: Potential Energies, Chain Statistics, and Model Building in *Topics in Nucleic Acid Structure*, Part 2, Chapter 1, Neidle, S., Ed., Macmillan, London, UK.

Olsthoorn, S. M., Bostelaar, L. J., DeRooij, J. F. M., Van Boom, J. H. and Altona, C. (1981). Circular Dichroism Study of Stacking Properties of Oligodeoxyadenylates and Polydeoxyadenylate, *Eur. J. Biochem.* **115**, 309–321.

Orbons, L. P. M., van der Marel, G. A., van Boom, J. H., and Altona, C. (1986). Hairpin and Duplex Formation of the DNA Octamer d(m5C-G-m5C-G-T-G-m5C-G) in Solution. An NMR Study, *Nucleic Acids Res.* **14**, 4187–4196.

Orbons, L. P. M., van der Marel, G. A., van Boom, J. H., and Altona, C. (1987). Conformational and Model-Building Studies of the Mismatched DNA Octamer d(m5C-G-m5C-G-T-G-m5C-G), *J. Biomol. Structure. Dyn.* **4**, 965–987.

Owen, R. J., Hill, L. R., and Lapage, S. P. (1969). Determination of DNA Base Compositions from Melting Profiles in Dilute Buffers, *Biopolymers* **7**, 503–516.

Oxender, D. L., Zurawski, G., and Yanofsky, C. (1979). Attenuation in the Escherichia coli Tryptophan Operon: Role of RNA Secondary Structure Involving the Tryptophan Codon Region, *Proc. Natl. Acad. Sci. USA* **76**, 5524–5528.

Owczarzy, R., Vallone, P. M., Gallo, F. J., Paner, T. M., Lane, M. J., and Benight, A. S. (1997). Predicting Sequence-Dependent Melting Stability of Short Duplex DNA Oligomers, *Biopolymers* **44**, 217–239.

Panayotatos, N. and Wells, R. D. (1981). Cruciform Structures in Supercoiled DNA, *Nature (London)* **289**, 466–470.

Papanicolaou, C., Gouy, M., and Ninio, J. (1984). An Energy Model that Predicts the Correct Folding of Both the tRNA and the 5S RNA Molecules, *Nucleic Acids Res.* **12**, 31–44.

Pardi, A., Martin, F. H., and Tinoco, I., Jr. (1981). Comparative Study of Ribonucleotide, Deoxyribonucleotide, and Hybrid Oligonucleotide Helices by Nuclear Magnetic Resonance, *Biochemistry* **20**, 3986–3996.

Patel, D. J., Kozlowski, S. A., Marky, L. A., Rice, J. A., Broka, C., Itakura, K. and Breslauer, K. J. (1982). Extra Adenosine Stacks into the Self-Complementary d(CGCAGAATTCGCG) Duplex in Solution, *Biochemistry* **21**, 445–451.

Patel, D. J., Shapiro, L., and Hare, D. (1987). DNA and RNA NMR Studies of Conformations and Dynamics in Solution, *Q. Rev. Biophys.* **20**, 35–112.

Peattie, D. A., Douthwaite, S., Garrett, R. A., and Noller, H. F. (1981). A "Bulged" Double Helix in a RNA-Protein Contact Site, *Proc. Natl. Acad. Sci. USA* **78**, 7331–7335.

Peritz, A. E., Kierzek, R., Sugimoto, N., and Turner, D. H. (1991). Internal loops in oligoribonucleotides: Symmetric loops are more stable than asymmetric loops, *Biochemistry* **30**, 6428–6436.

Petersheim, M. and Turner, D. H. (1983). Base-Stacking and Base-Pairing Contributions to Helix Stability: Thermodynamics of Double-Helix Formation with CCGG, CCGGp, CCGGAp, ACCGGp, CCGGUp, and ACCGGUp, *Biochemistry* **22**, 256–263.

Peterson, R. D., Bartel, D. P., Szostak, J. W., Horvath, S. J. and Feigon, J. (1994). ^{1}H NMR studies of the high-affinity Rev binding site of the Rev responsive element of HIV-1 mRNA: Base pairing in the core binding element, *Biochemistry* **33**, 5357–5366.

Pinnavaia, T. J., Marshall, C. L., Mettler, C. M., Fisk, C. L., Miles, H. T. and Becker, E. D. (1978). Alkali Metal Ion Specificity in the Solution Ordering of a Nucleotide, 5′ Guanosine Monophosphate, *J. Am. Chem. Soc.* **100**, 3625–3627.

Pleij, C. W. A., Rietveld, K., and Bosch, L. (1985). A New Principle of RNA Folding Based on Pseudo Knotting, *Nucleic Acids Res.* **13**, 1717–1731.

Pley, H. W., Flaherty, K. M., and McKay, D. B. (1994). Model for an RNA Tertiary Interaction from the Structure of an Intermolecular Complex between a GAAA Tetraloop and an RNA Helix, *Nature (London)* **372**, 111–113.

Podder, S. K. (1971). Co-operative Non-Enzymic Base Recognition: A Kinetic Study of Interaction Between GpGpGpC and GpCpCpC and of Self-Association of GpGpGpC, *Eur. J. Biochem.* **22**, 467–477.

Pohl, F. M. (1976). Polymorphism of a synthetic DNA in solution, *Nature (London)* **260**, 365–366.

Pohl, F. M. and Jovin, T. M. (1972). Salt-induced Co-operative Conformational Change of a Synthetic DNA: Equilibrium and Kinetic Studies with Poly(dG-dC), *J. Mol. Biol.* **67**, 375–396.

Poland, D. (1978). *Cooperative Equilibria in Physical Biochemistry,* Clarendon Press, Oxford, UK.

Pörschke, D. (1974). Thermodynamic and Kinetic Parameters of an Oligonucleotide Hairpin Helix, *Biophys. Chem.* **1**, 381–386.

Pörschke, D. (1978). Molecular States in Single-Stranded Adenylate Chains by Relaxation Analysis, *Biopolymers* **17**, 315–323.

Pörschke, D. and Eggers, F. (1972). Thermodynamics and Kinetics of Base-Stacking Interactions, *Eur. J. Biochem.* **26**, 490–498.

Pörschke, D. and Eigen, M. (1971). Cooperative Non-enzymic Base Recognition III. Kinetics of the Helix-Coil Transition of the Oligoribouridylic:Oligoriboadenylic Acid System and of Oligoriboadenylic Acid Alone at Acidic pH, *J. Mol. Biol.* **62**, 361–381.

Pörschke, D., Uhlenbeck, O. C., and Martin, F. H. (1973). Thermodynamics and Kinetics of the Helix-Coil Transition of Oligomers Containing GC Base Pairs, *Biopolymers* **12**, 1313–1335.

Powell, J. T., Richards, E. G., and Gratzer, W. B. (1972). The Nature of Stacking Equilibria in Polynucleotides, *Biopolymers* **11**, 235–250.

Prakash, G. and Kool, E. T. (1992). Structural Effects in the Recognition of DNA by Circular Oligonucleotides, *J. Am. Chem. Soc.* **114**, 3523–3527.

Praseuth, D., Perrouault, L., LeDoan, T., Chassignol, M., Thuong, N., and Helene, C. (1988). Sequence-specific binding and photocrosslinking of Z and B oligodeoxynucleotides to the major groove of DNA via triple-helix formation, *Proc. Natl. Acad. Sci. USA* **85**, 1349–1353.

Pyle, A. M. and Cech, T. R. (1991). Ribozyme Recognition of RNA by Tertiary Interactions with Specific Ribose 2′-OH Groups, *Nature (London)* **350**, 628–631.

Pyle, A. M., McSwiggen, J. A., and Cech, T. R. (1990). Direct Measurement of Oligonucleotide Substrate Binding to Wild-Type and Mutant Ribozymes from *Tetrahymena, Proc. Natl. Acad. Sci. USA* **87**, 8787–8191.

Pyle, A. M., Moran, S., Strobel, S. A., Chapman, T., Turner, D. H., and Cech, T. R. (1994). Replacement of the Conserved G • U with a G-C Pair at the Cleavage Site of the *Tetrahymena* Ribozyme Decreases Binding, Reactivity, and Fidelity, *Biochemistry* **33**, 13856–13863.

Pyle, A. M., Murphy, F. L. and Cech, T. R. (1992). RNA Substrate Binding Site in the Catalytic Core of the *Tetrahymena* Ribozyme, *Nature (London)* **358**, 123–128.

Quartin, R. S. and Wetmur, J. G. (1989). The Effect of Ionic Strength on the Hybridization of Oligodeoxynucleotides with Reduced Charge due to Methylphosphonate Linkages to Unmodified Oligodeoxynucleotides Containing the Complementary Sequence, *Biochemistry* **28**, 1040–1047.

Record, M. T., Jr. (1967). Electrostatic Effects on Polynucleotide Transitions. II. Behavior of Titrated Systems, *Biopolymers* **5**, 993–1008.

Record, M. T., Jr. (1975). Effects of Na^+ and Mg^{++} Ions on the Helix-Coil Transition of DNA, *Biopolymers* **14**, 2137–2158.

Record, M. T., Jr., Woodbury, C. P., and Lohman, T. M. (1976). Na^+ effects on transitions of DNA and polynucleotides of variable linear charge density, *Biopolymers* **15**, 893–915.

Record, M. T. and Zimm, B. H. (1972). Kinetics of the helix–coil transition in DNA, *Biopolymers* **11**, 1435–1484.

Rentzeperis, D., Rippe, K., Jovin, T. M., and Marky, L. A. (1992). Calorimetric Characterization of Parallel-Stranded DNA: Stability, Conformational Flexibility, and Ion Binding, *J. Am. Chem. Soc.* **114**, 5926–5928.

Richards, E. G., Flessel, C. P., and Fresco, J. R. (1963). Polynucleotides. IV. Molecular Properties and Conformation of Polyribouridylic Acid, *Biopolymers* **1**, 431–446.

Riesner, D., Maass, G., Thiebe, R., Philippsen, P., and Zachau, H. G. (1973). The Conformational Transitions in Yeast $tRNA^{Phe}$ as Studied with $tRNA^{Phe}$ Fragments, *Eur. J. Biochem.* **36**, 76–88.

Riesner, D. and Römer, R. (1973). Thermodynamics and Kinetics of Conformational Transitions in Oligonucleotides and tRNA, in *Physico-Chemical Properties of Nucleic Acids*, Vol. 2, Duchesne, J., Ed., Academic, New York, pp. 237–318.

Riley, M., Maling, B., and Chamberlin, M. J. (1966). Physical and Chemical Characterization of Two- and Three- stranded Adenine-Thymine and Adenine-Uracil Homopolymer complexes, *J. Mol. Biol.* **20**, 359–389.

Rippe, K. and Jovin, T. M. (1992). Parallel-Stranded Duplex DNA, *Methods Enzymol.* **211**, 199–220.

Rivas, E. and Eddy, S. (1999). A Dynamic Programming Algorithm for RNA Structure Prediction Including Pseudoknots, *J. Mol. Biol.* **285**, 2053–2068.

Robertus, J. D., Ladner, J. E., Finch, J. T., Rhodes, D., Brown, R. S., Clark, B. F. C., and Klug, A. (1974). Structure of Yeast Phenylalanine tRNA at 3A Resolution, *Nature (London)* **250**, 546–551.

Robinson, D. R. and Grant, M. E. (1966). The Effects of Aqueous Salt Solutions on the Activity Coefficients of Purine and Pyrimidine Bases and Their Relation to the Denaturation of Deoxyribonucleic Acid by Salts, *J. Biol. Chem.* **241**, 4030–4042.

Romaniuk, P. J., Lowary, P., Wu, H-N, Stormo, G., and Uhlenbeck, O. C. (1987). RNA Binding Site of R17 Coat Protein, *Biochemistry* **26**, 1563–1568.

Rommens, J., Kerem, B.-S., Greer, W., Chang, P., Tsui, L.-C., and Ray, P. (1990). Rapid Nonradioactive Detection of the Major Cystic Fibrosis Mutation, *Am. J. Hum. Genet.* **46** , 395–396.

Ross, P. D. and Scruggs, R. L. (1965). Heat of the Reaction Forming the Three-Stranded Poly(A+2U) Complex, *Biopolymers* **3**, 491–496.

Saenger, W. (1984). *Principles of Nucleic Acid Structure*, Chapters 9, 11, and 12, Springer-Verlag, New York.

SantaLucia, J., Jr., Kierzek, R., and Turner, D. H. (1991). Stabilities of Consecutive A•C, C•C, G•G, U•C, and U•U Mismatches in RNA Internal Loops: Evidence for Stable Hydrogen-Bonded U•U and C•C^+ Pairs, *Biochemistry* **30** , 8242–8251.

SantaLucia, J., Jr., Kierzek, R., and Turner, D. H. (1992). Context Dependence of Hydrogen Bond Free Energy Revealed by Substitutions in an RNA Hairpin, *Science* **256**, 217–219.

SantaLucia, J., Jr. and Turner, D. H. (1993). Structure of $(rGGCGAGCC)_2$ in solution from NMR and restrained molecular dynamics, *Biochemistry* **32**, 12612–12623.

SantaLucia, J., Jr. (1998). A Unified View of Polymer, Dumbbell, and Oligonucleotide DNA Nearest-neighbor Thermodynamics, *Proc. Natl. Acad. Sci. USA* **95**, 1460–1465.

Sassanfar, M. and Szostak, J. W. (1993). An RNA Motif that Binds ATP, *Nature,(London)* **364**, 550–563.

Schera, M., Shalon, D., Davis, R. W., and Brown, P. O. (1995). Quantitative Monitoring of Gene Expression Patterns with a Complementary DNA Microarray, *Science* **270**, 467–470.

Schroeder, S., Kim, J., and Turner, D. H. (1996). GA and UU Mismatches Can Stabilize RNA Internal Loops of Three Nucleotides, *Biochemistry* **35**, 16105–16109.

Schmitz, M. and Steger, G. (1996). Description of RNA Folding by "Simulated Annealing," *J. Mol. Biol.* **255**, 254–266.

Sen, D. and Gilbert, W. (1988). Formation of Parallel Four-Stranded Complexes by Guanine-rich Motifs in DNA and its Implications for Meiosis, *Nature (London)* **334**, 364–366.

Sen, D. and Gilbert, W. (1990). A Sodium-Potassium Switch in the Formation of Four-Stranded G4–DNA, *Nature (London)* **344**, 410–414.

Senior, M. M., Jones, R. A., and Breslauer, K. J. (1988a). Influence of Loop Residues on the Relative Stabilities of DNA Hairpin Structures, *Proc. Natl. Acad. Sci. USA* **85**, 6242–6246.

Senior, M., Jones, R. A., and Breslauer, K. J. (1988b). Influence of Dangling Thymidine Residues on the Stability and Structure of two DNA Duplexes, *Biochemistry* **27**, 3879–3885.

Serra, M. J., Axenson, T. J., and Turner, D. H. (1994). A Model for the Stabilities of RNA Hairpins Based on a Study of the Sequence Dependence of Stability for Hairpins of Six Nucleotides, *Biochemistry* **33**, 14289–14296.

Serra, M. J., Lyttle, M. H., Axenson, T. J., Schadt, C. A., and Turner, D. H. (1993). RNA hairpin loop stability depends on closing base pair, *Nucleic Acids Res.* **21**, 3845–3849.

Serra, M. J., Barnes, T. W., Betschart, K., Guiterrez, M. J., Sprouse, K. J., Riley, C. K., Stewart, L., and Temel, R. E. (1997). Improved Parameters for the Prediction of RNA Hairpin Stability, *Biochemistry* **36**, 4844–4851.

Serra, M. J. and Turner, D. H. (1995). Predicting the Thermodynamic Properties of RNA, *Methods Enzymol.* **259**, 242–261.

Shen, P., Zengel, J. M., and Lindahl, L. (1988). Secondary Structure of the Leader Transcript from the Escherichia coli S10 Ribosomal Protein Operon, *Nucleic Acids Res.* **16**, 8905–8924.

Simpkins, H. and Richards, E. G. (1967). Preparation and Properties of Oligouridylic Acids, *J. Mol. Biol.* **29**, 349–356.

Steger, G. (1994). Thermal Denaturation of Double-stranded Nucleic Acids: Prediction of Temperatures Critical for Gradient Gel Electrophoresis and Polymerase Chain Reaction, *Nucleic Acids Res.* **22**, 2760–2768.

Steger, G., Hofmann, H., Förtsch, J., Gross, H. J., Randles, J. W., Sänger, H. L., and Riesner, D. (1984). Conformational Transitions in Viroids and Virusoids: Comparison of Results from Energy Minimization Algorithm and from Experimental Data, *J. Biomolec. Struct. and Dynamics* **2**, 543–571.

Stevens, C. L. and Felsenfeld, G. (1964). The Conversion of Two-Stranded Poly (A+U) to Three-Strand Poly (A+2U) and Poly A by Heat, *Biopolymers* **2**, 293–314.

Strobel, S. A. and Cech, T. R. (1993). Tertiary Interactions with the Internal Guide Sequence Mediate Docking of the P1 Helix into the Catalytic Core of the *Tetrahymena* Ribozyme, *Biochemistry* **32**, 13593–13604.

Strobel, S. A., Moser, H. E., and Dervan, P. B. (1988). Double-Strand Cleavage of Genomic DNA at a Single Site by Triple-Helix Formation, *J. Am. Chem. Soc.* **110**, 7927–7929.

Strobel, S. A. and Cech, T. R. (1995). Minor Groove Recognition of the Conserved G•U Pair at the *Tetrahymena* Ribozyme Reaction Site, *Science* **267**, 675–679.

Studier, F. W. (1969). Effects of the Conformation of Single-stranded DNA on Renaturation and Aggregation, *J. Mol. Biol.* **41**, 199–209.

Sturtevant, J. M. (1987). Biochemical Applications of Differential Scanning Calorimetry, *Annu. Rev. Phys. Chem.* **38**, 463–488.

Sugimoto, N., Kierzek, R., and Turner, D. H. (1987a). Sequence Dependence for the Energetics of Dangling Ends and Terminal Base Pairs in Ribonucleic Acid, *Biochemistry* **26**, 4554–4558.

Sugimoto, N., Kierzek, R. and Turner, D. H. (1987b). Sequence Dependence for the Energetics of Terminal Mismatches in Ribooligonucleotides, *Biochemistry* **26**, 4559–4562.

Sugimoto, N., Kierzek, R. and Turner, D. H. (1988). Kinetics for Reaction of a Circularized Intervening Sequence with CU, UCU, CUCU, and CUCUCU: Mechanistic Implications from the Dependence on Temperature and on Oligomer and Mg^{2+} Concentrations, *Biochemistry* **27**, 6384–6392.

Sugimoto, N., Sasaki, M., Kierzek, R., and Turner, D. H. (1989a). Binding of a Fluorescent Oligonucleotide to a Circularized Intervening Sequence from Tetrahymena thermophila, *Chemistry Lett.* 2223–2226.

Sugimoto, N., Tomka, M., Kierzek, R., Bevilacqua, P. C. and Turner, D. H. (1989b). Effects of Substrate Structure on the Kinetics of Circle Opening Reactions of the Self-splicing Intervening Sequence from *Tetrahymena thermophila*: Evidence for Substrate and Mg^{2+} Binding Interactions, *Nucleic Acids Res.* **17**, 355–371.

Sugimoto, N., Nakano, S., Katoh, M., Matsumura, A., Nakamuta, H., Ohmichi, T., Yoneyama, M., and Sasaki, M. (1995). Thermodynamic parameters to predict stability of RNA/DNA hybrid duplexes, *Biochemistry* **34**, 11211–11216.

Sullivan, K. M. and Lilley, D. M. J. (1986). A Dominant Influence of Flanking Sequences on a Local Structural Transition in DNA, *Cell* **47**, 817–827.

Sundaralingam, M. (1969). Stereochemistry of Nucleic Acids and their Constituents. IV. Allowed and Preferred Conformations of Nucleosides, Nucleoside Mono-, Di, Tri, Tetraphosphates, Nucleic Acids and Polynucleotides, *Biopolymers* **7**, 821–860.

Sundaralingam, M. (1973). Conformation of Biological Molecules and Polymers, The Concept of a Conformationally 'Rigid' Nucleotide and its Significance in Polynucleotide Conformational Analysis, *Jerusalem Symp. Quantum Chem. Biochem.* **5**, 417–455.

Sundquist, W. I. and Klug, A. (1989). Telomeric DNA Dimerizes by Formation of Guanine Tetrads Between Hairpin Loops, *Nature (London)* **342**, 825–829.

Sutcliffe, J. G. (1978), Complete Nucleotide Sequence of the *Escherichia coli* plasmid pBR322, *Cold Spring Harbor Symp. Quant. Biol.* **43**, 77

Suurkuusk, J., Alvarez, J., Freire, E., and Biltonen, R. (1977). Calorimetric Determination of the Heat Capacity Changes Associated with the Conformational Transitions of Polyriboadenylic Acid and Polyribouridylic Acid, *Biopolymers* **16**, 2641–2652.

Szewczak, A. A., Moore, P. B., Chan, Y.-L., and Wool, I. G. (1993). The conformation of the sarcin/ricin loop from 28S ribosomal RNA, *Proc. Natl. Acad. Sci. USA* **90**, 9581–9585.

Tang, C. K. and Draper, D. E. (1989). An unusual mRNA pseudoknot structure is recognized by a protein translational repressor, *Cell* **57**, 531–536.

Tibanyenda, N., DeBruin, S. H., Haasnoot, C. A. G., Van Der Marel, G. A., Van Boom, J. H., and Hilbers, C. W. (1984). The Effect of Single Base-Pair Mismatches on the Duplex Stability of d(T-A-T-T-A-A-T-A-T-C-A-A-G-T-T-G) : d(C-A-A-C-T-T-G-A-T-A-T-T-A-A-T-A), *Eur. J. Biochem.* **139**, 19-27.

Tinoco, I., Jr., Uhlenbeck, O. C., and Levine, M. D. (1971). Estimation of Secondary Structure in Ribonucleic Acids, *Nature (London)* **230**, 362–367.

Topping, R. J., Stone, M. P., Brush, C. K., and Harris, T. M. (1988). Non-Watson–Crick Structures in Oligodeoxynucleotides: Self-Association of d(TpCpGpA) Stabilized at Acidic pH, *Biochemistry* **27**, 7216–7222.

Triggs-Raine, B. L. and Gravel, R. A. (1990). Diagnostic Heteroduplexes: Simple Detection of Carriers of a 4-bp Insertion Mutation in Tay-Sachs Disease, *Am. J. Hum. Genet.* **46** , 183–184.

Trulson, M. O., Cruz, P., Puglisi, J. D., Tinoco, I., Jr., and Mathies, R. A. (1987). Raman Spectroscopic Study of Left-Handed Z-RNA, *Biochemistry* **26**, 8624-8630.

Tucker, E. E., Lane, E. H., and Christian, D. S. (1981). Vapor pressure studies of hydrophobic interactions. Formation of benzene–benzene and cyclohexane–cyclohexanol dimers in dilute aqueous solution, *J. Solution Chem.* **10**, 1–20.

Tuerk, C., Gauss, P., Thermes, C., Groebe, D. R., Guild, N., Stormo, G., Gayle, M., d'Aubenton-Carafa, Y., Uhlenbeck, O. C., Tinoco, I., Jr., Brody, E. N., and Gold, L. (1988). CUUCGG Hairpins: Extraordinarily Stable RNA Secondary Structures Associated with Various Biochemical Processes, *Proc. Natl. Acad. Sci. USA* **85**, 1364–1368.

Turner, D. H. (1986). Temperature-Jump Methods, in *Investigations of Rates and Mechanisms of Reactions*, Vol. 6, Part 2, Chapter 3, Bernasconi, C. F., Ed., J. Wiley, New York.

Turner, D. H., Sugimoto, N., and Freier, S. M. (1988). RNA Structure Prediction, *Annu. Rev. Biophys. Biophys. Chem.* **17**, 167–192.

Turner, D. H., Sugimoto, N., Kierzek, R., and Dreiker, S. D. (1987). Free Energy Increments for Hydrogen Bonds In Nucleic Acid Base Pairs, *J. Am. Chem. Soc.* **109**, 3783–3785.

Uhlenbeck, O. C., Borer, P. N., Dengler, B. and Tinoco, I., Jr. (1973). Stability of RNA Hairpin Loops: A6-Cm-U6, *J. Mol. Biol.* **73**, 483–496.

van den Hoogen, Y. T., van Beuzekom, A. A., de Vroom, E., van der Marel, G. A., van Boom, J. H., and Altona, C. (1988a). Bulge-out Structures in the Single-stranded Trimer AUA and in the Duplex (CUGGUGCGG).(CCGCCCAG). A Model-Building and NMR Study, *Nucleic Acids Res.* **16**, 5013–5030.

van der Hoogen, Y. T., Erkelens, C., de Vroom, E., van der Marel, G. A., van Boom, J. H., and Altona, C. (1988b). Influence of Uracil on the Conformational Behavior of RNA Oligonucleotides in Solution, *Eur. J. Biochem.* **173**, 295–303.

Varani, G., Cheong, C., and Tinoco, I., Jr. (1991). Structure of an Unusually Stable RNA Hairpin, *Biochemistry* **30** , 3280–3289.

Voloshin, O. H., Mirkin, S. M., Lyamichev, V. I., Belotserkovskii, B. P., and Frank-Kamenetskii, M. D. (1988). Chemical Probing of Homopurine-homopyrimidine Mirror Repeats in Supercoiled DNA, *Nature (London)* **333**, 475–476.

Wada, A., Yubuki, S. and Husimi, Y. (1980). Fine Structure in the Thermal Denaturation of DNA High Temperature Resolution Spectrophotometric Studies, *Crit. Rev. Biochem.* **9**, 87–144.

Walter, A. E., Turner, D. H., Kim, J., Lyttle, M. H., Müller, P., Mathews, D. H., and Zuker, M. (1994). Coaxial Stacking of Helixes Enhances Binding of Oligoribonucleotides and Improves Predictions of RNA Folding, *Proc. Natl. Acad. Sci. USA* **91**, 9218–9222.

Walter, A. E. and Turner, D. H. (1994). Sequence Dependence of Stability for Coaxial Stacking of RNA Helixes with Watson–Crick Base Paired Interfaces, *Biochemistry* **33**, 12715–12719.

Wang, A. H.-J., Fujii, S., van Boom, J. H., and Rich, A. (1982). Right-handed and Left-handed Double-helical DNA: Structural Studies, *Cold Spring Harbor Symp. Quant. Biol.* **47**, 33–44.

Wartell, R. M. and Benight, A. S. (1985). Thermal Denaturation of DNA Molecules: A Comparison of Theory with Experiment, *Phys. Rep.* **126**, 67–107.

Weeks, K. M. and Crothers, D. M. (1993). Major Groove Accessibility of RNA, *Science* **261**, 1574–1577.

Wells, R. D., Collier, D. A., Hanvey, J. C., Shimizu, M., and Wohlrab, F. (1988). The chemistry and biology of unusual DNA structures adopted by oligopurine:oligopyrimidine sequences, *FASEB J.* **2**, 2939–2949.

Wetmur, J. G. (1991). DNA Probes: Applications of the Principles of Nucleic Acid Hybridization, *Critical Rev. Biochem. Mol. Biol.* **26** , 227–259.

Wetmur, J. G. and Davidson, N. (1968). Kinetics of Renaturation of DNA, *J. Mol. Biol.* **31**, 349–370.

Williams, A. L., Jr. and Tinoco, I., Jr. (1986). A Dynamic Programming Algorithm for Finding Alternative RNA Secondary Structures, *Nucleic Acids Res.* **14**, 299–315.

Williams, A. P., Longfellow, C. E., Freier, S. M., Kierzek, R., and Turner, D. H. (1989). Laser Temperature-Jump, Spectroscopic and Thermodynamic Study of Salt Effects on Duplex Formation by dGCATGC, *Biochemistry* **28**, 4283–4291.

Williamson, J. R. (1994). G-Quartet Structures in Telomeric DNA, *Annu. Rev. Biophys. Biomol. Struct.* **23**, 703–730.

Williamson, J. R., Raghuraman, M. K., and Cech, T. R., (1989). Monovalent Cation-Induced Structure of Telomeric DNA: The G-Quartet Model, *Cell* **59**, 871–880.

Wimberly, B., Varani, G., and Tinoco, I., Jr. (1993). The conformation of loop E of eukaryotic 5S ribosomal RNA, *Biochemistry* **32**, 1078–1087.

Woodson, S. A. and Crothers, D. M. (1987). Proton Nuclear Magnetic Resonance Studies on Bulge-Containing DNA Oligonucleotides from a Mutational Hot-Spot Sequence, *Biochemistry* **26**, 904–912.

Wu, M. J., McDowell, J. A., and Turner, D. H. (1995). A Periodic Table of Symmetric Tandem Mismatches in RNA, *Biochemistry* **34**, 3204–3211.

Wu, M., SantaLucia, J. Jr., and Turner, D. H. (1997). Solution Structure of $(rGGCAGGCC)_2$ by Two dimensional NMR and the Iterative Relaxation Matrix Approach, *Biochemistry* **36**, 4449–4460.

Wu, M. and Turner, D. H. (1996). Solution Structure of (rGCGGACGC), by Two-Dimensional NMR and the Iterative Relaxation Matrix Approach, *Biochemistry* **35**, 9677–9689.

Wyatt, J. R., Puglisi, J. D., and Tinoco, I., Jr. (1990). RNA Pseudoknots: Stability and Loop Size Requirements, *J. Mol. Biol.* **214** , 455–470.

Xia, T., McDowell, J. A., and Turner, D. H. (1997). Thermodynamics of Nonsymmetric Tandem Mismatches Adjacent to G•C Base Pairs in RNA, *Biochemistry* **36**, 12486–12497.

Xia, T., SantaLucia, J. Jr., Burkard, M. E., Kierzek, R., Schroeder, S. J., Jiao, X., Cox, C., and Turner, D. H. (1998). Thermodynamic Parameters for an Expanded Nearest-Neighbor Model for Formation of RNA Duplexes with Watson-Crick Base Pairs, *Biochemistry* **37**, 14719–14735.

Xia, T., Mathews, D. H., and Turner, D. H. (1999). Thermodynamics of RNA Secondary Structure Formation, in *Prebiotic Chemistry, Molecular Fossils, Nucleotides, and RNA*, Söll, D., Moore, P. B. and Nishimura, S., Ed., Elsevier Science Ltd., Oxford.

Yang, Y., Kochoyan, M., Burgstaller, P., Westhof, E., and Famulok, M. (1996). Structural Basis of Ligand Discrimination by Two Related RNA Aptamers Resolved by NMR Spectroscopy, *Science* **272**, 1343–1347.

Yoon, K., Hobbs, C. A., Koch, J., Sardaro, M., Kutny, R., and Weis, A. L. (1992). Elucidation of the Sequence-Specific Third-Strand Recognition of Four Watson–Crick Base Pairs in a Pyrimidine Triple Helix Motif: T•AT, C•GC, T•CG, and G•TA, *Proc. Natl. Acad. Sci. USA* **89**, 3840–3844.

Yuan, R. C., Steitz, J. A., Moore, P. B., and Crothers, D. M. (1979). The 3′ Terminus of 16S rRNA: Secondary Structure and Interaction with Ribosomal Protein S1, *Nucleic Acids Res.* **7**, 2399-2418.

Zimm, B. H. and Bragg, J. K. (1959). Theory of the Phase Transition Between Helix and Random Coil in Polypeptide Chains, *J. Chem. Phys.* **31**, 526–535.

Zimmerman, S. B., Cohen, G. H., and Davies, D. R. (1975). X-ray Fiber Diffraction and Model-building Study of Polyguanylic Acid and Polyinosinic Acid, *J. Mol. Biol.* **92**, 181–192.

Zuker, M., Jaeger, J. A., and Turner, D. H. (1991). A Comparison of Optimal and Suboptimal RNA Secondary Structures Predicted by Free Energy Minimization with Structures Determined by Phylogenetic Comparison, *Nucleic Acids Res.* **19**, 2707–2714.

Zuker, M. and Sankoff, D. (1984). RNA Secondary Structures and Their Prediction, *Bull. Math. Biol.* **46**, 591–521.

Zuker, M. and Stiegler, P. (1981). Optimal Computer Folding of Large RNA Sequences Using Thermodynamics and Auxiliary Information, *Nucleic Acids Res.* **9**, 133–148.

Zuker, M. (1989). On Finding All Suboptimal Foldings of an RNA Molecule, *Science* **244**, 48–52.

CHAPTER 9

Size and Shape of Nucleic Acids in Solution

1. CHARACTERIZATION AT THE MACROMOLECULAR LEVEL

The preceding chapters have mainly dealt with nucleic acids at the atomic or local base pair level. This chapter adopts a more coarse-grained approach, treating DNA and RNA as large polymers or macromolecules. Even at this level, there are important questions to be asked. For example, How big is a nucleic acid? How much genetic information does it contain? Can it be distinguished from similar molecules by its size? Is it a repeated sequence? Has it been cleaved or processed? What is its shape: linear, circular, supercoiled, branched, or looped? How much must it be compressed to package in a small volume, such as a nucleus or a virus capsid? How much energy does it take to bend or twist it, if such deformations are required to bind with other molecules? Can changes in size, shape, frictional resistance, or gel sieving give information on its binding to other molecules? What motions may nucleic acids be undergoing *in vivo*, and how may such motions influence function?

This chapter deals mainly with experimental methods (electrophoresis, hydrodynamics, light scattering, and microscopy) for determining the macromolecular structure of nucleic acids and their complexes with proteins. It begins with an overview of the structural models that are used to interpret experimental results, and ends with a summary of what has been learned about the bending and twisting of DNA in solution. Several books, which are useful as general references about these topics, are listed at the end of this chapter (Cantor and Schimmel, 1980; Tanford, 1961; Freifelder, 1982; Van Holde et al., 1998; Bloomfield et al., 1974; Saenger, 1984).

2. MODELS FOR MOLECULAR STRUCTURE

At the macromolecular level of resolution, structural models characterized by a few dimensional parameters must obscure most details except size and overall shape. A fundamental distinction can be made between rigid models, appropriate for small nucleic acids such as tRNA and short DNA restriction fragments (and the globular proteins that bind to nucleic acids); and flexible chain models, used for single-stranded polynucleotides and high molecular weight DNA. The wormlike chain model is required for DNA of intermediate length. These models, shown in Figure 9-1, are useful both be-

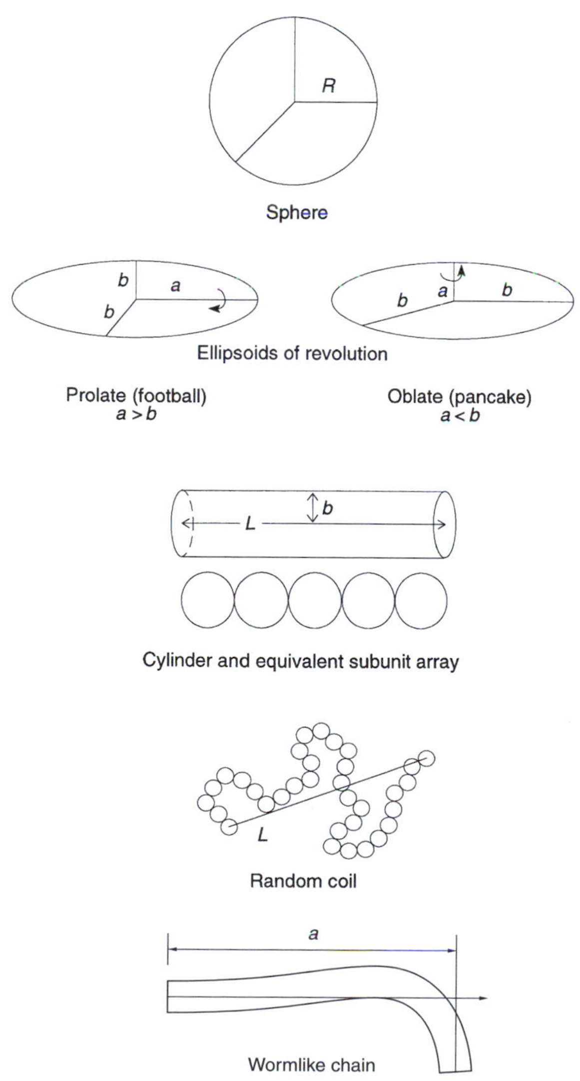

Figure 9-1
Simple structural models for biopolymers.

cause they represent nucleic acid structures with adequate realism, and because their hydrodynamic and scattering properties can be calculated with reasonable ease.

2.1 Rigid Models

The simplest model of a polymer is just a sphere, defined by its radius R. While this is an oversimplification of any real molecule, it is useful for order of magnitude estimates if molecular dimensions are not too asymmetric. This would be the case, for example, with tRNA, ribosomes, spherical viruses, and globular proteins.

Cylindrical rods are appropriate models for short fragments of double helical DNA, and for filamentous viruses. They may be characterized by their length L and radius b. Since analytical solutions for properties of cylinders are hard to obtain (because of their sharp ends), they are often replaced by equivalent ellipsoid or subunit assembly models. For example, for a cylinder of axial ratio $n = L/2b$, one may use n spheres of radius b. Subunit assemblies have proved very useful in modelling nucleoprotein complexes such as oligonucleosomes, chromatin fibers, and bacterial viruses such as T4 and λ. Arrays of spheres are arranged to reflect the outlines of the complex macromolecular geometry (Garcia de la Torre and Bloomfield, 1981).

2.2 Flexible Models

2.2.1 Random Coil

The simplest model for a flexible polymer is the random coil. It consists of N monomers, assumed to be spherical and of diameter (bond length) b. For obvious reasons, it is often called the pearl necklace model. The distance between the ends of the chain is L. This fluctuates statistically, so one can measure only an average value, such as the mean-square end-to-end distance $< L^2 >$. Random coils obey Gaussian chain statistics. Standard references (e.g. those cited at the beginning of this chapter) show that the probability that the distance between the beginning and end of a Gaussian chain lies between L and $L + dL$ is

$$P(L, N)dL = 4\pi \left(\frac{3}{2\pi b^2 N}\right)^{3/2} \exp\left(-\frac{3L^2}{2b^2N}\right) L^2 dL \tag{9-1}$$

The mean-square end-to-end distance is then calculated by multiplying L^2 by this probability and integrating over all L:

$$< L^2 >= \int_0^\infty L^2 P(L, N)dL \tag{9-2}$$

which yields

$$< L^2 >= b^2 N \tag{9-3}$$

The notable feature of this equation is that the mean-*square* distance is proportional to the *first* power of the number of bonds—a characteristic of Gaussian random walk processes.

Another measure of the size of a random coil is its radius of gyration or root-mean-square radius R_g, representing the root-mean-square (rms) distance of the segments from the center of mass of the coil. This may be shown to have the value

$$< R_g^2 >= \frac{1}{6} < L^2 >= \frac{b^2N}{6} \tag{9-4}$$

2.2.2 More Realistic Bond Models

The above model implies perfectly flexible, universal joints. It can be shown that the same Gaussian behavior occurs for more realistic chains, with b replaced by an effective bond length $b_{\text{eff}} = C_\infty b$, so long as interactions along the chain are short-ranged. The "∞" denotes a number of bonds large enough for random behavior to be achieved. The value of C_∞ depends on the nature of the chain: stiffer chains have larger C_∞. For example, for a chain with fixed bond angle θ and free rotation about the bond, $b_{\text{eff}} = b(1 + \cos\theta)/(1 - \cos\theta)$. With a C—C bond skeleton, where θ is the tetrahedral angle, $\cos\theta = \frac{1}{3}$, so $b_{\text{eff}} = 2b$ and $C_\theta = 2$. For more complex chains, such as the single-stranded polynucleotide backbone with six different bonds between each repeating unit (P—O, O—C5′, C5′—C4′, C4′—C3′, C3′—O, O—P), each with its own bond length, angle, and potential energy profile for internal rotation, one can also write $< L^2 >= b_{\text{eff}}^2 N = C_\infty v^2 N$, where N is now the number of nucleotides in the chain and v is the virtual bond length from P to P. Olson (1975) has calculated that $v = 5.63$ Å and $C_\infty = 11.1$ for the $C_{3'}$-*endo* sugar conformation, and $v = 6.69$ Å, $C_\infty = 24.1$ for the $C_{3'}$-*exo* conformation. The $C_{2'}$-*endo* is similar to $C_{3'}$-*exo*, and $C_{2'}$-*exo* to $C_{3'}$-*endo*. These large values of C_∞ indicate that single-stranded polynucleotides are locally rather stiff. A somewhat more realistic virtual bond model (Olson, 1980) takes into account three-bond correlations due to base stacking. This model is successful in reproducing the temperature dependence of chain dimensions in poly(rA) which is almost completely stacked at low T (Stannard and Felsenfeld, 1975) and progressively less stacked as T increases (Eisenberg and Felsenfeld, 1967).

2.2.3 Excluded Volume

If interactions are long ranged, as occurs when residues far from each other along the chain sequence interact physically (e.g., by excluded volume or polyelectrolyte repulsion), then the chain statistics are no longer Gaussian. In that case, Eq. 9-3 is replaced by

$$< L^2 >= b^2 N^{1+\varepsilon} \tag{9-5}$$

where $\varepsilon \approx \frac{1}{5}$ for nonpolyelectrolytes. For long, double-stranded DNA in 0.2 M salt, ε is about 0.1.

2.3 Wormlike Chain

An important model for double-stranded DNA is the wormlike chain. This chain represents behavior intermediate between the rigid rod and the random coil, thus taking into account the local stiffness but long-range flexibility of the double helix. The wormlike chain is defined by its contour length L (measured along the helix axis) and persistence length a. The persistence length is the average projection along the initial direction $< z >$ of the chain as L tends to infinity, as can be seen from the relation for chains of finite length (Bloomfield et al., 1974; Schellman, 1974):

$$< z >= a(1 - e^{-L/a}) \tag{9-6}$$

Another way to express this concept is that the average cosine of the angle θ between the tangents to the chain at its beginning and at L is

$$< \cos(\theta) >_L = e^{-L/a} \tag{9-7}$$

Detailed analysis of the wormlike coil model also shows that

$$< L^2 >= 2a(L - a + ae^{-L/a}) \tag{9-8}$$

If the contour length is much less than the persistence length, expansion of the exponential in Eq. 9-8 shows that $< L^2 >= L^2$, which is the behavior expected for a rigid rod. In the other limit, $L \gg a$, we find $< L^2 >= 2aL$. Since for a real chain, $L = Nb$, comparison with Eq. 9-3 for random coils shows that $2a$, twice the persistence length, is equivalent to the statistical segment length b_{eff}.

The persistence length is proportional to the bending stiffness of the chain. For DNA in 0.1 M NaCl, a is about 500 Å. Knowledge of the persistence length is important in understanding the energy required to bend DNA when it is packaged into a virus capsid, when it is coiled into a nucleosome, when it is complexed with binding proteins, and when it cyclizes into a closed circular molecule.

As an example of the use of these equations, consider a DNA molecule, such as that from bacteriophages T7 or λ, that contains about 40,000 bp. For B-DNA, with a contour length of 3.4 Å/bp, $L = 3.4 \times 40,000 = 136,000$ Å. With a persistence length $a = 500$ Å, Eqs. 9-8 and 9-6 give $R_g^0 \approx (2aL/3)^{1/2} \approx 6700$ Å, where the superscript 0 indicates neglect of excluded volume. The statistical segment length of DNA is $b_{\text{eff}} = 2a = 1000$ Å, so there are $N_{\text{eff}} = 136$ statistical segments in the phage DNA. Equation 9-5 indicates that we should multiply R_g^0 by $N_{\text{eff}}^{\varepsilon/2} \approx 1.28$ to get $R_g \approx 8600$ Å, which is a measure of the radius of the volume occupied by the free DNA. When the DNA is packed into a bacteriophage capsid, it must fit inside a protein shell whose radius is about 400 Å. The achievment of this 10^4-fold decrease in volume poses some interesting physical problems, which are discussed in Chapter 14.

2.4 Circular Molecules

To date, we have considered only linear chains. Intracellular DNA is often found to be circular. If the circular chain is nicked and considerably longer than the persistence length, then a flexible circular chain model is appropriate. For this type of chain, the end-to-end length $L = 0$ by definition. It can be shown that

$$< R_g^2 >_{\text{circular}} = \frac{1}{12} b^2 N = \frac{1}{2} < R_g^2 >_{\text{linear}} \tag{9-9}$$

Of particular interest in many molecular biology applications is the probability that the ends of a linear molecule will come close enough to cyclize (e.g., a plasmid with sticky-end restriction cuts). Jacobsen and Stockmayer (1950) showed, using Eq. 9-1 for a Gaussian chain, that this probability, denoted j, is

$$j dV = \left(\frac{3}{2\pi b^2 N} \right)^{3/2} dV \tag{9-10}$$

where dV is the volume element containing both ends of the chain. Note that j varies as $N^{-3/2}$, so the probability of cylization decreases rapidly with the length of the chain. Of course, it also decreases if the DNA is too short, for then the bending stiffness exacts too high an energetic cost; this behavior is not predicted by Eq. 9-10. Thus the cyclization probability reaches a maximum at intermediate DNA length, in the range of one or two persistence lengths. With such short DNA, j also depends on twisting rigidity, which is discussed in more detail in Section 14.2.6.

Covalently closed circular DNA (ccDNA) is generally found in a supercoiled state. It is discussed extensively in Chapter 10.

3. FRICTIONAL COEFFICIENTS FOR MODEL SHAPES

3.1 Translational and Rotational Friction Coefficients

Translational and rotational frictional coefficients have been calculated for the model structures shown in Figure 9-1. The translational frictional coefficient f_t is the proportionality constant between the velocity $\mathbf{v}$ with which a particle is moving, and the frictional force $\mathbf{F}_f$ which resists that motion:

$$\mathbf{F}_f = -f_t \mathbf{v} \tag{9-11}$$

where the minus sign arises because the frictional force opposes the particle motion. The motion is produced by an applied force (e.g., centrifugation or electrophoresis) or by a concentration gradient, as in diffusion. Likewise, the rotational frictional coefficient f_r

is the constant of proportionality between the angular velocity w of the molecule, and the frictional torque $\mathbf{T}_r$ resisting rotation:

$$\mathbf{T}_r = -f_r\omega \tag{9-12}$$

where the rotation may be produced by an applied electric or hydrodynamic flow field, or by diffusion of molecular axes. Later in this chapter we will discuss (Sections 5, 6, and 10) the experiments by which these frictional coefficients are measured, but it will be helpful now to have a sense of how they depend on molecular size and shape.

3.2 Spheres

The simplest structure is the sphere, of radius R and volume $\frac{4}{3}\pi R^3$. It has frictional coefficients

$$f_t = 6\pi\eta R \tag{9-13}$$

and

$$f_r = 8\pi\eta R^3 \tag{9-14}$$

where η is the solvent viscosity. The cgs unit for viscosity is the poise, (dyn s cm^{-3}), abbreviated P. The viscosity of water at 20°C is 0.01 P, or 1 cP (centipoise). Equation 9-13 is known as Stokes' law, and the hydrodynamic radius R is often called the Stokes radius. Note that translational friction varies only as the first power of the molecular size, and therefore is much less sensitive to molecular dimensions than is rotational friction, which varies as R^3 or molecular volume.

The hydrodynamic radius can be related to other molecular parameters. A polymer with molecular weight M and partial specific volume $\overline{v}_2$ has minimum molecular volume and radius $V_{\text{min}} = M\overline{v}_2/N_{\text{A}} = \frac{4}{3}\pi R^3_{\text{min}}$, where N_{A} is Avogadro's number so the minimum radius is

$$R_{\text{min}} = \left(\frac{3M\overline{v}_2}{4\pi N_{\text{A}}}\right)^{1/3} \tag{9-15}$$

Nucleic acids and proteins are highly hydrated, and at least the first hydration shell of the water moves hydrodynamically with the macromolecule. If δ_1 grams of water (specific volume v_1^0) are associated with 1 g of dry polymer, then $\overline{v}_2$ in Eq. 9-15 is replaced with the hydrated specific volume $\overline{v}_2 + \delta_1 v_1^0$ (Tanford, 1961) and the hydrated radius becomes

$$R_0 = R_{\text{min}}(1 + \delta_1 v_1^0/\overline{v}_2)^{1/3} \tag{9-16}$$

Values of δ_1 for biopolymers typically range from 0.2 to 0.5 g g^{-1} (Kuntz and Kauzmann, 1974), with nucleic acids at the upper end of this range. Taking $\delta_1 = 0.5$ g g^{-1}, $v_1^0 = 1.0$

$cm^3 g^{-1}$, and $\overline{v}_2 = 0.56$ $cm^3 g^{-1}$, we find the ratio $R_0/R_{min} = 1.24$. For globular proteins, the hydration correction often corresponds to augmenting the dry radius by about 2.8 Å, the diameter of a water molecule. For oligomeric B-DNA, the hydrodynamic diameter is 20.5 ± 1.0 Å (Eimer et al., 1990), not much different from the 20-Å phosphate–phosphate diameter determined by crystallography. The hydrodynamically immobilized waters of DNA thus appear to be mainly in the grooves, where they do not add substantially to the frictional resistance of the double helix.

3.3 Ellipsoids of Revolution and Cylinders

Other rigid models are perforce asymmetric, and therefore more complex. The ellipsoid of revolution model is widely used because its properties can be calculated exactly. This generalization of a sphere is obtained by rotating an ellipsoid about one of its axes. The axis about which rotation occurs is called the symmetry axis, and its half-length (analogous to the radius of a sphere) is a. The half-lengths of the two other axes are equal to b. The axial ratio $a/b = p$ and the volume $V = \left(\frac{4}{3}\right)\pi ab^2$, an obvious generalization of the result for a sphere. Oblate ellipsoids are pancake shaped, with $p < 1$. Prolate ellipsoids of revolution are football shaped, with $p > 1$. They are often used as models for cylindrical rods, giving reasonably accurate results if length and volume of the two objects are equated ($2a = L$, $b_{ell} = \sqrt{3/2}b_{cyl}$). However, since cylinders do not have gradually narrowing ends, their frictional resistance, especially to rotational motion, is somewhat different from that of ellipsoids. Therefore, numerically accurate equations for cylinders have been obtained.

Translational and rotational friction coefficients for ellipsoids and cylinders are listed in Table 9-1. They are expressed as ratios relative to the frictional coefficients of spheres of the same volume. That is,

$$F_t = f_t(\text{ellipsoid or cylinder})/6\pi\eta R_e \tag{9-17}$$

$$F_r = f_r(\text{ellipsoid or cylinder})/8\pi\eta R_e^3 \tag{9-18}$$

The equivalent radii R_e of the spheres of equal volume are given in the first row of the table. For ellipsoids, $q = 1/p$; for cylinders, $p = L/2b$, where $2b$ is the diameter.

In translation, the frictional coefficients are different for movement with the long axis parallel or perpendicular to the direction of motion. However, since what is almost invariably measured is the average over all orientations, only the average is given for F_t. In rotation, $F_{r,i}$ is the frictional coefficient for rotation around the ith axis ($i = a$ or b). Two sets of equations are given for cylinders in the last column of Table 9-1. Those labeled with subscript B (Broersma, 1960a,b) have been most commonly used, and hence are given for convenience. However, those labeled with subscript T (Tirado and Garcia de la Torre, 1979, 1980) are more accurate, and are to be preferred.

3.4 Random Coil

A random coil has on average a spherical domain, so it is not surprising that it behaves hydrodynamically like a sphere with effective hydrodynamic radius closely related

Table 9.1
Hydrodynamic Properties of Ellipsoids and Rods Relative to Spheres of the Same Volume

	Prolate Ellipsoid[a]	Oblate Ellipsoid[a]	Cylindrical Rod[b]
R_e	$(ab^2)^{1/3}$	$(ab^2)^{1/3}$	$(3/2p^2)^{1/3}(L/2)$
F_t	$\frac{(1-q^2)^{1/2}}{q^{2/3}\ln\left\{\frac{[1+(1-q^2)^{1/2}]}{q}\right\}}$	$\frac{(q^2-1)^{1/2}}{q^{2/3}\arctan(q^2-1)^{1/2}}$	$\frac{(2p^2/3)^{1/3}}{\ln p+\gamma}$ $\gamma_B = -0.037 + 5.8\left(\frac{1}{\ln 2p} - 0.358\right)^2$ $g_T = 0.312 + \frac{0.565}{p} + \frac{0.100}{p^2}$
$F_{r,a}$	$\frac{4(1-q^2)}{3(2-2q^{4/3}/F_t)}$	$\frac{4(1-q^2)}{3(2-2q^{4/3}/F_t)}$	$0.64\left(1 + \frac{0.677}{p} - \frac{0.183}{p^2}\right)$
$F_{r,b}$	$\frac{4(1-q^4)}{3q^2[2q^{-2/3}(2-q^2)/F_t-2]}$	$\frac{4(1-q^4)}{3q^2[2q^{-2/3}(2-q^2)/F_t-2]}$	$\frac{2p^2}{9(\ln p+\delta_a)}$ $\delta_{a,B} = -0.76 + 7.5\left(\frac{1}{\ln 2p} - 0.27\right)^2$ $\delta_{a,T} = -0.662 + \frac{0.917}{p} - \frac{0.050}{p^2}$

[a]Perrin (1934, 1936); Koenig (1975).

[b]Broersma (1960a,b); Newman et al. (1977); Tirado and Garcia de la Torre (1979, 1980); Garcia de la Torre and Bloomfield (1981).

to its radius of gyration (Kirkwood and Riseman, 1956). For coils without excluded volume,

$$f_t = 6\pi\eta\frac{3\sqrt{\pi}}{8}\left(\frac{b^2N}{6}\right)^{1/2} = 6\pi\eta(0.665 < R_g^2 >^{1/2}) \tag{9-19}$$

which shows that the effective hydrodynamic radius is about two thirds of the radius of gyration. A similar relation holds for coils with excluded volume, but the numerical factor varies slowly with the excluded volume parameter ε (see Eq. 9-5) (Bloomfield and Zimm, 1966).

Because of its flexibility, a random coil does not undergo a defined (rigid) rotational motion. The equivalent of a rotational relaxation time is the relaxation time τ_1 for the longest normal mode of internal motion of the coil. It is this time that dominates in a viscoelastic relaxation experiment, as discussed in Section 9.

3.5 Wormlike Coil

The frictional behavior of a wormlike coil varies with its length. If it is short relative to its persistence length, it behaves like a rigid rod; if it is long, it behaves like a random coil. There is a smooth transition between these two limits when the contour length is comparable to the persistence length. The dependence of the sedimentation coefficient of DNA on molecular weight, which displays this expected behavior, is discussed in Section 5. The rotational diffusion coefficient of a wormlike coil has been estimated by Monte Carlo computer simulation (Hagerman and Zimm, 1981). The results are complicated, and the interested reader is referred to the original paper for the detailed equations.

4. ELECTROPHORESIS

Gel electrophoresis has become the most common method of nucleic acid characterization. In molecular biology, it is used for nucleotide sequencing and for sequence similarity studies (Southern and Northern blots). Gel electrophoresis is also extensively used for DNA size determination, and to measure supercoiling distribution under various physiological and in vitro circumstances. It provided the first indication of the existence of bent DNA molecules, and it is increasingly used for quantitation of protein–nucleic acid complexes. Many of these uses are discussed in detail elsewhere in this book, especially in Section 14 and in Chapters 10 and 13. Here we develop some of the basic principles. An excellent review of the physical mechanisms of gel electrophoresis is that of Zimm and Levene (1992).

4.1 Free Particle Electrophoresis

Before embarking on a discussion of gel electrophoresis, it is useful to introduce some of the basic physical concepts by considering electrophoresis in solution, where there are no barriers to molecular motion. Consider the balance of electrical and frictional forces. A particle of charge Zq (q = proton charge = 1.6×10^{-19} coulombs) in a field of E V m^{-1} will experience an electrical force

$$F_{\mathrm{el}} = ZqE \tag{9-20}$$

(The above values are in SI units. In cgs units, $q = 4.8 \times 10^{-10}$ esu, and the field is measured in stat-V cm^{-1}, where one stat-V = 1/300 V.) After a very brief period of acceleration, the particle will reach a steady velocity v because of the retarding frictional force,

$$F_{\mathrm{fr}} = -f_{\mathrm{t}}v \tag{9-21}$$

where f_{t} is the translational frictional coefficient, and the minus sign indicates that the frictional force is directed in the opposite direction to the electrical force. In the steady state, the sum of the two forces is zero, allowing calculation of the velocity as

$v = ZqE/f_t$. The electrophoretic mobility μ is defined as the velocity per unit field, hence

$$\mu = v/E = Zq/f_t \tag{9-22}$$

This equation makes it appear a simple task, knowing the frictional coefficient (e.g., from diffusion measurements, as described in Section 6), to calculate the charge on a molecule or complex. Unfortunately, this is not the case. Our very simple derivation has assumed that the polyion undergoing electrophoresis is completely isolated from other ions. In any realistic situation, the solution will contain ions from the buffer, added salt, and so on. These ions will distribute themselves around the charged particle, forming an ion atmosphere of the Debye–Hückel type with the counterions tending close to the particle surface and the co-ions further away. The ion atmosphere will screen the field, lowering F_{el}. The tight association of counterions with a highly charged polyelectrolyte such as DNA, discussed in Chapter 11, amplifies this effect. Furthermore, the counterions will be drawn in the opposite direction from the polyion; the resulting asymmetric charge distribution will set up a local field opposing the applied field E. Both of these effects will retard the polyion motion, so μ will be reduced and the apparent Z will be lower than the true charge.

Elaborate calculations for spheres and cylinders (see Schellman and Stigter, 1977 for application to short DNA modeled as a rigid rod) have taken these effects into account. However, the results are very complicated and not generally useful. One striking and important result has been obtained, however, for random coil polyelectrolytes. Under a wide range of conditions, the electrical and frictional forces on each monomer in the polymer balance each other, so the polyion has essentially the same electrophoretic mobility as the monomer. This theoretical prediction has been verified experimentally for DNA (Bloomfield et al., 1974, pp. 396–399). The result demonstrates that electrophoresis in free solution is totally insensitive to DNA size. Fortunately, this conclusion does not hold for electrophoresis in gels, our next topic.

4.2 Gel Electrophoresis

In the gel electrophoresis experiment, a small amount of macromolecule in buffer is applied at the top of a gel of polyacrylamide or agarose that has been saturated with buffer, and an electric field is applied to drive the polyion toward the bottom of the gel. After a suitable time to allow separation of components, the gel is stained (typically with the fluorescent dye ethidium bromide in the case of DNA, or Coomassie Blue for proteins) and the band positions are visualized. Alternatively, the nucleic acid may be radiolabelled with ^{32}P, and the bands visualized by autoradiography. For large DNA molecules, agarose gels at a concentration of 1% or less are most frequently used; pore sizes are about 100 nm, comparable to a few persistence lengths of DNA. For small DNA fragments or proteins, polyacrylamide gels in the 2–20% concentration range are employed; pore sizes are 1–2 nm, not much larger than the DNA diameter. Two typical gel patterns are shown in Figure 9-2: an agarose sizing gel and a polyacrylamide sequencing gel.

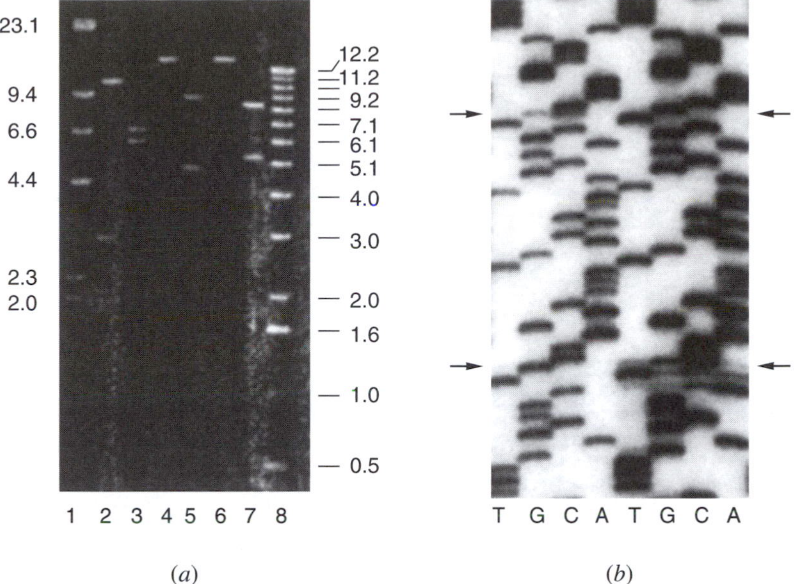

Figure 9-2
Typical gel electrophoresis patterns. (*a*) Agarose gel (0.8%) of restriction fragments of a genomic clone of a mouse opioid receptor, showing use of gel electrophoresis to determine DNA sizes. Lanes 1 and 8 are marker DNAs, whose lengths in kilobases (kb) are given on the left and right sides. Lanes 2–7 are digests with restriction enzymes KpnI, SacI, EcoRI, SalI, BamHI and SpeI. Electrophoresis was at 6 V cm^{-1} for 6 h; the gel was stained with ethidium bromide. (*b*) Portion of a 10% polyacrylamide sequencing gel of two clones of the TI plasmid from *Agrobacterium tumefaciens*. The arrows indicate two sequence differences between the clones: C to G (upper), and G to C (lower). [Courtesy Professor Anath Das, University of Minnesota.]

Gel electrophoresis is an empirical method for determining molecular weight, in the sense that standards of known molecular weight and homologous structure must be used for calibration. The data are then graphed in such a way as to give a linear plot over the range of sizes and mobilities of interest, and the molecular weights of the unknowns determined from those of the standards by interpolation. The two most common types of plot are log M versus μ_r (Fig. 9-3) and M versus $1/\mu_r$ (Fig. 9-4), where μ_r is the relative mobility. Note that the mobility–molecular weight dependence has a considerable sensitivity to the voltage gradient as well as to the gel concentration.

4.2.1 The Reptation Model

This behavior has been addressed theoretically with reasonable success (Zimm and Levene, 1992). The inverse dependence of μ on M (or contour length L) is explained by a simple theory (Lumpkin and Zimm, 1982), which assumes that the DNA moves with a wormlike motion with a friction coefficient proportional to its length; and that the

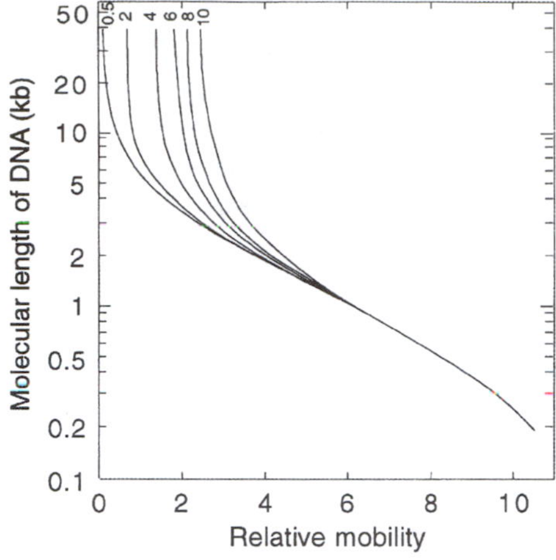

Figure 9-3
Molecular weight (kb) as a function of relative mobility for T7 DNA restriction fragments on 1.6% agarose gels. Curves are labeled with applied voltage gradients (V cm^{-1}). [Reprinted with permission from McDonell et al. (1977).]

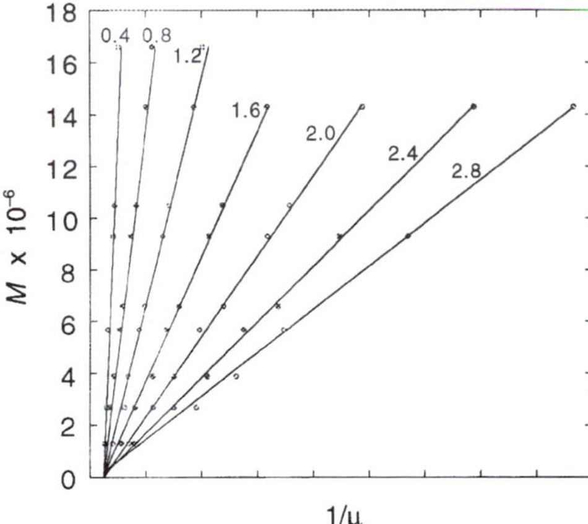

Figure 9-4
Molecular weight (Da) as a function of 1/relative mobility for λ DNA restriction fragments digested with *Hin* dIII and *Ava* I. Curves are labeled with % agarose. [Reprinted with permission from Southern, 1979.]

polymer conformation is described by Gaussian statistics without excluded volume. Since the DNA is stiff and moves in a wormlike fashion, it can be envisioned to move through some random "tube" in the gel formed by the gel fibers. This type of motion is known as "reptation" (Doi and Edwards, 1986). It has been directly observed by fluorescence microscopy of a single λ phage DNA molecule (Perkins et al., 1994a).

If Zq is the total effective charge on the polyion, Δs is an increment of length tangent to the polymer coil, and **E** is the electric field (voltage gradient) applied in the x direction, then the total tangential electric force on the polyion is

$$\frac{Zq}{L}\sum \mathbf{E}\cdot\Delta s = \frac{ZqE}{L}\sum \mathbf{i}\cdot\Delta s = \frac{ZqEh_x}{L} \tag{9-23}$$

where **i** is a unit vector along the x axis and h_x is the component of the polymer's end-to-end vector along that axis. These quantities are diagrammed in Figure 9-5.

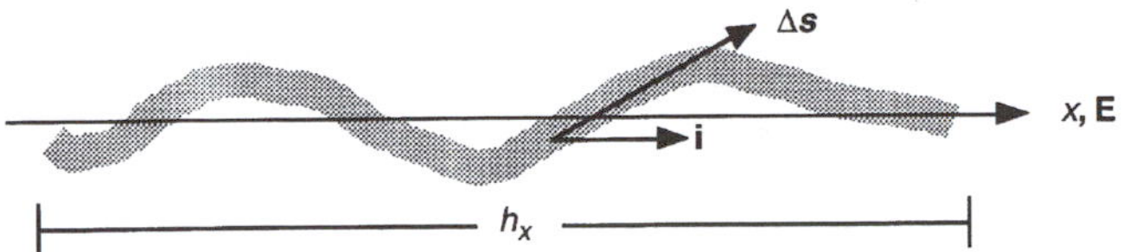

Figure 9-5
Wormlike chain motion in an electric field.

The electrical force is balanced by a frictional force:

$$f\dot{s} = ZqEh_x/L \tag{9-24}$$

where $\dot{s}$ is the mean velocity of the polymer along the tube. The component of the polymer's center-of-mass velocity along the x axis, $\dot{X}_{\text{cm}}$, is in the same proportion to $\dot{s}$ as h_x is to L:

$$\dot{X}_{\text{cm}} = \dot{s}h_x/L \tag{9-25}$$

Then combining Eqs. 9-23–9-25 and averaging over all chain conformations,

$$\mu = \frac{<\dot{X}_{\text{cm}}>}{E} = \frac{<h_x^2>Zq}{L^2 f} \tag{9-26}$$

The total charge Z is proportional to L, as is f according to the model, so Z/f is independent of L as noted above. Likewise, $<h_x^2>$ which varies like $<L^2>$, is proportional to L. Thus μ varies as $1/L$, or as the reciprocal of molecular weight.

Figure 9-3 shows that there is a significant dependence of mobility on applied field, even at quite low fields, which has practical significance, since it decreases the sensitivity of mobility to chain length. The physical basis of this phenomenon has been elucidated by Lumpkin et al. (1985). The tube through which the DNA moves becomes oriented because the field biases the direction of the leading end of the chain as it moves to extend the tube. The quantitative theory of this effect involves calculation of the orientation of an average chain (or tube) segment viewed as having an electric dipole moment. It predicts that the mobility should be the sum of two terms: The first, as before, is inversely proportional to chain length, and the second is independent of length but dependent on field. With reasonable choice of the parameters, the theory reproduces experimental results (Stellwagen, 1983; Hervet and Bean, 1987) quite well, as shown in Figure 9-6. A more refined theory (Duke et al., 1994) which takes into account the fluctuations of the chain end orientation in the field, accounts somewhat more accurately for the dependence of mobility on field strength at very low fields.

The reptation model of gel electrophoresis has some interesting consequences, since it implies that motion can occur relatively unimpeded only if all parts of the same molecule are in the same tube. Thus one predicts that circular DNA should have a lower mobility than linear DNA, even though it is more compact, since the likelihood

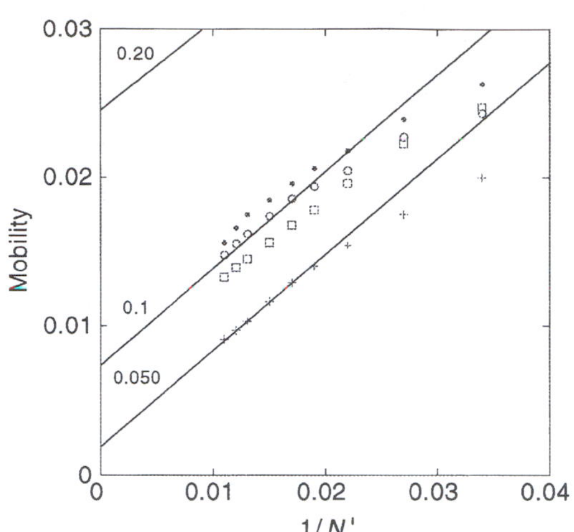

Figure 9-6
Fit of data of Stellwagen (1983) showing mobility (units of $cm^2\ V^{-1}\ h^{-1}$) as a function of chain length (N' = number of persistence lengths in chain) to theory of Lumpkin et al. (1985). The quantity Zq/f has been set to 18.3 $cm^2\ V^{-1}\ h^{-1}$ to fit the data at 0.64 V cm^{-1}, and the rms reptation tube length and DNA persistence length have both been set equal to $a = 67$ nm. Lines are labeled with reduced electric field $E' = aqE/2k_BT$. Experimental fields E (V cm^{-1}) are +, 0.64; □, 1.3; ○, 2.6; ●, 3.8. [Theory of gel electrophoresis of DNA, Lumpkin , O. J., Déjardin, P., and Zimm, B. H., *Biopolymers*, **24**, 1573–1593. Copyright ©1985. Reprinted by permission of John Wiley & Sons Inc.]

that chains on both sides of the circle are in the same tube is small. This decreased mobility appears to be the case under most circumstances, though it depends on gel concentration, molecular weight, and superhelix density (Mickel et al., 1977). Circular DNA may also get hung up on projections from the agarose gel matrix (Serwer and Hayes, 1987, 1989).

A similar consequence is that any disruption in the stiff double helix (e.g., by melting) should lower the mobility (Lerman et al., 1984; Abrams and Stanton, 1992). Such disruption may be envisioned to produce a branched polymer, and the likelihood that the new "ends" associated with the branches will enter the same tube as the main molecule is very small. Experimentally, this idea has been confirmed by performing gel electrophoresis on DNA molecules that have a low-melting (A,T rich) sequence in the middle, flanked by higher melting (G,C rich) sequences, in a gel that contains a gradient of a denaturant such as urea. As the DNA moves under the influence of the electrophoretic field from low-to-high denaturant, and enters the zone where the low-melting sequence is expected to undergo the helix–coil transition, its mobility decreases abruptly. This technique is very sensitive to DNA sequence, in favorable cases enabling separation of molecules that differ by single base pair substitutions.

When the DNA is shorter than the average pore size of the gel, the reptation model becomes inapplicable. In this regime, a sieving model in which the mobility is related to the probability that the molecule can move freely in the gel interstices without collision with the gel matrix (Ogston, 1958; Rodbard and Chrambach, 1970) is probably appropriate. In this regime, the logarithm of the mobility becomes a linear function of gel concentration.

4.2.2 Pulsed-Field Gradients for Separation of Very Large DNA Molecules

It is evident from Figure 9-3 that the gel electrophoretic mobility becomes independent of DNA size above about 10–20 kb. Above this size, the DNA random coil domain is very much larger than the average pore size in the gel, so the DNA must be extensively stretched out in a tube to undergo significant reptational motion in the field direction. Under these conditions, its frictional resistance is directly proportional to its length, so the ratio of electrical to frictional forces becomes independent of length. This independence stood as a major impediment to hopes of working on large, chromosome-size pieces of DNA until the development of pulsed-field gradient electrophoresis (PFGE) by Schwartz and Cantor (1984). Useful reviews of the physical basis and applications of this technique have been written by Cantor et al. (1988), Dawkins (1989), and Zimm and Levene (1992).

In PFGE, the electric field is applied alternately in two directions. When the field changes direction, the DNA molecule must try to reorient; molecules have a characteristic reorientation time that depends strongly on their length and their interaction with the gel matrix. Separation in PFGE depends on the ratio of the reorientation time to the time for which the field is applied before switching (the pulse time). If the pulse time is much faster than the reorientation time of all the molecules, then the molecules will be unable to follow. They will see just an average field and move as a group without separation. If the pulse time is much slower than the reorientation time of all the molecules, then all will be able to follow the field changes and again there will be no separation. In the intermediate case, the pulse and reorientation time are comparable for at least some of the molecules, which will then spend most of their time reorienting and will experience separation. By tuning the pulse time, the size range of the separation can be adjusted. It is now possible to separate chromosomal-sized pieces of DNA, up to about 10^7 bp, a several-hundred-fold increase in size over normal agarose gel electrophoresis.

As an example, Figure 9-7 shows the separation of the chromosomes of the yeast *Candida albicans*. The left panel shows a gel resulting from conditions that separate only the smaller chromosomes, up to about 2 Mb; the right panel shows separations of all chromosomes, up to 4.3 Mb. The pulse times in these experiments ranged from 1 to 15 min, and the total experiments took from 36 to 72 h. These very long times required to separate very large DNAs are characteristic of the technique.

The fundamental phenomenon of PFGE that allows separation is a minimum in mobility at a pulse time that is a function of DNA molecular weight. An example (Kobayashi et al., 1990) is shown in Figure 9-8 for field-inversion gel electrophoresis, developed by Carle et al. (1986), in which the pulsed fields are directly opposed at an angle of 180°. The field in the forward direction is applied three times longer than in the reverse. As Figure 9-8 shows, the minimum grows dramatically with the size of the DNA. This behavior cannot be explained by a reptational model. Several other theories have been developed that do account for the major observations (reviewed by Zimm and Levene, 1992). Each is highly approximate because of the very complex nature of the phenomenon, but they provide useful physical insight.

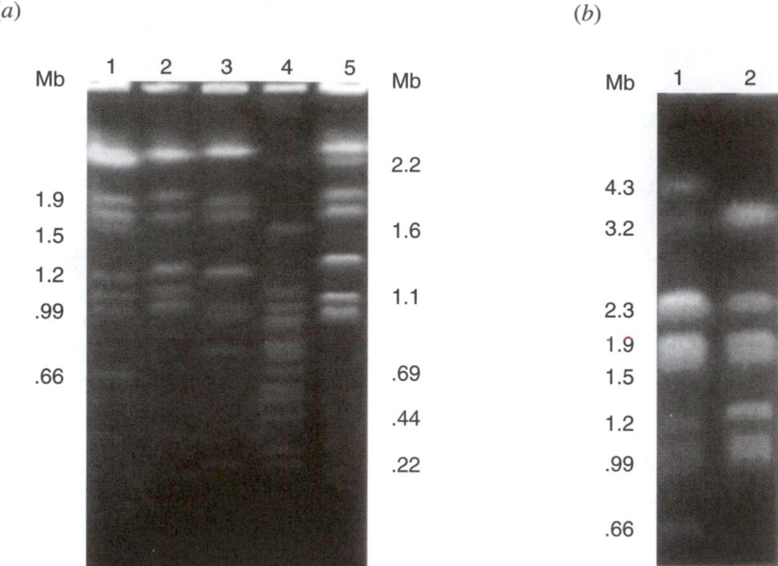

Figure 9-7
Pulsed-field gradient gel electrophoresis of yeast (*C. albicans*) chromosomes using the BioRad CHEF-DRII apparatus with a pulse angle of 120°. (*a*) Conditions for separation of smaller chromosomes: 60–300-s linear ramp, 36 h, 150 V, 0.9% agarose. The megabase (Mb) sizes of chromosomes are given at sides of gel for one *Candida* strain, lane 1, and for *Saccharomyces cerevisiae*, lane 4. (*b*) Conditions for separation of all chromosomes: 120–300-s linear pulse ramp, 24 h, followed by 420–900-s ramp, 48 h, 80 V, 0.6% agarose. Strains in lanes 1 and 2 are the same as in (*a*), lanes 1 and 2. [Courtesy of Dr. Bebe Magee, University of Minnesota.]

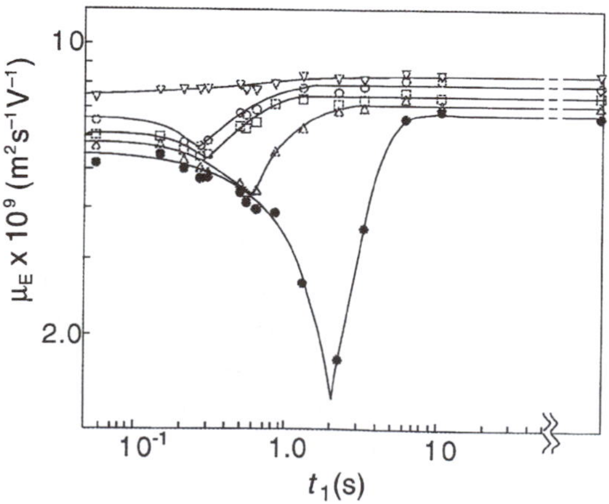

Figure 9-8
Field-inversion gel electrophoresis of DNA, as a function of the pulse time t_1 of the field in the forward direction. The pulse time in the reverse direction is $t_1/3$. The length of the DNA in kilobase pairs (kbp) is 6.56(▽), 9.42(○), 23.13(□), 48.50(△), and 166.0(●). [Reprinted with permission from Kobayashi T., Doe, M., Makens, Y., and Ogawa, M. (1990). *Macromolecules* **23**, 4480–4481. Copyright ©1990 American Chemical Society.]

One theory models the gel as a regular lattice of obstacles, and solves the dynamical equations by computer simulation (Deutsch and Madden, 1989). Results for a case in which the lattice spacing is comparable to the persistence length, and the energy due to the electric field is relatively high, are shown in Figure 9-9. The eight panels exemplify stages in a steady-state cycle characteristic of high field behavior. In Figure 9-9 (*a*), the chain is hooked around a gel fiber and is nearly fully extended. In Figure 9-9 (*b*), the chain has slid off the obstacle and begun to contract, the front more than the back. In Figure 9-9 (*c*), the front has curled into a rather dense coil which has lower mobility than the back. In Figure 9-9 (*d*), the entire molecule has been incorporated in the dense coil at the front, and is trapped between gel fibers. In Figure 9-9 (*e*) and (*f*), the coil unwinds, but remains hooked around several obstacles. In Figure 9-9 (*g*) and (*h*), the molecule returns to a conformation similar to (*a*), and the cycle begins again. Another model, which takes into account the random spacing of the gel fibers in agarose, is the "lakes-straits" model of Zimm (1991). The "lakes" are relatively large open volumes in the gel, in which several persistence lengths of the DNA can readily coil. They are connected by narrow "straits," which are narrow gates between fibers. A succession of alternating lakes and straits replaces the tube in the reptation model. Higher fields can force the chain to bunch up and overflow a lake by looping out through the fibers that form its boundary. Both of these models account for the mobility minimum phenomenon with plausible choices of parameters; and the cycling

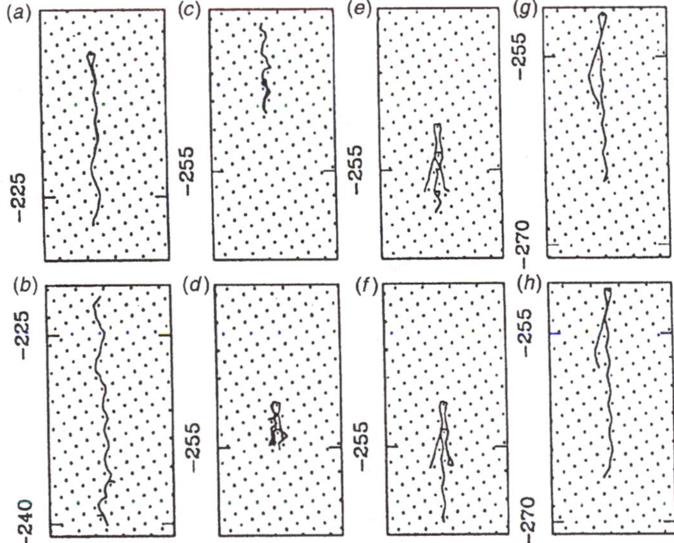

Figure 9-9
Simulated evolution of the conformation of a DNA molecule in a gel. In this simulation the lattice spacing is 2.1 times the persistence length a, the electrophoretic field E is downward, and the energy (relative to thermal) due to the field is $qEa/k_B T = 43$. The behavior displayed in these panels is described in the text. [Reproduced with permission from Deutsch and Madden, 1989.]

between extended and tightly bunched forms that they predict has been observed by fluorescence microscopy (Bustamante, 1991; Shi et al., 1995).

5. SEDIMENTATION VELOCITY

While gel electrophoresis is predominantly used for nucleic acid characterization, a wide range of hydrodynamic techniques that study molecular motion in dilute aqueous solution are well-established and are still valuable. These techniques are better understood theoretically, are uncomplicated by the poorly known properties of the gel phase, and reveal a wide variety of dynamic motions that are important in the functioning of nucleic acids.

Sedimentation is the most widely use of the various hydrodynamic techniques. Another term is centrifugation; and if the centrifuge operates at high speeds, it is called ultracentrifugation. We distinguish two types of sedimentation experiments: those in which the velocity of molecular motion is measured, and those in which the centrifuge runs until equilibrium is reached and one measures the unchanging concentration distribution. We first consider sedimentation velocity.

5.1 Basic Concepts and Equations of Centrifugation

The velocity with which a polymer moves in a centrifugal field is proportional to its molecular weight and buoyant density, and inversely proportional to its size. The constant of proportionality, the sedimentation coefficient, is a characteristic of the molecule and is often used to identify it, as when we speak of 5S RNA or 30S ribosomes. (S stands for the svedberg, the common unit of sedimentation defined below). Since high molecular weight, high-density molecules sediment more rapidly than small, light ones, sedimentation is a useful way to separate and purify molecules. Further, since molecular weight and size are often related for molecules of similar structure, such as nucleic acids, sedimentation can be used to determine molecular weight.

In a sedimentation velocity experiment, one measures the velocity of the macromolecular solute, dissolved in solvent, in response to a centrifugal field. It is conventional in a multicomponent system to denote water as component 1, polymer solute as component 2, and other small molecular solvent components (salts, buffer, etc.) as components 3, and so on. Consider a particle of mass m, spun with angular velocity ω at a distance r from the axis of rotation. The molecular weight of the polymer is M_2, so $m = M_2/N_A$, where N_A is Avogadro's number. If the specific volume (more rigorously, the partial specific volume) of the polymer is $\bar{v}_2$ cm^3 g^{-1}, and the density of the solvent is ρ g cm^{-3}, each polymer molecule displaces a mass of solvent $m\bar{v}_2\rho$, so its effective buoyant mass is $M_2(1 - \bar{v}_2\rho)/N_A$. The angular acceleration is $\omega^2 r$ (sometimes expressed as a multiple of g, the gravitational acceleration 980 cm s^{-2}), so the centrifugal force on the particle is

$$F_{\text{cent}} = (M_2/N_A)(1 - \bar{v}_2\rho)\omega^2 r \tag{9-27}$$

This produces a velocity $v = dr/dt$, which is resisted by a frictional force

$$F_{\mathrm{fr}} = -f_t v = -f_t(dr/dt) \tag{9-28}$$

In the steady state, the sum of these forces is zero, so one can define the sedimentation coefficient S as the velocity per unit acceleration:

$$S = \frac{dr/dt}{\omega^2 r} = \frac{M_2(1 - \bar{v}_2\rho)}{N_A f_t} \tag{9-29}$$

The first of these equalities allows determination of S in terms of measurable quantities. The second enables determination, from S and the buoyancy factor, of the quotient of molecular quantities M_2/f_t. (Note the similarity between Eq. 9-29 and Eq. 9-22 for electrophoresis. In both cases the numerator represents the response of the molecule to the driving field, and the denominator represents its frictional resistance to motion in the field.)

The frictional coefficient depends on polymer size and shape in a way that has been discussed in Section 3. It is useful, however, to list the following simple dependencies, obtained by considering the macromolecule as an equivalent sphere with Stokes radius R_h, thus with frictional coefficient $f_t = 6\pi\eta R_h$:

Shape	R_h	M_2/f_t
Sphere	$M^{1/3}$	$M^{2/3}$
Rod	$M/\ln M$	$\ln M$
Random coil	$M^{1/2}$	$M^{1/2}$

As an example of how these S versus M relations can shed light on nucleic acid structure, consider Figure 9-10, which shows the dependence of S on M for DNA in 0.2 M salt. The data are fit to the empirical equation (Crothers and Zimm, 1965)

$$S - 2.7 = 0.01517M^{0.445} \tag{9-30}$$

At low molecular weight, S increases only slightly with M (the rod limit of a wormlike coil), while at high M it increases faster, as expected in the random coil limit taking into account excluded volume and polyelectrolyte effects (see Eq. 9-5).

A theoretical equation (Gray et al., 1967) for the sedimentation coefficient of a linear wormlike coil, including excluded volume effects, has the same form as Eq. 9-30. Comparison betweeen theory and experiment led to the following structural parameters: persistence length $a = 450$ Å, hydrodynamic diameter $d = 27.2$ Å, and excluded volume parameter $\varepsilon = 0.11$. A somewhat more rigorous expression for wormlike coils without excluded volume (Yamakawa and Fujii 1973) gives similar numerical results. The sedimentation coefficient of relaxed circular wormlike coils with excluded volume (Gray et al., 1967) is of similar form to Eq. 9-30, but with a different numerical coefficient: $S - 2.7 = 0.01759M^{0.445}$. For typical relaxed circular

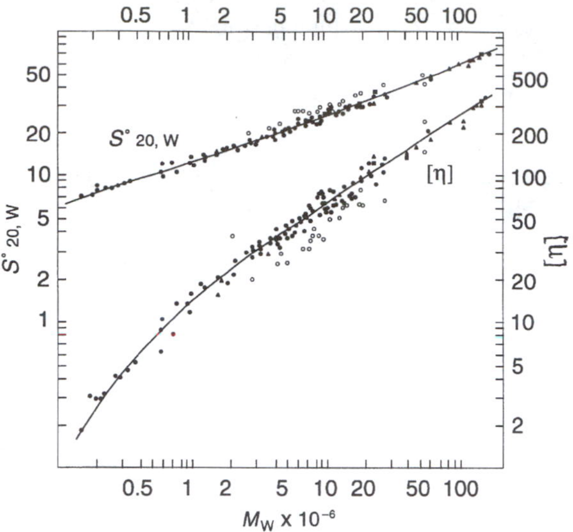

Figure 9-10
Dependence of sedimentation coefficient and intrinsic viscosity of DNA on molecular weight. The different symbols identify different sources of data listed in the original article. [Reprinted from Eigner, J. and Doty, P. The native, denatured and renatured states of deoxyribonucleic acid, *J. Mol. Biol.*, **12**, 549–580, Copyright ©1965, by permission of the publisher Academic Press Limited London.]

DNA, $S_{\text{linear}}/S_{\text{circular}} = 0.88 \pm 0.02$ in accord with experimental observations and with a theory for circular random coils (Bloomfield and Zimm, 1966).

The units of S are seconds (s). Generally, values of S are in the range 10^{-13}–10^{-10} s. To eliminate these small numbers, a sedimentation coefficient of 10^{-13} s is given the value of 1 S, after Svedberg, the inventor of the ultracentrifuge. The proper unit of angular velocity (ω) is radians per second (rad s^{-1}), but centrifuge speeds are most commonly stated in revolutions per minute (rpm). The conversion factor is 1 rpm = $60/2\pi$ rad s^{-1}. When doing calculations, it is important to work in consistent units.

The sedimentation coefficient is determined experimentally by measuring the position, as a function of time, of the band of solute material initially layered on top of the buffer solution, or of the midpoint of the boundary formed by redistribution of the components of the initially uniform solution in the centrifugal field. Rearrangement of Eq. 9-29 leads to $dr_m/r_m = S\omega^2 dt$, which when integrated yields

$$\ln(r_{\text{m}}/r_{\text{o}}) = S\omega^2 t \tag{9-31}$$

where r_{m} is to be interpreted as the midpoint of band or boundary at time t, and r_{o} is the starting position at $t = 0$. Thus S is obtained from the slope of a plot of ln r_{m} versus t, divided by ω^2.

An important use of Eq. 9-31 is to determine conditions for separation by differential centrifugation. Suppose that a solution contains two macromolecular components: a large, heavy, fast-sedimenting component with sedimentation coefficient S_1, and a smaller, slow-sedimenting component with sedimentation coefficient S_2. If r_m and r_o are the distances from the rotation axis of the bottom and meniscus of the centrifuge tube, ω can be chosen to be large enough to pellet the large component in an experimentally convenient time $t = 1/(S_1\omega^2)\ln(r_m/r_o)$; while ensuring that the position of the small component $r_{m,2} = r_o \exp[S_2\omega^2 t]$ is still close to r_o.

Various means are available to measure concentration distributions in sedimentation. In an analytical ultracentrifuge, and in some modern preparative instruments, the concentration across the cell can be monitored optically during the course of the run. For nucleic acids, which absorb light strongly at 260 nm, measurement of A_{260} as a function of r allows determination of the concentration distribution at quite low concentrations. For proteins, refractive index measurements using schlieren or Rayleigh interference fringe techniques are more common. The distribution can also be determined after a run is complete, by withdrawing and measuring solution from successive layers in the centrifuge tube. In this case, if the polymer has been radioactively labeled, or if it has some enzymatic or biological activity, very low concentrations can be detected.

There are numerous details that are important to the successful interpretation of sedimentation experiments. The band or boundary will not in general be perfectly sharp, but will be broadened by diffusion (an effect that increases with time) and by the presence of several solute components. The sedimentation coefficient is usually a decreasing function of concentration, often represented by the equation

$$S = S^\circ/(1 + k_s c) \tag{9-32}$$

so unless c is quite low, extrapolation of $1/S$ versus $1/c$ to $c = 0$ to obtain the infinite dilution S° may be necessary. Decrease of S with increasing c results from solvent backflow and from hydrodynamic interactions between molecules in concentrated solutions. The larger the molecule, the larger k_s. If S *increases* with increasing c, this is usually an indication of intermolecular association.

For highly charged macromolecules, such as nucleic acids, the sedimentation coefficient will be badly underestimated unless the experiment is conducted in a reasonable amount (0.1 *M* or so) of salt. This underestimation occurs because the centrifugal field tends to produce a separation of the heavy polyion from its light counterions. Since such a separation violates local electroneutrality, which is energetically very costly, the polyion sedimentation is markedly retarded. In the limit where there is no added salt, the apparent sedimentation coefficient will be $S_{true}/(Z + 1)$, where Z is the macroion charge.

Since S depends on f and ρ, both of which depend on solvent (f being proportional to solvent viscosity), it is usual to report S adjusted to standard conditions of infinite dilution at 20°C in water:

$$S^\circ_{20w} = S^\circ \left(\frac{\eta}{\eta_{20w}}\right) \frac{(1 - \bar{v}_2\rho)_{20w}}{(1 - \bar{v}_2\rho)} \tag{9-33}$$

This procedure depends, however, on the hydrodynamic size of the molecule being the same size in water as in the solvent of interest, which is not generally the case for flexible polyelectrolytes.

The sedimentation coefficient of DNA is a strong function of ionic strength as shown in Figure 9-11. Understanding this curve requires taking into account a number of effects. As the ionic strength goes down the DNA coil swells for two reasons. First, the persistence length increases as discussed in Chapter 11. Second, the DNA coil swells because of electrostatic repulsions between the phosphate groups, far distant from each other along the chain backbone. Both of these factors raise the frictional coefficient of the DNA and thereby lower its sedimentation coefficient. The change in salt concentration also has an effect on the buoyancy factor (see below). As Rinehart and Hearst (1972) have shown, the buoyancy factor increases as the sodium chloride concentration decreases. This somewhat compensates for the increase in frictional coefficient so that the dependence of S upon ionic strength is not as great as might have been expected.

Normally, one expects the sedimentation coefficient of high molecular weight DNA, like any random coil, to increase roughly as the square root of the molecular weight. A striking observation was made by Rubenstein and Leighton (1971), however, that when the DNA molecular weight reached several hundred million, the sedimentation coefficient no longer increased. Indeed, it appeared to decrease as the molecular weight rose still further. This effect was more pronounced at higher rotor speeds. This effect was explained theoretically by Zimm (1974), who used the bead and spring model of the DNA chain that we have already discussed, to show that the phenomenon arises from uneven frictional forces on the chain. On average, these forces are less near the center of the chain then they are at segments near the ends. The increased frictional drag causes the ends to lag behind, so average distances between segments are increased. This decreases the hydrodynamic shielding of one segment by another and increases the average friction, thereby decreasing the sedimentation coefficient.

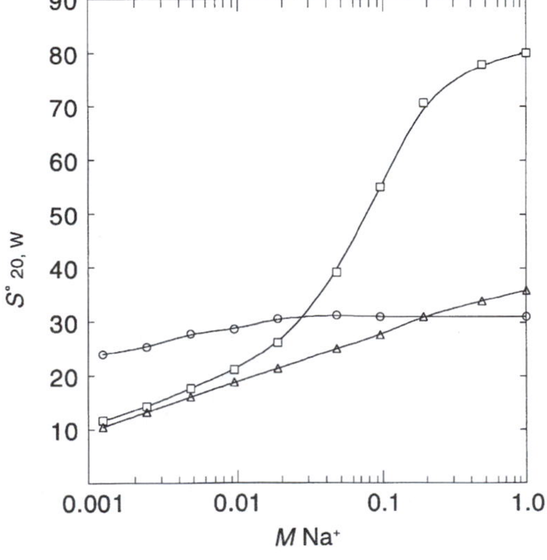

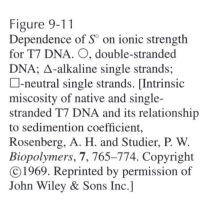

Figure 9-11
Dependence of S° on ionic strength for T7 DNA. ○, double-stranded DNA; Δ-alkaline single strands; □-neutral single strands. [Intrinsic miscosity of native and single-stranded T7 DNA and its relationship to sedimention coefficient, Rosenberg, A. H. and Studier, P. W. *Biopolymers*, **7**, 765–774. Copyright ©1969. Reprinted by permission of John Wiley & Sons Inc.]

5.2 Buoyancy Factor and Density Increment

In almost all biopolymer experiments, the solvent contains salt and buffer as well as water. Then the buoyancy factor $(1 - \overline{v}_2\rho)$ needs to be replaced by a more general factor, to take account of preferential solvation effects. As shown by Eisenberg (1976), this is $(\partial\rho/\partial c_2)_{\mu,T}$, where the subscript μ, T indicates that the density change is measured under conditions of constant temperature and chemical potentials of all solvent components. The buoyancy factor is determined by measuring the density of the solution as a function of polymer concentration, holding solvent chemical potentials constant by maintaining equilibrium with a large volume of solvent across a dialysis membrane. The buoyancy factor may also be written $(\partial\rho/\partial c_2)_{\mu,T} = (1 - \phi'\rho^\circ)$, where ϕ' is the apparent partial specific volume and ρ° is the solvent density. ($\overline{v}_2$ is ϕ' extrapolated to zero DNA concentration). Figure 9-12 shows ϕ' as a function of salt concentration for NaDNA in NaCl and for CsDNA in CsCl. It is evident that the buoyancy factor changes substantially with salt concentration. This figure also shows the salt dependence of the true partial specific volume at infinite dilution of DNA, $\overline{v}_2^\circ$, and the salt exclusion parameter $\Gamma_3 = (\partial w_3/\partial w_2)_{\mu,T}$ discussed in Chapter 11.

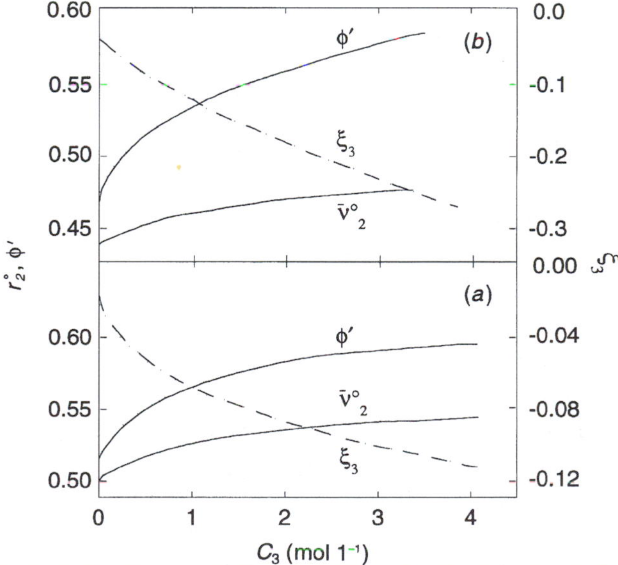

Figure 9-12
Apparent partial specific volume ϕ' as a function of salt concentration for (*a*) NaDNA in NaCl and (*b*) for CsDNA in CsCl. This figure also shows the salt dependence of the true partial specific volume at infinite dilution of DNA, v_2°, and the preferential solvation parameter ξ_3. [From Eisenberg, 1976.]

6. DIFFUSION

Diffusion is the expression of the random motion of molecules under the unceasing influence of collisions with their neighbors. This phenomenon is known as Brownian motion. It is important to understand diffusion if we are to have some concept of how molecules and particles may undergo random motion inside cells. The diffusion coefficient gives an indication of the size and shape of biological molecules. Diffusion also accounts for some aspects of band and boundary spreading in gel electrophoresis, chromatography, and sedimentation experiments. Perhaps the most interesting aspect of diffusion for biologists is that the rates of many reactions, for example the association of a repressor protein with an operator DNA sequence, are governed by the rate at which the reactants can diffuse together.

6.1 Fick's Laws of Diffusion

On the macroscopic level, diffusion is the equalizing of concentration differences between different parts of a system. If there is a concentration gradient $\partial c/\partial x$, diffusion will act to reduce the gradient to zero. This behavior is summarized by Fick's first and second laws of diffusion. If the concentration gradient is just in the x direction, then these laws are

$$J(x, t) = -D_t \left(\frac{\partial c_i}{\partial x} \right)_t \tag{9-34}$$

and

$$\left(\frac{\partial c_i}{\partial t} \right)_x = D_t \left(\frac{\partial^2 c_i}{\partial x^2} \right)_t \tag{9-35}$$

where $J(x, t)$ is the flux, or flow across unit area perpendicular to the flow direction per unit time, at x at time t, and D_t is the translational diffusion coefficient. If we take c_i in units of g cm^{-3}, and J is in units of g cm^{-2} s^{-1}, then we see that D_t has units of cm^2 s^{-1}, independent of the units of concentration used. Therefore, D_t has the same value whether the flow of matter is measured in grams, or moles, or molecules. It should also be noted that the units of J can be written (g cm^{-3})(cm s^{-1}), so that J is the product of concentration times velocity.

The basic solution of Fick's second law is obtained by considering that at $t = 0$, an infinitely sharp spike of material is formed at $x = 0$, with c_i initially equal to zero at all other points. It can be verified by direct substitution that the solution of Eq. 9-35 with these initial conditions is

$$c_i(x, t) = \frac{m_o}{\sqrt{4\pi Dt}} \exp \frac{-x^2}{4D_t t} \tag{9-36}$$

where m_o is the total amount of i in the system (i.e., $\int c_i(x, t)\, dx = m_o$). This Gaussian concentration distribution broadens with time according to its standard deviation $\sigma = \sqrt{2D_t t}$.

6.2 Diffusion and Random Walks

At the microscopic level, diffusion is the net result of purely random motions by individual particles, which is called Brownian motion. In a one-dimensional (1D) system, each molecule moves randomly to left or right; net flow occurs only because there are more molecules in the high-concentration region to move toward low c, than there are molecules in the low-c region to move toward high concentration. There is a strong connection between the random walk taken by a diffusing particle, and the motion of the end of a random coil polymer chain with respect to its beginning. For the latter, we previously noted (Eq. 9-3) that $< x^2 >= Nb^2$, where b is the bond length and N is the number of steps. By integrating the concentration distribution, Eq. 9-36, we find

$$< x^2 >= \int_{-\infty}^{\infty} x^2 c_i(x) dx = 2D_t t \tag{9-37}$$

which again shows the characteristic dependence of the mean-square displacement on the *first* power of the time (or the number of steps, if the number of diffusional steps per unit time is constant on average).

We have thus far considered only 1D diffusion for simplicity. It is not hard to show that the results equivalent to Eq. 9-37 are $< r^2 >= 4D_t t$ in two dimensions (2D), and $< r^2 >= 6D_t t$ in three dimensions (3D).

6.3 Measurement and Interpretation of Diffusion Coefficients

There are several standard ways of measuring the time evolution of concentration distributions, described in the standard references listed at the beginning of this chapter. For nucleic acids, nearly all modern diffusion measurements are made by dynamic light scattering, which is discussed in Section 12.

The relation between the diffusion and frictional coefficients was first derived by Einstein, who postulated an equivalent mechanical force F' acting on a single molecule. That force produces a velocity v, and the proportionality constant is the frictional coefficient f_t, so that $F' = f_t v$ (see Eq. 9-11). The force F' per molecule is then set equal to the thermodynamic force per mole (the gradient of the chemical potential) divided by N_A. Analysis developed in many standard texts then leads to

$$D_t = k_B T / f_t \tag{9-38}$$

where k_B is the Boltzmann constant (1.38×10^{-16} erg deg^{-1}) and T is the kelvin temperature. This equation is called the Einstein equation. It is strictly valid only at infinite dilution, but is generally adequate over the dilute concentration range in which biopolymers are normally studied.

The frictional coefficient depends on molecular size and shape. The simplest case is a sphere of radius R, for which Stokes' law gives

$$f_t = 6\pi\eta R \tag{9-39}$$

where η is the solvent viscosity. (In cgs units, 0.01002 P, or 1.002 cP, for water at 20°C.) Combination of these two equations gives

$$D_t = k_B T/6\pi\eta R \tag{9-40}$$

known as the Stokes–Einstein equation.

While Eqs. 9-39 and 9-40 strictly apply only to spheres, they are useful for other molecular structures as well, since frictional resistance may be characterized by an effective hydrodynamic radius R_h depending on size and shape. The parameter R_h is sometimes called the Stokes radius. The interpretation of translational and rotational frictional coefficients, in terms of the size and shape of molecular models of the sort displayed in Figure 9-1, has already been summarized in Table 9-1.

6.4 Molecular Weight from Sedimentation and Diffusion

The sedimentation and diffusion coefficients both depend inversely on friction coefficient f_t. By forming the quotient S/D_t, one eliminates f_t and obtains an expression for the molecular weight:

$$M = \frac{SRT}{D_t(1 - v_s\rho)} \tag{9-41}$$

where R is the gas constant (8.31×10^7 erg mol^{-1} deg^{-1}). Since S and D_t are readily measured, this is a useful way to determine the molecular weights of nucleic acids and well-defined nucleoprotein complexes such as viruses and nucleosomes.

6.5 Diffusion Controlled Reactions

Diffusion theory is very important in understanding the rates of bimolecular reactions in solution. If the activation energy of the reaction is fairly small, the rate may be limited by the diffusional encounter frequency of the reactants, undergoing the sort of motion shown schematically in Figure 9-13.

It is readily shown (e.g., Cantor and Schimmel, 1980, p. 920) that the bimolecular rate constant k_2 for the reaction of two spherical reactants with radii R_A, R_B and relative diffusion coefficients $D_{AB} = D_A + D_B$ is

$$k_2 = 4\pi D_{AB} R_{AB} N_{av}/1000 \tag{9-42}$$

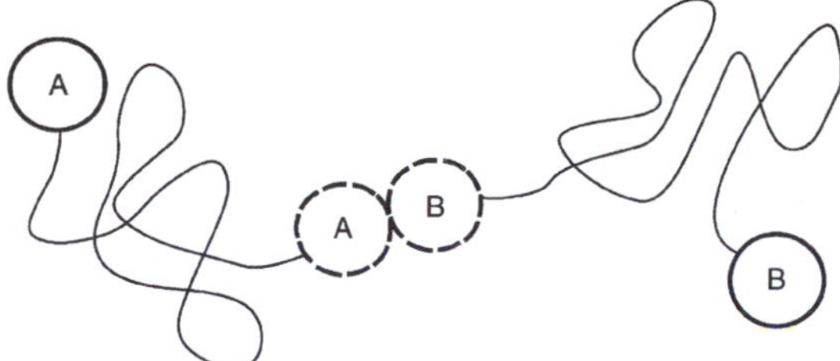

Figure 9-13
Diffusive motion of reactants A and B en route to a diffusion-controlled reaction.

For spherical molecules, the Stokes–Einstein equation is $D_i = kT/6\pi\eta R_i$. If we let $R_A/R_B = \alpha$, then

$$k_2 = \frac{2N_{av}kT}{3000\eta}\left(1 + \frac{1}{\alpha}\right)(1 + \alpha) \tag{9-43}$$

and if A and B are of equal size, $\alpha = 1$ so $k_2 = 8RT/3000\eta$ where the gas constant $R = N_{av}k = 8.31 \times 10^7$ ergs mol^{-1} deg^{-1}. For water at 20°C, this yields 6.5×10^9 L mol^{-1} sec^{-1}, a value that is relatively insensitive to molecular size.

There are numerous refinements to this simple theory that enable more accurate estimates of k_2 (Calef and Deutch, 1983; Berg and von Hippel, 1985; McCammon et al., 1986). One is taking account of nonuniform surface reactivity. If only a fraction of the surface of A and/or B is reactive, then k_2 must be reduced by a factor related to the product of the reactive fractions of each surface and to the rotational diffusion coefficients of the partners, which determine how rapidly they will be able to reorient to bring the reactive areas into juxtaposition. If the reactants are charged or if they interact by other intermolecular forces, their approach will produce attractions or repulsions that will accelerate or slow the reaction. Detailed computer calculations on enzyme–substrate interactions (Klapper et al., 1986) show that focusing of electric fields by the variable dielectric constant medium representing the polymer and solvent may produce effects that are considerably more complex than implied by a simple Coulombic interaction.

Other refinements of the theory include relaxation of the steady-state assumption and consideration of hydrodynamic interaction, in which solvent must be forced from between the two partners as they approach. A variant on simple diffusion control is gated diffusion, in which the binding site for a diffusing ligand opens and closes with some frequency comparable to the encounter frequency. Another variant is transport by segmental diffusion. An example would be a protein molecule bound to a nucleic acid that can be transported some distance by the motion of the DNA segment to which it is attached.

In nucleic acid research, much interest has been focused on diffusion-controlled reactions by the finding that the rate of combination of gene regulatory proteins (such as repressors and polymerases) with their target base sequences is faster than can be accounted for using the above rate constants (e.g., Riggs et al., 1970). This finding has led to the view, largely developed by Berg and von Hippel, that the kinetics of events such as repressor–promoter interaction are governed by 3D nonspecific binding to the DNA followed by effectively 1D sliding along the double helix to the target site. While the mathematic details of the application to DNA–protein interactions are very complicated (Berg et al., 1981; Berg, 1984), the basic results were put forward by Adam and Delbrück (1968). They showed that, for a process in i dimensions, the mean time of diffusion to the target is

$$\tau^{(i)} = \frac{b^2}{D^{(i)}} f^{(i)}\left(\frac{b}{a}\right) \tag{9-44}$$

where a is the target diameter and b is the diameter of the diffusion space. The first factor is the standard expression for the time to cover a root-mean-square distance b, and depends only weakly on dimension. The second factor, which Adams and Delbrück call the tracking factor, varies strongly with dimensionality. If $b \gg a$, then $f^{(i)}(b/a)$ is linear in b/a in 3D, logarithmic in 2D, and independent of b/a in 1D. This functional dependence produces a major reduction in search time in going from 3D to lower dimensionality.

7. SEDIMENTATION EQUILIBRIUM

7.1 Basic Equations

After a sufficiently long time, a solution subjected to a centrifugal field will reach an equilibrium in which there are no net flows of components, so concentrations become independent of time at each point in the cell. This condition is perhaps most readily thought of as a dynamic equilibrium between sedimentation and diffusion, in which the centrifugal force tending to push dense solute molecules toward the bottom of the cell is balanced by diffusion processes trying to equalize concentrations across the cell. A purely thermodynamic approach is also possible, in which the equation for sedimentation equilibrium is derived from the requirement that the total chemical potential for each component be constant across the cell. By either method, we find

$$c_i(r) = c_i(a) \exp\left[\frac{M_i(1 - \bar{v}_i\rho)\omega^2(r^2 - a^2)}{2RT}\right] \tag{9-45}$$

where $c_i(a)$ is the mass concentration of polymer component i at the reference point a in the cell.

In the simplest case of a single polymer solute (component 2), a plot of ln $c_2(r)$ versus r^2 will have a slope proportional to molecular weight M_2. In polydisperse mixtures, such a plot will be concave upward, since the largest molecules will be distributed preferentially toward the bottom of the cell. It can then be shown that the slope at r will yield the weight-average molecular weight at r:

$$< M_w(r) >= \frac{\sum_i c_i(r) M_i}{\sum_i c_i(r)} \tag{9-46}$$

7.2 Sedimentation Equilibrium in Associating Systems

An important use of sedimentation equilibrium in molecular biology is to determine whether DNA binding proteins are in monomeric or associated states under physiological conditions. Consider the simplest case of a monomer–dimer equilibrium of a protein P, $2P \rightleftharpoons P_2$, with equilibrium dissociation constant

$$K_d = \frac{[\mathrm{P}]^2}{[\mathrm{P}_2]} \tag{9-47}$$

If the total concentration of P in both forms is $[\mathrm{P}]_{\mathrm{tot}}$, so that

$$[\mathrm{P}]_{\mathrm{tot}} = [\mathrm{P}] + 2[\mathrm{P}_2] \tag{9-48}$$

then the concentration of monomer is

$$[\mathrm{P}]_1 = \frac{1}{2}\left(\frac{-K_\mathrm{d}}{2} + \sqrt{\frac{Kd^2}{4} + 2K_\mathrm{d}[P]_{\mathrm{tot}}}\right) \tag{9-49}$$

The mass and molar concentrations of monomer and dimer are related by $c_1 = M_1[\mathrm{P}]$ and $c_2 = 2M_1[\mathrm{P}_2]$, where M_1 is the molecular weight of monomer.

The equilibrium equations 9-47–9-49 must hold along with Eq. 9-45 at each point in the cell. The distribution of total protein, as affected by the dimerization equilibrium, can then be calculated and compared with experiment to obtain K_d. Figure 9-14 shows as an example the behavior of the RepA protein of plasmid P1, which mediates initiation of plasmid replication (DasGupta et al., 1993). The value of K_d which best fits the sedimentation equilibrium data, is about $2 \times 10^{-6} M$. Since the RepA concentration in DNA-binding reactions is in the nanomolar range, this shows that the protein is monomeric when it binds to DNA.

7.3 Buoyant Density Equilibrium

In nucleic acid work, the most frequent use of sedimentation equilibrium is a variation involving a density gradient formed by redistribution of a salt, usually CsCl, under the centrifugal field. If the density range spans the point of neutral buoyancy of the polymer, where $\bar{v}_2 = 1/\rho_o$, the polymer will tend to concentrate in a band around the position ρ_o where $\rho = \rho_o$ (see Fig. 9-15).

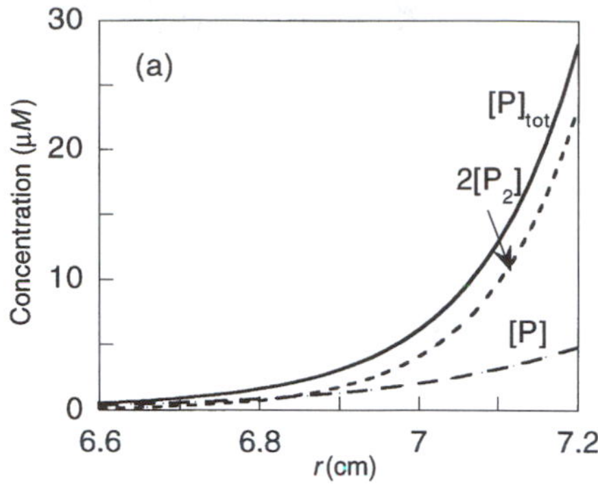

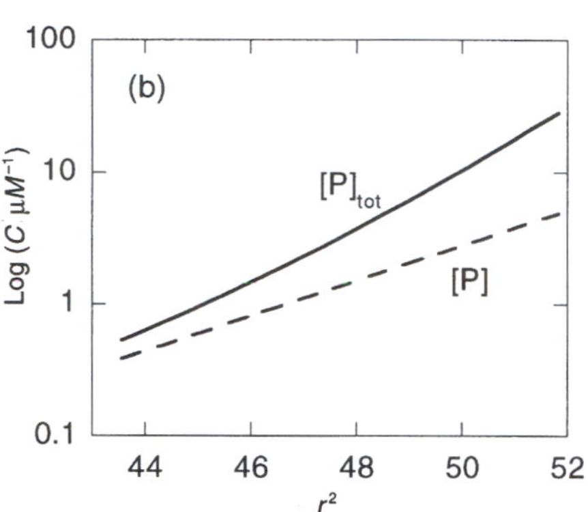

Figure 9-14
Sedimentation equilibrium behavior for the monomer–dimer equilibrium of protein RepA, using the data of DasGupta et al. (1993): monomer molecular weight 32,329 g mol^{-1}, partial specific volume 0.7215 mL g^{-1}, dissociation constant $K_d = 2.02 \times 10^{-6}$ mol L^{-1} at 5°C, rotation speed 12,000 rpm, loading concentration 9.33 μM. (*a*) Micromolar concentrations of total protein, monomer, and dimer (as μmol of monomer), as function of radial position. Note that the monomer dominates at low r, corresponding to low $[P]_{tot}$, while the dimer dominates at high r. (*b*) Plots of $\ln[P]_{tot}$ versus r^2 and $\ln[P]$ versus r^2. Note that the latter plot is linear, corresponding to a single molecular weight species, while the former plot is concave upward, corresponding to an increase in $< M_w >$ with increasing r^2 and $[P]_{tot}$.

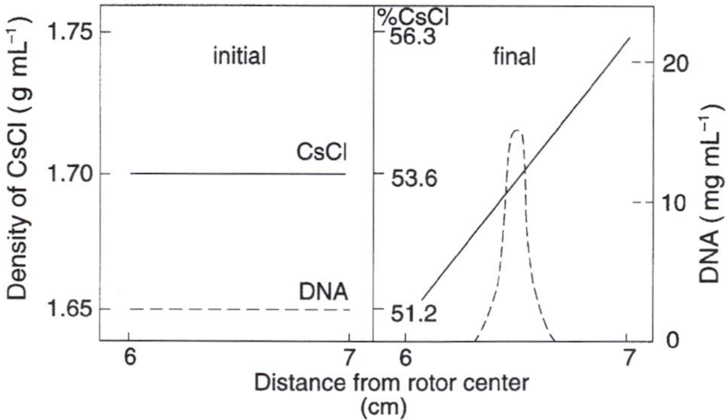

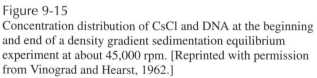

Figure 9-15
Concentration distribution of CsCl and DNA at the beginning and end of a density gradient sedimentation equilibrium experiment at about 45,000 rpm. [Reprinted with permission from Vinograd and Hearst, 1962.]

Even though the proper theory involves at least three components (water, polymer, and salt), we treat the system as if the solvent were a single component with variable density. We expand the density in a Taylor's series about $\rho_o : \rho = \rho_o + (\partial\rho/\partial r)(r - r_o) + \cdots$. Since $\overline{v}_2 = 1/\rho_o$ at neutral buoyancy, substitution in Eq. 9-45 yields

$$c_i(r) = c_i(0) \exp\left[-\frac{(r - r_o)^2}{2\sigma^2}\right] \tag{9-50}$$

which is a Gaussian distribution around ρ_o with standard deviation σ and variance σ^2

$$\sigma^2 = \frac{RT}{M_2\overline{v}_2(\partial\rho/\partial r)\omega^2 r_o} \tag{9-51}$$

The proper three-component theory can be shown to lead to equations with the same form, but with M_2 replaced by the solvated molecular weight $M_2(1 + \Gamma_1)$ and $\overline{v}_2$ by $(\overline{v}_2 + \Gamma_1\overline{v}_1)/(1 + \Gamma_1)$, where Γ_1 is the preferential solvent binding parameter—the grams of water "bound" per gram of polymer—discussed in Chapter 11.

Measurement of buoyant density can be used to determine GC content, since it has been shown by Schildkraut et al. (1962) that in CsCl gradients,

$$\rho_o = 1.660 + 0.098X_{\mathrm{GC}} \tag{9-52}$$

where X_{GC} is the mole fraction of GC base pairs. (This linear relation is violated for some synthetic sequences.) Cesium sulfate (Cs_2SO_4) is also used as a gradient former; it gives less linear but steeper gradients, which makes it useful to separate DNAs with widely different densities, as well as double-stranded RNAs and DNA–RNA hybrids. For details, see Szybalski (1968). Other uses of buoyant density equilibrium include

- Detection and measurement of glycosylation and alkylation.
- Separation of AT rich DNA on the basis of its affinity for $Hg^{(II)}$ and GC rich DNA on the basis of its affinity for $Ag^{(I)}$.
- Separation of native and denatured polynucleotides.
- Meselson–Stahl experiment demonstrating semiconservative replication in *Escherichia coli*.

Equation 9-51 suggests that measurement of the standard deviation of the bandwidth could be used to determine molecular weight. In fact, since $(\partial\rho/\partial r)$ varies as ω^2, the range of M varies as ω^4, making the technique potentially very useful over a very wide range of M. Vinograd and Hearst (1962) have demonstrated the basic validity of this approach for small proteins and DNAs. However, this approach fails for high molecular weight DNA. Schmid and Hearst (1969) have shown that the high concentration of DNA in the very narrow band (even though the solution as a whole is very dilute) make consideration of high-order virial coefficient corrections a necessity.

8. VISCOSITY

The viscosity of a fluid reflects its resistance to flow under an applied force. To move an object at a given velocity through molasses requires much more force than to move it through water. Therefore we say that molasses is more viscous than water. Viscosity is useful in the study of polymers because the addition of large molecules to a solvent increases its viscosity; the increase depends on the concentration, size, and structure of the polymer. Thus viscosity measurement, or viscometry, has been used to determine the molecular weights of nucleic acids, and the effects of solvents on their long-range structure and conformation. While viscometry is no longer a common method to study nucleic acids, it was one of the techniques that led, in the 1960s, to a fundamental characterization of the molecular weight and rigidity of DNA. Even now, the technique of viscoelastic recovery, discussed in Section 9.9, is one of the only physical methods available for determining the molecular weights of chromosome-sized DNA molecules.

Motion in one layer of a fluid causes motion in adjoining layers. To move layers with different relative velocities requires a force: the more viscous, the more force. The quantitative description of this relation was first enunciated by Newton. Suppose two parallel planes in the fluid each have area A, and are separated by distance h. If the upper plane moves with velocity v relative to the lower, then the shearing force F needed to maintain this velocity differential is given by

$$F = \eta A v / h \tag{9-53}$$

The coefficient of proportionality, η, is called the viscosity. It can be seen to have units of dyn cm^{-2}, or poise (P), in honor of the nineteenth century French physician Poiseuille, who investigated the flow of liquids in tubes. As noted earlier, water at 20°C has a viscosity of 0.01 P, or 1 cP.

Two things influence this force: the strength of the force between layers (greater in molasses than in water, and greater in water than in air), and the linear extent of the interactions. This finding explains why a cell extract is so viscous: The entangled network of macromolecules extends from one side of the container to the other, and the motion of one part of the fluid, for example, in pouring or stirring, is retarded by interactions with molecules far away (in regions with different velocities) mediated by intervening molecular interactions. Viscosity will decrease as polymer concentration decreases, since the probability of intermolecular interactions and the spans of intermolecular complexes will decrease.

This picture also predicts that bigger polymers, when dissolved in solution, will raise the viscosity more than smaller molecules, since their molecular domain has a greater extent. This concept was made quantitative by Einstein, who showed that for spheres occupying volume fraction ϕ in (dilute) solution, the solution viscosity relative to solvent viscosity η_o is

$$\eta = \eta_o(1 + \frac{5}{2}\phi) \tag{9-54}$$

The volume fraction (cm^3 polymer cm^{-3} solution) can be expressed as the product of the volume occupied by 1 g of hydrated polymer, v_h (cm^3 polymer g^{-1} polymer), and the weight concentration of polymer c_2 (g polymer cm^{-3} solution). Thus Eq. 9-54 can be rearranged to yield

$$\frac{\eta - \eta_o}{\eta_o c_2} = \frac{5}{2} v_h \tag{9-55}$$

It is customary to measure η at several concentrations and extrapolate to $c_2 = 0$, which yields the intrinsic viscosity $[\eta]$

$$[\eta] \equiv \lim_{c_2 \to 0} \frac{\eta - \eta_o}{\eta_o c_2} = \frac{5}{2} v_h \qquad \text{for spheres} \tag{9-56}$$

While this equation holds strictly only for spherical particles, it can be used to estimate the intrinsic viscosity for large, random coil molecules such as high molecular weight DNA ($[\eta]$ for DNA over a wide range of M is plotted in Fig. 9-10). A random coil behaves much like an effective sphere of radius R_g. Thus the volume occupied by 1 g of polynucleotide is $v_h = (4\pi/3)N_A \zeta^3 R_g^3$, where ζ is a factor of order unity. Thus we predict that

$$[\eta] = \frac{10\pi N_A}{3} \zeta^3 \frac{R_g^3}{M} \tag{9-57}$$

This relation is verified by theory and experiment, with $\zeta = 0.87$. We recall that for a random coil, R_g varies as $M^{1/2}$, so $[\eta]$ also varies as $M^{1/2}$. Plugging some typical numbers into this equation: If T4 phage DNA of $M = 110 \times 10^6$ were a random coil, $R_g \approx 10^4 Å = 10^{-4}$ cm. Then its intrinsic viscosity would be about 35,000 cm^3 g^{-1}. In a 1 μg mL^{-1} DNA solution, this corresponds to a viscosity 35 times that of water. In a lysate of phage-infected cells, the DNA concentration is probably considerably higher, and excluded volume and polyelectrolyte repulsion make R_g vary more rapidly than $M^{1/2}$, approximately as $M^{0.6}$. Thus one would expect that, if the DNA were unbroken, the viscosity of the lysate would be hundreds or thousands of times that of water. Indeed, one often observes gels of effectively infinite viscosity. Some of this is due to higher order concentration effects, and some to molecular entanglements that add to the purely hydrodynamic effects.

9. VISCOELASTICITY

9.1 Viscoelastic Relaxation

Viscometry is seldom used today to study DNA, because there are other more direct and readily interpretable ways to characterize short DNA; and because even the lowest shear gradients attainable cause distortion of large DNAs ($M_r > 10^8$), yielding unreliably low values of $[\eta]$. However, this very phenomenon of coil deformation in shear has been turned to advantage by Zimm and co-workers (Chapman et al., 1969; Klotz and Zimm, 1972; Dill and Zimm, 1980; Troll et al., 1980) in a related technique called viscoelastic relaxation or viscoelastic recovery. This technique can be used to measure the molecular weight of very large DNA, and in fact, as we shall see below, is most sensitive to the largest molecules in a population. It therefore has been used to explore the integrity of DNA in chromosomes, and to elucidate packaging mechanisms in bacteriophage.

The experimental method is rather simple in concept, though considerable delicacy is required in execution. A concentric cylinder apparatus is driven at an appropriately slow angular velocity for several turns. When the driving torque is turned off, the rotor does not simply stop; it actually turns backward, leading to the name viscoelastic recovery [see Fig. 9-16 (*a*)]. The backward angular displacement of the rotor decreases with time after cessation of the driving torque, in a manner that can be described as a superposition of exponential decays:

$$\theta(t) - \theta(0) = -\sum_{i=1}^{N} A_i \exp(-t/\tau_i) \tag{9-58}$$

where the τ_i values are relaxation times and the A_i values are amplitudes of the *i*th normal mode of motion of the polymer chain. Generally, the longest relaxation time is of most interest, since this is characteristic of the longest DNA chains in the system.

This behavior is based on the strong connection between a stretched polymer and an elastic spring. In the polymer, the restoring force is the greater randomness, or entropy, of the undistorted chain. The shear stress exerted by the turning rotor stretches the DNA molecules. When the torque is turned off, the chains try to maximize their entropy by returning to an undistorted ensemble of equilibrium conformations. In so doing, they drag solvent with them, setting up hydrodynamic forces that cause the rotor to rewind. These motions are opposed by viscous drag, causing their gradual decay (or relaxation) to zero. Since the largest molecules are the most distorted, they have the largest effect on the recovery. This combination of viscous and elastic forces leads to the term "viscoelasticity."

A theory of the springlike motion of high-polymer molecules, in the presence of viscous drag and thermal buffeting, shows that there is a spectrum of relaxation times. The longest, τ_1, is the most important in a viscoelastic relaxation experiment since it dominates the long-time part of the recovery behavior. The theory leads to the following expression for the longest relaxation time of a molecule of molecular weight M:

$$\tau_1 = BM[\eta]\eta_o/RT \tag{9-59}$$

where B is a constant and η_o is the viscosity of the solvent. This simple-appearing equation contains an important result. Since the intrinsic viscosity $[\eta]$, as shown above, varies as $M^{1/2}$ for a random coil, τ_1 varies as $M^{3/2}$. For polymers such as DNA with significant excluded volume or polyelectrolyte swelling of the coil, the exponent is even larger than three-halves. This is a stronger molecular weight dependence than nearly any other high-polymer technique. It therefore gives viscoelastic recovery a unique sensitivity to the largest molecules in a population.

As one example of the use of this technique, Figure 9-16 shows a viscoelastic relaxation curve for DNA isolated from giant heads of T2 bacteriophage. The capsids of these phage, grown in the presence of L-canavanine, are 12 times as long as normal T-even phage heads. This finding raises a question about the mechanism of DNA packaging within phage: Is the DNA packaged in one long piece, or in several pieces of normal length? The limiting slope of the semilog plot gives a longest relaxation time of 80.6 s. After extrapolation to zero DNA concentration and zero shear, the longest relaxation time in water at 25°C, $\tau^{\circ}_{25,w}$ was determined as 58 ± 6 s. According to the empirical equation $M = 1.56 \times 10^8\ (\tau^{\circ}_{25,w})^{0.6}$, the molecular weight of the largest molecules is 1.78×10^9, which is about 16 times normal. Thus the largest molecules in the population are proportional to the increased size of the head, implying a "headful stuffing" mechanism. However, analysis of the $t = 0$ intercept of Figure 9-16 shows that only 35% of the decay curve comes from these longest molecules, implying that many of the heads must have contained several shorter molecules.

To obtain unbroken chromosomal DNA molecules from *Drosophila*, Kavenoff and Zimm (1973) lysed cells in the viscoelastometer with detergent and pronase. The molecular weights for the largest DNA molecules ranged from 20 to 80 billion Da, and were proportional to the DNA contents of the chromosomes in the case of translocations or deletions. The results are consistent with a model of one continuous DNA molecule per chromosome, though this rests on a very long extrapolation of the empirical τ–M curve from samples of known size.

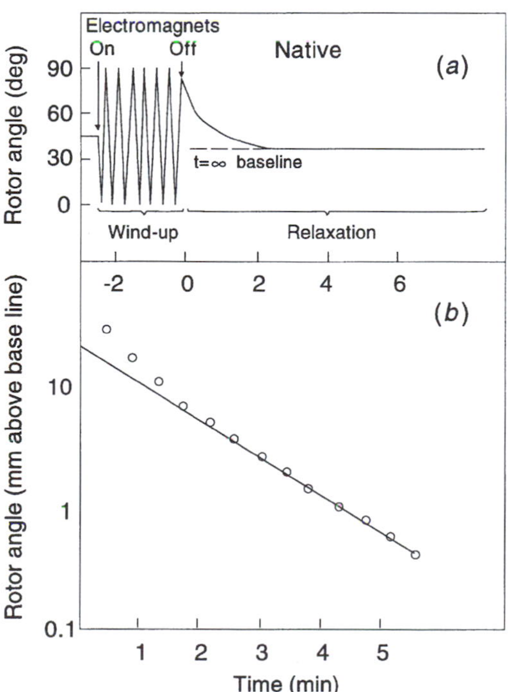

Figure 9-16
Viscoelastic recovery experiment on T2 DNA of 12 times normal length, at a concentration of 6.12 μg mL^{-1} DNA. (*a*) The angular displacement of a magnetically driven rotor increases steadily, for three revolutions (each maximum corresponds to a 90° rotation) in 2 min, until $t = 0$, when the torque is turned off. The displacement for $t > 0$ then decreases as the DNA relaxes and the rotor recoils in the opposite direction. (*b*) semilog plot of recovery phase to obtain longest relaxation time = 80.6 s. [Reprinted with permission from Uhlenhopp et al., 1974.]

9.2 Shearing Stress Breakage

We remarked above that passage of a large, flexible polymer such as DNA through a narrow capillary tube can severely stretch and distort it along the flow lines. Indeed, if the DNA is sufficiently long or the shear stress sufficient high, the molecule may be broken by the shear. The larger the molecule, the less the stress needed to break it. Normal laboratory fluid transfer operations, such as pipeting or pouring, are sufficient to break bacteriophage DNA down to fragments about 5–10 million molecular weight. In the early days of DNA research, before this phenomenon was understood, it was believed that a few million was the natural molecular weight of most DNA molecules. Only with carefully developed techniques, such as *in situ* cell lysis, is it possible to maintain integrity of chromosomal DNA molecules. As a practical matter, repeated passage through a hypodermic needle is a useful way to lower the viscosity of a DNA solution. Extensive sonication can lower the average size to about 200 bp or less, though the distribution is broad.

The mechanism of shear breakage is not well understood. It appears from the work of Yew and Davidson (1968) that shearing stress lowers the activation energy of the bond to be cleaved, thus accelerating the kinetics of a chemical cleavage event. In sonication or very strong vortexing, cavitation or small bubble formation seems to be involved in shear degradation.

10. ROTATIONAL DYNAMICS

Thanks largely to advances in lasers and electronics, which can measure processes on the nanosecond to microsecond time scale, methods that measure the rotational motion of nucleic acids have become important in characterizing their structures in solution. The uses to which rotational measurements have been put include determination of size and shape changes, measurements of persistence length of DNAs of different base composition and sequence, internal flexibility in tRNAs, and effects of ligand binding. As we noted earlier, rotation is very sensitive to small changes in size and shape because it depends approximately on the cube of the molecular length or radius. In contrast, translational techniques such as sedimentation and diffusion depend approximately on the first power of the linear dimensions. Detection of rotational motion generally depends on differences in optical properties of nucleic acid molecules along their principal molecular axes. This topic is discussed in Section 6.4 on linear dichroism.

10.1 Rotational Diffusion Coefficients and Relaxation Times

Under the influence of randomizing Brownian motion, a function $f(\theta, t)$ of the orientation of a molecular axis (Fig. 9-17) will exponentially decay, or relax, with a characteristic time constant: $f(\theta, t) = f(\theta, 0)e^{-t/\tau_r}$. This rotational relaxation time is denoted τ_r, and is inversely proportional to the rotational diffusion coefficient D_r, which has units of reciprocal seconds (s^{-1}). The relation between D_r and f_r, the rotational frictional coefficient (equal to the torque needed to rotate the molecule with unit angular velocity) is the same for rotation as for translation:

$$D_r = k_B T / f_r. \tag{9-60}$$

For a sphere of radius $R, f_r = 8\pi\eta R^3$ (Eq. 9-14).

There are two commonly defined rotational relaxation times, depending on the experiment being performed. The first, $\tau_r^{(1)}$, is obtained by experiments, such as dielectric relaxation, which directly measure the orientation, $f^{(1)}(\theta) = \cos(\theta)$, of the molecular axis with respect to an external field. The second type of relaxation time, $\tau_r^{(2)}$, is obtained in experiments (more common in contemporary practice) such as fluorescence anisotropy, electric dichroism, and NMR relaxation, which measure the orientation function $f^{(2)}(\theta) = \frac{1}{2}(3\cos^2\theta - 1)$. One can show that the relaxation time of axis a by rotational diffusion about axes b and c is $1/\tau_a^{(1)} = D_{r,b} + D_{r,c}$. For a sphere, which has three equivalent axes, $\tau_r^{(1)} = \frac{1}{2}D_r = f_r/2k_BT$, and $\tau_r^{(2)} = \frac{1}{6}D_r = f_r/6k_B$ Combining this with Eq. 9-60 and the volume $V = \frac{4}{3}\pi R^3$, we get the elegantly simple result

$$\tau_r^{(2)} = \eta V/kT \tag{9-61}$$

Since this result was first obtained by Debye, it is sometimes known as the Debye relaxation time. Results for molecular models of other shapes, such as ellipsoids of revolution and rigid rods, can be obtained from Table 9-1. In all cases, it is found that τ_r varies roughly according to V or R^3, where R is some characteristic linear dimension.

10.2 Photoselection: Fluorescence and Triplet State Anisotropy Decay

Techniques that measure rotational motion can be divided into two broad classes: those that orient molecules by application of an external field, and those that use polarized light to select molecules that are already oriented along the polarization direction. The most widely used of the photoselection techniques is fluorescence anisotropy decay, also known as fluorescence depolarization, which probes motions in the nanosecond time regime. Another recently developed technique is triplet state anisotropy decay, which is sensitive to microsecond motions. Our discussion will concentrate on fluorescence, since it is by far the more developed and widely applied. Both techniques are useful and important and these basic concepts and working equations are similar. A thorough discussion is given by Cantor and Schimmel (1980, pp. 454–465).

To understand photoselection techniques requires use of the concept of the electronic transition dipole moment, which was introduced in Chapter 6. Fluorescence involves first the absorption and then the emission of light. If the exciting light is vertically plane polarized with electric field **E** along the z-axis of Figure 9-17, then only those molecules with a component of their absorption transition dipole moment $\boldsymbol{\mu}_a$ in the same direction as **E** will be able to absorb. The amplitude of **E** along $\boldsymbol{\mu}_a$ is

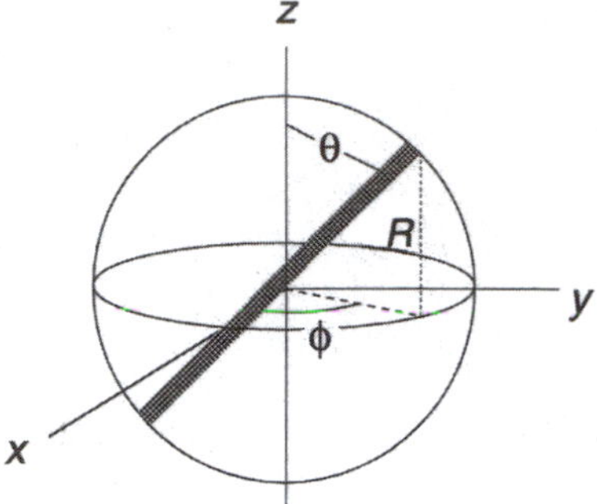

Figure 9-17
Diagram of spherical polar coordinate geometry for rotational dynamics experiments.

$\mathbf{E} \cdot \boldsymbol{\mu}_a = E\mu_a \cos\theta$, and the intensity of absorption, proportional to the square of the amplitude, varies as $\cos^2\theta$. Thus those molecules with $\cos\theta$ near 1 will be preferentially excited.

Assume initially that the system is rigid, so that the molecule does not rotate between excitation and emission, because it is very large or because the solution is effectively frozen. The fluorescence emission is polarized along the direction of the emission transition dipole moment μ_e. In the simplest and most common case, μ_e and μ_a are coincident. If the amplitude of the fluorescence electric field is **F**, then the component of the intensity of a single emission event in the direction of the excitation direction will be $(F\mu_e)^2 \cos^2\theta$. This term is multiplied by the probability that the molecule was excited in the first place, $\cos^2\theta$. Thus the total intensity of fluorescence along the excitation direction will be proportional to $F_\parallel = < \cos^4\theta >$, where the angular brackets denote an average over the whole population of fluorescent molecules. Likewise, the probability of emission perpendicular to the excitation direction will be proportional to $F_\perp = < \cos^2\theta \sin^2\theta \cos^2\phi >$ where ϕ is the azimuthal angle and $\sin\theta \cos\phi$ is the projection of μ_e on the x axis. The values of these averages are $F_\parallel = \frac{1}{5}$ and $F_\perp = \frac{1}{15}$.

Experimentally, one defines the fluorescence anisotropy as the difference between fluorescence intensity emitted parallel and perpendicular to the exciting field, normalized by the total emitted intensity (there are one parallel and two perpendicular directions of emission):

$$A = \frac{F_\parallel - F_\perp}{F_\parallel + 2F_\perp} \tag{9-62}$$

The maximum possible value of this quantity, assuming that the molecules are randomly oriented at excitation, have not rotated at all during the interval between excitation and fluorescence, and that the absorption and emission transition dipoles are coincident, is calculated as $A_o = \frac{2}{5}$ from the values of $F_\parallel$ and $F_\perp$ for a rigid system. In the more general case, μ_e and μ_a may not be the same, if intersystem crossing has occurred during excitation and vibrational relaxation. If the angle between μ_a and μ_e is ξ, the value of the anisotropy for a rigid system is calculated to be $A_o = (3\cos^2\xi - 1)/5$.

Now, consider the case in which the molecules rotate in the time between excitation and emission. This period is typically a few picoseconds to a few nanoseconds, during which the electronic excitation energy is redistributed within the molecule (primarily by transfer into vibrational modes) and/or transferred to surrounding solvent, and then emitted as a red-shifted photon as the molecule drops from excited singlet S_1 to ground state singlet S_0. The orientations of the molecules change as they rotate, so the fluorescence anisotropy (A) and polarization (P) will change. In the limiting case that rotation is very fast compared to emission, all initial orientation information will be lost, so A and P will decay to zero.

In the more general (and more useful) case, rotational relaxation time τ_r and fluorescence lifetime τ_F will be comparable. Then the parallel and perpendicular components of the emission will decay with time for two reasons: the fraction of molecules in the excited state decays according to $\exp(-t/\tau_F)$, and the orientation according to $\exp(-t/\tau_r)$. The result (Cantor and Schimmel, 1980, p. 461) is

$$A(t) = \left(\frac{2}{5}\right)\exp(-t/\tau_r)[(3\cos^2\xi - 1)/2] \tag{9-63}$$

Thus, the time-dependence of the anisotropy can be used to determine τ_r, and thereby provide information on molecular size and shape.

If the experiments are done in the steady-state rather than in a time-resolved manner, then one must average separately over the time dependence of components $F_\parallel$ and $F_\perp$ before computing the steady-state anisotropy $< A >$. The result is conveniently written in reciprocal form as

$$< A >^{-1} = A_o^{-1}(1 + \tau_F/\tau_r) = A_o^{-1}(1 + \tau_F kT/V\eta) \tag{9-64}$$

Another commonly used measure of anisotropy is the polarization P, defined as

$$P = \frac{F_\parallel - F_\perp}{F_\parallel + F_\perp} \tag{9-65}$$

which has maximum value $P_o = \frac{1}{2}$ for a rigid system. Algebraic rearrangement shows that

$$\left(\frac{1}{P} - \frac{1}{3}\right)^{-1} = \frac{3}{2}A \tag{9-66}$$

so the equation for the steady-state polarization is

$$\frac{1}{< P >} - \frac{1}{3} = \left(\frac{1}{P_o} - \frac{1}{3}\right)\left(1 + \frac{\tau_F}{\tau_r}\right) = \left(\frac{1}{P_o} - \frac{1}{3}\right)\left(1 + \frac{\tau_F kT}{V\eta}\right) \tag{9-67}$$

These are known as the Perrin equations (Perrin, 1934). Thus the molecular volume V can be obtained, if τ_F is known by independent measurement, by measuring $1/ < A >$

or $1/ < P >$ as a function of T/η. In these experiments, the viscosity is generally varied by changing the temperature (for water, η changes by 2% for each centigrade degree) or by adding an inert compound such as sucrose or glycerol to the solution.

The equations in this section are strictly valid only for rigid, spherical molecules. If the molecule is substantially nonspherical, the expressions become significantly more complicated, owing to the contributions from rotations about the nonequivalent axes. In the limit, as many as five relaxation times may enter into the decay of the anisotropy, though it is rare that more than two can be resolved experimentally. Another complexity arises if the chromophore has substantial independent mobility within the macromolecule. The measured rotational relaxation time will then be much shorter than that corresponding to the macromolecular dimensions. On the other hand, approximate equality of measured and calculated relaxation times indicates that the chromophore moves rigidly with the macromolecule. For example, τ_r obtained from anisotropy decay measurements of fluorescence from ethidium bromide bound intercalatively to $tRNA^{phe}$ of *E. coli* is 26 ns [Fig. 9-18(*a*)]. If the tRNA were a sphere, it should have $\tau_r = 12$ ns. Flexibility would yield a lower value. Instead, the higher value is consistent with a prolate ellipsoid model with an axial ratio of 3 or 4. By contrast, the anisotropy of the fluorescent Y base in the anticodon loop of $tRNA^{phe}$ decays with a longest time

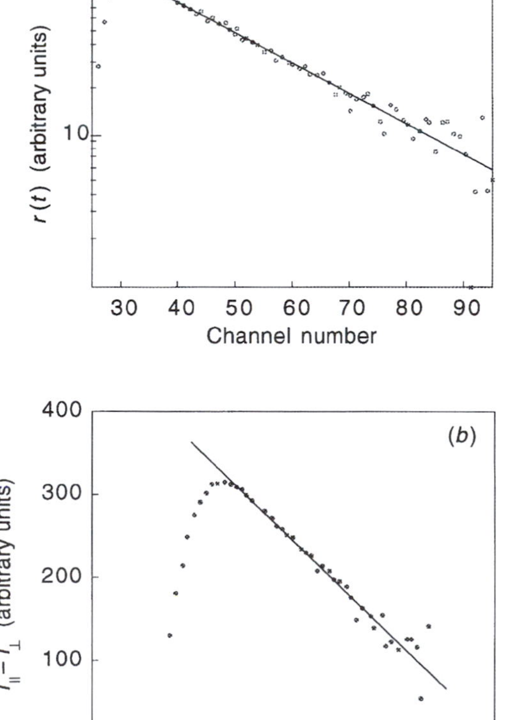

Figure 9-18
Fluorescence anisotropy decay of $tRNA^{phe}$: (*a*) Fluorescence from intercalated ethidium bromide, 1.20 ns/channel, relaxation time $\tau_D^{25,w} = 26.2$ ns [Reprinted with permission from Tao, T., Neeson, J. H., and Cantor, C. R., *Biochemistry*, **9**, 3514–3524. Copyright ©1970 American Chemical Society.]
(*b*) Fluorescence from Y base, 0.47 ns/channel, relaxation time $\tau_D = 9.2$ ns. [Reprinted with permission from Beardsley, K., Tao, T., and Cantor, C. R., *Biochemistry*, **9**, 3524–3532. Copyright ©1970 American Chemical Society.]

constant of about 9 ns [Fig 9-18(*b*)], indicating that the Y base is not rigidly attached to the tRNA.

Mobility of a chromophore within DNA has been used to determine the bending and twisting force constants of DNA. The initial observation was made by Wahl et al. (1970), who observed a 28-ns relaxation process in fluorescence depolarization experiments on a complex between DNA and ethidium, which has a fluorescence lifetime τ_F of 23 ns. Since ethidium is tightly bound to DNA by intercalation and since the relaxation time is too fast for rotation of the entire DNA molecule (calculated to be $\sim$ 1 ms), Wahl et al. (1970) concluded that the observed decay of the fluorescence anisotropy was due to internal rotatory Brownian motion in the DNA helix. This observation was put on a quantitative basis by Barkley and Zimm (1979) and Allison and Schurr (1979) who showed that the time-dependence of the anisotropy should be

$$A(t) = \frac{0.75\, e^{-\Gamma} + 0.45\, e^{-(\Gamma+\Delta)} + 0.4\, e^{-\Delta}}{3 + e^{-\Delta}} \tag{9-68}$$

This equation assumes that the absorption and emission transition moments are colinear and perpendicular to the local helix axis. The twisting decay function is $\Gamma(t) = 4kT(t/\pi C\rho)^{1/2}$, where C is the torsional stiffness and ρ is the frictional coefficient per unit length for rotation about the helix axis. The bending decay function is $\Delta(t) = B(t)t^{1/4}$, where $B(t)$ is a slowly varying function that depends on the bending stiffness. Except at very short times, nearly all the depolarization is due to twisting. A more complete theory, that takes into account local wobble of the chromophore as well as twisting and bending, fits the whole time course of the anisotropy decay somewhat better than the Barkley–Zimm theory (Schurr, 1984; Shibata et al., 1985). Numerical results of the two theories agree to within a factor of 2–4. The torsional stiffness parameter derived from these theories is about 2×10^{-19} erg cm, corresponding to an rms twisting fluctuation of about 6.9°–8° per bp (Schurr and Fujimoto, 1988). More details about twisting and bending energies are given at the end of this chapter.

Triplet anisotropy decay is another photoselection technique that has had fruitful application to DNA (Hogan et al., 1982). As with fluorescence anisotropy, an intercalating dye is used as a probe. However, in this case the dye is methylene blue rather than ethidium. Methylene blue has a high triplet yield. That is, it has a relatively high probability of being excited into a triplet state rather than an excited singlet state. Once in the triplet state, decay to the ground singlet is slow, since triplet–singlet transitions are forbidden by first-order quantum mechanical selection rules. The triplet lifetime of methylene blue is about 100 μs, compared with 23 ns for the singlet lifetime of intercalated ethidium. Thus triplet anisotropy decay techniques can measure motions over a time scale 1000 times larger than fluorescence methods allow.

The time dependence of the orientational distribution of photoselected molecules can be monitored by measuring either the polarization of emission from the triplet to the singlet state (phosphorescence) or the absorbance anisotropy of the depleted singlet state. A monitoring beam at the singlet transition frequency allows measurement of the decrease in absorbance that occurs when singlet chromophores undergo intersystem crossing to the triplet state. Hogan et al. (1982) chose the latter approach, because of its greater sensitivity. They used a series of DNA fragments from 65 to 600 bp long, and interpreted their results with the Barkley–Zimm (1979) theory. For the shorter fragments, less than or equal to 165 bp, DNA tumbles end-over-end as predicted for a rigid rod. For longer molecules, there is deviation from rigid rod behavior indicating segmental bending motions of the helix axis. At the earliest times, there is a rapid decay of anisotropy attributed to torsional motions of the helix with a time constant near 50 ns. Figure 9-19 schematically shows the time ranges over which these various types of motions are important as a function of DNA length.

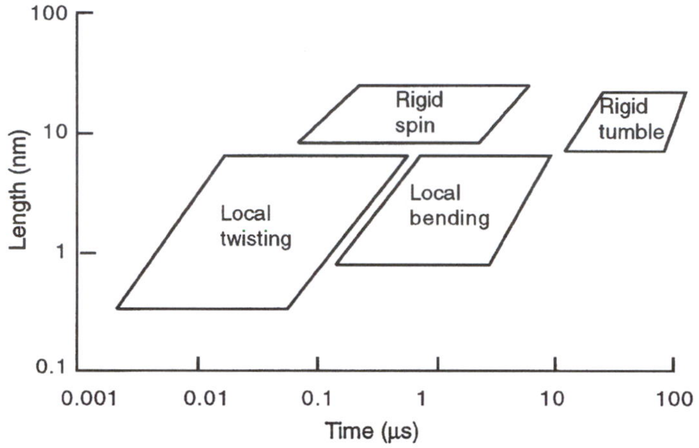

Figure 9-19
Time and length ranges for which different types of DNA rotational motion are important. [From Hogan et al. 1982.]

10.3 Transient Electric Dichroism and Birefringence

We turn now to the dynamics of orientation and disorientation in an applied field. In the early days of nucleic acid research, the most common means of orientation was a hydrodynamic shearing field. More recently, with the development of rapid electrical pulsers and rapid-response electronics, orientation in an electric field has become more common, though flow experiments are still useful. Regardless of the source of the orienting field, detection of orientation depends on differences in optical properties in different directions relative to the field. The difference in absorbance $\Delta A = A_{\parallel} - A_{\perp}$ is called dichroism and is discussed in Section 6.4. The analogous difference in refractive indexes, $\Delta n = n_{\parallel} - n_{\perp}$, is called birefringence. General reviews of the theory and methods of electric birefringence and dichroism are given by Fredericq and Houssier (1973) and by Kahn (1972) for birefringence. A review of applications to DNA and its complexes, with particular emphasis on ion atmosphere polarization effects, has been written by Charney (1988).

Chapter 6 considers the time-independent or steady state orientation in a field. In fact, the dichroism or birefringence in a typical experiment builds from zero after the field is turned on, reaches a steady plateau, and decays to zero after the field is turned off. (As a practical matter, the ionic strength is as low and the field pulse is as short as possible, to minimize joule heating and bulk electrophoresis.) This behavior, illustrated in Figure 9-20, is called transient dichroism or birefringence. It is expressed mathematically (Benoit, 1951; Tinoco, 1955) by

$$\frac{\Delta\varepsilon}{\Delta\varepsilon_{ss}} = 1 - \frac{3r}{2(r+1)}e^{-2D_r t} + \frac{r-2}{2(r+1)}e^{-6D_r t} \tag{9-69}$$

where $\Delta\varepsilon$ stands for dichroism or birefringence, depending on the experiment. (We will generally refer to dichroism for concreteness.) The parameter $\Delta\varepsilon_{ss}$ is the steady-state value, D_R is the rotational diffusion constant, and r gives the relative contributions of permanent and induced dipole moments to the rise kinetics. Specifically, we define dimensionless quantities $\beta = \mu E/kT$ and $\gamma = (\alpha_1 - \alpha_2)E^2/2kT$, where μ is the permanent dipole moment, E is the electric field, and α_1 and α_2 are the polarizabilities of the macromolecule in directions parallel and perpendicular to its principal hydrodynamic axis. The induced dipole moment is $(\alpha_1 - \alpha_2)E$. Then $r = P/Q = \beta^2/2\gamma$ with $P = (\mu/kT)^2$, $Q = (\alpha_1 - \alpha_2)/kT$.

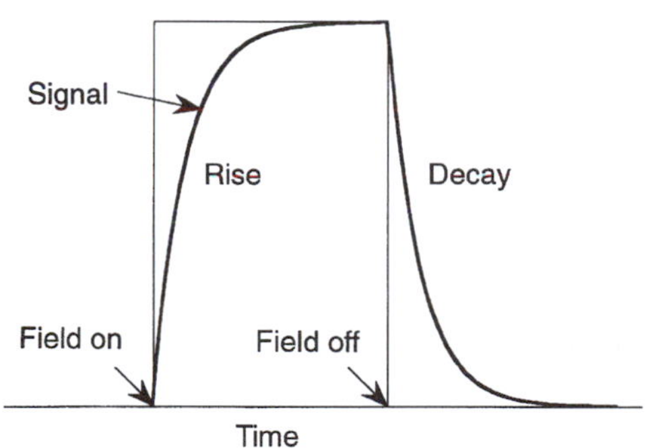

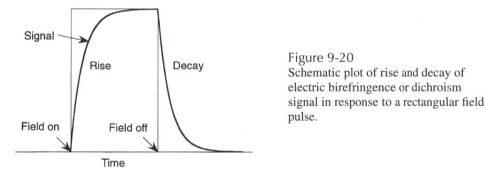

Figure 9-20
Schematic plot of rise and decay of electric birefringence or dichroism signal in response to a rectangular field pulse.

The course of the decay curve is given by $\Delta\varepsilon = \Delta\varepsilon_o e^{-6D_r t}$, where $\Delta\varepsilon_o$ is the dichroism at the moment the field is turned off. If $r = 0$ (pure induced moment) then the dichroism both rises and falls with time constant $6D_r$. If $r = 1$ (pure permanent moment), the dichroism rise is biexponential, with both $6D_r$ and $2D_r$ contributions; but the field-free decay constant is still $6D_r$. For this reason, the decay curve is more reliable for determining D_r, and molecular dimensions; while the rise curve is used mainly to obtain information about the mechanism of orientation.

The development of restriction enzyme technology and of preparative gel electrophoretic methods for isolating physical amounts of DNA has led to a number of careful studies in which the molecular properties of well-defined fragments have been characterized by electric birefringence and dichroism (Hogan et al., 1978; Stellwagen, 1981; Chen et al., 1982; Lewis et al., 1986). Figure 9-21 (Lewis et al., 1986) collects the

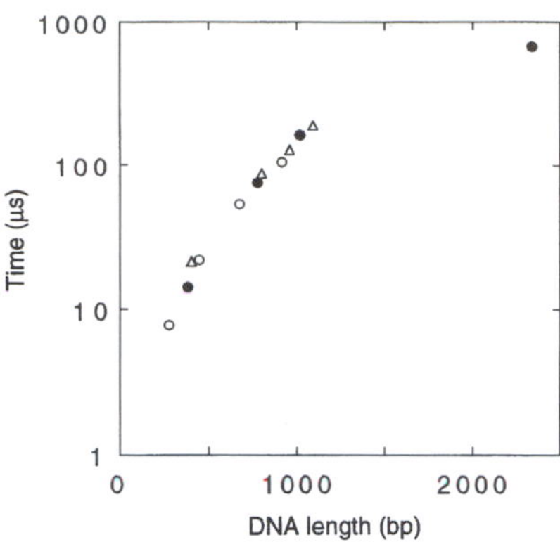

Figure 9-21
Decay time for the rotation of DNA restriction fragments as a function of DNA length. Data from Hagerman 1981 (○); Stellwagen, 1981 (△), and Lewis et al. 1986 (●). [Reprinted with permission from Lewis, R. J., Pecora, R., and Eden, D., *Macromolecules*, **19**, 134–139. Copyright ©1986 American Chemical Society.]

results of electric birefringence measurements from three research groups on B-form DNAs ranging from 80 to 2311 bp in length. The data all fall on the same curve, and there is good agreement with theoretical expectations for rigid rods at low molecular weight and with the Monte Carlo simulations of Hagerman and Zimm (1981) for larger molecules.

For poly[d(C-G)] fragments in different solvents, rotational correlation times agree with the lengths expected for the different conformations—2.8 Å/bp for A form and 3.7 Å/bp for Z form (Wu et al., 1981). Similar agreement between observed and expected increase in rise per base pair from 3.4 to 3.7 Å is found for the B–Z transition in 145 bp fragments of poly(dG-m^{5}dC) provoked by hexammine cobalt(III) (Chen et al., 1982).

Field-free decay of the electric dichroism has been used by Chen et al. (1985) to determine rotational relaxation times, and thereby end-to-end lengths, of four well-defined fragments of poly(dA-dT) ranging in size from 136 to 270 bp. Persistence lengths, calculated from the results of Hagerman and Zimm (1981) were in the range 200–250 Å. This is only about half the persistence length of naturally occurring, random sequence DNA.

Intercalation of drugs, such as ethidium and proflavine, increases the length of the DNA–drug complex, but the increase per intercalated base varies from 2.0–3.7 Å (Hogan et al., 1979). With the use of transient electrical birefringence, Sinden and Hagerman (1984) showed that interstrand psoralen cross-links of DNA extend the DNA, and do not cause kinking or bending. Earlier X-ray studies had been interpreted as indicating that psoralen cross links introduced significant bends in the DNA (Peckler, 1982). As discussed later in Section 9.14, sequence-dependent bending of DNA does occur naturally, and transient electric birefringence is an excellent method to characterize it (Hagerman, 1984).

Stellwagen (1981) used transient birefringence of fragments ranging from 80 to 4364 bp to examine in detail the orientation mechanism of DNA in electric fields. At sufficiently low-field strengths, the Kerr law is obeyed for all fragments. In that region, the rise of the birefringence is symmetrical with the decay for fragments less than or equal to 389 bp, indicating an induced dipole orientation mechanism. The induced polarizability is proportional to the square of the length of the DNA fragments, and inversely proportional to temperature, in accord with the theory of Hogan et al. (1978). In this theory, anisotropic ion flow produces an asymmetric ion atmosphere around the polyelectrolyte, resulting in an orienting torque. At high fields and with longer molecules, there is an indication of saturation behavior in the induced dipoles that may bear on the determination of base tilt from high-field extrapolations discussed in Section 6.4.

The induced dipole mechanism of DNA orientation is also confirmed in a reversing pulse electric birefringence study by Yamaoka and Matsuda (1980) of native calf thymus DNA ($M = 4.4 \times 10^6$) and sonicated fragments ($M = 1.24 \times 10^6$). In a reversing pulse experiment, the direction of the electric field is very rapidly switched. In principle, an induced dipole should adjust virtually instantaneously (since electrons and small ions move very rapidly compared to the macroion), with no reversal of sign of Δn, while a permanent dipole should exhibit a reversal. In practice, the relatively slow relaxation of the induced ion atmosphere polarization associated with long DNA molecules leads to a reversal even with an induced dipole mechanism. This phenomenon has been analyzed theoretically by Szabo et al. (1986), who assumed that ion atmosphere relaxation occurred only along the rod axis and could be characterized by a single relaxation time. For a 124 bp DNA fragment, with $D_r = 1.3 \times 10^5\ \mathrm{s}^{-1}$, the ion atmosphere relaxation time is calculated to be about 1.2×10^{-7} s. This relaxation time probably increases with DNA length.

Optical measurements have also been useful in determining orientation of DNA undergoing gel electrophoresis. Stellwagen (1985) measured the electric birefringence of DNA restriction fragments of three different sizes, 622, 1426, and 2936 bp, imbedded in agarose gels of different concentrations. She found that "the birefringence relaxation times observed in the gels are equal to the values observed in free solution, if the median pore diameter of the gel is larger than the effective hydrodynamic length of the DNA molecule in solution. However, if the median pore diameter is smaller than the apparent hydrodynamic length, the birefringence relaxation times increase markedly, becoming equal to the values expected for the birefringence relaxation of fully stretched DNA molecules. This apparent elongation indicates that end-on migration, or reptation is a likely mechanism for the electrophoresis of large DNA molecules in agarose gels." Similar conclusions were reached by Jonsson et al. (1988), who measured the linear dichroism spectra of 300–2319 bp DNA in 5% polyacrylamide and of 4361–23130 bp DNA in 1% agarose gels. In both cases, the fragments become preferentially oriented with the DNA helix axis parallel to the migration direction, and a considerable increase in orientation with length of DNA was observed. The gel electrophoretic orientation is high compared with dipole orientation in electric fields in free solution.

11. SCATTERING

11.1 Total Intensity Light Scattering

Light scattering has for many years been one of the standard methods for gaining information about the molecular weight, size, shape, and interactions of polymers. It was used in the early days of nucleic acid research to determine molecular weights of DNA molecules. While there are now more accurate and convenient ways to measure molecular weight, light scattering is still an important tool to determine DNA flexibility and its dependence on solvent conditions, and to assay condensation of DNA. The same principles that govern light scattering also apply to X-ray scattering, which has been used to determine the arrangement of DNA in bacteriophage particles, and to neutron scattering, which has elucidated the distribution of proteins in ribosomes.

11.1.1 Physical Basis

The basic theory of light scattering is presented in many standard references (e.g., Tanford, 1961; Bloomfield et al., 1974; Van Holde et al., 1998, Chapter 7). Light from a laser or intense thermal source is collimated into a narrow beam and directed onto the sample. The light has wavelength λ_0 in vacuo and $\lambda = \lambda_0/n$ in the sample, where n is the refractive index of the medium. The electric field of the light or X-rays causes the electrons of the illuminated sample to oscillate. These oscillating electrons in turn send out secondary, or scattered, radiation. A general expression for the intensity of the light scattered at angle θ, $I_s(\theta)$, relative to the incident intensity, I_o, is (Tanford, 1961)

$$I_s(\theta) = I_o \frac{4\pi^2 n^2 (\partial n/\partial c)^2 McP(\theta)}{N_A \lambda^4 r^2 (1 + 2BMc + 3Cc^2 + \cdots)} \tag{9-70}$$

This equation is appropriate for the common experimental setup in which the electric vector of the light is vertically polarized (say along the z axis), the beam is incident along the x axis, and the scattering is detected at angle θ in the x–y plane. If the beam is unpolarized, then $4\pi^2$ must be replaced by $2\pi^2(1 + \cos^2\theta)$. Equation 9-70 gives the scattering per unit volume of solution. The scattering volume is defined as the product of the length and the cross-sectional area of the "tube" of solution seen by the detector. A simple geometrical construction shows that this volume is proportional to $1/\sin\theta$.

Equation 9-70 shows how light scattering yields information on molecular weight M (or the weight average molecular weight $< M >_w$ if the sample is polydisperse), on interactions between polymer molecules (through the virial coefficients B and C), and on size and shape [through the scattering form factor $P(\theta)$]. The parameter $P(\theta)$ represents the ratio of scattering intensity of the actual macromolecule to that of a macromolecule of equal mass but very small size. The other factors in the equation are known or can be determined by independent measurements. The weight concentration c is usually expressed in grams per liter (g L^{-1}), and N_A is Avogadro's number. The distance from the scattering cell to the detector is r. The r^{-2} dependence reflects the fact that while the total scattered intensity remains constant, it passes through successive imaginary spherical surfaces of area $4\pi r^2$ surrounding the cell, so the intensity per unit area arriving at the detector falls off as r^{-2} · $\partial n/\partial c$ is the refractive index increment, measured at constant temperature, pressure, and chemical potential of diffusible solvent components. The $1/\lambda^4$ dependence is characteristic of Rayleigh scattering, that is, scattering from particles small compared to the wavelength of light.

Scattering of any sort of radiation occurs because of inhomogeneities in the sample. In X-ray diffraction from crystals, these inhomogeneities are the variations in electron density resulting from the periodic arrangement of atoms and molecules in the crystal lattice. In light scattering, the inhomogeneities arise from fluctuations in the refractive index of various microscopic portions of the solution. The refractive index fluctuation δn is related to the refractive index change accompanying a given concentration fluctuation δc, through the relation $\delta n = (\partial n/\partial c)\delta c$. The probability of a given concentration fluctuation is related to the bulk concentration c and to the solution nonideality—that is, the tendency of the macromolecules to cluster or to avoid each other. Positive virial coefficients $B, C \ldots$, represent a repulsive polymer–polymer interaction, hence a tendency for concentration fluctuations to be smaller than in ideal solutions, leading to lower scattering intensity. Negative virial coefficients represent attractive interactions, thus larger fluctuations and stronger scattering.

This equation is derived for a two-component system of polymer and simple solvent. Nucleic acids are almost invariably studied in the presence of added salt, so we have three or more components. Then a fluctuation in polynucleotide concentration will in general be correlated with a fluctuation in salt concentration, leading to scattering that is severely ionic strength dependent (in addition to the effects that ionic strength may have on the dimensions of the nucleic acid). It is therefore customary to work at relatively high-salt concentrations ($\sim 0.1M$) where polymer–salt interactions are constant. In order to take multicomponent effects properly into account, one must dialyze the nucleic acid solution against the buffer and subtract the scattering of the latter from that of the former (Eisenberg, 1976). Refractive index increments of DNA–salt solutions are tabulated by Cohen and Eisenberg (1968). In 0.2 M NaCl at a wavelength λ_o of 546 nm, the value is 0.166 $cm^3\ g^{-1}$ (Harpst, 1980).

The scattering form factor $P(\theta)$ must be taken into account when the polymer dimensions are significant with respect to the wavelength of light, about $\lambda/20$ or greater (~ 250 Å for visible light). Polymers smaller than this, such as tRNA, can be regarded as point scatterers. For such small particles, $P(\theta) = 1$. But larger particles scatter less light than might be expected from their mass, because light scattered from different parts of the same molecule is out of phase and interferes destructively. Only at $\theta = 0$ are all waves in phase, so $P(0) = 1$ for all molecules. Thus it is desirable to extrapolate light scattering measurements to $\theta = 0$ to obtain accurate molecular weight values for large polymers. It is shown in standard references that at low angles or for small particles such that $q < R_g >< 1$,

$$P(\theta) = 1 - \frac{1}{3}q^2 < R_g^2 > + \cdots \tag{9-71}$$

where $\mathbf{q}$ is the scattering vector, the difference between the wave vectors $\mathbf{k}_0$ and $\mathbf{k}_s$ of the incident and scattered beams. The magnitude of $\mathbf{q}$ is (see Fig. 9-22)

$$q = (4\pi n/\lambda_o)\sin(\theta/2) \tag{9-72}$$

Thus, we see that the dependence of $P(\theta)$ on q^2 [or $\sin^2(\theta/2)$] will give $< R_g^2 >$ with no assumptions about macromolecular structure.

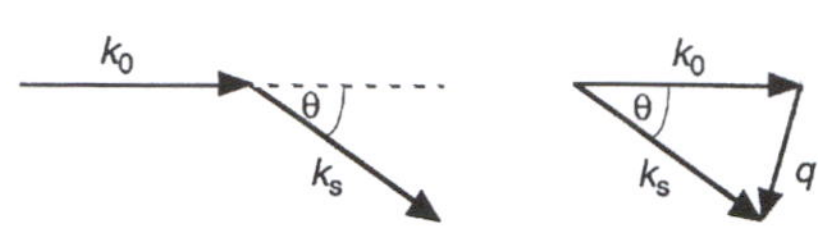

Figure 9-22
Diagram of scattering geometry. The incident and scattered wave vectors, $\mathbf{k}_o$ and $\mathbf{k}_s$, each have magnitude $2\pi/\lambda$. By trigonometry, the length of half the base of the triangle is $(2\pi/\lambda)\ \sin(\theta/2)$. The scattering vector $\mathbf{q}$, defined as the difference $\mathbf{k}_s - \mathbf{k}_o$, thus has magnitude $(4\pi/\lambda)\ \sin(\theta/2)$.

11.1.2 Data Analysis

If we define the Rayleigh ratio $R(\theta) = r^2 I_s(\theta)/I_o$, these equations can be rearranged in the compact form

$$\frac{K_c}{R(\theta)} = (1/ < M >_w + 2Bc \cdots)[1 + \frac{16\pi^2 n^2}{3\lambda_o^2} \sin^2\theta/2 < R^2 > + \cdots] \qquad (9\text{-}73)$$

where K is an optical constant $4\pi^2 n_o^2(\partial n/\partial c)^2/N_A\lambda^4$. (We have substituted n_o, the solvent refractive index, for n, the solution refractive index, an excellent approximation in dilute solutions.) Thus, if we extrapolate Kc/R_θ both to $c = 0$ and $\theta = 0$, we can get $< M >_w$, $< R_g^2 >$, and B. This simultaneous extrapolation may be done in a Zimm plot (Zimm, 1948), by plotting Kc/R_θ versus $\sin^2(\theta/2) + kc$, where k is an arbitrarily chosen constant of convenient magnitude.

Two such plots for scattering from solutions of ϕX 174 DNA are shown in Figure 9-23 (Sinsheimer, 1959). When we extrapolate points at a given concentration to $\theta = 0$ (i.e., to $\sin^2(\theta/2) = 0$), and then extrapolate these extrapolated points to $c = 0$, we get a line whose limiting slope is $2B/k$. In Figure 9-23(*a*), at low-salt concentration, this slope is positive. Thus $B > 0$, which indicates repulsion between the molecules due largely to polyelectrolyte effects. In Figure 9-23(*b*), at high-salt, $B < 0$, indicating intermolecular attraction. Extrapolating first to $c = 0$, and then to $\theta = 0$, we obtain a line whose limiting slope is proportional to $< R_g^2 >$. This slope is greater in low-salt concentration, demonstrating expansion of the chain due to intramolecular polyelectrolyte repulsions. Regardless of the order of extrapolation, the limiting lines extrapolate to the same intercept: $1/M$.

11.1.3 Structural Models

To use light scattering to distinquish between different models that may have the same radius of gyration, it is necessary to have expressions for $P(\theta)$ over the full range of angles. These expressions have been worked out rigorously only for a few cases (Tanford, 1961). For spheres of radius R (and mean-square radius $R_g^2 = \frac{3}{5}R^2$),

$$P(\theta) = \{(3/q^3)[\sin(qR) - qR\cos(qR)]\}^2 \qquad (9\text{-}74)$$

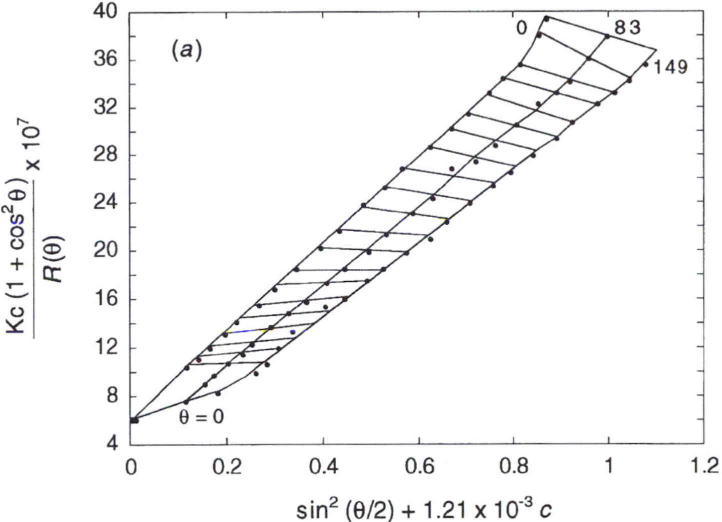

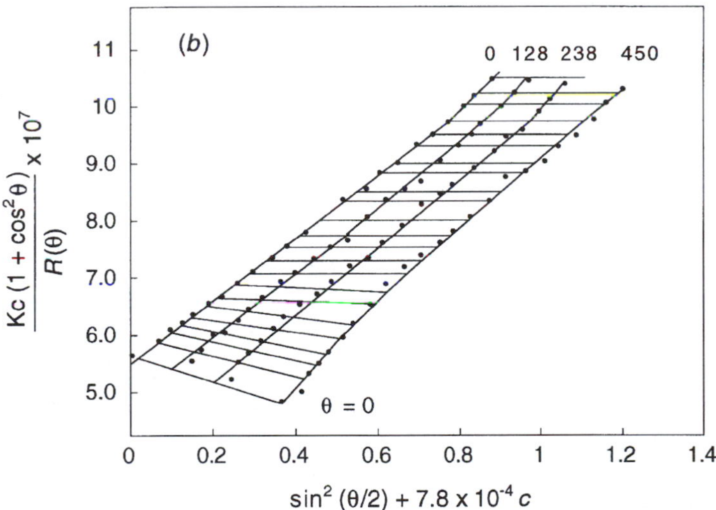

Figure 9-23
Zimm plot of light scattering of ϕX174 DNA in (*a*) 0.02 *M* NaCl and (*b*) 0.2 *M* NaCl. Curves are labeled with DNA concentrations in micrograms per milliliter. [Reprinted with permission from Sinsheimer, 1959.]

For rods of length L, with $R_g^2 = L^2/12$

$$P(\theta) = \frac{2}{qL}\int_0^{gL} \frac{\sin x}{x}\,dx - \left[\frac{\sin(qL/2)}{qL/2}\right]^2 \tag{9-75}$$

For Gaussian random coils, with $u = q^2 < R_g^2 >$

$$P(\theta) = (2/u^2)(e^{-u} + u - 1) \tag{9-76}$$

For wormlike coils, particularly those with excluded volume, the intramolecular segment distribution function is not known exactly. However, Sharp and Bloomfield (1968) obtained expressions for $P(\theta)$ as a function of persistence length and excluded volume, which appear to be quite accurate. (A minor error in this paper was corrected by Schmid et al. 1971.) Yamakawa and Fujii (1974b) used a more exact treatment of the dimensional statistics for short wormlike chains, but did not consider excluded volume. Thus their results in analyzing long DNA molecules give a persistence length that is too large, since all chain expansion is attributed to persistence length. The two theories agree well in the limit of no excluded volume. A later paper by Yoshizaki and Yamakawa (1980) calculates $P(\theta)$ for wormlike chains and helical wormlike chains (those with a structurally built-in helical path of the backbone) for the very wide range of wavelengths (q values) ranging from light through small-angle X-rays down to neutrons.

Figure 9-24 shows curves of $P(\theta)^{-1}$ versus $\sin^2(\theta/2)$ calculated for these various structural models applied to a molecule the size of T7 DNA, with $M = 25 \times 10^6$. The in vacuo wavelength λ_o was taken as 5460 Å. The spherical model has a radius of 500 Å, approximately that of a phage head. The random coil model assumes that the bond length is equal to the hydrated helix diameter, approximately 27 Å. The scattering curves for the most compact spherical and random coil structures show very little destructive interference, that for the straight rod shows a great deal, and the wormlike coil structures (both with 500-Å persistence length) are intermediate. The wormlike coil with excluded volume parameter $\varepsilon = 0.1$ is more expanded and has smaller $P(\theta)$, but it is evident that rather good data will be required to distinguish between similar values of ε.

11.1.4 High Molecular Weight DNA

The curvature for wormlike coils in Figure 9-24, extending to 10° and below, indicates the difficulties in applying light scattering to very large molecules. [A useful critique is given by Schmid et al. (1971)]. Some sample calculations are instructive. The Na salt of T7 DNA has a contour length $L = 25 \times 10^6/194 = 1.29 \times 10^5$ Å. With a persistence length a of 500 Å, the mean-square end-to-end distance $< L^2 > = 2aL = 1.29 \times 10^8$ Å^2, neglecting excluded volume effects. The mean-square radius is $< R_g^2 > = < L^2 > /6 = 2.15 \times 10^7$ Å^2. Typical light scattering instruments have a minimum detection angle $\theta_{min} = 30°$, so $\sin(\theta_{min}/2) = 0.259$. Thus for light with

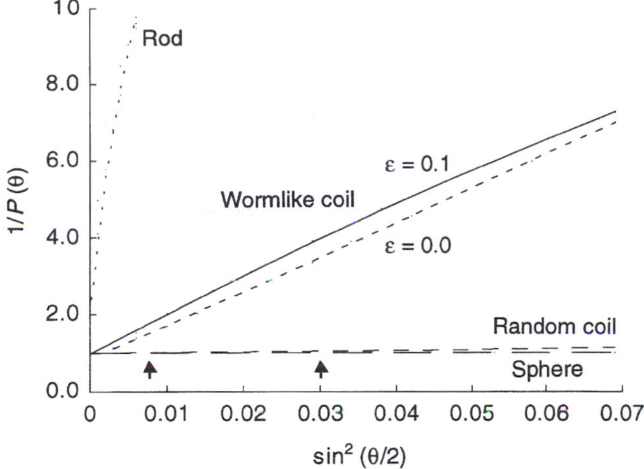

Figure 9-24
Reciprocal scattering form factor curves from 0° to 30° scattering angle for various models of T7 DNA ($M = 25 \times 10^6$) assuming in vacuo wavelength 546 nm. Wormlike coil calculations were performed according to the theory of Sharp and Bloomfield (1968) as corrected by Schmid et al. (1971). The arrows are at 10° and 20°.

wavelength in vacuo, $\lambda_o = 5460$ Å in aqueous solution with refractive index $n = 1.33$, $q_{\min} = (4\pi n/\lambda_o)\sin(\theta_{\min}/2) = 7.92 \times 10^{-4}$ Å^{-1}, and $q^2_{\min} < R^2_g > = 13.5$. This is much greater than unity, the condition for applicability of the expansion leading to Eq. 9-71. Taking the curve in Figure 9-24 and drawing the tangent to it at 30°, we find $1/P(0)_{\text{app}} \approx 3$. Thus M will be underestimated by a factor of 3. The limiting slope giving the radius of gyration will also be seriously in error. In order to get true limiting values for high molecular weight DNA, $\theta_{\min}$ must be substantially below 30°. Some modern light scattering instruments can go as low as $\theta_{\min} = 3°$. However, at these low angles, scattering by dust becomes a major problem, which may be removed by ultracentrifugation or ultrafiltration (Harpst et al.,1968; Krasna et al., 1970). Remarkably, millipore filtration does not break T7 DNA or even T2 DNA ($M = 120 \times 10^6$). It might have been thought that shear stresses in the narrow pores of the filter would break the very long molecules; but since the flow rate in any single pore is very small, high shear stresses are not developed.

Taking all these difficulties into account, Harpst (1980) analyzed data on T7 DNA in 0.195 M Na^+ previously obtained by Harpst et al. (1968) and Krasna et al. (1970). The analysis shows that linear extrapolations from the 10°–20° range of scattering angles gives incorrect values of M and $< R^2_g >$. Instead, a detailed fit of the data to calculated scattering curves, using the theory of Sharp and Bloomfield (1968), over the entire angular range is required. A proper fit can be obtained only when excluded volume effects are included. When this is done, M is found to be 25.5×10^6, $\varepsilon = 0.08$, and the persistence length is 600 Å, all in excellent agreement with other determinations. This paper confirms that light scattering is capable of yielding reliable results, but only if extraordinary care is taken in experiment and analysis. It probably defines the limit of feasible work with light scattering.

11.1.5 Turbidimetry

Light scattering may also be measured from the loss of beam intensity in an absorption spectrophotometer, a procedure known as turbidimetry. Just as with normal light scattering, particles larger than $\lambda/20$ cause destructive interference. The amount of destructive interference varies with wavelength. Thus, this correction must be applied for large particles: A scattering contribution to absorbance is often observed in aggregated nucleic acid samples. (Biochemists and molecular biologists will also be familiar with the use of turbidimetry to measure bacterial cell growth.) On the other hand, it can be used to estimate the molecular weights and dimensions of such particles. The basic theory for this effect was worked out by Doty and Steiner (1950); an excellent later treatment is given by Camerini-Otero et al. (1974).

The amount of light lost in an element of path length dx along the beam $-dI$ is proportional to dx and to the incident intensity I:

$$-dI = \tau I dx \tag{9-77}$$

The constant of proportionality is the turbidity, τ, which is related to concentration and molecular weight in just the same manner as regular light scattering. If we neglect concentration–dependent interactions,

$$\tau = HcMQ(\lambda) \tag{9-78}$$

H_o is an optical constant, given by $H(\lambda_o) = 32\pi^3 n_o^2(\partial n/\partial c)\mu^2/3N_A\lambda_o^4$. The parameter $Q(\lambda)$ is a function that depends on particle size and structure through the particle form factor $P(\theta)$ defined earlier.

Rearranging, we can write $HcQ/\tau = 1/M$, an equation analogous to $KcP(q)/R(\theta) = 1/M$ for angle-dependent light scattering. In both cases, to obtain M, Q or P must be extrapolated to unity or calculated from a knowledge of the structure. The extrapolation corresponding to q or $\theta \to 0$, in light scattering is $\lambda_o \to \infty$ in turbidity (since q and λ have a reciprocal relation). Examples of this procedure are given by Camerini-Otero et al. (1974). Tables of Q for common structures are given by Doty and Steiner (1950).

A useful relation that follows from the equations given above is

$$-\frac{d\log\tau}{d\log\lambda_o} = 4 - \frac{d\log Q}{d\log\lambda_o} = 4 - \beta \tag{9-79}$$

A log–log plot of turbidity versus wavelength gives β. This equation can be used to correct the spectra of strongly scattering samples, if determination of β can be made by turbidity measurements outside the absorption band (e.g., above 300 nm for nucleic acids).

11.2 X-Ray Scattering

X-ray scattering from solutions (as opposed to ordered crystals) obeys physical principles and equations very similar to light scattering. In both cases, the radiation is scattered from electrons. In X-ray scattering, there must be a difference in electron density ρ_e between macromolecule and solvent. The scattering intensity is proportional to $\Delta\rho_e^2$, just as it is to $(\partial n/dc)^2$ in light scattering. The dimensions of particles large enough to give rise to destructive interference is of order $\lambda/20$, or more precisely of order $q^{-1} = (\lambda/4\pi)/\sin(\theta/2)$. For scattering through large angles, for which $\sin(\theta/2)$ is not much less than unity, this indicates that relevant particle dimensions will be on the order of X-ray wavelengths, about 1.5 Å. However, special beam collimation techniques make it possible to work at very low scattering angles, down to millidegrees. In this range, $\sin(\theta/2) \approx \theta/2 \ll 1$, so considerably larger distances may be probed, on the order to tens to hundreds of angstroms. This is an important range, is not readily accessible to other scattering techniques.

As one example, the scattering pattern from a concentrated suspension of T4 bacteriophage capsids (Earnshaw and Harrison, 1977; Earnshaw et al., 1978) shows a broad maximum at a spacing of 27 Å, corresponding to the center-to-center spacing between adjacent helices. The maximum is modulated by a secondary pattern with a periodicity of about 600 Å, which is approximately the diameter of the phage head. This demonstrates that the DNA is wrapped in such a way as to maintain coherent order across the capsid.

11.3 Neutron Scattering

Neutrons, while particles and not electromagnetic waves, also show scattering and diffraction behavior similar to those of light and X-rays. Thermal neutrons have a wavelength of a few angstroms, similar to X-rays. (The wavelength λ may be calculated by combining three equations: the deBroglie relationship $\lambda = h/p$, where h is Planck's constant and p is the momentum; the relation between momentum and kinetic energy $\mathrm{KE} = \frac{1}{2}mv^2 = p^2/2m$ where v is the velocity and m is the particle mass; and the principle of equipartition of energy $KE = \frac{1}{2}kT$.) Neutron scattering, like X-ray scattering, can be measured over an angular range from millidegrees to degrees, giving access to structural detail in the range up to several hundred angstroms.

As with the other scattering methods, neutron scattering depends on contrast; that is, a difference in scattering power of sample and surroundings. However, contrast in neutron scattering can be manipulated by controlled hydrogen–deuterium substitution. These two isotopes have very different neutron scattering cross-sections. This fact has been cleverly utilized by Moore, Engelman, and their co-workers (Engelman et al., 1975; Moore and Engelman, 1979; Capel et al., 1988) to map out the distribution of proteins in the 30*S* ribosomal subunit of *E. coli.* This strategy takes advantage of the ability of *E. coli* to grow on media substituted with deuterium, and of biochemical techniques for purifying ribosomal proteins and RNA in large quantities and reconstituting ribosomes from the purified components. Thus it is possible to prepare ribosomes in which any pair of proteins is deuterium substituted, while all the other components contain hydrogen.

The scattering curve for each such substituted ribosome has three terms, two of which result from scattering by the individual proteins i and j. The third is the interference term, which is of the form $\sin(qd_{ij})/qd_{ij}$, where d_{ij} is the distance between the centroids of i and j. If a large number of such distances can be measured, the quaternary structure of the ribosome can be derived by triangulation. Since there are 21 proteins in the 30*S* subunit, this was a major undertaking, requiring 13 years for completion (Engelman et al., 1975; Capel et al., 1988). The protein distribution determined by neutron scattering agrees well with that obtained by other techniques, such as electron microscopy.

12. DYNAMIC LIGHT SCATTERING

12.1 Basic Principles

As scattering particles diffuse under the influence of Brownian motion, move in response to electrophoretic fields or cell motility, or undergo internal motions such as rotation and conformational changes, the electric field $E_s(t)$ and intensity $I_s(t) = |E_s(t)|^2$ scattered from them fluctuate with time. The time variation thus gives information on particle motions. There are three types of fluctuations to be considered in dynamic light scattering experiments.

1. *Occupation number fluctuations* are due to fluctuations in the number of particles N in the scattering volume. From Poisson statistics, we know that these fluctuations are inversely proportional to $\sqrt{N}$. To have fluctuations at the 1% level, N must be 10^4 or less in a scattering volume typically 1 mm on a side. This concentration is very low, hence only very strong scatterers, such as whole cells, will produce measurable occupation number fluctuations. These are not generally important for nucleic acid work. Further, since particles must diffuse a distance of about 1 mm to enter or exit the scattering volume, occupation number fluctuations are slow, typically on the order of seconds.

2. *Amplitude fluctuations* result from rotation and internal motions of molecules. In order to be observable, these motions must have a characteristic length that is not too small compared to λ. [This condition is equivalent to requiring that the total intensity form factor $P(\theta)$ be significantly < 1.] Such internal fluctuations are readily observed, if not always readily interpreted, in the dynamic scattering from large DNA molecules.

3. Most important, *phase fluctuations* arise from translation of molecule over a distance comparable to the wavelength of light λ. These fluctuations are rapid and observable from molecules over a very wide range of sizes. The characteristic time scale for phase fluctuations due to translational diffusion can be estimated from the Einstein equation for the mean-square random displacement of a molecule with diffusion coefficient D_t in time τ (see Eq. 9-37)

$$< [x(t+\tau) - x(t)]^2 >= 2D_t\tau \tag{9-80}$$

The distance traveled to produce a measurable change in the phase of the scattered light (i.e., a change in amplitude that is a significant fraction of the peak-to-trough amplitude) is on the order of $\lambda/10$. Thus, Eq 9-80 yields $\tau \approx (\lambda/10)^2/2D_t$. With λ_o about 5000 Å in vacuo, corresponding to about 3.8×10^{-5} cm in aqueous solution, and a typical D_t of 10^{-7} cm^2 s^{-1}, this gives $\tau \approx 7 \times 10^{-5}$s. The microsecond-to-millisecond range for dynamic light scattering fluctuations is observed for most biological macromolecules. Thus dynamic light scattering can be used for very rapid measurement of D_t.

The dynamic light scattering equations are derived in a more complete and systematic fashion in numerous sources (e.g., Berne and Pecora, 1976; Bloomfield and Lim, 1978; Schmitz, 1990; Chu, 1991). The important result is that the autocorrelation function $g^{(2)}(\tau)$ of the photocurrent resulting from the scattered light is

$$g^{(2)}(\tau) = 1 + \beta e^{-2q^2 D_t \tau} \tag{9-81}$$

where β is an instrumental constant. This equation showns that $-2q^2 D_t$ is the slope of a semilogarithmic plot of $g^{(2)}(\tau) - 1$ versus τ. Since q^2 is known, D_t can be determined.

If the sample is polydisperse, the apparent translational diffusion coefficient will be a weighted average of contributions from the various macromolecules in the solution. In this case, the plot of $\ln[g^{(2)}(\tau) - 1]$ versus τ will be somewhat curved. By fitting to a power series in τ, a procedure known as cumulant analysis (Koppel, 1972), it may be readily demonstrated that the apparent diffusion coefficient (the coefficient of the term linear in τ) is a z average in which the contribution of each species i is weighted according to its scattering power:

$$D_{t,app} = \langle D_t \rangle_Z = \frac{\sum_i D_{t,i} M_i c_i P_i(\theta)}{\sum_i M_i P_i(\theta)} \tag{9-82}$$

12.2 Applications of Dynamic Light Scattering

Numerous applications of dynamic light scattering to nucleic acids and nucleoproteins have been reviewed by Bloomfield (1985). Here we will give just a few examples for tRNA and DNA, which illustrate how molecular information can be obtained from the technique.

12.2.1 Transfer RNA

Ford and co-workers (Olson et al., 1976; Wang and Ford, 1981) have carried out a particularly instructive series of studies on tRNA. Figure 9-25(*a*) shows the dependence of D_t for tRNA as a function of RNA concentration for three different salt concentrations in the presence of 1 m*M* $MgCl_2$ (Olson et al., 1976). These data show that there is a strong concentration dependence of D_t, particularly in low-salt concentrations; and that the value of D_t extrapolated to [tRNA] = 0, D_o, has a maximum at about 0.1 *M* NaCl. The concentration dependence of D_t may be discussed in terms of a virial coefficient expansion.

The derivation of diffusive flux as a response to a chemical potential gradient leads to an expression for D_t of the form $D_t = D_o[1 + c(\partial \ln y/\partial c)]$. It may be shown that this is identical to the virial coefficient expansion in total intensity light scattering,

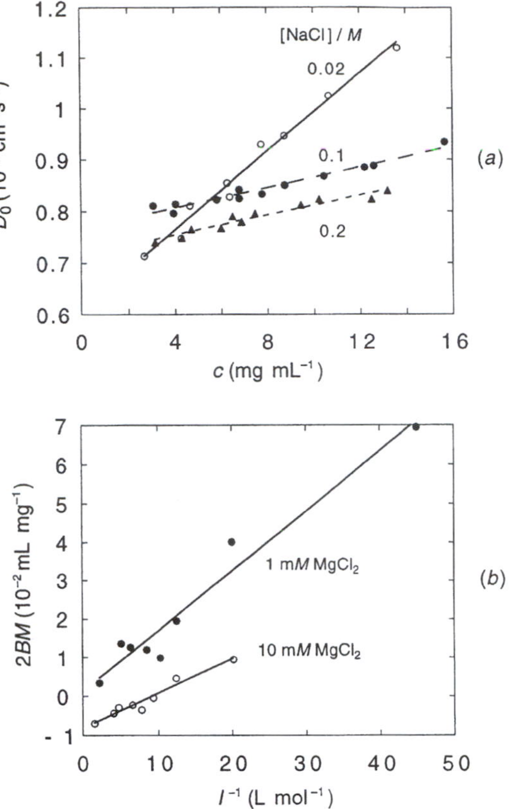

Figure 9-25
Dependence of the translational diffusion coefficient of tRNA on RNA concentration and ionic conditions. Top: diffusion coefficient in 1 m*M* $MgCl_2$ at the indicated NaCl concentrations. Bottom: Dependence of the diffusional virial coefficient of tRNA on the reciprocal of ionic strength at 1 m*M* and 10 m*M* $MgCl_2$. As discussed in the text, the slope of the line is proportional to the square of the charge on the tRNA. [Reprinted with permission from Olson et al. 1976.]

$D_t = D_o(1 + 2BMc + \cdots)$. There is also a hydrodynamic component to the diffusion coefficient, because of the concentration dependence of the frictional coefficient f in the expression $D_t = kT/f$. This equation has the form $(1 - \bar{v}_2 c)/(1 + k_f c + \cdots)$. The numerator gives the decrease in diffusion velocity due to flow against the solvent current produced by the motion of the other dissolved polymer molecules. The denominator gives the increase in f due to hydrodynamic interaction between solute molecules.

Thus one may write

$$D_t = D_o(1 - \bar{v}_2 c)\frac{1 + 2BMc + \cdots}{1 + k_f c + \cdots} = D_o(1 + 2B'Mc + \cdots) \qquad (9\text{-}83)$$

where $B' = B - (k_f + \bar{v}_2)/2M$ is the diffusional second virial coefficient. For an uncharged spherical molecule, $2BM = 8\bar{v}_2$ (Tanford, 1961). For a polyion of charge Z elementary units, in a solvent of ionic strength I, an additional term is added:

$1000v_1Z^2/2IM$. The frictional virial is (Batchelor, 1972) $k_f = 6.55v_2$. Thus the dependence of D_t on polyion concentration is predicted to be

$$D_t = D_o[1 + (0.45v_2 + 1000v_1Z^2/2IM)c + \cdots] \tag{9-84}$$

Figure 9-25(*b*) gives the slopes of D_t versus *c* curves, as a function of I^{-1}, in both 1 and 10 m*M* $MgCl_2$. From a knowledge of *M* and v_1, one obtains *Z*, which is −10 in 1 m*M* $MgCl_2$, and −8 in 10 m*M* $MgCl_2$.

Figure 9-26 shows the dependence of D_o on ionic strength in 1 m*M* and 10 m*M* $MgCl_2$. In the lower $MgCl_2$ solutions, there is a distinct maximum in D_o at about 0.1 *M* salt. This value indicates a more compact form of the tRNA, and is pictured in terms of a "block-and-hinge" motion connecting two regions of the molecule. This transition is not observed at higher $MgCl_2$ concentrations. The strong decrease in D_o at lower salt is ascribed to partial unfolding.

Dynamic light scattering has also been used to analyze the dimerization of two tRNA molecules, *E. coli* $tRNA^{Glu}$ and yeast $tRNA^{Phe}$, which interact through complementary anticodons (Wang et al., 1981). The diffusion coefficients of equimolar mixtures of the tRNAs as a function of concentration are shown at various temperatures in Figure 9-27. The theoretical lines for noninteracting monomers and pure dimer are also plotted. The equilibrium association constant *K* was obtained from the concentration dependence of the effective D_t. At 20°C, *K* varied from $1.0 \times 10^5\ M^{-1}$ without Mg^{2+} to $1.5 \times 10^6\ M^{-1}$ in 10 mM Mg^{2+}. Values of ΔH and ΔS, obtained from a van't Hoff plot of ln *K* versus $1/T$, were about −18 kcal mol^{-1} and −36 cal mol^{-1} deg^{-1}.

In an equilibrium mixture of this sort, the decay of the autocorrelation is a sum of exponentials, with contributions from each species with diffusion coefficient D_i

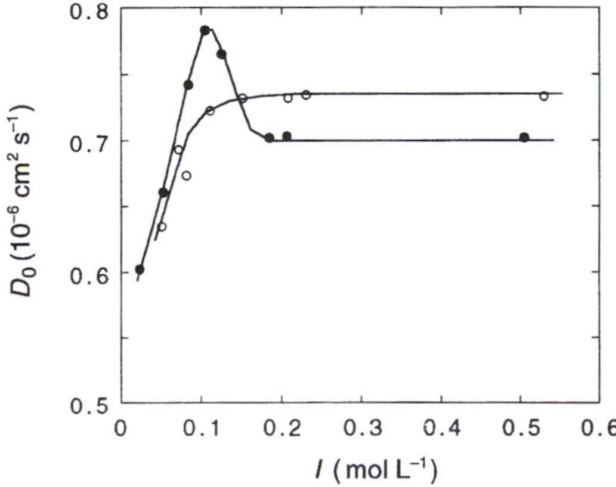

Figure 9-26
Dependence of the diffusion coefficient of tRNA on ionic strength, in (●) 1 m*M* and (x) 10 m*M* $MgCl_2$. [Reprinted with permission from Olson et al. 1976.]

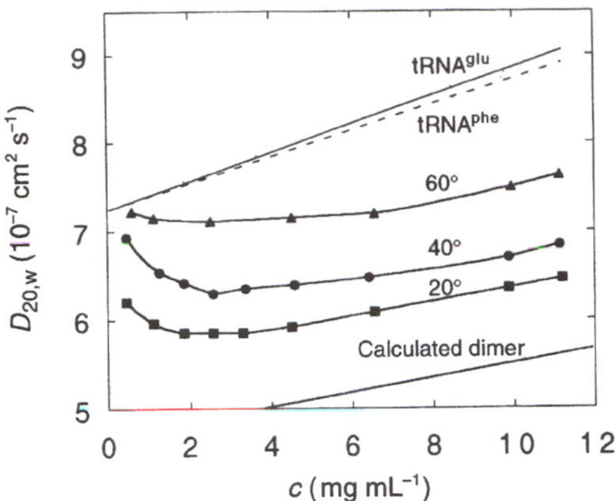

Figure 9-27
Diffusion coefficient as a function of tRNA concentration for tRNA^{phe} and tRNA^{Glu} monomers and equimolar mixtures, at the temperatures indicated. The line at the bottom is D versus c calculated for pure dimer. (65% of the sample is pure and capable of dimerization). [From Wang and Ford, 1981.]

weighted by its scattering power $M_i c_i P_i(\theta)$. In this case, we have two monomer tRNA species, and one dimer. It is a good approximation to assume that the molecular weights, concentrations, and diffusion coefficients of the two monomers are the same: M, c_1, and D_1. The dimer has parameters $2M$, c_2, and D_2. All species are small enough that $P_i = 1$. Then the z-averaged diffusion coefficient, Eq. 9-82, can be written

$$D_z = \frac{2Mc_1D_1 + 2Mc_2D_2}{2Mc_1 + 2Mc_2} \tag{9-85}$$

where the 2 in the first terms of numerator and denominator comes from the sum of the two monomers, and the 2 in the second terms comes from the doubled molecular weight of the dimer. The molar equilibrium constant for association is, when written in weight concentration units,

$$K = \frac{c_2/2M}{c_1^2/M^2} = \frac{M}{c}\frac{f}{(1-f)^2} \tag{9-86}$$

The second equality comes from setting $c_1 = (1-f)c$, $c_2 = 2fc$, where c is the total concentration of each monomer and f is the fraction dimerized. Thus $D_z = [(1-f)D_1 + 2fD_2]/(1+f)$, which may be rearranged to give

$$f = \frac{D_1 - D_z}{D_1 + D_z - 2D_2} \tag{9-87}$$

where f as a function of c is substituted in Eq. 9-86 to yield K, using the measured dependence of D_z on c, measuring the concentration dependence of D_1, and calculating the concentration dependence of D_2 from Eq. (9-84). The actual analysis (Wang et al., 1981) also took into account the fraction of unreactive species.

12.2.2 Short DNA Fragments

For short DNA fragments, dynamic light scattering and transient electric birefringence (or dichroism) enable measurement of both translational and rotational diffusion coefficients. Then application of hydrodynamic theories for rigid rods (Table 9-1) allows calculation of both the length and diameter of the cylinder. This strategy was applied by Mandelkern et al. (1981) to restriction fragments from the plasmid pBR322, ranging in length from 64 to 267 bp. Extrapolating the results to zero length to remove the effects of helix flexibility, they found a hydrodynamic diameter of 22–26 Å (the higher value according better with the field-free birefringence decay values), and a rise per base pair of 3.34 ± 0.1 Å. These values are in good agreement with the values obtained by fiber diffraction, taking into account a monolayer of water around the phosphates, and indicate that, in these respects at least, the structure of DNA in solution is not significantly different from that in the fiber. However, studies on very short synthetic oligomers, from 8–20 bp (Eimer et al., 1990); Eimer and Pecora, 1991), give a hydrodynamic diameter of 20 ± 1.5 Å, suggesting that the hydration does not extend significantly outside the DNA phosphates. The reason for this surprising result is not understood.

12.2.3 Plasmids

The linear dimensions of plasmids and other high molecular DNAs are comparable to or greater than the laser wavelengths used in dynamic light scattering experiments. Hence, internal motions of the DNA chain are reflected as well as translational diffusion. A schematic representation of D_{app} as a function of q^2 is shown in Figure 9-28. Since there is an inverse relation between q and the distance probed in light scattering experiments, the limiting value of D_{app} at small values of q, D_o, corresponds to translational diffusion of the molecular center of mass. The plateau at higher q reflects segmental diffusion within the DNA coil. Lin and Schurr (1978) have on the basis of model calculations, suggested that the value of q at the midpoint between the two regions should be related to the rms length b_G of the Gaussian subchain in a random

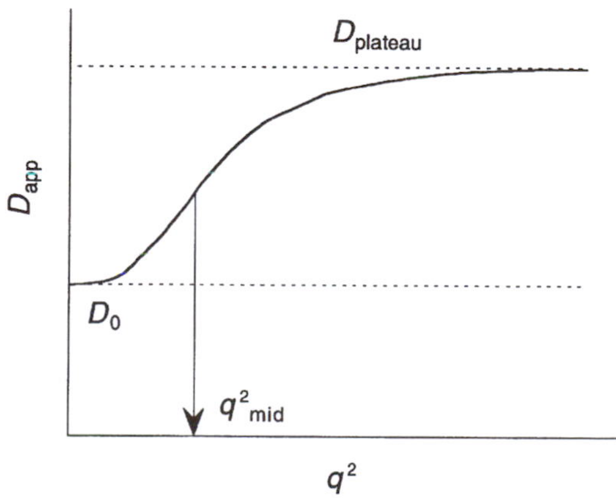

Figure 9-28
Apparent translational diffusion coefficient as a function of q^2, showing actual translation diffusion coefficient D_o at low q and contribution of internal modes to $D_{plateau}$ at high q. The value of q at the midpoint between D_o and $D_{plateau}$ is approximately related to the rms length b_G of a Gaussian subchain coil by $q^2_{mid}b^2_G \approx 8$ (Lin and Schurr, 1978).

coil representation of the DNA molecule. The relation is $q^2_{mid}b^2_G \approx 8$. They also show how the difference between D_o and the high-q plateau value, $D_{plat} - D_o$, can be equated to kT/f_{seg}, where f_{seg} is the segmental friction coefficient. The more flexible and mobile the segments, the smaller f_{seg}. Since the theory leading to these results is difficult and contains numerous approximations, precise quantitative results are not to be expected. However, a curve like Figure 9-28 can provide considerable qualitative or semiquantitative information about DNA flexibility, particularly on a comparative basis. For example, the segmental diffusion coefficient of a supercoiled plasmid is larger than for a linearized plasmid of the same length, but the amplitude of the segmental contribution is lower, due to the constraints of superhelicity. Because of the tightening of the coil, the translational diffusion coefficient increases with the number of superhelical turns (Langowski et al., 1992).

13. MICROSCOPY: VISUALIZATION OF SINGLE MOLECULES

The most direct way to obtain information about the structures of nucleic acids and their complexes is to look at them by microscopy. Modern microscopic techniques useful for visualizing nucleic acids include electron microscopy (EM); the family of scanning probe microscopies, especially scanning force microscopy (SFM); and high-resolution optical microscopy, especially fluorescence and confocal microscopy. The book by Slayter and Slayter (1992) provides an excellent overview of the basic principles of microscopy. Among the many general reviews of the applications of electron and other microscopies to nucleic acids and nucleoprotein complexes we may mention Brack (1981), Fisher and Williams (1979), Griffith and Christiansen (1979), Thresher and Griffith (1992,) and Morel (1995).

13.1 Electron Microscopy

13.1.1 Basic Aspects

Electron microscopy, with a resolution in modern instruments of 2–4 Å, will in principle allow visualization with resolution up to a single base pair. In practice, there are many difficulties in sample preparation that limit the information obtainable from EM, and the resolution or reproducibility is more typically 50–100 bp. However, it is still an enormously valuable technique. Some of its uses have been to determine molecular length and weight, to discriminate between double-stranded and single-stranded regions, to map regions of secondary structure in single-stranded molecules, and to identify circular molecules and replicative structures. Partial denaturation mapping gives an idea of the location of A-T rich tracts. Sequence complementarity between DNA and RNA molecules can be analyzed by a range of hybridization techniques. Transcription complexes can be visualized and promoter sites mapped by EM imaging of protein–nucleic acid complexes. Electron microscopy has also been important in elucidating the structures of ribosomes and chromatin.

In addition to the basic issues of magnification and resolution, there are two problems generic to all microscopy of biological specimens: contrast and maintenance of native structure. Large DNA molecules have an additional problem: spreading to enable an untangled view of the molecule. Good discussions of the physical chemical aspects of EM are given by Brack (1981), Freifelder (1982), and Cantor and Schimmel (1980, Chapter 10). Practical aspects are discussed by Thresher and Griffith (1992).

Contrast arises from differential scattering power of sample and surroundings. In EM, scattering of incident electrons is proportional to the square of the atomic number of the scattering atoms. Nucleic acids are mainly composed of light atoms, and so are the films on which they sit. Thus some means of contrast enhancement is generally needed. This enhancement is usually achieved by staining or shadowing. Positive staining is accomplished by binding heavy atoms, generally uranyl ions, to the nucleic acid. Negative staining (more properly, negative contrast) is achieved by embedding the molecules in an electron-opaque film of a heavy metal salt such as uranyl acetate. In this case, contrast is achieved between the continuous scattering background and the relatively nonscattering sample. Shadowing is performed by bombarding the sample with heavy atoms of platinum or tungsten evaporated from a hot wire source. These metal atoms pile up against the vertical sides of the molecules, like snow against a fence, producing very strong contrast. Shadowing may be done from one direction, in which case the length of the shadow cast on the "downwind" side of the molecule enables a measurement of its height. In nucleic acid work, it is more common to rotate the sample during shadowing, in which case the contrast arises from buildup of heavy atoms against vertical features.

Large DNAs are commonly spread with the surface spreading technique developed by Kleinschmidt (Kleinschmidt and Zahn 1959; Kleinschmidt, 1968), which uses a monolayer of the slightly basic protein cytochrome *c*. Smaller spreading agents cause less thickening of the apparent diameter of DNA, thus allowing visualization of finer details. For optimal high-resolution visualization of protein–DNA complexes, it is best to have no molecules at all coating the DNA. Thus it is necessary to treat a surface so that it will bind nucleic acids directly. This treatment is generally done by positively charging thin carbon films by submitting them to glow discharge in pentylamine vapor (Dubochet et al., 1971). These grids effectively adsorb nucleic acids and negatively charged proteins through a combination of electrostatic and hydrophobic forces.

13.1.2 Applications to Secondary Structure and Gene Structure Analysis

The panels of Figure 9-29 show examples of a variety of electron microscopic techniques used to visualize the secondary structure and loops in DNA molecules and complexes with proteins and RNA. Images such as these provide evidence regarding denaturation, binding sites, sequence homology, deletions and insertions, and similar structural features (Davis et al., 1971).

13.1.3 Physical Properties of DNA from Electron Microscopy

Electron microscopy can also be used for more physical, quantitative purposes. Lang et al. (1967, 1987) have made a careful study of the ability of EM to measure the lengths of double-stranded DNA and RNA molecules, and the variation of length with ionic strength of the hypophase. They find that the resolving power of the length measurements is about 40 nm or 143 bp, limited by length fluctuations among single molecules, not by EM. It would be expected to increase relatively as $(\text{bp})^{-1/2}$. The change in axial rise/bp h with the logarithm of ionic strength, $dh/d \log I$, is $-(4.54 \pm 0.28) \times 10^{-5} \mu\text{m/bp}$ for dsRNA, and $-(3.60\text{–}0.39) \times 10^{-5} \mu\text{m/bp}$ for dsDNA. The magnitude of the length change is least for NH_4Cl as the salt in the hypophase, and greatest for CsCl. These results are in line with the finding for closed circular dsDNA (Anderson and Bauer, 1978) that the winding angle ψ between adjacent base pairs of dsDNA in NH_4Cl changes according to $d\psi/d \log I = 0.168$ deg/bp, and the theoretical prediction (Manning, 1981) that winding of the double helix should be accompanied by a decrease in rise/bp of the DNA. The experimental ratio established by this work is $d\psi/dh = -4.67 \text{ deg nm}^{-1}$.

Detailed analysis of correlations in curvature along the lengths of adsorbed DNA molecules can, in principle, be used to determine the persistence length of DNA (Frontali et al., 1979). Measurements are made of angles θ between tangents to the coil at points separated by contour length l, and averages are taken over many points separated by l, and for several dozen molecules. Mathematical analysis gives $< \theta^2(l) >= l/a$, and $< \cos\theta(l) >= \exp(-l/2a)$. The agreement with solution measurements, as a function of ionic strength, is fairly good. The EM procedure is based on many more detailed measurements than are solution averages, but requires the validity of three assumptions, which may be open to some question. These assumptions are that the angle distribution is Gaussian, that the passage from 3D to 2D simply involves

Figure 9-29 (*facing page*).
Examples of electron microscopic techniques used to visualize the secondary structure and loops in DNA molecules and DNA complexes with proteins and RNA. (*a*) Double-stranded PM2 DNA and single-stranded fd phage DNA adsorbed onto pentylamine glow-discharged carbon films in the presence of 80% formamide, then rotary shadowed with platinum. Note the kinkier as well as thinner contour of the fd DNA. (*b*) Specific complex between λ repressor and λ operator, prepared by adsorption to pentylamine glow-discharged carbon film and stained with 0.5% aqueous uranyl formate. The DNA is positively stained and the repressor molecules appear negatively stained. The four subunits of the repressor tetramer can be clearly distinguished. [From Brack and Pirrotta, 1975.] (*c*) PM2 DNA molecule partially denatured with T4 gene 32 protein. The denatured single strands, since they are coated with protein, appear thicker than the dsDNA. [Reprinted with permission from Brack et al. 1975.] (*d*) Heteroduplex molecule between 1 DNA and λ polA (att-red), a phage carrying the *E. coli polA* gene as an insert. (*e*) *D*-loop molecules, formed by hybridization of PM2 restriction fragments HindIII f4 and f5 to linearized PM2 DNA. [Reprinted with permission from Brack et al. 1976.] (*f*) *R*-loops formed by hybridization between β-globin mRNA and a fragment of genomic DNA containing the *beta*globin gene. The loops contain the intervening sequences in DNA that are absent from the mRNA.

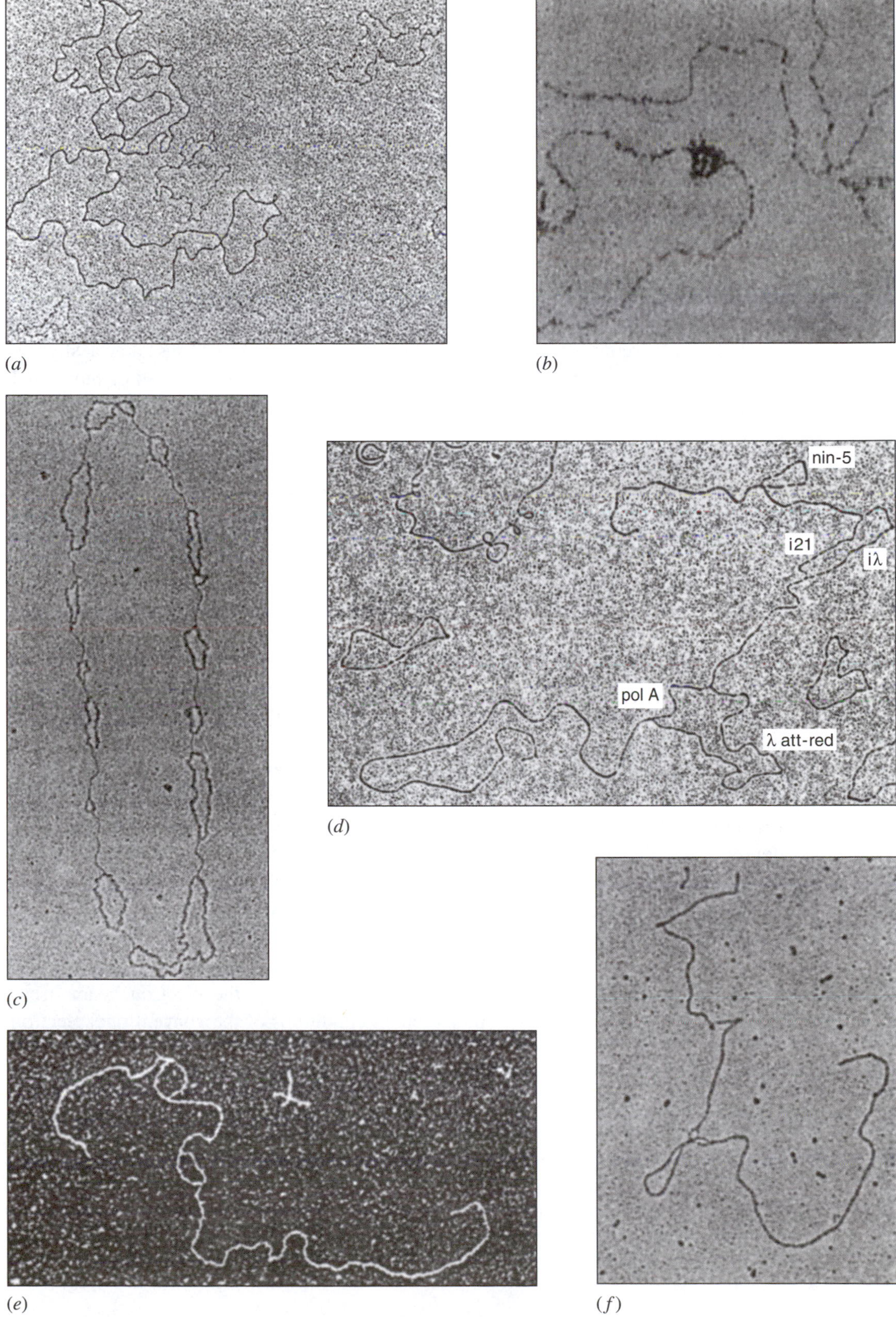

(a) (b) (c) (d) (e) (f)

blocking 1° of freedom, and that adsorption is at equilibrium. It appears that this procedure is valid between about 5 and 30 persistence lengths. Above that length, chain dimensions are influenced by intramolecular excluded volume (Bettini et al., 1980; Frontali, 1988).

13.1.4 Maintenance of Native Structure in Hydrated Samples: Cryoelectron Microscopy and Freeze Etching

Dehydration, staining and/or shadowing, and deposition on a 2D surface are all likely to distort the 3D structure that DNA molecules adopt in aqueous solution. These problems can be circumvented by cryoelectron microscopy (reviewed by Dubochet et al., 1992). In this technique, thin-film specimens are vitrified by very rapid cooling (on the order of 10^6K s^{-1}), which allows the water no time to crystallize. Contrast is elicited by phase contrast produced by strong defocusing. Micrographs taken at slightly different tilt angles can be combined as stereopairs to give 3D images of individual molecules. Computer image reconstruction yields representations that can be visualized more readily and analyzed quantitatively (Fig. 9-30).

Cryoelectron microscopy has been used to visualize supercoiled DNA molecules and their response to ionic strength (Adrian et al., 1990; Bednar et al., 1994). It was found that the superhelix is interwound rather than toroidal, that writhe increases and the superhelix diameter decreases with increasing salt, and that opposing segments appear to touch each other in a very tight superhelix at Mg^{2+} concentrations characteristic of *in vivo* levels. Supercoiling and ionic behavior are discussed in more detail in Chapters 10 and 11.

Freeze etching is a technique commonly used to examine samples from which the solvent water has not been completely removed. A sample is frozen, then cleaved by striking it with a knife. Solvent is sublimed from the cleaved surface, leaving nonvolatile macromolecules protruding from a frozen aqueous layer. Then a replica carbon film of the surface is made by bombarding with carbon atoms evaporated from a hot source. After the replica film is peeled off the surface, it is mounted and shadowed normally. Figure 9-31 shows a freeze–etch micrograph of calf thymus DNA, which has been condensed into a toroid by spermidine. The replica was shadowed from a single direction to highlight surface detail. Careful examination shows circumferential winding of the DNA strands, a result pertinent to understanding the mechanism of condensation. This quality of image has not been achieved by conventional sample preparation techniques. For more discussion of DNA condensation by cations, see Chapter 11.

The harsh chemical conditions, destructive action of the electron beam itself, and loss of water under the dry, high-vacuum conditions of the conventional electron microscope is particularly likely to lead to structural distortion of supramolecular complexes such as ribosomes, in which the forces stabilizing quaternary structure may be rather weak and delicately balanced. Conventionally dried samples may be quite shrunken and shrivelled compared with freeze-etched samples. Cantor and Schimmel (1980, Chapter 10) cite the case of freeze-dried ribosomes, which have dimensions of $170 \times 230 \times 250$ Å, corresponding to a volume of 5.1×10^6 Å^3. By contrast, normally air-dried ribosomes typically have dimensions of $160 \times 180 \times 200$ Å, resulting in a 38% smaller volume.

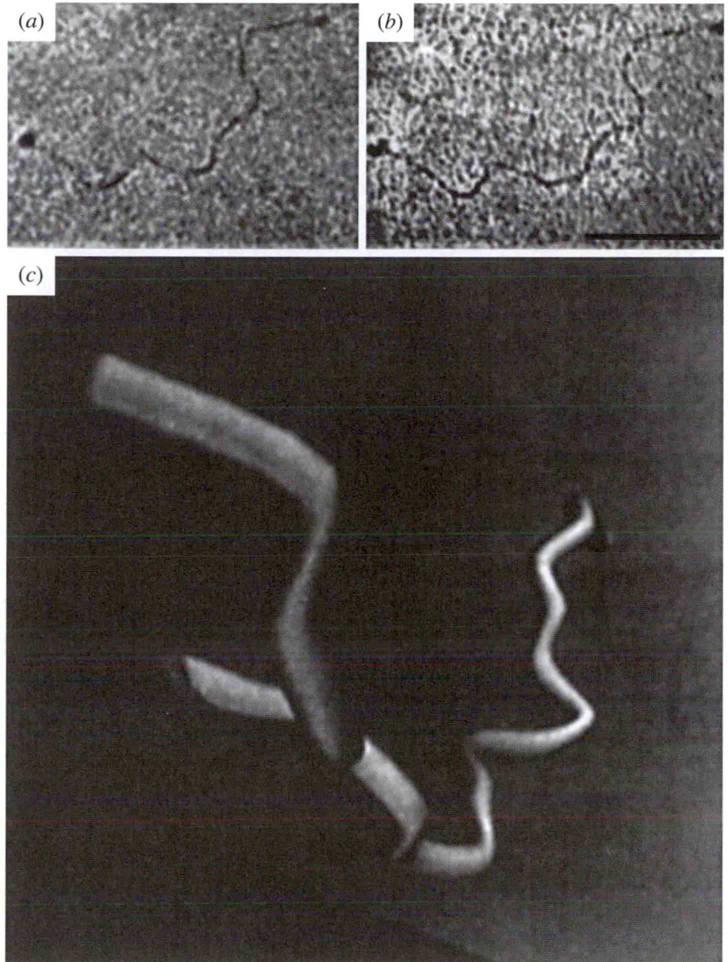

Figure 9-29
Cryoelectron microscopy of supercoiled plasmid pUC9 DNA. (*a* and *b*) Stereopair of images. (*c*) Computer reconstruction of 3D image. [Reproduced with permission from Dubochet et al., 1992.]

Figure 9-30
Circumferential wrapping of two DNA strands along the interior surface of a toroidal shaped spermidine–DNA complex. left 0° tilt view; right 10° tilt view. Only the torus-half farthest from the metal source is shown. Arrows depict the position of the higher lying DNA strand in both tilt views. [Reproduced with permission from Ruben et al. 1981.]

13.1.5 Scanning Transmission Electron Microscopy

Going beyond a visual image, scanning transmission electron microscopy (STEM) is useful for measuring the masses of macromolecular complexes such as protein–DNA complexes and virus structures (Hough et al., 1985; Thomas et al., 1994). Electrons are scattered elastically by the atoms in the specimen, yielding a signal proportional to the local mass density. This scattering enables measurement of the total mass of a particle, and the masses of its resolved domains. Scanning transmission electron microscopy has been shown to be capable of visualizing DNA without shadowing, staining, or other contrast enhancement (Hough et al., 1982), but it is most useful for quantitative characterization of protein–DNA interactions.

For example, Mastrangelo et al. (1985) used STEM to elucidate the binding of the large T antigen of SV40 at sites near the origin of replication. T-antigen binding to the origin is required for DNA replication; while concurrent binding at a nearby operator region causes maximum repression of early transcription. It is by this mechanism that T antigen shuts off its own production late in infection. The STEM mass footprinting (Fig. 9-32) shows that the monomer, dimer, and trimer states of the large T antigen all bind in the operator region, and monomer through tetramer in the origin region. Mass measurements were required to reach these conclusions, since the lengths and outlines of dimer and trimer are similar enough that they cannot be distinguished by geometry alone. Positional determination by STEM is accurate to about 5 bp.

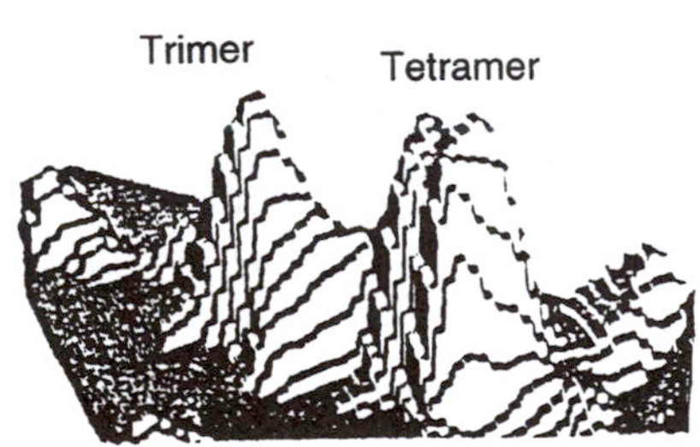

Figure 9-31
Perspective drawing of the STEM mass distributions for an SV40 T-antigen trimer in binding region I and a tandem tetramer in binding region II. Individual blocks making up the structure represent $1 \times 1\text{-nm}^2$ picture elements. Their height is proportional to the electron count for that element above a constant background. [Reproduced with permission from Mastrangelo et al., 1985.]

13.2 Scanning Probe Microscopy

13.2.1 General Features

The family of techniques collectively called scanning probe microscopy are not EM at all. The most prominent of these for biological applications are scanning tunneling microscopy (STM) and scanning force microscopy (SFM). Scanning force microscopy is also called atomic force microscopy (AFM). Instead of bombarding the sample with an electron beam, a very sharp tip (ideally, atomically sharp) is moved in a raster pattern within a few angstroms over the sample, and senses the height above the sample. In STM, the height-sensing mechanism is an electric current that depends exponentially on the distance between tip and conducting surface. In SFM, it is a response to the force (van der Waals, electrostatic, etc.) between tip and surface.

Changes in height and horizontal position are accomplished with a piezoelectric crystal. The expansion and contraction of the crystal are controlled by application of a voltage controlled by a feedback circuit that senses the tip displacement. Scanning probe microscopy has extraordinary resolution, in principle: an atomic diameter horizontally and a fraction of an angstrom vertically. In practice, the horizontal resolution is more often 10–50 Å with biological samples, similar to EM. Vertical displacements, while very sensitive to small changes, are not usually accurate for measuring heights.

The extreme sensitivity to distance between tip and sample gives these techniques excellent effective contrast, without need for staining or shadowing. They also have other advantages compared to EM. They can be performed in air or in physiological buffers. They do not subject the sample to degradation from a high-energy electron beam. Scanning probe microscopes are small; they weigh only a few kilograms and fit easily on a small table. Scanning force microscopy in particular promises the resolution of EM combined with the convenience of optical microscopy.

The book edited by Marti and Amrein (1993) gives a good survey of basic principles and techniques of scanning probe microscopies. The review by Engel (1991) emphasizes STM, while those by Bustamante et al. (1993, 1994) and Hansma and Ho (1994) treat SFM.

13.2.2 Scanning Tunneling Microscopy

Scanning tunneling microscopy is based on a nonclassical contrast principle, which is the variation of quantum mechanical tunneling current with distance from a conducting surface. In quantum mechanics, there is a finite probability of finding an electron above the surface of a sample, in what according to classical physics would be a forbidden region. The sample to be imaged constitutes one electrode; a very fine metal probe tip, poised about 10 Å above the sample and scanned across the surface in a raster pattern, constitutes the other. A voltage applied between the probe and the sample causes a tunneling current to flow across the gap. The tunneling current varies as $e^{-2\kappa d}$ where d is the distance between electrodes and κ is the inverse decay length for the wave functions in the barrier. For vacuum tunneling, κ is related to the work function ϕ by $\kappa = h/2\pi\sqrt{2m_e\phi}$, where m_e is the electron mass and h is Planck's constant (Marti and Amrein, 1993, Chapter 1). This exponential decay of tunneling current gives extraordinarily fine vertical resolution, better than 0.05 Å.

Initially, the STM technique was applied only to solid surfaces in high vacuum, but it has been found possible to image biological molecules in air or under water. Many of the early STM images of DNA were obtained on graphite substrates. It has since been realized that surface structures of graphite can mimic DNA (Clemmer and Beebe, 1991). Lindsay and Barris (1988) developed a method for producing stable absorbate patches of DNA on a gold surface under a buffer solution. Figure 9-33 shows an example of a double helical synthetic oligonucleotide prepared in that way. Although atomic resolution is within the power of STM, such resolution has been only rarely achieved to date (Bloomfield and Arscott, 1991). Numerous practical problems must be overcome before STM can become a standard technique of nucleic acid analysis.

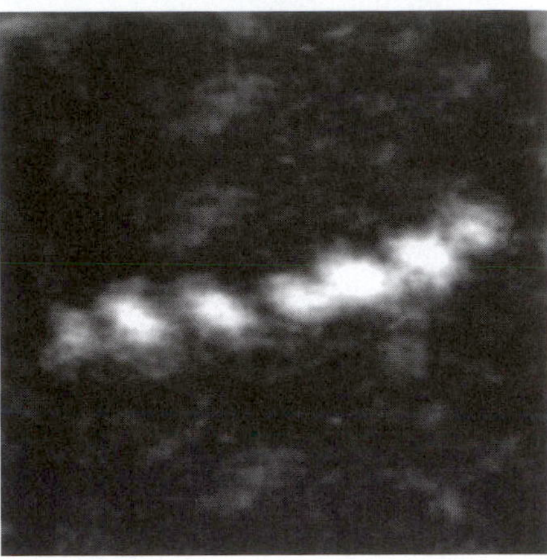

Figure 9-32
A 68 bp synthetic oligonucleotide with the λ O_R3 binding site in positions 19–35, imaged by STM in phosphate buffer on the (111) surface of a gold electrode after deposition under potentiometric control. Seven turns of the apparently right-handed double helix are clearly seen. [Reproduced with permission from Jing et al. 1993.]

The most serious is stable deposition of molecules on surfaces. Image formation with a nonconducting sample is also puzzling; it may involve actual mechanical contact between the probe tip and the substrate, leading to a distortion of electronic energy levels.

13.2.3 Atomic Force Microscopy

Unlike the STM, SFM can image insulating as well as conducting surfaces. The SFM of DNA is usually conducted on mica, which is atomically flat over large areas. Stable binding of double-stranded DNA and RNA to mica can be achieved by chemically modifying freshly cleaved mica with 3-aminopropyltriethoxy silane (Lyubchenko et al., 1992).

The SFM maps the topography of a surface by scanning the sample in a raster pattern beneath a fine tip at the end of a cantilever. As the tip encounters height changes on the surface, the cantilever begins to deflect. The deflection is detected by an "optical lever" arrangement/reflection of a laser off the cantilever onto a segmented photodiode, with the difference in output of the two segments proportional to the deflection. The sample is moved up or down to cancel the deflection, and the surface height is determined from the extent of vertical motion.

The SFMs have been most commonly operated in the constant force or contact mode, in which the tip moves relative to the surface like a needle across a phonograph record, showing the variations in sample height during scanning. While this is the simplest technique, it generates rather large lateral forces that can move or tear the sample or damage the tip. Tapping mode avoids these problems by oscillating the cantilever at very high frequency so that the tip is alternately brought into contact with the surface and lifted off. The force is always normal to the surface. This method provides high resolution while avoiding the artifacts due to dragging. Tapping is conducted in air by vibrating the cantilever. In liquids, the fluid damps the cantilever oscillation, so the fluid cell is oscillated instead. The DNA has been imaged in water and aqueous buffers, in propanol and butanol (which provide higher resolution) as well as in dry air.

A further advantage of working in liquid rather than in air is that strong capillary and electrostatic forces develop in air between the tip and the sample. These are minimized in liquids, leading to less sample distortion and allowing the use of smaller, sharper tips.

Applications of SFM to nucleic acids (reviewed by Hansma and Ho, 1994 and by Bustamante et al., 1994) include determination of the handedness of supercoiling in plasmids, length measurements, and cutting with the SFM tip at a selected location. Biotinylated DNA labeled with streptavidin complexes may be useful for high-resolution gene mapping, with resolution of a few dozen bases. At the other extreme of resolution, SFM has been used to image various stages in condensation of chromatin.

Imaging of DNA–protein complexes to show binding locations and DNA bending is a particularly promising application, since the sensitivity of SFM to height variations gives high effective contrast relative to EM. As an example (Erie et al., 1994), Figure 9-34 shows Cro protein from λ phage complexed to the operator sites O_R3, O_R2, and O_R1 of a 1 kbp fragment of λ DNA. The protein Cro is small, only 14.7 kDa; and three molecules, whose centers are only 23 bp (~ 7 nm) apart can be seen. The bend in the DNA induced by specific binding is also clearly seen. Interestingly, nonspecific binding by Cro also bends DNA, but the range of bending angles is much broader. The increasing ability to image in aqueous buffers, rather than in air, will enhance the physiological relevance of this technique even more. For example, the binding of E. coli RNA polymerase to DNA (Guthold et al., 1994) and the cleavage of DNA by DNase I (Bezanilla et al., 1994) have been observed in real time.

Figure 9-33
Contact SFM image of Cro–DNA complexes in which three λ Cro molecules (indicated by arrows) are bound specifically to the O_R3, O_R2, and O_R1 operator sites in a 1.0 kbp DNA fragment. The complexes were deposited in 20 m*M* tris acetate buffer, pH 7.5, 0.1 m*M* EDTA, 5 m*M* $MgCl_2$ onto freshly cleaved mica and imaged in air. The scan size is 250 × 250 nm. [Reprinted with permission from Erie, D. A. Yang, G., Schultz, H. C., and Bustamante, C. *Science* **266**, 1562–1566. Copyright ©1994 American Association for the Advancement of Science.]

13.3 Visualization and Manipulation of Individual DNA Molecules Undergoing Dynamic Motion

In EM and scanning probe microscopy, the molecules are generally adsorbed firmly to a surface. Therefore, although their structures can be discerned with increasing resolution, their dynamics are not observable. This limitation is being overcome with innovative light microscopy techniques (Beechem, 1994). The visualization of the dynamics of individual molecules avoids averaging over the motions of large numbers of molecules that will generally be out of phase with each other. This ability to observe single molecules is particularly important when systematic molecular motion, such as the translation of a DNA molecule through the active site of a processive enzyme, is being studied. On the other hand, any attempt to obtain thermodynamic properties from measurements of single molecules must take into account the large molecular and statistical fluctuations, and the need to average over a large number of observations.

Yanagida and co-workers (Morikawa and Yanagida, 1981; Matsumoto et al., 1981; Yanagida et al., 1983) developed a method to visualize individual long DNA molecules in solution under a fluorescent microscope connected to a highly sensitive video camera. The DNA is complexed with a fluorescent dye and trapped in a thin liquid layer between coverslips. Its fluorescent image is intensified and captured on video or optical disk for computer enhancement and analysis. Applications have included observation of DNA condensation and folding, digestion by DNase I, and motions during gel electrophoresis. The method has been reviewed in detail by Bustamante (1991).

The length of a long DNA molecule can be measured with reasonable accuracy by this technique, since it is greater than the spatial resolution of the microscope. The molecular width is generally less than the resolution, and appears to be greater than the true value. Thus the aspect ratio is distorted, and molecules appear to be relatively thicker than they actually are. This can complicate interpretation of condensation experiments, since a thick image cannot be unambiguously taken to mean that several molecules have aligned side by side. Surface effects in the thin films between the microscope coverslips must also be considered.

Another technique is to attach a micron-sized bead to one end of the DNA, and to use optical microscopy to observe the motion of the bead. The advantage of this procedure is that the bead can be manipulated by applying magnetic and hydrodynamic forces (Smith et al., 1992), or by using "optical tweezers," which move the bead by radiation pressure from an intense laser beam (Perkins et al., 1994a; Svoboda and Block, 1994). The fluorescence and bead techniques can be combined: binding a fluorescent dye to the DNA makes it possible to visualize the DNA while manipulating the bead.

If one end of the molecule is tethered to a surface, the response of individual molecules to well-defined stretching forces exerted on a bead at the other end can be measured (Smith et al., 1992). When interpreted in terms of wormlike coil statistics (Bustamante et al., 1994), the force-distance curve for dimers of λ-DNA agreed very well with the contour length (32.8 μm) and persistence length (53.4 nm) expected. A similar experiment stretched the DNA in hydrodynamic flow while a latex microsphere bead at one end of the DNA was held stationary by optical tweezers (Perkins et al., 1995). This setup avoided perturbations from the surface, but yielded virtually identical results.

Optical tweezers can be used to move a bead, attached to the end of DNA, over a well-defined path, and thereby to explore the motion of DNA in a variety of interesting situations. For example, Perkins et al. (1994a) followed the motion of a single, fluorescently labeled molecule of DNA in an entangled solution of unlabeled λ-phage DNA molecules, while they stretched it into various conformations having bends, kinks, and loops. The DNA obeyed several expectations of reptation theory; in particular, as it relaxed, it closely followed a path defined by its initial contour. Viscoelastic relaxation of individual molecules stretched to full extension by hydrodynamic flow was observed in the same apparatus (Perkins et al., 1994b). The longest relaxation time varied approximately as the 1.65 power of the molecular length, in good agreement with theoretical predictions and macroscopic relaxation measurements (see Section 9).

The tethered particle motion method has been put to ingenious use in studying the kinetics of transcript elongation by a single RNA polymerase molecule (Shafer et al., 1991; Yin et al., 1994). An RNA polymerase molecule is immobilized on a glass surface, and a microscopic particle is attached to one end of a DNA molecule bound to the polymerase in a stalled transcription complex. The DNA tethers the particle, which is visible by differential interference contrast light microscopy, to the immobilized enzyme. During the processive action of the polymerase, after the stalled complex has been restarted, the tether shortens or lengthens, depending on whether the particle is attached to the upstream or downstream end of the DNA. The magnitude of nanometer scale Brownian motion of the particle, detected and quantified by the extent of blurring of its image, increases as the length of the tether increases. Calibration with DNA molecules of known length allows determination of the tether length as a function of time, and thus of the rate of processing by the enzyme. A similar approach has been used to measure the rate of lactose repressor-mediated loop formation and breakdown in single DNA molecules (Finzi and Gelles, 1995). One end of the DNA is attached directly to a glass cover slip. Looping of the DNA by a tetrameric repressor molecule decreases the effective length of the tether and thereby reduces Brownian motion of the bead at the other end.

14. BENDING AND TWISTING OF DNA

Some of the most important information to come from the techniques described in this chapter concerns the elastic behavior of DNA. As structural and dynamic measurements have become more detailed and refined, we have come to realize that DNA often interacts with proteins and drugs not in its relaxed, equilibrium conformation, but rather in a structure in which it is bent and/or twisted away from equilibrium. This insight will be amplified in Chapter 13, but we develop some of the fundamentals in this section.

14.1 Energetics of Bending and Twisting

14.1.1 Definitions

Our aim is to calculate the cost, in free energy, of a particular amount of bending or twisting. The pertinent equations are entirely analogous to Hooke's law for a spring. This states that the force required to stretch the spring a length Δx beyond its equilibrium length is $F = k\Delta x$, where the force constant of the spring is k. If the force is integrated over the displacement, the work required to stretch the spring by Δx is $W = \frac{1}{2}k(\Delta x)^2$. Since the free energy change ΔG is the reversible work done in stretching the spring, we find $\Delta G = \frac{1}{2}k(\Delta x)^2$.

This very familiar squared dependence on deformation appears in almost identical form if an elastic rod such as DNA is bent or twisted (Fuller, 1971). Suppose a length L is bent, without twisting, along the arc of a circle of radius R_c. Then the free energy difference from its relaxed, unbent state is

$$\Delta G_{\text{bend}} = \frac{1}{2}B\left(\frac{1}{R_c}\right)^2 L \tag{9-88}$$

where B is the bending force constant. ΔG_{bend} is proportional to the length L undergoing the bending, and inversely to R_c^2. (Bending in a tight circle, of small R_c, requires more energy than bending in a large circle.) As another way to understand this equation, note that $L = \theta R_c$, where θ is the angle between the tangents to the beginning and ends of the rod. Thus

$$\Delta G_{\text{bend}} = \frac{1}{2}B\left(\frac{\theta^2}{L}\right) \tag{9-89}$$

Of course, it is unlikely that a molecule of DNA will bend exactly along the circumference of a circle. However, for any given small length, the bending free energy is proportional to the square of the curvature of that portion; the total ΔG_{bend} is obtained by summing these contributions along the chain.

Similarly, if the rod is twisted by ϕ radians while maintaining its axis straight, the free energy difference between the twisted and untwisted states is

$$\Delta G_{\text{twist}} = \frac{1}{2}C(\phi/L)^2 L \tag{9-90}$$

where C is the twisting force constant. ΔG_{twist} depends quadratically on the twist per unit length ϕ/L, and linearly on the total length L being twisted.

As discussed in detail below, B is generally about 2×10^{-19} erg cm, and C about 2.5×10^{-19} erg cm. It is instructive to consider specific examples of what these numbers mean. Suppose that 200 bp ($L = 680$ Å) of DNA are bent uniformly in a circle of radius $R_c = 55$ Å, corresponding roughly to the geometry of the DNA in a nucleosome. Then $\Delta G_{\text{bend}} = 2.25 \times 10^{-12}$ erg, or 55 times the thermal energy $k_B T$ (4.1×10^{-14} erg at 25°C). In more familiar units, this corresponds to a free energy of 32 kcal mol^{-1}, a very appreciable energy that must be compensated by binding interactions with histones (see Chapters 13 and 14). Next, consider a 20 bp length of DNA whose ends must twist by 36° (one-tenth of a turn) to accomodate a binding protein. This requires $\Delta G_{\text{twist}} = 7.3 \times 10^{-14}$ erg, 1.8 $k_B T$, or 1.0 kcal mol^{-1}.

The elastic properties of materials are often expressed in terms of their Young's modulus E for compression or extension. If the length of a circular rod of length L and radius R is changed by ΔL, the free energy change is $\Delta G = \frac{1}{2} EAL(\Delta L/L)^2$ where A is the cross-sectional area of the rod, πR^2. During bending, the outer part of the rod is stretched, the inner part compressed. Analysis leads to the relation between Young's modulus and bending force constant $B = EI$, where I is the moment of inertia of the rod about its axis, $\pi R^4/4$. A value of $B = 2 \times 10^{-19}$ erg cm implies $E = 2.5 \times 10^9$ dyne cm^{-2} (with $R = 10$ Å), and $\sigma = 0$. For comparison, E for many metals and alloys is in the range 10^{11}–10^{12} dyn cm^{-2}.

14.1.2 Relation to Persistence Length and RMS Fluctuations

Earlier in Section 9.2.3 we noted that DNA can be described as a wormlike coil, whose persistence length a characterizes its stiffness: the average correlation of the chain with its initial direction as it is subjected to random thermal buffeting. The bending force constant, on the other hand, measures the force required to bend the chain with a given radius of curvature. Although a describes a thermally averaged random behavior, and B a systematically applied force, they are both determined by the stiffness of the molecule, so there must be a relation between them. For an isotropic rod, the relation is obtained by performing a Boltzmann-weighted average over all possible angles θ, and considering displacements in both directions perpendicular to the backbone, rather than just in the plane of a circle (Landau and Lifshitz, 1958; Bloomfield et al., 1974, p. 164). The simple result is

$$a = B/k_B T \tag{9-91}$$

With $B = 2 \times 10^{-19}$ erg cm, and $k_B T = 4 \times 10^{-14}$ erg, $a = 5x10^{-6}$ cm, or 500 Å. Schellman (1980 a,b) has developed similar equations for more complex and realistic models of DNA, in which the isotropic bending potential is replaced with an anisotropic potential reflecting the helix symmetry, or in which abrupt chain interruptions occur due to breathing, kinks, or ligand binding.

The statistical mechanical treatment leading to Eq. 9-91 also provides a formula for the mean-square bending angle $< \theta_L^2 >$:

$$< \theta_L^2 > = 2Lk_B T/B \tag{9-92}$$

(This finding is obvious from Eq. 9-89 by letting the average $\Delta G_{\text{bend}} = k_B T$.) If we consider a length L of DNA equal to the distance b between base pairs, Eqs. 9-91 and 9-92 show that the mean-square bending angle $< \theta_b^2 >$ between adjacent base pairs is $2b/a$. With $b = 3.4$ Å, $a = 500$ Å, and $T = 298\ K$, this leads to $< \theta_b^2 >^{1/2} = 0.12$ rad, or 6.7°.

A similar analysis for twisting yields

$$< \phi_L^2 > = Lk_B T/C \tag{9-93}$$

With $C = 3.4 \times 10^{-19}$ erg cm, the average twist angle between adjacent bases $< \phi_b^2 >^{1/2}$ is 0.063 rad or 3.6°.

14.1.3 Static and Dynamic Bending

Our treatment has related the persistence length to the elastic energy needed to bend the DNA away from a straight path that is assumed to be of lowest energy. In fact, the fundamental definition of persistence length is in terms of the length over which the original directional correlation is lost. Such loss of correlation can occur either because of dynamic bending, as considered up to this point, or because of static bending due to sequence-dependent variations along the polymer chain. If the persistence lengths due to dynamic and static bending are a_d and a_s, then the observed persistence length will be (Trifonov et al., 1987; Schellman and Harvey, 1995)

$$a^{-1} = a_d^{-1} + a_s^{-1} \tag{9-94}$$

a_d and a_s may each be longer than 500 Å.

Perhaps supporting this idea, a theoretical analysis (Song et al., 1990; Song and Schurr, 1990) of time-dependent electrooptic experiments that probe bending and twisting motions (Diekmann et al., 1982; Pörschke et al., 1987) gives longer persistence lengths, (∼ 2100 Å). Song and Schurr (1990) speculate that there may be two persistence lengths, equilibrium and dynamic, and that the relation between them implied in Eq. 9-91 (*a* is an equilibrium quantity, *B* determines the response to time-dependent bending forces) may not hold on very short time scales. They suggest this may be because the potential energy for DNA bending is not strictly harmonic, but "instead is dimpled or scalloped, so that it exhibits several thermally accessible discrete minima separated by barriers." If these barriers are high enough that crossings are improbable in the 100 μs time scale of the transient electrooptic experiments, but not so high that equilibration cannot take place over longer times, then the apparent persistent length could be time dependent. The validity of this interesting idea remains to be seen, but it is not necessarily inconsistent with our limited but increasingly detailed knowledge of sequence-dependent structural variations along the helix.

Manning (1988) suggested that any locally stiff polymer has three persistence lengths: the bending persistence length discussed to this point, a persistence length for twisting (Wilcoxon and Schurr, 1983), and a persistence length for contraction/extension fluctuations. These have similar but not identical values, on the order of 100 bp. Considerations of all three types of deformations may be necessary to understand long-range, "action at a distance" phenomena such as propagation of structural distortions associated with B–Z transition sequences (Wartell et al., 1982) or the long-time persistence of several coexisting secondary structures in relaxed supercoiled DNAs (Song et al., 1990).

14.2 Experimental Evaluation of Persistence Length and Torsional Stiffness

The earliest studies of persistence length of DNA were done using hydrodynamic and light scattering techniques. Despite sample heterogeneity in most of these studies, approximations inherent in hydrodynamic theories, and the limited sensitivity of translational hydrodynamic and light scattering radius of gyration measurements to small changes in persistence length, work from that era yielded generally consistent results, which are in reasonable agreement with more recent measurements. It is important to realize that while short DNA is often described as "rodlike", it deviates measurably from the properties of a rod if it contains more than about 50–100 bp. This is demonstrated in Figure 9-35. The ratio z/L is given for wormlike chains by $(a/L)[1 - \exp(-L/a)]$. For a DNA molecule whose contour length L equals one persistence length ($\sim$ 500 Å or 150 bp), $z/L = 0.63$. The translational frictional coefficient ratio is relatively insensitive to L, since resistance lost in shortening is largely regained in thickening the molecular domain. However, the rotational frictional coefficient, depending on the cube of the effective length, decreases drastically relative to the rigid rod value as L increases.

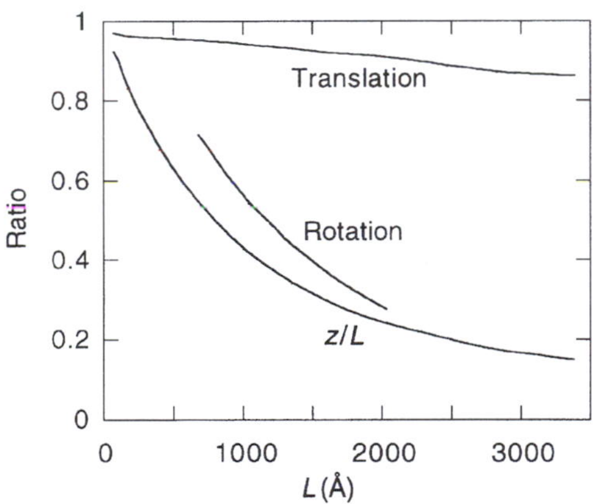

Figure 9-34
Dependence on DNA length (from 20 to 1000 bp) of three characteristic parameters, relative to rigid rods. Here z/L is the chain projection on initial axis direction. Translation is the translational friction coefficient, calculated from theories of Yamakawa and Fujii (1973) and Tirado and Garcia de la Torre (1979). Rotation is the rotational frictional coefficient of long axis, calculated from theories of Hagerman and Zimm (1981) and Tirado and Garcia de la Torre (1980). A persistence length of 500 Å has been assumed.

14.2.1 Hydrodynamics

Perhaps the first systematic determination of a comes from the sedimentation theory of wormlike chains developed by Hearst and Stockmayer (1962), applied to S versus M data of various workers. The characteristic feature of the data, reproduced by their theory, is a slower increase of log S with log M at low M than at high M. As noted in the discussion of Figure 9-10, this occurs because at low M, DNA behaves mainly as a rod, while at high M it behaves as a random coil. Transition between these regimes is determined by DNA stiffness. The Hearst–Stockmayer analysis was extended by

Gray et al. (1967) to incorporate an approximate treatment of excluded volume. They obtained $a = 450$ Å for DNA in 0.2 M salt, and a hydrodynamic diameter d of 27 Å. A more refined treatment of wormlike coils without excluded volume, applicable to relatively low molecular weight DNA, gave $a = 650$ Å and $d = 25 \pm 1$ Å (Yamakawa and Fujii, 1973) when applied to the same data. The persistence length may be slightly overestimated because of neglect of excluded volume.

Hagerman (1981) performed transient electric birefringence measurements of rotational diffusion for a series of DNA fragments from 104 to 910 bp. Assuming a rise/bp of 3.4 Å and a hydrodynamic diameter of 26 Å, he obtained a persistence length of about 500 Å under moderate to high salt conditions. Similar results were obtained by Elias and Eden (1981a,b) and Diekmann et al. (1982).

Combined use of translational and rotational diffusion can help to define more precisely the range of structure parameters. An example is given in Figure 9-36, which shows that measurements of D_t and D_r (which can be made with an accuracy of a few percent) on a series of fragments, can determine the persistence length to within 50 Å or less. Values of 25–27 Å for the hydrodynamic diameter seem reasonable if the distance between phosphates, 20 Å, is augmented by about a monolayer of water. A value of 24 ± 1.2 Å is obtained from fluorescence polarization anisotropy measurements on 43 and 69 bp restriction fragments (Wu et al., 1987). Eimer and Pecora (1991), however, have measured the translational and rotational diffusion coefficients of a series of short (8, 12, and 20 bp) B duplex oligonucleotides (which should be very well modeled as short cylinders), and found that the most consistent fit to the data is $d = 20 \pm 1.5$ Å. The reason for this discrepancy is not currently understood.

14.2.2 Light Scattering

The use of light scattering to measure the persistence length of DNA is best carried out with relatively small molecules, whose $< R_g >$ is less affected by intramolecular excluded volume and for which the angular extrapolations are less difficult. Godfrey and Eisenberg (1976) performed light scattering and hydrodynamic measurements on seven

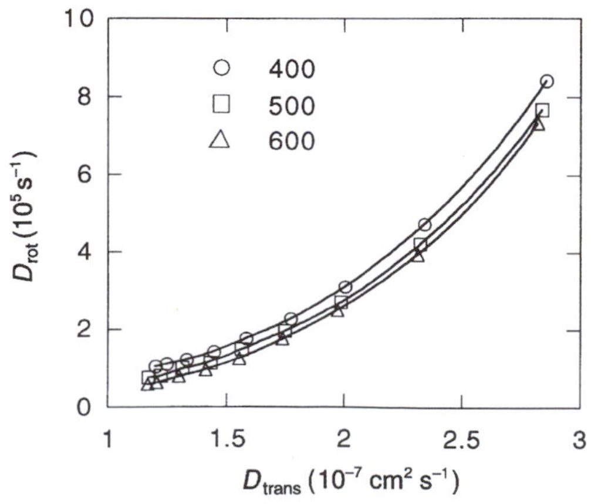

Figure 9-35
Sensitivity of combined translational and rotational diffusion measurements to persistence length. The legend shows the persistence length in angstrom. The points correspond to DNAs with 155 (upper right) to 525 (lower left) bp. All molecules were assumed to have a rise per residue of 3.4 Å and a hydrodynamic diameter of 26 Å.

narrow molecular weight fractions of calf thymus DNA in the range 0.3–1.3 × 10^6 in 0.2 M salt. Scattering measurements were made down to $\theta = 20°$, and limiting values of M and $< R^2 >$ were obtained after correction for optical anisotropy effects (negligible above 1 million). The dependence of $< R_g^2 >$ on M gave persistence length 540 ± 56 Å, invariant over the range of M examined. This value is in good agreement with persistence lengths determined on the same samples from sedimentation and viscosity measurements. The sensitivity of the angular dependence of light scattering to persistence length for a plasmid-size DNA is shown in Figure 9-37. Schmid et al. (1971) have written a thoughtful analysis of the limitations of light scattering for determining persistence length.

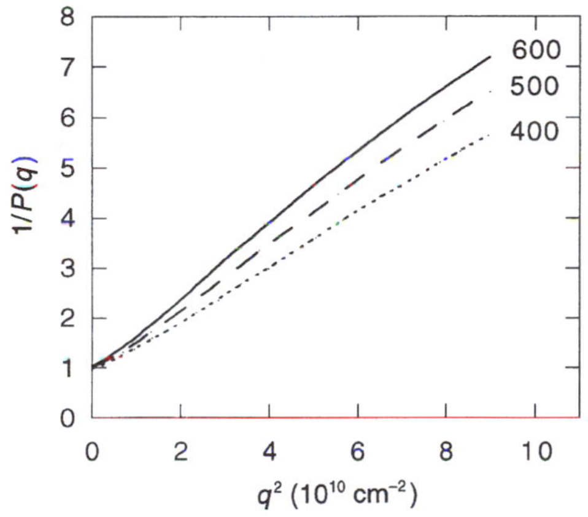

Figure 9-36
Reciprocal scattering function for a 3000 bp plasmid, as a function of $q^2 = [(4\pi n/\lambda)\sin(\theta/2)]^2$, with an Ar^+ ion laser irradiation ($\lambda = 4880$ Å) in aqueous solution ($n = 1.334$), for persistence lengths of 400, 500, and 600 Å. Calculation according to Sharp and Bloomfield (1968) as corrected by Schmid et al. (1971) for the case of no excluded volume.

14.2.3 Electron Microscopy

As noted in Section 13.1.3, measurement of chain curvature by EM can be a useful way to determine persistence length. It has the virtue of being insensitive to chain polydispersity and aggregation, which may complicate light scattering and hydrodynamic measurements. In principle, it allows determination of local bending flexibilities in different parts of a molecule. The major assumptions (Frontali et al., 1979) are that as the chain settles on the grid, it reequilibrates rather than just giving a 2D projection of a 3D conformation, and that the interactions with the surface do not change the flexibility. Even if these assumptions are not strictly valid, it is likely that stiffness comparisons between DNAs of different compositions or sequences will remain valid. In fact, measurements in 0.3–0.5 M ammonium acetate give persistence lengths for various T2 and T7 phage DNA fragments of 540 ± 50 Å, in good agreement with solution measurements under similar ionic conditions.

14.2.4 Single Molecule Stretching

Through the ingenious use of magnetic beads (Smith et al., 1992; Strick et al., 1996), micro-fibers (Cluzel et al., 1996), optical tweezers (Smith et al., 1996; Wang et al., 1997; Baumann et al., 1997), and hydrodynamic flow (Perkins et al., 1995), it has become possible to measure how the extension of individual DNA molecules varies with applied force in the piconewton range. An equation that adequately describes the extension x of a wormlike coil with contour length L_0 in response to a stretching force F is (Bustamante et al., 1994; Marko and Siggia, 1995)

$$\frac{Fa}{k_B T} = \frac{1}{4}\left(1 - \frac{x}{L_0}\right)^{-2} - \frac{1}{4} + \frac{x}{L_0} \tag{9-95}$$

This equation assumes that the DNA is of fixed length, and describes the elasticity of the DNA due the loss of entropy as the number of configurations of the chain is restricted upon extension.

Near full extension, Eq. 9-95 predicts that x approaches L_0 as $F^{-1/2}$. In this limit, it has been found (Smith et al., 1996) that DNA can be stretched, in a process that increases its internal energy or enthalpy, beyond the contour length L_0 defined by B-form geometry. An equation that describes this extensible regime is (Odijk, 1995)

$$\frac{x}{L_0} = 1 - \frac{1}{2}\left(\frac{k_B T}{Fa}\right)^{1/2} + \frac{F}{S} \tag{9-96}$$

where S is the elastic stretch modulus. By fitting x versus F measurements to these equations, it is possible to determine the persistence length and stretch modulus as a function of ionic conditions, obtaining results in good agreement with other types of measurements.

At extensions beyond those in which elastic behavior is obeyed, DNA undergoes a reversible overstretch transition to a form 1.7-fold longer than B-DNA (Cluzel et al., 1996; Smith et al., 1996). The orientations of bases and backbone in this unusual structure are presently unknown.

14.2.5 Twisting Stiffness by Electrooptic Measurements

Fujimoto and Schurr (1990) used time-resolved fluorescence polarization anisotropy of intercalated ethidium to obtain the torsional twisting constant C of the host DNA. Since depolarization occurs from a variety of motions, including rapid librations of the dye in its binding site, and bending as well as twisting of the DNA, a fair bit of complex theoretical interpretation is required before C can be extracted from the measurements. A minimum value of C can be set by assuming that the persistence length a is infinite, so that all depolarization is due to twisting. A maximum for C is obtained by assuming that a has its static value of 500 Å. In this case, C for most of the DNAs studied is in the range $(2.8 \pm 0.2) \times 10^{-19}$ erg cm, in reasonable agreement with the value obtained from ligation measurements described below. The two extreme C values differ by a factor of 1.9. An important point of this study is that the twisting force constant appears to be independent of DNA base composition, in contrast to the results of earlier studies, although it is not ruled out that particular sequences have C significantly different from the average.

14.2.6 Ligation of Short Fragments into Circles

The way to determine the bending and twisting force constants of DNA, that is least dependent on complicated and uncertain theory, is to measure the length-dependence of the ring closure probability for short circles. This approach was devised by Shore et al. (1981) and further developed by Shore and Baldwin (1983a,b). A useful review is by Crothers et al. (1992). Specifically, what is measured is the j factor, defined by Jacobson and Stockmayer (1950) as the ratio of equilibrium constants for cyclization (K_c) and linear bimolecular association (K_a) through cohesive ends. The cohesive ends are generated by restriction endonucleases, and the rates of cyclization or linear association are measured by gel electrophoresis. The j factor is given by Eq. 9-10 for an idealized flexible chain. For a real DNA molecule, the theory is much harder, though it has been worked out by Shimada and Yamakawa (1985). However, extraction of the desired elastic constants can be performed without a detailed theory.

The mechanism for the ring closing reaction is presumed to be

$$L \underset{k_{21}}{\overset{k_{12}}{\rightleftharpoons}} S \tag{9-97}$$

$$E + S \underset{k_{32}}{\overset{k_{23}}{\rightleftharpoons}} ES \overset{k_{34}}{\longrightarrow} E + P \tag{9-98}$$

where L is the linear form, S is the substrate (with ends in proximity but not yet covalently joined) for the ligase enzyme E, ES is the enzyme–substrate complex, and P is the ligated circular product. Under the low [E] experimental conditions employed, juxtaposition of the cohesive ends is in rapid equilibrium relative to the slower covalent closure reaction. Then steady-state solution of the kinetic equations 9-94 and 9-95 yields for the measured first-order rate constant k_1 of ligation of circles

$$k_1 = \frac{k_{34}[E_o]K_c(1 - f_s)}{K_m + [S]} \tag{9-99}$$

where $K_c = k_{12}/k_{21}$, K_m is the Michaelis constant $(k_{32} + k_{34})/k_{23}$, and f_s is the fraction of DNA in the S form.

The mechanism for linear dimerization is similar except that two L forms must come together to make an S, leading to an expression for the measured second-order rate constant

$$k_2 = \frac{k_{34}[E_o]K_a(1 - f_s)^2}{2(K_m + [S])} \tag{9-100}$$

(The factor of 2 arises because of the identical cohesive ends, so that linear molecule A reacts with both A and B.) Then, under conditions where $S \ll K_m$ so that $f_s < 0.03$, we obtain

$$j = K_c/K_a = k_1/2k_2 \tag{9-101}$$

Plots of the j factor as a function of DNA length are shown in Figure 9-38. Figure 9-38(a) shows that the variation in j is less than 10-fold over this wide size range, from about 1.6–29 persistence lengths. Molecules as short as two persistence lengths are still quite flexible: The increasing probability, with shorter chains, that the ends will be found in the same neighborhood (Eq. 9-10) is able to compensate for the somewhat higher bending energy of shorter chains. Below 240 bp, the closure probability decreases dramatically. The large vertical line at 250 bp shows the range of variation for the fragments in Figure 9-38(b). This figure shows an approximate 10 bp periodicity, suggesting that the variation of j with bp reflects the need to twist the ends into juxtaposition before ligation can occur.

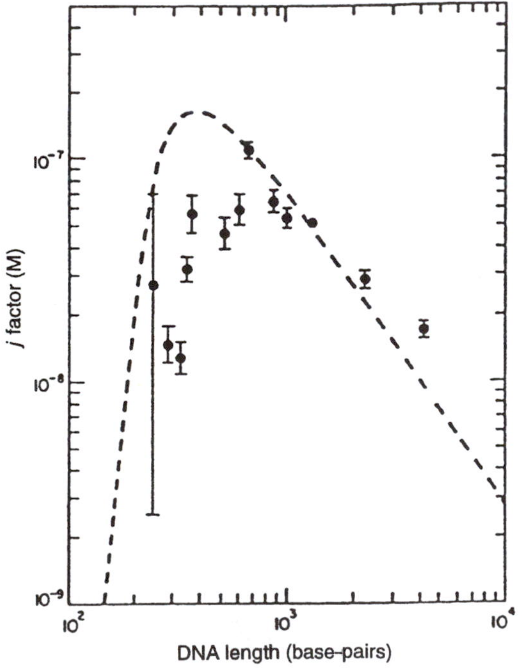

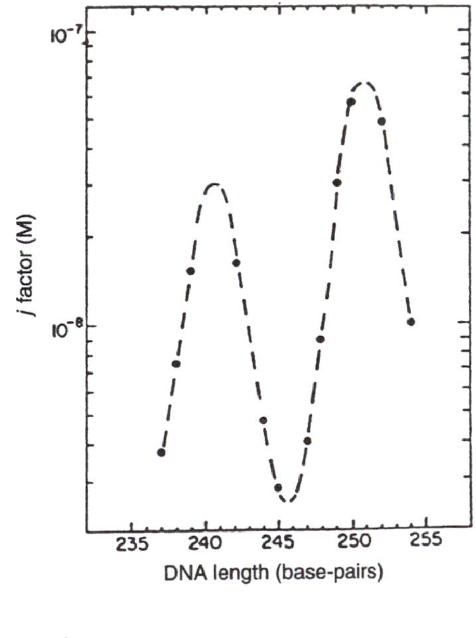

Figure 9-37
Cyclization probability j as a function of DNA length for a series of *Eco*RI restriction fragments ranging in length from (a) 250–4362 bp and (b) 237–254 bp. [Reprinted from Shore, D. and Baldwin, R. L., Twisting. I. Relation between Twist and Probability; II. Topoisomerism Analysis, *J. Mol. Biol.*, **170**, 957–981, 983–1078, Copyright ©1983, by permission of the publisher Academic Press Limited London.]

Detailed analysis (Shore and Baldwin, 1983) shows that a curve like Figure 9-38(*b*) can be used to determine the torsional force constant C so long as writhe is negligible (i.e., the circle is planar) and the DNA helix is continuous in nicked circles (i.e., twist is an integer). The first of these assumptions is validated by the calculations of Le Bret (1979), which show that the onset of writhe occurs only after about two or more turns of twist. The second assumption was proven by sealing experiments with nicked circles. Then C was calculated to be 2.4×10^{-19} erg-cm. However, the computer simulations of Levene and Crothers (1986a,b) indicate that there are substantial writhe fluctuations in small circles. Analysis of the Shore and Baldwin (1983) data by computer simulation (Levene and Crothers, 1986a,b; Crothers et al., 1992) yields $C = 3.4 \times 10^{-19}$ erg cm. A further assumption is that the DNA has no sequence-dependent bends or twists, so that j is not biased by some local feature and B and C are proper averages over the sequence (see below).

Using the experimental approach of Shore and Baldwin (1983) (but conducting both cyclization and linear association reactions in the same vessel) and the theory of Shimada and Yamakawa (1985), Taylor and Hagerman (1990) found $C = 2.0 \pm 0.2 \times 10^{-19}$ erg cm, and $a = 450 \pm 15$ Å (corresponding to $B = 1.8 \pm 0.06 \times 10^{-19}$ erg cm) independent of NaCl concentration above 10–20 m*M*. Both Shore and Baldwin (1983) and Taylor and Hagerman (1990) analyzed the distribution of topoisomers and found the helical repeat to be about 10.45 bp/turn in this ionic strength range. The C value found by Taylor and Hagerman (1990) is smaller than that found in other cyclization experiments. This range of C values indicates that there may be sequence-dependent variations in DNA torsional rigidity (Crothers et al., 1992).

To summarize, various techniques (ligation, hydrodynamics, light scattering, and electron microscopy) give generally similar results. It is important that different approaches be used, because each depends on different assumptions. The major remaining questions appear to be the molecular meaning of a longer dynamic persistence length, and the consequences of sequence-dependent variations in static geometry and flexibility.

14.3 Ionic Effects on Persistence Length and Bending

14.3.1 Ionic Strength Effects

As the ionic strength decreases, it is easy to imagine that the electrostatic repulsions between the phosphate groups, becoming less shielded by intervening salt, would lead to a stiffening of the wormlike coil. This concept has been developed theoretically (Skolnick and Fixman, 1977; Odijk, 1977) in the form

$$a = a_{\text{int}} + a_{\text{el}} \qquad (9\text{-}102)$$

where a, the total persistence length, is the sum of an intrinsic component a_{int} and an electrostatic component a_{el}. The intrinsic part a_{int} is the value at high salt concentration, and a_{el} has the form

$$a_{\text{el}} = \frac{1}{4\kappa^2 b_{\text{j}}} \qquad (9\text{-}103)$$

where $1/\kappa$ is the Debye–Hückel screening length and $b_j = e^2/\varepsilon k_B T$ is the Bjerrum length (7.1 Å in water at 25°C) (see Chapter 10). The parameter a_{int} is 450–500 Å; in order for electrostatic contributions to be noticeable, a_{el} must be an appreciable fraction of this. At 5 m*M* ionic strength, $1/\kappa = 42$ Å, so $a_{el} \approx 63$ Å. Thus it is only around 5 m*M* ionic strength or below that measurable electrostatic effects are expected. These predictions have been confirmed in detail for λ DNA by single molecule stretching experiments (Wang et al., 1997; Baumann et al., 1997) and for other DNAs by earlier studies (Rizzo and Schellman, 1981; Schurr and Schmitz, 1986; Hagerman, 1983). As noted by Hagerman (1983), *a* should not depend significantly on ionic strength under most physiologically relevant conditions, that is, for greater than 20–50 m*M* Na^+ or 0.2–0.5 m*M* Mg^{2+}. In fact, this appears to be true for the torsional force constant and the helical rise per base pair, as well (Taylor and Hagerman, 1990). However, studies at lower salt concentration are useful for probing the various contributions to DNA stiffness.

For short DNA, such as restriction fragments of a few hundred bp or less, the forces on the ends of the molecule are less than for the very long chains for which these equations are derived, so the stiffening at low salt concentration is less than predicted (Elias and Eden, 1981b; Hagerman, 1983).

An interesting challenge to the validity of macroscopic elasticity theory applied to DNA has emerged from single molecule stretching studies as a function of ionic strength. As noted in Section 14.1.1, the bending modulus *B*, and therefore the persistence length *a*, are predicted to be proportional to the Young's modulus *E*. The elastic stretch modulus *S* is also expected to be proportional to *E*. However, *a* increases with decreasing ionic strength, while *S* is found to decrease (Baumann et al., 1997). This suggests that different molecular mechanisms are operating. In low-salt, long-range Coulombic interactions increase *a*; while it may be that the decrease in *S* is due to local melting of A,T rich regions of the DNA.

14.3.2 Ion-induced Bending

While the above equations express the persistence length as a function of ionic strength, single molecule stretching experiments (Baumann et al., 1997) show that multivalent cations have quite a different effect than do monovalent ions. Ions such as Mg^{2+}, spermidine^{3+}, and spermine^{4+} lower the "high-salt" a_{int} in Eq. 9-102 significantly below the 450–500 Å observed in Na^+, to as low as 250 Å. It seems likely that this is due to transient bending of the negatively charged surface of the DNA toward the concentration of positive charge.

This bending appears to be one instance of the general prediction (Rich, 1978; Manning et al., 1989) that DNA will bend toward a surface neutralized by a localized, positively charged ligand. The bending occurs both because of Coulombic attraction between the DNA phosphates and the cation, and because of unbalanced Coulombic repulsions between phosphates. This prediction, advanced to explain folding of DNA in nucleoprotein complexes, has been verified by an experiment (Strauss and Maher 1994) in which six phosphates along the minor groove of a G,C rich sequence of DNA were replaced by neutral methylphosphonates, causing an approximately 20° bend toward the neutralized patch. In a similar study (Strauss et al., 1996), DNA bending was produced by tethering ammonium ions through hexamethylene linkers to one side of a DNA double helix.

14.4 Composition and Sequence Dependence of Bending and Twisting Constants

If DNA must bend and twist in order to bind with histones, regulatory proteins, and enzymes (Chapters 13 and 14), then it is important to know whether the bending and twisting free energies depend on base composition and sequence. Thomas and Bloomfield (1983) used total intensity scattering to show that the persistence length of poly(dG-dC)·poly(dG-dC) in the B form is about twice that of native DNA with random sequence and composition; and the Z form of this synthetic DNA, obtained at high salt concentration, is even stiffer. Chen et al. (1985) found by transient electric dichroism that the bending rigidity of alternating poly(dA-dT) sequences is less than that of random sequence DNA. Hogan and Austin (1987) used results from a variety of techniques (light scattering, pulsed birefringence, triplet anisotropy decay and triplet quenching) to conclude that the bending stiffness of DNA increases with increasing GC content. If it is assumed that the Poisson ratio $\sigma = B/C$ remains constant, then these results imply that the torsional force constant C also increases with %GC. However, Fujimoto and Schurr (1990) have found that there is no significant dependence of C on base composition. Sorting out these conflicting results will not be easy, but the thoughtful analysis of Fujimoto and Schurr inclines one currently to the position that the bending and twisting rigidity of DNA do not depend strongly on DNA base composition, though they may depend on local sequence.

14.5 Sequence-Directed Bending

14.5.1 Characterization of DNA Bending by Comparative Electrophoresis

The DNA helix axis is straight only under idealized conditions. We have already discussed the bending of DNA that is induced by thermal forces and characterized quantitatively by the persistence length. Some DNA molecules are intrinsically curved. Curvature results when certain base sequences are repeated in phase with the DNA helical repeat. In this arrangement the small bends of the repeating elements add coherently because they are all toward the same side of the helix and therefore in the same direction. For a review of this topic, see Crothers et al., 1990. Chapter 13 considers DNA bending by protein binding.

14.5.1(a) *Finding a Bending Locus* The fortuitous discovery of bent DNA (Marini et al., 1982) stemmed from the highly anomalous electrophoretic mobility of molecules in which the helix axis is systematically curved. Figure 9-39 (Wu and Crothers, 1984) illustrates the principle of comparative electrophoresis (reviewed by Crothers and Drak, 1992), in which the mobility is compared for molecules that are nearly identical in length and sequence, but which differ in shape because the helix is

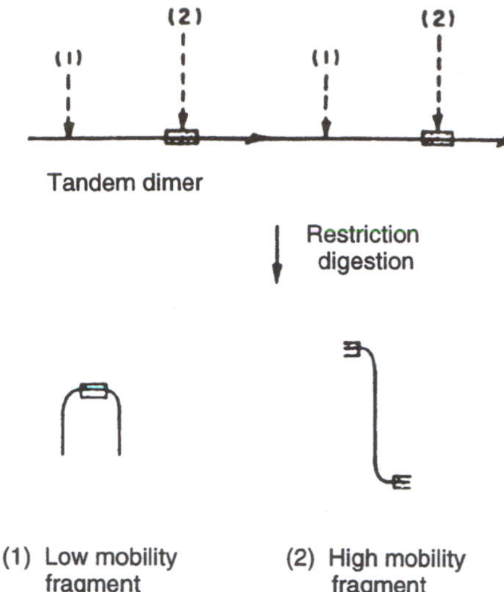

Figure 9-38
Construction of circularly permuted variants of a DNA sequence differing in conformation because of the location of a bending locus, shown by the box. The tandem dimer is cleaved with two different enzymes (1) and (2), yielding fragment (1), which is strongly bent and therefore of low gel mobility, and fragment (2), which is nearly linear because the bend is near the end, and is therefore of high mobility. [Reprinted with permission from *Nature*. Wu, H.-M. and Crothers, D. M. (1984). *Nature (London)* **308**, 509–513. Copyright ©1984 Macmillan Magazines Limited.]

bent. Tandemly repeated DNA fragments cut with a series of restriction enzymes that cleave only once per repeat can be used to create a corresponding set of molecules of circularly permuted sequence in which the bent segment is located near the middle or at the end of the fragment. Placing the bend near the center yields large overall curvature, whereas molecules with the bend close to the end are nearly straight in shape.

The electrophoretic consequence of increased overall curvature is a reduction of electrophoretic mobility. This observation can be qualitatively justified by the theory of Lumpkin and Zimm (1982), as summarized by Eq. 9-26. Molecules that differ only in shape have the same values of charge (Zq) and contour length (L); it can also be argued that they should have the same frictional coefficient (f) for interaction with the gel along the tube contour. However, bent molecules have a shorter average end-to-end distance, so $< h_x^2 >$ will be reduced relative to linear molecules. Hence, according to Eq. 9-26, the mobility μ will be reduced.

Note that this argument assumes that DNA molecules retain at least some of their solution curvature as they are drawn through the gel. This assumption may seem to contradict the simple model of Lumpkin and Zimm (1982), according to which the leading segment of the chain determines the tube path that must be followed by the rest of the molecule. However, Levene and Zimm (1989) have argued that the gel matrix should be viewed as elastic, which allows the shape of the tube to change in response to curvature of the DNA helix axis. The results of Drak and Crothers (1991), interpreted with Eq. 9-26, are consistent with the view that bent molecules are less curved in the gel than in solution, as would be expected from minimization of the overall energy of the gel matrix–DNA molecule system.

Figure 9-40(*a*) (Wu and Crothers, 1984) shows the experimental variation of electrophoretic mobility observed for circularly permuted sequences from kinetoplast DNA of the parasite *Leishmania tarentolae*. Extrapolation of the mobilities to their

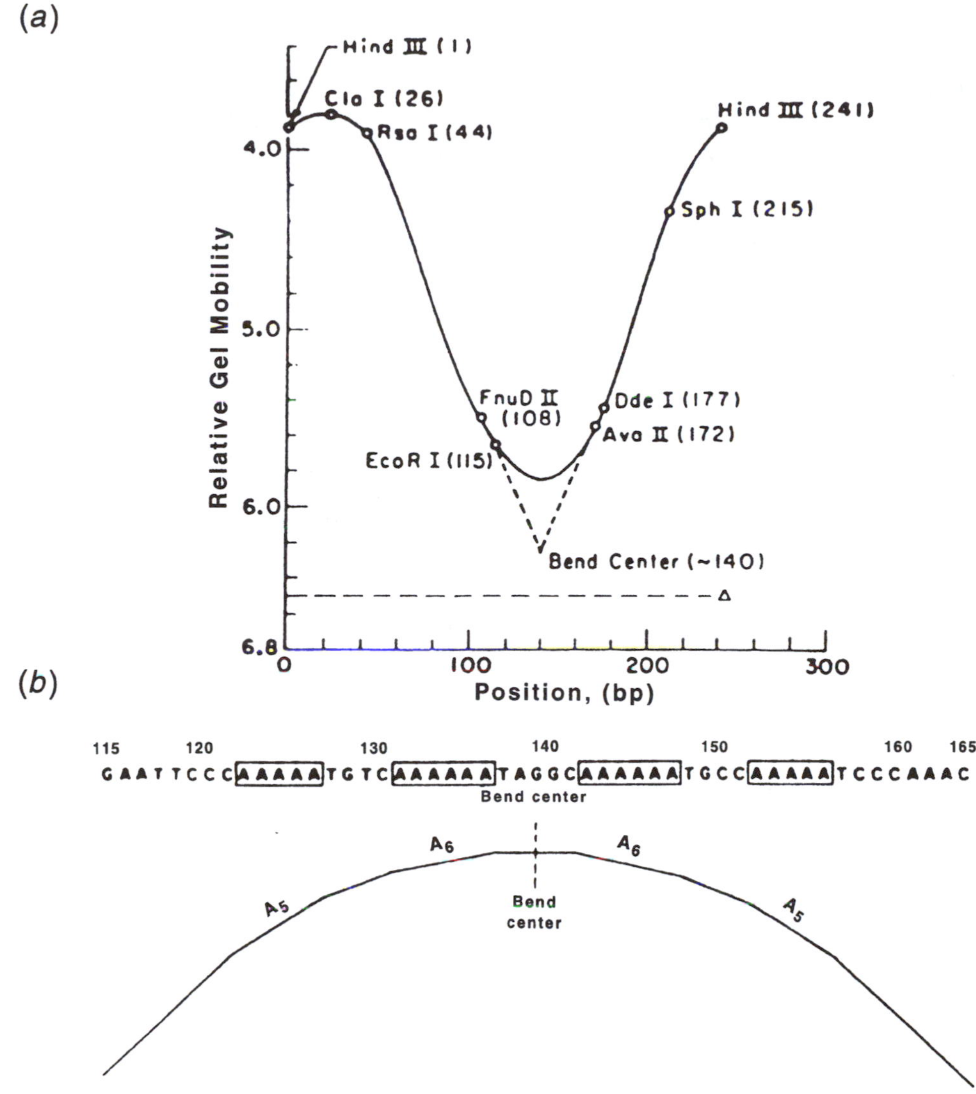

Figure 9-39
Identification of the bending locus in a curved DNA. (*a*) Experimental variation of the mobility of circularly permuted DNA sequences carrying a bending locus. The extrapolated minimum in the curve, corresponding to a maximum in the mobility, reflects positioning of the ends of the molecule at base pair 140. Hence, this site should correspond to the center of the bend. (*b*) DNA sequence at the bend locus, revealing A_{5-6} tracts repeated at 10 bp intervals. The curve represents the DNA helix axis, which is assumed to be deflected at each end of an A tract. [Reprinted with permission from *Nature*. Wu, H.-M. and Crothers, D. M. (1984). *Nature (London)* **308** 509–513. Copyright ©1984 Macmillan Magazines Limited.]

maximum (the minimum in the curve in the figure) shows that mobility is maximized when the end of the fragment is at about position 140 in the parent sequence. Figure 9-40(*b*) shows the distinctive sequence found at that locus. Tracts of homopolymeric $(dA \cdot dT)_n$ segments (called A tracts), each about one-half of a helical turn long ($n =$ 5–6), are repeated at 10 bp intervals, or nearly in phase with the DNA helical repeat.

14.5.1(b) *Phasing and Sequence Requirements for Bending* The requirement for phase match between A tracts and the DNA helical repeat in order to produce bending was confirmed by ligation of double helical oligomers of varying length (Hagerman, 1985; Koo et al., 1986). Figure 9-41 (Koo et al., 1986) shows how R_L, the ratio of apparent size from gel mobility to the real molecular size, varies with the

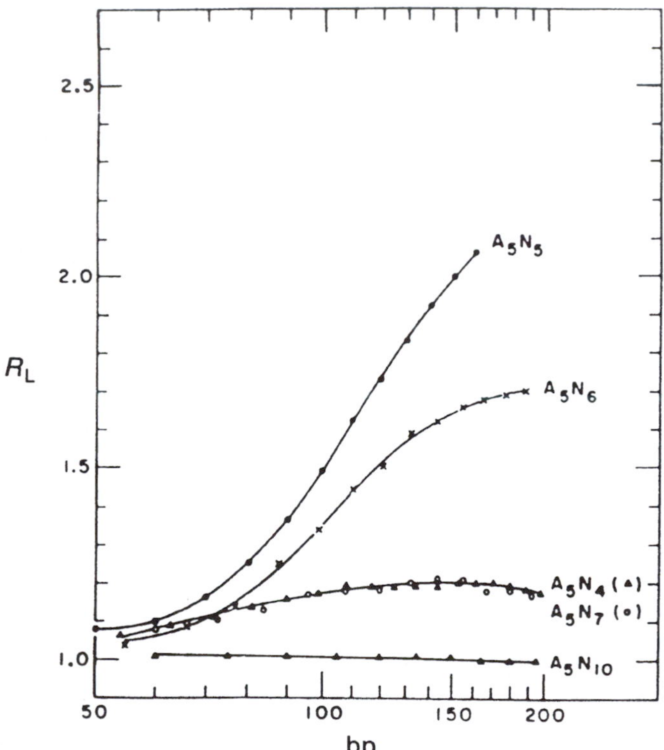

Figure 9-40
Dependence of the ratio R_L of apparent to real DNA chain length on DNA size for different phasings of A tract bending sequences. The anomaly in electrophoretic mobility is measured by deviations of R_L from 1: Note that these are maximal for molecules containing the repeated sequence A_5N_5, in which the A tracts are repeated at 10 bp intervals. The anomaly is smaller but still pronounced for repeats of A_5N_6, for which the 11 bp repeat is also close to the DNA helical repeat. However, A_5N_4 and A_5N_7 are nearly normal in mobility. The A_5N_{10} multimers are normal in electrophoretic mobility, indicating that A tracts separated by 1.5 helical turns do not cause DNA to bend. [Reprinted with permission from *Nature*. Koo, H. S., Wu, H.-M., and Crothers, D. M. (1986)].

number of repeats of sequences containing A tracts. Ligation of oligomers containing A_5N_5 or A_5N_6, corresponding to repeated A tracts every 10 or 11 bp, respectively, yields a series of molecules of decidedly anomalous electrophoretic mobility. The relative mobility R_L is largest, and hence most anomalous, for these two phasings, and much less so for repeats at 9 or 12 bp. Of particular note is the normal mobility of molecules in which the A tracts are repeated every 15 bp, or 1.5 helical turns. The lack of a mobility anomaly in this case shows that the A tract does not serve as an isotropically flexible hinge, since curvature from such a mechanism should also add coherently when the hinges are 1.5 turns apart. Evidently these molecules have a zigzag structure, which is not curved as judged by electrophoresis.

Koo et al. (1986) studied sequences in which A tracts of length 5 were interrupted by other nucleotides. Substitution of T, G, or C at the central position was found to eliminate bending nearly completely. Studies with other nucleotides (Koo and Crothers, 1987; Diekmann, 1987; Diekmann et al., 1987) show that the thymine methyl group is dispensable for the bending phenomenon. However, since an I•C pair can replace A•T with only moderate attenuation of bending, the guanine NH_2 group appears to play a role in preventing formation of the curved structure.

14.5.1(c) *The Helical Repeat in Bent Molcules by Comparative Electrophoresis* Figure 9-41 implies that the anomaly in electrophoretic mobility will be largest when the phasing of A tracts exactly matches the DNA helical repeat. Exact match yields planar molecules in which the helix axis follows a circular trajectory; inexact match produces structures that have either left or right handed writhe, and higher gel mobility. This principle can be used to determine the helical repeat, as illustrated in Figure 9-42. Oligomers containing three A tracts in 31 or 32 bp correspond to sequence repeats of 10.33 and 10.67, respectively; a sequence repeat of 10.5 results from two A tracts in 21 bp. Measurements of R_L for oligomers produced by ligation of this set, along with oligomers with one A tract in 10 or 11 bp, allows estimation of the sequence repeat which would optimize R_L, equivalent to the helical repeat. The result is 10.35 ± 0.05 bp /turn. This is smaller than the normal value for B-DNA in solution of around 10.5 bp/turn, presumably because poly(dA)•poly(dT) has a reduced helical repeat (Peck and Wang, 1981; Rhodes and Klug, 1981; Strauss et al., 1981).

14.5.1(d) *Bend Direction by Comparative Electrophoresis* Molecules containing two bends can form a set of rotational isomers that differ in the number of base pairs between the two bends. Figure 9-43 illustrates the cis and trans isomers for molecules containing an intrinsic A tract bend (the heavy solid line) and a bend induced by binding of a dimeric protein. If one knows the helical repeat of the DNA between reference coordinate frames that define the two bends, then identification of the cis isomer, the species with the lowest electrophoretic mobility, enables determination of the relative bend direction. For example, if two bends are toward the minor groove at their reference coordinate frames, an integral number of helical turns will separate the bends in the cis isomer. In contrast, if a half-integral number of turns separates the bends in the cis isomer, then the two bends must be toward opposite grooves.

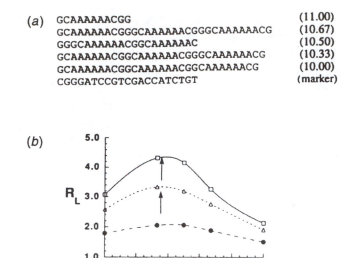

(*a*)

GCAAAAAACGG	(11.00)
GCAAAAAACGGGCAAAAAACGGGCAAAAAACG	(10.67)
GGGCAAAAAACGGCAAAAAAC	(10.50)
GCAAAAAACGGCAAAAAACGGGCAAAAAACG	(10.33)
GCAAAAAACGGCAAAAAACGGCAAAAAACG	(10.00)
CGGGATCCGTCGACCATCTGT	(marker)

(*b*)

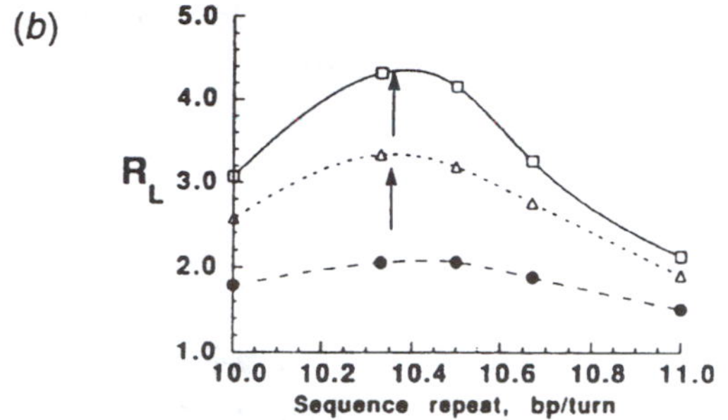

Figure 9-41
(*a*) Sequences of the oligonucleotides used in the ligation ladder experiment. The numbers in parentheses indicate the corresponding sequence repeat. Top strands only are shown (written in the 5′ → 3′ direction); bottom strands have the same length as their complementary top strand, but their ends are shifted by three bases, so that the duplexes have a 3-base 5′-end overhang for the ligation reaction. (*b*) Illustrative results of the ligation ladder experiments for multimers of the sequences in which contain 15 A tracts. The different polyacrylamide gel percentages used are •, 5%; Δ, 8%; and □, 12% (W/V). Note that the magnitude of the electrophoretic anomaly, measured by $R_L - 1$, increases with gel percentage, but the interpolated position of the optimal sequence repeat, indicted by the vertical arrow, does not change significantly with gel percentage. From this experiment it is concluded that the average DNA helical repeat in the sequences in panel *a*, taken equal to the sequence repeat which maximizes R_L, is 10.34 ± 0.04 bp/turn. (Drak and Crothers, 1991).

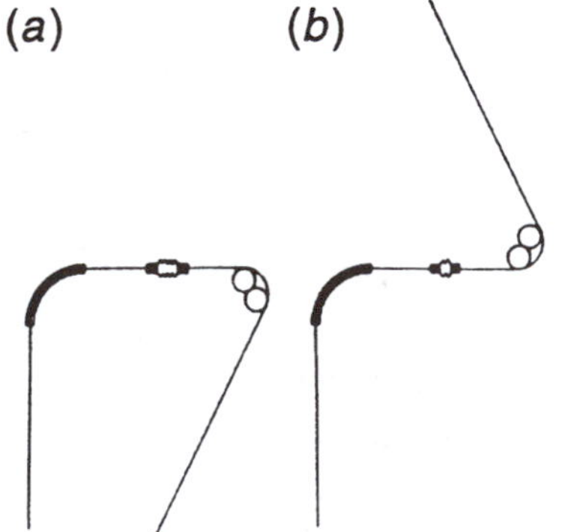

Figure 9-42
Diagram showing the cis (*a*) and trans (*b*) isomers of molecules containing an A tract bend (heavy line) and a protein-induced bend. The dimeric CAP protein is indicated by two circles. The isomers differ by one-half of a helical turn in length of the linker region between the two bends. [Reprinted with permission from *Nature*. Zinkel S. S. and Crothers, D. M. (1987). *Nature (London)* **328**, 178–181. Macmillan Magazines Limited.]

A simple symmetry argument stated by Koo et al. (1986) suggests that the bend induced by A tracts is toward either the major or the minor groove in a reference coordinate frame located at the center of the A tract. This suggestion is based on the observation that molecules containing alternating A_6 tracts and T_6 tracts on one of the complementary strands have essentially the same degree of curvature as molecules in which the A_6 tracts are all on one strand and the T_6 tracts on the other. The implication is that the operation of dyad rotation of an $A_6 \cdot T_6$ tract about an axis running between the major and minor grooves, thus interchanging the strands, does not appreciably affect the bend direction. Hence, the bend must be directed approximately along the dyad axis. Therefore its direction is either toward the major or the minor groove in a coordinate frame at or near the center of the A tract. The remaining ambiguity, whether toward the major or minor groove, was resolved by Zinkel and Crothers (1987) in favor of bending toward the minor groove, by combining A tract bends with the bend of DNA known to curve around CAP protein, as illustrated by the models in Figures 9-42 and 9-43. These results contained some indications that the A tract structure is not perfectly dyad symmetric about its central axis, but the sequences studied by Hagerman (1986) provided decisive evidence to that effect. Comparison of ligated repeats of the sequences $N_2A_4T_4$ and $N_2T_4A_4$ showed that the former are highly curved, whereas the latter are nearly normal in mobility, facts which are not compatible with full dyad symmetry of A tract structure. However, the observations can be accommodated by a slight change in the direction of the bend (Ulanovsky and Trifonov, 1987; Koo and Crothers, 1988). One way to visualize this is to retain the view that bending is toward the minor groove, but to shift and rotate the reference coordinate frame away from the center of the A tract by about one-half of a base pair toward its 3′-end. The consequence is to move the bends of the two A tracts closer together in A_4T_4 but further apart in T_4A_4:

```
N A A A A T T T T N     N T T T T A A A A N
      ↑   ↑                 ↑       ↑
```

where the arrows indicate the location of the reference frame in which the overall bend is directed toward the minor groove. In the illustration shown, the two bends in A_4T_4 are about 120° apart in relative helical phasing, so they have a substantial resultant and hence add together effectively. However, they are approximately 180° apart in T_4A_4, resulting in their mutual cancellation.

Once the A-tract bend direction is established, it can be used as a standard for characterizing other bends. One example is the bend produced by the bulge defect, in which there is an additional base on one strand. Rice and Crothers (1989) constructed molecules having A tracts interdigitated with bulge defect bends, and varied the phasing between them. The results implied bending of DNA away from the strand containing the extra adenine base, as expected for structures in which the base is stacked into the helix. Comparative electrophoresis studies of protein-induced bends is considered in Chapter 13.

14.5.2 Characterization of Bend Magnitude by Cyclization Kinetics

The theory of gel electrophoresis is not sufficiently developed to enable reliable determination of the A tract bend angle from mobility anomalies. Koo et al. (1990) addressed this problem by studying the kinetics of cyclization of molecules containing varying numbers of a 21 bp segment containing two A tracts. Comparison of Figures 9-38 and 9-44 demonstrates the dramatic effect of systematic curvature on the DNA cyclization rate. The peak j factor value in Figure 9-44 is three orders of magnitude larger than the peak value for normal DNA molecules in Figure 9-38. Furthermore, the optimum size for cyclization is reduced from about 500 bp to about 155 bp, a size for which cyclization of normal molecules is virtually undetectable.

Computer simulation of the properties of the set of bent DNA molecules enables estimation of the bend angle (reviewed by Crothers et al., 1992). The parameters affecting the rate of cyclization, in addition to the bend angle, include the persistence length, the helical repeat, and the torsional force constant C. The computer, supplied with estimates of these parameters, generates an ensemble of chains containing random bending and torsional fluctuations, along with appropriately placed systematic bends. This population is then examined for the density of states in which the two ends are in proximity, with the two end segments parallel to each other, and with the two terminal base pairs related by a twist angle that corresponds to the twist angle between base pairs in a double helix. Ensembles as large as 10^{10} chains are needed to get adequate statistical representation of those chains whose conformation is suited for ligation into circles. Figure 9-44 shows a comparison of simulated j values with the experimental results.

Because the molecules under study are ligated multimers of a 21 bp oligomer, they contain nearly integral numbers of helical turns. For this reason the simulation is not very sensitive to the torsional modulus C, which was taken equal to 3.4×10^{-19} erg cm (Levene and Crothers, 1986a,b). However, the results are quite sensitive to bend angle (panel a), helical repeat (panel b), and persistence length (panel c). As panel b shows, the best value of the helical repeat is about 10.35 bp/turn, the same as was determined for these molecules from the experiment in Figure 9-42. The best value for the bend angle is $18^\circ \pm 2^\circ$, and the persistence length is within 10% of the best value of 475 Å. Since this is the same within experimental error as the value that fits the cyclization kinetics of normal DNA using a similar calculation (Levene and Crothers, 1986a,b), we conclude that the stiffness of A tract DNA is probably not greatly different from that of normal DNA.

14.5.3 Other Methods for Characterization of DNA Bending

Rotational Dynamics. Electric dichroism was one of the methods used by Marini et al. (1982) in their search for the physical basis of the unusual gel electrophoretic properties of A tract containing molecules from *L. tarentolae*. By using a single exponential fit to characterize the dichroism decay, they found that the A tract molecules relaxed substantially more rapidly than control molecules of similar size. Hagerman (1984), using the more sensitive electric birefringence method, found that the terminal decay time revealed a significantly more rapid reorientation for the putatively curved molecules, although the difference is not as great as when the overall decay is measured.

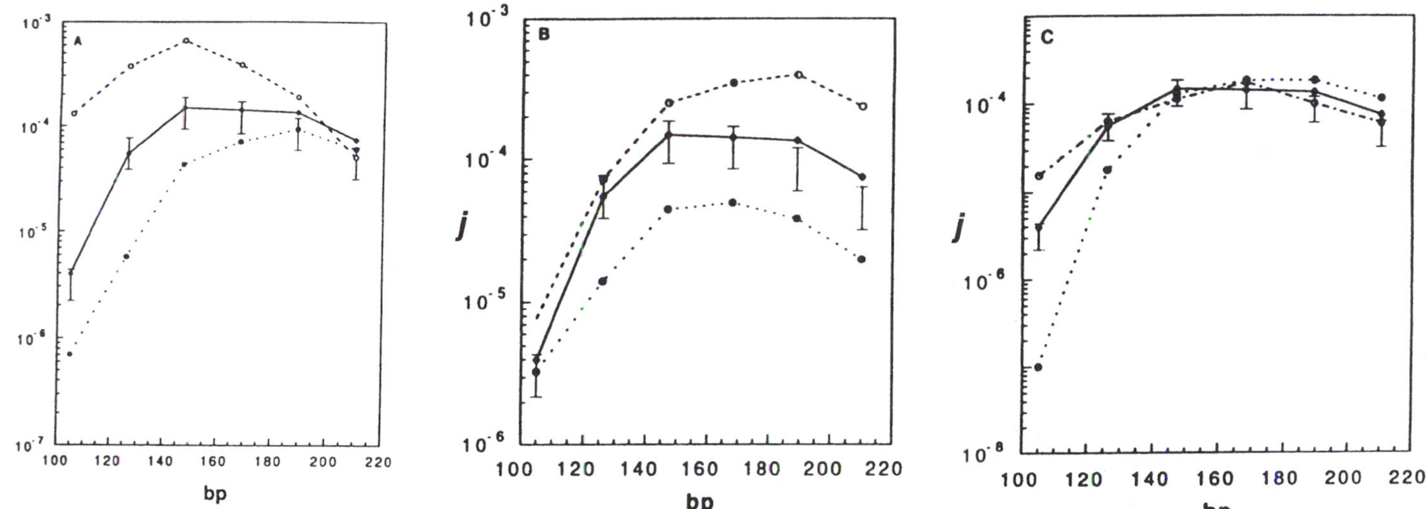

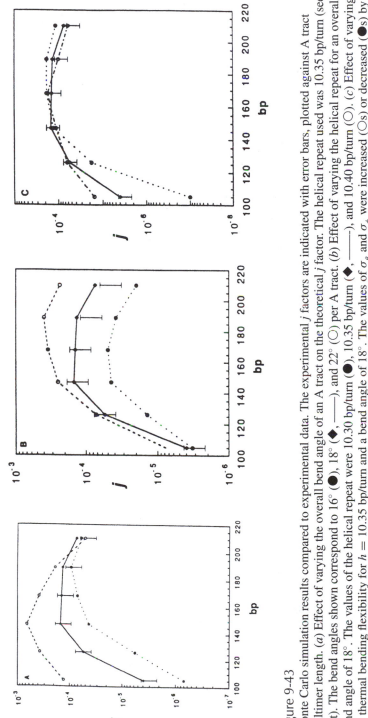

Figure 9-43
Monte Carlo simulation results compared to experimental data. The experimental *j* factors are indicated with error bars, plotted against A tract multimer length. (*a*) Effect of varying the overall bend angle of an A tract on the theoretical *j* factor. The helical repeat used was 10.35 bp/turn (see text). The bend angles shown correspond to 16° (●), 18° (◆, ——), and 22° (○) per A tract. (*b*) Effect of varying the helical repeat for an overall bend angle of 18°. The values of the helical repeat were 10.30 bp/turn (●), 10.35 bp/turn (◆, ——), and 10.40 bp/turn (○). (*c*) Effect of varying the thermal bending flexibility for $h = 10.35$ bp/turn and a bend angle of 18°. The values of σ_β and σ_ϕ were increased (○s) or decreased (●s) by 10%, corresponding to an increase or decrease in persistence length of 20% relative to the normal value of 475 Å (◇, ——). [Reprinted with permission from Koo, H.-S., Drak. J., Rice, J. A., and Crothers, D. M. (1990). *Biochemistry* **29**. 4227–4234. Copyright ©1990 American Chemical Society.]

A problem in all such measurements is to be sure that the molecules compared have the same values of persistence length and rise per base pair, parameters that strongly affect rotational dynamics. Levene et al. (1986) approached this problem by using circularly permuted varieties of the same sequence as the basis for comparison, reporting electric dichroism experiments on molecules with the four A-tract bend [Fig. 9-40(*b*)] at the center or at the end. Figure 9-45 shows the comparative decay curves on a logarithmic scale. The molecule with the bend at the center (SKRI 240) shows a larger and faster initial decay component; the terminal decay times differ by only about 20%.

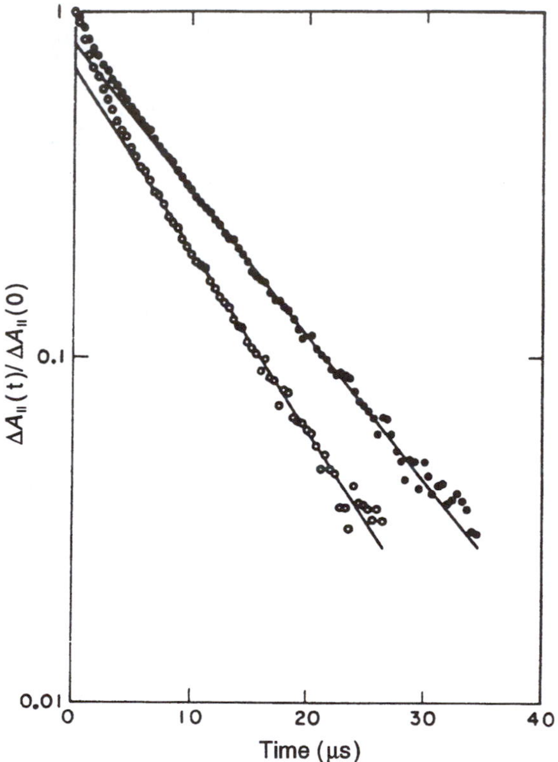

Figure 9-44
Semilogarithmic plot of the decay portion of the electric dichroism signal for SKRI 240 (O) and SK 240 (O) in 1.0 m*M* Tris-HCl and 0.025 m*M* $MgCl_2$, pH 8.0. Experimental conditions were $T = 6.0°C$, field strength $= 8.0$ kV cm^{-1}, and pulse width $= 12$ μs. The solid lines give the best least-squares fits to the longest relaxation times; these were $\tau_{SK} = 8.2\mu s$ and $\tau_{SKRI} = 10.2\mu s$. [Reprinted with permission from Levene, S. D., Wu, H.-M., and Crothers, D. M. (1986). *Biochemistry* **25**, 3988–3995. Copyright ©1986 American Chemical Society.]

The theoretical basis for this behavior is not yet resolved. It seems possible that the bent molecules might have a distribution of states, so that the terminal decay component reflects the portion of the population that is least bent. However, the value of the A-tract bend angle (18°), which Levene et al. (1986) deduced from application of hydrodynamic theory to the terminal decay, is the same as that found by Koo et al. (1990) from cyclization kinetic studies. Since the latter value characterizes the average of the population, it may be that the terminal dichroism/birefringence decay time also reflects the average properties of the population. Bending modes are a possible source of the large initial decay component.

14.5.3(a) *Electron Microscopy* As illustrated in Figure 9-46, electron microscopy (Griffith et al., 1986) reveals that molecules containing repeated A tracts, 18 times repeated in the example shown, are strongly curved. Since the molecules appear approximately circular, the average bend angle is 360/18 or about 20°. Théveny et al. (1988) showed that curved regions containing repeated A tracts could be localized in larger molecules using electron microscopy.

14.5.3(b) *Crystallography* Crystal structures have been reported for DNA molecules containing A tracts 5 or 6 bp long (Nelson et al., 1987; Coll et al., 1987; DiGabriele et al., 1989). All structures show A • T base pairs with a pronounced propeller twist, leading to the proposal of bifurcated hydrogen bonds. In such a structure the

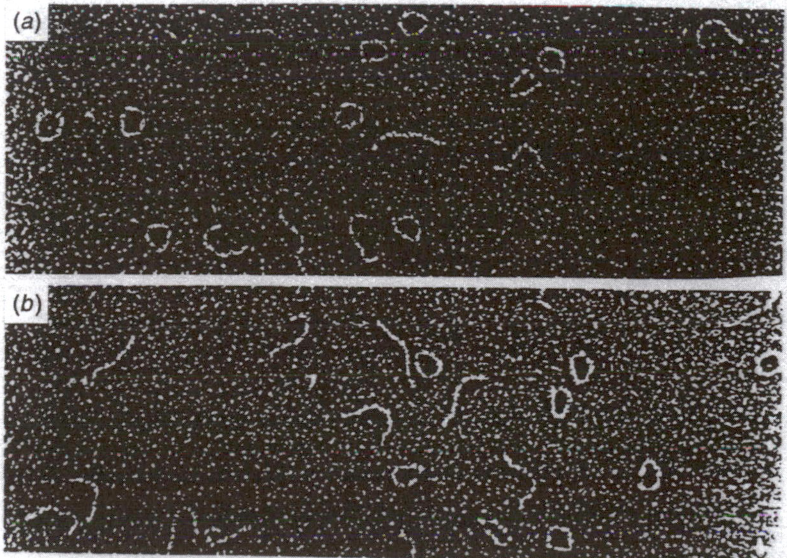

Figure 9-45
Visualization of the bent helix segment and the effect of distamycin binding. When the cloned 219 bp bent helix fragment was prepared for EM on ice as in Figure 9-1, more circular forms were present (*a*) than when the sample was mounted at room temperature (*b*). (Griffith et al., 1986).

C6 amino group of adenine, while remaining hydrogen bonded in the normal Watson–Crick sense, forms a second hydrogen bond to the O4 atom of the 3′ neighboring thymine. The result is a structure with narrow minor groove that should be relatively rigid. However, the importance of the secondary hydrogen bonds for intrinsic curvature is cast into doubt by the ability of I•C base pairs, which cannot participate in a bifurcated hydrogen-bond network, to substitute for A•T pairs with only minor loss of curvature (Koo and Crothers, 1987; Diekmann et al., 1987).

All crystallographic structures of A-tract containing molecules share the feature of propeller twisted base pairs and a relatively straight helix axis. Curvature in these molecules occurs in the regions around the A tracts. However, there is no consistent direction of curvature, and it has been concluded that the bends observed are the result of crystal packing forces (DiGabriele et al., 1989). Possibly the organic alcohols used for crystallization cause the A tract to shift to a structure that is not intrinsically bent (Sprous et al., 1995), a property found for other organic solvents by Marini et al. (1984).

14.5.3(c) *NMR Spectroscopy* The NMR measurements support the view that A tracts must exceed a minimum length of 3–4 bp before they begin to take on the structure characteristic of poly(dA)•poly(dT). This transition is illustrated for the A•T base pair imino proton chemical shifts in Figure 9-47 (Nadeau and Crothers, 1989). Imino protons near the 5′-end of the A tract are clearly distinguishable from those in long segments of poly(dA)•poly(dT), or those near the 3′-end. The study of proton exchange rates by Leroy et al. (1988) showed that exchange rates are markedly lengthened as A tracts reach and exceed 4 bp in length. So far it has not been possible to use NMR measurements to build a persuasive structural model for DNA bending induced by A tracts. However, there are several indications that minor groove width is quite narrow in the A tract region, probably even more so than in the crystal (Kintanar et al., 1987; Katahira et al., 1988; Nadeau and Crothers, 1989).

14.5.4 Models for DNA Bending

There are two general categories of models for describing DNA bending; the models are not necessarily mutually exclusive. According to the wedge model, curvature of DNA can result from addition of small wedge angles between successive DNA base pairs. Trifonov and Sussman (1980) developed this idea, and, along with Zhurkin et al. (1979), they showed how dinucleotide-specific roll and tilt deformations can in principle lead to systematic curvature when the same deformation is repeated in phase with the DNA helical repeat. In this view, curvature at A tracts results from repeated wedge angles, primarily due to roll, between successive A•T base pairs. When the tracts are about one-half of a helical turn long, the resultant of these wedge angles yields an overall deflection of the helix axis. When the roll between base pairs is in a direction to narrow the gap between the bases on the minor groove side, then the overall direction will be toward the minor groove near the center of the A tract, in agreement with experiment.

The junction model has its origins in model building studies of Selsing et al. (1979). They focused on the abrupt change in helix axis direction that occurs at the junction

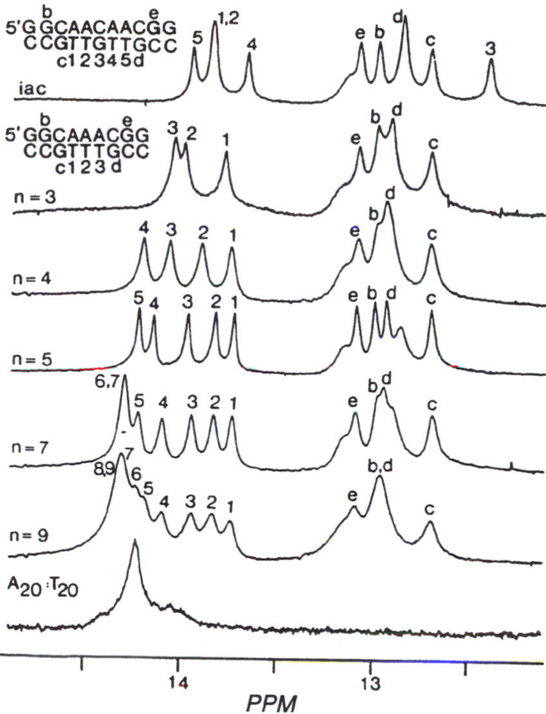

Figure 9-46
Imino proton spectra (500 MHz) of the various duplexes at 12°C. Spectra were subjected to 2-Hz exponential line broadening to improve signal to noise. The ThyH3 resonances are all numbered from left to right beginning at the 5′-end of the A tract, as shown for the A_3-tract duplex (Nadeau and Crothers, 1989).

between A- and B-form helices. Because the base pairs in the two forms have different inclinations relative to their respective helix axes, stacking the base pairs at the junction parallel to each other causes the two helix axes to be nonparallel. The locus at which the two axes meet is called a "junction bend" (Wu and Crothers, 1984; Levene and Crothers, 1983). When two junctions are one-half of a helical turn apart, the resulting bends are in the same direction, and overall curvature results. This concept is illustrated in Figure 9-48.

The two models can be used to describe the same geometry. For example, repeated roll angles between the base pairs in an A tract yields a helical structure in which the base pairs are inclined relative to the helix axis, meaning that the base pair long axis deviates from the canonical direction perpendicular to the helix axis (Fig. 9-48). Fiber diffraction studies of poly(dA)•poly(dT) yield a structure with this feature (Lipanov et al, 1990). Base pair inclination leads to junction bends at the intersection of the A tract helix axis with the adjacent segments of B-DNA. The roll and tilt components

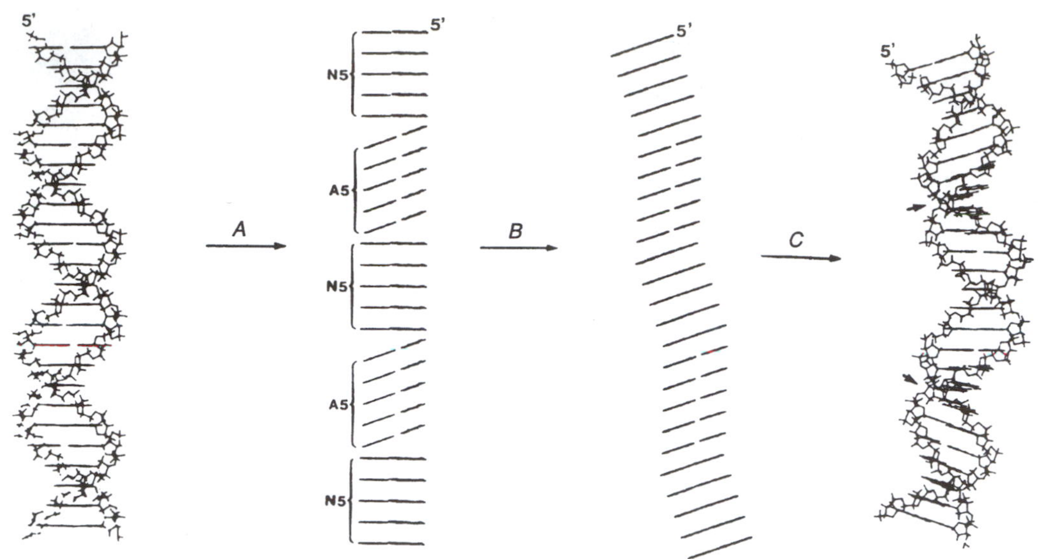

Figure 9-47
Schematic illustration of the A-tract induced bending of a DNA segment of sequence $N_5A5N_5A_5N_5$. In Step a, the B-form double helix on the left was unwound, its sugar–phosphate backbone removed (for purposes of clarity), and the base pairs within the A-tracts tilted or inclined relative to the helix axis in the direction characteristic of poly(dA)•poly(dT). (As drawn, the central figures represent views into the minor groove along the pseudo-dyad axis of each base pair. Had the backbones been shown, they would run lengthwise outside the base pairs, forming a ladderlike structure). In Step b, local helix axes reorient to facilitate base stacking at the junctions between the structurally dissimilar A5 and N5 regions; thus small bends in the helix axis arise from the inclination of the A-T pairs in combination with the requirement for favorable base stacking at the junctions. When these small local bends are positioned in phase with the helix repeat, large global curvature results. This can be seen in Step c where (1) the backbone was replaced, (2) 36° twists were applied about the local helix axes between each set of adjacent base pairs, and (3) the entire double helix was repositioned to put the overall bend in the plane of the page. Note that the direction of curvature produced by Steps a–c is geometrically equivalent to compression of the minor grooves at the centers of the A tracts (shown by the two small arrows); this is in accord with the bend direction deduced from comparative electrophoretic mobility studies (Koo and Crothers, 1986; Zinkel and Crothers, 1987; Rice and Crothers, 1989). In the figure on the extreme right, the bend magnitude is 20° per A tract (10°/junction), close to the value of 18°/A tract derived from the experiment (Koo et at., 1990); in the central two schematic figures, however, the bend magnitudes are twice those values for visual emphasis. [Reprinted with permission from Crothers et al., 1990.]

of the junction bends are described by Koo and Crothers (1988). The wedge model focuses instead on the gradual deflection of the local helix axis through the A tract. The view sometimes taken that these two descriptions necessarily imply a different structure is a mistaken one.

The wedge model has the virtue of generality, since it allows for curvature in regions of DNA that do not contain A tracts. There are several sets of parameters for describing deviations from ideal electrophoretic mobility, which are generally assumed to be a result of curvature. Parameter sets that allow calculation of the curvature of any DNA sequence have been published, as illustrated by the values given by Bolshoy et al. (1991) and De Santis et al. (1992). Testing of these theoretical values against accurately measured curvature magnitudes and directions remains a future objective.

References

Abrams, E. S. and Stanton, V. P., Jr. (1992). Use of denaturing gradient gel electrophoresis to study conformational transitions in nucleic acids. *Methods Enzymol.* **212**, 71-104.

Adam, G. and Delbrück, M. (1968). Reduction of dimensionality in biological processes, in *Structural Chemistry and Molecular Biology*, Rich A. and Davidson N., Eds., Freeman, San Francisco, CA pp. 198–215.

Adrian, M., Ten Heggeler-Bordier, B., Wahli, W., Stasiak, A. Z., Stasiak, A., and Dubochet, J. (1990). Direct visualization of supercoiled DNA molecules in solution. *EMBO J.* **9**, 4551–4554.

Anderson, P. and Bauer, W. (1978). Supercoiling in closed circular DNA: Dependence upon ion type and concentration, *Biochemistry* **17**, 594–601.

Barkley, M. D. and Zimm, B. H. (1979). Theory of twisting and bending of chain macromolecules: Analysis of the fluorescence depolarization of DNA, *J. Chem. Phys.* **70**, 2991–3007.

Batchelor, G. K. (1972). Sedimentation in a dilute suspension of spheres *J. Fluid Mech.* **52**, 245–268.

Baumann, C. G., Smith, S. B., Bloomfield, V. A., and Bustamante, C. (1997). Ionic effects on the elasticity of single DNA molecules, *Proc. Natl. Acad. Sci. USA* **94**, 6185–6190.

Beardsley, K., Tao, T., and Cantor, C. R. (1970). Studies on the conformation of the anticodon loop of phenylalanine transfer ribonucleic acid. Effect of environment on the fluorescence of the Y base, *Biochemistry* **9**, 3524–3532.

Bednar, J., Furrer, P., Stasiak, A., Dubochet, J., Egelman, E. and Bates, A. (1994). The twist, writhe and overall shape of supercoiled DNA change during counterion-induced transition from a loosely to a tightly interwound superhelix. Possible implications for DNA structure in vivo. *J. Mol. Biol.* **235**, 825–847.

Beechem, J. M. (1994). Single molecule spectroscopies and imaging techniques shed new light on the future of biophysics. *Biophys J.* **67**, 2133–2134.

Benoit, H. (1951). The Kerr effect demonstrated by dilute solutions of rigid macromolecules. *Ann. Phys.* **6**, 561–609.

Berg, O. G. (1984). Diffusion-controlled protein-DNA association: Influence of segmental diffusion of the DNA, *Biopolymers* **23**, 1869–1889.

Berg, O. G. and von Hippel, P. H. (1985). Diffusion-controlled macromolecular interactions. *Ann. Rev. Biophys. Chem.* **14**, 131–160.

Berg, O. G., Winter, R. B., and von Hippel, P. H. (1981). Diffusion-driven mechanisms of protein translocation on nucleic acids. 1. Models and theory, *Biochemistry* **20**, 6929–6948.

Berne, B. J. and Pecora, R. (1976). *Dynamic Light Scattering with Applications to Chemistry, Biology, and Physics*, Wiley-Interscience, New York, 376 pp.

Bettini, A., Pozzan, M. R., Valdevit, E., and Frontali, C. (1980). Microscopic Persistence Length of Native DNA: Its Relation to Average Molecular Dimensions, *Biopolymers* **19**, 1689–1694.

Bezanilla, M., Drake, B., Nudler, E., Kashlev, M., Hansma, P. K., and Hansma, H. G. (1994). Motion and enzymatic degradation of DNA in the atomic force microscope, *Biophys. J.* **67**, 2454–2459.

Bloomfield, V. A. (1985). Biological Applications, in *Dynamic Light Scattering: Applications of Photon Correlation Spectroscopy*, Pecora, R., Ed., Plenum, New York, pp. 381–388.

Bloomfield, V. A. and Arscott, P. G. (1991). Scanning tunneling microscopy of nucleic acids. *Nucleic Acids Mol. Biol.* **5**, 40–53.

Bloomfield, V. A., Crothers, D. M., and Tinoco, I., Jr. (1974). *Physical Chemistry of Nucleic Acids*, Chapter 5, Harper and Row, New York.

Bloomfield, V. A. and Lim, T. K. (1978). Quasi-elastic Laser Light Scattering, *Methods Enzymol.* **48F**, 415–494.

Bloomfield, V. A. and Zimm, B. H. (1966). Sedimentation, Viscosity, etc., of Dilute Solutions of Stiff Ring and Straight-Chain Polymers, *J. Chem. Phys.* **44**, 315–323.

Bolshoy, A., McNamara, P., Harrington, R. E., and Trifonov, E. N. (1991). Curved DNA without A-A: Experimental estimation of all 16 DNA wedge angles, *Proc. Natl. Acad. Sci. USA* **88**, 2312–2316.

Borochov, N. and Eisenberg, H. (1984). Conformation of LiDNA in Solutions of LiCl. *Biopolymers* **23**, 1757–1769.

Brack, C. (1981). DNA Electron Microscopy. *CRC Crit. Revs. Biochem.* **10**, 113–169.

Brack, C., Bickle, T. A., and Yuan, R. (1975). The relation of single-stranded regions in bacteriophage PM2 supercoiled DNA to the early melting sequences, *J. Mol. Biol.* **96**, 693–702.

Brack, C., Eberle, H., Bickle, T. A., and Yuan, R. (1976). A map of the sites of bacteriophage PM2 DNA for the restriction endonuclease HindIII and HpaII, *J. Mol. Biol.* **104**, 305–309.

Brack, C. and Pirrotta, V. (1975). Electron microscopic study of the repressor of bacteriophage λ and its interaction with operator DNA, *J. Mol. Biol.* **96**, 139–152.

Broersma, S. (1960a). Rotational diffusion constant of a cylindrical particle. *J. Chem. Phys.* **32**, 1626–1631.

Broersma, S. (1960b). Viscous force constant for a closed cylinder. *J. Chem. Phys.* **32**, 1632–1635.

Bustamante, C. (1991). Direct observation and manipulation of single DNA molecules using fluorescence microscopy, *Annu. Rev. Biophys. Biophys. Chem.* **20**, 415–446.

Bustamante, C., Erie, D. A., and Keller, D. (1994). Biochemical and structural applications of scanning force microscopy. *Curr. Opin. Struct. Biol.* **4**, 750–760.

Bustamante, C., Keller, D., and Yang, G. (1993). Scanning force microscopy of nucleic acids and nucleoprotein assemblies. *Curr. Opin. Struct. Biol.* **3**, 363–372.

Bustamante, C., Marko, J., Siggia, E., and Smith, S. (1994). Entropic elasticity of lambda-phage DNA. *Science* **265**, 1599–1600.

Calef, D. F. and Deutch, J. M. (1983). Diffusion-controlled reactions, *Annu. Rev. Phys. Chem.* **34**, 493–524.

Camerini-Otero, R. D., Franklin, R. M., and Day, L. A. (1974). Molecular weights, dispersion of refractive index increments, and dimensions from transmittance spectrophotometry. Bacteriophages R17, T7, and PM2, and Tobacco Mosaic Virus, *Biochemistry* **13**, 3763–3773.

Cantor, C. and Schimmel, P. (1980). Biophysical Chemistry (3 vols), Freeman, San Francisco, CA.

Cantor, C. R., Smith, C. L., and Mathew, M. K. (1988). Pulsed-field gel electrophoresis of vary large DNA molecules, *Annu. Rev. Biophys. Biophys. Chem.* **17**, 287–304.

Capel, M. S., Kjelgaard, M., Engelman, D. M., and Moore, P. B. (1988). Positions of S2, S13, S16, S17, S19 and S21 in the 30S Ribosomal Subunit of *Escherichia coli*. *J. Mol. Biol.* **200**, 65–87.

Carle, G. F., Frank, M., and Olson, M. V. (1986). Electrophoretic separations of large DNA molecules by periodic inversion of the electric field, *Science* **232**, 65–68.

Chapman, R. E., Jr., Klotz, L. C., Thompson, D. S., and Zimm, B. H. (1969). An instrument for measuring retardation times of deoxyribonucleic acid solutions. *Macromolecules* **2**, 637–643.

Charney, E. (1988). Electric linear dichroism and birefringence of biological polyelectrolytes, *Q. Rev. Biophys.* **21**, 1–60.

Chen, H. H., Charney, E., and Rau, D. C. (1982). Length changes in solution accompanying the B-Z transition of poly (dG-m^{5}dC). induced by $Co(NH_3)_6^{3+}$, *Nucleic Acids Res.* **10**, 3561–3571.

Chen, H. H., Rau, D. C., and Charney, E. (1985). The Flexibility of Alternating dA-dT Sequences, *J. Biomol. Struct. Dyn.* **2**, 709–719.

Chu, B. (1991). *Laser Light Scattering: Basic Principles and Practice*, 2nd ed., Academic, San Diego, CA.

Clemmer, C. R. and Beebe, T. P., Jr. (1991). Graphite: A mimic for DNA and other biomolecules in scanning tunneling microscope studies, *Science* **251**, 640–642.

Cluzel, P., Lebrun, A., Heller, C., Lavery, R., Viovy, J. L., Chatenay, D., and Caron, F. (1996). DNA: An extensible molecule, *Science* **271**, 792–794.

Cohen, G. and Eisenberg, H. (1968). Deoxyribonucleate solutions: Sedimentation in a density gradient, partial specific volumes, density and refractive index increments, and preferential interactions, *Biopolymers* **6**, 1077–1100.

Coll, M., Frederick, C. A., Wang, A. H.-J., and Rich, A. (1987). A bifurcated hydrogen-bonded conformation in the d(A · T) base pairs of the DNA dodecamer d(CGCAAATTTGCG) and its complex with distamycin, *Proc. Natl. Acad. Sci. USA* **84**, 8385–8389.

Crothers, D. M. and Drak, J. (1992). Global features of DNA structure by comparative gel electrophoresis. *Methods Enzymol.* **212B**, 46–71.

Crothers, D. M., Drak, J., Kahn, J. D., and Levene, S. D. (1992). DNA bending, flexibility, and helical repeat by cyclization kinetics, *Methods Enzymol.* **212B**, 3–29.

Crothers, D. M., Haran, T. E., and Nadeau, J. G. (1990). Intrinsically bent DNA. *J. Biol. Chem.* **265**, 7093–7096.

Crothers, D. M. and Zimm, B. H. (1965). Viscosity and Sedimentation of the DNA from Bacteriophages T2 And T7 and the Relation to Molecular Weight, *J. Mol. Biol.* **12**, 525–536.

DasGupta, S., Mukhopadhyay, G., Papp, P. P., Lewis, M. S., and Chattoraj, D. K. (1993). Activation of DNA binding by the monomeric form of the P1 replication initiator RepA by heat shock prteins DnaJ and DnaK, *J. Mol. Biol.* **232**, 23–34.

Davis, R. W., Simon, M., and Davidson, N. (1971). Electron microscope heteroduplex methods for mapping regions of base sequence homology in nucleic acids. *Methods Enzymol.* **21**, 413–428.

Dawkins, H. J. (1989). Large DNA separation using field alternation agar gel electrophoresis, *J. Chromatog.* **11**, 615–639.

De Santis, P., Palleschi, A., Savino, M., and Scipioni, A. (1992). Theoretical prediction of the gel electrophoretic retardation changes due to point mutations in a tract of SV40 DNA. *Biophys. Chem.* **42**, 147–152.

Deutsch, J. M and Madden, T. L. (1989). Theoretical studies of DNA during gel electrophoresis, *J. Chem. Phys.* **90**, 2476–2485.

Diekmann, S. (1987). DNA methylation can enhance or induce DNA curvature, *EMBO J.* **6**, 4213–4217.

Diekmann, S., Hillen, W., Morgenmeyer, B., Wells, R. D., and Pörschke, D. (1982). Orientational relaxation of DNA restriction fragments and the internal mobility of the double helix. *Biophys. Chem.* **15**, 263–270.

Diekmann, S., von Kitzing, E., McLaughlin, L., Ott, J., and Eckstein, F. (1987). The influence of exocyclic substituents of purine bases on DNA curvature, *Proc. Natl. Acad. USA* **84**, 8257–8261.

DiGabriele, A. D., Sanderson, M. R., and Steitz, T. A. (1989). Crystal lattice packing is important in determining the bend of a DNA dodecamer containing an adenine tract, *Proc. Natl. Acad. USA* **86**, 1816–1820.

Dill, K. A. and Zimm, B. H. (1980). Dynamics of polymer solutions. 1. Theory for an instrument; 2. The determination of molecular weight distribution by viscoelasticity. *Macromolecules* **13**, 426–432, 432–436.

Doi, M. and Edwards, S. F. (1986). *The Theory of Polymer Dynamics*, Oxford University Press, Oxford, UK, Chapter 6.

Doty, P. and Steiner, R. F. (1950). Light scattering and spectrophotometry of colloidal solutions, *J. Chem. Phys.* **18**, 1211–1220.

Drak, J. and Crothers, D. M. (1991). Helical repeat and chirality effects on DNA gel electrophoretic mobility, *Proc. Natl. Acad. Sci. USA* **88**, 3074–3078.

Dubochet, J., Adrian, M., Dustin, I., Furrer, P., and Stasiak, A. (1992). Cryoelectron microscopy of DNA molecules in solution. *Methods Enzymol.* **211**, 507–518.

Dubochet, J., Ducommun, M., Zollinger, M., and Kellenberger, E. (1971). A new preparation method for dark-field electron microscopy of biomacromolecules, *J. Ultrastruct. Res.* **35**, 147–167.

Duke, T., Viovy, J.-L., and Semenov, A. N. (1994). Electrophoretic mobility of DNA in gels. I. New biased reptation theory including fluctuations, *Biopolymers* **34**, 239–247.

Earnshaw, W. C. and Harrison, S. C. (1977). DNA Arrangement in Isometric Phage Heads. *Nature (London)* **268**, 598–602.

Earnshaw, W. C., King, J., Harrison, S. C., and Eiserling, F. A. (1978). The Structural Organization of DNA Packaged within the Heads of T4 Wild-Type, Isometric and Giant Bacteriophages, *Cell* **14**, 559–568.

Eigner, J. and Doty, P. (1965). The native, denatured, and renatured states of deoxyribonucleic acid. *J. Mol. Biol.* **12**, 549–580.

Eimer, W. and Pecora, R. (1991). Rotational and translational diffusion of short rodlike molecules in solution: Oligonucleotides, *J. Chem. Phys.* **94**, 2324–2329.

Eimer, W., Williamson, J., Boxer, S., and Pecora, R. (1990). Characterization of the overall and internal dynamics of short oligonucleotides by depolarized dynamic light scattering and NMR relaxation measurements. *Biochemistry* **29**, 799–811.

Eisenberg H. (1976). *Biological Macromolecules and Polyelectrolytes in Solution*, Clarendon Press, Oxford, UK.

Eisenberg, H. and Felsenfeld, G. (1967). Studies of the Temperature-dependent Conformation and Phase Separation of Polyriboadenylic Acid Solutions at Neutral pH. *J. Mol. Biol.* **30**, 17–37.

Elias, J. G. and Eden, D. (1981a). Transient Electric Birefringence Study of the Persistence Length and Electrical Polarizability of Restriction Fragments of DNA. *Macromolecules* **14**, 410–419.

Elias, J. G. and Eden, D. (1981b). Transient Electric Birefringence Study of the Length and Stiffness of Short DNA Restriction Fragments. *Biopolymers* **20**, 2369–2380

Engel, A. (1991). Biological applications of scanning probe microscopes, *Annu. Rev. Biophys. Biophys. Chem.* **20**, 79–108.

Engelman, D. M., Moore, P. B., and Schoenborn, B. P. (1975). Neutron scattering measurements of separation and shape of proteins in the 30 S ribosomal subunit of Escherichia coli: S2-S5, S5-S8, S3-S7. *Proc. Natl. Acad. Sci. USA* **72**, 3888–3892.

Erie, D. A., Yang, G., Schultz, H. C., and Bustamante, C. (1994). The role of DNA bending in protein recognition and specificity: Direct visualization of specific and non-specific Cro-DNA complexes. *Science* **266**, 1562–1566.

Finzi, L. and Gelles, J. (1995). Measurement of lactose repressor-mediated loop formation and breakdown in single DNA molecules. *Science* **267**, 378–380.

Fisher, H. W. and Williams, R. C. (1979). Electron Microscopic Visualization of Nucleic Acids and of Their Complexes with Proteins, *Annu. Rev. Biochem.* **48**, 649–679.

Fredericq, E. and Houssier, C. (1973). *Electric Dichroism and Electric Birefringence*, Clarendon Press, Oxford, UK.

Freifelder, D. (1982). *Physical Biochemistry: Applications to Biochemistry and Molecular Biology*, Chapter 3, 2nd ed., Freeman, San Francisco, CA.

Frontali, C. (1988). Excluded-Volume effect on the bidimensional conformation of DNA molecules adsorbed to protein films, *Biopolymers* **27**, 1329–1331.

Frontali, C., Dore, E., Ferrauto, A., Gratton, E., Bettini, A., Pozzan, M. R., and Valdevit, E. (1979). An Absolute Method for the Determination of the Persistence Length of Native DNA from Electron Micrographs, *Biopolymers* **18**, 1353–1373.

Fujimoto, B. S. and Schurr, J. M. (1990). Dependence of the torsional rigidity of DNA on base composition. *Nature (London)* **344**, 175–178.

Fuller, F. B. (1971). The writhing number of a space curve. *Proc. Natl. Acad. Sci. USA* **68**, 815–819.

García de la Torre, J. and Bloomfield, V. A. (1981). Hydrodynamic properties of complex, rigid, biological macromolecules: theory and applications, *Q. Rev. Biophys.* **14**, 81–139.

Godfrey, J. E. and Eisenberg, H. (1976). The Flexibility of Low Molecular Weight Double-Stranded DNA as a Function of Length. II. Light Scattering Measurements and the Estimation of Persistence Lengths from Light Scattering, Sedimentation and Viscosity. *Biophys. Chem.* **5**, 301–318.

Gray, H. B, Jr., Bloomfield, V. A., and Hearst, J. E. (1967). Sedimentation coefficients of linear and cyclic wormlike coils with excluded-volume effects, *J. Chem. Phys.* **46**, 1493–1498.

Griffith, J., Bleyman, M., Rauch, C. A., Kitchin, P. A., and Englund, P. T. (1986). Visualization of the bent helix in kinetoplast DNA by electron microscopy, *Cell* **46**, 717–724.

Griffith, J. D. and Christiansen, G. (1979). Electron microscope visualization of chromatin and other DNA-protein complexes, *Annu. Rev. Biophys. Bioeng.* **7**, 19–35.

Guthold, M., Bezanilla, M., Erie, D. A., Jenkins, B., Hansma, H. G., and Bustamante, C. (1994). Following the assembly of RNA polymerase–DNA complexes in aqueous solutions with the scanning force microscope, *Proc. Natl. Acad. Sci. USA* **91**, 12927–12931.

Hagerman, P. (1984). Evidence for the existence of stable curvature of DNA in solution, *Biochemistry* **81**, 4632–4636.

Hagerman, P. J. (1985). Sequence dependence of the curvature of DNA: A test of the phasing hypothesis, *Biochemistry* **24**, 7033–7037.

Hagerman, P. J. (1986). Sequence-directed curvature of DNA, *Nature (London)* **321**, 449–450.

Hagerman, P. J. (1981). Investigation of the Flexibility of DNA Using Transient Electric Birefringence, *Biopolymers* **20**, 1503–1535.

Hagerman, P. J. (1983). Electrostatic Contribution to the Stiffness of DNA Molecules of Finite Length, *Biopolymers* **22**, 811–814.

Hagerman, P. J. and Zimm, B. H. (1981). Monte Carlo Approach to the Analysis of the Rotational Diffusion of Wormlike Chains, *Biopolymers* **20**, 1481–1502.

Hansma, H. and Hoh, J. (1994). Biomolecular imaging with the atomic force microscope, *Annu. Rev. Biophys. Biomol. Struct.* **23**, 115–139.

Harpst, J. A. (1980). Analysis of Low Angle Light Scattering Results from T7 DNA. *Biophys. Chem.* **11**, 295–302.

Harpst, J. A., Krasna, A. I. and Zimm, B. H. (1968). Molecular Weight of T7 and Calf Thymus DNA by Low-Angle Light Scattering, *Biopolymers* **6**, 595–603.

Hearst, J. E. and Stockmayer, W. H. (1992). Sedimentation constants of broken chains and wormlike coils. *J. Chem. Phys.* **37**, 1425–1433.

Hervet, H. and Bean, C. P. (1987). Electrophoretic mobility of λ phage HIND III and HAE III DNA fragments in agarose gels: A detailed study. *Biopolymers* **26**, 727–742.

Hogan, M. E. and Austin, R. H. (1987). Importance of DNA Stiffness in Protein–DNA Binding Specificity. *Nature (London)* **329**, 263–266.

Hogan, M., Dattagupta, N., and Crothers, D. M. (1978). Transient Electric Dichroism of Rod-Like DNA Molecules. *Proc. Natl. Acad. Sci. USA* **75**, 195–199.

Hogan, M., Dattagupta, N., and Crothers, D. M. (1979). Transient Electric Dichroism Studies of the Structure of the DNA Complex with Intercalated Drugs. *Biochemistry* **18**, 280–288.

Hogan, M. , Wang, J., Austin, R. H., Monitto, C. L., and Hershkowitz, S. (1982). Molecular motion of DNA as measured by triplet anisotropy decay. *Proc. Natl. Acad. USA* **79**, 3518–3522.

Hough P. V., Mastrangelo, I. A., Wall, J. S., Hainfeld, J. F., Simon, M. N., and Manley, J. L. (1982). DNA–protein complexes spread on N2-discharged carbon film and characterized by molecular weight and its projected distribution. *J. Mol. Biol.* **160**, 375–386.

Hough, P. V. C, Mastrangelo, I. A., Wall, J. S., Hainfeld, J. F., Wilson, V. G., Ryder, K., and Tegtmeyer, P. (1985). STEM footprints and bound mass distributions for DNA control proteins. *Bio/Technol.* **3**, 549–553.

Jacobson, H. and Stockmayer, W. H. (1950). Intramolecular reaction in polycondensations. I. The theory of linear systems. *J. Chem. Phys.* **18**, 1600–1606.

Jing, T. W., Jeffrey, A. M., DeRose, J. A., Lyubchenko, Y. L., L. Shlyakhtenko, S., Harrington, R. E., Appella, E., Larsen, J., Vaught, A., Rekesh, D., Lu, F.-X., and Lindsay, S. M. (1993). Structure of hydrated oligonucleotides studied by *in situ* scanning tunneling microscopy, *Proc. Natl. Acad. Sci. USA* **90**, 8934–8938.

Jonsson, M., Akerman, B., and Norden, B. (1988). Orientation of DNA during Gel Electrophoresis Studied with Linear Dichroism Spectroscopy. *Biopolymers* **27**, 381–414.

Kahn, L. D. (1972). Electric Birefringence, *Methods Enzymol.* **26**, 323–337.

Katahira, M., Sugeta, H., Kyogoku, Y., Fujii, S., Fujisawa, R., and Tomita, K. (1988). One- and 2D NMR studies on the conformation of DNA containing the oligo(dA)oligo(dT) tract. *Nucluic Acids Res.* **16**, 8619–8631.

Kavenoff, R. and Zimm, B. H. (1973).Chromosome-Sized DNA Molecules from *Drosophila, Chromosoma* **41**, 1–27,

Kintanar, A., Klevit, R. E. and Reid, B. R. (1987). Two-dimensional NMR investigation of a bent DNA fragment: assignment of the proton resonances and preliminary structure analysis. *Nucl. Acids Res.* **15**, 5845–5862.

Kirkwood, J. G. and Riseman, J. (1956). in *Rheology*, Vol. 1, Eirich, F., ed. Academic, New York, p. 495.

Klapper, I., Hagstrom, R., Fine, R., Sharp, K., and Honig, B. (1986). Focusing of electric field in the active site of Cu–Zn superoxide dismutase: Effects of ionic strength and amino-acid modification, *Proteins: Structure, Function Genet.* **1**, 47–59.

Kleinschmidt, A. K. and Zahn, R. K. (1959). Ueber Deoxyribonucleinsäure-Molekeln in Protein-Mischfilmen. *Z. Naturforsch.* **14b**, 770–779.

Kleinschmidt, A. K. (1968). Monolayer technique in electron microscopy of nucleic acid molecules. *Methods Enzymol.* **12B**, 361–377.

Klotz, L. C. and Zimm, B. H. (1972). Retardation times of deoxyribonucleic acid solutions. II. Improvements in apparatus and theory. *Macromolecules* **5**, 471–481.

Kobayashi, T., Doi, M., Makino, Y., and Ogawa, M. (1990). Mobility minima in field-inversion gel electrophoresis. *Macromolecules* **23**, 4480–4481.

Koenig, S. (1975). Brownian motion of an ellipsoid. A correction to Perrin's results. *Biopolymers* **14**, 2421–2423.

Koo, H.-S. and Crothers, D. M. (1987). Chemical determinants of DNA bending at adenine-thymine tracts. *Biochemistry* **26**, 3745–3748.

Koo, H.-S. and Crothers, D. M. (1988). Calibration of DNA curvature and a unified description of sequence-directed DNA bending. *Proc. Natl. Acad. Sci. USA* **85**, 1763–1767.

Koo, H.-S., Drak, J., Rice, J. A., and Crothers, D. M. (1990). Determination of the extent of DNA bending by an adenine-thymine tract, *Biochemistry* **29**, 4227–4234.

Koo, H.-S., Wu, H.-M., and Crothers, D. M. (1986). DNA bending at adenine-thymine tracts. *Nature (London)* **320**, 501–506.

Koppel, D. E. (1972). Analysis of macromolecular polydispersity in intensity correlation spectroscopy: The method of cumulants. *J. Chem. Phys.* **57**, 4814–4820.

Krasna, A. I., Dawson, J. R., and Harpst, J. A. (1970). Characterization of acid-denatured DNA by low-angle light scattering, *Biopolymers* **9**, 1017–1028.

Kuntz, I. D., Jr. and Kauzmann, W. (1974). Hydration of proteins and polypeptides, *Adv. Protein Chem.* **28**, 239–345.

Landau, L. D. and Lifshitz, E. M. (1958). *Statistical Physics*, Pergamon, London, UK. p. 478.

Lang, D., Bujard, H., Wolff, B., and Russell, D. (1967). Electron microscopy of size and shape of viral DNA in solutions of different ionic strengths, *J. Mol. Biol.* **23**, 163–181.

Lang, D., Steely, H. T., Jr., Kao, C. Y., and Ktistakis, N. T. (1987). Length, mass, and denaturation of double-stranded RNA molecules compared with DNA. *Biochim. Biophys. Acta* **910**, 271–281.

Langowski, J., Kremer, W., and Kapp, U. (1992). Dynamic light scattering for study of solution conformation and dynamics of superhelical DNA. *Methods Enzymol.* **211**, 430–448.

Le Bret, M. (1979). Catastrophic variation of twist and writhing of circular DNAs with constraint? *Biopolymers* **18**, 1709–1725.

Lerman, L. S., Fischer, S. G., Hurley, I., Silverstein, K., and Lumelsky, N. (1984). Sequence-determined DNA separations, *Annu. Rev. Biophys. Bioeng.* **13**, 399–423.

Leroy, J.-L, Charreitier, E., Kochoyan, M., and Gueron, M. (1988). Evidence from base-pair kinetics for two types of adenine tract structures in solution: Their relation to DNA curvature, *Biochemistry* **27**, 8894–8898.

Levene, S. D. and Crothers, D. M. (1983). A computer graphics study of sequence-directed bending in DNA. *J. Biomol. Struct. Dyn.* **1**, 429–436.

Levene, S. D. and Crothers, D. M. (1986). Ring closure probabilities for DNA fragments by Monte Carlo simulation. *J. Mol. Biol.* **189**, 61–72.

Levene, S. D. and Crothers, D. M. (1986b). Topological distributions and the torsional rigidity of DNA: A Monte Carlo study of DNA circles. *J. Mol. Biol.* **189**, 73–83.

Levene, S. D., Wu, H.-M., and Crothers, D. M. (1986). Bending and flexibility of kinetoplast DNA, *Biochemistry* **25**, 3988–3995.

Levene, S. D. and Zimm, B. H. (1989). Understanding the anomalous electrophoresis of bent DNA molecules: A reptation model. *Science* **245**, 396–399.

Lewis, R. J., Pecora, R., and Eden, D. (1986). Transient Electric Birefringence Measurements of the Rotational and Internal Bending Modes in Monodisperse DNA Fragments, *Macromolecules* **19**, 134–139.

Lin, S. C. and Schurr, J. M., (1978). Dynamic Light Scattering Studies of Internal Motions in DNA. I. Applicability of the Rouse-Zimm Model. *Biopolymers* **17**, 425–461.

Lindsay, S. M. and Barris, B. (1988). Imaging deoxyribose nucleic acid molecules on a metal surface under water by scanning tunneling microscopy. *J. Vac. Sci. Technol.* **A 6**, 544–547.

Lipanov, A. A., Chuprina, V. P., Alexeev, D. G., and Skuratovskii, I. Ya. (1990). Bh-DNA: Variations of the Poly[d(A)].Poly[d(T)] structure within the framework of the fibre diffraction studies. *J. Biomol. Struct. Dyn.* **7**, 811–826.

Lumpkin, O. J., Déjardin, P., and Zimm, B. H. (1985). Theory of gel electrophoresis of DNA. *Biopolymers* **24**, 1573–1593.

Lumpkin, O. J. and Zimm, B. H. (1982). Mobility of DNA in gel electrophoresis. *Biopolymers* **21**, 2315–2316.

Lyubchenko, Y. L., Gall, A. A., Shlyakhtenko, L. S., Harrington, R. E., Jacobs, B. L., Oden, P. I., and Lindsay, S. M. (1992). Atomic force microscopy imaging of double stranded DNA and RNA, *J. Biomol. Struct. Dyn.* **10**, 589–606.

Mandelkern, M., Dattagupta, N., and Crothers, D. M. (1981). Conversion of B DNA between solution and fiber conformations. *Proc. Natl. Acad. Sci. USA* **78**, 4294–4298.

Manning, G. S. (1981). Theoretical evidence for the coupling of winding to compression in the solution conformation of duplex DNA. *Biopolymers* **20**, 2337–2350.

Manning, G. S. (1988). Three Persistence Lengths for a Stiff Polymer with an Application to DNA B–Z Junctions. *Biopolymers*. **27**, 1529–1542.

Manning, G., Ebralidse, K., Mirzabekov, A., and Rich, A. (1989). An estimate of the extent of folding of nucleosomal DNA by laterally asymmetric neutralization of phosphate groups. *J. Biomol. Struct. Dyn.* **6**, 877–89.

Marini, J., Levene, S. D., Crothers, D. M., and Englund, P. T. (1982). Bent helical structure in kinetoplast DNA. *Proc. Natl. Acad. Sci. USA* **79**, 7664–7668.

Marini, J. C., Effron, P. N., Goodman, T. C., singleton, c. K., Wells, R. D., Wartell, R. M., and Englund, P. T. (1984). Physical characterization of a kinetoplast DNA fragment with unusual properties. *J. Biol. Chem.* **259**, 8974–8979.

Marko, J. F. and Siggia, E. D. (1995). Stretching DNA. *Macromolecules* **28**, 8759–8770.

Marti, O. and Amrein, M., Eds. (1993). STM and SFM in Biology, Academic, San Diego, CA.

Mastrangelo, I. A., Hough, P. V., Wilson, V. G., Wall, J. S., Hainfeld, J. F., and Tegtmeyer, P. (1985). Monomers through trimers of large tumor antigen bind in region I and monomers through tetramers bind in region II of simian virus 40 origin of replication DNA as stable structures in solution. *Proc. Natl. Acad. Sci. USA* **82**, 3626–3630

Matsumoto S., Morikawa K., and Yanagida M. (1981). Light microscopic structure of DNA in solution studied by the 4′,6-diamidino-2-phenylindole staining method. *J. Mol. Biol.* **152**, 501–16.

McCammon, J. A., Northrup, S. H., and Allison, S. A. (1986). Diffusional dynamics of ligand-receptor association. *J. Phys. Chem.* **90**, 3901–3905.

McDonell, M. W., Simon, M. N., and Studier, F. W. (1977). Analysis of restriction fragments of T7 DNA and determination of molecular weights by electrophoresis in neutral and alkaline gels. *J. Mol. Biol.* **110**, 119–146.

Mickel, S., Arena, V., Jr., and Bauer, W. (1977). Physical properties and gel electrophoresis behavior of R12-derived plasmid DNAs. *Nucleic. Acids. Res.* **4**, 1465–1482.

Moore, P. B. and Engelman, D. M. (1979). On the feasibility and interpretation of intersubunit distance measurements using neutron scattering. *Methods. Enzymol.* **59**, 629–638.

Morel, G., Ed. (1995). *Visualization of Nucleic Acids*, CRC Press, Boca Raton, FL.

Morikawa K. and Yanagida M. (1981). Visualization of individual DNA molecules in solution by light microscopy: DAPI staining method. *J. Biochem.* **89**, 693–696.

Nadeau, J. G. and Crothers, D. M. (1989). Structural basis for DNA bending. *Proc. Nat'l. Acad. Sci. USA* **86**, 2622–2626.

Nelson, H. C. M., Finch, J. T., Luisi, B. F., and Klug, A. (1987). The structure of oligo(dA)ï-oligo(dT) tract and its biological implications. *Nature (London)* **330**, 221–226.

Newman, J., Swinney, H. L., and Day, L. A. (1977). Hydrodynamic properties and structure of fd virus. *J. Mol. Biol.* **116**, 593–606.

Odijk, T. (1977). Polyelectrolytes near the rod limit. *J. Polym. Sci., Polym. Phys. Ed.* **15**, 477–483.

Odijk, T. (1995). Stiff chains and filaments under tension. *Macromolecules* **28**, 7016–7018.

Ogston, A. G. (1958). The spaces in a uniform random suspension of fibers. *Trans. Faraday Soc.* **54**, 1754–1757.

Olson, T., Fournier, M. J., Langley, K. H., and Ford, N. C., Jr. (1976). Detection of a Major Conformational Change in Transfer Ribonucleic Acid by Laser Light Scattering. *J. Mol. Biol.* **102**, 193–203.

Olson, W. K. (1975). Configurational Statistics of Polynucleotide Chains. A Single Virtual Bond Treatment. *Macromolecules* **8**, 272–275.

Olson, W. K. (1980). Configurational Statistics of Polynucleotide Chains. An Updated Virtual Bond Model to Treat Effects of Base Stacking. *Macromolecules* **13**, 721–728.

Peck, L. and Wang, J. C. (1981). Sequence dependence of the helical repeat of DNA in solution. *Nature (London)* **292**, 375–378.

Peckler, S., Graves, B., Kanne, D., Rapoport, H., Hearst, J. E., and Kim, S.-H. (1982). Structure of a Psoralen-Thymine Monoadduct Formed in Photoreaction with DNA. *J. Mol. Biol.* **162**, 157–172.

Perkins, T. T., Quake, S., Smith, D., and Chu, S. (1994b). Relaxation of a single DNA molecule observed by optical microscopy. *Science*. **264**, 822–826.

Perkins, T. T., Smith, D. E., and Chu, S. (1994a). Direct observation of tube-like motion in a single polymer chain. *Science* **264**, 819–822.

Perkins, T. T., Smith, D. E., Larson, R. G., and Chu, S. (1995). Stretching of a single tethered polymer in a uniform flow. *Science* **268**, 83–87.

Perrin, F. (1934). Mouvement brownien d'un ellipsoide (I). Dispersion dielectrique pour des molecules ellipsoidales. *J. Phys. Rad.* **[7] 5**, 497–511.

Perrin, F. (1936). Mouvement brownien d'un ellipsoide (II). Rotation libre et depolarisation des fluorescences. Translation et diffusion de molecules ellipsoidales. *J. Phys. Rad.* **[7]7**, 1–11.

Porschke, D., Zacharias, W., and Wells, R. D. (1987). B-Z DNA Junctions Are Neither Highly Flexible nor Strongly Bent. *Biopolymers* **26**, 1971–1974.

Rhodes, D. and Klug, A. (1981). Sequence-dependent helical periodicity of DNA. *Nature (London)* **292**, 378–380.

Rice, J. A. and Crothers, D. M. (1989). DNA bending by the bulge defect. *Biochemistry* **28**, 4512–4516.

Rich, A. (1978). Localized positive charges can bend double helical nucleic acid. *Fed. Eur. Biochem. Soc.* **51**, 71–81.

Riggs, A. D., Bourgeois, J., and Cohn, M. (1970). The *lac* repressor-operator interaction. *J. Mol. Biol.* **53**, 401–417.

Rinehart, F. P. and Hearst, J. E. (1972). The ionic strength dependence of $S^{\circ}_{20,w}$ for DNA in NaCl. *Biopolymers* **11**, 1985–1987.

Rizzo, V. and Schellman, J. (1981). Flow dichroism of T7 DNA as a function of salt concentration. *Biopolymers* **20**, 2143–2163

Rodbard, D. and Chrambach, A. (1970). Unified theory for gel electrophoresis and gel filtration. *Proc. Natl. Acad. Sci. USA* **65**, 970–977.

Rosenberg, A. H. and Studier, F. W. (1969). Intrinsic viscosity of native and single-stranded T7 DNA and its relationship to sedimentation coefficient. *Biopolymers* **7**, 765–774.

Ruben, G. C., Marx, K. A., and Reynolds, T. C. (1981). Stereoscopic visualization of spermidine-DNA torus structure prepared by the freeze–fracture, deep-etch technique. *39th Ann. Proc. Electron Microsc. Soc. Am.* 438–439.

Rubenstein, I. and Leighton, S. B. (1971). The influence of rotor speed on the sedimentation behavior in sucrose gradients of high molecular weight DNAs. *Biophys. Soc. Abs. 209A.*

Saenger, W. (1984). *Principles of Nucleic Acid Structure*, Springer-Verlag, New York.

Schellman, J. A. (1974). Flexibility of DNA. *Biopolymers* **13**, 217–226.

Schellman, J. A. (1980a). The flexibility of DNA. I. Thermal fluctuations. *Biophys. Chem.* **11**, 321–328.

Schellman, J. A. (1980b). The flexibility of DNA. II. Spontaneous and ligand induced distortions. *Biophys. Chem.* **11**, 329–337.

Schellman, J. A. and Harvey, S. C. (1995). Static contributions to the persistence length of DNA and dynamic contributions to DNA curvature. *Biophys. Chem.* **55**, 95–114.

Schellman, J. A. and Stigter, D. (1977). Electrical Double Layer, Zeta Potential, and Electrophoretic Charge of Double-Stranded DNA. *Biopolymers* **16**, 1415–1434.

Schildkraut, C. L., Marmur, J., and Doty, P. (1962). Determination of the Base Composition of Deoxyribonucleic Acid from its Buoyant Density in CsCl. *J. Mol. Biol.* **4**, 430–443.

Schmid, C. W. and Hearst, J. E. (1969). Molecular weights of homogeneous coliphage DNA's from density-gradient sedimentation equilibrium. *J. Mol. Biol.* **44**, 143–160.

Schmid, C. W., Rinehart, F. P., and Hearst, J. E. (1971). Statistical Length of DNA from Light Scattering. *Biopolymers* **10**, 883–893.

Schmitz, K. S. (1990). *An Introduction to Dynamic Light Scattering by Macromolecules*, Academic, San Diego, CA.

Schurr, J. M. 1984. Rotational Diffusion of Deformable Macromolecules with Mean Local Cylindrical Symmetry. *Chem. Phys. (Netherlands)* **84**, 71–96.

Schurr, J. M. and Fujimoto, B. S. (1988). The Amplitude of Local Angular Motions of Intercalated Dyes and Bases in DNA. *Biopolymers* **27**, 1543–1569.

Schurr, J. M. and Schmitz, K. S. (1986). Dynamic Light Scattering Studies of Biopolymers: Effects of Charge, Shape, and Flexibility. *Annu. Rev. Phys. Chem.* **37**, 271–305.

Schwartz, D. C. and Cantor, C. R. (1984). Separation of yeast chromosome-sized DNAs by pulsed field gradient gel electrophoresis. *Cell* **37**, 67–75.

Selsing, E., Wells, R. D., Alden, C. J., and Arnott, S. (1979). Bent DNA: Visualization of a base-paired and stacked A–B conformational junction. *J. Biol. Chem.* **254**, 5417–5422.

Serwer, P. and Hayes, S. J. (1987). A Voltage Gradient-Induced Arrest of Circular DNA during Agarose Gel Electrophoresis. *Electrophoresis* **8**, 244–246.

Serwer, P. and Hayes, S. J. (1989). Atypical Sieving of Open Circular DNA During Pulsed Field Agarose Gel Electrophoresis. *Biochemistry* **28**, 5827–5832.

Schafer, D. A., Gelles, J., Sheetz, M., and Landick, R. (1991). Transcription by single molecules of RNA polymerase observed by light microscopy. *Nature (London)* **352**, 444–448.

Sharp, P. and Bloomfield, V. A. (1968). Light Scattering from Wormlike Chains with Excluded Volume Effects. *Biopolymers* **6**, 1201–1211.

Shi, X., Hammond, R. W., and Morris, M. D. (1995). DNA conformational dynamics in polymer solutions above and below the entanglement limit. *Anal. Chem.* **67**, 1132–1138.

Shibata, J. H., Fujimoto, B. S., and Schurr, J. M. (1985). Rotational Dynamics of DNA from 10^{-10} to 10^{-5} Seconds: Comparison of Theory with Optical Experiments. *Biopolymers* **24**, 1909–1930.

Shimada, J. and Yamakawa, H. (1985). Statistical mechanics of DNA topoisomers: The helical wormlike chain. *J. Mol. Biol.* **184**, 319–329.

Shore, D. and Baldwin, R. L. (1983). Energetics of DNA Twisting. I. Relation between Twist and Cyclization Probability; II. Topoisomer Analysis. *J. Mol. Biol.* **170**, 957–981, 983–1078.

Shore, D., Langowski, J., and Baldwin, R. L. 1981. DNA Flexibility Studied by Covalent Closure of Short Fragments into Circles. *Proc. Natl. Acad. Sci. USA* **78**, 4833–4837.

Sinden, R. R. and Hagerman, P. J. (1984). Interstrand Psoralen Cross-Links Do Not Introduce Appreciable Bends in DNA. *Biochemistry* **23**, 6299–6303.

Sinsheimer, R. L. (1959). A single-stranded deoxyribonucleic acid from bacteriophage ϕX174. *J. Mol. Biol.* **1**, 43–53.

Skolnick, J. and Fixman, M. (1977). Electrostatic persistence length of a wormlike polyelectrolyte. *Macromolecules* **10**, 944–948.

Slayter, E. M. and Slayter, H. S. (1992). *Light and Electron Microscopy*, Cambridge University Press, Cambridge UK.

Smith, S. B., Cui, Y. J., and Bustamante, C. (1996). Overstretching B-DNA: The elastic response of individual double-stranded and single-stranded DNA molecules. *Science* **271**, 795–799.

Smith, S. B., Finzi, L., and Bustamante, C. (1992). Direct mechanical measurements of the elasticity of single DNA molecules by using magnetic beads. *Science* **258**, 1122–1126.

Song, L., Allison, S. A., and Schurr, J. M. (1990). Normal mode theory for the Brownian dynamics of a weakly bending rod: Comparison with Brownian dynamics simulations. *Biopolymers*. **29**, 1773–1791.

Song, L., Fujimoto, B. S., Wu, P., Thomas, J. C., Shibata, J. H., and Schurr, J. M. (1990). Evidence for allosteric transitions in secondary structure induced by superhelical stress. *J. Mol. Biol.* **214**, 307–326.

Song, L. and Schurr, J. M. (1990). Dynamic bending rigidity of DNA. *Biopolymers*. **30**, 229–237.

Southern, E. M. (1979). Measurement of DNA length by gel electrophoresis. *Anal. Biochem.* **100**, 319–323.

Sprous, D., Zacharias, W., Wood, Z. A., and S. C. Harvey (1995). Dehydrating agents sharply reduce curvature in DNAs containing A tracts. *Nucleic Acids Res.* **23**, 1816–1821.

Stannard, B. and Felsenfeld, G. (1975). The Conformation of Polyriboadenylic Acid at Low Temperature and Neutral pH. A Single-Stranded Rodlike Structure. *Biopolymers* **14**, 299–307.

Stellwagen, N. C. (1981). Electric birefringence of restriction enzyme fragments of DNA: Optical factor and electric polarizability as a function of molecular weight. *Biopolymers* **20**, 399–434.

Stellwagen, N. C. (1983). Accurate molecular weight determinations of deoxyribonucleic acid restriction fragments on agarose gels. *Biochemistry* **22**, 6180–6185.

Stellwagen, N. C. (1985). Orientation of DNA Molecules In Agarose Gels by Pulsed Electric Fields. *J. Biomol. Struct. Dyn.* **3**, 299–314.

Strauss, F., Gaillard, C., and Prunell, A. (1981). Helical periodicity of DNA, Poly(dA)·Poly(dT) and Poly(dA-dT)·Poly(dA-dT) in solution. *Eur. J. Biochem.* **8**, 215–222.

Strauss, J. K. and Maher, L. J., III. (1994). DNA bending by asymmetric phosphate neutralization. *Science* **266**, 1829–1834.

Strauss, J. K., Roberts, C., Nelson, M. G., Switzer, C., and Maher, L. J., III. (1996). DNA bending by hexamethylene-tethered ammonium ions. *Proc. Natl. Acad. Sci. USA* **93**, 9515–9520.

Strick, T. R., Allemand, J. F., Bensimon, D., Bensimon, A., and Croquette, V. (1996). The elasticity of a single supercoiled DNA molecule. *Science* **271**, 1835–1837.

Svoboda, K. and Block, S. M. (1994). Biological applications of optical forces. *Annu. Rev. Biophys. Biomol. Struct.* **23**, 247–285.

Syzbalski, W. (1968). Use of cesium sulfate for equilibrium density gradient centrifugation. *Methods Enzymol.* **12**, 330–360.

Szabo, A., Haleem, M., and Eden, D. (1986). Theory of the transient electric birefringence of rod-like polyions: Coupling of rotational and counterion dynamics. *J. Chem. Phys.* **85**, 7472–7479.

Tanford, C. (1961). *Physical Chemistry of Macromolecules*, Chapter 5, Wiley, New York.

Tao, T., Nelson, J. H., and Cantor, C. R. (1970). Conformational studies on transfer ribonucleic acid. Fluorescence lifetime and nanosecond depolarization measurements on bound ethidium bromide. *Biochemistry* **9**, 3514–3524.

Taylor, W. H. and Hagerman, P. J. (1990). Application of the method of phage T4 DNA ligase-catalyzed ring-closure to the study of DNA structure. II. NaCl-dependence of DNA flexibility and helical repeat. *J. Mol. Biol.* **212**, 363–376.

Théveny, B., Coulaud, D., Le Bret, M., and Revet, B. (1988). Local variations of curvature and flexibility along DNA molecules analyzed from electron micrographs. *Structure Expression* **3**, 39–55.

Thomas, D., Schultz, P., Steven, A. C., and Wall, J. S. (1994). Mass analysis of biological macromolecular complexes by STEM. *Biol. Cell* **80**, 181–92.

Thomas, T. J. and Bloomfield, V. A. (1983). Chain Flexibility and Hydrodynamics of the B and Z Forms of Poly(dG-dC).Poly(dG-dC). *Nucleic Acids Res.* **11**, 1919–1929.

Thresher, R. and Griffith, J. (1992). Electron microscopic visualization of DNA and DNA-protein complexes as adjunct to biochemical studies. *Methods Enzymol.* **211**, 481–490.

Tinoco, I., Jr. (1955). The Dynamic Electrical Birefringence of Rigid Macromolecules. *J. Am. Chem. Soc.* **77**,4486–4489.

Tirado, M. M. and García de la Torre, J. (1979). Translational friction coefficients of rigid, symmetric top macromolecules. Application to circular cylinders. *J. Chem. Phys.* **71**, 2581–2588.

Tirado, M. M. and García de la Torre, J. (1980). Rotational diffusion of rigid, symmetric top macromolecules. Application to circular cylinders, *J. Chem. Phys.* **73**, 1986–1993.

Trifonov, E. N. and Sussman, J. L. (1980). The pitch of chromatin DNA is reflected in its nucleotide sequence. *Proc. Natl. Acad. Sci. USA* **77**, 3816–3820.

Trifonov, E. N., Tan, R. K.-Z., and Harvey, S. C. (1987). Static persistence length of DNA, in *Structure and Expression*, Olson, W. K., Sarma, M. H., Sarma, R. H., and Sundaralingam, M., Eds., Adenine Press, Schenectady, NY, pp. 243–253.

Troll, M., Dill, K. A., and Zimm, B. H. (1980). Dynamics of polymer solutions. 3. An instrument for stress relaxations on dilute solutions of large polymer molecules. *Macromolecules* **13**, 436–438.

Uhlenhopp, E. L., Zimm, B. H., and Cummings, D. J. (1974). Structural Aberrations in T-even Bacteriophage. VI. Molecular Weight of DNA from Giant Heads. *J. Mol. Biol.* **89**, 689–702,

Ulanovsky, L. E. and Trifonov, E. N. (1987). Estimation of wedge components in curved DNA. *Nature (London)* **326**, 720–722.

Van Holde, K. E., Johnson, W. C. and Ho, P. S. (1998). *Principles of Physical Biochemistry*, Prentice-Hall, Englewood Cliffs, NJ.

Vinograd, J. and Hearst, J. E. (1962). Equilibrium sedimentation of macromolecules and viruses in a density gradient. *Fortsch. Chem. Org. Naturstoffe* **20**, 372–422.

Wahl, P., Paoletti, J. and LePecq, J.-B. (1970). Decay of fluorescence emission anisotropy of the ethidium bromide–DNA complex: Evidence for an internal motion in DNA. *Proc. Natl. Acad. Sci. USA* **65** 417–421.

Wang, C.-C. and Ford, N. C. (1981). Laser Light-Scattering Analysis of the Dimerization of Transfer Ribonucleic Acids with Complementary Anticodons. *Biopolymers* **20**, 155–168.

Wang, M. D., Yin, H., Landick, R., Gelles, J., and Block, S. M. (1997). Stretching DNA With Optical Tweezers. *Biophys. J.* **72**, 1335–1346.

Wartell, R. M., Klysik, J., Hillen, W., and Wells, R. D. (1982). Junction between Z and B Conformations in a DNA Restriction Fragment: Evaluation by Raman Spectroscopy. *Proc. Natl. Acad. Sci. USA* **79**, 2549–2553.

Wilcoxon, J. and Schurr, J. M. (1983). Temperature Dependence of the Dynamic Light Scattering of Linear ϕ29 DNA. Implications for Spontaneous Opening of the Double Helix. *Biopolymers*. **22**, 2273–2321.

Wu, H.-M. and Crothers, D. M. (1984). The locus of sequence-directed and protein-induced DNA bending. *Nature (London)* **308**, 509–513.

Wu, H. M., Dattagupta, N., and Crothers, D. M. (1981). Solution structural studies of the A and Z forms of DNA. *Proc. Natl. Acad. Sci. USA* **78**, 6808–6811.

Wu, P., Fujimoto, B. S., and Schurr, J. M. (1987). Time-Resolved Fluorescence Polarization Anistropy of Short Restriction Fragments: The Friction Factor for Rotation of DNA about its Symmetry Axis. *Biopolymers*. **26**, 1463–1488.

Yamakawa, H. and Fujii, M. (1973). Translational friction coefficient of wormlike chains. *Macromolecules* **6**, 407–415.

Yamakawa, H. and Fujii, M. (1974b). Light Scattering from Wormlike Chains. Determination of the Shift Factor. *Macromolecules* **7**, 649–654.

Yamaoka, K. and Matsuda, K. (1980). Electric Dipole Moments of DNA in Aqueous Solutions as Studied by the Reversing-Pulse Electric Birefringence. *Macromolecules* **13**, 1558–60.

Yanagida M., Hiraoka Y., and Katsura I. (1983). Dynamic behaviors of DNA molecules in solution studied by fluorescence microscopy. *Cold Spring Harbor Symp. Quant. Biol.* **47**, 177–87.

Yew, F. F. H. and Davidson, N. (1968). Breakage by hydrodynamic shear of the bonds between cohered ends of λ-DNA molecules. *Biopolymers* **6**, 659–679.

Yin, H., Landick, R., and Gelles, J. (1994). Tethered particle motion method for studying transcript elongation by a single RNA polymerase molecule. *Biophys. J.* **67**, 2468–2478.

Yoshizaki, T. and Yamakawa, H. (1980). Scattering Functions of Wormlike and Helical Wormlike Chains. *Macromolecules* **13**, 1518–1525.

Zhurkin, V. B., Lysov, Y. P., and Ivanov, V. I. (1979). Anisotropic flexibility of DNA and the nucleosomal structure. *Nucleic Acids Res.* **6**, 1081–1096.

Zimm, B.H. (1948). Apparatus and methods for measurement and interpretation of the angular variation of light scattering; Preliminary results on polystyrene solutions, *J. Chem. Phys.* **16**, 1099–1116.

Zimm, B. H. (1974). Anomalies in sedimentation. IV. Decrease in sedimentation coefficients of chains at high field. *Biophys. Chem.* **1**, 279–291.

Zimm, B. H. (1991). 'Lakes-straits' model of field-inversion electrophoresis of DNA. *J. Chem. Phys.* **94**, 2187–2206; **95**, 3026.

Zimm, B. H. and Levene, S. D. (1992). Problems and prospects in the theory of gel electrophoresis of DNA. *Q. Rev. Biophys.* **25**, 171–204.

Zinkel, S. S. and Crothers, D. M. (1987). DNA bend direction by phase sensitive detection, *Nature (London)* **328**, 178–181.

CHAPTER 10

Supercoiled DNA

1. INTRODUCTION

The DNA within the cell is frequently found in circular rather than linear form. If the two ends are covalently joined, we refer to closed circular DNA, or ccDNA. Even if the ends are not covalently joined, segments may be held, for example, by protein binding in chromatin or bacterial nucleoids, in closed domains whose ends are prevented from rotating with respect to each other. Because of these topological constraints (i.e., constraints that depend only on the connectivity of the chain, not on its deformation in a continuous manner), the double helix axis may be forced to adopt a higher order curve in space. This leads to the term supercoiling; that is, the coiling of a coil. There are a number of excellent reviews on general aspects of supercoiled DNA, (e.g., Bauer,

1978; Bauer et al., 1980; Cozzarelli and Wang, 1990; Vologodskii, 1992; Bates and Maxwell, 1993).

The supercoiled conformation has a higher energy than relaxed DNA. Supercoiling therefore leads to enhanced reactivity toward proteins or reagents that modify duplex winding or that react with single-stranded DNA, and thus may be an important intracellular mechanism for regulating the reactions of DNA during transcription, replication, and recombination and repair. Conversely, use of ccDNA as a substrate for protein binding allows characterization of changes in DNA structure induced by the protein (White et al., 1992). Supercoiling is also key for DNA compaction. There have been extensive in vitro studies of how the stress of supercoiling is transferred to distortions of the double helix such as local denaturation, B–Z transition, or cruciform formation.

An important aspect of DNA loop formation, in general, is that it provides a mechanism for "action at a distance"; that is, for one DNA segment to influence another at considerable distance along the chain (Wang and Giaever, 1988). However, supercoiling involves different structural and energetic features than simple planar looping.

Analysis of supercoiling can be regarded as one example of the general field of "biochemical topology". The topological method uses changes in linking, catenation, and knot formation to deduce enzymatic mechanisms and DNA substrate structure. While consideration of these broader aspects of biochemical topology is not possible here, several reviews (Wasserman and Cozzarelli, 1986; Dröge and Cozzarelli, 1992) show how it has been used to analyze DNA replication and recombination, and the mechanisms of action of topoisomerases.

2. TOPOISOMERASES AND GYRASES

Changes in supercoiling are catalyzed by enzymes called topoisomerases (for reviews see Wang, 1985; Maxwell and Gellert, 1986; Wang, 1987; Wang, 1991). These enzymes can alter a fundamental topological property of ccDNA, the linking number Lk (defined as the number of times one strand of DNA is wound around the other), by breaking and rejoining backbone bonds. They can also form or resolve knots or catenanes (interlocked rings) in ccDNA, as is required when replication of ccDNA produces two catenated circles that must be separated. Topoisomerases are part of a more general class of DNA strand transferases, including other enzymes such as Int protein and resolvases, that carry out breakage and joining reactions during recombination.

2.1 Types of Topoisomerases

Two types of topoisomerases have been identified. Type I enzymes break and rejoin one strand at a time, catalyzing the passing of one strand through the other; they change Lk by ± 1. Type II enzymes break and rejoin both strands together, causing the crossing of dsDNA through a double-strand break and changing Lk by ± 2. Both types cause the relaxation of negatively supercoiled DNA (the kind most frequently found in cells; see below) toward its thermodynamic minimum. Prokaryotic topoisomerase II (DNA gyrase) can also use the energy of ATP hydrolysis to promote negative supercoiling.

It appears that in prokaryotes, the equilibrium degree of supercoiling results from a balance between the formation of negative supercoils by DNA gyrase and their removal by topoisomerase I. Eukaryotic topoisomerase II, while also requiring ATP, can only relax DNA. A third class of topoisomerase, called reverse gyrase, has been discovered in thermophilic archaebacteria (Kikuchi and Asai, 1984). This enzyme has the surprising property of increasing Lk above that of the most stable structure by coupling ATP hydrolysis to positive supercoiling, at temperatures above 55°C. This result is unexpected because no positively supercoiled DNA has ever been isolated from cells under normal conditions. It may be speculated that this has the function of stabilizing DNA against thermal denaturation by overtwisting the right-handed double helix.

2.2 Mechanisms of Topoisomerases

To change the linking number of ccDNA requires a sequence of binding of DNA to topoisomerase, backbone scission, formation of a transient DNA–protein link, turning of one strand about the other by passing one or both strands through the break, and backbone resealing. These processes are drawn schematically in Figure 10-1 for Type I

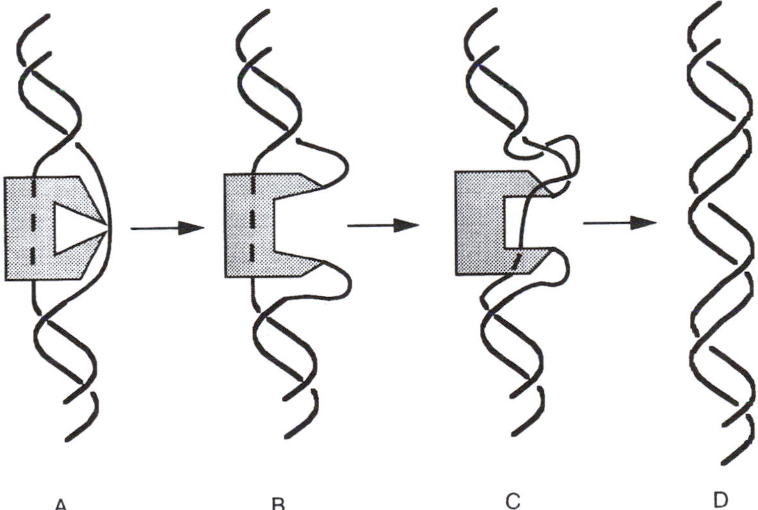

Figure 10-1
Strand passage model for prokaryotic topoisomerase I. In A, the enzyme has bound to the DNA and unwound the double helix. In B, the enzyme has nicked one strand, and bound to each broken end. In C, the unbroken strand has passed through the break. Between C and D the break has been ligated and the enzyme has dissociated. The net result is an increase of +1 in Lk: the DNA fragment in A has two right-handed turns, that in D has three. [After Dean et al., 1982.]

and in Figure 10-2 for Type 2. A three-dimensional (3D) crystal structure of the 67K N-terminal fragment of *Escherichia coli* DNA topoisomerase I (Lima et al., 1994) provides details that suggest how strand passage may occur. Structural analysis of DNA gyrase complexes by gel electrophoresis (Liu and Wang, 1978a), enzymatic digestion (Liu and Wang, 1978b), and electric dichroism (Rau et al., 1987) shows that about one turn of DNA wraps around the enzyme, with the ends close together.

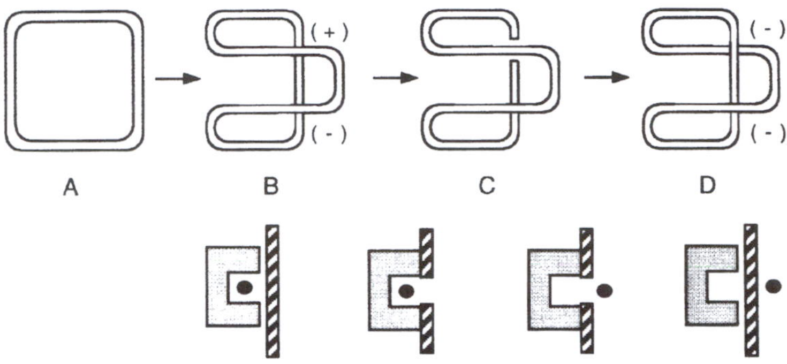

Figure 10-2
Sign inversion model of Brown and Cozzarelli (1979) for prokaryotic topoisomerase II (DNA gyrase), showing how Lk is changed by 2 by breaking of both strands. In going from A to B, the enzyme binds to and stabilizes the upper (+) crossing node. In C, the back double helix has been broken, and in D, the front duplex has been passed through the gap and the break resealed, giving a (−) node. In this case, Lk has changed from 0 to −2. The enzyme is actually tetrameric. Each end of the broken double helix is bound to one of the two A subunits of the tetramer; the two B subunits serve as ATPases.

3. GEOMETRY AND TOPOLOGY OF DNA SUPERCOILING

3.1 Fundamental Supercoiling Concepts

The definitions and concepts required for an adequate analysis of supercoiling are very clearly set forth in the chapter by Cozzarelli et al. (1990), whose treatment we follow closely. The lucid discussion by Crick (1976) is also recommended. There are two fundamentally different forms of DNA supercoiling (Bauer and Vinograd, 1968), illustrated in Figure 10-3. The left-hand panel shows a plectonemic, or interwound, supercoil, which is the characteristic state of ccDNA in solution. The right-hand drawing is of a solenoidal, or toroidal, supercoil, a form that is typified by the left-handed wrapping of DNA around the histone core in nucleosomes. (The ends of the nucleosomal DNA are prevented from relative movement by protein binding, rather than covalent closure. The protein is not shown in the figure.)

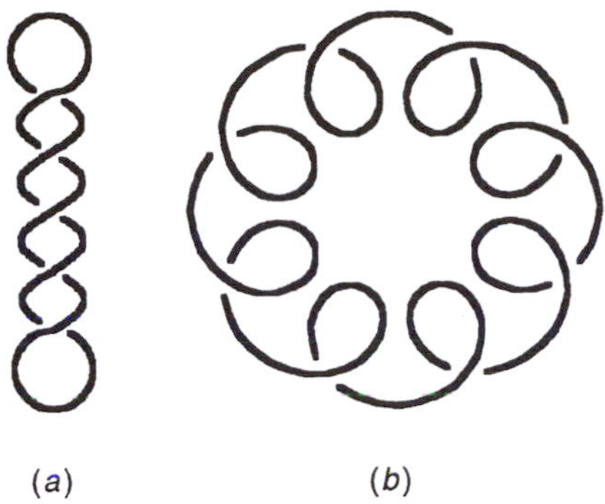

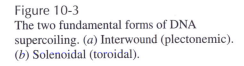

Figure 10-3
The two fundamental forms of DNA supercoiling. (*a*) Interwound (plectonemic). (*b*) Solenoidal (toroidal).

If the ends of double helical DNA are brought together and covalently joined or fixed relative to one another, the number of times one strand passes around the other becomes a topological constant that cannot be changed so long as no cuts are introduced in one or both chains. This constant is termed the linking number Lk. In order to separate the two single strands of DNA from one another during semiconservative replication, Lk must be reduced to zero. This reduction is accomplished by topoisomerases, which break and rejoin the strands.

While Lk must remain constant, it may be apportioned into the twisting of one strand about the other (Tw) or the writhing of the duplex axis in space (Wr). (In the older literature, quantities similar in meaning, but not always identical to Lk, Tw, and Wr, were often designated α, β, and τ respectively.) These quantities are connected by the fundamental relation

$$Lk = Tw + Wr \tag{10-1}$$

3.1.1 Linking Number

The linking number is a topological quantity, which remains unchanged so long as the DNA backbone remains continuous. In contrast, Tw and Wr are geometrical quantities, which may change as the shape of the DNA changes through bending, twisting, or kinking. We must now define these quantities with some care. To do so, we first consider three curves, W and C following the backbones of the Watson and Crick strands, and A following the axis down the center of the double helix. We establish the convention (which ignores the polarity of the phosphate backbone) that W, C, and A all point in the same direction.

The linking number describes the intertwining of W and C. However, it is more convenient to consider the Lk of C (or W) with the central axis A. These definitions of Lk are mathematically equivalent, since W or C can be continuously deformed into A without changing Lk. One way to define Lk is then to consider A as the perimeter of a surface, and to count the number of times C punctures that surface, as in Figure 10-4. (Think of A as an embroidery hoop, and C as the thread, but note that the thread wraps around the outside as well as the inside of the hoop.) A vector **n** is drawn normal to the surface, whose direction is determined by the direction of A and the right-hand rule.

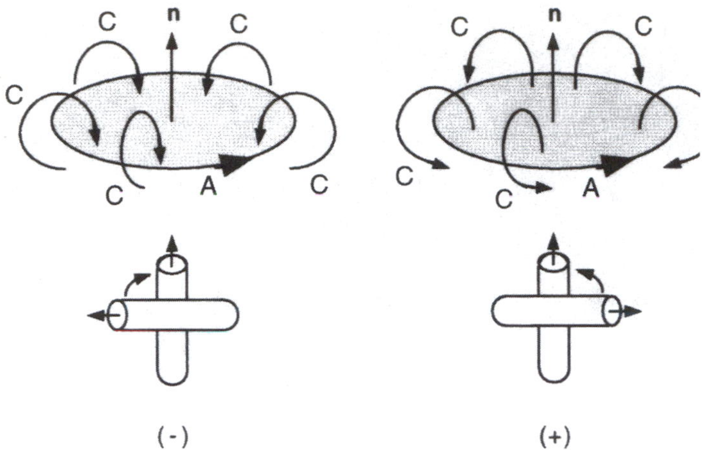

Figure 10-4
Two equivalent ways to compute Lk. If the C strand stitching the surface defined by the A axis goes in the opposite direction of the normal **n** to the surface, or if the upper strand is rotated clockwise to align with the lower, the contribution to Lk is negative.

If C crosses the surface in the same direction as **n**, +1 is associated with the puncture. If the crossing is in the opposite direction, the contribution is −1. Then, Lk is the sum of all these contributions. It is thus readily seen that Lk must be an integer, and that it is unchanged if the directions of both C and A are reversed.

Another way to determine Lk is to project C and A on a plane surface, and then consider the nodes where the two curves cross (the lower part of Fig. 10-4). The upper segment is rotated by an angle less than 180° to point in the same direction as the lower segment. If this rotation is clockwise, the node is given a value of $-\frac{1}{2}$; if counterclockwise, its value is $+\frac{1}{2}$. Then, Lk is the sum of the node values. There are two nodes associated with each puncture (the thread must go both under and over the hoop between successive punctures), which accounts for the factor $\frac{1}{2}$. To close the curve requires an even number of nodes (an equal number of overs and unders), so Lk is again shown to be an integer.

While each ccDNA molecule in a preparation has a fixed, integral value of Lk, the linking number will generally differ, in a relatively narrow distribution, from one molecule to another, and the average $< \mathrm{Lk} >$ will not usually be an integer. Molecules that differ only in a topological quantity such as Lk are called topoisomers.

3.1.2 Writhe

Writhe describes the tortuosity of the DNA axis A in space. It is calculated by counting the intersections of *A* with itself in a plane projection, using the sign convention defined for Lk . Each clockwise node contributes −1 to Wr, and each counterclockwise node contributes +1. However, in contrast to Lk, the measured $\mathrm{Wr_p}$ depends on the particular projection p. Then, the Wr defined in Eq. 10-1 is the average of Wr_p over all possible projections. This average is not always easy to compute, but some qualitative features may be readily deduced.

Consider, as a model for a piece of DNA, the familiar tightly coiled telephone handset cord, depicted in Figure 10-5. If it is unstretched, the successive helical coils are very close to one another. Therefore, almost every two-dimensional (2D) projection will show many crossovers, and the absolute value of Wr (whether positive or negative is immaterial here) is high. As the cord is increasingly stretched out, more and more views will not show nodes, and |Wr| decreases. In the limit of complete stretching, no projection shows an intersection, and $Wr = 0$. This result accords with our intuitive use of "writhing," which correlates it with the tortuosity of the path. A little more thought reveals that any planar piece of DNA has $Wr = 0$, since crossovers can be seen in projection only if some segment is out of the plane.

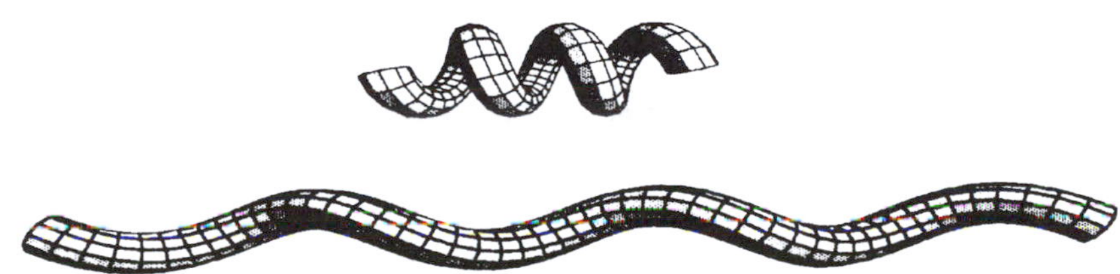

Figure 10-5
Replacement of Wr by twist as a tightly wound helix is stretched and ends are held in fixed orientations (Lk = constant).

3.1.3 Twist

Twist is often thought of as the number of times that the C strand turns around the axis A. More careful consideration shows that this is true only if the DNA is linear or planar. In this case, as we have seen above, $Wr = 0$, so according to Eq. 10-1, $Tw = Lk$. More generally, consider (Fig. 10-6) a cross-sectional plane *P* perpendicular to A at some point *a*. The plane will intersect C at some point *c*. Let $\mathbf{v}_{ac}$ be the unit vector from *a* to *c*. As *P* moves along A, $\mathbf{v}_{ac}$ will turn. Twist is defined as the number of 360° turns that $\mathbf{v}_{ac}$ makes around A. Note that the coordinate frame in which Tw is defined is independent of the surface on which *A* is wound. To get the total twist of a double helical DNA, one adds the rotation between successive base pairs, defined as the fraction of a turn projected in a plane perpendicular to the local helix axis.

Comparing the two parts of Figure 10-5, we see that for the unstretched cord, where Wr is a maximum, Tw is essentially zero. However, for the completely stretched helix, with $Wr = 0$, one edge of the cord twists many times around its axis. The intermediate state B is a left-handed solenoidal helix. One edge of the ribbon is A, and the other is C. The $\mathbf{v}_{ac}$ vector rotates in a left handed sense as one moves along

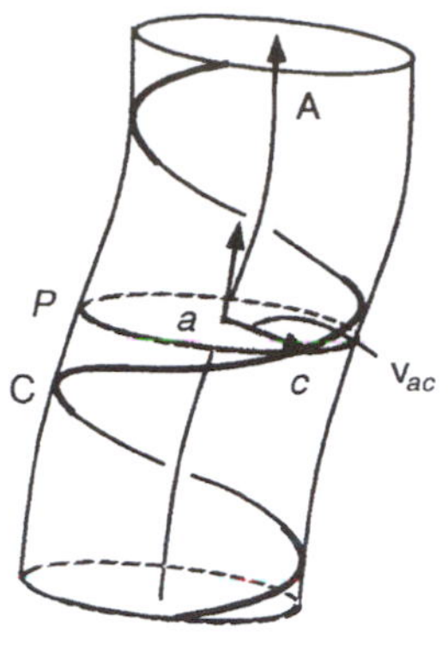

Figure 10-6
Diagram illustrating how to calculate Tw. See text for explanation. [Reprinted with permission from Cozzarelli et al., 1990, Fig. 8, p. 150].

the ribbon, corresponding to negative twist. Thus by stretching, we have gradually converted left-handed (negative) Wr into negative Tw. Throughout this conversion, Lk has remained constant.

3.1.4 Winding Number, Surface Twist and Surface Linking Number

While the three traditional parameters Lk, Wr, and Tw are sufficient to describe the major aspects of the topology and geometry of DNA, they are not necessarily the most informative or useful. In particular, it is helpful to have explicit descriptors for the two contributions to Tw: the winding of C relative to the normal to the surface, and the convolutions of the surface normal itself, which in turn depends on the shape of the surface. White et al. (1988) provided such descriptors for DNA wrapped about an actual or virtual surface. Such a surface might be, for example, the histone core of a nucleosome.

The essential features are illustrated in Figure 10-7. Let the surface be M, and require A (rather than the physical backbone of the DNA) to lie on M. We draw a unit vector **v** starting from A and extending along the normal to M. Both **v** and $\mathbf{v}_{ac}$ lie in a plane perpendicular to and cutting A. Then, define the winding number Φ as the number of times $\mathbf{v}_{ac}$ rotates past v as the circuit of A is traversed. The parameter Φ must be an integer for ccDNA. If there are N base pairs in the DNA, then $h = N/\Phi$ is the average helix repeat defined with respect to the surface M. For example, in an experiment in which the DNA lies on a surface or is wrapped around a protein, and is subject to chemical or enzymatic cleavage, the cleavage probability oscillates with period h (see Chapter 13). This shows how Φ can be measured experimentally.

State B of intermediate twist in Figure 10-5 is useful for visualizing the distinction between Tw and Φ. The surface on which the ribbon is wound is a cylinder. As one moves along the ribbon, the vector $\mathbf{v}_{ac}$ remains perpendicular to the surface normal, because both edges of the ribbon lie on the cylindrical surface. Hence Φ is zero. However, as we saw before, Tw is negative because of the left-handed rotation of $\mathbf{v}_{ac}$ in the plane perpendicular to A. In general, Tw is smaller (more negative) than Φ for a left-handed solenoidal helix, such as is found in nucleosomal core particles.

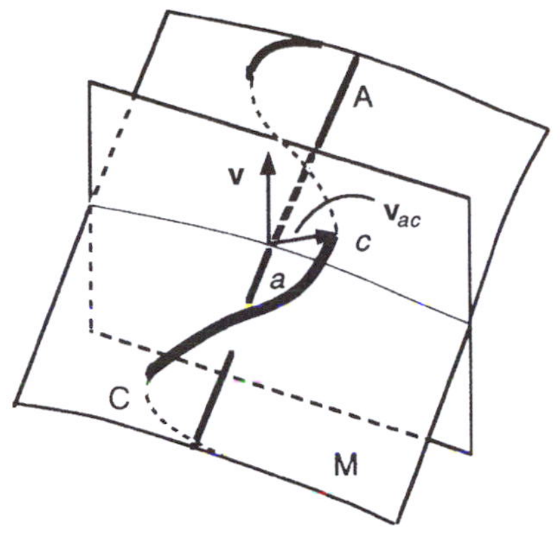

Figure 10-7
Vectors defining winding number, surface twist, and surface linking number (see text). [Reprinted with permission from White, J. H., Cozzarelli, N. R., and Bauer, W. R. (1988). *Science* **241**, 323–327. Copyright ©1988 American Association for the Advancement of Science.]

While Φ describes the winding of C around A on the surface, it gives no information about the shape of A itself, which is described by a second component of Tw, the surface twist STw. Surface twist is defined as the number of twists of the displacement curve A_ε about A. The displacement curve, diagrammed for three common surfaces in Figure 10-8, is the curve obtained by moving a distance ε away from A along the surface normal **v** in such a way that A_ε does not cross A. For example, think of A as the path followed by the feet of a walking person, and A_ε as that path followed by the head, if the body is always maintained perpendicular to the ground. It is then easy

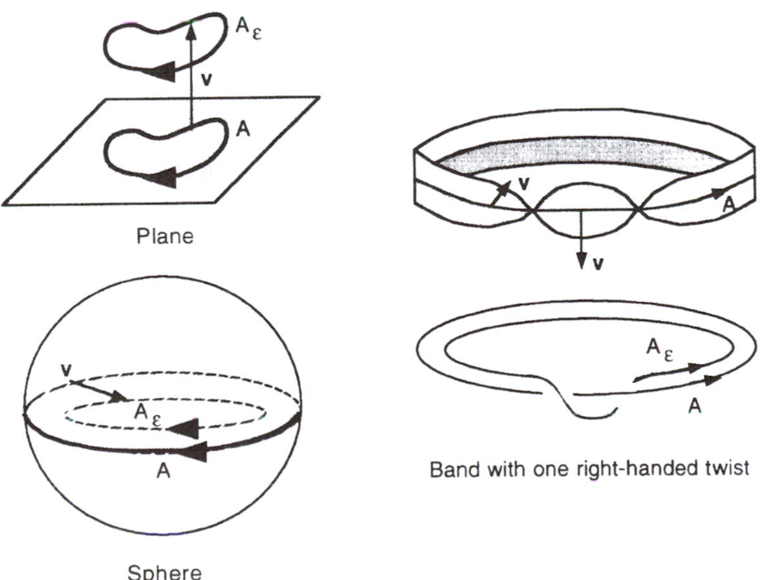

Figure 10-8
Illustrations of displacement curves and surface twisting and linking for simple models. [Adapted from White et al. 1988.]

to see that STw is always zero for paths on a plane or on the equator of a sphere, since there is no sideways motion of the head relative to the feet. On the other hand, a path that traverses peaks and valleys will lead to motions of the head side-to-side or back-and-forth, contributing to STw. The example shows a strip surface twisted once in a right-handed sense, yielding STw $= +1$.

To recapitulate, Φ gives information on helical repeat, and STw gives information on the overall superhelical shape. Together, these parameters give the total twist:

$$\text{Tw} = \Phi + \text{STw} \tag{10-2}$$

where STw and Wr describe different aspects of the overall shape. Their sum gives another important number, the surface linking number SLk:

$$\text{SLk} = \text{STw} + \text{Wr} \tag{10-3}$$

where SLk measures the linking number of A with A_ε. It must be integral, like Lk, but is invariant only under smooth deformations of the surface M. In a smooth deformation, the DNA axis must never leave the surface (though it may slide along it), and the direction of **v** along A must vary continuously. In contrast, Lk is invariant under continuous deformations, which require only that A remains intact. Smooth deformations are a subset of continuous deformations.

3.1.5 Analysis of DNA Binding to a Protein Surface

These concepts are pertinent to understanding the DNA structural changes accompanying the binding of a protein to DNA. A simple experiment, performed in many molecular biology laboratories, is to bind a protein to ccDNA that contains a specific binding site, add topoisomerase to relax the DNA, remove all proteins with detergent,and then to determine the new distribution of topoisomers by gel electrophoresis. The common result is that the binding protein has prevented the complete relaxation of the circular DNA. The residual ΔLk is due to both the rotation of the DNA around the protein, and the change in helical repeat of the DNA induced by the protein. Apportioning ΔLk between these two factors is most readily done using surface linking and twisting concepts (White et al., 1992).

For example, consider the conversion of minichromosomal DNA to a protein-free interwound supercoil. These two structures are diagrammed schematically in Figure 10-9: The minichromosome is wrapped solenoidally on a torus (left), and the free DNA is wrapped plectonemically on a capped cylinder (right). To convert from one to the other, the DNA would have to be lifted off the surface of the torus and rewound on the cylinder. The two shapes cannot be smoothly interconverted. The SLk for the minichromosome is $\pm n$, where n is the number of toroidal windings and the sign is negative for left-handed winding. In contrast, SLk $= 0$ and STw $= -$Wr for the interwound supercoil, because A_ε can be totally inside of A (i.e., inside the surface of the capped cylinder) and thus unlinked to it. It is a general result that any surface that can be deformed smoothly into a spheroid has SLk $= 0$.

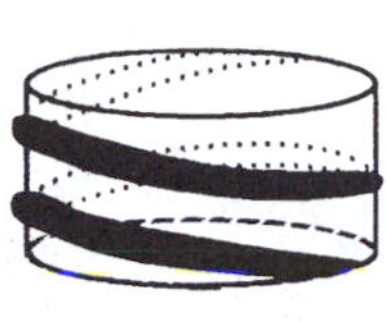
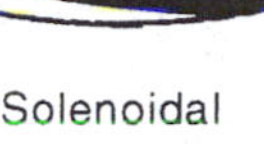

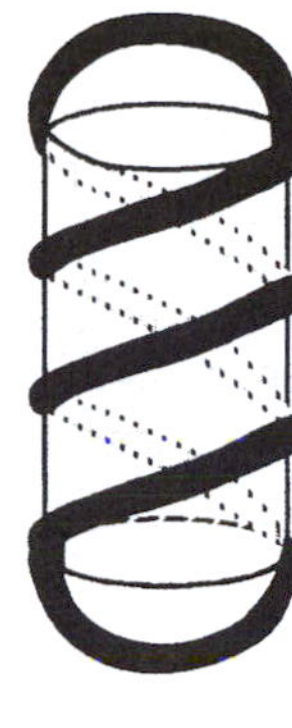

Figure 10-9
Topological change during conversion of toroidally wound minichromosomal DNA to interwound ccDNA. [Adapted from Cozzarelli et al. 1990.]

By combining Eqs. 10-1–10-3, we obtain the useful equation

$$\mathrm{Lk} = \mathrm{SLk} + \Phi \qquad (10\text{-}4)$$

which divides the Lk more clearly than Eq. 10-1 does into an axis shape component SLk and a helical twist component Φ. These components can be measured chemically or enzymatically, and SLk can be calculated from the DNA shape. The review by White et al. (1992) describes this analysis in some detail.

3.2 Description of Supercoiled DNA

In the previous sections, we defined the important topological and geometric variables. We now show how they are used to describe a particular plectonemically or solenoidally wound supercoiled DNA. The starting point is the relaxed state, which is defined operationally as the equilibrium state after nicking and religation under particular solvent conditions. It is denoted by subscript 0. The DNA will be planar on average, so $\mathrm{SLk}_0 = 0$. While there may be very small, compensating amounts of writhe and surface twist, it is conventional also to assume that Wr_0 and $\mathrm{STw}_0 = 0$. Therefore, in the relaxed state $\mathrm{Lk}_0 = \mathrm{Tw}_0 = \Phi_0 = \mathrm{N/h}_0$. Note, however, that the common assumption that Tw of supercoiled DNA equals N/h_0 would be true only if $\mathrm{STw} = 0$ and h was unchanged upon supercoiling. These conditions are not generally the case.

3.2.1 Specific Linking Difference (Superhelix Density)

Relaxed DNA can be supercoiled either by underwinding (reducing Lk below Lk_0), for example, through the action of DNA gyrase, or by wrapping the DNA around a protein complex. The underwound DNA forms a plectonemic supercoil characterized by the linking difference

$$\Delta\,\mathrm{Lk} = \mathrm{Lk} - \mathrm{Lk}_0 \qquad (10\text{-}5)$$

or the specific linking difference (sometimes imprecisely called superhelix density)

$$\sigma = \text{Lk}/\text{Lk}_0 - 1 \tag{10-6}$$

Both ΔLk and σ are generally negative, though positive supercoiling can also occur (Kikuchi and Asai, 1984). The wrapping of DNA around a protein complex, for example, in nucleosomes, does not in itself change Lk. This change occurs *in vivo* through the action of topoisomerases that remove supercoils so as to allow the DNA to lie smoothly on the protein surface. Experimental σ values for protein-wrapped and underwound free DNA are generally similar (values from −0.04 to −0.08 are typical), and the two forms can be readily interconverted by removal of protein.

3.2.2 Descriptors of Supercoiling

Cozzarelli et al. (1990) emphasize that no single variable is uniquely suited to describe supercoiling. The ΔLk has two major advantages. First, as described below, it is readily measured by intercalating dye binding or gel electrophoretic experiments. Second, the free energy of supercoiling is proportional to the square of ΔLk. This excess energy, as discussed later, may be transformed biologically into DNA conformational changes such as denaturation, B–Z transition, or cruciform formation. However, as a topological quantity, ΔLk is not directly related to the geometric shape of the DNA axis, which is so characteristic of the supercoiled state. Moreover, ΔLk depends on a careful and consistent definition of the relaxed reference state. For example, it is common to define Lk_0 as a constant, $N/10.5$, since the average helix repeat under physiological conditions is 10.5 bp/turn. On the other hand, if Lk_0 is defined as the equilibrium value of Lk after nicking and religation under particular solvent conditions, then it will change with temperature, ionic conditions, and ligand binding. This definition is more flexible and potentially more informative.

Consider, for example, the most common laboratory procedure for making negatively supercoiled DNA. Ethidium bromide intercalates between the bases of DNA, unwinding the double helix. If the drug binds to ccDNA, the negative twist is compensated by positive Wr to maintain constant linking number. The positive supercoils are then removed by topoisomerase action, and the drug is dialyzed away. If Lk_0 is defined as a constant, then ΔLk has a negative value once the topoisomerase acts. If Lk_0 is defined as a solution-dependent variable, $\Delta\text{Lk} < 0$ only after the ethidium is removed. Since topoisomerase treatment has relaxed the supercoiling, the ligand-bound DNA is planar, the ligand-free DNA is an interwound superhelix, and the latter definition seems most sensible. On the other hand, consider nucleosomal DNA wrapped around a histone core and then relaxed with topoisomerase. In this case, the variable definition of Lk_0 poorly reflects the geometry of the DNA (which is not planar) but provides a useful standard state for free energy changes produced by ligand binding or further bending.

The most direct and intuitive descriptor of supercoiling is n, which is just the number of superhelical turns about the superhelix axis. For solenoidal supercoiling, there is a simple relation between n and SLk: $\text{SLk} = +n$ for right-handed superhelices and SLk $= -n$ for left-handed superhelices. For ccDNA solenoidally wrapped on a torus, SLk is an integer, and therefore so is n. For plectonemic supercoiling, we consider the equivalent model of solenoidal wrapping about a capped cylinder: There are $n/2$ turns going up, and $n/2$ coming down. Thus there are n crossovers, or nodes, in an electron microscope view of plectonemically wound ccDNA. However, since SLk $= 0$ for interwound supercoils, there is no relation between SLk and n.

The third common and useful descriptor of supercoiling is Wr. Writhe is difficult to measure experimentally, but can be readily calculated for well-defined models (Fuller, 1971; Crick, 1976; White and Bauer, 1986). For a right-handed (negative) plectonemic supercoil with pitch angle γ (the angle between the tangent to the helix and the perpendicular to the helix axis) and n crossings when viewed perpendicular to the superhelix axis,

$$\mathrm{Wr} = -n \sin \gamma \tag{10-7}$$

For a left-handed (negative) solenoidal supercoil,

$$\mathrm{Wr} = -n + n \sin \gamma \tag{10-8}$$

Note the opposite handedness for these two types of negative supercoils. In a plectonemic supercoil, the helix axis doubles back, so the relative orientation of crossing strands is opposite to that in the solenoid. Thus to achieve the same topological sign, the strands must overlay in opposite sense, which in a supercoil can be done only by reversing handedness. Note also the opposite effects of maximally stretching the supercoils, so that γ approaches 90° and sin γ approaches 1. For the solenoid, this decreases Wr to 0 and increases Tw, as shown in Figure 10-5. For the plectonemic supercoil, the end-on views dominate in the averaging of Wr over all orientations. These are made up of crossings of distant segments with opposite orientations, leading in the limit to Wr $= -n$, the number of supercoil turns.

The parameters ΔLk, n, and Wr all contain different information about supercoiling. A supercoiled structure is best understood if all three are known, though this is rarely possible. The ΔLk is most readily measured and related to energetics, but gives no information about size and shape. The parameter n is intuitive and measurable by electron microscopy for plectonemic structures, but is not readily obtained otherwise. Writhe contains the most information on axis shape, but is the hardest to evaluate.

3.3 Measurement of Supercoiling Parameters

3.3.1 Measurement of ΔLk by Ethidium Intercalation

Three methods have been commonly used to measure ΔLk. The first takes advantage of changes in sedimentation properties accompanying the binding of ethidium bromide. The ethidium intercalates between the bases, unwinding the helix, causing an expansion of the hydrodynamic domain of the ccDNA and a decrease in its sedimentation coefficient. As more ethidium is added, the sedimentation coefficient finally reaches that of relaxed circular DNA, then increases again as positive supercoils are introduced (Fig. 10-10). If the number of ethidium molecules bound per base at the minimum is ν, the unwinding angle per ligand bound is ϕ, and the average helix repeat in the relaxed form is h_0, then

$$\sigma = -h_0(\phi/360)(2\nu) \tag{10-9}$$

According to most recent measurements, $h_0 = 10.5$ bp/turn, and $\phi = 26°$. The binding isotherm, hence ν, can also be measured by sedimentation equilibrium in a CsCl density gradient (Bauer and Vinograd, 1968).

3.3.2 Measurement of ΔLk by Band Counting Gel Electrophoresis

The second technique is band counting in gel electrophoresis, introduced by Keller and Wendel (1975). The gel electrophoretic mobility of ccDNA depends on the size of its coil domain, hence indirectly on ΔLk. In a population of molecules with a distribution of ΔLk, different molecules will electrophorese as individual bands distributed about $\Delta Lk_m \approx \langle \Delta Lk \rangle$. Two populations differing only in $\langle \Delta Lk \rangle$ migrating on the same gel will show bands of the same ΔLk that overlap, so differences in $\langle \Delta Lk \rangle$ can

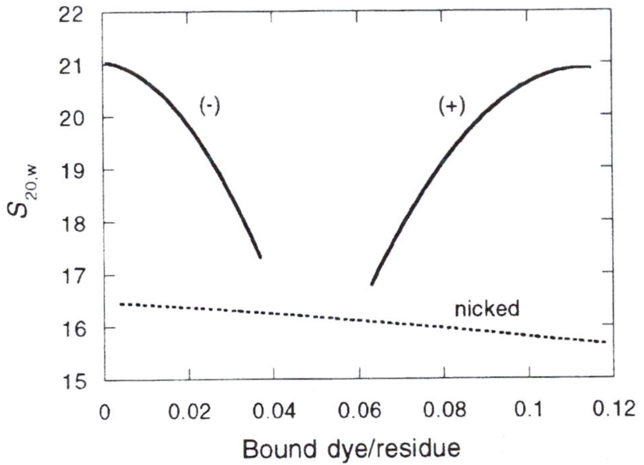

Figure 10-10
Dependence of sedimentation coefficient on ethidium bromide binding for closed circular and nicked circular SV40 DNA. [From data of Bauer and Vinograd, 1968.] The ligand intercalates between the bases, unwinding the helix, causing an expansion of the hydrodynamic domain of the ccDNA, and a decrease in its sedimentation coefficient. As more ethidium is added, the sedimentation coefficient approaches that of relaxed circular DNA, then increases again as positive supercoils are formed. The slight downward trend of the data from left to right is due to the lengthening of the DNA and decrease of its density upon intercalation of ligand, effects that occur with both closed and relaxed DNA.

be measured by band counting. An example is shown in Figure 10-11, demonstrating the difference in distribution of topoisomers as a function of temperature (Depew and Wang, 1975). Analysis of these band patterns showed that adjacent bands differ by $\Delta Lk = \pm 1$, and that the helix unwinds by about -0.012 deg $bp^{-1} {}^{\circ}C^{-1}$.

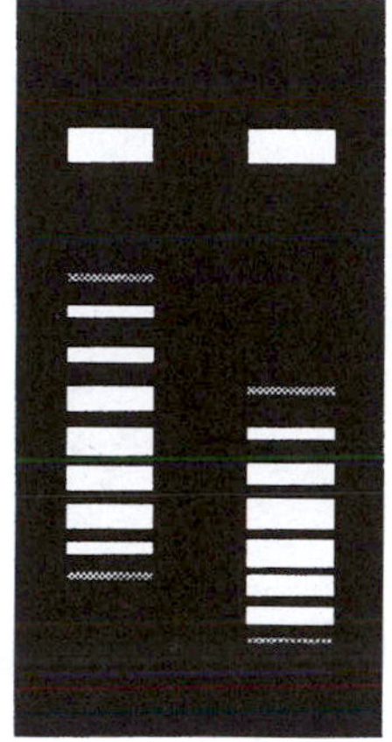

Figure 10-11
Schematic of band counting gel electrophoresis experiment to determine the change in winding angle per base pair, reflected in change in ΔLk distribution after incubation with ligase, as a function of temperature. Left and right lanes are for treatment with ligase at 21 and 14°C, respectively. [After Depew and Wang, 1975.]

3.3.3 Measurement of ΔLk by Two-Dimensional Gel Electrophoresis

The third technique is 2D gel electrophoresis (Lee et al., 1981; Bowater et al., 1992), which can expand the range of resolvable topoisomers and give information on structural transitions. The DNA solution is first electrophoresed in normal buffer in one dimension. The gel with DNA is then incubated for several hours in a dilute solution of an intercalating drug, turned 90°, and electrophoresed in the second dimension. The intercalator reduces the extent of supercoiling, and the 2D pattern allows following the migration of individual topoisomers.

As an example, consider Figure 10-12 (Wang et al., 1982) which demonstrates the abrupt change in plasmid writhing number and mobility accompanying a B–Z transition. The left panel shows a mixture of topoisomers, obtained by relaxing a plasmid with a topoisomerase in the presence of varying amounts of ethidium. The spots can be numbered from 1 to 28 in order of decreasing Lk; spot 0 is the nicked circle. The first-dimension pattern, from top to bottom, shows a decrease of positive supercoiling from 1 to 4, the relaxed state between 4 and 5, and increasingly negative supercoiling from 5 to 28. The second dimension, from left to right, shows the reduction of (negative) supercoiling by intercalation: topoisomers 1–14 are all positively supercoiled, and 16–28 are negatively supercoiled. The smooth curve shows that no abrupt transition has occurred as a result of intercalation and reduction of linking number. On the right is the same plasmid with a $d(GC)_{16}$ insert. The abrupt change in first-dimension migration between topoisomers 17′ and 18′ is immediately apparent. More quantitatively, it is noted that topoisomers 13′–16′ migrate approximately as fast as topoisomers

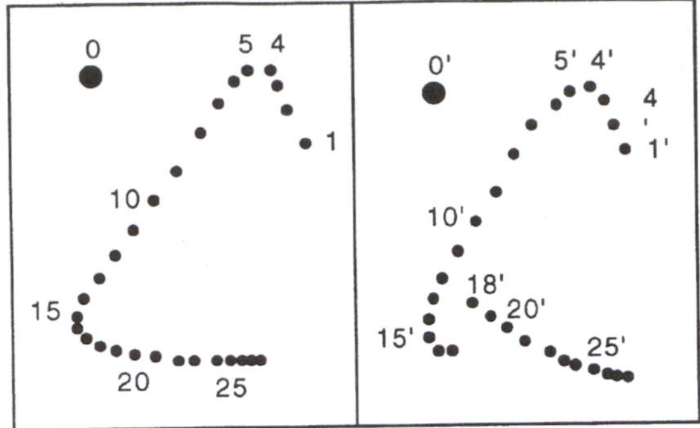

Figure 10-12
Two-dimensional gel electrophoresis analysis of a supercoiled plasmid. [After Wang et al. 1983], demonstrating the change in plasmid writhing number and mobility accompanying a B–Z transition. Direction of electrophoresis in the first dimension is downward; in the second dimension, to the right. See text for explanation.

indexed 6′ higher, 19′–22′ respectively, suggesting that they have the same Wr. Therefore, they differ in Tw by 6, since $(Tw_{14} - Tw_{20}) = (Lk_{14} - Lk_{20}) - (Wr_{14} - Wr_{20}) = 6 - 0$. This behavior is expected if the 32 bp GC insert flips from a 10.5 bp/turn B helix to a 12 bp/turn Z helix, since the difference in Tw should be $(32/10.5) + (32/12) = 5.7$.

3.3.4 Dependence of Sedimentation Coefficient on ΔLk

Shimada and Yamakawa (1988) combined hydrodynamic theory with Monte Carlo simulation to calculate the sedimentation coefficient (S) of ccDNA as a function of the difference in linking number, $\Delta Lk = Lk - \overline{Lk}$ between a given degree of supercoiling and the relaxed state. Their results show reasonably good agreement with the experimental data of Wang (1974) at low values of ΔLk. As $|\Delta Lk|$ increases from 0, S increases rather rapidly as the chain becomes more compact. At higher $|\Delta Lk|$, S exhibits a broad maximum, decreases somewhat, and then rises steadily. This behavior is attributed to successive transitions into interwound and branched forms. Similar behavior is obtained for S calculated from supercoiled conformations simulated by Monte Carlo methods (Vologodskii and Cozzarelli, 1994a).

3.3.5 Measurement of Crossovers by Electron Microscopy

As noted above, the other supercoiling parameter that is relatively easily measured, simply by counting intersections in electron micrographs, is *n*, the number of crossovers.

3.4 Characterization of Supercoiled DNA Molecules

3.4.1 Interwound (Plectonemic) Supercoils in Plasmids

Experimental values of ΔLk and n, along with measured dimensions of supercoiled molecules, can be used to determine all the other supercoiling parameters for ccDNA. A typical molecule is shown in Figure 10-13, along with a diagrammatic interpretation of the electron micrograph that shows branch points. Such branching, which has not been mentioned in our discussion up to now, is characteristic of plectonemic supercoils, and presumably represents one of the ways the molecule accommodates to superhelical stress. It should also be noted that branching, along with the doubling back innate to ccDNA, brings together distant sequences; this may be important in recombination, enhancer action, and so on. Vologodskii et al. (1992) calculated that the probability of juxtaposition of sites is about 100-fold higher in supercoiled than in relaxed circular DNA. The angular distribution between approaching sites is also made more asymmetric and is strongly peaked by supercoiling (Vologodskii and Cozzarelli, 1994a).

If a molecule like that shown in Figure 10-13 is straightened out and freed of non-topological intersections, it might look like Figure 10-14(a). Figure 10-14(b) depicts, for comparison, an equal length of nucleosomal DNA with the same superhelix density. We discuss the plectonemic structure first.

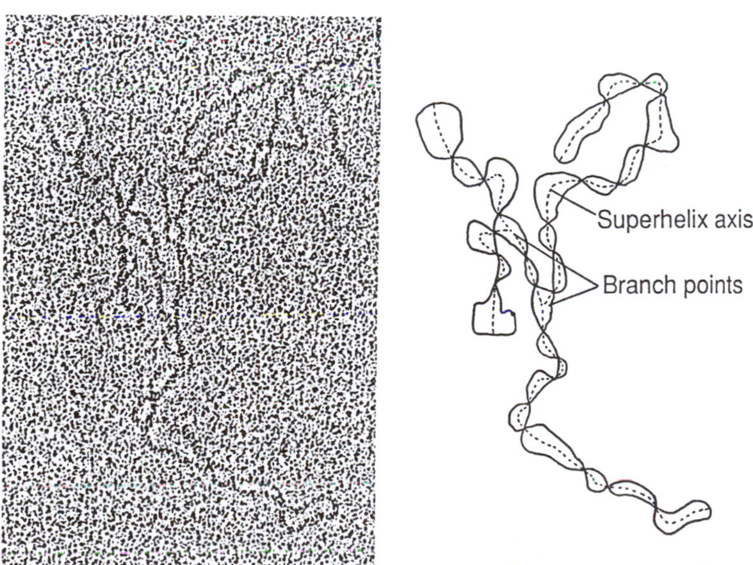

Figure 10-13
Electron micrograph and diagram of supercoiled pBR322 plasmid DNA. [Reprinted from *J. Mol. Biol.* **213**, 931–951 (1990). Boles, T. C., White, H. H., and Cozzarelli, N. R., Structure of plectonemically supercoiled DNA, Copyright ©1990, by permission of the publisher Academic Press Limited London.]

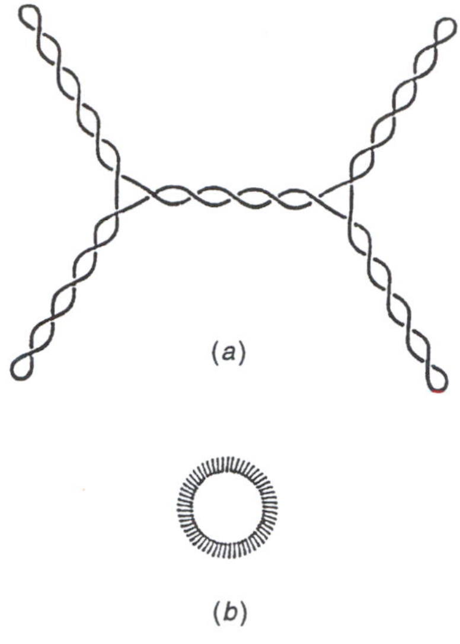

Figure 10-14
Schematic representation of plectonemic (a) and solenoidal (b) supercoiled DNA. The molecules are drawn to the same scale; both contain 4.6 kb of DNA with specific linking difference $\sigma = -0.06$. [With permission from Cozzarelli et al., 1990, Fig. 15, p. 169.]

The total length of the DNA is L (3.4 Åtimes N, the number of base pairs), and the superhelix has axis length l. In their study, Boles et al. (1990) found that l/L remained constant at 0.41 over the entire range of σ (-0.02 to -0.12) studied. This implies that plectonemically supercoiled DNA is very long and thin (the maximum l/L would be 0.5), and that as supercoiling increases, r decreases, and $\sin\gamma = 2l/\mathrm{L}$ remains constant at 55°.

For n superhelical turns of pitch p and radius r, the length of DNA involved in cylindrical winding is $2\pi n\sqrt{r^2 + (p/2\pi)^2}$. (This equation can readily be deduced from the Pythagorean theorem. The circumference of the helix is $2\pi r$, and the length of one turn projected along the axis is p. If one imagines unwinding one turn of the helix onto a flat surface, the hypotenuse of the right triangle is $\sqrt{4\pi^2 r^2 + p^2}$.) The length of DNA wrapped around E end caps of radius r is $E\pi r$. Thus

$$L = 2\pi n\sqrt{r^2 + (p/2\pi)^2} + E\pi r \tag{10-10}$$

It is also easy to see that

$$p = (l - Er)/2n \tag{10-11}$$

Boles et al. (1990) found that n and $\Delta\mathrm{Lk}$ are simply proportional: $n = -0.89\ \Delta\mathrm{Lk}$, or

$$n = -0.89\sigma \mathrm{Lk}_0 \tag{10-12}$$

Solution of these three equations for r as a function of σ shows that r drops rather sharply at first, for 150 Å at $\sigma = -0.02$ to 55 Å at $\sigma = -0.06$, and then more slowly to 50 Å at $\sigma = -0.08$.

Another relation between the supercoil parameters comes from Eq. 10-7, which becomes

$$\mathrm{Wr} = -n \sin \gamma = -0.82n = 0.73\Delta\mathrm{Lk} \qquad (10\text{-}13)$$

Thus n, ΔLk, and Wr are all proportional to each other. These equations enable a calculation of the change in twist relative to the relaxed form:

$$\Delta\mathrm{Tw} = \Delta\mathrm{Lk} - \mathrm{Wr} = 0.27\Delta\mathrm{Lk} \qquad (10\text{-}14)$$

Therefore, for every change of -1 in Lk, there is a drop of 0.73 in Wr and 0.27 in Tw. This constant proportion is a consequence of the constancy of the pitch angle γ, that is of l/L.

Finally, since SLk $= 0$ for plectonemic supercoiling,

$$\mathrm{STw} = -\mathrm{Wr} = -0.73\Delta\mathrm{Lk} \qquad (10\text{-}15)$$

which allows calculation of Φ and h as a function of ΔLk. The general equation (White et al., 1988)

$$h = \frac{h_0}{s - \frac{\mathrm{SLk}}{\mathrm{Lk}_0} + 1} \qquad (10\text{-}16)$$

which is derived readily from Eqs. 10-1 to 10-6, in this case becomes $h = h_0/(\sigma + 1)$.

Values for these parameters, which should be typical of all plectonemically supercoiled DNAs under similar ionic and temperature conditions, are shown in Figure 10-15 for 1 kb of DNA and for one superhelical turn. These values were determined at 37°C in 0.1 M ionic strength buffer with Na^+ as counterion. The average helix rotation angle per base pair has been found to vary by about -1.2×10^{-2} deg °C^{-1} (Depew and Wang, 1975). It also decreases with increasing salt concentration: -0.162° for a 10-fold increase in [Na^+], corresponding to $d\sigma/dp\mathrm{Na} = -0.00451$. There are also substantial changes in σ with counterion (Anderson and Bauer, 1978).

3.4.2 Solenoidal Supercoils in Nucleosomes

It is instructive to compare supercoiling parameters for plectonemically wound DNA with those for DNA wrapped solenoidally about core nucleosomes (Cozzarelli et al., 1990). Structural studies, discussed in detail in Chapter 14, show that 146 bp of DNA are wrapped 1.8 times in a left-handed helix around a histone octamer core with a pitch of 28 Å and radius of 43 Å (Richmond et al., 1984). The calculation, for a model linkerless case, assumes that the helix repeat is 10.0 bp/turns, compared with 10.5 bp/turn for free DNA. To facilitate comparison with the plectonemically supercoiled plasmid, the calculation is carried out for $N = 4600$ bp; Figure 10-14 shows the dramatic difference in size and shape between these two structures.

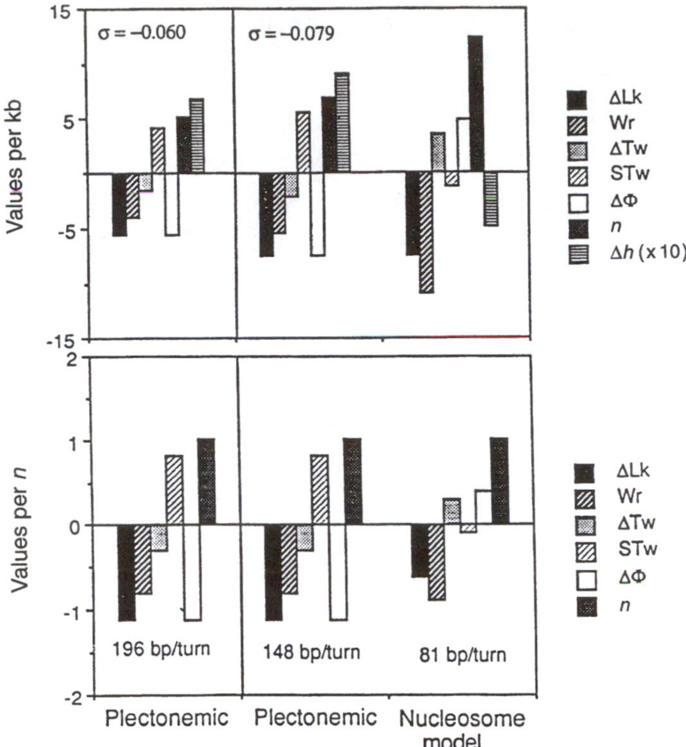

Figure 10-15
Supercoiling parameters as functions of specific Lk (upper) or bp/turn (lower) for plectonemic (left) and solenoidally wound nucleosomal (right) DNA. [With permission from Cozzarelli et al., 1990, Fig. 18, p. 176.]

Equation 10-4 gives ΔLk per solenoidal supercoil: $\Delta Lk = \mathrm{SLk} + \Delta\Phi = -n + \Delta\Phi$. There are $146/1.8 = 81$ bp/superhelical turn, so $n = 4600/81 = 57$ left-handed supercoils. With $h_{\text{nuc}} = 10.0$ and $h_0 = 10.5$ bp/turn, $\Delta\Phi = N(1/h_{\text{nuc}} - 1/h_0) = 0.39$ per supercoil, or 22 for the entire 4600 bp. Thus $\Delta\mathrm{Lk} = -57 + 22 = -35 = -0.61\ n$, or -1.1 per nucleosome.

The writhe is obtained from Eq. 10-8. Consideration of the helix geometry shows that $\sin\gamma = p/\sqrt{4\pi^2 r^2 + p^2}$. With $p = 28$ Å and $r = 43$ Å, $\sin\gamma = 0.10$ and $\gamma = 6.0°$, so

$$\mathrm{Wr} = -n + n\sin\gamma = -0.90n = 1.46\Delta\mathrm{Lk}$$

As in the plectonemic case, n, ΔLk, and Wr are all proportional to each other, but here Wr is a better estimator of n than is ΔLk. Use of Eqs. 10-3 and 10-1 gives the further results

$$\mathrm{STw} = \mathrm{SLk} - \mathrm{Wr} = -0.17\Delta\mathrm{Lk}$$

and

$$\Delta\mathrm{Tw} = -0.46\Delta\mathrm{Lk}$$

The specific linking difference, calculated for one turn, is $\sigma = \Delta\text{Lk}/\text{Lk}_0 = -0.61/(81/10.5) = -0.079$. These values are graphed on the right hand side of Figure 10-15. Comparing plectonemic and solenoidal supercoils at $\sigma = -0.079$, we see that the same ΔLk arises in two quite different ways. The plectonemic supercoil is right-handed, and has high values of pitch p and pitch angle γ. The solenoidal supercoil is left-handed, and has low values of p and γ. The small pitch leads to great compaction, which is key to chromosomal packing.

4. ENERGETICS OF TWISTING AND BENDING

The shape deformations of supercoiled DNA, relative to relaxed molecules, produce stresses that can be related to twisting and bending energies. Much information has been obtained from analysis of equilibrium distributions of topoisomers, determined by the banding counting/gel electrophoresis method. This technique may be applied to mixtures obtained by ligation of nicked circular DNA (Depew and Wang, 1975) or treatment of ccDNA with a topoisomerase (Pulleyblank et al., 1975). A schematic distribution of band intensities, proportional to relative topoisomer populations, is shown in Figure 10-16. It is immediately apparent that the distribution is a Gaussian bell-shaped curve. This curve is a consequence of the quadratic dependence of the free energy of supercoiling on the linking difference (Bauer and Vinograd, 1970), which governs the fluctuations of Lk about its most probable value Lk_0. We follow the analysis of Depew and Wang (1975).

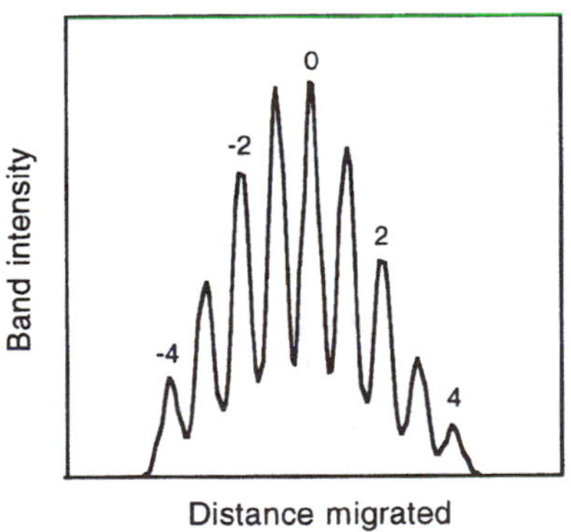

Figure 10-16
Simulated Gaussian distribution of topoisomers, calculated according to Eq. 10-19 with $K/RT = 0.11$ and $\omega = 0.39$; parameters were taken from the work of Depew and Wang (1975). Peaks are labeled with ΔLk value relative to most probable Lk_0.

4.1 Free Energy Differences between Topoisomers

It will generally be the case that the two ends across the nick of a relaxed circle will need to rotate by $2\pi\omega$ rad, or $\omega(< 0.5)$ turns, to be in register for ligation. The free energy for proper juxtaposition near the most probable linking number is

$$G(\text{Lk}_0) = K\omega^2 \tag{10-17}$$

where K is a constant for given solvent, temperature, and type of DNA molecule. For linking numbers different from Lk_0,

$$G(\mathrm{Lk}_0 + \Delta\mathrm{Lk}) = K(\Delta\mathrm{Lk} + \omega)^2 \tag{10-18}$$

The relative population of topoisomers is given by the Boltzmann distribution

$$\begin{aligned}\frac{N(\mathrm{Lk}_0 + \Delta\mathrm{Lk})}{N(\mathrm{Lk}_0)} &= \exp\left\{-\frac{[G(\mathrm{Lk}_0 + \Delta\mathrm{Lk}) - G(\mathrm{Lk}_0)]}{RT}\right\} \\ &= \exp\left\{-\frac{K[(\Delta\mathrm{Lk} + \omega)^2 - \omega^2]}{RT}\right\} \\ &= \exp\left[\frac{K\omega}{RT}\right]\exp\left\{-\frac{K(\Delta\mathrm{Lk} - \omega)^2}{RT}\right\}\end{aligned} \tag{10-19}$$

which is a Gaussian with standard deviation $\sqrt{RT/2K}$ centered at $\mathrm{Lk}_0 - \omega$. This equation can be analyzed to yield K and ω from experimental topoisomer distributions.

If ω is negligible relative to $\Delta\mathrm{Lk}$, then $\Delta G_{\Delta\mathrm{Lk}} = K(\Delta\mathrm{Lk})^2$ is the free energy difference between a molecule with Lk and relaxed DNA. This will depend on the number of base pairs N in the DNA: A larger DNA will be able to accomodate a given linking deficit at less energetic cost. In fact, the free energy difference per base pair $\Delta g_{\Delta\mathrm{Lk}}$ is expected to be approximately constant for fixed ΔLk/bp:

$$\Delta g_{\Delta\mathrm{Lk}} = \Delta G_{\Delta\mathrm{Lk}}/N = NK(\Delta\mathrm{Lk}/N)^2 \tag{10-20}$$

This equation implies that NK is also constant, which is confirmed experimentally for $N > 2000$ bp. To within about $\pm 10\%$, $NK = 1200RT$, or about 700 kcal at 20°C. With 10.5 bp/turn, and $\Delta\mathrm{Lk} = -1$ for every 20 turns ($\sigma \approx -0.05$), $\Delta g_{\Delta\mathrm{Lk}}$ is about 16 cal/bp, corresponding to 0.17 kcal/turn.

For shorter DNA, NK increases somewhat with decreasing N: for a 200 bp fragment, $NK = 3900RT$. This apparently reflects the decreased contribution of writhing fluctuations to the distribution in ΔLk (Le Bret, 1979; Shore and Baldwin, 1983b; Horowitz and Wang, 1984). According to the theoretical work of Le Bret (1979), writhing fluctuations appear abruptly at circle sizes above about 1000 bp. Monte Carlo simulations of small circles and determination of their Wr distribution support the idea that there is a narrow range of circle sizes over which Wr fluctuations increase sharply (Levene and Crothers, 1986). Consequently, torsional flexibility in small circles (< 500 bp) is dominated by fluctuations in twist, whereas the main contributor to torsional flexibility in large (> 2000 bp) circles is Wr. It follows that large circles are much more effective than small circles at absorbing unwinding stress that may accompany the binding of proteins or other ligands. The relation of these factors to the formation of very small closed circles, and the insight thereby obtained on bending and twisting force constants for DNA, has been discussed in Chapter 9.

The energy content of supercoiled DNA can be quite dramatic. Consider a typical plasmid, with 3000 bp, 300 helical turns, and $\sigma = -0.05$. Then $\Delta\mathrm{Lk} = -15$, and $\Delta G_{\Delta\mathrm{Lk}} = (3000)(700)(-15/3000)2 = 52.5\ \mathrm{kcal\ mol^{-1}}$. If ΔLk is reduced by one, to -14, $\Delta G_{\Delta \mathrm{Lk}}$ becomes 45.7 kcal mol^{-1}, a difference of 7 kcal mol^{-1} (similar to hydrolyis of an ATP molecule). Thus small reductions in supercoil density can provide substantial driving forces when coupled to other reactions.

4.2 DNA Shape as a Balance of Elastic and Interaction Energies

There has been considerable theoretical analysis of the equilibrium elastic behavior of DNA, particularly devoted to the question of whether the plectonemic or toroidal form is more stable for free ccDNA (Benham, 1979, 1983; LeBret, 1979; Calladine, 1980). It is now clear both from experimental and detailed computer calculations (e.g., Hao and Olson, 1989; Schlick and Olson, 1992; Vologodskii and Frank-Kamenetskii, 1992) that the toroidal form is unstable with respect to the plectonemic. Energy minimization and Monte Carlo calculations show a fairly sharp transition, the result of elastic instability, from a circle to a figure eight structure as ΔLk is increased to about 1.5–1.9. (The exact value depends on the ratio of twisting to bending force constants.) Subsequent increases in ΔLk produce corresponding increases in Wr (i.e. crossovers), as shown in Figure 10-17.

The range of shapes adopted by supercoils with a given value of ΔLk depends on the relative values of bending, twisting, and helix self-contact forces. Molecular dynamics simulations suggest that different shape families may have similar energies, leading to conformational transitions even near zero twist (Schlick et al., 1994). This may be the explanation of measurements (Shibata et al., 1984; Song et al., 1990)

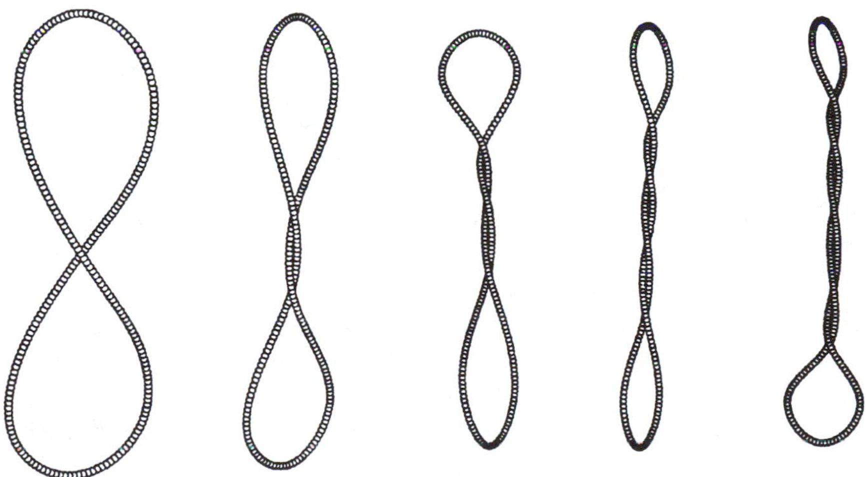

Figure 10-17
Energy-minimized interwound superhelical structures.From left to right, ΔLk = 2, Wr = 1.2; ΔLk = 3; Wr = 2.4; ΔLk = 4, Wr = 3.2; ΔLk = 5; Wr = 3.9; ΔLk = 6, Wr = 4.5. Each increase in ΔLk by one produces approximately one more superhelical turn. [Reprinted from Schlick, T. and Olson, W. K., Supercoiled DNA energetics and dynamics by computer simulation, *J. Mol. Biol.* **223**, 1089–1119, Copyright ©1992, by permission of the publisher Academic Press Limited London.]

indicating that there are two states of supercoiled plasmid DNA which, once formed, interconvert only very slowly, over periods of weeks or months. These states are distinguished by differences in torsional rigidity (related to helix secondary structure) and gel mobility (related to tertiary structure). While one of the states has normal properties, the other has a low torsion constant and does not convert completely to B-form DNA even upon relaxation of the superhelical constraints by linearization or relaxation by topoisomerase I.

The helix self-contact energy used in these calculations is often parameterized in terms of a hard cylinder radius, in which the actual radius of the DNA helix is increased by repulsive electrostatic forces (Stigter, 1977). Values of the effective radius obtained in this way are remarkably accurate in predicting the probability of knot formation during the random cyclization of DNA (Shaw and Wang, 1993; Rybenkov et al., 1993), a probability that is very sensitive to radius (Le Bret, 1980).

Computer simulation is one of the only ways of gaining information about the details of supercoil behavior. The molecules are too big for high-resolution techniques like crystallography and NMR; electrophoresis, hydrodynamic, and light scattering techniques only provide one or two parameters; and cryoelectron microscopy (Bednar et al., 1994) introduces poorly understood solvent and temperature perturbations. Computer modeling also enables simulation of sequence-specific effects such as intrinsic curvature, which have important effects on supercoiling transitions (reviewed in Vologodskii and Cozzarelli, 1994b).

5. STRESS-INDUCED REACTIONS

Supercoiling can produce changes in helical pitch and Wr, which affect binding of proteins and other ligands that require a precise fit to a particular DNA conformation. If the stresses become large enough and if sequences permit, supercoiling can also induce large-scale conformational changes, such as local denaturation, transition from B-DNA to Z-DNA or H-DNA, or cruciform formation (Hsieh and Wang, 1975; Benham, 1979). A fairly simple but useful theory has been constructed by Frank-Kamenetskii and Vologodskii (1984), which makes clear some of the basic concepts.

Consider superhelical ccDNA N bp long, carrying an artificial purine–pyrimidine insert n bp long that is capable of adopting the Z conformation. In general, there is competition with cruciform formation and melting, but the theory assumes for simplicity that only the B–Z transition is important. As the negative superhelicity σ increases, the insert adopts the Z form. This may occur in two stages. First, some fraction, m_0 bp, flips to Z at σ_0. Second, the remaining $n - m_0$ bp undergoes the B $\rightleftharpoons$Z transition as σ gradually becomes more negative. The flipping occurs when the free energy increase due to formation of an m bp Z segment is balanced by the decrease due to strain relaxation in superhelical DNA:

$$2F_{\mathrm{j}} * + m\Delta F_{\mathrm{BZ}} + \Delta G_m = 0 \tag{10-21}$$

The parameter F_j* is the junction free energy between B and Z regions, ΔF_{BZ} is the free energy change per bp for B–Z transition, and ΔG_m is the change in superhelix energy when m bp are converted from B to Z. This is

$$\Delta G_m = 10RTN[(\sigma + 1.8m/N)^2 - \sigma^2] \tag{10-22}$$

(Nordheim et al., 1982; Depew and Wang, 1975; Pulleyblank et al., 1975). The factor 1.8 appears because there are 10.5 bp/turn in the B form and 12 bp/turn in the Z form. Substituting Eq. 10-22 in Eq. 10-21 and solving for the minimal value of $-\sigma$ gives

$$\sigma_o = -(1/36RT)[\Delta F_{BZ} + 11.4\surd(2F_j * RT/N)] \tag{10-23}$$

and

$$m_o = 0.175\surd(2F_j * N/RT) \tag{10-24}$$

If $m_o > n$, the whole insert flips at some σ determined by Eqs. 10-21 and 10-22 with m replaced by n. For $m_o < n$ the flipping is followed by a gradual growth of the Z region with increasingly negative σ. For $m_o < m \leq n$,

$$\Delta F_{BZ} + 36RT(\sigma + 1.8m/N) = 0 \tag{10-25}$$

reflects the balance between the B–Z free energy increase and decrease in superhelical strain. These equations neglect the entropic terms that result from energetic equivalence of the m bp within n, but this is negligible for the $m_o \geq n$ case.

Use of Z-forming inserts with different lengths allows solving for F_j* and $\Delta F_{BZ.}$ The thermodynamic parameters for $(d(GC)_x \cdot d(GC)_x$ and $d(GT)_x \cdot d(AC)_x$ are similar and equal to $F_j = 4 - 5$ kcal^{-1} mol^{-1} of junctions and $\Delta F_{BZ} = 0.5 \pm 0.7$ kcal mol^{-1} of bp.

When the Z-forming sequence is self-complementary, the possibility for competing cruciform formation exists. Phase diagrams for some typical situations have been calculated by Anshelevich et al. (1988).

The kinetics of the B–Z transition induced by negative supercoiling has been examined by Peck et al. (1986). They employed a plasmid containing a $d(pCpG)_{16} \cdot d(pCpG)_{16}$ insert and used the binding of Z-DNA specific antibodies to follow the transition. The rate of the transition depends strongly on σ. With $\sigma = -0.09$, the transition was complete in less than 50 s. At $\sigma = -0.07$, the transition half-life was about 2 min, increasing about 10-fold at $\sigma = -0.05$.

5.1 Cruciform Formation

When a sequence in a plasmid is palindromic or self-complementary, superhelical stress may drive the sequence into paired stem–loop or hairpin structures called cruciforms (Murchie and Lilley, 1992; Kallenbach and Zhong, 1994). These structures, schematically depicted in Figure 10-18, were first demonstrated by Gellert et al., (1979), Lilley

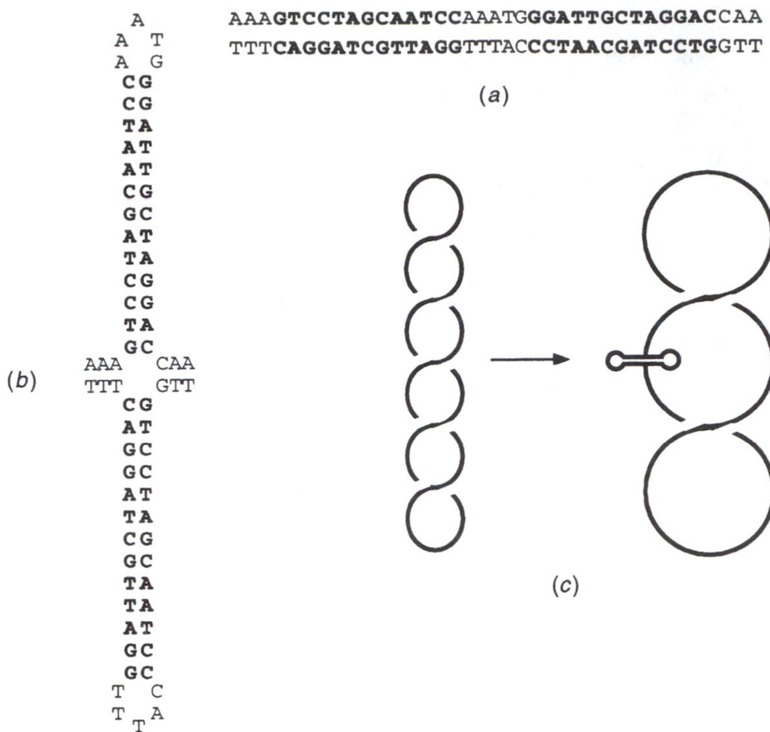

Figure 10-18
Formation of a cruciform from a self-complementary sequence. (*a*) Sequence of an inverted repeat in the *E. coli* plasmid ColE1, with self-complementary bases in bold face type. (*b*) The sequence rearranged into a cruciform. (*c*) Formation of a cruciform relaxes superhelical stress. [Adapted from Lilley, 1989.]

(1980), and Panayotatos and Wells (1981). A lucid and concise review, whose treatment we follow closely here, is by Lilley et al. (1987).

Cruciforms may be distinguished from normal double helical DNA by their different mobility in gel electrophoresis, and by their altered chemical and enzymatic reactivity in the single-strand loop and four-way junction regions. Because the four-way junction is formally equivalent to the Holliday (1964) junction proposed as a key intermediate structure in homologous genetic recombination, cruciforms have been intensively studied (Seeman and Kallenbach, 1994). While the free energy lost by disruption of base pairing in the original double helix is largely regained through base pairing in the hairpin, the unpaired bases in the loops still lead to a positive ΔG of cruciform formation of 13–18 kcal mol^{-1}; the lower value is observed with $(dA\text{-}dT)_n$ sequences. At least this amount of free energy must be released by reduction of superhelical stress for the cruciform to be thermodynamically stable. The loop size is normally between 4 and 6 bp. The four-way junction appears to be fully base paired.

While cruciform structures may be thermodynamically stable, their formation may be kinetically quite slow. A large number of base pairs must be disrupted before

they are reformed in the hairpin, which could lead to a very large activation energy and very slow cruciform extrusion kinetics. The initial studies by several groups of workers confirmed this expectation: extrusion times of several hours even at elevated temperatures were observed. However, a different inverted repeat sequence in ColE1 was found that gave much more rapid kinetics. Numerous artificial sequences have now been tested, and there appear to be two main classes of cruciforms, with quite different kinetic parameters and salt dependences. These have been termed S type (the more common, slower type) and C type (found naturally only with the ColE1 inverted repeat). Table 10-1 shows typical kinetic parameters for the two types.
The very high activation enthalpy for the C-type sequence means that an 8°C temperature rise causes a 2000-fold increase in rate.

These very different kinetic properties suggest two different mechanisms, which are diagrammed in Figure 10-19. The upper, C-type pathway involves a large-scale opening of the entire inverted repeat, which then presumably reforms the base pairs in the hairpin in a concerted process. This pathway explains the high activation enthalpy (since all the base pairs are broken at once) and the low-salt requirement (since low salt destabilizes the double helix). The lower, S-type pathway requires only a few base pairs to be disrupted at any given time, hence it has a much lower activation energy. Transition from the double helix to the hairpin occurs through a gradual process of proto-cruciform formation and branch migration. It has been found that the rate of S-type extrusion depends not just on the ionic strength, but on ionic size, suggesting that specific ion-binding cavities are formed by the geometry of the four-way junction (Sullivan and Lilley, 1987). The proposed S-type mechanism also suggests that extrusion kinetics will be very sensitive to base changes in the central region of the inverted repeat, but not in flanking sequences. In contrast, C-type extrusion should be relatively insensitive to sequence. These expectations are confirmed by experiment.

Whether an inverted repeat sequence undergoes a C- or S-type transition appears to depend on the sequence of flanking regions up to 100 bp away (Sullivan and Lilley, 1986). Flanking regions rich in A + T favor the C-type path, apparently due to long-range destabilization of the inverted repeat.

A third type of cruciform extrusion reaction has been observed in supercoiled DNA that contains $(A\text{-}T)_n$ inserts. This type has no detectable kinetic barrier and a very

Table 10.1
Kinetic Properties of S- and C-type Cruciform Extrusion[a]

Kinetic class	S-type	C-type
Example plasmid	pIRbke8	ColE1
NaCl optimum	50–60 mM	0–10 mM
$\Delta H^\dagger$ (kcal mol^{-1}	42	180
$\Delta S^\dagger$ at 37°C (cal mol^{-1} deg^{-1}	60	440

[a]From Lilley et al., 1987

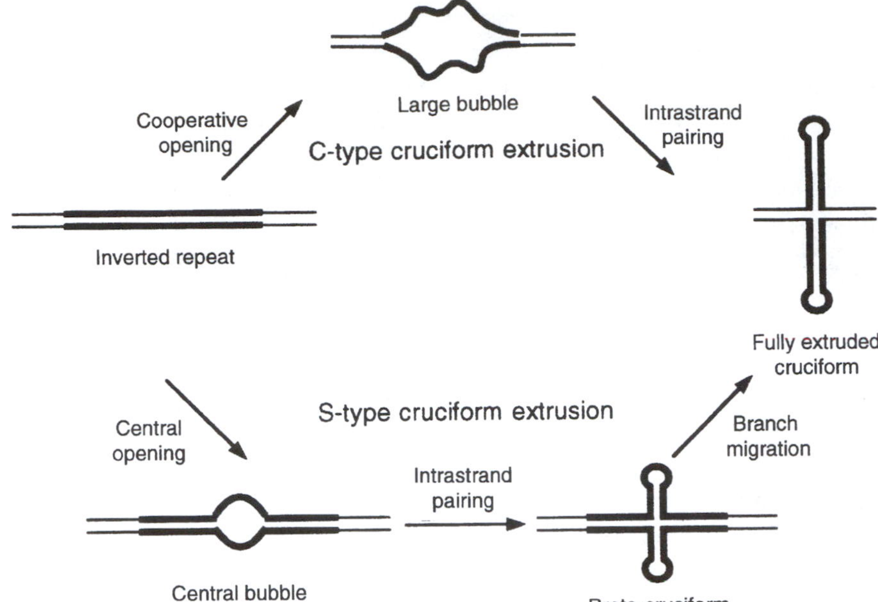

Figure 10-19
Both C- and S-type mechanisms of cruciform extrusion. [Reprinted with permission from Lilley, D. M. J. The inverted repeat as a recognisable structural feature in supercoiled DNA molecules, *Proc. Natl. Acad. Sci. USA* **77**, 6468–6472, 1988.]

low free energy of formation, which may result from perturbed helical structure of the torsionally stressed $(A\text{-}T)_n$.

5.2 H-DNA

H-DNA is an intramolecular triple helix formed with homopurine–homopyrimidine stretches and stabilized by supercoiling and hydrogen ions. It is reviewed by Frank-Kamenetskii (1992) and Mirkin and Frank-Kamenetskii (1994), and is discussed in Chapter 8.

References

Anderson, P. and Bauer, W. (1978). Supercoiling in Closed Circular DNA, Dependence upon Ion Type and Concentration, *Biochemistry* **17**, 594–601.

Anshelevich, V. V., Vologodskii, A. V., and Frank-Kamenetskii, M. D. (1988). A theoretical study of formation of DNA noncanonical structures under negative superhelical stress, *J. Biomolec. Struct. Dyn.* **6**, 247.

Bates, A. D. and Maxwell, A. (1993). *DNA Topology*, IRL Press, Oxford, UK.

Bauer, W. and Vinograd, J. (1968). The interaction of closed circular DNA with intercalative dyes. I. The superhelix density of SV40 DNA in the presence and absence of dye. *J. Mol. Biol.* **33**, 141–171.

Bauer, W. and Vinograd, J. (1970). Interaction of closed circular DNA with intercalative dyes. II. The free energy of superhelix formation in SV40 DNA, *J. Mol. Biol.* **47**, 419–435.

Bauer W. R. (1978). Structure and Reactions of Closed Duplex DNA. *Annu. Rev. Biophys. Bioeng.* **7**, 287–313

Bauer, W. R., Crick, F. H. C., and White, J. H. (1980). Supercoiled DNA, *Sci. Amer.* **243(1)**, 118–133.

Bednar, J., Furrer, P., Stasiak, A., Dubochet, J., Egelman, E., and Bates, A. (1994). The twist, writhe and overall shape of supercoiled DNA change during counterion-induced transition from a loosely to a tightly interwound superhelix. Possible implications for DNA structure *in vivo. J. Mol. Biol.* **235**, 825–847.

Benham, C. J. (1979). Torsional stress and local denaturation in supercoiled DNA. *Proc. Natl. Acad. Sci. USA* **76**, 3870–3874.

Boles, T. C., White, J. H., and Cozzarelli, N. R. (1990). Structure of plectonemically supercoiled DNA. *J. Mol. Biol.* **213**, 931–951.

Bowater, R., Aboul-ela, F., and Lilley, D. M. J. (1992). Two-dimensional gel electrophoresis of circular DNA topoisomers, *Methods Enzymol.* **212**, 105–120.

Brown, P. O. and Cozzarelli, N. R. (1979). A sign inversion mechanism for enzymatic supercoiling of DNA. *Science* **206**, 1081–1083.

Calladine, C. R. (1980). Toroidal Elastic Supercoiling of DNA. *Biopolymers* **19**, 1705–1713.

Cozzarelli, N. R., Boles, T. C., and White, J. H. (1990). A Primer on the Topology and Geometry of DNA Supercoiling, in *DNA Topology and Its Biological Effects*, Cozzarelli, N. R. and Wang, J. C., Eds., Cold Spring Harbor Laboratory Press, Cold Springs Harbor, NY.

Cozzarelli, N. R. and Wang, J. C. Eds., (1990). *DNA Topology and Its Biological Effects*, Cold Spring Harbor Laboratory Press, Cold Springs Harbor, NY, 480 pp.

Crick, F. H. C. (1976). Linking numbers and nucleosomes, *Proc. Natl. Acad. Sci. USA* **73**, 2639–2643.

Dean, F., Krasnow, M. A., Otter, R., Matzuk, M. M., Spengler, S. J., and Cozzarelli, N. R. (1982). Escherichia coli Type-1 topoisomerases, Identification, mechanism, and role in recombination, *Cold Spring Harbor Symp. Quant. Biol.* **47**, 769–777.

Depew, R. E. and Wang, J. C. (1975). Conformational fluctuations of DNA helix. *Proc. Natl. Acad. Sci. USA* **72**, 4275–4279.

Dröge, P. and Cozzarelli, N. R. (1992). The topological structures of DNA knots and catenanes, *Methods Enzymol.* **212**, 120–130.

Frank-Kamenetskii, M. D. (1992). Protonated DNA structures, *Methods Enzymol.* **211**, 180–191.

Frank-Kamenetskii, M. D., and Vologodskii, A. V. (1984). Thermodynamics of the B–Z transition in superhelical DNA, *Nature (London)* **307**, 481.

Fuller, F. B. (1971). The writhing number of a space curve. *Proc. Natl. Acad. Sci. USA* **68**, 815–819.

Gellert, M., Mizuuchi, K., O'Dea, M. H., Ohmori, H., and Tomizawa, J. (1979). DNA gyrase and DNA supercoiling. *Cold Spring Harbor Symp. Quant. Biol.* **43**, 35–40.

Hao, M.-H. and Olson, W. K. (1989). Global equilibrium configurations of supercoiled DNA. *Macromolecules* **22**, 3292–3303.

Holliday, R. (1964). A mechanism for gene conversion in fungi. *Genet. Res.* **5**, 282–304.

Horowitz, D. S. and Wang, J. C. (1984). Torsional Rigidity of DNA and Length Dependence of the Free Energy of DNA Supercoiling. *J. Mol. Biol.* **173**, 75–91.

Hsieh, T.-S. and Wang, J. C. (1975). Thermodynamic properties of superhelical DNAs. *Biochemistry* **14**, 527–535.

Kallenbach, N. and Zhong, M. (1994). DNA cruciforms. *Curr. Opin. Struct. Biol.* **4**, 365–371.

Keller, W. and Wendel, I. (1975). Stepwise relaxation of supercoiled SV40 DNA. *Cold Spring Harbor Symp. Quant. Biol.* **39**, 199–208.

Kikuchi, A. and Asai, A. (1984). Reverse gyrase—a topoisomerase which introduces positive superhelical turns into DNA, *Nature (London)* **309**, 677–681.

Le Bret, M. (1979). Catastrophic Variation of Twist and Writhing of Circular DNAs with Constraint?. *Biopolymers* **18**, 1709–1725.

Le Bret, M. (1980). Monte Carlo computation of supercoiling energy, the sedimentation constant, and the radius of gyration of unknotted and circular DNA. *Biopolymers* **19**, 619–637.

Lee, C.-H., Mizusawa, H., and Kakefuda, T. (1981). Unwinding of double-stranded DNA helix by dehydration. *Proc. Natl. Acad. Sci. USA* **78**, 2838–2842.

Levene, S. D. and Crothers, D. M. (1986). Topological Distributions and the Torsional Rigidity of DNA. A Monte Carlo Study of DNA Circles. *J. Mol. Biol.* **189**, 73–83.

Lilley, D. M. J. (1980). The inverted repeat as a recognisable structural feature in supercoiled DNA molecules. *Proc. Natl. Acad. Sci. USA* **77**, 6468–6472.

Lilley, D. M. J. (1988). DNA opens up—supercoiling and heavy breathing. *Trends in Genetics* **4**, 111–114.

Lilley, D. M. J. (1989). Structural isomerism in DNA, The formation of cruciform structures in supercoiled DNA molecules, *Chem. Soc. Revs.* **18**, 53–83.

Lilley, D. M. J., Sullivan, K. M., and Murchie, A. I. H. (1987). The extrusion of cruciform structures in supercoiled DNA—Kinetics and mechanism, in *Nucleic Acids and Molecular Biology*, Vol. 1, Eckstein, F. and Lilley, D. M. J., Eds., Springer- Verlag, Berlin, Germany, pp 126–137.

Lima, C. D., Wang, J. C., and Mondragón, A. (1994). Three-dimensional structure of the 67K N-terminal fragment of *E. coli* DNA topoisomerase I. *Nature (London)* **367**, 138–146.

Liu, L. F. and Wang, J. C. (1978a). *Micrococcus luteus* DNA gyrase, Active components and a model for its supercoiling of DNA. *Proc. Natl. Acad. Sci. USA* **75**, 2098–2102

Liu, L. F. and Wang, J. C. (1978b). DNA–DNA gyrase complex, The wrapping of the DNA duplex outside the enzyme. *Cell* **15**, 979–984.

Maxwell, A. and Gellert, M. (1986). Mechanistic Aspects of DNA Topoisomerases. *Adv. Protein Chem.* **38**, 69–107.

Mirkin, S. and Frank-Kamenetskii, M. (1994). H-DNA and related structures. *Annu. Rev. Biophys. Biomol. Struct.* **23**, 541–576.

Murchie, A. I. H. and D. M. J. Lilley. (1992). Supercoiled DNA and cruciform structures. *Methods Enzymol.* **211**, 158–180.

Nordheim, A., Lafer, E. M., Peck, L. J., Wang, J. C., Stollar, B. D., and Rich, A. (1982). Negatively supercoiled plasmids contain left-handed Z-DNA segments as detected by specific antibody binding. *Cell* **31**, 309–316.

Panayotatos, N. and Wells, R. D. (1981). Cruciform structures in supercoiled DNA. *Nature (London)* **289**, 466–470.

Peck, L. J., Wang, J. C., Nordheim, A., and Rich, A. (1986). Rate of B to Z Structural Transition of Supercoiled DNA. *J. Mol. Biol.* **190**, 125–127.

Pulleyblank, D. E., Shure, M., Tang, D., Vinograd, J., and Vosberg, H. P. (1975). Action of nicking-closing enzyme on supercoiled and nonsupercoiled closed circular DNA, Formation of a Boltzmann distribution of topological isomers. *Proc Natl Acad Sci USA* **72**, 4280–4284.

Rau, D. C., Gellert, M., Thoma, F., and Maxwell, A. (1987). The structure of the DNA gyrase-DNA complex as revealed by transient electric dichroism. *J. Mol. Biol.***193**, 555–569.

Richmond, T. J., Finch, J. T., Rushton, B., Rhodes, D., and Klug, A. (1984). Structure of the nucleosome core particle at 7 Å resolution. *Nature (London)* **311**, 532–537.

Rybenkov, V. V., Cozzarelli, N. R., and Vologodskii, A. V. (1993). Probability of DNA knotting and the effective diameter of the DNA double helix *Proc. Natl. Acad. Sci. USA* **90**, 5307–5311.

Schlick, T. and Olson, W. K. (1992). Supercoiled DNA energetics and dynamics by computer simulation. *J. Mol. Biol.* **223**, 1089–1119.

Schlick, T., Olson, W., Westcott, T., and Greenberg, J. (1994). On higher buckling transitions in supercoiled DNA. *Biopolymers* **34**, 565–597.

Seeman, N. and Kallenbach, N. (1994). DNA branched junctions. *Annu. Rev. Biophys. Biomol. Struct.* **23**, 53–86.

Shaw, S. Y. and J. C. Wang. (1993). Knotting of a DNA chain during ring closure, *Science* **260**, 533–536.

Shibata, J. H., Wilcoxon, J., Schurr, J. M., and Knauf, V. (1984). Structures and Dynamics of a Supercoiled DNA. *Biochemistry* **23**, 1188–1194.

Shimada, J. and Yamakawa, H. (1988). Sedimentation coefficients of DNA topoisomerases: The helical wormlike chain. *Biopolymers* **27**, 675–682.

Shore, D. and Baldwin, R. L. (1983b). Energetics of DNA Twisting II. Topoisomer Analysis. *J. Mol. Biol.* **170**, 983–1007.

Song, L., Fujimoto, B. S., Wu, P., Thomas, J. C., Shibata, J. H., and Schurr, J. M. (1990). Evidence for allosteric transitions in secondary structure induced by superhelical stress, *J. Mol. Biol.* **214**, 307–326.

Stigter, D. (1977). Interactions of highly charged colloidal cylinders with applications to double-stranded DNA. *Biopolymers* **16**, 1435–1448.

Sullivan, K. M. and Lilley, D. M. J. (1986). A dominant influence of flanking sequences on a local structural transition in DNA. *Cell* **47**, 817–827.

Sullivan, K. M. and Lilley, D. M. J. (1987). Influence of cation size and charge on the extrusion of a salt-dependent cruciform. *J. Mol. Biol.* **193**, 397–404.

Vologodskii, A. (1992). *Topology and Physics of Circular DNA*, CRC Press, Boca Raton, FL.

Vologodskii, A. V. and Cozzarelli, N. R. (1994a). Conformational and thermodynamic properties of supercoiled DNA. *Annu. Rev. Biophys. Biomol. Struct.* **23**, 609–643.

Vologodskii, A. and Cozzarelli, N. (1994b). Supercoiling, knotting, looping and other large-scale conformational properties of DNA. *Curr. Opin. Struct. Biol.* **4**, 372–375.

Vologodskii, A. V. and Frank-Kamenetskii, M. D. (1992). Modeling supercoiled DNA. *Methods Enzymol.* **211**, 467–480.

Vologodskii, A. V., Levene, S. D., Klenin, K. V., Frank-Kamenetskii, M., and Cozzarelli, N. R. (1992). Conformational and thermodynamic properties of supercoiled DNA. *J. Mol. Biol.* **227**, 1224–1243.

Wang, J. C. (1974). Interactions between twisted DNAs and entymes: The effects of superhelical turns. *J. Mol. Biol.* **87**, 797–816.

Wang, J.C. (1985). DNA Topoisomerases. *Annu. Rev. Biochem.* **54**, 665–697.

Wang, J. C. (1987). Recent studies of DNA topoisomerases. *Biochim. Biophys. Acta* **909**, 1–9.

Wang, J. C. (1991). DNA Topoisomerases, Why So Many? *J. Biol. Chem.* **266**, 6659–6662.

Wang, J. C. and Giaever, G. N. (1988). Action at a distance along a DNA. *Science* **240**, 300–304.

Wang, J. C., Peck, L. J., and Becherer, K. (1982). DNA Supercoiling and Its Effects on DNA Structure and Function. *Cold Spring Harbor Symp. Quant. Biol.* **47**, 85–91.

Wasserman, S. A. and Cozzarelli, N. R. (1986). Biochemical Topology, Applications to DNA Recombination and Replication. *Science* **232**, 951–960.

White, J. H. and Bauer, W. R. (1986). Calculation of the twist and writhe for representative models of DNA. *J. Mol. Biol.* **189**, 329–341.

White, J. H., Cozzarelli, N. R., and Bauer, W. R. (1988). Helical repeat and linking number of surface-wrapped DNA. *Science* **241**, 323–327.

White, J. H., Gallo, R. M., and Bauer, W. R. (1992). Closed circular DNA as a probe for protein-induced structural changes. *Trends Biochem. Sci.* **17**, 7–12.

CHAPTER 11

Interaction of Nucleic Acids with Water and Ions

Nucleic acids function through their interactions with other molecules. Chapters 12–14 will deal with the binding of ligands such as drugs and proteins to nucleic acids. In this chapter, we are concerned with species that are ubiquitous in all nucleic acid environments: water and ions. Aside from the effects of ions on the interactions of nucleic acids with other charged ligands, a major theme is that water and ions can profoundly influence the conformation, and therefore the properties, of DNA and RNA. Indeed, as Privé et al. (1991) note, "... DNA is built from five structural elements rather than three: the familiar bases, sugars, and phosphates, but also ordered waters and bound counterions." Section 8.5 discusses some of the structural transformations that have been observed as solvent conditions are varied. Studies under extreme conditions of high salt or high cosolvent concentration have been important in delineating conformational variability of the nucleic acids. In addition, such conditions may actually have biological relevance: some halophiles live in quite salty conditions, and the activity of water inside cells is significantly reduced by high concentrations of protein and nucleic acid. Such variability must always be considered when thinking about structure–function relations.

1. HYDRATION

There are many reasons for wanting to understand the molecular interactions between nucleic acids and water. Water is the major solvent component for nucleic acids in all biologically relevant situations: its concentration is about 55 *M* while most other species are less than 1 *M*. It was shown in early fiber diffraction studies that water content, or relative humidity, influences the DNA helix structure. The B form occurs at high relative humidity ($> 85\%$), the A form occurs at relative humidity between 75 and 80%), and a disordered form occurs between 55 and 75% relative humidity. It has since been found that base and salt composition influence these values. Thus changing interactions with water are a prototype of how interactions with other molecules can regulate nucleic acid structure. Infrared (IR) spectroscopic studies and, more recently, X-ray crystallographic structure determination have determined the relative strengths of interaction of water with various sites on nucleic acids, and have identified networks of hydrogen-bonded waters that appear to be important in stabilizing particular helix forms. Crystallographic and other physical chemical approaches to DNA hydration have been surveyed by Saenger (1984), Westhof (1988) and Berman (1991, 1994). Useful general reviews of biomolecular hydration have been written by Cooke and Kuntz (1974), Saenger (1987), and Westhof (1993).

Among the issues to be understood are specification of hydration sites and energetics of water binding, the role of hydration in control of helix geometry, the coupling of hydration to ion and ligand binding, and the importance of solvent structure in condensation and packaging of DNA in viruses and chromosomes. This last topic, which involves the role of hydration forces in repelling or attracting DNA double helices, will be discussed in Chapter 14.

1.1 Properties of Water

The peculiar properties of water, compared with other common solvents, are well known (Eisenberg and Kauzmann, 1969). The most important features of H_2O are its extensive hydrogen bonding capability and its large dipole moment. Both of these result from the partial positive changes on the hydrogen atoms and the partial negative charge associated with the nonbonded electron on the oxygen.

The hydrogen-bonding capability of water leads to its interactions with nucleic acid groups, such as O and N atoms that are hydrogen-bonding acceptors, and protons that are hydrogen-bonding donors. It also leads to extensive, though imperfect and transitory, hydrogen-bonded networks in water itself. Formation or disruption of such networks leads to a decrease or increase, respectively, in the entropy, S, thus affecting the thermodynamics of processes that affect water hydrogen bonding. For example, introduction of a hydrophobic residue into water leads to the formation of hydrogen-bonded water clusters, and to a negative ΔS. This hydrophobic behavior may be partially responsible for the low solubility of the nucleic acid bases in water, and consequently for their tendency to self-associate by stacking in aqueous solution (Crothers and Ratner, 1968). On the other hand, a "structure-breaking" species such as ClO_4^- disrupts water hydrogen bonding, thereby increasing the entropy of aqueous perchlorate solution.

The large dipole moment of water makes it a good solvent for ions. Cations interact favorably with the O, and anions interact with the H atoms. Since a high dipole moment leads to a high dielectric constant, ε, Coulombic interactions between ions, whose energy is $E = q_1 q_2/\varepsilon r$ (where q_1 and q_2 are the ion charges and r is their separation) will be weaker in water ($\varepsilon \approx 80$) than in organic solvents ($\varepsilon \approx 2$). In addition, most liquids with which nucleic acids interact are ionic or highly polar, and since nucleic acids themselves are highly charged, the properties of water as a solvent for ions are crucial.

1.2 Sites of Water Interaction with DNA: Gravimetric and IR Studies

The most direct way to study hydration of any molecule is to start with it completely dry, equilibrate the sample with water at controlled levels of water activity (relative humidity), and determine at each level the amount and location of bound water. This type of study was performed by Falk and co-workers (Falk et al., 1962, 1963; Falk, 1965) using gravimetric and IR spectroscopic techniques. The relative humidity was controlled by equilibrating the DNA sample in a thermostated enclosure with containers of saturated salt solution of known relative humidity.

To determine the amount of water bound, the DNA samples (in the form of fibers) were weighed repeatedly until equilibrium had been reached, as indicated by constant weight for 1 week. The results are shown in Figure 11-1 for NaDNA. These results were fit to a two-layer adsorption model, which enabled the determination of the number n of water molecules absorbed in the first layer, and the energy difference $E_1 - E_L$ between water adsorption to the first and successive layers. For NaDNA, $n = 2.2$ water molecules/nucleotide and $E_1 - E_L = 1.7$ kcal mol^{-1}. For LiDNA, $n = 2.0$ and $E_1 - E_L = 2.1$ kcal mol^{-1}. Similar hydration studies of purines, pyrimidines, nucleosides, and nucleotides showed that only the salts of the nucleotides that contained the ionic phosphate group formed stable hydrates over the relative humidity range 0–92% at room temperature. Thus the data on DNA were interpreted as formation of a stable dihydrate of the ionic phosphate group at very low relative humidity, followed at higher relative humidity by a series of higher hydrates involving the oxygen atoms of the sugars and base carbonyls and the nitrogen atoms of the heterocyclic bases. The small but reproducible hysteresis (excess water binding on desorption relative to adsorption) along with a deviation from the two-layer adsorption model above 80% relative humidity, were attributed to filling the "empty space" between DNA helices and the energy required to push the helices apart after the space had been filled.

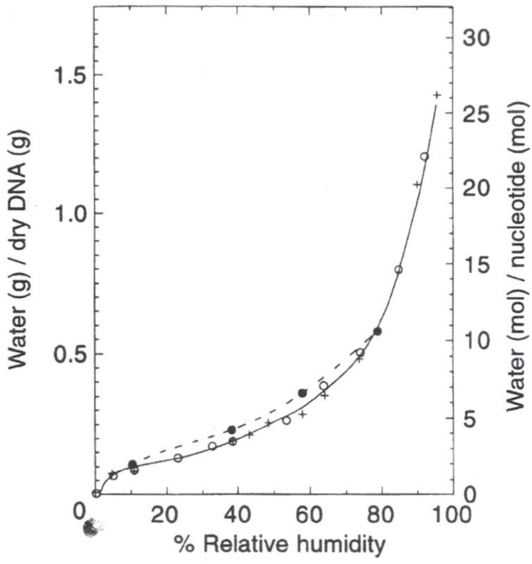

Figure 11-1
Adsorption and desorption of water by calf thymus and salmon sperm DNA. [Reprinted with permission from Falk, M., Hartman, K. A., Jr., and Lord, R. C., *J. Am. Chem. Soc.*, **84**, 3843–3846 (1962). Copyright ©1962 American Chemical Society.]

Further insight into the relative strengths of DNA hydration sites comes from IR spectroscopy of DNA films as a function of relative humidity. The frequency of the asymmetric OH stretch of the water near 3400 cm^{-1} increases with relative humidity in DNA films up to 65% relative humidity; by 92% relative humidity it has the value characteristic of bulk water. The lower frequencies are indicative of hydrogen bonding to the DNA. A strong band of 1240 cm^{-1} at 0% relative humidity shifts to 1220 cm^{-1} by 65% relative humidity; this band is assigned to the asymmetric PO_2 stretch. The lowered frequency is attributable to the higher reduced mass of the hydrated phosphate. Other bands showing shifts below 60% relative humidity are assigned to P–O and C–O stretching vibrations. Bands that do not begin to change until near 65% relative humidity are assigned to C=O and ring stretching vibrations as well as NH and NH_2 bending vibrations.

This sequence of events leads to the model shown in Figure 11-2. There is a primary hydration shell, consisting of water with different IR OH frequencies than liquid H_2O, which is immediately adjacent to DNA. There are 18–23 such water molecules per nucleotide in B-DNA. Two classes are distinguished within the primary hydration shell.

Class 1 contains 11–12 waters per nucleotide bound directly to DNA. About 6 of these are tightly bound to the sugar–phosphate chain: two most strongly bound to the ionic phosphates and four to the phosphodiester and furanose oxygen atoms. The other 4–6 Class 1 waters are less tightly bound to sites on the bases, such as C=O, heterocyclic N, NH, and NH_2.

Class 2 contains 8–9 additional waters per nucleotide with modified IR characteristics. These are hydrogen bonded to the Class 1 waters.

A secondary shell surrounds the primary hydration shell. Its properties are essentially those of liquid water.

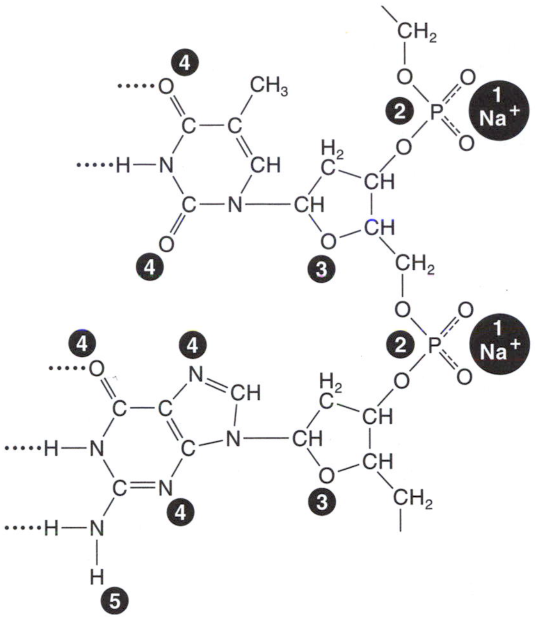

Figure 11-2
A schematic drawing of one strand of DNA containing T and G bases, indicating sites for possible adsorption as shaded areas. Reprinted with permission from Falk, M., Hartman, K. A., Jr., and Lord, R. C., *J. Am. Chem. Soc.*, **85**, 387–391 (1963). Copyright ©1962 American Chemical Society.]

1.3 Crystallographic Determination of Water Binding

A second way to look at the locations of bound water molecules is by X-ray crystallography (Berman, 1991, 1994). Before discussing specific results, some general remarks and cautions are in order. The DNA crystals are obtained from a solvent that contains, in addition to water, salts of cations that bind strongly to DNA (especially Mg^{2+} and spermine^{4+}), and sometimes alcohols such as 2-methyl-2,4-pentanediol (MPD7). At the level of resolution of some of the older structures, ~ 2.0 Å, it is difficult to distinguish Mg and spermine from H_2O; and high levels of MPD7 may distort apparent hydration occupancy statistics. Protons are not located, and specific hydrogen bonds are assigned on the basis of suitable O–O and O–N distances (typically 2.4–3.6 Å) and angular geometries. Modern, higher resolution structures have generally taken these factors into account.

1.3.1 B-DNA

In the first crystal structure determinations of B-form DNA (see Dickerson et al., 1982 for an overview), Dickerson et al. studied three forms of the self-complementary dodecamer d(CGCGAATTCGCG)$_2$: (1) the dodecamer at room temperature, (2) the dodecamer at 16K, and (3) the 9-bromo derivative d(CGCGAATT^{Br}CGCG)$_2$ at 7°C in MPD7. The hydration patterns in these structures are discussed in Drew and Dickerson (1981) and Kopka et al. (1983). They identified 114 solvent peaks in the MPD7 structure. This is 27% of all the water in the asymmetric unit, or an average of 4.75 solvent molecules per nucleotide; which should be compared with the 18–23 waters per nucleotide in the primary hydration shell detected by IR spectroscopy. All solvent peaks were treated as water, but some may actually be magnesium or spermine ions.

Water distances up to 3.5 Å were regarded as hydrogen-bonds, and those between 3.5 and 4.1 Å as looser DNA–water interactions. Visibility of solvent molecules is related to their crystallographic *B* values, which represent mean-square displacements due to thermal vibration and/or static disorder. The lower the *B* value, the less disorder. For each of the three structures, the order of mean *B* values for bound solvent locations was bases < sugars < phosphates.

The 16 K and MPD7 structures show extensive hydration (65 solvent peaks, an average of $3H_2O/P$) along the phosphate backbone. The hydration was not visible in the room temperature structure due to high *B* values. In the major groove, 19 other solvent molecules form a first hydration layer in contact with base edge N and O atoms and 36 more are found in upper hydration layers, tending to form strings or clusters spanning phosphate groups across the major groove. In the minor groove, a zigzag spine of hydration is found in all three structures.

The spine of hydration in the A + T rich center of the dodecamer is shown schematically in Figure 11-3. The first shell of hydration (Class 1) directly bridges adenine N3 and thymine O2. These are bridged in turn by a second layer of waters (Class 2) which gives approximately tetrahedral coordination to the first-shell water oxygens. The integrity of the regular zigzag spine disintegrates in passing from the A + T—containing center to the G + C—containing ends. This disintegration is due in part to disruption by N2 guanine amino groups, and in part to a change in geometry of the minor groove.

The minor groove hydration spine may be responsible for the stability of the B form of DNA containing only A-T and I-C base pairs (no N2 amino), and disruption of the spine may allow an easy transition to the A form of DNA with a high content of G-C base pairs.

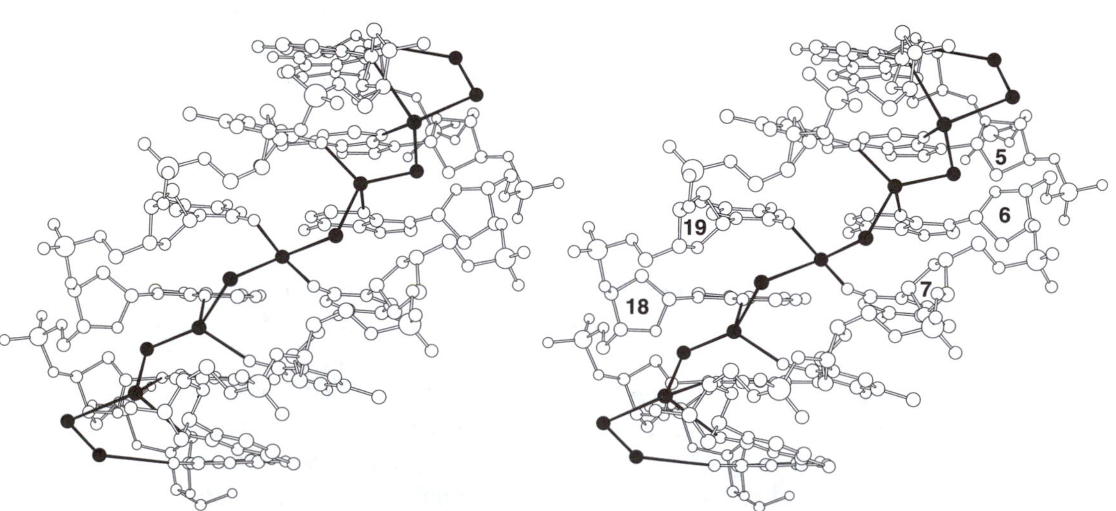

Figure 11-3
Stereodiagram of AT hydration spine. Reprinted with permission from Drew and Dickerson, 1981.

Hydration in upper levels (i.e., not in direct contact with the nucleic acid atoms) is sparse, and consists mainly of strings of water across the groove that have little contact with the spine below. Sugar O4′ (O1′ in the older nomenclature) atoms, which line the bottom of the minor groove, are closely associated with waters. This pattern may reflect the geometry of the minor groove rather than intrinsic affinity. The O3′ and O5′ atoms of phosphate groups along the backbone are the least hydrated, despite a lack of steric impediments.

Thus the apparent order of water binding strength in B-DNA, based on the percentage occupancy of hydration sites in the MPD7 structure, is minor groove spine (in A + T rich region) > free phosphate O > sugar O-4′ > major groove bases > phosphate O3′ > O5′. A clear discrepancy with the model proposed by Falk et al. is the weak crystallographic hydration of the esterified O3′ and O5′ phosphate oxygen atoms. However, this may reflect differences in the averaging over bound water populations by the IR and X-ray methods, rather than a real difference in preferred hydration sites.

A more recent set of B-DNA structures, obtained from three decamers of related sequence, has given higher resolution and suggested a wider range of hydration possibilities (Privé et al., 1987, 1991; Heinemann and Alings, 1989). These crystals are about 48% solvent, 52% DNA and have square rather than hexagonal packing. In the mismatch helix d(CCAAGATTGG), 52 non-Mg-cluster waters were localized, along with three octahedrally coordinated Mg clusters. In the d(CCAACGTTGG) helix, crystallography located 56 noncluster waters and 4 Mg complexes. The third decamer has the sequence d(CCAGGCCTGG). The difference between the effective 2.2 Å resolution of the Drew dodecamer and the 1.3–1.6 Å resolution of these decamers allows substantially more detailed structural definition.

One interesting observation from comparison of these three structures is that the width of the minor groove determines the hydration pattern, a conclusion confirmed by subsequent work (Quintana et al., 1992; Lipanov et al, 1993; Berman, 1994). The d(CCAACGTTGG) helix has a narrow minor groove in its central section, about 4.2 Å wide which is occupied by a distinct spine of hydration of the sort originally observed in the AT region of the dodecamer. The minor groove widens to about 6.7 Å toward the ends of the d(CCAACGTTGG) helix, and in this wide region is found two side-by-side ribbons of water molecules. This two-ribbon structure is also observed in the d(CCAAGATTGG) helix, which has a consistently wide minor groove ($\sim$ 7.2 Å). The d(CCAGGCCTGG) helix has a minor groove of intermediate width, and is occupied only sparsely by ordered waters. It seems that this intermediate width is not compatible with either hydration pattern.

The junction between the spine and ribbons in d(CCAACGTTGG) is bridged by a hydrated magnesium ion that is coordinated to both regions of structured water. Similarly, a hydrated Mg^{2+} bridges the ribbons at the break between molecules in the crystal, donating one pair of waters to each double ribbon.

The early studies of the spine of hydration in the AT rich region of the dodecamer, suggested that this water network was important in stabilizing B-DNA against a B-to-A transition (Drew and Dickerson, 1981; Kopka et al., 1983). With the correlation of high A,T content with a narrow minor groove, and the relative resistance of A,T rich DNA to undergoing the B–A transition, it has been speculated (Privé et al., 1991) that the spine confers greater stabilization of B-DNA than the ribbons, perhaps because of the greater isolation of the spine from bulk solvent.

1.3.2 A-DNA

Hydration has been analyzed crystallographically in two A-form deoxyoligonucleotides. The crystal structure of the octamer d(GG^{Br}UA^{Br}UACC)$_2$ has been refined to 1.7 Å by Kennard et al. (1986). Eighty four solvent molecules were located in the asymmetric unit, corresponding to 5.25 H_2O per nucleotide compared with 4.75 in the B-DNA dodecamer. The sugar–phosphate backbone and functional groups of the bases are highly hydrated. In the central ^{Br}UA^{Br}UA region of the major groove, there is a ribbon of water molecules, a closed circular network of four pentagons with shared edges (Fig. 11-4). These waters are linked to base O and N atoms and to the solvent chains connecting the O1 phosphate oxygen atoms on each strand. This hydration network is similar to the cluster of water pentagons observed by Neidle et al. (1980) in the crystal

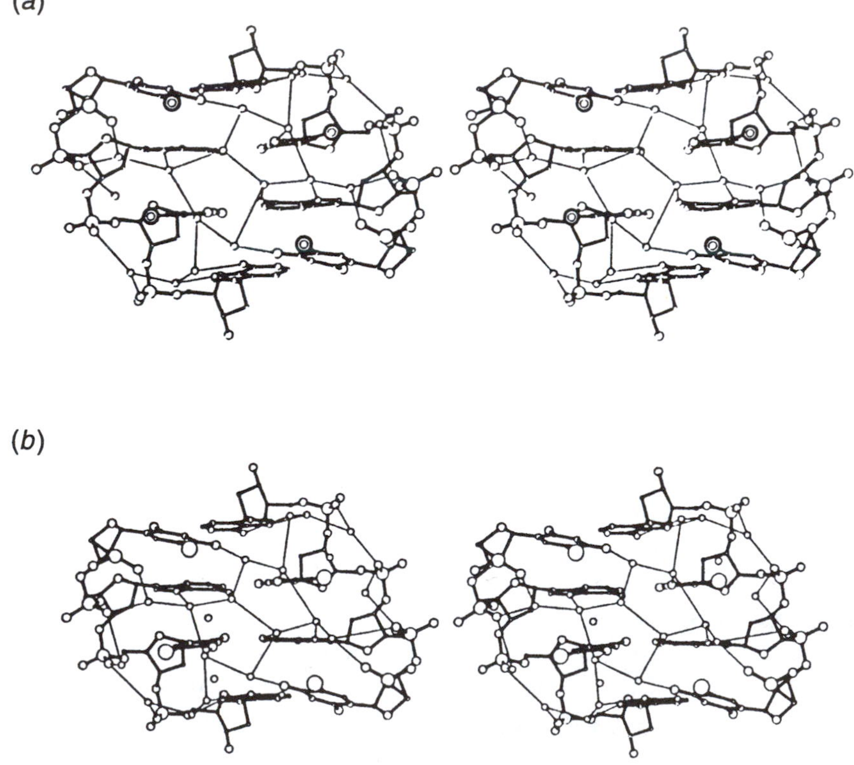

Figure 11-4
(*a*) Stereo diagram of the pentagonal arrangement of the water molecules in the central ^{Br}U-A-^{Br}U-A region of the major groove of the A-DNA octamer d(G-G-^{Br}U-A-^{Br}U-A-C-C). (*b*) The corresponding view in the native octamer d(G-G-T-A-T-A-C-C). [Reprinted with permission from Kennard et al., 1986.]

structure of the complex of d(CpG) with proflavin. The minor groove has a continuous water network in the central region and other networks at each end.

Conner et al. (1984) analyzed the hydration of the A-form tetramer ^{I}CCGG. They located 86 water molecules per tetramer, or 5.4 per nucleotide. Of these, 45.5 waters lie within a first hydration shell (3.5 Å from a DNA O or N). The trend of relative occupancy is free phosphate O > major groove N,O > backbone O3′ > minor groove N,O > backbone O5′ > sugar O4′. In contrast with B-DNA, O3′ atoms are frequently hydrated, because they are turned more toward the solvent environment. No ordered water structure comparable to the B-DNA minor groove hydration spine was observed. Many minor groove sites are blocked by crystal packing contacts.

A comparison of these two A-form structures shows that the tetramer has quite different hydration from the octamer, owing apparently to intermolecular packing disruptions. By comparing the B and A forms, Conner et al. (1984) propose that the spine of hydration stabilizes the deep, narrow minor groove of B-DNA. This hydration constrains the base pairs to the center of the helix. If the spine is disrupted, the minor groove opens and the base pairs can slide toward the surface of the helix, which is characteristic of A-family structures.

Another comparison between A and B-DNA patterns is illuminating. The N and O atoms in the major groove of B-DNA are highly hydrated, but no regular network is apparent. As pointed out by Privé et al. (1991), the roles of the major and minor grooves in B- and A-DNAs are reversed. In each case, the narrower groove (minor in B, major in A) shows a regular hydration network, perhaps because it is relatively inaccessible to bulk solvent. The wider groove, in contrast, is extensively but irregularly hydrated.

1.3.3 Z-DNA

The Z-form crystals often diffract better than the A or B form, so it has been possible to obtain considerable insight into the relative role of water and ions in stabilizing the Z conformation. Chevrier et al. (1986) studied crystals of the left-handed hexamer of d(^{5Br}C-G-^{5Br}C-G-5^{Br}C-G) grown at 18 and at 37°C. They find that the crystal grown at higher temperature has a more extensive hydration shell, particularly manifest in a spine of hydration running deep in the minor groove. This spine of hydration links exocyclic O2 atoms of the cytosines and bridges to the water molecules stabilizing the syn conformation of the guanine bases. A hydrated sodium ion bound to N7 of guanine was also observed, along with a string of waters bridging anionic phosphate oxygens. In all, 83 solvent atoms were localized in the 37°C structure refined to 1.4 Å, an average of 6.9 per nucleotide. This compares with approximately 100 water molecules and ions, or 8.3/nucleotide, identified by Gessner et al. (1985) in a 1.25-Å refinement of the Z form of d(CGCGCG)$_2$. In that structure, with the Z form stabilized with Mg^{2+}, three hydrogen bonds can be visualized between the base pairs and the octahedrally arranged waters that hydrate each magnesium ion.

The tendency of GC-containing polymers to adopt the Z conformation, and of AT-containing polymers to resist, is perhaps explained by the observation that the helical groove in d(^{m5}CGTA^{m5}CG)$_2$ is regularly filled with two water molecules per base pair, hydrogen-bonded to the bases (Wang et al., 1984). In helix segments containing AT base pairs, this regular occupancy is not seen, perhaps owing to solvent disorder.

1.3.4 Some Generalizations

The foregoing data present an apparent paradox. The A and Z forms of DNA are preferred to the B form under conditions of low water activity, but they have more water molecules per nucleotide that are located by crystallography. Since only a small fraction of the total waters are located in X-ray structures, there is no necessary contradiction. There may be fewer waters in the low-humidity forms, but held in more orderly structures. However, there is one crystallographic rationale for the adoption of non-B structures by DNA under low humidity conditions (Saenger et al., 1986).

From inspection of the backbone geometries of oligonucleotides, it is found that the free O atoms of adjacent phosphate groups average at least 6.6 Å apart. This distance is too far to be bridged by a single water molecule with a hydrogen-bonding distance of 3.5 Å, and the free O atoms are individually hydrated. In A- and Z-DNA, the free phosphate oxygens are as close as 5.3 and 4.4 Å respectively. Depending on the particular crystal form, 33–58% of the interphosphate gaps in A-DNA and 38–63% in Z-DNA, but none in B-DNA, are bridged by one water molecule. This hydration economy allows more efficient utilization of available water at low humidity, favoring the non-B geometries. Some bridging hydration structures are shown in Figure 11-5.

It has been observed (Westhof, 1988) that waters not involved in bridges, and which are located near nonpolar regions of grooves, tends to function like internal waters in proteins by forming networks of hydrogen-bonds to polar atoms that are too far away to form good hydrogen-bonds among themselves. Water molecules also tend to cluster around unusual bases in DNA and tRNA in such a way as to compensate for missing hydrogen bonds from unformed base pairs. A thorough survey of crystallographic coordinates (Schneider et al., 1992, 1993) shows that the positions of waters on the bases can be systematized in a way that relates DNA conformation and hydration.

1.4 Amount of Associated Water Viewed Nonspecifically

A variety of physical methods give indications of the amount of water held close to or perturbed by nucleic acid surfaces (Texter, 1978). They include hydrodynamics, calorimetry, NMR, and Brillouin scattering and they lack the specificity of spectroscopic or crystallographic methods, but are sensitive to different sorts of water–polynucleotide interactions, thus these methods help to give an overall picture. Since each method probes different aspects of hydration, exact agreement among them should not be expected, but a reasonable consistency is found.

1.4.1 Hydrodynamic Methods

The self-diffusion coefficient of water in a DNA solution, D, is reduced below its value in pure water, D_o, for two reasons. First, the large and slowly moving DNA obstructs the motion of the water molecules, which must take longer paths to move from one end of the diffusion apparatus to the other. Second, during the fraction of time that a water molecule is bound in a hydration layer of the DNA, it will move with

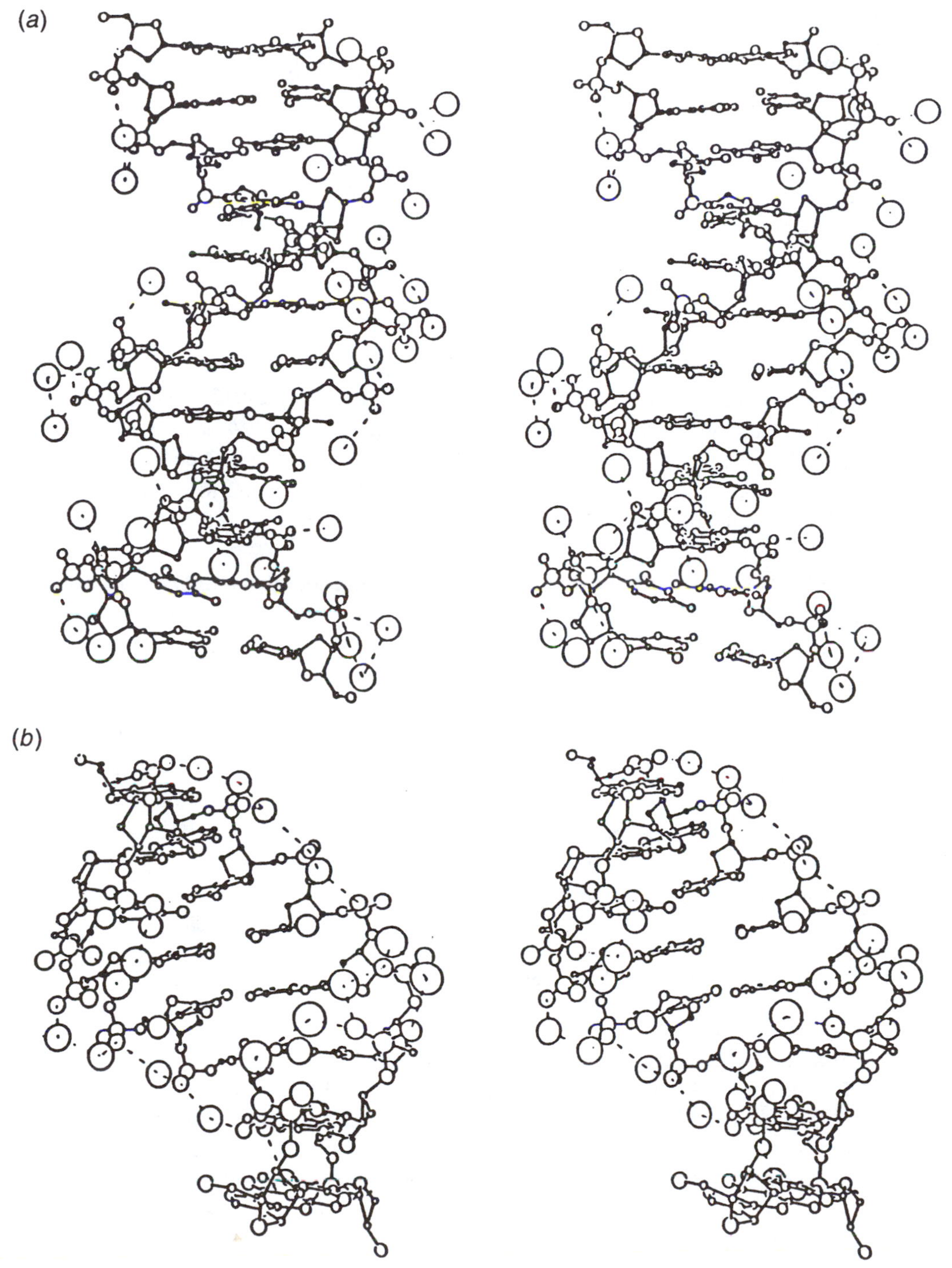

Figure 11-5
Hydration of phosphate groups in B-DNA (*a*) and A-DNA (lower), illustrating hydration economy in A-DNA. [Reprinted with permission from *Nature*, **324**, 385–388 (1986). Saenger W., Hunter, W. N., and Kennard, O. Copyright ©1986 Macmillan Magazines Limited.]

the diffusion coefficent of the DNA, which is negligible compared to D_o. Wang (1955) showed that these two effects could be combined in the approximate expression

$$D/D_o = 1 - [\alpha(V_p\rho_o + \Gamma') + \Gamma' + \Delta_1']w \tag{11-1}$$

where α is a geometrical factor taken as five thirds for an infinitely long, thin rod, V_p is the apparent specific volume of the dry DNA in solution, ρ_o is the density of pure water, Γ' is the hydration in gram H_2O/g DNA, w is the weight fraction of dry DNA, and

$$\Delta_1' = [\alpha\theta V_p\rho_o(1 - V_p\rho_o)w]/[1 - (1 - V_p\rho_o)w] \tag{11-2}$$

With an apparatus that allowed measurement of the diffusion of $H_2{}^{18}O$ through a capillary tube in DNA solutions with w ranging from 0 to 0.124, Γ' was determined to be 0.35 g/g, a value that did not depend significantly on the NaCl concentration. This value corresponds to about 6.5 H_2O per nucleotide, an amount lower than the 18–20 H_2O per nucleotide determined gravimetrically but higher than the amount localized crystallographically.

Analysis of the dependence of the sedimentation coefficient of DNA on molecular weight, as described in Chapter 9 (Gray et al., 1967), yields a hydrodynamic diameter of the DNA molecule of 26–27 Å, compared with the phosphate–phosphate diameter of 20 Å. The difference corresponds to one layer of hydrodynamically immobilized water on the outside of the helix, or about 14.5 H_2O per nucleotide. However, analysis by Kuntz and Kauzmann (1974) of the flow patterns near the surface of spheres whose radius is not very much larger than that of a water molecule casts some doubt on the applicability of conventional hydrodynamic theory, using "stick" boundary conditions, since much of the predicted flow disturbance occurs in a fraction of a water diameter. Also, as discussed in Chapter 9, hydrodynamic measurements on short, double helical oligonucleotides give a diameter closer to 20 Å.

1.4.2 *Unfrozen Water Near Nucleic Acid Surfaces*

Water near the surface of the double helix has different thermal and spectroscopic properties. Heat capacity studies of DNA in the presence of different amounts of water (Privalov and Mrevlishvili, 1967) show that 9–10 H_2O per nucleotide can be added without evidence of a liquid–solid phase transition as the temperature is dropped below 0°C. Some evidence of exothermic fusion was observed in samples containing more than 14 waters per nucleotide, and icelike freezing, at −1 to −4°C, was seen above 19 waters per nucleotide. Before the influence of the DNA on the water–ice transition is no longer observed, 20–30 waters per nucleotide must be added.

When comparing the OH stretching vibrational spectra of HDO in DNA, as a function of relative humidity, with that in liquid water, Falk et al. (1970) found no ice-like ordered water in the hydration shell of DNA. The distributions of hydrogen-bond strengths are similar for most of the water of hydration and for liquid water. When a film of partially deuterated DNA of low water content ~ 9 HDO per nucleotide, is slowly cooled to as low as −150°C, there is a continuous change in the OD stretching band of HDO, indicating no freezing of the adsorbed water. For a similar film with 14 HDO per nucleotide, there is an abrupt change in band shape between −10 and −20°C, indicative of freezing. Thus an inner hydration layer of 9–10 waters per nucleotide cannot crystallize, even when the surrounding water crystallizes to ice I. An intermediate layer of about 3 waters per nucleotide crystallizes only with difficulty. These IR spectra give evidence only of lack of ice-like order, not of mobility of water of hydration. Mobility can be assessed by other spectroscopic techniques.

When aqueous DNA solutions are cooled to −35°C, some of the water has an NMR spectrum narrower than that of ice, but broader than that of supercooled water at that temperature (Kuntz et al., 1969). This shows that the DNA decreases, but does not completely eliminate, the mobility of adjacent water molecules. Integration of the water peak shows that 0.59 g H_2O/g DNA, or 11 water molecules per nucleotide, is affected in this way.

Brillouin scattering is a method of using laser scattering to determine motions in the 10^{-9}–10^{-12} s time scale (1–1000 GHz). Tao et al. (1987) carried out measurements of films of NaDNA and LiDNA at 75% relative humidity (rh) (primary hydration) and 92% rh (secondary hydration). They observed a water rotational relaxation time of 4×10^{-11} s in the primary shell, and 2×10^{-12} s (close to that of pure water) in the secondary shell. Motion in both shells becomes slower as T is lowered from 300 to 140 K. The primary shell relaxation shows simple Arrhenius behavior with an activation energy of about 5 kcal mol^{-1} (compared with E_a of 4 kcal mol^{-1} in pure water). The secondary shell relaxation is more complex, with a larger, T-dependent E_a that may indicate cooperative behavior. The picosecond time scale is one in which significant short-range fluctuating motions can occur in the DNA. Perhaps coupling between DNA and hydration layer motions influences DNA dynamical behavior.

1.5 Theoretical Aspects of Hydration

Some theoretical approaches to nucleic acid structure have been described in Chapter 7. A major challenge to theory is to compute the hydration properties of biomolecules. An overview of the problems has been presented by Finney et al. (1985). One problem is the proper choice of potential functions, especially for water itself and for Coulombic interactions. Another is equilibration, the need to explore the astronomical number of configurations available to the system. Adequate sampling is made much more difficult by the large number of water molecules needed to solvate a single nucleic acid (1951 waters were used in the dodecamer simulation described below).

The crystal structure of the dodecamer d(CGCGAATTCGCG) shows a spine of hydration in the AATT region, with water bridges between N3 of adenine and O2 of thymine. In earlier theoretical work (Subramanian et al., 1988; Subramanian and Beveridge, 1989), the spine of hydration was predicted to extend beyond the AATT region, and only monodentate binding to the acceptor atoms, rather than bridging, was calculated. More recent calculations (Subramanian et al., 1990a) removed most of these discrepancies. The earlier calculations differed from the later in several subtle but significant ways: The Monte Carlo calculation was started from the canonical B form, the partial charges on the DNA were taken from AMBER 3.0, and the TIP4P water model was used. The newer calculation started from the experimentally determined crystal structure, used the GROMOS force field for DNA, and employed the SPC model for water. These changes led to excellent agreement with experiment. It appeared that the choice of potentials was more important than the choice of starting structure, showing that predicted hydration behavior is very sensitive to theoretical assumptions, and underlining the urgency of establishing reliable potentials (particularly for water and charged groups).

Another study (Subramanian et al., 1990b) compared the Monte Carlo predictions of hydration of r(GpC)2 with the crystallographically ordered water sites. This small oligonucleotide crystal is a stringent test, since it has been solved to $R < 15\%$ and resolution better than 1 Å, with a large number of waters located (Berman et al., 1988). The calculation combined the TIP4P water potential with the AMBER 3.0 force field. Theory and experiment were generally consistent, with two exceptions that are rationalized in the paper.

2. POLYELECTROLYTE BEHAVIOR

Nucleic acids are highly charged polyanions, with one negative charge per phosphate. They therefore interact strongly with ions, particularly cations, in the intracellular environment. Some typical values for the molalities of major cations and anions in eukaryotic and prokaryotic cells are given in Table 11-1. These values vary widely depending on cell type and growth conditions. However, this wide variability is suggestive of the great potential for regulation of biological processes by changes in ion concentrations.

A great deal of experimental and theoretical effort has been devoted to understanding the polyelectrolyte behavior of nucleic acids and its consequences for biologically relevant behavior. Such consequences include strong effects on the binding of charged ligands such as proteins and drugs, on the equilibria and rates of helix–coil and helix–helix transitions, and on the condensation and packaging of DNA and RNA. These consequences are the result of direct Coulombic interactions, and also of indirect effects such as the salt dependence of DNA flexibility and the influence of ions on water structure and hydration.

While we shall go into more detail later in this chapter, it is worthwhile to focus our attention on a few major points at the outset. There are three major types of ionic interactions that must be considered in nucleic acid solutions. Two of these are already familiar: specific ion binding (e.g., of a protein or drug to a specific site on a DNA molecule), which generally involves a substantial nonionic component; and interactions of the Debye–Hückel type involving the diffuse ion atmosphere. A third, which is essentially unique to highly charged linear polyelectrolytes such as DNA, is territorial binding of condensed counterions. Such ions are constrained to remain within a few angstroms of the DNA surface, but are free to move parallel to the double helix. Analysis of counterion condensation has led to an important basic concept: Reactions involving a net change in the number of bound ions are often strongly driven by the entropy change attending uptake or release of ions from the territorial binding domain.

Some useful monographs that treat polyelectrolyte behavior in a broader context are those by Oosawa (1971), Eisenberg (1976), Hara (1992), and Schmitz (1993).

Table 11.1
Intracellular Ionic Concentrations (m*M*)

Ions	Human Blood Plasma[a]	Skeletal Muscle[a]	Cultured Animal Cells[b]	*E. coli* Cytoplasm[cd]
Cations				
Na^+	150	14		
K+	5	150		230–930[d]
Mg^{2+}	0.9	8		23
Ca^{2+}	2.5	1		
$Putrescine^{2+}$			0.25–2.5	21
$Spermidine^{3+}$			1–2	6
$Spermine^{4+}$			1–2	
Anions				
Cl^-	105	16		Low
HCO_3^-	27	10		
$Glutamate^-$				30–260[d]
$Protein^-$	17	50		
Other	6	146		

[a]Metzler, 1977.

[b]Cohen, 1971. Polyamine levels vary widely in eukaryotes depending on tissue type and growth state.

[c]Kuhn and Kellenberger, 1985.

[d]Richey et al., 1987. Values for K^+ and $glutamate^-$ depend on extracellular osmolality.

2.1 Counterion Condensation

Counterion condensation theory provides a readily understood physical picture and a means for quantitative interpretation of the interactions between a polyion and its counterions (Manning, 1977, 1979). Like most theories, counterion condensation theory works with a simplified model. In reality, there are discrete charges on a polyion (one charge on each backbone phosphate), the polyion and small ions have finite radii, the polymer chain is somewhat flexible, and the molecular nature of water leads to dielectric saturation and structure-related effects.

2.1.1 Theory

Counterion condensation theory makes several idealizing approximations. The real polyelectrolyte chain, of length L with Z charged groups of valence Z_p (-1 for polynucleotides) is replaced by a continuous line charge with linear charge density $\beta = Z_p q/(L/Z) = Z_p q/b$ where q is the magnitude of the elementary charge (4.8×10^{-10}

esu or 1.6×10^{-19} C) and b is the linear charge spacing. Treating the polymer as a straight line or cylinder is satisfactory so long as the Debye–Hückel screening length $1/\kappa$ is less than the persistence length (Chapter 9). The parameter κ is defined by the equation

$$\kappa^2 = (4\pi q^2/\varepsilon kT) \sum z_i^2 n_i = (8\pi q^2/\varepsilon kT)I \tag{11-3}$$

where I is the ionic strength, $\frac{1}{2}\sum z_i^2 n_i$, and the sum runs over all ionic species i. Numerically, for water at 25°C, if I is expressed in molar units, $\kappa = 3.31 \times 10^7\ I^{1/2}\ \mathrm{cm}^{-1}$. (The ionic strength should include the contribution from the uncondensed counterions associated with the DNA. This contribution is generally important only at very low salt or high DNA concentration.) When the persistence length of DNA is about 500 Å, the approximation of treating DNA as a line charge will hold so long as the ionic strength is above 3.7×10^{-5} M. Further well-justified approximations are that interactions between polyions are neglected, and the dielectric constant ε is taken as that of pure bulk solvent (78.3 for water at 25°C).

The fundamental assumption of counterion condensation theory is that, if the charge density parameter

$$\xi = q^2/\varepsilon kTb = b_j/b \tag{11-4}$$

is greater than unity, sufficiently many counterions will "condense" on the polyion to lower the effective value of ξ to unity. The Bjerrum length is b_j, or the distance at which the Coulomb energy between two unit charges equals the thermal energy. These condensed counterions are territorially bound: confined close to the polyion backbone, but free to translate along it. The counterion condensation lowers the electrostatic potential of the polyion so that the uncondensed mobile ions may be treated in the Debye–Hückel approximation. These ideas are sketched in Figure 11-6. The main result is that, for $\xi > 1$, θ_N counterions of valence N are territorially bound per polyion

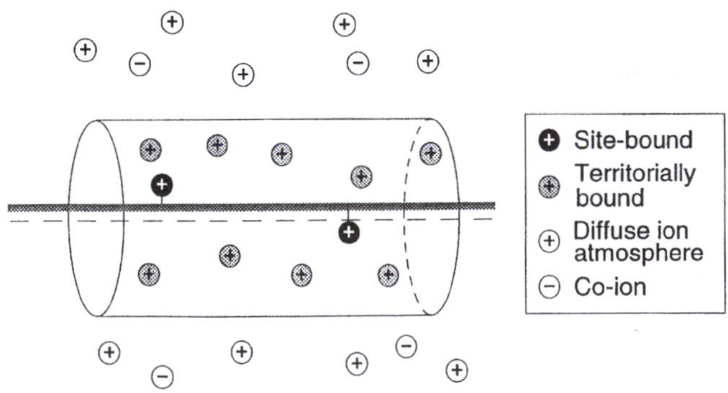

Figure 11-6
Types of ionic interactions in polyelectrolyte solutions, according to counterion condensation theory.

charge, so that the charge is reduced by the factor $r = N\theta_N = 1 - 1/N\xi$. The condensed ions are confined to a volume V_p, extending a radius Δx from the polyion surface, and have a local concentration c_b. For double-stranded DNA with $b = 1.7\,\AA$, $\xi = 4.2$, so strong counterion condensation occurs. Numerical results for ions of different valence are shown in Table 11-2.

Table 11.2
Counterion Condensation Parameters for Double-Stranded B-DNA

N	θ_N	r	V_p(cm^3/equiv)	Δx (Å)	$c_b(M)$
1	0.76	0.76	647	7.3	1.18
2	0.44	0.88	1121	11.2	0.39
3	0.31	0.92	1563	14.2	0.20
4	0.24	0.94	1995	16.8	0.12

2.1.2 Experimental Verification

Experimental measurements of ion behavior in nucleic acid solutions tend to verify the predictions of counterion condensation theory. Perhaps the most direct test is NMR studies of ^{23}Na quadrupolar relaxation rate in DNA solutions. The relaxation rate, which is determined through measurements of line-widths or of relaxation times T_1 and T_2, is sensitive to the electric field gradient at the position of the Na nucleus, and to the rotational relaxation time of the nuclear spin of the ion. Since different gradients and relaxation times will presumably be found close to the DNA phosphate backbone and in bulk solution, it should be possible to determine the fraction of counterions in each environment. Such studies (Bleam et al., 1980), show that Na^+ ions associated with DNA are not dehydrated or immobilized, confirming the basic applicability of long-range electrostatic theory. They further show, in corroboration of counterion condensation theory, that the fractional extent of neutralization θ_1 of DNA is independent of the bulk salt concentration over a wide range from 3 mM to 1.3 M. The value of θ_1 obtained was 0.75 ± 0.1, in very good agreement with the predicted 0.76. A similar conclusion is reached by Braunlin et al. (1987) and Padmanabhan et al. (1988). However, experiments reveal small but reproducible differences in the affinity of various monovalent cations for DNA, in the order $NH_4^+ > Cs^+ > K^+ > Li^+ > Na^+$, showing a nonelectrostatic component of the binding. Moreover, there are various complexities in the interpretation of quadrupolar relaxation that are not well understood, and it must be realized that the values of θ_1 are computed assuming the validity of the two-state (territorially bound/free) model.

Similar confirmation of the predictions of counterion condensation theory has been obtained for the divalent ion Mn^{2+} by NMR measurement of the proton relaxation enhancement of water (Granot and Kearns, 1982a,b). The extent of Mn^{2+} binding per DNA phosphate was measured as 0.43 ± 0.04, compared with predicted 0.44, and was independent of Mn concentration from 2.8×10^{-5} to $2.1 \times 10^{-3} M$.

The discontinuity at $\xi = 1$ predicted by counterion condensation theory is also noted with other polyions (where the charge density parameter can be changed) in thermodynamic measurements of ion activity coefficients, osmotic coefficients, and Donnan distribution coefficients; and in an abrupt change in electrophoretic mobility.

Counterion condensation theory in its standard form is derived for an infinitely long, uniformly spaced line of charges on the polyion. The requirement of infinite length is, in fact, the requirement that the polyion be much longer than the Debye length κ^{-1}. At an ionic strength of 0.1 *M* in water at 25°C, κ^{-1} is 9.6 Å; at 1 m*M*, 96 Å. Thus under physiological salt concentrations the basic assumption is well satisfied even for short oligonucleotides. It breaks down only at very low salt or for very short oligonucleotides. For example, Fenley et al. (1990a) calculated that in 0.01 *M* NaCl, the fractional condensation θ around a 20 bp oligomer is 0.757, compared with the limiting value of 0.764. For 10 bp, $\theta = 0.728$. Only for a very short a 5-mer oligonucleotide, does θ drop significantly, to 0.614. Deviations are greater at lower salt. For oligomers longer than 20 bp, θ approaches its limiting value as a linear function of 1/bp (see also Olmsted et al., 1989).

2.1.3 More Detailed Theories of Ionic Spatial Distribution

While counterion condensation theory allows simple and surprisingly accurate calculation of many polyelectrolyte phenomena, the physical model on which it rests appears rather oversimplified. Therefore, much theoretical effort has been devoted to exploring its validity and limitations. These efforts have been reviewed by Anderson and Record (1982, 1990). Poisson–Boltzmann (PB) theory, developed in Appendix A-2, removes one of the simplifications, attributing to DNA a finite cylindrical radius rather than a line charge; but it retains other assumptions, such as a continuous charge density and point charge small ions. Analytical solutions of the PB equation are not possible except in the limit of no added salt; but numerical solutions have been obtained, along with certain general characteristics of the equation (Schellman and Stigter, 1977; Stigter, 1977). Useful tables of numerical solutions have been given by Stigter (1975, 1982).

The PB calculations show that the gradient of counterion concentration near the surface is steep, but not discontinuous as in counterion condensation theory. The surface concentration of counterions is not invariant to bulk salt concentration, and integrating the PB counterion concentration out to the radius implied by V_p gives a lower concentration than predicted by counterion condensation theory (Zimm and Le Bret, 1983).

The PB theory itself is based on some significant approximations, notably neglect of both the finite size and spatial correlations of the small ions. These effects have been investigated both analytically and by grand canonical Monte Carlo (GCMC) simulations. It appears that these approximations roughly cancel each other, so the PB equation is valid up to about 0.1 *M*. A notable result of both PB and GCMC calculations is that, if the integration of counterion concentration is carried out to the radius at which the average electrostatic potential energy is $-k_B T$, rather than to the radius specified by V_p, the amount of counterion contained within that radius is independent of salt concentration. In fact, the GCMC value for the charge neutralization comes very close to the value predicted by counterion condensation theory (Lamm et al., 1994).

The GCMC calculations give important insight into end effects on ion distributions surrounding oligomeric DNA (Olmsted et al., 1989). Figure 11-7 (*a*) shows the surface concentration of monovalent counterions as a function of distance from one end of an *N*-mer, with *N* varying from 8 to 72. The interior concentration does not reach a limiting value until 18–20 phosphates in from one end. The average concentration is a linear function of $1/N$ for $N > 24$ [Fig. 11-7(*b*)], but the end effects do not become negligible until $N \gg 100$. We shall discuss the implications of these results for helix–coil transitions and ligand binding below.

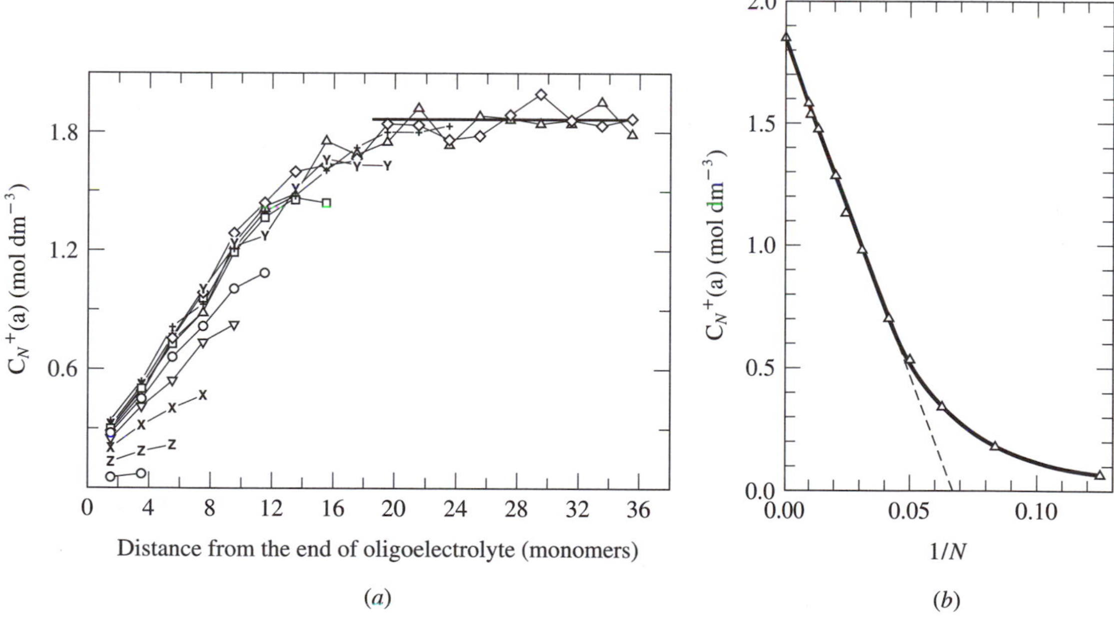

Figure 11-7
(a) Surface concentration of monovalent cation, modeled as a 3-Å sphere, as a function of distance (in nucleotides) from one end of an oligomeric B-DNA with (from bottom to top) $N = 8$, 12, 16, 20, 24, 32, 40, 48, 72, and 100 nucleotides. The solid line is the limiting value for an infinitely long DNA. Grand canonical Monte Carlo calculations were done with a mean ionic activity coefficient for the salt of $a_\pm = 1.76$m*M*, and a concentration of DNA nucleotides of $C_u = 2.49$m*M* for all *N*. (*b*) Average surface concentration of cations for the oligomers in (*a*), showing a linear dependence on $1/N$ for $N > 24$. [With permission from Olmsted et al. 1989.]

At salt concentrations higher than 0.1 M the situation becomes very complicated, and it is essential to take small ion effects propertly into account. Such a theory is necessary, for example, to understand the transition from B- to Z-DNA provoked by high salt (see below). However, passage from a continuum to a detailed molecular statistical theory should probably also take into account effects of ions and high electric field on the local molecular arrangement of water. Most calculations have treated water as a continuum with a dielectric constant ε equal to its bulk value; while the orientation of water dipoles around highly charged ions would be expected to lead to dielectric saturation effects, resulting in a reduction of ε. Saturation has been investigated by Jayaram et al. (1990), who showed that it has significant effects on the total energy of the DNA–water–counterion system, the internal energy of counterion binding, and the local counterion distribution. Counterion condensation is enhanced relative to the bulk dielectric model, but the net result is that the fraction of condensed counterions is predicted to be independent of salt concentration just as in the primitive counterion condensation model. Dielectric saturation also appears necessary to maintain this independence of θ on bulk salt in the counterion condensation model, if the line charge distribution is replaced with a double–helical array in which the point charges are placed at the DNA phosphate coordinates (Fenley et al., 1990a).

2.1.4 Interaction of Residual Charge with Ion Atmosphere

While simple counterion condensation theory seems to predict adequately the extent of counterion condensation, it does not consider the interaction of the residual polyion charge with the surrounding ion atmosphere. This has been done by Manning (1972) and Record (1975), who showed that this interaction is thermodynamically equivalent to the binding of an additional $(2N\xi)^{-1}$ of a counterion per polyion charge. One then defines a thermodynamic binding parameter Ψ, related to the condensation binding parameter θ by

$$N\Psi_N = N\theta_N + (2N\xi)^{-1} = 1 - (2N\xi)^{-1} \qquad (11\text{-}5)$$

With monovalent counterions, $\Psi = 0.88$ for double helical B-DNA and 0.70 for single-stranded DNA. Thermodynamic measurements such as helix–coil transitions and ligand binding are appropriately analyzed in terms of Ψ rather than θ.

2.1.5 Mixtures of Counterions

Many experiments on nucleic acids are conducted in mixtures of salts, for example, $MgCl_2$ and NaCl, where the aim is to understand the interaction of the higher valent counterion with the polynucleotide in the presence of an excess of univalent salt (Manning, 1978). Under these circumstances it is accurate to take the ionic strength equal to the salt concentration c_1, so that κ is determined just by the univalent ion concentration. A simple approach at low occupancy by N-valent counterions is then to assume that the total degree of neutralization remains constant, that is that binding of an N-valent ion displaces N univalent ions, or that

$$\xi^{-1} = (1 - \theta_1 - N\theta_N) \tag{11-6}$$

This assumption, which relates θ_1 and θ_N, makes it possible to minimize the free energy in terms of the single variable θ_N. An additional assumption of this one-variable theory is that V_P, the volume of the territorially bound region (see Table 11-2) is equal to that for 1:1 salts. Minimizing the total free energy with respect to θ_N leads to an expression for the equilibrium constant for the transfer of the N-valent ion from free solution to the territorially bound region:

$$K_N \equiv \theta_N / c_N \tag{11-7}$$

$$\log K_N = \log(V_P/1000e) + N \log[(1000e/V_P)(1 - \xi^{-1} - N\theta_N)c_1^{-1}] \tag{11-8}$$

This expression makes it particularly easy to predict the salt dependence of K_N:

$$d \log K_N / d \log c_1 = -N \tag{11-9}$$

As we shall see later, consideration of screening interactions between the residual charge density and the ion atmosphere reduce the slope from $-N$ to $-N\Psi_N$.

Manning's (1978) two-variable theory avoids these approximations and gives somewhat more accurate results, at the cost of more complicated calculations. Minimizing the free energy with respect to both θ_N and θ_1 gives the pair of equations (Wilson et al., 1980)

$$1 + \ln(1000\theta_1/c_1 V_{P1}) = -2\xi 1 - \theta_1 - N\theta_N) \ln(1 - e^{-kb}) \tag{11-10}$$

$$\ln(\theta_N/c_N) = \ln(V_{PN}/1000e) + N \ln[(1000e/V_{P1})c_1^{-1}\theta_1] \tag{11-11}$$

which may be solved iteratively. The predictions of this theory have been confirmed by gel electrophoresis of DNA in ion mixtures (Ma and Bloomfield, 1995).

2.2 Effects of Nonspecific Ion Interactions on Nucleic Acid Behavior

With this understanding of the interaction of counterions with the highly charged backbone of polynucleotides, we move to consider the effects of salt concentration on some important nucleic acid reactions, such as helix–coil transitions, helix–helix transitions, and binding of charged ligands. These are all phenomena that will be treated in more detail elsewhere in this book: Chapter 8 for transitions, Chapters 12 and 13 for binding. At this point, our aim is to understand how they may be influenced by salt concentration and conversely, how salt-dependence may give insight into mechanism.

2.2.1 General Principles

The fundamental insight is that the high local concentration of territorially bound counterions potentiates reactions that change the charge density of the polynucleotide backbone. The maintenance of such a high local concentration is entropically unfavorable, since an entropy gain of

$$\Delta S = R \ln(c_{\mathrm{loc}}/c_{\mathrm{bulk}}) \tag{11-12}$$

would be realized per mole of territorially bound ions released into bulk solution. Such release is prevented by the high electrostatic field at the surface of the nucleic acid; counterion condensation theory is derived by balancing these two factors. If a reaction changes the linear charge density of the backbone, condensed counterions will be released or bound, and the thermodynamics of the reaction is largely governed by the attendant entropy change.

In reality, the equilibria of nucleic acid reactions are influenced not only by counterions, but also by co-ions and water. Consider a very general type of reaction in which the polynucleotide helix H binds n_{L} moles of ligand L / mol bp to give H', with the simultaneous displacement of Δn_{M+} mol cation (counterion), Δn_{X}^{-} mol anion (coion), and Δn_{w} mol water,

$$H + n_{\mathrm{L}}L \rightleftharpoons H' + \Delta n_{\mathrm{M}_+}\mathrm{M}^+ + \Delta n_{X-}X^- + \Delta n_{\mathrm{w}}\mathrm{W} \tag{11-13}$$

The thermodynamic equilibrium constant for this reaction is

$$K = [a'_{\mathrm{H}}][a_{\mathrm{M+}}]^{\Delta n\mathrm{M}^+}[a_{\mathrm{X-}}]^{\Delta n_{\mathrm{X-}}}[a_{\mathrm{W}}]^{\Delta n_{\mathrm{W}}}/[a_{\mathrm{H}}][a_{\mathrm{L}}]^{n_{\mathrm{L}}} \tag{11-14}$$

where the a_{i} values are activities of species i. The observed equilibrium constant K_{obs} is

$$K_{\mathrm{obs}} = [\mathrm{H}']/[\mathrm{H}][a_{\mathrm{L}}]^{n_{\mathrm{L}}} \tag{11-15}$$

where we have replaced the polynucleotide activities with concentrations. Thus,

$$K_{\mathrm{obs}} = K/[a_{\mathrm{M^+}}]^{\Delta n_{\mathrm{M+}}}[a_{\mathrm{X^-}}]^{\Delta n_{\mathrm{X^-}}}[a_{\mathrm{W}}]^{\Delta n_{\mathrm{W}}} \tag{11-16}$$

The dependence of K_{obs} on counterion concentration can then be found by logarithmic differentiation:

$$\partial \ln K_{\mathrm{obs}}/\partial \ln a_{\mathrm{M+}} = -\Delta n_{\mathrm{M^+}} \tag{11-17}$$

Similar expressions are obtained for the dependence on co-ion and water, but these have been less well studied. A more rigorous treatment of this theory is presented in the review by Record et al. (1978).

2.2.2 Helix–Coil Transitions

The thermal helix–coil transition can be written in the form above with $n_L = 0$, since the transition is provoked by heat rather than ligand binding. The single-stranded coil has a lower charge density (larger charge spacing b) than the double-stranded helical B-form (Table 11-3), so counterions will be released in going from helix H to coil C (= H′). The entropy gain attending this release will be greater at lower salt concentration, so the helix will be stabilized relative to coil as the salt concentration increases.

The number of counterions released per cooperative unit denatured will be

$$\Delta n_{M^+} = n_{M^+,H} - n_{M^+,C} = N_u(\theta_H - \theta_C) = N_u \Delta\theta \tag{11-18}$$

where $n_{M^+,H}$ and $n_{M^+,C}$ are the number of condensed counterions per phosphate in the H and C forms, and N_u is the number of phosphates per cooperative melting unit. One must also take into account the residual screening interaction between the unneutralized charge density and the ion atmosphere. This leads to the replacement of $\Delta\theta$ by $\Delta\Psi$ (Eq. 11-5), or to

$$\Delta n_{M^+} = (N_u/2)(\xi_C^{-1} - \xi_H^{-1}) = [N_u/2b_j](b_C - b_H) \tag{11-19}$$

Thermodynamic manipulation of these equations (Record et al., 1978) finally leads to an equation for the dependence of the melting temperature T_m on the mean ionic activity $a_\pm$ of the salt:

$$S_\infty = dT_m/d \ln a_\pm = \alpha\beta\Delta\psi \tag{11-20a}$$

Table 11.3
Counterion Condensation Parameters of DNA and RNA with Univalent Cations

Polynucleotide	b (Å)	ξ	θ_1	V_P(cm^3 equiv−1)	Δx (Å)	c_b (M)
B- DNA	1.7	4.2	0.76	647	7.3	1.18
A-DNA	1.3	5.5	0.82	406	6.3	2.02
Z-DNA	1.9	3.8	0.73	778	7.8	0.94
ssDNA	4.3	1.7	0.40	2160	12.0	0.18
PolyA • poly(U)	1.6	4.5	0.78	583	7.1	1.33
Poly(A)	3.1	2.3	0.57	1600	12.0	0.35
Poly(U)	4.5	1.6	0.37	2200	11.0	0.17
Poly(A) • 2poly(U)	1.0	7.1	0.86	253	5.3	3.41

$\beta = RT_m^2/\Delta H^\circ_{obs}$, ΔH°_{obs} is the observed enthalpy of the transition per mole of nucleotides, and α is a correction factor for activity coefficients. The subscript on S_∞ indicates that Eq. (11-20) pertains to polynucleotides long enough that end effects are not important.

Experimentally, β is measured calorimetrically as $50 \pm 2°$ independent of salt concentration and temperature, and $dT_m/d\ln a_\pm = 8.9 \pm 0.2$ for [NaCl] between 2.5 and 100 m*M* (Fig. 11-8). The parameter α is about 0.95 in such solutions. Combining these values gives $\Delta\Psi = 0.19 \pm 0.02(\Delta\theta = 0.37 \pm 0.03)$, corresponding to the average charge spacing in the coil form, $b_C = 4.3 \pm 0.2$ Å. Such information on the conformation of single-stranded polynucleotides is difficult to obtain except from such salt-dependent melting studies.

For oligonucleotides, the lower ion concentrations near the ends have a significant effect on the ionic dependence of thermal melting. For an N-mer undergoing a transition to a form containing yN nucleotide units ($y = 1$ for hairpin helices and $y = \frac{1}{2}$ for dimer helices), $dT_m/d\ln a_\pm$ can be written (Olmsted et al., 1991)

$$S_{N\to yN} = -\frac{RT_m^2 N(2\Delta\Gamma_{N\to yN})}{M\Delta H^\circ_{obs}} \tag{11-20b}$$

M is the number of nucleotides that contribute to the enthalpy change upon denaturation. It can be approximated as $N - X$, where X is the number of unstacked bases in the helical form. The parameter Γ is the preferential ionic interaction parameter defined in the

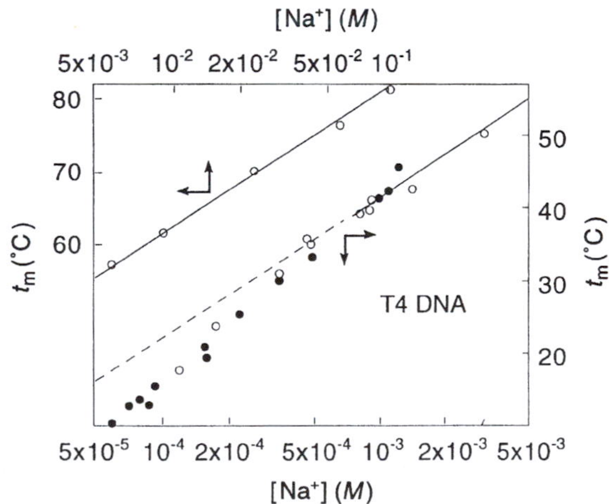

Figure 11-8
Dependence of T_m on Na^+ concentration for T4 DNA. [Effects of Na^+ and Mg^{++} ions on the helix–coil transition of DNA, Record, M. T., Jr., *Biopolymers*, **14**, 2137–2158. Copyright ©1975. Reprinted by permission of John Wiley & Sons Inc.]

Appendix Section A-1, and $\Delta\Gamma - N \rightarrow yN$ is the change in Γ during the transition. The parameter $2\Delta\Gamma_{N\rightarrow yN}$ is the difference in the average number of univalent counterions thermodynamically associated per oligomer charge, and is equal to $\Delta\Psi$ defined above.

Plots of Γ as a function of $1/N$ for B form and single-stranded DNA, calculated by the grand canonical Monte Carlo method, are shown in Figure 11-9. The convergence of the two plots as N decreases implies that the T_m of short hairpins (where $y = 1$ and both forms have the same N) should depend less on salt concentration than the T_m of long polynucleotides; this agrees with experiment. However, since N is halved in denaturation of a dimer duplex ($y = \frac{1}{2}$), behavior is more complex. In fact, the theory explains the surprising observation (Braunlin and Bloomfield, 1991) that dissociation of $d(GGAATTCC)_2$ has virtually the same salt dependence as that of polymeric DNA.

The salt-dependence of T_m can be used to estimate the number of unstacked bases in a hairpin (Olmsted et al., 1991). Equation (11-20b) leads to

$$\frac{S_{N\rightarrow yN}}{S_\infty} = \frac{N}{N-X}\frac{\Delta\Gamma_{N\rightarrow yN}}{\Delta\Gamma_\infty} \tag{11-21}$$

Figure 11-10 shows data on denaturation of $d(TA)_N$ hairpins (Elson et al., 1970), plotted according to this equation. The slope yields $X = 5.5 \pm 1.2$ bases, in good agreement with four unstacked bases in the loop and one at each end of the helix.

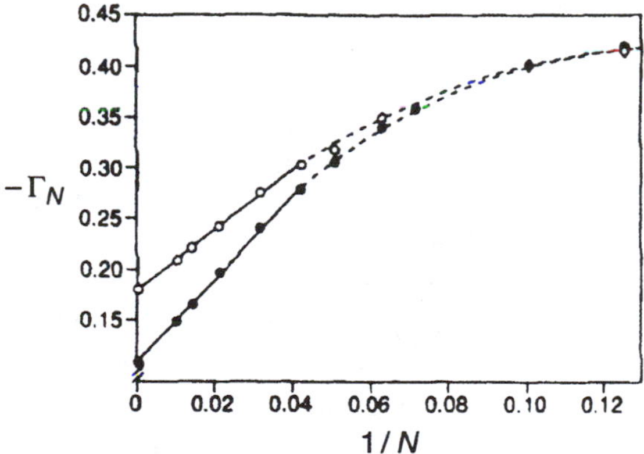

Figure 11-9
Dependence of preferential ionic activity coefficient on reciprocal of number of nucleotide charges, for B-DNA (●) and single-stranded DNA (○) at $a_\pm = 1.76\text{m}M$. [Importance of oligoelectrolyte and effects for the thermodynamics of conformational transitions of nucleic acid oligomers: A grand canonical Monte Carlo analysis, Olmsted, M. C., Anderson, C. F., and Record, M. T., Jr., *Biopolymers*, **31**, 1593–1604. Copyright ©1991. Reprinted by permission of John Wiley & Sons, Inc.].

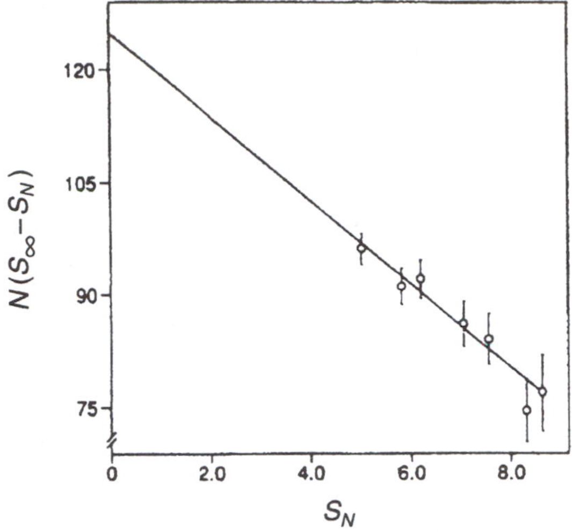

Figure 11-10
Analysis of data on melting of $d(TA)_N$ hairpin oligomers (Elson et al., 1970) according to Eq. 11-20c. The slope $= -X$, giving $X = 5.5 \pm 1.2$ bases. [Importance of oligoelectrolyte and effects for the thermodynamics of conformational transitions of nucleic acid oligomers: A grand canonical Monte Carlo analysis, Olmsted, M. C., Anderson, C. F., and Record, M. T., Jr., *Biopolymers*, **31**, 1593–1604. Copyright ©1991. Reprinted by permission of John Wiley & Sons, Inc.].

2.2.3 Helix–Helix Transitions

In principle, transitions between double helical conformations H and H', which differ in linear charge density, can be treated in exactly the same way. In practice, however, such transitions involve factors, such as hydration or binding of small amounts of multivalent cations, in addition to simple electrostatics. For example, Z-DNA is more extended than B-DNA (Table 11-3). The higher charge density of B-DNA suggests that the B form should be stabilized relative to the Z form by increasing salt. In reality, however, the transition from the B to the Z form of poly(dG-dC) is provoked by high salt. This behavior is inexplicable in terms of the simple forms of counterion condensation or PB theory, and requires a more sophisticated treatment.

Such a theory has been developed by Soumpasis et al. (1984, 1987). It treats the DNA phosphates as discrete charges of finite radius, fixed at positions specified by the equilibrium geometry of the helical form as determined by X-ray diffraction. The mobile ions are treated as spheres, whose radius is adjusted for best fit with experiment. The solvent is treated as a continuum, with dielectric constant equal to that of bulk water. Modern statistical mechanical techniques were applied to this so-called "restricted primitive model," which has allowed successful treatment of ionic solutions at concentrations well above the realm of validity of the PB or counterion condensation theories, through proper consideration of ionic correlations and excluded volume effects. Figure 11-11 shows calculated free energies, in units of thermal energy (kT), of dodecamers of B, Z, and A forms of DNA up to 5 M univalent salt. The theory successfully predicts the B-Z_I transition in d(G-C)•d(G-C) (that is, $G_Z < G_B$) above 2.2 M salt, with a reasonable hydrated ionic radius of 4.9 Å. It also predicts that A-form DNA will be more stable than B-form DNA above 1.8 M salt, a transition for which there is evidence in poly(d^{n2}A-T) and in poly(dG)•poly(dC). However, it also predicts that the Z_{II} form is more stable than B, and the C conformation the most stable of all (calculation not shown) at low salt concentrations. These predictions are clearly

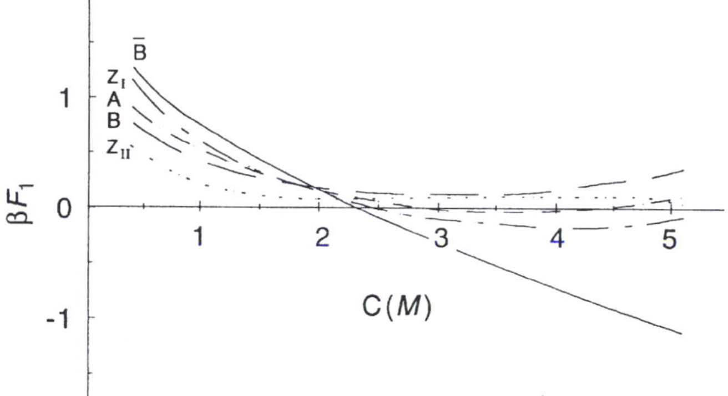

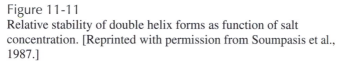

Figure 11-11
Relative stability of double helix forms as function of salt concentration. [Reprinted with permission from Soumpasis et al., 1987.]

in disagreement with experiment. It is likely that one root of these disagreements is an inadequate treatment of hydration. Until realistic treatments at high salt of both ionic behavior and hydration are combined in the same theory, such theories must be regarded with skepticism.

Similar results are obtained from a counterion condensation calculation with realistic helical geometry of the phosphate charges, if the constant dielectric model is used (Fenley et al., 1990b). However, if dielectric saturation is considered (which, as shown above, is necessary to obtain agreement with experiment for the helical model), the electrostatic behavior of Z_I and Z_{II} become quite different. This result again emphasizes that water structure near the helix (which is responsible for dielectric saturation) and the interactions of such water with condensed or bound counterions must be better understood before electrostatic theories of helix stability can be considered reliable.

2.2.4 Ligand Binding

Equation (11-13) applies directly to the binding of a ligand to a nucleic acid. If the ligand is cationic with charge Z^+, it will displace Z univalent counterions. The residual screening interaction with the diffuse ion atmosphere then leads to $\Delta n_{M^+} = Z\Phi$ and to

$$\partial \ln K_{obs}/\partial \ln a_{M^+} = -Z\Psi \tag{11-22}$$

Figure 11-12 shows the salt dependence of the binding constants of putrescine^{2+}, spermidine^{3+} and spermine^{4+} to DNA (Braunlin et al., 1982). The slopes are 1.7, 2.5, and 3.3 respectively, leading to Ψ values of 0.85, 0.83, and 0.82. These values are in good agreement with the expected 0.88.

These effects can be very large in protein–nucleic acid complexes. Figure 11-13 shows the salt dependence of the binding constant of the *lac* repressor to DNA. The log–log plot has a slope of 10. The value of $\Psi = 0.88$ corresponds to the displacement of 11

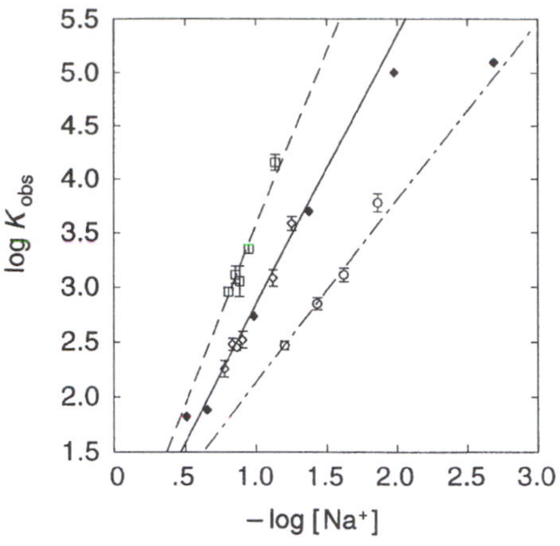

Figure 11-12
Salt dependence of binding of putrescine^{2+} (□), spermidine^{3+} (◇,◆), and spermine^{4+} (○) to DNA. [Equilibrium dialysis of polyamine binding to DNA, Braunlin, W. H., Strick, T. J., and Record, M. T., Jr., *Biopolymers*, **21**, 1301–1314. Copyright ©1982. Reprinted by permission of John Wiley & Sons, Inc.]

sodium ions. Since two groups on the repressor must be protonated for binding to occur, and since protonation and Na^+ release are coupled, de Haseth et al. (1977) conclude that 12 ± 2 phosphates are involved in ionic interactions with each protein molecule. Such results suggest that relatively small variations in intracellular ion concentrations may have large effects on DNA–protein binding. As seen from Figure 11-13, an increase in NaCl from 0.10 to 0.16 *M* decreases K_{obs} by two orders of magnitude.

The binding of a highly charged cationic ligand to DNA drastically changes the counterion distribution not only at the binding site, but also for an extended distance to either side. An example of the predicted effect is shown by the GCMC calculations of Olmsted et al. (1995) in Figure 11-14, in which a molecule with eight positive charges is bound to the center of a DNA with 72 nucleotides. The reduction of the counterion concentration extends about 24 bases to either side of the center of the binding site. For longer polynucleotides, interior binding of an oligocationic ligand effectively divides

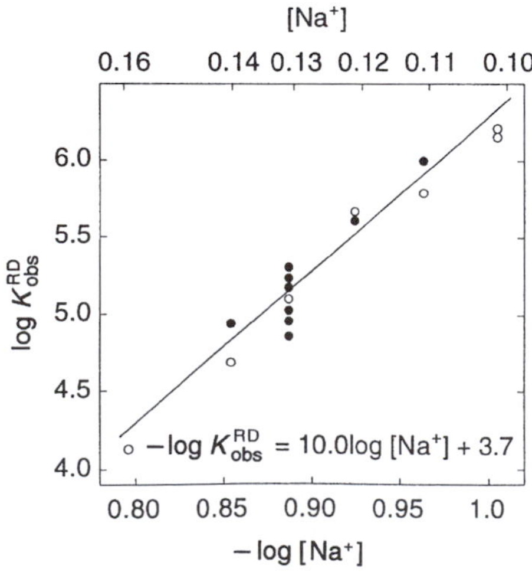

Figure 11-13
Salt dependence of the nonspecific binding of *lac* repressor to DNA. The open and filled circles represent two different preparations, See original article for details. [Reprinted with permission from deHaseth, P. L., Lohman, T. M., and Record, M. T., Jr., *Biochemistry*, **16**, 4785–4790 (1977). Copyright ©1977 American Chemical Society.]

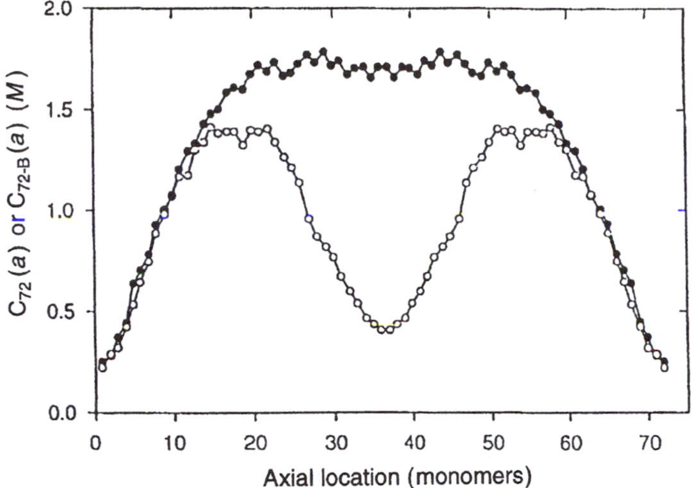

Figure 11-14
The GCMC calculations of the concentration of monovalent counterion along the surface of B-DNA containing 72 bases uncomplexed (●) and with the 8 central charges eliminated to simulate binding of cationic ligand bearing 8 positive charges (○). [Reprinted with permission from Olmsted et al., 1995.]

the DNA into two separate domains as far as counterion binding is concerned. The predicted effect on K_{obs} is shown in Figure 11-15 as a function of DNA length. The salt dependence is small for very short oligomers, but passes through a maximum at intermediate N before declining somewhat to the infinitely long polynucleotide value predicted by Eq. 11-22.

The strong effects of ionic concentrations on protein binding have interesting implications for ionic regulation of gene expression, which have been explored by Record and co-workers (Record et al., 1985; Leirmo et al., 1987). They note that in *E. coli*, the dominant ions are K^+ and glutamate$^-$, and that levels of these ions can vary widely in response to external osmotic conditions (Table 11-1). (*In vitro* experiments show that there is a specific anion effect on binding, in addition to the general electrostatic effects we have been considering: Protein–DNA interactions are considerably stronger in glutamate compared to chloride.) However, the *in vivo* formation of functional protein–operator DNA complexes is only weakly dependent on intracellular ion concentrations, suggesting the existence of compensation mechanisms to buffer against large environmental changes (Richey et al., 1987). One possibility, based on a thermodynamic model of the *lac* operon, is that a large number of nonspecific repressor binding sites moderates the salt sensitivity of binding at the operator site. Another possibility is that macromolecular crowding effects in the concentrated solutions within the cell (Berg, 1990; Garner et al., 1990) may compensate for the activity coefficient changes associated with ion displacement.

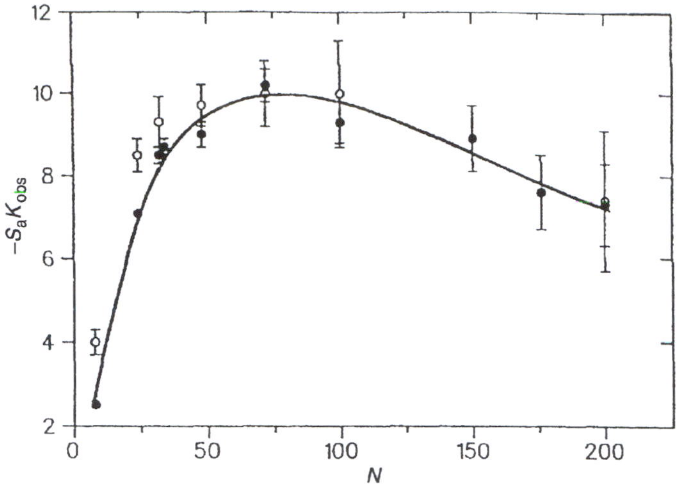

Figure 11-15
GCMC calculations of $-S_aK_{obs} = -dK_{obs}/d\ln a_{\pm}$ for binding of L^{8+} to the center of B-DNA containing N nucleotides, at $a_{\pm} = 1.76mM$ (●) and 12.3 mM (○). The line is drawn to guide the eye. [Reprinted with permission from Olmsted et al., 1995.]

3. HYDRATION IN IONIC SOLUTIONS

We have been considering water and ionic interactions with nucleic acids separately. In fact, they are inextricably interconnected. This interdependence is particularly important in concentrated salt solutions, such as those employed in density gradient experiments for characterization and separation of nucleic acids on the basis of their densities, since the high concentration of salt can markedly influence the water activity. As we shall see, this affects the interpretation of sedimentation coefficients or molecular weights obtained from such experiments. Furthermore, high salt concentration often provokes transitions among polynucleotide conformations, such as the B–Z or B–A transitions of DNA. A question of great mechanistic interest is whether such a transitions is the direct result of ionic interactions, or of the effect of salt on hydration of the DNA.

3.1 Thermodynamic Treatment

The thermodynamic measure of hydration can be written

$$\Gamma_1 = (\partial m_1/\partial m_2)_{\mu_1} \tag{11-23}$$

That is, Γ_1 is the number of moles of water (component 1) that must be added per mole of component 2 (nucleic acid) in order to keep the chemical potential of water constant. It represents, in rough physical terms, the amount of water "bound" by 1 mol of nucleic acid, which thus must be "replenished" in the bulk solution. On a mass instead of a molar basis, one defines the hydration

$$\Gamma_1' = (M_1/M_2)\Gamma_1 \qquad (11\text{-}24)$$

as the number of grams of water that are "bound" per gram of nucleic acid.

Thermodynamically, one cannot distinguish between water "binding" and salt "exclusion," since in a three-component system, according to the Gibbs–Duhem equation, there are only two independent chemical potentials at a given composition. If Γ_3' is the thermodynamic binding parameter for component 3 (salt) on a gram/gram basis, then it is readily shown (Cohen and Eisenberg, 1968) that

$$\Gamma_3' = -\Gamma_1' \frac{(1 - \rho^\circ v_1)}{(1 - \rho^\circ \bar{v}_3)} \qquad (11\text{-}25)$$

where ρ° is the density of the solvent mixture (water + salt) in the absence of polymer, and v_i is the partial specific volume of component i .

Perhaps the most conceptually straightforward experimental measure of preferential hydration is isopiestic distillation. Two vessels, containing identical amounts of water and salt, are placed in a thermostatted container, and a known mass of nucleic acid is added to one of the vessels. Water will distill from the vessel without the nucleic acid to the vessel with the nucleic acid, until the water activities in the two vessels are equal at equilbrium. The mass of water transferred, per gram of nucleic acid added, equals Γ_1'. The parameter Γ_1' can also be determined by high-precision determination of the density change of a water–nucleic acid–salt solution, and by buoyant density determination in sedimentation equilibrium (Hearst and Vinograd, 1961a).

3.2 Results and Molecular Interpretation

These procedures have been used to determine the preferential hydration of DNA in various salts. Typical results are shown in Figure 11-16 for various Cs and Li salts (Hearst and Vinograd, 1961b). Data for Na and K salts fall on the same curve. Strikingly, the hydration depends only on the water activity, not on the specific nature of the cations and anions. Note the similarity, despite the differences of experimental technique and definition, to the gravimetric hydration in Figure 11-1. Tunis and Hearst (1968) have shown, from an analysis of the buoyant density of DNA as a function of base composition, that in CsCl a mole of A-T base pairs binds about 2 mol of water more than a mole of G-C base pairs. It has been speculated that this higher hydration is the reason that AT rich DNA does not undergo a B–A or B–Z transition as water activity is reduced.

Wolf and Hanlon (1975) have combined hydration data with circular dichroism (CD) measurements of the fraction of B-form content of DNA as a function of salt concentration, to develop an interesting model of the role of hydration on the conformation of DNA. As seen in Figure 11-17, the fraction of B relative to (A + C) conformation decreases linearly as Γ goes from fully hydrated (60–80 mol H_2O/mole nucleotides) to about 12–14 moles H_2O/mole nucleotides. Below this critical range of G, the B

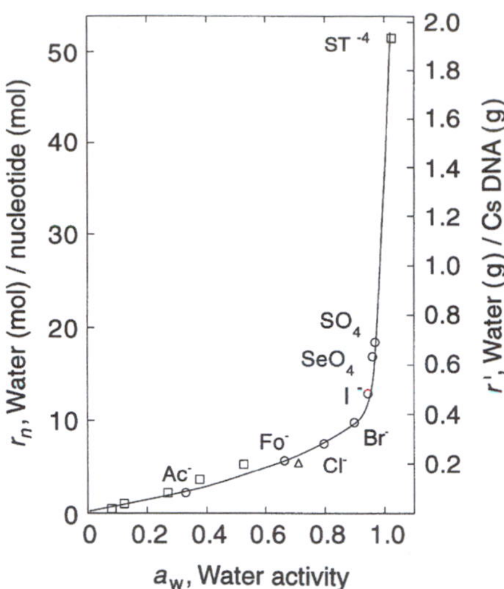

Figure 11-16
The net hydration of DNA from bacteriophage T4 as a function of water activity, for various Cs and Li salts. [With permission from Hearst and Vinograd, 1961b.]

content declines more rapidly. Linear extrapolation suggests that the fully B form must have at least 18 mol H_2O/mol nucleotides, while alkali metal salts of DNA that are completely in the C and/or A conformation have about 4 mol H_2O/mol nucleotides. The ammonium salt retains about 7 mol mol^{-1} in the C or A forms. Connecting these observations with the IR hydration site assignments of Falk et al. (1963) suggests that the entire primary hydration shell of about 18 mol H_2O/mol nucleotides (those water molecules with modified IR properties) is required to maintain the B conformation.

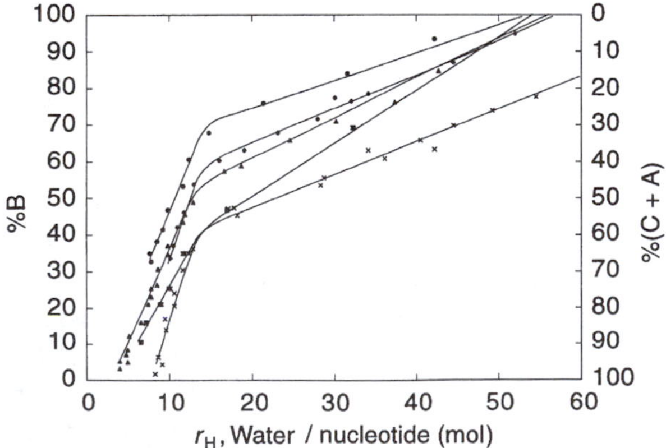

Figure 11-17
Fractional B content of calf thymus DNA as a function of net hydration from the data of Hearst and coworkers. [Reprinted with permission from Wolf, B. and Hanlon, S., *Biochemistry*, **14**, 1661–1670 (1975). Copyright ©1975 American Chemical Society.]

As Class 2 waters (those hydrogen bonded to the Class 1 waters, which in turn are hydrogen-bonded directly to the DNA) are removed, the B conformation is gradually converted to C and A. When all the Class 2 waters are gone, at about 12 moles H_2O/mol nucleotides, removal of the Class 1 waters directly hydrogen-bonded to the DNA causes a more abrupt disruption of the B structure. The transition is complete when only the most tightly held waters, those hydrogen-bonded to the phosphates, remain. Wolf and Hanlon also point out that ionic interactions, of both the Donnan and binding types, must be included for a complete interpretation.

4. SPECIFIC BINDING OF METAL IONS AND POLYAMINES

In addition to the delocalized association of ions with nucleic acids that we have considered up to now, there are important specific interactions, at particular binding sites, that affect nucleic acid structure and which may be used for various analytical and preparative purposes. In this section, we consider the structural and thermodynamic consequences of some of these interactions. Related material is contained in Chapter 12, on complexes with drugs, and in Chapter 3, on covalent reactions produced by metal complexes.

4.1 Metal Ion Binding to Nucleosides and Nucleotides

Studies on metal ion interactions with the nucleic acid bases have identified preferred sites of interaction for different classes of metals, determined thermodynamic binding parameters, and elucidated the structures of some complexes. Some useful reviews of metal ion-monomer interactions include Eichhorn (1973a), Daune (1974), Martin and Mariam (1979), and Marzilli (1981). An insightful review of broader biomolecular scope is Tam and Williams (1985).

The metal ions whose interactions with nucleic acids have been most thoroughly studied are classified according to the periodic table as

Alkali metals: Li, Na, K, Rb, Cs

Alkaline earth: Mg, Ca, Sr, Ba

Transition metals:

Divalent: Mn, Co, Ni, Pd, Pt, Cu, Zn, Cd, Hg

Other: Ag^{I}; Ru^{III}, Co^{III}, Au^{III}; Os^{VI}

The sites to which these metal cations may bind are the electronegative oxygen and nitrogen atoms of the nucleic acid constitutents. Their binding characteristics are described briefly as follows:

Phosphate oxygen atoms: These oxygen atoms bind with all metal ions through salt linkages. Although specific complexes are found in crystals, solution studies indicate that binding is generally territorial in the sense of counterion condensation theory.

Sugar hydroxyl groups: Alkali metals and alkaline earths, but not transition metals, bind with the lone-pair electrons of the oxygen, which enter the metal coordination sphere. These tend to be the weakest sites.

Base ring nitrogen atoms: These nitrogen atoms can interact with all three types of metal ions, but binding is strongest and most specific with d orbitals of transition metals. The N7 position of guanine is a preferred site when the glycosidic bond has blocked N9 of purines and N1 of pyrimidines.

Exocyclic base keto groups: Direct metal binding can occur at the O2 of cytosine or O2, O4 of thymidine/uracil, but generally not to O6 of guanine, since N7 is preferred.

Exocyclic base amino groups: There is no binding to these groups since the lone-pair electrons needed for metal binding are delocalized over the heterocylic ring.

Some metal ions can bind to more than one site, forming cyclic chelate compounds. A summary of coordination sites is given in Table 11-4.

Thermodynamic measurements of binding gives information on the extent of complexation between metal ions and nucleic acids under particular solution conditions, and also provides insight into the factors governing complex formation. Table 11-5 displays binding constants K and standard-state thermodynamic values (Gibbs free energy ΔG°, enthalpy ΔH°, and entropy ΔS°) for complex formation of metal ions with AMP^{2-}, ADP^{3-}, and ATP^{4-}. These nucleotides have been thoroughly studied because of the importance of ATP in energy metabolism. Given the high physiological concentration of Mg^{2+} ($\sim$ 10 mM), ADP and ATP exist mainly as the Mg complexes. The predominence of these complexes has implications for the mechanisms of action of the many enzymes that utilize hydrolyze or synthesize ATP.

Table 11-5 shows that the stability of complexes increases with increasing ligand charge, as would be expected for ion pair formation. The small enthalpies show that association is mainly entropy driven. (The large ΔH for Mg is due to its abnormally

Table 11.4
Summary of Metal Ion Coordination Sites with the Phosphate, Base, and Ribose Moieties of Nucleotides and Nucleic Acids

Site	Metal Ions
Phosphate	Li^+, Na^+, K^+, Rb^+, Cs^+, Mg^{2+}, Ca^{2+}, Sr^{2+}, Ba^{2+}, trivalent lanthanides
Phosphate and ribose	B^{III}, UO_2^{2+}
Phosphate and base[a]	$Co^{2+} = Ni^{2+}$, Mn^{2+}, Zn^{2+}, Cd^{2+}, Pb^{2+}, Cu^{2+}
Ribose and base	Co^{3+}
Base	Ag^+, Hg^{2+}

[a]Increasing affinity for base relative to phosphate from left to right. [Reprinted with permission from Izatt, R. M., Christensen, J. J., and Rytting, J. H., *Chem. Rev.* **71**, 439–481 (1971). Copyright American Chemical Society.]

Table 11.5
Thermodynamic Parameters for Complex Formation of AMP, ADP, and ATP with Divalent Metal Ions at 25°C in 0.1 M KNO_3[a]

	Metal Ion	Log *K*	$\Delta G°$ kcal mol^{-1}	$\Delta H°$ kcal mol^{-1}	$\Delta S°$ cal/mol deg^{-1}
5′-AMP^{2-}	Ba^{II}	1.73	−2.36	−2.0	1.2
	Sr^{II}	1.79	−2.44	−1.4	4.4
	Ca^{II}	1.85	−2.52	−0.6	6.4
	Mg^{II}	1.97	−2.68	3.4	20.4
	Co^{II}	2.53	−3.60	−1.1	8.4
	Mn^{II}	2.40	−3.27	−1.0	7.6
	Zn^{II}	2.72	−3.70	−1.2	8.2
	Nie^{II}	2.84	−3.87	−1.0	9.6
	Cu^{II}	3.18	−4.33	−2.0	8.0
5′-ADP^{3-}	Ba^{II}	2.36	−3.22	−2.9	1.1
	Sr^{II}	2.54	−3.46	−2.7	5.5
	Ca^{II}	2.86	−3.90	−1.2	9.1
	Mg^{II}	3.17	−4.32	3.6	26.6
	Co^{II}	4.20	−5.72	−2.0	12.5
	Mn^{II}	4.16	−5.67	−2.4	11.0
	Zn^{II}	4.28	−5.83	−2.0	12.5
	Ni^{II}	4.50	−6.13	−1.9	14.1
	Cu^{II}	5.90	−8.04	−4.1	13.0
5′-ATP^{4-}	Ba^{II}	3.29	−4.59	−3.9	2
	Sr^{II}	3.54	−4.83	−3.0	6
	Ca^{II}	3.97	−5.40	−0.9	12
	Mg^{II}	4.22	−5.80	2.6	27.5
	Co^{II}	4.66	−6.40	−2.2	14
	Mn^{II}	4.78	−6.50	3.0	12
	Zn^{II}	4.85	−6.60	−2.7	13
	Ni^{II}	5.02	−6.90	2.5	15
	Cu^{II}	6.13	−8.40	−4.3	14

[a]From Khan and Martell, (1966, 1967).

high heat of hydration.) The entropy of complex formation varies inversely with the ionic radius of the cation, presumably reflecting the greater water release upon ion pair formation by the smaller cations, which have higher charge densities.

The crystal structures of metal complexes with nucleosides and nucleotides show considerable diversity (Table 8-3, pp. 204–205, in Saenger, 1984). One useful generalization is that the N7 of purines accomodates metals with sixfold coordination, while the N3 of pyrimidines requires lower coordination numbers. The NMR as well as X-ray evidence suggests that some ions, such as Mn^{2+}, can bind simultaneously to a ring nitrogen and a phosphate oxygen, forming a cyclic chelate.

4.2 Metal Ion Binding to Polynucleotides

Metal ions have major effects on double helix stability and structure (Eichhorn, 1981; Sigel, 1993). The most widely studied consequence of metal ions binding to polynucleotides has been the effect on helix stability as reflected in the thermal helix–coil transition. Ions will raise the transition temperature (T_m) if they bind more strongly to the double–helical form than to the single-stranded form; such binding may be specific or nonspecific. Monovalent cations, except for Ag^+, generally appear to operate by the nonspecific polyelectrolyte mechanisms discussed in Section 11.2.

In contrast, divalent cations show evidence of site-binding. Increasing concentrations of alkali metal cations, which bind to phosphates, increase T_m, as shown in Figure 11-18 (Dove and Davidson, 1962). The stabilization is much greater with divalent than with monovalent cations. For example, $3 \times 10^{-4} M$ Mg^{2+} stabilizes DNA as effectively as 5×10^{-2} M Na^+. This appears to be due both to the increased effects of counterion condensation and residual ion atmosphere screening for divalent relative to univalent

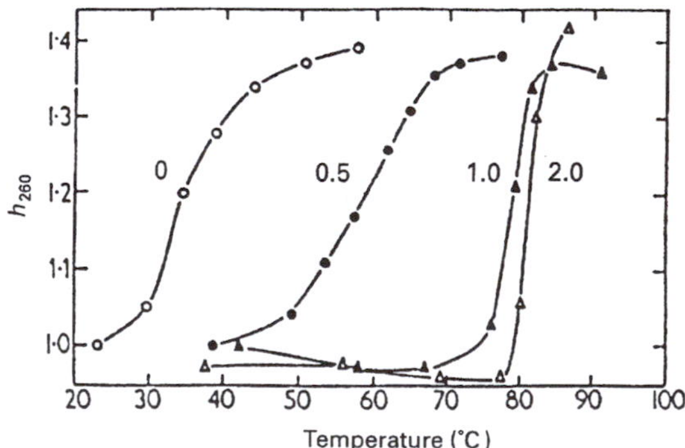

Figure 11-18
Effect of Mg^{2+} on thermal denaturation of *Bacillus megaterium* DNA, at pH 7 and ionic strength $3 \times 10^{-4} M$. Curves are labeled with the number of equivalents of Mg^{2+} per equivalent of DNA phosphate. [Reprinted with permission from Dove and Davidson, 1962.]

cations, and to site binding of Mg^{2+} ions to the phosphates in the double-stranded form. The broadening of the transition at half-saturation, $r = 0.5$, is also attributable to site-binding.

In addition to stabilizing electrostatic effects, alkaline earths and transition metals can destabilize the double helix, as shown in Figure 11-19 (Eichhorn and Shin, 1968). The destabilization occurs by metal ion binding to the bases, so the relative importance of destabilization to stabilization is in the order of preference of base to phosphate:

$$\text{Mg}^{\text{II}} < \text{Co}^{\text{II}} < \text{Ni}^{\text{II}} < \text{Mn}^{\text{II}} < \text{Zn}^{\text{II}} < \text{Cd}^{\text{II}} < \text{Cu}^{\text{II}}$$

Zn^{I}, Cd^{II}, and Cu^{II} strongly depress T_m, but also aid full renaturation, evidently by cross-linking the bases and preventing extensive strand separation. In contrast, Co^{II}, Ni^{II}, and Mn^{II} permit only partial renaturation. The disruption of base stacking and pairing by these ions also leads to DNA aggregation (Knoll et al., 1988).

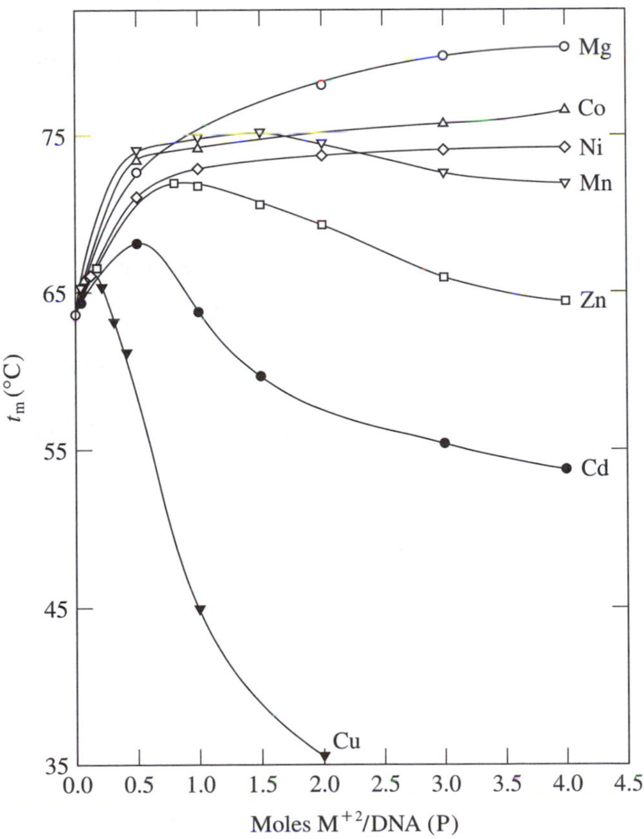

Figure 11-19
Variations of T_m for solutions of DNA as a function of divalent metal ion concentration. [Reprinted with permission from Eichhorn, G. L. and Shin, Y. A., *J. Am. Chem. Soc.*, **90**, 7323–7328 (1968). Copyright ©American chemical Society.]

Both Ag^{I} and Hg^{II} behave similarly to Cu^{II}, but are even more extreme since they appear to bind only to the bases, and produce an ordered structure in which the ion apparently is held between the two helical strands. A fuller ranking would therefore be

$$Zn^{II} < Co^{II} < Ni^{II} < Mn^{II} < Zn^{II} < Cd^{II} < Cu^{II} < Ag^{II} < Hg^{II}$$

There is base specificity in these phenomena, with most metals manifesting a general preference for purines and for G,C rich DNA, owing to the N7 of guanine; but Hg^{II} binds particularly tightly to A,T rich DNA, because of a strong affinity for the N3 of thymine. As discussed in Chapter 9, this can serve as the basis for density-gradient separation of DNA on the basis of G,C content.

Raman spectroscopy and pH measurements (Duguid et al., 1993; Duguid et al., 1995) show that the transition metals provoke a variety of structural changes, including decreases in backbone order, base unstacking, distortion of glycosyl torsion angles, rupture of hydrogen bonds, and metal binding to purine N7 and pyrimidine N3 atoms.

X-ray structures of metal–nucleotide complexes generally show close approach of the metal to the P, N, or O ligand, indicating that the ligand has displaced the water of hydration of the ion, forming an inner-sphere complex. In contrast, the success of counterion condensation theory with nucleic acids suggests that territorially bound ions retain their water of hydration, interacting mainly electrostatically. Nuclear magnetic resonance has provided evidence on the extent of inner-sphere complex formation with DNA and the cation binding environments (Braunlin, 1995). The paramagnetic ions Mn^{2+} and Co^{2+} affect the relaxation rates of the phosphorus atoms in ^{31}P NMR experiments. Relaxation rate measurements can be interpreted in terms of average internuclear distances, since the interaction varies as R^{-6}, and then in terms of the fraction of directly coordinated metal ion. Over a range from 20 to 160 mM, only $15 \pm 5\%$ of the bound Mn^{2+} and Co^{2+} form inner-sphere complexes with the phosphates (Granot et al., 1982). The paramagnetic ions also affect the NMR relaxation properties of the water and base protons, so the effects of polynucleotide binding on proton relaxation enhancement give information about metal ion interactions with the bases and water as well as with the phosphates. The number of waters coordinated to Mn^{2+} and Co^{2+} are little changed on average upon binding to DNA (Granot and Kearns, 1982a). In studies with the polyribonucleotide poly(rI)•poly(rC), the relaxation behavior of the H at the 6 position of C (cytosine) is consistent with less than 10% occupancy of phosphate inner-sphere binding sites. However, there is substantial broadening of the H at the 8 position of I (inosine) imino resonance by Mn^{2+}, indicative of some inner-sphere complexing at N7 (Chang and Kearns, 1986). This finding is consistent with the results of NMR dispersion measurements of water proton relaxation rates in Mn-DNA solutions (Kennedy and Bryant, 1986). Such measurements, in which the strength of the applied magnetic field is varied, are able to detect some rotationally immobilized Mn^{2+} on the DNA; inner-sphere complexes with the G or C bases are thought to be the most likely causes of such irrotational binding.

There is also clear NMR evidence that Mg^{2+} forms site-bound complexes with the DNA phosphates, in which at least one water from the Mg hydration sphere is lost. This evidence comes from measurements of the NMR spectrum and relaxation rates of ^{25}Mg (Rose et al., 1980; Berggren et al., 1992), which show slow exchange between site-bound and solution environments. A striking additional finding (Rose et al., 1982) is that even high concentrations of competing ions such as Na^{+}, Ca^{2+}, Hg^{2+}, Zn^{2+} and Co^{2+} are ineffective in displacing all of the $^{25}Mg^{2+}$ from its site-bound environment. The ^{43}Ca NMR (Braunlin et al., 1992) shows that there are at least two classes of binding environments, that site binding correlates with GC content, and that Ca^{2+} binding is very sensitive to local DNA structure.

4.3 Binding of Metal–Ligand Complexes

Up to this point, we have considered only aqua complexes of metal ions. However, there are some metal–ligand complexes whose interaction with nucleic acids is of considerable interest. Perhaps the most prominent of these, because of its antitumor activity, is cis-dichlorodiammine platinum(II) (see Chapter 12). It has been striking that the cis isomer is a very effective anticancer agent, while the trans isomer is inactive. A variety of studies have shown that the cis compound forms an adduct with DNA in which it loses the two chloride ions and forms two Pt–N bonds to the N7 atoms of adjacent guanines on the same strand. The crystal structure of the complex is shown in Figure 11-20 (Sherman et al., 1985). The Pt forms a square planar complex with two ammonia ligands and the two N7 atoms of guanine. The base stacking of the guanines is completely disrupted in this complex. Assuming the same structure existed in DNA, this would cause a localized disruption of the double helix, which might inhibit replication. However, it is speculated that the disruption might be sufficiently localized to elude detection by repair enzymes. Model building indicates that the trans isomer is unable to form an intrastrand GpG cross-link because of steric constraints. It does form a variety of other intrastrand cross-links, but these may cause greater perturbations and be more susceptible to repair mechanisms.

The ammonia ligands of the complex hexammine cobalt(III), $[Co(NH_3)_6]^{3+}$, exchange extremely slowly with water. $[Co(NH_3)_6]^{3+}$ has therefore been useful as a substitutionally inert inorganic analogue of naturally occurring multivalent cations such as the polyamine spermidine^{3+}, in studies of the condensation behavior of DNA, as detailed in Chapter 14. We consider below the structural basis of its stabilization of the Z form of poly(dG-dC). In solutions of oligonucleotides such as d(GGCCGGCC) with contiguous guanines, $[Co(NH_3)_6]^{3+}$ induces CD of an A-DNA character, ^{59}Co NMR shows that tumbling of the $[Co(NH_3)_6]^{3+}$ is greatly slowed, and very large upfield ^{59}Co chemical shifts are observed (Xu et al., 1993b). The H NMR spectrum of another oligo with adjacent guanines, d(CCCCGGGG), changes in the presence of $[Co(NH_3)_6]^{3+}$, and an A-form CD spectrum is observed (Xu et al., 1993c). Such site binding and induction of a B–A transition are not seen in oligos without GG sequences.

Chiral complexes of ruthenium, zinc, and cobalt have been developed that can recognize local structural features, related to helicity, along the DNA strands (Barton, 1986). A simple example, the tris(phenanthroline) metal complex, is shown in Figure 11-21(*a*). One of the phenanthroline (phen) rings can intercalate between the DNA base pairs. The overall energy of binding is then governed by the steric interactions of the other two rings with the helical groove. For binding to B-DNA, the alignment of the Δ

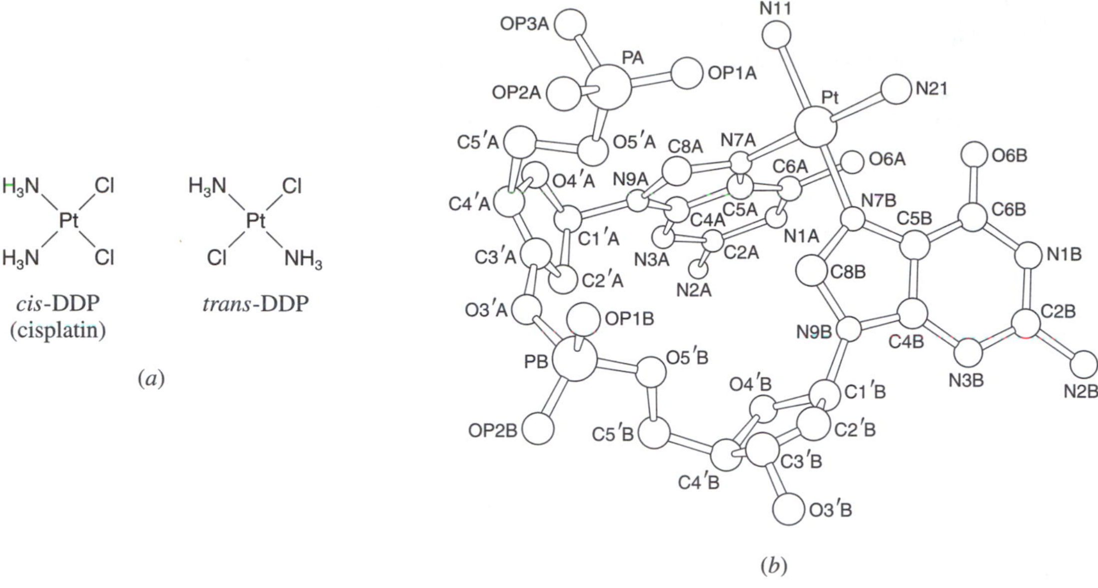

Figure 11-20
(*a*) Square planar complexes of *cis*- and *trans*-diamminedichloroplatinum(II) (DDP).
(*b*) Molecular structure of a *cis* – $[Pt(NH_3)_2\{d(pGpG)\}]$ adduct. [Reprinted with permission from Sherman, S. E., Gibson, D., Wang, A. H.-J., and Lippard, S. J. (1985). *Science* **230**, 412–417. Copyright ©1986 American Association for the Advancement of Science.]

(right) isomer is favorable, while that of the Λ-(left) isomer leads to steric clashes with the backbone phosphates (Figure 11-21b). Thus it is found that $[\Delta\ Ru(phen)_3]^{2+}$ binds more strongly to calf thymus DNA than does the Λ isomer, a phenomenon termed enantiomeric selectivity or enantiospecificity.

It might be expected that $[\Lambda\text{-}Ru(phen)_3]^{2+}$ would show equal selectivity for Z-DNA, but this is not the case, since Z-DNA is not just a left-handed version of B-DNA. It will be recalled that Z-DNA has a very wide and shallow major groove, and a very deep, sharp minor groove, in contrast to the better defined and more nearly equivalent major and minor grooves of B-DNA. Adequate enantiospecificity is obtained, however, with bulkier complexes. The most extensively used is the complex of ruthenium with 4,7-diphenylphenanthroline(DIP). The structure of $[Ru(DIP)_3^{2+}$ is shown in Figure 11-22. Barton (1986) has shown that spectroscopic assays, utilizing luminescence quenching or hypochromism of the charge-transfer band of the $[Ru(DIP)_3]^{2+}$, can be used to demonstrate enantioselective binding. Such assays, shown in Figure 11-23, demonstrate that the B helix binds only the Δ-enantiomer; while there is only a slight stereoselectivity in binding the Λ enantiomer to the Z helix. These conclusions are consistent with those arrived at by model building.

These comparative enantiomeric selectivities may be used to probe whether an unknown DNA helix is right or left handed. If only the Δ isomer binds, the helix is probably right handed and B-like. If the Λ isomer binds, but only slightly more

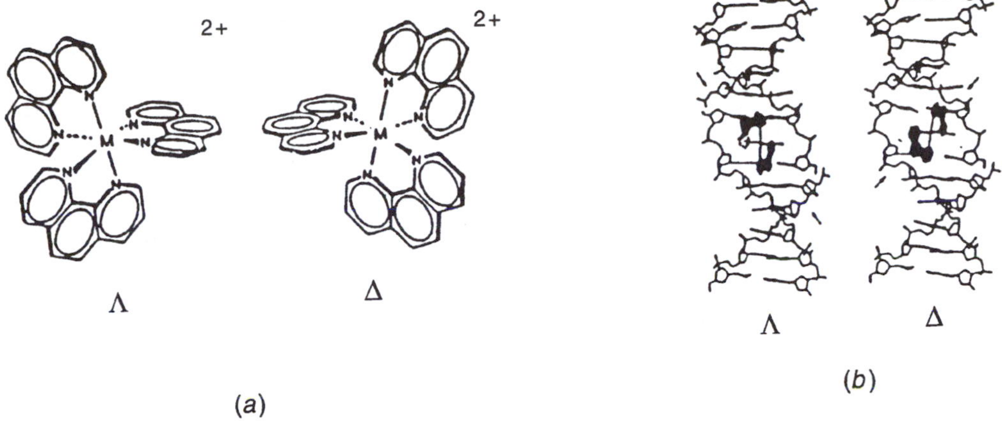

Figure 11-21
(*a*) The Λ (left) and Δ (right) enantiomers of a tris(phenanthroline) metal complex. (*b*) Models of the intercalation of the Λ and Δ enantiomers in a B-DNA helix. [Reprinted with permission from Barton, J. K. (1986). *Science* **233**, 727–734. Copyright ©1986 American Association for the Advancement of Science.]

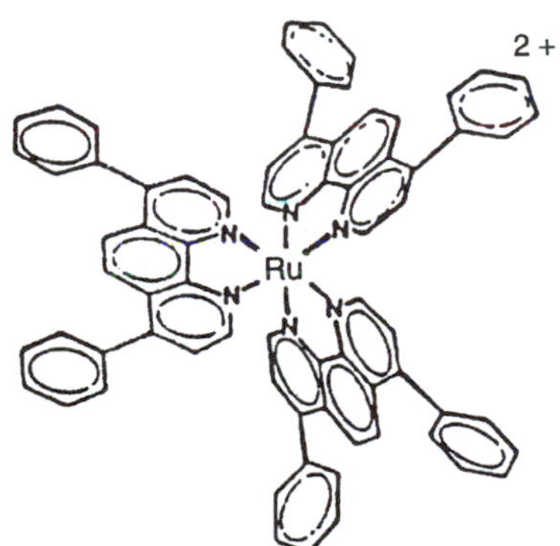

Figure 11-22
The structure of $[Ru(DIP)_3]^{2+}$. [Reprinted with permission from Barton, J. K. (1986). *Science*, **233**, 727–734. Copyright ©1986 American Association for the Advancement of Science.]

strongly than the Δ isomer, then the helix is probably in the Z conformation. A stronger preference for the Λ isomer would indicate a left-handed, but non-Z, conformation. There is no detectable binding of these complexes to single-stranded DNA or to double-stranded RNA.

Enantiospecific binding may be coupled with reactivity to specifically cleave left-handed regions in a plasmid DNA (Barton and Rafael, 1985). With OH radical chemistry, presumably similar to that discussed in Chapter 3 for Fe–EDTA complexes, $[\Lambda\text{-}Co(DIP)_3]^{3+}$ oxidatively cleaves the sugar–phosphate backbone. The cleavage, which requires photoactivation, can serve to localize those regions in which Z-DNA occurs in an otherwise right-handed helix.

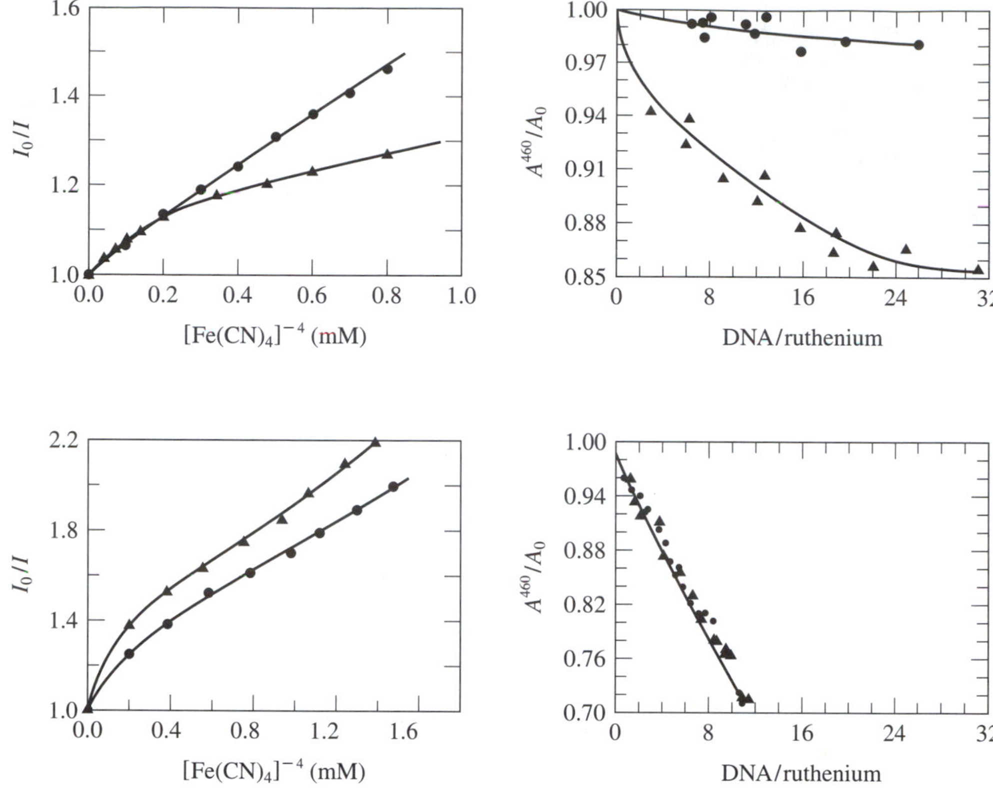

Figure 11-23
Spectroscopic assays for DNA helical handedness using Λ (●) and Δ (Δ) $[Ru(DIP)_3]^{2+}$ enantiomers bound to B-DNA (*a*) and Z-DNA (*b*). Left: Stern–Vollmer plot of luminescence quenching by ferrocyanide. Right: Hypochromism in the 460 nm charge-transfer band. [Reprinted with permission from Barton, J. K. (1986). *Science* **233**, 727–734. Copyright ©1986 American Association for the Advancement of Science.]

4.4 Effect of Metal Binding on Polynucleotide Structure

In addition to effects on helix–coil melting behavior, divalent and multivalent metal ions cause transitions between B and Z forms of polydeoxynucleotides of suitable alternating purine–pyrimidine sequence. The Z form of poly(dG-dC) was first observed at very high-salt concentrations. However, it has since been observed that the B–Z transition can occur in millimolar salt concentrations in the presence of low concentrations of multivalent cations (Behe and Felsenfeld, 1981; Behe et al., 1985; Woisard et al., 1985).

The Z-form of $d(CGCGCG)_2$ is stabilized 10^4 times more effectively by $[Co(NH_3)_6]^{3+}$ than by Mg^{2+} (Gessner et al., 1985). The structural basis of this stabilization, and of the relative effectiveness of different cations, has been established crystallographically (Fig. 11-24). In a 1.25 Å X-ray structure, it was seen that the NH_3 groups of the $[Co(NH_3)_6]^{3+}$ ion donate five hydrogen-bonds directly to one guanine N7 and two O6 atoms as well as to two backbone phosphate oxygen atoms. The B conformation does not have a geometry that allows such extensive hydrogen bonding. The phosphate group P9 in Figure 11-24(*b*) has rotated down to form a Z_{II} conformation that allows formation of the two hydrogen-bonds. In contrast, the Mg^{2+} complexes have only three hydrogen-bonds, between the base pairs and the octahedrally arranged waters that hydrate the magnesium ion. (In solution, the Mg^{2+} waters of hydration also presumably can exchange readily with solvent water, further weakening the interaction with DNA.) Similar structures are found for complexes of $[Co(NH_3)_6]^{3+}$ with $d(CGTACGTACG)_2$ in the Z form (Brennan et al., 1986), emphasizing the specificity of ligand binding to guanine residues. In both these crystals, the cations bind between

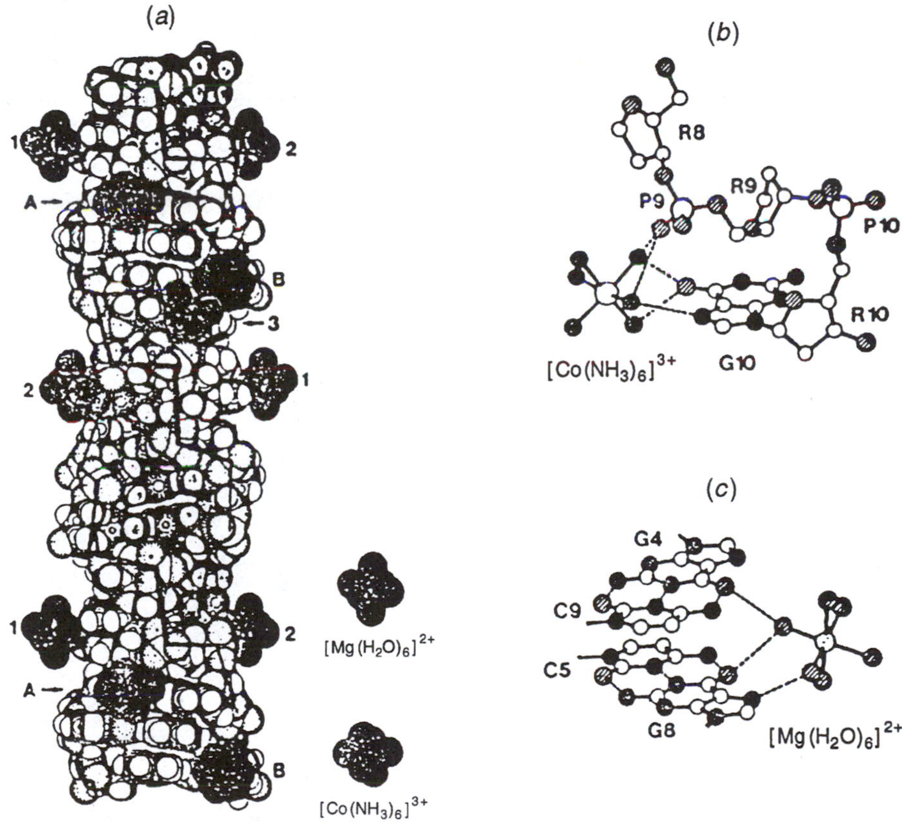

Figure 11-24
(*a*) The Z-DNA with its attached ions in a $[Mg/Co(NH_3)_6]^{3+}$ crystal. The complexes largely bridge between two adjacent duplexes in the crystal; $[Mg(H_2O)_6]^{2+}$ complexes 1 and 2 and $[Co(NH_3)_6]^{3+}$ complexes A and B show the two different ways in which the same complex interacts with the bridged DNA molecules. The Mg complex 3 is not intermolecular and binds only to one duplex. (*b*) Coordination of cobalt hexammine to Z-DNA (site A). (*c*) Coordination of a magnesium–water octahedral complex with Z-DNA (site 2). [Reprinted with permission from Gessner, R. V., Quigley, G. J., Wang, A. H.-J., van der Marel, G. A., van Boom, J. H., and Rick, A., *Biochemistry*, **24**, 237–240 (1985). Copyright ©1985 American Chemical Society.]

two different DNA helices in the lattice. This binding may help to explain the great effectiveness of $[Co(NH_3)_6]^{3+}$ in condensing DNA (see Chapter 14).

Considerable evidence has accumulated on the role of cations in stabilizing certain structural features of tRNA. Teeter et al. (1981) have summarized solution evidence that indicates 4–6 strong binding sites for cations in tRNA, in addition to the weaker interactions associated with charge neutralization. These sites, located more precisely by crystallographic studies, are shown in Figure 11-25. They correspond to two basic types: electronegative pockets and electronegative clefts. Pockets, which are formed by high concentrations of negative charge produced at a point by a sharp bend or loop in the backbone, or by juxtaposition of Ns and Os on adjacent bases, are occupied by metals, which are themselves essentially point charges. Thus metal ions binding in these pockets tend to stabilize bends or loops in the tRNA arms or in single stranded regions such as the variable loop and the anticodon loop. The dominant intracellular divalent ion Mg^{2+}, as the hexahydrate $[Mg(H_2O)_6]^{2+}$, occupies pockets formed solely by phosphates or by phosphates and base atoms, a behavior similar to that exhibited by Mg^{2+} with DNA. A third kind of pocket, formed by base atoms alone, may be occupied by second-row transition metals.

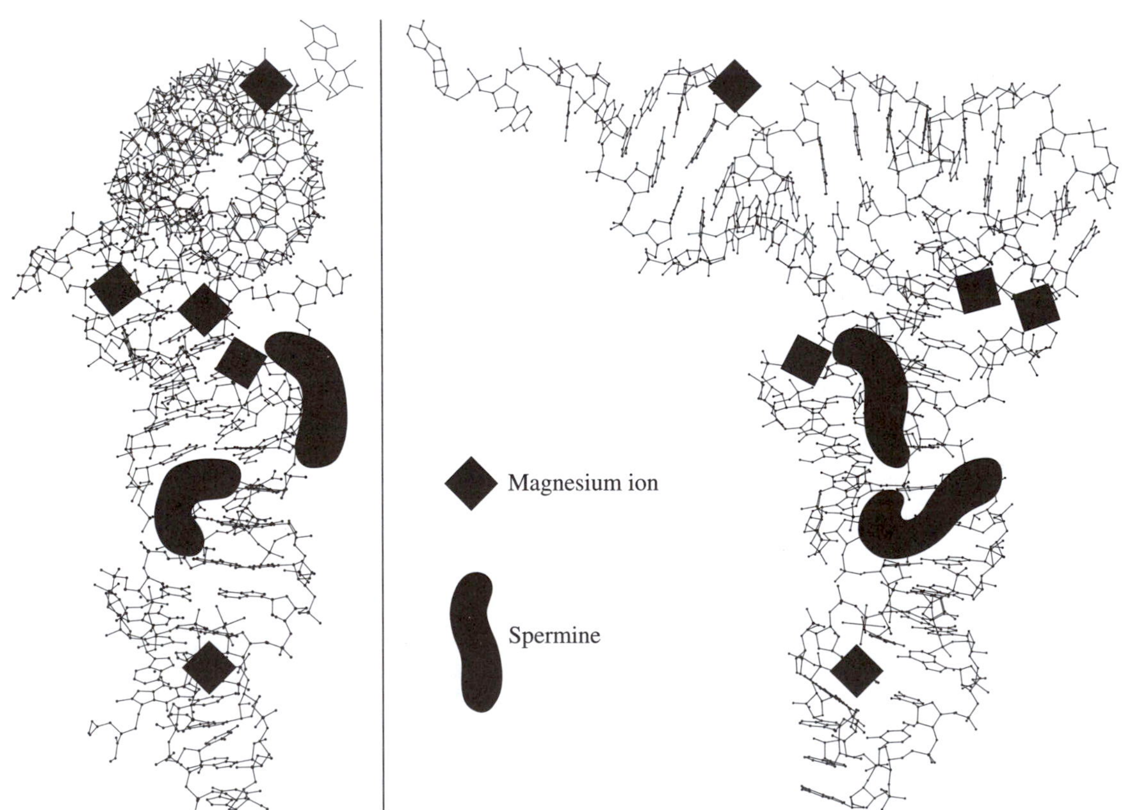

Figure 11-25
Two views of Mg^{2+} and spermine^{4+} binding sites to yeast tRNA^{phe}. [Reprinted with permission from Teeter et al., 1981.]

As a first-row transition metal, Co^{2+} binds to a phosphate-base electronegative pocket of $tRNA^{phe}$ at guanine residue 15. On the other hand $[Co(NH_3)_6]^{3+}$, with additional hydrogen-bonding possibilities, binds through cis ammine ligands to N7 and O6 positions of adjacent purine bases in the major groove, as well as to phosphate oxygens in neighboring strands (Hingerty et al., 1982). This multipoint binding increases the stability of the tRNA structure.

The most notable specific structural effect of monovalent cations on nucleic acid structure is the stabilization of G quartets, especially the strong preference for K^+ over Na^+ (Hardin et al., 1991; Kang et al., 1992; Smith and Feigon, 1992). This selectivity appears to depend on the size of the cation that binds in the cavity formed by the Hoogsteen base pairing of the G bases. Quadrupolar NMR (Xu et al., 1993a) demonstrates that both $^{39}K^+$ and $^{23}Na^+$, when specifically bound, tumble slowly on a time scale characteristic of the oligonucleotide $d(T_2G_4T)$. However, rapid exchange with atmospherically bound cations, which rotate much more rapidly, shows that the lifetime of the complex is less than 1 ms.

4.5 Polyamine Binding to Nucleic Acids

All cells contain polyamines. The three naturally occurring polyamines are putrescine^{2+} (1,4-diaminobutane), spermidine^{3+}, and spermine^{4+}:

Putrescine $^+H_3N{-}(CH_2)_4{-}NH_3^+$

Spermidine $^+H_3N{-}(CH_2)_4{-}NH_2^+{-}(CH_2)_3{-}NH_3^+$

Spermine $^+H_3N{-}(CH_2)_3{-}NH_2^+{-}(CH_2)_4{-}NH_2^+{-}(CH_2)_3{-}NH_3^+$

Putrescine and spermidine are found in prokaryotes; spermidine and spermine are found in eukaryotes. Polyamine metabolism is highly regulated in a way that is strongly coupled to cell growth and proliferation, so that intracellular levels can vary by several orders of magnitude depending on cell growth rate (Tabor and Tabor, 1984). While polyamines interact strongly with any polyanionic compounds, such as phospholipids in membranes, their interactions with nucleic acids have been the major focus as a possible mechanism for their effects on cell growth rates.

Such interactions are manifest *in vitro* through the ability of polyamines to raise the helix–coil transition temperature of DNA (Tsuboi, 1964), to provoke the B–Z transition (Behe and Felsenfeld, 1981) , and to condense DNA into toroidal or rodlike particles (Gosule and Schellman, 1976), a process discussed in more detail in Chapter 14. These phenomena depend substantially on the structure of the polyamine as well as its charge (Thomas and Bloomfield, 1984, 1985), suggesting that specific as well as general electrostatic interactions are involved.

An early molecular model for the binding of spermidine and spermine to DNA (Liquori et al., 1967) was based on the crystal structure and energy refinement of spermine tetrahydrochloride. A plausible sequence of dihedral angles, with trans states for the C–C bonds and mainly gauche states for the C–N bonds, gives a conformation that allows the polyamine to lie nicely in the minor groove of the DNA. All four basic groups of spermine can form hydrogen bonds with the phosphates, two on one chain and two on the other (Suwalsky et al., 1969). In this model, there is no need to distort the DNA structure to accomodate polyamine binding.

Direct crystallographic determination of the structure of spermine complexes with tRNA (Quigley et al., 1978) and with a synthetic DNA dodecamer (Drew and Dickerson, 1981) have shown rather more complex results. Binding tends to be in the major, rather than the minor, groove, and distortion of the nucleic acid upon binding is frequent. Phenylalanyl tRNA has two localized spermines (Fig. 11-25). One spermine is in the major groove at one end of the anticodon stem. The other is near the variable loop, curling around phosphate 10 in a region where the backbone turns sharply. Both of these binding sites are electronegative clefts, formed when rows of backbone atoms approach closely (Teeter et al., 1981), and in both, the spermine stabilizes some distortion of the RNA structure. In DNA as well, spermine asymmetrically bridges the major groove.

Feuerstein et al. (1986) used molecular mechanics to calculate minimized energy structures for $d(A\text{-}T)_5 \cdot d(A\text{-}T)_5$, $d(G\text{-}C)_5 \cdot d(G\text{-}C)_5$, and their complexes with spermine. They found that for $d(A\text{-}T)_5 \cdot d(A\text{-}T)_5$ the total energy of the major groove model is greater than 70 kcal mol^{-1} more stable than the minor groove model, and the binding energy is 120 kcal mol^{-1} more favorable. Although these calculations were done in vacuo, without water as part of the system, the magnitude of the difference between the major and minor groove models is large enough to be convincing. With $d(G\text{-}C)_5 \cdot d(G\text{-}C)_5$, the energy-minimized calculation predicts a 25° bend in the oligomer produced by spermine binding (Fig. 11-26). This is reminiscent of the bend observed in $tRNA^{phe}$. Closer examination shows that base stacking is not disrupted, and that the bend occurs over several base pairs with no evidence of abrupt kinking.

While crystallography and computer simulation indicate that polyamine binding occurs at well-defined sites and can strongly influence nucleic acid structure, solution measurements are consistent with a looser, more general interaction. We have already mentioned (see Fig. 11-12) that the ionic strength dependence of polyamine binding to DNA obeys the predictions of counterion condensation theory, which assumes the free translation along the polyion backbone characteristic of territorial binding. Calorimetry shows that the enthalpy of binding of spermine to DNA is 0 cal mol^{-1} P, compared with -300 cal mol^{-1} P for poly(L-lysine) and +350 cal mol^{-1} P for Mg^{2+} (Ross and Shapiro, 1974). Thus polyamine binding is entropy driven, a characteristic of purely electrostatic interactions without significant dehydration or structural changes. An even more direct view of solution events in the interaction of spermine with d(CGCGAATTCGCG) is obtained from NMR studies (Wemmer et al. , 1985). The narrow resonance lines of the spermine protons are not broadened in the complex under conditions where most of the spermine is bound, and very weak positive nuclear Overhauser effects NOEs are observed between spermine protons. Both of these findings indicate rapid spermine rotation, essentially equivalent to that of free spermine. The NMR spectrum of the DNA in the complex shows no changes beyond those produced by simple salts. Thus despite the tight binding, $K = 10^6$ under the conditions used, the spermine appears to move independently of the DNA on the time-scale probed by NMR. The challenge in reconciling the solution and X-ray results is obvious.

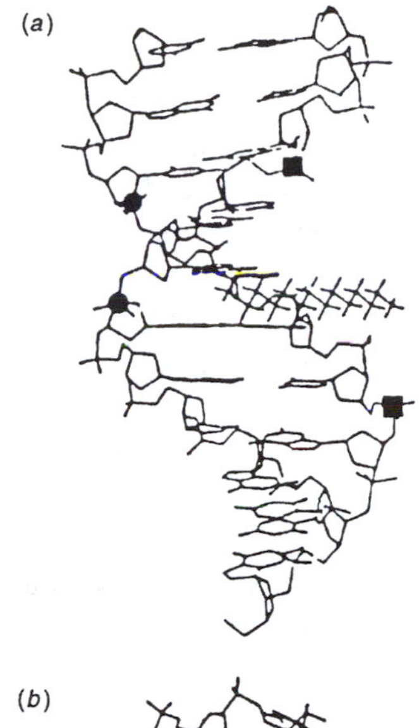

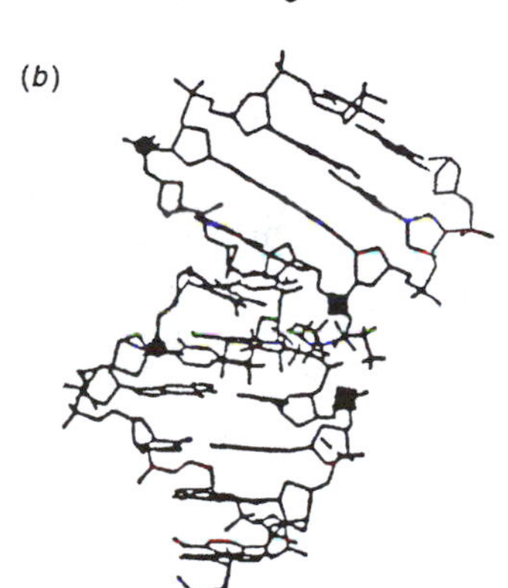

Figure 11-26
Views into the major groove of $d(G\text{-}C)_5 \cdot d(G\text{-}C)_5$ with spermine in place before (*a*) and after (*b*) energy minimization. Note the decrease in distance across the major groove (■) and the increase across the minor groove (●) after energy minimization was performed. Squares and circles represent the same points on the helix, and are included for comparison between (*a*) and (*b*). [With permission from Feuerstein et al., 1986.]

5. MIXED AQUEOUS–NONAQUEOUS SOLVENTS

While water is the normal biological milieu, mixed solvents are often used in physical or preparative work on nucleic acids. In particular, ethanol is often used to precipitate DNA. We will discuss ethanol-mediated condensation later, in Chapter 14, but deal now with the effects of alcohols on DNA secondary structure. Addition of nonaqueous solvents means loss of water for hydration. Since helix geometry depends on hydration, we expect and observe changes in helix geometry.

5.1 A-DNA

DNA undergoes a transition from the B to the A conformation between 60 and 80% (w/w) ethanol. When these solutions are precipitated by adding salt, the secondary conformation does not change, although the condensates have quite different morphologies (Gray et al., 1979).

5.2 Z-DNA

Ethanol at a concentration of 60% (v/v) induces the transition of poly[d(G-C)] from B to Z. The presence of small amounts of divalent cations markedly reduces the ethanol needed: 20% at 0.4 m*M* $MgCl_2$ and 10% at 4 m*M* $MgCl_2$. The left-handed Z* DNA produced in these mixtures differs from normal Z DNA formed at high-salt or high ethanol concentrations in that it is highly aggregated, supports binding of intercalating drugs, and serves as a template for *Escherichia coli* RNA polymerase. In the absence of ethanol, submillimolar concentrations of divalent metal ions induce the B–Z transition only at elevated temperature; 20% (v/v) ethanol lowers the activation energy so that the interconversion takes place at room temperature (van de Sande and Jovin, 1982; van de Sande et al., 1982).

With the methylated polymer poly(dG-^{m5}dC)·poly(dG-^{m5}dC), only 20% ethanol is required to produce the Z-form, even in the absence of divalent ions (Behe and Felsenfeld, 1981).

5.3 P-DNA

DNA in methanol up to 95% (v/v) undergoes a transition to a form that has been identified as having 10.2 bp/turn, compared with 10.4 bp/turn in aqueous solution. When the temperature is raised, or methanol is added (e.g., to a volume composition of 47.5:5:47.5% MeOH/H_2O/EtOH), the DNA undergoes a cooperative transition to the P form, which has CD and UV absorption spectra resembling denatured DNA. However, it is not denatured in the sense of strand separation, since addition of water causes immediate reversion to the B form. The CD spectra up to 190 nm shows that the P form has no base stacking interactions (Zehfus and Johnson, 1981), and IR spectroscopy shows that there is little or no hydrogen bonding. Electron microscopy reveals that the P form has a condensed tertiary structure (Zehfus and Johnson, 1984). This collapsed structure occurs when about 90% of the DNA charge is neutralized, just as for DNA in aqueous solution condensed by multivalent cations (see Chapter 14).

APPENDIX

A.1 Donnan Effect

Before considering various sophisticated mechanisms of interaction between ions and nucleic acids, we should recall an elementary, but very important, cause of ion redistribution in membrane equilibrium experiments on polyelectrolyte solutions: the Donnan effect. Interpretation of preferential hydration of nucleic acids involves a salt redistribution component, which has a Donnan contribution without any specific ion binding. Redistribution of ions at low-salt concentrations can also severely affect binding, light scattering, and hydrodynamic experiments. The Donnan effect is the result of two basic laws of equilibrium: electroneutrality and the equality of chemical potentials of diffusible components.

Consider two solutions in equilibrium across a semipermeable membrane. The solution on one side contains solvent (component 1), nucleic acid M_ZP (component 2), which dissociates into Z mol of a univalent cation M^+ and 1 mol of polyion P^{Z-}, and a uni-univalent salt MX (component 3) with counterion M^+ in common with the polymer. The solution on the other side [denoted by a prime (′)] contains just solvent and salt. Concentrations will be measured in molality (m) (mol kg^{-1} solvent).

Electroneutrality requires

$$m'_{M^+} = m'_{X^-} \tag{11-A.1}$$

on the side without the nucleic acid, and

$$m_{M^+} = Zm_P + m_{X^-} \qquad \text{or} \qquad m_{M^+} = m_u + m_{X^-} \tag{11-A.2}$$

on the side with the nucleic acid. The equivalent concentration of charged groups (nucleotides) is $m_u = Zm_P$.

At equilibrium, the chemical potential of diffusible components (water and salt) must be equal on both sides of the membrane. The general expression for the chemical potential μ_i of component i is

$$\mu_i = \mu_i^\circ + RT \ln a_i \tag{11-A.3}$$

where a_i is the activity. For a salt MX that can dissociate into M^+ and X^-,

$$a_{MX} = a_{M^+} a_{X^-} \tag{11-A.4}$$

Thus the condition of membrane equibrium for the salt is $\mu_3 = \mu'_3$, or

$$a_{M^+} a_{X^-} = a'_{M^+} a'_{X^-} = a_3'^2 \tag{11-A.5}$$

Ignoring activity coefficients, thus equating activities to molalities,

$$m_{M^+} m_{X^-} = m'_{M^+} m'_{X^-} \tag{11-A.5}$$

The extent of ion redistribution is obtained by combining these equations to get

$$m_{X^-}(m_u + m_{X^-}) = m_{X^-}'^2 \tag{11-A.7}$$

and solving the quadratic to yield

$$m_{X^-} = -\tfrac{1}{2} m_u + m'_{X^-}(1 + m_u^2/4m_{X^-}'^2)^{1/2} \tag{11-A.8}$$

Define the distribution parameter Γ representing the decrease in molality of anionic co-ion (or increase of cationic counterion) on the nucleic acid side, per mole of nucleotide,

$$\Gamma = (m_{X^-} - m'_{X^-})/m_u = (m_{X^+} - m'_{X^+})/m_u - 1 \tag{11-A.9}$$

Expansion of the square root gives

$$\Gamma = -\tfrac{1}{2} + m_u/8m'_{X^-} - m_u^3/128m'^3_{X^-} + \cdots \tag{11-A.10}$$

In the limit of infinitely dilute polymer, $m_u \to 0$, $\Gamma \to -\frac{1}{2}$: salt component 3 is "rejected" from the polymer compartment. Γ approaches this limit as the polyion concentration decreases or the salt concentration increases. For NaDNA, at weight concentration c mg ml^{-1} and an average nucleotide molecular weight (including the Na) of 330, Table 11-4 may be constructed.

According to counterion condensation theory, a fraction r of the polyion charge is neutralized by condensed counterions, so the effective equivalent molality of polyion units to be used in Equation 11-A.10 is $(1 - r)m_u$. For univalent counterions, $r = 0.76$.

A.2 Poisson-Boltzmann Equation

Most analyses of the behavior of ionic solutions begin with a combination of electrostatics and statistical mechanics known as the Poisson–Boltzmann equation. The key electrostatic quantity is the electrostatic potential ψ. The electrical work, or free energy change, of introducing an infinitesimal unit of charge dq into a region of potential ψ is $dG_{el} = \psi dq$.

In a region of solution not containing charges, ψ is obtained by solution of Laplace's equation

$$\nabla^2\psi = 0 \tag{11-A.11}$$

while in a region containing charge of density ρ, ψ is obtained from the Poisson equation

$$\nabla^2\psi = -4\pi\rho/\varepsilon \tag{11-A.1}$$

where ε is the dielectric constant and ∇^2 is the Laplacian operator. In Cartesian coordinates,

$$\nabla^2 = \frac{\partial^2}{\partial x^2} + \frac{\partial^2}{\partial y^2} + \frac{\partial^2}{\partial z^2} \tag{11-A.13}$$

To determine the charge density $\rho(\mathbf{r})$ at some position $\mathbf{r}$, one now invokes the Boltzmann distribution law. This law states that the ratio of the concentration of an ion of charge q_i at $\mathbf{r}$, n_i ($\mathbf{r}$), to its concentration at a position of zero reference potential, the bulk concentration n_i (number/cm^3), is given by

$$n_i(\mathbf{r})/n_i = \exp[-G_{el}(\mathbf{r})/kT] \tag{11-A.14}$$

where $G_{el}(\mathbf{r})$ is the electrostatic work required to bring the charge from the reference state to $\mathbf{r}$. Replacing $G_{el}(\mathbf{r})$ by $q_i\psi(\mathbf{r})$ and letting $q_i = z_i q$ where q is the proton charge and z_i is the valence, one obtains

$$\rho(\mathbf{r}) = q\sum z_i n_i(\mathbf{r}) = q\sum z_i n_i \exp[-z_i q\psi(\mathbf{r})/kT] \tag{11-A.15}$$

where the sum is over all positive and negative small ions (but not polyions). Combining this with the Poisson equation yields the PB equation

$$\nabla^2\psi(\mathbf{r}) = -(4\pi q/e)\sum z_i n_i \exp[-z_i q\psi(\mathbf{r})/kT] \tag{11-A.16}$$

This equation is now just for the potential [since $\rho(\mathbf{r})$ has been eliminated] which is to be solved applying the appropriate geometry and boundary conditions. Since the PB equation is nonlinear in ψ, it is very difficult to solve analytically except under special circumstances. A common approximation is to linearize the PB equation by expanding the exponential and keeping only the leading terms (using $e^x \approx 1 + x + \cdots$):

$$\sum z_i n_i \exp[-z_i q\psi(\mathbf{r})/kT] = \sum z_i n_i + [(q/kT)\sum z_i^2 n_i]\psi(\mathbf{r}) + \ldots \tag{11-A.17}$$

The first term on the right-handed side is equal to zero because of electroneutrality. Substituting the remaining second term in Eq. 11-A.16) gives the linearized Poisson-Boltzmann equation

$$\nabla^2\psi(\mathbf{r}) = -[(4\pi q^2/\varepsilon kT)\sum z_i^2 n_i]\psi(\mathbf{r}) = -\kappa^2\psi(\mathbf{r}) \tag{11-A.18}$$

where

$$\kappa^2 = (4\pi q^2/\varepsilon kT)\sum z_i^2 n_i = (8\pi q^2/\varepsilon kT)I \tag{11-A.19}$$

and I is the ionic strength. Numerically, for water at 25°C, if I is expressed in molar units, $\kappa = 3.31 \times 10^7 I^{1/2}\text{cm}^{-1}$.

The linearized PB equation forms the basis of the Debye–Hückel theory for solutions of spherical ions. In spherical polar coordinates (r, θ, ϕ) the Laplacian of ψ is

$$\nabla^2\psi = \frac{1}{r^2}\frac{\partial}{\partial r}\left(r^2\frac{\partial\psi}{\partial r}\right) + \frac{1}{r^2\sin\theta\partial\theta}\frac{\partial}{\partial\theta}\left(\sin\theta\frac{\partial\psi}{\partial\theta}\right) + \frac{1}{r^2\sin\theta}\left(\frac{\partial^2\psi}{\partial\phi^2}\right) \tag{11-A.20}$$

If we consider the potential surrounding a spherical ion, symmetry guarantees that it will have no angular dependence, so the linearized PB equation in the charged region around a spherical ion is

$$\frac{1}{r^2}\frac{\partial}{\partial r}\left(r^2\frac{\partial\psi}{\partial r}\right) = -\kappa^2\psi \tag{11-A.21}$$

which may be written in the form

$$\frac{d^2(r\psi)}{dr^2} = -\kappa^2 r\psi \tag{11-A.22}$$

It is easily verified that this has the solution

$$\psi = Ae^{-\kappa r}/r + Be^{\kappa r}/r \tag{11-A.23}$$

The coefficient B of the positive exponential term must equal zero, since the potential must go to zero as r goes to ∞. Further analysis, detailed in Tanford (1961), shows that $A = Zq/\varepsilon$ in the region outside the central ion of charge Zq. Thus the effect of the added salt, in the Debye–Hückel approximation, is to screen the direct Coulomb potential by the exponentially decaying function $e^{-\kappa r}$. The resulting screened Coulomb potential

$$\psi = (Zq/\varepsilon r)e^{-\kappa r} \tag{11-A.24}$$

has a much shorter range than the Coulomb potential. The distance κ^{-1} over which the potential decays is called the Debye–Hückel length. While the linearized PB equation can be solved analytically, it is valid only when the electrostatic potential energy $q\psi$ at the central ion surface is small compared to the thermal energy kT. It is rarely small for highly charged ions such as nucleic acids.

Much effort has been devoted to solutions of the PB equation for cylindrical polyions. In cylindrical coordinates (r, z, θ), if it is assumed that the charge distribution and potential are uniform so there is no axial (z) or angular (θ) dependence, the Laplacian of the potential is

$$\nabla^2\psi = \frac{1}{r}\frac{\partial}{\partial r}\left(r\frac{\partial\psi}{\partial r}\right) \tag{11-A.25}$$

The nonlinear PB equation for cylindrical polyelectrolytes can be solved exactly only if there is no added salt, so that the only ions contributing to the charge density are the associated counterions. A useful model, which can be solved numerically in the more common case of added salt, is the cylindrical cell model. This model assumes parallel rods whose centers are a distance R apart. Symmetry then requires that $d\psi/dr = 0$ at $r = R/2$, while the boundary conditions at the surface of the cylinder, $r = b$, are

$$\psi_1(b) = \psi_2(b) \tag{11-A.26}$$

and

$$\varepsilon_1\left(\frac{\partial\psi_1}{\partial r}\right)_b-\varepsilon_2\left(\frac{\partial\psi_2}{\partial r}\right)_b=-4\pi\sigma \tag{11-A.26b}$$

where 1 and 2 refer to nucleic acid and solvent, and σ is the surface charge density, provide the other equations needed to solve the PB equation numerically. Such solutions have been obtained over a wide range of conditions by Stigter (1975). Their relation to counterion condensation theory is discussed by Stigter (1978) and by Anderson and Record (1980).

A.3 Counterion Condensation Theory

The assumptions described in the body of this chapter allow derivation of useful working equations. Along the backbone of the polyion, charge–charge interactions are of the screened Debye–Hückel form $r^{-1}\exp^{-\kappa r}$ where κ and I are defined in Eq. 11-A.19. We sum interactions over all pairs of polyion charges, to compute G_{el}, the electrostatic free energy of the system:

$$G_{\mathrm{el}}=-n_{\mathrm{p}}RT\xi\ln(1-e^{-\kappa b})\approx-n_{\mathrm{p}}RT\xi[\ln(\kappa b)+O(\kappa)] \tag{11-A.27}$$

where n_p is total number of moles of charged groups on the polyions and O(κ) denotes terms that have roughly the same magnitude as κ. We keep just the ln(κb) term, since that is the term characteristic of polyelectrolytes. The parameter θ_N is the number of territorially bound counterions of valence N per charged group on the polyion. Each charged group has its charge lowered from q to q_{net}:

$$q_{\mathrm{net}}/q=1-N_{qN} \tag{11-A.28}$$

Replacing q by q_{net} in the expression for G_{el}:

$$G_{\mathrm{el}}=-n_{\mathrm{p}}RT(1-N_{q_N})^2\xi\ln(\kappa b) \tag{11-A.29}$$

or, in terms of electrostatic free energy per polyion site in units of thermal energy,

$$g_{\mathrm{el}}=G_{\mathrm{el}}/n_{\mathrm{p}}RT=(1-N_{q_N})^2\xi\ln(\kappa b) \tag{11-A.30}$$

We now consider other contributions to the total free energy of the system. These are the interactions of bound (b) and free (f) counterions with their immediate environment, exclusive of ionic interactions (μ° terms), and translational mixing ($RT\ln c$ terms). Then the chemical potentials for free and bound ions are

$$\mu_{\mathrm{f}}=\mu_{\mathrm{f}}^\circ+RT\ln c_{\mathrm{f}} \tag{11-A.31}$$

$$\mu_{\mathrm{b}}=\mu_b^\circ+RT+RT\ln\gamma_b c_b \tag{11-A.32}$$

where the activity coefficient γ_b is given by

$$RT \ln \gamma_b = 2RTN(1 - N_{q_N})\xi \ln(\kappa b) \quad (11\text{-A.33})$$

The local concentration of territorially bound ions is

$$c_b = 1000 q_N / V_\mathrm{p} \quad (11\text{-A.34})$$

where V_p is the volume of "bound" region in cubic centimeters per mole of polyion equivalent.

To evaluate the unknown parameters, we treat counterion condensation as a binding reaction

$$M_{\mathrm{N}^+} \text{ (free) } = M_{\mathrm{N}^+} \text{ (bound)} \quad (11\text{-A.35})$$

with reaction affinity

$$\Delta\mu = \mu_\mathrm{b} - \mu_\mathrm{f} = 0 \quad (11\text{-A.36})$$

at equilibrium. This procedure is equivalent to minimizing G with respect to θ_N at constant c_b and V_p. In the limit of infinite dilution, this enables unique solutions for θ_N, c_b, and V_p:

$$\theta_N = N^{-1}(1 - 1/N\xi) \quad (11\text{-A.37})$$
$$V_p = 4\pi e N_\mathrm{a} N' n^{-1}(\nu + \nu')(\xi - 1/N)b^3 \quad (11\text{-A.38})$$
$$c_b = 1000\theta_N / V_\mathrm{p} \quad (11\text{-A.39})$$

where ν and ν' are the number of counterions and co-ions in the formula for the salt, and N and N' are the valences of counterion and co-ion.

References

Anderson, C. F. and Record, M. T., Jr. (1980). The relationship between the Poisson–Boltzmann model and the condensation hypothesis: An analysis based on the low salt form of the Donnan coefficient. *Biophys. Chem.* **11**, 353–360.

Anderson C. F. and M. T. Record, Jr. (1982). Polyelectrolyte theories and their applications to DNA. *Annu. Rev. Phys. Chem.* **33**, 191–222.

Anderson, C. F. and Record, M. T., Jr. (1990). Ion distributions around DNA and other cylindrical polyions: Theoretical descriptions and physical implications. *Annu. Rev. Biophys. Biophys. Chem.* **19**, 423–465.

Barton, J. K. (1986). Metals and DNA: Molecular left-handed complements. *Science* **233**, 727–734.

Barton, J. K. and Raphael, A. L. (1985). Site-specific cleavage of left-handed DNA in pBR322 by Λ-tris(diphenylphenanthroline)cobalt(III). *Proc. Natl. Acad. Sci. USA* **82**, 6460–6464.

Behe, M. and Felsenfeld, G. (1981). Effects of methylation on a synthetic polynucleotide: The B–Z transition in poly(dG-m^{5}dC).poly(dG-m^{5}dC). *Proc. Natl. Acad. Sci. USA* **78**, 1619–1623.

Behe, M. J., Felsenfeld, G., Szu, S. C., and Charney, E. (1985). Temperature-dependent conformational transitions in poly(dG-dC) and poly(dG-m^{5}dC). *Biopolymers* **24**, 289–300.

Berg, O. G. (1990). The influence of macromolecular crowding on the thermodynamic activity constants of spherical and dumbbell-shaped molecules. *Biopolymers* **30**, 1027–1037.

Berggren, E., Nordenskiöld, L., and Braunlin, W. H. (1992). Interpretation of ^{25}Mg spin relaxation in Mg-DNA solutions: Temperature variation and chemical exchange effects. *Biopolymers* **32**, 1339–1350.

Berman, H. M. (1991). Hydration of DNA. *Curr. Opin. Struct. Biol.* **1**, 423–427.

Berman, H. M. (1994) Hydration of DNA: take 2. *Curr. Opin. Struct. Biol.* **4**, 345–350.

Berman, H. M., Sowri, A., Ginell, S., and Beveridge, D. (1988). A Systematic Study of Patterns of Hydration in Nucleic Acids: (I) Guanine and Cytosine. *J. Biomol. Struct. Dyn.* **5**, 1101–1110.

Bleam, M. L., Anderson, C. F., and Record, M. T., Jr. (1980). Relative binding affinities of monovalent cations for double-stranded DNA. *Proc. Natl. Acad. Sci. USA* **77**, 3085–3089.

Braunlin, W. H. (1995). NMR studies of cation binding environments on nucleic acids, in *Advances in Biophysical Chemistry*, Vol. 5, Allen Bush, D., Ed., JAI Press, Greenwich, CT.

Braunlin, W. H., Anderson, C. F., and Record, M. T., Jr. (1987). Competitive Interactions of $Co(NH_3)_6^{3+}$ and Na^+ with Helical B-DNA Probed by ^{59}Co and ^{23}Na NMR. *Biochemistry* **26**, 7724–7731.

Braunlin, W. H. and Bloomfield, V. A. (1991). ^{1}H NMR study of the base-pairing reactions of d(GGAATTCC): Salt effects on the equilibria and kinetics of strand association. *Biochemistry* **30**, 754–758.

Braunlin, W. H., Drakenberg, T., and Nordenskiöld, L. (1992). Ca^{2+} binding environments on natural and synthetic polymeric DNA's. *J. Biomol. Struct. Dyn.* **10**, 333–343.

Braunlin, W. H., Strick, T. J., and Record, M. T. Jr. (1982). Equilibrium dialysis studies of polyamine binding to DNA. *Biopolymers* **21**, 1301–1314.

Brennan, R. G., Westhof, E., and Sundaralingam, M. (1986). Structure of a Z-DNA with two different backbone chain conformations. Stabilization of the decadeoxyoligonucleotide d(CGTACGTACG) by $[Co(NH_3)_6]^{3+}$ binding to the guanine. *J. Biomol. Struct. Dyn.* **3**, 649–665.

Chang, D-K and Kearns, D. R. (1986). Distribution of Mn^{2+} ions around poly(rI) • poly(rC). *Biopolymers* **25**, 1283–1297.

Chevrier, B., Dock, A. C., Hartmann, B., Leng, M., Moras, D., Thuong, M. T., and Westhof, E. (1986). Solvation of the left-handed hexamer d(5BrC-G-5BrC-G-5BrC-G) in crystals grown at two temperatures. *J. Mol. Biol.* **188**, 707–719.

Cohen, G. and Eisenberg, H. (1968). Deoxyribonucleate solutions: Sedimentation in a density gradient, partial specific volumes, density and refractive index increments, and preferential interactions. *Biopolymers* **6**, 1077–1100.

Cohen, S. S. (1971). *Introduction to the Polyamines*, Prentice-Hall, New York, p. 29.

Conner, B. N., Yoon, C., Dickerson, J. L., and Dickerson, R. E. (1984). Helix geometry and hydration in an A-DNA tetramer: I-CCGG *J. Mol. Biol.* **174**, 663–695.

Cooke, R. and Kuntz, I. D. (1974). The properties of water in biological systems. *Annu. Rev. Biophys. Bioeng.* **3**, 95–126.

Crothers, D. M. and Ratner, D. I. (1968). Thermodynamic studies of a model system for hydrophobic bonding. *Biochemistry* **7**, 1823–1827.

Daune, M. (1974). Interactions of Metal Ions with Nucleic Acids, Chapter 1, Sigel, H., Ed., in *Metal Ions in Biological Systems*, Vol 3: High Molecular Complexes, Marcel Dekker, New York, pp. 1–43.

deHaseth, P. L., Lohman, T. M., and Record, M. T., Jr. (1977) Nonspecific interaction of *lac* repressor with DNA: An association reaction driven by counterion release. *Biochemistry* **16**, 4783–4790.

Dickerson, R. E., Drew, H. R., Conner, B. N., Wing, R. M., Fratini, A. V., and Kopka, M. L. (1982). The anatomy of A-,B-, and Z-DNA. *Science* **216**, 475–485.

Dove, W. F. and Davidson, N. (1962). Cation effects on the denaturation of DNA. *J. Mol. Biol.* **5**, 467–478.

Drew, H. R. and Dickerson, R. E. (1981). Structure of a B-DNA dodecamer. III. Geometry of Hydration. *J. Mol. Biol.* **151**, 535–556.

Duguid, J., Bloomfield, V. A., Benevides, J., and Thomas, G. J., Jr. (1993). Raman spectroscopy of DNA–metal complexes. I. Interactions and conformational effects of the divalent cations: Mg, Ca, Sr, Ba, Mn, Co, Ni, Cu, Pd, and Cd *Biophys. J.* **65**, 1916–1928.

Duguid, J. G., Bloomfield, V. A., Benevides, J. M., and Thomas, G. J., Jr. (1995). Raman spectroscopy of DNA-metal complexes. II. The thermal denaturation of DNA in the presence of Sr^{2+}, Ba^{2+}, Mg^{2+}, Ca^{2+}, Mn^{2+}, Co^{2+}, Ni^{2+} and Cd^{2+}. *Biophys. J.* **69**, 2623–2641.

Eichhorn, G. L. (1981). The effect of metal ions on the structure and function of nucleic acids, Chapter 1, in *Advances in Inorganic Biochemistry*, Vol 3, Metal Ions in Genetic Information Transfer, Eichhorn, G. L. and Marzilli, L. G., Eds., Elsevier/North Holland The Netherlands, pp. 1–46.

Eichhorn, G. L. and Shin, Y. A. (1968) Interaction of metal ions with polynucleotides and related compounds. XII. The relative effect of various metal ions on DNA helicity. *J. Am. Chem. Soc.* **90**, 7323–7328.

Eisenberg, D. and Kauzmann, W. (1969). *The Structure and Properties of Water*, University Press, Oxford, UK. p. 296.

Eisenberg, H. (1976). Biological Macromolecules and Polyelectrolytes in Solution. Oxford University Press, London, UK.

Elson, e. L., Scheffler, J. E., and Baldwin, R. L. (1970). Helix formation by d(TA) oligomers. *J. Mol. Biol.* **54**, 401–415.

Falk, M. (1965). Hydration of purines, pyrimidines, nucleosides, and nucleotides, *Can. J . Chem.* **43**, 314–318.

Falk, M., Hartman, K. A., Jr., and Lord, R. C., (1962). Hydration of deoxyribonucleic acid. I. A gravimetric study. *J. Am. Chem. Soc.* **84**, 3843–3846.

Falk, M., Hartman, K. A., Jr., and Lord, R. C. (1963). Hydration of deoxyribonucleic acid. II. An infrared study. *J. Am. Chem. Soc.* **85**, 387–391.

Falk, M., Poole, A. G., and Goymour, C. G., (1970). Infrared study of the state of water in the hydration shell of DNA. *Can. J. Chem.* **48**, 1536–1542.

Fenley, M. O., Manning, G. S., and Olson, W. K. (1990a). Approach to the limit of counterion condensation. *Biopolymers* **30**, 1191–1203.

Fenley, M. O., Manning, G. S., and Olson, W. K. (1990b). A numerical counterion condensation analysis of the B–Z transition of DNA. *Biopolymers* **30**, 1205–1213.

Feuerstein, B. G., Pattabiraman, N., and Marton, L. J. (1986). Spermine-DNA interactions: A theoretical study. *Proc. Natl. Acad. Sci. USA* **83**, 5948–5952.

Finney, J. L., Goodfellow, J. M., Howell, P. L., and Vovelle, F. (1985) Computer Simulation of Aqueous Biomolecular Systems. *J. Biomol. Struct. Dyn.* **3**, 599–622.

Garner, M. M., Cayley, D. S., and Record, M. T., Jr. (1990). Calculation of macromolecular crowding effects on protein–DNA interactions *in vivo*. *Biophys. J.* **57**, 62a.

Gessner, R. V., Quigley, G. J., Wang, A. H.-J. van der Marel, G. A., van Boom, J. H., and Rich, A. (1985). Structural basis for stabilization of Z-DNA by cobalt hexaammine and magnesium cations. *Biochemistry* **24**, 237–240

Gosule, L. C. and Schellman, J. A. (1976). Compact form of DNA induced by spermine. *Nature (London)* **259**, 333–335.

Granot, J. and Kearns, D. R. (1982a). Interactions of DNA with Divalent Metal Ions. II. Proton Relaxation Enhancement Studies. *Biopolymers* **21**, 203–218.

Granot, J. and Kearns, D. R. (1982b). Interactions of DNA with Divalent Metal Ions. III. Extent of Metal Binding: Experiment and Theory. *Biopolymers* **21**, 219–232.

Granot, J., Feigon, J., and Kearns, D. R. (1982). Interactions of DNA with Divalent Metal Ions. I. 31P-NMR Studies. *Biopolymers* **21**, 181–201.

Gray, D. M., Edmondson, S. P., Lang, D., and Vaughn, M. (1979). The circular dichroism and X-ray diffraction of DNA condensed from ethanolic solutions. *Nucleic Acids Res.* **6**, 2089–2107.

Gray, H. B., Jr, Bloomfield, V. A., and Hearst, J. E. (1967). Sedimentation coefficients of linear and cyclic wormlike chains with excluded volume effects. *J. Chem. Phys.* **46**, 1493–1498.

Hara, M., ed. (1992). *Polyelectrolytes: Science and Technology*, Marcel-Dekker, New York.

Hardin, C. C., Henderson, E., Watson, T., and Prosser, J. K. (1991). Monovalent cation induced structural transitions in telomeric DNAs: G-DNA folding intermediates. *Biochemistry* **30**, 4460–4472.

Hearst, J. E. and Vinograd, J. (1961a). A three-component theory of sedimentation equilibrium in a density gradient. *Proc. Nat. Acad. Sci. USA* **47**, 999–1004.

Hearst, J. E. and Vinograd, J. (1961b). The net hydration of T-4 bacteriophage deoxyribonucleic acid and the effect of hydration on buoyant behavior in a density gradient at equilibrium in the ultracentrifuge. *Proc. Nat. Acad. Sci. USA* **47**, 1005–1014.

Heinemann, U. and Alings, C. (1989). Crystallographic study of one turn of GC-rich B-DNA. *J. Mol. Biol.* **210**, 369–381.

Hingerty, B. E., Brown, R. S., and Klug, A. (1982). Stabilization of the tertiary structure of yeast phenylalanine tRNA by $Co(NH_3)_6^{3+}$: X-ray evidence for hydrogen bonding to pairs of guanine bases in the major groove. *Biochim. Biophys. Acta* **697**, 78–82.

Izatt, R. M., Christensen, J. J., and Rytting, J. H. (1971). Sites and thermodynamic quantitites associated with proton and metal ion interaction with ribonucleic acid, deoxyribonucleic acid, and their constituent bases, nucleosides, and nucleotides. *Chem. Revs.* **71**, 439–481.

Jayaram, B., Swaminathan, S., Beveridge, D. L., Sharp, K., and Honig, B. (1990). Monte Carlo simulation studies on the structure of the counterion atmosphere of B-DNA. Variations on the primitive dielectric model. *Macromolecules*. **23**, 3156–3165.

Kang, C., Zhang, X., Ratliff, R., Moyzis, R., and Rich, A. (1992). Crystal structure of four-stranded *Oxytricha* telomeric DNA. *Nature (London)* **356**, 126–131.

Kennard, O., Cruse, W. B. T., Nachman, J., Prange, T., Shakked, Z., and Rabinovich, D. (1986). Ordered water structure in an A-DNA octamer at 1.7 Å resolution, *J. Biomol. Struct. Dyn.* **3**, 623–647.

Kennedy, S. D. and Bryant, R. G. (1986). Manganese-deoxyribonucleic acid binding modes. *Biophys. J.* **50**, 669–676.

Khan, M. M. T. and Martell, A. E. (1966). Thermodynamic quantities associated with the interaction of adenosine triphosphate with metal ions. *J. Am. Chem. Soc.* **88**, 668–671.

Khan, M. M. T. and Martell, A. E. (1967). Thermodynamic quantities associated with the interaction of adenosinediphosphoric and adenosinemonophosphoric acids with metal ions. *J. Am. Chem. Soc.* **89**, 5585–5590.

Knoll, D. A., Fried, M. G., and Bloomfield, V. A. (1988). Heat-induced DNA aggregation in the presence of divalent metal salts, in *Structure and Expression: DNA and Its Drug Complexes*, Sarma, R. H., and Sarma, M. H., Ed., Adenine Press, Albany, NY, pp. 123–145.

Kopka, M. L., Fratini, A. V., Drew, H. R., and Dickerson, R. E. (1983). Ordered water structure around a B-DNA dodecamer. A quantitative study. *J. Mol. Biol.* **163**, 129–146.

Kuhn, A. and Kellenberger, E. (1985). Productive phage infection in Escherichia coli with reduced internal levels of the major cations. *J. Bacteriol.* **163**, 906–912.

Kuntz, I. D., Jr., and Kauzmann, W. (1974). Hydration of proteins and polypeptides. *Adv. Protein Chem.* **28**, 239–345.

Kuntz, I. D. Jr., Brassfield, T. S., Law, G. D., and Purcell, G. V. (1969). Hydration of macromolecules. *Science* **163**, 1329–1331.

Lamm, G., Wong, L., and Pack, G. (1994). Monte Carlo and Poisson–Boltzmann calculations of the fraction of counterions bound to DNA. *Biopolymers* **34**, 227–237.

Leirmo, S., Harrison, C., Cayley, D. S., Burgess, R. R., and M. Record, T., Jr. (1987). Replacement of KCl by K glutamate dramatically enhances protein–DNA interactions in vitro. *Biochemistry* **26**, 2095–2101.

Li, A. Z., Qi, L. J., Shih, H. H., and Marx, K. A. (1996). Trivalent counterion condensation on DNA measured by pulse gel electrophoresis. *Biopolymers* **38**, 367–376.

Lipanov, A., Kopka, M. L., Kaczor-Grzeskowiak, M., Quintana, J., and Dickerson, R. E. (1993). Structure of the B-DNA decamer C-C-A-A-C-I-T-T-G-G in two different space groups: Conformational flexibility of B-DNA. *Biochemistry* **32**, 1373–1389.

Liquori, A.M., L. Costantino, V. Crescenzi, V. Elia, E. Giglio, R. Puliti, M. de Santis Savino and V. Vitagliano (1967) Complexes between DNA and polyamines: a molecular model, *J. Mol. Biol.* 24, 113–122.

Ma, C. and Bloomfield, V. A. (1995). Gel electrophoresis measurement of counterion condensation on DNA. *Biopolymers* **35**, 211–216.

Manning, G. S. (1972) On the application of polyelectrolyte "limiting laws" to the helix-coil transition of DNA. I. Excess univalent cations. *Biopolymers*. **11**, 937–949.

Manning, G. S. (1977) Limiting laws and counterion condensation in polyelectrolyte solutions. IV. The approach to the limit and the extraordinary stability of the charge fraction. *Biophys. Chem.* **7**, 95–102.

Manning, G. S. (1978) The molecular theory of polyelectrolyte solutions with applications to the electrostatic properties of polynucleotides. *Q. Rev. Biophys.* **11**, 179–246.

Manning, G. S. (1979) Counterion binding in polyelectrolyte theory. *Accts. Chem. Res.* **12**, 443–449.

Martin, R. B. and Mariam, Y. H. (1979). Interactions between metal ions and nucleic bases, nucleosides, and nucleotides in solution, Sigel H, Ed., in *Metal Ions in Biological Systems*, Vol 8, Chapter 2, Nucleotides and Derivatives: Their Ligating Ambivalency, pp. 57–124 Marcel-Dekker, New York.

Marzilli, L. G. (1981). Metal complexes of nucleic acid derivatives and nucleotides: Binding sites and structures, Eichhorn, G.L. and Marzilli, L.G., Eds., in *Advances in Inorganic Biochemistry*, Vol 3: Chapter 2, Metal Ions in Genetic Information Transfer, Elsevier/North-Holland, The Netherlands, pp. 47–85.

Metzler, D. E. (1977). *Biochemistry: The Chemical Reactions of Living Cells*, Academic, New York. p. 269.

Neidle, S., Berman, H. M., and Shieh, H. S. (1980). Highly structured water network in crystals of a deoxynucleoside–drug complex, *Nature (London)* **288**, 129–133.

Olmsted, M. C., Anderson, C. F., and Record, M. T., Jr. (1989). Monte Carlo Description of Oligoelectrolyte Properties of DNA Oligomers: Range of the End Effect and the Approach of Molecular and Thermodynamic Properties to the Polyelectrolyte Limits. *Proc. Natl. Acad. Sci. USA* **86**, 7766–7770.

Olmsted, M. C., Anderson, C. F., and Record, M. T., Jr. (1991). Importance of oligoelectrolyte end effects for the thermodynamics of conformational transitions of nucleic acid oligomers: A grand canonical Monte Carlo analysis, *Biopolymers* **31**, 1593–1604.

Olmsted, M. C., Bond, J. P., Anderson, C. F., and Record, M. T., Jr. (1995). Grand canonical Monte Carlo molecular and thermodynamic predictions of ion effects on binding of an oligocation (L8+) to the center of DNA oligomers, *Biophys. J.* **68**, 634–647.

Oosawa, F. (1971). *Polyelectrolytes*, Marcel-Dekker, New York, 160 pp.

Padmanabhan, S., Richey, B., Anderson, C. F., and Record, M. T., Jr. (1988). Interaction of an *N*-Methylated Polyamine Analogue, Hexamethonium(2+), with NaDNA: Quantitative ^{14}N and ^{23}Na NMR Relaxation Rate Studies of the Cation-Exchange Process. *Biochemistry* **27**, 4367–4376.

Privalov, P. L. and Mrevlishvili, G. M. (1967). Macromolecule hydration in native and denatured state. *Biofizika* **12**, 22–29.

Privé, G. G., Yanagi, K., and Dickerson, R. E. (1991). Structure of the B-DNA decamer C-C-A-A-C-G-T-T-G-G and comparison with isomorphous decamers C-C-A-A-G-A-T-T-G-G and C-C-A-G-G-C-C-T-G-G. *J. Mol. Biol.* **217**, 177–199.

Privé, G. G., Heinemann, U., Chandrasegaran, S., Kan, L.-S. M., Kopka, L., and Dickerson, R. E. (1987). Helix Geometry, Hydration, and G.A Mismatch in a B-DNA Decamer. *Science*. **238**, 498–504.

Quigley, G. J., Teeter, M. M., and Rich, A. (1978). Structural analysis of spermine and magnesium ion binding to yeast phenylalanine transfer RNA. *Proc. Natl. Acad. Sci. USA* **75**, 64–68.

Quintana, J. R., Grzeskowiak, K., Yanagi, K., and Dickerson, R. E. (1992). Structure of a B-DNA decamer with a central T-A step: C-G-A-T-T-A-A-T-C-G. *J. Mol. Biol.* **225**, 379–395.

Record, M. T., Jr. (1975). Effects of Na^+ and Mg^{++} ions on the helix-coil transition of DNA. *Biopolymers* **14**, 2137–2158.

Record, M. T., Jr., Anderson, C. F., and Lohman, T. M. (1978). Thermodynamic analysis of ion effects on the binding and conformational equilibria of proteins and nucleic acids: the roles of ion association or release, screening, and ion effects on water activity. *Q. Revs. Biophys.* **11**, 103–178.

Record, M. T., Jr, Anderson, C. F., Mills, P., Mossing, M., and Roe, J.-H. (1985). Ions as regulators of protein-nucleic acid interactions in vitro and in vivo, *Adv. Biophys.* **20**, 109–135.

Richey, B., Cayley, D. S., Mossing, M. C., Kolka, C., Anderson, C. F., Farrar, T. C., and Record, M. T., Jr. (1987). Variability of the intracellular ionic environment of E coli: Differences between in vitro and in vivo effects of ion concentrations on protein-DNA interactions and gene expression. *J. Biol. Chem.* **262**, 7157–7164.

Rose, D. M., Bleam, M. L., Record, M. T., Jr., and Bryant, R. G. (1980). ^{25}Mg NMR in DNA solutions: Dominance of site binding effects. *Proc. Natl. Acad. Sci. USA* **77**, 6289–6292.

Rose, D. M., Polnaszek, C. F., and Bryant, R. G. (1982). ^{25}Mg NMR investigations of the magnesium ion-DNA interaction. *Biopolymers* **21**, 653–664.

Ross, P. D. and Shapiro, J. T. (1974). Heat of interaction of DNA with polylysine, spermine, and Mg^{++}. *Biopolymers* **13**, 415–416.

Saenger, W. (1984) Water and Nucleic Acids, in *Principles of Nucleic Acid Structure*, Chapter 17, Springer-Verlag, Berlin, pp. 368-384.

Saenger, W. (1987). Structure and Dynamics of Water Surrounding Biomolecules. *Annu. Rev. Biophys. Biophys. Chem.* **16**, 93–114.

Saenger, W., Hunter, W. N., and Kennard, O. (1986). DNA conformation is determined by economics in the hydration of phosphate groups. *Nature (London)* **324**, 385–388.

Schellman, J. A. and Stigter, D. (1977). Electrical Double Layer, Zeta Potential, and Electrophoretic Charge of Double-Stranded DNA. *Biopolymers*. **16**, 1415–1434.

Schmitz, K. S. (1993). *Macroions in Solution and Colloidal Suspension*, VCH, New York.

Schneider, B., Cohen, D., and Berman, H. M. (1992). Hydration of DNA bases: Analysis of crystallographic data. *Biopolymers* **32**, 725–750.

Schneider, B., Cohen, D. M., Schleifer, L., Srinivasan, A. R., Olson, W. K., and Berman, H. M. (1993). A systematic method for studying the spatial distribution of water molecules around nucleic acid bases. *Biophys. J.* **65**, 2291–2303.

Sherman, S. E., Gibson, D., Wang, A. H.-J., and Lippard, S. J. (1985). X-ray structure of the major adduct of the anticancer drug cisplatin with DNA: *cis*-$[Pt(NH_3)_2\{d(pGpG)\}]$. *Science* **230**, 412–417.

Sigel, H. (1993). Interaction of metal ions with nucleotides and nucleic acids and their constituents. *Chem. Soc. Rev.* **22**, 255-267.

Smith, F. W. and Feigon, J., (1992). Quadruplex structure of *Oxytricha* telomeric DNA oligonucleotides. *Nature (London)* **356**, 164–168.

Soumpasis, D.-M. (1984). Statistical Mechanics of the B–Z Transition of DNA: Contribution of Diffuse Ionic Interactions. *Proc. Natl. Acad. Sci. USA* **81**, 5116-5120.

Soumpasis, D. M., Wiechen, J., and Jovin, T. M. (1987). Relative stabilities and transitions of DNA conformations in 1:1 electrolytes: A theoretical study. *J. Biomol. Struct. Dyn.* **4**, 535–551.

Stigter, D. (1975). The charged colloidal cylinder with a Gouy double layer. *J. Colloid Interface Sci.* **53**, 296–306.

Stigter, D. (1977). Interactions of Highly Charged Colloidal Cylinders with Applications to Double-Stranded DNA. *Biopolymers* **16**, 1435–1448.

Stigter, D. (1978). A comparison of Manning's polyelectrolyte theory with the cylindrical Gouy model. *J. Phys .Chem* .**82**, 1603–1606.

Stigter, D. (1982) Coil Expansion in Polyelectrolyte Solutions. *Macromolecules* **15**, 635–641.

Subramanian, P. S. and Beveridge, D. L. (1989). A theoretical study of the aqueous hydration of canonical B d(CGCGAATTCGCG): Monte Carlo simulation and comparison with crystallographic ordered water sites. *J. Biomol. Struct. Dyn.* **6**, 1093–1122.

Subramanian, P. S., Pitchumani, S., Beveridge, D. L., and Berman, H. M. (1990b). A Monte Carlo simulation study of the aqueous hydration of $r(GpC)_2$: Comparison with crystallographic ordered water sites. *Biopolymers* **29**, 771–783.

Subramanian, P. S., Ravishanker, G., and Beveridge, D. L. (1988). Theoretical Considerations on the "Spine of Hydration" in the Minor Groove of d(CGCGAATTCGCG) • d(GCGCTTAAGCGC): Monte Carlo Computer Simulation. *Proc. Natl. Acad. Sci. USA* **85**, 1836–1840.

Subramanian, P. S., Swaminathan, S., and Beveridge, D. L. (1990a). Theoretical account of the 'spine of hydration' in the minor groove of duplex d(CGCGAATTCGCG). *J. Biomol. Struct. Dyn.* **7**, 1161–1165.

Suwalsky, M., Traub, W., Shmueli, U., and Subirana, J. (1969). An X-ray study of the interaction of DNA with spermine. *J. Mol. Biol.* **42**, 363–373.

Tabor, C. W. and Tabor, H. (1984). Polyamines. *Annu. Rev. Biochem.* **53**, 749–790.

Tam, S.-C. and Williams, R. J. P. (1985). Electrostatics and Biological Systems. *Structure Bonding* **63**, 105.

Tanford, C. (1961). *Physical Chemistry of Macromolecules*, Chapter 7, Wiley, New York.

Tao, N. J., Lindsay, S. M., and Rupprecht, A. (1987). The dynamics of the DNA hydration shell at gigahertz frequencies. *Biopolymers* **26**, 171–188.

Teeter , M. M., Quigley, G. J., and Rich, A. (1981). The binding of metals to tRNA. *Adv. Inorg. Biochem.* **3**, 233–272.

Texter, J. (1978). Nucleic acid—water interactions. *Prog. Biophys. Mol. Biol.* **33**, 83–97.

Thomas, T. J. and Bloomfield, V. A. (1984). Ionic and structural effects on the thermal helix-coil transition of DNA complexed with natural and synthetic polyamines. *Biopolymers* **23**, 1295–1306.

Thomas T. J., Bloomfield, V. A., and Canellakis, N. (1985). Differential effects on the B to Z transition of poly(dG-me5dC).poly(dG-me5dC) produced by N1 and N8-acetyl spermidine. *Biopolymers* **24**, 724–729.

Tsuboi, M. (1964). The melting temperature of nucleic acid in solution. *Bull. Chem. Soc. Jpn.* **37**, 1514–1522.

Tunis, M.-J. B. and Hearst, J. E. (1968). On the hydration of DNA. II. Base composition dependence of the net hydration of DNA. *Biopolymers* **6**, 1345–1353.

van de Sande, J. H. and Jovin, T. M. (1982). Z* DNA, the left-handed helical form of poly[d(G-C)] in $MgCl_2$-ethanol, is biologically active. *EMBO J.* **1**, 115–120.

van de Sande, J. H., McIntosh, L. P., and Jovin, T. M. (1982). Mn^{2+} and other transition metals at low concentration induce the right-to-left helical transformation of poly[d(G-C)]. *EMBO J.* **1**, 777–782.

Wang, A. H.-J., Hakoshima, T., van der Marel, G. van Boom, J. H., and Rich, A. (1984). AT base pairs are less stable than GC base pairs in Z-DNA: The crystal structure of $d(m^5CGTAm^5CG)_2$. *Cell* **37**, 321–331

Wang, J. H. (1955) The hydration of desoxyribonucleic acid. *J. Am. Chem. Soc.* **77**, 258–260.

Wemmer, D. E., Srivenugopal, K. S., Reid, B. R., and Morris, D. R. (1985). NMR Studies of Polyamine Binding to a Defined DNA Sequence. *J. Mol. Biol.* **185**, 457-459.

Westhof, E. (1988). Water: An Integral Part of Nucleic Acid Structure. *Annu. Rev. Biophys. Biophys. Chem.* **17**, 125–144.

Westhof, E. (1993) *Water and Biological Macromolecules*, CRC Press, Boca Raton, FL.

Wilson, R. W., Rau, D. C., and Bloomfield, V. A. (1980). Comparison of Polyelectrolyte Theories of the Binding of Cations to DNA. *Biophys. J.* **30**, 317–326.

Woisard A, Fazakerley, G. V., and Guschlbauer, W. (1985). Z-DNA is formed by poly(dC-dG) and poly(dm^5-dC) at micro or nanomolar concentrations of some zinc(II) and copper(II) complexes. *J. Biomol. Struct. Dynam.* **2**, 1205–1220.

Wolf, B. and Hanlon, S. (1975). Structural transitions of deoxyribonucleic acid in aqueous electrolyte solutions. II. The role of hydration. *Biochemistry* **14**, 1661–1670.

Xu, Q., Deng, H., and Braunlin, W.H. (1993a) Selective localization and rotational immobilization of univalent cations on quadruplex DNA. *Biochemistry* **32**, 13130–13137.

Xu, Q., Jampani, S. R. B., and Braunlin, W. H. (1993b). Rotational dynamics of hexaamminecobalt(III) bound to oligomeric DNA: Correlation with cation-induced structural transitions. *Biochemistry* **32**, 11754–11760.

Xu, Q., Shoemaker, R. K., and Braunlin, W. H. (1993c). Induction of B-A transitions of deoxyoligonucleotides by multivalent cations in dilute aqueous solution. *Biophys. J.* **65**, 1039–1049.

Zehfus, M. H. and Johnson, W. C., Jr. (1981). Properties of P-form DNA as revealed by circular dichroism *Biopolymers* **20**, 1589–1603.

Zehfus, M. H. and Johnson, W. C., Jr. (1984). Conformation of P-form DNA. *Biopolymers* **23**, 1269–1281.

Zimm, B. H. and Le Bret. M. (1983). Counter-ion condensation and system dimensionality. *J. Biomol. Struct. Dyn.* **1**, 461–471.

CHAPTER 12

Interaction and Reaction with Drugs[1]

Many natural compounds interact strongly and specifically with nucleic acids, particularly DNA. These drugs range from potent antitumor and antibiotic compounds of considerable importance in medicine, to agents such as ethidium, which is widely used as a fluorescent stain for DNA. Illustrative structures are given in Figure 12-1. Drugs that react covalently with DNA are considered separately in Section 6.

Study of the structure, binding specificity, and dynamics of drug–DNA complexes has dual objectives: to elucidate the properties of the drug, and to probe the interactive capability of the "host" nucleic acid in this host–guest interaction. Structure is of vital importance, since it is on this basis that one seeks to understand the spectroscopic properties, binding specificity, hydrodynamic, and dynamic characteristics. Drug–DNA

[1]Dedicated by DMC to Werner Müller who taught me much about this field.

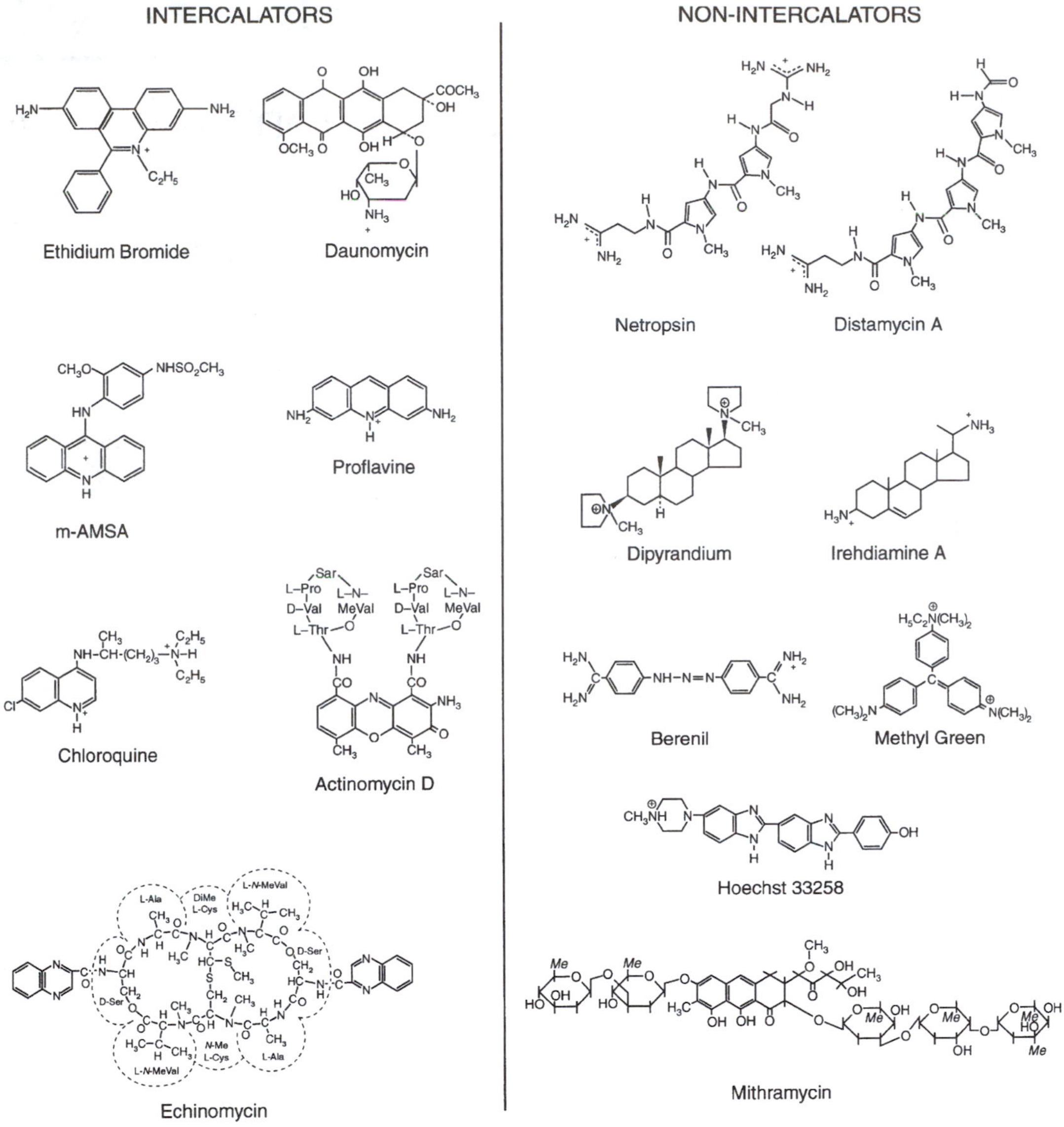

INTERCALATORS

Ethidium Bromide

Daunomycin

m-AMSA

Proflavine

Chloroquine

Actinomycin D

Echinomycin

NON-INTERCALATORS

Netropsin

Distamycin A

Dipyrandium

Irehdiamine A

Berenil

Methyl Green

Hoechst 33258

Mithramycin

Figure 12-1
Structures of some DNA binding compounds.

complexes have been studied for several decades by a large variety of methods. We will focus on the major issues, with selective examples chosen from the large body of literature.

1. MODES OF INTERACTION: INTERCALATION AND GROOVE BINDING

There are two major classes of DNA binding compounds: those that bind by intercalation, and those that attach to DNA by some other mode. In intercalation, first proposed as a structural model by Lerman (1961), a flat, usually heteroaromatic polycyclic ring is inserted between two base pairs in the double helix. Stacking parallel to its neighboring base pairs, the intercalated residue lengthens the DNA double helix by approximately 1 bp spacing, and also produces substantial unwinding of the helix. These properties provide a set of diagnostic criteria for intercalators, enabling direct experimental test of the binding mode for specific drugs, as discussed in Section 4. The properties of intercalating compounds have been extensively reviewed (Berman and Young, 1981; Waring, 1981; Wilson and Jones, 1981, 1982; Dougherty and Pilbrow, 1984, Neidle and Abraham, 1984; Zimmerman, 1986; Wang, 1987).

The final proof of the intercalation model has come from X-ray crystallographic analysis of drug–oligonucleotide complexes (see Chapter 4). Figure 12-2 shows views of the intercalated complexes of ethidium and daunomycin with DNA. Note that for ethidium, the long axis of the drug parallels the long axis of the base pair, but that for daunomycin, the drug long axis points between the major and minor grooves. In both cases, the nonplanar substituents attached to the intercalated "chromophore" (so-called because these compounds are generally colored) protrude into the minor groove. Positioning in the minor groove is also found for the cyclic peptide side chains of actinomycin (Sobell, 1974). In contrast, most protein–DNA interactions target the DNA major groove rather than the minor groove (see Chapter 13). In general, the DNA minor groove is too small to allow access by protein structural motifs such as the α-helix, but it is able to accommodate small drug molecules and their side chains. However, the exclusion of protein interactions from the minor groove is by no means absolute; for example, a peptide loop from DNAase I interacts with its substrate there (Suck and Oeffner, 1986)

Groove binding drugs, reviewed in detail by Zimmer and Wähnert (1986), generally interact with the DNA minor groove. Figure 12-3 shows the results of an NMR study of the complex between berenil and a double helical DNA oligonucleotide (Lane et al., 1991). The drug prefers binding to DNA sequences containing mainly A•T pairs, a specificity that has been ascribed to hydrophobic contact between adenine C2 hydrogen atoms and the phenyl rings in the drug (Brown et al., 1990). This interaction is similar to the proposed hydrophobic basis for the specificity of binding netropsin to the DNA minor groove in A•T rich regions, in which the pyrrole rings of the drug molecule provide the hydrophobic surface (Kopka et al., 1985). Hydrogen bonds are also formed between the drug amide NH groups and the adenine N3 and thymine O2 atoms in the case of netropsin. Both berenil and netropsin have curved shapes, which match the helical curvature of the DNA minor groove. A number of other compounds bind to the DNA minor groove, including some, such as chromomycin, which are specific for G•C containing sequences (Gao and Patel, 1989).

2. THEORETICAL DESCRIPTION OF BINDING EQUILIBRIA

Drugs that bind to DNA noncovalently generally do so reversibly, so the interaction can be described by classical thermodynamics. Here we consider theoretical models that describe the equilibrium, to be followed in Section 12.3 by consideration of experimental methods that probe the strength of binding and the nature of the preferred binding sites. Underlying the complexity of this problem is the variability of local DNA structure and interaction potential due to local sequence fluctuations. Consequently,

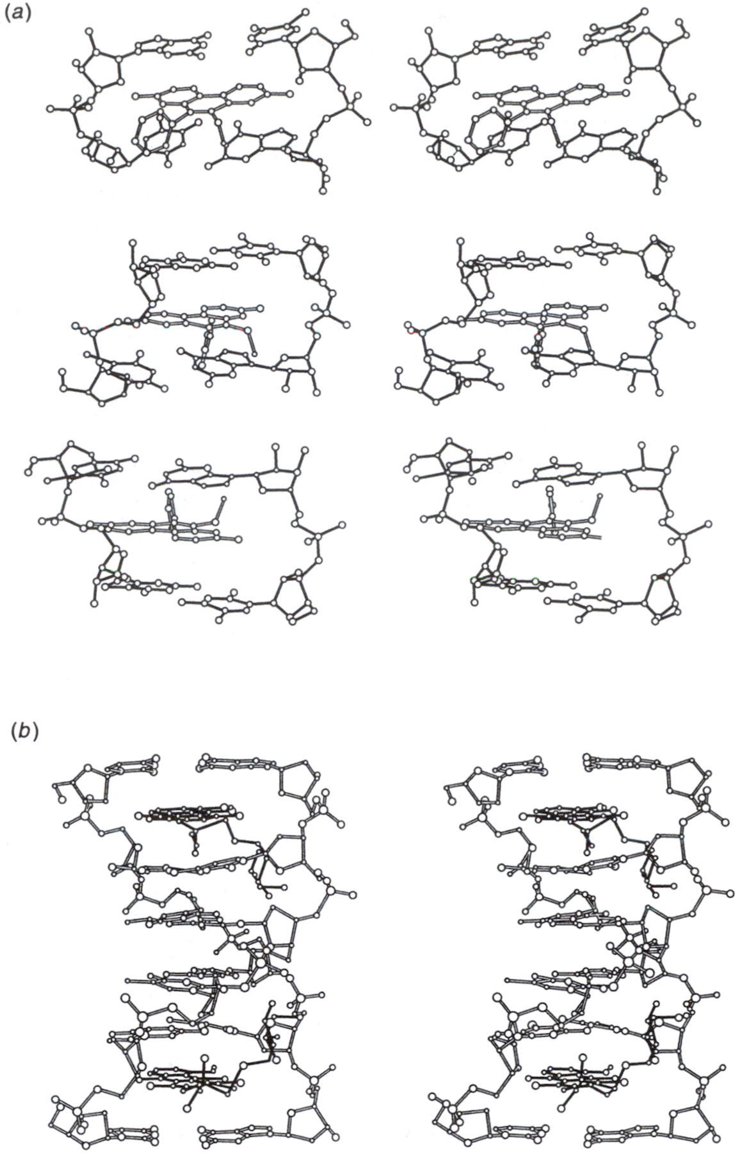

Figure 12-2
(*a*) Stereopairs of intercalative drug binding, as visualized in the ethidium/iodo CpG crystal structure. The two upper views are from the minor groove, while the lower view is from the major groove. [Reprinted from Aspen, SC., Tsai, G.-C., and Sobell, H. M., Visualization of drug–nuclei interactions at atomic resolution II Structure of an ethidium/denucleotide monophosphate crystalline comples, theduim: 5-isododocytidylyl (3′–5′) guanosine, *J. Mol. Biol.*, **114**, 317–331, copyright ©1977, by permission of the publisher Academic Press Limited London.] (*b*) Structure of the daunomycin complex with d(CGATCG), showing two drugs bound to the hexanucleotide. The upper drug is viewed from the major groove, and the lower from the minor groove. [Reprinted with permission from Wang, A. H.-J., Ughetto, G., Quigley, G. J., and Rich, A., *Biochemistry* **26**, 1152–1163 (1987).]

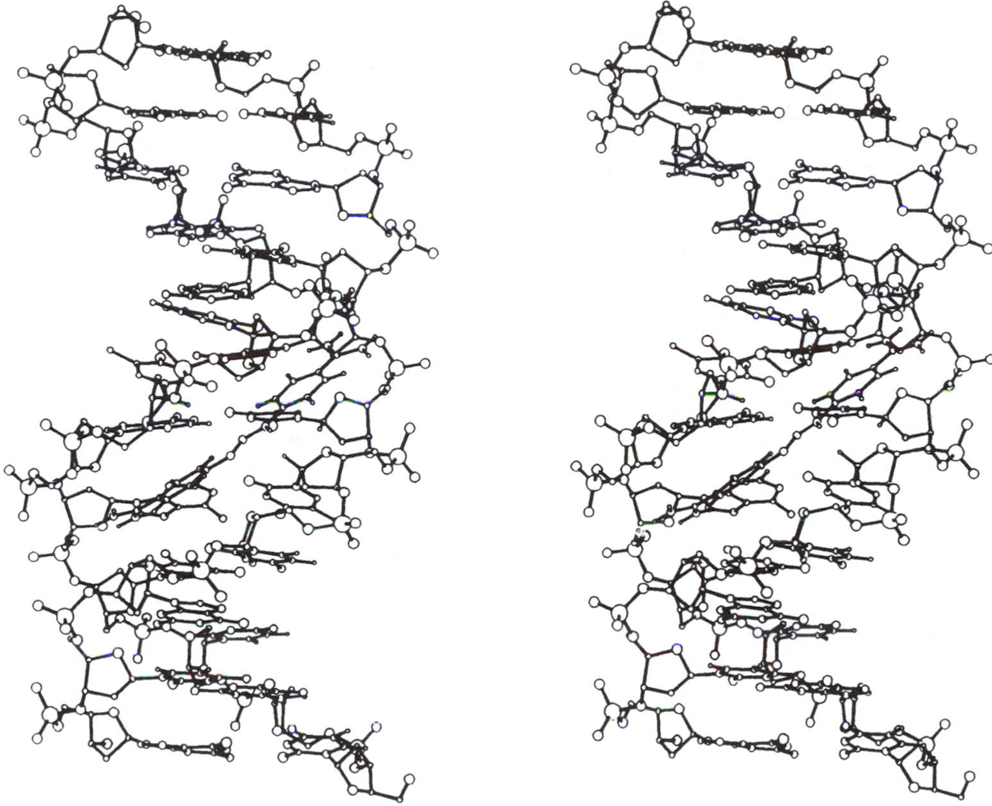

Figure 12-3
Stereoview of the refined NMR structural model for the binding of berenil to the dodecameric DNA sequence d(CGCGAATTCGCG). [Reprinted with permission from Lane, A. N., Jenkins, T. C., Brown, T., and Neidle, S., *Biochemistry* **30**, 1372–1385 (1991). Copyright ©1991 American Chemical Society.]

DNA is distinctly nonuniform from the perspective of a drug molecule seeking its natural binding site.

2.1 Equilibrium Binding Models

2.1.1 Independent Binding Sites Model

The simplest thermodynamic model for drug–DNA interaction is that devised by Scatchard (1949). The Scatchard model considers that a specific group of n base pairs constitutes a binding site; we call $B_{ap} = 1/n$ the apparent number of binding sites per base pair, where n is the number of base pairs occupied by the bound drug. In order to move a bound molecule to a new site, it must be shifted by some integral multiple of n base pairs (Fig. 12-4). The total number of potential binding sites per DNA molecule in this model is N/n, where N is the total number of base pairs in the molecule.

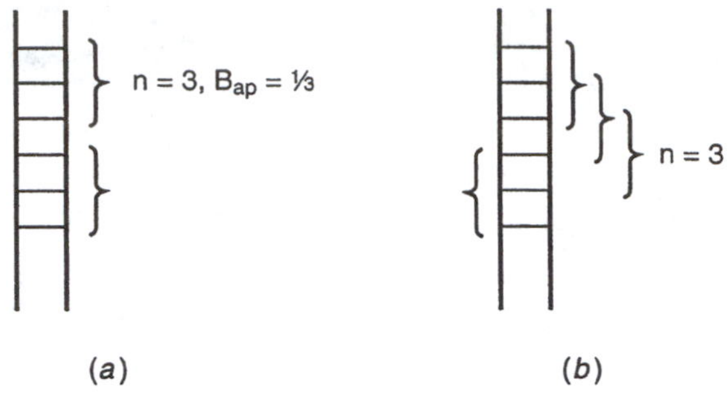

Figure 12-4
(*a*) *Scatchard model* for binding to DNA when $B_{ap} = \frac{1}{3}$. Six base pairs provide two binding sites, because binding sites do not overlap. (*b*) *Neighbor exclusion model*. Because the binding sites overlap, 6 bp provide four *potential* binding sites. However, occupancy of one site prevents the bases in that site from being used again in another site.

2.1.2 The Neighbor Exclusion Model

This somewhat more realistic model is closely related to the Scatchard model. Once again, a bound drug molecule occupies n base pairs, but now the binding sites overlap: The center of the bound molecule can be shifted 1 bp at a time in occupying new binding sites, assuming that that region of the DNA is not occluded by other bound drugs (Fig. 12-4). At saturation, the number of drug molecules bound per DNA is N/n, as for the Scatchard model, but the number of potential binding sites is approximately $N (N \gg n)$ when the occupancy ratio r of bound drug molecules per base pair is low. This difference is responsible for the contrasting shapes of the two binding isotherms discussed below. Neighbor exclusion effects were first noted by Cairns (1962), when it was observed that proflavine binding saturated at one drug for every 2 bp, even though in principle one might expect one intercalation site for each base pair.

The only genuinely realistic model for drug–DNA interactions is one that incorporates not only neighbor exclusion, but also the sequence preference or specificity characteristic of the drug (Crothers, 1968). Computational methods for predicting the corresponding binding isotherms have been available for more than 20 years, but it is only now that sufficient experimental information about drug sequence preference is becoming available to make it possible to assign values to the local binding constants that are needed to calculate an overall binding isotherm and compare it with experiment.

2.2 The Scatchard Binding Isotherm

We define K_{ap} as the apparent binding constant for the reaction

$$\text{DNA} + \text{drug} \rightleftharpoons \text{Complex}$$

and let C_N° be the concentration of DNA base pairs, C_B the concentration of bound drug, C_F the concentration (more precisely, the activity) of free drug, and $r = C_B/C_N^\circ$ the ratio of bound drug to base pairs. Then the concentration of free DNA binding sites is $(B_{ap} - r)C_N^\circ$, and

$$K_{ap} = \frac{[\text{complex}]}{[\text{free drug}]\,[\text{free sites}]} = \frac{rC_N^0}{C_F(B_{ap} - r)C_N^0} \tag{12-1}$$

which can be rearranged to

$$\frac{r}{C_F} = K_{ap}(B_{ap} - r) \tag{12-2}$$

This form gives rise to a Scatchard plot: The variation of r/C_F with r should be linear, with slope $-K_{ap}$ and x-axis intercept at $r = B_{ap}$; the intercept on the y-axis is $K_{ap}B_{ap}$.

Cases in which the drug aggregates or self-associates can lead to gross errors in binding constant and exclusion range if account is not taken of the effect. The parameter C_F in these equations is the activity of the drug, or, to a first approximation, the concentration of the monomeric form of the free drug. If the drug aggregates, C_F can be much smaller than the total free drug concentration C_{F*} and serious errors can result from replacing C_F by C_{F*} The consequence of assuming $C_F = C_{F*}$ is an artefactual decline of r/C_{F*} as r increases, leading to an underestimate of B_{ap} and overestimate of K_{ap}.

2.3 The Neighbor Exclusion Binding Isotherm

The excluded site problem has been solved by a variety of statistical mechanical methods (Crothers, 1968; Zasedatelev et al., 1971; McGhee and von Hippel, 1974). The simplest general approach to problems of this kind is the method of sequence generating functions, assuming that the DNA molecule is long enough so that end effects are unimportant. Dattagupta et al. (1980) describe briefly how such calculations can be done, including cases in which binding is accompanied by a DNA structural change. The neighbor exclusion binding isotherm, first written in this form by McGhee and von Hippel (1974) is

$$\frac{r}{C_F} = K(1 - nr)\left[\frac{1 - nr}{1 - (n-1)r}\right]^{n-1} \tag{12-3}$$

Recall that n is the number of base pairs covered by a bound drug, or alternatively the minimum distance in base pairs between the center of adjacent bound drug molecules. The equilibrium binding constant for all sites is assumed to be K; note that in making this assumption we ignore the sequence dependence of drug binding affinity. Comparison of Eq. 12-3 to an equation analogous to Eq. 12-1 for the binding constant in terms of concentrations of free drug, complex, and free binding sites allows one to obtain an expression for the concentration of free binding sites C_S in the neighbor exclusion model:

$$C_S = f(r)C_N^0 \qquad (12\text{-}4)$$

in which

$$f(r) = \frac{(1 - nr)^n}{[1 - (n-1)r]^{n-1}} \qquad (12\text{-}5)$$

Figure 12-5 illustrates the agreement between experiment and the neighbor exclusion model for the binding of ethidium to DNA. Excellent agreement is found for a model in which $n = 2$, according to which at least 2 bp must intervene between adjacent intercalated ethidium residues (Bauer and Vinograd, 1970; Bresloff and Crothers, 1975).

Unlike the Scatchard model, the neighbor exclusion model predicts a curved isotherm when r/C_F is plotted against r. The initial linear part of the isotherm when r is small has slope $-K(2n-1)$, and r/C_F approaches zero asymptotically at $r = 1/n$. Application of the Scatchard model to the linear region of an isotherm more properly described by the neighbor exclusion model would lead to the conclusion that $B_{ap} = 1/(2n-1)$, yielding an apparent site size of $2n - 1$ instead of the actual value n.

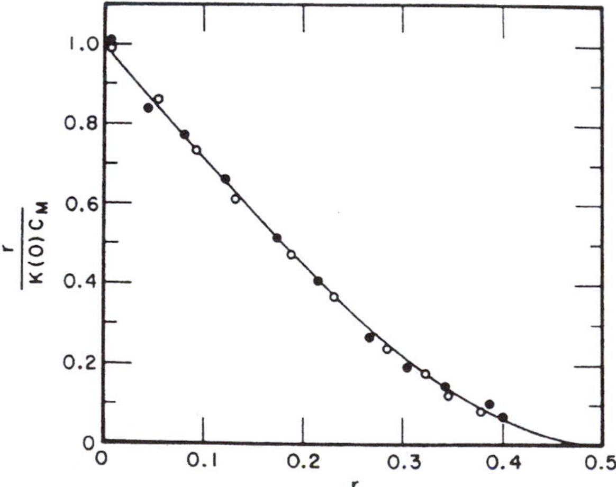

Figure 12-5
Equilibrium dialysis isotherm for ethidium binding to *Micrococcus luteus* (○) and calf thymus (●) DNA, compared with the isotherm calculated using the neighbor exclusion model, with $n = 2$. Isotherms were normalized so that the intercept is 1. [Reprinted with permission from Bresloff, J. L. and Crothers, D. M., *Biochemistry*, **20**, 3547–3554 (1981), Figure 1. Copyright ©1981 American Chemical society.]

The intercept of r/C_F at $r = 0$ is K instead of $K_{ap}B_{ap}$ for the Scatchard isotherm, from which we conclude that the parameters for the two models are related by $K_{ap} = K/B_{ap}$. Since $B_{ap} \leq 1$, the Scatchard model binding constant K_{ap} is generally greater than the neighbor exclusion binding constant K; the reason for this is that the number of potential binding sites in the neighbor exclusion model is larger by $n = 1/B_{ap}$ than in the Scatchard model. With a larger number of sites, it takes a correspondingly smaller binding constant to achieve a given level of binding.

It has become common in using Eq. 12-3 to allow nonintegral values of n in using computational methods to fit the neighbor exclusion isotherm to the data. This tactic is of dubious utility, since the equation was derived using a lattice model for which, by definition, the sites are separated by integral numbers of base pairs. Nonintegral values of n imply that the model is not adequate to describe the data accurately. Contributing factors might include partial exclusion effects at distances near n, and may especially reflect neglect of the role of sequence preference in binding.

The origins of neighbor exclusion effects by simple intercalators such as ethidium are not well understood. A neighbor exclusion range of 2 is common for intercalators, but binding by some RNA homopolymers, for example, poly (A)•poly (U), are characterized by $n = 3$, as well as by modest cooperativity in binding of ethidium ions that are separated by the minimum distance (Bresloff and Crothers, 1981). The intercalator tilorone exhibits an exclusion range of 3 in binding to several synthetic DNA polynucleotides (Sturm et al., 1981). Possible contributors to neighbor exclusion include constraints on sugar pucker at the intercalation site, electrostatic repulsion between bound drugs, and other mechanisms for transmitting long-range interactions. It is likely that the apparent site size is also influenced by preference for some binding sites over others, as discussed in the Section 12.2.4.

Friedman and Manning (1984) examined the potential role of ionic effects in binding of ligands to DNA. They considered the influence of both drug charge and the effect on average charge spacing when DNA is lengthened by intercalation, using the counterion condensation theory described in Chapter 11. The effects can be understood qualitatively in terms of the influence the bound ligand has on the average charge spacing along the helix axis. Binding of a ligand carrying one or more positive charges increases the average charge spacing and hence decreases the linear charge density. A similar consequence results from intercalative binding even if the ligand is uncharged, since DNA length increases, and the phosphate charge remains unchanged. This increase in charge spacing decreases the electrostatic stress on the system that results from repulsion between the polymer charges. In addition, the increased charge separation leads to dissociation of "territorially bound" or condensed counterions. Consequently binding becomes weaker as the salt concentration increases, because the entropy gain upon counterion dissociation is smaller (see Eq. 10-13). This finding is true even if the ligand is uncharged, as found, for example, for actinomycin. However, the influence of salt concentration on binding strength is not as large in this instance as it is for a charged ligand (Müller and Crothers, 1968).

Friedman and Manning (1984) also pointed out that the electrostatic effects have an anticooperative character, in that the decrease in electrostatic stress on the system lessens as the degree of binding rises. The result is that the favorable electrostatic contribution to the binding free energy diminishes as the degree of binding increases, which dictates a decrease in binding constant, assuming other factors to be unchanged. Neighbor exclusion also reflects anticooperative binding, since the number of accessible binding sites drops as the lattice becomes more occluded. All anticooperative binding effects lead to similar curvature of the binding isotherm as exhibited by the data in Figure 12-5. It is not possible on the basis of the simple shape of the binding isotherm to distinguish site exclusion based on steric factors from these electrostatic effects, or from the influence of sequence specificity discussed in the Section 12.2.4.

2.4 Inclusion of Sequence Specificity

Binding sites of different affinity are readily incorporated into the Scatchard model for drug binding. If there are two different sites of substantially different affinity, the Scatchard isotherm consists of a superposition of two isotherms of the form given by Eq. 12-2. The slope of the plot of r/C_F against r will be greater at small r than when r is large, because the sites that are being filled at small r have greater affinity. If the sites are characterized by binding constants K_1 and K_2 and apparent site sizes B_1 and B_2, then the limiting slope as $r \to 0$ can be shown to be

$$\lim_{r \to 0} \left[\frac{\partial (r/C_F)}{\partial r} \right] = -\frac{K_1 + K_2\alpha}{1 + \alpha} \tag{12-6}$$

where $\alpha = K_2B_2/K_1B_1$. When the first binding site is much stronger than the second, $\alpha \to 0$, and the initial slope is $-K_1$. The intercept on the y axis is $K_1B_1 + K_2B_2$, and the limiting slope as $r \to (B_1 + B_2)$ is $-K_2$. The overall direction of curvature expected (Fig. 12-6) is similar to that of the neighbor-exclusion isotherm, Figure 12-5. Distinguishing multiple sites from neighbor exclusion as the source of curvature in the isotherm is a difficult experimental problem.

The matrix method was used by Crothers (1968) to calculate binding isotherms that include both neighbor exclusion and sequence specificity, in an attempt to distinguish possible models for binding of actinomycin to DNA (Müller and Crothers, 1968). Figure 12-7 illustrates the effect of requiring that the binding site contain at least one G•C pair, with a neighbor exclusion range of $n = 4$. For the upper curve, the DNA is assumed to contain only G•C pairs, while the assumed G•C content decreases for the lower curves. Note that the intercept on the r/C_F axis decreases as the G•C content decreases; the limiting form is

$$\lim_{r \to 0} \left[\frac{r}{KC_F} \right] = B_0 \tag{12-7}$$

where B_0 is the number of potential binding sites per base pair. In the case illustrated in the figure, binding does not occur at potential intercalation sites flanked by two A•T pairs; the fraction of these is $(1 - \beta)^2$, where β is the fractional G•C content, and the sequence is assumed to be random. Hence, one expects $B_0 = 1 - (1 - \beta)^2 = 2\beta - \beta^2$, as is consistent with the intercept values on the vertical axis.

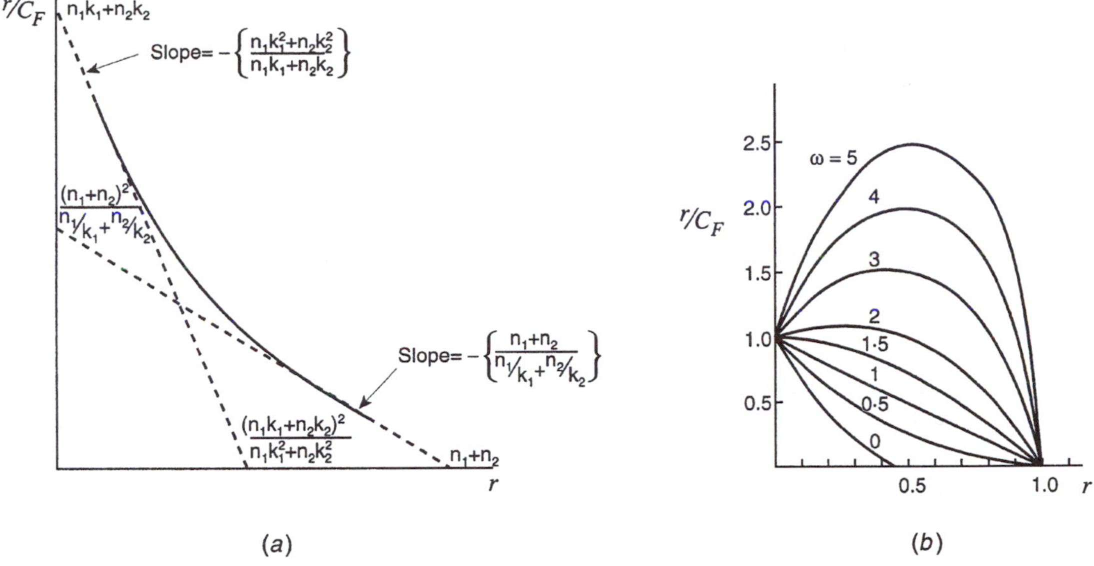

Figure 12-6
(*a*) Schematic curve of r/C_F versus r for two independent binding modes, 1 and 2, showing the intercepts and limiting slopes. [Reprinted with permission from Dougherty G. and Pigram, W. J., *CRC Crit. Rev. Biochem.* **12**, 103–132 (1982). Copyright ©1982 American Chemical Society. See Fig. 4.] (*b*) Theoretical plots for the cooperative binding of a ligand of $n = 1$ for $K = 1\ M^{-1}$ with ω ranging from 0 to 5 [Reprinted from McGhee, J. D. and von Hippel, P. H., Theoretical aspects of DNA–protein interactions: Co-operative and non-co-operative binding of large ligands to a one-dimensional homogeneous lattice, *J. Mol. Biol.*, **86**, 469–489, copyright ©1974, by permission of the publisher Academic Press Limited London.]

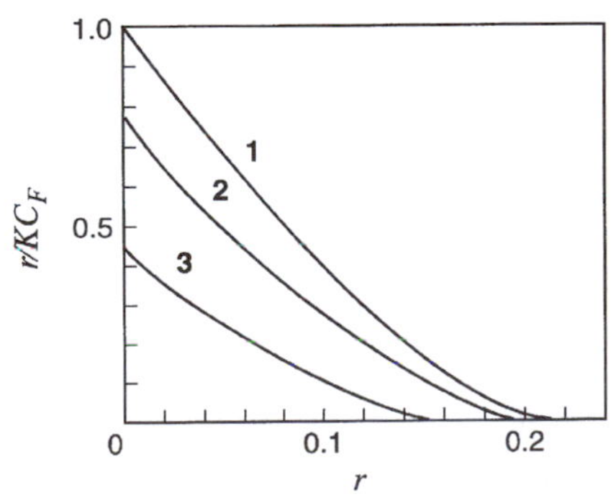

Figure 12-7
Calculated binding isotherms for a model in which a small molecule intercalates between 2 bp when at least one of them is G•C. In addition, there must be at least 4 bp between adjacent bound monomers. The plot shows r/KC_F versus r, where K is the intrinsic binding constant, r is the ratio of bound monomers to base pairs, and C_F is the concentration of drug free in solution. The curves are for different base compositions: The fractional G•C compositions are (1) 1.0; (2) 0.505; (3) 0 0.257. The base sequence is assumed to be random (Crothers, 1968). [Calculation of binding isotherms for heterogeneous polymers, Crothers, D. M., *Biopolymers*, **6**, 575–584. Copyright ©1968. Reprinted by permission of John Wiley & Sons, Inc.]

There is no a priori way to discern from a single binding isotherm what relative influence neighbor exclusion, electrostatic factors, and sequence preference have on the apparent binding site size. For example, curve 3 in Figure 12-7 can readily be fitted to a neighbor exclusion binding isotherm with a smaller value of n than for curve 1. However, in the model it is the requirement for a G•C pair that increases the apparent binding site size.

Some progress can be made on this problem by studying binding as a function of base composition. For example, Müller and Crothers (1968) were able to fit the binding isotherm for actinomycin with several different DNAs to a model in which the neighbor exclusion range is $n = 6$, and a binding site must contain at least one G•C pair. The data are clearly inconsistent with a simple requirement for GpC, which is widely regarded as the binding target for actinomycin on the basis of the strong binding to that sequence (Krugh et al., 1975; Krugh and Reinhardt, 1975; Kastrup et al., 1978; Aivasashvilli and Beabealishvilli; 1983; Phillips and Crothers, 1986). The detailed footprinting studies of Goodisman et al. (1992) indicate that the strongest binding sites contain a GpC sequence flanked by A•T pairs; the strongest binding sequence among the 14 observed had the sequence TGCT. It is only now becoming possible, on the basis of data like those collected by Goodisman et al. (1992), to begin to predict the overall thermodynamics of actinomycin binding from the microscopic binding properties at specific sequences.

Another drug for which progress has been made in understanding the thermodynamics of binding to the various kinds of binding sites is tilorone. For this drug, it was possible to generate a good approximation to the binding isotherm for a DNA on the basis of the binding properties observed for a series of synthetic polynucleotides (Sturm et al., 1981).

2.5 Influence of Cooperativity

Cooperativity results when a drug molecule bound close to another drug attaches more tightly than it would if bound alone. The general result is curvature of the Scatchard plot in the opposite sense from that produced due to neighbor exclusion or binding site heterogeneity. In some cases the isotherm can have a strongly positive slope at small or intermediate values of r. Figure 12-6(b) illustrates the effect of increasing the value of the cooperativity parameter ω, which is the factor by which the equilibrium binding constant is increased when binding occurs adjacent to an already bound ligand, on the shape of the neighbor exclusion isotherm.

An interesting source of cooperativity occurs when drug binding shifts the nucleic acid conformation. For example, Figure 12-8 shows the isotherm found for binding of ethidium to a mixture of poly(dA) and poly(rU) (Bresloff and Crothers, 1981), which is normally triple helical. Since ethidium generally binds more strongly to the double helix (Waring, 1974), addition of the drug switches the conformation from a triple to a double helix. The broken lines show the binding isotherms inferred for the double (upper dashed curve) and triple (lower dotted curve) helices. All theoretical isotherms were calculated using the method of sequence generating functions, which is readily adapted to problems of this kind (Dattagupta et al., 1980).

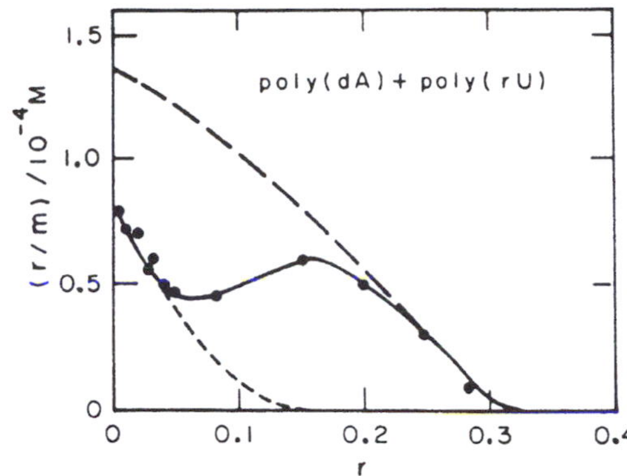

Figure 12-8
Ethidium binding to the 1:1 mixture of poly(dA) + poly(rU), showing the binding isotherm inferred for the triple helix (- - -) and double helix (– – –). The solid line shows the calculated binding, allowing for a switch between the two forms. [Reprinted with permission from Bresloff, J. L. and Crothers, D. M., *Biochemistry* **20**, 3547–3553 (1981), Figure 7. Copyright ©1981 American Chemical Society.]

The contrasting isotherm curvatures produced by neighbor exclusion and sequence specificity on one hand and cooperativity on the other means that the system is underdetermined if only the overall equilibrium binding isotherm is measured. For example, the excellent agreement between neighbor exclusion isotherm and experimental data for ethidium (Fig. 12-5) may result from fortuitous cancellation of the effects of cooperativity and sequence preference in the binding of ethidium to DNA. Additional information, for example, from footprinting and binding to specific sequences, is needed before binding thermodynamics can be fully understood.

3. EXPERIMENTAL STUDIES OF BINDING EQUILIBRIA

Early studies of drug–nucleic acid complexes focused on bulk properties, using primarily optical methods to characterize the interaction. In more recent years, as defined DNA sequences became available for study, methods such as footprinting and affinity cleavage have been used to address the problem of sequence variability of the interaction. An important objective in the years ahead is to refine knowledge of specific binding constants to a sufficient degree to enable prediction of the drug binding properties of a large DNA molecule or a bulk sample. Only this achievement can signal full understanding of the equilibrium binding properties of this important class of drugs.

3.1 Dialysis and Phase Partition

Methods that rely on partition of drug between two phases, one that contains DNA and another that does not, bring the experimentalist as close as practically possible to rigorous thermodynamic measurement of overall DNA binding affinity. In particular, optical methods generally require that assumptions be made concerning the uniformity of the optical properties of bound drug, to which a sceptical view should be taken in view of the heterogeneity of the binding sites. Footprinting methods deal qualitatively with site heterogeneity in a satisfying way, but a rigorous quantitative treatment requires special care (Shubsda et al., 1994). These methods should be viewed as complementary to measurement of the overall affinity, which should, in principle, be predictable from the binding constants for specific sites. The partition methods are also not without their problems; in particular, one should be aware of the possible importance of electrostatic potential differences between the phases, such as occurs in dialysis equilibrium at low salt concentrations (Chapter 11). A negative potential in the DNA phase will increase the positively charged drug concentration there by a mechanism that is unrelated to its specific binding.

Simple dialysis equilibrium and phase partition (Waring et al., 1975; Davanloo and Crothers, 1976) methods require little additional explanation: The concentration of drug in the aqueous phase that lacks DNA is C_F, and the concentration in the DNA containing phase is $C_B + C_F$. From these values, it is simple to calculate a binding isotherm. Often the most difficult point experimentally is to measure the very low drug concentration in the phase that lacks DNA. The solvent phase partition method has an advantage in this regard: One can choose a solvent in which the drug is more soluble than it is in water, thereby increasing the tendency to partition in favor of the organic solvent. A disadvantage of this experimental approach is the unknown effect of dissolution of a small amount of the organic component in the aqueous phase.

Competition dialysis between DNAs of different composition or sequence provides an important tool for studying the binding specificity of drugs. Müller and co-worker (Müller and Crothers, 1975; Müller et al., 1975) used this approach to obtain the dependence of drug binding on DNA base composition in the limit of small degrees of binding r. Generalizing Eq. 12-6 to a model in which there are multiple binding sites j, each of affinity K_j and frequency B_j per base pair, leads to the result

$$\lim_{r \to 0}\left[\frac{r}{C_F}\right] = \sum_j B_j K_j \equiv \sigma \tag{12-8}$$

from which we conclude that the intercept of a Scatchard plot on the vertical axis depends on the sum σ of products of binding site affinity constants K_j and their frequencies per base pair B_i. For example, if the binding site i appears once per 20 bp, then $B_i = 0.05$. If the nature of the binding site is known or postulated, one can predict the variation of this intercept as the DNA sequence or base composition is changed.

It may sometimes be difficult experimentally to determine r/C_F in the limit of zero r because values of C_F become very small and inaccurate. However, if two different DNA preparations a and b are placed on opposite sides of the membrane, the binding values r_a and r_b can be determined with good precision (although several days may be required for equilibration), and their ratio is readily extrapolated to zero r values. Because the two samples are in dialysis equilibrium, C_F is the same for samples a and b, and

$$\frac{r_a}{r_b} = \frac{\sigma_a}{\sigma_b} \tag{12-9}$$

The limiting value of this ratio as r approaches zero is called α_{ab}, or

$$\alpha_{ab} = \lim_{r \to 0}\left[\frac{\sigma_a}{\sigma_b}\right] \tag{12-10}$$

This approach has the advantage that because α_{ab} is evaluated at zero r, the influence of neighbor exclusion and cooperativity are eliminated.

Selection of a pair of nucleic acids a and b should yield a value of α_{ab} which is fixed by the corresponding values of σ_a and σ_b given by equation 12-8. Comparison of measured values with those based on assumed models for the binding site allows one to eliminate models that disagree with the experimental results. Early applications of this approach (Müller and Crothers, 1975) used bacterial DNAs of differing average G•C content, and assumed a statistically random base pair sequence. For example, when a is *M. luteus* DNA (72% G•C) and b is *Bacillus subtilis* DNA (44% G•C), then a model in which intercalation can only occur adjacent to one side of a G•C pair predicts an α value that is the ratio of the G•C contents, or $\alpha_{ab} = 1.63$. If the site instead requires an A•T pair, then α_{ab} is predicted to be 0.50. A model in which intercalation occurs only between two G•C pairs has a predicted α_{ab} value of 2.66, and if two A•T pairs are required, $\alpha_{ab} = 0.25$. Table 12-1 gives reported α_{ab} values for this choice of competing DNAs for some common DNA-binding ligands. With the exception of berenil, which is highly A•T specific, and the G•C-specific ligands actinomycin and methylene blue, most α_{ab} values fall within the range of 0.5–1.63, which implies relatively nonselective binding. Note in addition that the α_{ab} value for actinomycin (1.75) is much less than would be predicted if the binding site had a simple requirement for the sequence GpC ($\alpha_{ab} = 2.66$). With the current availability of DNA molecules of defined sequence,

Table 12.1
Specificity Ratios $\alpha_a b$ for some DNA Binding Ligands

Ligand	α_{ab}^{a}
Berenil	0.17
Ellipticine	0.99
β-Rhodomycin	1.00
Chloroquine	1.10
Quinacrine	1.17
Ethidium	1.23
Proflavine	1.26
Propidium	1.35
Acridine orange	1.57
Actinomycin D	1.75
Methylene blue	1.84

[a]From Müller and Crothers (1975). The parameter α_{ab} is defined by Eqs. 12-6 and 12-8, and refers to measured values when DNA *a* is *M. luteus* and DNA *b* is *B. subtilis*. Increasing values of α_{ab} correspond to increasing selectivity for G•C versus A•T.

partition dialysis should be a useful method for accurate study of the relative affinities of different binding sites (Woodson and Crothers, 1988)

3.2 Optical Methods

Absorbance and fluorescence spectroscopy are particularly convenient methods for characterization of DNA binding ligands. Their use requires that there be a detectable difference between the absorption or fluorescence properties of the ligand when free and DNA bound. Many DNA binding drugs show a pronounced change in their absorption spectra when DNA binding occurs. Figure 12-9 shows spectra for ethidium bromide in the presence of increasing amounts of DNA. The absorption maximum shifts progressively to the red, and there is a reduction in the peak absorbance.

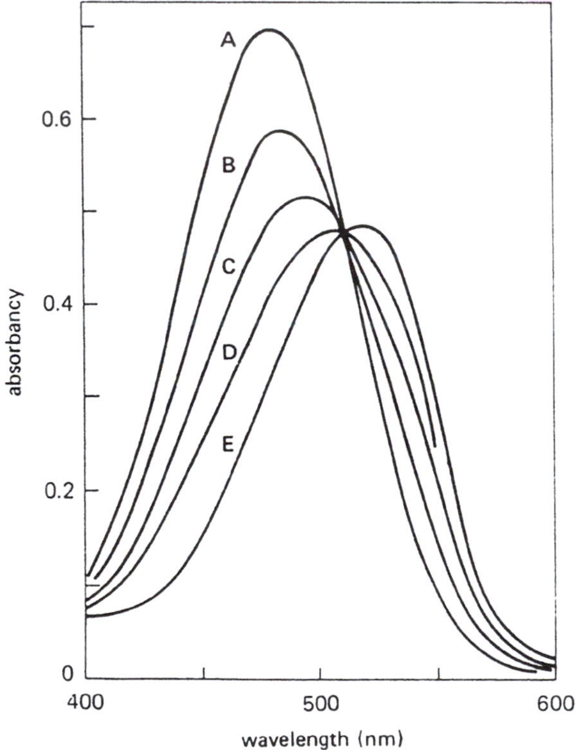

Figure 12-9
Shift of the absorption spectrum of ethidium bromide in the presence of DNA. (*a*) spectrum of the drug alone in buffer. (*b–e*) With increasing concentrations of bacteriophage T2 DNA added. Note the isosbestic point above about 500 nm. [Reprinted with permission from *Nature*. Waring, M. J., *Nature (London)* **219**, 1320–1325 (1968). Copyright ©1968 Macmillan Magazine Limited.]

All of the absorbance spectra in Figure 12-9 intersect at a common point called the isosbestic. The most common circumstance under which an isosbestic point is obtained is when there are only two forms of the drug, in this case free in solution and intercalated. Then the absorbance in a 1-cm pathlength cell (or optical density) A is given by

$$A = \varepsilon_F C_F + \varepsilon_B C_B \tag{12-11}$$

where ε_F and ε_B are the extinction coefficients of the free and bound forms of the drug, respectively, and C_F and C_B are the corresponding concentrations. At the isosbestic wavelength ε_F and ε_B are equal, and the absorbance does not depend on how the fixed total drug concentration is partitioned between free and bound. It should not, however, be assumed that because there is an isosbestic wavelength, only two forms of the drug can be present. For example, there could be a mixture of free dye with two forms of bound dye having different spectra, but if the two bound forms are always present in a constant ratio, an isosbestic wavelength will still be observed.

When systematic absorbance spectra fail to show an isosbestic point, then there must be more than two physical states of the dye. An example is shown in Figure 12-10

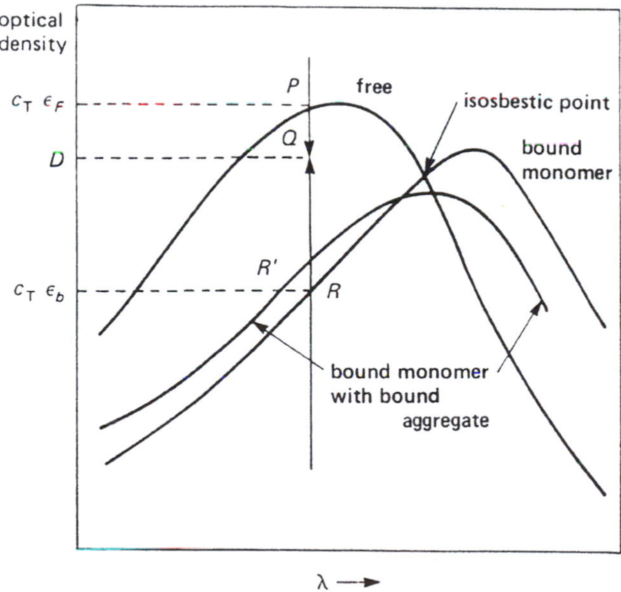

Figure 12-10
Spectra of proflavine, free and bound as monomer. The spectrum of proflavine bound both as monomer and as aggregate is schematic only, though based on experimental observations. [The interaction of aminoacridines with nucleic acids, Blake, A. and Peacocke, A. R., *Biopolymers*, Copyright ©1968. Reprinted by permission of John Wiley & Sons, Inc.]

for binding of proflavine to DNA. When the DNA concentration is high, the many sites available for intercalation cause most of the drug to be bound in that form, with the remainder free in solution. However, at lower DNA concentration the dye becomes crowded on the polymer and the spectrum shifts in a different way, causing the isosbestic to disappear. This effect is thought to be due to binding of the dye by a different mechanism, namely stacking on the outside of the double helix, mediated by electrostatic forces and the tendency of the drug to self-associate.

When a dye intercalates between the DNA base pairs, the visible absorption band is usually shifted to the red, with an accompanying reduction in maximum extinction coefficient. This phenomenon presumably arises from interaction between the drug chromophore and the electron system of the base pairs. However, such spectral shifts should not be regarded as diagnostic for intercalation. For example, a nonintercalating derivative of proflavine shows a red shift of the absorption spectrum upon binding (Müller et al., 1973). Large changes in absorbance are often observed when dyes stack together on the surface of a polyelectrolyte such as DNA. Frequently, there is a blue shift in the absorbance maximum, in contrast to the red shift on intercalation, but again these changes should not be regarded as decisively diagnostic.

Optical measurements are well suited for quantitative characterization of binding equilibria when the optical properties of the complex are independent of the degree of binding r. In such cases, the change in extinction coefficient $\Delta\varepsilon = \varepsilon_B - \varepsilon_F$ upon binding can be obtained by extrapolation of the optical properties to infinite nucleic acid concentration. For this purpose, one can use the linear form of the neighbor exclusion binding isotherm, valid when r is much less than $1/n$,

$$\frac{r}{C_F} = K[1 - (2n - 1)r] \qquad (12\text{-}12)$$

To relate the limiting properties of the binding isotherm at small r to the optical properties, we define the measured apparent extinction coefficient $\varepsilon_{ap} = A/C_T$, where C_T is the total concentration of drug, and let $\Delta\varepsilon_{ap} = \varepsilon_{ap} - \varepsilon_F$. In the limit of large DNA concentration at fixed C_T, $\Delta\varepsilon_{ap}$ approaches $\Delta\varepsilon$. If the extinction coefficient of the bound drug is independent of r, the fraction of the drug that is bound is $\Delta\varepsilon_{ap}/\Delta\varepsilon$. Consequently,

$$r = \frac{\Delta\varepsilon_{ap} C_T}{\Delta\varepsilon C_N^0} \qquad (12\text{-}13)$$

where C_N^0 is the concentration of DNA base pairs. Using this result, substituting in Eq. 12-12, and expanding to first order in $\Delta\varepsilon - \Delta\varepsilon_{ap}$ leads to the result

$$\Delta\varepsilon_{ap} = \Delta\varepsilon - \frac{\Delta\varepsilon_{ap}}{K\left[C_N^0 - (2n - 1)\, C_T\right]} \qquad (12\text{-}14)$$

Hence a plot of $\Delta\varepsilon_{ap}$ against $\Delta\varepsilon_{ap}/[C_N^0 - (2n - 1)C_T]$ will intercept the vertical axis at $\Delta\varepsilon_{ap} = \Delta\varepsilon$ when C_N^0 approaches infinity. The calculation requires an estimate of n, but since C_T is much smaller than C_N^0 in the range of interest, the term involving $2n - 1$ has little effect on the plot. The slope of the linear plot is $1/K$. Once $\Delta\varepsilon$ is established, Eq. 12-13 can be used to calculate r from the measured value of $\Delta\varepsilon_{ap}$; the binding isotherm showing r/C_F as a function of r can be compared with theoretical curves such as that predicted by the neighbor exclusion model.

A similar approach can be used to deduce the difference in fluorescence properties of free and bound drug. The fluorescence analogue to Eq. (12-11) is

$$F = i_F C_F + i_B C_B \tag{12-15}$$

where i_F and i_B are the molar fluorescence analogues to the molar extinction coefficients of free and bound ligand, respectively, and F is the fluorescence intensity. Thus the fluorescence analogue to Eq. (12-14) is

$$\Delta i_{ap} = i_B - i_F - \frac{i_{ap} - i_F}{K[C_N^0 - (2n-1)C_T]} \tag{12-16}$$

where Δi_{ap} is the observed fluorescence minus that of the free drug, divided by the total drug concentration, C_T. Once $i_B - i_F$ is known, the measured fluorescence can be used with the analogue of Eq. (12-13) to calculate the degree of binding and hence the binding isotherm.

The *model independent* analysis of Bujalowski and Lohman (1987) (reviewed by Lohman and Mascotti, 1992), can be used to determine a binding isotherm without resorting to the assumption that the spectral or fluorescence change induced upon drug binding is independent of degree of binding r. The experiment requires measuring the binding equilibrium at a set of different total concentrations. Specifically, this can be done by titrating the nucleic acid into different fixed total concentrations of a fluorescent ligand, whose fluorescence we assume to be quenched on binding. Let the fractional fluorescence quenching Q_j in bound state j be

$$Q_j = \frac{F_F - F_j}{F_F} \tag{12-17}$$

Then the total fractional fluorescence quenching Q_{obs} is

$$Q_{obs} = \frac{i_F - i_{obs}}{i_F} = \sum_j Q_j r_j \tag{12-18}$$

where r_j is the fractional occupancy of sites of kind j, i_F is the fluorescence intensity for the free drug F_F divided by its concentration, and i_{obs} is the observed fluorescence of the drug-nucleic acid mixture divided by the total drug concentration. The key assumption in the method is that a given fluorescence quenching Q_{obs} always has the same set of site occupancies r_j. This assumption requires that there be no concentration-dependent aggregation of the nucleic acid or drug. Use is then made of the equation

$$C_T = C_F + C_N^0 \sum_j r_j = C_F + C_N^0 r_T \tag{12-19}$$

where r_T is the total concentration of bound ligand divided by the total nucleic acid concentration, by graphically interpolating the fluorescence quenching data to select different combinations of total nucleic acid concentration C_N^0 and total ligand concentration C_T that have the same value of Q_{obs} and hence r. Then, according to equation (12-19), a plot of the values of C_T against the corresponding values of C_N^0 has slope r and intercept C_F.

The procedure is illustrated in Figure 12-11, giving the results for interaction of a fluorescent peptide with poly(U) (Lohman and Mascotti, 1992). Panel (*a*) shows the quenching curves obtained by titrating a fixed total ligand concentration with poly(U). Points at fixed Q_{obs} in panel (*a*) are then transferred to panel (*b*) to give a series of linear plots of C_T against C_N^0, each at fixed Q_{obs} and hence fixed r. The slope of each line gives the value of r, and the intercept is C_F, according to Eq. (12-19). These values are then used to construct the Scatchard plot, as shown in panel (*c*).

When a discrete number of reacting components are present in a complex equilibrium, the method of *chemometric analysis* described by Kubista et al. (1993) can be used to resolve the observed absorption spectrum into spectral components for the individual species. The equilibrium constants characteristic of the system are also obtained. The technique requires measurement of the full ligand absorption spectrum at varying concentration of nucleic acid. If, for example, the nucleic acid has two binding sites of different binding constants, then the analysis yields absorbance spectra for the free and bound forms, along with the individual binding constants.

3.3 Footprinting and Affinity Cleavage

Phase partition and optical methods characterize average or bulk drug binding properties of a DNA sample, whereas the footprinting and affinity cleavage techniques rely on DNA sequencing technology to assess binding at particular sites on the DNA molecule. This feature offers considerable advantage in the effort to understand the DNA sequence preference of drug binding.

Naturally occurring substances that bind to and cleave DNA include the bleomycins, neocarzinostatin, and the enediyne compounds, discussed in Section 12.6.2. Synthetic agents that mimic this function, also considered below, include the 1,10 phenanthroline–$Cu^{(I)}$ complex (where phen = 1, 10-phenanthroline), and $Fe^{(II)}$ (methidiumpropyl edta), or Fe (MPE) (where edta = ethylenediaminetetraacetic acid). The latter, developed and exploited by Dervan and co-workers (Hertzberg and Dervan, 1982), has been particularly useful for studying the site of binding of other drugs because Fe (MPE) is relatively nonspecific in its DNA binding, and hence produces comparatively uniform cleavage of naked DNA. Added drugs that are specific in their binding compete locally

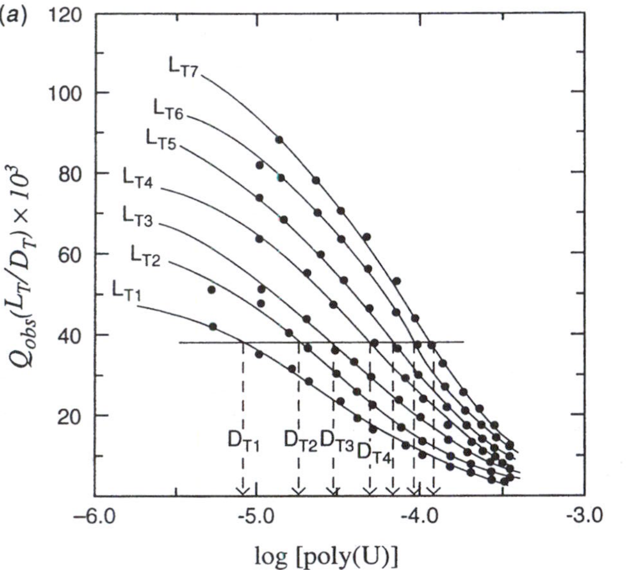

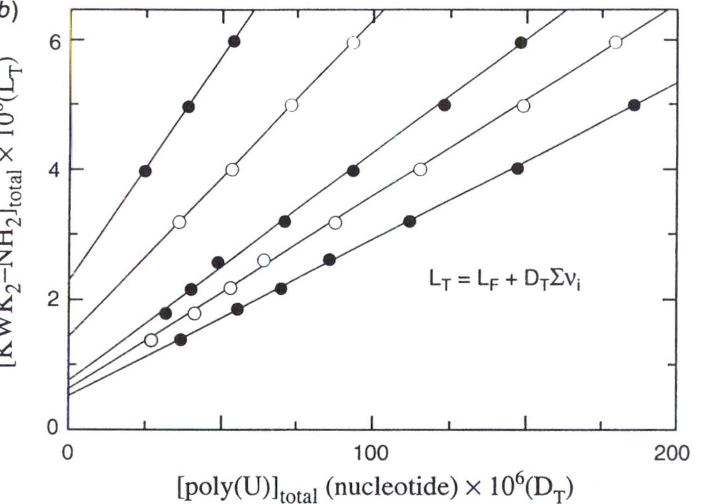

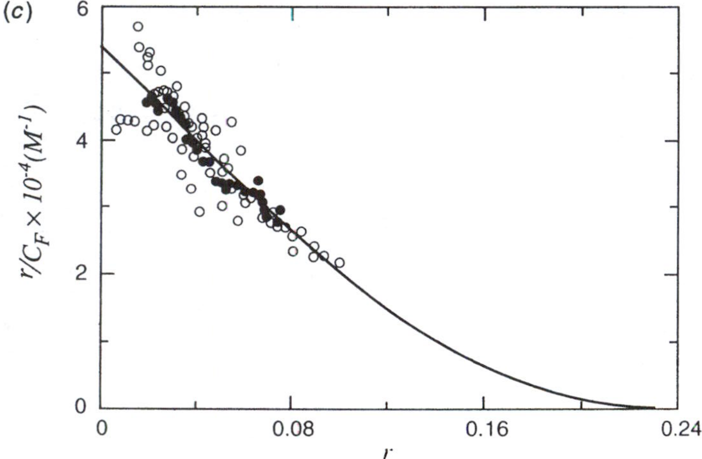

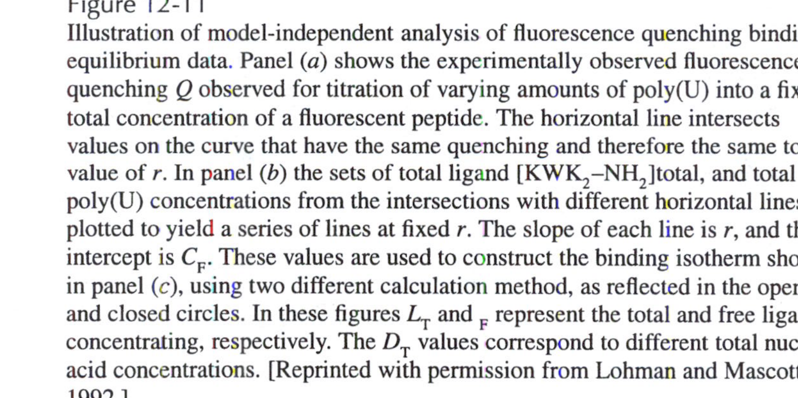

Figure 12-11
Illustration of model-independent analysis of fluorescence quenching binding equilibrium data. Panel (*a*) shows the experimentally observed fluorescence quenching Q observed for titration of varying amounts of poly(U) into a fixed total concentration of a fluorescent peptide. The horizontal line intersects values on the curve that have the same quenching and therefore the same total value of r. In panel (*b*) the sets of total ligand [KWK_2–NH_2]total, and total poly(U) concentrations from the intersections with different horizontal lines are plotted to yield a series of lines at fixed r. The slope of each line is r, and the intercept is C_F. These values are used to construct the binding isotherm shown in panel (*c*), using two different calculation method, as reflected in the open and closed circles. In these figures L_T and $_F$ represent the total and free ligand concentrating, respectively. The D_T values correspond to different total nucleic acid concentrations. [Reprinted with permission from Lohman and Mascotti 1992.]

with Fe(MPE) binding, reducing the cleavage around the site of high affinity as a consequence. Comparative studies with inhibition of enzymatic cleavage by DNAase I show reasonable agreement between the binding sequences detected by the two methods. Figure 12-12 shows protection against the two DNA cleavage agents by actinomycin D, chromomycin A_3 and distamycin A. DNAase I protection, shown by the horizontal bars, generally reports a larger apparent binding site size than does Fe (MPE), where drug-induced protection is indicated by the histograms. This difference probably reflects the macromolecular nature of DNAase I as a probe, compared to the smaller DNA binding agent Fe (MPE). The reader is referred to the review by Shubsda et al. (1994) for a discussion of the special steps needed to interpret footprinting data such as those in Figure 12-12 in a quantitative way.

Phillips and co-workers (Phillips and Crothers, 1986; White and Phillips, 1989) developed an alternative method for mapping drug binding sites in which one measures drug-induced inhibition of the elongation step in RNA polymerase-catalyzed mRNA synthesis. Drug binding typically causes a transient accumulation of RNA species whose growth is inhibited by a drug molecule bound at a specific site on the DNA. In some cases transcription termination is induced by the bound drug (White and Phillips, 1988). The data can be analyzed to obtain both the relative occupancy of binding sites and the rate at which drug dissociates from each site in the presence of RNA polymerase.

Figure 12-13 shows the actinomycin binding sites found by transcriptional footprinting compared with data on the same DNA fragment obtained from DNAase I protection. Transcriptional footprinting, especially when studied bidirectionally (White and Phillips, 1989), shows exceptional precision in locating the drug binding site. However, the only sites identified by this technique are those from which the drug dissociates slowly in the presence of RNA polymerase. In the case of actinomycin, for example, DNAase I footprinting identifies binding sites that lack a GpC sequence, which are not found by transcriptional analysis. Sites containing the GpC sequence are unusual in that the dissociation rate in the presence of polymerase is substantially slower than from naked DNA (White and Phillips, 1988).

Table 12-2 summarizes some binding specificities of several DNA binding drugs. In many cases, the agreement among the various methods is good, but in some instances, such as daunomycin, there remains disagreement over the nature of the strongest site. It is likely in such circumstances that a number of sequences are nearly equivalent in binding strength.

3.4 Calorimetric Studies of Drug Binding

Direct measurement of the heat of drug binding provides a clear view of the relative importance of enthalpy and entropy contributions to the free energy of binding. Figure 12-14 illustrates measurement of the heat evolved upon mixing netropsin and a solution of a polynucleotide of alternating sequence. A total heat production of about 0.1 mcal can be measured with good precision.

Systematic calorimetric experiments by Breslauer et al. (1987), some of which are summarized in Table 12-3, demonstrate that while there is no fixed rule for the contribution of entropic and enthalpic contributions, the binding reaction is generally

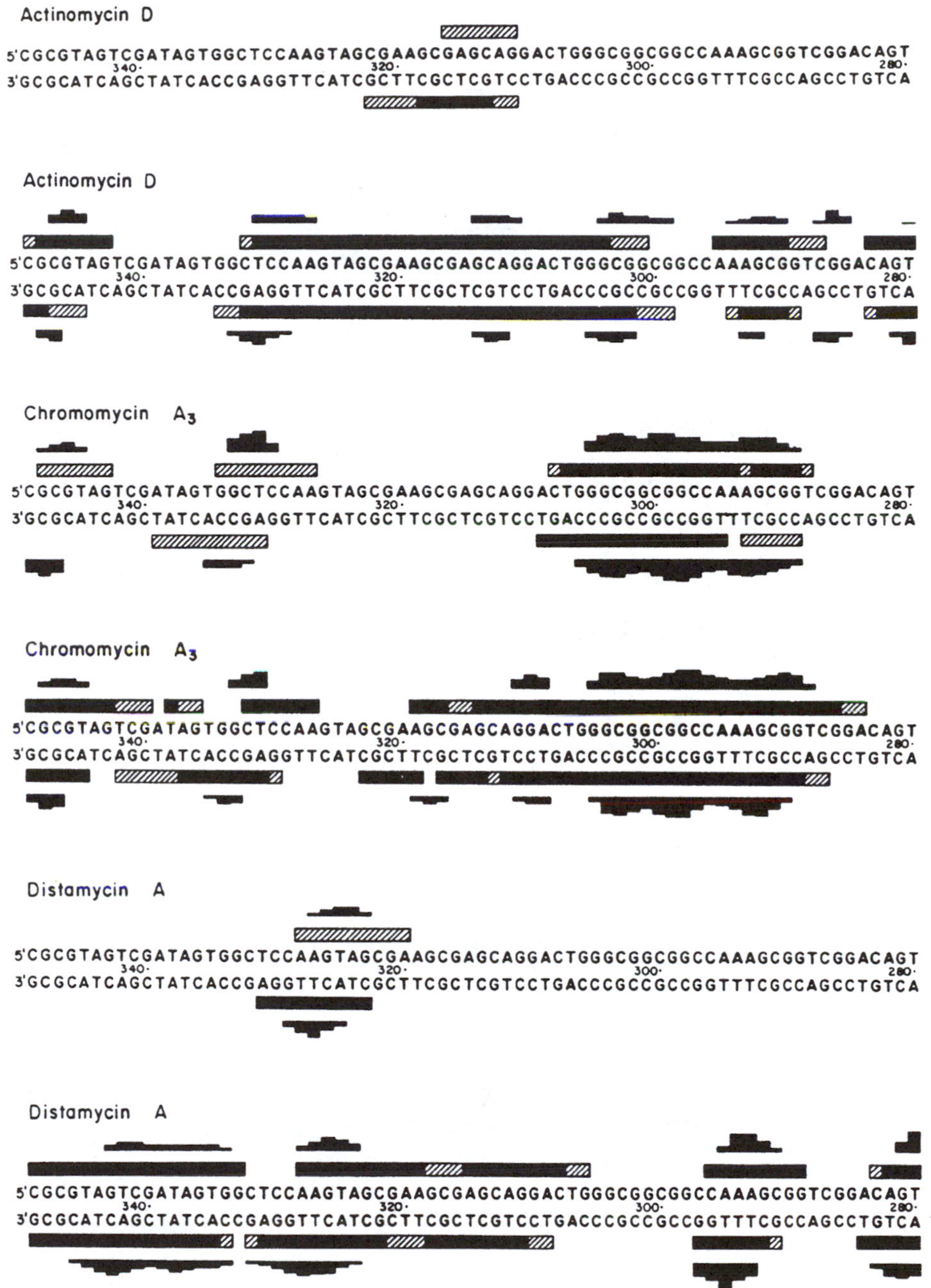

Figure 12-12
The Fe(MPE) and DNase I footprints on both strands of 70 nucleotides of a 381 bp DNA fragment. The DNase I footprints are shown as light and dark bars due to partial and complete cleavage inhibition, respectively. The Fe(MPE) footprints are shown as histograms. Two binding densities are shown for each inhibiting drug; top is 0.06 drug/bp; bottom is 0.25 drug/bp.

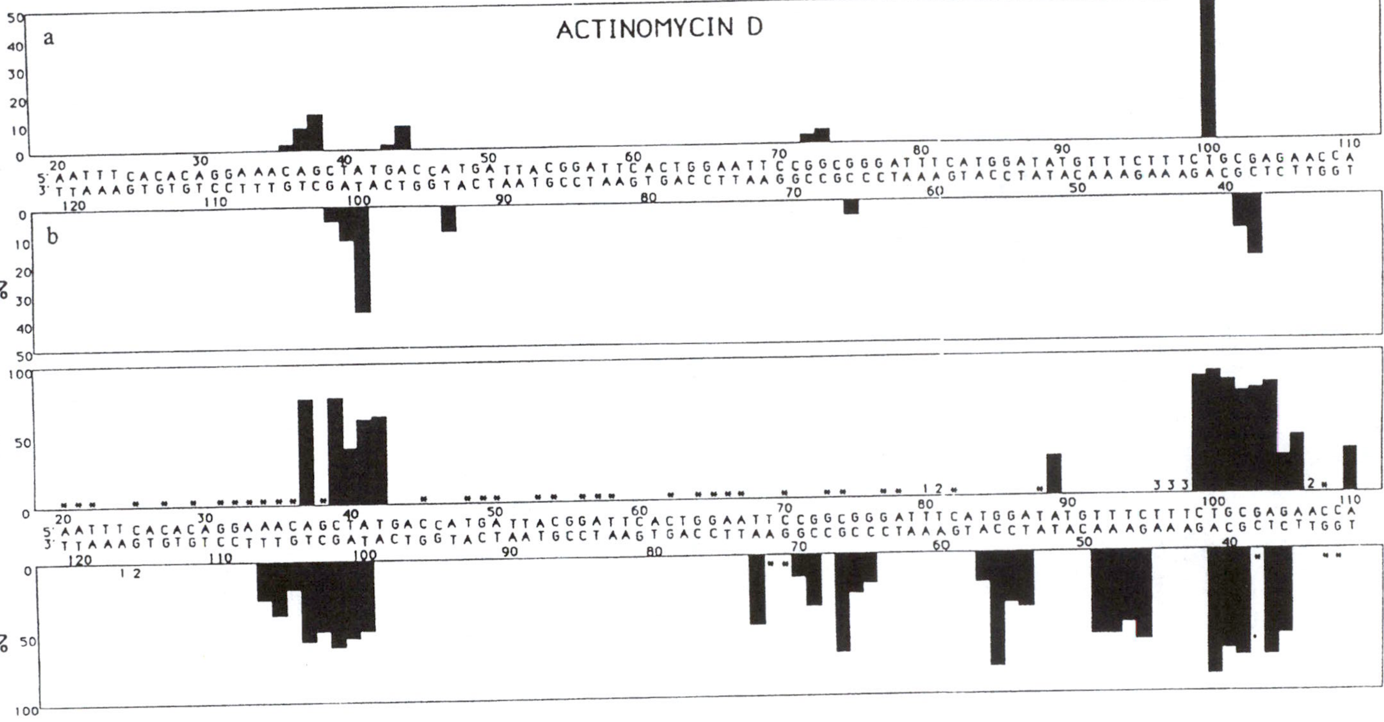

Figure 12-13
Bidirectional transcription footprints and DNase I footprints. The upper panel shows the mole percent of blocked transcripts arising from the UV5 promoter (*a*) and the N25 promoter (*b*). The lower panel for the two strands shows the decreased DNase I cutting efficiency at each site. Asterisks show sites that were not cut by DNase I and did not yield a corresponding band in the control lane. [Reprinted with permission from White, R. J. and and Phillips, D. R., **28**, 6259–6269 (1989). Copyright ©1989 American Chemical Society.]

Table 12.2
Reported Binding Site Preferences for DNA Binding Drugs

Method	Sequence	Reference
Actinomycin		
Fe(MPE) cleavage	One or more G•C pairs	Van Dyke et al. (1982)
DNAase I cleavage	GpC preferred	Lane et al. (1983)
Transcription	GpC, strong sites are AGCT and TGCT	White and Phillips (1989)
Bleomycin		
Affinity cleavage	Cleavage at GpT, GpC	McLean et al. (1989)
Chromomycin		
Fe(MPE) cleavage	Two contiguous G•C pairs, GGG,AGC>GCC,CCG>AGC, TCC>GTC	Van Dyke and Dervan (1983)
Distamycin		
Fe(MPE) cleavage	A•T rich, minimum 4 bp protected	Van Dyke et al. (1982)
Affinity cleavage	A•T rich; a strong site is AATTT	Taylor et al. (1984)
Daunomycin		
DNAase I cleavage	Adjacent G•C pairs, flanked by A•T	Chaires et al. (1987)
Transcription	CA > GC,CG,CT,TC,CC,AC > AA,AT,TA	Skorobogaty et al. (1988)
Fluorescence	CGTACG ≥ TAGCTG ≅ TCATACC, CGCGCG	Roche et al. (1994)
Echinomycin		
Fe(MPE) cleavage	CpG; two strong sites are ACGT and TCGT	Van Dyke and Dervan (1983)
Chemical footprinting	CpG	Low et al. (1984); McLean and Waring (1988)
Transcription	CpG, strong sites: CCGG,ACGG,GCGG	White and Phillips (1989)
Ethidium		
DNAase I cleavage	Prefers alternating pur-pyr, avoids A_n	Fox and Waring (1986)
Netropsin		
DNAase I cleavage	A•T rich	Lane et al. (1983)
Nogalamycin		
DNAase I	Alternating pur-pyr, especially TG, GT	Fox and Waring (1986)
Transcription	CA	White and Phillips (1989)

exothermic (negative ΔH°). Exceptions to this rule include actinomycin, for which the reaction heat is nearly zero, and binding is entropically driven. In the case of the steroidal diamines, the binding process is endothermic; the explanation given earlier (Dattagupta et al., 1978) for this finding remains plausible: binding of the diamine may cause substantial DNA kinking and base pair unstacking, which is a sufficiently endothermic process to make the net heat of binding positive.

A consistent feature of the ΔH° values in Table 12-3 is their relatively greater endothermic character when the binding target is poly(dA)•poly(dT) rather than

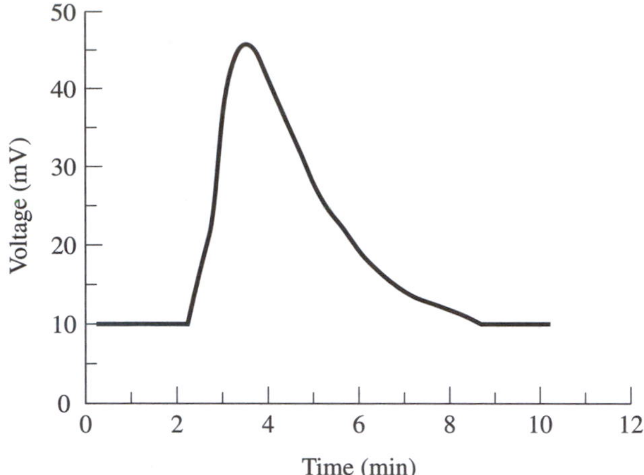

Figure 12-14
Typical calorimetric heat burst curve. This curve was produced upon mixing a solution of Netropsin (0.757 mM) and a solution of poly[d(I-C)]•poly[d(I-C)] at a phosphate/drug ratio of 10:1. The area under then curve corresponds to a total heat production of 1.18 millicalorie (1 cal = 4.184 J). [From Breslauer et al., 1987.]

any other double helix. Comparison of the heats of binding of the intercalators ethidium and daunomycin and the minor groove binders distamycin and netropsin to the dA•dT homopolymer as drug receptor, relative to the heats of binding to the alternating copolymer poly[d(A-T)]•poly[d(A-T)], shows that reaction with the latter target is more exothermic by an average of roughly 10 kcal mol^{-1}. This finding may mean that drug binding to the homopolymer requires an endothermic change in which the special structure of poly(dA)•poly(dT) reverts to a more canonical B-DNA conformation in order to accommodate the bound drug. This phenomenon is probably related to the endothermic process that causes DNA bending at $dA_n \cdot dT_n$ tracts to disappear at elevated temperatures (Crothers et al., 1990; see Chapter 9).

The overall binding strength of the drugs examined, characterized by the value of ΔG°, varies much less than does the reaction enthalpy from one drug or binding target to another. This results from the tendency of the $T\Delta S^\circ$ term to vary in a way that compensates the enthalpy changes ΔH° in the expression for the Gibbs free energy change, $\Delta G^\circ = \Delta H^\circ - T\Delta S^\circ$. This "enthalpy–entropy compensation," is a commonly observed phenomenon. In several cases, illustrated in Table 12-3, drug binding is primarily entropically rather than enthalpically driven, meaning that the $-T\Delta S$ term is more negative than ΔH. Examples include ethidium and daunomycin binding to poly(dA)•poly(dT), along with actinomycin binding to DNA samples.

Table 12.3
Thermodynamic parameters for binding drugs to DNA[a b]

Drug	DNA	$\Delta H°$	$\Delta S°$	$\Delta G°$
Ethidium[b]	Poly[d(A-T)] · poly[d(A-T)]	−10.0	−3	−9.1
	Poly(dA) · poly(dT)	−1.2	18	−7.2
	Poly[d(I-C)] · poly[d(I-C)]	−9.2	1	−9.3
	Salmon testes DNA	−12.4	−10	−9.5
Daunomycin[b]	Poly[d(A-T)] · poly[d(A-T)]	−8.9	2	−9.4
	Poly(dA) · poly(dT)	−2.1	21	−8.4
	Poly[d(G-C)] · poly[d(G-C)]	−10.4	−5	−9.0
	Salmon testes DNA	−9.9	−3	−9.0
Actinomycin[c]	Salmon testes DNA	0.9	31	−7.6
Distamycin[b]	poly[d(A-T)] · poly[d(A-T)]	−18.5	−20	−12.6
	Poly(dA) · poly(dT)	−4.2	24	−11.4
	$d(GCGAATTCGC)_2$	−15.8	−16	−11.5
Netropsin[b]	Poly[d(A-T)] · poly[d(A-T)]	−11.2	5	−12.7
	Poly(dA) · poly(dT)	−2.2	33	−12.2
	Poly[d(I-C)] · poly[d(I-C)]	−9.9	4	−11.1
Steroidal diamines:				
Dipyrandium[d]	Poly[d(A-T)] · poly[d(A-T)]	4.2	36	−6.5
Irehdiamine[e]	*M. luteus* DNA	22		
	Calf thymus DNA	12		

[a]Energies are expressed in kilocalories per mole (kcal/mol), and entropy values in cal K^{-1} mol^{-1} (eu).

[b]Breslauer et al. (1987).

[c]Marky et al. (1983a).

[d]Marky et al. (1983b).

[e]Dattagupta et al. (1978). These results were obtained from the difference of forward and reverse activation energies for binding, rather than from calorimetry.

3.5 Effect of Ligand Binding on DNA Melting

Any ligand that binds differently to the double helix and single stranded forms of DNA will perturb the helix–coil melting transition. We can view this as a simple equilibrium process,

$$\text{helix } (r_h \text{ ligands bound}) \rightleftharpoons \text{coil } (r_c \text{ ligands bound}) + (r_h - r_c) \text{ ligands} \qquad (12\text{-}20)$$

where r_h and r_c are the extent of drug binding per base pair in helix and coil forms respectively. If r_h is greater than r_c, then addition of the drug should drive the equilibrium toward the helix form, and stabilize the DNA against melting. Similarly, drugs can destabilize DNA by binding to a greater extent to the coil form than to the helix. Ligands that bind preferentially to triple helices are also known (Mergny et al., 1992). Figure 12-15 shows differential melting curves for poly[d(A-T)] • poly[d(A-T)] when increasing amounts of distamycin are added. Small amounts of added drug yield biphasic melting curves, which result because part of the double helix can melt readily as long as binding sites remain available to accommodate the drug displaced by the unfolding process. The second melting transition reflects disruption of the duplex regions and release of the drug to free solution rather than to alternative binding sites.

For a preliminary quantitative characterization of this process, we note that the equilibrium constant expression for the melting–dissociation process depends on the concentration of ligand according to

$$K = \frac{[\text{coil}]}{[\text{helix}]}[\text{ligand}]^{(r_h - r_c)} = K'[\text{ligand}]^{(r_h - r_c)} \tag{12-21}$$

where K' is the equilibrium constant for the helix–coil transition. Consider the case in which the ligand concentration is in large excess over the binding site concentration. Then the ligand concentration is effectively constant during the melting–dissociation process, and the melting curve will be monophasic, as illustrated in Figure 12-15. At

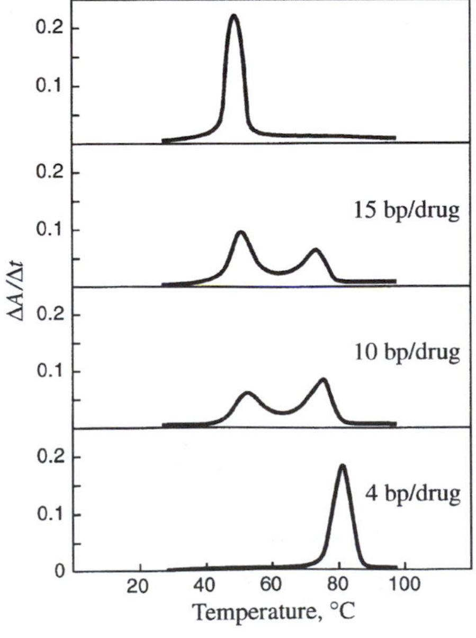

Figure 12-15
Differential melting curves ($\Delta A/\Delta t$) versus temperature t for the helix–coil transition of poly[d(A-T)] in the absence and presence of distamycin A at the indicated DNA/drug ratio. [From Breslauer et al., 1987.]

the temperature T_m which corresponds to the peak of the differential melting curve, K' is a constant, independent of ligand concentration, and

$$\frac{\partial \ln K}{\partial(1/T_m)} = \frac{-\Delta H_m^\circ}{R} = (r_h - r_c)\frac{\partial \ln [\text{ligand}]}{\partial(1/T_m)} \tag{12-22}$$

where ΔH_m° is the heat of the process in Eq. 12-20, including the heat of melting and of dissociating the bound drug. According to Eq. 12-22, a plot of $1/T_m$ against the logarithm of the ligand concentration should be linear (assuming ΔH_m° does not depend strongly on temperature), with slope given by $-R(r_h - r_c)/\Delta H_m^\circ$. This plot is useful for circumstances in which the ligand is in excess, so that monophasic melting curves with a well defined T_m result. Examples are the binding of monovalent ions to DNA (Chapter 11), or binding of drugs under conditions of near saturation of the binding sites at the limiting values of r. Integration of Eq. 12-22 to find the shift in $(1/T_m)$ yields an integral of the form

$$\Delta\left(\frac{1}{T_m}\right) = \frac{-R}{\Delta H_m^\circ}\int (r_h - r_c)d \ln [\text{ligand}]$$

More detailed statistical mechanical theories of DNA melting with coupled binding equilibria have been presented. For example, McGhee (1976) illustrated biphasic calculated transition curves using a simplified model for DNA melting that neglects the contributions of loop entropy, heterogeneity of base pair stability, and sequence dependence of ligand binding strength. A full theory that takes all of these factors into account has not yet been developed, in part because the experimental information necessary to test the theory, particularly the quantitative sequence dependence of binding, is not generally available.

The approach taken by Crothers (1971) can be used to determine the expected temperature difference, with and without added ligand, between transition curve points of a given fractional helix content θ. If $\theta = 0.5$ as at T_m, then this is equivalent to asking for the influence of added ligand on the transition midpoint temperature. The helix–coil transition in the presence of ligand is determined by a parameter s', which is the equilibrium constant for growth of a helical segment at the expense of adjacent unpaired bases when binding and melting are coupled. Assuming that factors such as cooperativity of melting and the entropy of denatured loops are not affected by the binding equilibrium, θ is a function only of s', which in turn is a function of the external variables temperature T and free ligand concentration C_F, or activity a.

The basis of the theory is calculation of the temperature increase required to reduce $\ln s'$ enough to compensate for the increase in $\ln s'$ that results from ligand binding, assuming that the drug binds preferentially to the helical form. The result of taking $\Delta \ln s' = 0$ due to offsetting changes of temperature and ligand binding is (Crothers, 1971)

$$\frac{1}{T_\theta} - \frac{1}{T'_\theta} = \frac{R}{\Delta H_m^\circ}\int_0^a (r_h - r_c)d \ln a' \tag{12-23}$$

This equation shows how the reciprocal temperature at a defined transition point, $1/T'_\theta$, should vary with the difference in degree of binding to helix and coil at free ligand or drug activity a; the integral on the right-hand side should be evaluated at T'_θ; T_θ is the transition temperature characterized by fractional helix content θ in absence of added drug.

Equation 12-23 can be developed further if assumptions are made about the binding model. Using the Scatchard model, inserting $a = C_\mathrm{F}$ in Eq. 12-2, substituting for $d \ln a'$ in Eq. 12-23 and integrating yields (Crothers, 1971)

$$\frac{1}{T_\theta} - \frac{1}{T'_\theta} = \frac{R}{\Delta H^\circ_\mathrm{m}} \left[\frac{(1 - r_\mathrm{c}/B_\mathrm{c})^{B_\mathrm{c}}}{(1 - r_\mathrm{h}/B_\mathrm{h})^{B_\mathrm{h}}} \right] = \frac{R}{\Delta H^\circ_\mathrm{m}} \left[\frac{(1 + K_\mathrm{h} a)^{B_\mathrm{h}}}{(1 + K_\mathrm{c} a)^{B_\mathrm{c}}} \right] \tag{12-24}$$

in which B_h and B_c are the apparent site sizes for binding to helix and coil, respectively, and K_h and K_c are the corresponding apparent binding constants for the Scatchard model. The activity a can usually be replaced by the free ligand concentration C_F.

Alternatively, Eq. 12-3 for the neighbor exclusion model can be used to calculate $d \ln a$; insertion into Eq. 12-23 and integration yields

$$\frac{1}{T_\theta} - \frac{1}{T'_\theta} = \frac{R}{\Delta H^\circ_\mathrm{m}} \left[\ln \frac{1 - (n_\mathrm{h} - 1) r_\mathrm{h}}{1 - n_\mathrm{h} r_\mathrm{h}} - \ln \frac{1 - (n_\mathrm{c} - 1) r_\mathrm{c}}{1 - n_\mathrm{c} r_\mathrm{c}} \right] \tag{12-25}$$

in which n_h and n_c are the binding site sizes for the neighbor exclusion model.

In the limit of small r, both Eqs. 12-24 and 12-25 reduce to

$$\frac{1}{T_\theta} - \frac{1}{T'_\theta} = \frac{R}{\Delta H^\circ_\mathrm{m}} (r_\mathrm{h} - r_\mathrm{c}) \tag{12-26}$$

which provides a useful plot of experimental data when binding is relatively weak so that much of the drug is unbound. In this case, transition curves are generally monophasic, although usually broadened from the curve observed without drug, and T_m can be unambiguously identified.

Another important limit is that in which binding to the double helix is saturated ($K_\mathrm{h} a \gg 1$), but binding to the denatured form can be neglected ($K_c a \ll 1$). In this case both Eqs. 12-24 and 12-25 reduce to

$$\frac{1}{T_\theta} - \frac{1}{T'_\theta} = \frac{R}{\Delta H^\circ_\mathrm{m}} \ln (K_\mathrm{h} a)^{B_\mathrm{h}} \tag{12-27}$$

where B_h is replaced by the equivalent $1/n_\mathrm{h}$ in the neighbor exclusion model. This approach is useful for interpreting the behavior of the second maximum in biphasic melting transitions such as those seen in Figure 12-15, since the second transition corresponds effectively to melting of the drug-saturated helix.

4. SOLUTION STUDIES OF THE STRUCTURE OF DRUG-DNA COMPLEXES

A number of global structural features of DNA complexes can be investigated by "low resolution" solution methods, although determination of the structure at atomic resolution requires X-ray or NMR methods, which are considered in Chapters 4 and 5, respectively. The first question that is generally addressed is whether or not the compound is an intercalator. Experimental tests include the ability to unwind and lengthen the DNA double helix while preserving orientation of the heteroaromatic ring system parallel to the DNA base pairs. For nonintercalating compounds, a key question is the locus of the interaction, which is often found to be the DNA minor groove. Additional characteristics that are accessible from solution structural studies include bending and changes in the stiffness of the double helix.

4.1 Measurement of the Extent of DNA Unwinding

Unwinding of DNA by drug binding causes relaxation of negative superhelical turns, which can be detected by a number of assays. Any DNA for which the superhelical density is known can be used for this purpose. Essential for the assay is measurement of the degree of drug binding r, not just the amount added per nucleotide or base pair. Figure 12-16 illustrates the use of sedimentation velocity to detect the conformational relaxation of closed circular DNA upon drug binding to characterize unwinding by ethidium, daunomycin, and actinomycin (Waring, 1971). The minimum in the sedimentation coefficient of the closed circular form reflects formation of the relaxed species, whose sedimentation coefficient S is indistinguishable from that of the nicked form. The relative ability of a given drug to unwind DNA is measured by the molar ratio of bound drug to DNA base pairs at the sedimentation coefficient minimum. As shown in Figure 12-16, ethidium relaxes ϕX174 DNA at an r value of 0.04 molecules per DNA nucleotide, whereas daunomycin requires more than twice as much drug, or about 0.09 per nucleotide.

Conversion of these figures to total unwinding angles requires a reference or comparison compound. Comparative studies of unwinding by alkaline denaturation and drug binding (Wang, 1974) established a value of 26° for the unwinding by ethidium, to which other drugs can then be compared. Table 12-4 gives some unwinding values for DNA binding agents. There is considerable variation in the extent of drug-induced unwinding. Note that echinomycin, called a bis-intercalating compound because both chromophores (Fig. 12-1) intercalate into the double helix, unwinds DNA by roughly twice the amount found for other intercalators, as expected for bis intercalation. All known intercalators unwind DNA to some extent, providing an important criterion for distinguishing these compounds from reversibly binding nonintercalators, which generally change the DNA twist by smaller amounts than are reflected in Table 12-4. However, compounds that react covalently with DNA can unwind DNA significantly. Examples include *N*-acetoxy-2-acetylaminofluorene (AAAF), unwinding angle 22° (Drinkwater et al., 1978), and *cis*-diamminedichloroplatinum(II), unwinding angle 11°–60°, depending on degree of superhelicity (Scovell and Collart, 1985)

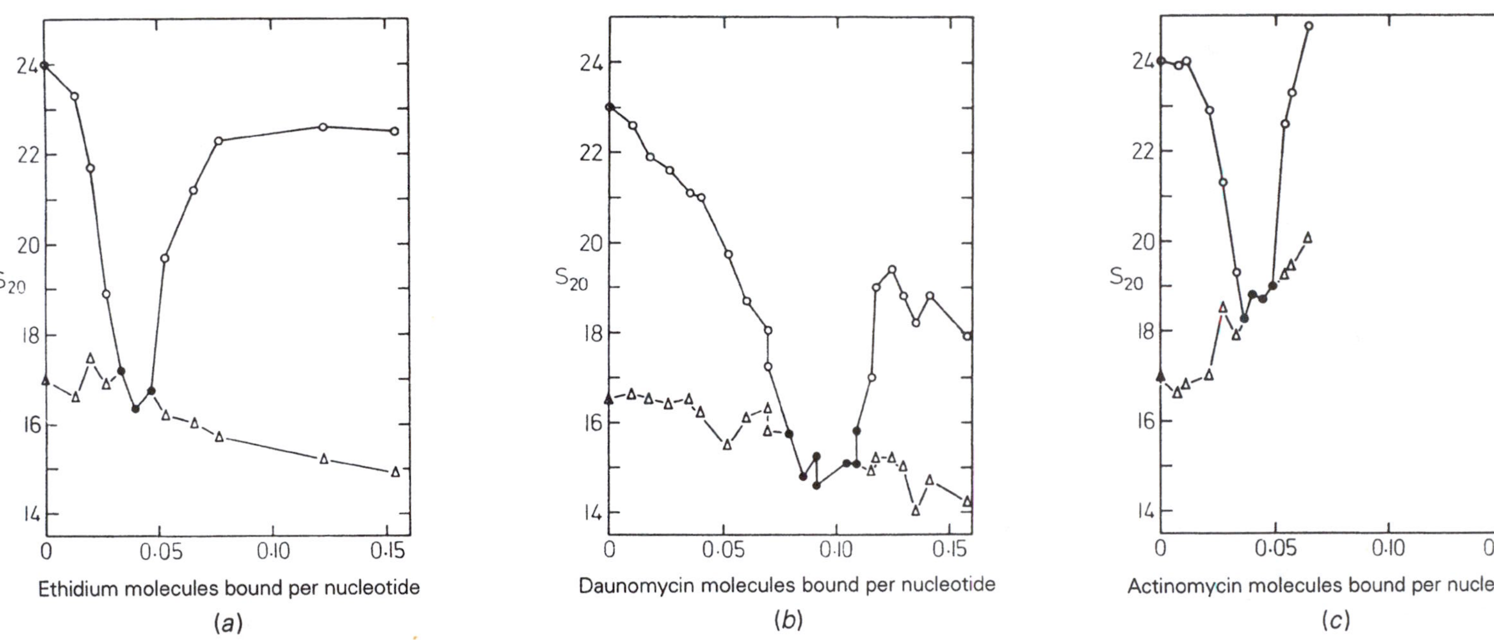

Figure 12-16
Effects of intercalating drugs on the sedimentation coefficient S_{20} of bacteriophage ϕX174 DNA. (*a*) Ethidium, (*b*) daunomycin, (*c*) actinomycin D. The buffer was 0.05 *M* tris-HCl, pH 7.9. The abscissa shows the average level of drug binding to the two DNA components, closed and nicked circles, present together in the sample. Symbols: (○= S_{20} of closed circles. (△ = S_{20} of nicked circles.
● = weight-average S_{20} from single boundary formed by closed and nicked circles sedimenting together. [From Waring, 1971.]

Table 12.4
Unwinding Angles of Several DNA Intercalators[a]

Drug	Unwinding Angle
Ethidium	26°
Proflavine	17°
Daunomycin	11°
Actinomycin	26°
Echinomycin	48°

[a]Tabulated by Neidle and Abraham (1986).

4.2 Hydrodynamic Studies of Structural Changes

Drugs that bind to DNA can alter not only the helical twist, but also the contour length and stiffness (both bending and torsional), and can, as an additional complication, induce systematic bends into the DNA double helix. Early use of hydrodynamic methods to study intercalating compounds focused on the length change expected to accompany the process. Early observations by Lerman (1961) showed that acridines produce an increase in the viscosity and a decrease in the sedimentation coefficient of DNA molecules. These effects could, in principle, be the result of an increase in molecular length, stiffness, or both, but length increase seems to be the dominant source of the change in hydrodynamic parameters of high molecular weight DNA samples to which most intercalators are bound; actinomycin is an important exception to this general rule.

Short DNA molecules, of dimension comparable to or less than the persistence length (Chapter 9), provide a means to distinguish an increase in length from an increase in stiffness or persistence length, since the latter has little effect on the hydrodynamic properties of a molecule which is already near the rodlike limit. This principle was recognized by Müller and Crothers (1968) in studies of actinomycin–DNA complexes, and Cohen and Eisenberg (1969) in experiments on acridine–DNA complexes; Mauss et al. (1967) had earlier used low molecular weight DNA samples in analogous light scattering experiments for the same reason. Figure 12-17 shows viscosity data for the actinomycin–DNA complex, from which it can be seen that the intrinsic viscosity rises as drug is added to a low molecular weight sample of DNA ($M = 10^5$, ~ 150 bp), but declines when the molecular weight is large. Müller and Crothers (1968) took the behavior of the low molecular weight sample as representative of the rigid rod limit, and, in combination with sedimentation data to eliminate the influence of helix diameter changes, calculated a length increment per actinomycin bound of about 4.5 Å. The decline in viscosity for high molecular weight samples was ascribed to favorable chain–chain interactions induced by actinomycin, causing a decrease in the excluded volume of the polymer chain. This interpretation was favored over changes in persistence length because there was no evidence that the percentage decrease in viscosity was approaching a plateau as molecular weight increased. (It can be seen by comparing Eqs. 9-3 and 9-5 that in the high molecular weight limit a decrease in the statistical segment length b should yield a percentage decrease in the intrinsic viscosity that is independent of N, whereas a change in the excluded volume parameter ε yields a percentage change that continues to increase with N.)

Reinert (1981, 1983) focused on quantitative interpretation of viscosity changes for molecules of different molecular weight at a constant degree of drug binding. Since changes in persistence length and contour length affect the viscosity differently depending on molecular weight, comparison of DNA molecules with substantially different sizes enables calculation of changes in the two parameters. Figure 12-18 shows the result of application of Reinert's analysis (Reinert, 1981) to the data of Figure 12-17. A length increase slightly greater than 1 bp spacing per bound actinomycin is verified for all molecular weights. This expected consequence of intercalation is accompanied by a gradual decrease in persistence length, reaching about 30% decline at the minimum of the curve. Since this analysis quantitatively accounts for the data of Figure 12-17, Reinert's interpretation of the viscosity decrease at higher molecular weight in terms of a persistence length decrease is to be preferred over that given by Müller and Crothers (1968). However, it remains possible that changes in excluded volume contribute to viscosity changes upon drug binding to high molecular weight DNAs.

A persistence length change can result either from a change in DNA flexibility or from a bend or kink induced by drug binding. Reinert (1981, 1983) interpreted the broad minimum in persistence length as a function of degree of drug binding seen in Figure 12-18 in terms of a drug-induced kink accompanied by a stiffening of the DNA molecule. The results of analysis of viscosity changes for binding of several drugs led

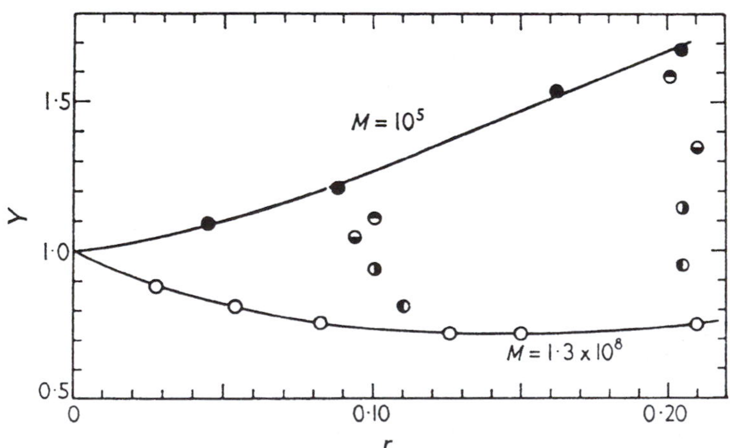

Figure 12-17
Ratio Y of intrinsic viscosity of complex to that of DNA plotted against the degree of binding r for several DNA molecular weights, ranging from 1×10^5 (~ 150 bp) to 1.3×10^8 (~ 200 kbp). [Reprinted from Müller, W. and Crothers, D. M., Studies of the binding of actinomycin and related compounds to DNA, *J. Mol. Biol.* **35**, 251–290, Copyright ©1968, by permission of the publisher Academic Press Limited London.]

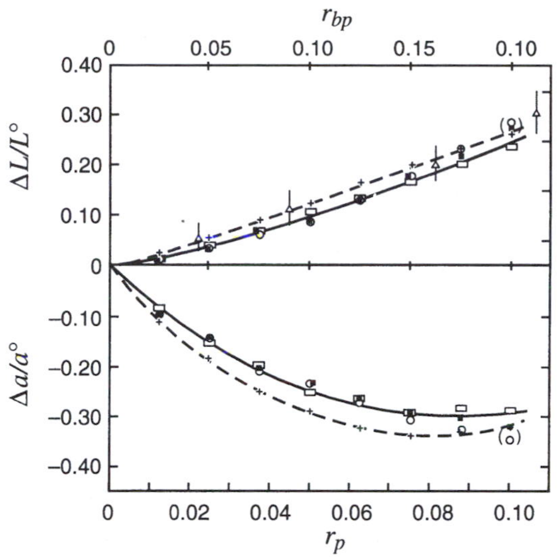

Figure 12-18
Calculated relative change of DNA persistence length and contour length, $\Delta a/a^0$ and $\Delta L/L^0$, as a function of the ratio r_p, the ratio of bound drug molecules to DNA phosphate. The experimental points correspond to the data in Figure 12-17. [Reprinted from Reinert, K. E. Aspects of specific DNA–protein interaction, local bending of DNA molecules by in-register binding of the oligiopeptide antibiotic distamycin, *Biophys. Chem.* **13**, 1–14, Copyright ©1981 with kind permission of Elsevier Science-NL., Amsterdam, The Netherlands].

to the parameters for lengthening, kinking, and stiffening collected in Table 12-5. In future work, it should be possible to sharpen the distinction between altered flexibility and induced bending by study of DNA sequences designed such that drug binding sequences are arrayed in specific phasings relative to the DNA helical repeat, as done for sequence directed bending effects (see Chapter 9).

Table 12.5
Lengthening, Bending and Stiffening of DNA by Drugs[a]

Drug	ΔL/drug[b] (nm)	γ[c] (deg)	ΔA/drug[d] (nm)
Anthracyclines			
Adriamycin	0.40	10.5	0.34
Daunomycin	0.39	11.4	0.46
Aclacinomycin	0.36	9.9	0.29
Proflavine	0.35	6.7	0.08

[a]From Reinert (1983).

[b]The contour length increase per bound drug bound is ΔL/drug.

[c]The parameter γ is the bend angle induced by binding of one drug molecule.

[d]ΔA/drug is the effective length of a DNA segment from which all flexibility is removed by binding one drug molecule.

4.3 Linear Dichroism Studies of DNA Complexes

The technique of linear dichroism (LD) (see Chapter 9) can be used to determine the orientation of bound drug molecules relative to the DNA helix axis. This subject has been reviewed in detail by Nordén et al. (1992). The drug must have one or more optical transition moments that do not overlap with the DNA absorbance, and one must know the orientation of the transition moment within the drug molecule. The technique is based on orienting DNA either by flow or in an electric field. The reduced LD is defined by

$$LD^r = \frac{A_\parallel - A_\perp}{A_{\text{iso}}} \tag{12-28}$$

where $A_\perp$ and $A_\parallel$ are the absorbances for light polarized perpendicular and parallel, respectively, to the DNA axis, which is parallel to the electric field direction or to the flow direction. The isotropic absorbance, A_{iso}, is given by

$$A_{\text{iso}} = \frac{1}{3}(A_\perp + 2A_\parallel) \tag{12-29}$$

For a planar intercalated aromatic chromophore in which the plane of the drug is perpendicular to the DNA axis, one expects $A_\parallel$ to be zero in the limit of perfect orientation (infinite flow rate or electric field), because the optical transition moments lie in the plane of the base. In this case, combining Eqs. (12-28) and (12-29) gives $LD^r_\infty = -\frac{3}{2}$. In the other limit the chromophore direction lies along the DNA axis, and $A_\perp$ is zero, so $LD^r_\infty = +3$. In general, if the transition moment lies at an angle α to the DNA axis, the LD at perfect orientation (LD^r_∞) is given by

$$LD^r_\infty = \frac{3}{2}(3\cos^2\alpha - 1) \tag{12-30}$$

The LD is zero when $3\cos^2\alpha - 1 = 0$, corresponding to orientation angle $\alpha = 54.7°$.

The orientation angle expected for transition moments aligned along the DNA minor groove is about 45°, which should provide a positive dichroism signal. On the other hand, intercalated chromophores that have transition moments in the plane of the ring should have negative dichroism. This distinction provides an important diagnostic for intercalation as opposed to groove binding (Nordén et al., 1992). Intercalating compounds that have been examined are characterized by a substantial negative LD signal (Hogan et al., 1979).

The exact orientation angles calculated from Eq. (12-30) are subject to uncertainty because of the problem of extrapolating to perfect orientation. Some authors (Nordén et al., 1992) prefer to assume that the DNA bases are within a few degrees of the 90° angle to the helix axis as required by the canonical B-DNA structure, and then determine drug orientation angles from the ratio of drug/DNA linear dichroism. However, dynamic motion of the DNA bases can be expected to increase the component of their transition moment along the helix axis from fluctuations such as propeller twisting, and indeed the extrapolated electric dichroism for small DNA restriction fragments implies base orientation angles in the range of 73° (Hogan et al., 1978). However, larger DNA molecules display reduced LD approaching the limiting value of $-\frac{3}{2}$ (Dieckmann et al., 1982; Lee and Charney, 1982). The origin of this distinction between large and small DNA molecules is unknown; possibilities include more effective straightening of large molecules in electric fields, and distortion of DNA structure to remove base tilt and propeller twist fluctuations by the large longitudinal stress that large molecules experience in high electric fields. These unknowns lead to uncertainties of about 15° in absolute orientation angles determined by LD methods. However, the relative angles are more accurate than that.

The length increase that accompanies intercalation leads to an increase in the rotational correlation time τ. For small DNA molecules, less than or equal to 150 bp, the length increase can be estimated from the dependence of τ on L^3. The results indicate an increase of roughly 1 bp spacing for each ligand intercalated (Hogan et al., 1979).

4.4 Circular Dichroism and Ligand Complex Structure

The transition moment of a DNA bound ligand has complex interactions with the helically organized multiple transition moments of the DNA bases. For this reason, it is difficult to calculate the circular dichroism (CD) of drugs bound to DNA and interacting with the chiral transition moment environment. The problem is made even more difficult because the polarization direction and wavelength maxima of the base electric dipole transition moments are not yet fully established. However, progress has been made on this problem in recent years. For example, Lyng et al. (1991, 1992) calculated the expected CD for groove bound and intercalated drug molecules when bound to specific DNA sequences. A general result is that the CD signal should be much larger for a groove bound ligand than for an intercalated species. Furthermore, the CD expected for minor groove bound molecules is positive when the transition moment is directed along the groove. Agreement between experiment and calculated spectra is qualitatively correct, but not quantitatively accurate.

5. KINETICS OF LIGAND BINDING

It has been recognized for many years that there are striking differences between the kinetic properties of DNA binding ligands, and that these differences may have important consequences for drug action (Müller and Crothers, 1968). Noncovalently bound ligands can differ by as much as six orders of magnitude in the lifetime of their DNA complexes, from submilliseconds to thousands of seconds. Bimolecular association rate constants can also vary widely, for example from about $10^4\ M^{-1}s^{-1}$ for actinomycin D (Müller and Crothers, 1968), or about $10^7\ M^{-1}s^{-1}$ for a typical intercalator such as proflavin (Li and Crothers, 1969), to values approaching diffusion limitation ($\sim 10^9\ M^{-1}s^{-1}$) for groove binding ligands (Müller et al., 1973). Differences in the lifetime of the complexes can have important consequences for inhibition of enzymes such as RNA and DNA polymerase. which move processively along the double helix (Müller and Crothers, 1968; White and Phillips, 1988).

Even among simple intercalators one finds substantial differences in the kinetic properties. Two main kinetic classes have been characterized. Proflavine illustrates the class that binds by a simple two-step process (Li and Crothers, 1969)

$$P + \text{DNA} \underset{k_{-1}}{\overset{k_1}{\rightleftharpoons}} (P)_{\text{out}} \underset{k_{-2}}{\overset{k_2}{\rightleftharpoons}} (P)_{\text{in}}$$

in which the first step yields a transiently stable "outside" bound complex, after which intercalation occurs in the second step. A reaction mechanism of this kind has two relaxation times τ, which can be measured by the temperature jump method, and are given by

$$\frac{1}{\tau_1} = k_1 \left(\bar{C}_{\text{D}} + \bar{C}_{\text{F}}\right) + k_{-1} \qquad \text{(12-31a)}$$

and

$$\frac{1}{\tau_2} = k_{-2} + \frac{k_2 \left(\bar{C}_{\text{D}} + \bar{C}_{\text{F}}\right)}{1/K_1 + \left(\bar{C}_{\text{D}} + \bar{C}_{\text{F}}\right)} \qquad \text{(12-31b)}$$

Equation 12-31a predicts a linear dependence of the faster relaxation rate $1/\tau_1$ on the sum of the equilibrium concentrations of free DNA binding sites and free drug, $\bar{C}_{\text{D}} + \bar{C}_{\text{F}}$, as is observed experimentally. Equation 12-31b predicts that the slower relaxation will reach a plateau value given by $k_2 + k_{-2}$ when the concentration becomes large. This behavior is illustrated in Figure 12-19 for proflavine binding to calf thymus DNA. The intercept at zero concentration gives k_{-2}, and the concentration $\bar{C}_{\text{D}} + \bar{C}_{\text{F}}$ required for $1/\tau_2$ to rise halfway to the plateau value from the intercept value is $1/K_1$, or the inverse of the equilibrium constant for the first step. Illustrative values for the conditions of Figure 12-19 (0.2 M Na^+, 10°C) are $k_1 = 1 \times 10^7 M^{-1}\ \text{s}^{-1}$, $k_{-1} = 3.5 \times 10^3 s^{-1}$, $k_2 = 1.5 \times 10^3 s^{-1}$, $k_{-2} = 1.1 \times 10^2 s^{-1}$, $K_1 = 2.8 \times 10^3 M^{-1}$, and $K_2 = 13.6$. The dissociation time of the complex is the reciprocal of the dissociation rate constant.

The first step has been interpreted as formation of an "outside" or groove-bound complex, which is followed by the intercalation step. Evidence favoring groove binding in the first step includes the strong influence of glucosylation such as in T2 phage DNA on the energetic and spectroscopic properties of the first but not the second step (Li and Crothers, 1969). Furthermore, the weak binding characterized by K_1 is characteristic of the external binding mode known to occur at high concentrations of added drug (Peacock and Skerrett, 1956). On this basis, one can estimate that the time required to remove proflavine from its intercalative binding site is about 10 ms, whereas the time for dissociation of the outside bound complex is about 300 μs. The lifetime

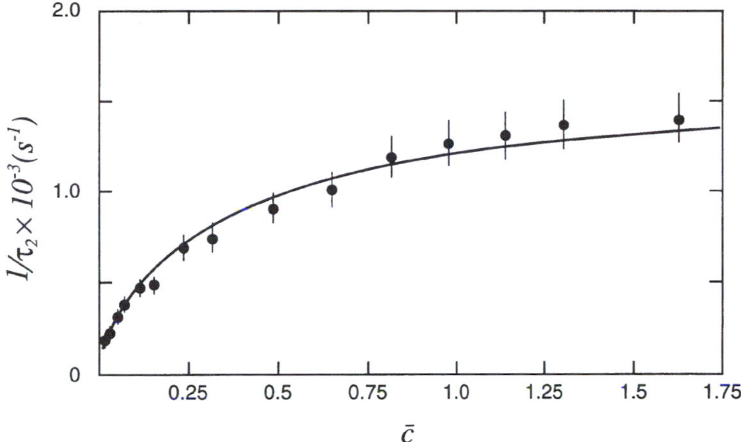

Figure 12-19
Variation of the slower relaxation time τ_2 with the sum of the concentrations of free DNA binding sites and free proflavine. The solid curve was calculated from Eq. 12-31b, with adjustment for best fit of the kinetic constants in the two-step binding mechanism. [Reprinted from Li, H. J. and Crothers, D. M. Relaxation studies of the proflavine–DNA complex: The kinetics of an intercalation reaction, *J. Mol. Biol.*, **39**, 461–477, Copyright ©1969, by permission of the publisher Academic Press Limited London].

of an outside-bound complex of 2,7-di-*t*-butyl-proflavine, a derivative rendered non-intercalating by virtue of substitution with bulky *t*-butyl side chains, is about 250 μs (Müller et al., 1973).

The lifetime of bound drugs can be studied directly by mixing the complex with low concentrations of a detergent (SDS-sodium dodecyl sulfate), and observing the time-dependent optical change that accompanies dissociation (Müller and Crothers, 1968). The dissociation time for intercalated actinomycin can range up to many thousands of seconds (Müller and Crothers, 1968; White and Phillips, 1988), whereas the lifetime of other important drugs may lie between the values for proflavine and actinomycin. For example, the lifetime of bound daunomycin is about 1 s (reviewed by Chaires, 1990; Chaires et al., 1985; Roche et al, 1994).

The other major kinetic class of DNA binding ligands is that in which the relaxation rates for the drug–DNA mixture continue to rise as concentration increases, without the plateau character seen in Figure 12-19 (Bresloff and Crothers, 1975; Wakelin and Waring, 1980). This behavior is illustrated for an ethidium analogue in Figure 12-20. When only a simple bimolecular reaction step is involved in binding, it is expected that the relaxation rate should continue to rise, but when there are two or more relaxation times, there must be more than one bound form of the drug, and, in simple models, the conversion between bound species should become limited by a first-order reaction step, such as drug dissociation or movement on the DNA. This limit leads to independence of concentration for one of the relaxation times, as seen in Figure 12-19.

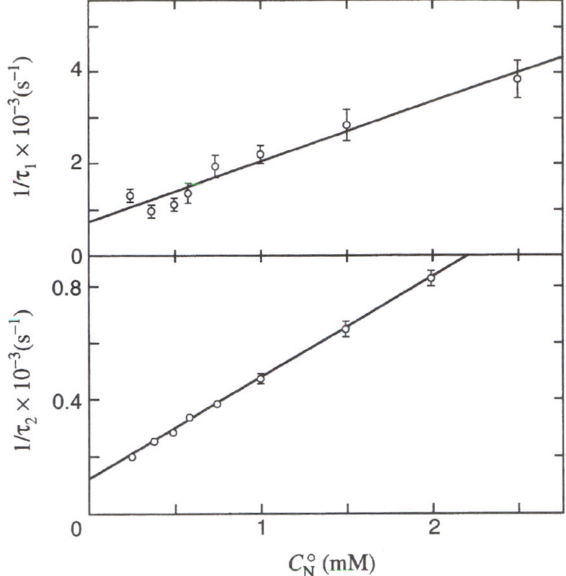

Figure 12-20
Continuous increase of the reciprocal relaxation times τ_1 and τ_2 with DNA concentration for binding the ethidium analogue carboxy dimidium. The nucleic acid concentration extends well beyond the range required to bind most of the ligand present, so simple two-step binding mechanisms would yield a plateau value for the slower of the two relaxation times. The results require catalysis by DNA of the equilibration over multiple binding sites. [Reprinted from Wakelin, L. P. and Waring, M. J. Kinetics of drug–DNA interaction. Dependence of the binding mechanism on structure of the ligand, *J. Mol. Biol.*, **144**, 183–214, Copyright ©1980, by permission of the publisher Academic Press Limited London.]

Failure of drugs such as ethidium to show a plateau for the slower relaxation rate as concentration increases requires a more complicated reaction mechanism. Interpretation has focused on direct transfer of the drug from one DNA molecule to another, a process that should retain a concentration dependence for equilibration over the bound sites (Bresloff and Crothers, 1975; Wakelin and Waring, 1980). The multiple sites could, for example, include drug bound in different sequence contexts. It would not be surprising if these complexes differed in kinetic and thermodynamic properties, creating a multiplicity of sites with a multiplicity of relaxation times. The complexity of the problem is emphasized by the fact that not all experimental approaches reveal a multiplicity of relaxation times for ethidium binding, depending on the perturbation and detection methods (Macgregor et al., 1985). The most recent experiments continue to favor some kind of direct transfer of the drug from one kind of binding site to another (Meyer-Almes and Pörshcke, 1993).

6. NATURAL PRODUCTS THAT REACT COVALENTLY WITH DNA

A number of naturally occurring compounds have been discovered that react with DNA to produce covalent adducts. In some cases these have proven to be important antitumor compounds. The DNA sites of reactivity are varied, ranging from minor groove functionalities including the guanine amino group and adenine N3 to the deoxyribose sugar. The modes of reaction include alkylation of the bases and radical cleavage of the DNA backbone. In the following sections we consider illustrative examples of these compounds.

6.1 DNA Minor Groove Alkylating Agents: Mitomycin, (+)-CC-1065, and Benzo[a]pyrene Diol Epoxide

The DNA minor groove provides an hospitable environment for hydrophobic ligands. Nature has evolved toxic compounds that target reactive functionalities to the minor groove on this basis. The primary sites of reaction are the nucleophilic guanine 2-amino group and adenine N3 ring nitrogen. Such reactions also have profound biological consequences, as illustrated by the carcinogenic diol epoxides, which are metabolic products of benzo[a]pyrene.

The mitomycins, which are clinically useful antitumor compounds, are DNA alkylating agents that can cross-link the two DNA strands through guanine amino groups at a CpG sequence. Reductive activation is required to produce the reactive species. This class of compounds has been reviewed by Tomasz (1994). As shown in Figure 12-21, reactions involve opening the aziridine ring to form an adduct at the 1 position of the drug. The other reactive functionality is at C10, where the carbamate moiety serves as the leaving group for the entering DNA nucleophile, resulting in a cross-link. Monoadducts also form, especially at G-containing sequences other than CpG. These products result from reaction at the aziridine ring.

Another DNA alkylating agent that reacts from the minor groove is (+)-CC-1065, but in this case the reactive nucleophile is adenine N3. This cytotoxic compound, produced by *Streptomyces zelensis*, has a distinct specificity for sequences such as PuNTTA and for A tracts (Reynolds et al., 1985). Figure 12-22 (Lin et al., 1991) shows the curved overall structure of the drug, which fits into the DNA minor groove, thus bringing the reactive cyclopropyl ring into apposition with N3 of adenine in a target sequence.

Benzo[a]pyrene is an environmental pollutant that is metabolized in mammalian cells to highly carcinogenic diol epoxides (BPDE). Of these, the (+)-*anti*-BPDE isomer [Figure 12-23(*a*)] is highly tumorigenic. Upon binding in the DNA minor groove the epoxide reacts by trans addition of the guanine amino group to yield the adduct shown in Figure 12-23(*b*). Studies of the adduct by NMR (Cosman et al.. 1992) show that the aromatic ring remains positioned in the minor groove, with its long axis directed in the 5′ direction along the modified strand.

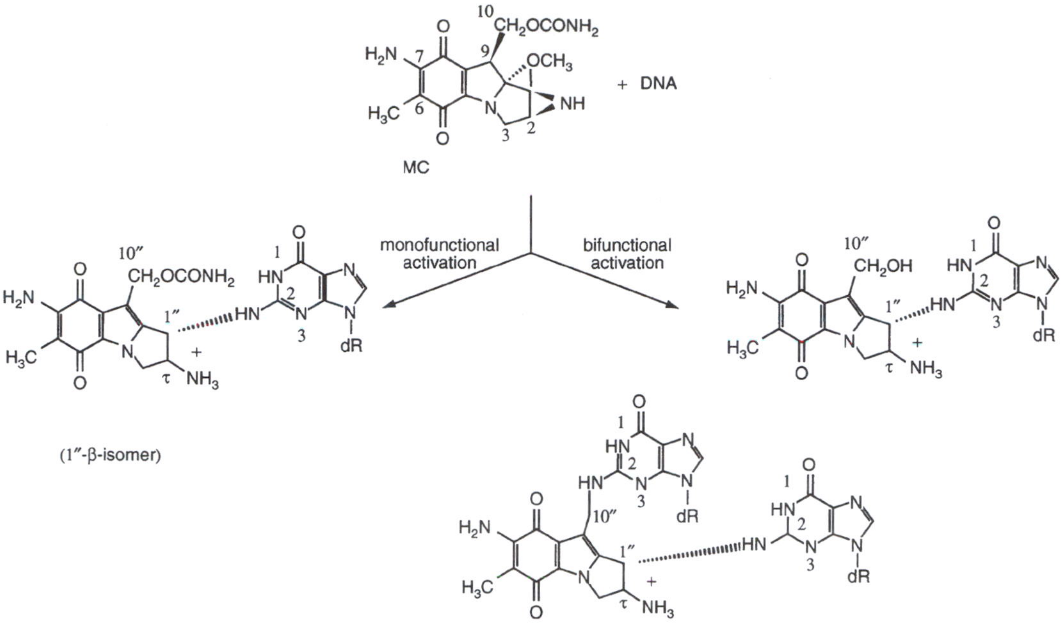

Figure 12-21
Adducts of mitomycin C and DNA formed under reductive activation. One or two electron ($1e^-$ or $2e^-$) reduction of the quinone in mitomycin yields an electrophilic species that can react to cross-link DNA. The initial site of reaction is the 1-position in the aziridine ring, and the guanine 2-amino group is the nucleophile. A cross-link results when a second properly placed guanine 2-amino group in a CpG sequence reacts at the 10 position to displace the carbamate group. [From Tomasz, 1994.]

6.2 Radical Mediated Cleavage from the Minor Groove: Bleomycin and the Enediynes

Several natural products can cause radical-mediated cleavage of the DNA backbone by abstracting hydrogen atoms from the deoxyribose component of the phosphodiester chain. The site attacked is typically the 1'-, 4'- or 5'-position. The origin of the hydrogen abstracting species can vary from oxidation–reduction chemistry as in the Fe–bleomycins, to photochemical, as in the Co–bleomycins, to formation of C–C bonds, as in the Bergman cycloaromatiztion reaction in the enediyne series of compounds. These agents have in common an affinity for the DNA minor groove, coupled with the ability to generate reactive radical forms of the deoxyribose ring by hydrogen abstraction. In the presence of molecular oxygen, oxidative cleavage of the chain results, by a process that culminates in β-elimination (see Chapter 3).

The bleomycins (BLM), reviewed by McGall and Stubbe (1988), are glycopeptide antibiotics with clinical antitumor utility. They have the capacity to chelate transition metals. Both $Fe^{(II)}$ and $Fe^{(III)}$ are thought to be bound in the active complex, for which a proposed chelation model is shown in Figure 12-24 (McGall and Stubbe, 1988). The

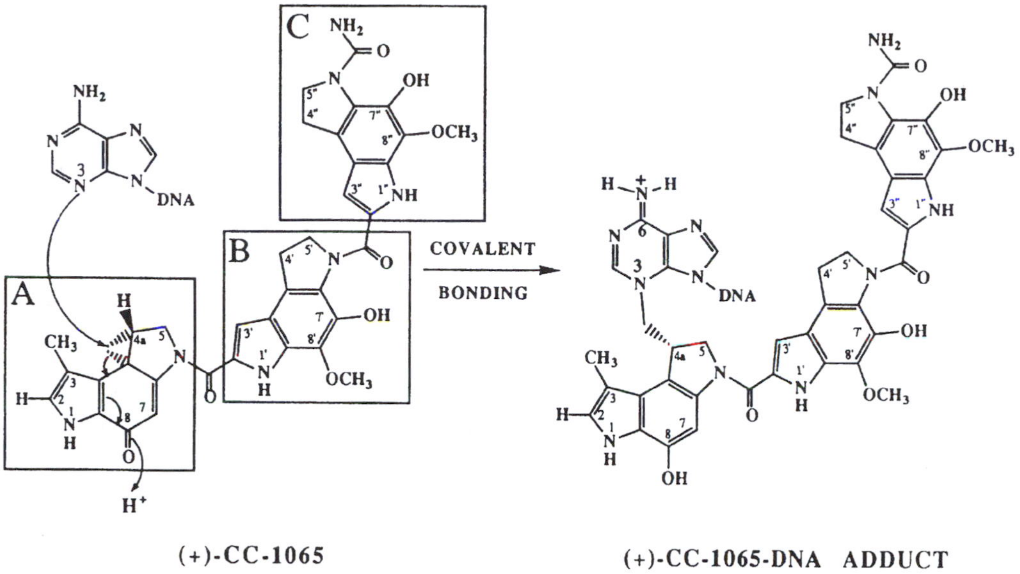

Figure 12-22
Reaction of (+)-CC-1065 with double-stranded DNA at N3 of adenine to form the (+)-CC-1065-DNA adduct. The covalently modified adenine is in the doubly protonated 6-amino form. [Reprinted with permission from Lin, C. H., Beale, J. M., and Hurley, L. H., *Biochemistry* **30**, 3597–3602 (1991). Copyright ©1991 American Chemical Society.]

structure consists of a metal binding portion coupled via a flexible peptide chain to a bithiazole group. Two dimensional NMR studies of peroxycobalt(III) bleomycin and cobalt(III) BLM show a chelation pattern similar to that in Figure 12-24 (Xu et al., 1994). Whereas the iron BLM require molecular oxygen and a reductant to generate the hydrogen abstracting species, the cobalt bleomycins are capable of light-activated cleavage of DNA at specific sequences (Nightingale and Fox, 1994). The production of a hydrogen scavenging species, called "activated BLM," requires reaction with O_2, followed by a one-electron reduction:

$$\text{Fe}^{(\text{II})}\text{—BLM} \overset{\text{O}_2}{\rightleftharpoons} \text{Fe}^{(\text{II})}\text{—BLM-O}_2 \overset{1e^-}{\rightleftharpoons} \text{“Activated BLM”}$$

The exact structure of the activated complex is still uncertain, but its properties are consistent with the peroxide species $Fe^{(III)}$-OOH, or with $Fe^{(III)}(:\ddot{O}:)$, in which the oxygen is at the oxidation level of atomic oxygen, or with $Fe(V) = O$ (McGall and Stubbe, 1988). Unlike the $Fe^{(II)}$-edta mediated cleavage of DNA, freely diffusing OH radical is not the hydrogen abstracting species in DNA cleavage by bleomycin.

Studies of the effect of deuterium substitution at specific positions in deoxyribose on the kinetics of chain cleavage can be used to identify the hydrogen that is abstracted in the rate-limiting step (Kozarich et al., 1989). Substitution at the 4′- position is found to produce a kinetic isotope effect (a reduction in the rate) of several-fold, indicating

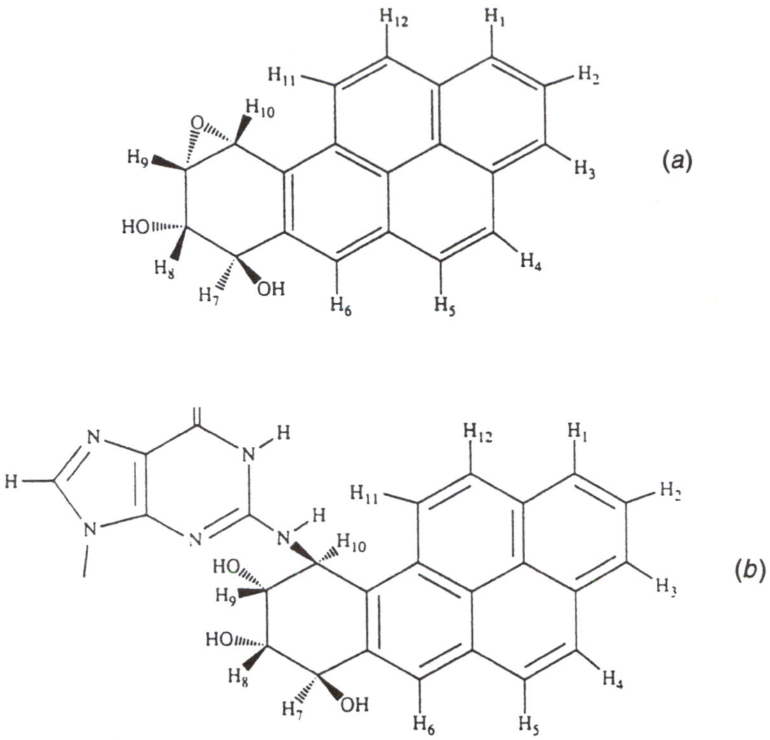

(*a*)

(*b*)

Figure 12-23
(*a*) Structure of the highly carcinogenic (+)-*anti*-benzo[a]pyrene diol epoxide. (*b*) Adduct of the diol epoxide in (*a*) with the guanine 2-amino group. [From Cosman et al., 1992.]

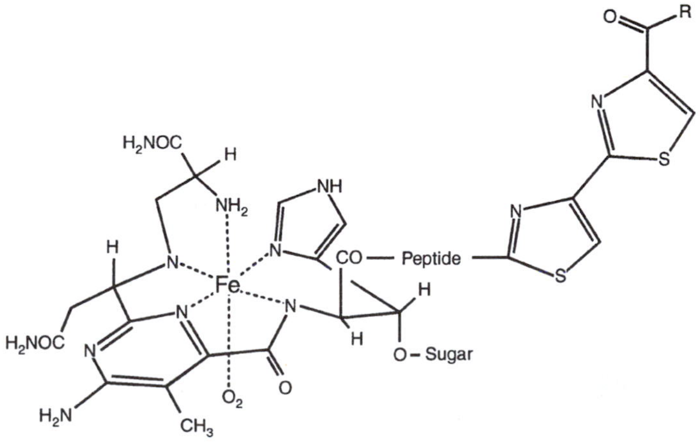

Figure 12-24
Proposed structure for $Fe^{(II)} \cdot \cdot O_2$BLM. [From McGall and Stubbe 1988.]

that activated BLM reacts by abstracting a hydrogen from the 4′- position of the deoxyribose ring (McGall and Stubbe, 1988; Worth et al., 1993). This process leads to chain cleavage as shown in Figure 12-25, with products dependent on whether or not molecular oxygen is present in the medium. Note that the phosphate residue at the 3′- end of the chain is modified by a glycolate, which affects electrophoretic mobility relative to the Maxam–Gilbert reaction standards.

Sequence specificity of BLM cleavage seems to reside in the metal binding domain, whereas the bithiazole moiety is needed for overall DNA binding affinity (Carter et al., 1990a; Kane et al., 1994). Bleomycin can also cleave RNA in a sequence specific way (Carter et al., 1990b)

The enediynes (reviewed by Nicolaou and Dai, 1991; Nicolaou et al., 1993) also cause radical mediated DNA cleavage based on minor groove binding, but hydrogen abstraction relies on an entirely different chemistry from that seen with bleomycin. Figure 12-26 (Nicolaou and Dai, 1991) shows the structures of calicheamicin, esperamicin, dynemicin, and the neocarzinostatin chromophore. Neocarzinostatin consists of a protein–chromophore complex, whose structure has been determined by X-ray crystallography (Kim et al., 1993). The exceptional toxicity of these compounds is based on their ability to cause double strand breaks in DNA at subnanomolar concentration levels.

The enediynes share the chemical feature of two triple bonds separated by a conjugated double bond, embedded in a 9- or 10-membered ring. Such structures are capable of undergoing a process known as the Bergman cycloaromatization reaction (Jones and Bergman, 1972), shown in simple form in Figure 12-27(*a*). If the two carbon atoms at the ends of the triple bond approach each other closely enough, a new bond is formed between them, generated by one electron from each of the triple bonds. The result is the aromatic biradical shown in the figure.

The reactive process in the calicheamicin chromophore is shown in Figure 12-27(*b*). Reduction of the trisulfide releases a thiolate, which is ideally positioned to attack the adjacent α, β unsaturated ketone. Conversion of the sp^2 hybridization at the bridgehead double bond to sp^3 has the effect of shortening, by a few tenths of an angstrom, the distance between the carbon atoms at the ends of the triple bond. This displacement greatly enhances the rate of the cycloaromatization reaction, thus generating a biradical, which is capable of abstracting a hydrogen atom from each of the two DNA strands. The result can be double-strand cleavage of DNA, a cataclysmic event for the cell.

The calicheamicin chromophore lacking the complex carbohydrate side chain (Fig. 12-26) is capable of causing nonsequence specific DNA strand breaks when present at about 0.1-m*M* concentration, but only a small fraction of the cleavage events result in double strand breakage (Drak et al., 1991). In contrast, calicheamicin γ_1^1 at 1-n*M* concentration cleaves a substantial fraction of added superhelical DNA, and a large fraction of the cuts are double-strand cleavage events. Furthermore, there is a pronounced sequence preference, which targets runs of pyrimidines such as TCCT. The enhanced affinity, sequence specificity, and specialization for double-strand cleavage are contributed by the complex carbohydrate, an aryl tetrasaccharide, which is attached to the aglycon chromophore. The carbohydrate moiety has unusual chemical features for a natural product, such as an N–O linkage between sugar residues, and an iodobenzene subunit.

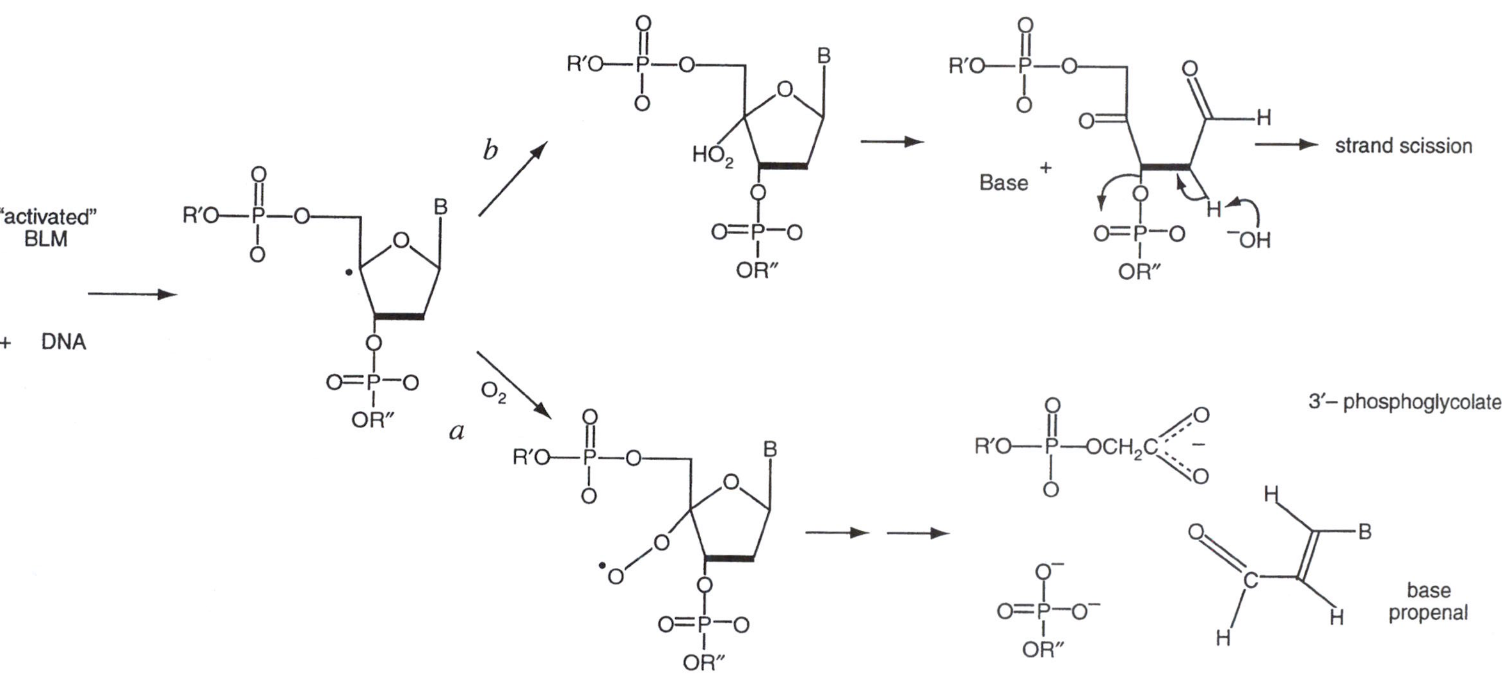

Figure 12-25
Illustration of the reactions leading to chain cleavage following hydrogen abstraction. Removal of H4′ by activated BLM leads, via a series of steps, to a β-elimination reaction that cleaves the chain. [From McGall and Stubbe 1988.]

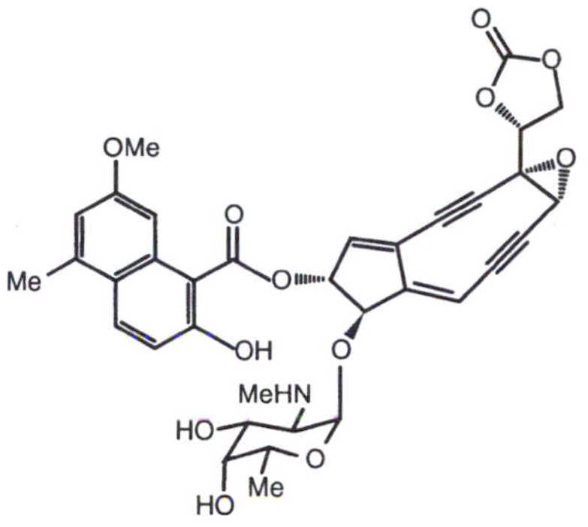

Figure 12-26
Structures of the enediyne antibiotics. Neocarzinostatin chromophore (**1**), calicheamicin γ_1^1 (**2**), esperamicin A_1 (**3**), and dynemicin A (**4**). [Reprinted with permission from Nicolau and Dai, 1991.]

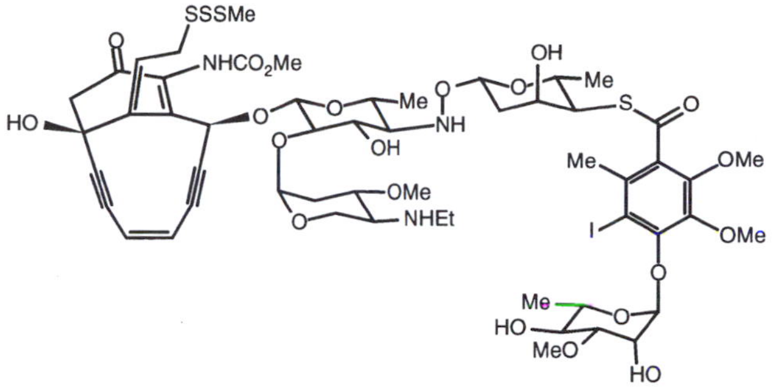

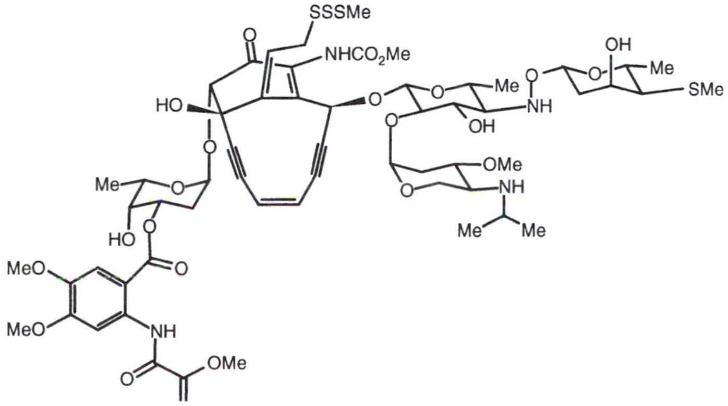

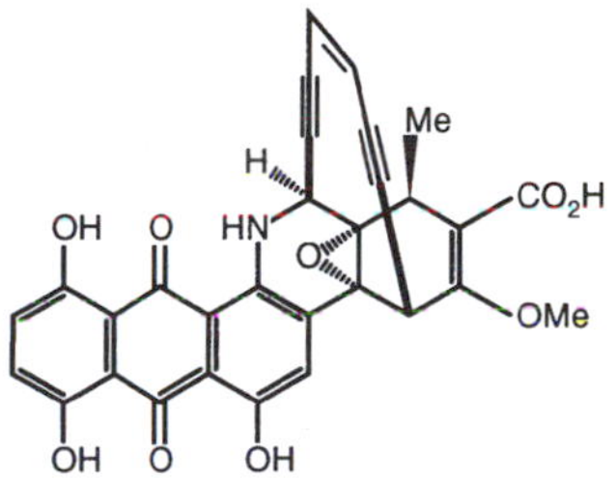

The function of the carbohydrate side chain is to steer the radical-generating enediyne component to the minor groove. Footprinting studies show that it binds strongly to the DNA minor groove in a sequence-specific way (Aiyar et al., 1992). The NMR structural studies of Paloma et al. (1994) and Walker et al. (1994) show that calicheamicin fits snugly into the DNA minor groove, with the reactive aglycon moiety brought into appropriate position to abstract the H4′ hydrogen from one strand and the H5″ (pro-S) hydrogen from the other strand. The model is consistent with the observed stagger of three nucleotides in the cutting events on the two strands (Fig. 12-28) and with the experimental sites of hydrogen abstraction (De Voss et al., 1990; Hangeland et al., 1992). Following hydrogen abstraction, reaction with oxygen leads eventually to double-stranded chain cleavage. Hence, the biological activity of these compounds is comprehensively explained on the basis of the structural and chemical features of their interactions and reactions with DNA.

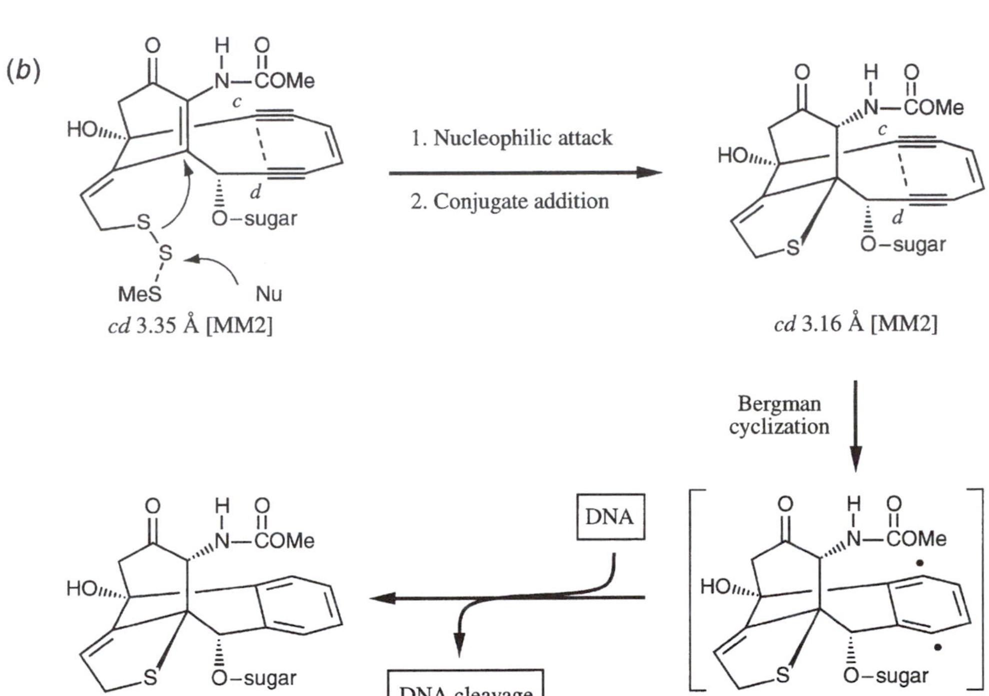

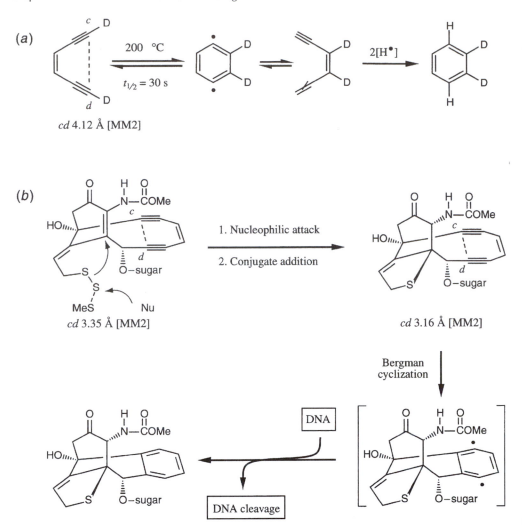

Figure 12-27
(*a*) The Bergman cycloaromatizion reaction. The distance between the centers *c* and *d*, critical for the rate of the process, is calculated to be 4.12 Å. (*b*) The internally triggered Bergman cyclization reaction in the calicheamicin chromophore. [Reprinted with permission from Nicolau and Dai 1991.]

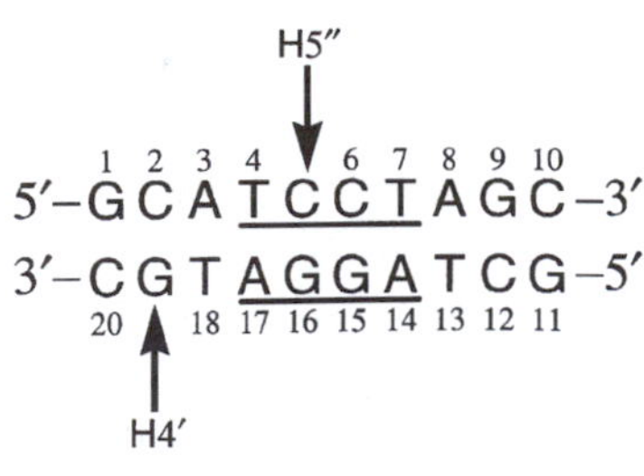

Figure 12-28
Typical sequence subject to double-strand cleavage by calicheamicin, showing the sequence recognized (underlined) and the sites of hydrogen abstraction: H5″ from the pyrimidine run, and H4′ from the purine strand, displaced in the 3′ direction from the purine tract recognition sequence. [Reprinted with permission from Paloma, L. G., Smith, J. A., Chazin, W. J., and Nicolaou, K. C., *J. Am. Chem. Soc.* **116**, 3697–3708 (1994). Copyright ©1994 American Chemical Society.]

6.3 Electrophilic Attack from the Major Groove: Pt Antitumor Compounds and the Carcinogenic Aflatoxins

The DNA major groove is not without nucleophilic sites, for example, the N7 ring nitrogen of A and G, which react with the electrophilic antitumor drug cisplatin to form inter- and intra-strand cross-linked adducts. The reaction of these inorganic ligands (reviewed by Lepre and Lippard, 1990) with the DNA major groove provides an interesting contrast to the behavior of natural products that react with the DNA minor groove by virtue of their ability to bind there.

Figure 12-29 illustrates four platinum(II) compounds that are active as antitumor agents, and two that are not. The original discovery of *cis*-DDP, or diamminedichloroplatinum(II), stemmed from the observation that the products of electrolysis at a Pt electrode inhibited cell division in *E. coli* (Rosenberg, et al., 1965). The active compounds subsequently discovered generally retain the inert cis ammine functions chelated with Pt, with labile cis ligands at the other two positions in square planar complexes. The trans isomers are inactive as antitumor agents, even though they are reactive with DNA. Selective *in vivo* repair or inactivation of trans-DDP adducts may be responsible for this difference, since the preferred reaction products differ for the cis and trans isomers. For example, trans-DDP is able to form interstrand cross-links between guanine and complementary cytosine residues (Brabec and Leng, 1993) but *cis*-DDP does not.

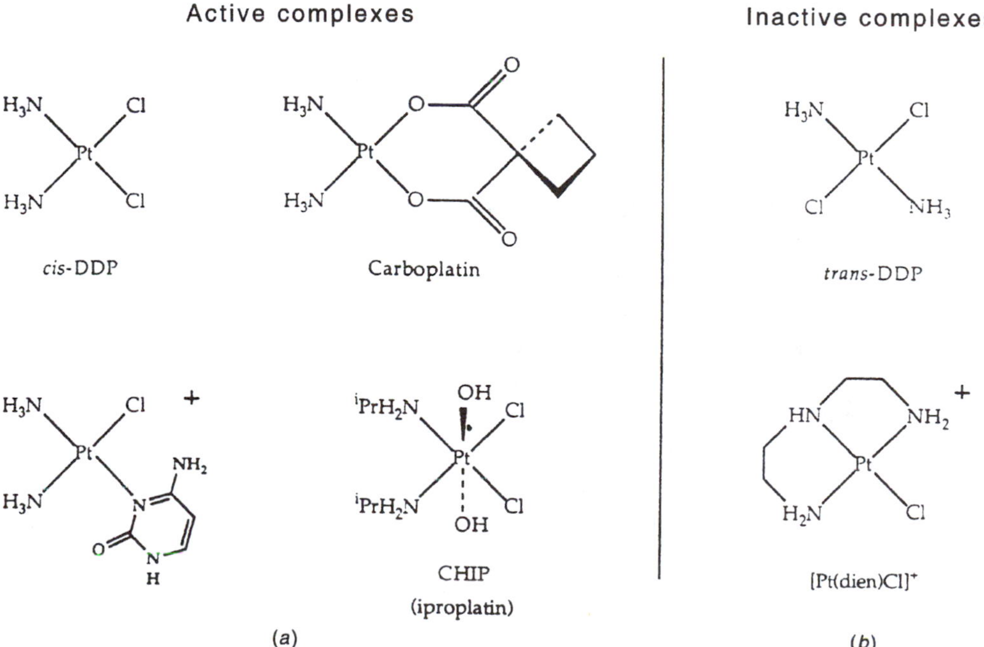

Figure 12-29
Structural diagrams of antitumor active (*a*) and inactive (*b*) platinum complexes. [From Lepre and Lippard 1990.]

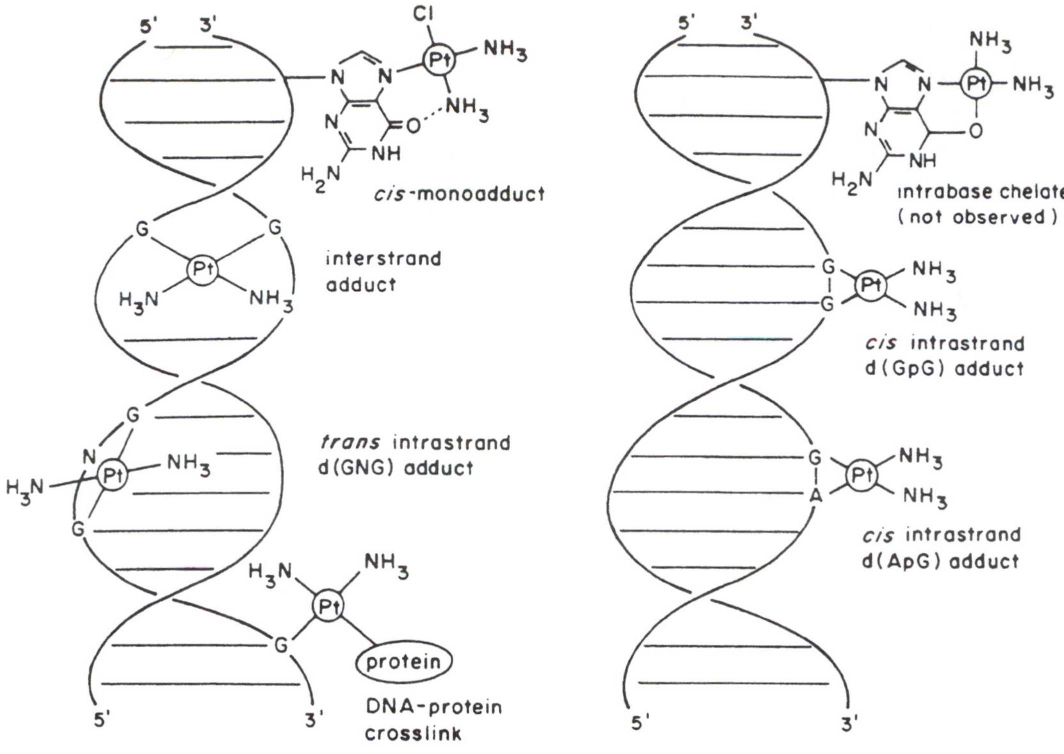

Figure 12-30
Schematic representation of possible platinum–DNA binding modes. [From Lepre and Lippard 1990.]

Figure 12-30 illustrates the main reaction products seen for reaction of *cis*-DDP with double stranded DNA. In some cases the local structures of the adducts are known at atomic resolution. For example, the principal adduct of *cis*-DDP with DNA, in which $Pt^{(II)}$ is reacted with the N7 positions of successive guanines in the sequence GpG shows formation of bonds at the cis liganding positions of Pt (Sherman et al., 1985). One predicts that the adduct should produce a strong distortion of double helical DNA, since stacking between the adjacent guanine rings is lost, yielding a sharp kink toward the major groove to accommodate $Pt^{(II)}$ binding to adjacent guanines.

Comparative electrophoresis measurements (reviewed by Crothers and Drak, 1992) are in agreement with this prediction, and reveal helix unwinding by adduct formation (Rice et al., 1988; Bellon and Lippard, 1990; Bellon et al., 1991) The bend, about 35°–40° in magnitude, is directed toward the major groove, as judged by comparison with A-tract bends. These conclusions are in agreement with the crystal structure of the adduct of *cis*-$[Pt(NH_3)_2Cl_2]$ with a double helical dodecamer of sequence dCCTCTG*G*TCTCC, where the asterisks denote the site of $Pt^{(II)}$ adduct formation at guanine N7 (Takahara et al., 1995). The structure is shown in Figure 12-31. There is no

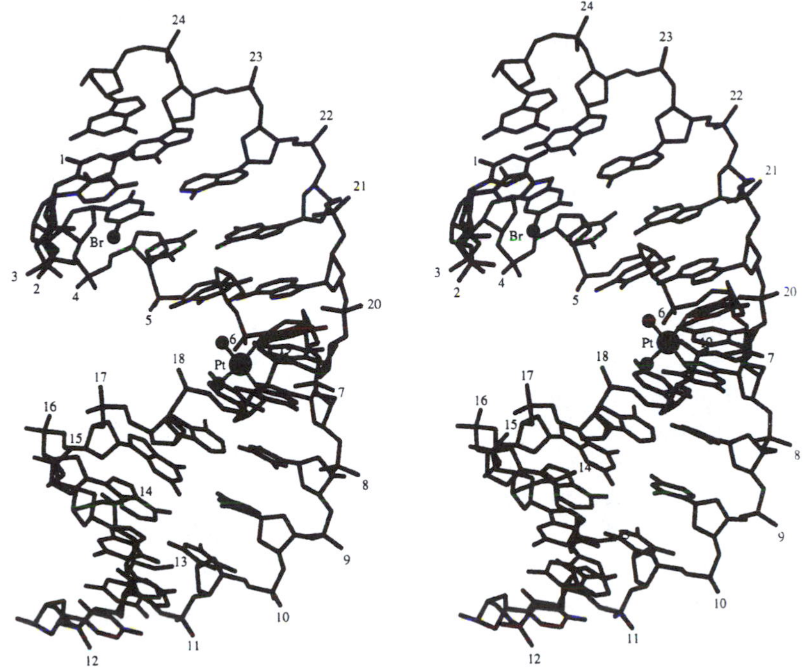

Figure 12-31
Stereoimage of one of the two duplexes in the unit cell for the complex of the double helix of dCCTCTG*G*TCTCC with cisplatin. [Reprinted with permission from *Nature*. Takahara, P. M., Rosenzweig, A. C., Frederick, C. A., and Leppard, S. T., *Nature (London)* **377**, 649–652 (1995). Copyright ©1995 Macmillan Magazines Limited.]

disruption of Watson–Crick hydrogen bonding, and there is a strong kink of 39°–55° toward the major groove. The roll angle at the GpG step is about 26°.

The aflatoxins are among the most carcinogenic of naturally occurring compounds. Produced by the fungus *Aspergillus flavus,* they are natural food contaminants. Aflatoxin B_1 is a highly substituted coumarin derivative, which, like benzo[a]pyrene, is metabolized to an epoxide which is thought to be the reactive agent for DNA modification. However, unlike the benzo[a]pyrene diol epoxide, which targets the minor groove, aflatoxins react with guanine N7 in the major groove, producing an adduct that is responsible for mutagenic and carcinogenic consequences (Muench et al., 1983; Misra et al., 1983; Sambamurti et al., 1988).

7. SYNTHETIC DNA BINDING AGENTS

Synthetic agents capable of binding to specific DNA sequences offer the opportunity to mimic the activity of gene regulatory proteins, to possible pharmaceutical advantage. Major problems facing such a technology include designing agents with sequence specificity, and delivering them in a therapeutically effective way. We begin here with consideration of agents that report on their own binding position by inducing cleavage reactions as a consequence of binding. This technology has important applications for the study of sequence-specific binding agents.

7.1 DNA Affinity Cleavage Agents

Cleavage of DNA by a reactive group carried to the DNA surface by a bound ligand is an example of affinity-based chemistry. This general technology, which evolved in nature through generation of compounds such as bleomycin, the enediynes, and spectronigrin (Cone et al., 1976), has provided powerful applications through the design of compounds to analyze sequence-dependent effects in DNA–ligand interactions. The method relies on inferring the binding site from the site of chain cleavage. One of the first reagents of this kind was 1,10-phenanthroline–$Cu^{(I)}$, which, together with H_2O_2, causes radical-mediated cleavage of DNA (Sigman et al., 1979; Pope et al., 1982; reviewed by Sigman and Spassky, 1989). The planar phenanthroline binds to DNA by intercalation, and oxidation–reduction chemistry of the bound copper generates species that cleave DNA.

Dervan and co-workers developed a series of affinity cleavage agents that have been highly informative on the nature and specificity of DNA binding interactions (reviewed by Dervan, 1986). The prototype of this series of compounds (Hertzberg and Dervan, 1982) is (methidiumpropyl-edta)(Fe^{II} (Fig. 12-32), which links the intercalative affinity of methidium (an analogue of ethidium) together with the OH radical-generating capacity of edta-$Fe^{(II)}$ in the presence of O_2 and a reducing agent. A number of affinity cleavage agents for drugs such as distamycin (Taylor et al., 1984) have followed in this series. In addition, the edta-$Fe^{(II)}$ chemistry has been tethered to DNA binding peptides, such as the DNA binding domain of Hin recombinase (Sluka et al., 1987).

The technology is capable of identifying the site of binding, and whether the edta–Fe portion is bound in the major or minor groove. Figure 12-33 illustrates the results. The 52-amino acid peptide corresponding to the Hin recombinase DNA binding domain was derivatized with edta at its amino terminus. The figure shows both protection of

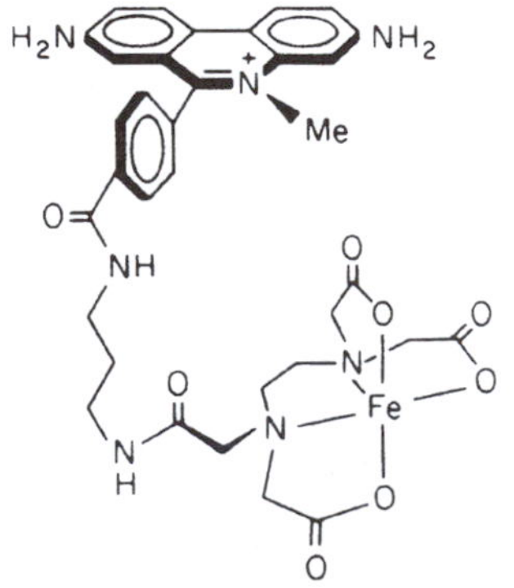

H2N NH2 N+ Me O NH O O O N O N Fe O N H O O

Figure 12-32
Structure of MPE–Fe. [Reprinted with permission from Hertzberg, R. P. and Dervan, P. B. *J. Am. Chem. Soc.* **104**, 313–315 (1982). Copyright ©American Chemical Society.]

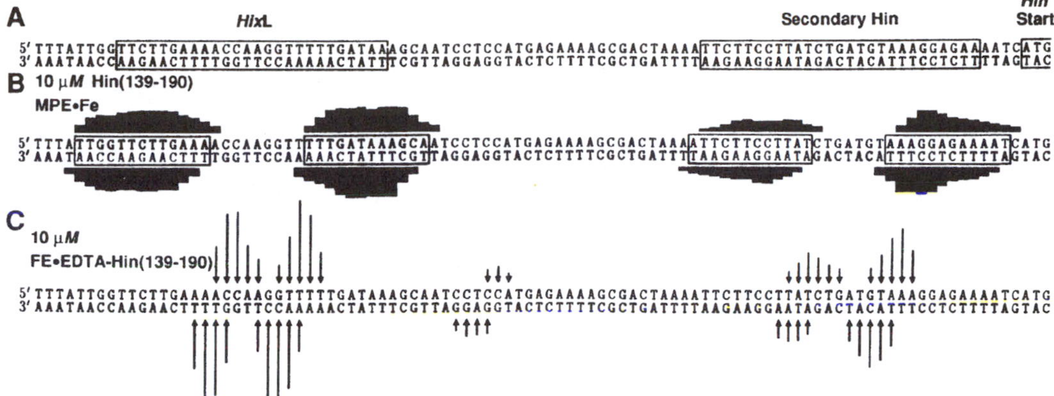

Figure 12-33
Interaction of the DNA binding domain of Hin recombinase with the *Hix*L DNA sequence, probed by edta-Fe complexes. (*a*) Sequence of the *Hix*L and secondary Hin sites. (*b*) Protection from MPE-Fe cleavage by interaction with the DNA binding domain. (*c*) Cleavage generated by a derivative of the DNA binding peptide having an edta-Fe residue attached at the amino terminus. [Reprinted with permission from Sluka, J. P., Horvath, S. J., Bruist, M. F., simon, M. I., and Dervan, P. B., *Science* **238**, 1129–1132 (1987). Copyright ©*American Association for the Advancement of Science*.]

the *Hix*L recombination site from cleavage by MPE-Fe (panel *b*), and the sites of DNA cleavage generated by the bound peptide derivatized with edta. The cleavage pattern shows that the amino terminal portion of the peptide is bound near the symmetry axis of the DNA binding site. Furthermore, the peptide amino terminus must be positioned in the DNA minor groove. This conclusion can be inferred from the pattern of stagger of the DNA cleavage sites. As shown in Figure 12-34, the residues on the two strands closest to a metal positioned in the minor groove are displaced to the 3′- side of the local dyad axis at the binding site. In contrast, when the metal or other cleavage ligand is placed in the major groove, the expected sites of cutting are displaced toward the 5′- direction from the symmetry axis. These two patterns are called minor and major groove stagger, respectively. Minor groove stagger is characteristic of cleavage by DNAse I, for example (see Chapter 13). The broad applicability of the affinity-based reaction technology can also be illustrated by the use of triple helix forming molecules to target specific sequences, followed by photoreaction with psoralen to produce a covalent linkage (Takasugi et al., 1991).

7.2 Sequence-Specific DNA Binding Agents

It has been a long-standing objective of the field of small molecule–DNA interactions to generate molecular structures that are capable of recognizing DNA sequences large enough to be unique in a realistic biological context, such as a viral infection, or even a eukaryotic regulatory sequence. An early effort in this direction was carried out by

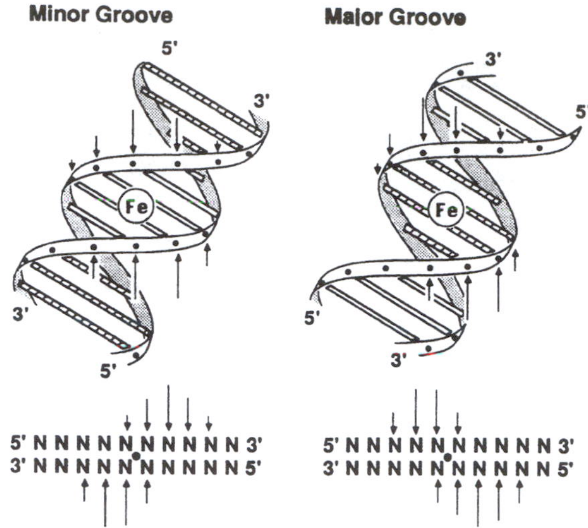

Figure 12-34
Cleavage patterns produced by a diffusible oxidant generated by edta–Fe localized in the minor and major grooves of right-handed DNA. The edges of the bases are shown as open and cross-hatched bars for the minor and major grooves, respectively. The "minor groove stagger" shown below, in which the cutting sites are displaced in the 3′ direction from the site of origin of the radicals, is typical of minor groove binding and DNA cleaving ligands. [Reprinted Science with permission from Sluka, J. P., Horvath, S. J., Bruist, M. F., Simon, M. I., and Dervan, P. B, *Scinece* **238**, 1129–1132 (1987). Copyright ©1987 *American Association for the Advancement of Science*.]

Müller and co-workers. Through study of the DNA composition dependence of drug binding affinity (see Section 12.2.4), they were able to identify specific ligands with a high specificity for G•C or A•T rich sequences (Müller and Crothers, 1975; Müller et al., 1975; Müller and Gautier, 1975). Then, in a strategy which, in today's terminology might be called template-directed combinatorial synthesis, polymeric forms of these mixed drugs were generated by polymerization on a specific viral DNA template. An example is shown in Figure 12-35 (Kosturko et al., 1979). Phenyl neutral red (PNR) is an intercalating derivative with high selectivity for G•C rich sequences, with an α value (Section 12.2.4) that indicates a preference for adjacent G•C pairs (Müller et al., 1975). Malachite green is a triphenylmethane dye with a pronounced specificity for A•T rich regions (Müller and Gautier, 1975). The acryl derivatives of these compounds were combined with a charged but otherwise non-DNA binding acrylamide, and polymerized by a radical-induced mechanism in the presence of DNA. Some evidence for specificity in rebinding to the DNA used as template in the polymerization could be detected (Kosturko et al., 1979).

Synthetic compounds based on the natural ligands netropsin and distamycin (Fig. 12-1) have been actively investigated as DNA sequence-specific binding agents (reviewed by Dervan, 1986). These drugs recognize the DNA minor groove by a combination of hydrogen bonding and hydrophobic interactions, as illustrated for netropsin in Figure 12-36 (Kopka et al., 1985). The successive amide NH groups lie between

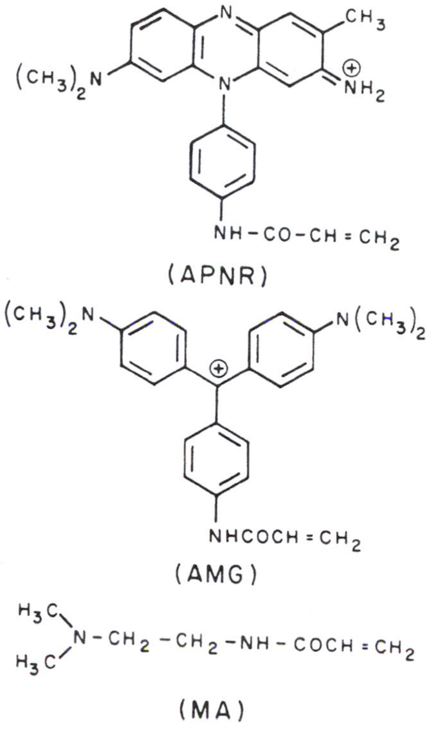

Figure 12-35
Chemical structures of three monomers used in template-directed synthesis of sequence-specific DNA binding compounds by Kosturko et al. (1979). The acryl derivatives of phenyl neutral red (APNR), a G • C specific intercalating ligand, malachite green (AMG), an A • T specific groove binding compound, and methylacrylamine (MA), a weakly binding charged acrylamide, are polymerized by a radical chain mechanism in the presence of DNA of defined sequence. [Reprinted with permission from Kosturko, L. D., Dattagupta, N., and Crothers, D. M., *Biochemistry* **18**, 5751–5756 (1979). Copyright ©American Chemical Society.]

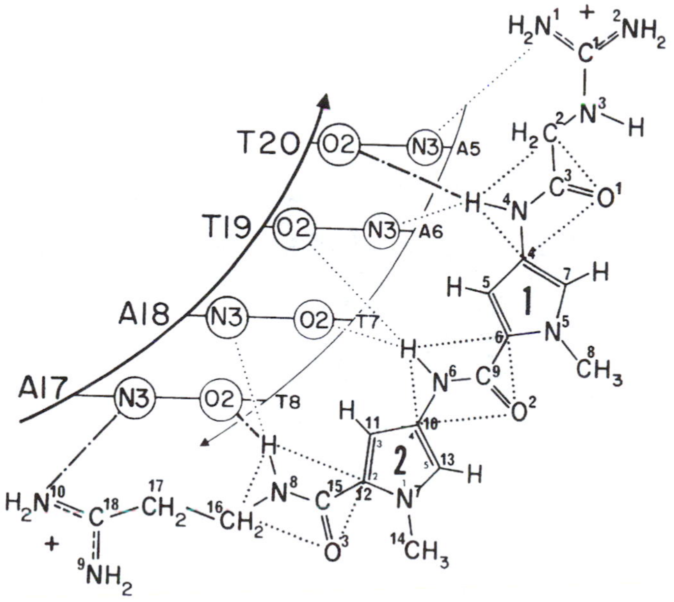

Figure 12-36
Diagrammatic representation of netropsin binding to DNA, as deduced from the X-ray structure (Kopka et al., 1985). The DNA is shown by a minor groove ladder, with adenine N3 and thymine O2 atoms indicated. Dot–dashed lines indicate N-to-N or N-to-O distances short enough to be standard hydrogen bonds, whereas dotted lines indicate distances of 3.2 Å or more. [From Kopka et al., (1985).]

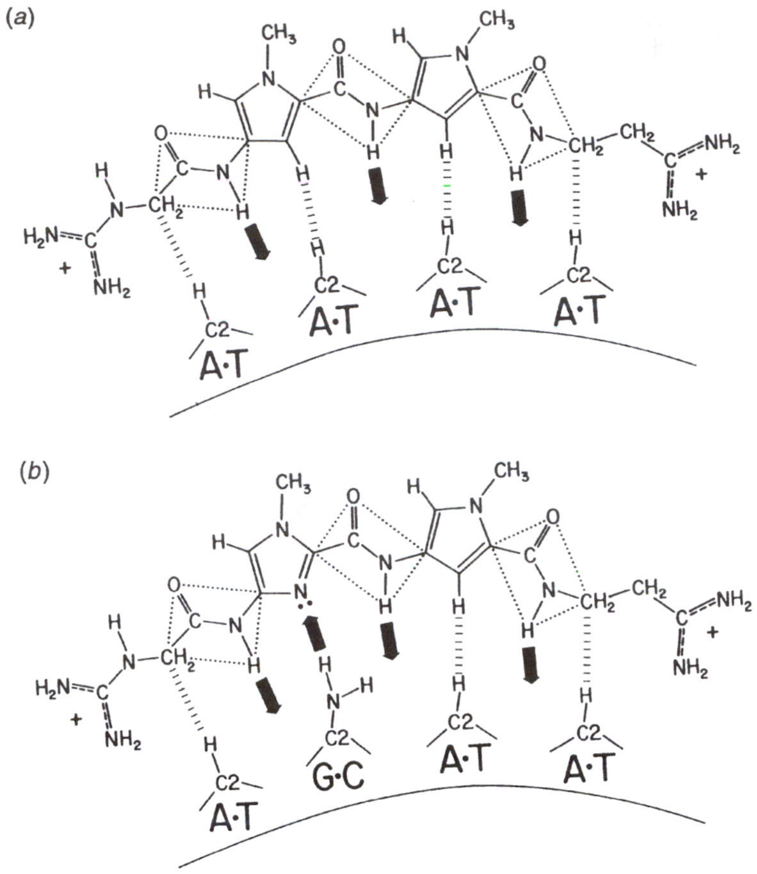

(a)

(b)

Figure 12-37
Schematization of bonding and base specificity in (*a*) netropsin and (*b*) in a proposed lexitropsin molecule designed so that it might be capable of recognizing and binding to G • C base pairs. Heavy black arrows indicate hydrogen bonds, in a donor H to acceptor direction. Bands of short parallel lines mark nonbonded van der Waals contacts. The floor of the DNA minor groove is at the bottom. Note that replacement of one methylpyrrole by methylimidazole provides both room for the NH_2 group of guanine, and the opportunity for still another stabilizing hydrogen bond. [From Kopka et al., (1985).]

the base pair planes and form bifurcated hydrogen bonds to successive thymine carbonyl groups. The successive methylpyrrole ring hydrogens make hydrophobic contact with the adenine H2 hydrogen, as shown schematically in Figure 12-37(*a*). The Dervan group has synthesized compounds which extend this ladder of contacts, and are capable of recognizing larger A • T tracts (Dervan, 1986).

Replacement of a methylpyrrole by another residue such as methylindole [Fig. 12-37(*b*)]) provides a potential hydrogen-bonding contact with a guanine amino group, and therefore could alter the sequence specificity for binding. Such molecules, dubbed *lexitropsins* by Dickerson and by Lown, have been extensively studied by the Lown group. For example, synthetic lexitropsins based on substitution of one or more imidazoles in the distamycin framework have been synthesized, and characterized for binding and inhibition of prokaryotic topoisomerase activity (Burckhardt et al., 1993). Enhanced preference for G•C-containing sites was indeed observed. Substitution of a thiazole for methylpyrrole shifts the binding site from AATT to CAAT, as characterized by NMR spectroscopy (Kumar et al., 1991). The use and potential of these compounds in antiviral drug development has been reviewed by Lown (1992). Given the variety of natural products that bind to DNA and have biological activity as a consequence, there is reason to be optimistic about the prospects for future development of potent synthetic compounds based on properties designed for binding to specific DNA sequences.

References

Aivasashvilli, V. A. and Beabealashvilli, R. Sh. (1983). Sequence-specific inhibition of RNA elongation by actinomycin D. *FEBS Lett.*, 124–128.

Aiyar, J., Danishefsky, S. J. and Crothers, D. M. (1992). Interaction of the aryl tetrasaccharide domain of calicheamicin γ_1^1 with DNA: influence on aglycon and methidiumpropyl-EDTA•Iron(II)-mediated DNA cleavage. *J. Am. Chem Soc.* **114**, 7552–7554.

Bauer, W. and Vinograd, J. (1970). Interaction of closed circular DNA with intercalative dyes. II. The free energy of superhelix formation in SV40 DNA. *J. Mol. Biol.* **47**, 419–435.

Bellon, S. F., Coleman, J. H. and Lippard, S. J. (1991). DNA unwinding produced by site-specific intrastrand cross-links of the antitumor drug *cis*- diamminedichloroplatinum(II). *Biochemistry* **30**, 8026–8035.

Bellon, S. and Lippard, S. (1990). Bending studies of DNA site-specifically modified by cisplatin, trans-diamminedichloroplatinum(II) and *cis*-[Pt(NH_3)$_2$(N3-cytosine)Cl$^+$. *Biophys. Chem.* **35**, 179–188.

Berman, H. M. and Young, P. R. (1981). The interaction of intercalating drugs with nucleic acids. *Annu. Rev. Biophys. Bioeng.* **10**, 87–114.

Blake, A. and Peacock, A. R. (1968). The interaction of aminoacridines with nucleic acids, *Biopolymers* **6**, 1225–1253.

Brabek, V. and Leng, M. (1993). DNA interstrand cross-links of trans diamminedichloroplatinum(II) are preferentially formed between guanine and complementary cytosine residues. *Proc. Natl. Acad. Sci. USA* **90**, 5345–5349.

Breslauer, K. J., Remeta, D.P., Chou, W.-Y., Ferrante, R., Curry, J., Zaunczkowski, D., Snyder, J. G., and Marky, L. A. (1987). Enthalpy–entropy compensations in drug–DNA binding studies, *Proc. Natl. Acad. Sci. USA* **84**, 8922–8926.

Bresloff, J. and Crothers, D. M. (1975). Ethidium reaction kinetics: Demonstration of direct ligand transfer between DNA binding sites, *J. Mol. Biol.* **95**, 103–123.

Bresloff, J. L. and Crothers, D. M. (1981). Equilibrium studies of Ethidium-polynucleotide interactions, *Biochemistry* **20**, 3547–3553.

Brown, D. G., Sanderson, M. R., Skelly, J. V., Jenkins, T. C., Abrown, T., Garman, E., Stuart, D. I. and Neidle, S. (1990). Crystal structure of a berenil-dodecanucleotide complex: the role of water in sequence-specific ligand binding. *EMBO J.* **9**, 1329–1334.

Bujalowski, W. and Lohman, T. M. (1987). A general method of analysis of ligand–macromolecule equilibria using a spectroscopic signal from the ligand to monitor binding. Application to *Escherichia coli* single-strand binding protein–nucleic acid interactions, *Biochemistry* **26**, 3099–3106.

Burckhardt, G., Luck, G., Store, K., Zinner, C. and Lown, J. W. (1993). Binding to DNA of selected lexitropisins and effects on prokaryotic topoisomerase activity, *Biochem. Biophys. Acta* **25**, 266–272.

Carter, B. J., de Vroom, E., Long, E. C., van der Marel , G. A., van Boom, J. H., and Hecht, S. M. (1990b). Site specific cleavage of RNA by Fe(II)•bleomycin, *Proc. Natl. Acad. Sci. USA* **87**, 9373–9377.

Carter, B. J., Murty, V. S., Reddy, K. S., Wang, S. N. and Hecht, S. M. (1990a). A role for the metal binding domain in determining the DNA sequence selectivity of Fe-bleomycin. *J. Biol. Chem.* **265**, 4193–4196.

Cairns, J. (1962). The application of autoradiography to the study of viruses, *Cold Spring Harbor Symp. Quant. Biol.* **27**, 311–318.

Chaires, J. B. (1990) Biophysical chemistry of the daunomycin-DNA interaction, *Biophys. Chem.* **35**, 191–292.

Chaires, J. B., Dattagupta, N. and Crothers, D. M. (1985). Kinetics of the daunomycin-DNA interaction, *Biochemistry* **24**, 260–267.

Chaires, J. B., Fox, K. R., Herrera, J. E., Britt, M. and Waring, M. J. (1987). Site and sequence specificity of the daunomycin–DNA interaction, *Biochemistry* **26**, 8227–8236.

Cohen, G. and Eisenberg, H. (1969). Viscosity and sedimentation study of sonicated DNA–proflavine complexes, *Biopolymers* **8**, 45–55.

Cone, R., Hasan, S. K., Lown, J. W. and Morgan, A. R. (1976). The mechanism of degradation of DNA by streptonigrin. *Can. J. Biochem.* **254**, 219–23.

Cosman, M., de Los Santos, Carlos, Fiala, R., Hingerty, B. E., Singh, S. B., Ibanez, V., Margulis, L. A., Live, D., Geacintov, N. E., Broyde, S. and Patel, D. J. (1992). Solution conformation of the major adduct between the carcinogen (+)-*anti*-benzo[a]pyrene diol epoxide and DNA, *Proc. Natl. Acad. Sci. USA* **89**, 1914–1918.

Crothers, D. M. (1968). Calculation of binding isotherms for heterogeneous polymers, *Biopolymers* **6**, 575–584.

Crothers, D. M. (1971). Statistical thermodynamics of nucleic acid melting transitions with coupled binding equilibria, *Biopolymers* **10**, 2147–2160.

Crothers, D. M., Haran, T. E., and Nadeau, J. G. (1990). Intrinsically bent DNA, *J. Biol. Chem.* 265, 7093–7096.

Crothers, D. M. and Drak, J. (1992). Global features of DNA structure by comparative gel electrophoresis, *Methods Enzymol.* **212**, 46–71.

Dattagupta, N., Hogan, M., and Crothers, D. M. (1978). Does Irehdiamine kink DNA? *Proc. Natl. Acad. Sci. USA* **75** 4286–4290.

Dattagupta, N., Hogan, M., and Crothers, D. M. (1980). Interaction of netropsin and distamycin with DNA: electric dichroism study, *Biochemistry* **19,** 5998–6005.

Davanloo, P. and Crothers, D. M. (1976). Phase Partition studies of actinomycin–nucleotide complexes, *Biochemistry* **15**, 4433–4438.

Dervan, P. (1986). Design of sequence-specific DNA binding molecules, *Science* **232**, 464–471.

De Voss, J. J., Townsend, C. A., Ding, D. W., Morton, G. O., Ellestad, G. A., Zein, N., Tabor, A. B., and Schreiber, S. L. (1990). Site-specific atom transfer from DNA to a bound ligand defines the geometry of a calicheamicin γ_1^1 complex, *J. Am. Chem. Soc.* **112**, 9669–9770.

Diekmann, S., Hillen, W., Jung, M., Wells, R. D., and Pörschke, D. (1982). Electric properties and structure of DNA restriction fragments from measurements of the electric dichroism, *Biophys. Chem.* **15**, 157–167.

Dougherty, G. and Pigram, W. J. (1982). Spectroscopic analysis of drug–nucleic acid interactions, *CRC Crit. Rev. Biochem.* **12**, 103–132.

Dougherty, G. and Pilbrow, M. R. (1984). Physico-chemical probes of intercalation, Int. *J. Biochem.* **16**, 1179–1192.

Drak, J., Iwasawa, N., Danishefsky, S., and Crothers, D. M. (1991). The carbohydrate domain of calicheamicin γ_1^1 determines its sequence specificity for DNA cleavage, *Proc. Natl. Acad. Sci. USA* **88**, 7464–7468.

Drinkwater, N. R., Miller, J.A., Miller, E. C., and Yang, N.-C. (1978). Covalent intercalative binding to DNA in relation to the mutagenicity of hydrocarbon epoxides and *N*-acetoxy-2-acetylaminofluorene, *Cancer Res.* **38**, 3247–3255.

Fox, K. R. and M. J. Waring (1986). Nucleotide Sequence Binding Preferences at Nogalamycin Investigated by DNase I Footprinting, *Biochemistry* **25**, 4349–4356.

Fox, K. R. and Waring, M. J. (1987). Footprinting at low temperatures: evidence that ethidium and other simple intercalators can discriminate between different nucleotide sequences. *Nucleic Acids Res.* **15**, 491–507.

Friedman, R. A. G. and Manning, G. S. (1984). Polyelectrolyte effects on site binding equilibria with application to the intercalation of drugs into DNA, *Biopolymers* **23**, 2671–2714.

Gao, X. L. and Patel, D. J. (1989). Antitumor drug–DNA Complexes: NMR studies of echinomycin and chromomycin complexes, *Q. Rev. Biophys.* **22**, 93–138.

Goodisman, J., Rehfus, R., Ward, B. and Dabrowiak, J. C. (1992). Site-specific binding constants for actinomycin D on DNA determined from footprinting studies, *Biochemistry* **31**, 1046–1058.

Hangeland, J. J., De Voss, J. J., Heath, J. A., Townsend, C. A., Ding, W., Ashcroft, J. S., and Ellestad, G. A. (1992). Specific abstraction of the 5′(S)- and 4′-deoxyribosyl hydrogen atoms from DNA by calicheamicin γ_1^1, *J. Am. Chem. Soc.* **114**, 9200–9202.

Hertzberg, R. P. and Dervan, P. B. (1982). Cleavage of double helical DNA by (methidiumpropyl-EDTA) iron(II), *J. Am. Chem. Soc.* **104**, 313–315.

Hogan, M., Dattagupta, N., and Crothers, D. M. (1978). Transient electric dichroism of rod-like DNA molecules, *Proc. Natl. Acad. Sci. USA* **75**, 195–199.

Hogan, M., Dattagupta, N., and Crothers, D. M. (1979). Transient electric dichroism of the structure of the DNA complex with intercalated drugs, *Biochemistry* **18**, 280–288.

Jain, S. C., Tsai, C.-C., and Sobell, H. M. (1977). Visualization of drug–nucleic interactions at atomic resolution II Structure of an ethidium/dinucleoside monophosphate crystalline complex, ethidium:5-iodocytidylyl (3′-5′) guanosine, *J. Mol. Biol.* **114**, 317–331.

Jones, R. R. and Bergman, R. G.(1972). *p*-benzyne. Generation as an intermediate in a thermal isomerization reaction and trapping evidence for the 1,4 benzenediyl structure, *J. Am. Chem. Soc.* **94**, 660–662.

Kastrup, R. V., Young, M. A., and Krugh, T. R. (1978). Ethidium complexes with self-complementary deoxytetranucleotides. Demonstration and discussion of sequence preferences in the intercalative binding of ethidium bromide, *Biochemistry* **17**, 4855–4865.

Kane, S. A., Natrajan, A., and Hecht, S. M. (1994). On the role of the bithiazole moiety in sequence-selective DNA cleavage by Fe • bleomycin, *J. Biol. Chem.* **269**, 10899–10804.

Kim, K.-H., Kwon, B.-M., Myers, A. G., and Rees, D. C. (1993). *Science* **262**, 1042–1046.

Kopka, M. L., Yoon, C., Goodsell, D. Pjura, P., and Dickerson, R. E. (1985). The molecular origin of DNA-drug specificity in netropsin and distamycin, *Proc. Natl. Acad. Sci. USA* **82**, 1376–1380.

Kosturko, L. D., Dattagupta, N., and Crothers, D. M. (1979). Selective repression of transcription by base sequence specific synthetic polymers, *Biochemistry* **18**, 5751–5756

Kozarich, J. W., Worth, L., Jr., Frank, B. L., Christner, D. F., Vanderwall, D. E., and Stubbe, J. (1989). Sequence-specific isotope effects on the cleavage of DNA by bleomycin, *Science* **245**, 1396–1399.

Krugh, T. R. and Chen, Y. C. (1975). Actinomycin D–deoxynucleotide complexes as models for the actinomycin D–DNA complex. The use of nuclear magnetic resonance to determine the stoichiometry and geometry of the complexes, *Biochemistry* **14**, 4912–4922.

Krugh, T. R. and Reinhardt, C. G. (1975). Evidence for sequence preferences in the intercalative binding of ethidium bromide to dinucleoside monophosphates, *J. Mol. Biol.* **97**, 133-162.

Kubista, M., Sjöback, and Albinsson, B. (1993). Determination of equilibrium constants by chemometric analysis of spectroscopic data. *Analyt. Chem.* **65**, 994–998.

Kumar, S., Bathini, Y., Joseph, T., Pon, T. R., and Lown, J. W. (1991). Structural and dynamic aspects of non-intercalative binding of a thiazole-lexitropin to the decadeoxy ribonucleotide d-$[CGCAATTGCG]_2$:An ^{1}H-NMR and molecular modeling study, *J. Biomol. Struct. Dyn.* **9**, 1–21.

Lane, M. J., Dabrowiak, J. C., and Vournakis, J. N. (1983). Sequence specificity of actinomycin D and Netropsin binding to pBR322 DNA analyzed by protection from DNase I *Proc. Natl. Acad. Sci. USA* **80**, 3260–3264.

Lane, A. N., Jenkins, T. C., Brown, T., and Neidle, S. (1991). Interaction of berenil with the *Eco*RI dodecamer $d(CGCGAATTCGCG)_2$ in solution studied by NMR, *Biochemistry* **30**, 1372–1385.

Lee, C. H. and Charney, E. (1982). Solution conformation of DNA, *J. Mol. Biol.* **161**, 289–303.

Lepre, C. A. and Lippard, S. J. (1990). Interaction of platinum antitumor compounds with DNA in *Nucleic Acids and Molecular Biology* **4**, Eckstein, F. and Lilley, D. M. J. Eds., Springer-Verlag, Berlin, Heidelberg, pp.9–38.

Lerman, L. S. (1961). Structural considerations in the interaction of DNA and acridines, *J. Mol. Biol.* **3**, 18–30.

Li, H. J. and Crothers, D. M. (1969). Relaxation studies of the proflavine–DNA complex: the kinetics of an intercalation reaction, *J. Mol. Biol.* **39**, 461–477.

Lin, C. H., Beale, J. M., and Hurley, L. H. (1991). Structure of the (+)-CC-1065-DNA adduct: critical role of ordered water molecules and implications for involvement of phosphate catalysis in the covalent reaction, *Biochemistry* **30**, 3597–3602.

Lohman, T. M. and Mascotti, D. P. (1992). Nonspecific ligand–DNA equilibrium binding parameters determined by fluorescence methods, *Methods Enzymol.* **212**, 424–458.

Low, C. M. L., Drew, H. R., and Waring, M. J. (1984). Sequence-specific binding of echinomycin to DNA: evidence for conformational changes affecting flanking sequences *Nucleic Acids Res.* **12**, 4865–4879.

Lown, J. W. (1992). Lexitropsins in antiviral drug development, *Antiviral Res.* **17**, 179–196.

Lyng, R., Rodger, A., and Nordén (1991). The CD of ligand–DNA systems. I. Poly(dG-dC) B-DNA, *Biopolymers* **31**, 1709–1720.

Lyng, R., Rodger, A., and Nordén, B. (1992). The CD of ligand-DNA systems. 2. Poly (dA-dT) B-DNA, *Biopolymers* **32**, 1201–1214

Macgregor, R. B., Clegg, R. M., and Jovin, T. M. (1985) Pressure-jump study of the kinetics of ethidium bromide binding to DNA, *Biochemistry* **24**, 5503–5510.

McGall, G. H. and Stubbe, J. (1988). Mechanistic studies of bleomycin-mediated DNA cleavage using isotope labeling, in *Nucleic Acids and Molecular Biology* 2, eds. Eckstein, F. and Lilley, D. M. J. Eds., Springer-Verlag, Berlin, Heidelberg, pp. 85–104.

McGhee, J. D. and von Hippel, P. H. (1976). Theoretical calculations of the helix-coil transition of DNA in the presence of large, cooperatively binding ligands, *Biopolymers* **15** 1345–1375.

McLean, M. J., Dar, A., and Waring, M. J. (1989). Differences between sites of binding to DNA and strand cleavage for complexes of Bleomycin with iron or cobalt. *J. Mol. Recog.* **1**, 184–192.

McLean, M. J. and Waring, M. J. (1988). Chemical probes reveal no evidence of Hoogsteen base pairing in complexes formed between Echinomycin and DNA in solution, *J. Mol. Recog*. **1**, 138–151.

Marky, L. A., Snyder, J. G., Remeta, D. P., and Breslauer, K. J. (1983a). Thermodynamcics of drug–DNA interactions, *J. Biomol. Struct. and Dyn*. **1**, 487–507.

Marky, L. A., Snyder, J. G., and Breslauer, K. J. (1983b). Calorimetric and spectroscopic investigation of drug–DNA interactions: II. Dipyrandium binding to poly d(A-T). *Nucleic Acids Res.* **11**, 5701–5715.

Mergny, J. L., Duval-Valentin, G., Nguyen, C. H., Perrouault, L., Faucon, B., Rougée, M., Montenay-Garestier, T., Bisagni, E. and Hélène, C. (1992). Triple helix-specific ligands. *Science* **256**, 1681–1684.

Meyer-Almes, F. J., and Pörshcke, D. (1993). Mechanism of intercalation into the DNA double helix by ethidium. *Biochemistry* **32**, 4246–4253

Mauss, Y., Chambron, Y. M., Daune, M., and Benoit, H. (1967). Morphological study by light-scattering of the complex formed by DNA and Proflavin. *J. Mol. biol.* **27**, 579–589.

Misra, R. P., Muench, K. F., and Humayun, M. Z. (1983). Covalent and noncovalent interactions of aflatoxin with defined DNA sequences, *Biochemistry* **22**, 3351–3359

Muench, K. F., Misra, R. P., and Humayun, M. Z. (1983). Sequence specificity in aflatoxin B_1 DNA interactions, *Proc. Natl. Acad. Sci. USA* **80**, 6–10.

Müller, W., Bünemann, H., and Dattagupta, N. (1975). Interactions of heteroaromatic compounds with nucleic acids. 2. Influence of substituents on the base and sequence specificity of intercalating ligands, *Eur. J. Biochem.* **54**, 279–291.

Müller, W. and Crothers, D. M. (1968). Studies of the binding of actinomycin and related compounds to DNA. *J. Mol. Biol*. **35**, 251–290.

Müller, W., Crothers, D. M., and Waring, M. (1973). A non-intercalating proflavine derivative, Eur. *J. Biochem*. **39**, 223–234.

Müller, W. and Crothers, D. M. (1975). Interaction of heteroaromatic compounds with nucleic acids 1. The influence of heteroatoms and polarizability on the base specificity of intercalating ligands, *Eur. J. Biochem*. **54**, 267–277.

Müller, W. and Gautier, F. (1975). Interactions of heteroaromatic compounds with nucleic acids. A•T-specific non-intercalating DNA ligands, *Eur. J. Biochem.* **54**, 385–94.

Neidle, S. and Abraham, Z. (1984). Structural and sequence dependent aspects of drug intercalation into nucleic acids, *CRC Crit. Rev. Biochem*. **17**, 73–121.

Nicoloau, K. C. and Dai, W.-M. (1991). Chemistry and biology of the enediyne anticancer antibiotics, *Angew. Chem. Int. Ed. Engl.* **30**, 1387–1416.

Nicoloau, K. C., Smith, A. L., and Yue, E. W. (1993). Chemistry and biology of natural and designed enediynes. *Proc. Natl. Acad. Sci. USA* **90**, 5881–5888.

Nightingale, K. P. and Fox, K. R. (1994). Light activated cleavage of DNA by cobalt-bleomycin, *Eur. J. Biochem.* **220**, 173–181.

Nordén, B., Kubista, M. and Kurucsev, T. (1992). Linear dichroism spectroscopy of nucleic acids, *Q. Rev. Biophys.* **25**, 51–170.

Paloma, L. G., Smith, J. A., Chazin, W. J., and Nicolaou, K. C. (1994). Interaction of calicheamicin with duplex DNA: role of the oligosaccharide domain and identification of multiple binding modes. *J. Am. Chem. Soc.* **116**, 3697–3708.

Peacocke, A. R. and Skerrett, J. N. H. (1956). The interaction of aminoacridines with nucleic acids, *Trans. Faraday Soc.* **52**, 261–279.

Phillips, D. R. and Crothers, D. M. (1986). Kinetics and sequence specificity of drug–DNA interactions: An in vitro transcription assay, *Biochemistry* **26**, 7355–7362.

Pope, L. M., Reich, K. A., Graham, D. R. and Sigman, D. S. (1982). Products of DNA cleavage by the 1,10-phenanthroline–copper complex, *J. Biol. Chem.* **257**, 12121–12128.

Reinert, K. E. (1981). Aspects of specific DNA–protein interaction; local bending of DNA molecules by in-register binding of the oligopeptide antibiotic distamycin, *Biophys. Chem.* **13**, 1–14.

Reinert, K. E. (1983). Anthracycline-binding induced DNA stiffening, bending and elongation; stereo-chemical inplications from viscometric investigations, *Nucl Acids Res*. **11**, 3411–3430.

Reynolds, V. L., Molineux, I. J., Kaplan, D. J., Swenson, D. H., and Hurley, L. H. (1985). Reaction of the antitumor antibiotic CC-1065 with DNA. Location of the site of thermally induced strand breakage and analysis of DNA sequence specificity, *Biochemistry* **24**, 6228–6237.

Roche, C, J., Thomson, J. A., and Crothers, D. M. (1994). Site selectivity of daunomycin, *Biochemistry* **33**, 926–935.

Rice, J. A., Crothers, D. M., Pinto, A. L., and Lippard, S. J. (1988). The major adduct of the antitumor drug *cis*-diamminedichloroplatinum(II) with DNA bends the duplex by ~40° toward the major groove, *Proc. Natl. Acad. Sci. USA* **85**, 4158–4161.

Rosenberg, B., VanCamp, L., and Krigas, T. (1965). Inhibition of cell division in *E. coli* by electrolysis products from a platinum electrode, *Nature*, **205**, 698–699.

Sambamurti, K., Callahan, J., Luo, X., Perkins, C. P., Jacobsen, J. S., and Humayun, M. Z. (1988). Mechanisms of mutagenesis by a bulky DNA lesion at the guanine N7 position, *Genetics* **120**, 863–873.

Scatchard, G. (1949). The attraction of proteins for small molecules and ions, *Ann. N. Y. Acad. Sci.* **51**, 660.

Scovell, W. M. and Collart, F. (1985). Unwinding of supercoiled DNA by cis- and trans-diamminedichloroplatinum(II): influence of the torsional strain on DNA unwinding1, *Nucl Acids Res.* **3**, 2881–2895.

Sherman, S. E., Gibson, D., Wang, A. H. J., and Lippard, S. J. (1985). Crystal and molecular structure of *cis*-[Pt(NH_3)$_2${d(pGpG)}], the principal adduct formed by *cis*-diamminedichloroplatinum(II) with DNA. *J. Am. Chem. Soc.* **110**, 1520–1524.

Shubsda, M., Kishikawa, H., Goodisman, J., and Dabrowiak, J. (1994). Quantitative footprinting analysis, *J. Mol. Recognit.* **7**, 133–139.

Sigman, D. S., Graham, D. R., Aurora, V., and Stern, A. M. (1979). Oxygen-dependent cleavage of DNA by the 1,10-phenanthroline • cuprous complex. Inhibition of *Escherichia coli* DNA polymerase I, *J.Biol. Chem.* **254**, 12269–12272.

Sigman, D. S. and Spassky, A. (1989). Dnase activity of 1,10-phenanthroline-copper ion, in *Nucleic Acids and Molecular Biology* 3, Eckstein, F. and Lilley, D. M.J., Eds., Springer-Verlag, Berlin, Heidelberg, pp.13–27.

Skorobogaty, A., White, R. J., Phillips, D. R., and Reiss, J. A. (1988). The 5'-CA DNA sequence preference of daunomycin, *FEBS Lett.* **227**, 103–106.

Sluka, J. P., Horvath, S. J., Bruist, M. F., Simon, M. I., and Dervan, P. B. (1987). Synthesis of a sequence-specific DNA-cleaving peptide, *Science* **238**, 1129–1132.

Sobell, H. M. (1974). How actinomycin binds to DNA, *Sci. Am.* **231**, 82–91.

Sturm, J., Schreiber, L., and Daune, M. (1981). Binding of ligands to a one-dimensional heterogeneous lattice. II. Intercalation of tilorone with DNA and polynucletides, *Biopolymers* **20**, 765–785.

Suck, D. and Oefner, C. (1986). Structure of DNase I at 2.0 A resolution suggests a mechanism for binding to and cutting DNA, *Nature (London)* **321**, 620–625.

Takahara, P. M., Rosenzweig, A. C., Frederick, C. A., and Lippard, S. J. (1995). Crystal structure of double stranded DNA containing the major adduct of the anticancer drug cisplatin, *Nature (London)* **377**, 649–652.

Takasugi, M., Guendouz, A., Chassignol, M., Decout, J. L., Lhomme, J., Thuong, N. T., and Hélène, C. (1991). Sequence-specific photo-induced cross-linking of the two strands of double-helical DNA by a psoralen covalently linked to a triple helix forming oligonucleotide, *Proc. Natl. Acad. Sci. USA* **88**, 5602–5606.

Taylor, J. S., Schultz, P. G., and Dervan, P. B. (1984). Sequence specific cleavage of DNA by distamycin–EDTA • Fe(II) and EDTA–distamycin • Fe(II), *Tetrahedron* **40**, 457–465.

Tomasz, M. (1994). The mitomycins: natural cross-linkers of DNA, in *Molecular Aspects of Anticancer Drug–DNA Interactions*, Vol 2, Neidle, S. and Waring, M., Eds., Macmillan, New York.

Van Dyke, M. M., and Dervan, P. B. (1983). Methidiumpropyl-EDTA-Fe(II) and DNAaseI footprinting reports different small molecule binding site sizes on DNA, *Nucleic Acids Res.* **11**, 5555–5567.

Van Dyke, M. M., Hertzberg, R. P., and Dervan, P. B. (1982). Map of distamycin, netropsin, and actinomycin binding sites on heterogeneous DNA: DNA cleavage-inhibition patterns with methidiumproply–EDTA • Fe(II), *Proc. Natl. Acad. Sci. USA* **79**, 5470–5474.

Van Dyke, M. M., and Dervan, P. B. (1984). Echinomycin Binding Sites on DNA, *Science* **225**, 1122–1127.

Wakelin, L. P. and Waring, M. J.)1980). Kinetics of drug–DNA interaction. Dependence of the binding mechanism on structure of the ligand, *J. Mol. Biol.* **144**, 183–214.

Walker, S., Gange, D., Gupta, V., and Kahne, D. (1994). Analysis of hydroxylamine glycosidic linkages: structural consequences of the NO bond in calicheamicin, *J. Am. Chem. Soc.* **116**, 3197–3206.

Wang, A. H. J. (1987). Interactions between antitumor drugs and DNA, *Nucleic Acids Mol. Biol.* **1**, 53-69.

Wang, A. H.-J., Ughetto, G., Quigley, G. J., and Rich, A. (1987). Interactions between an anthracycline antibiotic and DNA: molecular structure of daunomycin complexed to d(CpGpT • ApCpG) at 1.2-A resolution, *Biochemistry* **26**, 1152–1163.

Wang, J. C. (1974). The degree of unwinding of the DNA helix by ethidium. I. Titration of twisted PM2 DNA molecules in alkaline cesium chloride density gradients, *J. Mol. Biol.* **89**, 783–801.

Waring (1968). Drugs which affect the structure and function of DNA, *Nature (London)***219**, 1320–1325.

Waring, M. (1971). Binding of Drugs to supercoiled circular DNA: evidence for and against intercalation, *Progress Molec. Subcell. Biol.* **2**, 216–231.

Waring, M. J. (1974). Stabilization of two-stranded ribopolymer helices and destabilization of a three-stranded helix by ethidium bromide, *Biochem. J.* **143**, 483–486.

Waring, M. J. (1981). DNA modification and cancer, *Annu. Rev. Biochem.* **50**, 159–192.

Waring, M. J., Wakelin, L. P. G., and Lee, J. S. (1975). A solvent-partitian method for measuring the binding of drugs to DNA: Application to the quinoxaline antibiotics echinomycin and triostin A. *Biochem. Biophys. Acta* **407**, 200–212.

White, R. J. and Phillips, D. R. (1988). Transcriptional analysis of multisite drug–DNA dissociation kinetics: delayed termination of transcription by Actinomycin D, *Biochemistry* **27**, 9122–9132.

White, R. J. and Phillips, D. R. (1989). Bidirectional transcriptionl footprinting of DNA binding ligands, *Biochemistry* **28**, 6259–6269.

Wilson, W. D. and Jones, R. L. (1981). Intercalating drugs: DNA binding and molecular pharmacology, *Adv. Pharmacol. Chemother* **18**, 177–222.

Wilson, W. D. and Jones, R. L. (1982). Intercalation in biological systems. in *Intercalation Chemistry*, Whittingham, M. S. and Jacobson, A. J., Eds., Academic, New York, pp. 445–501.

Woodson, S. A. and Crothers, D. M. (1988). Binding of 9-aminoacridine to bulged-base DNA oligomers from a frame-shift hot spot *Biochemistry* **27**, 8904–8914.

Worth, L, Jr., Frank, B. L., Christner, D. F., Absalon, M. J., Stubbe, J., and Kozarich, J. W. (1993). Isotope effects on the cleavage of DNA by bleomycin: mechanism and modulation, *Biochemistry* **32**, 2601–2609.

Xu, R. X., Neettesheim, D., Otvos, J.D., and Petering, D. H. (1994). NMR determination of the structures of peroxycobalt(III) bleomycin and cobalt(III) bleomycin, products of the aerobic oxidation of cobalt(II) bleomycin by dioxygen, *Biochemistry* **33**, 907–916.

Zasedateler, A. A., Gurskii, G. V., and Vol'Renstein, M. V. (1971).Theory of one-dimensional adsorption. I. Adsorption of small molecules on a homopolymer. *Mol. Biol. (Moscow)*, **5**, 194–198.

Zimmer, C. and Wähnert, U. (1986). Nonintercalating DNA-binding ligands: Specificity of the interaction and their use as tools in the biophysical biochemical and biological investigations of the genetic material, *Prog. Biophys. Mol. Biol.* **47**, 31–112.

Zimmerman, H. W. (1986). Physicochemical and cytochemical investigations on the binding of ethidium and acridine dyes to DNA and to organelles in living cells, *Angew. Chem. Intl. Ed. Engl.* **25**, 115–196.

COLOR PLATES

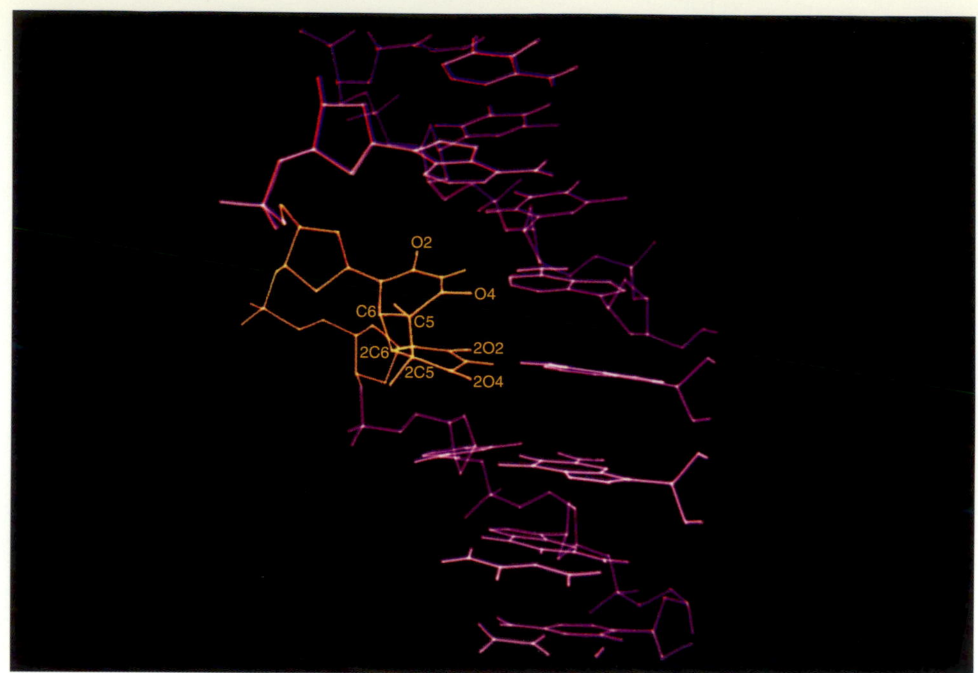

Figure 7-12
The central 7 bp in d(CGCGAAT[ ]TCGCG) • d(CGCGAATTCGCG), where T[ ]T refers to a thymine photodimer structure. [With permission from Rao et al., 1984.]

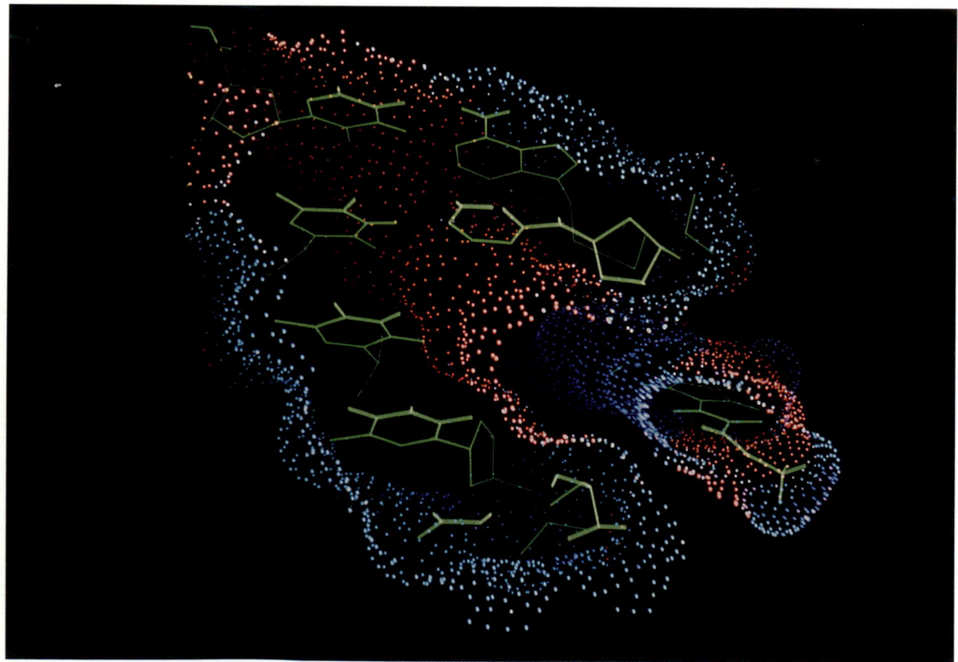

Figure 7-13
Electrostatic potential surfaces on netropsin and $dA_6 \cdot dT_6$. [With permission from Weiner et al., 1982.]

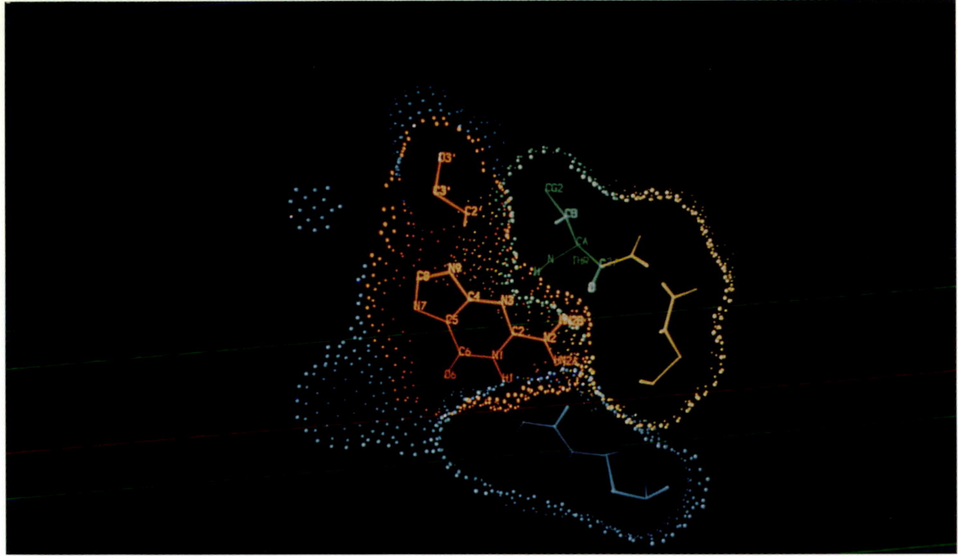

Figure 7-14
Illustration of hydrogen bond complementarity in a complex with actinomycin intercalated into a GC sequence. [See Lybrand et al., 1986]. The key G residue is in orange, the rest of the DNA is in blue and the actinomycin is in yellow and green. Note the key C=O hydrogen bond from actinomycin to the N2—H2B of the guanine.

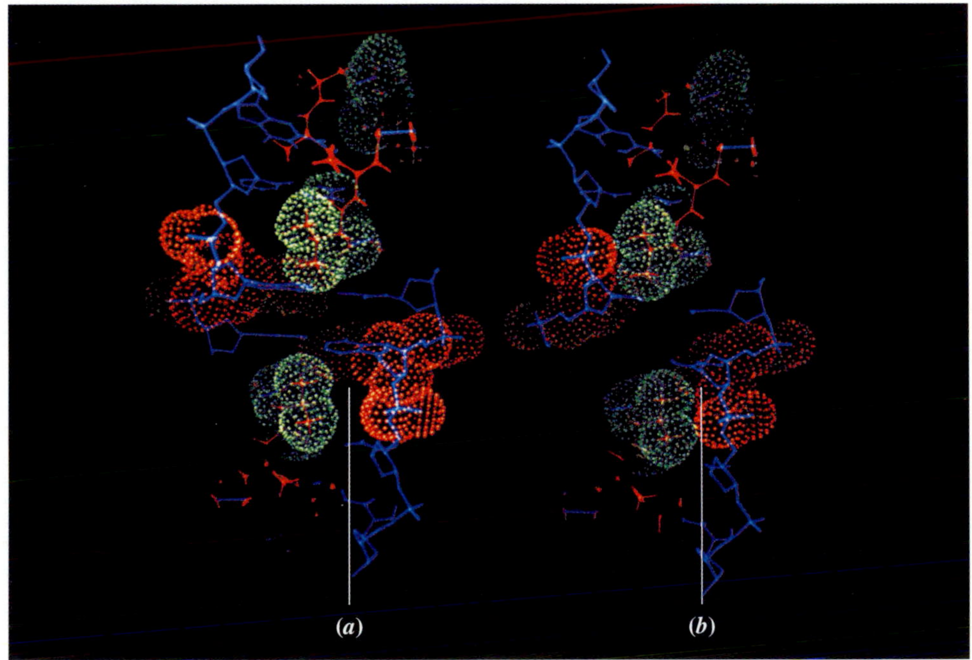

Figure 7-15
Hydrogen-bond complementarity in triostin–DNA interactions when the chromophore intercalates in a GC sequence. A comparison of the van der Waals complementarity of triostin intercalating into $d(CGATCG)_2$ when the central base pair is Watson–Crick (*a*) and Hoogsteen (*b*). [With permission from Singh et al., 1986.]

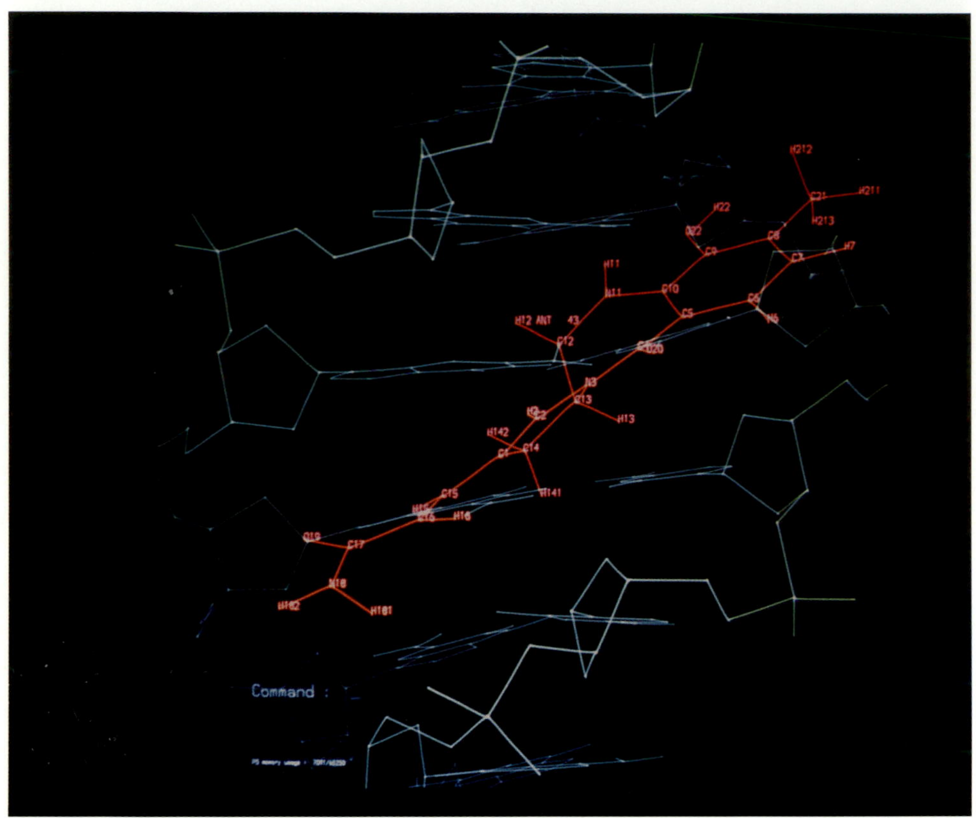

Figure 7-16
Structure of the lowest energy complex of anthramycin with $d(ATGCAT)_2$. [With permission from Rao et al., 1986a.]

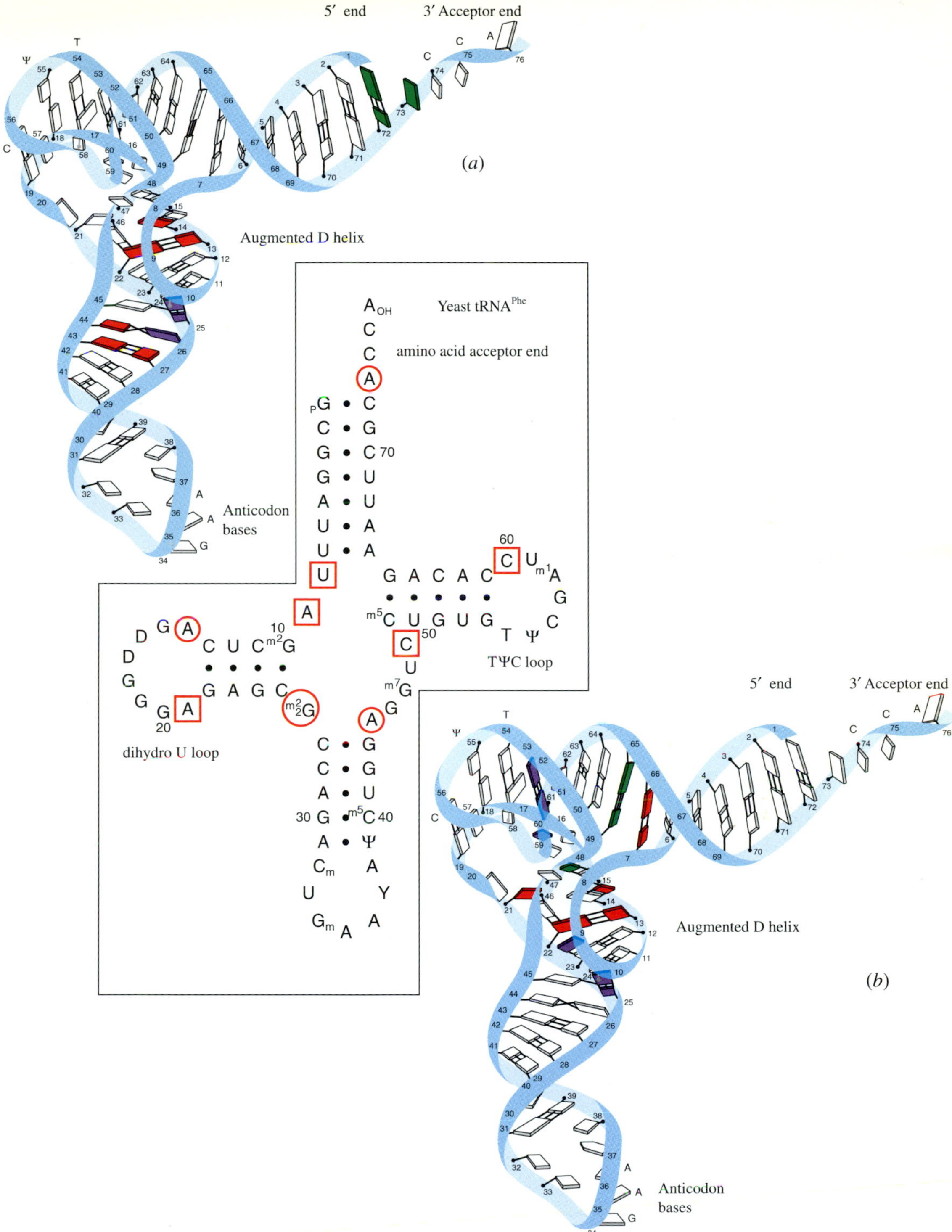

Figure 8-17
Three-dimensional structure of yeast phenylalanine tRNA showing correlations with free energy increments for stacking. (*a*) The weakly stacking single-strand sequence 16DD17 is unstacked. Strongly stacking unpaired bases and their adjacent base pairs are colored. The stacked, unpaired terminal nucleotides are A14, m_2^2 G26, A44, and A73. (*b*) The strongly stacking single-strand sequences 35AA36 and 74CC75 are stacked. Weakly stacking unpaired nucleotides adjacent to terminal base pairs are colored. The unpaired terminal nucleotides not stacked on their adjacent base pairs are U8, A9, A21, C48, and C60. [Adapted from Cantor and Schimmel (1980), redrawn from illustration Copyright © by Irving Geis.]

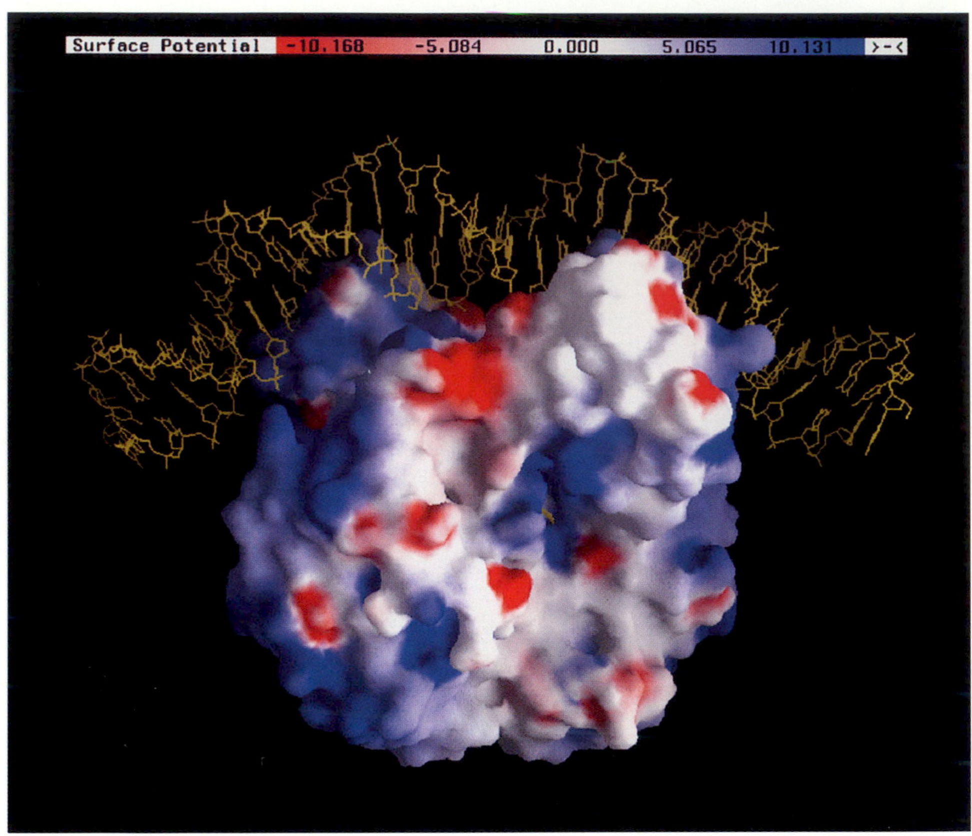

Figure 13-4
Representation of the electrostatic surface of CAP shown bound with DNA from the orthorhombic crystal form, made using the program GRASP. The DNA interacts with a strongly electopositive (blue) region of the protein (electronegative regions are red). On the lower left is an extended region of positive potential on the large domain of CAP that has been postulated to interact with a longer fragment of DNA. [From Passner, 1996. See also Warwicker et al., 1987. Figure courtesy of T. A. Steitz and J. Passner.]

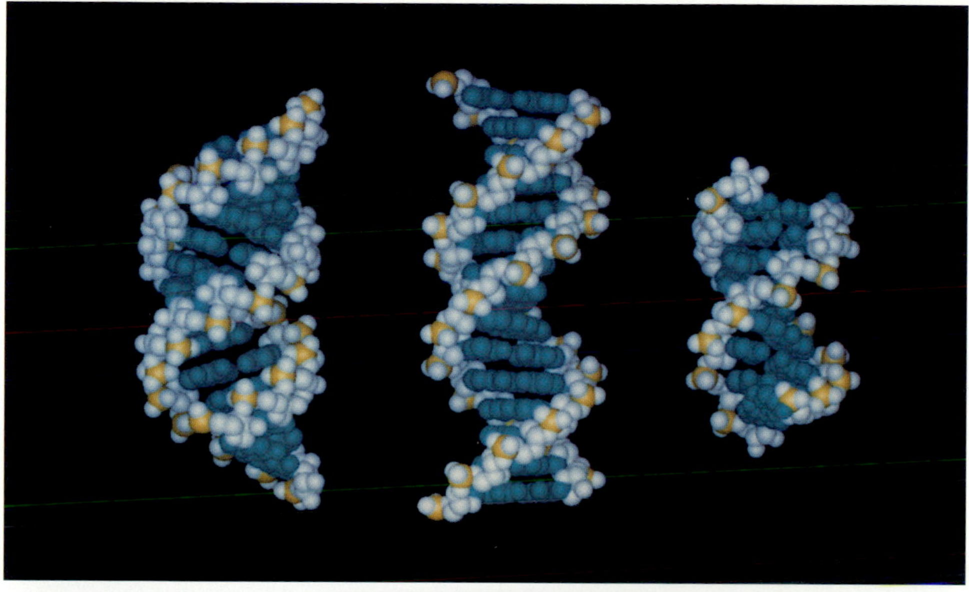

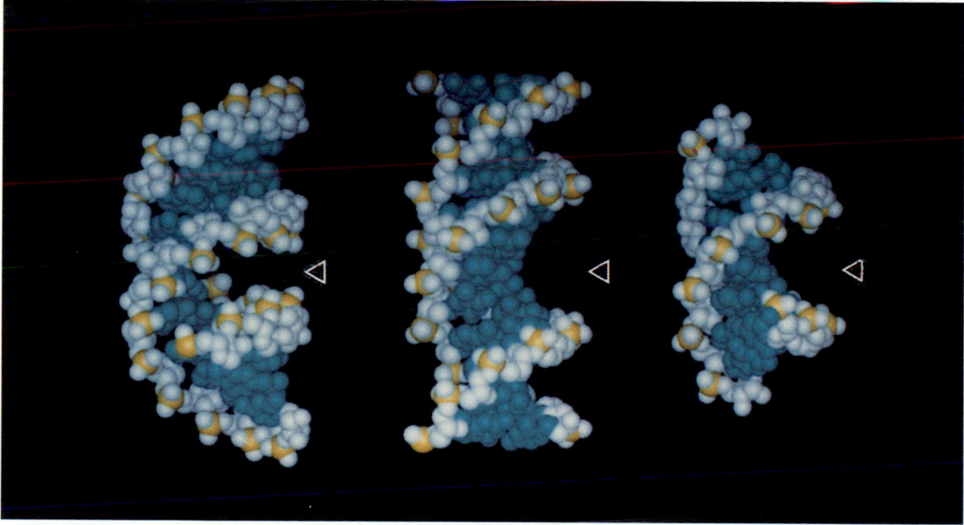

Figure 13-6
Structures of A- and B-form DNA in space-filling representation showing differences in major and minor groove widths and shapes. On the left is shown A-RNA, in the middle B-DNA, and on the right is an A-RNA helix having two extra bases (a "bulge" loop) on one strand, but stacked in a continuation of the right-handed helix. In the upper panel the helix axes are parallel to the page, whereas in the lower panel the helix axes have been tilted up by 32° to show the groove shapes. Note the deep and narrow major groove of RNA (left), and the wide major groove of DNA, (center), and the widened RNA major groove that results when extra bases are added to one strand (right). The bases are colored blue, the phosphoros atoms are yellow, and all other atoms are white. The edges of the bases are relatively accessible from the major groove of B-DNA and the minor or shallow groove of A-RNA. [From Weeks and Crothers, 1991, adapted from Steitz, 1990.]

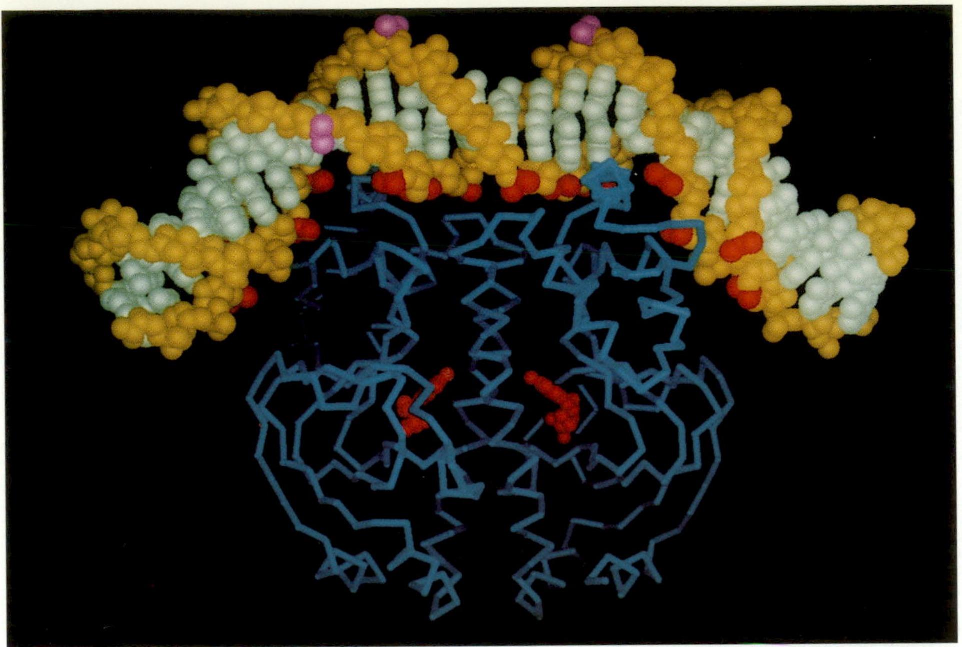

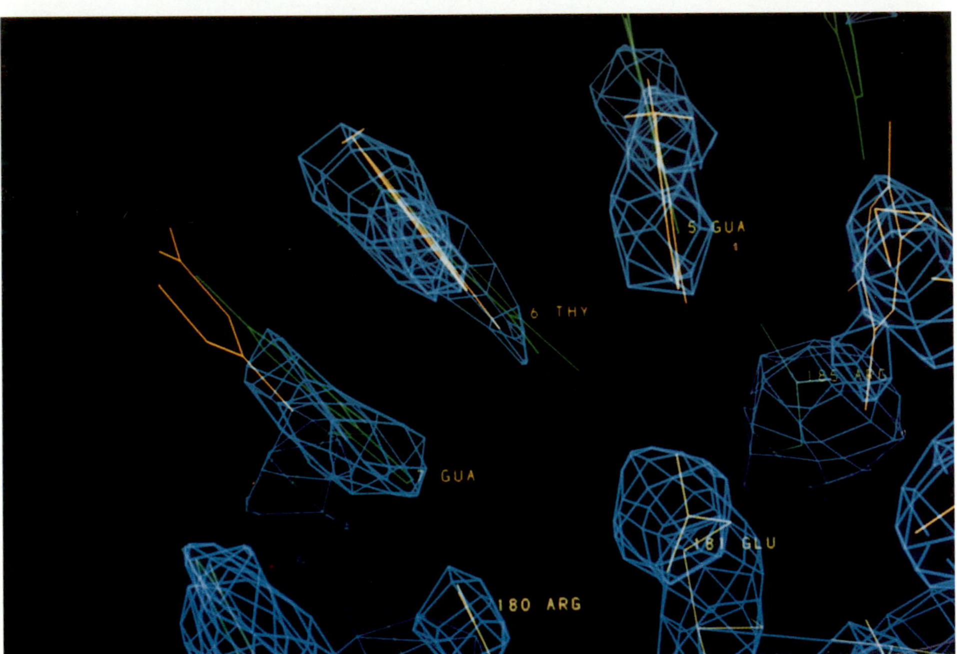

Figure 13-31
(*a*) Structure of the CAP–DNA complex. The protein is represented as an α-carbon backbone trace (blue) and the DNA (yellow and white) and cAMP (red) as space-filling models. The DNA phosphates whose ethylation interferes with DNA binding to CAP are shown in red, and phosphates that are hypersensitive to DNaseI are shown in pink. The second helix of the HTH motif is parallel to the base planes (perpendicular to the page) and enters approximately halfway into the major groove at the DNA kink. (*b*) Details of the structure at the kink between T6 and G5. [Reprinted with permission from Schultz, S. C., Shields, G. C., and Steitz, T. A. (1991). *Science* **253**, 1001–1007. Copyright ©1991 American Association for the Advancement of Science.]

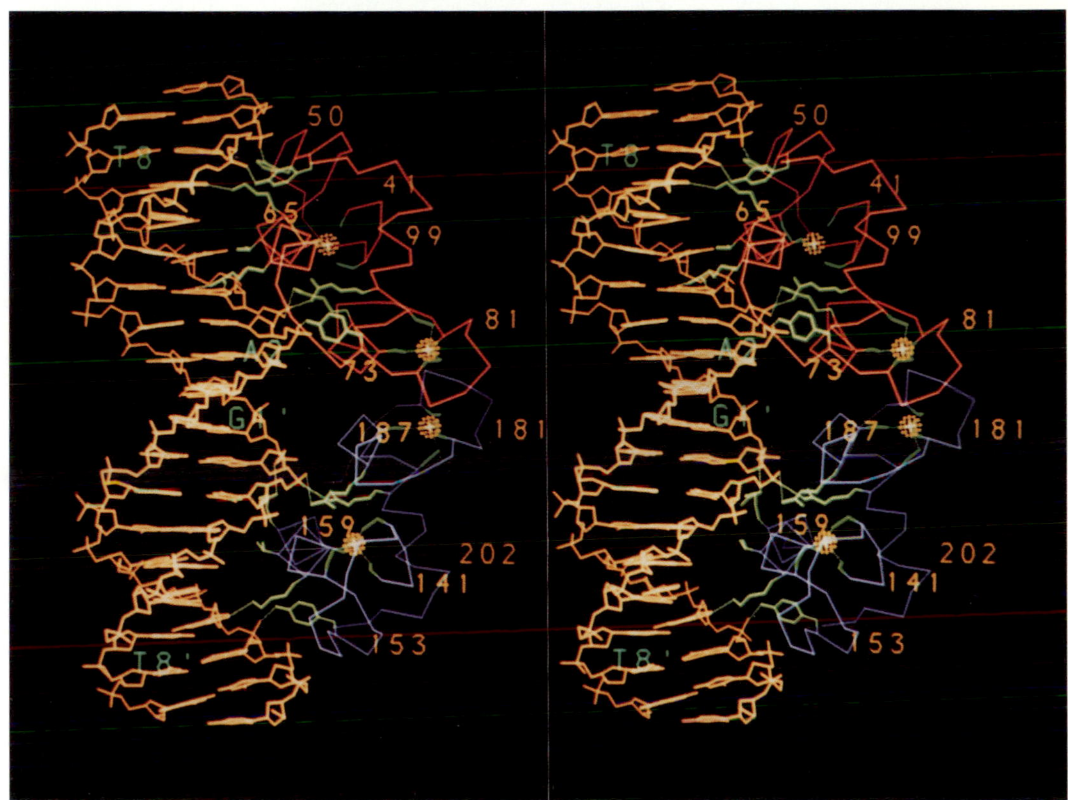

Figure 13-38
Stereoview of the complex of the rat glucocorticoid receptor DNA binding domain with a DNA sequence to which one domain binds in a specific mode, and the other in a non-specific manner. [Reprinted with permission from Luisi, B. F., Xu, a. X., Otwenowski, Z., Freedman, L. P. Yamamato, K. R., and Sigler, P. B. (1991). *Nature (London)* **352**, 497–505. Copyright ©1991 Macmillan Magazines Limited.] Only the α-carbon tracing of the protein is shown for clarity, except for the side chains interacting with the bases and phosphates, and the cysteine residues coordinating the Zn ions, which are shown in green. The subunit making specific contacts is in red, and the nonspecific subunit is in blue. Dashed lines indicate direct hydrogen bonds between the protein and DNA.

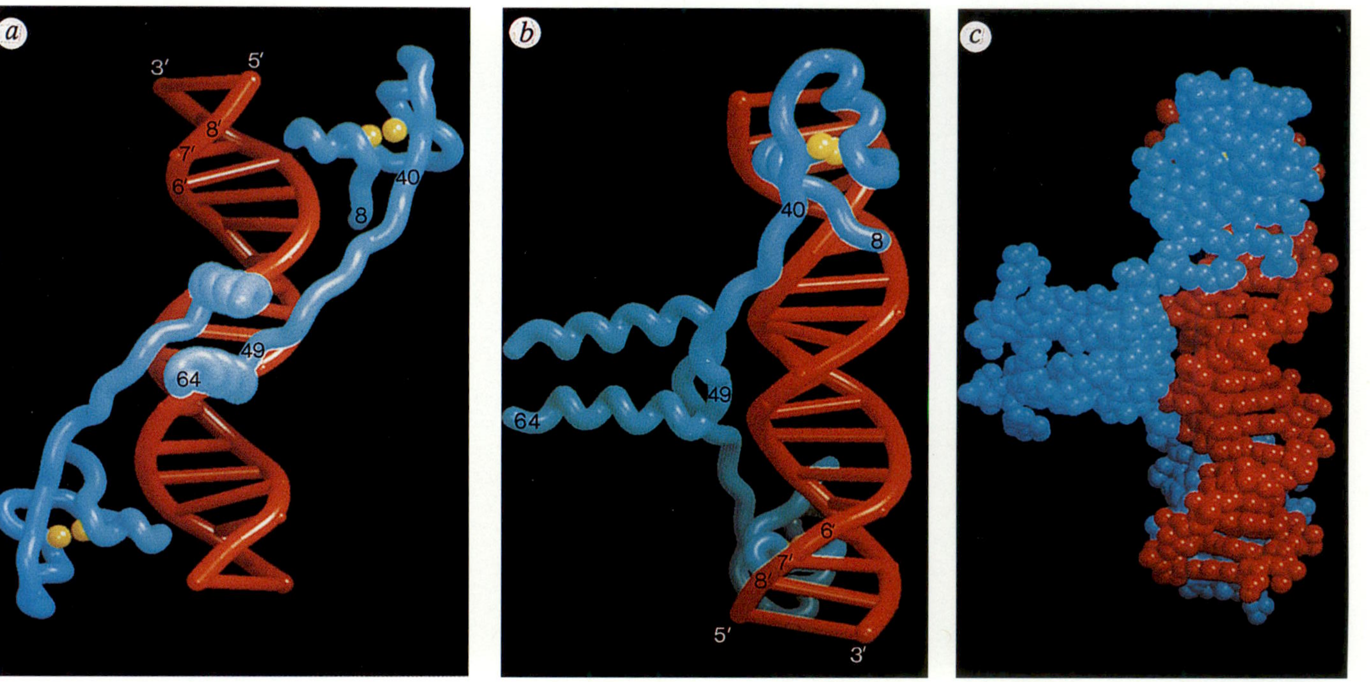

Figure 13-40
Structure of the GAL4–DNA complex. (*a*) View approximately along the twofold axis of the complex. Amino acid sequence numbers at the borders of the three protein modules (DNA recognition module, residues 8–40; linker region, residues 41–49; and dimerization element, residues 50–64) are shown on one subunit DNA (red) is shown as a helical ladder, using C3′ positions. The positions of bound metal ions are shown as yellow balls. (*b*) View approximately perpendicular to the twofold axis of the complex. (*c*) Space-filling model of the complex, from the same perspective as (*b*). [Reprinted with permission from Marmorstein, R., Casey, M., Ptashne, M., and Harrison, S. C. (1992). *Nature (London)* **356**, 408–414. Copyright ©1992 Macmillan Magazines Limited. See fig. 3.]

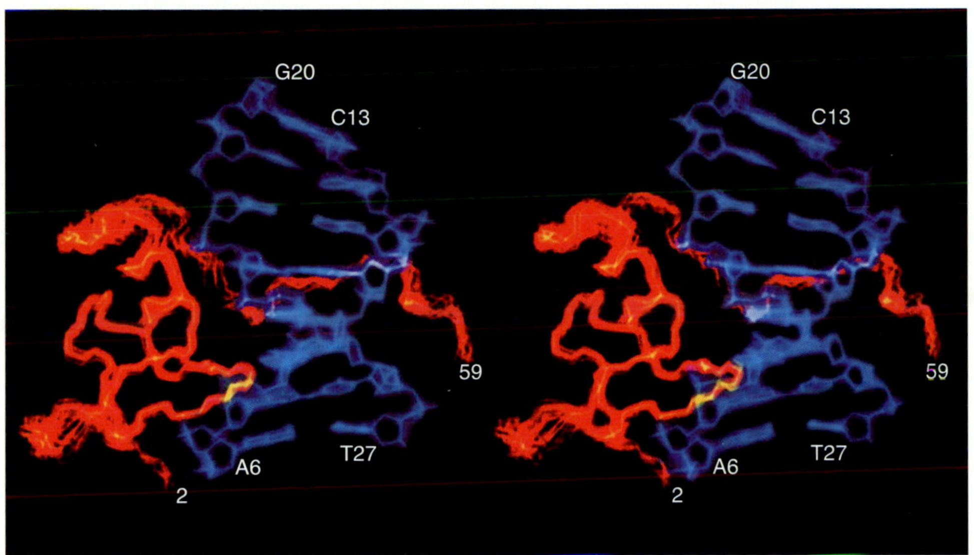

Figure 13-42
Stereoview of the complex of the GATA-1 DNA binding domain with DNA. The backbone (N, $C\alpha$, and C atoms) of the peptide is shown in red, and all non-hydrogen atoms of the DNA are shown in blue. [Reprinted with permission from Omichinski, J. G., Clore, G. M., Schaad, O., Felsenfeld, G., Trainor, C., Appella, E., Stahl, S. J., and Gronenbrn, A. M. (1993). *Science* **261**, 438–446. Copyright ©1993 American Association for the Advancement of Science. See Fig. 3.]

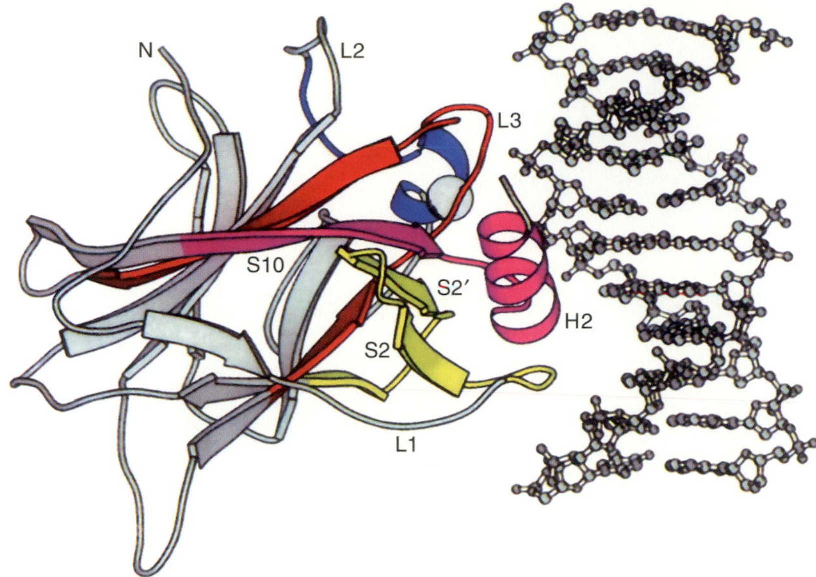

Figure 13-43
Schematic representation of the structure of the complex of the core domain of tumor suppressor protein p53 with DNA. The loop–sheet–helix motif consists of loop L1, β-sheet S2 and S2′, and helix H2. These fit into the DNA major groove, while loop L3, stabilized by the coordinated Zn ion, interacts with the adjacent minor groove. [Reprinted with permission from Cho, Y., Gorina, S., Jeffrey, P. D., and Pavletich, N. P. (1994). *Science* **265**, 346–355. Copyright ©American Association for the Advancement of Science. See Fig. 4B.]

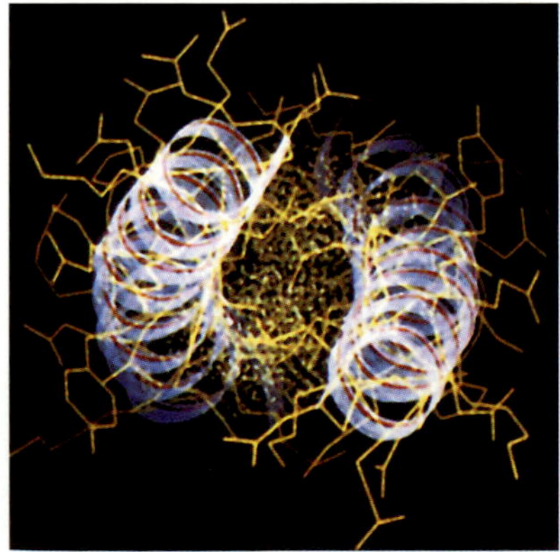

Figure 13-45
View of the parallel coiled-coil dimer along the superhelix axis from the NH_2 terminus. The main chain is highlighted with a ribbon, and the reduced van der Waals surfaces of the side chains at positions **a** and **d** are stippled in yellow. The helices are curved, and the overall superhelical twist is about 90°. [Reprinted with permission from O'Shea, E. K. Klemm, J. D., Kim, P. S., and Alber, T. (1991). *Science* **254**, 539–544. Copyright ©1991 American Association for the Advancement of Science. See Fig. 5.]

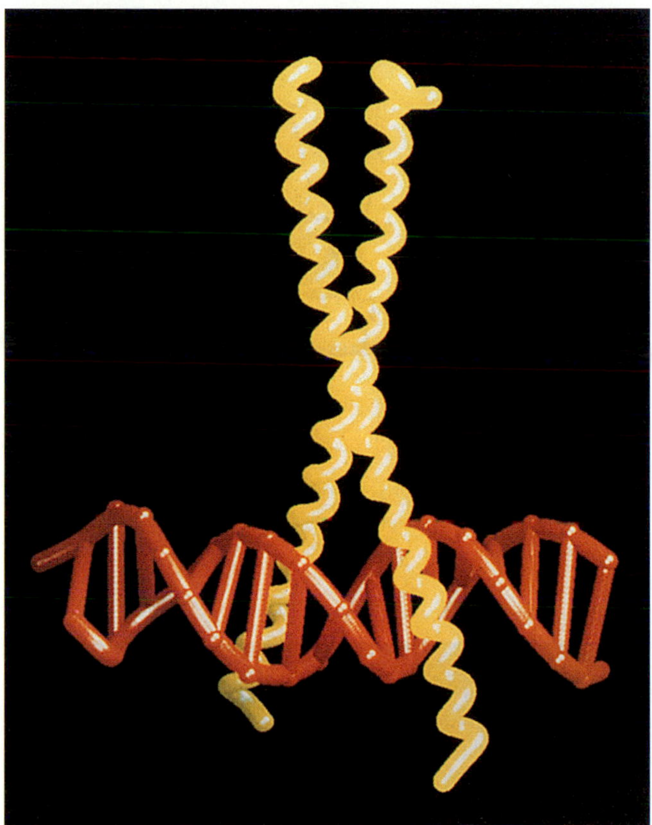

Figure 13-47
Structure of the GCN4 bZIP–DNA complex. The bZIP dimer (yellow) binds in the major groove of the DNA (red). Each bZIP protomer is a smoothly curved, continuous α helix. The carboxy terminal residues of the monomers pack together as a coiled coil, which gradually diverges to follow the major groove of either DNA half-site. The DNA in the complex is straight, and its conformation is in the B form across the region contacted by the protein. [From Ellenberger et al., 1992, Fig. 3]

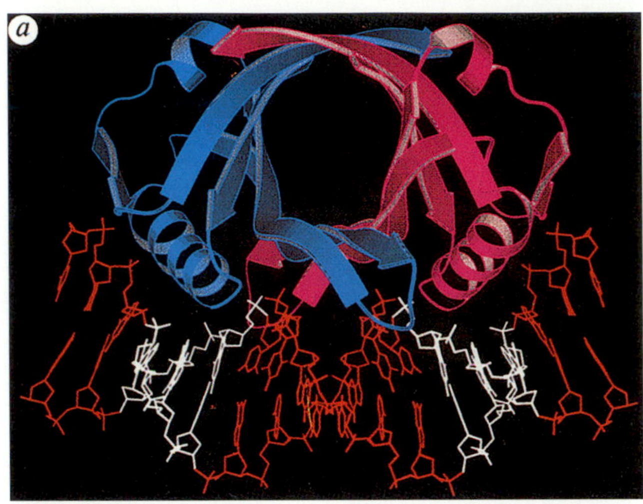

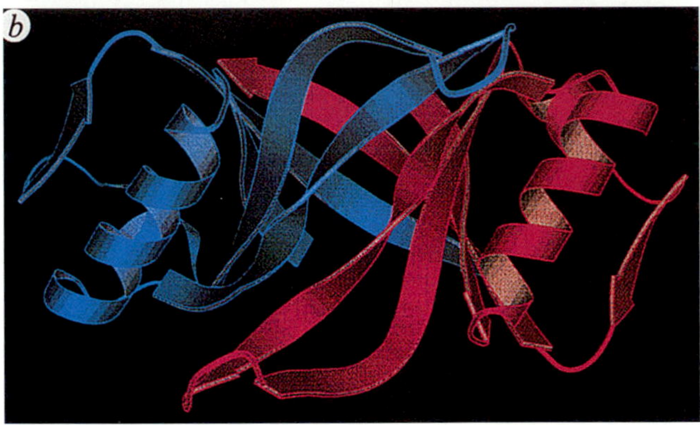

Figure 13-49
Schematic drawing of the E2–DNA complex. (*a*) Ribbon drawing of E2 residues 326–410 bound to its DNA target and viewed down the barrel axis of the protein. The two subunits of the protein are in blue and magenta, nucleotides in the conserved identity element are in white. (*b*) Ribbon view of the protein fragment alone, viewed roughly along the molecular dyad axis. Recognition α helices face the viewer. [Reprinted with permission from Hegde, R. S., Grossman, S. R., Laimins, L. A., and Sigler, P. B. (1992). *Nature* **359**, 505–512. Copyright ©1992 Macmillan Magazine Limited. See Fig. 2.]

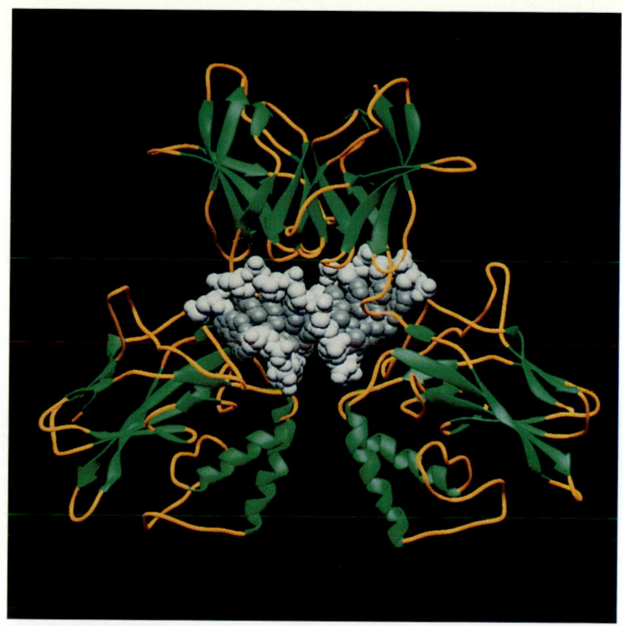

Figure 13-50
Schematic view of the X-ray structure of the complex of homodimeric NF-κB with DNA. This motif features interaction of protein loops with the DNA major groove. [Reprinted with permission from Ghosh, G., Van Duyne, G., Ghosh, S., and Sigler, P. B. (1995). *Nature* **373**, 303–310. Copyright ©1995 Macmillan Magazine Limited. See Fig. 2b.]

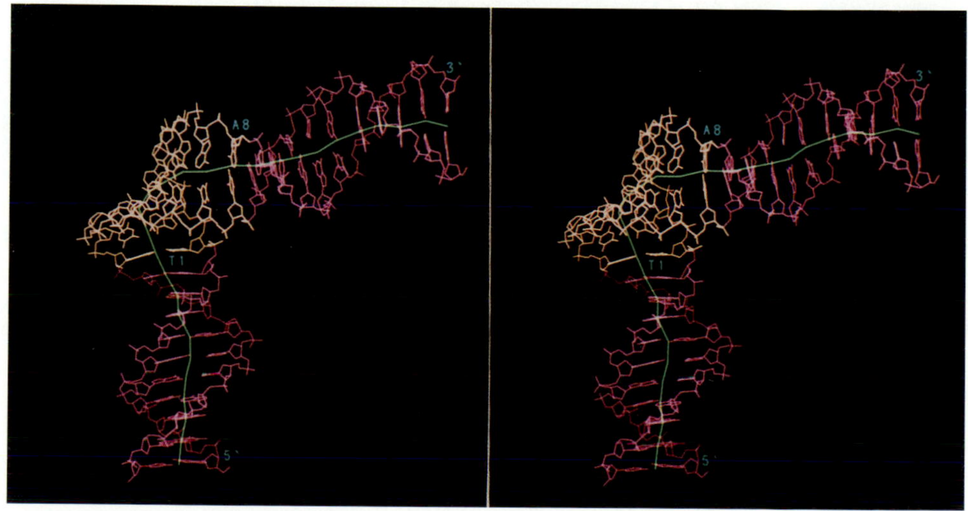

Figure 13-56
Stereoview of the structure of DNA as it is bent in the structure of the complex with TBP. The widened minor groove is visible in the upper left corner of the figure. [Reprinted with permission from Kim, Y. Geiger, J. H., Hahn, S., and Sigler, P. B. (1993a). *Nature* **365**, 512–520. Copyright ©1993 Macmillan Magazine Limited. See Fig. 2d.]

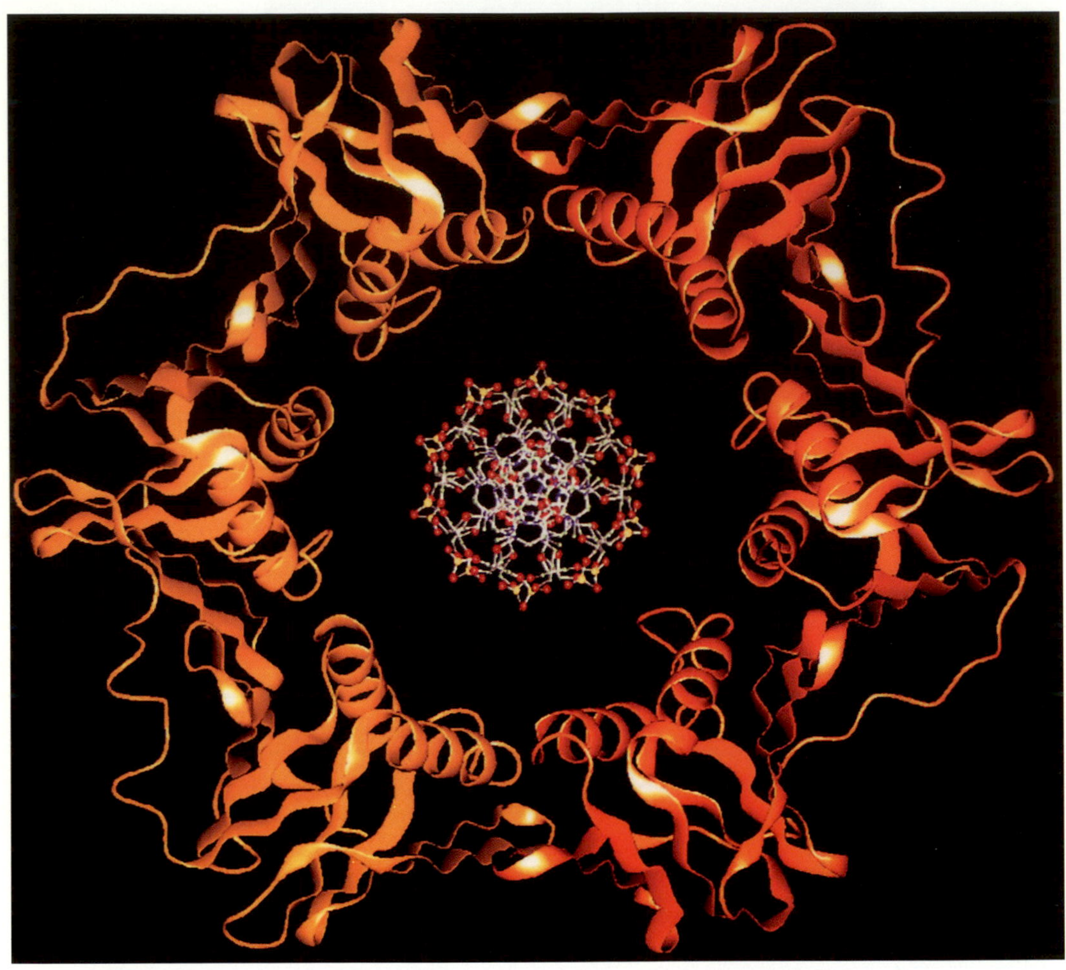

Figure 13-63
Ribbon representation of the polypeptide chain of a DNA polymerase III subunit dimer, looking down the axis of the ring. The α helices are shown as spirals and the β sheets as flat ribbons. The two monomers are colored yellow and red. A standard model for B-form DNA is in the middle of the structure. The DNA structure is hypothetical, and is placed in the geometric center of the β-subunit ring with the helix axis aligned along the twofold rotation axis of the ring. (Fig. 1 of Kong et al., 1992).

1. BIOLOGICAL ROLE OF PROTEIN-NUCLEIC ACID INTERACTIONS

Protein–nucleic acid interactions are ubiquitous in nature. Nearly all of the functions of nucleic acids are carried out in conjunction with participating proteins. This wide variety of interactions can generally be divided into three main categories: regulatory, enzymatic, and structural. Regulatory interactions often require recognition of a specific nucleic acid sequence or structure, thereby setting in motion a cascade of events that leads to gene expression or nucleic acid replication. Examples include gene regulatory proteins interacting with DNA to affect RNA production, translational control proteins interacting with mRNA to affect protein synthesis, or DNA binding proteins interacting with a replication origin to initiate the process of DNA synthesis. Regulatory effects can be either positive or negative, depending on whether the subsequent events are activated or repressed. These phenomena provide much of the subject matter of molecular genetics, and study of the underlying interaction mechanisms is required for full understanding of that subject.

Current research has revealed both complexity and underlying patterns in the interaction of gene regulatory proteins with DNA. These characteristics are evident in the mechanisms used for control of gene expression in eukaryotes. Figure 13-1 (Mitchell and Tjian, 1989) illustrates the general categories of DNA binding proteins that are found in such systems. Several interactions of the type shown are often used for the control of an individual gene, covering hundreds or even thousands of base pairs

CHAPTER 13

Protein–Nucleic Acid Interactions

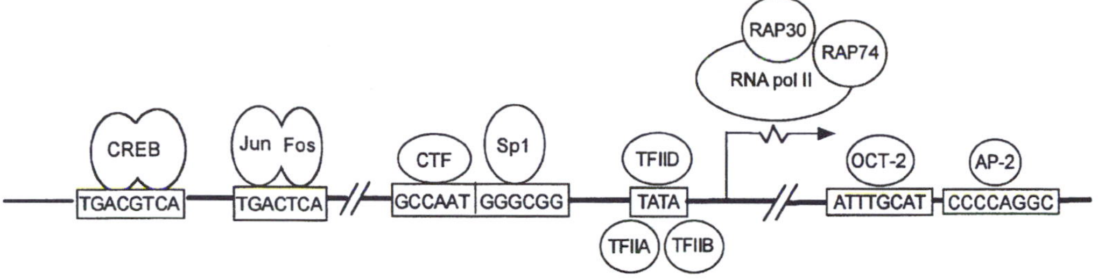

Figure 13-1
Features of the transcriptional control region for a mammalian protein-coding gene, showing a hypothetical array of cis elements that constitute the promoter and enhancer regions of a gene transcribed by pol II. Proteins that associate at these control regions are symbolically represented and include pol II, TFIIA, TFIIB, TFIID, RAP30, and RAP74 of the general transcriptional machinery, and various DNA binding proteins (CREB, Jun, Fos, CTF, Spl, OCT-2, and AP-2) that active through specific sequence elements. The transcription initiation site is indicated by an arrow, The figure is not meant to imply that all the DNA binding factors must be bound simultaneously as depicted here in order to initiate transcription. (from Mitchell and Tjian, 1989)

in the chromosome. Prokaryotes are less profligate in the use of DNA sequences for control of gene expression, with 40–100 bp segments commonly used for this purpose.

Genetic regulation in higher animals is now under intensive study. Control of transcription initiation generally involves a combination of ubiquitous control elements, along with tissue-specific factors that regulate production of the proteins that are produced in a specific cell type. Figure 13-2 illustrates the combination of ubiquitous and specific factors used to control β-globin synthesis in chicken erythroid cells (β-globin is a subunit of hemoglobin). The promoter, which is the DNA segment flanking the 5′-end of the coding sequence, covers about 200 bp in this case (Jackson et al., 1989); six binding sites for regulatory proteins have been identified in this region, some are specific for erythroid cells and some are not. The enhancer, a gene regulatory element defined by its ability to act at a distance and from either end of the gene sequence, begins about 300 bp to the 3′ side of the end of the coding sequence, and covers about 100 bp (Evans et al., 1988; Felsenfeld et al., 1989). The protein factors that interact with the β-globin enhancer are largely erythroid specific.

In considering the gene expression system that is presented as a linear array in Figure 13-2, it should be recalled that much of the DNA is folded by binding to histone and other chromosomal proteins, and that the entire domain probably has a specific higher order structure, which undoubtedly affects its function. A challenging problem for the future is understanding the relationship between chromatin structure and genetic function.

Enzymatic interactions with DNA include the polymerases required for transcription and replication. These interactions also have a regulatory component, since their initiation occurs at specific sites. Prokaryotic RNA polymerases are able to initiate transcription at many defined promoter sites without auxillary factors, although other

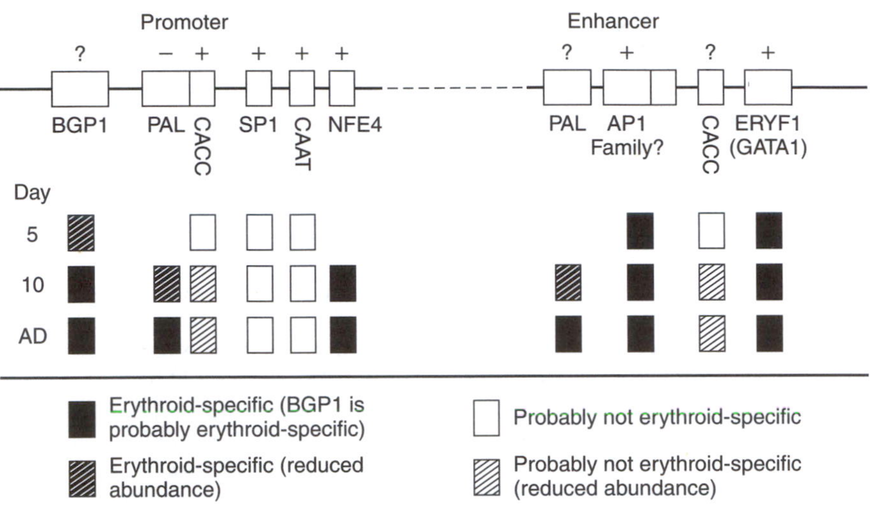

Figure 13-2
Combination of tissue-specific and ubiquitous regulatory proteins and their DNA binding sites in the adult chicken β-globin gene. The promoter is at the left-hand side of the diagram, and the enhancer element on the right. The + and − signs above the designated DNA binding sites indicate whether the element activates or inhibits globin production. The boxes below the DNA diagram indicate the level of production and erythroid specificity of the corresponding protein factor during development of the chicken from embryo to adult (AD). [Figure courtesy of Gary Felsenfeld.]

promoters may be very weak without an activator protein. Bacterial polymerases use the σ subunit to regulate initiation, then discard it when the enzyme enters the elongation mode, which, except for the requirement for specific termination, is less dependent on DNA sequence. Other proteins that act enzymatically on DNA include the topoisomerases and recombinases; these often have a regulatory component, since specific sequences may be targeted. A number of enzymes degrade DNA and RNA, such as restriction endonucleases, and a wide variety of endonucleases and exonucleases (see Chapter 3). Both RNA processing and mRNA-directed protein synthesis are essential enzymatic processes in which RNA can share the catalytic activity. Understanding all of these processes in detail will require many years of future research.

Protein–nucleic acid interactions can also have a primarily structural role. DNA is packaged into chromosomes by histones and other chromosomal proteins, reducing its physical extension by many thousands of times so that it can be stored compactly in the nucleus. Protein–RNA interactions are vital for the architecture of ribosomes and the ribonucleoprotein particles involved in RNA processing.

The field of protein–nucleic acid interactions is a main focus of current research in molecular genetics. Given the size and pace of the field, it is not possible to consider all aspects systematically here. Rather, we set forth the underlying principles that are best established at this time.

2. ELEMENTARY ENERGETICS AND SPECIFICITY OF INTERACTION

Nucleic acids are hydrated polyelectrolytes that have both charged and hydrogen bond donating and accepting groups exposed on their surfaces in the regular double helical form. If the helical structure is disrupted, the opportunities for specific recognition by hydrogen bonding are further increased. When a protein binds to DNA or RNA, ions and water are displaced by the formation of electrostatic and hydrogen-bonding linkages between the two partners. The net free energy of interaction is a difference that reflects the loss of one set of interactions and formation of another. Bulk entropy effects are also important, since it is more favorable to displace ions or water when the activities of these components of the solvent are lowered. It may in addition be necessary to take account of changes in structure that accompany binding, such as bending or melting of the double helix. While beyond the capabilities of present theory even if the structure of the complex is known, full computation of the equilibrium constant of a protein–nucleic acid interaction is a challenging goal for future research.

The interactions that stabilize protein binding can be divided into the simple categories electrostatic, hydrogen bonding, and hydrophobic, classifications that should be regarded as useful more for qualitative understanding than for quantitative rigor.

2.1 Electrostatic Effects

Electrostatic attraction results from the favorable interaction between unlike charges. Both DNA and RNA are negatively charged polyelectrolytes, and are strongly attracted to molecules that have multiple positive charges. When these polycations bind, positive ions of smaller charge that had been bound to the nucleic acid are displaced, resulting in the exchange of one set of electrostatic interactions for another. Since the bound small ions are released to lower concentration in the solvent, an entropy increase results. Furthermore, in addition to interaction of charged groups on the protein with the DNA or RNA phosphates, the positive end of polar groups on the protein, including hydrogen-bond donors, can interact with the phosphates. These interactions can be included either among the electrostatic or the hydrogen-bonding categories. Because of the variable electrostatic potential around both DNA and protein, as well as the complications introduced by the exchange of solvent ion interactions for those with protein, calculation and understanding of electrostatic effects is a formidable problem. For a review of the application of classical electrostatic calculations to nucleic acids and proteins, see Honig and Nicholls (1995).

The electrostatic influence on protein–nucleic acid interactions due to positive ion displacement is detected phenomenologically by a dependence of the binding constant on the concentration of salt added to the solvent. As discussed in detail in Chapter 11, binding of positively charged ligands to DNA or RNA results in release of the bound counterions to bulk solution:

$$\text{Protein} + \text{DNA}(r_D M^+) \rightleftharpoons \text{Protein-DNA}(r_C M^+) + \Delta r M^+ \qquad (13\text{-}1)$$

where $\Delta r = r_D - r_C$ is the difference in ion binding between naked DNA and the complex. Since the counterion enters the reaction with stoichiometric coefficient Δr, ion concentration affects the overall total equilibrium constant K_T according to

$$K_T = \frac{[\text{Complex}]}{[\text{Protein}][\text{DNA}]}[\text{M}^+]^{\Delta r} = K_{\text{obs}}[M^+]^{\Delta r} \tag{13-2}$$

where K_{obs} is the observed binding constant as normally defined (independent of buffer components), and we have replaced the ion activity a_M^+ by its concentration $[\text{M}^+]$. Since K_T cannot depend on the ion concentration,

$$\frac{\partial \ln K_T}{\partial \ln[\text{M}^+]} = 0 \tag{13-3}$$

therefore

$$\frac{\partial \ln K_{\text{obs}}}{\partial \ln[\text{M}^+]} = -\Delta r \tag{13-4}$$

Figure 11.13 shows a plot of ln K_{obs} versus sodium ion concentration for *lac* repressor interacting nonspecifically with DNA. The slope gives $\Delta r = 10$.

Interpretation of this result requires knowledge of the number of counterions associated with naked DNA; following the theory in Chapter 11, this number is expected to be about 0.88 per phosphate. Therefore, release of 10 Na^+ requires involvement of about 11 phosphates in salt linkages. Consideration of other complicating factors led deHaseth et al. (1977) to conclude that the nonspecific binding of *lac* repressor to DNA involves the formation of 12 ± 2 salt linkages to DNA phosphates.

The entropy change that accompanies this ion release is an important contributor to the overall free energy change of protein–nucleic acid binding (Record et al. 1976, 1978). A rough estimate of the entropy change per mole of ions can be made using the equation $\Delta S = R \ln (C_b/C_f)$, where C_b and C_f are the ion concentrations when bound to DNA and free in solution, respectively. Using an estimate of $C_b \sim 2M$ (LeBret and Zimm, 1984; Matthews and Richards, 1984), and taking $C_f = 50$ mM, the predicted entropy increase is about 7.4 cal K^{-1} mol^{-1} of ions. Using the well-studied example of λ repressor (Sauer et al., 1990), for which the estimated number of Na^+ ions released is $\Delta r \sim 5$, one calculates an entropy contribution to the free energy of ion release $[\Delta G = -\Delta r T \Delta S]$ of about -11 kcal mol^{-1} of protein bound at 22°C. The total binding free energy under these conditions is about -17 kcal mol^{-1}, confirming that the entropy of ion release is an important component of the overall thermodynamic balance.

Calculated net DNA–protein electrostatic free energies of interaction are also comparable in magnitude to the estimated free energy contributed by the entropy of ion release. For example, Matthew and Ohlendorf (1985) estimated that the electrostatic free energy change upon binding λ *cro* repressor to its operator site is about -17 kcal mol^{-1} at 0.1 M salt concentration.

The Coulombic free energy for the interaction of charges q_1 and q_2 in a medium of dielectric constant ε is

$$\Delta G_{el} = \frac{q_1 q_2}{\varepsilon r} \quad (13\text{-}5)$$

The Coulombic energy becomes very large when ions approach each other closely. For example, the energy of dissociating a crystal of KCl into its component gaseous ions is about 160 kcal mol^{-1}. The dielectric constant term in the denominator of Eq. 13-5 reduces the magnitude of these energies, but when ions approach each other closely in aqueous medium, the water molecules between them become fully aligned, and the effective dielectric constant is strongly reduced from its value in bulk solution (which is ~ 75 at physiological temperature). Consequently, one expects the Coulombic free energy of separating a protein with multiple positively charged groups from its negatively charged DNA binding site to be much larger than the observed binding free energy.

The reason that the contribution of the Coulombic free energy term does not dominate over the entropy of ion release is that binding involves displacement of counterions, with their Coulombic interactions, by charged groups on the protein, which form a new set of Coulombic interactions. A good electrostatic fit between protein and DNA can be expected to contribute a somewhat more favorable Coulombic energy than can be provided by the more mobile counterions, unless there is great entropy sacrifice by the latter. However, the factor in the free energy due to the advantage of fixing the charges by the protein is not large, judging from the calculated energy values.

Electrostatic interactions do not require free charges; they can result as well from electric dipoles. A protein folded into an α helix has a substantial dipole moment because of the alignment of the C=O· · ·H–N hydrogen bonds approximately parallel to the helix axis. The result is a substantial net positive displacement charge on the amino terminal end of the helix. Lockart and Kim (1992) used an optical probe whose absorption spectrum is dependent on the electric field to derive a calculated electric field of 3.4×10^7 V cm^{-1} at the probe site. The magnitude of the local field was independent of the length of the helix between 21 and 41 residues, and was not influenced by a nearby arginine side chain, whose electrostatic effects were quenched by the high dielectric constant of solvent water (Lockart and Kim, 1993). As discussed in Section 13.6, helices frequently have their (positive) amino termini directed toward the negatively charged DNA molecule in protein–DNA complexes. It is also apparent from structural studies that electrostatic effects can involve polar side chains.

Studies of the detailed structure of protein–DNA complexes reveals extensive interactions with phosphate oxygens. More often than not these involve polar groups on the protein rather than charged amino acid side chains. Figure 13-3 (Jordan and Pabo, 1988; Sauer et al., 1990) illustrates this principle, showing the contacts between λ repressor and the sugar–phosphate backbone chain in one-half of the operator site, which is the name given the specific binding site when the protein is a repressor. Two positively charged side chains, Lys-26 and Lys-19, are close enough to the DNA to be in electrostatic contact. Neutral but hydrogen-bond donating side chains from Tyr-22, Gln-33, Asn-52, Asn-58, and Asn-61 contact the phosphate oxygens. Additional

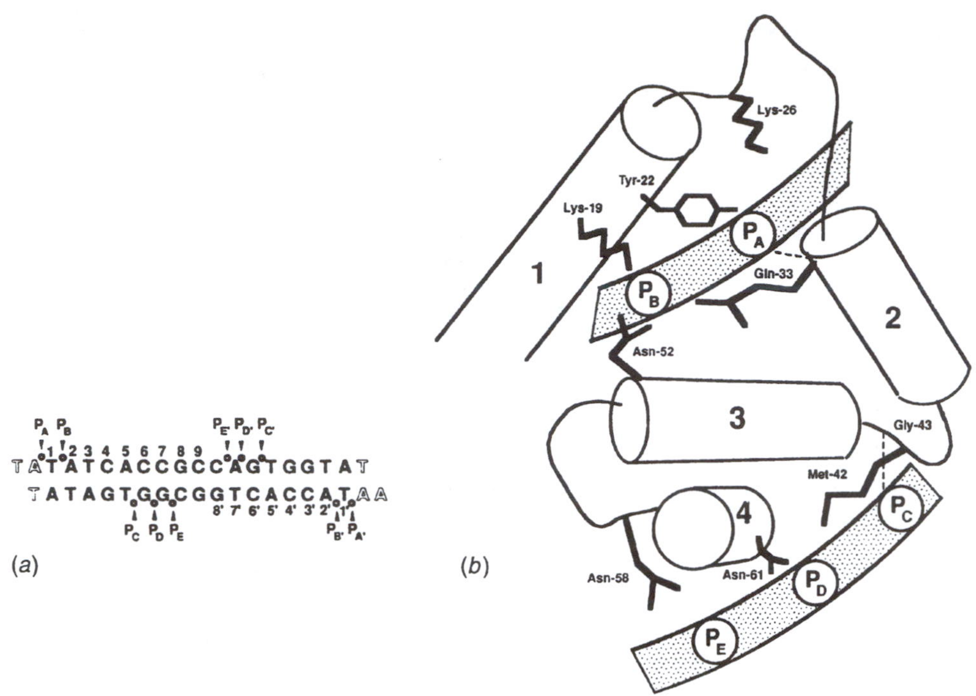

Figure 13-3
(*a*) Operator site used for cocrystallization of repressor. Sequence of the 17 bp site O_L is shown in bold face letters. The left half of the operator site, with base pairs numbered 1–9, matches the consensus sequence for the operator half-sites (see Fig. 13-8). The approximate twofold axis goes through base pair 9, and base pairs in the nonconsensus half of O_L1 are numbered $1'-8'$. Circles mark phosphates that ethylation interference experiments (Section 9.3) had implicated as contacts. These are labeled P_A to P_E in the consensus half-site and $P_{A'}$ to $P_{E'}$ in the nonconsensus half. (*b*) Contacts with the phosphate backbone. Phosphate contacts that λ repressor makes in the consensus half-site are shown. Broken lines indicate hydrogen bonds with the main-chain NH. [Reprinted with permission from Sauer et al., 1990.]

hydrogen-bonding contacts are made by the peptide backbone NH groups of Gln-33 and Gly-43. It is also noted (Jordan and Pabo, 1988) that the α-helix dipole moment (with positive pole at the amino terminal end of the helix) of helix 2 produces a favorable dipole–charge interaction with phosphate P_A. Finally, there is contact between the side chain of Met-42 and the deoxyribose sugar between phosphates P_C and P_D.

Clearly, the electrostatic framework of the protein is more complex than envisioned in simple earlier models in which charged protein side chains contact phosphates and displace bound counterions. A problem for the future is correlation of thermodynamic values for ion release (Δr in Eq. 13-4) with the full and partial charges that are seen to contact the DNA phosphates in detailed views of the structure of protein–DNA complexes, such as known for λ repressor.

A general approach to electrostatic interactions is provided by computation of the electrostatic potential around the surface of the protein. (The electrostatic potential is equal to the potential energy of a unit positive charge, and is a function of the distances from each of the charges on the protein, the local dielectric constant, etc. Its accurate calculation is a formidable problem.) Figure 13-4 (see color plate) illustrates such a calculation for the gene activating protein CAP (for catabolite activator protein; also called CRP, for cAMP receptor protein). Blue regions in the diagram indicate a positive electrostatic potential, to which negative charges are attracted. Hence, the DNA–phosphate backbone is expected to be drawn to the blue zones, which define a roughly semicircular ramp of net positive charge. As a consequence, DNA is bent around the protein, as discussed in more detail in Section 13.4.

One way to test the ability of theory to calculate electrostatic effects in DNA–ligand interactions is to compute the difference in free energy of binding for a protonated ligand and the unprotonated, neutral species (Misra and Honig, 1995). The difference, which results from the favorable interaction of the protonic charge with the negative electrostatic potential at the DNA binding site, is related to the shift in pK_a of the ligand upon binding. Misra and Honig (1995) show that the experimental variation of pK_a for a DNA intercalating ligand is accurately predicted by the theory. This relatively simple example provides a basis for optimism about accuracy of the theory for more complicated cases.

Of considerable interest is the extent to which the sugar–phosphate backbone contacts can provide specificity in electrostatic interactions. The preferred position of phosphates in a double helix depends on base sequence because of variations in helical twist and other structural parameters such as base pair roll and tilt (see Chapter 4). Furthermore, as we see below in Sections 13.4 and 13.5, DNA is often strongly distorted by protein binding. Since the ability to undergo such conformational shifts can be expected to be dependent on base sequence, these interactions are potential sources of specificity, even though the chemical groupings in the DNA sugar–phosphate backbone are invariant.

2.2 Hydrogen Bonding: Donors and Acceptors in DNA and RNA Duplexes

Nucleic acids, particularly DNA, are often found in a structurally monotonous double helical form, which must nonetheless be recognized by specifically binding proteins. Formation of a double helix blocks some of the hydrogen-bonding positions that could be used to distinguish one base sequence from another. The remaining hydrogen-bond donors and acceptors must be approached from the major or minor grooves, as illustrated in Figure 13-5 (Seeman et al., 1976; Steitz, 1990). As can be seen from the figure, the recognition features are more distinctive when viewed from the major groove: The pattern for a G•C pair reads acceptor (GN7), acceptor (GO6), donor (CN4), whereas that for an A•T pair is acceptor (AN7), donor (AN6), acceptor (TO4), and nonpolar (T-CH3). Since these are different from each other, and do not read the same forward and backward, the pattern can in principle be used to distinguish all four base pairs (Seeman et al., 1976). On the other hand, the view from the minor groove is less distinctive: acceptor, donor, acceptor for G•C and acceptor, acceptor for A•T pairs. Each of these patterns reads the same forward and backward, and therefore it may not be an easy matter to distinguish G•C from C•G or A•T from T•A base pairs by recognition in the minor groove.

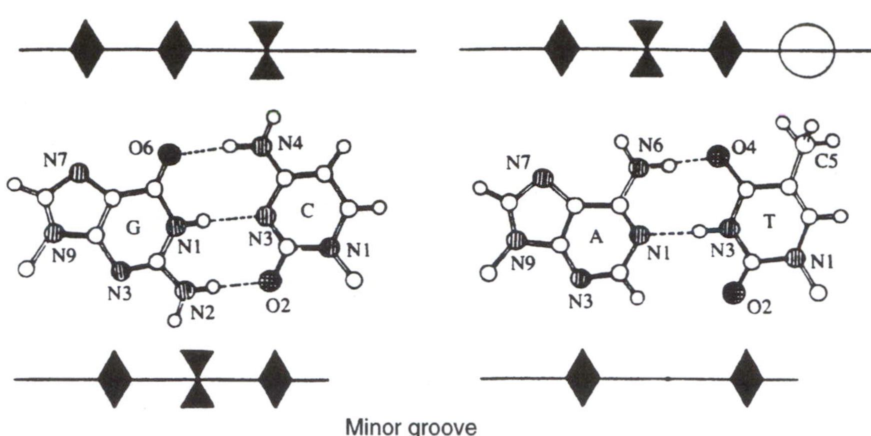

Figure 13-5
The hydrogen-bond donors and acceptors presented by Watson–Crick base pairs to the major groove and the minor groove. The symbols for hydrogen bond donors (▼▲) and acceptors (◆) show a varied pattern presented by the base pairs to the major and a poor information array in the minor groove. While it is possible to distinguish among AT, TA, GC, and CG in the major groove, functional groups in the minor groove allow only easy discrimination between AT and GC containing base pairs; ⊖ is a methyl group. [Reprinted with the permission of Cambridge University Press. Steitz, T. A. 1990 *Q. Rev. Biophys.* **23**, 205–280. Copyright ©1990.]

The difference between DNA B-form and the RNA A-form geometry has important consequences for the accesibility of the major and minor grooves. Figure 13-6 (see color plate) (Steitz, 1990) shows views of the two helices at an angle that enables one to sight along the grooves and judge their width (see the right-hand view in each case). In the B form, the major groove is wide enough to accommodate such protein sturctural motifs as an α helix or an antiparallel β ribbon, as described in Section 13.6. In contrast, the major groove in RNA is deep and narrow, too much so to provide easy access for protein recognition. The minor groove in DNA, while sometimes wide enough (depending on the sequence) to accommodate amino acid side chains, as in the case of the complex with DNAse I (Suck et al., 1988), is generally too narrow to accept an α helix. On the other hand, the RNA minor groove is wide and shallow enough to provide sites for specific recognition, as is found for the complex of tRNA^{Gln} bound to the corresponding aminoacyl synthetase (Rould et al., 1989).

Inaccesibility of the RNA major groove applies to long helices. Near helix ends the major groove functionalities are accessible, a principle that is exploited in the interaction of aspartyl tRNA synthetase with its cognate tRNA (Ruff et al., 1991). The RNA major groove is widened adjacent to "bulged" nucleotides (with one or more extra bases on one strand), as revealed by enhanced chemical reactivity of purine N7 (Weeks and Crothers, 1991). The effect can be generalized to asymmetric loops, in which the number of noncomplementary bases on the two strands are not equal (Weeks and Crothers, 1993). Chemical reactivity is less enhanced adjacent to small symmetric loops, which are also less destablizing for the double helix, presumably because helical stacking tends to be retained on both strands through the loop.

2.3 Hydrogen Bonding: Protein Donors and Acceptors

A number of polar amino acid side chains can serve as hydrogen-bond donors and acceptors to facilitate protein recognition of nucleic acid sequences. Table 13-1, prepared from original papers and the data reviewed by Steitz (1990) and Sauer et al. (1990), shows illustrative amino acid side chain–nucleotide base hydrogen-bonding

Table 13.1
The DNA Major Groove Recognition Interactions with Polar Amino Acid Side Chains Observed in Crystallographic Studies of Sequence-Specific Complexes[a]

Amino Acid	DNA Residue					
	AN6 (donor)	AN7 (acceptor)	TO4 (acceptor)	GN7 (acceptor)	GO6 (acceptor)	CN4 (donor)
Gln (—C(=O)—NH_2)	1,2	1,2	1	1	1	
Glu (—C(=O)—OH)	12					1,7,12
Ser (—OH)				2		
Lys (—NH_3^+)		9^b	9^b	$2,3^b,6,8,9,10$	$2,6^b,8,9,10$	
Asn (—C(=O)—NH_2)	4	4,10	11	2		10,11
Arg $(\text{NH—C(NH}_2\text{)=NH}_2^+)$	3^b	3^b	7	3,5,6,7,11,12	3,5,6,7,11	
Thr (—OH)	$3^b,8$	$3^b,8$		2		
His (N—H)				5,11		
Cys (—SH)	10			10		

[a]Structures in which the interaction occurs are designated by numbers; prepared from data reviewed by Steitz (1990) and Sauer et al. (1990) and more recent papers. Key to structures: 1 = 434 repressor (Aggarval et al., 1988); 2 = λ (cI) repressor (Jordan and Pabo, 1988; Clarke et al., 1991); 3 = trp repressor (Otwinowski et al., 1988); 4 = engrailed homeodomain (Kissinger et al., 1991); 5 = Zif268 Zn finger (Pavletich and Pabo, 1991); 6 = gluocorticoid receptor (Luisi et al., 1991); 7 = CAP (Schultz et al., 1991); 8 = *met* repressor (Somers and Phillips, 1992); 10 = papillomavirus E2 (Hedge et al., 1992); 11 = GCN4 (Ellenberger et al., 1992); 12 = Max (Ferré-D'Amaré et al., 1993).

[b]Via a bridging water molecule.

interactions that have been verified by crystallographic determination. Structures include complexes of DNA with 434 (Aggarwal et al., 1988) and λ (Jordan and Pabo, 1988; Clarke et al., 1991) bacteriophage repressors, with *Escherechia coli trp* repressor (Otwinowski et al., 1988), with the engrailed homeodomain (Kissinger et al., 1990), with Zif268 Zn finger (Pavletich and Pabo, 1991), with glucocorticoid receptor (Luisi et al., 1991), with *E. coli* CAP protein (Schultz et al., 1991), with GAL4 (Marmorstein et al., 1992), with *met* repressor (Sommers and Phillips, 1992), with papillomavirus E2 (Hedge et al., 1992), with GCN4 (Ellenberger et al., 1992), and with the eukaryotic transcription factor Max (Ferre-D'Amare et al., 1993).

Given the rapidly growing nature of this field, one can expand Table 13.1 accordingly, but some simple principles are evident from this early sample. First, there is clearly no correspondence between amino acid and nucleotide base, since a given nucleotide can be recognized by more than one side chain, and a given side chain can hydrogen bond to more than one kind of nucleotide. Also, amino acid side chains, such as glutamine, that have both hydrogen-bond donating and accepting capability can function in either mode. Positively charged side chains, arginine and lysine, can function as hydrogen bond donors; they need not be relegated to the simple role of electrostatic interaction with phosphate. Furthermore, bidentate interactions can occur, such as the hydrogen bonding of the arginine side chain simultaneously to the N7 and O6 of guanine as originally proposed by Seeman et al. (1976); their proposals for bidentate major groove binding of asparagine or glutamine to adenine, and arginine to guanine, are illustrated in Figure 13-7. Bonding can also occur to the hydrated form of DNA. For example, a water molecule serves to bridge between the amino acid side chain and a hydrogen bond acceptor on DNA in several cases in the complex with *trp* repressor (Otwinowski et al., 1988).

Even though there is no general amino acid-base interaction code for protein–DNA interactions, combination with additional stereochemical constraints could lead to better rationalization of the interactions. Suzuki and Yagi (1994) show that combination

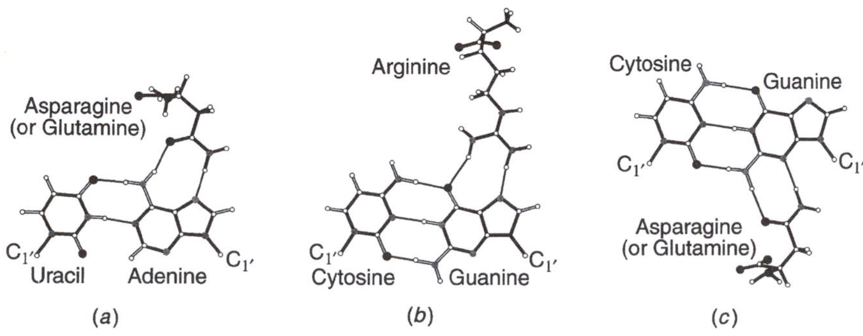

Figure 13-7
Proposed models for bonding of (*a*) glutamine (or asparagine) to a U • A pair in the major groove; (*b*) bonding of arginine to a C • G pair in the major groove; (*c*) bonding of asparagine (or glutamine) to a C • G pair in the minor groove [from Seeman et al., 1976].

of chemical rules, which set out the possible pairings between amino acids and the bases, together with stereochemical rules, which specify the position of amino acids in particular families of DNA binding protiens, leads to successful rationalization of the data. They were able to use the rules to predict the location of a specific protein binding site in a DNA sequence

Crystallographic work also indicates that hydrogen bonding interactions are not restricted to amino acid side chains. Clarke et al. (1991) found that the peptide carbonyl of a lysine residue at the amino terminal end of λ repressor interacts with the exocyclic amino group of a cytosine in the major groove. Similar interactions have since been reported; for example, in GAL4 the main chain carbonyl of a lysine also accepts a hydrogen bond from cytosine (Marmorstein et al., 1992).

2.4 Hydrophobic Interactions

It is a common if nonrigorous practice to lump together, under the heading of "hydrophobic interactions," those free energetic contributions that are not accounted for by considering the effects of charges (electrostatics) and dipoles (including hydrogen bonding). Broadly viewed, hydrophobic effects reflect the tendency of nonpolar groups to associate in aqueous solution rather than to expose themselves to maximal contact with the solvent water. This tendency results from the same factors that make nonpolar substances insoluble in water. The effect reflects the net free energy of multiple changes in the solution, including the magnitude of the van der Waals or dispersion interaction between the nonpolar molecules, plus the hydrogen bonding and dispersion interactions between the water molecules, compared to the dispersion interactions between water and the nonpolar groups. Special interactions between adjacent water molecules at their interface with the nonpolar groups are also important. It should be recognized that nonpolar aromatic residues are capable of hydrogen bonding, at least with water (Suzuki et al., 1992). Recent theoretical advances are moving toward a better understanding of the hydrophobic effect based on the properties of neat water (Hummer et al., 1996; Berne, 1996).

A simple view of a hydrophobic association is a merging of two hydrated nonpolar domains to form one larger hydrated domain:

$$2 \text{ Nonpolar (aq)} \rightarrow \text{NONPOLAR (aq)} + \text{bulk water} \tag{13-6}$$

The total volume of the nonpolar region remains about constant, but the surface/volume ratio decreases upon association, resulting in release of some water from its special structural and thermodynamic status at the surface; the released water rejoins the bulk form.

The large change in surface area exposed to water has important thermodynamic consequences. Water molecules at the surface are more ordered (ice-like) than those in bulk solution, so a reduction in hydrophobic surface area upon complex formation is accompanied by a positive ΔS, along with a positive value for ΔH corresponding to the heat of "melting." However, the value of ΔS for the hydrophobic effect is strongly temperature dependent, extrapolating to zero at about 386 K (Baldwin, 1986).

This, along with other factors, has led to a focus on a better constant for the process, namely, the large negative heat capacity change ΔC_{assoc} that accompanies hydrophobic associations (Spolar and Record, 1994; see also the commentary by von Hippel, 1994).

The heat capacity change ΔC_{assoc} is equal to the difference in heat capacity of the products and reactants in the association process,

$$\Delta C_{assoc} = C(\text{complex}) - [C(\text{protein}) + C(\text{DNA})] \tag{13-7}$$

The negative sign of ΔC_{assoc} can be rationalized by considering the role of the ordered water at the hydrophobic surface. The heat taken up by the gradual melting of the specially structured water as temperature is raised makes a positive contribution to the overall heat capacity. (The extrapolated temperature $T = 386$ K, at which ΔS becomes zero, reflects the completion of this melting process.) Since the combination of separate protein and DNA molecules has more surface water than does the complex, the heat capacity of protein plus DNA is larger than the heat capacity of the complex (considering just the hydrophobic effect), and the corresponding ΔC_{assoc} is negative.

Spolar and Record (1994) give an equation that relates the heat capacity change to the change in the nonpolar surface area exposed to water upon complex formation, ΔA_{np} (in Å^2) and a similarly defined change in the water-accessible polar surface area, ΔA_p,

$$\Delta C_{assoc} = (0.32 \pm 0.04)\, \Delta A_{np} - (0.14 \pm 0.04)\, \Delta A_p \quad (\text{in cal mol}^{-1}\ \text{K}^{-1}) \tag{13-8}$$

Since nonpolar surface is buried in a hydrophobic association process, ΔA_{np} is negative as is ΔC_{assoc}. Note that the second term in Eq. 13-8 has the oppositie sign from the first, corresponding to the different nature of water at a polar surface. Since ΔA_p is usually quite small, ΔC_{assoc} is dominated by the first term, corresponding to burial of the nonpolar surface. Spolar and Record (1994) apply these considerations to protein–DNA complexes for which data allowing calculation of ΔC_{assoc} are available. They conclude that large negative heat capacity changes, of the order of -0.5 to -3 kcal mol^{-1} K^{-1}, are characteristic of specific complex formation. Furthermore, the effect is frequently too large to be accounted for on the basis of rigid body association between the two partners. They infer that local protein folding accompanies complex formation in these cases, contributing additional burial of hydrophobic surface. An example showing the difference in the hydrophobic effect for specific and nonspecific complex formation is provided by Foguel and Silva (1994) for the *Arc* repressor–DNA complex.

In contrast to protein hydrophobic surfaces, a nucleic acid double helix exposes mainly polar groups to the aqueous surface, which interact favorably with water. Exceptions include, in the major groove, the pyrimidine 5,6 double bond, purine C(8)-H, and thymine methyl group, and in the minor groove adenine C(2)-H. The deoxyribose C(2')-H2 can also be considered to be in this category. Contacts between these sites and nonpolar side chains on bound proteins can be classified as hydrophobic.

Interactions of this kind reported so far involve the thymine methyl group. For example, a hydrophobic pocket formed by the side chains of Thr and Gln interacts with the thymine methyl group in the structure of 434 repressor (Aggarwal et al.,1988). For the λ repressor, Jordan and Pabo (1988) and Sauer et al. (1990) list among the hydrophobic interactions the contact between thymine methyl groups and the γ- carbon of Gln-22, and with pockets formed by the β- carbon of Ala-49, Gly-48, and the δ-carbon of Ile-54 in one case and by Gly-46 and the β- carbon of Ser-45 in another. The thymine methyl group also makes hydrophobic contacts with Ile in a homeodomain–DNA complex (Kissinger et al., 1990), with Val in the glucocorticoid receptor-DNA complex (Luisi et al., 1991), and with Phe in the papillomavirus E2 protein–DNA complex (Hegde et al., 1992).

Stacking of aromatic amino acids on the nucleic acid bases provides a relatively stereospecific interaction of the hydrophobic class. This kind of interaction is common in RNA–protein interactions, because many of the RNA nucleotides are in nonhelical conformations, and hence available for stacking with protein side chains (for a review, see Mattaj and Nagai, 1995). Examples at high structural resolution can be found in the complex of the U1A splicesomal protein complex with an RNA hairpin (Oubridge et al., 1994), in which stacking of a tyrosine side chain on cytosine and phenylalanine on adenine are seen. In these cases, however, the stereospecificity of the interaction is strongly governed by an extensive hydrogen-bonding network involving the affected bases. An example in which stacking with nonspecific single-stranded DNA bases is implied by the structure is the single stranded DNA binding protein T4gp32 (Shamoo et al., 1995).

2.5 The Thermodynamics of Specificity

One of the goals of studying protein–nucleic acid interactions is to understand the thermodynamic basis for the observed interaction specificity. Structural studies can reveal contacts between amino acid side chains and the nucleic acid bases or backbone, but cannot specify the magnitude of the free energy contribution made by that interaction. The missing data can be provided by mutagenesis of the protein or nucleic acid, followed by study of the binding affinity and specificity relative to the wild-type interaction.

Extensive work of this kind has been reported for the λ repressor interacting with its operator sites (Sauer et al., 1990). Table 13-2 summarizes the effect of mutations in the protein on the free energy of binding, relative to values for the wild-type protein. Values are given both for binding to the operator site and to nonoperator DNA. The table shows that all categories of interactions, including polar interactions with the phosphate backbone, hydrogen bonds to the bases, and hydrophobic effects can have similar dramatic effects on binding affinity. The binding free energy to operator DNA is reduced by 1.6 to more than 6.2 kcal mol^{-1} (corresponding to reduction in binding constant by an amount ranging from 16-fold to more than 10,000-fold) as a result of mutations at the critical amino acids. The effects on nonspecific binding are much smaller, indicating that nonoperator binding does not use the same contacts as found in the specific operator complex. This difference identifies the tabulated interactions as important for specificity, since their loss decreases the difference in binding affinity between operator and nonoperator DNA. Note that the thermodynamics of all of the mutations that have

Table 13.2
Thermodynamic Effects of Mutations in the λ Repressor on Operator Binding[a]

Mutant	Wild-Type Interaction	ΔG Relative to Wild-Type, kcal mol^{-1}	
		Operator $\delta\Delta G$	Nonoperator $\delta\Delta G$
Tyr-22 → Phe	Backbone phosphate	1.6	
Gln-33 → Tyr	Backbone phosphate	1.8	0
Gln-33 → Ser	Backbone phosphate	4.5	0.1
Asn-52 → Asp	Backbone phosphate	>6.2	>2.8
Gly-43 → Glu	Backbone phosphate	>6.2	0.5
Gln-44 → Ser	Hydrogen Bond to A(2)N7,–NH_2	3.7	0.2
Gln-44 → Tyr	Hydrogen Bond to A(2)N7,–NH_2	3.7	0.2
Gln-44 → Leu	Hydrogen Bond to A(2)N7,–NH_2	>6.2	0.2
Ser-45 → Leu	Hydrogen Bond to G(4)N7	>6.2	−0.1
Asn-55 → Lys	Hydrogen Bond to G(6)N7,Lys-4	>4.1	−3.1
Lys-4 → Gln	Hydrogen Bond to G(6)06,Asn-55	5.2	0.5
Ala-49 → Val	Hydrophobic T(3)–CH_3[b]	4.6	0.2
Ala-49 → Asp	Hydrophobic T(3)–CH_3[b]	>6.2	−0.1

[a]Prepared from Jordan and Pabo (1988) and data reviewed by Sauer et al. (1990). Necleotide pairs are numbered as in Figure 13-8.

[b]Contact listed by Jordan and Pabo (1988), but not in the tabulation of Sauer et al (1990).

large effects on the binding free energy can be qualitatively explained in terms of the protein–DNA contacts observed in the crystal structure.

The effect of mutations in the operator site can also be qualitatively understood in terms of the structure of the complex. Figure 13-8 (Sauer et al., 1990) shows the influence of single base pair changes in the operator site on binding affinity. The critical interactions for recognition are hydrogen-bonds between A-2 and Gln-44, G-4 and Ser-45, G-6 and Asn-55 and Lys-4, and G-8 and Thr-2. Hydrophobic contacts are also observed for thymines 1, 3, and 5. Change of any of these bases causes a moderate to severe reduction in binding affinity for the wild type repressor (Sauer et al., 1990).

2.6 The Role of Nucleic Acid Distortion and Distortability in Specificity

The local structural parameters of DNA such as helical twist, roll, and tilt, depend on sequence. A protein might be designed to interact with a specific DNA structural variant that is characteristic of a specific sequence. The interaction can therefore be sequence-specific, without necessarily requiring stereospecific contact between protein residues and functional groups on the bases responsible for binding specificity. Suppose, for

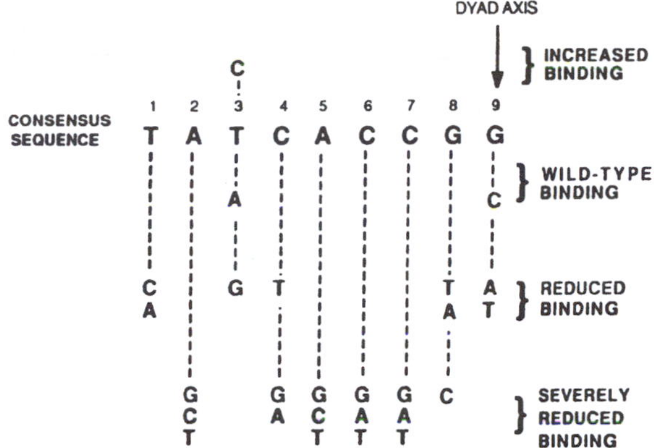

Figure 13-8
Effects of symmetric λ operator mutations. Starting with a symmetrized consensus sequence, symmetric changes were introduced into each operator half-site and repressor binding was assayed *in vivo* [Reprinted with permission from Sauer et al., 1990].

example, that a protein interacts with DNA deoxyribose–phosphate backbone residues that are separated by 34 Å and one helical turn. The only sequences permitted between these two sites would be those that provide the appropriate total helical twist and rise per turn. These considerations apply to cases such as recognition of *E. coli* promoters, for which the spacing and helical twist between the -10 and -35 regions of the promoter affect promoter strength (Mulligan et al., 1984).

Interaction of *trp* repressor with its target DNA sequence illustrates the importance of DNA distortions, or intrinsic deviations from canonical B-DNA structue, in establishing binding specificity (Shakked et al., 1994). The structure of DNA in the complex is intermediate between B and A forms, as judged by variables such as X displacement, which is the perpendicular distance from the vector between the C6 and C8 atom of a base pair to the helix axis (see Chapter 4). The free DNA crystallized without protein is also intermediate between the two forms, although somewhat closer to the B form than the complexed DNA.

Binding of phage 434 repressor to its DNA operator target provides another illustration of the importance of DNA distortion, as well as distortability, effects in binding specificity. Tight contacts at the edges of the binding site for the dimeric protein result in overtwisting of the central 4 bp. The sequence of these four bases has a significant effect on binding affinity, even though the structure of the complex shows that they do not contact the protein (Koudelka et al., 1987, 1988). The importance of rigidity of DNA in this effect is documented by the higher binding affinity of a DNA operator having a single-strand nick at the center of the binding site. Protein rigidity is also important: A repressor with a mutation that results in a relaxed dimer interface is less sensitive in its binding affinity to the sequence of the central 4 bp. The results can be understood in terms of a simple model in which local overtwisting at the operator center is required for binding. The energetic cost of this deformation depends on base sequence, which therefore modulates the binding affinity. Hogan and Austin (1987) drew attention to the possible role of bending as opposed to twisting deformation of the operator DNA.

The experiments of Paolella et al. (1994) further illustrate the role of DNA shape and distortability on binding affinity and specificity. Some proteins of the bZIP family can distinguish between the two related target sequences ATGACTCAT (the 9 bp AP-1 site) and ATGACGTCAT (the 10 bp dyad symmetric CRE or ATF site), and some cannot. For example, the yeast transcription factor GCN4 binds equally well to the two sites, but the cAMP response element binding (CREB) protein binds 50 fold better to the CRE site than to the AP-1 site. Gel electrophoresis experiments showed that the CRE site has an intrinsic bend toward the major groove in a coordinate frame at the center of the site, whereas the AP-1 site is straight. When GCN4 binds, it leaves the CRE and AP-1 DNA structures unchanged, showing that in this case the protein is able to adapt to differing DNA structures. However, CREB binding introduces a tendency to bend toward the minor groove in a coordinate frame at the center of the site. The result for the CRE site is to make the DNA nearly straight, yielding a high-affinity complex. The complex of CREB with the AP-1 site requires bending the DNA site toward the minor groove. A plausible interpretation of the weak binding affinity of CREB for this site is that the required DNA distortion energy for bending toward the minor groove is much larger than for the CRE site, where all that has to happen is removal of the anomalous curvature and return of DNA to its normal straight form. It is likely to be a general principle that proteins that discriminate rigorously between close sequence variants (such as restriction enzymes) will turn out to be more rigid and intolerant of DNA structural variations than proteins that accept sequence variations more readily.

The importance of DNA bendability is illustrated by the influence on binding affinity of DNA sequence changes in the binding site for *E. coli* CAP protein (Gartenberg and Crothers, 1988; Dalma-Weiszhausz et al., 1991). Alteration of the sequence at sites where bends are thought to occur, but where the protein does not contact the bases (Schultz et al., 1991) changes the binding affinity by up to two orders of magnitude. This system is discussed in more detail in Section 13.4.3.

A high-resolution view of the energetic cost of a localized bend or kink is provided by studies of the the interaction of *Eco*RI endonuclease with its DNA target sequence (Lesser et al., 1993). Specific functional groups that had been identified as participants in hydrogen bonds by structural studies of the complex were deleted through substitution of purine analogues. For five of six such functional group deletions, the observed penalty in binding free energy was about 1.5 kcal mol^{-1}. However, deletion of an adenine N6 amino group at the inner A in the binding site increased the (negative) binding free energy by -1.0 kcal mol^{-1}, in spite of the deletion of a base-protein hydrogen bond at that position. This effect was attributed to easier formation of the DNA kink, shich narrows the major groove at that position. Deletion of the N6 amino group reduces the steric clash in the major groove, providing a favorable free energy contribution that outweighs by 1.0 kcal mol^{-1} the unfavorable effect of loss of the hydrogen bond.

3. DNA SEQUENCE MOTIFS

3.1 Inherent Symmetries of Target DNA Sequences

Proteins bind to a restricted domain on a DNA molecule that is defined by its base sequence. The size of the sequence that is important for *recognition* can vary from a few base pairs up to several turns of the double helix. It is important to distinguish the recognition domain of the DNA from the flanking double helical regions that may provide additional nonspecific contacts (usually with the backbone), which are needed for adequate overall binding *affinity*, but which are relatively unimportant for the specificity of binding.

An increase in the size of the DNA region recognized is advantageous if the objective is improved ability to discriminate against incorrect sequences, since a larger number of specific hydrogen-bonding contacts increases the difference in the binding affinity compared to that for bulk DNA. The simplest way to increase the size of the recognized region is to repeat it and engage in interaction with an additional identical protein subunit. Amplification of affinity and specificity requires, however, that the protein subunits interact favorably when bound (either by direct contact or through another protein) so that they bind better to the repeated sequence than to a single copy of it.

There are two simple kinds of repeat of a double helical nucleotide sequence: direct (or tandem) and inverted. As illustrated in Figure 13-9(*a*), direct repeat implies that the sequence occurs a second time with the same $5' \rightarrow 3'$ polarity in the same DNA strand, whereas in an inverted repeat the sequence is repeated $5' \rightarrow 3'$ on the other strand [Fig. 13-9(*b*)]. An inverted repeat gives the binding site dyad symmetry, meaning that it is

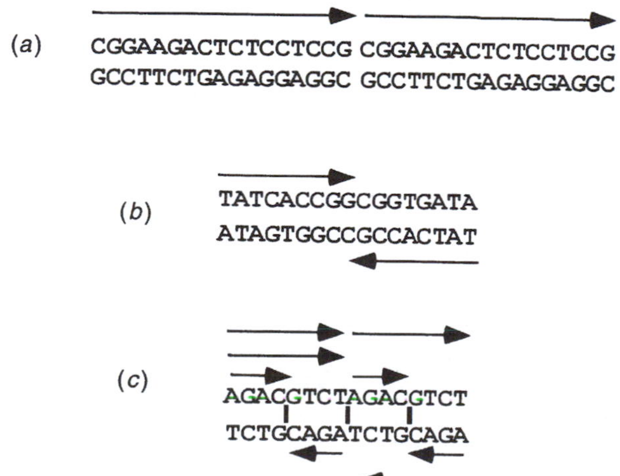

Figure 13-9
Idealized symmetries in target DNA sequences. Direct repeats are indicated by → and inverted repeats by →←; local dyad axes are shown by |. (*a*) Direct repeat: Yeast GAL4 binding sites Giniger and Ptashne (1988). (*b*) Inverted repeat: λ repressor binding sete (Sauer et al. (1990). (*c*) Direct repeat of dyad symmetric elements: *met* repressor binding sites (Phillips et al.(1989).

left unchanged by rotation through 180° about an axis that runs from the major to the minor groove at the center of the sequence. Sequences with inverted repeat symmetry are sometimes incorrectly referred to as palindromic, which means a line of text that reads the same forward and backward [*Madam I'm Adam*]. A true palindrome requires that the $5' \rightarrow 3'$ sequence be repeated $3' \rightarrow 5'$ on the same strand. Since there is no simple way to associate protein subunits with palindromic symmetry of DNA, its use as a structural principle seems unlikely.

Whether the target sequence is repeated in tandem or with dyad symmetry has a profound effect on the expected nature of the protein complex. A tandemly repeated DNA sequence motif can continue indefinitely: A protein monomer that interacts with the first repeat of the motif will find its "right" side in interaction with the "left" side of its neighbor, whose "right" side could in turn be in interaction with the "left" side of a third protein, and so on. This arrangement produces cooperative binding of the proteins, assuming favorable interactions at the interfaces between bound subunits. In such a circumstance, occupancy of the first site is thermodynamically more difficult than binding to subsequent sites in the chain; once binding is initiated, it tends toward full occupancy over a relatively narrow range of concentration of the protein.

Another way that tandemly bound proteins can act synergistically is illustrated in Figure 13-10. Yeast GAL4 proteins bound to tandem repeats of a 17-mer sequence (see Fig. 13-10) interact with a (hypothetical) target protein, whose binding affinity is enhanced by contacting more than one bound GAL4.

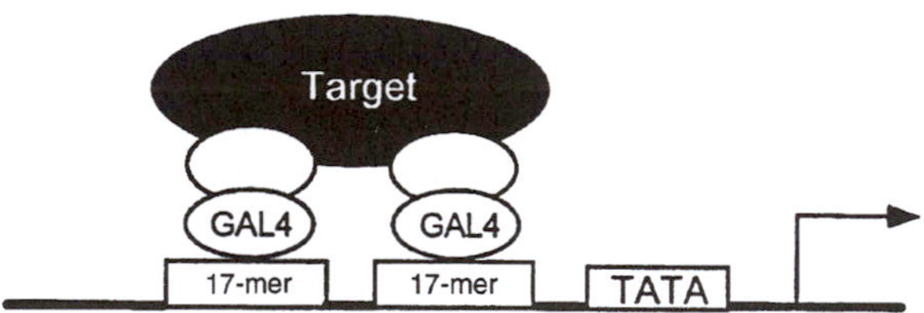

Figure 13-10
Schematic representation of a proposed model for multiple activators simultaneously contacting a common target. Two molecules of GAL4 derivative, bound to adjacent sites on DNA, simultaneously contact an unidentified target protein. The interaction energies add, leading to an exponential increase in the affinity of the target for multiple activators; the exponential increase in affinity generates a synergistic increase in transcription. More than 2, and possibly as many as 5–10 molecules of certain activators may simultaneously interact with the target. The target is shown as a single molecule although there may be multiple targets that interact with one another. [Reprinted with permission from Carey, M., Young, Sun, L., Green, M. R., and Ptashne, M. (1990). *Nature (London)* **345**, 361–364. Copyright ©1990 Macmillan Magazine Limited.]

In a site with dyad symmetry, on the other hand, the protein subunit that is bound to the inverted repeat must be rotated through 180° because of the dyad symmetry of the site. This rotation brings the two "right" sides of the proteins into contact, an interaction motif that cannot be repeated by either of the partners. Consequently, dyad symmetric sites are usually occupied by dimeric proteins whose dyad or *C*2 symmetry corresponds to that of the DNA sequence target.

An arrangement with combined symmetry is that in which dyad symmetric interaction sites are tandemly repeated [Fig. 13-9(*c*)]. Because of the local dyad axis in each site, direct repeat of the motif yields an overall dyad symmetry axis at the center of the set of repeats. A binding site such as that for *met* repressor in Figure 13-9(*c*) could in principle be occupied by a single protein dimer that utilizes the overall dyad axis of the site, or by two tandemly repeated dimers, each of which takes advantage of the local dyad symmetry axis. This principle is illustrated in Figure 13-11 for an idealized operator sequence for binding the *trp* repressor. It is in general not possible to discriminate between the two possibilities on the basis of the DNA sequence alone, and additional experiments such as determination of the binding stoichiometry of proteins that bind to sites like those of *trp* repressor (Carey, 1988; Carey, 1991a; Haran et al., 1992) and the structure of the complex (Otwinowski et al., 1988; Lawson and Carey, 1993) are required.

The two binding modes in Figure 13-11 place quite different requirements on the relative positioning of the DNA recognition elements on the two protein subunits that make up a dimer. In order to take advantage of the two local dyad symmetry axes as in Figure 13-11(*b*), the two subunits in the dimer must recognize bases that are immediately adjacent. Alternatively, if a single protein utilizes the larger dyad symmetry of the site as in Figure 13-11(*a*), it can take advantage of the two symmetrically related target sequences that are separated by a full helical turn. We will see in Section 6 that these

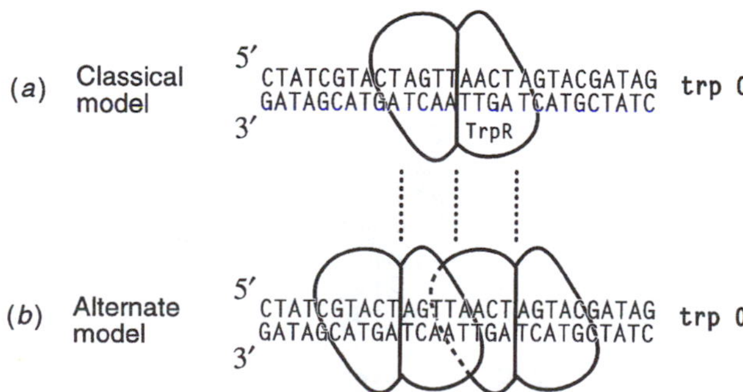

Figure 13-11
Two models of *trp* repressor binding. In the accepted model, only one *trp* repressor dimer can bind one operator. The second model suggests that if there are two axes of symmetry, two *trp* repressor dimers can bind to one operator [Staacke, D., Walter, B., Kisters-Worke, B., Wilchen-Bermann, B., and Müller-Hull, B. 1990). *EMBO J.* **9**, 1963–1967 by permission of Oxford University Press.]

two possibilities are reflected in the structures of dimeric protein–DNA complexes interacting by the antiparallel β ribbon and bZIP motifs on one hand [Fig. 13-11(*b*)] and the helix–turn–helix [Fig. 13-11(*a*)] motif on the other.

It is commonly observed that the dyad symmetry of inverted repeat binding sites is imperfect. Figure 13-12 (Sauer et al., 1990) illustrates the sequences of tandemly repeated bacteriophage λ operator sites. The string 5′-TATCACCGC-3′ (the dyad symmetry axis corresponds to the position of the C3′) is called the consensus sequence because it reflects the most probable base at each position among the set of natural operator sequences (see Section 13.3.2); a dyad symmetric repeat of the consensus usually has a very high binding affinity. Boxes in the figure enclose those base pairs that deviate from the consensus sequence. The most strongly conserved positions are the A • T base pair at position 2 (counting from the 5′-end) and the C • G at position 4, both of which are found in all the natural operator sites; the C • G pair at position 6 is conserved in 11 of the 12 operator half-site sequences in Figure 13-12. Note from Table 13-2 that bases in all three of these pairs make important hydrogen-bonding contacts to the repressor in the crystal structure. Other bases obviously contribute to specificity, however, because the binding affinity varies by a factor of 50 among the natural set of operators, generally decreasing as the number of deviations from the consensus increases.

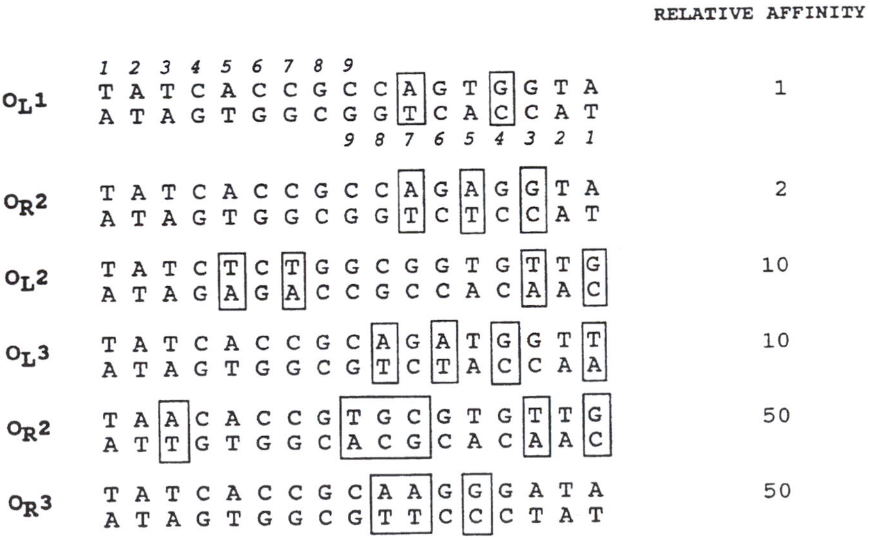

	Sequence	RELATIVE AFFINITY
	1 2 3 4 5 6 7 8 9	
O_L1	T A T C A C C G C C A G T G G T A A T A G T G G C G G T C A C C A T	1
	9 8 7 6 5 4 3 2 1	
O_R2	T A T C A C C G C C A G A G G T A A T A G T G G C G G T C T C C A T	2
O_L2	T A T C T C T G G C G G T G T T G A T A G A G A C C G C C A C A A C	10
O_L3	T A T C A C C G C A G A T G G T T A T A G T G G C G T C T A C C A A	10
O_R2	T A A C A C C G T G C G T G T T G A T T G T G G C A C G C A C A A C	50
O_R3	T A T C A C C G C A A G G G A T A A T A G T G G C G T T C C C T A T	50

Figure 13-12
Operator site sequences. The six operator sites found in the phage DNA are arranged according to their relative affinity for λ repressor. Relative affinities are expressed as the relative amount of repressor dimer needed (in the absence of cooperative effects) to give equivalent protection in DNase protection experiments. Base pairs that do not match the consensus sequence are enclosed in boxes. [Reprinted with permission from Sauer et al., 1990].

The reasons for deviations from the consensus sequence include the necessity for making subtle regulatory distinctions between binding the same protein to a variety of positions on the DNA. In addition, there may be a selective disadvantage to perfectly dyad symmetric sequences, which can form cruciform structures (Chapter 10) under superhelical tension *in vivo*, possibly increasing the rate of their excision from the genome.

3.2 Significance of Consensus

Usage of the concept of a "consensus" sequence is not uniform. For example, Staacke et al. (1990), in analyzing six sequences of *trp* repressor binding sites, find that at one position in the operator the base T occurs three times, G occurs twice, and C once. This finding is interpreted as a consensus for T. On the other hand, examination of 10 binding sites for NF-κB by Lenardo and Baltimore (1989) reveals occurrence of eight T and two C residues at one position, which they interpret as a consensus for pyrimidine.

Quantitative treatments can make the concept of consensus more precise (reviewed by Stormo, 1991). One main issue is the extent to which the observations deviate from expectation based on random sampling. For example, suppose that six DNA sequences are uncorrelated, and all four bases occur in essentially equal proportions in the genome. What is the probability that when the sequences are aligned and a given position compared, 3 will be of one kind, 2 of another, and 1 of a third kind, as in the observations of Staacke et al. (1990). It is clearly less likely that random sampling will produce 8 T and 2 C residues in 10 trials, as in the data set of Lenardo and Baltimore. How do the probabilities of these particular sampling outcomes compare with all other possibilities?

Schneider et al. (1986) and Berg and von Hippel (1987, 1988) address this problem in terms of information theory. One bit of information is sufficient to discriminate between two equally likely possibilities. Since a DNA sequence contains four possible bases at each position (we will take the bases to be equally probable), it takes two bits of information to specify the base at each position. For example, 00 specifies A, 01 specifies C, 10 specifies G, and 11 specifies T. Therefore the uncertainty H, which is the amount of information needed to specify the sequence, is two bits when we have no knowledge of the base at a particular position. The general formula for H is (Shannon and Weaver, 1949)

$$H = \sum_{i=1}^{M} P_i \log_2 P_i \tag{13-9}$$

in which P_i is the intrinsic probability of symbol i, and there are a total of M symbols. Note that use of the base 2 logarithm results in $H = 2$ for four symbols each of probability $P = \frac{1}{4}$. If the natural logarithm is used in this formula (Berg and von Hippel, 1987, 1988), the units of H are no longer bits.

Now suppose that we have a set of N sequences that are known to share a common function in binding the same regulatory protein. Alignment of a large number of such sequences shows that the base at a particular position is always A. Then, if another such sequence is found, it is very likely that it will also have A at the corresponding site, and the uncertainty H is reduced: when $P_A = 1$ and all other $P_i = 0$, then Eq. 13-9 gives $H = 0$.

Schneider et al. (1986) define a parameter R_{sequence} as the information gained about the sequence at each position L by aligning the sequences on the basis of their shared property:

$$R_{\text{sequence}}(L) = H_{\text{genome}} - H_{\text{sequence}}(L) \tag{13-10}$$

where H for the genome is calculated using Eq. 13-9 and the probabilities P_i of the bases as they occur in the genome as a whole. A first approximation for H_{sequence} results from using Eq. 13-9, and setting the probability for each base equal to its fractional occurrence at position L in the set of sequences examined. Note that the maximum possible value for R_{sequence} is 2, which is the case when H_{sequence} is zero. This results when the consensus sequence is invariant. Schneider et al. (1986) describe the correction needed in the calculation of H_{sequence} because of the limited size of the set of related sequences; this correction results in values of R_{sequence} which are always less than 2.

Figure 13-13 (Schneider et al., 1986) illustrates values of R_{sequence} at various positions L in and around *Hinc*II restriction sites in bacteriophage T7 DNA. All such sites are of sequence GTCGAC or GTTAAC. At the fully conserved GT and AC sites, R_{sequence} is nearly 2 when the sample size is 61 sequences [Fig. 13-13(*a*)], but is smaller, with larger error bars, when only 17 sequences are compared [Fig. 13-13(*b*)].

Berg and von Hippel (1988) suggest an alternative correction for small sample size in calculating R_{sequence}, with

$$R_{\text{sequence}} = \sum_{i=A}^{T} \frac{n_i(L) + 1}{N + 4} \log_2 \left[4 \frac{n_i(L) + 1.5}{N + 4.5} \right] \tag{13-11}$$

where n_i is the number of times base A, C, G, or T occurs at position L in the set of N sequences. This approach is numerically simpler than the computational method described by Schneider et al. (1986) for taking account of small sample size.

In a different attack on the problem, one can examine the relative probability P_j of random sampling outcomes when the base compositions of N uncorrelated sequences are compared. For example, suppose that six sequences for a putative DNA binding site are compared and the four bases are assumed to have equal and uncorrelated probability of appearing in the set of sequences at position L. The nine possible distributions of numbers of bases of each kind that add to six are summarized in Table 13-3. The most probable distribution ($P_1 = 0.3516$) is that in which there are three bases of one kind, two of another, one of a third kind, and none of the fourth kind. The least probable distribution ($P_9 = 0.0098$) is that in which random selection produces all six bases of the same kind.

Table 13.3
Probabilities of Random Sampling Outcomes on a Set of Six Sequences[a]

Distribution	P_j	Σ	α
3,2,1,0	0.3515	0.0000	0.000
2,2,1,1	0.2637	0.3516	0.188
3,1,1,1	0.1172	0.6152	0.415
2,2,2,0	0.0879	0.8203	0.745
4,1,1,0	0.0879	0.8203	0.746
4,2,0,0	0.0439	0.9082	1.037
3,3,0,0	0.0293	0.9521	1.320
5,1,0,0	0.0176	0.9814	1.732
6,0,0,0	0.0098	0.9902	3.010

[a]The distribution refers to the number of bases of a given kind that are identical when the sum of the number of bases is $N = 6$. Probabilities are calculated from Eq. (13-12), and $\sum$ (Eq. 13-13) is the sum of P_j values for all other distributions that are at least as likely as P_j'. The parameter α is defined by Eq. (13-14). The average value of α for the distribution is $\bar{\alpha} = 0.347$, and the square root of the variance is $\sigma = 0.408$.

These numerical results are based on the equation for the probability $P_j\{n_i\}$ of a particular distribution $\{n_i\}$

$$P_j\{n_i\} = \frac{N!}{\Pi n_i!}\frac{4!}{L!}\frac{1}{\Pi q_k!}4^{-N} \tag{13-12}$$

in which P_j is the probability that random selection of N bases will produce n_1 of one kind, n_2 of another, and so on. The parameter L is the number of bases for which $n_i = 0$, and q_k is the number of non-zero values of n_i which have the same value x_k. (The symbol $\Pi n_i!$ means $n_1 \cdot n_2 \cdot n_3 \ldots$, and L factorial, $L!$, is $L \cdot (L-1) \cdot (L-2) \cdots$) For example, in the distribution 2, 2, 1, 1 for $N = 6$ (Table 13-3), $L = 0$ and $q_1 = 2$ for $x_1 = 1$, and $q_2 = 2$ for $x_2 = 2$, while for the distribution 2, 2, 2, 0, $L = 1$ and $q_1 = 3$. The term $N!/\Pi n_i!$ in Eq. 13-12 is the standard expression for the number of ways of arranging N symbols if there are n_1 of type 1, n_2 of type 2, and so on; the rest of the factorial terms correct for the different ways of combining the symbols, and 4^{-N} normalizes the distribution so that the sum of the probabilities P_j is 1. As an example, consider the distribution 3, 3, 0, 0 (Table 13-3), for which $N = 6$, $L = 2$, $n_1 = 3$, $n_2 = 3$, $n_3 = 0$, $n_4 = 0$, and $q_1 = 2$. Then

$$P\{3,\ 3,\ 0,\ 0\} = \frac{6!}{3!3!0!0!}\frac{4!}{2!}\frac{1}{2!}4^{-6} = 0.0293$$

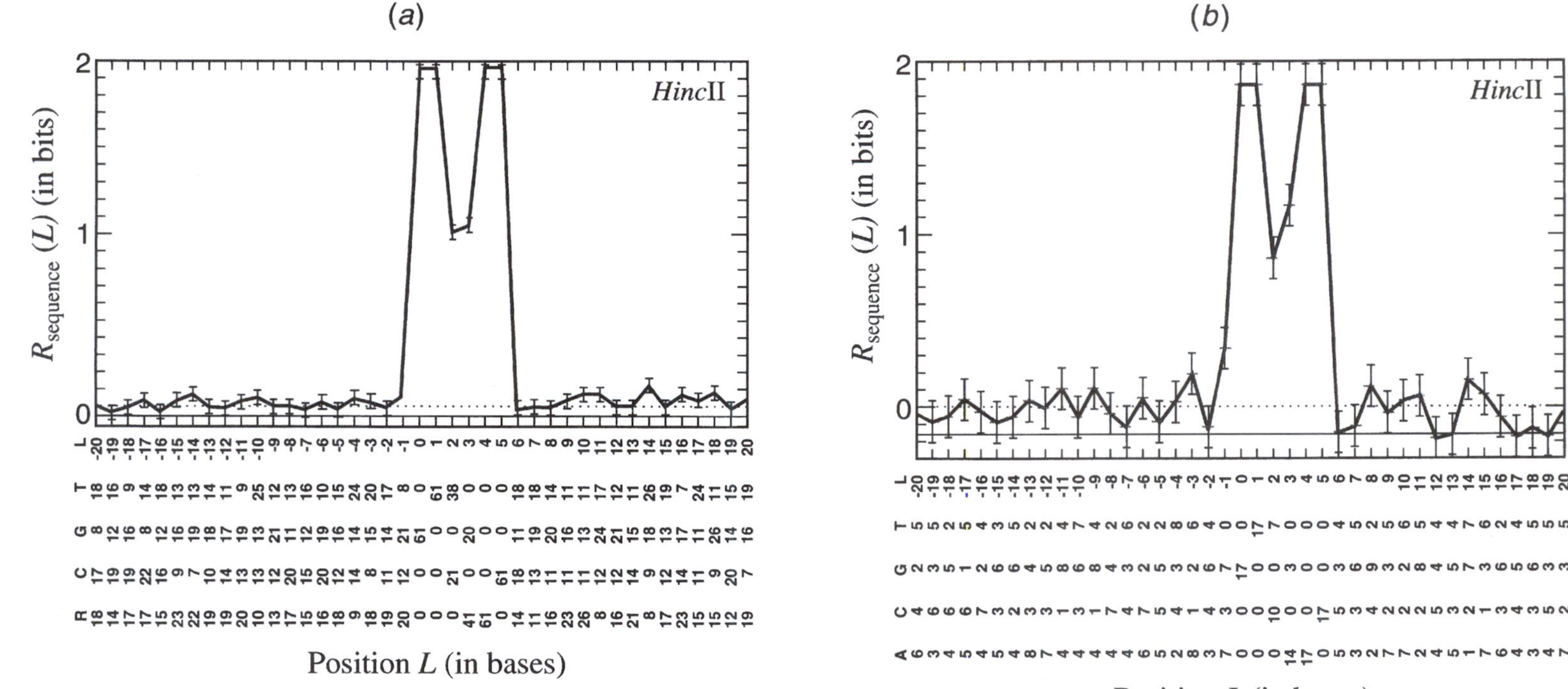

Figure 13-13
Information content, $R_{\text{sequence}}(L)$ in bits per base, at various positions (L) in and around H*inc*II sites (G-T-(T/C)-(A/G)-A-C). The numbers of bases of each kind at each position are given. The sites were obtained starting at the left end of the bacteriophage T7 DNA sequence, and only one orientation of each site was used. The left-most base in each site (G) was placed at position 0 in each case, and the sequence examined for 20 nucleotides in each direction from this base. The continuous lines are the zero, without sampling error correction. The broken lines are the zero, when the sampling error correction is made. The bars show one standard deviation above or below $R_{\text{sequence}}(L)$. Part (*a*) has 61 sequences, and part (*b*) has 17 sequences. [Reprinted from Schneider, T. D., Stormo, G. D., and Gold, L., Information Content of Bending Sites on Nucleotides Sequences, *J. Mol. Biol.* **188**, 415–431, Copyright ©1986, by permission of the publisher Academic Press Limited London.]

The value of the probability P_j is not directly useful, since even the most probable distributions of n_i values will have small P_j values when N becomes large. A useful quantity in this regard is the sum $\sum$ of the probabilities P_j' of all other distributions j' that are at least as probable as the distribution j actually observed:

$$\sum(P_j) = \sum_{P_j' \leq P_j} P_j' \tag{13-13}$$

If $\sum$ is near 1 for an observed distribution, then a large number of relatively probable distributions have been avoided in the set of observed sequences. For example, the distribution 6, 0, 0, 0 (Table 13-3) has $\sum = 0.9902$, meaning that more than 99% of the time one expects the more probable distributions than this one to be observed.

Since $\sum$ can approach very close to 1 when N is large, a logarithmic functional representation is useful; let α be defined by

$$\alpha(P_j) = -\log_{10}[1 - \sum(P_j')] \tag{13-14}$$

Then $10^{-\alpha}$ is the total probability that by chance the distribution described by P_j, or any of the even less probable distributions, would have been observed. The calculated values of α for a given N are approximately proportional to the values of R_{sequence} (Fig. 13-13). However, α is an open-ended quantity, whereas R_{sequence} has an upper limit of 2. These two approaches to the problem are complementary, with the utility of α found in its intutitve relationship to probabilities.

Figure 13-14 shows calculated α values for the λ operator sequences shown in Figure 13-12. The parameter α is 5 or greater for the distributions of bases observed for the set of sequences at positions 2, 4, and 6, which are the base pairs identified in the crystal structure as showing specific interactions with the repressor protein (Otwinowski et al., 1988). The statistical analysis in Figure 13-14 shows that the extent of conservation of bases observed at those positions would be expected once in 10^5 times or less when random trials are made. On the other hand, the distribution 6, 3, 3, 0 found for base pair 3 (Fig. 13-14) has an α value (1.03), which does not differ from average expectation (0.42) enough to be significant. Hence one cannot legitimately derive a "consensus" base for that position. By the same logic, the so-called consensus for T in the distribution 3, 2, 1, 0, which Table 13-3 shows to be the most probable distribution when $N = 6$, has $\alpha = 0$. Since any base other than T is equally likely to be in the majority in such a distribution, these results cannot be used to deduce a statistically significant consensus at that position.

3.3 Deviations from Consensus

Promoters for transcription initiation and RNA splice junctions are examples of sets of sequences that share a common function, yet have considerable variation of sequence within the set. In many cases, these variations serve a regulatory role by modulating function. The objective of quantitative analysis of this problem is to assign each member of the set a "score," which measures the degree of homology with the consensus. The

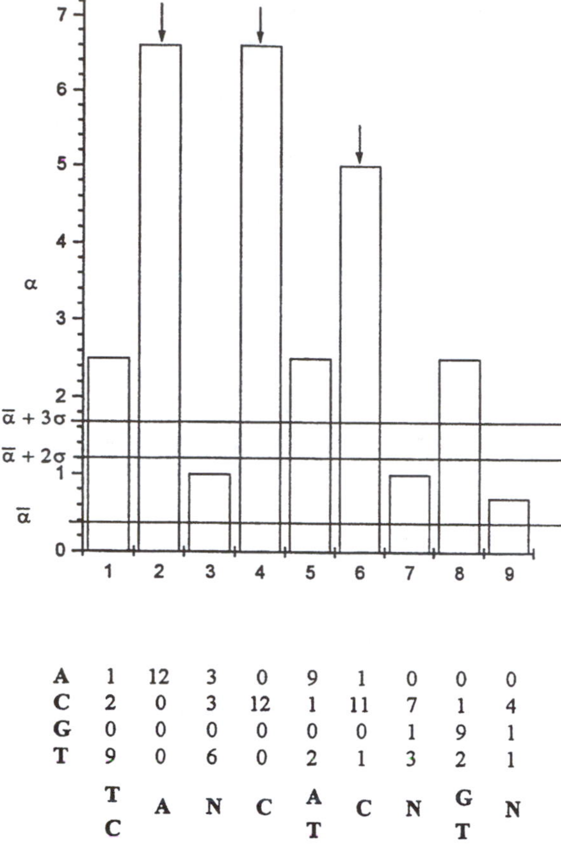

Figure 13-14
Variation of the α index for the operator sequences in Figure 13.12. The number of bases of each kind at each position is indicated, along with the consensus sequence derived. Only α values that deviate by at least two standard deviations σ from the mean (see corresponding line) are included in the consensus. The upper base indicates the preferred base if the consensus is divided. Arrows indicate the bases identified by the crystallographic structure to be in hydrogen-bonding interaction with the protein. The pseudodyad axis of the sequence runs through base 9 as numbered here and in Figure 13.12.

score can then be examined for correlation with function. An important additional use of such scores is to search DNA sequences of unknown function for homology to sequences of known function.

The variety of possible solutions of this problem is illustrated by the approaches of Staden (1984) and Mulligan et al. (1984), who provide plausible equations for calculating the homology score for a sequence. Since the treatments are ad hoc, there is no definitive reason to prefer one over the other. First, suppose that there are M positions in the sequence that contribute to the score, numbered 1, 2, ... m, $m+1, \ldots M$. Furthermore, suppose that a set of N sequences is compared in order to establish the consensus. Let the frequency of occurrence of A, T, G and C at position m in the set of N sequences be $f_{m,A}, f_{m,T}$, and so on. Then, according to Staden (1984) the homology score for any individual sequence is the sum

$$\text{Score} = \sum_{m=1}^{M} \ln f_{m,J} \tag{13-15}$$

here J refers to the base that actually occurs at position m in the sequence of interest. If $f_{m,J}$ is zero, which can happen only if the sequence of interest is not a member of the set used to establish the consensus, then $f_{m,J}$ is set equal to $1/N$. Note that all scores are negative according to Eq. 13-15, coming closer to zero as the degree of homology increases.

The score according to Mulligan et al. (1984) is based on summing the ratios of observed frequencies to the standard deviation in a random selection model. For example, suppose that base J occurs in the consensus set at position m a total of $Nf_{m,J}$ times. The random expectation value is $N/4$, with a standard deviation of $(N/4)^{1/2}$. Hence, the ratio of the observed value to the random expectation standard deviation is $Nf_{m,J}/(N/4)^{1/2} = 2N^{1/2}f_{m,J}$. On this basis, the score for the whole sequence is

$$\text{Score} = 2\sqrt{N}\sum_{m=1}^{M} f_{m,J} \tag{13-16}$$

which is a linear function of the frequencies, rather than logarithmic as in Eq. 13-15. Mulligan et al. (1984) further define a homology score for prokaryotic promoters that increments the score defined in Eq. 13-16 by another score to account for the influence of spacing between the promoter elements at −10 and −35. This result is corrected by subtraction of the score expected for a random base sequence. Figure 13-15 shows the

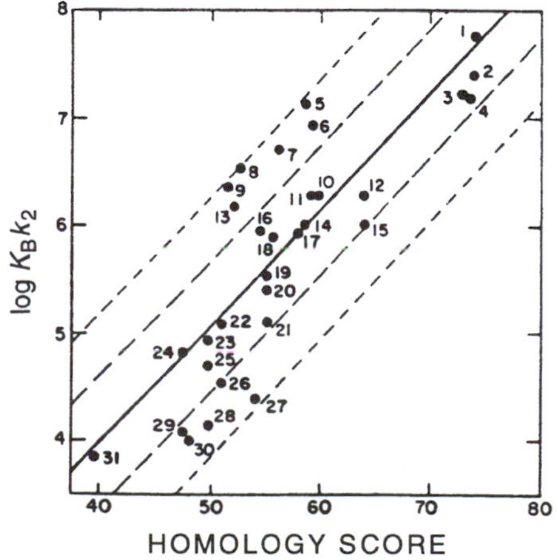

Figure 13-15
Correlation between log K_Bk_2 and the homology score calculated for a variety of *E. coli* promoters, with a correlation coefficient of 0.83. The dashed lines are drawn one (__) and two (- - -) standard deviations from the best-fit line. Twenty promoters fall within one standard deviation and eleven are between one and two standard deviations. [From Mulligan et al., 1984.]

correlation between homology score and promoter activity, the latter measured by the logarithm of the product of K_B, the equilibrium constant for binding polymerase to the promoter, times k_2, the rate constant for isomerization to the active "open" complex. For general treatments of the sequence analysis problem, see Karlin and Brendel (1992) and Karlin and Altschul (1993).

3.4 Categories of Target DNA Sequences

Table 13-4 illustrates some DNA sequences that have been observed to interact with eukaryotic regulatory proteins. Among higher organisms, the sequences and proteins used to control gene expression are reused at many promoters. Some of the sequences have evident dyad symmetry elements, indicated by opposing arrows in the table, and some do not. Some target sequences cover large numbers of base pairs, as illustrated by the binding site for the transcription factor TFIIIA in the 5S RNA gene of *Xenopus borealis*. In such cases, it is likely that some bases are more important than others for recognition by the protein. For example, the recognition site for TFIIIA is shown in groups of 11 bases; the fourth and fifth bases in these groups are G 10 out of 12 times. This arrangement has only a small probability of occurring at random, and one could conclude that it is likely that these bases are involved in the recognition process, even without knowledge of the structure of the protein–DNA complex.

4. TWISTING AND BENDING OF DNA SEQUENCES IN PROTEIN COMPLEXES

4.1 Binding Energetics for Proteins that Unwind DNA

The DNA molecule must be unwound to initiate vital cellular processes such as transcription and replication. Because of the strong stacking energy between the DNA bases, there is a considerable energetic cost to opening a denaturation bubble. Opening the first base pair requires breaking two stacking interactions; as the denatured bubble grows, only one additional stacking interaction is lost for each base pair opened. A recent estimate of the free energy required to initiate denaturation is 10.2 kcal mol^{-1} (Bauer and Benham, 1993). Strong protein–DNA interactions in enzymes such as RNA polymerase can be used to overcome the thermodynamic barrier to the initiation of unwinding.

In closed circular molecules, the binding energy of proteins that unwind DNA is strongly coupled to the degree of superhelicity. Characterization of these interactions is a powerful way to study protein-induced structural changes of DNA (reviewed by White et al., 1992). Binding a DNA unwinding protein to a superhelical molecule is enhanced by the superhelical stress, leading to an increased binding constant for the protein. The magnitude of the effect depends on the linking deficit $\Delta Lk = Lk - Lk_0$ (see Chapter 10), or the specific linking difference (superhelix density) σ. A recent

Table 13.4
Some Eukaryotic Protein-binding DNA Sequences

Protein source	Functional role	Target sequence	Reference
GR (mammalian cells)	Transcription	$GGTACAN_3TGTTCT$ $CCATGTN_3ACAAGA$	Mitchell and Tjian (1989)
Spl (mammalian cells)	Transcription	GGGCGG CCCGCC	Mitchell and Tjian (1989)
CTNF/NF-I (mammalian cells)	Transcription	GCCAAT CGGTTA	Mitchell and Tjian (1989)
c-Jur + Fos (mammalian cells)	Transcription	TGACTCA ACTGAGT	Mitchell and Tjian (1989)
AP-2 (mammalian cells)	Transcription	CCCCAGGC GGGGTCCG	Mitchell and Tjian (1989)
CREB (mammalian cells)	Transcription	TGACGTCA ACTGCAGT	Mitchell and Tjian (1989)
OCT-1 (mammalian cells)	Transcription	ATTTGCAT TAAACGTA	Mitchell and Tjian (1989)
SRF (mammalian cells)	Transcription	GATGTCCATATTAGGACATC CTACAGGTATAATCCTGTAG	Mitchell and Tjian (1989)
C/EBP (mammalian cells)	Transcription	TGTGGAAAT ACACCTTTA	Mitchell and Tjian (1989)
NF-kB (mammalian cells)	Transcription	GGGRNTYYCC CCCYNARRGG	Lenardo and Baltimore (1989)
GAL4 (yeast)	Transcription	CGGAAGACTCTCCTCCG GCCTTCTGAGAGGAGGC	Giniger and Ptashne (1988)
GCN4 (yeast)	Transcription	ATGACTCAT TACTGAGTA	Oliphant *et al.* (1989)
TFIIIA (*X. borealis*) (showing one strand only, in groups of 11)	Transcription	TCGGAAGCCAA- GCAGGGTCGGG- CCTGGTTAGTA- CCTGGATGGCA- GACCGCCTGGG	Rhodes and Klug (1986)
NF-I (animal virus)	Replication	$TTGGCN_5GCCAA$ $AACCGN_5CGGTT$	Chalberg and Kelly (1989)

estimate of the free energy $\Delta G_{\Delta Lk}$ of superhelical molecules containing N base pairs ($N \geq 2000$) at 37°C (Bauer and Benham, 1993; see also Section 10.4) is

$$\Delta G_{\Delta Lk} = \frac{740}{N}(\Delta Lk)^2 \quad \text{(in kcal mol}^{-1}\text{)} \tag{13-17}$$

By using the definition of $\sigma = \Delta Lk/Lk_0$ from Eq. 10-6 and $Lk_0 = N/h_0$, we can convert Eq. 13-17 to

$$\Delta G_{\Delta Lk} = \left(\frac{740}{h_0}\right)\left(\frac{\Delta Lk}{N/h_0}\right)^2 N = 6.7\sigma^2 N \quad \text{(in kcal mol}^{-1}\text{)} \tag{13-18}$$

where h_0 = 10.5 bp/turn has been used for the numerical conversion. Note that the total superhelix free energy scales with the number of base pairs and with the square of the superhelix density. For typical values of $\sigma = -0.05$ and $N = 3000$, the free energy from Eq. 13-18 is 50.3 kcal mol^{-1}.

There is a change in superhelix free energy upon protein binding because the protein neutralizes part of the super helical stress. Equation 13-18 gives,

$$\delta\Delta G_{\Delta Lk} = 6.7\left(\sigma_f^2 - \sigma_i^2\right)N \tag{13-19}$$

where the subscripts f and i denote the values of σ after and before protein binding, respectively. When the change in σ, $\delta\sigma = \sigma_f - \sigma_i$, is small compared to σ, Eq. 13-19 can be approximated by

$$\delta\Delta G_{\Delta Lk} = -141\sigma_i\delta Lk \quad \text{(in kcal mol}^{-1}\text{)} \tag{13-20}$$

where δLk is the change in local winding (either twist or writhe) induced by the protein. For a protein that unwinds DNA by one turn ($\delta Lk = -1$) binding to a typical superhelical plasmid DNA with $\sigma = -0.05$, the change in superhelix free energy given by Eq. 13-20 is -7 kcal mol^{-1}. At 37°C, this contributes a factor of 1.2×10^5 to the binding constant. Hence, there is a strong coupling between a protein's binding affinity and the presence of torsional stress in its target DNA. Methods for measuring the change in winding upon ligand binding are discussed in Chapter 10.

4.2 Bent DNA in the Nucleosome: Energetics and Variation in Bendability

Double helical DNA is a stiff polymer chain, which must nonetheless be packaged into a much smaller volume than would be occupied by chromosome-sized pieces of DNA in solution. The first step in this condensation process is bending of DNA around the core histone octamer to form a nucleosome (see Chapter 14; for reviews see Travers and Klug, 1987; Travers, 1989). The energy cost of this process is substantial. By using the wormlike chain model, with persistence length a (Chapter 9), the bending energy per mole for a DNA molecule with uniform curvature (see Eq. 9-89) is

$$\Delta G_{\text{bend}}^{(M)} = \frac{N_{\text{A}}B}{2L}(\Delta\theta)^2 = \frac{aRT}{2L}(\Delta\theta)^2 \tag{13-21}$$

where $\Delta\theta$ is the total curvature in a DNA segment of contour length L. For example, a typical nucleosome contains 146 bp of DNA and about 1.75 superhelical turns, so $\Delta\theta = 1.75 \times 2\pi = 11.0$ rad. With $a = 150$ bp and $RT = 0.6$ kcal mol^{-1}, we find from Eq. 13-21 that $\Delta G_{\text{bend}}^{(M)} = 37$ kcal mol^{-1}. This unfavorable contribution to the energy of nucleosome formation must be compensated by favorable interactions, probably largely electrostatic, between DNA and the positively charged core histone proteins.

Since DNA is a helical polymer, models in which the curvature is uniform are unrealistic. It is predicted on the basis of conformational energy calculations (Zhurkin et al., 1979) that the energy cost of bending is less when it results from roll between the base pairs (Chapter 4) as compared to tilt. Hence, one expects the curvature to be greatest near the positions where the major and minor grooves of DNA face inward toward the core histones. This prediction is supported by structural studies of nucleosomal core particles (Richmond et al., 1984).

Because of its variable base sequence, DNA is not a uniform helical polymer. Hence, the energy of packaging can be reduced if relatively bendable sequences are placed at sites of maximum curvature, as illustrated in Figure 13-16 (Trifonov and Sussman, 1980). This figure shows that in order to produce the nearly planar curve characteristic of DNA wound in a nucleosome, the bendable sites (which, in the model illustrated in Fig. 13-16, correspond to positions of tilt between the base pairs) must be placed at a spacing very close to the intrinsic helical repeat of linear DNA; a small phase deviation between the two periodicities yields a curve with left- or right-handed superhelical writhe (wr), depending on the sign of the phase difference. For example, the left-handed writhe characteristic of DNA in a nucleosome results when bends are incorporated every 10.17 bp in a DNA molecule in which the average twist (tw) between the base pairs (in the external reference frame) corresponds to a DNA helical repeat of 10.31 bp (Travers and Klug, 1987). If the bends were incorporated at some spacing

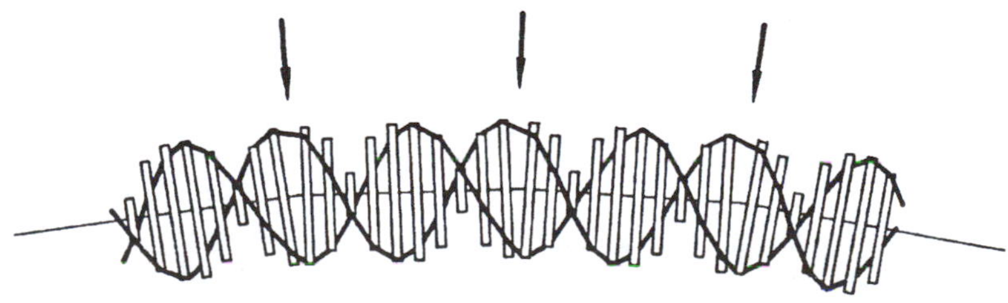

Figure 13-16
Schematic illustration of the unidirectional bending of a DNA molecule by the regular insertion of a nonparallel set of adjacent base pairs (arrows). [From Trifonov and Sussman, 1980.]

larger than 10.31 bp, a right-handed DNA superhelix would result. (See Section 13-4.3 for further discussion of this topic.)

If the packaging energy of DNA is significantly sequence-dependent, it should be possible to detect bendable sites by analysis of sequence periodicities in DNA molecules that are selected for strong nucleosome incorporation affinity. This result was found by Satchwell et al. (1986), who determined the sequence of 177 different DNA molecules isolated and cloned from chicken erythrocyte nucleosomal core particles. They analyzed their results primarily in terms of the occurrence of dinucleotides, with the objective of determining whether particular dinucleotides have a preference for bending toward the major or minor DNA grooves. Since nucleosomes have an approximate twofold symmetry axis, it is not possible to distinguish a given dinucleotide from its complement, which would appear instead in the sequence if the DNA were rotated 180° about the dyad axis.

Figure 13-17 shows the periodic variation of the occurrence of AA or TT dinucleotides in the set of core particle DNA sequences of length from 142 to 149 bp. The sequences are aligned so that the DNA minor groove faces the core histones at positions $6 + 10.2n$ (n is an integer). These positions correspond closely to the periodic maxima in the distribution of AA + TT dinculeotides, with the exception of the anomalous

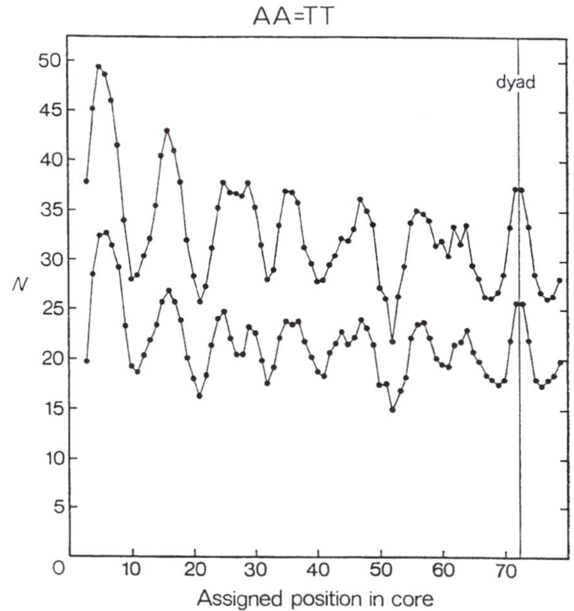

Figure 13-17
Variation in the occurrence of AA + TT dinucleotides (N) versus position in the sequence. Data are presented as a running three-bond average of occurrence averaged about base step 72.5 (the dyad axis). The upper curve corresponds to the whole set of 177 sequences, and the lower corresponds to a selected set of 117 sequences of length 144–146 bp.

region around the dyad axis (position 72.5), where there is a maximum in the AA/TT distribution function, but the DNA major groove faces inward toward the protein core. Accordingly, there is a preference for these dinucleotides to appear where DNA bends toward the minor groove, except for the nucleosome dyad axis. Special features of the structure may account for the anomaly at the dyad axis, but one should keep in mind that an important unknown factor in these experiments is the possible role of sequence preferences in protecting the core particle from micrococcal nuclease digestion in the original preparation. Similar plots for other dinucleotides show, for example, that GC is preferred at positions $1 + 10.2n$, where the major groove faces the core histones, again with the exception of the region around the dyad axis.

Fourier analysis of the periodic dinucleotide distribution functions yields corresponding amplitudes, periods, and phase angles ϕ . These findings are summarized in Table 13-5 (Satchwell et al., 1986), for the region of the sequence that excludes 23 bp around the dyad axis where the behavior appears to be anomalous. The phase angle is set to zero at position 1, so that dinucleotides with phase angles near zero are preferred when the major groove faces the core histones, and dinucleotides with

Table 13.5
Fourier Analysis of Variations in the Occurrence of Dinucleotides in Nucleosomal DNA[a]

Dinucleotide (bend toward major groove)	Fractional Variation in Occurrence	Phase Angle (deg)	Period (bp)
GC	0.27	25	10.15
CG	0.15	−51	10.42
GG/CC	0.12	25	10.15
TG/CA	0.08	−6	10.26
GT/AC	0.07	−69	10.47
AG/AC	0.05	−57	10.36
GA/TC		(no significant signal)	
AT	0.06	147	10.26
TA	0.13	174	10.26
AA/TT	0.20	172	10.26
(bend toward minor groove)			

[a]The frational occurrence is defined as the normalized amplitude per period divided by 5.0, which is the sum of $\cos^2\theta$ over 1 period of 10 steps, where $\theta = 0°, 36°, 72°$, and so on. The phase origin is set at step 1.0. Thus a phase angle of 0° at this portion, for a period of 10.21 bp, yields a phase angle of 0° also at position 72.5, which is coincident with an axis of ten-fold or dyad symmetry in the protein–DNA complex (Richmond et al., 1984). the dinucleotides are listed in order of tendency to bend in the direction of the major groove at the top of the list, to tendency to bend toward the minor groove at the bottom.

phase angles near 180° are preferred when the minor groove faces inward. The dinucleotides for which the largest periodic amplitude is seen are the A•T-rich sequences AA/TT $\phi = 172°$) and TA ($\phi = 174°$) and the G•C-rich sequences GC ($\phi = 25°$), CG ($\phi = -51°$), and GG/CC ($\phi = 25°$). From these observations, there emerges the generalization that DNA sequences of mixed-base composition bend most easily when the A•T-rich sequences are placed where the bend is toward the minor groove, and the G•C rich sequences are located where the major groove faces inward.

This general finding was confirmed and the underlying energetics placed on a more quantitative basis by the experiments of Shrader and Crothers (1989), who made synthetic DNA sequences based on the principle of *anisotropic flexibility*, according to which DNA prefers to bend toward the minor groove at A•T rich sequences, and toward the major groove at G•C rich sequences. For example, the 20 bp sequence (called GT hereafter) *n* times repeated:

$$(\mathrm{GT})_n = (\mathrm{T\underline{CGG}TG\underline{\underline{TTA}}GA\underline{GCC}TG\underline{\underline{TAA}}C})_n$$

contains G•C rich (underlined) and A•T rich (double underline) segments alternating every 5 bp. Multimers of this sequence with $n = 5$ or 6 incorporate into nucleosomes *in vitro* with greater affinity than any then-known natural nucleosome positioning sequence, such as those from amphibian 5S RNA genes. Figure 13-18 (Shrader and Crothers, 1989) shows relative free energies of nucleosome formation, deduced from competitive reconstitution experiments. The natural 5S RNA gene sequences are about 1.5 kcal mol^{-1} more favorable than bulk nucleosomal DNA in nucleosome formation, but five repeats of the anisotropically flexible GT sequence above are more favorable by another 1.5 kcal mol^{-1}. Relative exclusion of DNA from nucleosomes might be expected if the rules for anisotropic flexibility are deliberately broken. Wang and Griffith (1996) found that sequences of the form $[(\mathrm{G/C})_3\mathrm{NN}]_n$, in which there is no sytematic phasing of sequences capable of preferred bending toward the minor groove, are indeed weak in their bindng to core histones.

However, the energy advantage of these anisotropically flexible sequences is not large when expressed as the free energy per bend. A nucleosomal core particle contains 14 helical turns of DNA, and therefore about 28 loci, where the DNA flexes toward the major or minor groove as it winds around the core histones. Hence, the approximate 3 kcal mol^{-1} per nucleosome preferential binding energy of the most tightly binding sequences amounts to about 100 cal mol^{-1} per DNA bend. Recalling that the average thermal energy *RT* is about 600 cal mol^{-1} at physiological temperature, we conclude that there seems to be remarkably little variation of DNA packaging energy with base sequence.

When the $(\mathrm{GT})_n$ sequence is studied *in vivo*, it is found to be positioned in the linker region between nucleosomes (Tanaka et al., 1992). This unanticipated result may reflect curvature in the linker DNA, or some other unknown feature of the chromatin assembly process.

Optimal *in vitro* binding of anisotropically flexible DNA should occur when the sites of preferred bending are phased so that they just match the helical repeat of DNA in the core particle. Figure 13-19 (Shrader and Crothers, 1990) shows how the affinity of such sequences for core histones varies when the sequence phasing of the

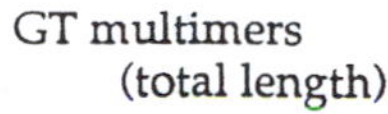

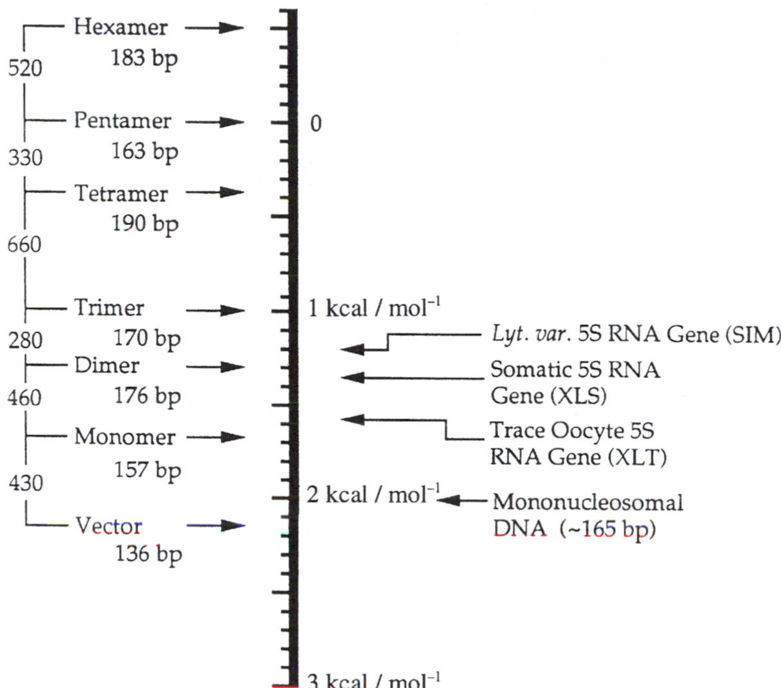

Figure 13-18
Nucleosome binding as a function of the length of anisotropically flexible DNA. The free energies of reconstitution of three natural nucleosome positioning sequences are compared with designed sequences containing various lengths of flexible DNA. Sequences that reconstitute most favorably are at the top of the diagram. The binding of the best natural sequences can be mimicked with approximately 40 bp of repetitive DNA. Additionally, the binding free energy increment for each additional flexible region is quite constant. The total length of each multimer (vector DNA plus the indicated number of 20 bp, GT oligonucleotides) is given below the multimer name. The difference in binding free energy between fragments is given at the extreme left. These free energies are all relative to the GT pentamer. [From Shrader and Crothers, 1989.]

anisotropically flexible sites is altered. The free energy minimum occurs at about 10.1-10.2 bp/turn, which is in agreement with other estimates of the helical repeat of DNA in the nucleosome.

4.3 Topological Consequences of Periodically Repeated DNA Bends

Regularly repeated small bends of the DNA double helix cause a marked curvature of the molecule when the helical repeat h of the bend is nearly equal to the intrinsic DNA helical repeat h'. We define the latter quantity as the number of base pairs required for one full rotation ($\Delta\text{Tw} = 1$) of the vector **ac** (see Fig. 10-6), directed from the helix

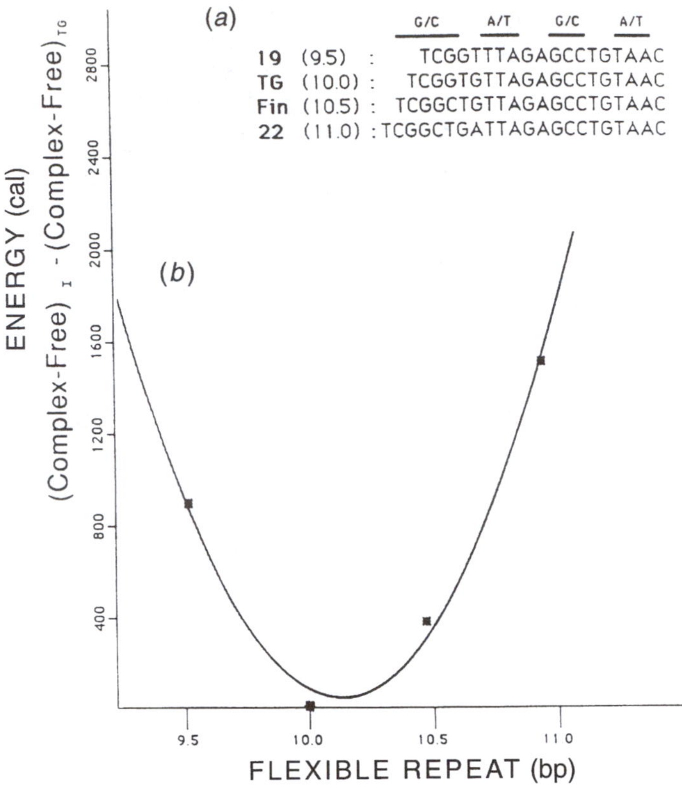

Figure 13-19

Optimal phasing of flexible regions. (*a*) Oligonucleotides used to determine the optimal phasing of flexible (A/T or G/C rich) regions for nucleosome reconstitution. Only the top strand of each molecule is shown. All oligonucleotides were double stranded with 4 bp asymmetric Ava 1 overhangs. (*b*) Free energy of nucleosome reconstitution plotted as a function of flexible repeat. Flexible repeat is defined as one half the length of the oligonucleotide monomer that makes up the central 95–110 bp of each fragment. The curve represents a least-squares fit of a parabola to the data. Free energies are relative to the free energy for nucleosome reconstitution of a fragment containing five copies of the TG oligonucleotide. The minimum of the curve at about 10.1–10.2 bp sequence repeat should coincide with the preferred helical repeat of binding sites on DNA in nucleosomes. [Reprinted from Shrader, T. E. and Crothers, D. M., Effects of DNA Sequences and Histone–Histone Interaction on Nucleosome Placement, *J. Mol. Biol.* **216**, 69–84, Copyright ©1990, by permission of the publisher Academic Press Limited, London.]

axis to the DNA backbone, in the plane perpendicular to the helix axis (White et al., 1988) The repeat h' is sometimes called the helical repeat in the "laboratory frame," because it takes no account of the shape of the surface on which the DNA is wound (see the discussion by Klug and Travers, 1989). The quantity h, called the helical repeat in the "local frame," is defined as the average spacing in base pairs between passages of the vector **ac** in Figure 10.7 past the vector **v**, which is normal to the surface. In a superhelix, h is the number of base pairs between equivalent positions in the superhelix. It can be measured, for example, by determination of the number of base pairs between successive maxima or minima in hydroxyl radical cleavage of DNA in the superhelix. When the DNA axis follows a writhing path, the vector **v** rotates in the **ac** plane, with the consequence that the number of passages of the rotating vector past the surface normal is not equal to the number of rotations about the helix axis. Hence, in general h' need not equal h. If the two repeats are exactly equal everywhere, the DNA helix axis lies in a plane [Fig. 13-20(a)]. However, if there is an appreciable systematic mismatch in phase, then the resulting curve is a superhelix. As illustrated in Figure 13-20, it is a consequence of the right-handed geometry of the DNA helix that repetition of DNA bends at a period h less than the intrinsic helical repeat of the linear form results in a left-handed solenoidal superhelix [Fig. 13.20(c)]. When the repetition period is greater than the intrinsic helical repeat, a right handed solenoid results [Fig. 13-20(b)].

Nucleosomal DNA provides a good example of this principle. The optimum phasing of bends in nucleosomal DNA, $h \approx 10.17$ bp/turn according to Figure 13-19 and other data (Satchwell et al., 1986), is smaller than the intrinsic helical repeat h', defined by the number of base pairs required for $\Delta\text{Tw} = 1$ in B-DNA. Viewed from the surface of the nucleosome, DNA has the same bend conformation every 10.17 bp, so the helical repeat of this structure is 10.17 bp. Since this is smaller than the intrinsic helical repeat h', the DNA follows a left handed superhelical path. It is important to realize that the helical repeat h of DNA depends on the surface on which it is wound, even if the twist is unchanged. Since the helical repeat h measures the distance in base pairs between equivalent sites on the surface, it is relatively easy to determine experimentally,

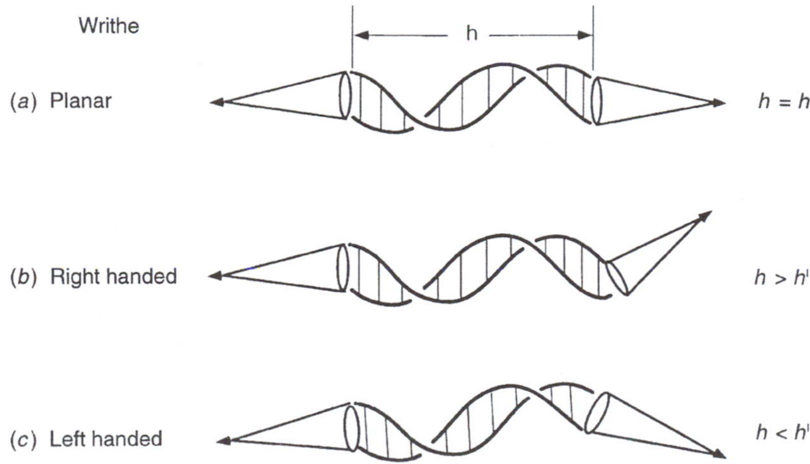

Figure 13-20
Dependence of the hand of superhelical writhe on the difference between the intrinsic helical repeat h' of the planar DNA molecule and the helical repeat h in the superhelix. When the two helical repeats match (a) the curve is planar, with components directed away from the observer. (b) When the helical repeat h' is less than h, the bend is introduced before completion of a full helical turn of the helix, and therefore it has a component directed above the plane, yielding a right-handed writhe. Similarly, (c) when h' is greater than h, the out-of-plane component is directed downward, and the writhe is left handed.

for example, by measuring the modulation of chemical or enzymatic cleavage rates. However, in the absence of a detailed molecular structure, the intrinsic helical repeat h' can only be deduced indirectly from the helical repeat h and the estimated pitch of the superhelix. We use Eq. 10.2, which relates the twist Tw, the winding number Φ relative to the surface, and the surface twist STw,

$$\mathrm{Tw} = \Phi + \mathrm{STw} \tag{13-22}$$

Consider the changes in these variables when a linear DNA is wrapped into one turn of a left-handed solenoidal superhelix of pitch $2\pi p$ (p is negative for a left handed solenoid) and radius r. Our objective is to calculate the intrinsic helical repeat h' that is implied by measured values of h, p, and r, assuming that the introduction of repeated bends is not accompanied by changes in twist.

Substituting $\Phi = N/h$, where N is the number of base pairs in one turn of the superhelix, and replacing Tw by N/h', we obtain from Eq. 13-22

$$\frac{1}{h'} - \frac{1}{h} = \frac{\mathrm{STw}}{N} \tag{13-23}$$

The surface twist of one turn of a solenoidal superhelix (White and Bauer, 1989;) is Bauer and White, 1990

$$\mathrm{STw} = \sin\alpha \tag{13-24}$$

where α is the superhelical pitch angle, with $\sin\alpha = p/[p^2 + r^2]^{1/2}$. Hence,

$$\frac{1}{h'} - \frac{1}{h} = \frac{\sin\alpha}{N} \tag{13-25}$$

Inserting a pitch of $2\pi p = -2.8$ nm, a nucleosome radius r of 4.3 nm, $N = 80$ bp, and a consensus estimate of $h = 10.17$ bp/turn yields $\sin\alpha = -0.103$ and $h' = 10.31$ bp/turn. This value for the intrinsic helical repeat of DNA on the nucleosome is significantly smaller than the average helical repeat of linear DNA in free solution, for which h_o is about 10.5–10.6 bp/turn. Hence, these simple geometrical considerations lead to the conclusion that the twist of free DNA must be significantly increased upon incorporation into nucleosomes: one must first wind up free DNA from an average helical repeat $h_o = 10.5$–10.6 bp/turn to $h' = 10.31$ bp/turn in order to bend it into a superhelix with the observed helical repeat and dimensions of DNA in the nucleosome. Bending and writhing to form the superhelix, without further change in twist, reduces the helical repeat from 10.31 to 10.17 bp/turn.

The major assumption underlying this conclusion is that DNA follows a reasonably regular superhelical path on the core particle (see White et al., 1989). The validity of this assumption is not fully settled, since the results of Zivanovitch et al. (1988), who studied the topological consequences of nucleosome formation on small DNA circles, lead to the conclusion that there is no net twist change associated with wrapping DNA into nucleosomal core particles.

We now summarize the estimated values of the topological changes that accompany bending of DNA around the core histones to form a nucleosome (see the discussion by White and Bauer, 1989). Values given for the linking number changes ΔLk and ΔSLk, which can refer only to closed curves, can be thought of as the average increment in linking provided in a closed circular DNA (ccDNA) by a single nucleosome, which is assumed to contain 1.8 left handed solenoidal turns in 146 bp of DNA. The experimentally accessible quantities are $\Delta\Phi$ and STw:

$$\Delta\Phi = \frac{146}{10.17} - \frac{146}{10.55} = 0.5 \tag{13.26a}$$

where 10.17 and 10.55 are, respectively, the estimated DNA helical repeat values for the nucleosome and for free DNA; Φ and the other topological quantities are measured in number of turns. Also,

$$\text{STw} = 1.8 \sin\alpha = -0.2 \tag{13.26b}$$

using the parameters of the superhelix described earlier. The total twist change can be derived from these parameters:

$$\text{Tw} = \Delta\phi + \text{STw} = 0.3 \tag{13.26c}$$

The writhe induced is, from Eq. 10-8 (with $\gamma = -\alpha$)

$$\text{Wr} = -1.8(1 + \sin\alpha) = -1.6 \tag{13.26d}$$

Hence, the increment in surface linking number is

$$\Delta\text{SLk} = \text{STw} + \text{Wr} = -1.8 \tag{13.26e}$$

as expected for 1.8 turns of a solenoidal superhelix, and the net linking number increment per nucleosome is

$$\Delta\text{Lk} = \text{Tw} + \text{Wr} = +\text{SLk} = -1.3 \tag{13.26f}$$

The uncertainty in the derived quantities is about ± 0.2, stemming from experimental uncertainty in the helical repeat values needed to calculate $\Delta\Phi$.

The overall consequence of this calculation is that the estimated decrease in Lk is only a little greater than 1/nucleosome, even though DNA makes nearly two superhelical turns around the core particle. The overwinding of DNA implied by a positive value of ΔTw partially compensates the negative superhelical turns. Experimentally determined values for ΔLk are in reasonable agreement with the calculation (Stein, 1980).

4.4 Bent DNA in Regulatory Protein Complexes

A number of DNA binding proteins have been found to bend DNA (see Section 13.9 for a discussion of experimental methods for detecting this phenomenon). The prototypical example of DNA bending by a gene regulatory protein is the *E. coli* catabolite activator protein (CAP, also called the cAMP receptor protein, or CRP), which causes a DNA bend at the site of binding (Wu and Crothers, 1984). Crystallographic analysis of the structure of the complex with DNA reveals two sharp kinks at TG dinucleotide steps, each one-half of a helical turn from the axis of approximate dyad symmetry of the binding site (Steitz, 1990; Schultz et al., 1991; see Section 13.6) where the DNA minor groove faces the protein. The kinks one-half of a helical turn away result in bends toward the major grooves, which face inward at those sites.

In addition to the primary DNA bends resulting from kinks, there is substantial evidence favoring additional DNA bending over an expanded region of the DNA. Mapping of the binding site (Section 13.9) indicates that it spans at least 30 bp. As aligned by de Crombrugghe et al. (1984), the consensus sequence (one strand only) for a variety of CAP binding sites over a 30 bp domain is

```
 M        m        M        m        M       m       M
−15      −10      −5        ↓        5      10      15
        AA                         CAC       T
NNNN        NTGTGANNNNNNNN             AN      NNNNN
        TT                         TGA       A
```

where M and m, respectively, show approximate positions where the major and minor DNA grooves face the protein, the arrow marks the dyad axis, and the upper base is the more common one when the consensus is divided. When there is no statistically significant consensus the symbol N is used, meaning that any nucleotide appears to be allowed. In addition to conservation of the TGTGA motif (or its complement, TCACA), the binding sites show a strong tendency toward A • T rich sequences one helical turn on each side of the dyad axis. Since the minor groove faces inward there, these sequences may be preferred because they enhance DNA bending at those loci.

For reasons discussed in Section 13.9, the electrophoretic mobility of protein–DNA complexes is strongly affected by the extent of DNA bending. Consequently, it was possible for Gartenberg and Crothers (1988) to examine the effect of DNA sequence changes on the bending induced by CAP protein. Since there are base-specific contacts within the TGTGA motif, attention was focused on the flanking regions, from 8 to 22 bp downstream of the dyad axis in the *lac* promoter. Mutations were localized, so that the influence of changing dinucleotide elements in the sequence could be assessed. Figure 13-21 (Crothers et al., 1991) summarizes the results, showing the particular dinucleotide sequence at each position that was found to give minimal bending [curve in Fig. 13-21(*b*)] or maximal bending [curve in Fig. 13-21(*a*)]. Note that the lower curve has a strong minimum (corresponding to high electrophoretic mobility) at the dinucleotide position that is 10 steps removed from the dyad axis, where the minor groove faces the protein. The dinucleotide GC produces the least bending when positioned there, whereas AT produces maximal bending (giving a local maximum). At the locus centered 16 nucleotides (1.5 helical turns) from the dyad axis

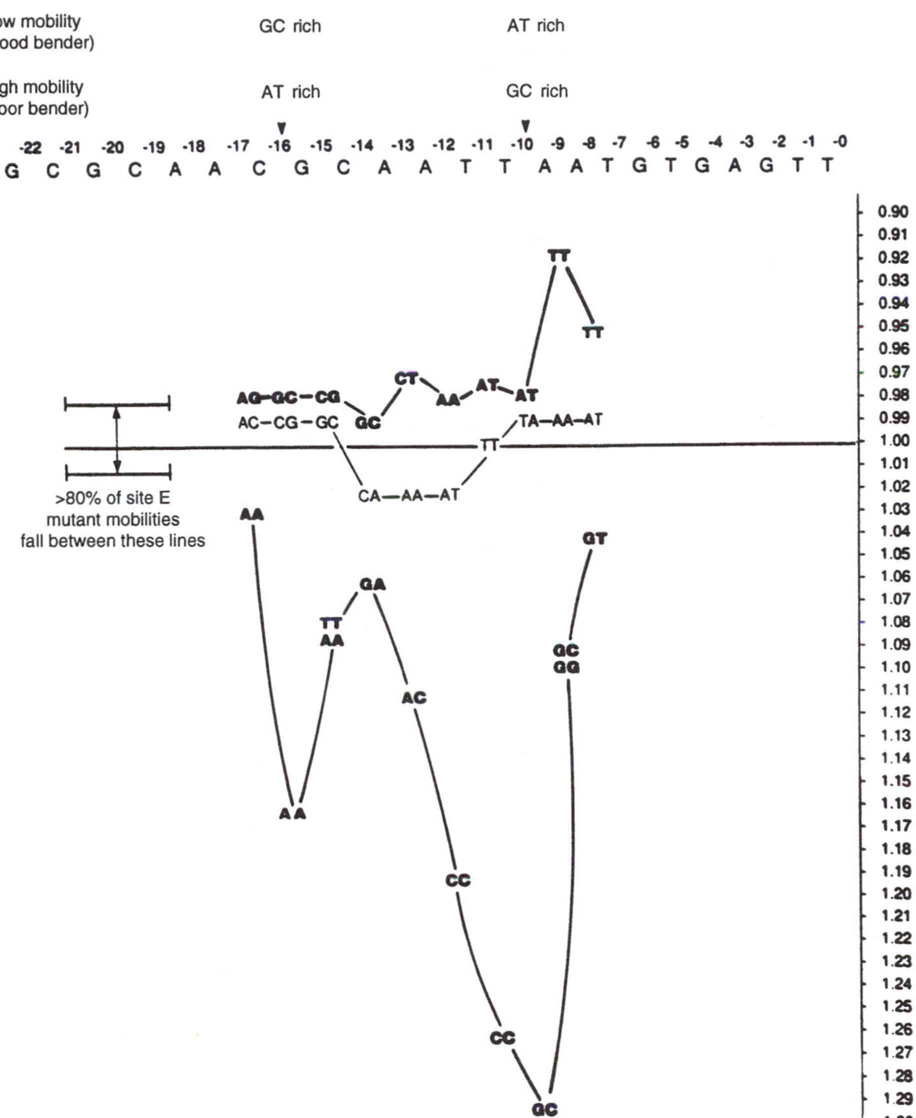

Figure 13-21
Sequence dependence of the mobility of CAP-DNA complexes. The dinucleotide shown is found to confer the largest (*b*) and smallest (*a*) mobility on the CAP–DNA complex. The high-mobility sequences show extremes at positions −10 and again at −16, sites, where the DNA minor and major grooves, respectively, face the protein. This observation implies maximum sensitivity of bending to sequence at the positions where the DNA bends by roll alternately into the minor and major grooves. Note that at positions −10 and −11, the worst benders (greatest mobility) are the G/C rich dinucleotides GC and CC, whereas the best benders are AT and TA. Near the −16 locus AA is the worst bender and GC is the best. The wild-type sequence is shown near the middle line in the figure. [Reprinted with permission from Crothers et al., 1991.]

there is a second minimum, but now the worst bender is AA and the best is GC. These results are in general agreement with the anisotropic flexibility model deduced from the bendability of DNA on nucleosomes (Satchwell et al., 1986).

The transition of MerR from repressor to activator of the *mer*R gene provides an informative illustration of the role of DNA bending distortion in modulating gene activity (Ansari et al., 1995). In its repressor form, MerR binds between the -10 and -35 promoter elements and bends the DNA towards itself. Upon binding $Hg^{(II)}$ ion, DNA bending by MerR is relaxed and promoter unwinding between -10 and -35 improves polymerase contacts at those sites. The combined action of unbending and untwisting activates the promoter.

4.5 Thermodynamics of DNA Bending by Proteins

Because of the energy required to bend DNA, the binding affinity of regulatory proteins that bend DNA is coupled to the overall structure of the target DNA. To study this effect experimentally, DNA binding sites can be incorporated into small DNA circles, which also contain A tracts that bend DNA in a preferential direction. The presence of A tracts fixes the rotational setting of DNA in the circle in order to minimize its bending energy. Consequently, protein binding sites can be inserted so that, depending on their phasing relative to the A tracts, their binding sites are bent by the circularization constraint in a direction favorable or unfavorable for protein binding (Kahn and Crothers, 1992). For example, incorporating the binding site for CAP protein at optimal phasing in small A tract-containing DNA circles results in an increase by about two orders of magnitude over the affinity for linear DNA (Kahn and Crothers, 1992). An effect of similar magnitude was seen for the TATA binding protein upon prebending of the TATA element (Parvin et al., 1995).

One of the interesting questions concerning the forces that cause DNA bending is the possible role of DNA charge neutralization on one side of the DNA helix in causing DNA to bend. Mirzabekov and Rich (1979) proposed that the positively charged lysine and arginine side chains that contact one face of the DNA bent around nucleosome cores help DNA to bend by reducing the phosphate–phosphate repulsion on one side of the double helix. This conjecture, supported by theoretical work by Manning et al. (1989) has also received experimental support. Strauss and Maher (1994) (see also the commentary by Crothers, 1994) incorporated "neutral patches," consisting of methylphosphonate residues as replacements for internucleotidic phosphate, at intervals separated by an integral number of helical turns along the double helix. A neutral patch consisted of three methylphosphonate residues on each strand, directly opposite each other across the minor groove. By altering the phasing of the neutral patches relatively to similarly repeated A tracts, they could judge the direction and magnitude of bending. As predicted, the bend is toward the neutral patch, narrowing the minor groove. In an electrophoresis medium containing multivalent ions, the bend was approximately 10° per neutral patch. Thus, by analogy, the forces that induce DNA bending include the reduction of electrostatic repulsion between phosphate residues that are neutralized by close apposition of positively charged residues from protein side chains.

5. LOOPING OF DNA SEQUENCES

Early studies of gene activation in eukaryotes revealed the existence of essential DNA sequences far removed from the site of transcription initiation (Benoist and Chambon, 1981; Banerji et al., 1981; Moreau et al., 1981). In the intervening years, it has become apparent that DNA can be looped or bent to bring distant sites into apposition as a result of interaction between the proteins bound to two or more sites (reviewed by Adhya, 1989). Figure 13-22, showing proposed models for this process for DNA recombination (Johnson and Glasgow, 1987) and bacterial gene activation (Su et al., 1990), illustrates the generality and potential complexity of the phenomenon. Recent work shows that light microscopy can be used to detect loop formation in individual DNA molecules (Finzi and Gelles, 1995). It seems likely that many DNA regulatory and recombinational (Kim and Landy, 1992) complexes will involve well-defined tertiary folding of DNA into a topology that allows multiple protein–protein contacts in the complex. This prospect gives impetus to study of the general problem of DNA flexibility and its dependence on sequence.

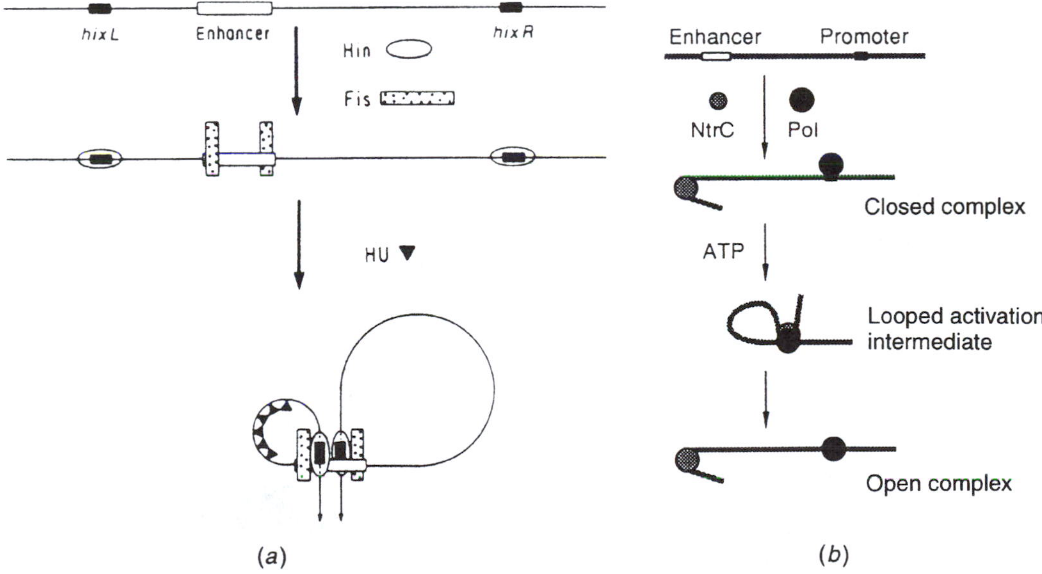

Figure 13-22
Illustration for models of folding DNA into loops to form active complexes. (*a*) Interaction of the Hin recombinase with the *hix*L and *hix*R DNA sites, and Fis protein with the enhancer site, followed by formation of a looped recombination complex, assisted by the HU protein. [Reprinted with permission from Johnson, R. C. and Glasgow, A. C. (1987). *Nature (London)* **329**, 462–465. Copyright ©1987. Macmillan Magazines Limited.] (*b*) Interaction of the prokaryotic enhancer-binding protein NtrC with RNA polymerase, mediated by ATP and DNA loop formation, resulting in formation of open complex, which is competent for RNA transcription (Su et al., 1990).

5.1 Phase Dependence of the Looping Interaction

Early evidence (Dunn et al., 1984) for the functional importance of DNA looping came from *in vivo* studies of the dependence of repression at the arabinose BAD promoter on the spacing between two sites for binding the AraC protein. Figure 13-23(*a*) shows the structure of the promoter region; for purposes of the present discussion, one only needs to know that the gene is active when an AraC dimer is bound to the *AraI* site, and inactive when the dimer bridges between the *araO*$_2$ site and the *araI* site (Lobell and Schleif, 1990), as shown in Figure 13-23(*b*). Addition of arabinose shifts this

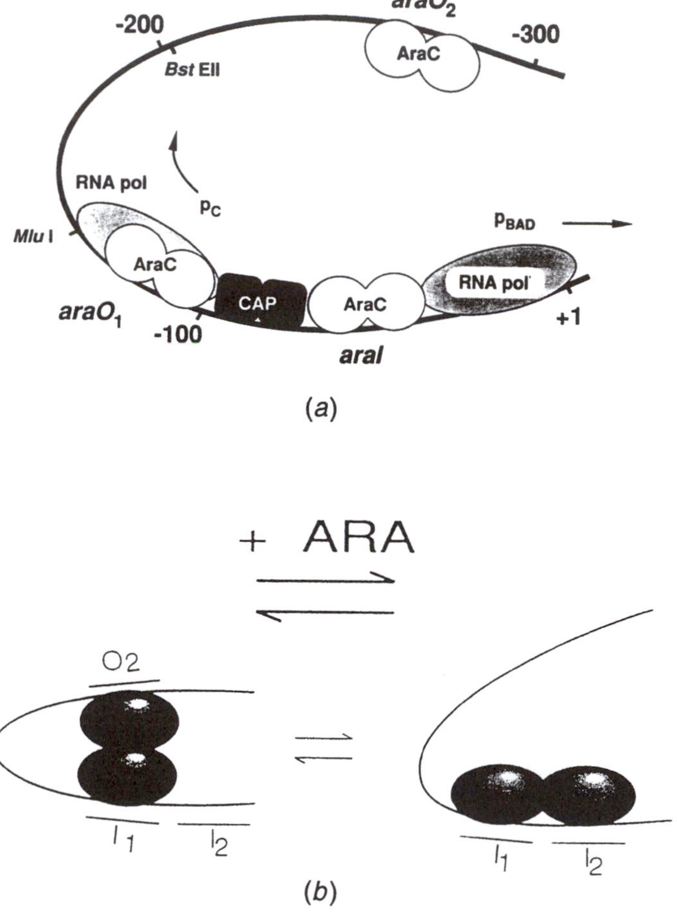

Figure 13-23
(*a*) Protein binding sites in the L-arabinose *araCBAD* regulatory region, showing the binding of RNA polymerase, araC, and CAP proteins to their regulatory sites (Lee and Schleif, 1989). (*b*) Active (right) and inactive (left) conformations of the regulatory region in (*a*), showing loop formation between araO$_2$ and half of *ara*I in the inactive (repressed) complex. Addition of arabinose switches the conformation to the active form, in which araC binds as a dimer to both half sites of *ara*I. [Reprinted with permission from Lobell, R. B. and Schleif, R. F. (1990). *Science* **250**, 528–532. Copyright ©American Association for the Advancement of Science.]

equilibrium to the active form, resulting in transcription of the *B*, *A*, and *D* genes needed for utilization of arabinose. As shown by Dunn et al. (1984), the DNA spacing between $AraO_2$ and *AraI* must be close to an integral multiple of the DNA helical repeat in order for repression to occur. The basis for this finding is illustrated in Figure 13-24, showing that half-integral multiples of the helical repeat prevent proteins bound to the two sites from touching each other. Takahashi et al. (1986) found a similar requirement for integral helical phasing between an enhancer site and the SV40 promoter. This requirement has subsequently been found in a number of other systems, confirming the generality of the phenomenon.

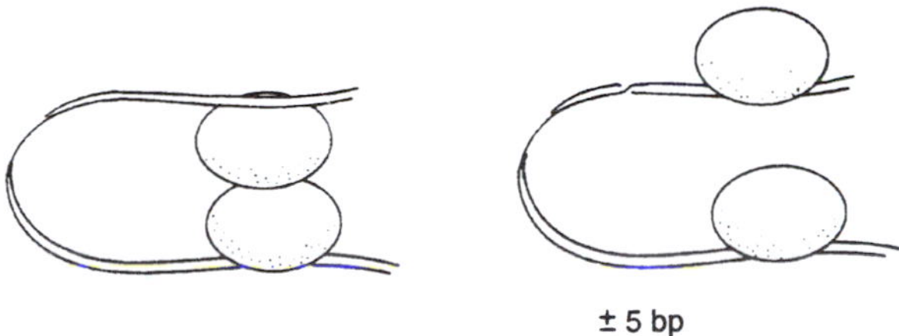

Figure 13-24
Model showing how one-half of a helical turn can interfere with loop formation (Dunn et al., 1984).

Experiments *in vitro* have verified the presence of protein-induced DNA loops, and confirmed their dependence on the phasing between the two protein binding sites. For example, Hochschild and Ptashne (1986) found cooperative binding of λ repressor to DNA molecules when there are two binding sites separated by an approximately integral multiple of 10.5 bp. These loops can be seen directly by electron microscopy (Griffith et al., 1986). Bridging of two *lac* operator sites by a single *lac* repressor tetramer was reported by Kramer et al. (1987), who showed that the looped complex is an electrophoretically distinguishable species.

5.2 DNA Size Dependence of the Looping Interaction: Implications for Protein Flexibility

Lee and Schleif (1989) measured *in vivo* repression of the *AraBAD* promoter by araC protein-induced looping between $AraO_2$ and AraI, using an artificial construct in which the CRP (CAP) site [Fig. 13-23(*a*)] was removed. This experiment allowed the looping interaction to be examined and detected down to 32 bp between the sites, as shown in Figure 13-25. No attenuation of the ability to repress and by inference to form a loop was found. Similarly, Hochschild and Ptashne (1986) found that the extent of cooperativity in binding of *lac* repressor was remarkably constant over the range examined, between about 2 and 6 helical turns separation.

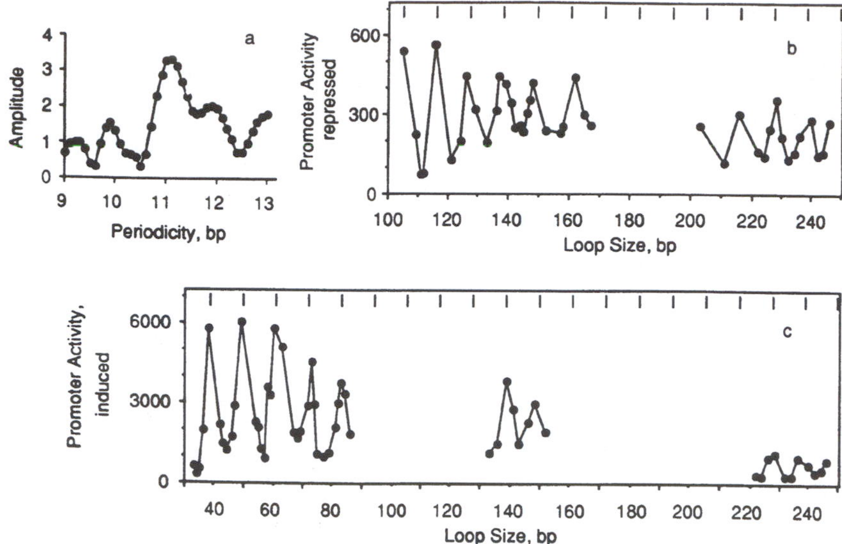

Figure 13-25
Variation of promoter activity in two different regulatory constructs ((*b*) and (*c*)) at the *araBAD* promoter, showing periodic repression and failure of repression as the loop size is changed. (*a*) The frequency spectrum, peaking at a value of 11.1 bp/turn (Lee and Schleif, 1989).

Simple models for DNA looping by proteins (Fig. 13-24) picture the complex with roughly a semicircle of DNA in the loop. The optimum DNA size for such an interaction should be about one-half of the optimum size for DNA cyclization (see Section 9.8), or about $0.5 \times 500 = 250$ bp (Shore and Baldwin, 1983a,b). Because of the stiffness of DNA, it is expected that the free energy cost of bending DNA into a semicircle should rise steeply as the molecule becomes appreciably smaller than 200 bp. For example, Eq. 13-15 predicts, with persistence length $a = 150$ bp and $RT = 0.6$ kcal mol^{-1}, that a bend angle of $180° = \pi$ rad has an energy cost of 2.2 kcal mol^{-1} at 200 bp, 4.5 kcal mol^{-1} at 100 bp, and 8.9 kcal mol^{-1} at 50 bp. If such a bend were actually required for the araC protein and *lac* repressor loop formation, the energy variation should have been clearly evident in the DNA length dependence of the interaction. Since the predicted variation was not found, it seems likely that the looping interaction between the proteins does not generally require a 180° bend from the DNA, especially when the sites are close together. If bound proteins can interact successfully, independent of the angle between the DNA molecules to which they are bound, there must be a flexible linkage at some point in the protein between the DNA binding domain and the protein–protein interaction interface (Lee and Schleif, 1989). Analysis of the crystal structure of repressor core tetramer (Friedman et al., 1995) provides a plausible explanation for the postulated protein flexibility in terms of partial unfolding of an extended coil domain.

5.3 Topology of DNA in Regulatory Loops

Fourier transformation (FT) of the periodic variation of repression in Figure 13-25 yields a periodicity of 11.1 bp, which is the helical repeat h of DNA in the loop. Assuming a shape like a plectonemicly wound supercoil, Eq. 10.16 takes the form $h = h_o/(1 + \sigma)$, which is readily rearranged to

$$\sigma = \frac{h_o - h}{h} \quad (13\text{-}27)$$

With $h = 11.1$, and $h_o = 10.55$ for relaxed DNA, we calculate $\sigma = -0.05$, comparable to the superhelix density in isolated plasmid DNAs. Note the important difference between the helical repeat in negatively superhelical DNAs of solenoidal and plectonemic shapes: h is less than h_o for the former, whereas h is greater than h_o for the latter. This difference arises because the superhelix path is locally right handed for negatively superhelical plectonemic (interwound) superhelices and left handed for solenoidal superhelices.

Another important point to keep in mind is the sharply reduced torsional flexibility of small circles (see Section 10.4). This principle also applies to small DNA loops that are torsionally constrained, for example by a tightly bound bridging protein such as *lac* repressor whose binding to two distant sites forms the loop. Small ($<$ 500 bp) loops of this kind are highly resistant to the kind of unwinding required to initiate transcription or replication. Theoretical considerations (Bauer et al., 1993) predict that introduction of intrinsic bends into small circles (or loops) increases the probability of writhe fluctuations, and hence should increase the overall torsional flexibility of the loop.

6. STRUCTURAL MOTIFS IN SEQUENCE SPECIFIC PROTEIN-DNA INTERACTIONS

A striking feature of specific protein–DNA interactions is the repeated use made of a limited set of protein structural motifs in the recognition process. These structures, considered in detail below, serve to present standard protein structural elements such as α helix and β sheet in a conformation that enables them to interact with the DNA bases, primarily although not exclusively in the major groove. Early models for protein–DNA interactions (Zubay and Doty, 1959; Sung and Dixon, 1970; Adler et al., 1972; Warrent and Kim, 1978), reviewed by Steitz (1990), proposed placing an α helix in the DNA major groove in an orientation parallel to the groove. Subsequent crystallographic work has verified the widespread use of the α helix in recognition of the major groove, but has shown that many orientations of the helix axis relative to the groove are possible. In general (reviewed by Steitz, 1990; Harrison and Aggarwal, 1990; Harrison, 1991; Pabo and Sauer, 1992), sequence-specific interactions involving the α helix come primarily from amino acid side chains that emanate from the amino end of α helices that penetrate the DNA major groove. This orientation places the positive end of the α helix dipole (see Section 13.2.1) in the major groove. As we detail below, several known protein structural motifs for DNA recognition involve major groove–α-helix interactions. Early structural models also proposed that two antiparallel β strands could interact with DNA in the major and minor grooves (Carter and Kraut, 1974; Church et al., 1977). This general motif is used for DNA recognition by the *arc* and *met* repressors (Breg et al., 1990), by the TATA binding protein, and for nonspecific DNA binding by HU protein (Tanaka et al., 1984).

6.1 The Helix–Turn–Helix and Related Motifs

The helix–turn–helix (HTH) motif, the first of the specific DNA binding motifs to be characterized (Steitz et al., 1982; Anderson et al., 1982; Weber et al., 1982), consists of two adjacent α helices separated by a non-helical turn of four amino acids. Figure 13-26 (Steitz et al., 1982) shows the location of the corresponding helices in CAP and λCro protein; identification of this motif was a consequence of its appearance in the DNA binding region of these two proteins (Anderson et al., 1981; McKay and Steitz, 1981). Note that helices 1, 2, and 3 in Cro are equivalent to helices D, E, and F in CAP. Helices 3 and F, which penetrate the major groove, have different tilt angles in the plane of the page. However, the spatial relationship between helices 2 and 3 (E and F) is conserved.

Examination of the sequences of these and other DNA binding proteins revealed sequence homologies (Steitz et al., 1982; Matthews et al., 1982; Sauer et al., 1982; Weber et al., 1982; Ohlendorf and Matthew, 1983; Laughon and Scott, 1984; Shepherd et al., 1984; reviewed by Pabo and Sauer, 1984; Sauer et al., 1990; Steitz, 1990), which enabled tentative assignment of other DNA binding proteins to this class, even though their structures had not yet been determined. Figure 13-27(*a*) shows the amino acid sequence of the HTH region of four proteins whose structure verifies their categorization in this class. The boxed residues are those most highly conserved.

Figure 13-27(*b*) provides some indication of the pattern and variability of amino acid sequence at the critical boxed residues in the HTH motif. Following Sauer et al.

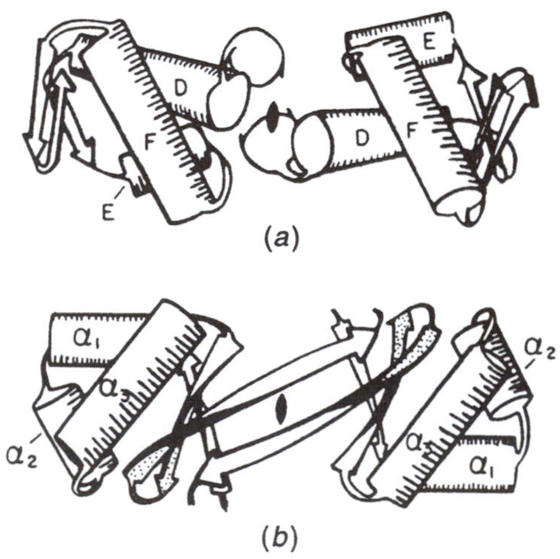

Figure 13-26
Schematic drawings comparing the backbone conformations in the DNA binding domain of two HTH motif proteins, CAP (*a*) and Cro repressor (*b*). The equivalent DNA binding helices F and α_3 are toward the viewer, and have clearly different tilt angles (Steitz et al., 1982).

λ repressor																			
Gln	Glu	Ser	Val	Ala	Asp	Lys	Met	Gly	Met	Gly	Gln	Ser	Gly	Val	Gly	Ala	Leu	Phe	Asn
λ cro																			
Gln	Thr	Lys	Thr	Ala	Lys	Asp	Leu	Gly	Val	Tyr	Gln	Ser	Ala	Ile	Asn	Lys	Ala	Ile	His
CAP			172	173			176	177	178					183			186		
Arg	Gln	Glu	Ile	Gly	Gln	Ile	Val	Gly	Cys	Ser	Arg	Glu	Thr	Val	Gly	Arg	Ile	Leu	Lys
Engrailed																			
Arg	Gln	Gln	Leu	Ser	Ser	Glu	Leu	Gly	Leu	Asn	Glu	Ala	Gln	Ile	Lys	Ile	Trp	Phe	Gln

··· ———Helix———→|———Turn———→|———Helix———→ ···

(*a*)

Residue:	172	173	176	177	178	183	186
Phage proteins	Leu-9 Val-3 Thr-2	Ala-14 Gly-1	Leu-4 Val-4 Ala-2 Thr-2 Phe-2	Gly-12	Val-6 Ile-3 Leu-2 Thr-2	Ile-8 Val-4 Leu-3	Trp-6 Leu-3 Ile-2 Ala-2 Tyr-2
Bacteria	Leu-6 Ile-6 Val-5 Ala-2	Ala-11 Gly-6	Leu-9 Phe-4 Val-2 Ala-2	Gly-13 Asn-2	Val-8 Leu-3 Ile-3	Ile-8 Val-7 Leu-3	Ile-5 Leu-3 Val-3 His-2 Glu-2
Eukaryotes	Leu-8 Ile-8 Met-2	Ala-14 Ser-4	Leu-14 Thr-2	Ser-4 Gly-3 Cys-3 Asn-3 Gln-2 Ala-2	Leu-17	Ile-12 Val-7	Trp-19
Summary	a Hydrophobic	b (2x) Ala, Gly	c Hydrophobic	d Often small (Gly)	e (4x) Hydrophobic	f (2x) Hydrophobic	g Hydrophobic

(*b*)

Figure 13-27
Sequence homologies in HTH proteins. (*a*) Illustrative sequences; the boxed residues are the most highly conserved. (*b*) Amino acid homology at the boxed residues, numbered according to the CAP sequence above. The number following each amino acid gives the number of times the residues occur within the family of proteins. Includes only amino acids that occur more than once. [Prepared from the sequences tabulated by Sauer et al. (1990).]

(1990), Figure 13-27(*b*) is divided into the categories of phage, bacterial, and eukaryotic proteins. When large numbers of sequences from diverse biological origin are compared, the sequence conservation is less dramatic than was inferred from comparison of the few bacterial and phage sequences that served to define the class originally. However, the sequence pattern clearly calls for hydrophobic residues at positions (CAP numbering, as in Fig. 13-27) 172, 176, 178, 183, and 186. Residue 173 is predominantly Ala or Gly, and residue 177, in the middle of the "turn" region, is usually a small amino acid, with Gly predominating.

The physical reason for conservation of the hydrophobic amino acids in the HTH motif is only indirectly related to DNA recognition. Hydrophobic packing between these side chains is responsible for stabilizing the geometric relationship between the two helices in the HTH motif, thus providing structural integrity to the α-helical recognition element. Figure 13-28 illustrates the relationship between these side chains

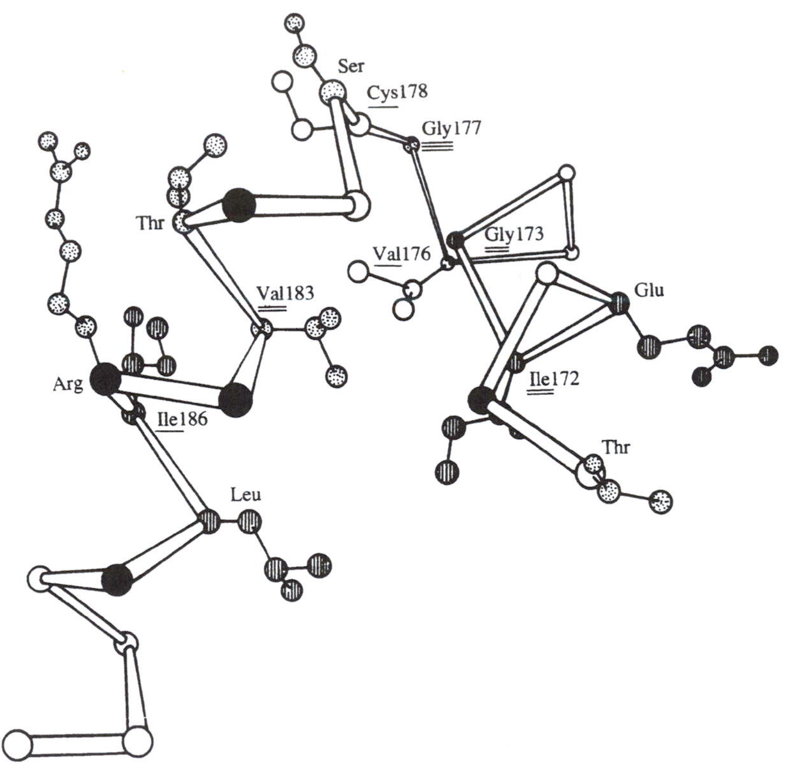

Figure 13-28
The HTH motif showing the positions of conserved hydrophobic residues that form the contacts between the two helices and the conserved residues at the bend between the helices. The sequence and numbers are those of CAP. Underlining increases with the extent of sequence conservation which was apparent from the prokaryotic DNA binding proteins that were known at the time. [Weber, I. T., McKay, D. B., and Steitz, T. A. (1982). *Nucleic Acids Res.* **10**, 5085–5102 by permission of Oxford University Press.]

in CAP protein. Residue 177 (or equivalent) in the loop appeared from early studies of prokaryotic proteins to be predominantly Gly, but the sequence in this region in the eukaryotic homeodomain proteins is much more divergent.

It should be recognized that the sequence conservation of this group of proteins is sufficiently weak that misclassification can result if only the sequence is known. If, however, it is known that a particular sequence resides in a DNA binding domain and the pattern shown in Figure 13-27 is seen, then the chances are quite good that the protein indeed contains an HTH motif.

Whereas the steric relationship between the two helices in the HTH motif is conserved, the geometric relationship between the two-helix motif and the DNA double helix is not. Figures 13-29 and 13-30 give a sense of the variability. In some cases, such as Cro and the *engrailed* homeodomain protein [Figs. 13-29(*a*) and 13-30], helix 3 is approximately parallel to the major groove, but it is not so in CAP and *trp* repressor [Fig. 13-29(*b*) and (*c*)]. Another interesting variation is the extended length of helix 3 in the homeodomain proteins (Fig. 13-30), which significantly alters the geometric relationship between the two-helix motif and the DNA double helix (Kissinger et al., 1990). Heterodimeric homeodomain proteins can display specificities not found in either subunit alone, as illustrated by binding of the yeast proteins a1 and $\alpha 2$, whose protein–protein contact is mediated by a C-terminal 22 amino acid tail on $\alpha 2$ (Stark and Johnson, 1994). In summary, the common feature of the HTH class is a conserved two-helix structure, which can be variably oriented with respect to the DNA major groove, creating recognition patterns that are specific to the individual proteins.

6.1.1 The POU Domain

Several eukaryotic DNA binding homeobox proteins share an additional region of conserved sequence (Herr and Sturm, 1988), called the POU-specific box; the entire conserved region including the adjacent homeobox subdomain is called the POU domain. Both the homeobox and POU domains can contribute to DNA binding (Sturm and Herr, 1988). These transcriptional regulatory proteins are cell type-specific, and are found in a variety of cells of varying degrees of differentiation, including cells producing immunoglobulins (Clerc et al., 1988; Ko et al., 1988; Scheidereit et al., 1988; Muller et al., 1988), cells of the nervous system (Treacy et al., 1991; Dick et al., 1991), and pituitary cells (Ingraham et al., 1988). Mutations in these proteins can affect cell lineage and differentiation (Finney et al., 1988).

The solution structure of the Oct-1 POU domain has been determined by NMR methods (Assa-Munt et al., 1993; Dekker et al., 1993). There are four α helices surrounding a conserved hydrophobic core. The POU domain is structurally similar to the monomeric DNA binding domains of the bacteriophage λ and 434 repressors. The proteins exhibit superimposable HTH motifs, except that in the POU domain, the first helix and the linker to the second helix of the motif are extended. These conserved structural features strongly suggest a model for the DNA complex that is analogous to binding by the prokaryotic repressors.

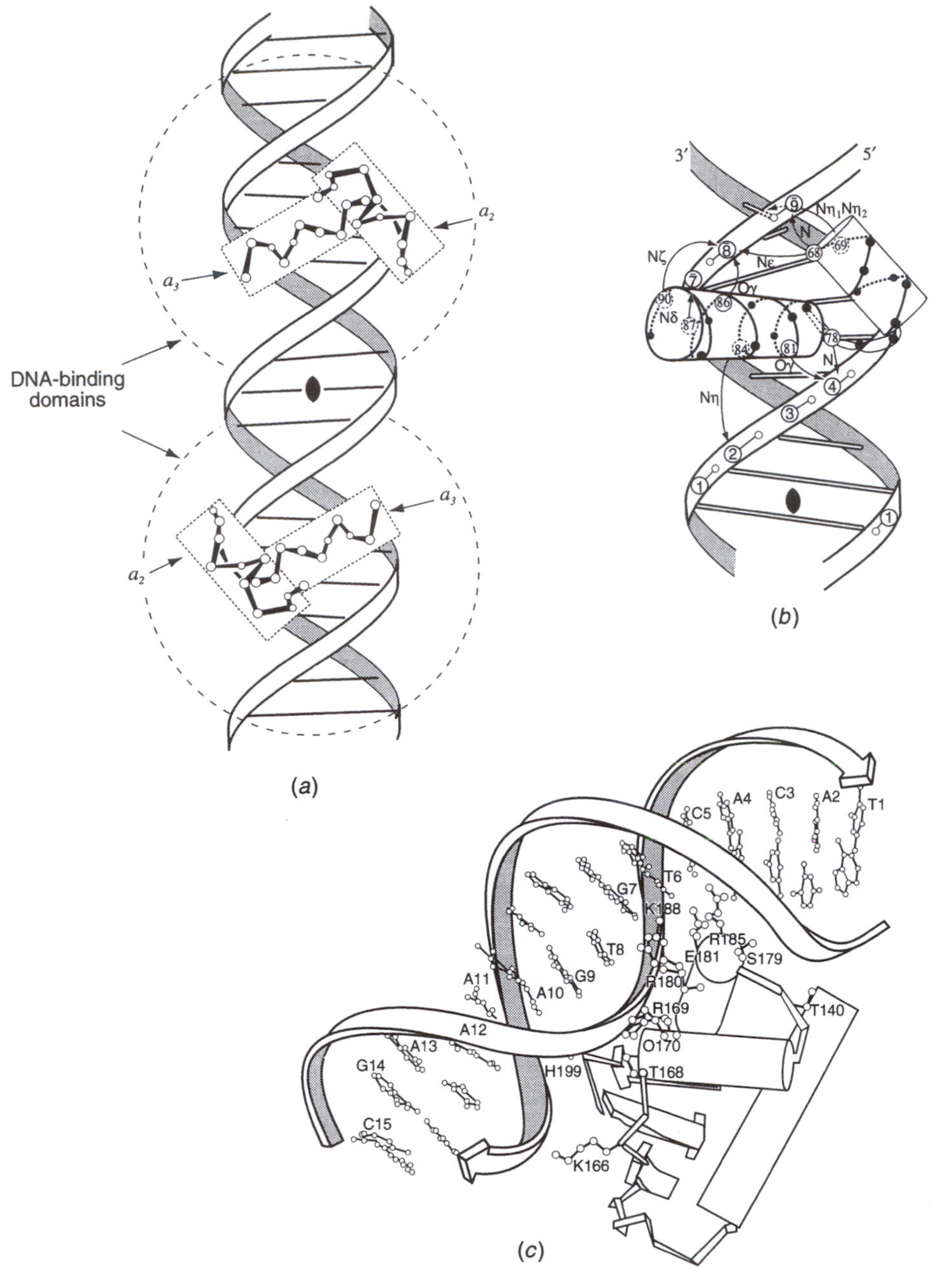

Figure 13-29
Positioning of the HTH motif relative to the DNA major groove in three prokaryotic complexes. (a) Cro repressor. [With permission from the Annual Review of Biochemistry, Volume 53, Copyright ©1984, by Annual Reviews, Inc.] (*b*) T*rp* repressor. [Reprinted with permission from Otwinowski, Z., Schevitz, R. W., Zlaz, R.-G., Lawson, C. L., Joachemiak, A., Marmostein, R.Q., Luise, B. F., and Sig.er, P. B. (1988). *Nature (London)* **335**, 321–329. Copyright ©1988 Macmillan Magazines Limited.] (c) CAP. [Reprinted with permission from Schultz, S. C., Shields, G. C., and Steitz, T. A. (1991). *Science* **253**, 1001–1007. Copyright ©1991 American Association for the Advancement of Science.]

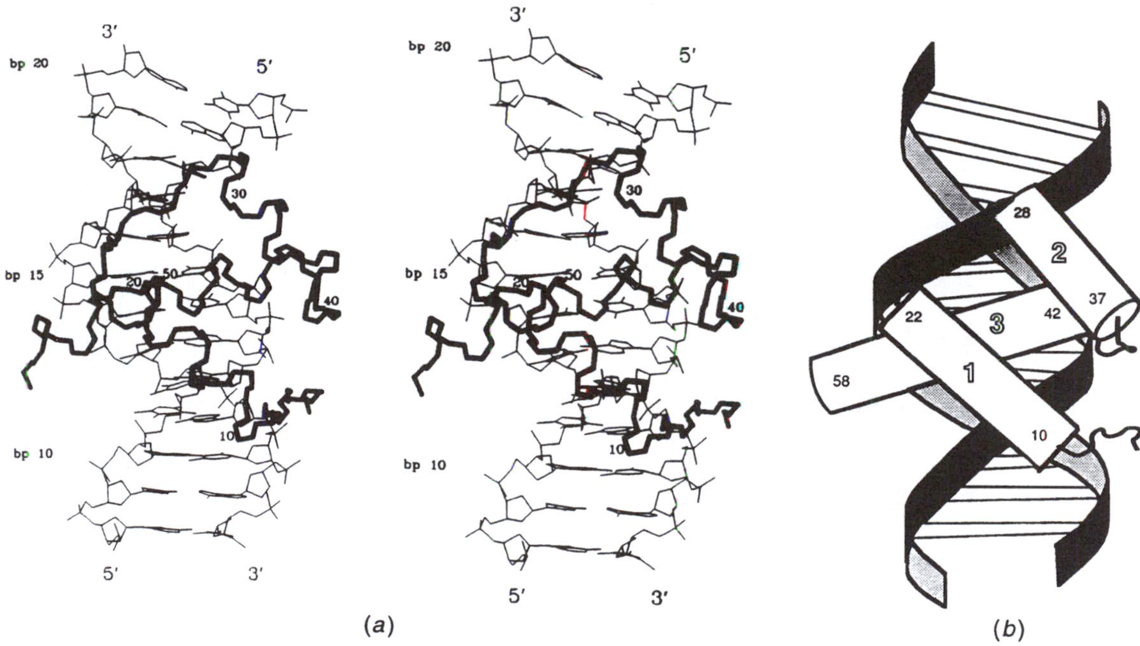

Figure 13-30
Structure of the DNA binding domain of a eukaryotic HTH motif protein, the homeodomain from the *engrailed* protein, in complex with DNA. (*a*) Stereo-diagram showing how the helices and the N-terminal arm are arranged in the complex. Only backbone atoms are shown for the protein. (*b*) Sketch identifying the helices of the HTH domain. [From Kissinger et al., 1990.]

6.1.2 DNA Binding Domains of Histone H5 and the HNF-3γ Proteins

The "linker" histones H1 and H5 are important for folding polynucleosomes into the more compact 30-nm fiber. These proteins are organized into a globular domain flanked by basic amino and carboxy terminal arms that are unstructured in solution. Ramakrishnan et al. (1993) determined the structure of the globular domain of histone H5 (GH5). They observed three α helices, which could be superimposed on helices D, E, and F in the DNA binding domain of *E. coli* CAP protein. This motif has been called the "winged helix" because of the anatomical resemblance of two extended loops on the monomer to butterfly wings. The relationship of this structure to the HTH motif has been reviewed by Brennan (1993) asnd Schwabe and Travers (1993). The motif is further represented in the eukaryotic transcription factor HNF-3γ whose structure in a DNA complex was determined by Clark et al. (1993). Binding to the DNA major groove involves the three-helix motif. Helices 1 and 2 are oriented with their amino termini directed toward the phosphate backbone, and helix 3 lies in the major groove. A long loop in the carboxy terminal domain reaches around into the minor groove, and features a bidentate interaction of arginine with thymine carbonyl O2. As is also true of GH5, HNF-3γ binds as a monomer, but again the three helix motif in the protein is homologous structurally (although not in sequence) with CAP.

6.1.3 The ets Domain

An illustrative case in which the structural homology to the HTH class of DNA binding proteins could not be predicted on the basis of sequence homology is provided by the *ets* domain proteins. This family of eukaryotic transcription activators shares a common DNA binding domain of about 85 amino acids. Its structure is homologous to *E. coli* CAP protein (Liang et al., 1994a,b).

6.1.4 DNA Bending by the HTH Motif

An important variation within this class is the response of DNA structure to the binding event. Curvature of DNA is induced by several proteins, but the most dramatic example is provided by CAP (Schultz et al., 1991; see color plate for Fig. 13-31), which induces a 90° DNA bend, according to the crystal structure. This is largely a result of two approximate 40° kinks that open the roll angle between base pairs 5 and 6 from the dyad axis, corresponding to the position of the second TG in the TGTGA motif [Fig. 13-31(*b*)]. The consequence is a distinct bend that narrows the major groove and curves the DNA toward the protein. A small additional bend ($\sim 10°$) can also be seen between base pairs 10 and 11, resulting from roll between the bases and propeller twist of base pair 11. The DNA bend in the complex with the structurally homologous protein HNF-3γ, at about 13° per subunit, is much less than in the CAP complex.

6.2 Zinc Finger Proteins: The Cys_2His_2Zn Motif in TFIIIA and Related Proteins

6.2.1 The General Categories of DNA Binding Zn Proteins

A number of DNA binding proteins bind transition metals (reviewed by O'Halloran, 1993). In some cases the protein serves a regulatory role to sense the presence of the metal ion and activate appropriate genes. In the case of the general class of DNA binding proteins that contain zinc, often called Zn finger proteins, the metal plays a structural role in stabilizing the DNA binding conformation. The DNA binding proteins that also bind zinc are very common in eukaryotes. Four distinct classes of protein–DNA complexes have so far been characterized structurally, and there are at least two additional classes on the basis of sequence pattern. The first of this general class to be discovered was TFIIIA from Xenopus oocytes, which is required for transcription of 5S RNA genes by RNA polymerase III (Engelke et al., 1980). Subsequent determination of the sequence of this protein (Brown et al., 1985; Miller et al., 1985) revealed tandem repeats of a motif containing two cysteines and two histidines per Zn atom (Fig. 13.32), which serve as binding sites for essential Zn atoms (reviewed by Rhodes and Klug, 1988; Berg, 1990). Another important Class I Zn finger protein is the mammalian transcription factor Sp1 (Kadonaga et al., 1987)

Analogous Zn binding motifs have since been found. In two of these motifs, two Zn atoms are chelated by eight (Cys_8Zn_2) and six (Cys_6Zn_2) cysteines (Fig. 13-32), whereas in another motif Zn is chelated by four cysteines, in the Cys_4Zn class. In the p53 tumor supressor protein, the coordination is Cys_3HisZn. The Cys_8Zn_2

Class I Cys_2His_2Zn Finger (TFIIIA)

X_3-Cys-X_{2-4}-Cys-X_{12}-His-X_{3-4}-His-X_4

Class II Cys_8Zn_2 (Glucocorticoid receptor)

X_2-Cys-X_2-Cys-X_{13}-Cys-X_2-Cys-X_5-Cys-X_5-Cys-X_9-Cys-X_2-Cys-X

Class III Cys_6Zn_2 (Gal4)

X_3-Cys-X_2-Cys-X_6-Cys-X_6-Cys-X_2-Cys-X_6-Cys-X_3

Class IV Cys_4Zn (GATA-1)

-Cys-X_2-Cys-X_{17}-Cys-X_2-Cys-

Class V Cys_7HisZn_2 [LIM(Michelsen et al., 1993), see also RING (Lovering et al., 1993)]

-Cys-X_2-Cys-X_{17}-His-X_2-Cys-X_2-Cys-X_2-Cys-X_{17}-Cys-X_2-Cys-

Figure 13-32
Cysteine and histidine repeats in Zn-containing DNA binding regulatory proteins. The X is an amino acid with varying degrees of conservation. These features of the sequence are suppressed in favor of displaying the pattern of conservation of Cys and His residues (see Valee et al., 1991; Omichinshi et al., 1993).

grouping includes proteins of the steroid receptor class, for example, the mammalian glucocorticoid receptor, and the Cys_6Zn_2 group includes yeast GAL4 protein and related fungal transcription factors. A Cys_4Zn chelation (Omichinski et al., 1993a,b) is found in the GATA-1 erythroid transcription factor (see Fig. 13-2). An NMR structure, featuring a β sheet motif, for a Cys_4Zn module from eukaryotic transcriptional elongation factor TFIIS was reported by Qian et al. (1993),.

One new class that is likely to be structurally distinct (Fig. 13-32) is called the LIM or RING motif. It has the composition Cys_7HisZn_2 (Michelsen et al., 1993; Lovering et al., 1993), and is found in a diverse collection of proteins that includes transcription factors. The conserved sequence in the RING motif is similar to, but not identical with the LIM motif. It is not yet clear whether they are separate classes since structure and its diversity in these classes are unknown at present. Future work may require their reclassification.

The term "zinc finger" was coined to describe tandemly repeated domains folded in part by coordination of the conserved Cys and His residues to Zn (Miller et al., 1985). Structural models envisioned contact of the individual fingers of this motif with DNA. There has been a tendency to lump all of the Zn-containing regulatory proteins into the Zn finger category, but this generic categorization does not take proper account of the distinctly different structures that have been determined for the different motifs shown in Figure 13-32. Vallee et al. (1991) have suggested the terms "Zinc twist" and "Zinc cluster" for proteins of the glucocorticoid receptor and Gal4 class, respectively (see Sections 13.6.3 and 13.6.4).

6.2.2 Cys_2His_2Zn Finger Proteins

Pavletich and Pabo (1991) determined the structure of the complex between DNA and a truncated protein containing three zinc fingers from the mouse protein Zif268. Figure 13-33(*a*) shows the folding of the zinc finger domain, which consists of an antiparallel β sheet hairpin structure followed by an α helix. This pattern was successfully predicted by Berg (1988) on the basis of homology with metal binding sites in other proteins [Fig. 13-33(*b*)]. The structure of the Zn-binding domain has also been determined in solution by NMR methods (Parraga et al., 1988; Lee et al., 1989). The tetrahedral Zn coordination site is created by two Cys residues provided by the β sheet (C7 and C12) and two His residues near the carboxy terminus of the α helix (H25 and H29), where the residue numbers are given for the first finger. In this frame, the β sheet runs from residue 5 to 16, the α helix from 19 to 30, and the next β sheet starts at residue 35. Hence, the Zn coordination site serves to hold the α helix and β sheet together. Successive fingers are highly homologous in structure, with packing between the α helix and the β sheet further stabilized by a hydrophobic core provided by conserved residues Phe-16, Leu-22, and His-25, along with other less conserved hydrophobic residues [Fig. 13-33(*a*)].

Figure 13-34(*a*) shows the interaction of three connected fingers with double helical DNA. Consistent with the pattern established for other DNA binding proteins, the amino terminus of each α helix lies in the major groove, and amino acid side chains from the helix and the nearby turn contact bases on their major groove faces [Fig. 13-34(*b*)]. For example, in the first finger, Arg-18 and Arg-24 hydrogen bond with G in

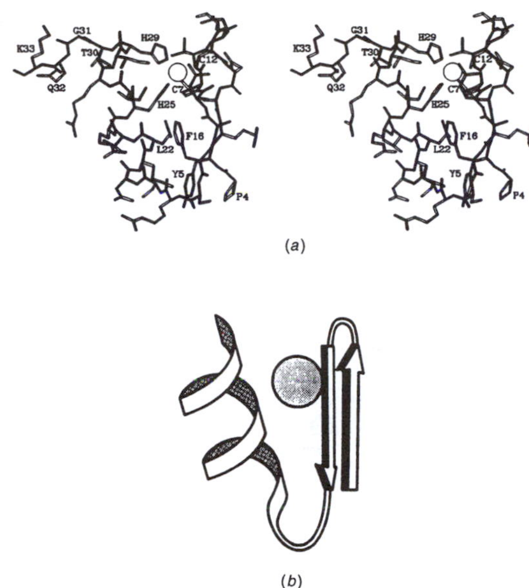

Figure 13-33
Structural features of Zn fingers. (*a*) Stereodiagram of the structure of finger 1 from the complex of Zif268 with DNA. [Reprinted with permission from Pavletich, N. P. and Pabo, C. O. (1991). *Science* **252**, 809–817. Copyright ©1991 American Association for the Advancement of Science.] The zinc binding site, shown by a sphere, is formed by residues C7, C12, H25, and H29. The β sheet is viewed from the side on the right hand side of the image, and the α helix runs from lower right to upper right. Conserved residues in the hydrophobic core between the αa helix and β sheet are also identified. (*b*) Schematic diagram of the structure of the Zn finger as predicted by Berg (1988) on the basis of homology to other metal binding proteins. The figure is oriented approximately as in (*a*).

the major groove. In some cases, the Arg-G hydrogen-bonding pattern is stabilized by an additional bond between Arg and an Asp residue from the α helix, as illustrated in Figure 13-35(*c*). Phosphate contacts in the first finger are provided by Arg-3, Arg-14, and His-25. Note that these phosphate contacts result from a residue preceding the β sheet, one in the β sheet, and one in the α helix.

As Figure 13-34(*a*) emphasizes, each finger interacts with a three nucleotide segment of the G rich strand in the binding site. The polypeptide strand winds around the DNA duplex and is antiparallel to the G rich DNA strand, meaning that it is oriented from amino-to-carboxy terminal in the $3' \rightarrow 5'$ direction along the primary interacting DNA strand. The helical winding of the protein is characterized by a rotation angle per finger of about 96° (3° × 32°/bp) about the DNA axis, and a translation of about 10 Å (3 × 3.3 Å/bp) along the axis, values close to expectation for double helical DNA. Hence the structure leads one to expect little DNA unwinding or bending in the complex.

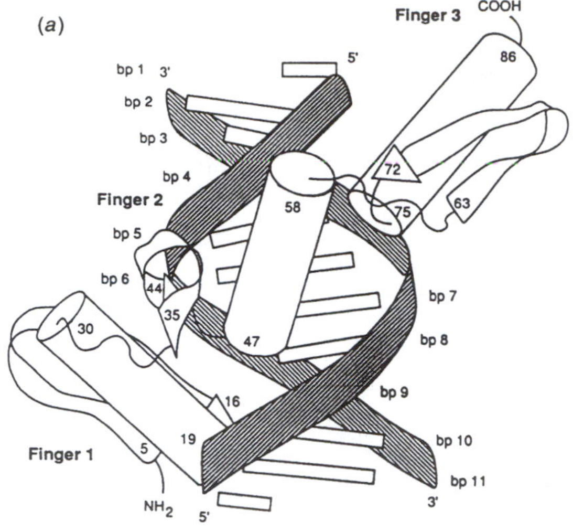

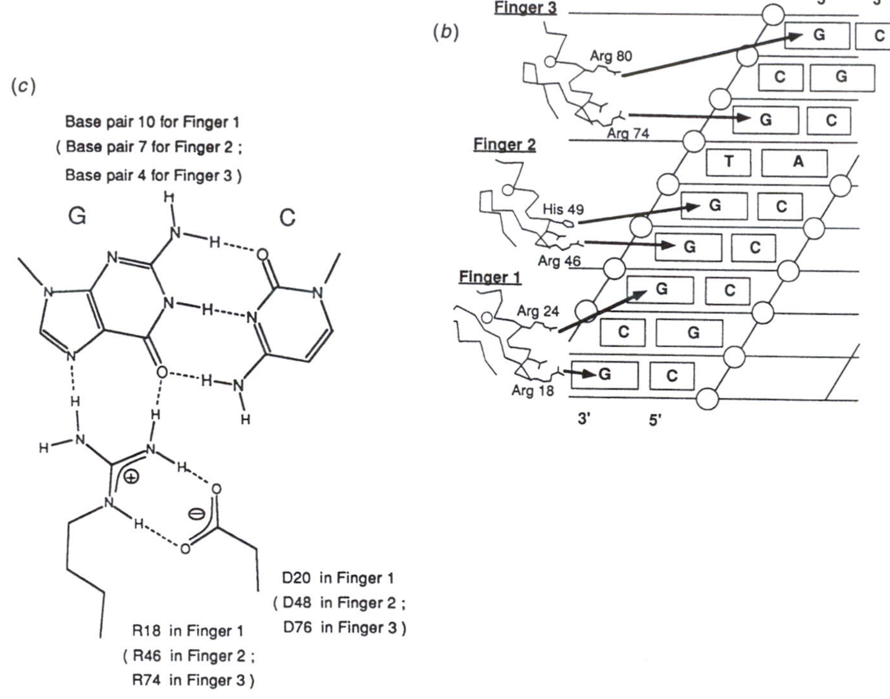

Figure 13-34
Structural features of the Zif268-DNA complex. [Reprinted with permission from Pavletich, N. P. and Pabo, C. O. (1991). *Science* **252**, 809–817. Copyright ©1991 American Association for the Advancement of Science.] (*a*) Sketch showing the relation of the Zn fingers with respect to each other and with respect to the DNA. The starting and ending residues of each α helix and β sheet are indicated, together with the base pair numbers. (*b*) Sketch summarizing the base contacts made by the Zif268 peptide. The DNA is represented as a cylindrical projection. (*c*) Drawing of the Asp–Arg–guanine interaction present in all three fingers.

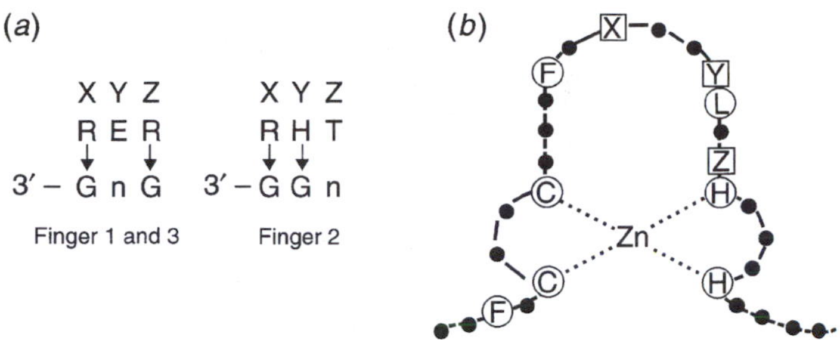

Figure 13-35
(*a*) Pattern of contacts between side chain and bases in the Zif268–DNA complex.
(*b*) Identification of residues X, Y, and Z in the Cys_2His_2 Class I Zn finger motif. Circled residues are conserved, and X, Y, and Z are boxed. [Reprinted with permission from Klevit, R. E. (1991). *Science* **253**, 1367. Copyright ©1991 American Association for the Advancement of Science.]

A conserved feature of the trinucleotide DNA sequence is G in the 3′ position. In contacting the TGG sequence, the second finger uses different contacts than are found for fingers 1 and 3, both of which interact with the sequence GCG in Figure 13-34(*b*). Figure 13-35 (Klevit, 1991) shows the recognition pattern schematically. Residue X, which just precedes the α helix, and residues Y and Z on the outer face of the α helix, interact in an "antiparallel" way with the nucleotide triplet GnG or nGG. Base contacts are observed with G when X, Y, or Z are Arg (R) or His (H). Generalization and extension of this pattern for Sp1 Zn fingers and variants were reported by Desjarlais and Berg (1992) and Kriwacki et al. (1992). Additional structures have been reported by Pavletich and Pabo (1993) and Fairall et al. (1993). It remains to be seen whether the coding patterns deduced are further conserved for other Class I Zn finger binding sequences, which show even less trinucleotide sequence conservation over the full target sequence. In many cases, there may be less regularity in the packing of fingers around DNA. However, the basic pattern of wrapping the string of fingers around the DNA duplex in the major groove is likely to be common to the Class I Zn finger proteins.

Cys_2His_2Zn Zn fingers have an additional distinction that sets them apart from the other classes. Since the fingers can be viewed as a linear array, the DNA sequence motif utilized is effectively a tandem repeat, Figure 13-9(*a*). In the case of Zn fingers, however, the repeat is a trinucleotide, and there are sequence variations from one repeat to another. There is in principle no upper limit on the number of repeated finger motifs in this arrangement. More complex binding patterns are also possible, as is thought to be the case for TFIIIA, for which it is proposed that fingers 1–3, 5, and 6–9 bind in the major groove, but fingers 4 and 6 span the minor groove (Clemens et al., 1992).

6.3 The Cys_8Zn_2 Motif in Mammalian Glucocorticoid Receptor

The glucocorticoid receptor protein is a member of a larger family of nuclear receptors. Upon binding hormone, the protein is translocated from the cytoplasm to the nucleus where it binds to specific DNA sequences and modulates transcription of the target genes (Chandler et al., 1983; Ponta et al., 1985). Members of the family have conserved domains for DNA binding, ligand binding, and transcription activation. The DNA binding domain has a conserved Cys_8Zn_2 motif, in which eight appropriately spaced Cys residues (see Fig. 13-32) bind two Zn^{2+} ions, which are crucial to the globular folding of the domain.

The structure of the DNA binding domain of the rat glucocorticoid receptor protein complexed with double helical DNA, reported by Luisi et al. (1991), is in good general agreement with the solution structure of the protein alone as determined by NMR methods (Hard et al., 1990). A similar structure was deduced from NMR measurements on the DNA binding domain of the estrogen receptor (Schwabe et al., 1990). A crystal structure of the estrogen receptor bound to DNA was reported by Schwabe et al. (1993). Since the estrogen and glucocorticoid receptor proteins differ in only a few base pairs in their DNA targets, comparison of the two DNA complexes provides insight into mechanisms of binding specificity (Schwabe et al., 1993). Nuclear magnetic resonance structural characterization has also been reported for the DNA binding domain of a closely related protein, the retinoid X receptor (Lee et al., 1993).

Figure 13-36 (Luisi et al., 1991) summarizes the amino acid sequence and overall structural organization of the two-Zn domain, which can be further subdivided into one-Zn "modules." The folded two-Zn domain is globular and dimerizes upon binding to approximately dyad symmetric double helical DNA. Figure 13-36 shows the amino acids involved in dimerization, DNA contact, and DNA sequence discrimination. The domain contains extensive α-helical segments, which are indicated by boxed regions in the sequence.

Figures 13-37 and 13-38 (see color plate) summarize the structure of the complex. The primary base-specific contacts are made in the major groove by amino acids in the α-helical segment that runs from Cys-457 through Glu-469 (see Fig. 13-36). These include, as summarized in Figure 13-37(*c*), hydrogen bonding of Arg-466 to G4, hydrophobic contact of Val-462 with T5, direct hydrogen bonding between Lys-461 and G7, and water-mediated hydrogen bonds between Lys-461 and G7 and T6. There are, in addition, a number of contacts with the phosphate backbone that may contribute to specificity because of the influence of DNA sequence on the shape of the phosphodiester chain. A common feature of the Class I and II Zn finger proteins is the use of one or more Zn ions coordinated to peptide side chains to stabilize a globular folding of the protein, thereby presenting an α-helical domain for interaction with the DNA major groove.

In addition to the estrogen and glucocorticoid receptor proteins that bind to their dyad symmetric DNA targets as homodimers, there are a number of Cys_8Zn_2 proteins that bind as heterodimers to tandemly repeated sequence elements. The structural basis for the dimer interface was reported by Rastinejad et al. (1995) for the 9-*cis*-retinoic acid receptor and thyroid hormone receptor bound to DNA.

6.4 The Cys_6Zn_2 Motif in GAL4

Activation of the genes coding for galactose metabolism in yeast requires DNA binding of the transcription factor GAL4 (Oshima, 1981). The 62 NH_2-terminal amino acids

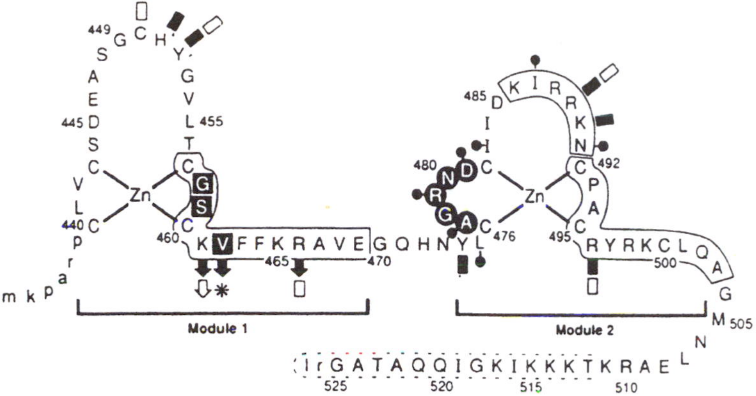

Figure 13-36
Amino acid sequence and general structural organization of the glucocorticoid DNA binding domain used for crystallographic analysis. [Reprinted with permission from Luisi, B. F., Xu, a. X., Otwenowski, Z., Freedman, L. P. Yamamato, K. R., and Sigler, P. B. (1991). *Nature (London)* **352**, 497–505. Copyright ©1991 Macmillan Magazines Limited.] Two tetrahedrally coordinated Zn atoms link the conserved cysteines as shown. Dimerization contacts are mediated by amino acids marked by solid dots. Residues that make phosphate contacts in specific and nonspecific binding modes are marked by solid and open rectangles, respectively, while specific and nonspecific base contacts are indicated by solid and open arrows, respectively. Three amino acids that direct the discrimination of glucocorticoid and estrogen response elements are indicated by white lettering in solid boxes, and those that discriminate between estrogen and thyroid response elements are shown in white on solid disks.

of this protein comprise a DNA and Zn^{2+} binding domain (Keegan et al., 1986; Pan and Coleman, 1989, 1990a,b), which contains the conserved Cys_6 sequence shown in Figure 13-32. The structure of this Zn finger protein clearly distinguishes it from other motifs. Transcription factors isolated from fungi (Salmeron and Johnston, 1986; Pfeifer et al., 1989) share homology with this motif, as shown in Figure 13-39(*a*), which compares the amino acid sequence of the DNA binding domain of GAL4 with several other fungal transcription factors. One of these, LAC9, is sufficiently homologous to GAL4 to bind the same DNA sequence, but PPR1 binds to a different sequence.

The solution structure of a peptide incorporating residues 7–49 (Kraulis et al., 1992) and 1–65 (Baleja et al., 1992) of GAL4, which provide the Zn and DNA binding domain, has been solved by NMR methods. As shown in Figure 13-39(*b*), residues approximately 9–40 form a compact globular domain containing two α helices connected by a loop. Each Zn atom is tetrahedrally coordinated by four S atoms from Cys residues. Cysteine residues 11 and 28 act to bridge the Zn atoms in an $-S-Zn^{(II)}-S-$ bimetal thiolate. The Zn and S atoms, and the two α helices, are related by an approximate two fold axis of symmetry.

Crystallographic methods were used by Marmorstein et al. (1992) to determine the structure of a 65 amino terminal fragment of GAL4 complexed with a dyad symmetric 17 bp sequence containing the specific binding site. The metal binding domain occupies residues 1–40, with a structure very similar to that found for the protein alone in

affect specificity (Kim et al., 1993). Also, Cuenoud and Schepartz (1993) constructed molecules in which the natural dimerization motif was replaced by stereochemically defined metal ion complexes, and observed alterations in specificity.

6.7.3 The Helix–Loop–Helix Motif

The helix–loop–helix (HLH) motif, which contains two amphipathic α helices connected by a peptide loop, is shared by such evolutionarily divergent proteins as the myc family, the *daughterless* protein from Drosophila, and the E12 and E47 proteins, which bind to the κE2 sequence in the immunoglobulin kappa chain enhancer (Murre et al., 1989). The role of HLH proteins in neurogenesis and myogenesis has been reviewed by Jan and Jan (1993). Clear homology also exists with the nuclear matrix lamin proteins (Murre et al., 1989). A number of additional members of this family have been reported (see Zhang et al., 1991). Helical wheel diagrams for the two 12 and 13 amino acid sequences in the HLH region, folded as α helices, show strong sequence conservation on their hydrophobic faces (Murre et al., 1989). The amino acid sequence between the helices is quite variable, but makes frequent use of the residues Gly, Pro, Asn, Asp, and Ser, which appear in the loop motif (Lesczynski and Rose, 1986), often found on the exterior of proteins. For example, the sequence proposed for the loop in c-myc is Pro-Glu-Leu-Glu-Asn-Asn-Glu. Helix–loop–helix proteins bind to DNA as dimers (Murre et al., 1989).

The DNA binding HLH proteins contain basic and leucine zipper domains in an overall sequential organization *Basic domain–helix 1–loop–helix 2-leucine zipper,* and are accordingly called bHLH-ZIP proteins. Crystallographic analysis of the structure of a truncated form of the transcription factor Max (Max 22–113) with a 22 bp DNA fragment shows that the *basic domain–helix* 1 motif forms a continuous α helix, as does the *helix 2-leucine zipper* region (Ferre D'Amare et al., 1993). As Figure 13-48 shows, the DNA binding and leucine zipper coiled-coil interactions are very similar to those in the bZIP family (Fig. 13-47), except that in Max the four helices in the dimer come together in the center to form a four helix bundle structure stabilized by hydrophobic packing. Based on these structural considerations, one expects that bHLH-ZIP proteins will generally have a stronger dimer interface than is characteristic of the bZIP class. Modification of the structure of the dimer interface is required to explain the binding of SREBP-1, a bHLH-ZIP protein that controls transcription of the low-density lipoprotien receptor gene (Yokoyama et al., 1993), because the binding sites are tandemly repeated, rather than dyad symmetric as found for other members of the class.

6.8 E2 and NFκB Transcription Factors

6.8.1 E2 Protein

The variety of ways in which protein α helices can be presented for interaction with the DNA major groove has continued to grow. Hegde et al. (1992) solved the crystal structure of an 84 amino acid peptide containing the DNA binding domain

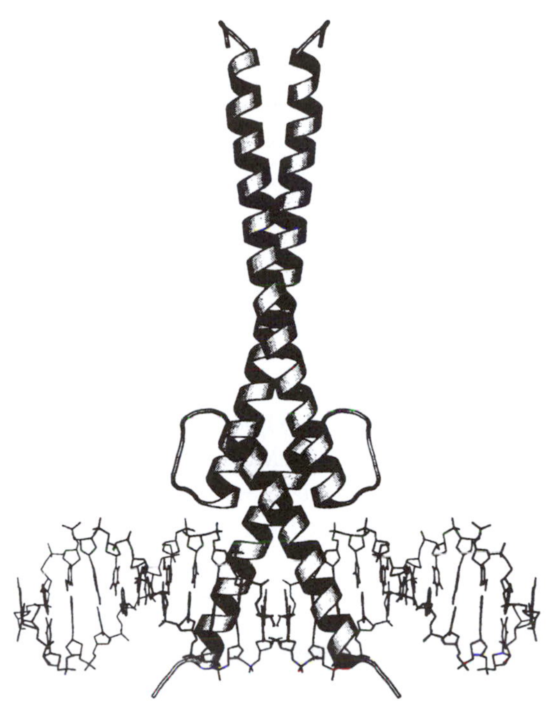

Figure 13-48
Cartoon representation of the structure of the Max (22–113) b/HLH/Z–DNA complex. Each subunit consists of two continuous α helices interrupted by a loop. The lower helix, which interacts with DNA in the major groove, includes the basic region and helix 1, and the upper helix contains helix 2 and the leucine zipper motif. The leucine zipper regions at the carboxy terminus come together to form a coiled coil. The four helix bundle is found in the middle of the molecule, adjacent to the loop. [Reprinted with permission from Ferre-DAmare, A. R., Prendergast, G. C., Ziff, E. B., and Burley, S. K. (1993), *Nature (London)* **363**, 38–45. Copyright ©1993 Macmillan Magazine Limited. See Fig. 3.]

of papillomavirus-1 E2 protein interacting with a 16 bp DNA oligonucleotide. The E2 protein is a transcriptional regulator that is required, in conjunction with the E1 protein, for the initiation of viral replication. Specific DNA binding is mediated through a carboxy-terminal 85 amino acid domain that is sufficient for dimerization as well. The DNA binding/dimerization domain is connected to a 160 amino acid amino terminal activation domain through a proline-rich linker.

The E2 protein is found to be a dyad-symmetric eight stranded antiparallel barrel made up of two identical "half-barrel" subunits as shown in Figure 13-49 (see color plate) (Hegde et al., 1992). Each subunit has two helices that make crossover connections on the outside of the barrel. The larger of these two helices, which connects β strands 1 and 2, serves as the recognition helix. The recognition helices from the two subunits interact with successive major grooves in the DNA at the dyad symmetric sequence G<u>ACC</u>GACGTC<u>GGT</u>C (Fig. 13-49), where the primary identity and contact bases are underlined.

The DNA is smoothly bent to a radius of curvature of about 45 Å (Fig. 13-49). The straightest segments are the identity elements where the DNA major groove interacts with the protein α helix. This pattern is in contrast to the DNA bending pattern of CAP protein (Fig. 13-31), where the main focus of bending is a 45° kink at the TG step in the major groove contacted by the protein. This difference appears to reflect the orientation of the recognition α helix parallel to the major groove in E2, thus tending to keep it wedged open. In contrast, the α helix in CAP is almost perpendicular to the DNA and provides a fulcrum for bending the DNA.

Distinctive features of the E2–DNA interface include the observation that 3 of the 4 bp involved in base discrimination participate in interactions that straddle 2 bp. For example, Asn contacts the N7 of one adenine and the cytosine amino group of the adjacent base pair. Also, Cys is reported for the first time as hydrogen-bond donor–acceptor in specific protein–DNA interacions (see Table 13-1).

6.8.2 NFκB Protein

The structure of a truncated version of NFκB transcription factor bound to DNA was reported by Ghosh et al. (1995) and Müller et al. (1995); see also the commentary by Baltimore and Beg (1995). A distinctive feature of the structure is the use of loops interacting with the major groove and phosphate backbone to achieve binding specificity (see color plate for Fig. 13-50). As pointed out by Müller et al. (1995) there is some similarity to the core domain of p53 (see Section 13.6.6). The protein has a relatively high β-strand content, and the loops are positioned in the sturcture to connect the β strands.

6.9 Antiparallel β Strand–Major Groove Interaction: *Arc* and *MetJ* Repressors

6.9.1 The Met and Arc Repressors

All of the DNA binding motifs discussed so far involve primarily interaction of α helical segments and adjacent residues with the DNA major groove. This feature is not, however, a prerequisite for DNA–protein interaction. Two proteins that share a distinctive recognition motif are the *Met J* (Rafferty et al., 1989) and *Arc* repressors (Breg et al., 1990); the *Met J* repressor structure was determined by crystallography, and the *Arc* repressor by NMR.

In addition to a shared structural homology, the *Arc* and *Met J* repressors have in common the property of binding to tandemly repeated sequences. An example is the so-called "Met-box" (Phillips et al., 1989). The symmetry of these sites, illustrated in Figure 13-51, would in principle permit binding of a single repressor dimer with its dyad axis coincident with the DNA dyad axis between the two AGACGTCT sequences (Alignment one in Fig. 13-51). However, the AGACGTCT sequence is self-complementary, and hence has an internal dyad axis. This symmetry provides the alternative possibility that two dimers may bind to the sequence, each with its dyad axis coincident with the internal dyad axis (Alignment two). An HTH motif protein would be expected to bind by Alignment one, since this arrangement would place the two DNA recognition elements in successive major grooves. The structure of the *Met J* repressor (Rafferty et al., 1989) revealed a dimer with an apparent HTH motif, as well as an antiparallel sheet motif, leaving initial ambiguity concerning the mode of interaction with DNA.

The crystal structure of the met repressor–operator complex illustrated recognition of the DNA major groove by a double-stranded antiparallel β ribbon, corresponding to Alignment two. (Somers and Phillips, 1992; Kim, 1992). Figure 13-52 shows a stereoview of a truncated part of the protein, revealing the mode of interaction of the β

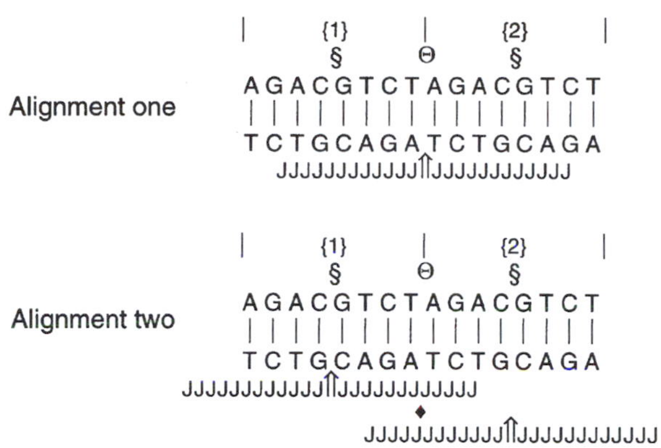

Figure 13-51
Two possible alignments of *MetJ* repressor on a tandem repeat of two met-box operators. In Alignment one, a single repressor dimer binds to the tandem site. The protein dyad axis coincides with the overall dyad axis of the DNA sequence. In the second Alignment, two repressor dimers occupy the tandem sites, with coincidcnce between the dimer dyad axes and the internal dyad axis of the Met-box sequence AGACGTCT. Alignment one is characteristic of HTH motif proteins, whereas Alignment two is suited for interaction of two dimeric antiparallel β sheets with the major groove. [Reprinted with permission from Phillips, S. E. V., Mainfield, I., Parsons, I., davidson, B. E., Rafferty, J. B., Somers, W. S., Margarita, D., Cohen, G. N., Saint-Gerona, D., and Stokley, P. G. (1989). *Nature* **341**, 711–715. Copyright ©Macmillan Magazine Limited.]

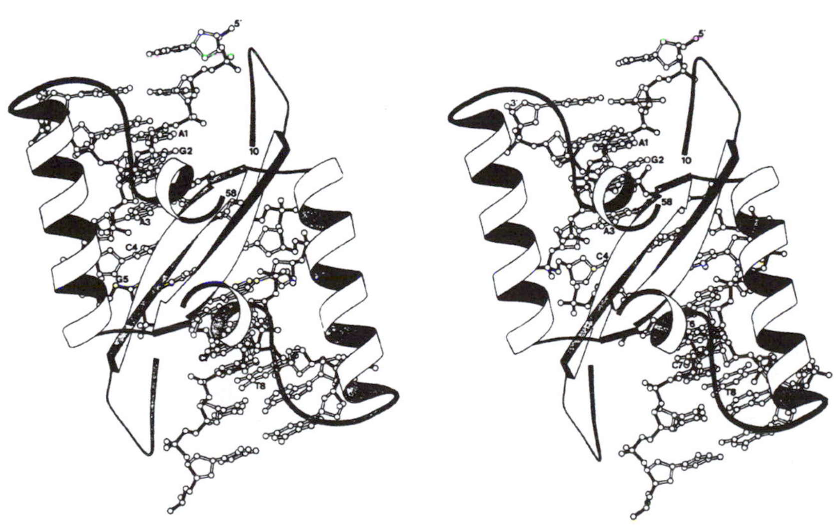

Figure 13-52
Stereodiagram of a ribbon drawing of the binding of β strands in the met repressor to the major groove of DNA. The protein structure is truncated to facilitate viewing of the binding interface. The two protein subunits in the dimer are distinguished by their shading. [Reprinted with permission from Somers, W. S. and Phillips, S. E. V. (1992). *Nature* **359**, 387–393. Copyright ©1992 Macmillan Magazine Limited. See Fig. 2.]

strands with the major groove. Note that each subunit contributes one of the two β strands, so that the β ribbon is part of a dimer interface. Lys and Thr side chains make direct hydrogen bonds to base pairs (see Table 13-1), and there are nummerous contacts to the phosphate backbone. Tandem binding of these dimers is explained by another dimerization interface provided by association between the α helices of two dimers spaced along the DNA. The homologous Arc repressor also utilizes a double-stranded antiparallel β ribbon to interact with the major groove (Raumann et al., 1994). The structure of a replication terminator protein complexed with DNA (Kamada et al., 1996) shows another example of an antiparallel β strand interacting with the DNA major groove.

6.10 Minor Groove Binding/DNA Bending Proteins

All of the protein–DNA interaction motifs considered so far have interaction with the major groove as a primary feature. There is a general class of proteins, whose mode of interaction with DNA is beginning to be understood, in which the primary contact and recognition site is the DNA minor groove. These proteins also share the feature of introducing a large bend into the DNA helix. We group these proteins together here, even though they represent several non-homologous structural classes (see the commentary by Crothers, 1993).

6.10.1 HU and IHF (Integration Host Factor)

These are structurally homologous prokaryotic proteins that are known to bend DNA strongly. The IHF protein is a sequence-specific DNA binding protein that assists formation of nucleoprotein complexes by overcoming the stiffness of DNA through induced bends. An example is found in phage λ site-specific recombination. Nash and co-workers (Goodman and Nash, 1989; Goodman et al., 1992) showed in a series of experiments that other DNA bending proteins, or even intrinsically bent DNA, can replace this activity under appropriate conditions. The HU protein differs from IHF in that it does not recognize specific sequences.

The crystal structure of HU protein (Tanaka et al., 1984) shows an antiparallel β-strand motif, through which DNA interaction is thought to occur (Yang and Nash, 1989; White *etal*., 1989). Involvement of the minor groove is indicated by the fact that methylation of adenine N3 (minor groove) in the specific IHF binding site generally interferes with binding, whereas guanine N7 (major groove) methylation does not (Yang and Nash, 1989). Strong DNA bending by IHF was demonstrated by gel electrophoresis methods (see Section 13.9). These facts have been incorporated into models in which the IHF/HU strands interact with the DNA minor groove, and the DNA is bent around the protein (Yang and Nash, 1989; White et al., 1989). The crystal structure (Rice et al., 1996) reveals a bend of about 160°, and extensive contacts with the minor groove and phosphodiester backbone.

6.10.2 HMG Box Proteins

These are eukaroytic proteins with homology to the “high-mobility group” chromosomal proteins HMG1 and HMG2 (reviewed by Johns, 1982). Like HU, HMG1 and HMG2 are relatively abundant, nonspecific DNA binding proteins. They bind to A • T rich sequences *in vitro* and prefer single- over double-stranded DNA (Triezenberg et al., 1988). As first recognized by Tjian and co-workers for the protein hUBF (Jantzen et al., 1990), certain sequence specific DNA binding proteins contain a region about 85 amino acids in length (the HMG box), which is homologous to two such sequences found in both HMG1 and HMG2. Figure 13-53 shows the homology of the A and B HMG boxes in HMG1 and HMG2 with HMG boxes in several sequence specific proteins, some of which also contain more than one HMG box. This same motif has also been reported in the products of genes in the sex-determining region of both mouse and human Y chromosomes (Gubbay et al., 1990; Sinclair et al., 1990). Analogous proteins have been found in yeast (Kolodrubetz and Burgum, 1990).

LEF-1 is a lymphoid-specific HMG domain protein that regulates T-cell receptor enhancer function (Travis et al., 1991). With the use of an electrophoresis assay (see Section 13.9) Grossschedl and co-workers (Giese et al., 1992) showed that the HMG domain of LEF-1 bends DNA strongly. Furthermore, it can substitute for IHF in a phage λ recombination assay that reflects DNA bending, if a LEF-1 binding site replaces an appropriately positioned IHF site. Subsequently, Paull et al. (1993) showed that the

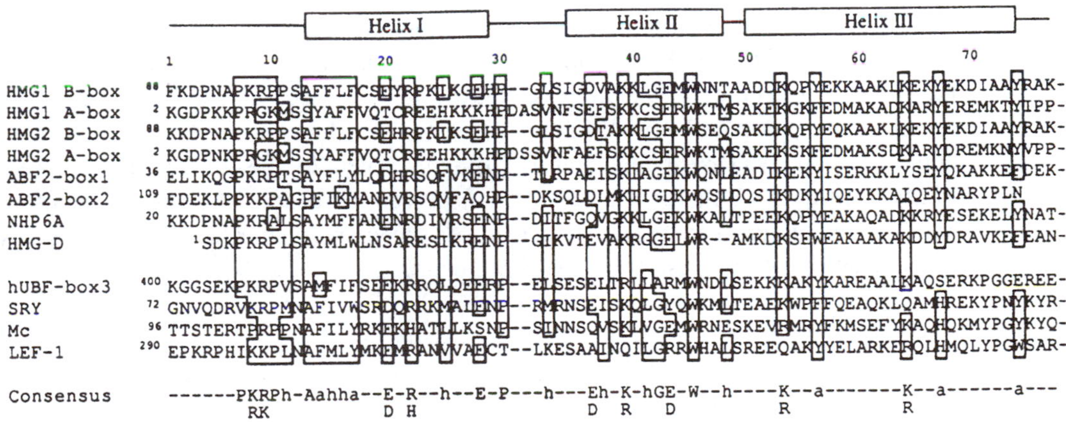

Figure 13-53
Protein sequence alignment of representative HMG boxes. The top eight represent HMG1- and HMG-like proteins that bind DNA nonspecifically, and the bottom four are members of the sequence-specific HMG box transcription factors. Boxed residues in the sequence are identical or conserved in at least 8 of the 12 sequences; the consensus sequence is shown below. Aromatic residues (a) include Y, F, W, and H; hydrophobics (h) include V, I, A, L, and F. The positions of the three helices determined by NMR (see Fig. 13.53) are indicated. [Weir, H. W., Karulis, P. J., Hill, c. S., Raine, A. R. C., Laue, E. D., and Thomas, J. O. (1993). *EMBO J.* **12**, 1311–1319 by permission of Oxford University Press.]

nonspecific DNA binding proteins HMG1 and HMG2 could replace the bacterial HU protein in a Hin recombination assay. LEF-1 also binds to the DNA minor groove, as indicated by methylation interference experiments, and by the fact that exchange of A • T pairs in the binding site by I • C does not inhibit binding (Giese et al., 1992). (Recall that A • T and I • C pairs have the same functional groups in the minor groove.) Thus the prokaryotic HU/IHF and the eukaryotic HMG1,2/HMG box proteins are similar in overall binding and DNA bending properties. They appear to function both as sequence specific (IHF, hUBF, SRY, LEF-1, etc.) and nonspecific (HU, HMG1, and HMG2) DNA bending elements that assist in the formation of protein–DNA complexes by counteracting the intrinsic stiffness of DNA. These proteins also bind strongly to DNA substrates that contain sharp bend angles, such as cruciform structures (Bianchi et al., 1989; Ferrari et al., 1992; Lilly, 1992) and Pt-modified DNA (Pil and Lippard, 1992).

However, HU and the HMG domain are not structurally homologous. Figure 13-54 shows the structure of the HMG domain as determined by NMR methods (Weir et al., 1993). There are three α helices, also marked in Figure 13-54. Note that the structure is L-shaped, with an angle of about 80° between the two arms. A number of the conserved amino acid residues are clustered at the junction of the two arms. There are a substantial number of positively charged R and K residues on the surface of the domain.

How this structure docks onto and bends DNA was revealed in the NMR structure of the complex determined by Werner et al. (1995). The concave surface of the L structure follows the DNA minor groove, which is enlarged to accommodate the interaction. At the local level this is achieved by roll between the base pairs, analogous to the DNA distortion induced by the TATA binding protein TBP, discussed below (see the commentary by Travers, 1995). Interesting features of the interaction interface include the insertion of the hydrophobic patch on the inner surface of the L into the DNA minor groove. As pointed out by Travers (1995) the reduced local dielectric constant could contribute to DNA bending away from the protein by the converse of the effect of asymmetric DNA charge neutralization discussed in Section 13.4.5. There is also a partial intercalation of an Ile side chain between successive adenines, which contributes to the positive roll that results in the overall bend.

6.10.3 HMGI(Y)

Another distinct class of "high-mobility group" proteins that seems to belong to the minor groove binding/DNA bending category is represented by HMGI and the closely analogous protein HMGY (Thanos and Maniatis, 1992; Du et al., 1993). [The HMGI(Y) proteins are not homologous in sequence to HMG1 and HMG2.] Specific binding sites for HMGI are found in the enhancer–promoter DNA sequence in the human interferon-β gene, and HMGI is required for NFκB-dependent virus induction of the gene (Thanos and Maniatis, 1992). The HMGI binding sites are phased at integral multiples of DNA helical turns, so that DNA bending at those sites should produce a nearly planar curved shape.

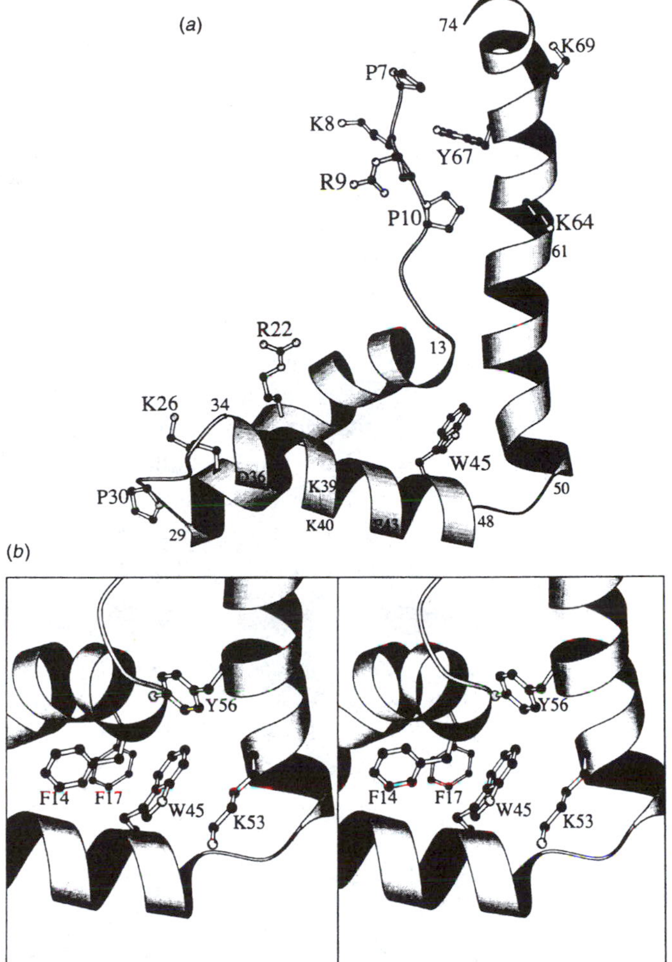

Figure 13-54
Schematic representation of an NMR structure of the B-domain HMG box, showing the location of some of the conserved residues (Fig. 13.53). (*a*) The overall structure. (*b*) A stereoview of the cluster of conserved residues at the junction of the two arms of the structure.

6.10.4 TATA Binding Protein, TBP

The TATA binding protein, so-called because it recognizes the 7 bp sequence TATAa/tAa/t centered at about position -28 relative to the transcriptional start site, is a universal eukaryotic transcription factor (reviewed by Sharp, 1992). The protein contains a highly conserved 180 amino acid carboxy terminal portion, which incorporates two homologous but not identical repeats of 66 amino acids flanking a basic region. TBP interacts with the TATA element in the minor groove (Lee et al., 1991), and bends promoter DNA (Horikoshi et al., 1992).

Solution of the crystal structure of TBP revealed a novel fold, in which the two repeated sequences form an α/β structure with approximate intramolecular dyad symmetry (Nikolov et al., 1992). The repeated domains form a "saddle"-shaped structure that interacts with DNA. The DNA binding surface is a curved, antiparallel β sheet of 10 strands, 5 from each repeat, illustrated in the stereodiagram in Figure 13-55.

A dramatic and unpredicted mode of DNA distortion was found at the TATA element when the crystal structure of TBP complexed with DNA was solved (Kim et al, 1993a; Kim et al., 1993b). As stated by Kim et al. (1993b), "binding of the saddle-shaped protein induces a conformational change in the DNA, inducing sharp kinks at either end of the sequence TATAAAAG. Between the kinks the right-handed double helix is smoothly curved and partially unwound, presenting a widened minor groove to TBPs concave, antiparallel β-sheet. Side-chain/base interactions are restricted to the minor groove, and include hydrogen bonds, van der Waals contacts and phenylalanine-base stacking interactions." Kim et al. (1993a) have a similar view ". . . the 8 bp of the TATA box bind to the concave surface of TBP by bending towards the major groove with unprecedented severity. This produces a wide open, underwound, shallow minor groove which forms a primarily hydrophobic interface with the entire under-surface of the TBP saddle. The severe bend and a positive writhe radically alter the trajectory of the flanking B-form DNA." Bending of the DNA, to a structure illustrated in Figure 13-56 (see color plate), is primarily the consequence of a series of steps with large positive roll, which act to widen the minor groove.

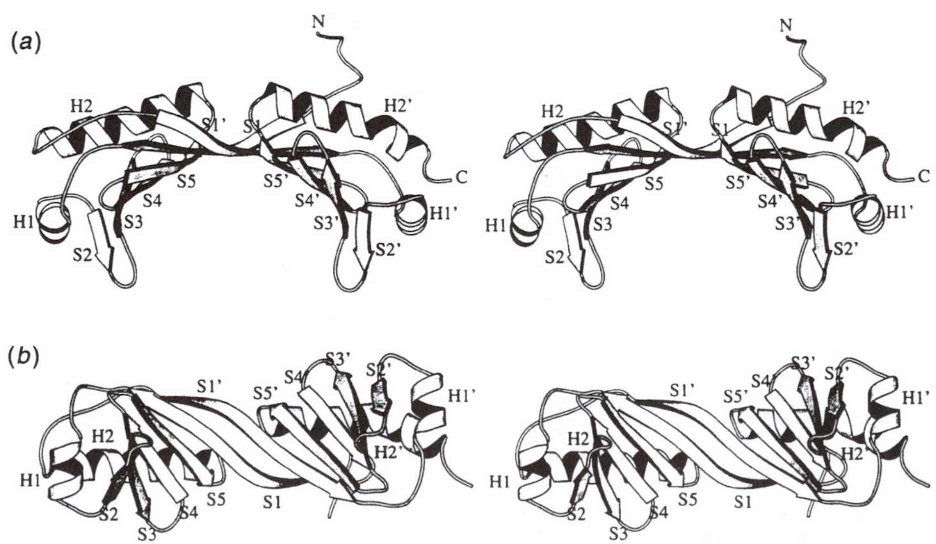

Figure 13-55
Stereodiagrams of the structure of TBP from *Aravidopsis thaliana*. (*a*) View perpendicular to the intramolecular twofold symmetry axis that relates the two homologous repeats. (*b*) View along the twofold axis, looking into the concave surface formed by the antiparallel β-sheet structure. The DNA binds to this surface through its minor groove (see Figure 13.55).

6.10.5 α-Helices Binding in the Minor Groove: The PurR Protein–DNA Complex

The *E. coli* PurR protein regulates purine biosynthesis. In its DNA binding, it combines an HTH motif with the binding of α helices in the DNA minor groove (Schumacher et al., 1994). The interaction of the HTH motif with the DNA major groove is highly homologous with λ repressor. Two dyad symmetry related α helices interact with the minor groove between the two sites of major groove–HTH interaction. Partial intercalation of Leu side chains helps to roll open the minor groove at the centrally located CpG step, kinking the DNA by about 45° in a direction away from the protein.

7. MOTIFS IN RNA-PROTEIN INTERACTIONS

Progress in characterizing the structure of RNA–protein complexes has been slower than for DNA complexes. With a few exceptions, the primary structures determined so far by crystallographic methods are complexes of tRNA synthetases with their cognate tRNAs, considered in Section 13.8. However, given the recent improvements in both enzymatic and chemical methods for large scale RNA synthesis, more rapid progress can be expected in the years ahead. We focus here on a few general principles, including especially the difference in interaction motifs that is dictated by the relative inaccessibility of the RNA major groove in a perfect duplex. According to Figure 13-5, nucleic acid double helices are best recognized from the major groove side, but as figure 13-6 shows, the major groove in an RNA double helix is deep and narrow. This inaccessibility prevents access of an α helix to the major groove, which, as we have seen, is the most common motif for recognizing a DNA double helix. Hence, the general expectation is that RNA recognition should make extensive use of single stranded regions, hairpin loops, and helices containing defects that serve to make the major groove side of the double helix more accessible (Weeks and Crothers, 1991, 1993). For a review of the subject of RNA binding proteins, see Burd and Dreyfus (1994) and Draper (1995). Some of the interactions described here do not yet satisfy the criterion for a "motif," which requires multiple observations of a similar structure in different systems. However, at the early stage of the effort to describe RNA–protein interactions, it seems likely that many observations will find subsequent repetition.

7.1 The RNP Domain: An RNA Binding Domain Containing the RNP1 and RNP2 Consensus Sequence Motifs

Many eukaryotic nuclear and cytoplasmic RNAs contain one or more repeats of an approximately 90 amino acid sequence called the "RNP motif," "RNP domain," or "RNA binding domain." These sequences were first identified as repeats in poly(A) binding proteins from yeast (Adam et al., 1986; Sachs et al., 1986). Since that time the motif has been observed in a variety of cells, including higher eukaryotes; RNAs bound include hnRNA, mRNA, and snRNA (reviewed by Dreyfus et al., 1988; Bandziulis et al., 1989; Mattaj, 1989). The ubiquitous appearance of this motif makes it likely that it is an evolutionarily ancient mechanism for specific RNA binding.

Nagai et al. (1990) used crystallographic methods to determine the structure of a 95 amino acid peptide, which contains the RNA binding domain of the U1A protein. This protein binds to stem–loop II of human U1 snRNA, recognizing primarily the loop sequence AUUGCACUCC. A related protein containing a similar RNP motif is U2A′, which recognizes the homologous sequence (U)AUUGCAGUAC(CU) in the loop of stem-loop IV of human U2 snRNA. Interchange of the underlined bases exchanges the binding specificities of the two proteins (Scherly et al., 1990). Hence the domain is able to bind to RNAs sharing common structural (stem–loop) and sequence (AUUGCA) features, but is also able to distinguish between closely related sequences.

The structure of the U1A protein RNP domain (Nagai et al., 1990) illustrated in Figure 13-57 serves as a basis for understanding the conservation that has been observed in the set of these sequences. The domain contains two α helices and a four-stranded

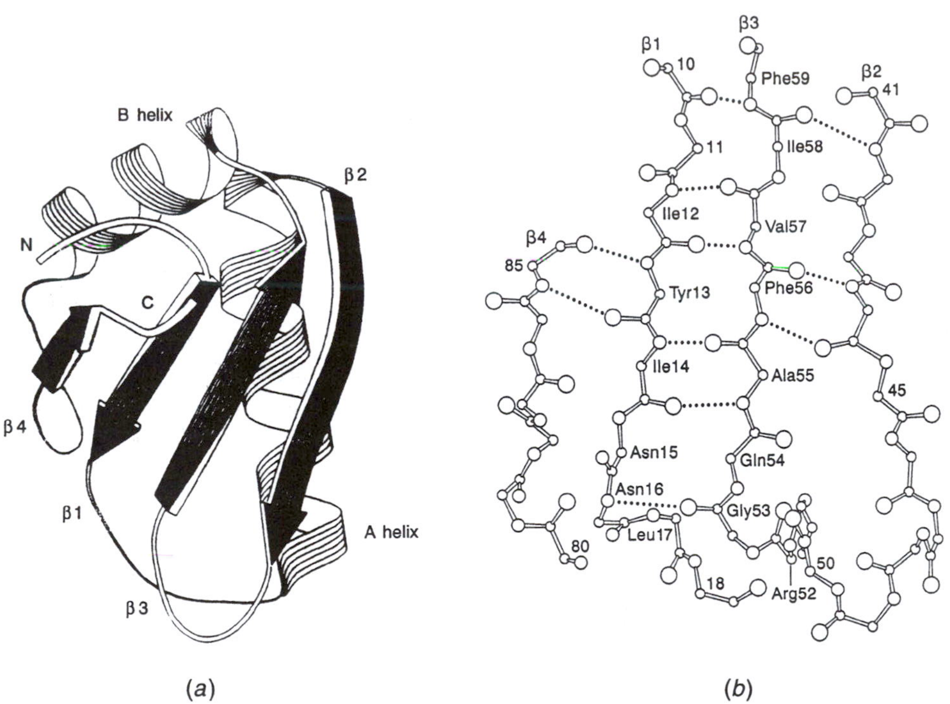

Figure 13-57
(a) Schematic ribbon representation of the RNP domain of the U1A protein. The structure contains a four-stranded β sheet (arrows) and two α helices. [Reprinted with permission from Nagai, K., Outridge, Jessen, T. H., Li, J. and Evans P. R. (1990). *Nature* **348**, 515–520. Copyright ©1990 Macmillan Magazine Limited. See Fig. 2a.] (b) Hydrogen-bonding network of the four-stranded β sheet. RNP2 (residues 12–17) and RNP1 (residues 52–59) lie side by side in the two strands $\beta 1$ and $\beta 3$ in the middle of the β sheet. [Reprinted with permission from Nagai, K., Outridge, Jessen, T. H., Li, J. and Evans P. R. (1990). *Nature* **348**, 515–520. Copyright ©1990 Macmillan Magazine Limited. See Fig. ca.]

antiparallel β sheet [Fig. 13-57(*a*)]. The two α helices are on the same side of the β sheet. Two regions of the sequence, called RNP1 and RNP2 (Dreyfus et al/, 1988), show particularly striking conservation (Fig. 13-58). These are located primarily on β strands 1 and 3, which are hydrogen bonded to each other in the β sheet, as shown in Figure 13-57(*b*). Specific interactions that stabilize this structure include, for example, a hydrogen bond between the side chains of Gln-54 from RNP1 and Tyr-13 from RNP2. Hence, the RNP1 and RNP2 motifs constitute a conserved core for the overall RNA binding domain.

The crystal structure of the RNP domain of U1A bound to an RNA hairpin was reported by Oubridge et al. (1994). The 10 nucleotide RNA loop binds to the surface of the β sheet as an open structure, and the AUUGCAC sequence of the loop interacts extensively with the conserved RNP1 and RNP2 motifs, The interactions include stacking of RNA bases with aromatic amino acid side chains and many direct and water-mediated hydrogen bonds. An NMR structure of U1A protein with the AUUAGCAC sequence contained in an asymmetric internal loop (Allain et al., 1996) shows very similar interaction features, in spite of the contrasting setting of the single-stranded region.

A single-stranded DNA binding protein that contains seqences similar to RNP1 and RNP2 has been shown to consist of a five stranded β barrel; three of the β strands share structural homology with β strands 1–3 in Figure 13-56 (Schnuchel et al., 1993; Schindelin et al., 1993).

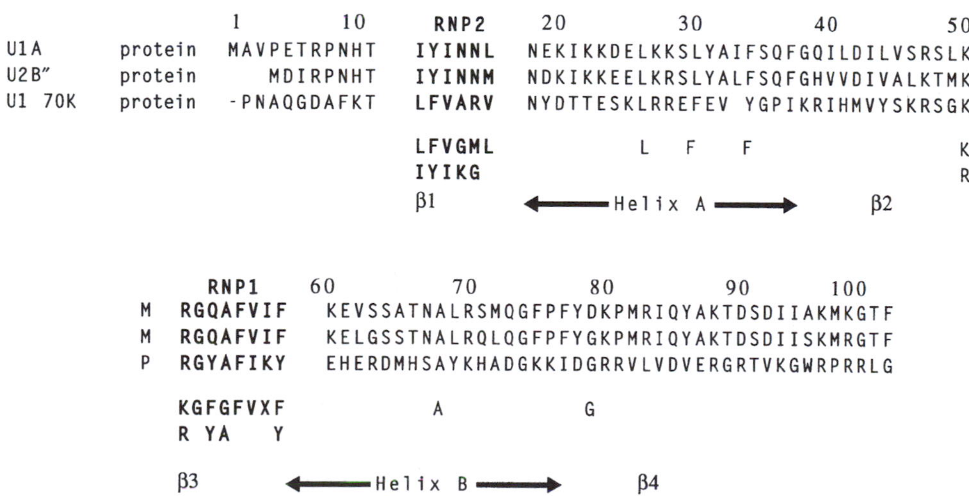

Figure 13-58
Amino acid sequences of the RNA binding domain of the U1A, U2B″ and U1 70K proteins. Consensus sequences found in more than 20 RNA binding proteins are shown below at positions where consensus could be established. [Reprinted with permission from Nagai, K., Outridge, Jessen, T. H., Li, J. and Evans P. R. (1990). *Nature* **348**, 515–520. Copyright ©1990 Macmillan Magazine Limited. See Fig. 1.]

7.2 Bacterial Virus Coat Protein-RNA Interaction

Several RNA viruses that infect *E. coli,* such as R17, f2, GA, Qβ, and SP, contain a major coat protein that binds with high specificity to a small region of the RNA genome. The nature of the protein–RNA recognition process in these related systems has been intensively studied, particularly by Uhlenbeck and his co-workers (reviewed by Witherell et al., 1991). The overall thermodynamics indicate moderately strong binding for a protein-nucleic acid complex ($\sim 10^9\ M^{-1}$). The electrostatic component inferred from the ionic strength of binding is relatively weak, corresponding to about five ion pairs (Witherell et al., 1991). Binding specificity, estimated to be above a 10^6-fold preference for the specific sequence over bulk RNA, is comparable to prokaryotic repressors, but below the cutting specificity of restriction enzymes. The structure of a coat protein–RNA complex has been solved by Vallegard et al. (1994).

Given the specificity and nonionic character of the interaction, it is perhaps surprising to find how few specific bases are required for recognition, as opposed to the general requirement for a defined stem–loop structure. Figure 13-59 shows the wild-type binding site sequences, along with the consensus sequence deduced from mutational studies (Witherall et al., 1991), for four different viral coat proteins. In the figure, a filled circle indicates bases that can be changed without affecting interaction strength; the larger the circle, the greater the number of base changes that were tested. Bases written in

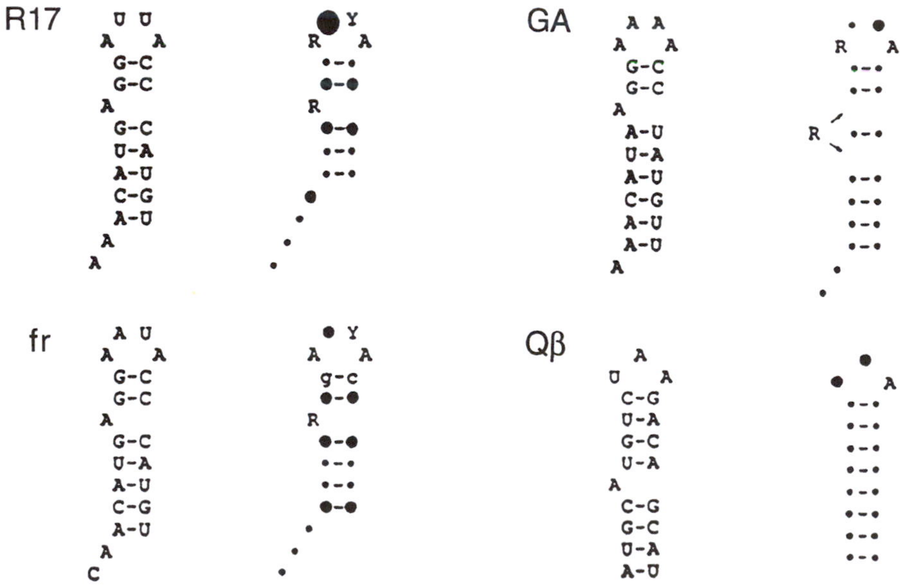

Figure 13-59
Natural target sequences for four RNA virus coat proteins, together with their binding site sequence and size requirements. Filled circles indicate positions where the base, or base pairs if the circles are connected by a line, can be changed without a major effect on binding. These diagrams are truncated to indicate the minimum binding site size for full affinity [Reprinted with permission from Witherell et al., 1991. See Fig. 6.]

lower case indicate that no sequence changes were made in the experiments. When the circles are connected by a line to indicate base pairing, base complementarity must be conserved.

The R17 coat protein appears to be among the most demanding with respect to sequence, but even in that case the constraints are limited: Binding requires a bulged purine nucleotide 2 bp away from a loop of four, whose sequence can be any member of the set RNYA (Fig. 13-59), where R is purine and Y is pyrimidine. Phage fr has similar requirements, except that the first base in the loop must be A. In the case of phage GA, the purine bulge can be either 2 or 3 bp away from the loop. The only requirement for Qβ coat protein is A in the third position of a loop of 3. The much lower information content of these binding sites (see Section 13.3.2) compared to bacterial regulatory protein DNA binding sites may reflect the comparatively low information content of the RNA virus genome, which means that the number of sites that must be distinguished is small.

It is important to recognize that lack of sequence specificity at an interaction site does not mean that there is no contact between protein and RNA there. For example, removal of the 5′ terminal nucleotide (A) from the R17 target sequence shown in Figure 13-59 reduces binding by about 1000-fold (Witherell et al., 1991). Given the lack of dependence on sequence at that site, the required contact is likely to be electrostatic. The R17 RNA target sequence does not have to be double helical at the base of the stem–loop structure, since the 3′ terminal GU can be removed without loss of binding strength (Witherell et al., 1991).

7.3 HIV Tat Protein-TAR RNA Interaction

The HIV transactivating protein Tat acts by binding to a segment of the viral RNA transcript near its 5′ end, thereby activating transcription by a mechanism that remains controversial. The Tat-responsive part of the RNA is called TAR. The primary features of this system for the present discussion include the small size of both RNA and protein segments that are responsible for "specific" binding. However, binding specificity, as measured by relative affinities of specific and nonspecific binding sites, is much less pronounced than for RNA viral coat proteins and the DNA–protein interaction systems considered earlier.

Tat is a small protein of 86 amino acids, of which only a much smaller number are vital for RNA binding. As shown by Weeks et al. (1990), a 38 amino acid carboxy terminal fragment of Tat binds TAR RNA in clear preference over DNA and nonspecific RNAs. Even a peptide as short as 14 amino acids, spanning the basic region of the protein beginning at Gly 48, is capable of specific recognition. Calnan et al. (1991) continued this reductionism further by showing that a 9 amino acid basic peptide consisting of arginine and lysine can mimic specific binding, as long as there is at least one arginine in the position 51–54 range.

Chemical interference and mutagenesis studies (Weeks et al., 1990; Weeks and Crothers, 1991) revealed the interaction target on TAR shown in Figure 13-60. Nucleotides N in the loop and base bairs N–N can be changed without altering the binding affinity of Tat peptides. The binding site requires only a U at the 5′ end of a bulge loop of at least two nucleotides, and the adjacent G•C and A•U base pairs.

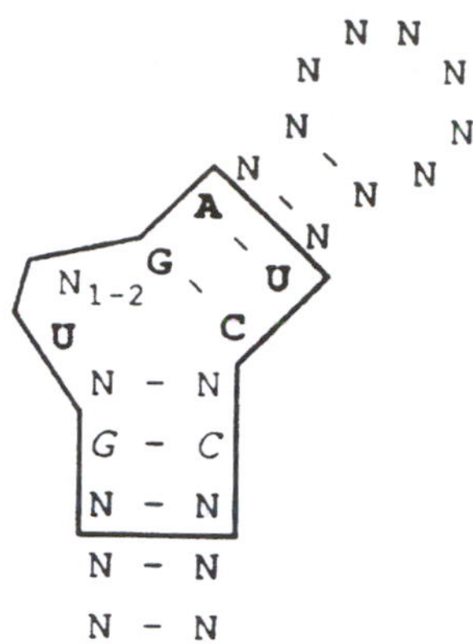

Figure 13-60
Summary of RNA structure and sequence determinants for specific TAR RNA binding by Tat peptides, for example, Tfr38. The contact site defined by chemical modification-based interference is enclosed by a solid line. Residues essential for specific interaction with Tfr38 are in boldface. A base pair whose modification moderately affects binding is italicized. The N indicates positions where base pairs or loop nucleotides can be changed without affecting complex stability (Fig. 5 of Weeks and Crothers, 1991).

The NMR experiments of Williamson, Frankel, and co-workers (Puglisi et al., 1992, 1993) elucidated the sequence requirements in the binding site in Figure 13-60 and the requirement for arginine in the peptide. Figure 13-61 shows the model derived from the NMR studies. The G residue in the conserved G • C pair is hydrogen bonded to arginine through its carbonyl and N7 in the major groove, the same pattern that has been identified for a number of protein–DNA interaction systems (see Table 13-1). Ionic interactions with phosphate help to stabilize the interaction, analogous to the

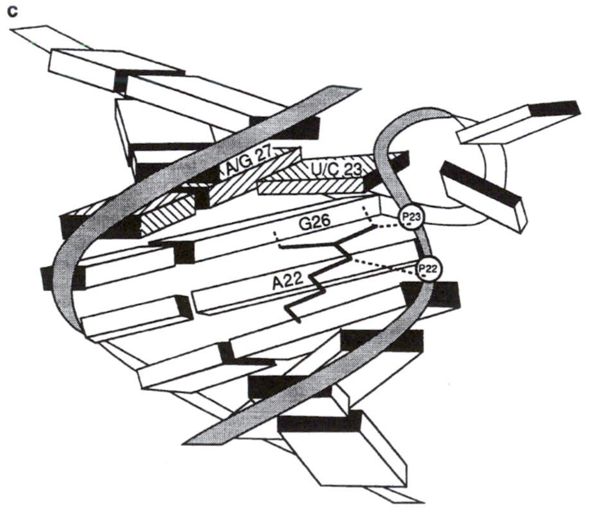

Figure 13-61
Schematic drawing of the structural model derived by Puglisi et al. (1993) from NMR measurements on the interaction of argininamide with TAR RNA. The guanidinium side chain is hydrogen bonded to G26, which is stacked on A22. Ionic interactions to phosphates 22 and 23 stabilize the position of the guanidinium group. Either U or C 23 in the bulge loop forms a base triple with A or G at position 27. This structure is induced by the binding of argininamide (Fig. 2 of Puglisi et al., 1993).

interaction of Asp or Glu with the arginine guanidinium group in the Cys_2His_2 Zn finger proteins (Pavletich et al., 1991). The conserved U in the bulge loop forms a base triple with the conserved A•U pair. This feature of the model was confirmed by showing that exchange of the A•U pair for G•C required a change of the bulge loop U to C (Puglisi et al., 1993). At slightly acidic pH, the structure is then stabilized by a C^+•G•C base triple.

Tat peptides and R17 coat protein interacting with their RNA targets show roughly equal sequence requirements in their preferred sites. However, the binding specificity of Tat peptides, measured by the comparative affinity for nonspecific sites, is only about 10–100-fold, depending on the nature of the competing nonspecific RNA (Weeks and Crothers, 1991), compared to 10^6 or more for R17 coat protein. This difference may reflect the limited set of contacts that are possible between small Tat peptides and their RNA target.

Weeks and Crothers (1991, 1993) compared the accesibility of purines in models for the TAR sequence to reaction with diethyl pyrocarbonate in the major groove. Reactivity is greatly increased adjacent to a bulge loop, and increases as the size of the loop is increased from 1 to 3. This observation led to the suggestion that bulge loops can facilitate RNA recognition by opening the major groove sufficiently to allow access by amino acid side chains. The model shown in Figure 13-6 (color plate) shows how this might work sterically: stacking of the bases in the loop continues the helix on the strand containing extra bases. Widening of the major groove results, as can be seen from the view sighting along the major groove in the lower right perspective in Figure 13-6 (color plate).

8. STRUCTURAL MOTIFS IN ENZYMES THAT ACT ON NUCLEIC ACIDS

Structural characterization of enzymes that act on DNA, particularly in complex with their nucleic acid substrates, has greatly expanded our knowledge of the chemical and physical principles underlying nucleic acid function. In this section, we review a selection of these systems, with an organizational focus on the underlying chemical concepts.

8.1 The Chemistry of Nucleic Acid Cleavage: Enhanced Formation of a Pentacoordinate Phosphorus Intermediate by Metaloenzymes

Reactions in which the internucleotidic bonds in RNA or DNA are formed or broken are members of the general class of phosphoryl-transfer reactions:

$$R_1{-}O{-}\overset{\displaystyle O}{\overset{\|}{\underset{\underset{\displaystyle O^-}{|}}{P}}}{-}O{-}R_2 \;+\; R_3{-}OH \;\longrightarrow\; R_1{-}O{-}\overset{\displaystyle O}{\overset{\|}{\underset{\underset{\displaystyle O^-}{|}}{P}}}{-}O{-}R_3 \;+\; R_2{-}OH$$

in which the phosphoryl group R_1OPO^- is transferred from R_2O^- to R_3O^-. In this reaction, R–OH can be, for example, a deoxyribose sugar, pyrophosphate, or water, depending on whether the chain is being formed or cleaved. Reactions of this kind generally proceed through a pentacoordinated intermediate which has a trigonal bipyramidal structure (reviewed by Westheimer, 1968). The reaction proceeds by attack of a nucleophile (RO^-) on tetrahedrally coordinated phosphorus, with the entering and leaving groups in line on the apices of the bipyramidal structure. The expected consequence is inversion of configuration at phosphorus (Gupta and Benkovic, 1984).

Formation and/or productive breakdown of the pentacoordinate phosphorus intermediate is a likely limiting step in phosphoryl transfer reactions. The process can therefore be accelerated by agents that make R_3–OH a better nucleophile, or that stabilize the leaving group oxyanion R_2–O^-. In model systems, divalent metal ions can accelerate nucleophilic displacement reactions at phosphorus (Herschlag and Jencks, 1987).

These chemical principles are seen clearly at work in the structures of the $3' \rightarrow 5'$ exonucleolytic site of DNA polymerase I (Beese and Steitz, 1991) and alkaline phosphatase (Kim and Wyckoff, 1991). Figure 13-62(*a*) shows the inferred structure around the proposed pentacoordinate intermediate for the DNA polymerase I exonuclease site (Beese and Steitz, 1991). The attacking nucleophile, OH^- in this case, is at one apex of the bipyramid, and the leaving group, deoxyribose 3′-OH, is at the other. Two metal ions (Mg^{2+}, Mn^{2+} or Zn^{2+}, but not Ca^{2+}) are positioned so that they aid in formation and productive breakdown of this complex: metal A stabilizes the negative charge on the attacking OH^- and metal B stabilizes the negative charge on the departing oxyanion. In addition, both metals help to stabilize the pentacoordinated oxygens at phosphorus. Amino acid side chains play a vital role in proper positioning of the metal ions, and may also contribute to catalysis in a more direct way, for example, by a hydrogen bond from Tyr-497 to the OH^- nucleophile to stabilize its charge and orient the lone electron pair appropriately for nucleophilic attack on phosphorus (Beese and Steitz, 1991). The ribonuclease H domain of HIV-1 reverse transcriptase appears to have a similar two-metal structure at the active site (Davies et al., 1991).

Figure 13-62(*b*) shows in schematic form the analogous pentacoordinate intermediate inferred from the structure of alkaline phosphatase (Kim and Wyckoff, 1991). In this case, the attacking nucleophile is the OH group of Ser 102, leading to transient linkage of phosphate to the enzyme. Subsequent attack by OH^- (from H_2O) leads to the production of enzyme and free phosphate.

Steitz and Steitz (1993) proposed that phosphoryl-transfer reactions catalyzed by RNA might proceed by an analogous mechanism, except that RNA phosphate and possibly other RNA functional groups serve to hold the two metal ions in the appropriate position for catalysis. An interesting example of a metalloribozyme that uses Pb^{2+} and Mg^{2+} is provided by Pan and Uhlenbeck (1992).

8.2 Maintenance of DNA Polymerase Processivity: The Sliding Clamp

Processive enzymes add nucleotides to the growing polymer without dissociation. DNA polymerase III is distinguished by its ability to replicate long stretches of DNA

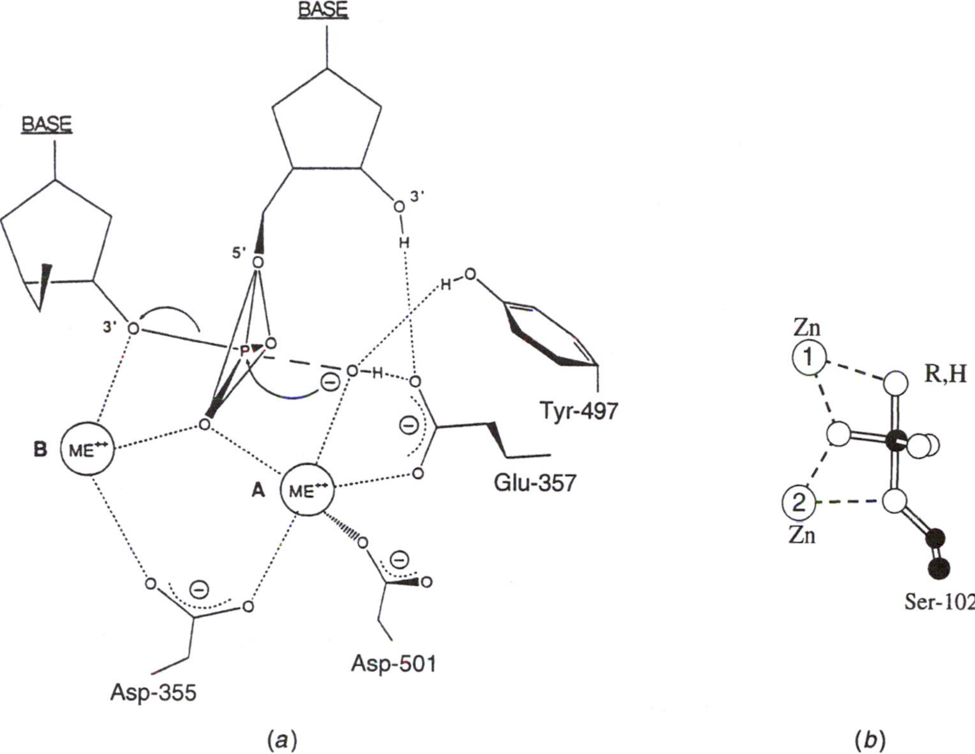

Figure 13-62
Pentacoordinate phosphorus intermediates proposed for two metalloenzymes. (*a*) The active site in the 3′ → 5′ exonuclease site of the large proteolytic (Klenow) fragment of *E. coli* DNA polymerase I. The OH- is the nucleophile, and deoxyribose 3′–O$^-$ directly opposite in the bipyramid is the leaving group. Both the nucleophile and the leaving group are stabilized by divalent metal ions A and B (Fig. 10 of Beese and Steitz, 1991) (*b*) Schematic view of the corresponding structure in alkaline phosphatase. In this case the two in-line substituents are the OH group from ser 102 and either the nucleoside or OH$^-$. [Reprinted from Kim, E. U. and Wyckoff, H. W., Reaction Mechanism of Alkaline Phosphatase Based on Crystal Structures, *J. Mol. Biol.* **218**, 449–464, Copyright ©1991, by permission of the publisher Academic Press London. See Fig. 6.]

without dissociating, and is hence a strongly processive enzyme. The β subunit of the polymerase is vital for this function. Kong et al. (1992) elucidated the structural basis for this property when they solved the structure of the protein. A dimer of the β subunit forms a ring-shaped structure lined by 12 α helices that can encircle duplex DNA (see color plate for Fig. 13-63). The structure is highly symmetrical, with each monomer containing three domains of identical topology. Once locked around DNA by a process that requires ATP, the circular structure can slide along the double helix, preventing dissociation of polymerase.

8.3 DNase I: Binding to and Distortion of the Minor Groove

Suck et al. (1998) solved the structure of DNase I with a nicked DNA oligonucleotide. This enzyme cleaves DNA in a relatively nonspecific way, although it is generally observed that sequences with narrow minor grooves are poorly cut. An unusual feature of the complex is interaction of a tyrosine and two arginine side chains with the DNA minor groove, resulting in substantial conformational changes of the DNA. The tyrosine residue stacks with the deoxyribose ring of a thymine, and flips the C4′–C5′ and C5′–O5′ torsional angles from the usual gauche/trans to trans/gauche, producing a large roll angle (13°) between the T and adjacent C. The two argine residues hydrogen bond to pyrmidine carbonyl oxygen atoms. Overall, there is a bend of about 21° *away* from the enzyme, which is in a direction to widen the minor groove at the interaction site to about 15 Å, compared to the canonical value of 12 Å. This result rationalizes the tendency of DNase I to avoid double helices with narrow minor grooves, since such sequences would presumably be more difficult to distort into the shape found in the complex (Suck et al., 1988).

8.4 Aminoacyl tRNA Synthetase–tRNA Complexes

A critical linkage in the process of genetic coding is the assignment of specific amino acids to a base triplet codon. As originally hypothesized by Crick (1957) in his adaptor hypothesis, this linkage is accomplished by tRNA molecules that contain an anticodon complementary to the codon. Attachment of a specific amino acid to the 2′- or 3′-OH of the tRNA requires recognition of the tRNA, amino acid, and ATP by an enzyme called an aminoacyl-tRNA synthetase, of which there is usually (at least) one for each amino acid. The ultimate "chicken and egg" problem is to understand how this crucial reaction, which now involves an elaborate interaction between protein and RNA, could have gotten started before there was a coding system for the production of proteins. The general problem of recognition of tRNAs by their cognate synthetases has been reviewed by Saks et al. (1994).

8.4.1 Two Classes of Aminoacyl-tRNA Synthetases

It is now evident that tRNA synthetases fall into two major classes (Eriani et al., 1990), possibly implying two independent origins for these evolutionarily ancient enzymes. Class I enzymes, of which there are 10 (Table 13-6), contain a "Rossman fold" motif, which consists of a central β-sheet flanked by α helices and serves in general as a nucleotide binding domain. In Class I aminoacyl-tRNA synthetases the nucleotide binding domain contains the characteristic sequence motifs His-Ile-Gly-His (HIGH) and Lys-Met-Ser-Lys-Ser (KMSKS). Class II synthetases, (Table 13-6) lack these motifs, but share instead one or more of the three sequence motifs identified by Eriani et al. (1990). It is also notable that the position of the acyl linkage, whether at the 2′- (Class I) or the 3′-OH (Class II), shows a strong correlation with synthetase class (Table 13-6).

Table 13.6
Partition of Aminoacyl Synthetases into Classes I and II on the Basis of the Presence or Absence of Protein Sequence Motifs.[a]

Class II Synthetases					
Motif 3 Only		Motifs 1, 2, 3		Class I Synthetases HIGH + KMSKS	
Gly (α2β2)	3′—OH				
Ala (α4)	3′—OH				
		Pro (α2)	3′—OH		
		Ser (α2)	3′—OH		
		Thr (α2)	3′—OH		
		Asp (α2)	??		
		Asn (α2)	3′—OH		
				Glu (α)	2′—OH
				Gln (α)	?
		His (α2)	3′—OH		
		Lys (α2)	3′—OH		
				Arg (α)	2′—OH
				Val (α)	2′—OH
				Ile (α)	2′—OH
				Leu (α)	2′—OH
				Met (α2)	2′—OH
		Phe (α2β2)	2′—OH	Tyr (α2)	??
				Trp (α2)	?

[a]The subunit structure is indicated, along with the position of aminoacylation. [Reprinted with permission from Eriani, G., Delarue, M., Poch, O., Gangloss, J., and Moras, D. (1990). *Nature (London)*, **347**, 203–206. Copyright ©1990 Macmillan Magazines Limited].

8.4.2 Class I Synthetase-tRNA Recognition

Crystallographic methods have so far been successful in the determination of the structure of several tRNA–synthetase complexes, for example, glutaminyl tRNA synthetase (GlnRS), a Class I enzyme, with $tRNA^{Gln}$ (Rould et al., 1989), AspRS (Class II) with $tRNA^{Asp}$ (Ruff et al., 1991), and serRS (Class II) with $tRNA^{Ser}$ (Biou et al., 1994). These structures are rich in the detailed view they provide of RNA–protein interactions.

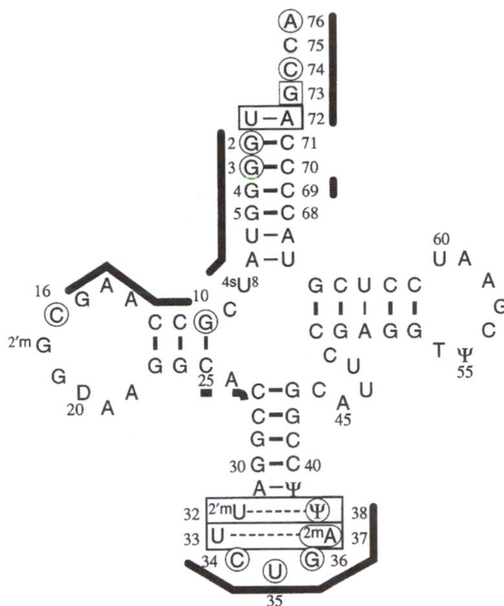

Figure 13-64
Schematic summary of protein–RNA and RNA–RNA interactions seen in the refined GlnRS–tRNA^{Gln} crystal structure. Bases directly and specifically recognized by the enzyme are circled, whereas bases that interact with the protein through a single water molecule are in boldface. Nucleotides that facilitate the tRNA in assuming the conformation necessary for binding the protein are boxed. Interactions between the enzyme and sugar–phosphate backbone are indicated by thick dark lines. [Reprinted with permission from Rould, M. A., Perona, J. J., and Steitz, T. A. (1991). *Nature* **352**, 213–218. Copyright ©1991 Macmillan Magazine Limited. See Fig. 7.]

Figure 13-64 summarizes the numerous interaction sites on tRNA^{Gln} observed in the complex with GlnRS. The spatially widespread character of the contacts arises because the surfaces of the two macromolecules have a large degree of overlap, as shown in Figure 13-65. Rould et al. (1991) summarized the principal features of the interaction in terms of three general ways in which nucleotides function as identity elements for recognition by GlnRS:

1. There are direct and water-mediated contacts with the protein in the minor groove, for example, with the amino group of G2, G3, and G10. A well-ordered water network between the minor grooves and the protein enforces the requirement for base pairs at those guanines.

2. Identity element nucleotides can form RNA structures required for enzyme binding that other nucleotides do not favor. This principle is illustrated by Figure 13-66(*a*), showing the structure at the acceptor stem. The U1 • A72 base pair is broken, and G73 (the so-called discriminator base) folds back to stack on C75 and A76 and make a hydrogen bond between its amino group and the 5′ phosphate of A72. Another example is the formation of nonstandard ${}^{2'm}$U32 • Ψ38 and U33 • ^{2m}A37 pairs in the acceptor stem loop.

3. Single-stranded bases can interact with recognition pockets on the protein, as illustrated in Figure 13-66(*b*), which shows schematically the structure of the protein pocket interacting with the anticodon loop. Bases C34, U35, and G36 are in direct contact with the protein. Note that the nonstandard pairs at positions 32 and 33 in the adjacent part of the anticodon loop are essential for the anticodon conformation which is recognized. This example helps explain why identity elements do not have to be in direct contact with the protein. Another nucleotide that interacts directly with the protein is C16.

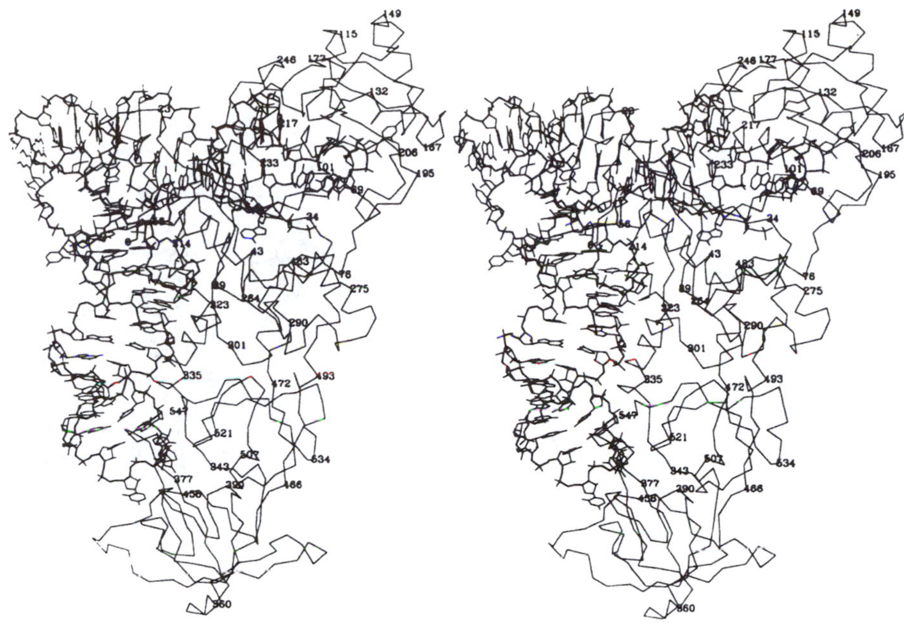

Figure 13-65
Stereodrawing of the α-carbon backbone of the GlnRS with its cognate tRNA and ATP bound. [Reprinted with permission from Rould, M. A., Perona, J. J., Söll, D., and Steitz, T. A. (1989). *Science* **246**, 1135–1142. Copyright ©1989 American Association for the Advancement of Science. See Fig. 4.]

8.4.3 Class II Synthetase-tRNA Interaction

The Class II synthetase–tRNA interaction is not homologous with the corresponding Class I complex (Ruff et al., 1991); even the orientation between synthetase and tRNA is different. Opposite sides of the tRNA molecule are recognized by AspRS and GlnRS: $tRNA^{Asp}$ interacts using its variable-loop and the major groove side of the acceptor stem, whereas in $tRNA^{Gln}$ the D-loop side and minor groove side of the acceptor stem face the protein.

In Class II synthetases the Rossman dinucleotide binding motif is replaced by a new motif, shown schematically in Figure 13-67(*a*), which consists of a large β sheet surrounded by α helices. The sequence motifs identified by Eriani *etal.* (1991) are located in this domain. This region of the molecule interacts with the acceptor stem of the tRNA, as shown in Figure 13-67(*b*). It is likely that the basic features of this interaction are conserved in other Class II synthetase-tRNA interactions.

AspRS has important recognition interactions with $tRNA^{Asp}$ in the acceptor stem. However, unlike GlnRS, in this case the 3′ acceptor stem continues in a regular helical stack, and the protein forms contacts with the first base pair and discriminator base in the major groove. This interaction is permitted sterically because the major groove is accessible for the first few base pairs from the end of a double helix (Weeks and Crothers, 1993).

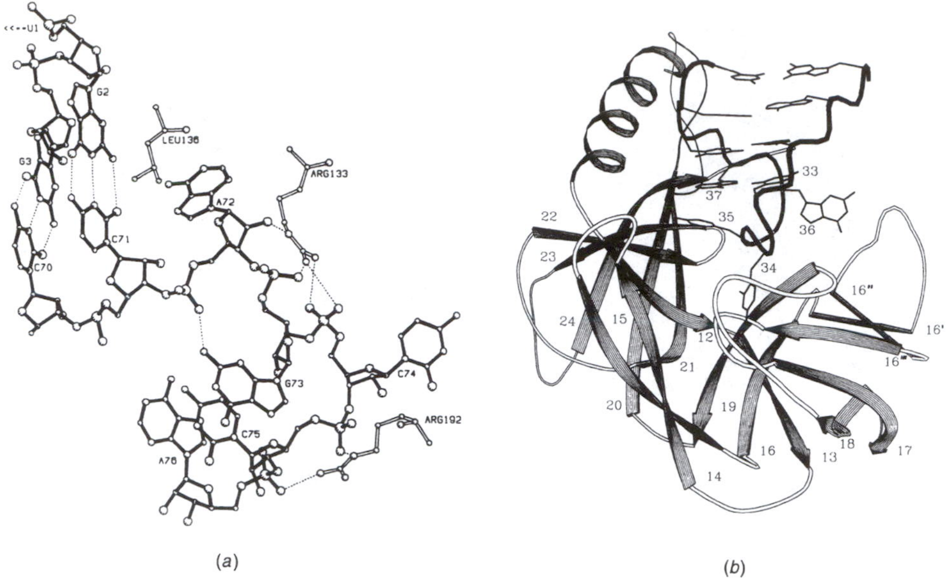

Figure 13-66
Some details of the interaction of GlnRS with its cognate tRNA. (*a*) Structure of the acceptor stem, showing the disruption of the U1 • A72 base pair, and the folding back of G73 to form a hydrogen bond between its amino group and the phosphate backbone. [Reprinted with permission from Rould, M. A., Perona, J. J., Söll, D., and Steitz, T. A. (1989). *Science* **246**, 1135–1142. Copyright ©1989 American Association for the Advancement of Science. See Fig. 7.] (*b*) Schematic diagram of interactions at the anticodon loop. Nucleotides 34,35 and 36 interact specifically with the protein, and the nonstandard pairs at residues 32 • 38 and 33 • 37 assist in forming the requisite conformation of the anticodon loop. The protein in this region (residues 340–544) consists of two β-barrel domains. [Reprinted with permission from Rould, M. A., Perona, J. J., and Steitz, T. A. (1991). *Science (London)* **352**, 313–318. Copyright ©1991 Macmillan Magazine Limited. See Fig. 2.]

8.5 Structures of DNA Topoisomerases

Topoisomerases are essential enzymes that allow DNA to be broken and resealed, usually to solve topological problems. Type I topoisomerases open one strand and allow local rotation about the remaining backbone linkage. In superhelical DNAs, this action causes changes in the DNA linking number in units of ± 1. The consequence can be relaxation of superhelicity, for example. In type II topoisomerases both DNA strands are broken at once, allowing passage of one duplex through another. The consequence in superhelical DNAs is change of the linking number in units of ± 2. Because of the ability to pass one strand through the other, DNA can also be decatenated by these enzymes, allowing resolution of DNA tangles.

Structures of protein domains of both kinds of topoisomerases without DNA have been reported. Fragments of *E. coli* topoisomerase I (Lima et al., 1994) and yeast topoisomerase II (Berger et al., 1996) reveal a large central cavity. These structures provide attractive models for topoisomerase function. For example, for yeast topoisomerase II the model is described (Berger et al., 1996) as "an ATP-modulated clamp with two sets

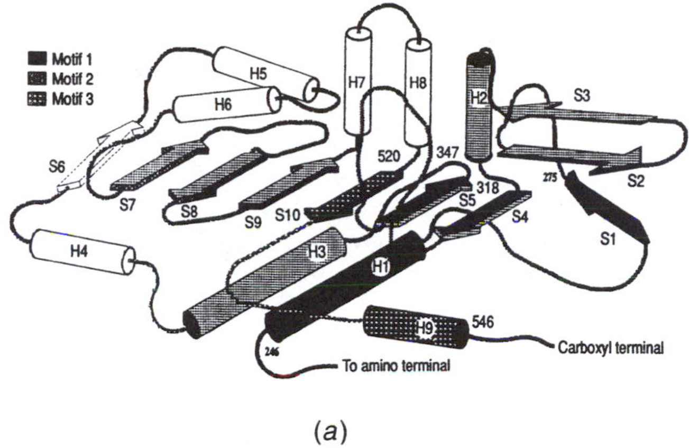

(*a*)

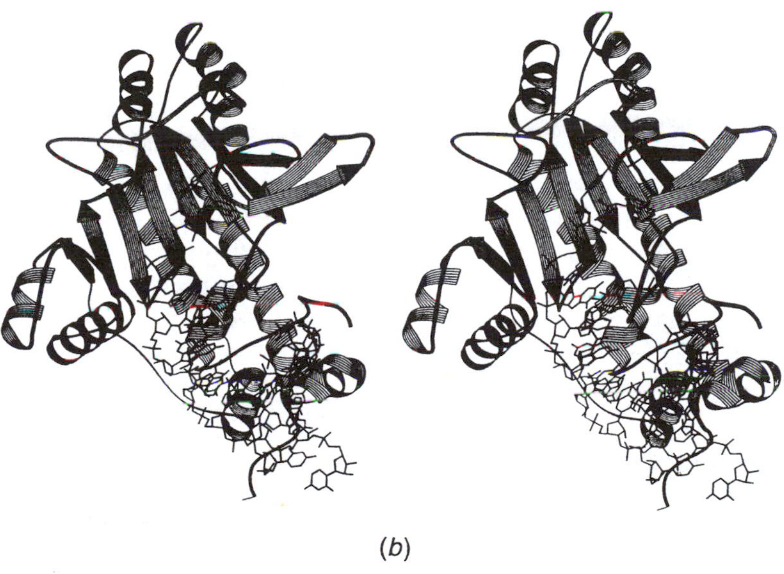

(*b*)

Figure 13-67
Topology and acceptor stem binding of the carboxyl-terminal domain of AspRS. (*a*) The three conserved motifs characteristic of Class II synthetases (Eriani et al., 1990) are shaded as shown: motif 1 (residues 258–275) includes α-helix H1 and β-strand S1; motif 2 (residues 315–349) includes β-strands S4 and S5; and motif 3 (residues 517–549) includes β-strand S10 and helix H9. All of the shaded areas are thought to be topologically invariant in Class II synthetases. [Reprinted with permission from Ruff, M., Krishnaswamy, S., Boeglin, M., Poterszman, A., Mitschler, A., Podjarny, A., Rees, B., and Thierny, J. C. (1991). *Science* **252**, 1682–1689. Copyright ©1991 American Association for the Advancement of Science. See Fig. 7.] (b) Ribbon steroview of the carboxyl terminal domain of one subunit of AspRS (residues 219–557) and the acceptor end of the interacting cognate tRNA. This view emphasizes the active site cavity built around a six-stranded antiparallel β sheet. [Reprinted with permission from Ruff, M., Krishnaswamy, S., Boeglin, M., Poterszman, A., Mitschler, A., Podjarny, A., Rees, B., and Thierny, J. C. (1991). *Science* **252**, 1682–1689. Copyright ©1991 American Association for the Advancement of Science. See Fig. 6.]

of jaws at opposite ends, connected by multiple joints. An enzyme with bound DNA (in the central hole) can admit a second DNA duplex through one set of jaws, transport it through the cleaved first duplex, and expel it through the other set of jaws."

9. SOLUTION CHARACTERIZATION OF BINDING INTERACTIONS

Understanding the physical basis for protein–nucleic acid interactions requires characterization of these complexes in solution. Among the questions to be answered are the binding affinity and its dependence on such environmental parameters as ionic strength and temperature. Variation of affinity with sequence changes in either nucleic acid or protein is of particular interest for comparison with predictions based on contacts observed in structural studies. Footprinting experiments, both chemical and enzymatic, define the locus of binding, knowledge of which is a prerequisite for structural studies of the complex. Interference methods, in which reaction of specific base or phosphate groups prevents complex formation, help to define molecular contacts. All of these methods depend heavily on gel electrophoresis, both of the nucleic acid and its complex with protein. Spectroscopic methods, such as NMR, are well suited to detect complex formation, but their relative insensitivity (fractional m*M* concentrations are required for NMR) make them ill suited to study binding equilibria. Fluorescence spectroscopy, among the most sensitive of spectroscopic techniques, can be used in the nanomolar concentration range, and so is suited to measurements of partially associated systems with dissociation constants as small as 10^{-9} *M*.

9.1 Filter Binding for Detection of Complexes

The filter binding method for detection of protein–nucleic acid complexes was for more than a decade the mainstay of solution studies of binding interactions. In early classic experiments this approach was used to detect binding of mRNA to ribosomes (Nirenberg and Leder, 1964), binding of RNA polymerase to DNA (Jones and Berg, 1966), and binding of tRNA to aminoacyl tRNA synthetase (Yarus and Berg, 1967). Bourgeois and co-workers, with a focus on *lac* repressor (Riggs et al., 1970b), developed the approach to a quantitative method for characterizing regulatory protein–DNA interactions. More recently, the introduction of a double filter approach by Wong and Lohman (1993), which allows determination of the amounts of protein-bound and free DNA simultaneously, has improved the quantitative reliability of this method.

This method is based on the fact that double-stranded nucleic acids do not adhere to nitrocellulose filter, although proteins often do. When a nucleic acid is bound to a protein, the complex is usually bound to the filter. Hence, with a ^{32}P label on DNA, the assay for formation of a complex reduces to detection of radioactivity on the filter. This approach, like the gel electrophoresis methods considered in Section 13.9.2, is a pseudoequilibrium method: Because the assay is not instantaneous, it is necessary to assume that the equilibrium is not perturbed during the time course of the assay itself, during which it is necessary to wash the filter to remove nonspecifically associated DNA.

Even though some features of the filter binding process remain mysterious, for example, the fact that often even in large protein excess not all the nucleic acid can be induced to bind to the filter, this method remains arguably the most accurate general technique for determining *absolute* binding constants in protein–nucleic acid interactions. An illustrative application is that of Ebright et al. (1989) to binding of *E. coli* CAP protein to DNA. Earlier efforts on this system used DNA fragments of several hundred base pairs, and were hampered by relatively poor retention of the specific complex on the filter and high-nonspecific binding activity. Ebright et al. (1989) showed that the system is well behaved when 40 bp fragments are used. The CAP concentrations measured varied from picomolar (pM) to micromolar (μM); this large range of accessible concentrations is characteristic of the filter binding method. The technique can also be used at high-salt concentrations, which tend to cause problems in the gel electrophoresis method (see below). Extensive application of the filter binding method to RNA–protein interactions have been reported by Uhlenbeck and co-workers (see Witherell et al., 1991).

9.2 Gel Electrophoresis Methods

It has been known for many years that nucleoprotein particles such as RNA polymerase–DNA complexes (Chelm and Geiduschek, 1979) can be observed as discrete bands in nondenaturing gel electrophoresis. Early experiments on related systems also showed, for example, that DNA complexes with histone fractions could be detected by electrophoresis (Klevan et al., 1978), and that the electrophoretic mobility of the nucleosome is affected by binding of HMG proteins (Sandeen et al., 1980). In 1981, Garner and Revzin (1981) and Fried and Crothers (1981) found that regulatory proteins such as *lac* repressor bound to DNA can be detected as discrete bands in electrophoresis. In low ionic strength electrophoresis buffers it is even possible to detect weak complexes such as *lac* repressor bound nonspecifically to DNA. An excess of nonspecific competitor DNA can be used to abolish this effect (Fried and Crothers, 1981). In general, protein–nucleic acid complexes are stabilized in the gel by the low ionic strength of the electrophoresis buffer, and also by a caging or molecular sequestration effect (Fried and Liu, 1994), which increases the local concentration of protein around its binding target.

The "gel retardation" or "mobility shift" assay has become standard for qualitative characterization of protein–nucleic acid interactions. It has a distinct advantage in that it can be used to assay complex mixtures: Even if more than one DNA binding activity is present, the complexes can often be distinguished on the basis of mobility. Furthermore, the slab gel format allows multiple assays to be performed simultaneously.

9.2.1 Absolute Binding Constants by Gel Electrophoresis

Like filter binding, gel electrophoresis analysis of protein–DNA complexes is a pseudoequilibrium method. The potential artefacts include processes that occur during the approximately 15–30 s required while the complex enters the gel and separates from any free protein or nucleic acid that may be present. During this process, the ionic conditions surrounding the sample are also switched from loading buffer to that characteristic of the gel running buffer, usually lowering the ionic strength in the process. If, for example, free protein is present, the lowered salt concentration may induce it to bind to free DNA sites, thus perturbing the equilibrium. It is also possible that the process of entry into the gel can lead to artifactual dissociation of protein–nucleic acid complexes. (Instability of the complex in the gel itself can usually be recognized from spreading of the band corresponding to the complex, assuming that the mobilities of free protein and DNA are not equal.) In many cases, the composite of these artifacts makes the gel method less accurate than filter binding for determination of absolute binding constants. Nevertheless, the gel method is capable of producing reasonable estimates for binding constants, and has had a number of successful applications (Carey, 1991b; Liu-Johnson et al, 1986; Weeks and Crothers, 1992). On the other hand, serious underestimation of peptide–RNA equilibrium binding constants has also been documented (Hall and Stump, 1992; Long and Crothers, 1995).

9.2.2 Relative Binding Constants by Gel Electrophoresis

It is in the determination of relative binding affinity of different DNA sequences that the gel electrophoresis method excels. Properly done, the experiment is carried out under conditions of limiting protein and at high enough total concentration to assure that essentially all of the protein present is bound to one or another of the competing DNA fragments present in the mixture. This condition avoids any rebinding artifacts in the process of entry into the gel. In the simplest case, there are two competing DNA fragments, and the size of these is adjusted so that four species can be observed on the gel: both of the free DNA bands and both complexes.

Figure 13-68 shows an example, in which a wild-type CAP binding sequence ATT*AAT (where the asterisk marks the internucleotide position 10 nucleotides from the binding site dyad axis) is compared with a sequence mutated to ATC*TGT. Since the mutation converts an A•T rich sequence to G•C rich at a position where the minor groove is compressed by bending (Section 13.4), one expects a reduction in binding affinity. The gel shows clearly that wild-type DNA is converted to complex at substantially lower protein concentration than is the case for the mutant sequence.

Quantitative analysis of the gel is done by determining the counts in each band, and calculating the relative binding constant (K_{rel}) from the ratio of binding constants:

$$K_{rel} = \frac{K_{mut}}{K_{wt}} \tag{13-28}$$

where

$$K_{rel} = \frac{\text{(counts in mutant complex)/(counts in mutant DNA)}}{\text{(counts in wildtype complex)/(counts in wildtype DNA)}} \tag{13-29}$$

In the example shown in Figure 13-67, K_{rel} is 0.20, meaning that binding to the mutant sequence is five times weaker than to the wild type. A similar electrophoretic analysis

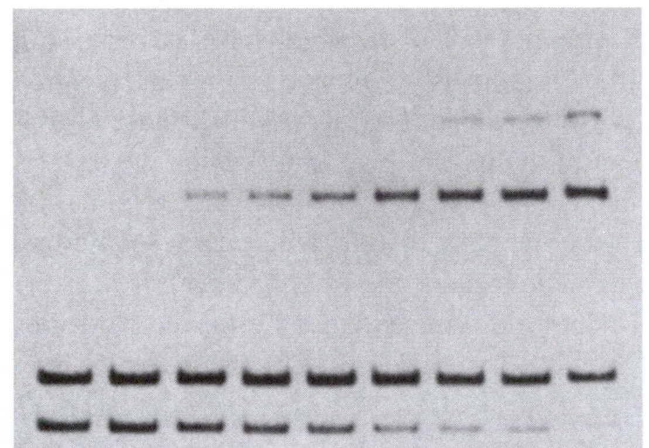

Figure 13-68
Gel electrophoresis assay for determination of the relative binding affinity of DNA fragments that contain mutant CAP binding sites. The mutant binding site fragment (24A) was mixed with a shorter wild-type (wt) fragment) and titrated with limiting concentrations of CAP protein. Both the free DNA bands (lower) and complexes (upper) are resolved by electrophoresis. [Dalma-Weiszhausz, D. D., Gartenberg, M. R., and Crothers, D. M. (1991). *Nucleic Acid Res.* **19**, 611–616 by permission of Oxford University Press.]

of relative binding to wild type and mutant TAR RNA sequences by Tat peptides has been reported by Weeks and Crothers (1991, 1992).

9.2.3 Electrophoretic Detection of Protein-Induced DNA Bending

In addition to their utility for assaying the extent of protein–DNA complex formation, electrophoretic measurements can give information about change of DNA shape when a protein is bound (reviewed by Crothers et al., 1991). These techniques are able to define the position and direction in space of an induced bend, and to provide an estimate of its magnitude by comparison with standard bends. There are two main electrophoretic approaches to detecting induced bends, one of which, the circular permutation method, is designed to identify the position of the center of the bend in the DNA molecule. In the other method, a standard bend of known direction is placed at variable phasing relative to an unknown bend by inserting variable length DNA linkers between the two. This approach can determine the direction of the bend relative to a defined coordinate frame in the DNA molecule.

The logic of the circular permutation experiment is the same as that discussed in Section 9.3. According to the simple theory given by Lumpkin and Zimm (1982), the mobility of a DNA chain of fixed contour length should vary with the mean-square end-to-end distance (Eq. 9-17). It was found empirically (Wu and Crothers, 1984) that the mobility of a DNA molecule having a globular protein, such as *lac* repressor, bound is approximately independent of the position of the binding site, as long as the protein does not induce DNA bending. [More recently, however, evidence has emerged (Gartenberg et al., 1990) indicating that this approximation may break down when the bound protein is large and asymmetric, making it advisable in suspicious cases to check for the presence of a bend by other methods as well.] Circular permutation of the sequence provides molecules having the induced bend near the center (short end-to-end distance and low mobility) and near the end (long end-to-end distance and high mobility).

Figure 13-69 shows a plot of the electrophoretic data by which the bend center can be identified. The mobilities of the various fragments are plotted against the position of the DNA end in the numbering system of the parent fragment. The distance moved down the gel is a maximum (a minimum in the plot in Fig. 13-69) when the bend is near the end. The position of the bend center is estimated by extrapolating the lines on both sides of the minimum to their intersection. Note that the wild-type CAP binding sequence has a smaller mobility than the mutant sequence when the bend is near the center. This result implies a larger bend in the wild-type sequence. Further complexities

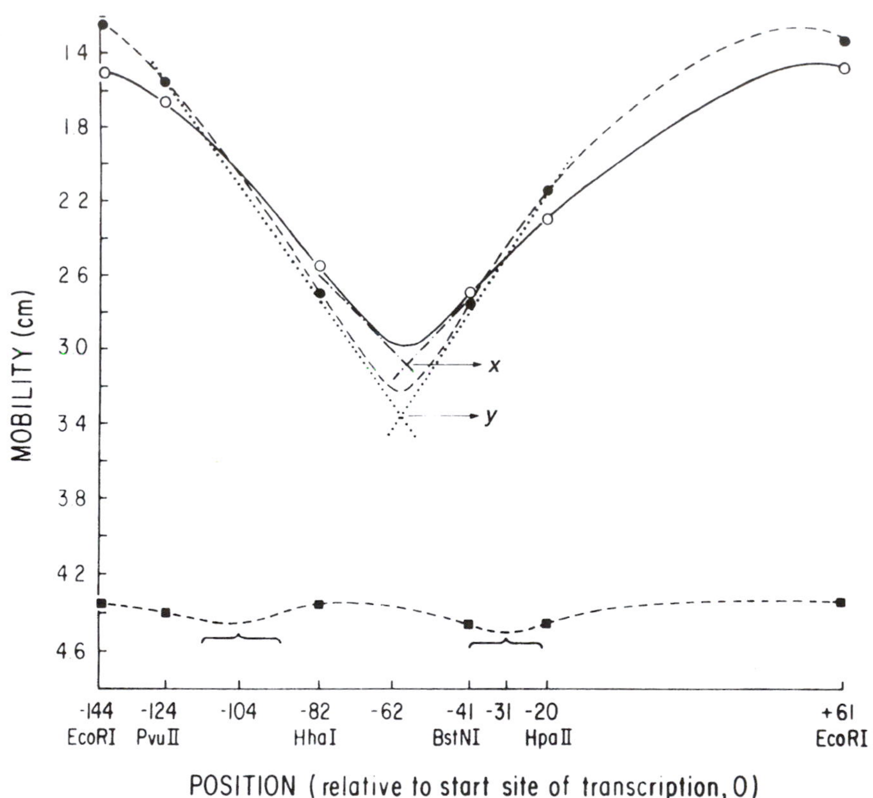

Figure 13-69
Mapping the bending locus. Filled circles show the mobility of a complex formed between CAP and a 203 bp DNA fragment containing a wild-type CAP site; open circles indicate the complex of CAP with a weaker binding and bending mutant sequence (sy203). Squares show the mobility of the naked DNAs, which run identically. The center of sequence symmetry is between -61 and -62. The bend center in the sy203 fragment, labeled *x*, is at about -58, and the estimated center of the wt203 bend (*y*) is at about -60. The regions of the naked DNA where a bend is suspected, are indicated by brackets (Fig. 8 of Liu-Johnson et al., 1986).

of this plot are discussed by Liu-Johnson et al. (1986) and Crothers et al. (1991); see also Crothers and Drak (1992).

Figure 13-70 illustrates the conceptual basis for the variable phasing method (Zinkel and Crothers, 1987). Molecules having an integral number of helical turns between the centers of two bends are shaped like the "cis" isomers in Figure 13-70(*a*), whereas the "trans" isomer [Fig. 13-70(*b*)] results when a half-integral number of turns separates the two centers. The cis isomer is identified experimentally by its minimum electrophoretic mobility. In order to deduce the direction of bending at a coordinate frame a fixed number of base pairs away, it is necessary to know the helical repeat between the two bends, which requires an experiment to determine whether or not DNA is unwound by protein binding. Topological relaxation of superhelical DNA in the presence of bound protein, followed by two-dimensional gel electrophoresis to assay the topoisomer distribution (Chapter 10), can be used for this purpose.

9.2.4 Estimation of Bend Magnitude by Electrophoresis

Two methods have been proposed for estimating relative bend magnitude from comparative electrophoresis. Both depend on an independent estimate of the bend angle induced by A tracts (see Koo et al., 1990). Thompson and Landy (1988) provide a calibration equation for the magnitude of the positional variation of mobility from circular permutation plots like that in Figure 13-69. Zinkel and Crothers (1990) used the logic of separating the electrophoretic effect of a centered bend from the retardation effect of protein binding at the molecular end. They constructed a set of calibration molecules that have A-tract bends of variable magnitude in the center, and a protein binding site at the end. The comparison molecule has the protein-induced bend in the center. The experiment matches the mobility of the comparison molecule to a value interpolated between members of the calibration set. In the specific case studied, it was concluded that a CAP bend is equivalent to 5.6 phased A-tract bends, or about 100°.

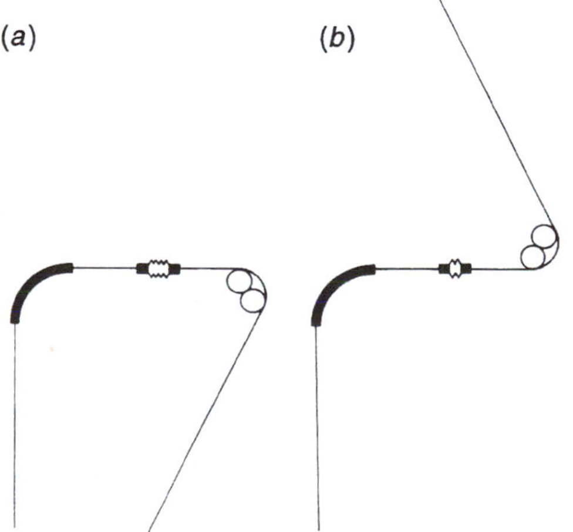

Figure 13-70
Logic of the phase-sensitive experiment for determination of bend direction. Shown are *cis* and *trans* isomers of DNA constructs containing both A-tract (heavy line) and CAP-induced bends. Phasing between the two bends is controlled by the length of the variable linker region in the center of the molecule. [Reprinted with permission from Zinkel, S. and Crothers, D. M. (1987). *Nature* **328**, 178–181. Copyright ©1987 Macmillan Magazine Limited. See Fig. 1.]

9.3 Footprinting and Interference Methods

9.3.1 *Enzymatic Footprinting*

Binding a protein to a nucleic acid affects the ability of enzymes and chemical reagents to digest or modify the DNA or RNA. If the protein is bound to a specific site, then the modified reactivity is localized to the region of the binding site. Comparison of the reactivity with and without bound protein leaves a "footprint" in the region of binding. The origins of this technology can be found in DNAse I digestion of chromatin, which revealed a polynucleosomal structure (Noll, 1974) and DNAse I digestion of specific protein–DNA complexes (Galas and Schmitz, 1978). For reviews, see Tullius (1989) and Hochschild (1991).

Footprinting experiments can be highly informative, both with respect to quantitative evaluation of binding affinity, the location of the binding site, and detection of structural changes. Figure 13-71 (Hochschild and Ptashne, 1986) illustrates the potential. Phage λ repressor binds cooperatively to DNA, and in so doing is able to induce DNA loops with substantial DNA curvature. Figure 13-71 shows the DNAse I digestion pattern that results when two operator sites are present, separated by 6 helical turns (right half of the figure). The extent of protection increases with the concentration of added protein (increasing from left to right). Increasing protection from digestion is observed in parallel at the two operator sites, O_R1 and O_R1^m. Quantitation of these effects enables determination of the binding constant (Hochschild, 1991; Tullius, 1989*b*). Note that when the binding site O_R1^m is isolated on the DNA (right half of Fig. 13-71), protection, and therefore binding, is much weaker than when it is coupled with O_R1. It can be concluded that occupancy of the two operator sites is a cooperative process when they are separated by six turns of DNA.

The figure also shows that the DNA between the two sites develops a pattern of alternating protection and sensitivity with a period equal to the DNA helical repeat, which is a consequence of DNA bending that allows the proteins bound to the two sites to touch each other. The minor groove is widened when it faces outward on the loop, and narrowed when it faces inward. Since DNAse I prefers a widened major groove (see Section 13.8.3), the enhanced cutting sites correspond to positions where the minor groove faces outward.

9.3.2 *Chemical Footprinting and Interference*

The distinction between chemical footprinting (or modification protection) and interference experiments is illustrated in Figure 13-72 (Wissman and Hillen, 1991). In the modification protection experiment [Fig. 13-72(*a*)], the protein is bound to DNA before the chemical modification reaction is carried out; a control DNA sample is subjected to the same reaction without bound protein. The DNAs are then cleaved at modified bases, and the extent of reaction is compared with and without bound protein. The modification interference experiment differs in that the DNA is modified before interaction with protein [Fig. 13-72(*b*)]. Separation of bound and unbound forms of DNA and evaluation of their modification patterns reveals the residues that are critical for complex formation.

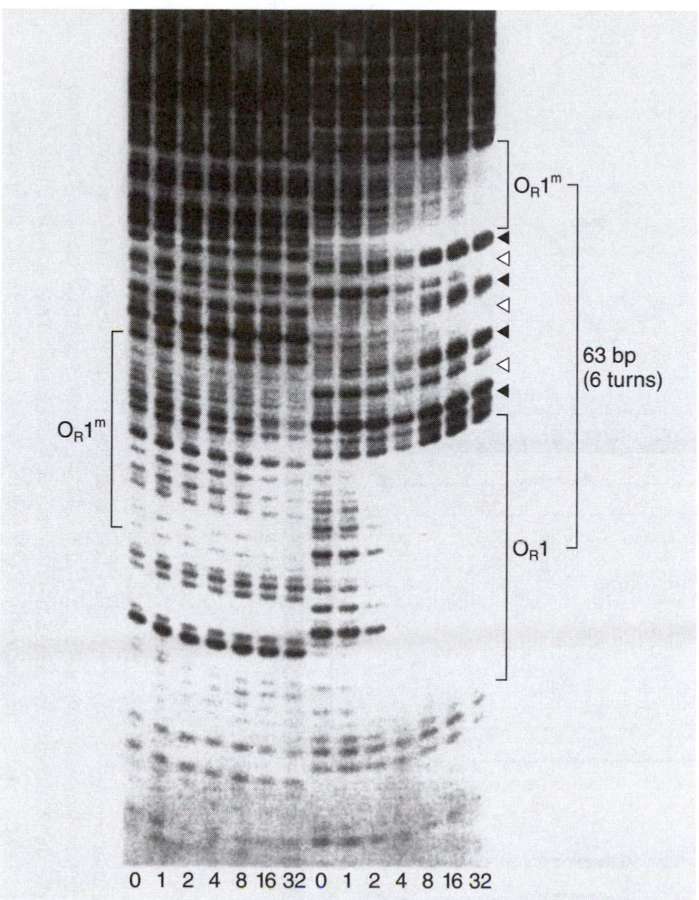

Figure 13-71
DNase I footprinting assay showing that λ repressor binds cooperatively to operator sites separated by six turns of the DNA helix. The binding of repressor to operator O_R1^m is measured in the absence (left panel) and presence (right panel) of a high affinity site O_R1^m six turns away on the same DNA template. Note the alternating sites of hypersensitivity (filled triangles) and protection (open triangles) through the region of the DNA loop between the two protein binding sites. [From Hochschild and Ptashne, 1986.]

The products of three common reactions used in chemical footprinting–interference measurements on DNA are illustrated for guanine in Figure 13-73 (reviewed by Wissmann and Hillen, 1991). Methylation of DNA at guanine N7 (illustrated) and adenine N3 results from treatment with dimethyl sulfate. Upon treatment with base at elevated temperature, these adducts lead to chain cleavage, as observed in DNA chemical sequencing reactions. Ethylation of DNA phosphates by *N*-ethyl-*N*-nitrosourea removes a phsophate charge, and interferes with DNA electrostatic interactions. Carbethoxylation of DNA purines results from reaction at N7 with diethyl pyrocarbonate in the

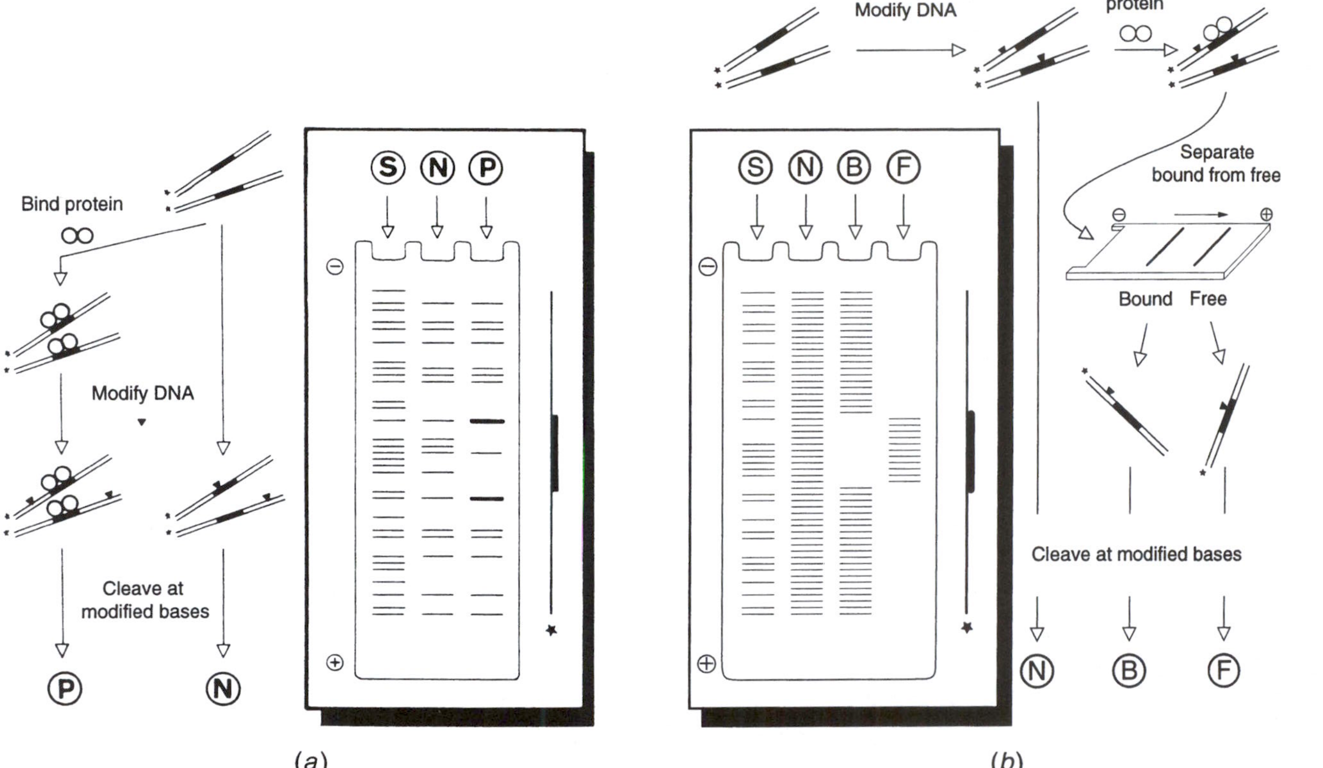

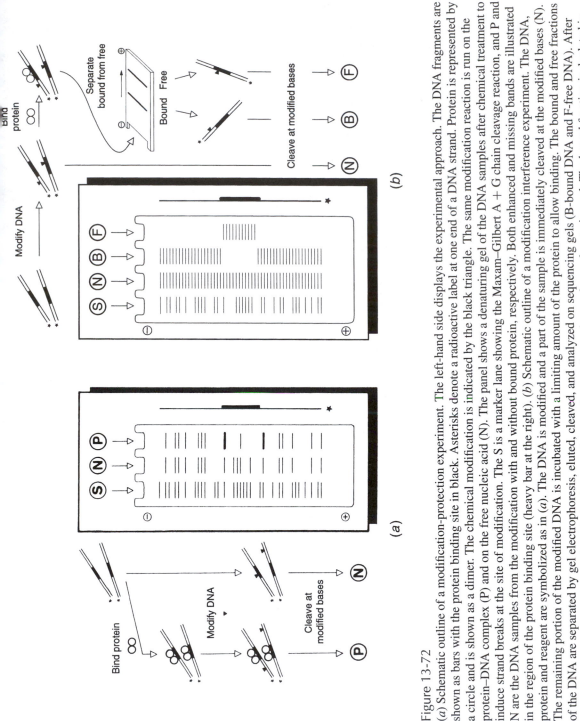

Figure 13-72
(*a*) Schematic outline of a modification-protection experiment. The left-hand side displays the experimental approach. The DNA fragments are shown as bars with the protein binding site in black. Asterisks denote a radioactive label at one end of a DNA strand. Protein is represented by a circle and is shown as a dimer. The chemical modification is indicated by the black triangle. The same modification reaction is run on the protein–DNA complex (P) and on the free nucleic acid (N). The panel shows a denaturing gel of the DNA samples after chemical treatment to induce strand breaks at the site of modification. The S is a marker lane showing the Maxam–Gilbert A + G chain cleavage reaction, and P and N are the DNA samples from the modification with and without bound protein, respectively. Both enhanced and missing bands are illustrated in the region of the protein binding site (heavy bar at the right). (*b*) Schematic outline of a modification interference experiment. The DNA, protein and reagent are symbolized as in (*a*). The DNA is modified and a part of the sample is immediately cleaved at the modified bases (N). The remaining portion of the modified DNA is incubated with a limiting amount of the protein to allow binding. The bound and free fractions of the DNA are separated by gel electrophoresis, eluted, cleaved, and analyzed on sequencing gels (B-bound DNA and F-free DNA). After chemical treatment as in a) to induce chain cleavage, the samples are electrophoresed on a denaturing gel. The bound fraction is depleted in modification sites within the binding site, and the free fraction is correspondingly enhanced. [Reprinted with permission from Wissman and

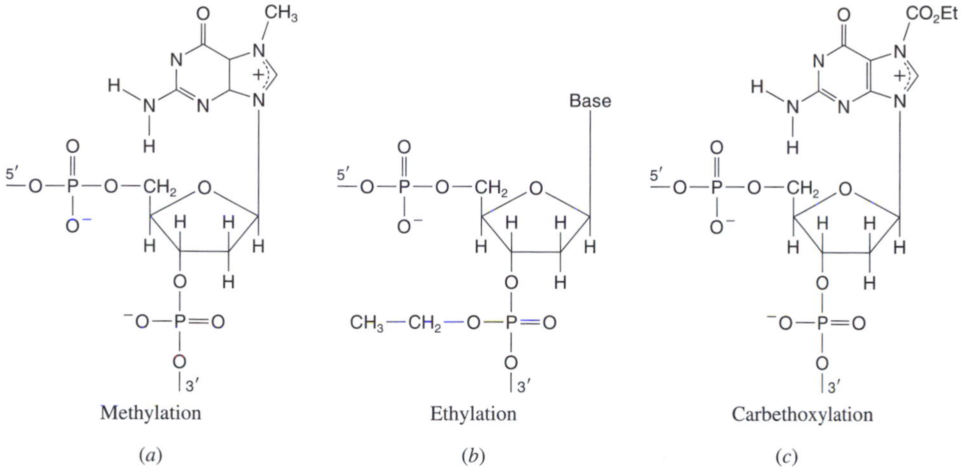

Methylation

(*a*)

Ethylation

(*b*)

Carbethoxylation

(*c*)

Figure 13-73
Chemical structures of modified guanine nucleotides resulting from some commonly used chemical footprinting agents. (*a*) Reaction with dimethyl sulfate yields methylation of G at N7 and; A is methylated at N3. (*b*) Reaction with *N*-ethyl-*N*-nitrosourea ethylates phosphate residues. (*c*) Reaction with diethyl pyrocarbonate causes carbethoxylation of purine N7. [Reprinted with permission from Wissman and Hillen, 1991.]

major groove. Ethylated and carbethoxylated nucleotides also require further chemical treatment to induce strand breaks.

Hydroxyl radical is an important reagent for both protection and interference experiments (reviewed by Tullius, 1989a; Dixon et al., 1991; Price and Tullius, 1992). The radical is generated from molecular oxygen or hydrogen peroxide plus iron-edta (ethylenediaminetetracetic acid = edta) in the Fenton reaction,

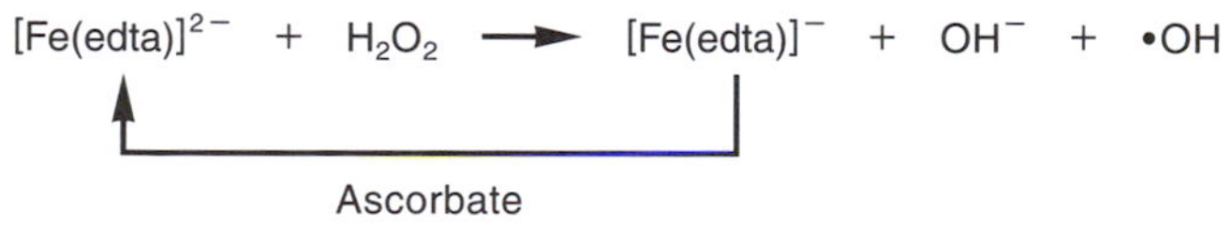

$$[\mathrm{Fe(edta)}]^{2-} + \mathrm{H_2O_2} \longrightarrow [\mathrm{Fe(edta)}]^{-} + \mathrm{OH}^{-} + \bullet\mathrm{OH}$$

Ascorbate

where ascorbate serves to regererate $Fe^{(II)}$-edta from $Fe^{(III)}$-edta, so that only catalytic amounts of iron-edta are needed for the procedure. Another reagent that generates oxidative cleavage is 1,10-phenanthroline–copper, which, unlike iron–edta, binds to DNA (reviewed by Sigman et al., 1991).

Hydroxyl radical abstracts hydrogen from deoxyribose in the DNA minor groove. It is thought that the major degradation pathway involves attack at H1′ (Fig. 13-74), which is followed by loss of the base and two β elimination reactions. The result is 3′ and 5′ chain ends, whose mobility is the same as that of the Maxam–Gilbert DNA sequencing reaction products. A minor pathway *b* involving H4′ abstraction leads to a 3′-end modified by phosphoglycolate, which by virtue of its additional charge migrates on a gel slightly faster than the corresponding phosphate.

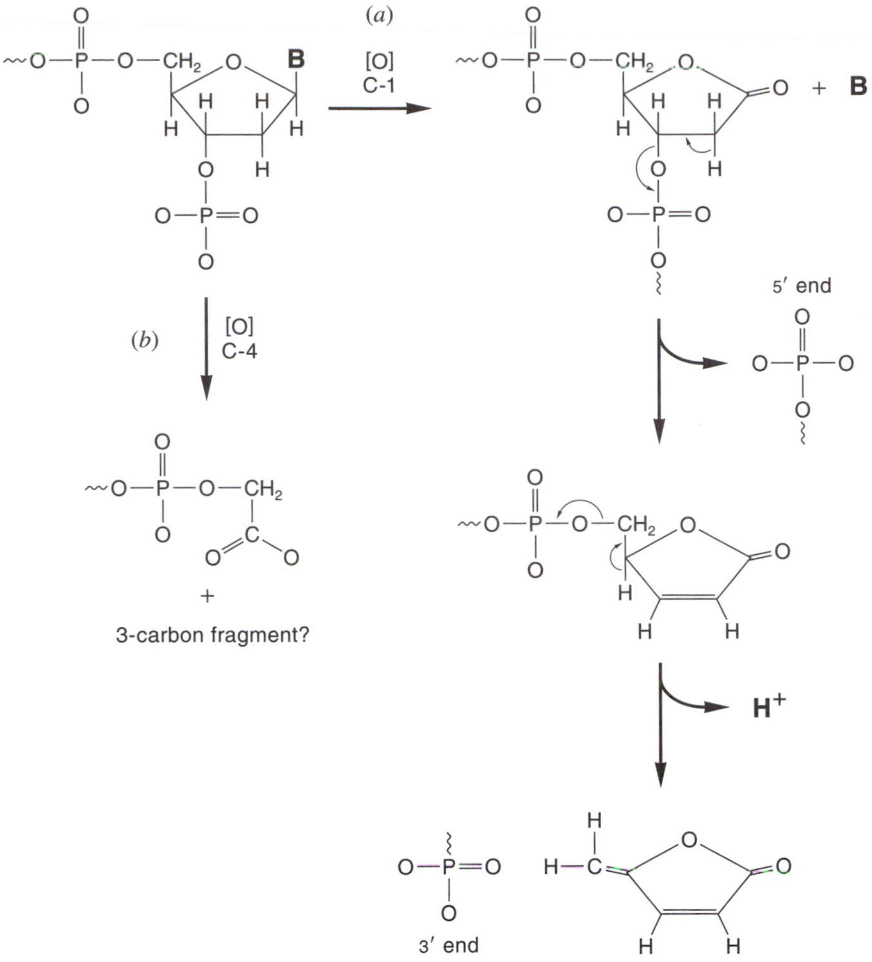

Figure 13-74
Postulated reaction mechanism for the scission of DNA by OH radical generated by 1,10-phenanthroline–copper complex. Reaction pathway (*a*) accounts for greater than 70% of the scission events at any sequence position. The minor pathway (*b*) yields chains modified at their 3′-ends, leading to altered electrophoretic mobility. [Reprinted with permission from Sigman et al., 1991.]

A number of other reagents, summarized in Table 13-7 (Lilley, 1992), are available for chemical footprinting/interference measurements on DNA and RNA. Some of these chemical approaches are suitable for footprinting analysis inside cells (reviewed by Sasse-Dwight and Gralla, 1991).

Hydroxyl radical attack leads to elimination of a nucleoside unit without the requirement for further chemical treatment. This principle provides the basis for the "missing nucleoside experiment", in which DNA reacted with OH radical is subsequently allowed to interact with limiting amounts of protein (reviewed by Dixon et al.,

Table 13.7
Chemical Probes Used to Study DNA Structure[a]

Probe	Comments
Haloacetaldehydes	A > C base, etheno aduct
Osmium tetroxide	T base, 5,6-diester adduct; cleaved with piperidine
Permanganate	T base, diol product; cleaved with piperidine
Diethyl pyrocarbonate (DEP)	A > C base; carbethoxylation; cleaved with piperidine
Formaldehyde	Cross-linking
Glyoxal	G base, etheno adduct
Glycidaldehyde	G base
Bisulfite	C base; deamination to dU
Hydroxylamine	C base
Dimethyl sulfate (DMS)	G base; N7 methylation; cleaved with peperidine
Methylene blue	Photooxidation of G base
Singlet oxygen	Cleavage reported at DNA kinks
Ethylnitrosourea	Phosphate ethylation
Carbodiimide	Single-strand-specific T and G base modification
Ozone	Strand scission at (A + T)-rich sequences
Psoralens	Supercoiling-dependent cross-linking
Edta–Fe	Deoxyribose cleavage by HO• generated uniformly
MPE–Fe	Binding of complex to DNA; strand scission
Cu(*o*-phen)	Attack of deoxyribose leads to backbone scission
Transition ions	Binding or cleavage as function of ion stereochemistry
Uranyl ion	Photooxidation of deoxyribose

[a]From Lilley (1992).

1991). The protein-bound and free DNA components are then separated by nondenaturing gel electrophoresis, and each is subsequently analyzed on a denaturing gel to detect the presence of chain interruptions. Missing nucleosides at critical interaction sites result in less frequent chain breaks in protein-bound DNA compared to free DNA. The ratio of band intensities can be used to estimate the free energy of interaction contributed by particular nucleoside units.

Another technique for mapping the thermodynamic contributions of nucleotides in the binding site begins with primer extension along a single stranded template that includes the full binding site sequence (Liu-Johnson et al., 1986). Dideoxynucleotides, as in the Sanger sequencing technique, provide stops at individual nucleotides. The partly polymerized sample is then reacted with limiting amounts of protein, and the free DNA and protein–DNA complexes are separated by nondenaturing electrophoresis.

Primer extended samples containing a DNA duplex that is too short to reconstitute the full binding site are systematically excluded from the protein–DNA complex, to an extent that reflects the sum of the free energy contributions of the missing contacts. The increment in the binding free energy per nucleotide added can be determined from the increase in relative band intensity for each residue across the binding site.

9.4 Affinity Cleavage Methods

In affinity cleavage methods, the reagent that cuts the nucleic acid chain is attached covalently to a ligand that binds sequence specifically. Extensive development of reagents of this kind, beginning with (methidiumpropyl-edta)iron(II) or MPE-iron (Hertzberg and Dervan, 1982), has been reported by Dervan and co-workers (reviewed by Dervan, 1988). In MPE-iron the chelating agent edta is coupled to the intercalating agent MPE (Chapter 12). Reduction to the $Fe^{(II)}$ form of the DNA bound ligand results in generation of OH radical from molecular O_2, leading to DNA strand cleavage by hydrogen abstraction, as discussed in Section 13.9.3. Since MPE-iron binds DNA in a relatively nonspecific manner, footprinting of bound proteins can be carried out by observing DNA sequences that are protected from cleavage by addition of proteins.

The iron cleaving reagent can also be coupled to a specifically binding ligand such as a bound protein. Figure 13-75 illustrates the principle of this experiment. A 52 amino acid protein, based on the sequence-specific DNA binding domain of Hin recombinase, was derivatized with edta at the amino terminus. The HTH domain of the protein presumably interacts with the major groove, but the Fe-edta induced cleavage results from interaction of the protein amino terminus with the DNA minor groove [Fig. 13-75(*a*)]. Note that cutting by $Fe^{(II)}$ at the nearest residues in the minor groove yields chains that are not of identical length. Reading one chain in the 5′ to 3′ direction, it reaches its closest approach to the iron atom after the complementary strand does. The result is an offset or stagger of the chain ends [Fig. 13-75(*b*)] in which the 5′-end extends beyond the 3′-end. The opposite stagger is diagnostic for cleavage reactions in the major groove.

10. KINETICS AND THE SEARCH MECHANISM

Protein–nucleic acid complexes are static only in a resting state, or perhaps transiently, between the steps in an active process that leads to gene expression. The simplest view of this process is that the protein moves by simple diffusion through solution until it encounters a specific binding site, to which it then attaches. While much remains to be learned about the mechanism of the complex series of events that culminates in gene expression, we know enough now to state that the simplest view is indeed too simple.

Early measurements of the rate of reaction of *lac* repressor with operator DNA (Riggs et al., 1970a) revealed a bimolecular rate constant of about $5 \times 10^{10} M^{-1} s^{-1}$. As discussed below, this value is roughly 1000 times larger than expected for a "diffusion-limited" reaction of two macromolecules, namely, one that occurs as fast as they can find each other by simple three-dimensional diffusive motion. One way to speed up the reaction is to have the protein encounter the DNA molecule far from the operator

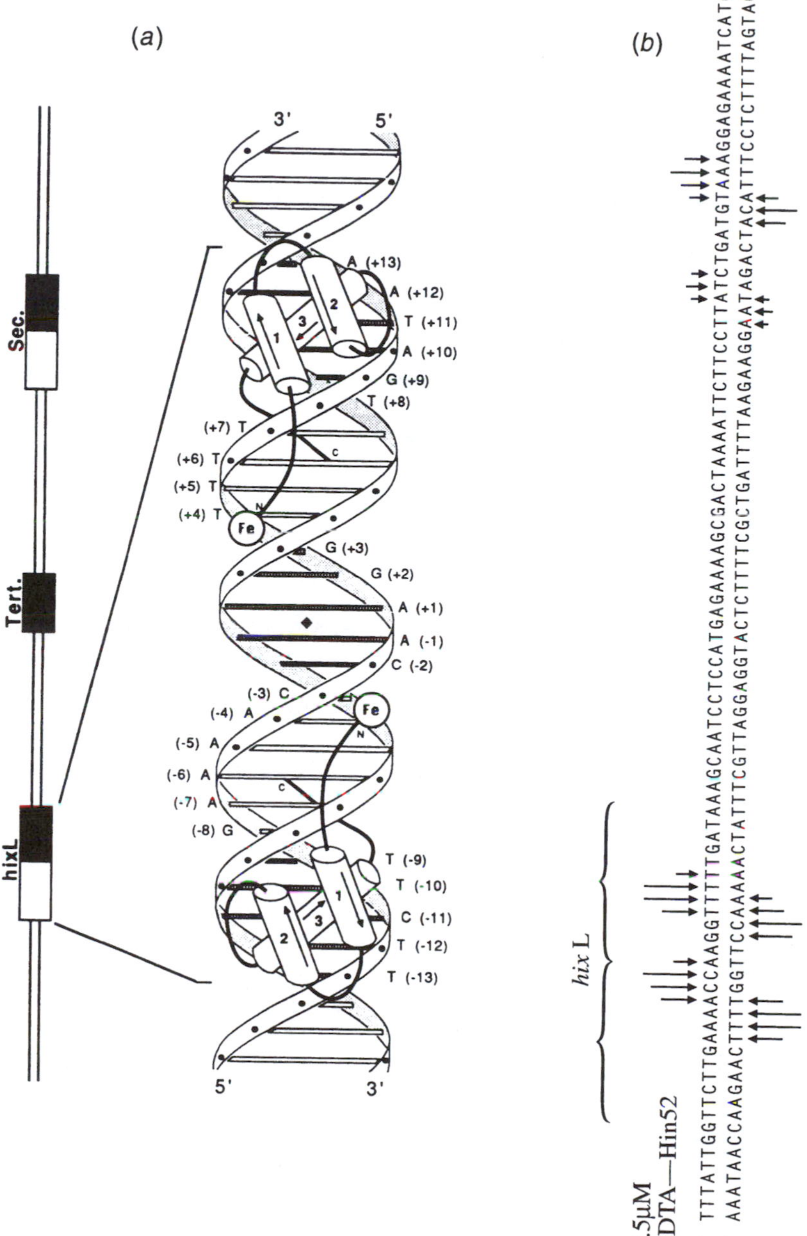

Figure 13-75
(*a*) Schematic model of the complex of the DNA binding domain of Hin recombinase with the *hix*L binding sequence, showing attachment of Fe to the amino terminus through edta. (*b*) The DNA cleavage pattern observed for the 52 amino acid fragment of Hin recombinase. The stagger of the cutting sites in which the $5' \rightarrow 3'$ strand is longer that the $3' \rightarrow 5'$ strand is consistent with minor groove binding of the amino terminal region of the protein. [Reprinted with permission from Sluka, J. P., Horvath, S. J., Glasgow, A. C., Simon, M. I., and Dervan, P. D. (1990). *Biochemistry* **29**, 6551–6561. Copyright ©1990 American Chemical Society.]

site, and then move along it by one-dimensional diffusion (Adam and Delbruck, 1968; Richter and Eigen, 1974). This mechanism reduces the dimensionality of the search process, and leads to the idea of facilitated target location (reviewed by von Hippel and Berg, 1989).

10.1 Rate Constant of a Diffusion-Limited Reaction

The upper limit on the rate with which two particles A and B can react depends on their diffusion constants D_A and D_B, their distance apart, α, when capture occurs, on the fraction κ of their surfaces that is reactive, and on a factor f that reflects effects due to electrostatic attraction or repulsion as the particles encounter each other. Consideration of an equation originally derived by Smoluchowski leads to the result that the association rate constant is then given by

$$k_1 = \frac{4\pi\kappa\alpha f(D_A + D_B)N_A}{1000} \tag{13-30}$$

where N_A is Avogadro's number, and the factor of 1000 normalizes the units of k_1 to $M^{-1}s^{-1}$. Estimation of the values of quantities in this equation (von Hippel and Berg, 1989), including D_A for a repressor approximately equal to D_B for DNA approximately equal to 5×10^{-7} cm^{2}s^{-2}, $\alpha = 50$ Å, $\kappa = 0.05$, and $f = 1$ leads to a value of about 10^8 $M^{-1}s^{-1}$ or less for k_1, compared to measured values that can approach 10^{11} $M^{-1}s^{-1}$. Consequently, simple diffusion cannot explain the rapidity with which the reaction occurs.

10.2 Sliding and Transfer Processes

Reaction of a sequence-specific protein with a large DNA molecule is complicated by the ability of the protein to bind nonspecifically to the large number of such sites present on the molecule. The nature of the contributing events is illustrated in Figure 13-76 for repressor protein (von Hippel and Berg, 1989). The protein can, in principle, bind to a DNA chain nonspecifically, slide along it, transfer between segments, and dissociate either transiently or completely. In general, the ability to bind nonspecifically enhances the rate of reaction because the protein is held in the vicinity of its ultimate target. However, in the case of a large molecule, or in the presence of separate nonoperator DNA molecules, a nonspecifically bound protein can become trapped in a DNA domain far from the operator site, and may react slowly for that reason.

10.3 The Two-Step Reaction Mechanism

von Hippel, Berg, and co-workers have treated the complex problem outlined in Figure 13-76 in detail, both theoretically and experimentally (Berg et al., 1981; Winter et al., 1981). For purposes of simplicity, we reduce the problem here to one in which the DNA molecules are less than a few persistence lengths long ($<$1000 bp), thus ignoring

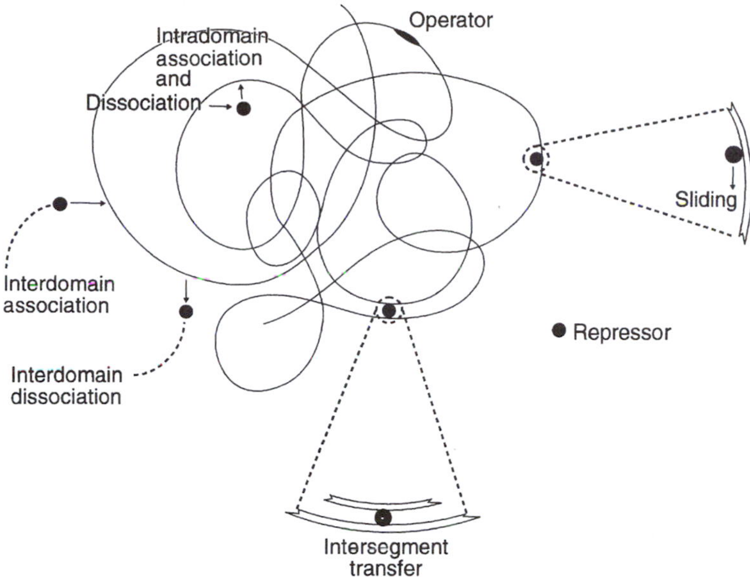

Figure 13-76
Schematic view of *lac* repressor interacting with a large operator-containing DNA molecule in dilute solution. The (upper) expanded view shows repressor bound to a segment of nonoperator DNA, on which it can either "slide" or engage in intradomain dissociation–association processes in seeking its specific (operator) target site. The (lower) expanded view shows a repressor molecule double bound to two DNA segments; this corresponds to the intermediate state in the intersegment or direct transfer process. [Reprinted with permission from von Hippel and Berg, 1989. See Fig. 1.]

problems such as what is called "domain trapping." In this case, one can understand the process in terms of a two-step "bind-and-slide" reaction mechanism:

$$\mathrm{R} + \mathrm{D{-}O} \underset{k_{-1}}{\overset{\text{Bind}\atop k_1}{\rightleftharpoons}} \mathrm{R{\bullet}D{-}O} \underset{k_{-2}}{\overset{\text{Slide}\atop k_2}{\rightleftharpoons}} \mathrm{D{-}O{\bullet}R}$$

In the first of these steps, the protein (R) attaches to the operator DNA (D) nonspecifically, and then in the second step it translocates to the operator (O) and forms the specific complex. Transfer of the protein from one DNA segment to another may also contribute to the "sliding" process.

Since all base pair units in principle provide equivalent nonspecific binding sites, the concentration units for k_1 are expressed in terms of the molar concentration of base pairs. As discussed above, the upper limit for k_1 is about $10^8\ M^{-1}$. However, k_a, the overall bimolecular rate constant for forming the O • R complex, is expressed in terms of the molar concentration of operator DNA. Since many binding events as measured by k_1 contribute to operator capture as measured by k_a, k_a can be much larger than k_1. For example, if a molecule containing M base pairs is short enough so that all captured repressors form the operator complex ($k_2 \gg k_{-1}$), then $k_a = Mk_1$. More generally, Berg et al. (1981) showed that

$$k_a = 2k_1[D_1/k_{-1}\ell^2]^{1/2}\tanh(k_{-1}M^2\ell^2/D_1)^{1/2} \tag{13-31}$$

in which ℓ is the length of a base pair ($\sim$ 3.4 Å) and D_1 is the diffusion constant for the sliding process. The ratio $R = D_1/k_{-1}\ell^2$ measures the relative rates of diffusion (D_1/ℓ^2) and dissociation (k_{-1}), and can be thought of as the effective target size on the DNA, or the average sliding distance in base pairs. In the limit when $R \gg M$, then Eq. 13-28 predicts that $k_a = Mk_1$, as expected. However, it should be realized that k_1 is not independent of M, because the diffusion constant of the DNA molecule (D_B) affects k_1, unless the DNA has a much smaller diffusion coefficient than characterizes the repressor.

For strong specific compared to nonspecific binding ($k_2 \gg k_{-2}$) and reasonably short DNA molecules, the number ($\sim M$) of nonspecific sites is small enough so that they do not contribute appreciably to the overall binding constant. Hence, with the equilibrium constant for binding independent of M, and $K = k_a/k_d$, the overall dissociation rate constant k_d must also depend on M. This process can be thought of in terms of the mechanism (Fig. 13-76) as reflecting a rapid equilibrium (relative to the rate of dissociation) between the specific complex D–O•R and the nonspecifically bound states O–D•R. In this case, the rate of dissociation is just k_{-1} times the concentration of states O–D•R. When the sliding range R is large compared to M, then the concentration of nonspecifically bound states is proportional to the degeneracy M of that state, and the overall dissociation rate should be proportional to $k_{-1}M$. The corresponding equation of Berg et al. (1981) is

$$k_d = \frac{k_1}{K_{RO}(k_{-1}\ell^2/(4D_1)]^{1/2}\coth(k_{-1}M^2\ell^2/D_1)^{1/2} + K_{RD}} \tag{13-32}$$

in which K_{RD} is the nonspecific binding constant, and K_{RO} is the specific binding constant. The value of k_d predicted by Eq. 13-32 reaches a plateau as M approaches ∞, and is a function of the sliding range R.

These ideas have been tested for EcoRI endonuclease binding by Jack et al. (1982) and for *Cro* repressor by Kim et al. (1987). These investigators showed that the association and dissociation rates of proteins that bind to specific DNA sequences depend on the size of the DNA in which the site is embedded. This finding clearly shows that flanking nonspecific DNA can participate in the mechanism of protein binding. Figure 13-77 shows the length dependence for the dissociation rates. As the DNA size increases, there are more sites to serve the role of nonspecific binding, so their population increases, as does the overall rate of dissociation. The approach of the curve to a plateau value, at which point one exceeds the range of the sliding reaction by which the flanking DNA sites are accessed, allows determination of the sliding range R. The results for EcoRI indicate a range of 1300 bp (in 100 m*M* Tris buffer); for *Cro* repressor the corresponding value is 300 bp (10 m*M* Tris + 100 m*M* KCl). Given the fact that the sliding range should increase as ionic strength decreases, because of the reduced value

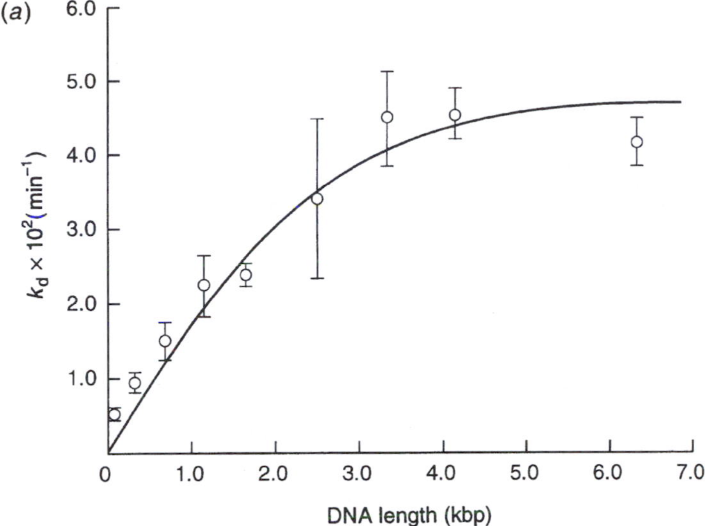

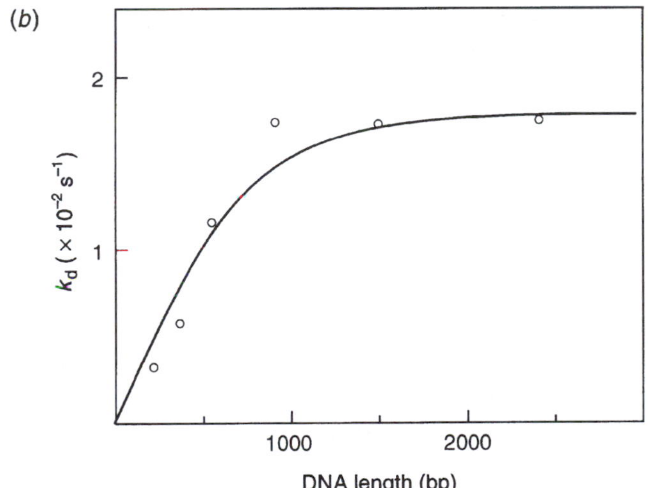

Figure 13-77
The DNA length dependence of the dissociation rate for (*a*) EcoRI and (*b*) *Cro* repressor. In (*a*) a sliding length of $R = 1300$ bp was deduced (100 m*M* Tris buffer, pH 7.6). (Fig. 1 of Jack et al., 1982). In (*b*) a value of $R = 300$ bp was deduced (10 m*M* Tris, 100 m*M* KCl). [Reprinted with permission from Kim, J. G., Takeda, Y., Matthews, B. W., and Anderson, W. F., Kinetic Studies on Cro Repressor–Operator DNA Interaction, *J. Mol. Biol.* **196**, 149–158, Copyright ©1987, by permission of the publisher Academic Press London. See Fig. 9]

of k_{-1}, these results are in reasonable agreement in spite of having been measured for very different proteins.

10.4 Need for a Three Step Mechanism: Bind, Slide, and Transfer/Isomerize

Since *lac* repressor, a potentially bifunctional tetramer, is able to form bridged complexes in solution (Krämer et al., 1987), it is of particular interest to examine the role of direct or intersegment transfer events in that case. The experiments of Ruusala and Crothers (1992), which were directed at this objective, revealed the presence of a metastable intermediate and transfer events in the search–capture process. The experimental strategy compared the relative reaction rates of two restriction fragments (Fig. 13-78) of nearly equal size (~200 bp) and therefore diffusion constant, but differing by a factor of 6 in the number of operator sites present on the fragment. If the sliding range is large compared to the fragment size, the fragments should react at equal rates, but if the sliding range is small compared to the distance between sites on the hexaoperator, then it should have a sixfold advantage over the monooperator fragment.

The result found experimentally depends on the concentration of the system. At low operator–repressor concentrations, and 50-m*M* salt concentration, the two fragments reacted at equal rates, leading to the conclusion that the sliding range is greater than 200 bp. At 150 m*M* salt concentration the hexaoperator had a 1.8-fold advantage, from which the equations of Berg et al. (1981) allow one to deduce a sliding length of 100 bp, in good agreement with the earlier estimate of Winter et al. (1981).

However, in the limit of high concentration, the hexaoperator had an advantage of about 3.5-fold in reaction rate over the monooperator fragment, even at 50 m*M* salt concentration. The experimental results imply that there is a metastable intermediate bound at operator sites, but which has not yet isomerized to the final state. In the presence of operator concentrations above a few nanomolar, this state reacts faster with an additional operator than it can isomerize by a unimolecular pathway to the final complex. It was concluded that the lifetime of the metastable state is at least 1 s. Furthermore, repressor in this state is readily transferred to other operator sites, with

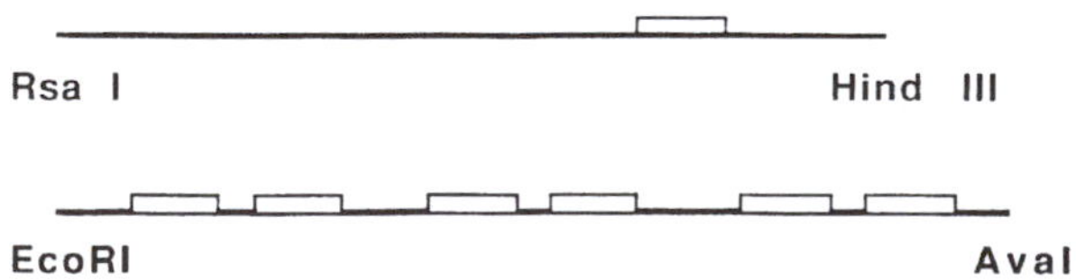

Figure 13-78
Operator fragments used in the experiment of Ruusala and Crothers (1992) to determine relative reaction rates of a fragment containing a single operator (upper) and a fragment of nearly the same length containing six operator sites (lower).

probability 0.5 once the bridged complex is formed, which accounts for the advantage enjoyed by the hexaoperator at high concentrations (Ruusala and Crothers, 1992).

The role of transfer events can also be seen in repressor dissociation kinetics (Fried and Crothers, 1984; Ruusala and Crothers, 1992). Figure 13-79 illustrates the catalysis of the dissociation rate of a specific repressor–operator complex by excess operator DNA. These results can be understood if one postulates a mechanism in which repressor can bridge two operators and transfer its tight binding affinity back and forth between the two operators in a process that is more rapid than the unimolecular dissociation of the R–O complex. Hence, the pathway for operator-catalyzed repressor dissociation involves formation of a bridged complex with an attacking operator, followed by transfer of the primary affinity to the attacking operator and finally by dissociation of the now weakened bond to the original operator:

$$\text{R–O} + \text{O}' \rightleftharpoons \text{O}' \sim \text{R–O} \rightleftharpoons \text{O}'\text{–R} \sim \text{O} \rightleftharpoons \text{R–O}' + \text{O}$$

These results point out the potential role of sccondary or distant DNA sequences in affecting the rates of isomerization processes in protein–DNA complexes. Much work remains to be done to resolve the detailed reaction mechanisms involved in protein–nucleic acid interactions.

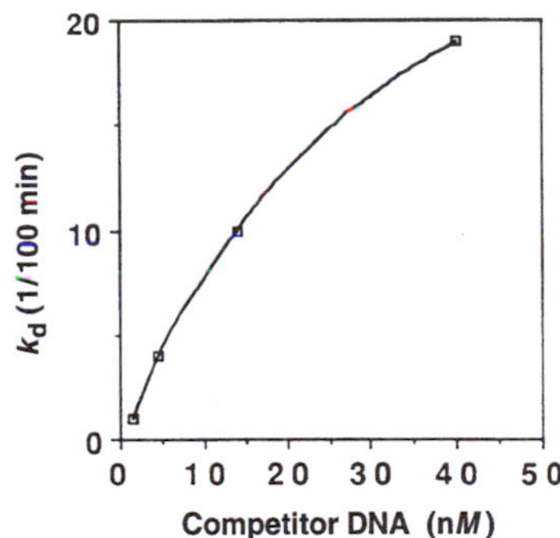

Figure 13-79
Catalysis of the rate of dissociation of *lac* repressor from its operator by added operator DNA. For all experimental conditions, added operator was in large excess over the starting complex; 50 mM salt concentration (From Ruusala and Crothers, 1992).

References

Ackers, G. K., Johnson, A. D., and Shea, M. A. (1982). Quantitative Model for Gene Regulation by λ Phage Repressor, *Proc. Natl. Acad. Sci. USA* **79**, 1129–1133.

Adam, G. and Delbrück (1968). Reduction of Dimensionality in Biological Diffusion Processes, in *Structural Chemistry and Molecular Biology*, Rich, A. and Davidson, N. Eds., Freeman, NY, pp. 198–215.

Adam, S. A., Nakagawa, T., Swanson, M. S., Woodruff, T. K., and Dreyfuss, G. (1986). mRNA Polyadenylate-Binding Protein: Gene isolation and Sequencing and Identification of a Ribonucleoprotein Consensus Sequence, *Mol. Cell Biol.* **6**, 2932–2943.

Adhya, S. (1989). Multipartite Genetic Control Elements: Communication by DNA Loop, *Annu. Rev. Genet.* **23**, 227–250.

Adler, K., Beyreuther, K., Fanning, E., Geisler, N., Gronenborn, B., Klemm, A., Müller-Hill, S. S., Pfahl, M., and Schmitz, A. (1972). How lac Repressor Binds to DNA, *Science (London)* **237**, 322–326.

Aggarwal, A. K., Rodgers, D. W., Drottar, M., Ptashne, M., and Harrison, S. C. (1988). Recognition of a DNA Operator by the Repressor of Phage 434: A View at High Resolution, *Science* **242**, 99–107.

Agre, P., Johnson, P. F., and McKnight, S. (1989). Cognate DNA Binding Specifity Retained After Leucine Zipper Exchange Between GCN4 and C/EBP, *Science* **246**, 922–926.

Allain, F. H.-T, Gubser, C. C., Howse, P. W. A., Nagai, K., Neuhaus, D., and Varani, G. (1996). Specificity of Ribonucleoprotein Interaction Determined by RNA Folding During Complex Formation, *Science (London)* **380**, 646–650.

Anderson, W. F., Ohlendorf, D. H., Takeda, Y., and Matthews, B. W. (1981). Structure of the Repressor from Bacteriophage and Its Interaction with DNA, *Science (London)* **290**, 754–758.

Anderson, W. F., Takeda, Y., Ohlendorf, D. H., and Matthews, B. W. (1982). Proposed α-helical Supersecondary Structure Associated with Protein–DNA Recognition, J. Mol. Biol. **159**, 745–751.

Ansari, A. Z., Bradner, J. E., and O'Halloran, T. V. (1995) DNA-Bend Modulation in a Repressor-to-Activator Switching Mechanism, *Science (London)* **374**, 371–375.

Assa-Munt, N., Mortishire-Smith, R.J., Aurora, R., Herr, W., and Wright, P.E. (1993). The Solution Structure of the Oct-I POU-Specific Domain Reveals a Striking Similarity to the Bacteriophage λ Repressor DNA–Binding Domain, *Cell* **73**, 193–205.

Baldwin, R. L. (1986). Temperature Dependence of the Hydrophobic Interaction in Protein Folding. *Proc. Natl. Acad. Sci. USA* **83**, 8069–8072.

Baleja, J. D., Marmorstein, R., Harrison, S. C., and Wagner, G. (1992). Solution Structure of the DNA–Binding Domain of Cd_2-GAL4 from *S. cerevisiae, Nature (London)* **356**, 450–453.

Baltimore, D. and Beg, A. A. (1995). A Butterfly Flutters By, *Science* **373** 287–288.

Bandziulis, R. J., Swanson, M. S., and Dreyfuss, G. (1989) RNA-binding Proteins as Developmental Regulators, *Genes and Develop.* **3**, 431–437.

Banerji, J., Rusconi, S., and Schaffner, W. (1981). Expression of a β-globin Gene is Enhanced by Remote SV40 DNA Sequences, *Cell* **27**, 299–308.

Baranger, A. M., Palmer, C. R., Hamm, M. K., Gieber, H. A., Brauweller, A., Nyborg, J. K., and Schepartz, A. (1995). Mechanism of DNA-binding Enhancement by the Human T-Cell Leukaemia Virus Transactivator Tax, *Science (London)* **376**, 606–608.

Bauer, W. R. and Benham, C. J. (1993). The Free Energy, Enthalpy and Entropy of Native and Partially Denatured Closed Circular DNA, *J. Mol. Biol.* **234**, 1184–1196.

Bauer, W. R., Lund, R. A., and White, J. H.(1993). Twist and Writhe of a DNA Loop Containing Intrinsic Bends, *Proc. Natl. Acad. Sci. USA* **90**, 833–837.

Bauer, W. R. and White, J. H. (1990). Surface Linking and Helical Repeat of Protein-Wrapped DNA, in *Nucleic Acids and Molecular Biology* 4, Eds. Eckstein, F. and Lilley, D. M. J., Springer-Verlag, Berlin, Heidelberg, New York, London, Paris, Tokyo, Hong Kong, Barcelona, pp. 39–54.

Beese, L. S. and Steitz, T. A. (1991). Structural Basis for the 3′ - 5′ Exonuclease Activity of *Escherichia Coli* DNA Polymerase 1: A Two Metal Ion Mechanism, *The EMBO Journal* **10**, 25–33.

Benoist, C. and Chambon, P. (1981). In Vivo Sequence Requirements of the SV40 Early Promotor Region, *Science (London)* **290**, 304–310.

Berg, J.M. (1988). Proposed Structure for the Zinc-binding Domains from Transcription Factor IIIA and Related Proteins, *Proc. Natl. Acad. Sci. USA* **85**, 99–102.

Berg, J. M. (1990). Zinc Finger Domains: Hypotheses and Current Knowledge, *Annu. Rev. Biophys. Chem.* **19**, 405–421.

Berg, O. G. and von Hippel, P. H. (1987). Selection of DNA Binding Sites by Regulatory Proteins. Statistical Mechanical Theory and Application to Operators and Promoters, *J. Mol. Biol.* **193** 723–750.

Berg, O. G. and von Hippel, P. H. (1988). Selection of DNA Binding Sites by Regulatory Proteins. II. The Binding Specificity of Cyclic AMP Receptor Protein to Recognition Sites, *J. Mol. Biol.* **200** 709–723.

Berg, O. G., Winter, R. B., and von Hippel, P. H. (1981) Diffusion-Driven Mechanisms of Protein Translocation on Nucleic Acids. 1. Models and Theory, *Biochemistry* **20**, 6929–6948.

Berger, J. M., Gamblin, S. J., Harrison, S. C., and Wang, J. C. (1996). Structure and Mechanism of DNA Topoisomerase II, *Science (London)* **379**, 225–232.

Berne, B. (1996). Inferring the Hydrophobic Interaction from the Properties of Neat Water, *Proc. Natl. Acad. Sci. USA* **93**, 8800–8803.

Bianchi, M.E., Beltrame, M., and Paonessa, G. (1989). Specific Recognition of Cruciform DNA by Nuclear Protein HMG1, *Science* **243**, 1056–1059.

Biou, V., Yaremchuk, A., Tukalo, M., and Cusack, S. (1994). The 2.9 Å Crystal Structure of *T. thermophilus* Seryl-tRNA Synthetase Complexed with $tRNA^{Ser}$, *Science* **263**, 1404–1410.

Breg, J. N., van Opheusden, H. J., Burgering, M. J. M., Boelens, R. and Kaptein, R. (1990). Structure of Arc Repressor in Solution: Evidence for a Family of β-sheet DNA–binding Proteins, *Science (London)* **346**, 586–589.

Brennan, R. G. (1993). The Winged Helix DNA-Binding Motif: Another Helix–Turn–Helix Takeoff, *Cell* **74**, 773–776.

Brown, R. S., Sander, C., and Argos, P. (1985). The Primary Structure of Transcription Factor TFIIIA Has 12 Consecutive Repeats, *FEBS Lett.* **186**, 271–276.

Burd, C. G. and Dreyfus, G. (1994). Conserved Structures and Diversity of Functions of RNA–Binding Proteins, *Science* **265**, 615–621.

Calnan, B. J., Tidor, B., Biancalana, S., Hudson, D., and Frankel, A. D. (1991). Arginine-Mediated RNA Recognition: The Arginine Fork, *Science* **252**, 1167–1171.

Carey, J. (1988). Gel Retardation at Low pH Resolves Trp Repressor-DNA Complexes for Quantitative Study, *Proc. Natl. Acad. Sci. USA* **85**, 975–979.

Carey, J. (1991a). How Does trp Repressor Bind to Its Operator?, *J. Biol. Chem.* **266**, 24509–24513.

Carey, J. (1991b). Gel Retardation, *Method Enzymol.* **208**, 103–117.

Carey, M., Young-Sun, L., Green, M. R., and Ptashne, M. (1990). A Mechanism for Synergistic Activation of a Mammalian Gene by GAL4 Derivatives, *Science (London)* **345**, 261–364.

Carter, C. W. and Kraut, J. (1974). A Proposed Model for Interaction of Polypeptides with RNA, *Proc. Natl. Acad. Sci. USA* **71**, 283–287.

Chalberg, M. D. and Kelly, T. J. (1989), Animal Virus DNA Replication, *Annu. Rev. Biochem.* **58**, 671–717.

Chandler, V. L., Maler, B. A., and Yamamoto, K. R. (1983). DNA Sequences Bound Specifically by Glucocorticoid Receptor in Vitro Render a Heterologous Promoter Hormone Responsive in Vivo, *Cell* **33**, 489–499.

Chelin, B. J. and Geiduschek, E. P. (1979a). Gel Electrophoretic Separation of Transcription Complexes: An Assay for RNA Polymerase Selectivity and a Method for Promoter Mapping, *Nucleic Acids Res.* **7**, 1951–1867.

Cho, Y., Gorina, S., Jeffrey, P. D., and Pavletich, N. P. (1994) Crystal Structure of a p53 Tumor Supressor-DNA Complex: Understanding Tumorigenic Mutations, *Science* **265**, 346–355.

Church, G. M., Sussman, J. L., and Kim, S.-H. (1977). Secondary Structure Complementarity between DNA and Proteins, *Proc. Natl. Acad. Sci. USA* **86**, 423–425.

Clark, N. D., Beamer, L. J., Goldberg, H. R., Berkower, C. and Pabo, C. O. (1991). The DNA Binding Arm of Lamda Repressor: Critical Contacts from a Flexible Region, *Science* **254**, 267–270.

Clark, K. L., Halay, E. D., Lai, E., and Burley, S. K. (1993) Co-Crystal Structure of the HNF3/*fork head* DNA-recognition Motif Resembles Histone H5, *Science (London)* **364**, 412–420.

Clemons, K. R., Liao, X. Wolf, V., Wright, P. E., and Gottesfeld, J. M. (1992). Definition of the Binding Sites of Individual Zinc Fingers in the Transcription Factor IIIA-5S RNA Gene Complex, *Proc. Natl. Acad. Sci. USA* **89**, 10822–10826.

Clerc, R. G., Corcoran, L. M., Baltimore, D., Sharp, P. A., Ingraham, H. A., Rosenfeld, M. G., Finney, M., Ruvkun, G., and Horvitz, H. R. (1988b). The POU Domain: A Large Conserved Region in the Mammalian pit-I, oct-1, oct-2 and *Caenorhabditis elegans unc-86, Genes Develop.* **2**, 1513–1516.

Crick, F. H. C. (1957). Discussion Comment after the Paper by M. H. F. Wilkins, "Molecular Structure of DNA and Nucleoprotein and Possible Implications in Protein Synthesis," Biochemical Society Symposium, Crook, E. M., Ed., Vol. 14, Cambridge University Press, Cambridge, UK, pp. 13–18.

Crick, F. H. C. (1953). The Packing of α-Helices: Simple Coiled-Coils, *Acta Crystallo.* **6**, 689–697.

Crothers, D. M. (1993). Architectural Elements in Nucleoprotein Complexes, *Curr. Biol.* **3**, 675–676.

Crothers, D. M. (1994). Upsetting the Balance of Forces in DNA, *Science* **266**, 1819–1820.

Crothers, D. M. and Drak, J. (1992). Global Features of DNA Structure by Comparative Gel Electrophoresis, *Methods Enzymol.* **212**, 46–71.

Crothers, D. M., Gartenberg, M. R., and Shrader, T. E. (1991). DNA Bending in Protein-DNA Complexes, *Methods. Enzymol.* **208,** 118–146.

Cuenoud, B. and Schepartz, A. (1993). Design of a Metallo-bZIP Protein that Discriminates between CRE and AP1 Target Sites: Selection against AP1, *Proc. Natl. Acad. Sci. USA* **90**, 1154–1159.

Cusack, S., Berthet, Colominas, C., Hartlein, Nassar, N., and Leberman, R. (1990). A Second Class of Synthetase Structure Revealed by X-ray Analysis of *Escherichia coli* seryl-tRNA Synthetase at 2.5 Å, *Science (London)* **347**, 249–255.

Dalma-Weiszhausz, D. D., Gartenberg, M. R., and Crothers, D. M. (1991). Sequence-Dependent Contribution of Distal Binding Domains to CAP Protein-DNA Binding Affinity *Nucleic Acid Res.* **19**, 611–616.

Davies, J. F. II, Hostomska, Z., Hostomsky, Z., Jordan, S. R., and Matthews, D. A. (1991). Crystal Structure of the Ribonuclease H Domain of HIV-1 Reverse Transcriptase, *Science* **252**, 88–95.

de Crombrugghe, B., Buy, S., and Buc, H. (1984). Cyclic AMP Receptor Protein: Role in Transcription Activation, *Science* **224**, 831–838.

de Haseth, P. L., Lohman, T. M., and Record, M. T., Jr. (1977). Non Specific Interaction of lac Repressor with DNA: An Association Reaction Driven by Counterion Release, *Biochemistry* **16**, 4783–4790.

Dekker, N. Cox, M., Boelens, R., Verrijzer, C.P., van der Vliet and Kaptein, R. (1993). Solution Structure of the POU-Specific DNA–binding domain of Oct-1, *Science (London)* **362**, 852–855.

Dervan, P. (1988). Sequence Specific Recognition of Double Helical DNA. A Synthetic Approach, *Nucleic Acids and Molecular Biology*, Eckstein, F. and Lilley, D. M. J., Ed. Springer-Verlag, Berlin, Heidelberg, New York, London, Paris, Tokyo, Hong Kong, Barcelona, pp. 49–64.

Desjarlais, J. R. and Berg, J. M. (1992). Toward Rules Relating Zinc Finger Protein Sequences and DNA Binding Site Preferences, *Proc. Natl. Acad. Sci. USA* **89**, 7345–7349.

Dick, T., Yang, X., Yeo, S., and Chia, W. (1991). Two Closely Linked Drosophila POU Domain Genes are Expressed in Neuroblasts and Sensory Elements, *Proc. Natl. Acad. Sci. USA* **88,** 7645–7649.

Dixon, W. J., Hayes, J. J., Levin, J. R., Weidner, M. F., Dombrowski, B. A., and Tullius, T. D. (1991). Hydroxyl Radical Footprinting, *Methods Enzymol.* **208**, 380–413.

Draper, D. E. (1995). Protein–RNA Recognition, *Annu. Rev. Biochem.* **64**, 593–620.

Dreyfuss, G., Swanson, M. S., and Pinol-Roma, S. (1988) Heterogeneous Nuclear Ribonucleprotein Particles and the Pathway of mRNA Formation, *TIBS* **13**, 86–91.

Du, W., Thanos, D. and Maniatis, T. (1993). Mechanisms of Transcriptional Synergism between Distinct Virus Inducible Enhancer Elements, *Cell* **74**, 887–898.

Dunn, T. M., Hahn, St., Ogden, S., and Schleif, R. F. (1984). An Operator at -280 Base Pairs that is Required for Repression of araBAD Operon Promoter: Addition of DNA Helical Turns between the Operator and Promoter Cyclically Hinders Repression, *Proc. Natl. Acad. Sci. USA* **81**, 5017–5020.

Ebright, R. H., Ebright, Y. W., and Gunasekera, A. (1989) Consensus DNA Site for the *Escherichia coli* Catabolite Gene Activator Protein (CAP): CAP Exhibits a 450-fold Higher Affinity for the Common DNA Site than for the *E. coli* Lac DNA Site, *Nucleric Acids Res.* **17**, 10295–10350.

Engelke, D. R. (1980). Specific Interaction of a Purified Transcription Factor with an Internal Control Region from 5S RNA Genes, *Cell* **9**, 717–728.

Ellenberger, T. E., Brandl, C. J., Struhl, K., and Harrison, S. C. (1992). The GCN4 Basic Region Leucine Zipper Binds DNA as a Dimer of Uninterrupted α-Helices: Crystal Structure of the Protein-DNA Complex, *Cell* **71**, 1223–1237.

Eriani, G., Delarue, M., Poch, O., Gangloss, J., and Moras, D. (1990). Partition of tRNA Synthetases into Two Classes Based on Mutually Exclusive Sets of Sequence Motifs, *Science (London)* **347,** 203–206.

Evans, T. Reitman, M., and Felsenfeld, G. (1988). An Erythrocyte-Specific DNA–Binding Factor Recognizes a Regulatory Sequence Common to All Chicken Globin Genes, *Proc. Natl. Acad. Sci. USA* **85**, 5976–5980.

Fairall, L., Schwabe, J. W. R., Chapman, L., Finch, J. T., and Rhodes, D. (1993). The Crystal Structure of a Two Zinc-Finger Peptide Reveals an Extension to the Rules for Zinc-Finger/DNA Recognition, *Science (London)* **366**, 483–487.

Felsenfeld, G., Evans, T., Jackson, P. D., Knezetic, J., Lewis, C., Nickol, J., and Reitman, M. (1989). Regulatory Protein Action in the Neighborhood of Chicken Globin Genes During Development. *Prog. Clin. Biol. Res.* **316A** 73–87.

Ferrari, S., Harley, V. R., Pontiggia, A., Goodfellow, P. N., Lovell-Badge, R., and Bianchi, M. E. (1992). SRY, Like HMG1, Recognizes Sharp Angles in DNA, *EMBO J.* **11**, 4497–4506.

Ferre-D'Amare, A. R., Prendergast, G. C, Ziff, E. B., and Burley, S. K (1993). Recognition by Max of its Cognate DNA through a Dimeric b/HLH/Z Domain, *Science (London)* **363**, 38–45.

Finney, M., Ruvkun, and Horvitz, H. R. (1988). The *C. elegans* Cell Lineage and Differentiation Gene unc-86 Encodes a Protein with a Homeodomain and Extended Similarity to Transcription Factors, *Cell* **55**, 757–769.

Finzi, L. and Gelles, J. (1995). Measurement of Lactose Repressor-Mediated Loop Formation and Breakdown in Single DNA Molecules, *Science* **267**, 378–380.

Foguel, D. and Silva, J. K. (1994). Cold Denaturation of a Repressor-Operator Copmplex: The Role of Entropy in Protein-DNA Recognition, *Proc. Natl. Acad. Sci. USA* **91**, 8244–8247.

Fried, M. and Crothers, D. M. (1981). Equilibria and Kinetics of lac Repressor-Operator Interactions by Polyacrylamide Gel Electrophoresis, *Nucluic Acids Res.* **9**, 6505–6525.

Fried, M. G. and Crothers, D. M. (1984). Kinetics and Mechanism in the Reaction of Gene Regulatory Proteins with DNA, *J. Mol. Biol.* **172**, 263–282.

Fried, M. G. and Liu, G. (1994). Molecular Sequestration Stabilizes CAP–DNA Complexes during Polyacrylamide Gel Electrophoresis, *Nucl. Acids Res.* **22**, 5054–5059.

Friedmann, A., Fischmann, T. O., and Steitz, T. A. (1995). Crystal Structure of *lac* Repressor Core Tetramer and Its Implications for DNA Looping, *Science* **268**, 1721–1727.

Galas, D. J. and Schmitz, A. (1978). DNAse Footprinting: A Simple Method for the Detection of Protein-DNA Binding Specificity, *Nucluic Acids Res.* **5**, 3157–3170.

Garner, M. M. and Revzin, A. (1981). A Gel Electrophoresis Method for Quantifying the Binding of Proteins to Specific DNA Regions: Application to Components of the *Escherichia coli* Lactose Operon Regulatory System, *Nucleic Acids. Res.* **9,** 3047–3060.

Gartenberg, M. R., Ampe, C., Steitz, T. A., and Crothers, D. M. (1990). Molecular Characterization of the GCN4–DNA Complex, *Proc. Natl. Acad. Sci. USA* **87**, 6034–6038.

Gartenberg, M. R. and Crothers, D. M. (1988). DNA Sequence Determinants of CAP-Induced Bending and Protein Binding Affinity, *Science (London)* **333**, 824–829.

Ghosh, G., Van Duyne, G., Ghosh, S., and Sigler, P. B. (1995) Structure of NFκB p50 Homodimer Bound to a κB Site, *Science (London)* **373**, 303–310.

Giese, K., Cox, J., and Grosschedl, R. (1992). The HMG Domain of Lymphoid Enhancer Factor 1 Bends DNA and Facilitates Assembly of Functional Nucleoprotein Structures, *Cell* **69**, 185–195.

Giniger, E. and Ptashne, M. (1988). Cooperative DNA Binding of the Yeast Transcriptional Activator GAL4, *Proc. Natl. Acad. Sci. USA* **85**, 382–386.

Goodman, S. D. and Nash, H. A. (1989). Functional Replacement of a Protein-Induced Bend in a DNA Recombination Site, *Science* **341**, 251–254.

Goodman, S. D., Nicholson, S. C., and Nash, H. A. (1992). Deformation of DNA during Sitespecific Recombination of Bacteriophage λ: Replacement of IHF Protein by HU Protein Or Sequence-Directed Bends, *Proc. Natl. Acad. Sci. USA* **89**, 11910–11914.

Griffith, J., Hochschild, A., and Ptashne, M. (1986). DNA Loops Induced by Cooperative Binding of λ Repressor, *Science* **322**, 750–752.

Gubbay, J., Collignon, Koopman, P., Capel, B., Economou, A., Munsterberg, A., Vivian, N., Goodfellow, P., and Lovell-Badge, R. (1990). A Gene Mapping to the Sex-Determining Region of the Mouse Y Chromosome is a Member of a Novel Family of Embryonically Expressed Genes, *Science (London)* **346**, 245–250.

Gupta, A. P. and Benkovic, S. J. (1984). Sterochemical Course of the $3' \rightarrow 5'$-Exonuclease Activity of DNA Polymerase I, *Biochemistry* **23**, 5874–5881.

Hall, K. B. and Stump, W. T. (1992). Interaction of N-terminal Domain of U1A Protein with an RNA Stem/loop, *Nucleic Acids Res.* **20**, 4283–4290.

Haran, T. E., Joachimiak, A., and Sigler, P. B. (1992). The DNA Target of the *trp* Repressor, *EMBO J.* **11**, 3021–3030.

Harbury, P. B., Zhang, T., Kim, P., and Alber, T. (1993). A Switch Between Two-, Three- and Four-Stranded Coiled Coils in GCN4 Leucine Zipper Mutants, *Science* **262**, 1401–1407.

Hard, T., Kellenbach, E., Boelens, R., Maler, B. A., Dahlman, K., Freedman, L. P., Carlstedt-Duke, J., Yamamoto, K. R., Gustafsson, J. A., and Kaptein, R. (1990). Solution Structure of the Glucocorticoid Receptor DNA Binding Domain, *Science* **249**, 157–160.

Harrison, S. C. (1991). A Structural Taxonomy of DNA-binding Domains, *Science (London)* **353**, 715–719.

Harrison, S. C. and Aggarwal, A. K. (1990). DNA Recognition by Proteins with the Helix–Turn–Helix Motif, *Annu. Rev. Biochem.* **59**, 933–969.

Hegde, R. S., Grossman, S. R., Laimins, L. A., and Sigler, P. B. (1992). Crystal Structure at 1.7 Å of the Bovine Papilomavirus-1 E2 DNA Binding Domain Bound to its DNA Target, *Science (London)* **359**, 505–512.

Herr, W. and Sturm, R.A. (1988). The POU Domain: A Large Conserved Region in the Mammalian *pit-1, oct-1, oct-2*, and *Caenorhabditis elegans unc-86* Gene Products, *Genes Develop.* **2**, 1513–1516.

Herschlag, D. and Jencks, W. P. (1987). The Effect of Divalent Metal Ions on the Rate and Transition-State Structure of Phosphoryl-Transfer Reactions, *J. Am. Chem. Soc.* **109**, 4665–4674.

Hertzberg, R.P. and Dervan, P.B. (1982). Cleavage of Double Helical DNA by (Methidiumpropyl-edta). iron(II), *J. Am. Chem. Soc.* **104**, 313.

Hochschild, A. (1991). Detecting Cooperative Protein-DNA Interactions and DNA Loop Formation by Footprinting, *Methods Enzymol.* **208**, 343–361.

Hochschild, A., and Ptashne, M. (1986). Cooperative Binding of λ Repressors to Sites Separated by Integral Turns of the DNA Helix, *Cell* **44**, 681–687.

Hogan M. E. and Austin R.H. (1987). Importance Of DNA Stiffness In Protein-DNA Binding Specificity, *Science (London)* **329**, 263–266.

Honig, B. and Nicholls, A. (1995). Classical Electrostatics in Biology and Chemistry, *Science* **268**, 1144–1149.

Horikoshi, M., Betuccioli, C., Takada, R., Wang, J., Yamamoto, T and Roeder, R.G. (1992). Transcription Factor TFIID Induces DNA Bending upon Binding to the TATA Element, *Proc. Natl. Acad. Sci. USA* **89**, 1060–1064.

Hummer, G., Garde, S., Garcia, A. E., Pohorille, A., and Pratt, L. E. (1996). An Information Model of Hydrophobic Interactions, *Proc. Natl. Acad. Sci. USA* **93**, 8951–8955.

Ingraham, H. A., Chen, R., Mangalam, H. J., Elsholtz, H. P., Flynn, S. E., Lin, C. R., Simmons, D. M., Swanson, L., and Rosenfeld, M.G. (1988). A Tissue-Specific Transcription Factor Containing a Homoedomain Specifies a Pituitary Phenotype, *Cell* **55**, 519–528.

Jack, W. E., Terry, B. J., and Modrich, P. (1982). Involvement of Outside DNA Sequences in the Major Kinetic Path by which EcoRl Endonuclease Locates and Cleaves its Recognition Sequence, *Proc. Natl. Acad. Sci. USA* **79**, 4010–4014.

Jackson, P. D., Evans, T., Nickol, J. M., and Felsenfeld, G. (1989). Developmental Modulation of Protein Binding to β-globin Gene Regulatory Sites within Chicken Erythrocyte Nuclei, *Genes and Development* **3**, 1860–1873.

Jan, Y. H. and Jan, , L. Y. (1993). HLH Proteins, Fly Neurogenesis, and Vertebrate Myogenesis, *Cell* **75**, 827–830.

Jantzen, H.-M., Admon, A., Bell, S. P., and Tijan, R. (1990) Nucleolar Transcription Factor hUBF contains a DNA-binding Motif with Homology to HMG Proteins, *Science (London)* **344**, 830–836.

Johns, E. W., Ed. (1982). *The HMG Chromosomal Proteins*, Academic, New York.

Johnson, R. C. and Glasgow, A. C. (1987). Spatial Relationship of the Fis Binding Sites for Hin Recombinational Enhancer Activity, *Science (London)* **329**, 462–465.

Jones, O. W. and Berg, P.(1966). Studies on the Binding of RNA Polymerase to Polynucleotides, *J. Mol. Biol.* **22**, 199–209.

Jordan, S. R. and Pabo, C. O. (1988). Structure of the Lambda Complex at 2.5 Å Resolution: Details of the Repressor-Operator Interactions, *Science* **242**, 893–899.

Kadonaga, J. T., Camer, K. R., Masiarz, F. R., and Tijan, R. (1987) Isolation of cDNA Encoding Transcription Factor Spl and Functional Analysis of the DNA Binding Domain, *Cell* **51**, 1059–1090.

Kahn, J. D. and Crothers, D. M. (1992). Protein-Induced Bending and DNA Cyclization, *Proc. Natl. Acad. Sci. USA* **89**, 6343–6347.

Kamada, K., Horiuchi, T., Ohsumi, K., Shimamoto, N and Morikawa, K. (1996). Structure of a Replication–terminator Protein Complexed with DNA, *Science (London)* **383**, 598–603.

Karlin, S. and Altschul, S.F. (1993). Applications and Statistics for Multiple High-Scoring Segments in Molecular Sequences, *Proc. Natl. Acad. Sci. USA* **90**, 5873–5877.

Karlin, S. and Brendel, V. (1992). Chance and Statistical Significance in Protein and DNA Sequence Analysis, *Science* **257**, 39–49.

Keegan, L., Gill, G., and Ptashne, M. (1986). Separation of DNA Binding from the Transcription-Activating Function of a Eukaryotic Regulatory Protein, *Science* **231**, 699–704.

Kim, E. U. and Wyckoff, H. W. (1991). Reaction Mechanism of Alkaline Phosphatase Based on Crystal Structures, *J. Mol. Biol.* **218**, 449–464.

Kim, S.-H. (1992). β Ribbon: A New DNA Recognition Motif, *Science* **255**, 1217–1218.

Kim, S. and Landy, A. (1992). Lambda Int Protein Bridges Between Higher Order Complexes at Two Distant Chromosomal Loci *att*L and *att*R, *Science* **256**, 198–203.

Kim, Y., Geiger, J. H., Hahn, S., and Sigler, P. B. (1993). Crystal Structure of a Yeast TBP/TATA–Box Complex, *Science (London)* **365**, 512–520.

Kim, J. G., Takeda, Y., Matthews, B. W., and Anderson, W. F. (1987). Kinetic Studies on Cro Repressor-Operator DNA Interaction, *J. Mol. Biol.* **196**, 149–158.

Kim, J. L., Nikolov, D. B., and Burley, S. K. (1993b). Co-Crystal Structure of TBP Recognizing the Minor Groove of a TATA Element, *Science (London)* **365**, 520–511.

Kim, J., Tzamarias, D., Ellenberger, T., Harrison, S.C., and Struhl, K. (1993). Adaptability at the Protein–DNA Interface Is an Important Aspect of Sequence Recognition by bZIP Proteins, *Proc. Natl. Acad. Sci. USA* **90**, 4513–4517.

Kissinger, C. R., Liu, B., Martin-Blanco, E., Komberg, T., and Pabo, C. O. (1990). Crystal Structure of an Engrailed Homeodomain-DNA Complex at 2.8 Å Resolution: A Framework for Understanding Homeodomain–DNA Interactions, *Cell* **63**, 579–590.

Klevit, R. E. (1991). Recognition of DNA by Cys 2, His2 Zinc Fingers, *Science* **253**, 1367.

Klevan, L., Dattagupta, N., Hogan, M., and Crothers, D. M. (1978). Physical Studies of Nuclesome Assembly, *Biochemistry* **17**, 4533–4540.

Klug, A. and Travers, A. A. (1989). The Helical Repeat of Nucleosome-Wrapped DNA, *Cell* **56**, 9–11.

Ko, H.-S., Fast, P., McBride, W., and Staudt, L. M. (1988). A Human Protein Specific for the Immunoglobulin Octamer DNA Motif Contains a Functional Homeobox Domain, *Cell* **55**, 135–144.

Kolodrubetz, D. and Burgum, A. (1990). Duplicated *NHP6* Genes of *Saccharomyces cerevisiae* Encode Protein Homologous to Bovine High Mobility Group Protein 1, *J. Biol. Chem.* **265,** 3234–3239.

Koo, H.-S., Drak, J., Rice, J. A., and Crothers, D. M. (1990) Determination of the Extent of DNA Bending by an Adenine-Thymine Tract, *Biochemistry* **29**, 4227–4234.

Kong, X-P., Onrust, R., O'Donnell, M., and Kuriyan, J. (1992) Three-Dimensional Structure of the β Subunit of *E. Coli* DNA Polymerase III Holoenzyme: A Sliding DNA Clamp, *Cell* **69**, 425–437.

Koudelka G. B., Harrison S. C., and Ptashne M. (1987). Effect of Non-Contacted Bases on the Affinity of 434 Operator for 434 Repressor and Cro, *Science (London)* **326**, 886–8888.

Koudelka G. B., Harbury P., Harrison S. C., and Ptashne M. (1988) DNA Twisting and the Affinity of Bacteriophage 434 Operator For Bacteriophage 434 Repressor, *Proc. Natl. Acad. Sci. USA* **85**, 4633–4637.

Krämer, H., Niemolle, M., Amouyal, M., Revet, B., von Wilcken-Bergmann and Müller-Hill, B. (1987). lac Repressor Forms Loops with Linear DNA Carrying Two Suitably Spaced lac Operators, *EMBO J.* **6**, 1481–1491.

Kraulis, P. J., Raine, A. R. C., Gadhavi, P. L., and Laue, E. D. (1992). Structure of the DNA binding Domain of Zinc GAL4, *Science* **356**, 448–450.

Kriwacki, R.W., Schultz, S.C., Steitz, T.A., and Caradonna, J. P. (1992). Sequence-Specific Recognition of DNA by Zinc-Finger Peptides Derived from the Transcription Factor SP1, *Proc. Natl. Acad. Sci. USA* **89**, 9759–9763.

Landschulz, W. H., Johnson, P. F., and McKnight, S. L. (1989). The Leucine Zipper: A Hypothetical Structure Common to a New Class of DNA Binding Proteins, *Science* **240**, 1759–1764.

Laughon, A. and Scott, M. P. (1984). Sequence of a *Drosphila* Segmentation Gene; Protein Structure Homology with DNA-binding Proteins, *Science (London)* **310**, 25–31.

Lawson, C. L. and Carey, J. (1993). Tandem Binding in Crystals of a *trp* Repressor/Operator Half-Site Complex, *Science (London)* **366**, 178–182.

LeBret, M. and Zimm, B.H. (1984). Distribution of Counterions around a Cylindircal Polyelectrolyte and Manning's Condensation Theory, *Bioplymers* **23**, 287–312.

Lee, D.-H and Schlief, R. F. (1989). *In vivo* DNA Loops in araBAD: Size Limits and Helical Repeat, *Proc. Natl. Acad. Sci. USA* **86**, 476–480.

Lee, D. K., Horikoshi, M., and Roeder R. G. (1991). Interaction of TFIID in the Minor Groove of the TATA Element, *Cell* **67,** 1241–1250.

Lee, M. S., Gippert, G. P., Soman, K. V., Case, D. A., and Wright, P. E. (1989). Three-Dimensional Solution Structure of a Single Zinc Finger DNA-Binding Domain, *Science* **245**, 635–637.

Lee, M. S., Kliewer, S. A., Provencal, J., Wright, P. E., and Evans, R. M. (1993). Structure of the Retinoid X Receptor α DNA Binding Domain: A Helix Required for Homodimeric DNA Binding, *Science* **260**, 1117–1121.

Lenardo, M. J. and Baltimore, D. (1989). NF-KB: A Pleiotropic Mediator of Inducible and Tissue-Specific Gene Control, *Cell* **58**, 227–229.

Lesser, D. R., Kurpiewske, M. R., Waters, T., Connolly, B. A., and Jen-Jacobson, L. (1993). Facilitated Distortion of the DNA Site Enhances *Eco*RI Endonuclease-DNA Recognition, *Proc. Natl. Acad. Sci. USA* **90**, 7548–7552.

Leszczynski, J. F. and Rose, G. D. (1986). Loops in Globular Proteins: A Novel Category of Secondary Structure, *Science* **234**, 849–855.

Liang, H., Mao, X., Olejniczak, T.,Nettesheim, D. G., Yu, L., Meadows, R. P., Thompson, C. B., and Fesik, S. W. (1994b). Solution Structure of the *ets* Domain of Fli-1 when Bound to DNA, *Structural Biol.* **1**, 871–875.

Liang, H., Olejniczak, T., Mao, X., Nettesheim, D. G., Yu, L., Thompson, C. B., and Fesik, S. W. (1994a). The Secondary Structure of the *Ets* Domain of Human Fli-1 Resembles that of the Helix–Turn–Helix Binding Motif of the *Escherichia Coli* Catabolite Gene Activator Protein, *Proc. Natl. Acad. Sci. U. S. A.* **91**, 11655–11659.

Lilley, D. M. J. (1992). Probes of DNA Structure, *Methods. Enzymol.* 212, 133–139.

Lima, C. D., Wang J. C., and Mondragon A. (1994) Three-dimensional Structure of the 67K N-terminal Fragment of E. coli DNA Topoisomerase I. *Science (London)* **367**, 138–146.

Liu-Johnson, H.-N., Gartenberg, M. R., and Crothers, D. M. (1986). The DNA Binding and Bending Angle of *E. Coli* CAP Protein, *Cell* **47**, 995–1005.

Lobell, R. B. and Schleif, R. F. (1990). DNA Looping and Unlooping by AraC Protein, *Science* **250**, 528–532.

Lockhart, D. J. and Kim, P. S. (1992). Intenal Stark Effect Measurement of the Electric Field at the Amino Terminus of an α Helix, *Science* **257**, 947–951.

Lockhart, D. J. and Kim, P. S. (1993). Electrostatic Screening of Charge and Dipole Interactions with the Helix Backbone, *Science* **260**, 198–202.

Long, K. S. and Crothers, D. M. (1995). Interaction of the Human Immunodeficiency Virus Type I Tat-Derived Peptides with TAR RNA, *Biochemistry* **35**, 8885–8895.

Lovering, R., Hanson, I. M., Borden, K. L. B., Martin, S., O'Reilly, N. J., Evan, G. I., Rahman, D., Pappin, D. J. C., Trowsdale, J., and Freemont, P. S. (1993). Identification and Preliminary Characterization of a Protein Motif Related to the Zinc Finger, *Proc. Natl. Acad. Sci. USA* **90**, 2112–2116.

Luisi, B. F., Xu, W. X., Otwinowski, Z., Freedman, L. P., Yamamoto, K. R., and Sigler, P. B. (1991). Crystallographic Analysis of the Interaction of the Glucocorticoid Receptor with DNA, *Nature* **352**, 497–505.

Lumb, K. J. and Kim, P. S. (1995). Measurement of Interhelical Electrostatic Interactions in the GCN4 Leucine Zipper, *Science* **268**, 436–439.

Lumpkin, O. J. and Zimm, B. H. (1982). Mobility of DNA in Gel Electrophoresis, *Biopolymers* **21**, 2315–2316.

Manning, G. S. (1989). An Estimate of the Extent of Folding of Nucleosomal DNA by Laterally Asymmetric Neutralization of Phosphate Groups, *J. Biomol. Struct. Dynam.* **6**, 877–889.

Marmorstein, R., Carey, M., Ptashne, M., and Harrison, S. C. (1992). DNA Recognition by GAL4: Structure of a Protein-DNA Complex, *Science (London)* **356**, 408–414.

Mattaj, 1. W. (1989). A Binding Consensus: RNA–Protein Interactions in Splicing, snRNPs, and Sex, *Cell* **57,** 1–3.

Mattaj, I. W. and Nagai, K. (1995). Recruiting Proteins to the RNA World, *Science Struct. Biol.* **2**, 518–522.

Matthews, B. W., Ohlendorf, D. H., Anderson, W. F., and Takeda, Y. (1982). Structure of the DNA-binding Region of lac Repressor Inferred from its Homology with cro Repressor, *Proc. Natl. Acad. Sci. USA* **79**, 1428.

Matthew, J. B. and Ohlendorf, D. H.(1985). Electrostatic Deformation of DNA by a DNA Binding Protein, *J. Biol. Chem.* **260**, 5860–5862.

Matthew, J. B. and Richards, F. M. (1984). Differential Electrostatic Stabilization of A-, B-, and Z- Forms of DNA, *Biopolymers* **23**, 2743–2759.

McKay, D. B. and Steitz, T. A. (1981). Structure of Catabolite Gene Activator Protein at 2.9 Å Resolution Suggests Binding to Left-handed B-DNA, *Science (London)* **290**, 744–749.

Michelsen, J. W., Schmeichel, K. L., Beckerle, M. C., and Winge, D. R. (1993). The LIM Motif Defines a Specific Zinc-Binding Protein Domain, *Proc. Natl. Acad. Sci. USA* **90**, 4404–4408.

Miller, J., McLachian, A. D., and Klug, A. (1985). Repetitive Zinc-binding Domains in the Protein Transcription Factor IIIA from *Xenopus* Oocytes, *EMBO J.* **4**, 1609–1614.

Mirzabekov, A. D. and Rich, A. (1979). Asymmetric Lateral Distribution of Unshielded Phosphate Groups in Nucleosomal DNA and its Role in DNA Bending. *Proc. Natl. Acad. Sci. USA* **76**, 1118–1121.

Misra, B. K. and Honig, B. (1995). On the Magnitude of the Electrostatic Contribution to Ligand-DNA Interactions, *Proc. Natl. Acad. Sci. USA* **92** 4691–4695.

Mitchell, P. J. and Tjian, R. (1989). Transcriptional Regulation in Mammalian Cells by Sequence-Specific DNA Binding Proteins, *Science* **245**, 371–378.

Moreau, P., Hen, R., Wasylyk, B., Everett, R., Gaub, M. P., and Chambon, P. (1981). The SV40 72 Base Pair Repeat has a Striking Effect on Gene Expression both in SV40 and other Chimeric Recombinants, *Nucleic Acids Res.* **9**, 6047–6068.

Müller, C. W., Rey, F. A., Sodeoka, M., Verdine, G., and Harrison, S. C. (1995). Structure of the NFκB Homodimer Bound to DNA, *Science (London)* **373**, 311–317.

Muller, M. M., Ruppert, S., Schaffner, W., and Matthias, P. (1988). A Cloned Octamer Transcription Factor Stimulates Transcription from Lymphoid-Specific Promoters in Non-B Cells, *Science (London)* **336**, 544–551.

Mulligan, M. E., Hawley, D. K., Entriken, R., and McClure, W. R. (1984). *Esherichia coli* Promoter Sequences Predict *in vitro* RNA Polymerase Selectivity, *Nucleic Acids Res.* **46**, 123–132.

Murre, C., McCaw, P. S., and Baltimore, D. (1989). A New DNA Binding and Dimerization Motif in Immunoglobulin Enhancer Binding, *daughterless, MyoD, and myc* Protein, *Cell* **56**, 777–783.

Nagai, K., Oubridge, Jessen, T. H., Li, J., and Evans, P. R. (1990). Crystal Structure of the RNA-binding Domain of the Ul Small Nuclear Ribonucleoprotein A, *Science (London)* **348**, 515–520.

Nikolov, D. B., Hu, S.-H., Lin, J., Gasch, A., Hoffmann, A., Horikoshi, M., Chua, N.-H, Roeder, R. G., and Burley, S. K. (1992). Crystal Structure of TFIID TATA-box Binding Protein, *Science* **360**, 40–46.

Nirenberg, M. and Leder, P. (1964). RNA Codewords of Protein Synthesis, *Science* **145**, 1399–1407).

Noll, M. (1974). Internal Structure of the Chromatin Subunit, *Nucleic Acids Res.* **1**, 1573–1579.

O'Halloran, T. V. (1993). Transition Metals in Control of Gene Expression, *Science* **261**, 715–725.

Ohlendorf, D. H. and Matthew, B. W. (1983). Structural Studies of Protein-nucleic Acid Interactions, *Annu. Rev. Biophys. Bioeng.* **12**, 259–284.

Oliphant, A. R., Brandl, C. J., and Struhl, K. (1989). Defining the Sequence Specificity of DNA–Binding Proteins by Selecting Binding Sites from Random-Sequence Oligonucleotides: Analysis of Yeast GCN4 Protein, *Mol. Cell Biol.* **9**, 2944–2949.

Omichinski, J. G., Clore, G. M., Schaad, O., Felsenfeld, G., Trainor, C.,, Appella, E., Stahl, S. J., and Gronenbrn, A. M. (1993a). NMR Structure of a Specific DNA Complex of Zn-Containing DNA Binding Domain of GATA-1, *Science* **261**, 438–446.

Omichinski, J. G., Trainor, C., Evans, T., Gronenbom, A. M., Clore, G. M., and Felsenfeld, G. (1993b). A Small Single"Finger" Peptide from the Erythroid Transcription Factor GATA-1 Binds Specifically to DNA as a Zinc or Iron Complex, *Proc. Natl. Acad. Sci. USA* **90**, 1676–1680.

O'Neil, K. T., Hoess, R. H., and DeGrado, W. F. (1990). Design of DNA–Binding Peptides Based on the Leucine Zipper Motif, *Science* **249**, 774–778.

O'Shea, E. K., Klemm, J. D., Kim, P. S., and Alber, T. (1991). X-ray Structure of the GCN4 Leucine Zipper, a Two-Stranded, Parallel Coiled Coil, *Science* **254**, 539–544.

O'Shea, E. K., Rutkowski, R., and Kim, P. S. (1989a). Evidence that the Leucine Zipper is a Coiled Coil, *Science* **243**, 538–542.

O'Shea, E. K., Rutkowski, R., and Kim, P. S. (1992). Mechanism of Specificity in the Fos-Jun Oncoprotein Heterodimer, *Cell* **68**, 699–708.

O'Shea, E. K., Rutokowski, R., and Stafford, W. F. Ill (1989b) Preferential Heterodimer Formation by Isolated Leucine Zippers from Fos and Jun, *Science* **245**, 646–648.

Oshima, Y. (1981). *The Molecular Biology of the Yeast Saccharomyces, Metabolism and Gene Expression*, Strathem, J. N., Jones, E. W., and Broach, J. R. Eds., Cold Spring Harbor Laboratory, Cold Spring Harbor, New York, pp. 159–180.

Otwinowski, Z., Schevitz, R. W., Zhang, R. -G., Lawson, C. L., Joachimiak, A., Marmostein, R. Q., Luisi, B. F., and Sigler, P. B.(1988). Crystal Structure of *trp* Repressor/Operator Complex at Atomic Resolution, *Science* **335**, 321–329.

Oubridge, C., Ito, N., Evans, P. R., Teo, C.-H., and Nagai, K. (1994). Crystal Structure at 1.92 Å Resolution of the RNA-Binding Domain of the U1A Splicesomal Protein Complexed with an RNA Hairpin, *Science (London)* **372**, 432–438.

Pabo, C. O. and Sauer, R. T. (1984). Protein-DNA Recognition, *Annu. Rev. Biochem.* **53**, 293–321.

Pabo, C. O. and Sauer, R. T. (1992). Transcription Factors: Structural Families and Principles of DNA Recognition, *Annu. Rev. Biochem.* **61**, 1053–1095.

Pan, T. and Coleman, J. E. (1989). Structure and Function of the Zn(II) Binding Site within the DNA-binding Domain of the GAL4 Transcription Factor, *Proc. Natl. Acad. Sci. USA* **86,** 3145–3149.

Pan, T. and Coleman, J. E. (1990a). GAL4 Transcription Factor is not a "Zinc Finger" but Forms a $Zn(II)_2$-CYS_6 Binuclear Cluster, *Proc. Natl. Acad. Sci. USA* **87**, 2077-2081.

Pan, T. and Coleman, J. E. (1990b). The DNA Binding Domain of GAL4 Forms a Binuclear Metal Ion Complex, *Biochemistry* **29,** 3023–3029.

Pan, T. and Uhlenbeck, O. C. (1992). A Small Metalloribozyme with a Two-Step Mechanism, *Science (London)* **358**, 560–563.

Paolella, D. N., Palmer, C. R., and Schepartz, A. (1994). DNA Targets of Certain bZIP Proteins Distinguished by an Intrinsic Bend, *Science* **264**, 1130–1133

Parraga, G., Horvath, S. J., Eisen, A., Taylor, W. E, Hood, L., Young, E. T., and Klevit, R. E. (1988). Zinc-Dependent Structure of a Single-Finger Domain of Yeast ADRI, *Science* **241**, 1489–1492.

Parvin, J. D., McCormick, R. J., Sharp, P. A., and Fisher, D. E. (1995). Pre-Bending of a Promoter Sequence Enhances Affinity for the TATA-Binding Fator, *Science (London)* **373**, 724–727.

Passner, J. (1997). Structural Studies Involving the *Escherichia coli* Catabolite Gene Activator Protein, Ph.D Thesis, Yale University, New Haven, CT.

Paull, T. T., Haykinson, M. J., and Johnson, R. C. (1993). The Non-Specific DNA Binding and Bending Proteins HMG1 and HMG2 Promote The Assembly Of Complex Nucleoprotein Structures, *Genes Develop.* **7**, 1521–34.

Pauling, L. and Corey, R. B. (1953). Compound Helical Configurations of Polypeptide Chains. Structure of Proteins of the α-Keratin Type, *Science (London)* **171**, 59–61.

Pavletich, N. P. and Pabo, C. O. (1991). Zinc Finger-DNA Recognition: Crystal Structure of a Zif268–DNA Complex at 2.1 Å, *Science* **252**, 809–817.

Pavletich, N. P. and Pabo, C. O. (1993). Crystal Structure of a Five-Finger GLI–DNA Complex: New Perspectives on Zinc Fingers, *Science* **261**, 1701–1707.

Perini, G., Wagner, S., and Green, M. R. (1995). Recognition of bZIP Proteins by the Human T-Cell Leukaemia Virus Transactivator TAX, *Science (London)* **376**, 602–605.

Pfeifer, K., Kim, K.-S., Kogan, S., and Guarente, L. (1989). Functional Dissection and Sequence of Yeast HAP1 Activator, *Cell* **56**, 291–301.

Phillips, S. E. V., Mainfield, I., Parsons, I., Davidson, B. E., Rafferty, J. B., Somers, W. S., Margarita, D., Cohen, G. N., Saint-Girons, I., and Stokley, P. G. (1989). Cooperative Tandem Binding of met Repressor of *Escherichia coli, Nature* **341**, 711–715.

Pil, P. M. and Lippard, S. J. (1992). Specific Binding of Chromosomal Protein HMG1 to DNA Damaged by the Anticancer Drug Cisplatin, *Science* **256**, 234–237.

Ponta, H., Kennedy, N., Skorch, P., Hynes, N. E., and Groner, B. (1985). Hormonal Response Region in the Mouse Mammary Tumor Virus Long Terminal Repeat Can Be Dissociated from the Proviral Promoter and Has Enhancer Properties, *Proc. Natl. Acad. Sci. USA* **82**, 1020–1024.

Price, M. A. and Tullius, T. D. (1992). Using Hydroxyl Radical to Probe DNA Structure, *Methods Enzymol.* **212**, 194–219.

Puglisi, J. D., Chen, L., Frankel, A. D., and Williamson, J. R. (1993). Role of RNA Structure in Arginine Recognition of TAR RNA, *Proc. Natl. Acad. Sci. USA* **90**, 3680–3684.

Puglisi, J. D., Ran, R., Calnan, B. J., Frankel, A. D., and Williamson, J. R. (1992). Conformation of the TAR RNA–Arginine Complex by NMR Spectroscopy, *Science* **257**, 7680.

Qian, X., Jeon, C., Yoon, H., Agarwal, K., and Weiss, M. A. (1993). Structure of a New Nucleic-Acid-Binding Motif in Eukaryotic Transcriptional Elongation Factor TFIIS, *Science (London)* **365**, 277–279.

Rafferty, J. B., Somers, W. S., Saint-Girons, I., and Phillips, S. E. V. (1989). Three-dimensional Crystal Structures of *Escherichia coli met* Repressor with and without Corepressor, *Science (London)* **341**, 705–710.

Ramakrishnan, V., Finch, J. T., Graziano, V., Lee, P. L., and Sweet, R. M. (1993). Crystal Structure of Globular Domain of Histone H5 and its Implications for Nucleosome Binding, *Science* **362**, 219–223.

Rasmussen, R., Benvegnu, D., O'Shea, E. K., Kim, P. S., and Alber, T. (1991). X-ray Scattering Indicates that the Leucine Zipper Is a Coiled Coil, *Proc. Natl. Acad. Sci. USA* **88**, 561–564.

Rastinejad, F., Perlman, T., Evans, R. M., and Sigler, P. B. (1995). Structural Determinants of Nuclear Receptor Assembly on DNA Direct Repeats, *Science (London)* **375**, 203–211.

Raumann, B. E., Rould, M. A., Pabo, C. O., and Sauer, R. T. (1994). DNA Recognition by β-sheets in the Arc Repressor-Operator Crystal Structure, *Science (London)* **367**, 754–757.

Record, M. T., Jr., Anderson, C. F., and Lohman, T. M. (1978) Thermodynamic Analysis of Ion Effects on the Binding and Conformational Equilibria of Proteins and Nucleic Acids: The Roles of Ion Association or Release, Screening, and Ion Effects on Water Activity, *Q. Rev. Biophys.* **11**, 103–178.

Record, M. T., Jr., Lohman, T. M., and de Haseth, P. L.(1976). Ion Effects on Ligand-Nucleic Acid Interactions, *J. Mol. Biol.* **107**, 145–158.

Rhodes, D. and Klug, A. (1986). An Underlying Repeat in Some Transcriptional Control Sequences Corresponding to Half a Double Helical Turn of DNA, *Cell* **46**, 123–132.

Rhodes, D. and Klug, A. (1988). "Zinc Fingers": A Novel Motif for Nucleic Acid Binding, in *Nucleic Acids and Molecular Biology* **2**, Eckstein, F. and Lilley, D. M. J., Eds., Springer-Verlag, Berlin, Heidelberg, New York, London, Paris, Tokyo, Hong Kong, Barcellona, pp. 149–166.

Rice, P. A., Yang, S., Mizuuchi, K. and Nash, H. A. (1996). Crystal Structure of an IHF-DNA Complex: a Protein-Induced DNA U-turn, *Cell* **87**, 1295–1306.

Richmond, T. J., Finch, J. T., Rushton, B., Rhodes, D., and Klug, A. (1984). Structure of the Nucleosome Core Particle at 7 Å Resolution, *Science (London)* **311**, 532–537.

Richter, P. H. and Eigen, M. (1974). Diffusion Controlled Reaction Rates in Spheroidal Geometry. Application to Repressor–Operator Association and Membrane Bound Enzymes. *Biophys. Chem.* **2**, 255–263.

Riggs, A. D., Bourgeois, S., and Cohn, M. (1970a). The lac Repressor–Operator Interaction. III. Kinetic Studies, *J. Mol. Biol.* **53**, 401–417.

Riggs, A. D., Suzuki, H., and Bourgeois, S. (1970b). lac Repressor–Operator Interaction, *J. Mol. Biol.* **48**, 67–83.

Rould, M. A., Perona, J. J., and Steitz, T. A. (1991). Structural Basis of Anticodon Loop Recognition by Glutaminyl-tRNA Synthetase, *Science (London)* **352**, 213–218.

Rould, M. A., Perona, J. J., Söll, D., and Steitz, T. A. (1989). Structure of *E. coli* Glutaminyl-tRNA Synthetase Complexed with $tRNA^{Gln}$ and ATP at 2.8 Å Resolution, *Science* **246**, 1135–1142.

Ruff, M., Krishnaswamy, S., Boeglin, M., Poterszman, A., Mitschler, A., Podjarny, A., Rees, B., Thierry, J. C., and Moras, D. (1991). Class II Aminoacyl Transfer RNA Synthetases: Crystal Structure of Yeast Aspartyl-tRNA Synthetase Complexed with $tRNA^{Asp}$, *Science* **252**, 1682–1689.

Ruusala, T. and Crothers, D. M. (1992). Sliding and Intermolecular Transfer of the *lac* Repressor: Kinetic Perturbation of a Reaction Intermediate by a Distant Sequence, *Proc. Natl. Acad. Sci. USA* **89**, 4903–4907.

Sachs, A. B., Bond, M. W., and Kornberg, R. D. (1986). A Single Gene from Yeast for Both Nuclear and Cytoplasmic Polyadenylate-Binding Proteins: Domain Structure and Expression, *Cell* **45**, 827–835.

Saks, M. E., Sampson, J. R., and Abelson, J. N. (1994). The Transfer RNA Identity Problem: A Search for Rules, *Science* **263**, 191–197.

Salmeron, J. M., Jr. and Johnston, S. A. (1986). Analysis of *Kluyveromyces lactis* postive Regulatory Gene LAC9 Reveals Functional Homology to, but Sequence Divergence from, the *Saccharomyces Cerevisiae* GAL4 Gene, *Nucleic Acids Res.* **14** 7767–7781.

Sandeen, G., Wood, W. I., and Felsenfeld, G. (1980). The Interaction of High Mobility Proteins HMG14 and 17 with Nucleosomes, *Nucleic Acids Res.* **8**, 3757–3778.

Sasse-Dwight, S. and Gralla, J. (1991). Footprinting Protein–DNA Complexes *in Vivo, Methods Enzymol.* **208**, 146–168.

Satchwell, S. C., Drew, H. R., and Travers, A. A. (1986) Sequence Periodicities in Chicken Nucleosome Core DNA, *J. Mol. Biol.* **191**, 659–675.

Sauer, R. T., Jordan, S. R., and Pabo, C. O. (1990). Lamda Repressor: A Model System for Understanding Protein–DNA Interactions and Protein Stability, *Adv. Prot. Chem.* **40**, 1–61.

Sauer, R. T., Yocum, R. R., Doolittle, R. F., Lewis, M., and Pabo, C. O. (1982). Homology among DNA-binding Proteins Suggests Use of a Conserved Super-secondary Structure, *Science* **298**, 447–451.

Scheidereit, C., Cromlish, J. A., Gerster, T., Kawakami, K., Balmaceda, C-G., Currie, R. A., and Roeder, R. G. (1988). A Human Lymphoid-specific Transcription Factor that Activates Immunoglobulin Genes is a Homeobox Protein, *Science* **336**, 551–557.

Scherly, D., Boelens, W., Dathan, N. A, van Venrooij, W. J., and Mattaj, 1. W. (1990). Major Determinants of the Specificity of Interaction Between Small Nuclear Ribonucleoproteins UIA and U2B" and their Cognate RNAS, *Science (London)* **345**, 502–506.

Schindelin, H., Marahlel, M.A., and Heinemann, U. (1993). Universal Nucleic Acid-Binding Domain Revealed by Crystal Structure of the *B. Subtilis* Major Cold-Shock Protein, *Science (London)* **364**, 164–168.

Schneider, T. D., Stormo, G. D., and Gold, L. (1986). Information Content of Binding Sites on Nucleotide Sequences, *J. Mol. Biol.* **188**, 415–431.

Schnuchel, A., Wiltscheck, R., Czisch, M., Herrier, M., Willimsky, G., Graumann, P., Marahlel, M.A., and Holak, T.A. (1993). Structure in Solution of the Major Cold-Shock Protein from *Bacillus subtilis, Nature (London)* **364**, 169–171.

Schultz, S. C., Shields, G. C., and Steitz, T. A. (1991). Crystal Structure of a CAP-DNA Complex: The DNA is Bent by 90°, *Science* **253**, 1001–1997.

Schumacher, M. A., Choi, K. Y., Zalkin, H., and Brennan, R. G. (1994). Crystal Structure of LacI Member, PurR, Bound to DNA: Minor Groove Binding by α Helices, *Science* **266**, 763–770.

Schwabe, J. W., Chapman, L, Finch, J. T., and Rhodes, D. (1993) The Crystal Structure of the Estrogen Receptor DNA-Binding Domain Bound to DNA: How Receptors Discriminate between their Response Elements, *Cell* **75**, 567–578.

Schwabe, J. W. R., Neuhaus, D., and Rhodes, D. (1990). Solution Structure Of The DNA Binding Domain Of The Oestrogen Receptor, *Science (London)* **348**, 458–461.

Schwabe, J. W. and Travers, A. A. (1993). What is Evolution Playing At?, *Current Biol.* **3**, 628–630.

Seeman, N. C., Rosenberg, J. M., Rich, A. (1976) Sequence-specific Recognition of Double Helical Nucleic Acids by Proteins, *Proc. Natl. Acad. Sci. USA* **73**, 804–808.

Shakked, Z., Guzikevich-Guerstein, G., Frolow, F., Rabinowich, D., Joachimiak, A., and Sigler, P. B. (1994). Determinants of Repressor/Operator Recognition from the Structure of the *Trp* Operator Binding Site, *Science (London)* **369**, 469–473.

Shamoo, Y., Friedman, A. M., Parsons, M. R., Konigsberg, W. H. and Steitz, T. A. (1995). Crystal Structure of a Replication Fork Single-Stranded DNA Binding Protein (T4gp32) Complexed to DNA, *Science (London)* **376**, 362–366.

Shannon, C. E. and Weaver, W. (1949). The Mathematical Theory of Communication, University of Illinois Press, Urbana, IL.

Sharp, P.A. (1992). TATA-Binding Protein Is a Classless Factor, *Cell* **68**, 819–821.

Shepherd, J. C. W., McGinnis, W., Carrasco, A. E., DeRoberts, E. M., and Gehring, W. J. (1984). Fly and Frog Homeo Domains Show Homologies with Yeast Mating Type Regulatory Proteins, *Science* **310**, 70–71.

Shore, D. and Baldwin, R. L. (1983a). Energetics of DNA Twisting. I. Relation between Twist and Cyclization Probability, *J. Mol. Biol.* **170**, 957–981.

Shrader, T. E., and Crothers, D. M. (1989). Artificial Nucleosome Positioning Sequences, *Proc. Natl. Acad. Sci. USA* **86,** 7418–7422.

Shrader, T. E. and Crothers, D. M. (1990). Effects of DNA Sequence and Histone–Histone Interactions on Nucleosome Placement, *J. Mol. Biol.* **216**, 69–84.

Sigman, D. S., Kuwabara, M. D., Chen, C.-H. B., and Bruice, T. W. (1991). Nuclease Activity of 1, 10-Phenanthroline-Copper in Study of Protein–DNA Interactions, *Methods Enzymol.* **208**, 414–457.

Sinclair, A. H., Erta, P., Palmer, M. S., Hawkins, J. R., Griffiths, B. L., Smith, M. J., Foster, J. W., Frischauf, A.-M., Lovell-Badge, R., and Goodfellow, P. N. (1990). A Gene from the Human Sex-determining Region Encodes a Protein with Homology to a Conserved DNA-Binding Motif, *Science (London)* **346**, 240–244.

Sluka, J. P., Horvath, S. J., Glasgow, A. C., Simon, M. I., and Dervan, P. D. (1990). Importance of Minor-Groove Contacts for Recognition of DNA by the Binding Domain of Hin Recombinase, *Biochemistry* **29**, 6551–6561.

Somers, W. S. and Phillips, S. E. V. (1992). Crystal Structure of the Met Repressor-operator Complex at 2.8 Å Resolution Reveals DNA Recognition by β-strands, *Science (London)* **359**, 387–393.

Sorger, P. K. and Nelson, H. C. M. (1989). Trimerization of a Yeast Transcriptional Activator via a Coiled-Coil Motif, *Cell* **59**, 807–813.

Spolar, R. S. and Record, M. T., Jr. (1994). Coupling of Local Folding to Site-Specific Binding of Proteins to DNA, *Science* **263**, 777–784.

Staacke, D., Walter, B., Kisters-Woike, B., Wilcken-Bermannn, B. v., and Müller-Hill, B. (1990). How Trp Repressor Binds to Its Operator, *EMBO J.* **9**, 1963–1967.

Staden, R. (1984). Computer Methods to Locate Signals in Nucleic Acid Sequences, *Nucleic Acids Res.* **12**, 505–519.

Stark, M. R. and Johnson, A. D. (1994). Interaction between two Homeodomain Proteins Is Specified by a Short C-Terminal Tail, *Science (London)* **371**, 429–432.

Stein, A. (1980). DNA Wrapping in Nucleosomes: The Linking Number Problem Re-examined. *Nucleic Acids Res.* **80**, 4803–4820.

Steitz, T. A. (1990). Structural Studies of Protein-Nucleic Acid Interaction: the Sources of Sequence-Specific Binding, *Quart. Rev. Biophys.* **23**, 205–280.

Steitz, T. A., Ohlendorf, D. H., McKay, D. B., Anderson, W. F. and Matthews, B. W. (1982). Structural Similarity in the DNA Binding Domains of Catabolite Gene Activator and Cro Repressor Proteins, *Proc. Natl. Acad. Sci. USA* **79**, 3097–3100.

Steitz, T. A. and Steitz, J. A. (1993). A General Two-Metal-Ion Mechanism for Catalytic RNA, *Proc. Natl. Acad. Sci. USA* **90**, 6498–6502.

Stormo, G. D. (1991). Probing Information Content of DNA-Binding Sites, *Methods Enzymol.* **208**, 458–468.

Strauss, J. K. and Maher, J., III (1994). DNA Bending by Asymmetric Phosphate Neutralization, *Science* **266**, 1829–1834.

Sturm, R. A. and Herr, W. (1988). The POU Domain is a Bipartite DNA-binding Structure, *Science (London)* **336**, 601–604.

Su, W., Porter, S. Kuster, S., and Echols, H. (1990). DNA Looping and Enhancer Activity: Association between DNA–Bound Ntr-C activator and RNA Polymerase at the Bacterial GlnA Promoter, *Proc. Natl. Acad. Sci. USA* **87**, 5504–5508.

Suck, D., Lahm, A., and Oefner, C. (1988). Structure Refined to 2 Å of Nicked DNA Octanucleotide Complex with DNAase I, *Science (London)***332**, 620–625.

Sung, M. T. and Dixon, G. H. (1970). Modification of Histones During Spermiogenesis in Trout: A Molecular Mechanism of Altering Histone Binding to DNA, *Proc. Natl. Acad. Sci. USA* **67,** 1616–1623.

Suzuki, M. and Yagi, N. (1994). DNA Recognition Code of Transcription Factors in the Helix–turn–helix, Probe Helix, Hormone Receptor, and Zinc Finger Families, *Proc. Natl. Acad. Sci. USA* **91,** 12357–12361.

Suzuki, M. (1989). SPXX, A Frequent Sequence Motif in Gene Regulatory Proteins, *J. Mol. Biol.* **207**, 61–84.

Suzuki, S., Green, P.G., Bumgarner, R.E., Dasgupta, S., Goddard III, W.A., and Blake, G.A. (1992). Benzene Forms Hydrogen Bonds with Water, *Science* **257**, 942–945.

Takahashi, K., Vigneron, M., Matthes, H., Wildeman, A., Zenke, M., and Chambon (1986). Requirement of Stereospecific Alignments for Initiation from the Simian Virus 40 Early Promoter, *Science* **319**, 121–126.

Talanian, R. B., McKnight, C. J., and Kim, P. S. (1990) Sequence-Specific DNA Binding to a Short Peptide Dimer, *Science* **249**, 769–771.

Tanaka, I., Appelt, K., Dijk, J., White, S. W., and Wilson, K. S. (1984). 3 Å Resolution Structure of a Protein with Histone-Like Properties in Prokaryotes, *Science (London)* **310**, 376–381.

Tanaka, S., Zatchej, M., and Thoma, F. (1992). Artificial Nucleosome Positioning Sequences Tested in Yeast Minichromosomes: a Strong Rotational Setting is not Sufficient to Position Nucleosomes in vivo. *EMBO J* **11**, 1187–1193.

Thanos, D. and Maniatis, T. (1992). The High Mobility Group Protein HMG I(Y). is Required for NF-(B-Dependent Virus Induction of the Human IFN-β Gene, *Cell* **71**, 777–789.

Thompson, J. F. and Landy, A. (1988). Empirical Estimation of Protein-induced DNA Bending Angles: Applications to λ Site-specific Recombination Complexes, *Nucleic. Acid Res.* **16**, 9687–9705.

Travers, A. A. and Klug, R. R. S. (1987). The Bending of DNA in Nucleosomes and Its Wider Implications, *Phil. Trans. R. Soc. Lond. B* **317**, 537–561.

Travis, A., Amsterdam, A., Belanger, and Grosschedl (1991). LEF-1, A Gene Encoding A Lymphoid-Specific Protein, with an HMG Domain, Regulates T-Cell Receptor Enhancer Function, *Genes Develop.* **5**, 880–894

Travers, A. A. (1989). DNA Conformation and Protein Binding, *Annu. Rev. Biochem.* **58**, 427–452.

Travers, A. A. (1995). Reading the Minor Groove, *Structural Biol.* **2**, 615–618.

Treacy, M. N., He, X., and Rosenfeld, M. G. (1991). I-POU: A POU-Domain Protein that Inhibits Neuron-specific Gene Activation, *Science (London)* **350**, 577–584.

Triezenberg, S. J., LaMarco, K. L., and McKnight, S. L. (1988). Evidence of DNA: Protein Interactions that Mediate HSV-1 Immediate Early Gene Activation by VP16, *Genes Develop.* **2**, 730–742.

Trifonov, E. N. and Sussman, J. L. (1980). The Pitch of Chromatin DNA is Reflected in Its Nucleotide Sequence, *Proc. Natl. Acad. Sci. USA* **77**, 3816–3820.

Tullius, T. D. (1989a). Structural Studies of DNA through Cleavage by the Hydroxyl Radical, in *Nucleic Acids and Molecular Biology* **4**, Eckstein, F. and Lilley, D. M. J. Eds., Springer-Verlag, Berlin, Heidelberg, New York, London, Paris, Tokyo, Hong Kong, Barcellona.

Tullius, T. D. (1989b). Physical Studies of Protein–DNA Complexes by Footprinting, *Annu. Rev. Biophys. Biophys. Chem.* **18**, 213–237.

Turner, R. and Tijan, R. (1989). Leucine Repeats and an Adjacent DNA Binding Domain Mediate the Formation of Functional cFos-cJun Heterodimers, *Science* **243**, 1689–1694.

Valegard, K., Murray, J. B., Stockley, P. G., Stonehouse, N. J. and Liljas, A. (1994). Crystal Structure of an RNA Bacteriophage Coat Protein–Operator Complex *Science (London)* **371**, 623–626.

Vallee, B. L., Coleman, J. E., and Auld, D. S. (1991). Zinc Fingers, Zinc Clusters, and Zinc Twists in DNA-binding Protein Domains, *Science* **88b**, 999–1003.

von Hippel, P. H. (1994). Protein-DNA Recognition: New Perspectives and Underlying Themes, *Science* **263**, 769–780.

von Hippel, P. H. and Berg, O. G. (1989). Facilitated Target Location in Biological Systems, *J. Biol. Chem.* **264,** 675–678.

Wang, Y.-H. and Griffith, H. D. (1996). The $[G/C)_3NN]_n$ Motif: A Common DNA Repeat that Excludes Nucleosomes, *Proc. Natl. Acad. Sci. USA* **93**, 8863–8867.

Warrant, R. W. and Kim, S.-H. (1978). α-Helix-double Helix Interaction Shown in the Structure of a Protamine-transfer RNA Complex and a Nucleoprotamine Model, *Science (London)* **271,** 130–135.

Warwicker, J., Engelman, B. P., and Steitz, T. A. (1987) Electrostatic Calculations and Model-Building Suggest That DNA Bound to CAP Is Sharply Bent, *Proteins: Structure, Function Gene.* **2**, 283–289.

Weber, I. T., McKay, D. B., and Steitz, T. A. (1982). Two Helix DNA Binding Motif of CAP Found in lac Repressor and gal Repressor, *Nucleic. Acids Res.* **10**, 5085–5102.

Weber, I. T. and Steitz, T. A. (1987). Structure of Catabolite Gene Activator Protein and Cyclic AMP Refined at 2.5 Å Resolution, *J. Mol. Biol.* **198**, 311-326.

Weeks, K. M., Ampe, C., Schultz, S. C., Steitz, T. A., and Crothers, D. M. (1990). Fragments of the HIV-1 Tat Protein Specifically Bind TAR RNA, *Science* **249**, 1281–1285.

Weeks, K. M. and Crothers, D. M. (1991). RNA Recognition by Tat-Derived Peptides: Interaction in the Major Groove?, *Cell* **66**, 577–588.

Weeks, K. M. and Crothers, D. M. (1992). RNA Binding Assays for the Tat-Derived Peptides: Implications for Specificity, *Biochemistry* **31**, 10281–10287.

Weeks, K. M and Crothers, D. M. (1993). Major Groove Accessibility of RNA *Science* **261**, 1574–1577.

Weir, H. M., Kraulis, P. J., Hill, C. S., Raine, A. R. C., Laue, E. D. and Thomas, J. O. (1993). Structure of the HMG Box Motif in the B-domain of HMGI, *EMBO J.* **12**, 1311–1319.

Weiss, M. A., Ellenberger, T., Wobbe, C. R., Lee, J. P., Harrison, S. C., and Struhl, K. (1990). Folding Transition in the DNA-binding Domain of GCN4 on specific Binding to DNA, *Science (London)* **347**, 575–578.

Werner, M. H., Huth, J. R., Gronenborn, A. M., and Clore, G. M. (1995), Molecular Basis of Human 46X, Y Sex Reversal Revealed from the Three-Dimensional Solution Structure of the Human SRY-DNA Complex, *Cell* **81**, 705–714.

Westheimer, F. H. (1968). Pseudorotation in the Hydrolysis of Phosphate Esters, *Acid. Chem. Res.* **1**, 70–79.

White, J. H. and Bauer, W. R. (1989). The Helical Repeat of Nucleosome Wrapped DNA, *Cell* **56**, 9–11.

White, J. H., Cozzarelli, N. R., and Bauer, W. R. (1988). Helical Repeat and Linking Number of Surface Wrapped DNA, *Science* **241**, 323–327.

White, J. H., Gallo, R. M., and Bauer, W. R. (1989). Effect of Nucleosome Distortion on the Linking Deficiency in Relaxed Minichromosomes, *J. Mol. Biol.* **207**, 193–199.

White, J. H., Gallo, R. M., and Bauer, W. R. (1992). Closed Circular DNA as a Probe for Protein-Induced Structural Changes, *TIBS* **17**, 7–12.

White, S. W., Krzysztof, A., Wilson, K. S., and Tanaka, 1. (1989). A Protein Structural Motif that Bends DNA, *Proteins: Structure, Function and Genet.* **5**, 281–288.

Winter, R. B., Berg, O. G., and von Hippel, P. H. (1981). Diffusion-Driven Mechanisms of Protein Translocation on Nucleic Acids. 3. The *Escherichia coli lac* Repressor–Operator Interaction: Kinetic Measurements and Conclusions, *Biochemistry* **20**, 6961–6977.

Wissman, A. and Hiller, W. (1991). DNA Contacts Probed by Modification Protection and Interference Studies, *Methods. Enzymol.* **208**, 365–379.

Witherell, G. W., Gott, J. M., and Uhlenbeck, O. C. (1991). Specific Interaction between RNA Phage Coat Proteins and RNA, *Prog. Nucleic Acid Res. Mol. Biol.* **40**, 185–220.

Wong I. and Lohman T. M. (1993), A Double-Filter Method for Nitrocellulose-Filter Binding: Application to Protein-Nucleic Acid Interactions, *Proc. Natl. Acad. Sci. USA* **90**, 5428–5432.

Wu, H. -M. and Crothers, D. M. (1984). The Locus of Sequence-Directed and Protein-Induced DNA Bending, *Science (London)* **308**, 509–513.

Yang, C.-C. and Nash, H. A. (1989). The Interaction of *E. Coli* IHF Protein with Its Specific Binding Sites, *Cell* **57**, 869–880.

Yarus, M. and Berg, P. (1967). Recognition of tRNA by Aminoacyl tRNA Synthetases, J. *Mol. Biol.* **28**, 479-490.

Yokoyama, C., Wang, X., Briggs, M. R., Admon, A., Wu, J., Hua, X., Goldstein, J. L., and Brown, M, S. (1993). SREBP-1, a Basic–Helix–Loop–Helix–Leucine Zipper Protein That Controls Transcription of the Low Density Lipoprotein Receptor Gene, *Cell* **75**, 187–197.

Zhang, Y., Babin, J., Feldhaus, A. L., Singh, H., Sharp, P. A. and Binou, M. (1991). HTF4: A New Human Helix–Loop Helix Protein, *Nucleic. Acids Res.* **19**, 4555.

Zhurkin, V. B., Lysov, Y. P., and Ivanov, V. L. (1979). Anisotroic Flexiblibity of DNA and the Nucleosomal Structure, *Nucleic. Acids Res.* **6**, 1081–1096.

Zinkel, S. S. and Crothers, D. M. (1987). DNA Bend Direction by Phase-Sensitive Detection, *Science (London)* **328**, 178–181.

Zinkel, S. S. and Crothers, D. M. (1990). Comparative Gel Electrophoresis Measurement of the DNA Bend Angle Induced by the Catabolite Activator Protein, *Biopolymers* **29**, 29–38.

Zivanovic, Y., Goulet, I., Revet, B., Le Bret, M., and Prunell, A. (1988). Chromatin Reconstituton On Small DNA Rings. II. DNA Supercoiling on the Nucleosome, *J. Mol. Biol.* **200**, 267–285.

Zubay, G. and Doty, P. J. (1959). The Isolation and Properties of Dexoyribonucleoprotein Particles Containing Single Nucleic Acid Molecules, *J. Mol. Biol.* **7**, 1–20.

CHAPTER 14

Higher Order Structure

1. DNA CONDENSATION IN VITRO

1.1 Biological Relevance

The DNA molecule is most commonly studied by physical techniques in dilute solution, where it is a wormlike or random coil with a very low fractional volume occupancy of its domain (e.g., 4×10^{-5} for T4 DNA). However, in biological systems such as cells and viruses, DNA is typically very tightly packaged, with a fractional volume occupancy that may approach 0.5. Figure 14-1, showing the enormous increase in volume of the

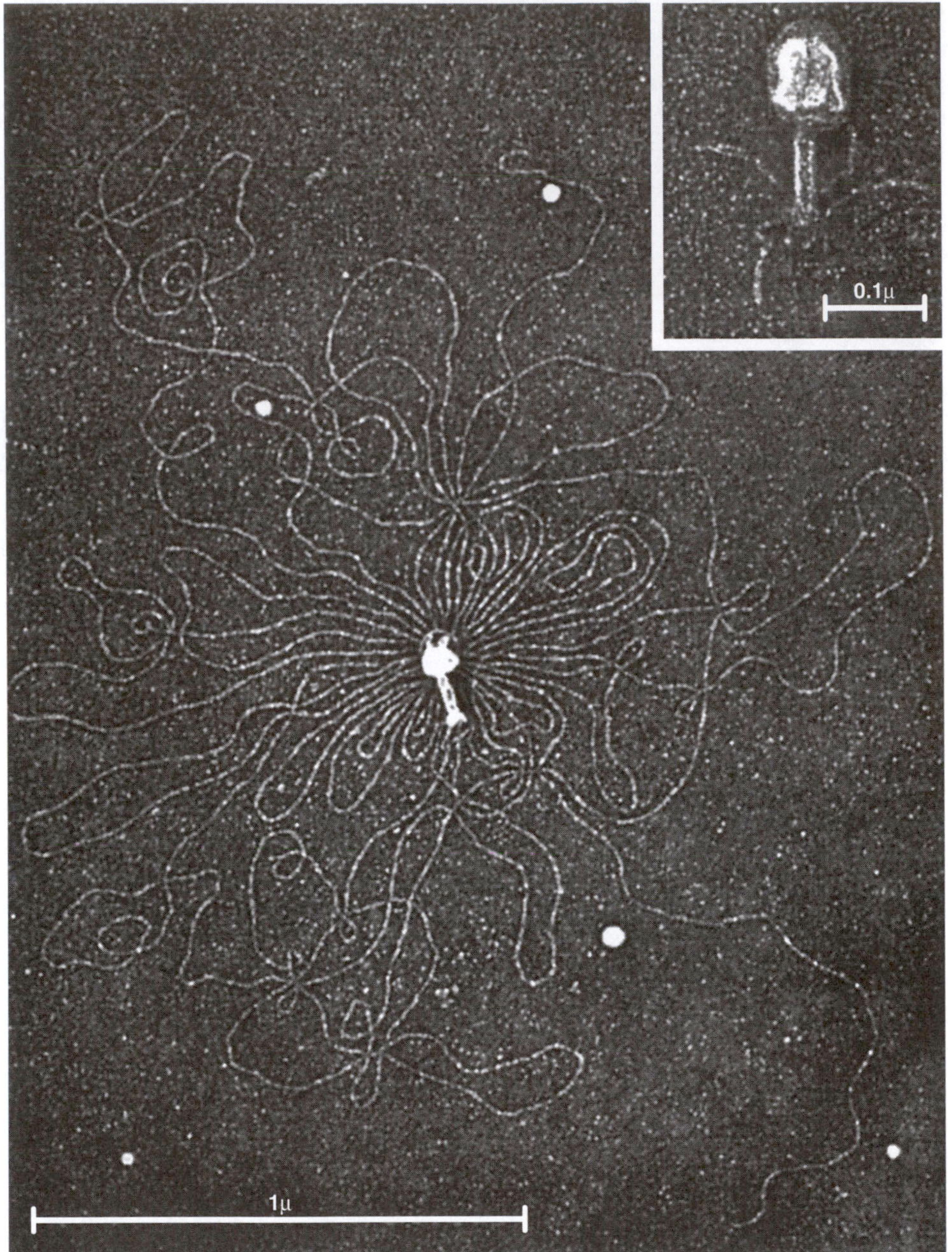

Figure 14-1
Deoxyribonucleic acid released by osmotic shock from T2 bacteriophage. [Reprinted with permission from Kleinschmidt et al. (1962).]

DNA coil, compared with its tight packaging within the phage capsid, dramatizes the challenge of understanding the reverse process of packaging and condensation.

We define condensation as a decrease in the volume occupied by a DNA molecule from the large domain dilutely occupied by a wormlike random coil, to a compact state in which the volume fractions of solvent and DNA are comparable. In the condensed state, DNA helices may be separated by just one or two layers of water. While condensation of single molecules has been observed, it is more common that several molecules are incorporated into the condensed structure. Thus condensation is difficult to distinguish rigorously from aggregation or precipitation. We generally reserve the term condensation for situations in which the aggregate is of finite size and orderly morphology. A succinct review of various aspects of DNA condensation is found in Bloomfield (1996).

Model systems that can produce condensation of DNA in solution are of great interest for understanding the packaging of DNA and RNA in viruses, and the functional arrangement of DNA in eukaryotic nuclei and prokaryotic nucleoids. There is evidence that DNA condensation is required for efficient catenation and recombination (Krasnow and Cozzarelli, 1982), and the rate of DNA renaturation is greatly accelerated by DNA condensation (Sikorav and Church, 1991), in both cases presumably because the concentration of reactants in a confined search space is enhanced. Changes in intracellular DNA compaction produced by varying polyamine levels affect susceptibility to radiation damage (Hung et al., 1983). The DNA condensed with cationic liposomes has been shown to be an efficient agent for transfection of eukaryotic cells (Gershon et al., 1993), with promise in gene therapy. In this section, we shall consider mainly *in vitro* models for condensation of the sort that might be encountered in virus packaging and arrangement in prokaryotic cells (Section 14.2). In Section 14.3, the even more complex topic of DNA packaging in chromatin will be discussed.

1.2 Agents that Cause Condensation *In Vitro*

Many different types of substances cause condensation of DNA. In all cases, the condensing agent appears to work either by decreasing repulsions between DNA segments (e.g., by neutralization of phosphate charge by cations) or by making DNA–solvent interactions less favorable (e.g., by adding ethanol, which is a poorer solvent than water for DNA, or by adding another polymer that excludes volume, to the DNA). In addition, multivalent cations may in some cases cause localized bending or distortion of the DNA, which can also facilitate condensation.

1.2.1 Multivalent Cations

In aqueous solutions, condensation is caused by cations of valence 3 or greater: the naturally occurring polyamines spermidine^{3+} and spermine^{4+} (Gosule and Schellman, 1976; Chattoraj et al., 1978), the inorganic cation $Co(NH_3)_6^{3+}$ (Widom and Baldwin, 1980, 1983a), cationic polypeptides such as polylysine (Laemmli, 1975), and basic proteins such as histones H1 and H5 (Hsiang and Cole, 1977; Garcia-Ramírez and Subirana, 1994). Divalent metal cations do not provoke condensation in water at room temperatures except under special circumstances (Ma and Bloomfield, 1994), but they will do so at somewhat elevated temperatures (Knoll et al., 1988; Rau and Parsegian, 1992b) or in water–methanol mixtures (Wilson and Bloomfield, 1979; Votavova et al., 1986).

1.2.2 Alcohols

At high concentrations, ethanol is widely used as a precipitant in the purification of DNA, but under properly controlled conditions it can produce particles of defined morphology (Lang, 1973; Eickbush and Moudrianakis, 1976). While 80% (v/v) ethanol is normally used to precipitate DNA, as little as 15–20% will cause condensation if $Co(NH_3)_6^{3+}$ is also added to a solution at low ionic strength (Arscott et al., 1995). Methanol and isopropanol behave similarly. The P form of DNA (Zehfus and Johnson, 1981, 1984), formed in ethanol–methanol solutions with low water content, is also condensed; its bases are neither paired nor stacked.

1.2.3 Neutral and Anionic Polymers

Even neutral polymers, such as polyethylene glycol, at high concentrations and in the presence of adequate concentrations of salt can provoke DNA condensation (Lerman, 1971). The resulting structure has been termed Ψ-DNA, or psi-DNA, the acronym for *P*olymer-and-Salt-*I*nduced, which describes the condensation process. Psi-DNA is also produced by anionic polymers, such as polyaspartate, polyglutamate, and the anionic peptides found in the capsid of bacteriophage T4 (Laemmli et al., 1974). As will be discussed in Section 1.3.4, Ψ-DNA is characterized by its unusual circular dichroism (CD) spectrum, which derives from its liquid crystalline structure.

1.2.4 Cationic Liposomes

When DNA is condensed with cationic liposomes, the complex becomes a very efficient agent for transfection of eukaryotic cells, presumably because of two factors. First, the condensed state of the DNA protects it from nucleases and allows it to pass more easily through small openings; and second, the lipid coating on the DNA increases its permeability through cell membranes. Gershon et al. (1993) suggest that cationic liposomes initially form clusters along the uncondensed DNA. At a critical density, these clusters coalesce by DNA-induced membrane fusion, and the DNA condenses to a form completely encapsulated by lipid.

A somewhat related but even more complicated DNA packaging system involves positively charged micelles and flexible negatively charged polypeptides or single-stranded RNA (Ghirlando et al., 1992). The DNA is partly embedded in a micellar scaffold, and partly condensed into tightly packed chiral structures. It is suggested that "the DNA induces the elongation of the micelles into rodlike aggregates, forming a closely packed matrix in which the DNA molecules are immobilized. In contrast, the flexible anionic polymers stabilize clusters of spherical micelles, which are proposed to effect a capping of the rodlike micelles, thus arresting their elongation and creating surfactant-free segments of the DNA that are able to converge and collapse."

1.3 Structures of Condensed DNA

1.3.1 Toroids

When condensation is induced by addition of polyamines or $Co(NH_3)_6^{3+}$ to very dilute ($\sim 1 \mu g\, mL^{-1}$) DNA solutions at low ionic strength, toroids and rods are produced [Fig. 14-2(c and d)]. The toroids have drawn a great deal of attention, since they are similar in size and appearance to DNA bundles [Fig. 14-2(a and b)] produced by gentle lysis of T2 and T7 phage heads (Klimenko et al., 1967; Richards et al, 1973).

Perhaps the most thoroughly studied type of condensation is that provoked by multivalent cations. Interest in this area is particularly intense since Gosule and Schellman (1976) first showed that toroidal particles are formed upon treatment of T7 DNA with spermidine^{3+}, a naturally occurring polyamine that is found inside phage heads and bacterial cells. These particles resemble some of those obtained by gentle lysis of phage particles. The condensation by spermidine occurs abruptly once a critical concentration of the cation is reached. Chattoraj et al. (1978) showed that condensation occurs with other phage DNAs, and the particles are of similar size regardless of DNA length, indicating that several small DNA molecules are incorporated in a single particle. In addition to toroids, rods and globules are observed. The CD spectrum of the condensed DNA is not significantly different from that of B-DNA.

The morphology of spermidine-condensed calf thymus DNA particles was studied by freeze fracture electron microscopy by Marx and Ruben (1983). They showed that torus-shaped condensates exist under the hydrated conditions of the freeze fracture experiment . High contrast replicas of the spermidine–DNA toruses showed circumferential wrapping of fibers that were the proper size to be individual DNA double helices (Fig. 9-31). In other studies (Marx and Reynolds, 1982), an arithmetic ladder series of enzymatic digestion patterns was also interpreted to support circumferential wrapping.

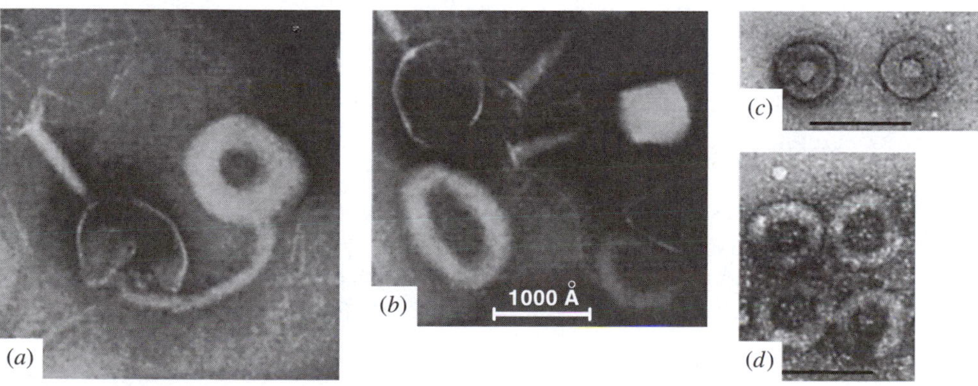

Figure 14-2
((*a*) and (*b*)) Toroidal DNA bundles produced by gentle lysis of T2 phage capsids. [Reprinted with permission from Klimenko et al., 1967.] ((*c*) and (*d*)) Toroidal particles of T7 phage DNA condensed with spermidine^{3+}. [By permission from Chattoraj et al., 1978.] In all cases the scale bar is 1000 Å.

It was supposed that the enzyme enters at a particularly accessible site at the toroid surface, thus making cleavages at an integral number of turns of the DNA.

1.3.2 Rods

Small numbers of rods are sometimes seen in electron micrographs of mainly toroidal condensates. However, under some conditions, especially in the presence of high concentrations of alcohols, rods become the predominant form. Lang (1973) and Lang et al. (1976) showed that the particle morphology depends on the ethanol concentration. Rodlike structures are formed with uniform diameters and three distinct length classes. The particles become shorter and thicker with increasing ethanol concentration. Eickbush and Moudrianakis (1976) found similar rodlike particles, but observed that the morphology can vary substantially depending on the salt concentration and type. In particular, addition of spermidine gives rise to a high proportion of toroids. Arscott et al. (1995) used mixtures of water with methanol, ethanol, and isopropanol to vary the solvent dielectric constant ε from 80 to 50, and provoked condensation of pUC18 plasmids by $Co(NH_3)_6^{3+}$. A high proportion of rods occur in the range of ε from 65 to 70, with mainly toroids at lower alcohol (higher ε) and fibers at higher alcohol; this behavior is independent of the species of alcohol. Rods become shorter as well as more numerous as ε decreases. Arscott et al. (1995) speculated that the combined effects of solvent and $Co(NH_3)_6^{3+}$ locally destabilize the double helix, permitting DNA foldbacks that lead to rodlike condensates. Condensation with permethylated spermidine also produces a large proportion of rodlike particles (Plum et al., 1990), perhaps due to the more hydrophobic character of this ligand.

1.3.3 Fibers and Platelets

As the concentration of alcohol is raised to high levels, discrete toroids and rods are replaced by more extensively aggregated structures. Precipitation of short sonicated fragments of salmon sperm DNA by ethanol leads to hexagonal platelet crystals about 150 Å thick (Giannoni et al., 1969). These appear to be similar to synthetic polymer crystals, which crystallize by lamellar chain folding. In the presence of $Co(NH_3)_6^{3+}$ and in sufficient alcohol to lower the dielectric constant below 65, DNA collapses into a network of multistranded fibers (Arscott et al., 1995). Circular dichroism spectroscopy shows that the DNA has undergone a transition from the B to the A conformation, indicating that ethanol and $Co(NH_3)_6^{3+}$ synergistically promote the B–A transition. The A-DNA strongly self-adheres and rapidly aggregates into fibrous networks, not allowing time for more compact and orderly condensates to form.

1.3.4 Ψ-DNA

A different type of condensation, discovered by Lerman (1971), is produced by neutral or acidic polymers at high salt. Poly(ethylene oxide) (PEO) is the polymer most commonly used to provoke Ψ condensation. The collapse is apparently caused primarily by repulsive interactions between DNA and added polymer, rather than by attraction between DNA segments. The salt neutralizes repulsive DNA–DNA electrostatic repulsions. Sedimentation and electron microscopic (EM) studies show that, at low concentrations, the DNA adopts a compact structure comparable to the size of phage heads. As with cation-induced condensation, Ψ condensation is abrupt and cooperative.

The CD spectrum of Ψ-DNA is strikingly different from normal B-form DNA, as seen in Figure 14-3 (Jordan et al., 1972). The magnitude is much larger, and the spectrum is strongly nonconservative. Achievement of the long-range order resulting in a fully developed Ψ spectrum is a slow process. X-ray diffraction measurements (Jordan et al., 1972; Maniatis et al., 1974) show spacings characteristic of normal B-form DNA. The helices are predominantly parallel, compatible with a folded chain structure of the compact Ψ state, an arrangement similar to the usual mode of crystallization of simple linear polymers. A folded chain structure is also suggested by the observation (Laemmli, 1975) that PEO-collapsed DNA contains sites, spaced 200–400 bp apart and presumably located at each fold, which are susceptible to cleavage by a single-strand specific endonuclease.

Keller, Bustamante, and co-workers (Keller and Bustamante, 1986a,b; Kim et al., 1986) developed a theory that explains the Ψ-type CD spectrum, showing that it arises from a long-range chiral structure and delocalization of electronic excitations throughout the particle. The magnitude of the spectrum is determined by the volume, the chromophore density, and the pitch of the aggregate, while the shape is determined mainly by the pitch and handedness. Aggregates with opposite handedness give mirror-image CD spectra.

Highly acidic polypeptides also cause Ψ condensation. (Laemmli, 1975). These include synthetic polyanions such as poly(glutamic acid) and poly(aspartic acid), as well as the highly acidic internal peptides found in the mature T4 bacteriophage head. Repulsive interactions with the internal peptides may help to stabilize the intraphage DNA in its highly compacted form.

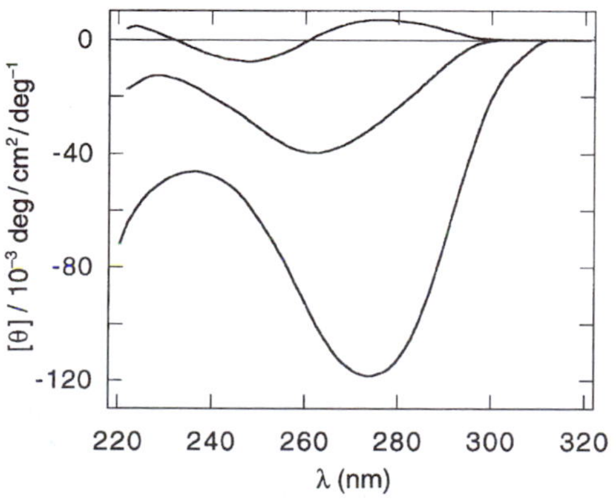

Figure 14-3
Circular dichroism spectrum of T7 DNA, showing progression from B-DNA to Ψ-DNA spectrum. From top to bottom, the PEO concentration and time after mixing are 80.9 mg mL^{-1}, 0.7 h; 126.5 mg mL^{-1}, 0.7 h; 126.5 mg mL^{-1}, 48 h. All solutions contained 22 μM P-DNA and 0.26 M Na^{+}. [After Jordan et al., 1972.]

1.3.5 Liquid Crystals

Liquid crystalline arrays may be formed at the very high concentrations characteristic of condensed DNA, since a parallel arrays of rods can accomodate many more molecules per unit volume than a randomly oriented "brush-heap." If the DNA is short and linear, such as nucleosomal DNA of about one persistence length, the liquid crystalline phase forms abruptly and spontaneously upon slight increase in concentration or decrease in temperature (Rill et al., 1983; Rill, 1986); the critical concentration is near 200 mg mL^{-1}. At least three liquid crystal phases can be distinguished as the DNA concentration is varied from 160 to 290 mg mL^{-1} (Strzelecka et al., 1988; Strzelecka and Rill, 1990). The phase diagram depends on the length and diameter of the DNA molecules. The effective diameter depends in turn on the ionic strength, since the phosphate–phosphate diameter is augmented by the thickness of the ion atmosphere, which increases with decreasing ionic strength (see Chapter 11).

Short DNA fragments precipitated from dilute solution by spermidine, form a liquid crystalline phase in equilibrium with a concentrated isotropic phase (Sikorav et al., 1994). The liquid crystalline ordering occurs in the presence of attractive interactions, in contrast to that seen at very high concentrations in the presence of monovalent salt.

The effect of charge on the isotropic-nematic liquid crystal transition in polyelectrolytes has a subtle complication (Stroobants et al., 1986). In solutions of neutral rods, the transition is purely entropic, driven by the greater translational entropy of aligned rods even though they have lost rotation entropy by enforced parallelism. If the rods are charged, the parallel orientation has the greatest electrostatic repulsion, so there is a tendency for the rods to remain twisted and higher concentrations are required. On the other hand, attractive forces, if they exist, are also probably maximized in the parallel orientation.

Liquid crystalline domains of DNA can also be seen in electron micrographs of dinoflagellate chromosomes (Rill et al., 1989), suggesting that this is a pertinent mode of organization *in vivo* when the DNA is uncomplexed by chromosomal proteins. Liquid crystals also appear in bacteria with high copy number plasmids, as determined by *in vivo* X-ray scattering experiments (Reich et al., 1994a). The structure of the liquid crystalline phase is determined by the supercoiling density and handedness of the plasmids, rather than by environmental factors as is the case for linear DNA molecules. Accordingly Reich et al. (1994b) have suggested "that supercoiling-regulated liquid crystallinity represents an effective packaging mode of nucleosome-free, topologically constrained DNA molecules in living systems."

1.4 Size Distributions, Thermodynamics, and Kinetics

1.4.1 Sizes of Condensed Particles

A striking experimental result is that collapsed particles have approximately the same range of sizes, regardless of the molecular weight of the DNA from which they are formed (Chattoraj et al., 1978; Widom and Baldwin, 1980; Arscott et al., 1990; Bloomfield, 1991). Toroidal condensates of high molecular weight DNA from bacteriophage generally contain only one DNA molecule; while condensates of smaller

DNAs, such as calf thymus or pUC plasmids, contain several molecules, but have virtually identical size and morphology. A typical toroid has an inner radius R_i of 150–200 Å, and an outer-radius R_o of 350–500 Å. This value corresponds to an average of about 40,000 bp of B-DNA in a typical particle. Equally striking, rods formed in aqueous buffers tend to have about the same volume and proportions as toroids; that is, length = circumference = $\pi(R_o + R_i) \approx 1800$ Å and diameter = thickness = $(R_o - R_i) \approx 300$ Å. A linearized plasmid containing 2700 bp wraps approximately five times around the circumference of a toroid, or back and forth along the length of a rod.

The inner radii of toroids is probably determined in large measure by the persistence length of DNA: Too small a radius exacts too high a cost in bending energy, while too large a radius leads to less than optimal overlap of mutually attractive DNA surfaces. See Section 14.1.5 for a more detailed discussion. It has been calculated that torsional elastic free energy causes a decrease in the equilibrium radius of toroids formed from closed circular DNA, compared to linear or relaxed circular DNA (Grosberg and Zhestkov, 1985), a prediction confirmed by the experiments of Arscott et al. (1990).

1.4.2 Thermodynamics of Condensation

Any mechanism that attempts to explain DNA condensation must account for the regulation of condensate size. This can be approached by an equilibrium thermodynamics argument (Bloomfield, 1991) based on considering the multimolecular condensation of plasmid molecules, represented as a monomer–n-mer equilibrium $n\mathrm{D}_1 \rightleftharpoons \mathrm{D}_n$. Expressing the equilibrium constant in mole fraction units X automatically takes account of the translational entropy changes in the association process (Tanford, 1974), so that the equilibrium constant K reflects the free energy of interactions within the condensed particle:

$$K = \frac{\mathrm{X}_n}{\mathrm{X}_1^n} = \frac{(1-f_u)(\mathrm{X}_1^0)^{1-n}}{nf_u^n} \tag{14-1}$$

Here f_u is the fraction of uncondensed DNA and X_1^0 is the initial mole fraction of DNA molecules. Under typical conditions, the plasmid is about 3000 bp long and has a concentration of 10 μg mL^{-1}, so that $\mathrm{X}_1^0 \approx 10^{-10}$. The parameter f_u is near 0.5, and the number of plasmid molecules in the condensate is $n \approx 10$. Using $\Delta G^\circ = -RT \ln K$, leads to $n^{-1}\Delta G^\circ/RT \approx -200$, or about -0.007 RT per mole of base pairs. Since in a hexagonally packed array of cylinders the average number of nearest neighbors is about five (six in the interior and four at the surface), the estimated ΔG° per mole of interactions is only about -0.0014 RT. Although this is only a crude estimate, it shows that condensation involves delicate free energy balances, each very small but multiplied by tens of thousands of interactions between base pairs. The total free energy change is large enough to stabilize the condensed particle under suitable solution conditions, but not so large that condensation is irreversible.

In order for a complex of associating molecules to be stable but not grow indefinitely large, the free energy of association must vary nonmonotonically with n, the degree of association (Tanford, 1974). For small n, ΔG_n° will be strongly negative, to favor association. For large n, ΔG_n° will eventually become positive, disfavoring further growth of the particle. It might be, for example, that the attractive free energy varies like $-\alpha n$, while the repulsive free energy varies like β_n^2 (where α and β are positive constants), so $\Delta G_n^\circ = -\alpha n + \beta n^2$. If $\alpha \approx n^* \beta$, the maximum in the size distribution will be near n^*. With a suitable choice of parameters, it is possible to approximately reproduce the size distribution observed experimentally (Bloomfield, 1991). The likely sources of attractive and repulsive interactions are discussed in Section 1.4.3.

The very small energy of pairwise interaction explains the observation (Widom and Baldwin, 1980) that DNA shorter than about 400 bp does not form ordered, toroidal condensates. Very short DNA molecules cannot nucleate stable aggregates since they cannot develop adequate overlap, either internally through circle formation, or intermolecularly through side-by-side interactions. Since each pairwise interaction generates less than 0.01 $k_B T$ per base pair, several hundred base pairs of overlap per molecule are the minimum to stabilize the nucleus against thermal disruption (Bloomfield, 1991). Short DNA molecules will aggregate under stronger condensing conditions, but orderly condensates are not formed.

1.4.3 Kinetics and Reversibility of Condensation

Kinetic as well as thermodynamic factors are important in condensation. Widom and Baldwin (1980) showed that condensation of λ DNA by $Co(NH_3)_6^{3+}$, at DNA concentrations above 1 μM phosphate, takes minutes to hours, and becomes slower as the DNA concentration is increased. Decondensation by Na^+ or Mg^{2+} occurs in seconds to minutes with a rate independent of [DNA]. Thus condensation is a readily reversible process. Widom and Baldwin suggest that intermolecular DNA contacts compete with, and slow down, intramolecular condensation. Electron microscopic study of condensation of linear pUC9 plasmids (1350 and 2700 bp) (Arscott et al., 1990) shows that upon prolonged incubation, the toroids grow and aggregate while maintaining their identity. These observations suggest that DNA condensation in vitro is a complex process whose thermodynamic and kinetic components are yet to be fully unraveled.

1.5 Forces and Mechanisms

1.5.1 Overview

The forces and mechanisms underlying DNA condensation are not well understood at the molecular level. A general survey of the physical chemical issues is presented by Gosule and Schellman (1976). They point out that a substantial thermodynamic driving force would be necessary to package viral DNA under physiological salt conditions. For example, the domain of T7 DNA as a wormlike coil free in solution occupies about 10,000 times the volume of the T7 phage head. Thus to compact the DNA would require a large pressure, to counteract the entropy loss of the compressed coil. An additional, and even larger, pressure would be required to overcome the electrostatic repulsion between the DNA segments as they approach; and yet more energy must be expended to bend or kink the DNA into a suitably small volume. Presumably, high-salt concentrations or addition of multivalent cations act, at least in part, to reduce unfavorable electrostatic interactions; while addition of neutral or acidic polymer raises the energy of the uncondensed DNA, making the condensed state relatively more favorable.

A complete understanding of DNA condensation must include the thermodynamic and kinetic factors governing collapse, as well as specification of intermolecular interactions. While we are far from this goal at this time what follows is an attempt to summarize some features that appear to be understood.

1.5.2 Cross-Linking

When one tries to understand the forces that may be important in stabilizing the condensed form, the most obvious is cross-linking by the condensing ligand. There are several pieces of evidence favoring this idea. Suwalsky et al. (1969) used X-ray diffraction to measure the distance between DNA fibers in DNA–spermine complexes, as a function of relative humidity. The distance increases with hydration, but reaches an upper limit consistent with spermine cross-linking. Allison et al. (1981) condensed DNA with homologs of spermidine, in which the butyl moiety is replaced by longer pentyl through octyl groups. Aggregation became a more prevalent occurrence relative to condensation as the length of the end chain increased, suggesting an important role for cross-linking. Schellman and Parthasarathy (1984) used spermidine analogues of the structure [NH_3^+–$(CH_2)_3$–NH_2^+–$(CH_2)_n$–NH_3^+ with $n = 3,4,5$, and 8] and aliphatic diamines [NH_3^+-$(CH_2)_n$–NH_3^+ with $n = 2,3,4,$and 6] to collapse DNA. They found that the interhelical spacing varied systematically with the length of the methylene bridge, and that the ionic strength of the solution had no effect on the spacing. They therefore suggested that the arrangement of DNA in the complexes is determined by the structure of the polycation, not by long-range electrostatic repulsive and attractive forces.

It must be noted, however, that in B-DNA crystals where spermine or $Co(NH_3)_6^{3+}$ has been localized, it has never been found in a cross-bridging location. Instead, it is generally associated with some loop or distortion of a double helix. This accords with the theoretical prediction by Feuerstein et al. (1986) that the minimum energy structure of a spermine–DNA complex is one in which the spermine causes a bending of the major groove in which it is docked. However, in Z-DNA spermine does bridge between the helices (Egli et al., 1991). Furthermore, the spacing between DNA helices condensed by Mn^{2+} ions in methanol–water solvent is 32 Å, much too far to be bridged by the divalent ion (Rau and Parsegian, 1992a). Thus the bulk of the evidence indicates that cross-linking by cations is not a general mechanism for B-DNA condensation.

1.5.3 Polymer Theory and Mixing Entropy

A general phase-transition theory for polymer collapse has been developed by Post and Zimm (1979). This theory contains an interaction parameter χ that contains the details of the differential interaction between DNA and its solvent milieu, but which cannot be calculated reliably from molecular theory. The parameter $\chi = 0.5$ corresponds to an ideal solvent. As χ increases, solvent–segment contacts become more unfavorable, leading to phase separation. The theory predicts, in accord with observations (Post and Zimm, 1982b), that stiff-chain polymers such as DNA will collapse abruptly, while more flexible polymers will collapse gradually as solvent quality decreases. A refinement of the theory (Post and Zimm, 1982a) shows that monomolecular condensation occurs only at very low DNA concentrations ($< 1\mu g\ mL^{-1}$), while multimolecular aggregation will be expected at most experimentally realizable concentrations. Figure 14-4 is a phase diagram showing regions of expanded and collapsed single coils, and aggregated molecules, for DNAs of three molecular weights. Collapse and aggregation depend on both χ and molecular weight. As the molecular weight increases, collapse occurs at lower values of χ, and the aggregated state dominates except at very low DNA concentration.

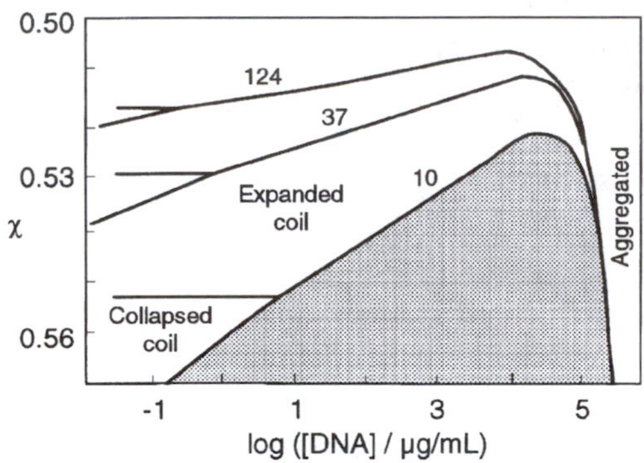

Figure 14-4
Phase diagram showing regions of expanded and collapsed single coils, and aggregated molecules, for DNAs of molecular weights 10, 37, and 124 million. [Post, C. B. and Zimm, B. H. (1982). Theory of DNA Condensation Collapse versus Aggregation. *Biopolymers* **21**, 2123–2137. Copyright ©1982. Reprinted by permission of John Wiley & Sons, Inc.]

A simple estimate of entropy loss due to the demixing of DNA and solvent upon condensation is $\Delta S_{mix} = -RL/a$, where R is the gas constant, L is the contour length, and a is the persistence length (Riemer and Bloomfield, 1978). Neglecting the complications of stiffness and χ value considered by Post and Zimm, this yields

$$\Delta G_{mix} = -T\Delta S_{mix} = RTL/a \tag{14-2}$$

This value will be used, along with simple estimates of other contributions to condensation free energy, in Table 14-1.

1.5.4 Bending

If a length L of DNA with persistence length a is bent in a path with radius of curvature R_c, the bending free energy is

$$\Delta G_{bend} = RTaL/2R_c^2 \tag{14-3}$$

Inside a condensed particle, R_c will vary from point to point, being less at the inside of a toroid than at the periphery, so a mean value must be used (Riemer and Bloomfield,

1978). A greater complication arises from local sequence-directed bending (Reich et al., 1992), or from bending due to site binding of multivalent cations, in which case one may even have $\Delta G_{\text{bend}} < 0$ (Marquet and Houssier, 1991). However, Eq. 14-3 provides an order of magnitude estimate of the energetic cost of bending; it is used in this way in Table 14-1. Some ways by which multivalent cations might affect DNA bending are discussed below.

1.5.5 Ionic Interactions

1.5.5(a) *Partial Charge Neutralization of DNA by Condensing Cations.* Electrostatic forces are clearly important in condensation. The laws of physics require that something must be done to neutralize the strong repulsive interactions as negatively charged DNA segments approach closely. Ionic effects on condensation were investigated systematically by Wilson and Bloomfield (1979), who found a striking regularity that has since been confirmed under a wide variety of circumstances: approximately 90% of the DNA charge must be neutralized for condensation to occur. Note that full charge neutralization does not occur; approximately 10% of the DNA charge remains.

Wilson and Bloomfield (1979) studied condensation of T7 bacteriophage DNA with spermidine, spermine, and other multivalent cations by light scattering, which enabled convenient detection of condensation under a wide range of ionic conditions. Diffusion coefficient measurements by dynamic light scattering indicated that condensates in solution are of a size similar to those observed by EM. The critical concentration of spermidine increases with increasing salt. The amount of binding of spermidine and

Table 14.1
Contributions to $\Delta G/RT$ per Mole of Toroidally Condensed DNA in the Presence of Z_2-Valent Cations[a]

Cation Valence Z_2	$\Delta G_{\text{mix}}/RT$ Eq. 14-2	$\Delta G_{\text{bend}}/RT$ Eq. 14-3	$\Delta G_{\text{elec}}/RT$ Eq. 14-7	$\Delta G_{\text{fluct}}/RT$ Eq. 14-8	$\Delta G_{\text{tot}}/RT$	$\Delta G_{\text{tot}}/RT$ per P-DNA
+1	312	306	96,600	−9,650	87,600	0.97
+2	312	306	24,100	−14,600	10,100	0.11
+3	312	306	10,700	−15,800	−4,410	−0.049
+4	312	306	6,040	−16,200	−9,520	−0.106

[a]Calculations assume aqueous solution at 20°C with dielectric constant $\varepsilon = 80$, 45,000 bp of DNA with persistence length $a = 50$ nm, toroids with mean radius of curvature $R_c = 35$ nm, thickness $R_o - R_i = 20$ nm, and interhelical distance $X_{\text{cond}} = 2.75$ nm. To convert to kilojoules per mole (kJ mol^{-1}), multiply the numbers in the table by 2.436.

simple salt to the DNA was calculated using the counterion condensation theory (Manning, 1978) discussed in Chapter 10, modified to consider two cations of different valence. The working equations are

$$1 + \ln \frac{1000\theta_1}{c_1 V_{P1}} = -2Z_1\xi(1 - Z_1\theta_1 - Z_2\theta_2)\ln(1 - e^{-\kappa b}) \tag{14-4}$$

and

$$\ln \frac{\theta_2}{c_2} = \ln \frac{V_{P2}}{1000e} + \frac{Z_2}{Z_1} \ln \frac{1000\theta_1 e}{c_1 V_{P1}} \tag{14-5}$$

where Z_1, c_1, V_{P1}, and θ_1 are the valence, molar concentration, molar territorial binding volume (see Appendix 11-A.3), and fraction of DNA phosphate binding sites occupied, for cations of type 1 (the lower valent ion). The parameters Z_2, c_2, V_{P_2}, and θ_2 are the corresponding quantities for the higher valent ion; ξ is the counterion condensation parameter, the ratio of charge spacing b to Bjerrum length $q^2/\varepsilon kT$; and κ is the Debye–Hückel screening parameter. Iterative numerical solution of these equations for θ_1 and θ_2 allows calculation of the total fraction of DNA phosphate charges neutralized,

$$r = Z_1\theta_1 + Z_2\theta_2 \tag{14-6}$$

The charge neutralization predicted by these equations has been directly verified by gel electrophoresis of plasmids in buffers containing various amounts of multivalent cations (Ma and Bloomfield, 1995).

Experiments show that DNA condensation occurs when an essentially constant fraction, $r \approx 0.89$–0.90, of the DNA charge is neutralized by counterion condensation. The same rule holds with Mg^{2+} as the lower valent ion, with spermine^{4+} as the higher valent ion, and in a wide range of water–cosolvent mixtures (Wilson and Bloomfield, 1979; Bloomfield et al., 1994; Arscott et al., 1995; Flock et al., 1995, 1996). It was also confirmed by Benbasat (1984) with phage ϕW14 DNA, which has variable and lower charge density owing to the incorporation of the positively charged base α-putrescinylthymine^{2+}. Widom and Baldwin (1980, 1983a) found that $Co(NH_3)_6^{3+}$ is a fivefold more efficient condensing agent than spermidine, despite equal binding of these isovalent compounds (Plum and Bloomfield, 1988), with collapse occurring at about 85% charge neutralization. However, the dependence of the critical $Co(NH_3)_6^{3+}$ concentration on added salt confirms the ion exchange behavior underlying Eqs. 14-4 and 14-5. Yen et al. (1983) directly demonstrated by electrophoretic light scattering that the condensation of λ-DNA by spermidine^{3+} or spermine^{4+} is accompanied by a decrease in charge density to approximately the $1 - r = 0.1$ level.

All of these results indicate that DNA condensation in the presence of multivalent cations is determined by the total charge neutralization of the DNA, rather than by the binding of the multivalent cation per se, which can be as low as $\theta_2 = 0.1$ for Mg^{2+} – spermine^{4+} solutions.

1.5.5(b) *Condensation and Catenation in Three-Ion Systems.* The biological importance of this result is seen from the work of Krasnow and Cozzarelli (1982) who investigated the catenation of DNA rings by topoisomerases, a process that requires polyvalent cations such as spermidine. They showed that this is due to compaction of the DNA into aggregates where the high local DNA concentration favors the catenated state. The same critical concentration of spermidine or $Co(NH_3)_6^{3+}$ induced both aggregation and catenation, and the critical concentrations for both increased equally when competing mono- or divalent cations were added. An extension of the counterion condensation equations 14-4–14-6 to accomodate three cations of different valence (+1, +2, +3), as required for topoisomerase reactions, showed that DNA rings cooperatively aggregated when about 90% of DNA charge was neutralized, just as with the results above for two cations. A phase diagram showing the predicted ionic conditions for aggregation is given in Figure 14-5.

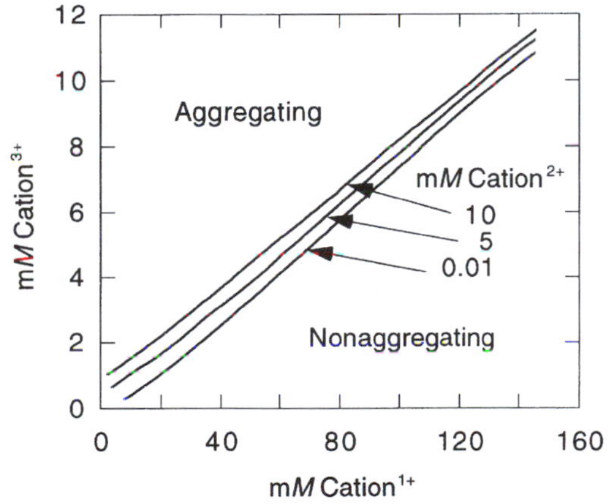

Figure 14-5
Predicted ionic concentrations for DNA aggregation in buffers containing +1, +2, and +3 cations. [After Krasnow and Cozzarelli, 1982.]

1.5.5(c) *Kinetics of Ion Binding and Unbinding.* The kinetics of multivalent cation binding appears to play an important role in regulating the rate of DNA condensation (Pörschke, 1984). The time for intramolecular condensation of λ phage DNA by spermine is in the millisecond range, when the reactant concentrations are micromolar. An induction period for intramolecular condensation is attributed to binding of spermine below the critical concentration, with the rate-limiting step being the binding of the critical amount, which is entropically disfavored in an excluded-site model. The DNA condensation itself occurs in the submillisecond range, indicating high flexibility and fast bending motions.

The DNA can be rapidly decondensed by application of a strong electric field pulse, probably by a dissociation field effect that reduces spermine binding below the critical level (Pörschke, 1985). This effect is similar to decondensation by adding salt, which also displaces multivalent cations (Widom and Baldwin, 1980).

1.5.5(d) *Repulsive Coulombic Interactions.* Although the charge of the DNA is reduced to about 10% of its original value under the ionic conditions required for condensation, it is not reduced to zero. In fact, the amount of remaining charge is impressively large. If a condensed DNA particle contains 40,000 bp, it has 80,000 negative phosphates, or about 8000 elementary charges after 90% neutralization. This much charge concentrated in a particle a few hundred Å in radius will generate a powerful repulsive force. The free energy change ΔG_{elec} due to electrostatic repulsions in going from uncondensed to condensed DNA has been estimated (Bloomfield and Riemer, 1978; Marquet and Houssier, 1991) from the results of Oosawa (1971) to be

$$\Delta G_{elec} = \frac{n_{tot} k_B T}{2\xi Z_2^2} \ln \frac{V_{uncond}}{V_{cond}} \tag{14-7}$$

where n_{tot} is the total number of DNA phosphate charges before neutralization, and Z_2 is the valence of the condensing cation. The uncondensed molecule is assumed to be a sphere of radius equal to its radius of gyration R_g (see Eqs. 9-4 and 9-8) and volume $V_{uncond} = (4\pi/3)R_g^3 = (4\pi/3)(aL/3)^{3/2}$. The volume of the condensed particle is $V_{cond} = 2\pi R_c \times \pi[(R_o - R_i)/2]^2$ (see Section 14.1.4.1). Equation 14-7 is used to estimate electrostatic repulsion in Table 14-1.

Direct measurements of the electrostatic repulsive force as a function of distance between aligned DNA molecules (discussed in Section 1.5.6) show that it cannot be explained just by a simple application of screened Debye–Hückel interactions (Podgornik et al., 1989). Instead, fluctuations of the somewhat flexible DNA helices expand the range of the repulsive interactions. Odijk (1993b) has developed a mean-field theory that explains these data fairly well, using the persistence length to estimate the elastic bending properties of DNA that govern the undulations.

1.5.5(e) *Attractive Ionic Interactions.* Substantial reduction of the repulsive electrostatic interaction between DNA helices does not in itself account for the attractive interaction stabilizing the condensed form. One possible source of attraction is the London dispersion force arising from the mutual polarization of electron clouds. Bloomfield et al. (1980) analyzed the balance between polyelectrolyte repulsive forces and attractive forces due to London dispersion interactions. They concluded that dispersion forces would have to be 2–5 times larger than normal, in order to cause DNA collapse when the repulsions have been reduced by the presence of multivalent counterions.

A conceptually similar mechanism, suggested by Oosawa (1971), is induced dipole interactions between the fluctuating ion atmospheres surrounding rodlike macroions such as DNA. Marquet and Houssier (1991) used the Oosawa theory to obtain the simple equation for the attractive fluctuation free energy

$$\Delta G_{fluct} = -\frac{3Lk_B T}{X_{cond}} \frac{(\theta_2 Z_2^2 \xi)^2}{(1 + \theta_2 Z_2^2 \xi)^2} \tag{14-8}$$

where θ_2 is the fractional charge neutralization from cation of charge Z_2, and X_{cond} is the center-to-center interhelix distance in the condensed DNA. Combination of ΔG_{fluct}

with the other sources of free energy in Table 14-1 shows that a net attraction can be attained in water with cations of charge +3 or greater, in accord with experiment. ΔG_{tot} for divalent cations in 50% methanol-water is very slightly negative, indicating stable condensates consistent with the results of Wilson and Bloomfield (1979). However, Eq. 14-8 neglects the effects of excess monovalent counterions and of Debye–Hückel screening, which will tend to reduce the attractive free energy.

A somewhat different attraction mechanism, which utilizes the concept of correlated ionic fluctuations, has been developed by Rouzina and Bloomfield (1996). This theory emphasizes the pseudo-two-dimensional (2D) character of the counterion distribution very close to the highly charged DNA surface. Coulombic repulsion between these surface-adsorbed but mobile ions leads to a 2D ionic lattice. When two DNA molecules with ionic surface lattices approach each closely, the lattices adjust in complementary fashion, positive charge opposite negative, leading to a net attraction. The magnitude of this attraction is determined by the surface charge density and the solution dielectric constant. For B-DNA in water, the attraction is calculated (in agreement with experiment) to be stable with respect to disruptive thermal motions of the ions so long as the counterions have charge +3 or greater.

1.5.5(f) *Bending and Buckling Induced by Multivalent Cations.* Multivalent cations may cause attraction between DNA helices not just by correlations between fluctuating ion atmospheres, but also by inducing DNA bending or buckling. At spermine and $Co(NH_3)_6^{3+}$ concentrations slightly below those needed for condensation, and also in the presence of Mg^{2+}, electrooptical measurements show that rotational relaxation times of DNA decrease (Pörschke, 1986; Marquet et al., 1985, 1987). The apparent persistence length in the presence of spermine and Mg^{2+} is nearly halved, a result affirmed in more detail by single-molecule stretching experiments (Baumann et al., 1997a), so the radius of curvature is comparable to that of condensed toroids; and even stronger bending is provoked by $Co(NH_3)_6^{3+}$. Work with DNA of different composition shows that binding of spermine to A-T regions induces bending, while binding to poly(dG-dC) stiffens the polynucleotide (Marquet and Houssier, 1988).

The mechanism of reduction in stiffness by multivalent cations is unclear. The NMR evidence (reviewed in Chapter 11) indicates that these ions are mainly bound to B-DNA in the mobile fashion predicted by counterion condensation theory, rather than by site binding, although hexammine cobalt shows some preference for G,C rich sequences (Braunlin and Xu, 1992; Xu et al., 1993a) and may induce a B–A transition (Xu et al., 1993b). It might take only a small amount of site binding-induced bending, at susceptible sequences, to enhance the local curvature of DNA. Such bending could occur either through specific interactions, (e.g., hydrogen-bonding), or through a nonspecific charge neutralization process. As discussed in more detail in Chapter 9, neutralization of phosphate charges on one side of the double helix causes a bending into the neutral surface (Strauss and Maher, 1994; Manning et al., 1989). Whether transient directional bending can occur if ion binding is largely mobile and symmetrical around the DNA remains to be seen.

Manning (1980, 1985, 1989) has elaborated an interesting alternative mechanism; that is, neutralization of the stiff double helix makes it elastically unstable, like a long thin column under compression. The compressed segment buckles outward, and then its most stable shape is strongly curved or even folded, allowing attractive forces to come into play to maintain a collapsed state. The "buckling persistence length" is estimated to be about 50 bp, which is in reasonable agreement with the inner radius of toroidal condensates (Arscott et al., 1990); Odijk (1993a) has arrived at a similar estimate of minimum toroidal radius from considerations of the behavior of a semiflexible polymer confined in a tightly curved "tube" (formed in this case from neighboring polymer segments). However, single molecule stretching experiments on λ-DNA in the presence of condensing concentrations of multivalent cations do not show any evidence of an abrupt buckling transition (Baumann et al., 1997b).

1.5.6 Hydration Forces

An additional force affecting the interaction of DNA helices is the hydration force, due to reconfiguration of water between macromolecular surfaces.The hydration force appears to be a general force in biomolecular systems (Leikin et al., 1993). In addition to DNA, it is also observed with charged, zwitterionic, and neutral phospholipid bilayers, neutral and charged polysaccharides, protein assemblies such as collagen fibers and hemoglobin, and voltage-gated transmembrane ion channels. It can be either attractive or repulsive. Repulsive forces are short-ranged, exponentially decaying, independent of ionic strength, and similar in behavior for all types of molecular systems examined. These similarities suggest a common underlying force, probably involving polarization of water by polar groups on the surface.

The concept of hydration forces developed largely from the direct measurement of osmotic stress as a function of separation between macromolecules. The DNA in an ordered fiber is equilibrated with a reservoir containing water and ions that can exchange, and a large inert polymer (e.g., polyethylene glycol, PEG) that cannot exchange. The distance between DNA molecules is measured from the Bragg spacing in X-ray diffraction, assuming hexagonal packing. The osmotic pressure is controlled by the polymer concentration, giving osmotic stress versus distance curves of the sort shown in Figure 14-6. Since stress is force per unit area, a measured osmotic stress is readily converted to a force between surfaces. Techniques for osmotic stress measurements are described in detail by Parsegian et al. (1986).

Force–distance measurements of DNA parallel helices as a function of salt concentration and valence show a surprising result: All ionic influences nearly disappear when the separation between DNA surfaces is 5–15 Å (Rau et al., 1984). This behavior is evident in the upper left hand (repulsive) regions of Figures 14-6–14-8, measured under very different ionic conditions. The repulsive pressure decays exponentially, with a characteristic length of 2.5–3.5 Å. Similar behavior is seen even within intact bacteria (Reich et al., 1995). At interaxial distances of 26–30 Å typical of DNA packing in bacteriophage, the packing energy per base is 0.1–0.4 kcal mol^{-1}, similar to the values in Table 14-1, and corresponding to a DNA pressure of 1.2–5.5 $\times$ 10^7 dyn cm^2.

Hydration forces can under suitable conditions be attractive as well as repulsive, as seen in Figures 14-6–14-8 where the osmotic pressure–distance curve exhibits a discontinuity or drops abruptly to zero. These curves have several properties that are not easily explicable by more traditional forces (Rau and Parsegian, 1992 a,b). They

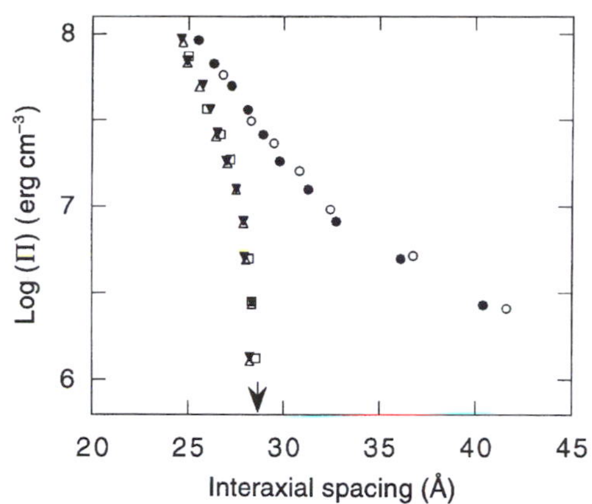

Figure 14-6
Osmotic pressure–distance relations for DNA at 20°C in 0.25 *M* NaCl +1 m*M* EDTA (○), 0.50 *M* NaCl +1 m*M* EDTA (●), 1 m*M* $Co(NH_3)_6Cl_3$ (□), 20 m*M* $Co(NH_3)_6Cl_3$ + 0.25*M* NaCl (Δ), and 100 m*M* $Co(NH_3)_6Cl_3$ + 0.25*M* NaCl (∇). The osmotic stress was applied by PEG, and all buffers contained 10 m*M* Tris to maintain pH at 7.5. The arrow indicates the equilibrium center-to-center distance (28.3 Å) between spontaneously assembled helices in the absence of applied stress. [Reprinted with permission from Rau and Parsegian, 1992a.]

can extend 8–10 Å surface-surface, do not depend on ionic strength or composition, are two orders of magnitude greater than the van der Waals prediction, and decay exponentially over characteristic distance 1.4–1.5 Å independent of the condensing ligand. Parameters for forces between DNA molecules in solutions of monovalent cations have been tabulated by Podgornik et al. (1994).

This type of behavior is explained by a theory that postulates the rearrangement of surface-bound water by condensing ligands, to create regions of hydration attraction, or water bridging, between helices. The behavior of the condensing ligands, in turn, is influenced by the surface lattice of DNA phosphates and other polar groups. Leikin et al. (1991) state "Hydration force magnitudes depend on the strength of surface water ordering, while the decay length and sign, attraction or repulsion, depend on the mutual structuring of water on the two surfaces. Attraction results from a complementary ordering, while repulsion is due to symmetrical structuring."

Hydration forces are very difficult to predict from first principles, since they depend on summation of hundreds or thousands of very weak intermolecular interactions. However, some general statements can be made using an order parameter formalism (Leikin et at, 1991). This predicts a repulsive hydration pressure between two similar, homogeneous planar surfaces separated by distance h:

$$P_{\text{rep}}^{\text{homo}} = \frac{R}{\sinh^2(h/2\lambda_w)} \approx 4Re^{-h/\lambda_w} \tag{14-9}$$

and an attractive pressure between two complementary surfaces

$$P_{\text{attr}}^{\text{homo}} = -\frac{A}{\cosh^2(h/2\lambda^w)} \approx -4Ae^{-h/\lambda_w} \tag{14-10}$$

where R and A are coefficients and λ_w is the water correlation length (about 4–5 Å). This value is observed for Na^+ or tetramethylammonium$^+$ as counterions.

If the surfaces themselves have a lattice structure, of periodicity α, then the homogeneous pressure is replaced by an inhomogenous pressure

$$P_{\text{rep}}^{\text{inhom}} = \frac{\alpha < \sigma^2 >}{\sinh^2\{h[\lambda_{\text{w}}^{-2} + (2\pi/\alpha)^2]^{1/2}\}} \tag{14-11}$$

proportional to the mean-square value of the surface "charge" density σ. For B-DNA, with $a = 34$ Å and $\lambda_{\text{w}} = 4$–5 Å, the effective correlation length

$$\lambda = \frac{1}{2}[\lambda_{\text{w}}^{-2} + (2\pi/\alpha)^2]^{-1/2} \tag{14-12}$$

is in the range 1.6–1.8 Å, compared with the measured 1.3 Å for DNA condensed with $Co(NH_3)_6^{3+}$ or Mn^{2+}.

The hydration force can also be analyzed thermodynamically (Leikin et al., 1991; Rau and Parsegian, 1992a). It shows that attraction is entropy driven by rearrangement or release of water, which depends on counterion binding. Attraction is enhanced by increasing bulk water entropy with chaotropic anions such as perchlorate. Thermodynamic analysis begins with the standard relation between free energy, temperature, and pressure changes, but with osmotic pressure Π replacing external pressure P:

$$dG = -SdT + Vd\Pi \tag{14-13}$$

Differentiating both sides with respect to T and Π gives the Maxwell relation

$$(\partial S/\partial \Pi)_{\text{T}} = -(\partial V/\partial T)_{\Pi} \tag{14-14}$$

This relation can be integrated to give the entropy change accompanying a change of osmotic stress

$$\Delta S = -\int (\partial V(T, \Pi)/\partial T)_{\Pi} d\Pi \tag{14-15}$$

In the region of abrupt change due to packing transition, one uses a variant of the Clausius–Clapeyron equation,

$$\Delta S = -(d\Pi_{\text{t}}/dT_{\text{t}})\Delta V \tag{14-16}$$

and the total entropy change as a function of helix spacing (which is in turn related to water volume) is a sum of the continuous and transition contributions.

The total free energy change equals the work done on the system by changing the array spacing at constant osmotic pressure

$$W = \Delta G = \int \Pi dV \tag{14-17}$$

Then the enthalpy change accompanying the change of DNA hydration is

$$\Delta H = W + T\Delta S \tag{14-18}$$

When this analysis is applied to DNA in aqueous $MnCl_2$ solutions (Leikin et al., 1991; Rau and Parsegian, 1992b) using data of the sort shown in Figure 14-8, ΔS and ΔH are positive and increase exponentially as the helices approach, with $\lambda \approx 4$ Å. They show no discontinuity in slope or magnitude associated with the collapse transition. The parameters $T\Delta S$ and ΔH are of nearly equal magnitude; ΔG is about 0.1 of either, in the range 0.1–0.001 k_BT depending on interhelix distance. The exponential decay length decreases with increasing T, but is independent of $MnCl_2$ concentration at 10 mM and above. The parameter $T\Delta S$ for transfer of a single water molecule from the DNA containing phase into the bulk $MnCl_2$ solution is 0.3–3 cal mol^{-1}. Although this is a very small number, it becomes significant when summed over all the waters associated with DNA. The small values of the thermodynamic quantities for MnDNA arise from an almost equal mix of attractive and repulsive contributions. A transition

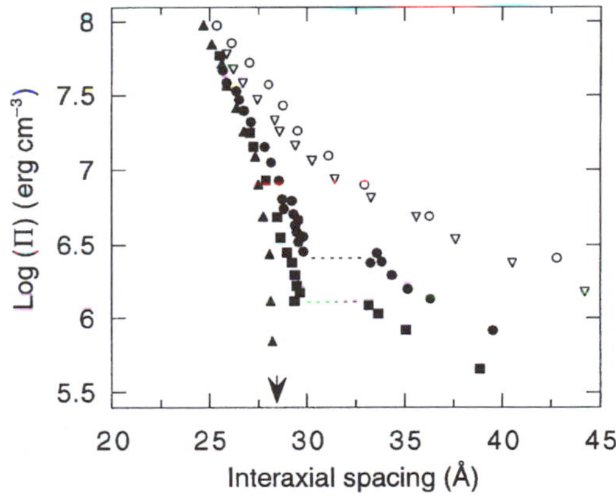

Figure 14-7
Osmotic pressure–distance relations for DNA at 20°C in 0.25 M NaCl, 10 mM Tris buffer, pH 7 at several concentrations of $Co(NH_3)_6Cl_3$: 0 mM (○), 2 mM (▽), 8 mM (●), 12 mM (■) and 20 mM (△). Spontaneous precipitation occurs at 17 mM. The arrow indicates the equilibrium spacing in the spontaneously assembled hexagonal array of DNA helices. [Reprinted with permission from Rau and Parsegian, 1992a.]

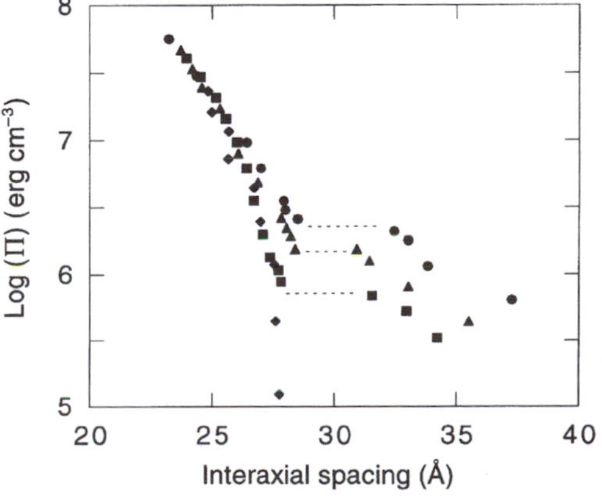

Figure 14-8
Osmotic pressure–distance relations for DNA in 50 mM $MnCl_2$ + 10mM Tris, showing transition behavior as a function of temperature. Spontaneous precipitation with no applied osmotic stress occurs between 40 and 45°C. [Reproduced with permission from Rau and Parsegian, 1992b.]

from repulsion to attraction may require only small rearrangements of bound Mn^{2+} ions. On the other hand, application of the same sort of analysis to DNA condensation by $Co(NH_3)_6^{3+}$ using the data in Figure 14-7 (Rau and Parsegian, 1992a) shows that a change of 0.20 ions bound per bp accompanies the spontaneous ($\Pi = 0$) condensation transition.

1.5.7 Other Long-Range Forces

A very surprising aggregation of mononucleosomal (160 bp) DNA is observed in concentrated (60–80 mg mL^{-1}) but isotropic solutions (Wissenburg et al., 1994, 1995). The aggregates are detected by both light scattering and EM, and appear to contain several hundred DNA molecules—in the same range of 10^4-10^5 bp of DNA as observed in DNA condensed by multivalent cations. The intermolecular forces leading to the formation of these aggregates are quite unclear, but may be related to an attractive long-range force that leads to a large third virial coefficient, although the second virial coefficient (reflecting pairwise interactions) is essentially unaffected (Odijk, 1994).

1.5.8 Secondary Structure Changes

An intriguing possibility is that condensed DNA structures are stabilized not (or not only) by rather nonspecific intermolecular forces, but by interaction between helical segments that have been perturbed from the normal B conformation. There is a range of circumstantial evidence supporting this hypothesis. Much of it arises from studies of Z-forming polymers, particularly poly[d(G-C)] and poly[d(G-me^{5}C)]. Van de Sande and Jovin (1982) showed that the left-handed form of poly[d(G-C)] (designated Z*-DNA) produced by 0.4 m*M* $MgCl_2$ and 20% EtOH, or 4 m*M* $MgCl_2$ and 10% EtOH, sedimented readily out of solution at low speed, indicative of condensation and aggregation. Detailed studies (Revet et al., 1983; Castleman and Erlanger, 1983; Castleman et al., 1984; Thomas and Bloomfield, 1985) have shown that collapse/aggregation and B–Z transition generally occur together for poly[d(G-C)] and poly[d(G-me^{5}C)] in a wide variety of conditions. Electron microscopy often shows close parallel association of DNA strands under condensing conditions. Some helical modifications observed with poly[d(G-C)], such as four-stranded hairpins and loops built up by the sticking together of two segments of DNA, have also been seen with natural sequence plasmid DNAs (Revet et al., 1983).

Another piece of evidence that Z-DNA favors condensation was provided by Ma et al. (1995) who inserted $d(CG)_n$ sequences ($n = 12$ or 20) into plasmids and provoked condensation with $Co(NH_3)_6^{3+}$. Plasmids with longer $d(CG)_n$ inserts condense more extensively at natural superhelical densities; and enzymatic and chemical probing showed that the inserts convert from B to Z form in the presence of $Co(NH_3)_6^{3+}$ under the conditions of condensation. The Z conformation may enhance DNA condensation through the higher flexibility of secondary structure around the $d(CG)_n$ insert region, or through greater exposure of bases to solvent in the Z form. Although the inserts are relatively short, conformational perturbations modulated by condensing ions and dehydrating agents may extend beyond the sequence proper (Reich et al., 1991, 1993).

Further insight into the mechanism of aggregation comes from use of transition metal ions that prefer to bind to the bases, in contrast to Mg^{2+}, which preferentially binds to phosphate. An aggregating, left-handed form of poly[d(G-C)] can be produced in the absence of EtOH by Mn^{2+}, Ni^{2+}, and Co^{2+} (Van de Sande et al., 1982). In an informative violation of the "90% charge neutralization rule" described in Section 1.5.5 (a), Mn^{2+} can cause condensation in water of supercoiled but not linear DNA (Ma and Bloomfield, 1994). Since neither $MgCl_2$ nor NaCl reverses the condensation, the condensation mechanism does not appear to be electrostatic. Instead, the supercoiled MnDNA is more extensively digested than the linear form by S1 nuclease, suggesting that supercoiling cooperates with Mn^{2+} in stabilizing helix distortions. Supercoiling also provides a "pressure" that brings opposite sides of the circle together, enhancing lateral association of DNA strands.

The ability of transition metals to aggregate natural sequence calf thymus DNA at elevated temperatures has been extensively studied by Knoll et al. (1988). There is a strong correlation between the effectiveness of the divalent cations in producing aggregation, their effectiveness in producing apparent thermal melting of DNA (Eichhorn and Shin, 1968), and their strength of binding to the bases, particularly N7 of guanine. When present at M^{2+}:DNA phosphate ratios near unity, Cu^{2+}, Zn^{2+}, Ni^{2+}, Cd^{2+}, Mn^{2+} and Co^{2+} produce a hyperchromic effect resembling that accompanying the double-single strand thermal transition as T is raised. However, the optical change is immediately reversed when the metal ions are complexed with EDTA. This behavior is in contrast to normal thermal melting, where strand separation makes renaturation very slow and often irreversible. Thus it appears that the hyperchromic change associated with disruption of base stacking occurs without extensive unwinding of the strands of the double helix. Aggregation occurs at higher M^{2+}/DNA phosphate ratios, about 30:1, than the apparent melting transition, but also does not involve extensive formation of single-stranded regions, as evidenced by lack of digestibility by S1 nuclease. Multiwavelength spectroscopy demonstrates that the aggregated DNA is not melted.

All of these results are compatible with a theory (Knoll et al., 1988) of the ionic strength dependence of the transition from B-form helix to a conformation whose bases are accessible to site binding by metal ions. Disruptions of B-DNA structure consistent with such a conformation (or a range of disrupted conformations) have been found by Raman spectroscopy of DNA–transition metal complexes (Duguid et al., 1993). This theory predicts the existence of an intermediate range of salt concentrations over which this conformation is more stable, provided that its charge spacing is not much greater than that of the B form. A small change in spacing would be the case if the conformation were not extensively unwound. The theory also predicts that the B form should again become more stable at the high-salt concentrations at which the aggregation experiments were conducted.

A similar picture is presented by the unusual P form of DNA, formed in low-water–high-alcohol solutions, which is condensed and whose bases are neither paired nor stacked (Zehfus and Johnson, 1981, 1984). The P form is produced in a cooperative transition when the temperature or EtOH concentration of a 95% MeOH solution is increased. Collapse occurs when about 90% of the DNA charge is neutralized by counterion condensation. The B $\rightarrow$ P transition is not reversed on cooling or adding MeOH, but is instantaneously reversed when water is added, indicating no extensive strand separation in the P form. Similar unstacked but rapidly renatured DNA is observed at lower MeOH in the presence of divalent ions (Votavová et al., 1986).

The condensation behavior of DNA appears to depend on sequence-dependent secondary structure. Poly(dA)–(dT) adopts the Ψ^+ conformation in the presence of NaCl and PEG (Chaires, 1989), opposite from other DNAs, which adopt the Ψ- conformation under these conditions. Reiterated short A_N-tracts ($N < 3$) allow formation of long-range chiral order when the DNA is condensed in ethanol–water mixtures, but longer tracts do not; while phased A tracts of the sort leading to sequence-directed bending produce very small toroids (Reich et al., 1992). These variations are presumably related to sequence-dependent helix structure, flexibility, and curvature.

1.5.9 Summary

An overview of all these results indicates that DNA condensation arises from the interplay of several forces. The entropy loss upon collapse of the expanded wormlike coil exacts a free energy toll. Stiffness sets limits on tight curvature. Electrostatic repulsions must be overcome by high-salt concentrations or by the correlated fluctuations of territorially bound multivalent cations. Hydration must be adjusted to allow a cooperative accommodation of the water structure surrounding surface groups on the DNA helices as they approach. Dispersion forces make a small but perhaps not negligible attractive contribution. Repulsive excluded volume interactions with other polymers in the solution can force the DNA segments closer together. Additional attractive free energy may be provided by bridging through condensing ligands, and/or by interhelical binding between bases whose normal intraduplex pairing and stacking have been disrupted by interactions with solvent or ligand. However, in cation-induced condensation of random sequence DNA, only a small fraction of the base pairs can be disrupted since the CD spectrum is very similar to that of B-form DNA.

Probably the dominant contributions come from Coulombic and hydration forces. While these forces are sometimes discussed as if one excludes the other, this is unlikely to be the case. The interpretation of the insensitivity to ionic conditions of osmotic stress measurements at close approach of DNA surfaces, as showing that Coulombic interactions are insignificant, is based on mean-field electrostatic theories of the Debye–Hückel and Poisson–Boltzmann type. Proper accounting for ionic effects demands consideration of ionic fluctuations and correlations, which are very difficult to handle by analytical theory and which have largely been studied by elaborate computer simulations (Nilsson et al., 1991; Lyubartsev and Nordenskjöld, 1995). Such studies tend to show behavior, including attractive interactions mediated by multivalent cations, that could not have been predicted by mean-field theories. On the other hand, the evidence is very strong that water structure is a crucial, but until recently often unappreciated, determinant of interactions between polar and charged surfaces of biological macromolecules. Since both hydration and ionic forces at close range appear to depend on the effect of apposing surface lattices on the fluctuating correlations of ions and water molecules between them, the two types of forces may be very difficult to disentangle.

2. DNA PACKAGING IN BACTERIOPHAGE AND BACTERIA

2.1 Double-Stranded DNA Packaging in Bacteriophage

2.1.1 Common Features of Double-Stranded DNA Packaging

The *in vitro* studies presumably explain many of the forces operating in allowing phage DNA to be packaged into virus capsids. However, they do not explain the detailed biochemical mechanism. Fortunately, there is a good deal of evidence, surveyed by Earnshaw and Casjens (1980) and Casjens and Hendrix (1987), that all dsDNA phage use essentially the same packaging mechanism, thus conclusions on one should apply to all. Many studies have identified the following common features.

1. The DNA is inserted into a preformed protein container, the mature prohead. The immature prohead differs morphologically and chemically from the mature prohead. Immature proheads contain scaffolding proteins, or assembly cores, which are absent from complete phage. The prohead expands at or near the time of DNA packaging, serving either to produce an osmotic pressure differential (Serwer, 1975, 1980) or to regulate the capacity of the inner surface of the capsid to interact with DNA (Earnshaw and Casjens, 1980).
2. Proheads and completed phage particles contain 5–30 copies of a protein species near the portal vertex (that vertex to which the tail will join), which take part in initiation of prohead assembly, DNA packaging, and the injection of the DNA from the phage.
3. In order for packaging to proceed correctly, an *in vitro* system requires proheads, a correct and uncut or catenated DNA substrate, two accessory "nonstructural" phage-coded proteins, polyamines, and ATP. As can be understood from our preceding discussion, the polyamines probably serve to neutralize electrostatic repulsions and perhaps to cross-link or perturb DNA helices near their binding sites. The ATP is the source of free energy for the packaging machinery. The nonstructural proteins serve to recognize the correct DNA. For example, in λ phage, the gene A protein (gpA) specifically binds near the left end of λ-DNA, which is the end that first enters the prohead. The gpA–DNA complex is then somehow recognized as the appropriate substrate for packing into proheads. The rate of packaging in T7, bacteriophage probably typical of most dsDNA phage, is 28 ± 6 kbp min^{-1} for the last 20–50% of the DNA packaged (Son et al., 1993).
4. The packing density and spacing of DNA in phage heads is extremely constant, with the exception of T4, which has glucosylated hydroxymethyl cytosine residues, and deletion mutants of λ. There is about 1.5 g of water for each g of DNA. A "terminase" endonuclease, associated with the portal vertex of the capsid, somehow recognizes a completely packed phage head as its proper substrate, and cleaves the concatemeric DNA into a proper linear "headful" when full packing density is achieved (Streisinger et al., 1967). We do not

currently understand how the terminase recognizes that the head is full, or how it determines the correct place on the DNA to make the terminating cut. Mutations in the portal protein of phage P22 lead to variation in the length and packing density of the DNA, suggesting that the packing density is not determined solely by the coat protein shell or the DNA itself (Casjens et al., 1992).

2.1.2 Structure and Arrangement of Packaged DNA

Raman spectroscopy shows that intraphage DNA has normal B-form geometry, despite its tight packing and winding and a notably different electrostatic environment of the phosphates relative to aqueous solution (Aubrey et al., 1992). X-ray diffraction and EM show a close-packed, apparently orderly winding. In isometric heads, such as those of P22 and λ phages, X-ray patterns (Earnshaw and Harrison, 1977; see Chapter 9) show that the DNA is wound tightly into locally parallel layers concentric with the capsid shell. In wild-type λ, the DNA is tightly and uniformly packed with helix centers separated by 27 Å; in deletion mutants, the local interhelix distance increases proportionately to the decrease in amount of DNA. In giant T4 phage, with long prolate heads, the DNA forms a coil with axis perpendicular to the phage tail (Earnshaw et al., 1978). The spacing decreases as packaging proceeds (Earnshaw and Casjens, 1980). The constant final interhelix distance or packing density, independent of capsid volume, is a strong verification of the headful packing model.

Much attention has been given to the way in which the DNA is wound within the capsid. Is it wound parallel, perpendicular, or at an angle to the long axis of the phage? Is the end that first enters held near the inner surface of the capsid shell, or at the center? Do the DNA strands wind smoothly in concentric layers, or do they run predominantly straight with sharp bends at the ends of the capsid? Are DNA molecules in each phage particle oriented the same way, or is the orientation only weakly determined? How is the DNA arranged so as to be able to eject rapidly without tangling, and without requiring turning of the whole bundle within the capsid? How is ATP hydrolysis coupled to packaging? Current experimental evidence has not provided clear answers to any of these questions, but there are many interesting and thoughtful proposals.

Most packing models agree that, in order to avoid knots and entanglements, the first end of the DNA to enter must be tethered near the entry vertex. The other end of the DNA, which is the first to leave, also is located near this vertex or protrudes into the tail tube, so that ejection involves unwinding from the interior of the coil. Thus, the entire DNA is not required to rotate with respect to the inner surface of the capsid.

Three proposed packaging geometries are shown in Figure 14-9. One of the earliest proposals was that by Richards et al. (1973) based on the EM of five different DNA bacteriophages. They proposed two alternative models, both consistent with the data available at the time. In the first, the DNA is wound like a ball of yarn. In the second, the DNA is wound like a spool, whose axis is perpendicular to the axis of the phage particle. In this model, there are a greater number of turns in the central region than at the two ends. An equally orderly, but somewhat different packing model, has been proposed for T4 phage by Black et al. (1985) on the basis of ion etching experiments.

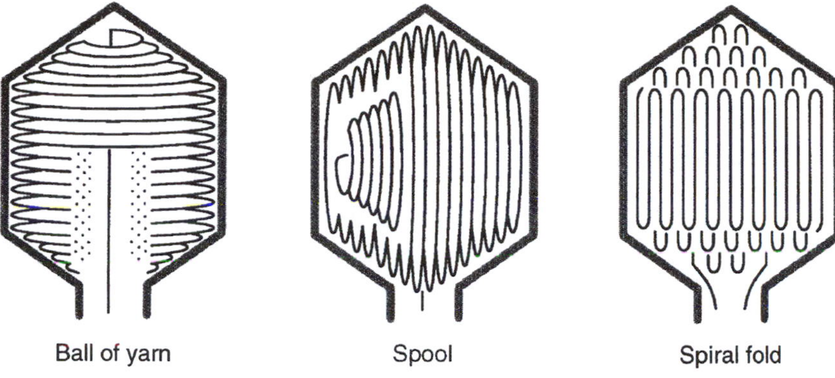

Figure 14-9
Three models for packing of DNA in phage heads.

This technique progressively erodes virus components from outside to inside. The end of the DNA that first enters the prohead is eroded most slowly, suggesting that it is near the center of the capsid. The results are consistent with a spiral-fold model with DNA strands running parallel to the long axis of the phage, folding back sharply at the top and bottom of the capsid. The folds are arranged radially about the long axis in spirally organized shells. The spool and spiral-fold packaging models are, respectively, related to the toroidal and rodlike particles formed in *in vitro* condensation experiments.

The giant T4 head results (Earnshaw et al., 1978) and electric dichroism studies on three isometric phages (Kosturko et al., 1979) show a preferred axis for the DNA coil. These studies support the spool or spiral-fold model, rather than the ball-of-yarn-model, in which the winding axis would be random.

However, photochemical cross-linking studies designed to determine the protein and DNA neighbors of various pieces of λ DNA, which has a nonpermuted sequence, suggest a more random structure. Widom and Baldwin (1983b) used UV irradiation of λ phage containing 5-bromodeoxyuridine in place of thymine to probe the loci on the DNA which are in contact with the capsid. Electron microscopy and density gradient analysis of the results showed capsid–DNA contacts distributed nearly randomly over the entire genome. They also used the psoralen derivative AMT (Chapter 3) to produce DNA–DNA cross-links after intercalation into the intact λ phage. The EM examination of the denatured DNA showed different patterns of cross-links in each molecule, with no consistent pattern. Haas et al. (1982) and Welsh and Cantor (1987) obtained similar results with bis-psoralen as a cross-linking agent. Cross-links were located by their appearance in a restriction digest as X-shaped molecular features. Significant cross-linking frequencies were found between all possible pairs of restriction fragments, consistent with a random packaging model. The T7 DNA packaging concentrates the last end packaged near the inner surface of the T7 capsid, but no part of the DNA is excluded from contact with the capsid surface (Serwer et al., 1992). Overall, these studies show that the positioning of local sequences of DNA inside viruses is variable from particle to particle.

2.1.3 Hypothetical Packaging Mechanisms

We examine two mechanisms that reconcile generally orderly close packing with particle–particle variability. The first of these (Harrison,1983), in Figure 14-10 (left), is a variant of the spool model. To begin packaging, the leading end of the DNA is attached to the capsid near the entry vertex. The DNA enters along the axis and adds turns initially to the far end of the spool. At an intermediate stage, turns are uniformly distributed throughout the spool of intermediate density. To package a complete headful, further tightening must occur to accomodate the remainder of the DNA. Such tightening can happen in a different way from one head to the next: layers can add near the axis or around the outside.

Spooling or toroidal packing involves a series of unidirectional loops, requiring supertwists of the DNA within the capsid. The second model (Serwer, 1986), in Figure 14-10 (right), avoids toroidal supertwisting by postulating that DNA is packaged in successive segments that alternate in winding direction and do not complete closed 360° loops. The kinks are compatible with structures proposed for drug intercalation sites or nucleosomes, and may account for high-affinity ethidium binding sites in packaged T7 DNA. This model is reminiscent of the spiral-fold model (Black et al., 1985).

Whether one of these, or some other, packaging model is correct, it seems likely that the orderly arrangement of the DNA begins at the early stages of packaging, governed by both the stiffness and intersegmental forces of the DNA and the geometry of the capsid (Earnshaw and Casjens, 1980). This mechanism is akin to what Lepault et al. (1987) have termed "constrained nematic crystallization."

2.1.4 Packaging in Bacillus Subtilis Phage ϕ29

The most detailed biochemical studies on a well-defined, highly efficient packaging system are those by Anderson and co workers (Guo et al., 1987 a–c), using the *B. subtilis* ϕ29 phage. They exemplify some aspects of biochemical mechanism that must be important for a full understanding of the packaging process. As noted above, all phage systems appear to require proheads, DNA, and two accessory proteins for packaging. In ϕ29, the accessory proteins have been identified as the products of genes 3 and 16; they are called gp3 and gp16. The ϕ29 phage is unusual in that gp3 is covalently bound to the 5′-termini of the DNA, and that the prohead (but not the mature virion) contains a 120 base RNA species that is essential for packaging (Reid et al., 1994a,b). Thus the *in vitro* packaging system contains purified proheads with RNA, DNA–gp3, and gp16. In the first step, gp16 binds to, and is modified by, the prohead. This complex then binds to DNA–gp3, causing a second modification of gp16, which enables it to bind ATP and hydrolyze. DNA–gp3 aggregates are produced, reminiscent of the covalently linked concatemeric DNA required for efficient packaging in other *in vivo* systems. Kinetic analysis shows that these initiation steps are rate limiting, and that subsequent translocation of DNA–gp3 into the prohead is a very rapid process. The packaging is accompanied by gp16-catalyzed ATP hydrolysis, with approximately one molecule of ATP required for 2 bp, an amount equivalent to 9000 molecules per virion.

Adenosine triphosphate is required for packaging, presumably to provide energy for driving the DNA into the head against a thermodynamic gradient. One idea is that

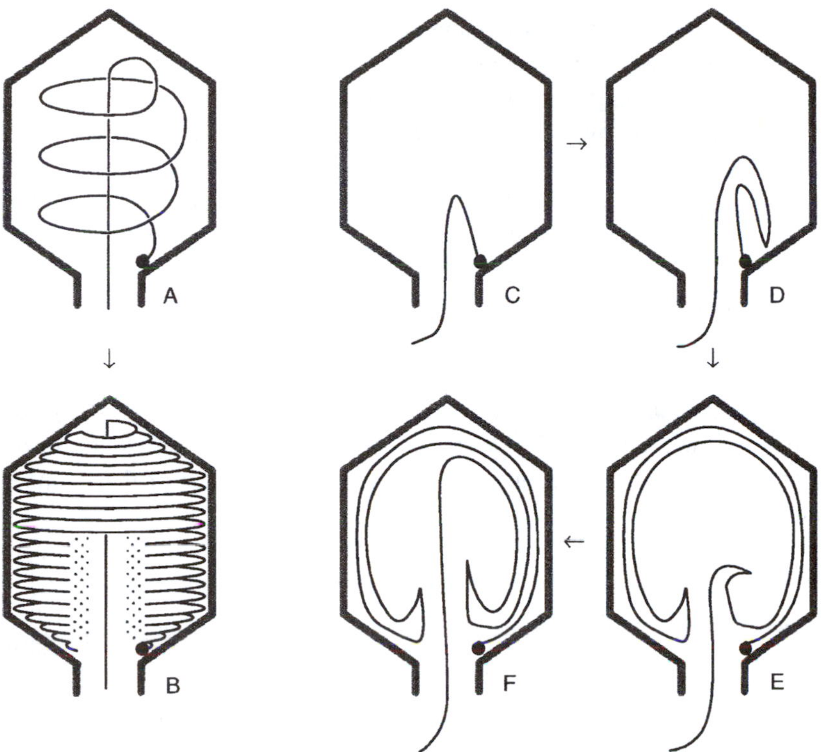

Figure 14-10
Two possible mechanisms of DNA packaging. Left: Spool model (Harrison, 1983). (A) Early stages and (B) fully packaged. Right: Nontoroidal kink model for bacteriophage T7 (Serwer, 1986). (C) A 180° fold forms as the first DNA enters the T7 capsid through its internal cylinder, creating two segments. (D) The segments elongate as DNA enters the capsid. (E) After the segments further elongate and bend into 360° arcs, a second-fold forms. (F) A third segment enters the capsid.

DNA is drawn into the capsid by an osmotic pressure gradient, and that ATP hydrolysis drives the osmotic pump (Serwer, 1988). However, most speculation centers on some kind of mechanical motor, perhaps involving relative motion of the capsid, portal vertex protein complex, and concatenated DNA (Casjens and Hendrix, 1987). For example, supercoiled plasmid DNA has been shown to wrap around the outside of the isolated ϕ29 head–tail connector (Turnquist et al., 1992). This wrapping is hypothesized to be crucial for initiation of DNA packaging: DNA to be packaged would be supercoiled, wrapped around the connector, linearized, and translocated by rotation of the connector relative to the viral capsid with the aid of ATP hydrolysis.

2.2 Packaging of Single-Stranded DNA in Filamentous Phages

The filamentous phage are nucleoprotein complexes about 10^4 Å in length and 60–80 Å in diameter, containing a circular, single-stranded DNA molecule. The best known is

M13, which is widely used as a cloning vehicle. It is one of a closely related group generically designated Ff. Other related groups, on which much physical work has been done, are Pf1, Pf3, and Xf. The phage infects its bacterial host by adsorbing to a pilus. After penetration, the DNA becomes paired with a complementary strand in the cytoplasm. Subsequent replication produces a pool of positive strand circles, complexed with virus-coded ssDNA binding protein. Concurrently, transcription produces five virus structural proteins, of which at least the major coat and adsorption proteins, and perhaps the other three, are sequestered in the cell membrane prior to assembly. Virus assembly is initiated by the interaction of the DNA hairpin in the gpV complex with two cap proteins, presumably in the membrane. The major coat protein molecules then diffuse laterally to the assembly site, replacing the single-strand binding proteins to form a helical protein coat as the DNA is extruded from the cytoplasm, through the membrane, and into solution. The lengths of the virus particles are determined by the size of the circular DNA molecules, which pass from one end of the filament to the other and back again.

Since the phage cannot be crystallized, and since diffraction data on oriented fibers gives information mainly about the protein, information on DNA structure and its interaction with the protein subunits has been obtained mainly from EM, from hydrodynamics and light scattering, and from molecular modeling. Data on the four groups are summarized in Table 14-2.

The most striking thing about this table is that, for Ff and Pf3, the number of nucleotides per protein subunit is nonintegral, seemingly outside of experimental error. Further, even though Xf and Pf1 have almost identical numbers of nucleotides, the latter has twice as many subunits, and is twice as long, as the former. Detailed model building has led to the proposed DNA structures shown in Figure 14-11. A variety of evidence leads to the proposal that Ff and Xf have bases in, but highly tilted; while Pf1 and Pf3 have bases out. In Pf3, the central channel is large enough to accomodate neutralizing cations from solution. In the narrower Pf1, the paraxial phosphates are neutralized by

Table 14.2
Physical Parameters of Filamentous Bacteriophages[a]

	Ff	Xf	Pf1	Pf3
Total number of nucleotides	6408	7420 ± 240	7352 ± 100	5833
Total major protein subunits	2750 ± 100	3500 ± 220	7370 ± 150	2500 ± 160
Nucleotides/subunit	2.33 ± 0.08	2.12 ± 0.15	1.00 ± 0.03	2.33 ± 0.15
Length, (nm)	880 ± 30	977 ± 35	1960 ± 70	683 ± 30
Diameter of DNA region (nm)	2.2 ± 0.2	2.2 ± 0.2	1.6	2.2 ± 0.2
Nucleotide axial separation/chain (nm)	0.28	0.26	0.53	0.23
Subunit axial separation (nm)	0.32	0.28	0.27	0.27

[a]Adapted from Day et al., 1988.

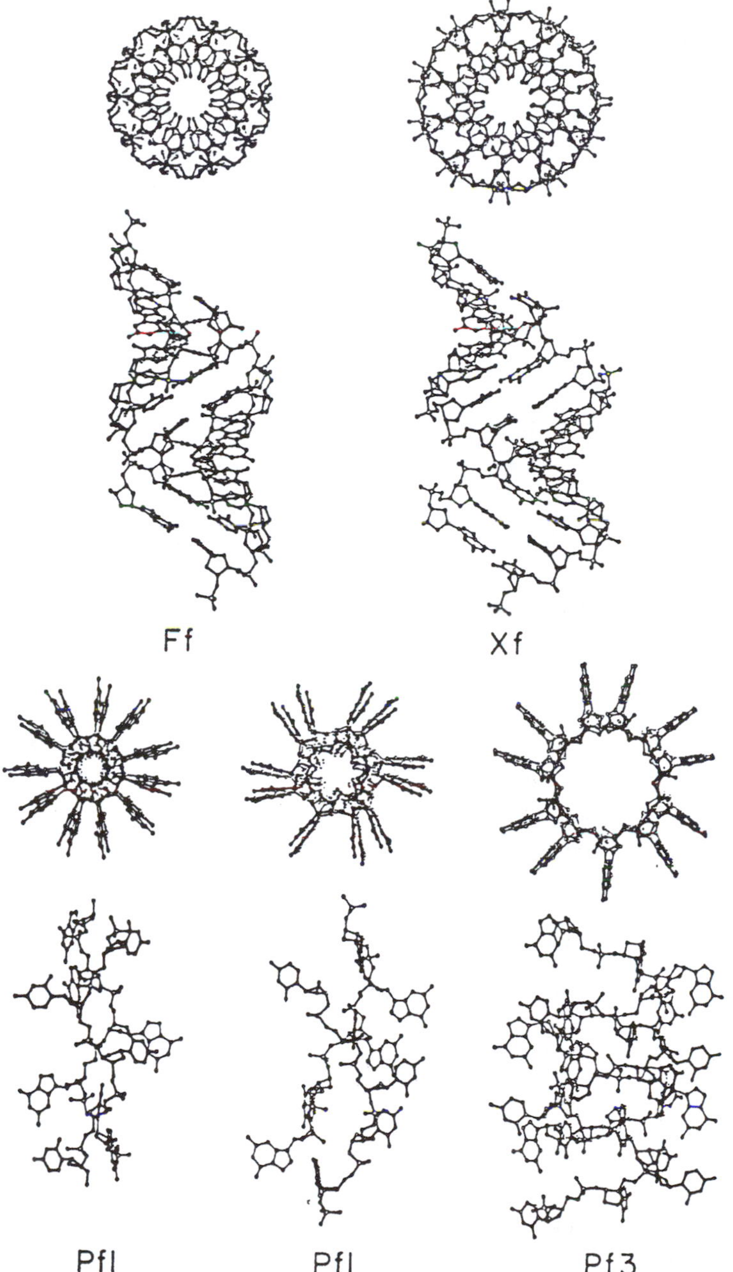

Figure 14-11
End-on and sideways views of stereochemically feasible DNA models for filamentous viruses. These structures satisfy all known constraints, but the constraints are insufficient to specify unique structures, Day et al. (1988). [Reproduced with permission from the Annual Review of Biophysics and Biophysical Chemistry, Vol. 17, ©1988, by Annual Reviews Inc.]

lysine and arginine side chains (Liu and Day, 1994). Thus some intriguing variations are hidden in these apparently monotonous structures.

Marzec and Day (1994) developed a model that treats the symmetry matching problem in structures made of two interacting coaxial helices of point charges, and applied it to the electrostatic interactions between DNA and protein in filamentous viruses. They find that coaxial helices with optimally related symmetries can lock into "spatial resonance configurations" that maximize their interaction. The model accounts for experimental nucleotide/subunit ratios and other aspects of DNA–protein interfaces in these viruses.

2.3 DNA Condensation in Bacteria

Very little is known about the detailed organization of DNA in the bacterial nucleoid. It has been demonstrated (for reviews see Pettijohn, 1988; Schmid, 1988) that there are two general levels of organization. The first is a long-range structure of about 43 independently supercoiled chromosomal domains, averaging about 100 kbp each. The second is a short-range structure, about 60–120 bp, with numerous DNA-binding proteins and a negative supercoil density $\sigma \approx -0.05$ to -0.08. Although bacterial DNA often appears to be confined to particular regions of the cell, the nucleoids of bacteria do not have a degree of DNA condensation typical, for example, of bacteriophage DNA, eukaryotic nucleosomes, or *in vitro* condensates (Kellenberger et al., 1986; Kellenberger, 1987). The DNA of dinoflagellates, however, does show significant compaction and, in fact, appears to adopt a liquid crystalline state. Robinow and Kellenberger (1994) concluded from a survey of evidence from a wide variety of microscopic and fixation procedures that the nucleoid of growing bacterial cells is in a dynamic state, with part of the chromatin extending far out from the bulk of the nucleoid in order to be transcribed in the cytoplasm. This extension presumably enables it to reach the maximum number of ribosomes.

3. DNA PACKAGING IN EUKARYOTES

3.1 Overview of Chromatin Structure

There are several orders of magnitude more DNA in eukaryotic cells than in viruses or prokaryotes. This means that even more efficient and highly regulated methods must be adopted by eukaryotic cells to package their DNA so as to compress it within the dimensions of the cell while making it available for cellular processes such as transcription, replication, recombination, and cell division. The fundamental structure for accomplishing this is, of course, the chromosome. Human cells, for example, have about 3.9×10^9 bp in 23 haploid chromosomes. Each eukaryotic chromosome contains a single linear DNA molecule, as first determined directly by viscoelastic retardation experiments (Kavenoff and Zimm, 1973). Thus there are about 1.7×10^8 bp in the average human chromosomal DNA molecule, which would be about 5.8 cm long if fully extended. Chromosomes also contain a great deal of protein. In fact, the mass of a chromosome is approximately $\frac{1}{3}$ DNA and $\frac{2}{3}$ protein. The protein is roughly 50%

highly basic histones and 50% non-histone proteins, of which the most abundant is DNA topoisomerase II, required for chromosomal condensation (Earnshaw, 1991).

To fit inside the nucleus, the DNA is compressed by a linear factor of about 10^4-10^5. This compaction is achieved by complexing the DNA with proteins; the complex is called chromatin. The standard monograph on the structure of chromatin and its relation to function, summarizing a large amount of experimental information and giving an interesting historical perspective, is van Holde (1989). The book by Wolffe (1992) is a concise and readable treatment with a more biological emphasis. Some other general references to the chemistry and biology of chromosomes and chromatin include Adolph (1988, 1990), Wassarman and Kornberg (1989), and the proceedings of the 1993 Cold Spring Harbor Symposium on *DNA and Chromosomes* that celebrated the 40th anniversary of the discovery of the structure of the double helix.

Perhaps the two key insights into chromatin structure were (1) the proposal by Kornberg (1974) that chromatin is composed of repeating units of two each of the histones H2A, H2B, H3, and H4, and 200 bp of DNA; and (2) the EM visualization by Olins and Olins (1974) of chromatin as "beads on a string," providing a structural model for the arrangement of the repeating units. Although many of the details of these two papers have been modified and extended, they provided a conceptual framework on which most subsequent work has built.

A diagram of the various stages of packaging of DNA into chromosomes is shown in Figure 14-12. The lowest level structures, about which most is known, are the nucleosome, the nucleosome filament, and the 30-nm chromatin filament. The review by Widom (1989) presents a good introduction to these structures.

The fundamental structural entity in chromatin is the nucleosome. Nucleosomes have two turns of DNA wrapped around a histone octamer core containing two copies each of the histones H2A, H2A, H3, and H4. In higher eukaryotes (but not in yeast) a linker histone H1 (or the related H5 in avian erythrocytes) binds to the outside of the nucleosome and leads to formation of the 30-nm filament.

Nuclease digestion of chromatin yields an initial population of DNA fragments whose length is an approximate multiple of 166–246 bp, depending on the organism and cell type. Further digestion leads to particles called chromatosomes (Simpson, 1978), which contain a fixed length, approximately 166 bp, of DNA along with histone H1 and two copies each of H2A, H2A, H3, and H4. The DNA that has been digested away, from 0 to 80 bp in length, is called linker DNA. It connects neighboring nucleosomes. A final stage of nuclease digestion yields the nucleosome core particle. This contains 146 bp of DNA (see Table 6-4 of van Holde, 1989) and the histone octamer core. Histone H1 has been lost, along with 20 bp of DNA. Nucleosome core particles have a common, conserved structure in all eukaryotes. At low resolution they appear as disks about 11 nm in diameter and 5.7 nm high, with about 1.75 turns of DNA wrapped around the octamer core (Richmond et al., 1984).

In low-salt solution, chromatin unfolds into a nucleosome filament that looks like beads on a string; it consists of nucleosomes connected by linker DNA. At higher, physiological salt concentrations, the filament becomes more compact, folding into a 30-nm diameter filament, perhaps a helix with six nucleosomes per turn, which is thought to be the biologically relevant form. Striations with a periodicity of 11 nm

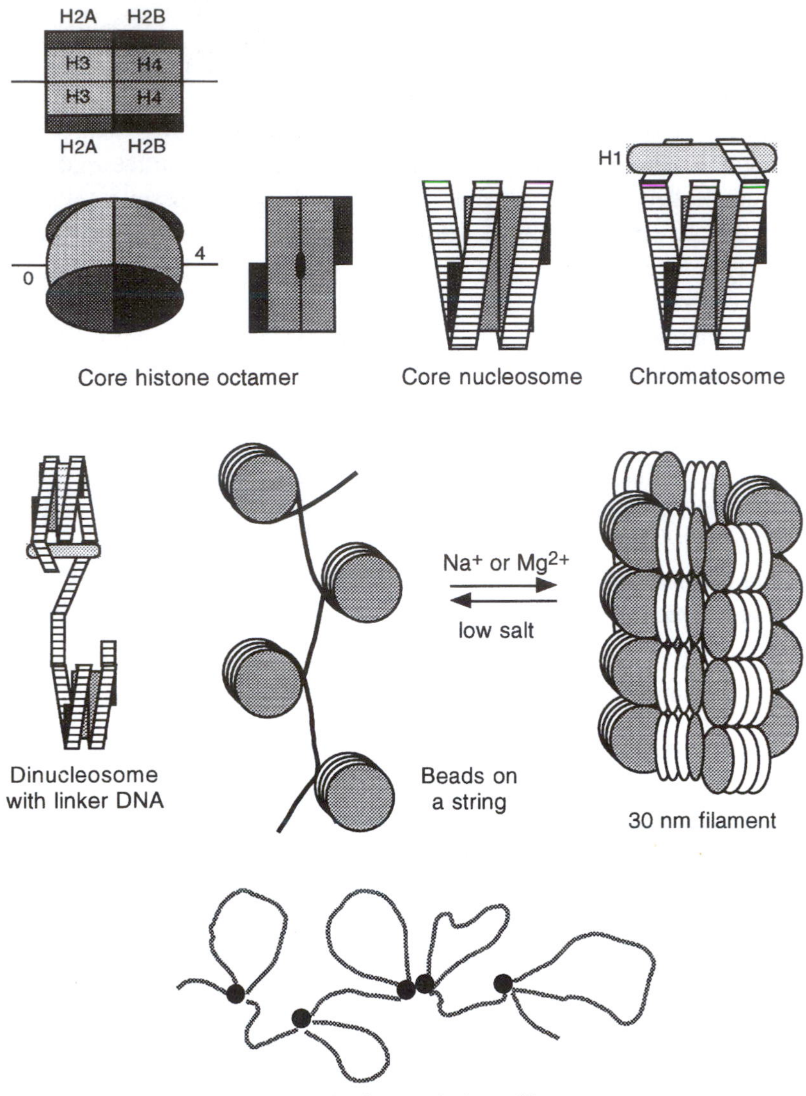

Figure 14-12
Schematic diagram of levels of chromatin structure. In the upper left we find top, front, and side views of the core histone octamer, with gray-scale coding of histones indicated in the top view. The pseudo dyad lies in the plane of the paper in top and front views, running from the entry point of the DNA (turn 0) to turns ± 4. The side view looks down the pseudodyad axis from position 4. Dinucleosome, "beads on a string", and 30nm filament are drawn to half scale of the nucleosome structures on the line above. The loops of 30nm filament attached to the nuclear matrix are drawn at another two orders of magnitude reduction in scale.

are visible along the length of the 30-nm filament. This spacing corresponding to the diameter of the nucleosome core particles whose flat faces are oriented approximately parallel (within 20°) to the long axis of the filament.

Very little is firmly known about the structure of chromatin above the level of the 30 nm filament. The higher level structure of mitotic chromosomes, which is chromatin in its most condensed state, is very difficult to study because it is too small for light microscopy but too large for conventional transmission EM. Proteins responsible for folding above the level of the 30 nm filament will probably be present in trace amounts, two or three orders of magnitude lower than individual histones, and therefore very difficult to isolate and identify (Widom, 1991).

The predominant model (Earnshaw, 1991) is that chromosomal DNA of higher eukaryotes, already organized in the 30-nm filament, is tethered in loops of about 50–100 kbp to the chromosomal scaffold or nuclear matrix. These loops do not appear to stem from a central continuous axis, but instead are attached at irregular points to the scaffold (Wolffe, 1992; Zlatanova and van Holde, 1992). Among the proteins that have been identified as involved in the nuclear matrix are the lamins (Franke, 1987), topoisomerase II, and components of the centromere and telomeres. The loop structure is probably regulated locally by topoisomerase II (Zlatanova and van Holde, 1992).

3.2 Nucleosome Cores

Our knowledge of the structure of nucleosome cores is not quite at the atomic level of resolution (Wolffe, 1992; Ramakrishnan, 1994). It is based on X-ray crystallography, neutron diffraction, and enzymatic and chemical cleavage. These techniques show that DNA is on the outside of the nucleosome, wrapped in 1.75 left-handed superhelical turns around the histone octamer. The nucleosome is nearly symmetric, with the center of the dyad axis of the nucleosome coinciding with the center of the nucleosomal DNA.

A 7-Å resolution X-ray structure of the nucleosome core particle (Richmond et al., 1984) shows the shape of the DNA. It is distorted B form, with average pitch 28 Å. The DNA is not bent uniformly, but shows sharp bends or kinks at ±1 and ±4 double-helical turns from the DNA center. These bends are not due to alcohol in the crystallization buffer (Struck et al., 1992). They may be related to DNA sequence and/or to contacts with histone H3 at either side of the bends. The 7-Å structure shows clusters of electron density corresponding to individual histones. These have long rods of electron density with dimensions characteristic of helices. The X-ray structure combined with cross-linking studies shows the location of each histone within the octamer core. From either end, they are in the order H2A–H2B, H4–H3, H3–H4, H2B–H2A. The central 80 bp turn of DNA interacts with the tetramer $(H3–H4)_2$. This central DNA–protein complex has dyad symmetry. The remaining approximately 20 bp at each end interact mainly with one H2A–H2B heterodimer.

The DNA in solution has 10.5 bp/turn. Nucleosomal DNA appears to have 10.0 bp/turn toward the ends, but 10.7 bp/turn near the center. The juncture between these regions presumably accounts for the distortion. The average twist of nucleosomal DNA is 10.2 bp/turn. This accounts, at least in part, for the "linking number paradox" (Klug and Lutter, 1981): DNA coiled around a nucleosome core should have 1.75 superhelical turns, but only 1 turn is observed after relaxation by topoisomerase. The overwinding of DNA from 10.5 to 10.2 bp/turn can account for 0.4 turns of this discepancy. Monte Carlo computer simulation of ccDNA partially wrapped around a core of proteins shows that the DNA domain collapses suddenly upon a small increase in wrapping of the DNA around the core (Zhang et al., 1994).

The 3.1-Å crystal structure of the histone octamer (Arents et al., 1991), a reinterpretation of an earlier structure (Burlingame et al., 1985; Wang et al., 1994) is in accord with the lower resolution structure (Klug et al., 1980; Richmond et al., 1984). The octamer is a three-part structure, with a central (H3–H4)2 tetramer flanked by two H2A–H2B dimers. It has a diameter of 65 Å, with a length that varies from a maximum of 60 Å to a minimum of 10 Å. Depending on orientation, it therefore appears to be a flat disk or a wedge. The disk is the planar projection of a left-handed protein superhelix with pitch about 28 Å, which is the path along which the DNA runs in the nucleosome.

Despite only weak sequence similarity, each of the histone molecules has a similar three-dimensional (3D) structure: rather elongated, with a long central helix bordered at each end with a loop and a shorter helix. This structure has been termed the "histone-fold." There is extensive interdigitation in a "handshake motif" between pairs of associated histones H3–H4 and H2A–H2B. The protein superhelix on the surface of the octamer arises from the spiral array of these dimers.

Core histones all have a globular domain and charged tails with large numbers of lysine and arginine residues. Posttranslational modifications such as acetylation and phosphorylation take place on the tails. The core histone tails apparently do not organize DNA (a role taken by the globular domains); but those of H4, at least, are implicated in binding to DNA (Hong et al., 1993). Electrostatic interactions of the DNA phosphates with arginines in the globular regions of the core histones, especially H3 and H4, appear to be most important in organizing nucleosomal DNA structure. The sequences of H3 and H4 are among the most highly conserved of all known proteins, attesting to their fundamental importance in the functional organization of DNA. The core octamer structure includes only about 70% of the amino acids in the histones; the amino-terminal tails and some of the carboxy terminus of H2A are missing. Presumably these regions are disordered and are therefore not visible in the crystal structure.

Histones are removed from DNA by high salt, implying that the major interactions are electrostatic. First, H2A and H2B dissociate, followed by H3 and H4. These results, along with chemical cross-linking, show that H2A and H2B form a stable dimer, while H3 and H4 form a stable tetramer $(H3–H4)_2$ (Kornberg and Thomas, 1974). In solution, the octamer is in equilibrium with the $(H3–H4)_2$ tetramer and H2A-H2B dimer (Eickbush and Moudrianakis, 1978). Mixtures of H3 and H4 alone tend to form the tetramer under physiological conditions, though H3–H4 dimer can also be detected. The formation of small amounts of higher oligomer is prevented by addition of H2A–H2B dimer, which apparently acts as a "molecular cap" in regulating the assembly pathway toward the formation of tripartite octamers (Baxevanis et al., 1991). Under physiological salt conditions, the time for dissociation of octamers into H2A–H2B heterodimers and $(H3–H4)_2$ tetramers is faster than 1 s (Feng et al., 1993), arguing against a simple displacement of octamers from DNA during read-through by RNA or DNA polymerase.

3.3 Linker Histones and Linker DNA

By careful nuclease digestion, Simpson (1978) isolated the chromatosome, consisting of the core octamer of histones, one molecule of H1, and about 160 bp of DNA. Thermal denaturation studies showed that H1 stabilizes (and therefore interacts with) not only with the linker DNA but also the core DNA. The additional DNA in the chromatosome beyond the ends of the core particle probably does not continue the superhelix trajectory to complete two full turns. This path would be obstructed by the H2A molecules. Instead, these ends make an angle of 50°-60° with the dyad axis. The H1 molecule appears to sit on the surface of the chromatosome at the points where the DNA enters and exits, causing the entry and exit points to be juxtaposed.

The H1 molecule has a globular central domain that contacts the DNA and protects it from nuclease digestion. It also has extended, basic N- and C-terminal arms that probably interact with the linker DNA (Hartman et al., 1977; Aviles et al., 1978). These arms are highly charged. They are required for chromatin condensation (Thoma et al., 1983), and contain sites for phosphorylation, which is correlated with condensation during the cell cycle (Bradbury, 1992). The arms are extended and disordered in solution, but presumably adopt a more defined structure when they bind to chromatin, perhaps an α-helix that lies in the major groove. More detailed ideas about this structure are reviewed by Ramakrishnan (1994).

The crystal structure of the globular domain of H5, termed GH5, has been solved to 2.5-Å resolution (Ramakrishnan et al., 1993). It consists of a three-helix bundle with a β-turn at the C-terminus abutting the helical core. This is not a canonical helix–turn–helix (HTH) type of DNA binding protein, but rather a variant on this theme. The structure is very similar to that of the transcription factor HNF-3γ (Clark et al., 1993b). This protein has two "wings" that bind DNA, with a DNA binding helix in the middle. One of these wings corresponds to the amino terminal arm in GH5, which adopts several conformations in the crystal; the other corresponds to the C-terminal arm that is highly disordered and not modeled in GH5. Mapping the HNF-3γ structure onto GH5 provides a model for DNA binding by the latter, and explains the importance of several key amino acids in sequence conservation and cross-linking.

Linker histones H1 and H5 bind preferentially to nucleosomal DNA associated with an octamer of core histones, rather than to linear, uncomplexed duplex DNA (Hayes and Wolffe, 1993). For this preferential binding to occur, there must be free linker DNA on either side of the core. Binding of one linker histone to the core protects an additional 20 bp of linker DNA from micrococcal nuclease digestion. This additional DNA is asymmetrically distributed, with 15 bp on one side of the core and 5 bp on the other. Cross-linking studies (Hayes et al., 1994) show that there is strong interaction between GH5 and the nucleosome at a single site on one side of the dyad axis. This site also contacts the core histones, and association of GH5 changes the core histone–DNA interactions.

Further studies will undoubtedly elucidate some of the more subtle features of the linker histones. We note, for example, that several studies (Thoma et al., 1983; Clark and Thomas, 1988; Leuba et al., 1993) show that the behavior of H1 and H5, while similar, are not identical. Further, although yeast does not have proteins corresponding to H1, the folding and packaging behavior of yeast chromatin is similar to that of higher eukaryotes (Lowary and Widom, 1989), implying that it has a protein or protein domain that serves the functions of H1.

3.4 Nucleosome Positioning

Folding of the DNA in chromatin is often thought of in terms of compaction and generalized reduction of access to the DNA. However, linker DNA is more accessible to micrococcal nuclease than core DNA. If this difference in accessibility is maintained for trans-acting transcription factors, then the precise location of a nucleosome on a stretch of DNA might be critical in regulating the activity of that DNA sequence. For this to occur, the nucleosome must be reproducibly positioned—that is, has the same location and orientation with respect to the DNA—in all cells of a given population. Nucleosome positioning is reviewed by Simpson (1991) and Wolffe (1994).

Translational positioning is defined by the beginning and ending of the nucleosome contacts along the DNA. Rotational positioning is defined by the orientation of the DNA surface toward or away from the histone core. Nucleases and chemical probes have been used to determine both of these features. The 5S RNA genes exhibits especially strong positioning, and have been used in many studies of the effect (Simpson, 1991). This DNA has a periodic modulation in minor groove width that occurs every turn of the helix. Hydroxyl radical footprinting shows that these modulations are maintained and exaggerated when the DNA is bound to the nucleosome (Hayes et al., 1990).

Variation of minor groove width, with a periodicity of the helix repeat, leads to curvature of the DNA (see Chapter 9). The narrow minor grooves will naturally bend in toward the histone core, and the wide minor grooves will face out, thus lessening the free energy requirements for DNA bending around the core. which must occur in any case (Drew and Travers, 1985). This idea has been tested with synthetic sequences of the form $(A/T)_3NN(G/C)_3NN$, in which segments consisting exclusively of A and T or G and C, separated by 2 bp, are repeated with a 10 bp period (Shrader and Crothers, 1989, 1990). These repeated motifs form nucleosomes even better than natural positioning sequences. However, the free energy differences for nucleosome formation, between the best sequences and bulk DNA, are small, only about 100 cal mol^{-1}, when normalized to the number of bends.

Although differences between positioning sequences are small, there is evidence that the interactions holding the DNA in a defined translational and rotation position relative to the histone octamer are very strong, since positioning is maintained between 0 and 75°C, and between 0 and 0.8 *M* salt (Bashkin et al., 1993). This implies that the energy of bending DNA around the nucleosome is independent of salt concentration and temperature in this range, and shows that a great deal of energy is needed to displace DNA from previously established contacts with histones in the nucleosome core.

There is considerable evidence that nucleosome positioning depends on interactions with the trans-acting proteins during the assembly process, as well as on DNA sequence (Wolffe, 1994). These trans-acting factors can serve as "bookends" for arrays of nucleosomes, or can promote the assembly of adjacent nucleosomes.

3.5 Influence of Nucleosomes on Transcription

It was initially thought that incorporation of DNA into chromatin that had been compacted by histones would generally repress transcription, by occluding access of transcription factors and polymerases to the DNA. However, it has turned out that this is not the case (Workman and Buchman, 1993; Lu et al., 1994; Paranjape et al., 1994; Wolffe, 1994). Both transcription and replication can proceed past a nucleosome, though initiation is blocked or impeded by a nucleosome (Lorch et al., 1987; Bonne-Andrea et al., 1990). Instead, the detailed folding of DNA in positioned nucleosomes has an important role in determining the access of transcription factors to regulatory DNA sequences, and in regulating the activation and repression of transcription. This regulation can occur by at least three mechanisms (Wolffe, 1994).

First, transcription factors bind primarily to exposed DNA, that is, sequences that are in linker DNA or are oriented toward the solution in positioned nucleosomes. Conversely, DNA sequences that are turned inward toward the histone core by rotational positioning are likely to be repressed.

Second, in positioned nucleosomes there are well-defined contacts between DNA and specific histones. Posttranslational modification or dissociation of these histones can modulate access to the DNA without dissociation of the entire nucleosome. Examples come from the interaction of transcription factor TFIIIA with the 5S rRNA genes (Simpson, 1991). The strongly basic amino-terminal tails of the core histones probably lie in the major groove of the DNA and impede binding of TFIIIA. When these tails are acetylated, their interaction with the DNA is weakened and TFIIIA can bind (Lee et al., 1993). The TFIIIA binding also competes with binding of the H2A–H2B dimer; the rRNA genes are activated by displacement of H2A–H2B. Similar effects arise from the phosphorylation of the basic tails of H1 and H2B (Hill et al., 1991). There are also numerous instances in which linker histone H1 is displaced by trans-acting regulatory factors.

Third, the wrapping of one or two turns of DNA around a nucleosome forms loops that may bring regulatory elements into close proximity. For example, two recognition elements for the glucocorticoid receptor (GR) in the mouse mammary tumor virus long terminal repeat are separated by 92 bp. They are brought near to each other by wrapping around the surface of the histone core, so that they can bind cooperatively to the GR dimer (reviewed by Wolffe, 1994). If the regulatory sequences are located in the linker regions, looping brings together elements separated by approximately 160 bp.

The general picture that emerges is that DNA, histones, and transcription factors are intimately linked in determining chromatin structure, and proper regulation of gene expression can occur only when the framework is correctly assembled. Studies of histone–DNA interactions alone, or of transcription factor–DNA interactions alone, tell only part of the story.

A key question in transcription is whether the histone core octamer remains bound to the DNA in a nucleosome during passage of RNA polymerase, or whether it dissociates and then rebinds either randomly or to a specific site (Clark et al., 1993a). This choice is connected to the changes in supercoiling produced by binding of histones and by passage of the transcription complex. Clark and Felsenfeld (1991) studied the binding of histone core octamers to positively supercoiled DNA. The CD spectra and chemical reactivity of the proteins and DNA shows no distortions from the "classical" structures involving negative supercoils. Furthermore, reconstituted octamer–DNA complexes with equal numbers of positive or negative supercoils have similar sedimentation coefficients and therefore similar degrees of compaction, showing that positive supercoiling is not accompanied by significant unfolding of the complex (Clark et al.,

1993a). However, since histone binding induces positive superhelical turns, binding to positively supercoiled DNA increases superhelical stress, while binding to negatively supercoiled DNA reduces stress. Free energy considerations therefore dictate that octamers should bind more strongly to negative than to positive supercoils, a prediction borne out in an equilibrium exchange experiment.

Since RNA polymerase produces positive supercoils in front of the transcription complex and negative supercoils behind, octamer binding to DNA should be destabilized in front of the transcription complex. The histone core would then be transferred to a nearby piece of acceptor DNA (Clark and Felsenfeld, 1992). In an *in vitro* system, it has been demonstrated that the octamer can "step around" a transcribing polymerase without leaving the DNA: The DNA ahead of the polymerase uncoils from the octamer as the DNA behind coils around it (Studitsky et al., 1994).

Nucleosome cores can be displaced by direct binding of a transcription factor, as was demonstrated (Workman and Kingston, 1992) by addition of the DNA binding domain of the yeast GAL4 protein to nucleosome cores. This ternary complex of GAL4-AH, core histone proteins and DNA was unstable in the presence of nonspecific competitor DNA: It disproportionated into either the original nucleosome core and GAL4, or GAL4 bound to naked DNA. The interaction of GAL4-AH with nucleosomes containing multiple binding sites shows a complex, cooperative character that depends on the N-termini of the core histones (Vettese-Dadey et al., 1994). The histone amino termini, when present, inhibit binding of GAL4-AH especially near the center of the nucleosome core. Binding in this region is increased by the presence of additional GAL4-AH molecules bound to more accessible positions near the ends of the core. This cooperative effect is abolished if the amino termini of the core histones are removed by tryptic digestion.

The N-termini of the core histones can also affect the repression of transcription factor by linker histone H1 (Juan et al., 1994). The H1 molecule has different effects on the binding of transcription factors USF and GAL4-AH, repressing the former but having little effect on the latter. If the amino-terminal tails of the core histones are removed, the H1-mediated repression of USF binding is alleviated; if the tails are acetylated, repression is partly alleviated.

Binding of transcription factors to nucleosomes can also be stimulated by the histone binding protein, nucleoplasmin (Chen et al., 1994). In the presence of the GAL4-AH, a fragment of GAL4, which contains the DNA binding and dimerization domains, nucleoplasmin removes H2A and H2B from the nucleosome. This facilitates the subsequent displacement of the $(H3–H4)_2$ tetramers onto competing DNA.

3.6 The 30-nm Filament

While many studies of the effects of nucleosomes on transcription have employed small DNA fragments bound to only one or a few nucleosomes, a full understanding of transcription and replication *in vivo* must include highly polymerized chromatin. If the salt concentration is near physiological, virtually all of the nucleosome filament folds into a 30-nm chromatin filament. A great deal of effort has been made to understand the structure of this filament, "because this is the folded state in which most of the chromatin is maintained during most of the cell cycle. It must be unfolded to allow transcription or replication, and it must be further folded to allow meiosis or mitosis" (Widom, 1989).

The appearance of chromatin fragments in the EM depends on ionic conditions (Thoma et al., 1979). At low salt concentration, an extended zigzag fiber of "beads on a string" is seen. At intermediate ionic strength, a flat ribbon about 25 nm wide is observed. At 0.1 *M* NaCl, near physiological ionic strength, a 30-nm fiber with 11-nm striations is seen, similar to that observed in preparations of cell nuclei. Therefore, the 30-nm fiber is generally considered to be the structure adopted by native chromatin.

Folding of the nucleosome filament into the 30-nm filament appears to be a continuous process, rather than a two-state transition with defined endpoints. Compaction can occur in moderate concentrations of salt, due to screening of DNA charges by monovalent cations, and reduction of effective linker DNA charge by counterion condensation by multivalent cations (Hansen et al., 1989; Clark and Kimura, 1990; Yao et al., 1991). Linker histones H1 and H5 are involved in higher order folding, but are not essential for compaction. If linker histones are present, the concentration of salt needed for compaction is lowered since the basic groups on the linker histone lower the effective charge of the linker DNA. The linker histones promote more orderly packaging of chromatin, however (Thoma and Köller, 1977), perhaps by selecting a single compact structure from the large family of compact states permitted by salt alone (Yao et al., 1991). This selection may be accomplished by binding the entering and exiting stretches of DNA close together at the surface of the core (Worcel et al., 1978). The globular domains of the linker histones are seen by a variety of techniques to be in the interior (Graziano et al., 1994; Leuba et al., 1993) but the arms appear to be partially exposed (Leuba et al., 1993).

There are several models (reviewed by Wolffe, 1992 and Pruss et al., 1995) of how the histones and DNA are folded into the 30-nm fiber. Perhaps the most widely accepted is a solenoidal winding model, with approximately six nucleosomes per helical turn and a helical pitch of 11 nm (Finch and Klug, 1976). Since the width of the nucleosomal disk is 11 nm, this suggests that the nucleosomes are stacked with their long axes parallel to the fiber axis, a suggestion confirmed by transient electric birefringence (McGhee et al., 1980, 1983). Folding into the 30 nm fiber appears to be stabilized by cooperative interactions between linker histones arrayed down the middle of the fiber, an idea supported by a variety of binding, chemical cross-linking, and enzymatic accessibility studies.

The simple solenoid model leaves unspecified the path of the linker DNA. In the "coiled linker" model (Felsenfeld and McGhee, 1986), the linker DNA is coiled between neighboring nucleosomes. This proposal is consistent with cryo-EM and small-angle X-ray measurements which show that the diameter of the chromatin fiber is variable and increases with increasing linker length in species that have long linkers (Athey et al., 1990; Williams and Langmore, 1991). It is also consistent with the demonstration (Yao et al., 1991) that short linker DNA in dinucleosomes has substantial flexibility under the ionic conditions that stabilize chromatin folding.

Studies of nucleosome repeat lengths (Widom, 1992; Yao et al., 1993) show that they are roughly quantized by integral multiples of the helical twist of DNA. Since DNA in the nucleosome core has essentially a constant length of 146 bp, this implies that linker DNA lengths are quantized. Widom (1992) postulates that this arises from the constraints of higher order chromatin structure.

References

Adolph, K. W., Ed. (1988). *Chromosomes and Chromatin* CRC Press, Boca Raton, FL.

Adolph, K. W., Ed. (1990). *Chromosomes: Eukaryotic, Prokaryotic and Viral*, CRC Press, Boca Raton, FL.

Allison, S. A., Herr, J. C., and Schurr, J. M. (1981). Structure of Viral ϕ29 DNA Condensed by Simple Triamines, A Light-Scattering and Electron-Microscopy Study, *Biopolymers* **20**, 469–488.

Arents, G., Burlingame, R. W., Wang, B. C., Love, W. E., and Moudrianakis, E. N. (1991). The nucleosomal core histone octamer at 3.1 A resolution: a tripartite protein assembly and a left-handed superhelix. *Proc. Nat. Acad. Sci. USA* **88**, 10148–10152.

Arscott, P. G., Li, A.-Z., and Bloomfield, V. A. (1990). Condensation of DNA by trivalent cations. 1. Effects of DNA length and topology on the size and shape of condensed particles. *Biopolymers* **30**, 619–630.

Arscott, P. G., Ma, C., Wenner, J., and Bloomfield, V. A. (1995). DNA Condensation by Cobalt Hexaammine(III) in Alcohol–Water Mixtures: Dielectric Constant and Other Solvent Effects. *Biopolymers* **36**, 345–365.

Athey, B. D., Smith, M. F., Rankert, D. A., Williams, S. P., and Langmore, J. P. (1990). The diameters of frozen-hydrated chromatin fibers increase with DNA linker length: evidence in support of variable diameter models for chromatin. *J. Cell Biol.* **111**, 795–806.

Aubrey, K. L., Casjens, S. R., and Thomas, G. J., Jr. (1992). Secondary structure and interactions of the packaged dsDNA genome of bacteriophage P22 investigated by Raman difference spectroscopy. *Biochemistry* **31**, 11835–11842.

Aviles, F. J., Chapman, G. E., Kneale, G. G., Crane-Robinson, C. and Bradbury, E. M. (1978). The conformation of histone H5. Isolation and characterisation of the globular segment. *Eur. J. Biochem.* **88**, 363–371.

Bashkin, J., Hayes, J. J., Tullius, T. D., and Wolffe, A. P. (1993). Structure of DNA in a nucleosome core at high salt concentration and at high temperature. *Biochemistry* **32**, 1895–1898.

Baumann, C. G., Smith, S. B., Bloomfield, V. A., and Bustamante, C. (1997a). Ionic effects on the elasticity of single DNA molecules, *Proc. Natl. Acad. Sci. USA* **94**, 6185–6190.

Baumann, C. G., Smith, S. B., Bloomfield, V. A., and Bustamante, C. (1997b). Manuscript in preparation.

Baxevanis A. D. Godfrey J. E., and Moudrianakis E. N. (1991). Associative behavior of the histone $(H3–H4)_2$ tetramer: dependence on ionic environment. *Biochemistry* **30(36)**, 8817–23.

Benbasat, J. A. (1984). Condensation of Bacteriophage ϕW14 DNA of Varying Charge Densities by Trivalent Counterions. *Biochemistry* **23**, 3609–3619.

Black, L., Newcomb, W., Boring, J., and Brown, J. (1985). Ion etching of bacteriophage T4: Support for a spiral-fold model of packaged DNA. *Proc. Natl. Acad. Sci. USA* **82**, 7960–7964.

Bloomfield, V. A. (1991). Condensation of DNA by multivalent cations: Considerations on mechanism. *Biopolymers* **31**, 1471–1481.

Bloomfield, V. A. (1996). DNA condensation. *Curr. Opin. Struct. Biol.* **6**, 334-341.

Bloomfield, V. A., Ma, C., and Arscott, P. G. (1994). Role of Multivalent Cations in Condensation of DNA, in *Macro-Ion Characterization: From Dilute Solutions to Complex Fluids*, Schmitz, K. S., Ed., American Chemical Society, Washington, DC., pp. 195–209.

Bloomfield, V. A., Wilson, R. W., and Rau, D. C. (1980). Polyelectrolyte Effects in DNA Condensation by Polyamines, *Biophys. Chem.* **11**, 339–343.

Bonne-Andrea, C., Wong, M. L., and Alberts, B. M. (1990) In vitro replication through nucleosomes without histone displacement. *Nature (London)* **343**, 719–726.

Bradbury, E. M. (1992). Reversible histone modifications and the chromosome cell cycle, *BioEssays* **14**, 9–16.

Braunlin, W. H. and Xu, Q. (1992). Hexaamminecobalt(III) binding environments on double-helical DNA. *Biopolymers* **32**, 1703–1711.

Burlingame, R. W., Love, W. E., Wang, B. C., Hamlin, R., Nguyen, H. X., and Moudrianakis, E. N. (1985). Crystallographic structure of the octameric histone core of the nucleosome at a resolution of 3.3 A. *Science* **228**, 546–553.

Casjens, S. and Hendrix, R. (1987). Control mechanisms in dsDNA bacteriophage assembly, in Calendar, R. Ed., in *The Bacteriophages*, Vol. 1, Chapter 2, Plenum, New York, pp. 15–91.

Casjens, S., Wyckoff, E., Hayden, M., Sampson, L., Eppler, K., Randall, S., Moreno, E. T., and Serwer, P. (1992). Bacteriophage P22 portal protein is part of the gauge that regulates packing density of intravirion DNA. *J. Mol. Biol.* **224**, 1055–1074.

Castleman, H. and Erlanger, B. F. (1983). Electron microscopy of "Z-DNA." *Cold Spring Harbor Symp. Quant. Biol.* **47**, 133–141.

Castleman, H., Specthrie, L., Makowski, L., and Erlanger, B.F. (1984). Electronmicroscopy and circular dichroism of the dynamics of the formation and dissolution of supramolecular forms of Z-DNA. *J. Biomol. Struct. Dynam.* **2**, 271–283.

Chaires, J. B. (1989). Unusual condensation behavior of poly(dA)-poly(dT), *Biopolymers* **28**, 1645–1650.

Chattoraj, D. K., Gosule, L. C., and Schellman, J. A. (1978). DNA Condensation with Polyamines. II. Electron Microscopic Studies, *J. Mol. Biol.* **121**, 327–337.

Chen, H., Li, B., and Workman, J. L. (1994). A histone-binding protein, nucleoplasmin, stimulates transcription factor binding to nucleosomes and factor-induced nucleosome disassembly. *EMBO J.* **13**, 380–390.

Clark, D. J. and Felsenfeld, G. (1991). Formation of nucleosomes on positively supercoiled DNA. *EMBO J.* **10**, 387–395.

Clark, D. J. and Felsenfeld, G. (1992). A nucleosome core is transferred out of the path of a transcribing polymerase. *Cell* **71**, 11–22.

Clark, D. J., Ghirlando, R., Felsenfeld, G., and Eisenberg, H. (1993a). Effect of positive supercoiling on DNA compaction by nucleosome cores. *J. Mol. Biol.* **234**, 297–301.

Clark, D. J. and Kimura, T. (1990). Electrostatic mechanism of chromatin folding. *J. Mol. Biol.* **211**, 883–896.

Clark, D. J. and Thomas, J. O. (1988). Differences in the binding of H1 variants to DNA. Cooperativity and linker-length related distribution. *Eur. J. Biochem.* **178**, 225–233.

Clark, K. L., Halay, D. E., Lai, E., and Burley, S. K. (1993b). Co-crystal structure of the HNF-3γ/fork head DNA-recognition motif resembles histone H5. *Nature (London)* **364**, 412–420.

Day, L. A., Marzec, C. J., Reisberg, S. A., and Casadevall, A. (1988). DNA packing in filamentous bacteriophages. *Annu. Rev. Biophys. Biophys. Chem.* **17**, 509–539.

Drew, H. R. and Travers, A. A. (1985). DNA bending and its relation to nucleosome positioning. *J. Mol. Biol.* **186**, 773–790.

Duguid, J., Bloomfield, V. A., Benevides, J., and Thomas, G. J., Jr. (1993). Raman spectroscopy of DNA-metal complexes. I. Interactions and conformational effects of the divalent cations: Mg, Ca, Sr, Ba, Mn, Co, Ni, Cu, Pd, and Cd. *Biophys. J.* **65**, 1916–1928.

Earnshaw, W. C. (1991). Large scale chromosome structure and organization. *Curr. Opin. Struct. Biol.* **1**, 237–244.

Earnshaw, W. C. and Casjens, S. R. (1980) DNA packaging by the double-stranded DNA bacteriophages. *Cell* **21**, 319–331.

Earnshaw W. C. and Harrison, S. C. (1977). DNA arrangement in isometric phage heads. *Nature (London)* **268**, 598–602.

Earnshaw, W. C., King, J., Harrison, S. C., and Eiserling, F. A. (1978). The structural organization of DNA packaged within the heads of T4 wild-type, isometric and giant bacteriophages. *Cell* **14**, 559–568.

Egli, M., Williams, L. D., Gao, Q., and Rich, A. (1991). Structure of the pure-spermine form of Z-DNA (magnesium free) at 1-Å resolution. *Biochemistry*. **30**, 11388–11402.

Eichhorn, G. L. and Shin, Y. A. (1968). Interaction of metal ions with polynucleotides and related compounds. XII. The relative effect of various metal ions on DNA helicity *J. Am. Chem. Soc.* **90**, 7323–7328.

Eickbush, T. H. and Moudrianakis, E. N. (1976). The Compaction of DNA Helices into Either Continuous Supercoils or Folded-Fiber Rods and Toroids, *Cell* **13**, 295–306.

Eickbush, T. H. and Moudrianakis, E. N. (1978). The histone core complex: An octamer assembled by two sets of protein–protein interactions. *Biochemistry* **17**, 4955–4964.

Felsenfeld, G. and McGhee, J. D. (1986). Structure of the 30 nm chromatin fiber. *Cell* **44**, 375–377.

Feng, H. P., Scherl, D. S., and Widom, J. (1993). Lifetime of the histone octamer studied by continuous-flow quasielastic light scattering: test of a model for nucleosome transcription. *Biochemistry*. **32**, 7824–7831.

Feuerstein, B. G., Pattabiraman, N., and Marton, L. J. (1986). Spermine-DNA Interactions: A Theoretical Study. *Proc. Natl. Acad. Sci. USA* **83**, 5948–5952.

Finch, J. T., and Klug, A. (1976). Solenoidal model for superstructure in chromatin. *Proc. Natl. Acad. Sci. USA* **73**, 1897–1901.

Flock, S., Labarbe, R., and Houssier, C. (1995). Osmotic effectors and DNA structure: Effect of glycine on precipitation of DNA by multivalent cations. *J. Biomol. Struct. Dyn.* **13**, 87–102.

Flock, S., Labarbe, R., and Houssier, C. (1996). Dielectric constant and ionic strength effects on DNA precipitation. *Biophys. J.* **70**, 1456–1465.

Franke, W. W. (1987) Nuclear lamins and cytoplasmic intermediate filament proteins: a growing multigene family. *Cell* **48**, 3–4.

Garcia-Ramírez, M. and Subirana, J. (1994). Condensation of DNA by basic proteins does not depend on protein composition. *Biopolymers* **34**, 285–292.

Gershon, H., Ghirlando, R., Guttman, S. B., and Minsky, A. (1993). Mode of formation and structural features of DNA-cationic liposome complexes used for transfection. *Biochemistry* **32**, 7143–7151.

Ghirlando, R., Wachtel, E. J., Arad, T., and Minsky, A. (1992). DNA packaging induced by micellar aggregates: a novel in vitro DNA condensation system. *Biochemistry* **31**, 7110–7119.

Giannoni, G., Padden, F. J., Jr., and Keith, H. D. (1969). Crystallization of DNA from Dilute Solution. *Proc. Natl. Acad. Sci. USA* **62**, 964–971.

Gosule, L. C. and Schellman, J. A. (1976) Compact Form of DNA Induced by Spermidine. *Nature (London)* **259**, 333–335.

Graziano, V., Gerchman, S. E., Schneider, D. K., and Ramakrishnan, V. (1994). Histone H1 is located in the interior of the chromatin 30-nm filament. *Nature (London)* **368**, 351–354.

Grosberg, A. Y. and Zhestkov, A. V. (1985). On the toroidal condensed state of closed circular DNA. *J. Biomol. Struct. Dyn.* **3**, 515–520.

Guo, P., Erickson, S., and Anderson, D. (1987a) A small viral RNA is required for in vitro packaging of bacteriophage ϕ29 DNA. *Science* **236**, 690–694.

Guo, P., Peterson, C., and Anderson, D. (1987b). Initiation events in in vitro packaging of bacteriophage ϕ29 DNAγ p3. *J. Mol. Biol.* **197**, 219–228.

Guo, P., Peterson, C., and Anderson, D. (1987c). Prohead and DNA–gp3-dependent ATPase activity of the DNA packaging protein gp16 of bacteriophage ϕ29. *J. Mol. Biol.* **197**, 229–236.

Haas, R., Murphy, R. F., and Cantor, C. R. (1982). Testing models of the arrangement of DNA inside λ by cross-linking the packaged DNA. *J. Mol. Biol.* **159**, 71–92.

Hansen, J. C., Ausio, J., Stanik, V. H., and van Holde, K. E. (1989). Homogeneous reconstituted oligonucleosomes, evidence for salt-dependent folding in the absence of histone H1. *Biochemistry* **28**, 9129–9136.

Harrison, S. (1983). Packaging of DNA into bacteriophage heads: A model. *J. Mol. Biol.* **171**, 577–580.

Hartman, P. G., Chapman, G. E., Moss, T., and Bradbury, E. M. (1977). Studies on the role and mode of operation of the very-lysine-rich histone H1 in eukaryote chromatin. The three structural regions of the histone H1 molecule. *Eur. J. Biochem.* **77**, 45–51.

Hayes, J. J., Pruss, D., and Wolffe, A. P. (1994). Contacts of the globular domain of histone H5 and core histones with DNA in a "chromatosome." *Proc. Nat. Acad. Sci. USA* **91**, 817–821.

Hayes, J. J., Tullius, T. D., and Wolffe, A. P. (1990). The structure of DNA in a nucleosome. *Proc. Natl. Acad. Sci. USA* **87**, 7405–7409.

Hayes, J. J. and Wolffe, A. P. (1993). Preferential and asymmetric interaction of linker histones with 5S DNA in the nucleosome. *Proc. Nat. Acad. Sci. USA* **90**, 6415–6419.

Hill, C. S., Rimmer, J. M., Green, B. N., Finch, J. T., and Thomas, J. O. (1991). Histone–DNA interactions and their modulation by phosphorylation of -Ser-Pro-X-Lys/Arg- motifs. *EMBO J.* **10**, 1939–1948.

Hong, L., Schroth, G. P., Matthews, H. R., Yau, P., and Bradbury, E. M. (1993). Studies of the DNA binding properties of histone H4 amino terminus. Thermal denaturation studies reveal that acetylation markedly reduces the binding constant of the H4 "tail" to DNA. *J. Biol. Chem.* **268**, 305–314.

Hsiang, M. W. and Cole, R. D. (1977). Structure of Histone H1–DNA Complex, Effect of Histone H1 on DNA Condensation. *Proc. Natl. Acad. Sci. USA* **74**, 4852–4856.

Hung, D. T., Marton, L. J., Deen, D. F., and Shafer, R. H. (1983). Depletion of intracellular polyamines may alter DNA conformation in 9L rat brain tumor cells. *Science* **221**, 368–370.

Jordan, C. F., Lerman, L. S., and Venable, J. H., Jr. (1972). Structure and circular dichroism of DNA in concentrated polymer solutions. *Nature New Biol.* **236**, 67–70.

Juan, L. J., Utley, R. T., Adams, C. C., Vettese-Dadey, M., and Workman, J. L. (1994). Differential repression of transcription factor binding by histone H1 is regulated by the core histone amino termini. *EMBO J.* **13**, 6031–6040.

Kavenoff, R. and Zimm, B. H. (1973). Chromosome-sized DNA molecules from Drosophila. *Chromosoma* **41**, 1–27.

Kellenberger, E. (1987). The compactness of cellular plasmas; in particular, chromatin compactness in relation to function. *Trends Biochem. Sci.* **12**, 105–107.

Kellenberger, E., Carlemalm, E., Sechaud, J., Ryter, A., and De Haller, G. (1986). Considerations on the condensation and the degree of compactness in non-eukaryotic DNA-containing plasmas, in *Bacterial Chromatin*, Gualerzi, C. O., and Pon, C. L., Eds., Springer-Verlag, Berlin, Germany, pp. 11–25.

Keller, D. and Bustamante, C. (1986a). Theory of the interaction of light with large inhomogeneous molecular aggregates. I. Absorption. *J. Chem. Phys.* **84**, 2961–2971.

Keller, D. and Bustamante, C. (1986b). Theory of the interaction of light with large inhomogeneous molecular aggregates. II. Psi-type circular dichroism. *J. Chem. Phys.* **84**, 2972–2980.

Kim, M.-H., Ulibarri, L., Keller, D., Maestre, M. F., and Bustamante, C. (1986). Theory of the interaction of light with large inhomogeneous molecular aggregates. III. Calculations. *J. Chem. Phys.* **84**, 2981–2989.

Kleinschmidt, A. K., Lang, D., Jacherts, D., and Zahn, R. K. (1962). Preparation and length measurements of the total deoxyribonucleic acid content of T2 bacteriophages. *Biochem. Biophs. Acta* **61**, 857–864.

Klimenko, S. M., Tikhchonenko, T. I., and Andreev, V. M. (1967). Packing of DNA in the Head of Bacteriophage T2. *J. Mol. Biol.* **23**, 523–533.

Klug, A. and Lutter, L. C. (1981). The helical periodicity of DNA on the nucleosome. *Nucleic Acids Res.* **9**, 4267–4283.

Klug, A., Rhodes, D., Smith, J., Finch, J. T., and Thomas, J. O. (1980). A low resolution structure for the histone core of the nucleosome. *Nature (London)* **287**, 509–516.

Knoll, D. A., Fried, M. G., and Bloomfield, V. A. (1988). Heat-induced DNA aggregation in the presence of divalent metal salts, in *Structure and Expression Volume 2, DNA and Its Drug Complexes*, (Proceedings of the 5th Conversation Biomolecular Stereodynam) Sarma, M. H. and Sarma, R. H., Eds., Adenine Press, Schenectady, NY. 123–145.

Kornberg, R. (1974). Chromatin structure: A repeating unit of histones and DNA. *Science* **184**, 868–871.

Kornberg, R. and Thomas, J. O. (1974). Chromatin structure: oligomers of histones. *Science* **184**, 865–868.

Kosturko, L. D., Hogan, M., and Dattagupta, N. (1979). Structure of DNA within three isometric bacteriophages. *Cell* **16**, 515–522.

Krasnow, M. A. and Cozzarelli, N. R. (1982). Catenation of DNA Rings by Topoisomerases. Mechanism of Control by Spermidine. *J. Biol. Chem.* **257**, 2687–2693.

Laemmli, U. K. (1975). Characterization of DNA Condensates Induced by Poly(ethylene oxide) and Polylysine. *Proc. Natl. Acad. Sci. USA* **72**, 4288–4292.

Laemmli, U. K., Paulson, J. R., and Hitchins, V. (1974). Maturation of the head of bacteriophage T4. V. A possible DNA packaging mechanism: in vitro cleavage of the head proteins and the structure of the core of the polyhead. *J. Supramol. Struc.* **2**, 276–301.

Lang, D. (1973). Regular Superstructures of Purified DNA In Ethanolic Solutions. *J. Mol. Biol.* **78**, 247–254.

Lang, D., Taylor, T. N., Dobyan, D. C., and Gray, D. M. (1976). Dehydrated Circular DNA, Electron Microscopy of Ethanol- Condensed Molecules. *J. Mol. Biol.* **106**, 97–107.

Lee, D. Y., Hayes, J. J., Pruss, D., and Wolffe, A. P. (1993). A positive role for histone acetylation in transcription factor access to nucleosomal DNA. *Cell* **72**, 73–84.

Leikin, S., Rau, D. C., and Parsegian, V. A. (1991). Measured entropy and enthalpy of hydration as a function of distance between DNA double helices. *Phys. Rev. A* **44**, 5272–5278.

Leikin, S., Parsegian, V. A., Rau, D. C., and Rand, R. P. (1993). Hydration forces. *Annu. Rev. Phys. Chem.* **44**, 369–395.

Lepault, J., Dubochet, J., Baschong, W., and Kellenberger, E. (1987). Organization of double-stranded DNA in bacteriophages, a study by cryo-electron microscopy of vitrified samples. *EMBO J.* **6**, 1507–1512.

Lerman, L. S. (1971). A Transition to a Compact Form of DNA in Polymer Solutions. *Proc. Natl. Acad. Sci. USA* **68**, 1886–1890.

Leuba, S. H., Zlatanova, J., and van Holde, K. (1993). On the location of histones H1 and H5 in the chromatin fiber. Studies with immobilized trypsin and chymotrypsin. *J. Mol. Biol.* **229**, 917–929.

Liu, D. J. and Day, L. A. (1994). Pf1 virus structure: helical coat protein and DNA with paraxial phosphates. *Science*. **265**, 671–674, 1994.

Lorch, Y., LaPointe, J. W., and Kornberg, R. D. (1987). Nucleosomes inhibit the initiation of transcription but allow chain elongation with the displacement of histones. *Cell* **49**, 203–210.

Lowary, P. T. and Widom, J. (1989). Higher-order structure of Saccharomyces cerevisiae chromatin. *Proc. Natl. Acad. Sci. USA* **86**, 8266–8270.

Lu, Q., Wallrath, L. L., and Elgin, S. C. (1994). Nucleosome positioning and gene regulation. *J. Cell. Biochem.* **55**, 83–92, 1994.

Lyubartsev, A. P. and Nordenskjöld, L. (1995). Monte Carlo simulation study of ion distribution and osmotic pressure in hexagonally oriented DNA. *J. Phys. Chem.* **99**, 10373–10382.

Ma, C. and Bloomfield, V. A. (1994). Condensation of Supercoiled DNA Induced by $MnCl_2$, *Biophys. J.* **67**, 1678–1681.

Ma, C. and Bloomfield, V. A. (1995). Gel electrophoresis measurement of counterion condensation on DNA. *Biopolymers* **35**, 211–216.

Ma, C., L. Sun and Bloomfield, V. A. (1995). Condensation of plasmids enhanced by Z-DNA conformation of $d(CG)_n$ inserts. *Biochemistry* **34**, 3521–3528.

Maniatis, T., Venable, J. H., and Lerman, L. S. (1974). The Structure of Ψ DNA. *J. Mol. Biol.* **84**, 37–64.

Manning, G. S. (1978). The molecular theory of polyelectrolyte solutions with applications to the electrostatic properties of polynucleotides. *Q. Rev. Biophys.* **11**, 179–246.

Manning, G. S. (1980). Thermodynamic stability theory for DNA doughnut shapes induced by charge neutralization. *Biopolymers* **19**, 37–59.

Manning, G. S. (1985). Packaged DNA: An elastic model *Cell Biophys.* **7**, 57–89.

Manning, G. S. (1989) Self-attraction and natural curvature in null DNA. *J. Biomol. Struct. Dyn.* **7**, 41–61.

Manning, G., Ebralidse, K., Mirzabekov, A., and Rich, A. (1989). An estimate of the extent of folding of nucleosomal DNA by laterally asymmetric neutralization of phosphate groups. *J. Biomol. Struct. Dyn.* **6**, 877–89.

Marquet, R. and C. Houssier. (1988). Different binding modes of spermine to A-T and G-C base pairs modulate the bending and stiffening of the DNA double helix. *J. Biomol. Struct. Dyn.* **6**, 235–246.

Marquet, R. and Houssier, C. (1991). Thermodynamics of cation-induced DNA condensation. *J. Biomol. Struct. Dyn.* **9**, 159–167.

Marquet, R., Houssier, C., and Fredericq, E. (1985). An electro-optical study of the mechanisms of DNA condensation induced by spermine, *Biochim. Biophys. Acta* **825**, 365–374.

Marquet, R., Wyart, A., and Houssier, C. (1987). Influence of DNA length on spermine-induced condensation. Importance of the bending and stiffening of DNA. *Biochim. Biophys. Acta* **909**, 165–172.

Marx, K. A. and Reynolds, T. C. (1982). Spermidine-Condensed ϕX-174 DNA Cleavage by Micrococcal Nuclease: Torus Cleavage Model and Evidence for Unidirectional Circumferential DNA Wrapping. *Proc. Natl. Acad. Sci. USA* **79**, 6484–6488.

Marx, K. A. and Ruben, G. C. (1983). Evidence for Hydrated Spermidine-Calf Thymus DNA Toruses Organized by Circumferential DNA Wrapping. *Nucleic Acids Res.* **11**, 1839–1854.

Marzec, C. J. and Day, L. A. (1994). An electrostatic spatial resonance model for coaxial helical structures with applications to the filamentous bacteriophages. *Biophys. J.* **67**, 2205–2222.

McGhee, J. D., Nickol, J. M., Felsenfeld, G., and Rau, D. C. (1983). Higher order structure of chromatin: Orientation of nucleosomes within the 30 nm chromatin solenoid is independent of species and spacer length. *Cell* **33**, 831–841.

McGhee, J. D., Rau, D. C., and Felsenfeld, G. (1980). Orientation of the nucleosome within the higher order structure of chromatin. *Cell* **22**, 87–96.

Nilsson, L. G., Guldbrand, L., and Nordenskjöld, L. (1991). Evaluation of the electrostatic osmotic pressure in an infinite system of hexagonally oriented DNA molecules. A Monte Carlo simulation. *Mol. Phys.* **72**, 177–192.

Odijk, T. (1993a). Physics of tightly curved semiflexible polymer chains. *Macromolecules* **26**, 6897–6902.

Odijk, T. (1993b). Undulation-enhanced electrostatic forces in hexagonal polyelectrolyte gels. *Biophys. Chem.* **46**, 69–75.

Odijk, T. (1994). Long-range attractions in polyelectrolyte solutions. *Macromolecules* **27**, 4998–5003.

Olins, A. L. and Olins. D. E. (1974). Spheroid chromatin units (v bodies). *Science* **183**, 330–332.

Ooosawa, F. (1971). *Polyelectrolytes*, Marcel-Dekker, New York, Chapter 9.

Paranjape, S. M., Kamakaka, R. T., and Kadonaga, J. T. (1994). Role of chromatin structure in the regulation of transcription by RNA polymerase II. *Annu. Rev. Biochem.* **63**, 265–297.

Parsegian, A. V., Rand, R. P., Fuller, N. L., and Rau, D. C. (1986). Osmotic stress for the direct measurement of intermolecular forces. *Methods Enzymol.* **127**, 400–416.

Pettijohn, D. E. (1988). Histone-like proteins and bacterial chromosome structure. *J. Biol. Chem.* **263**, 12793–12796.

Plum, G. E., Arscott, P. G., and Bloomfield, V. A. (1990). Condensation of DNA by trivalent cations. 2. Effect of cation structure. *Biopolymers* **30**, 631–643.

Plum, G. E. and Bloomfield, V. A. (1988). Equilibrium Dialysis Study of Binding of Hexammine Cobalt (III) to DNA. *Biopolymers* **27**, 1045–1051.

Podgornik, R., Rau, D. C., and Parsegian, V. A. (1989). The action of interhelical forces on the organization of DNA double helices: Fluctuation-enhanced decay of electrostatic double-layer and hydration forces. *Macromolecules* **22**, 1780–1786.

Podgornik, R., Rau, D. C., and Parsegian, V. A. (1994). Parametrization of direct and soft steric-undulatory forces between DNA double helical polyelectrolytes in solutions of several different anions and cations, *Biophys. J.* **66**, 962–971.

Pörschke, D. (1984). Dynamics of DNA condensation. *Biochemistry* **23**, 4821–4828.

Pörschke, D. (1985). Short electric-field pulses convert DNA from "condensed" to "free" conformation. *Biopolymers* **24**, 1981–1993.

Pörschke, D. (1986). Structure and dynamics of double helices in solution: Modes of DNA bending. *J. Biomol. Struct. Dyn.* **4**, 373–389.

Post, C. B. and Zimm, B. H. (1979). Internal Condensation of a Single DNA Molecule. *Biopolymers* **18**, 1487–1501.

Post, C. B. and Zimm, B. H. (1982a). Theory of DNA Condensation: Collapse versus Aggregation. *Biopolymers* **21**, 2123–2137.

Post, C. B. and Zimm, B. H. (1982b). Light Scattering Study of DNA Condensation, Competition Between Collapse and Aggregation. *Biopolymers* **21**, 2139–2160.

Pruss, D. J., Hayes, J., and Wolffe, A. P. (1995). Nucleosomal Anatomy—Where Are The Histones. *Bioessays* **17**, 161–170.

Ramakrishnan, V. (1994) Histone structure. *Curr. Opin. Struct. Biol.* **4**, 44–50.

Ramakrishnan, V., Finch, J. T., Graziano, V., Lee, P. L., and Sweet, R. M. (1993). Crystal structure of globular domain of histone H5 and its implications for nucleosome binding. *Nature (London)* **362**, 219–223.

Rau, D. C., Lee, B. K., and Parsegian, V. A. (1984). Measurement of the repulsive force between polyelectrolyte molecules in ionic solution: Hydration forces between parallel DNA double helices. *Proc. Natl. Acad. Sci. USA* **81**, 2621–2625.

Rau, D. C. and Parsegian, V. A. (1992a). Direct measurement of the intermolecular forces between counterion-condensed DNA double helices. Evidence for long range attractive hydration forces. *Biophys. J.* **61**, 246–259.

Rau, D. C. and Parsegian, V. A. (1992b). Direct measurement of temperature-dependent solvation forces between DNA double helices. *Biophys. J.* **61**, 260–271.

Reich, Z., Friedman, P., Levin-Zaidman, S., and Minsky, A. (1993) Effects of adenine tracts on the B–Z transition. Fine tuning of DNA conformational transition processes. *J. Biol. Chem.* **268**, 8261–8266.

Reich, Z., Ghirlando, R., and Minsky, A. (1991). Secondary conformational polymorphism of nucleic acids as a possible functional link between cellular parameters and DNA packaging processes. *Biochemistry* **30**, 7828–7836.

Reich, Z., Ghirlando, R., and Minsky, A. (1992). Nucleic acids packaging processes: effects of adenine tracts and sequence-dependent curvature. *J. Biomol. Struct. Dyn.* **9**, 1097–1109.

Reich, Z., Levin-Zaidman, S., Gutman, S. B., Arad, T., and Minsky, A. (1994b). Supercoiling-regulated liquid-crystalline packaging of topologically-constrained, nucleosome-free DNA molecules. *Biochemistry* **33**, 14177–14184.

Reich, Z., Wachtel, E. J., and Minsky, A. (1994a). Liquid-crystalline mesophases of plasmid DNA in bacteria. *Science* **264**, 1460–1463.

Reich, Z., Wachtel, E. J., and Minsky, A. (1995). In vivo quantitative characterization of intermolecular interactions. *J. Biol. Chem.* **270**, 7045–7046.

Reid, R. J., Bodley, J. W., and Anderson, D. (1994a). Characterization of the prohead-pRNA interaction of bacteriophage phi 29. *J. Biol. Chem.* **269**, 5157–5162.

Reid, R. J., Bodley, J. W., and Anderson, D. (1994b). Identification of bacteriophage phi 29 prohead RNA domains necessary for in vitro DNA–gp3 packaging. *J. Biol. Chem.* **269**, 9084–9089.

Revet, B., Delain, E., Dante, R., and Niveleau, A. (1983). Three dimensional association of double-stranded helices are produced in conditions for Z-DNA formation. *J. Biomol. Struct. Dyn.* **1**, 857–871.

Richards, K. E., Williams, R. C., and Calendar, R. (1973). Mode of DNA packing within bacteriophage heads. *J. Mol. Biol.* **78**, 255–259.

Richmond, T. J., Finch, J. T., Rushton, B., Rhodes, D., and Klug, A. (1984). Structure of the Nucleosome Core Particle at 7 Å Resolution. *Nature (London)* **311**, 532–537.

Riemer, S. C. and Bloomfield, V. A. (1978). Packaging of DNA in bacteriophage heads, Some considerations on energetics. *Biopolymers* **17**, 785–794.

Rill, R. L. (1986). Liquid crystalline phases in concentrated aqueous solutions of Na^+ DNA. *Proc. Natl. Acad. Sci. USA* **83**, 342–346.

Rill R. L., Hilliard, P. R., Jr., and Levy, G. C. (1983). Spontaneous ordering of DNA. Effects of intermolecular interactions on DNA motional dynamics monitored by ^{13}C and ^{31}P nuclear magnetic resonance spectroscopy. *J. Biol. Chem.* **258**, 250–2566.

Rill, R. L., Livolant, F., Aldrich, H. C., and Davidson, M. W. (1989). Electron microscopy of liquid crystalline DNA: direct evidence for cholesteric-like organization of DNA in dinoflagellate chromosomes. *Chromosoma* **98**, 280–286.

Robinow, C. and Kellenberger, E. (1994). The bacterial nucleoid revisited. *Microbiol. Revs.* **58**, 211–232.

Rouzina, I. and Bloomfield, V. A. (1996). Macroion attraction due to electrostatic correlation between screening counterions. 1. Mobile surface-adsorbed ions and diffuse ion cloud. *J. Phys. Chem.* **100**, 9977–9989.

Schellman, J. A. and Parthasarathy, N. (1984). X-ray diffraction studies on cation-collapsed DNA. *J. Mol. Biol.* **175**, 313–329.

Schmid, M. B. (1988) Structure and function of the bacterial chromosome. *Trends Biochem. Sci.* **13**, 131–135.

Serwer, P. (1975). Buoyant density sedimentation of macromolecules in sodium iothalamate density gradients. *J. Mol. Biol.* **92**, 433–448.

Serwer, P. (1980). A metrizamide-impermeable capsid in the DNA packaging pathway of bacteriophage T7. *J. Mol. Biol.* **138**, 65–91.

Serwer, P. (1986). Arrangement of double-stranded DNA packaged in bacteriophage capsids. An alternative model. *J. Mol. Biol.* **190**, 509–512.

Serwer, P. (1988). The source of energy for bacteriophage DNA packaging, An osmotic pump explains the data. *Biopolymers* **27**, 165–169.

Serwer, P., Hayes, S. J., and Watson, R. H. (1992). Conformation of DNA packaged in bacteriophage T7. Analysis by use of ultraviolet light-induced DNA-capsid cross-linking. *J. Mol. Biol.* **223**, 999–1011.

Shrader, T. E. and Crothers, D. M. (1989). Artificial nucleosome positioning sequences. *Proc. Natl. Acad. Sci. USA* **86**, 7418–7422.

Shrader, T. E. and Crothers, D. M. (1990). Effects of DNA sequence and histone-histone interactions on nucleosome placement. *J. Mol. Biol.* **216**, 69–84.

Sikorav, J.-L. and Church, G. M. (1991). Complementary recognition in condensed DNA: Accelerated DNA renaturation. *J. Mol. Biol.* **222**, 1085–1108.

Sikorav, J. L., Pelta, J., and Livolant, F. (1994). A liquid crystalline phase in spermidine-condensed DNA. *Biophys. J.* **67**, 1387–1392.

Simpson, R. T. (1978). Structure of the chromatosome, a chromatin core particle containing 160 base pairs of DNA and all the histones. *Biochemistry* **17**, 5524–5531.

Simpson, R. T. (1991). Nucleosome positioning: Occurrence, mechanisms, and functional consequences. *Prog. Nucleic Acid Res. Mol. Biol.* **40**, 143–184.

Son, M., Watson, R. H., and Serwer, P. (1993). The direction and rate of bacteriophage T7 DNA packaging in vitro. *Virology* **196**, 282–289.

Strauss, J. K. and Maher, L. J., III. (1994). DNA bending by asymmetric phosphate neutralization. *Science*.**266: 1829–1834.**

Streisinger, G., J. Emrich and M.M. Stahl (1967) Chromosome structure in phage T4. III. Terminal redundancy and length determination, Proc. Natl. Acad. Sci. USA 57, 292–295.

Stroobants, A., Lekkerkerker, H. N. W., and Odijk, T. (1986). Effect of electrostatic interaction on the liquid crystal phase transition in solutions of rodlike polyelectrolytes. *Macromolecules* **19**, 2232–2238.

Struck, M. M., Klug, A., and Richmond, T. J. (1992). Comparison of X-ray structures of the nucleosome core particle in two different hydration states. *J. Mol. Biol.* **224**, 253–264.

Strzelecka, T. E., Davidson, M. W., and Rill, R. L. (1988). Multiple liquid crystal phases of DNA at high concentrations. *Nature (London)* **331**, 457–460.

Strzelecka, T. E. and Rill, R. L. (1990). Phase transitions of concentrated DNA solutions in low concentrations of 1:1 supporting electrolyte. *Biopolymers* **30**, 57–71.

Studitsky, V. M., Clark, D. J., and Felsenfeld, G. (1994). A histone octamer can step around a transcribing polymerase without leaving the template. *Cell* **76**, 371–382.

Suwalsky, M., Traub, W., Shmueli, U., and Subirana, J. A. (1969). An X-ray study of the interaction of DNA with spermine. *J. Mol. Biol.* **42**, 363–373.

Tanford, C. (1974). Thermodynamics of micelle formation: Prediction of micelle size and size distribution. *Proc. Natl. Acad. Sci. USA* **71**, 1811–1815.

Thoma, F. and Köller, T. (1977). Influence of histone H1 on chromatin structure. *Cell* **12**, 101–107.

Thoma, F., Köller, T., and Klug, A. (1979). Involvement of histone H1 in the organization of the nucleosome and of the salt-dependent superstructures of chromatin. *J. Cell Biol.* **83(2 Pt 1)**, 403–427.

Thoma, F., Losa, R., and Köller, T. (1983). Involvement of the domains of histones H1 and H5 in the structural organization of soluble chromatin. *J. Mol. Biol.* **167**, 619–640.

Thomas, T. J. and Bloomfield, V. A. (1985). Toroidal Condensation of Z DNA and Identification of an Intermediate in the B to Z Transition of Poly(dG-m5dC) • Poly(dG-m5dC). *Biochemistry* **24**, 713–719.

Turnquist, S., Simon, M., Egelman, E., and Anderson, D. (1992). Supercoiled DNA wraps around the bacteriophage phi 29 head-tail connector. *Proc. Nat. Acad. Sci. USA* **89**, 10479–10483.

van Holde, K. E. (1989). *Chromatin*, Springer Verlag, New York.

van de Sande, J. H. and Jovin, T. M. (1982). Z* DNA, the left-handed helical form of poly[d(G-C)] in $MgCl_2$-ethanol, is biologically active. *EMBO J.* **1**, 115–120.

van de Sande, J. H., McIntosh, L. P., and Jovin, T. M. (1982). Mn^{2+} and other transition metals at low concentration induce the right-to-left helical transformation of poly[d(G-C)]. *EMBO J.* **1**, 777–782.

Vettese-Dadey M., Walter, P., Chen, H., Juan, L. J., and Workman, J. L. (1994). Role of the histone amino termini in facilitated binding of a transcription factor, GAL4-AH, to nucleosome cores. *Mol. Cell. Biol.* **14**, 970–981.

Votavová, H., Kucerová, D., Felsberg, J., and Sponar, J. (1986). Changes in Conformation, Stability and Condensation of DNA by Univalent and Divalent Cations in Methanol–Water Mixtures. *J. Biomol. Struct. Dyn.* **4**, 477–489.

Wang, B. C., Rose, J., Arents, G., and Moudrianakis, E. N. (1994). The octameric histone core of the nucleosome. Structural issues resolved. *J. Mol. Biol.* **236**, 179–188.

Wassarman, P. M. and Kornberg, R. D. (1989). *Nuecleosomes Methods Enzymol.* **170**, 683 pp.

Welsh, J. and Cantor, C. R. (1987). Studies on the arrangement of DNA inside viruses using a breakable bis-psoralen crosslinker, *J. Mol. Biol.* **198**, 63–71.

Widom, J. (1989). Toward a Unified Model of Chromatin Folding. *Annu. Rev. Biophys. Biophys. Chem.* **18**, 365–395.

Widom, J. (1991). Nucleosomes and chromatin. *Curr. Opin. Struct. Biol.* **1**, 245–250.

Widom J. (1992) A relationship between the helical twist of DNA and the ordered positioning of nucleosomes in all eukaryotic cells. *Proc. Nat Acad. Sci. USA* **89**, 1095–1099.

Widom, J. and Baldwin, R. L. (1980). Cation-Induced Toroidal Condensation of DNA. Studies with $Co^{3+}(NH_3)_6$. *J. Mol. Biol.* **144**, 431–453.

Widom, J. and Baldwin, R. L. (1983). Monomolecular Condensation of Lambda-DNA Induced by Cobalt Hexammine. *Biopolymers* **22**, 1595 1620.

Widom, J. and Baldwin, R. L. (1983b). Tests of spool models for DNA in phage lambda. *J. Mol. Biol.* **171**, 419–437.

Williams, S. P. and Langmore, J. P. (1991). Small angle X-ray scattering of chromatin. Radius and mass per unit length depend on linker length. *Biophys. J.* **59**, 606–618.

Wilson, R. W. and Bloomfield, V. A. (1979). Counterion-Induced Condensation of Deoxyribonucleic Acid. A Light-Scattering Study. *Biochemistry* **18**, 2192–2196.

Wissenburg, P., Odijk, T., Cirkel, P., and Mandel, M. (1994). Multimolecular aggregation in concentrated isotropic solutions of mononucleosomal DNA in 1 M sodium chloride. *Macromolecules* **27**, 306–308.

Wissenburg, P., Odijk, T., Cirkel, P., and Mandel, M. (1995). Multimolecular aggregation of mononucleosomal DNA in concentrated isotropic solutions. *Macromolecules* **28**, 2315–2328.

Wolffe, A. P. (1992), *Chromatin: Structure and Function*, Academic, San Diego, CA.

Wolffe, A. P. (1994) Nucleosome positioning and modification: chromatin structures that potentiate transcription. *Trends Biochem. Sci.* **19**, 240–244.

Worcel, A., Han, S., and Wong, M. L. (1978). Assembly of newly replicated chromatin. *Cell* **15**, 969–977.

Workman, J. L. and Buchman, A. R. (1993). Multiple functions of nucleosomes and regulatory factors in transcription. *Trends Biochem. Sci.* **18**, 90–95.

Workman, J. L. and Kingston, R. E. (1992). Nucleosome core displacement in vitro via a metastable transcription factor-nucleosome complex. *Science* **258**, 1780–1784.

Xu, Q., Jampani, S. R. B., and Braunlin, W. H. (1993a). Rotational dynamics of hexaamminecobalt(III) bound to oligomeric DNA: Correlation with cation-induced structural transitions. *Biochemistry* **32**, 11754–11760.

Xu, Q., Shoemaker, R. K., and Braunlin, W. H. (1993b). Induction of B–A transitions of deoxyoligonucleotides by multivalent cations in dilute aqueous solution. *Biophys. J.* **65**, 1039–1049.

Yao, J., Lowary, P. T., and Widom, J. (1991). Linker DNA bending induced by the core histones of chromatin. *Biochemistry* **30**, 8408–8414.

Yao, J., Lowary, P. T., and Widom, J. (1993). Twist constraints on linker DNA in the 30-nm chromatin fiber: implications for nucleosome phasing. *Proc. Nat. Acad. Sci. USA* **90**, 9364–9368.

Yen, W. S., Rhee, K. W., and Ware, B. R. (1983). Condensation of polyamines onto nucleic acids. *J. Phys. Chem.* **87**, 2148–2152.

Zehfus, M. H. and Johnson, W. C., Jr. (1981). Properties of P-form DNA as revealed by circular dichroism. *Biopolymers* **20**, 1589–1603.

Zehfus, M. H. and Johnson, W. C., Jr. (1984). Conformation of P-form DNA. *Biopolymers* **23**, 1269–1281.

Zhang, P., Tobias, I., and Olson, W. K. (1994). Computer simulation of protein-induced structural changes in closed circular DNA. *J. Mol. Biol.* **242**, 271–290.

Zlatanova, J. S., and van Holde, K. E. (1992). Chromatin loops and transcriptional regulation. *Crit. Revs. Eukaryotic Gene Expr.* **2**, 211–224.

Index

A

B

C

D

F

G

N

Q

R

S

T

U

V

W

X

Y–Z